CALCULUS
SINGLE VARIABLE

CARL V. LUTZER • H.T. GOODWILL

WILEY

John Wiley & Sons, Inc

VICE PRESIDENT & PUBLISHER	Laurie Rosatone
ACQUISITIONS EDITOR	David Dietz
MARKETING MANAGER	Jonathan Cottrell
SENIOR DEVELOPMENT EDITOR	Mary O'Sullivan
PROJECT EDITOR	Jennifer Brady
SENIOR DESIGNER	Jim O'Shea
MEDIA EDITOR	Melissa Edwards
PRODUCTION MANAGER	Micheline Frederick
SENIOR PRODUCTION EDITOR	Ken Santor

This book was set in LaTex by the authors and printed and bound by Courier Kendallville. The cover was printed by Courier Kendallville.

This book is printed on acid free paper.

Library of Congress Cataloging in Publication Data:

ISBN 13   978-0470-17930-7

Printed in the United States of America

10 9 8 7 6 5 4 3 2 1

CONTENTS

CONTENTS

CONTENTS

TO THE STUDENT

During your study of calculus this year, you will likely have questions about why the mathematics works the way it does, where it arises in the real world, and how it's used by professionals. Based on our experiences with our own students, we've tried to anticipate and answer as many of those questions as possible so that formulas and techniques will make sense to you. This is one part of a larger goal. We wrote this text to help you develop new skills and broaden your intuition, and to give you an idea of how mathematics is used to both describe and change our world. But in order to achieve that goal, we need your help. Specifically, we need you to read, so in the short space that we have here, we want to offer some advice about reading technical content.

Effective reading strategies

The most important thing we can tell you is that learning technical content takes time, and often requires substantial mental effort, so first and foremost, be patient.

Many people find that a scan-then-dig strategy works well when reading technical writing. Your first pass through a section is meant only to familiarize you with the general organization of topics and terms in the section so that you have a mental "road map" of the learning that you'll do. The second pass is where you begin to dig into the main ideas and develop the relevant skills.

Note our use of the word "dig." Learning technical material, like the physical act of digging, is an active endeavor that often requires significant effort, even for professionals. Recently when reading a book about applications of physics to the medical sciences, one of us found himself saying, "That doesn't seem right. That would mean the relationship is something like …" after which there was a lot of writing on scrap paper, a few occurrences of, "Oh …" and finally, "Now I get it."

The point is that in order to really learn something and get good at, you have to do it. Keep a pencil and paper handy when you read so you can check calculations (that will help you get used to the ways that various terms interact and develop a basic understanding of what they mean), and work through the "Try It Yourself" examples that we provide so that your reading is an active, not a passive endeavor. If your answer disagrees with ours, and you can't figure out why, we've posted full solutions online.

Effective practice

When taking a quiz or exam, and more importantly, later, when you're actually using calculus to solve problems in the real world, you'll have to think creatively about how to combine the ideas and techniques that you learn here. You won't develop that creative mathematical thinking by watching someone else do it. You have to practice it yourself. We've already mentioned this kind of thing as part of effective reading strategies, but here we mean to offer advice about making homework exercises effective learning tools.

Specifically, many students find that having a solutions manual is helpful, and we have worked with Dr. Justin Young to provide a good one for this text. Our advice is to use it as a stepping stone, rather than a crutch. Each time you complete an exercise by reading part or all of its solution in the manual, *try to do a similar exercise on your own*. Otherwise, someone else is doing the creative "heavy lifting" for you, and you may end up in the situation of saying, "I don't understand why I got all of the homework exercises right but did poorly on the exam."

In closing

Our last piece of advice is to talk about what you're learning. The best way to solidify and synthesize ideas is often to articulate them to other people. And if "other people" includes your teacher, he or she can help you by hearing and responding to the way that you describe ideas and techniques.

PREFACE

This project started when a former student spoke to us about calculus texts. He said, "The texts that are out there are good reference books, but they don't actually *teach* you anything. If I couldn't get it from the teacher, there's no way I could learn it from the text." After a review of the most popular texts on the market today, we understood what he meant. While correct and concise, these texts are written in a style that's difficult to parse for readers who are unaccustomed to technical writing, and they often neglect to provide motivation of their content at a practical level that speaks to modern students. The result is that both students and teachers tend to *skim* these texts rather than actually *read* them. Although they make great reference books in later years, during the actual learning process these texts function as little more than very heavy exercise sets.

A calculus text should spark curiosity in its readers and it should hold students' attention. It should motivate the material in relevant and interesting ways, and answer the spectrum of students' questions from "How do we do this?" to "Why do we do this?" to "How can I make sense of this?" As a complement to their classroom experience, it should engage students in the discussion and discovery of ideas, develop their intuition, and help them learn about the power, beauty, and versatility of calculus. This book is our effort to provide such a text.

Engaging students

This text is written primarily for students who are interested in physical and medical sciences, mathematics, engineering, or computer science, but also provides inlets to the subject and exercises for students whose interests lie in business or social science. Most students who take calculus are interested in applications, so we often use applications in our examples and in the preliminary discussion of a new topic.

For example, the idea of a limit is introduced using a thought experiment in which we track the position of a planet that passes between us and the sun; the concept of the derivative is introduced using clinical data about the radius of an artery; derivatives of trigonometric functions are motivated by a vibrating guitar string; Riemann sums are introduced to calculate the net displacement of an accelerating vehicle based on data from a radar gun; the idea of a sequence is motivated by concentrations of medication in a patient's bloodstream; and the list goes on.

We recognize that the applications must be presented at an elementary level, with broad generalizations, so that people who are new to a subject can quickly get a rough understanding without being drawn away from the mathematical topic at hand. At the same time, we have found that students feel a connection to the subject when it is presented in context, and that they appreciate a glimpse of the rich tapestry that lies beyond what they are learning at the moment.

Calling on students' experience

Part of engaging students is helping them learn by connecting the ideas of calculus to their experience and observations. Doing so provides them with a way to understand the material in terms of familiar facets of life, so calculus makes sense to them rather than being a random assortment of facts and formulas. Here are some examples that we use to accomplish this:

Newton's method: We begin our discussion of Newton's method by asking the question, "Suppose a model rocket is coming back down, after its flight. It's descending at 6 feet per second, and is 30 feet above the ground. How much longer before it lands?" Students always respond with the expected answer of five seconds based on their experience with motion. Pedagogically, what's important about their answer is that they've assumed a *constant* rate of change. That's Newton's method in a nutshell.

Mean Value Theorem: We begin by asking you to think about a race against a friend. The friend promises to run at a steady, constant rate from start to finish, and though you may speed up and slow down at will, by some fluke you tie. Could you have been running faster than your friend at each and every moment, or slower the whole time? No, because one of you would have won the race uncontested. If you start and finish together, there must have been some time at which you were running at the same speed. This discussion opens the door to the Mean Value Theorem.

Substitution: When we apply the substitution technique to definite integrals, we begin by determining a change in altitude in two different ways—one calculation based on time, and the other based on position. This allows us to help students view the substitution technique as a change in our point of view, from initial and terminal *times* to starting and stopping *locations*.

By drawing on students' experience with the world, we help them internalize the ideas of calculus, solidify their understanding, and develop their intuition.

Our approach to writing

On the one hand, it is important that students see mathematics done rigorously and well. On the other, we are aware that our readers have little or no experience reading technical documents. For this reason, we have chosen to write in a colloquial style during the development of ideas, so that by the time students read the formal language of a theorem, they already know what it says.

We also use color to help students follow longer or more complicated calculations, and to reinforce similarities across examples. We make common use of underbraces to provide guidance as students read equations, and often include margin notes to answer common questions without interrupting the flow of the presentation.

Promoting active reading

Some years ago, we attended a seminar in which Dr. P.K. Imbrie of Purdue University spoke about ways of engaging students, even in large classes. He suggested breaking up lectures with small "try it yourself" moments that keep students actively involved in the learning process. There are other ways, of course, but Dr. Imbrie's idea made a lot of sense to us as authors: students who are inexperienced at learning mathematics from a book often read in a passive way, as if it's a novel, but learning mathematics is an active endeavor. To promote active learning while students read, we often include a "Try It Yourself" example immediately after one whose solution is fully laid out. The answers to these Try It Yourself examples are provided in the text, and if students need to see a full solution they can access it online. These examples act as pedagogical "stepping stones," half-way between examples and exercises, allowing students to try something for themselves but with a virtual safety net.

Reading v. Spelling in mathematics

For many students, a mathematical equation is just a collection of symbols that they are expected to manipulate in order to arrive at an answer, but they see no meaning in the equation itself. So starting early on, we teach students how to interpret the equations they see, and to assess the ones they write themselves so that mathematics becomes a communication of ideas.

We consider this skill to be a kind of quantitative literacy, closely related to English literacy. When reading, we don't view each symbol (letter) separately, but look at groups of symbols to derive meaning. For example, the following sentence would be confusing if read letter by letter (try it, out loud):

> "T-h-e-n-e-x-t-t-i-m-e-I-s-l-e-e-p-,-i-t-w-o-n-'-t-b-e-o-n-m-y-c-a-l-c-u-l-u-s-b-o-o-k-.-It-d-i-d-n-'-t-h-e-l-p-l-a-s-t-t-i-m-e-a-n-d-n-o-w-I-h-a-v-e-a-c-r-i-c-k-i-n-m-y-n-e-c-k-."

Instead, we group the letters and read,

> "The next time I sleep, it won't be on my calculus book. It didn't help last time and now I have a crick in my neck."

Each grouping (word) conveys a concept, and together they communicate an idea. We help students move beyond spelling to reading mathematics by demonstrating this skill in the presentation and by including exercises that ask students to make the transition from English to mathematical notation and back.

Additionally, we often use dimensional analysis to check that calculations of physical objects and phenomena make sense. When done at an elementary level, this helps students internalize the meaning of the difference quotient, which has units such as meters per second, and understand why Riemann sums approximate the net change in, for example, position (when summands have the form rate X time). Dimensional analysis also helps students to remember formulas such as those for cylindrical shells and washers, which are used when calculating volumes of solids of revolution.

Helping students avoid common errors

We have found that students often benefit from learning about common mistakes, so we pause to discuss them at the end of each section. For example, students sometimes forget whether the recursion formula in Newton's method requires the derivative in the numerator or the denominator; and they sometimes make use of the *same letter* for the substitution as the original variable of an integral, which invariably leads to errors. So throughout the book we discuss ways to remember formulas and avoid common errors.

Exercises and projects

Since different schools require different levels of rigor, we've designed the exercise sets to include various levels of difficulty. We've also sorted them according to the particular skills that are required, and we've separated skill exercises from those that develop concepts or demonstrate relevance by way of application. At the end of each chapter you'll find a Chapter Review, and projects that have been distilled from those we've assigned to teams of students in our own classes through the years. Some focus on purely mathematical ideas, while others ask students to use their skill and understanding in context.

Helping students learn the "art" of calculus — decision making

When working with integration or infinite series, students have to decide which technique is appropriate to each problem. Knowing *when* to use the ratio test, for example, is just as important as knowing *how* to use it, and making that initial decision is itself a skill. In order to help students develop that skill, we include designated groups of exercises called Mixed Practice sets that require students to use techniques from previous sections on integration techniques and convergence tests, respectively. This helps students learn to analyze problems and decide on appropriate methods at the same time that they are learning the mechanics that each choice entails.

Proofs

Proofs lie at the heart of deeper mathematical understanding, but first-year students tend to find formal proofs unsatisfactory as answers to the question, "Why?" Indeed, from their point of view, formal proofs obfuscate facts more often than they illuminate them. With this in mind, we frequently precede proofs with informal discussions that focus on the salient ideas students are about to see, and put them in a context that allows students to intuit the relationship between quantities. This often makes the formal proofs easier for students to follow, and leaves them more satisfied with the proof and the role it plays.

Technology

Generally speaking, today's students are technologically savvy. They are familiar with chat rooms, have pages on Facebook, maintain blogs, send emails and text messages, and are typically comfortable interacting online. It's important to understand that their mathematical education, like the rest of their lives, has been affected by the presence of technology. Many of today's students begin a math problem by reaching for their graphing calculator because they've been trained to think in terms of graphs, which have always been readily accessible. Using computational technology is as normal to them as driving a car.

In principle, we believe that computational technology is a powerful tool that can help expedite the problem solving process, but only if the user knows what to look for—much in the same way that your car's maximum speed is irrelevant if you're lost. So we tend to focus on ideas and concepts, and bring in technology where we believe it's appropriate.

We also believe that numerical techniques should not be discussed and then discarded. They are important tools for engineers, scientists, and applied mathematicians, so we regularly ask students to use them to approximate quantities that are otherwise impossible for them to calculate. To facilitate students' use of these algorithms, we have provided instructions for implementing them on a generic spreadsheet program (a tool that is often used by engineers and scientists, but isn't as common in mathematics education) in several places throughout the book.

In closing

In a discussion with Dr. George Thurston, he said (we paraphrase),

> There seems to be a certain momentum that draws the teaching community away from the original motivation and understanding of a topic—away from the *why* of things, toward the what and how of it.

This was our own experience learning calculus, and based on conversations with students, it is the experience that many people still have today. In this book we strive to focus students' attention not only on *what's* true but also on *why* it's true, so they can both *do* the calculus and *understand* the calculus.

We've worked to produce a text that is interesting to students by providing some context to the subject, and by using students' intuition and life experience as a framework in which to present ideas. Simultaneously, we've focused on making the text accessible to students by supporting their reading with a healthy sprinkling of margin notes, and by couching much of the presentation in an informal style that's comfortable for many students, while preserving the rigor of the mathematics itself. We hope that after digging through this text a little, you'll find ways that it can complement your classroom teaching, and that you'll ask your students to crack the cover, and honestly try reading it.

Since this is a preliminary edition, we welcome your feedback—especially ideas about other ways to help students. You can email your comments to us at lutzer@wiley.com. We look forward to hearing from you. Thanks!

IN-TEXT FEATURES

- **Chapter Openers.** Each chapter sets the stage with a unique, real-world example from a varied assortment of topics.

- **Try It Yourself.** Many of the examples that are worked out in the text are followed immediately by a similar example that students are asked to work out for themselves. An *answer* is provided in the text, and the full solution is provided at the book's companion website and in *WileyPLUS*.

- **You Should Know/You Should be Able to.** Appearing at the end of each section a boxed lists that serve as checklists for students to make sure they understand key concepts and can demonstrate key skills.

- **Exercises.** Each section includes relevant skill-development exercises, and exercises devoted to applications from the sciences, business, economics, engineering, physics, and mathematics.

- **Projects and Applications.** These are chapter-ending cumulative reviews that require students to apply the concepts and demonstrate the skills they learned in the chapter.

DIGITAL RESOURCES

Wiley PLUS is an innovative, research-based, online environment for effective teaching and learning.

What do students receive with *WileyPLUS*?

A Research-based Design. *WileyPLUS* provides an online environment that integrates relevant resources, including the entire digital textbook, in an easy-to-navigate framework that helps students study more effectively.

- WileyPLUS adds structure by organizing textbook content into smaller, more manageable "chunks".

- Related media, examples, and sample practice items reinforce the learning objectives.

- Innovative features such as calendars, visual progress tracking and self-evaluation tools improve time management and strengthen areas of weakness.

One-on-one Engagement. With *WileyPLUS* Lutzer & Goodwill/ Calculus students receive 24/7 access to resources that promote positive learning outcomes. Students engage with related examples (in various media) and sample practice items, including:

- **Algorithmically generated exercises,** with randomized numeric values and an answer-entry palette for symbolic notation, are provided on-line through *WileyPLUS*. Students can click on "help" buttons for hints, link to the relevant section of the text, show their work or query their instructor using a white board, or see a step-by-step solution (depending on instructor-selecting settings).

- **Applets** are available on-line through *WileyPLUS*. We have found that there are specific times in the teaching of calculus when students benefit from exploring a given topic with a Java applet, Maple worksheet, or spreadsheet.

Measurable Outcomes. Throughout each study session, students can assess their progress and gain immediate feedback. *WileyPLUS* provides precise reporting of strengths and weaknesses, as well as individualized quizzes, so that students are confident they are spending their time on the right things. With *WileyPLUS*, students always know the exact outcome of their efforts.

What do instructors receive with *WileyPLUS* ?

WileyPLUS provides reliable, customizable resources that reinforce course goals inside and outside of the classroom as well as visibility into individual student progress. Pre-created materials and activities help instructors optimize their time:

Customizable Course Plan: *WileyPLUS* comes with a pre-created Course Plan designed by a subject matter expert uniquely for this course. Simple drag-and-drop tools make it easy to assign the course plan as-is or modify it to reflect your course syllabus.

Pre-created Activity Types Include:

- ◇ Assignments with algorithmic exercises

- ◇ Readings and resources

- ◇ Print Tests

Course Materials and Assessment Content:

- ◇ Lecture Notes PowerPoint Slides

- ◇ Classroom Response System (Clicker) Questions

- ◇ Instructor's Solutions Manual

- ◇ Gradable Reading Assignment Questions (embedded with online text)

- ◇ Assignments: auto-graded end-of-chapter problems coded algorithmically with randomized numeric values. Instructor sets availability of student hints, links to text, whiteboard/"Show Work" feature, and worked-out solution

- ◇ Test Bank

Gradebook: *WileyPLUS* provides instant access to reports on trends in class performance, student use of course materials and progress towards learning objectives, helping inform decisions and drive classroom discussions.

WileyPLUS. Learn More.www.wileyplus.com. Powered by proven technology and built on a foundation of cognitive research, *WileyPLUS* has enriched the education of millions of students, in over 20 countries around the world.

ACKNOWLEDGEMENTS

In addition to those listed below who patiently waded through draft chapters and provided useful feedback, we wish to thank the staff at John Wiley & Sons who expertly guided us through the process from prospectus to publication, especially David Dietz, Mary O'Sullivan, and Leslie Lahr. We also wish to express our gratitude to Joseph DeLorenzo, Ray Hodges, Jeffery McBeth, David Ross, and George Thurston for their support and contributions to this effort.

Darry Andrews, *Ohio State University*

Scott Barnett, *Henry Ford Community College*

William Basener, *Rochester Institute of Technology*

Phil Bowers, *Florida State University*

David Bradley, *University of Maine*

Mark Bridger, *Northeastern University*

Ray Cannon, *Baylor University*

Kristen Chatas, *Washtenaw Community College*

Min Chen, *Purdue University*

David Collingwood, *University of Washington*

Randy Combs, *West Texas A&M University*

Ted Dobson, *Mississippi State University*

Michael Dorff, *Brigham Young University*

Gregory Dresden, *Washington & Lee University*

David Finn, *Rose–Hulman Institute of Technology*

Ralph Grimaldi, *Rose–Hulman Institute of Technology*

Huiqiang Jiang, *University of Pittsburgh*

Joe Lackey, *New Mexico State University*

Hong-Jian Lai, *West Virginia University*

Glenn Ledder, *University of Nebraska–Lincoln*

Sergey Lvin, *University of Maine*

Vania Masconi, *Ball State University*

Philip McCartney, *Northern Kentucky University*

Jie Miao, *Arkansas State University*

Mona Mocanasu, *Metropolitan State College of Denver*

Aaron Montgomery, *Central Washington University*

Karen Mortensen, *University of Illinois at Urbana-Champaign*

John Orr, *University of Nebraska–Lincoln*

Emma Previato, *Boston University*

Dennis Reissig, *Suffolk County Community College*

Peter Roper, *University of Utah*

Amit Savkar, *University of Connecticut*

Chris Storm, *Adelphi University*

Barrett Walls, *Georgia Perimeter College*

Scott Wilde, *Baylor University*

Brock Williams, *Texas Tech University*

Robin Wilson, *California State Polytechnic University, Pomona*

Tzu-Yi Alan Yang, *Columbus State Community College*

Last, but certainly not least, we take a moment to thank our families. They have sacrificed a great deal of time with us so that we could pursue this effort to help others, and have done so lovingly without complaint. Without their sacrifice and support this project could never have come to fruition.

--Carl Lutzer and Tim Goodwill
  September, 2010

PHOTO CREDITS

Icons

Icon   page 5 © Evgeny Rannev/iStockphoto; Icon page 18 © Purestock / SuperStock; Icon page 43 ©kemie/iStockphoto.

Chapter 1

UN 1.1 page 4 Photo Researchers Inc.; Figure 2.2 page 12, Carl Lutzer; UN 1.2 page 5, © Tom England/iStockphoto; Figure 5.6 page 34, Carl Lutzer; UN 1.3 page 35, © music Alan King/Alamy; Figure 5.16 page 40, Carl Lutzer; UN 1.4 page 49, Nick Wass/©AP/Wide World Photos; UN 1.6 page 83, © Janne Ahvo/iStockphoto; UN 1.7 page 83, © Frank van den Bergh/iStockphoto.

Chapter 2

UN 2.1 page 89,© Michiel de Boer/iStockphoto; UN 2.2 page 107, © cloki/ iStockphoto; UN 2.3 page 137, "Phil Walter/Getty Images, Inc.".

Chapter 3

UN page 160, © Jan Kratochvila/iStockphoto; UN 3.2 page 167, © Greg Nicholas/ iStockphoto; UN 3.3 page 169, © Neil Sullivan/iStockphoto; UN 3.4 page 171, © Kris Hanke/iStockphoto; UN 6.1 page 202, © Royalty-Free/CORBIS; UN 6.2 page 202, moodboard RF/Photolibrary; UN 3.5 page 211, © Pogorelov Vladimir/iStockphoto; UN 3.6 page 211, © Shannon Stent/iStockphoto; Figure 7.1 page 213, Copyright © Yoav Levy /Phototake; Figure 7.5 page 219,"VGL/amanaimagesRF /Getty Images, Inc."; Figure 7.7 page 219, © Glyn Jones/©Corbis; Figure 7.7 page 219, © Frank Leung/iStockphoto; Figure 7.8 page 219, "VGL/amanaimagesRF /Getty Images, Inc."; UN 3.7 page 226, © Cristian Andrei Matei/iStockphoto; UN 3.8 page 233, © Mike Norton/iStockphoto.

Chapter 4

Figure 1.1 page 236, Bill Aron /PhotoEdit; Figure 1.3 page 238, Carl Lutzer; Figure 1.9 page 245, ©Planetpix/Alamy; UN 4.1 page 251, „David Madison / Getty Images, Inc."; Figure 2.1 page 251, ©Mooneydriver/iStockphoto; Figure 2.2 page 252, © Lawrence Manning/©Corbis; Figure 2.2 page 252, © brentmelissa/ iStockphoto; Figure 2.2 page 252, © Mark Ross/iStockphoto; Figure 2.3 page 253, ©Mooneydriver/iStockphoto; Figure 2.12 page 258, Photo by Rachel Goldstein; Figure 3.3 page 263, ©Mark Ross/iStockphoto; Figure 3.4 page 264, SMC Images /Getty Images, Inc.; Figure 3.8 page 270, ©Pogorelov Vladimir/iStockphoto; Figure 7.1 page 301, © oversnap/iStockphoto; Figure 7.1 page 301, „Joe Michl / Getty Images, Inc.";   Figure 7.1 page 301, Ralph A Clevenger/ Flirt Collection/ Photolibrary; UN 4.3 page 306, „rubberball /Getty Images, Inc."; Figure 7.12 page 307, ©peter zelei/iStockphoto; Figure 7.2 page 310, ©Mlenny Photograhy | Alexander Hafemann/iStockphoto; Figure 9.4 page 322, „SSPL via Getty Images, Inc."; Figure 9.5 page 322, ©Mary Evans Picture Library/The Image Works; Figure 7.14 page 308, Bela Szandelszky/©AP/Wide World Photos; Figure 8.2 page 315, Carl Lutzer.

Chapter 5

UN 5.1 page 325, © www.gerardbrown.co.uk/Alamy; UN 5.2 page 335, Imagestate

Chapter 1
Tools of the Trade

You learned at an early age that planets orbit the sun, and our sun orbits the center of the Milky Way because of a force called *gravity*. Many people today find that fact unremarkable, but for most of human history celestial motion was something of a mystery. It wasn't until the early 1600s that our modern understanding of the universe really began to take shape. That's when the telescope was invented, Galileo Galilei saw the moons of Jupiter, and Johannes Kepler published his finding that planets move around the sun in elliptical orbits. After that, people knew *what* happened, and turned their attention to *why*. Men like Robert Hooke, Edmund Halley, and Christopher Wren struggled with the problem until 1684, when Halley paid a visit to Isaac Newton in Cambridge.

Halley asked if a force that obeyed an *inverse square law* would result in the orbits that Kepler had described. Newton's ready response was, *yes*. In fact, he'd already made the calculations ... though he had misplaced them (see [29]). Newton promised to find or redevelop the calculations, and three years later, he published his famous *Philosophiae Naturalis Principia Mathematica* in which he introduced the calculus. Since that time, it has become not only a tool for making calculations, but a way of understanding and talking about the world we live in.

> It's okay if the term *inverse square law* is unfamiliar to you. We'll talk about it in more detail in a moment.

This chapter is intended as a review of the basic functions that you will see in calculus—tools of the trade, as it were. The point of such a review is not to rehash the mechanics of how these functions work, but rather to help you begin thinking about mathematics in a new way.

- Whereas algebra and precalculus traditionally focus on the *value* of a function, calculus is fundamentally about how functions *change*, so our discussion will often come back to the topic of change.

- We'll emphasize equations as a communication of ideas rather than as a collection of symbols to be manipulated. Part of this emphasis will involve a technique called *dimensional analysis*, which provides deeper understanding of functions and change based on the units involved in a problem.

- While there are a lot of mathematical ideas that are intriguing, or beautiful, or just plain fun, humans have been driven by the need to describe and understand the world we live in. So our discussion will often include situations in which particular types of functions or important ideas arise.

As you read, you'll see that examples in the presentation are often followed by "try it yourself" examples, to which an answer is provided here but not the full solution. Our hope is that you'll try these examples, and *learn by doing* as you go. If you find that you need help, you can find a full solution on-line.

1.1 Power Functions

In the chapter opener, Halley asked Newton about a force that obeyed an inverse square law. What he had in mind was a force, F, that changes with distance like

$$F = \frac{\text{(some constant)}}{r^2}, \tag{1.1}$$

where r is the radial distance (i.e., straight-line distance) between the sun and a planet. This equation tells us that F gets smaller when r increases. Alternatively, you could say that r gets smaller as F increases. The important thing to understand is that the quantities F and r change *together*.

As often as possible, we design notation to help us remember what variables mean. In this case, F reminds us of the word *force*, and r reminds of the word *radial*.

In equation (1.1) we wrote the numerator as "some constant," so that we could focus on the relationship between F and r. More specifically, the relationship between distance and gravitational force is

$$F = \frac{Gm_1m_2}{r^2}, \tag{1.2}$$

where

> G is called the *gravitational constant*. Its value doesn't depend on the objects, the distance between them, or anything else. It is the same from situation to situation, from moment to moment.
>
> m_1 and m_2 are the masses of the objects involved.

Notice how subscripts are used to distinguish between the masses. This is common practice when you have more than one occurence of the same *kind* of quantity.

Although F and r change together, we typically think of gravitational attraction as depending on distance. When we think of the relationship in this way, we write F as a function of r:

$$F(r) = \frac{Gm_1m_2}{r^2}.$$

The number r is called the **argument** of the function F. We could also describe this function by writing

$$\underset{\substack{\smile \\ \text{The argument} \\ \text{of the function}}}{\overset{\substack{\frown \\ \text{The name of} \\ \text{the function}}}{F} \quad : \quad r \quad \longmapsto \quad \underset{\substack{\smile \\ \text{The output of} \\ \text{the function}}}{Gm_1m_2r^{-2}}.} \tag{1.3}$$

The "mapping arrow" in this notation ($\longmapsto$) emphasizes the idea that a function takes an input, *does something with it,* and arrives at an output. By way of analogy, a function is like a verb that connects two nouns (the input and the output)—it's an *action*.

As you might guess from the name of this section, F is an example of a **power function**, which takes an input and raises it to a specified (nonzero) power. You can multiply the resulting value by a constant, and it's still called a power function.

The specified power could be 1, in which case we usually don't write it.

Power functions and arrow notation

Example 1.1. The circumference and area of a circle whose radius is r are given by the power functions

These formulas are derived in Appendix A.

$$A(r) = \pi r^2 \quad \text{and} \quad C(r) = 2\pi r.$$

Write these functions with the "arrow" notation that you see in (1.3).

Solution: $A : r \longmapsto \pi r^2$ and $C : r \longmapsto 2\pi r$. ∎

Power functions are a great place to begin studying calculus because they change in a straightforward way. We'll begin with the simplest kind of exponents: integers.

▷ Parity of an integer exponent

The word **parity** refers to whether an integer is even or odd. When a power function has an even exponent, the function value changes in the same way whether we move left or right from $x = 0$ (see the graph of $f_2(x) = x^2$ in Figure 1.1). By contrast, power functions with odd exponents change in *opposite* ways when we move left or right from $x = 0$ (see the graph of $f_3(x) = x^3$ in Figure 1.1).

> Since f_1 and f_2 are the same kind of thing—power functions with positive, integer exponents—we're using the same letter to denote them, and distinguishing between them with subscripts, much as we did with masses a moment ago. Note that the subscripts have been chosen according to the exponent.

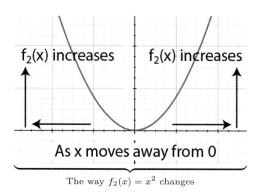
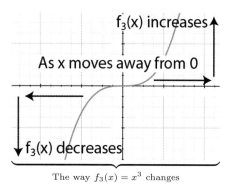

The way $f_2(x) = x^2$ changes   The way $f_3(x) = x^3$ changes

Figure 1.1: The way that a power function changes depends on the parity of its exponent.

Note: Calculus is about *change*, so we're talking in those terms. We could also use the language of geometry, and focus on symmetry: *When the exponent of a power function is even, its graph is symmetric about the y-axis. When its exponent is odd, its graph is symmetric about the origin.*

▷ Polarity of an integer exponent

The word **polarity** refers to whether a number is positive or negative. In the case of power functions, the polarity of the exponent determines whether the magnitude (i.e. size) of the function value grows or shrinks as we move away from $x = 0$.

When the exponent is positive, the size of the function value *grows* when the size of the argument grows, as seen in Figure 1.1. However, when the exponent is negative, the size of the function value *shrinks* as the size of the argument grows. For example, consider the functions

$$g_1(x) = x^{-1} \quad \text{and} \quad g_2(x) = x^{-2}.$$

As you see in Tables 1 and 2, the numbers $g_1(x)$ and $g_2(x)$ *decrease* as x moves farther away from zero, and the function values grow in size when x heads toward zero. You can also see this happen in Figure 1.2, which depicts the graphs of g_1 and g_2.

x	$g_1(x)$	$g_2(x)$
1	1.0000	1.0000
3	0.3333	0.1111
5	0.2000	0.0400
7	0.1429	0.0204
9	0.1111	0.0123
11	0.0909	0.0083
13	0.0769	0.0059
15	0.0667	0.0044
17	0.0588	0.0035

Table 1

x	$g_1(x)$	$g_2(x)$
1	1	1
.5	2	4
.25	4	16
.125	8	64
.0625	16	256
.0312	32	1024
.0156	64	4096

Table 2

Note: When the exponent of a power function is negative and x is near zero, small changes in the argument (x) cause *large* changes in the function value (y).

▷ Magnitude

The word **magnitude** means the "size" or "extent" of something. When it comes to power functions, those whose exponents have larger magnitude tend to "do things faster." For example, in Figure 1.2 you an see that $g_2(x)$ tends to zero faster than $g_1(x)$ as $|x|$ gets large. And as $|x|$ approaches zero, $g_2(x)$ grows faster than $g_1(x)$.

> Later, in our study of *improper integrals* and *infinite series*, this simple idea will be very important.

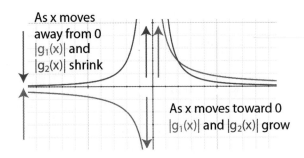

Figure 1.2: The graphs of power functions $y = g_1(x)$ and $y = g_2(x)$.

❖ Fractional Exponents

In the chapter opener, we mentioned Kepler's discovery that the planets of our solar system orbit the sun along elliptical trajectories. Another of his discoveries was that the time it takes a planet to complete one orbit around the sun, say T (measured in Earth-years), depends on the shape of its orbit according to

$$T(a) = a^{3/2}, \tag{1.4}$$

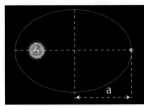

Recall, the semi-major axis of an ellipse is half of the major axis (i.e., the longer one).

where a denotes the length of the semi-major axis of the planet's orbit, measured in *astronomical units*. When a power function has a fractional exponent, $\frac{p}{q}$ (in lowest terms), parity and polarity still determine the function's behavior.

Parity: When q is odd, the parity of p affects behavior in exactly the same way it does for integer exponents (see p. 3). When q is even, as in $x^{1/2}$ or $x^{-5/6}$, negative values of x aren't in the domain of the function (see Figure 1.3), so the parity of p is irrelevant.

Polarity: The polarity of the exponent affects behavior in exactly the same way it does for integer exponents (see p. 3).

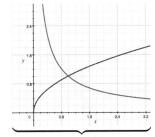
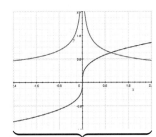

The graphs of $y = x^{1/2}$ and $y = x^{-5/6}$ The graphs of $y = x^{1/3}$ and $y = x^{-6/17}$

Figure 1.3: Graphs of various power functions with rational exponents.

How do fractional exponents affect a power function?

Example 1.2. Discuss the behavior of $h(x) = x^{2/3}$.

Solution: The exponent in $h(x)$ is positive, so $|h(x)|$ will grow as $|x|$ does, and will head toward zero as $|x|$ does. The denominator of the exponent is odd, so the domain of h includes $x < 0$. Since the numerator is even, the value of $h(x)$ will change in the same way whether we move right or left of zero. ∎

Try It Yourself: How do fractional exponents affect a power function?

Example 1.3. Without using a graphing utility, discuss the behavior of $j(x) = -3x^{9/5}$, and $k(x) = x^{-13/8}$.

Full Solution On-line

Answer: Discussion of $j(x)$ and $k(x)$ is found on-line.   ∎

✦ Introduction to Dimensional Analysis

An equation that says "kilograms = meters" is nonsensical, so it's good practice to check that both sides of an equation describe the same kind of quantity (mass, length, force, etc.) before you begin working with it. The act of checking is called **dimensional analysis,** and is done by accounting for the units that are associated with each term in the equation. Comparing units can also be a handy way of remembering formulas.

Dimensional analysis

Example 1.4. Use dimensional analysis to determine which of πr^2 and $2\pi r$ describes the area, and which the circumference of a circle.

Solution: Suppose we measure length in meters (m). Then

$$area \text{ has units of square meters (m}^2\text{), and}$$
$$\underbrace{circumference}_{\text{This is a length}} \text{ has units of meters (m).}$$

The radius, which is a length, will also have units of meters (m). Now let's look at the formulas:

Units can help you reason your way to/through a formula when you get mixed up.

$$\underbrace{\overset{\uparrow}{\pi r^2}}_{\substack{\text{Two factors of } r, \text{ each with} \\ \text{units of meters. This formula} \\ \text{leaves us with units of m}^2.}} \qquad \underbrace{\overset{\uparrow}{2\pi r}}_{\substack{\text{One factor of } r. \text{ This formula} \\ \text{leaves us with units of m.}}}$$

Based on the units involved, we know that $A \neq 2\pi r$ because the left-hand side has units of m^2 but the right-hand side has units of m. So it must be that $A = \pi r^2$ and $C = 2\pi r$.   ∎

Try It Yourself: Dimensional analysis

Example 1.5. Use dimensional analysis to determine which of $\frac{4}{3}\pi r^3$ and $4\pi r^2$ describes the surface area, and which the volume of a sphere.

Full Solution On-line

Answer: See solution on-line.   ∎

When a factor doesn't change the *kind* of thing that's being described, such as the number π in $A = \pi r^2$ and $C = 2\pi r$, we say it's **dimensionless.** Not all constant factors are dimensionless, as seen in the next example.

Constants with units

Example 1.6. Suppose we measure force in newtons, denoted by N. When mass is measured in kilograms (kg), and distance in meters (m), what are the units associated with the *gravitational constant* in equation (1.2)?

When we use these units, the numerical value of G is 6.673×10^{-11}. If we measure mass and distance differently, the units and value of G will change accordingly (not because gravity has changed, but because we've changed the way we're measuring its effect).

Solution: The "=" in equation (1.2) means more than just numerical equality. From a physical point of view, it means that they are the *same kind of thing.* The left-hand side is a force, so the right-hand side must be also, and must have units of force. If we write the units of G as $\square$ (meaning "temporarily unknown"), equation (1.2) has the form

$$F = G\frac{m_1 m_2}{r^2} \quad \rightarrow \quad \text{force } = \square\frac{(\text{mass})(\text{mass})}{(\text{distance})^2} \quad \rightarrow \quad \text{N} = \square\frac{(\text{kg})^2}{(\text{m})^2}.$$

By "solving" for $\square$, we find that G must have units of $(\text{N})(\text{m}^2)/(\text{kg})^2$. Note that without the units of G, equation (1.2) would say that force = $(\text{mass/distance})^2$, which is nonsensical. The factor of G on the right-hand side of the equation changes the *kind* of thing that you see on the right-hand side.   ∎

> You might feel uneasy about this kind of "algebraic" work with units, but it's the same thing you do intuitively when you say that you travel 10 kilometers by driving at 5 km/hr for 2 hours (rate × time = distance).

❖ Common Difficulties with Power Functions

One of the most common mistakes, in one form or another, is distributing exponents over sums and differences. For example, you know that

$$(a+b)^2 \neq a^2 + b^2 \quad \text{since} \quad (a+b)^2 = (a+b)(a+b) = a^2 + 2ab + b^2.$$

The mistake of distributing powers over sums is more common when the power is negative or fractional. Always keep in mind that

Exponents don't distribute over sums

$$\frac{1}{a+b} \neq \frac{1}{a} + \frac{1}{b} \qquad \text{because} \qquad (a+b)^{-1} \neq a^{-1} + b^{-1},$$

$$\sqrt{a+b} \neq \sqrt{a} + \sqrt{b} \qquad \text{because} \qquad (a+b)^{1/2} \neq a^{1/2} + b^{1/2}.$$

Exponents don't distribute over sums

You should know

- the terms *argument, parity,* and *polarity;*

- the units on the left- and right-hand sides of an equation must be the same (e.g., an equation that says *meters = kilograms* can't possibly be right).

> This second bullet is particularly important because it's at the heart of a skill that many beginners lack: the ability to tell when something is wrong.

You should be able to

- explain how the parity of an integer exponent is related to the symmetry of a power function;

- explain how the polarity of an exponent is related to the way a power function changes;

- determine the behavior of a power function;

- discern the parity and polarity of a power function's exponent based on a graph;

❖ 1.1 Skill Exercises

Use your understanding of parity, polarity, rational, and integer exponents to match the graphs in Figure 1.4 with the functions in #1–4.

1. $f(x) = x^{1/6}$　　　2. $f(x) = x^{-2}$　　　3. $f(x) = x^{-13/3}$　　　4. $f(x) = x^{3/7}$

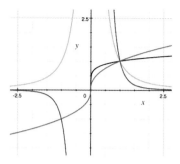

Figure 1.4: For #1–4.

In each of 5–10 use the parity, polarity, and magnitude that you see in the exponent to compare and contrast the way that $f(x)$ and $g(x)$ change when x moves away from zero. Write your answer in complete sentences, and avoid using pronouns.

5. $f(x) = x^{4/3}$, $g(x) = x^{-4/3}$　　　8. $f(x) = x^{7/8}$, $g(x) = x^{-7/8}$

6. $f(x) = x^{5/3}$, $g(x) = x^{-5/3}$　　　9. $f(x) = x^{6/11}$, $g(x) = x^{7/11}$

7. $f(x) = x^{5/4}$, $g(x) = x^{-5/4}$　　　10. $f(x) = x^{8/13}$, $g(x) = x^{7/13}$

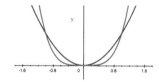

Figure 1.5: For #11.

11. Which graph in Figure 1.5 is $y = t^2$ and which is $y = t^4$? How can you tell? Answer in complete sentences (avoid using pronouns).

12. Which graph in Figure 1.6 is $y = t^3$ and which is $y = t^5$? How can you tell? Answer in complete sentences (avoid using pronouns).

13. If an object is $h(t)$ meters above sea level after t seconds:

    (a) Write the equation that says, "the object was 23.4 meters above sea level after 7 seconds."

    (b) Write the inequality that says, "the object was below sea level after t seconds.

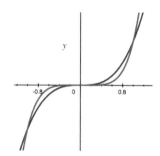

Figure 1.6: For #12.

14. In business, the *revenue function* relates the number of items sold (x) and the amount of money earned from sales (R). Suppose $R(x) = 0.25x$. The number $R(x)$ has units of dollars.

    (a) What are the units of the variable x?

    (b) Based on part (a), what units must the number 0.25 have?

    (c) Interpret those units using plain, simple language. What does the number 0.25 mean to us in this context?

15. When we model the behavior of something that swings back and forth under the force of gravity (e.g., the arm of a pendulum), we find that $T = 2\pi\sqrt{L/g}$, where L denotes the length of the pendulum, T denotes the period of its motion, and g is the acceleration due to gravity (with units of $\frac{m}{sec^2}$). Perform a dimensional analysis to verify that the equation makes sense.

16. A company's customer service center processes $q(t)$ calls per month (in hundreds) t months after the national debut of a product line.

    (a) In a complete sentence, interpret the equation $q(4) = 2.5$ in terms of months and the number of calls per month.

    (b) Write an equation that says, "The number of calls per month 3 months after the debut is larger than the number of calls-per-month 7 months after the debut."

    (c) Write an equation that says, "The number of calls per month t_1 months after the debut is larger than the number of calls-per-month t_2 months after the debut."

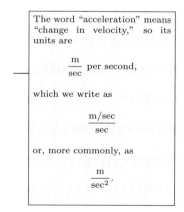

The word "acceleration" means "change in velocity," so its units are

$$\frac{m}{sec} \text{ per second,}$$

which we write as

$$\frac{m/sec}{sec}$$

or, more commonly, as

$$\frac{m}{sec^2}.$$

17. A famous equation in modern physics is $m = m_0/\sqrt{1 - \left(\frac{s}{c}\right)^2}$, where m and m_0 denote mass, measured in kilograms, and c is the speed of light, measured in m/sec. What have to be the dimensions of s in order for this equation to make sense?

18. The *momentum* of an object is denoted by p, and is related to both its mass (m) and its velocity (v) by $p = mv$ (see Appendix E). What are the units of momentum?

19. Coulomb's force law says that the force between two ions, F, is related to their charges (q_1 and q_2) and the distance between them, r, according to $F = kq_1q_2/r^2$, where k is a constant. If force is measured in newtons, charge in coulombs, and distance in meters, what have to be the units of k in order for the equation to make sense?

20. Add together functions you know in order to make a function whose graph is like the one in Figure 1.7.

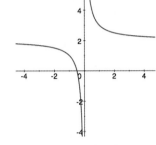

Figure 1.7: For #20.

Physics

❖ 1.1 Concept and Application Exercises

21. Equation (1.2) formulates gravitational force according to an inverse square law. The gravitational *potential* function, $U(r)$, is slightly different:

$$F(r) = \frac{Gm_1m_2}{r^2} \quad \text{and} \quad U(r) = -\frac{Gm_1m_2}{r}.$$

Answer the following without using a graphing utility:

(a) Which function values gets smaller (in magnitude) faster as $r > 0$ grows large?

(b) Which function values gets larger (in magnitude) faster as $r > 0$ shrinks toward zero?

22. A flywheel is often used to store energy in the form of rotational motion. Suppose that the disk shown in the margin is spinning, and that it makes 8π radians per second.

Mechanical Engineering

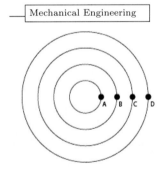

(a) Point A is 1 cm from the center. How far does the point travel in 1 second? *(Hint: what's the circumference of the circle around which it's traveling?)*

(b) Points B, C and D are each 1 cm farther from the center. Determine how far each travels in 1 second.

(c) Write an equation that relates a point's distance from the center, r, to the distance it travels in one second.

(d) How does your formula change if the wheel rotates at ω radians per second, instead of 8π?

23. When an object is rotating freely, a quantity called angular momentum, denoted by L, is conserved (i.e., is *constant*). In simple situations, the formula for angular momentum is $L = mr^2\omega$, where m is the object's mass, r is its distance to the axis of rotation, and ω is its angular velocity (in radians per second).

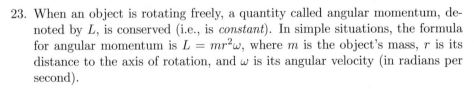
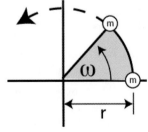

(a) What does this formula say happens to ω when the object moves farther away from the center of the rotation (assuming no change in mass)?

Figure 1.8: The formula in #23 is really the *magnitude* of the angular momentum in a special case.

(b) Suppose we decrease angular momentum by applying torque ("rotational force") to the system. If ω remains constant, what must happen?

24. The *Ideal Gas Law* describes the relationship among pressure (P), volume (V), the number of molecules in a gas (N), and the temperature (T) as $PV = Nk_BT$, where k_B is a number called *Boltzmann's constant*. When distance is measured in meters, force in newtons, and temperature in kelvins (so $T \geq 0$), Boltzmann's constant is $k_B = 1.3806503 \times 10^{-23}$. Suppose $N = 10^{23}$. Then

$$P = \frac{1.3806503\,T}{V}. \tag{1.5}$$

When temperature is held constant, equation (1.5) defines pressure as a function of volume. The graph of P is called an *isotherm*.

The prefix "iso" means "same," and the ending "therm" refers to thermal energy.

(a) Draw isotherms for two different temperatures.

(b) Label the curves as T_0 and T_1, so that $T_0 < T_1$.

(c) Pick a point on the T_1-isotherm, somewhere near the left side, and call it A. Then pick a point on the T_0-isotherm, somewhere near the right side, and call it B.

(d) Label the V-coordinates of A and B as V_1 and V_0, respectively ($V_1 < V_0$), and mark them on the V-axis.

The word *adiabatic* means that no heat flows into or out of the system, perhaps because compression (or expansion) happens too quickly, or because of insulation or isolation.

(e) During an *adiabatic compression*, the point (V, P) moves from B to A along a smooth curve called an *adiabat*. Connect A to B by such a curve. It should look a lot like a parabola that opens up, or like a slightly tilted isotherm.

25. The exact relationship between pressure and volume in an *adiabatic* change of an ideal gas (see #24) is $P = cV^{-(d+2)/d}$, where c is a constant that depends on initial relationship between pressure and volume, and d is the number of so-called "degrees of freedom" for the gas. (Near room temperature, the number d is 3 for a monatomic gas, and 5 for a diatomic gas.)

(a) For a fixed value of c, which adiabat (monatomic or diatomic) heads to zero faster for large values of V?

(b) For a fixed value of c, which adiabat (monatomic or diatomic) is steeper when V is near zero?

The phrase *same initial conditions* means "the same number of particles, same initial starting temperature, same initial pressure, and same initial volume (which we'll assume to be very small)."

(c) Let's interpret your finding from part (b) in terms of the real world: helium is a monatomic gas, and hydrogen is a diatomic gas. Suppose these two gases are held separately with the same initial conditions. According to the Ideal Gas Law, which of them will have a greater temperature increase during an adiabatic compression?

1.2   Polynomials

A **polynomial** function is a sum of power functions whose exponents are non-negative integers. The largest power is called the **degree**, and the summand with that power is called the **leading term**. For example,

$$f(t) = 4 + 2t + 10t^2 + 2t^7 \quad \text{and} \quad g(t) = -4t^3 + t$$

are polynomial functions of degree 7 and 3, respectively; and the leading terms are $2t^7$ and $-4t^3$, respectively.

✧ Linear Polynomials

Polynomial functions of degree 1 are called **linear functions** and in Chapter 3 you will see that they are very important in our study of calculus because the *rate at which they change is constant*. In the physical world, they often appear when two quantities are proportional. For example, you know that an elastic object (such as a spring) extends farther when you pull with greater force. If we denote your force by F and the object's extension beyond its natural length by x ($x < 0$ means compression), the simplest relationship in which larger force results in greater extension is

$$x = \frac{1}{k}F, \quad \text{or equivalently} \quad F = kx,$$

where k is a positive number called the **spring constant**. This is a famous and important equation called **Hooke's Law**. Instead of thinking of extension as the consequence of a given force, we can say that F is the force required to hold the spring at a given extension, x. Writing this in function notation, $F(x) = kx$.

> We'll make the phrase "rate of change" technically precise in Chapter 3. For now, the following examples and Figure 2.1 provide numerical and graphical descriptions of the meaning.

> Notice that for a given force, the extension of the object depends on k. When k is large, the spring is "stiff," and the extension is small.

Linear functions and change

Example 2.1. Discuss how the force in Hooke's Law varies as we increase the extension beyond x_0.

Solution: Let's begin by investigating the change in F that occurs when we increase x from x_0 to $x_0 + 1$. Using Hooke's Law for F, we see that

$$\Delta F \;\; = \;\; F(x_0 + 1) - F(x_0) = k(x_0 + 1) - kx_0 = k.$$

Δ is the upper-case Greek letter delta, and is often used to mean "change in"

> A subscript of 0 often indicates an initial (starting) value of some kind. Sometimes it's also used to indicate that a particular variable has been fixed for the purposes of discussion, which is what's happening here.

Similarly, the change in F that results from increasing the extension from x_0 to $x_0 + 0.12$ is

$$\Delta F = F(x_0 + 0.12) - F(x_0) = 0.12k.$$

And in general, extending the spring by an additional Δx requires F to increase by $\Delta F = k\Delta x$. Note that ΔF does not depend on x_0. This is what we mean by saying that the rate of change in the linear function F is *constant*. ∎

Try It Yourself: Linear functions and change

Example 2.2. Suppose $y(t) = 4t$. Show that raising t from t_0 to $t_0 + \Delta t$ increases y by $4\Delta t$, regardless of t_0.

Answer: See on-line solution. ∎

When there is some kind of initial or resting value, it's often true that the graph of a linear function no longer passes through the origin. For example, as the heart

Full Solution On-line

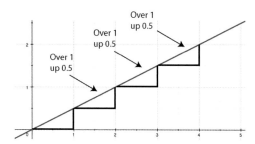

Figure 2.1: When a function is linear, increasing its argument by 1 always results in the same change in value, no matter where we start.

fills with blood its volume changes in response to internal pressure. If we denote the internal pressure by P and the heart's volume by V, and model the elastic properties of the heart wall with Hooke's Law, we see

$$V = \underbrace{V_0}_{} + \underbrace{\frac{P}{k}}_{} \quad \text{where } k > 0 \text{ is some constant.}$$

This is the initial volume, when   This is the change in volume,
there is no excess internal pressure.   due to internal pressure.

Linear functions and change

Example 2.3. Discuss how the heart's volume varies as the internal pressure increases beyond P_0.

Solution: Since we're thinking of volume as a function of internal pressure, let's write $V(P) = V_0 + P/k$. Then by increasing the pressure from P_0 to $P_0 + 1$, the heart's volume increases by

$$\Delta V = V(P_0 + 1) - V(P_0) = \left(V_0 + \frac{P_0 + 1}{k} \right) - \left(V_0 + \frac{P_0}{k} \right) = \frac{1}{k}.$$

Moreover, if we increase the internal pressure by some fraction of 1, the heart's volume will change by exactly that fraction of $1/k$. For example, the change in V that occurs when pressure is increased from P_0 to $P_0 + 0.27$ is

$$\Delta V = V(P_0 + 0.27) - V(P_0) = 0.27 \left(\frac{1}{k} \right).$$

Note that ΔV does not depend on P_0. The current pressure has nothing to do with the response of V to a change in pressure. This is what we mean by saying that the rate of change in the linear function V is constant.   ∎

Try It Yourself: Linear functions and change

Example 2.4. Suppose $x(t) = 3 - 14t$. Show that x changes by $-14\Delta t$ when t rises from t_0 to $t_0 + \Delta t$, regardless of t_0.

Answer: See on-line solution.   ∎

Full Solution
On-line

❖ Nonlinear Polynomials

Polynomial functions whose degree is larger than 1 are said to be **nonlinear**. They occur naturally in a variety of applications, from micro-mechanical systems to human biology to imaging science.

Nonlinear polynomials and imaging

Example 2.5. In a black-and-white photo each picture element (pixel) has an associated intensity, $x \in [0, 1]$, where $x = 0$ means black and $x = 1$ means white. Suppose we change the intensity at each pixel from x to $f(x) = 3x - 5x^2 + 4x^3 - x^4$. Graph the function f and explain how its application changes the original image.

> There would be no change at all if we had used $f(x) = x$ instead.

Solution: The graph of f is shown in Figure 2.2 with the line $y = x$. Note that $f(x) > x$ except at $x = 0$ and $x = 1$, so almost all the pixels are becoming lighter. Further, you can see that $f(x) \approx x$ when $x > 0.85$ but that $f(x)$ is substantially larger than x when $x \in (0.1, 0.7)$, so the effect on pixels with dark and mid-range intensities is more pronounced than it is on lighter pixels. ∎

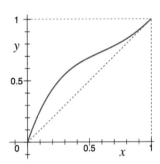

Figure 2.2: (left) The graphs of $y = x$ and $y = f(x)$ for Example 2.5; (middle) original image; (right) image filtered by the nonlinear polynomial $f(x)$.

When working with a polynomial function, it's often helpful to know that its qualitative behavior is determined by its leading term when the argument is large. This fact is discussed further in the next example.

Large-scale behavior of polynomials

Example 2.6. Discuss $f(t) = 4 + 2t + 10t^2 + 2t^7$ when $|t|$ is large.

Solution: The graph of f is shown in Figure 2.3, and you can see that its behavior is similar to $y = 2t^7$ when $|t|$ is large. To understand why this happens, think in terms of percentages. Suppose you and a friend want to buy a \$20 pizza, but he only has \$8. If you chip in \$12, you've contributed

$$\frac{\left(\text{your contribution}\right)}{\left(\text{total value}\right)} = \frac{12}{12 + 8} = 0.6 \quad \text{(i.e., 60\%) of the money.}$$

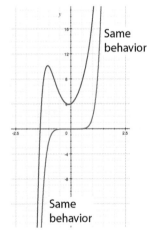

Figure 2.3: The graphs of $y = f(t)$ and $y = 2t^7$.

Similarly, we can determine what percentage of $f(t)$ comes from the leading term by calculating

$$\frac{\left(\text{leading term}\right)}{\left(\text{total value}\right)} = \frac{2t^7}{4 + 2t + 10t^2 + 2t^7}. \tag{2.1}$$

It's easier to interpret (2.1) if we factor $2t^7$ out of the denominator:

$$\frac{\left(\text{leading term}\right)}{\left(\text{total value}\right)} = \frac{2t^7}{2t^7 \left(\frac{4}{2t^7} + \frac{2t}{2t^7} + \frac{10t^2}{2t^7} + 1\right)} = \frac{1}{\frac{2}{t^7} + \frac{1}{t^6} + \frac{5}{t^5} + 1}.$$

Now when t is very large, the fraction of $f(t)$ that comes from $2t^7$ is

$$\frac{\left(\text{leading term}\right)}{\left(\text{total value}\right)} = \frac{1}{\underbrace{\frac{2}{t^7} + \frac{1}{t^6} + \frac{5}{t^5}}_{} + 1} \approx 1 \quad \text{(i.e., 100\%)},$$

These fractions are extremely small when t is extremely big.

which is why $2t^7$ dominates the behavior of $f(t)$ when $|t|$ is large.   ■

If we want to discuss the way a polynomial function changes when its argument is small, a factored form is extremely helpful.

Behavior of polynomials near their roots

Example 2.7. Suppose $f(t) = 2(t-1)^3(t-2)(t-3)^2$. Discuss its behavior (a) when $|t|$ is large, and (b) when t is near a root of f.

Solution: (a) If you were to multiply the factors together you'd see a 6th-degree polynomial whose leading term is $2t^6$. Therefore, $f(t)$ will behave like $2t^6$ when $|t|$ is large (see Figure 2.4).

(b) Let's begin by discussing what happens when $t \approx 1$. We know that $f(1) = 0$ because of the $(t-1)^3$ factor. The key to understanding the behavior of f *near* time $t = 1$ is recognizing that the other factors play no role in forcing the function value toward zero. Specifically, $(t-2) \approx -1$ and $(t-3)^2 \approx 4$ when t is near 1, so $f(t) \approx -8(t-1)^3$. Because $f(t)$ behaves like a cubic near $t = 1$, its graph will "snake" through the t-axis there, just like the graph of a cubic function. We can perform the same kind of qualitative analysis near the other roots ...

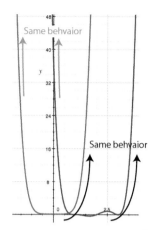

Figure 2.4: The curves $y = 2t^6$ and $y = f(t)$.

t	$f(t) = 2(t-1)^3(t-2)(t-3)^2$		Behavior
≈ 1	$\approx 2(t-1)^3(-1)(4)$	$= -8(t-1)^3$	cubic
≈ 2	$\approx 2(1)(t-2)(1)$	$= 2(t-2)$	linear
≈ 3	$\approx 2(8)(1)(t-3)^2$	$= 16(t-3)^2$	quadratic

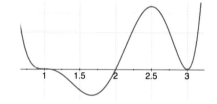

Figure 2.5: Zooming in on the graph of $f(t)$ in Example 2.7.

Since $f(t)$ is like a first-degree power function near $t = 2$, the graph of f will "crash" through the t-axis there, just as a line would. And since $f(t)$ is like a quadratic near $t = 3$, the graph of f will rebound off of the t-axis just like a parabola at its vertex.   ■

Try It Yourself: Polynomials in factored form

Example 2.8. Use the technique of Example 2.7 to analyze the behavior of $g(t) = t - 4t^3$, both when $|t|$ is large, and near its roots.

Answer: See on-line solution.   ■

Full Solution On-line

You should know

- the terms *polynomial, degree,* and *leading term;*
- that linear functions have a constant rate of change.

You should be able to

- discuss a polynomial's behavior when its argument is large;
- calculate and discuss change in linear functions.

❖ 1.2 Skill Exercises

Use your understanding of parity to determine the behavior of the polynomials in #1–4 when $|t|$ becomes *very large*. Explain in complete sentences and sketch a rough graph of each function on a large scale.

1. $f(t) = 3t^5 - 7t^2$

2. $g(t) = \frac{3}{5}t^{11} + 3t^8 - \frac{6}{5}t^7$

3. $k(t) = -8t^6 + 9t^3$

4. $h(t) = 4t^{10} - 13t^5 - 300t + 1$

In each of #5–10, sketch a rough graph of the function based on its behavior near its roots, the sign of its leading term, and the degree of the polynomial (i.e., once the terms are multiplied together).

5. $f(t) = (t-1)(t-2)(t+3)$

6. $f(t) = (t-1)(t+2)(3-t)$

7. $f(t) = (t-1)^2(2.5-t)(3.5-t)$

8. $f(t) = (t-1.5)(2-t)(3.5-t)^2$

9. $f(t) = (t+4)^3(1.5-t)(3.5-t)^2$

10. $f(t) = (t+4)(2+t)^3(1-t)^2$

11. The technique used in Example 2.7 can be used in some nontrivial cases.

   (a) Use the technique in example Example 2.7 to graph the function $f(t) = 4t^3 + 19t^2 - 5t$. (*Hint: begin by factoring this formula.*)

   (b) Based on part (a), graph the function $f(t) = 4t^3 + 19t^2 - 5t + 1$.

Use the technique of #11 to graph the functions in #12–14.

12. $f(t) = 10t^3 - 22t^2 + 4t + 1$

13. $f(t) = t^3 - 12t^2 + 36t - 2$

14. $f(t) = 4t^4 - 7t^3 - 2t^2 + 3$

Determine/design polynomial functions whose graphs could be the ones shown in #15–18.

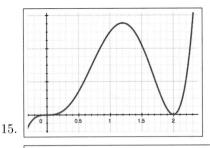

15.

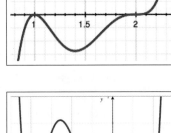

17.

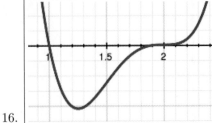

16.

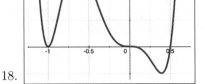

18.

19. Scratch out one branch of the graph in Figure 2.6, and connect the remaining two so that the graph could be the graph of a polynomial with odd degree.

20. Scratch out one branch of the graph in Figure 2.6, and connect the remaining two so that the graph could be the graph of a polynomial with even degree.

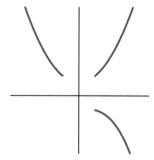

Figure 2.6: For #19–20.

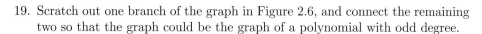

❖ 1.2 Concept and Application Exercises

21. In business, a *demand equation* relates the number of items that can be sold (x) to the item's sale price (p), and we use it to write p as a function of x. Suppose $p(x) = -0.15x + 75$.

    (a) Graph $p(x)$. Determine the values of the y- and x-intercepts.

    (b) The variable x has units of "items sold," and $p(x)$ has units of dollars. What are the units of the number -0.15?

    (c) Where do we see the number -0.15 in the graph of p?

    (d) What is the "real-world" meaning of the number 75?

    (e) What is the "real-world" meaning of the x-intercept?

    (f) What's the "real-world" domain of p?

22. In business, the *revenue function* relates the number of items sold, x, to the amount of money earned from sales, R. Suppose $R(x) = (\# \; sold)(price) = (x)(-1.35x + 125) = -1.35x^2 + 125x$. | Business |

    (a) The variable x has dimensions of "units sold." What are the dimensions associated with 125?

    (b) What does the number 125 mean to us in this context?

23. Suppose y denotes the height of a moving object (measured in feet), and t denotes time (measured in seconds). If these two variables are related according to $y = 3t + 7 \ldots$ | Kinetics |

    (a) Sketch the graph of the line $y = 3t + 7$.

    (b) Given what you know about y (that it is measured in feet), what are the units associated with the number 7? What are the units associated with the number 3?

    (c) Calculate the height at time, $t = 1$, and at time $t = 2$.

    (d) Explain the difference between the values found in part (c) both graphically (based on your sketch) and physically (based on the units).

24. When neglecting air resistance, the height of an object that's been tossed into the air (denoted by H) is governed by the equation $H = at^2 + v_0 t + h_0$. The | Kinetics |
left-hand side represents height, so it has units of meters. In order for the equation to make sense, the right-hand side must also have units of meters. If t is measured in seconds, what units have to be associated with the constants h_0, v_0 and a in order for the formula to make sense?

25. An object's height above ground is described by $s(t) = -4.9t^2 + 40t + 18$ when | Kinetics |
it is propelled upward with an initial velocity of 40 meters/second from an initial height of 18 meters. What's the domain of s? (Some values of t are after impact!)

26. The Ideal Gas Law (p. 9, #24) is sometimes written as $P = 8.315T\rho$, where T is the temperature of the gas (in kelvins), and ρ is the average number of mols of gas per cubic meter. It's an inconvenient fact that many real gases do not obey the Ideal Gas Law, but we can make our gas model more flexible by extending it to read

$$P = 8.315T(\rho + B\rho^2 + C\rho^3 + \cdots), \qquad (2.2)$$

where B and C depend on temperature, but not on density. Equation (2.2) is called the *virial expansion* of the Ideal Gas Law, the number B is called the second *virial coefficient,* and C is called the third (the first virial coefficient is 1). When ρ is very small, ρ^2 is minuscule and ρ^3 is negligible, so let's approximate equation (2.2) by

$$P = 8.315T(\rho + B\rho^2). \qquad (2.3)$$

(a) What have to be the units of B in order for (2.3) to make sense?

(b) Graph P as a function of ρ when $T = 200$ K. What does your graph tell you about the pressure when $0 < \rho < 1/35$?

(c) A negative pressure means that the gas is attracted to itself, so the volume decreases. Explain what happens to the density as the volume decreases ($\rho = (\# \text{ mols})/V$).

(d) When ρ changes, so does the pressure P; and when P changes, so does V. The change in V affects ρ ... In short, all three change together. If ρ begins in $(0, 1/35)$, where will it end up (provided that we keep T constant)?

27. Some important problems in mechanics and electromagnetics are solved using **Legendre polynomials**, the first three of which are $P_0(x) = 1$, $P_1(x) = x$ and $P_2(x) = 0.5(3x^2 - 1)$.

(a) Show that $2P_2(x) = 3xP_1(x) - P_0(x)$.

In general, the Legendre polynomials are related to each other according to the formula

$$nP_n(x) = (2n - 1)xP_{n-1}(x) - (n - 1)P_{n-2}(x). \qquad (2.4)$$

(b) Verify that the formula in part (a) is just (2.4) when $n = 2$.

(c) Use (2.4) to find formulas for $P_4(x)$ and $P_5(x)$.

28. The **Chebyshev polynomials** are used to approximate functions because they minimize a kind of error called *Runge's phenomenon*. There are two "kinds" of Chebyshev polynomials, and like the Legendre polynomials (see #27), the polynomials in each category are related to each other according to a recurrence relation. For the Chebyshev polynomials of the *first kind,* $T_0(x) = 1$, $T_1(x) = x$, and

$$T_n(x) \quad = \quad 2xT_{n-1}(x) - T_{n-2}(x). \qquad (2.5)$$

(a) Use (2.5) to find formulas for $T_2(x)$, and $T_3(x)$.

(b) Use (2.5) to show that $T_n(1) = 1$, regardless of n.

29. Consider the interface of a gas and a liquid, much like the air-blood interface in your lungs. Gas at higher pressure drives more molecules into the liquid than gas at low pressure, so we might expect the concentration of gas molecules in

Physical Chemistry

A mol is a group of 6.0221415×10^{23} of something—in this case, gas molecules.

Here are some measured values of the second virial coefficient for nitrogen gas, N_2:

T	B
100	-100
200	-35
400	9.0

Data from [11, p.9]

Negative values of B indicate pairwise attractions between gas molecules (see [23]). You see this reflected in the pressure calculations of this exercise.

The letter "g" in *Legendre* is soft, like the "g" in the English word *massage*. Dictionaries often denote this sound by the letters ZH. Using this convention, the name *Legendre* is pronounced "Luh-ZH-ON-drah."

An equation that defines the "next" of something by the "previous" one is called a **recurrence relation**. We'll talk about them more in Chapter 8.

The name *Runge* is pronounced "RUN-guh."

the liquid, c, to be proportional to the pressure, P. Said mathematically, $c = \sigma P$, or in the notation of functions: $c(P) = \sigma P$, where σ is a positive number called the *solubility*. Show that raising gas pressure from P_0 to $P_0 + \Delta P$ increases $c(P)$ by $\sigma \Delta P$, regardless of P_0.

> The symbol σ is the lower-case Greek letter sigma, pronounced "SIG-mah."

30. In 1887 a German physicist named Heinrich Rudolf Hertz discovered that shining light on a metal surface causes electrons to be ejected from the metal, a phenomenon now called the *photoelectric effect*. Amazingly, the *kinetic energy* of an ejected electron depends not on the intensity of the light we use, but only its *frequency*, ν. Experiments tell us that kinetic energy, K, depends on frequency according to $K = h\nu - W$, where $h = 6.63 \times 10^{-34}$ is a number called Planck's constant, and W is the energy that binds the electron to its atom (which depends on the metal). Show that raising the frequency of the light from ν_0 to $\nu_0 + \Delta \nu$ increases the kinetic energy of the ejected electrons by exactly $h\Delta\nu$, regardless of ν_0.

> The symbol ν is a lower-case Greek letter pronounced "noo."

In each of #31–34 describe what happens when we apply $f(x)$ to an image, as in Example 2.5. (You might find it helpful to graph $y = f(x)$ and $y = x$ on the same axes, as in Figure 2.2.)

> Imaging

31. $f(x) = 1 - x$

32. $f(x) = 0.2 + 0.8x$

33. $f(x) = x^2$

34. $f(x) = 4(x - 0.5)^3 + 0.5$.

1.3   Rational Functions

Just as a rational number is a *ratio* of integers, a **rational function** is a *ratio* of polynomials. The following example shows how rational functions arise naturally when we wrap the entire real number line onto a circle without overlapping.

Rational functions (the stereographic projection)

Example 3.1. Imagine that we set a circle on the real number line at $t = 0$ (see Figure 3.1), and all of the points on the number line are attracted to the "north pole" of the circle, which is at $(0, 2)$. As points leap off the number line under the force of that attraction, they travel along straight paths toward the north pole until they collide with the circle, where they are stuck forever. If P is the point at which the number t collides with the circle during its trip toward the north pole, we say that P is the **stereographic projection** of t onto the circle. (a) Show that the Cartesian coordinates of P are rational functions of t, and (b) discuss the domain and range of each.

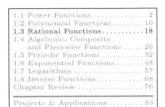

"North Pole"
(0,2)

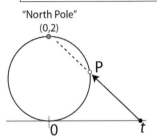

Figure 3.1: Diagram of the stereographic projection.

Solution: (a) To determine the precise location of P, consider the line that passes through $(0, 2)$ and $(t, 0)$. It has a slope of $\frac{\text{rise}}{\text{run}} = \frac{2-0}{0-t}$, so its equation is

$$y = \left(-\frac{2}{t}\right) x + 2, \tag{3.1}$$

which is the slope-intercept formula for a line, $y = mx + b$. (Notice that the slope depends upon the particular value of t. Whatever number that is, think of it as being fixed for the duration of our calculation.) On the other hand, the equation of the unit circle with center at $(0, 1)$ is

$$(x - 0)^2 + (y - 1)^2 = 1. \tag{3.2}$$

We can find the intersection of this line and circle by substituting equation (3.1) into equation (3.2) to get

$$x^2 + \left(1 - \frac{2x}{t}\right)^2 = 1.$$

Expanding the square on the left-hand side reveals

$$x^2 + \left(1 - \frac{4x}{t} + \frac{4x^2}{t^2}\right) = 1 \tag{3.3}$$

$$\left(1 + \frac{4}{t^2}\right) x^2 - \frac{4x}{t} = 0 \tag{3.4}$$

$$x\left[\left(1 + \frac{4}{t^2}\right) x - \frac{4}{t}\right] = 0. \tag{3.5}$$

(0,2)

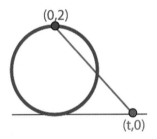

(t,0)

Figure 3.2: The line described by equation (3.1) and the circle described by equation (3.2).

Don't just read it. Work it out! How does equation (3.4) follow from (3.3), or (3.5) from (3.4)?

Since $x = 0$ corresponds to the "north pole" of the circle, which is not the point P, we conclude from this equation that

$$0 = \left(1 + \frac{4}{t^2}\right) x - \frac{4}{t}.$$

Equation (3.5) has the form $ab = 0$, so either a or b must be zero.

To solve this equation for x, let's add $4/t$ to both sides and then multiply by t^2. So doing yields

$$4 = (t^2 + 4)x,$$

after which dividing by $(t^2 + 4)$ brings us to

$$x_p = \frac{4t}{t^2 + 4}. \tag{3.6}$$

The subscript of p reminds us that this is the x-coordinate of the point P.

Since P is on the line described by equation (3.1), its y-coordinate must be

$$y_p = -\frac{2}{t}x_p + 2 = -\frac{2}{t}\left(\frac{4t}{t^2+4}\right) + 2 = \frac{2t^2}{t^2+4}. \tag{3.7}$$

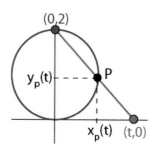

Figure 3.3: The stereographic projection.

Equations (3.6) and (3.7) describe the relationship between t, and x and y, respectively. Writing x and y as functions of t, we have

$$x_p(t) = \frac{4t}{t^2+4} \quad \text{and} \quad y_p(t) = \frac{2t^2}{t^2+4}.$$

(b) Since every number t has a projection onto the circle, the domain of both x_p and y_p is all of $\mathbb{R}$. The points on the circle have x-coordinates that range from -1 to 1, so the range of x_p is $[-1, 1]$. Similarly, the points on the circle have y-coordinates that range from 0 to 2, but no value of t is projected onto the "north pole" of the circle, so $y_p(t)$ *will never be 2*. Therefore, the range of y is $[0, 2)$. ∎

> At a purely mechanical level, we can substitute any value of t into our formulas and they make sense, so the domain is all of $\mathbb{R}$.

Because the behavior of a polynomial is dominated by its leading term when $|t|$ is large, the behavior of a rational function at large $|t|$ is easy to predict.

When the numerator has a smaller degree than the denominator

Example 3.2. Describe the behavior of $x_p(t) = \frac{4t}{t^2+4}$ when t is large.

Solution: Since polynomials are dominated by their leading terms when $|t|$ is large, let's factor a t^2 out of the denominator to see

$$x_p(t) = \frac{4t}{t^2+4} = \underbrace{\frac{4t}{\left(1+\frac{4}{t^2}\right)t^2} \approx \frac{4t}{(1)t^2}}_{\text{for } big \text{ values of } t} = \frac{4}{t} = 4t^{-1}.$$

> The value of $4/t^2$ is nearly zero when t is large, so $1 + \frac{4}{t^2} \approx 1$.

This technique reduces our analysis to what we know of power functions. Since the exponent of our power function is negative, the function values will tend toward zero as t gets large (see Figure 3.4). ∎

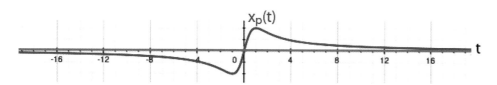

Figure 3.4: The graphs of $x_p(t) = \frac{4t}{t^2+1}$ and $y = 0$.

Because the value of $x_p(t)$ is virtually constant when $|t|$ is large, the graph of $x_p(t)$ approaches a horizontal line. That line is called a **horizontal asymptote** of the graph. From the standpoint of calculus, the importance of a horizontal asymptote has to do with *change* or, rather, lack thereof. In the example above, the function values are virtually constant when t is large, which makes sense because large values of t are all projected onto the circle near the north pole (so they all have roughly the same x-coordinate).

> Note that we talk about an asymptote *of the graph*, rather than talking about an asymptote of the function. That's because a function is an action; and while it doesn't make any sense to say that an action approaches a line, it *does* make sense to say that a curve approaches a line.

When the numerator and denominator have the same degree

Example 3.3. Describe the behavior of $y_p(t) = \frac{2t^2}{t^2+4}$ when t is large.

Solution: As in Example 3.2, the leading terms dominate numerator and denominator, so we know that $y_p(t) \approx \frac{2t^2}{t^2} = 2$ when $|t|$ is large. Consequently, the graph will have a horizontal asymptote at $y = 2$. ∎

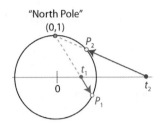

"North Pole"
(0,1)

Figure 3.5: For Example 3.4.

Full Solution
On-line

Try It Yourself: Another stereographic projection

Example 3.4. Suppose the unit circle from Example 3.1 is placed with its center at the origin, and points on the number line that are outside the circle are attracted to the "north pole," while points inside the circle are repelled from it (see Figure 3.5). Determine the formulas for the coordinates of the stereographic projection, $x_p(t)$ and $y_p(t)$. Then discuss their domain, range, and behavior for large $|t|$.

Answer: $x_p(t) = \frac{2t}{t^2+1}$ and $y_p(t) = \frac{t^2-1}{t^2+1}$ (full solution on-line). ∎

When the degree of a rational function's numerator is *exactly* 1 larger than the degree of its denominator, something special happens.

The degree of the numerator is 1 larger than the degree of the denominator

Example 3.5. Discuss the behavior of $f(t) = \frac{8t^3+14}{5t^2+t+4}$ when t is large.

Solution: The leading terms in the numerator and denominator are $8t^3$ and $5t^2$, respectively. When we factor them out, we see that

$$f(t) = \frac{\left(1 + \frac{14}{8t^3}\right)8t^3}{\left(1 + \frac{1}{5t^2} + \frac{4}{5t^2}\right)5t^2} \approx \frac{(1)8t^3}{(1)5t^2} = \frac{8t}{5}$$

when $|t|$ is very large. This tells us that when $|t|$ is large, the graph of $f(t)$ looks like a line with a slope of 8/5. If we want more specifics, we can use polynomial division to rewrite

$$f(t) = \frac{8}{5}t - \frac{8}{25} + \frac{1}{25}\frac{382 - 152t}{5t^2 + t + 4}. \tag{3.8}$$

The last term is a rational function whose numerator has a smaller degree than its denominator, so its contribution is negligible when $|t|$ is large. That means

$$f(t) \approx \frac{8}{5}t - \frac{8}{25}$$

when $|t|$ is large, so the graph of f approaches the line $y = \frac{8}{5}t - \frac{8}{25}$ as $|t|$ grows large (see Figure 3.6). ∎

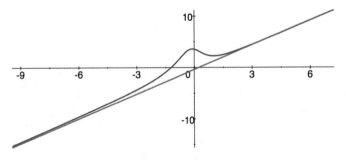

Figure 3.6: The curve $y = f(t)$ and its oblique asymptote $y = \frac{8}{5}t - \frac{8}{25}$ from Example 3.5. The graph of f does approach the line to the left of zero, but it takes longer, so you don't see it as clearly.

In Example 3.5 the graph of f approaches the line $y = \frac{8}{5}t - \frac{8}{25}$ as $|t|$ grows large. Because this is an oblique line, we refer to it as an **oblique asymptote** of the graph of f. Like horizontal asymptotes, the importance of oblique asymptotes lies in what they tell us about the way that a function is changing. When the graph of f has an oblique asymptote, it looks roughly linear at large $|t|$, and linear functions have a constant rate of change.

Try It Yourself: An oblique asymptote

Example 3.6. Use polynomial division to determine the oblique asymptote for the graph of $f(t) = \frac{4t^3 + t}{t^2 + t + 2}$.

Answer: $y = 4t - 4$ ■

Full Solution On-line

When the numerator has a larger degree than the denominator

Example 3.7. Describe $f(t) = \frac{7t^4 + 8t^3 + 3}{3t^2 + t + 12}$ when $|t|$ is large.

Solution: The leading terms tell us that $f(t) \approx \frac{7}{3}t^2$ when t is large. That is, the function grows quadratically when $|t|$ is large, so this graph has neither horizontal nor oblique asymptotes.   ■

Division by zero can lead to asymptotes

Example 3.8. Discuss the important behaviors of $f(t) = \frac{3t^2 + 4t - 2}{5t - 10}$.

Solution: Polynomial division shows us that $f(t) \approx \frac{3}{5}t + 4$ when t is large (i.e., the graph will have an oblique asymptote of $y = \frac{3}{5}t + 4$). But this function also has interesting behavior near $t = 2$. The numerator remains near 18 as t gets closer and closer to 2 (check!), but the value of the denominator heads toward zero. This causes the function value to "blow up" (see Table 3.1).

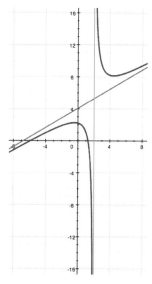

Figure 3.7: Figure for Example 3.8.

t	numerator	denominator	$f(t)$
1.99	$17.8403 \approx 18$	-0.05	$\approx -18/0.05 = -360$
1.999	$17.984 \approx 18$	-0.005	$\approx -18/0.005 = -3600$
1.9999	$17.9984 \approx 18$	-0.0005	$\approx -18/0.0005 = -36000$
1.99999	$17.99984 \approx 18$	-0.00005	$\approx -18/0.00005 = -360000$
2	18	0	UNDEFINED
2.00001	$18.00016 \approx 18$	0.00005	$\approx 18/0.00005 = 360000$
2.0001	$18.0016 \approx 18$	0.0005	$\approx 18/0.0005 = 36000$
2.001	$18.0160 \approx 18$	0.005	$\approx 18/0.005 = 3600$
2.01	$18.1603 \approx 18$	0.05	$\approx 18/0.05 = 360$

Table 3.1: As t gets near 2, the numerator stays near 18 but the denominator heads to zero.

Consequently, the graph of f looks more and more like a vertical line as t comes closer and closer to 2 (see Figure 3.7).   ■

In Example 3.8 the graph of f approaches the vertical line $t = 2$ (in a "parallel," not a "transverse" way) as the number t gets closer to 2, so we say that the line $t = 2$ is a **vertical asymptote** of the graph of f. The importance of vertical asymptotes, as with the other asymptotes, is what they tell us about the way that a function's value changes. In the previous example, when t is close to 2, small changes in t result in *large* changes in function value.

Try It Yourself: Locating vertical asymptotes

Example 3.9. Suppose $f(t) = \frac{2t^2-t-6}{3t^2-5t-2}$. (a) Determine times at which the denominator of $f(t)$ is zero. (b) By checking values of t near the roots from part (a), show that the graph of f has a vertical asymptote at one but not the other.

Full Solution
On-line

Answer: The graph of f has a vertical asymptote at $t = -1/3$ but not at $t = 2$. ∎

❖ Common Difficulties with Rational Functions

Students often assume that a vertical asymptote will occur wherever the denominator has a root (as in Example 3.8). In fact, while there *might* be a vertical asymptote, there doesn't *have* to be (as in Example 3.9).

Also, students often believe that a function's graph cannot cross its asymptotes, but look back to Example 3.5 on p. 20. Specifically, let's consider equation (3.8), which describes $f(t)$. The third term on the right-hand side is negative when t is large, so the graph of f is below the oblique asymptote; but the term is positive when t is small, so the graph of f is above the asymptote.

> **You should know**
>
> - the terms *rational function* and *asymptote*.

> **You should be able to**
>
> - use the the "large $|t|$" behavior of polynomials to determine the "large $|t|$" behavior of rational functions;
>
> - use a leading-term analysis to determine the slope of an oblique asymptote to the graph of a rational function;
>
> - recognize where graphs of rational functions may have vertical asymptotes;
>
> - determine the equations of asymptotes to graphs of rational functions.

❖ 1.3 Skill Exercises

In each of #1–6, determine the behavior of $f(t)$ for large $|t|$.

1. $f(t) = \frac{6t^4}{5t^3+8}$
2. $f(t) = \frac{7t^6}{9t^6+10}$
3. $f(t) = \frac{3t^4}{9t^2+1}$
4. $f(t) = \frac{t^5}{3t^2-4}$
5. $f(t) = \frac{7t^2}{7t^2+8}$
6. $f(t) = \frac{13t^6}{5t^5+9}$

Sketch a graph of each function in #7–12 (without a graphing utility). Pay special attention to $t = 0, \pm1$, and when t is very large.

7. $f(t) = \frac{5t-1}{t^2}$
8. $f(t) = \frac{5t^2+6}{1+t^2}$
9. $f(t) = \frac{5t^2+t+5}{1+t^2}$
10. $f(t) = \frac{6t^2+5}{1+t^2}$
11. $f(t) = \frac{6t^2+5}{1-t^2}$
12. $f(t) = \frac{6t^2}{t^2-1}$

In each of #13–24, without using polynomial division, (a) determine whether the graph of the given function has an asymptote, and if it does (b) find the slope of the asymptote.

13. $f(t) = \frac{2t-4}{t+7}$ 17. $f(t) = \frac{t^2+3}{6t^2+2t+1}$ 21. $f(t) = \frac{t^6+t^5}{1-8t^5}$

14. $f(t) = \frac{8t+3}{8t+4}$ 18. $f(t) = \frac{t^4+t-3}{8t^4+3t^2+7}$ 22. $f(t) = \frac{6t^9+2}{t^8+t^6+1}$

15. $f(t) = \frac{5t^2-1}{t+6}$ 19. $f(t) = \frac{7t-t^3}{1+t^2}$ 23. $f(t) = \frac{16t^3+2}{t^8-4t^7}$

16. $f(t) = \frac{3t^2+6t}{1+t}$ 20. $f(t) = \frac{6t^8+5}{1+t^3}$ 24. $f(t) = \frac{6t^9+2}{t^8-4t^7}$

❖ 1.3 Concept and Application Exercises

25. Design a rational function whose rate of change is nearly constant for large values of t.

26. Design a rational function whose graph has no asymptotes.

27. Suppose $f(t)$ is a rational function whose numerator and denominator have the same degree. How does $f(t)$ behave for large values of t? Support your answer with an example, and an analysis of your example that includes the techniques of this section.

28. Suppose $f(t)$ is a rational function in which the degree of the numerator is larger than the degree of the denominator. How does $f(t)$ behave for large values of t? Support your answer with an example, and an analysis similar to those in the examples in this section.

29. Can the graph of a rational function have only one horizontal asymptote? Two? Three?

30. Determine/design the formula for a rational function whose graph looks like the one shown below.

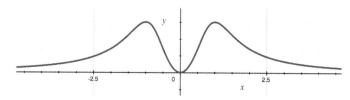

31. In our discussion on p. 18, we found formulas for the coordinates of the stereographic projection of the number t onto the unit circle, $(x_p(t), y_p(t))$. The graphs of $x_p(t)$ and $y_p(t)$ are shown below, over the positive t-axis. In complete sentences, discuss the "story" these graphs tell about the way the projection moves on the circle as t varies. In particular ...

   (a) What's going on when the graphs intersect? Does it make sense that they *should* intersect?

   (b) What happens to the x-coordinate of the projection as t continues to grow? How is that reflected in the graph below? Does it make sense that it should happen that way?

   (c) Based on the geometry, will $y_p(t)$ ever reach 2?

   (d) Why is the north pole called *the point at infinity*?

   (e) Based on your understanding of the projection, what happens to $x_p(t)$ and $y_p(t)$ when t is negative? (Try sketching their graphs without a graphing utility.)

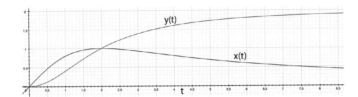

32. In our discussion on p. 18, we found formulas for the coordinates of the stereographic projection of the number t onto the unit circle that's centered at $(0, 1)$.

    (a) Find analogous formulas when the center of the unit circle is at $(0, b)$, where $-1 < b < 1$.

    (b) Find analogous formulas when the center of the circle is at $(0, 1)$, but the radius of the circle is $r > 0$.

    (c) Find analogous formulas when the radius of the circle is $r > 0$, and the center of the circle is at $(0, b)$, where $-r \leq b \leq r$.

    (d) Discuss how change in b and r affect the domain and range of $x_p(t)$ and $y_p(t)$, and their behavior for large values of $|t|$.

33. When you run a business, you have to store inventory. If we're talking about a *lot* of it, you'll have to rent space from a warehouse. You can rent *less* space if you keep less inventory on hand, but then you'll have to restock the warehouse more frequently and that costs money. In this exercise, you're going to develop a formula for the yearly cost of warehousing your inventory in terms of the number of calls that you make to your supplier in order to restock the warehouse.

         | Modeling in Business |

    We'll make two assumptions: (1) the warehouse owners don't rent you space by the second, so it doesn't matter that your warehousing needs diminish as you sell your inventory—you pay for the warehousing space that you need when your warehoused inventory is full, and (2) each time you place an order with your supplier, you pay a surcharge (much like a "handling" fee).

    (a) Your projections indicate that you'll sell a total of 25,000 units in the year. Assuming that you always restock with the same number of units, how many are brought to the warehouse if you make n calls to your supplier?

    (b) Suppose that a single unit of inventory requires 0.5 square meters of warehouse space. How much warehouse space will you need to rent (see your answer from part (a))?

    (c) If it costs $3.00 to rent a square meter of warehouse space for a month, how much will you spend on the warehouse each month? How much will you spend each year? I'll denote the cost per year by $C_w(n)$.

    (d) If you incur a surcharge of $40.00 for each call to your supplier, what's the cost of restocking the warehouse n times in a year? I'll denote this number by $C_s(n)$.

    (e) If we denote the total cost of warehousing inventory by C_T, we have $C_T(n) = C_w(n) + C_s(n)$. Write down a formula for $C_T(n)$ and perform a dimensional analysis to make sure it is giving you the right kind of information.

34. Myoglobin is an oxygen storage protein that is found in muscle tissue in many species (see [31, p. 278]). The relationship between the *partial pressure* of

         | Biology |

oxygen, denoted here by x, and the fraction of myoglobin that has successfully bound an O_2 molecule, y, is

$$y = \frac{x}{K + x},$$

where K is a number called the *apparent dissociation constant*.

(a) If we treat y as a function of x, what is the domain of $y(x)$?

(b) What happens to the fraction of oxygen-bound myoglobin, y, as the partial pressure of oxygen increases?

(c) What happens to the fraction of oxygen-bound myoglobin, y, as the partial pressure of oxygen nears zero?

(d) Explain how this makes sense intuitively.

35. The interaction of atoms in a noble gas can be described in terms of the *Lennard-Jones potential* (see [11, p. 241]). In the case of argon, this potential function is

Physical Chemistry

$$u(x) = K\frac{1759.37 - x^6}{x^{12}},$$

where x is the distance between the centers of two gas molecules, measured in angstroms, and K has a numerical value of 70.3748752. Without using a graphing utility, predict what happens to the value of $u(x)$ as two molecules of argon gas (a) get far away, and (b) come close together.

1.4   Algebraic, Composite, and Piecewise Functions

A function is called **algebraic** when we can calculate its value by using a finite number of the *algebraic* operations: addition, subtraction, multiplication, division, and root-taking. All of the functions we've discussed heretofore have been algebraic. You can add and subtract these functions, or multiply and divide them by each other, and the resulting function is still called algebraic. You can also *compose* them, which we talk about below.

When a function is *not algebraic,* we say it's **transcendental**. Some of the most important transcendental functions are discussed in second half of this chapter.

> If a function is *not* algebraic, its value cannot be calculated by a *finite* number of algebraic operations. We'll see what that means in some detail when we study *Taylor series* in Chapter 8.

❖ Composition of Functions

On p. 2 we characterized a function as an *action,* because it takes a number (from the domain) and *produces* another (in the range). When we perform two or more actions in succession, we refer to the overall process as a **composite** function.

Designing a composite function

Example 4.1. Suppose $f(t) = 3t - 2$ and $g(t) = \sqrt{t}$. If we always use $f(t)$ then $g(t)$, in that order, determine (a) the output associated with $t = 9$, and (b) a general formula for the composite function.

Solution: (a) We always use f first, so let's calculate $f(9) = 25$. Then we apply g, the square root, and see that $g(25) = 5$.

(b) In general, we determine $f(t)$ and then calculate the the square root of that number, $\sqrt{f(t)}$. Since $f(t) = 3t - 2$, this composite function is described by the formula $\sqrt{3t - 2}$. ∎

Try It Yourself: Designing a composite function

Example 4.2. Suppose $f(t) = 3t - 2$ and $g(t) = \sqrt{t}$. If we always use $g(t)$ then $f(t)$, in that order, determine (a) the output associated with $t = 9$, and (b) a general formula for the composite function.

Answer: (a) 7, and (b) $3\sqrt{t} - 2$. ∎

Full Solution On-line

When we use the function g after the function f, as in Example 4.1, we refer to the composite function as "g composed with f" or as "g of f," and denote it by $g \circ f$. As a mnemonic device, it might help you to think of the symbol $\circ$ as an abbreviation of the word "of." Then at a purely semantic level,

$$g \circ f = g \text{ of } f = g(f).$$

In Example 4.1 g was the square-root function, so $g(f) = \sqrt{f}$. Then, since the value of f is calculated as $3t - 2$ in that example,

$$\underbrace{(g \circ f)(t)}_{\text{the value of the composite function at time } t} = \sqrt{3t - 2}. \tag{4.1}$$

Similarly, when we use the function f after the function g, as in Example 4.2, we refer to the composite function as "f composed with g" or "f of g" and denote it by $f \circ g$. Using the functions from Example 4.2,

$$\underbrace{(f \circ g)(t)}_{\text{the value of the composite function at time } t} = 3\sqrt{t} - 2. \tag{4.2}$$

It's important to note that $f \circ g$ is different from $g \circ f$, which you can easily see in equations (4.1) and (4.2). But more importantly, that makes sense when you think about f and g as actions, because there are many things you do each day for which the outcome depends on the order in which you perform actions.

> Sometimes, the order matters: putting on socks then shoes is very different from putting on shoes then socks.

Try It Yourself: Reading composition notation

Example 4.3. Suppose $f(t) = 1/(3 + 5t)$ and $g(t) = \sqrt{10 - \sqrt{t}}$. Determine a general formula for (a) the function $(f \circ g)(t)$ and (b) the function $(g \circ f)(t)$.

Full Solution
On-line

Answer: (a) $\dfrac{1}{3 + 5\sqrt{10 - \sqrt{t}}}$, and (b) $\sqrt{10 - \sqrt{\dfrac{1}{3 + 5t}}}$ ∎

Of course, we can perform more than two actions in a row.

An algebraic function in sighting theory

Example 4.4. In the mid-1940s Bernard Koopman developed the modern theory of sighting around the experiences of fighter pilots who were searching for ships in the open ocean. Based on some simple geometry (see the End Notes), he asserted that the likelihood of a pilot spotting an object depends on the across-the-water distance between them, r, according to a formula of the form

$$\ell = \frac{1}{(r^2 + 1)^{3/2}}.$$

> We've suppressed constants such as the area of the object and the altitude of the plane for the sake of simplicity. Those aspects of the scenario are included in the End Notes.

Write ℓ as the composition of (a) two functions, and (b) three functions.

Solution: Our goal is to take a relatively involved computation and write it as the cumulative effect of simpler steps. Your choice of steps depends on what you think is simple. With that said, it's often best to ask what the last step in a calculation would be. For example, suppose you were using a scientific calculator to determine ℓ when $r = 10$. First you'd calculate $10^2 + 1 = 101$. Then you'd raise that to the 1.5 power, and the very last thing you'd do is hit the "reciprocal" key.

(a) Based on the description above, in which the last step is to calculate a reciprocal, let's characterize ℓ as $(f \circ g)(r)$ where f is the "reciprocal" function. Then

$$f \circ g = f(g) = \frac{1}{g}, \quad \text{which is supposed to be} \quad \frac{1}{(r^2 + 1)^{3/2}}.$$

That means $g(r)$ must be $(r^2 + 1)^{3/2}$.

(b) To write ℓ as the composition of *three* functions, we think of g as the cumulative effect of two simpler steps. We have

$$g(r) = (r^2 + 1)^{3/2} = \sqrt{(r^2 + 1)^3},$$

so the last thing we'd do when calculating $g(r)$ is hit the square-root key on the calculator. Therefore, let's write $g(r)$ as the composition $(h \circ k)(r)$, where h is the square-root function. Then

$$(h \circ k) = h(k) = \sqrt{k}, \quad \text{which is supposed to be} \quad \sqrt{(r^2 + 1)^3},$$

so $k(r)$ must be $(r^2 + 1)^3$. That is, we have $\ell = (f \circ h \circ k)(r)$ where

$$f(t) = \frac{1}{t}, \quad h(t) = \sqrt{t} \quad \text{and} \quad k(t) = (t^2 + 1)^3.$$ ∎

❖ Piecewise Defined Functions

Sometimes the formula describing the function $f(t)$ changes from one interval to the next. In this case, $f(t)$ is called a **piecewise defined** function.

A piecewise linear function

Example 4.5. Based on the graph of f shown in Figure 4.1, write a formula for calculating $f(t)$.

Solution: When $t < 0$, the graph of this function is linear and has a slope of -1. Specifically, it's the line $y = -t$. The behavior of the function changes when $t > 0$. The graph is still linear, but has a slope of 1. Specifically, it's the line $y = t$. Since the behavior is different when $t \leq 0$ than it is when $t > 0$, we split our formulation into two parts, writing

$$f(t) = \left\{ \begin{array}{ll} -t & \text{when } t < 0 \\ t & \text{when } t \geq 0 \end{array} \right.$$

You might recognize this as the **absolute value** function, which is perhaps the most important piecewise defined function.   ■

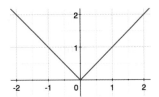

Figure 4.1: The graph of f for Example 4.5

Piecewise defined functions and the human heart

Example 4.6. Suppose the heart has just beat, and the voltage on its AV node is x. By the time the next signal from the pacemaker arrives, the voltage on the AV node will have decayed to a fraction of that, say mx where $m \in (0, 1)$. If mx is less than a so-called *critical voltage*, the heart will beat. Otherwise it continues to relax. If the critical voltage is denoted by V_c, we describe the voltage on the AV node at this moment of decision (when the signal from the pacemaker arrives) by

The term AV is an abbreviation for *atrioventricular*.

The AV node acts as a kind of "trigger lock," which prevents the heart muscle from contracting if it hasn't had enough time to relax since the last contraction.

$$y = \left\{ \begin{array}{ll} mx + b & \text{if } x \in [0, x_c] \\ mx & \text{if } x > x_c \end{array} \right. , \quad \text{where } b > 0 \text{ and } x_c = \frac{V_c}{m}.$$

Graph y as a function of x when $m = 0.5$, $b = 2$, and $V_c = 3$.

Solution: We do this in two steps, as shown in Figure 4.2. (1) Graph both $y = mx$ and $y = mx + b$ without regard for where they apply. (2) Trace over these graphs on the appropriate intervals.   ■

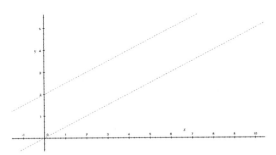
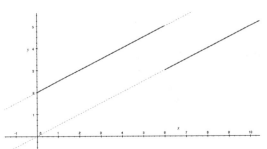

Figure 4.2: (left) Step 1 of graphing a piecewise defined function is to sketch the graphs of all the pieces; (right) step 2 is to darken the graphs over the relevant intervals.

You should know

- the terms *algebraic, transcendental, composite, compose, piecewise;*

- the notation $f \circ g$ and $(f \circ g)(t)$;

You should be able to

* formulate $(f \circ g)(t)$ based on formulas for $f(t)$ and $g(t)$;

* write a composite function as the composition of simpler functions;

* graph piecewise defined functions.

❖ 1.4 Skill Exercises

In #1–6 you should (a) determine a formula for $(f \circ g)(t)$, and (b) $(g \circ f)(t)$.

1. $f(t) = 2t + 4$, $g(t) = 8t - 1$ 4. $f(t) = \sqrt{9 - t}$, $g(t) = t^3$

2. $f(t) = -3t + 17$, $g(t) = 2t + 1$ 5. $f(t) = \frac{t+1}{t^2+2}$, $g(t) = 3t^2 + 13$

3. $f(t) = 8/t$, $g(t) = \sqrt{3t + 4}$ 6. $f(t) = t^3 - 1$, $g(t) = \frac{2t^2+t}{t+8}$

In each of #7–12 find g so that $y(t) = (f \circ g)(t)$.

7. $y(t) = \sqrt{t^6 + 2}$ where $f(t) = \sqrt{t}$ 9. $y(t) = \sqrt{3t^4 + 1}$ where $f(t) = t^3$

8. $y(t) = \frac{t^2+3}{t^2+4}$ where $f(t) = \frac{t+5}{t+6}$ 10. $y(t) = \frac{t^2+2t+3}{3t^2+6t+7}$ where $f(t) = \frac{t+1}{t+4}$

In each of #11–14 find f so that $y(t) = (f \circ g)(t)$.

11. $y(t) = \frac{t+3}{t+4}$ where $g(t) = t + 2$ 13. $y(t) = \frac{81}{t^2+4}$ where $g(t) = \sqrt{t + 3}$

12. $y(t) = \sqrt{t + 1}$ where $g(t) = (t + 4)^3$ 14. $y(t) = t^2 + 8t$ where $g(t) = t^3 + 9$

In each of #15–22 sketch a graph of the piecewise defined function.

15. $f(t) = \begin{cases} 3t & \text{if } t < 0 \\ 0.5t & \text{if } t \geq 0 \end{cases}$ 19. $f(t) = \begin{cases} 3 & \text{if } t < 1 \\ 2 - t^2 & \text{if } t \geq 1 \end{cases}$

16. $g(t) = \begin{cases} -2t & \text{if } t < 0 \\ 3t & \text{if } t \geq 0 \end{cases}$ 20. $g(t) = \begin{cases} 4 - 2t & \text{if } t < 3 \\ -2 & \text{if } t \geq 3 \end{cases}$

17. $f(t) = \begin{cases} 3 - t & \text{if } t \leq 1 \\ t - 2 & \text{if } t > 1 \end{cases}$ 21. $f(t) = \begin{cases} -t^2 & \text{if } t < 0 \\ t^2 & \text{if } t \geq 0 \end{cases}$

18. $g(t) = \begin{cases} 5 + t & \text{if } t \leq 2 \\ 10 - t & \text{if } t > 2 \end{cases}$ 22. $g(t) = \begin{cases} (t + 1)^2 & \text{if } t \leq -1 \\ 4 - t^2 & \text{if } t > -1 \end{cases}$

In exercises #23–26 you should graph the function $f \circ g$.

23. f from #17 and $g(t) = t^2$ 25. f from #17, and g from #22

24. f from #18 and $g(t) = 6 - t^2$ 26. f from #18, and g from #20

❖ 1.4 Concept and Application Exercises

Exercises 27–32 focus on the **unit step function**, defined by

$$u(t) = \begin{cases} 0 & \text{if } t < 0 \\ 1 & \text{if } t \geq 0 \end{cases}$$

Specifically, you'll see in these exercises the way that we use $u(t)$ as a kind of mathematical "switch."

27. Graph the function $u(t)$.

28. Graph the function $f(t) = u(t)t^2$.

29. Graph the function $f(t) = u(t)(t + 5)$.

30. Graph the function $f(t) = 1 - u(t)$.

31. Graph the function $f(t) = (1 - u(t))t^3$.

32. Graph the function $f(t) = (1 - u(t))(3 - t)$.

Exercises 33–36 concern the *greatest integer* function (colloquially known as the "floor" function), which is denoted by $\lfloor t \rfloor$ and defined as

$$\lfloor t \rfloor = \text{the greatest of the integers whose value is at most } t.$$

33. Determine (a) $\lfloor \pi \rfloor$, (b) $\lfloor \sqrt{2} \rfloor$, and (c) $\lfloor 7 \rfloor$.

34. Determine (a) $\lfloor 0.7 \rfloor$, (b) $\lfloor -3.1 \rfloor$, and (c) $\lfloor -9 \rfloor$.

35. Suppose $f(t) = t^2$ and $g(t) = \lfloor t \rfloor$.

    (a) Graph $(g \circ f)(t)$ and locate the first six breaks in the graph at $t > 0$.

    (b) As t grows, what happens to the distance between the breaks that you found in part (a)? Why?

36. Suppose $f(t) = t^2$ and $g(t) = \lfloor t \rfloor$.

    (a) Graph $(f \circ g)(t)$ and locate the first six breaks in the graph at $t > 0$.

    (b) As t grows, what happens to the distance between the breaks that you found in part (a)? Why?

37. Suppose $c(x)$ is the cost (in dollars) of buying x liters of aniseed oil. Graph the function c when

$$c(x) = \begin{cases} 0.2898x & \text{if } x \in [0, 0.5) \\ 0.07932x & \text{if } x \in [0.5, 1) \\ 0.06339x & \text{if } x \in [1, 4) \\ 0.0513975x & \text{if } x \geq 4 \end{cases}$$

38. If we assume that the Earth has a uniform density, ρ, the planet's gravitational pull on a mass of m kilograms that's r meters from its center is

$$F_g(r) = \begin{cases} \frac{4}{3}\pi\rho G m r & \text{if } r \leq R \\ \frac{4\pi R^3 \rho G m}{3r^2} & \text{if } r \geq R \end{cases}$$

where R is the radius of the Earth (in meters) and G is the gravitational constant (discussed on p. 2).

    (a) This description of F_g seems to indicate that both pieces of the function apply when $r = R$. Verify that they have the same value.

    (b) Sketch a graph of F_g and mark the location of R on the horizontal axis. (The basic structure of the graph doesn't depend on the particular values of m, G and R.)

39. Suppose your cup of hot chocolate is sitting on the table. When you push on the cup with a force of f newtons, the table offers resistance via the friction force, F, according to

$$F(f) = \begin{cases} f & \text{if } f \le 3 \\ 2 & \text{if } f > 3 \end{cases}$$

   (a) Sketch a graph of $F(f)$.

   (b) Based on this graph, explain what happens to the cup as you slowly increase the amount of force that you apply to it.

40. Various holding tanks are shown below. If fluid is pumped into the tanks at a constant rate, sketch a graph that shows the depth of the fluid in each tank as a function of time, $d(t)$. (Tank (d) is filled by pumping fluid into the *left* well of the tank.)

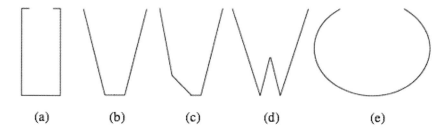

    (a)   (b)   (c)   (d)   (e)

41. Find piecewise defined functions whose graphs look like your sketches from parts (b), (c), and (d) of #40.

Exercises #42–47 investigate the domain of composite functions.

42. Suppose $t = 3$ is in the domain of $f \circ g$. Explain why $t = 3$ must be in the domain of g.

43. Suppose $t = 7$ is in the domain of $f \circ g$. What must be true about $g(7)$?

44. Suppose $f(t) = \sqrt{t}$ and $g(t) = 13 - 2t^2$. Determine (a) the domain of $f \circ g$, and (b) the domain of $g \circ f$.

45. Suppose $f(t) = 1/t$ and $g(t) = t^2 + t - 10$. Determine (a) the domain of $f \circ g$, and (b) the domain of $g \circ f$.

46. Suppose $f(t) = \sqrt{t} + 1$ and the domain of g is the interval $[6, 9]$. What is the domain of $g \circ f$?

47. Suppose $f(t) = \sqrt{1 - t^2}$ and the domain of $f \circ g$ is all of $(-\infty, \infty)$. What do you know about the domain and range of g?

In each of #48–51 describe what happens when we apply $f(x)$ to an image, as in Example 2.5 in Section 1.2. (You might find it helpful to graph $y = f(x)$ and $y = x$ on the same axes, as in Figure 2.2.)

                                                                                | Imaging |

48. $f(x) = \begin{cases} 0.6x + 0.2 & \text{if } x < 0.5 \\ x & \text{if } x \ge 0.5 \end{cases}$

50. $f(x) = \begin{cases} 0 & \text{if } x < 0.2 \\ \frac{5}{3}\left(x - \frac{1}{5}\right) & \text{if } x \in [0.2, 0.8] \\ 1 & \text{if } x > 0.8 \end{cases}$

49. $f(x) = \begin{cases} x & \text{if } x < 0.5 \\ 0.6x + 0.5 & \text{if } x \ge 0.5 \end{cases}$

51. $f(x) = \begin{cases} x & \text{if } x < 0.2 \\ -\frac{5}{3}\left(x - \frac{4}{5}\right) + \frac{4}{5} & \text{if } x \in [0.2, 0.8] \\ x & \text{if } x > 0.8 \end{cases}$

1.5   Periodic Functions & Transformations

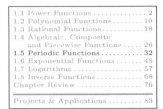
There are a lot of processes and phenomena that repeat themselves on a regular or roughly regular basis: the beating of your heart, the distance of a piston from the top of its cylinder, the length of daylight, etc. In day-to-day language, we say these things are *periodic*. We make the idea of periodicity mathematically precise by saying that a function f is **periodic** if there is some number $\lambda > 0$ for which

$$f(t) = f(t - \lambda) \quad \text{at every } t. \tag{5.1}$$

> The symbol λ is the lower-case Greek letter "lambda."

If t is measured in seconds, for example, equation (5.1) says that the function value at time t is the same as it was λ seconds earlier. The smallest λ that makes equation (5.1) true is called the **period** of f.

Numerical consequences of periodicity

Example 5.1. Suppose the period of f is $\lambda = 6$. Show $f(2) = f(20)$.

Solution: We know from (5.1) that $f(20) = f(14) = f(8) = f(2)$.   ∎

Graphical consequences of periodicity

Example 5.2. Find the period of f, whose graph is in Figure 5.1.

Solution: Equation (5.1) says the graph of f has the same height at time t as it did λ seconds earlier, so we look for repetition in the graph. Notice that we can slide the graph of f by 4, 8, 12 … units horizontally and it will look exactly the same. The smallest of these numbers is 4, so that's the period of f.   ∎

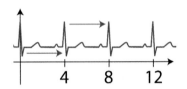

Figure 5.1: The graph of f from Example 5.2.

The most important periodic functions are indisputably sine, cosine, and tangent. Though you're familiar with these functions, they're important enough that we should pause for a brief review (for further review, see Appendix A).

The unit circle: As you see in Figure 5.2, a point on the unit circle that's separated from $(1,0)$ by θ radians has coordinates $(\cos(\theta), \sin(\theta))$. For this reason, sine and cosine are often called the **circular functions** and have a natural period of 2π (i.e., $\sin(\theta + 2\pi) = \sin(\theta)$).

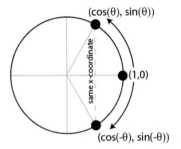
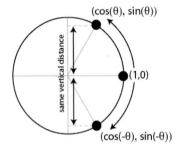
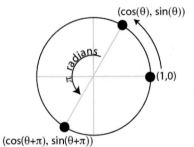

Figure 5.2: (left) Since cosine is the x-coordinate of a point on the unit circle, we have $\cos(-\theta) = \cos(\theta)$; (middle) since sine is the y-coordinate of a point on the unit circle, we have $\sin(-\theta) = -\sin(\theta)$; (right) since $\tan(\theta)$ is the slope of a line through the origin, tangent has a period of π.

In Appendix A you'll see that a similar triangles argument leads us to the fact that $\tan(\theta) = \frac{\sin(\theta)}{\cos(\theta)}$, which is exactly the slope of the line that connects the origin to the point at $(\cos(\theta), \sin(\theta))$. Because that line has the same slope as the one connecting the origin to $(\cos(\theta + \pi), \sin(\theta + \pi))$, the tangent has a period of π (see the right-hand image of Figure 5.2).

Cosine is even: Moving θ radians away from $(1,0)$ along the unit circle gives us the same x-coordinate whether clockwise or counter-clockwise (see Figure 5.2). For this reason, $\cos(-\theta) = \cos(\theta)$. Consequently, when we graph the cosine function, its left and right "halves" are mirror images of each other (see Figure 5.3). Because the graph of a power function with an *even* exponent also exhibits this symmetry, we say that the cosine (and any other function that has this quality) is an **even** function.

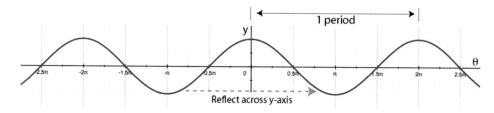

Figure 5.3: The graph of $y = \cos(\theta)$.

Sine is odd: When moving counter-clockwise away from $(1,0)$, around the unit circle, the y-coordinate is initially positive; but when moving in a clockwise direction it is initially negative (though with the same magnitude, see Figure 5.2). For this reason, $\sin(-\theta) = -\sin(\theta)$. Consequently, when we graph the sine function we see that the left and right "halves" are reflections of each through the origin (see Figure 5.4). Because the graph of a power function with an *odd* exponent also exhibits this behavior, we say that the sine (and any other function that has this quality) is an **odd** function.

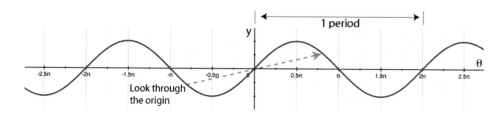

Figure 5.4: The graph of $y = \sin(\theta)$.

Tangent is odd: The graph of the tangent in Figure 5.5 shows that the left and right halves are reflections through the origin, just as we see in the graph of sine. We can also verify that the tangent is an odd function by using what we know of the sine and cosine:

$$\tan(-\theta) = \frac{\sin(-\theta)}{\cos(-\theta)} = \frac{-\sin(\theta)}{\cos(\theta)} = -\frac{\sin(\theta)}{\cos(\theta)} = -\tan(\theta).$$

Consequently, when we graph the tangent function (see Figure 5.5) the left and right "halves" are reflections of each through the origin. Note also that the graph of the tangent has vertical asymptotes at $\theta = \pm\pi/2$, at $\pm 3\pi/2$ and, in fact, at all odd multiples of $\pi/2$. Fundamentally, this happens for the same reason it did in Example 3.8 (see p. 21)—the denominator, $\cos(\theta)$, approaches zero at these values of θ, but the numerator does not.

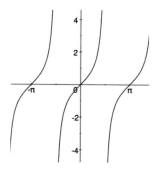

Figure 5.5: The tangent is also an odd function.

❖ Transformations and Sound Waves

You hear sound when your ear detects a periodic change in air pressure. If we could take a photograph of the pressure near your ear, we'd see periodic highs and lows, so we describe the sound wave with a cosine.

- The *volume* of the sound you hear is determined by the the *amplitude* of the cosine wave, which measures the difference between the highs and lows (see below).

- The *tone* is determined by the number of high-low cycles that arrive at your ear each second. We call this number the **frequency** of the wave, and it's affected by two things: the length of a high-low cycle (see Figure 5.6), and how quickly the pressure wave is moving (more on this below).

1 high-low cycle |←→|

Figure 5.6: A traveling sound wave.

▷ Scaling a function affects its amplitude

The function $\cos(x)$ varies at most 1 unit away from its average value of 0, so we say $\cos(x)$ has an **amplitude.** of 1. To describe waves with different volumes, we use cosines with different amplitudes, which is achieved by **scaling** the cosine.

> In a mathematical context, the word *scale* means "to make larger or smaller by multiplication." For example, when we scale $\cos(x)$ by 3 we get $3\cos(x)$, and the scale factor is 3.

Scaling a function changes the amplitude

Example 5.3. Determine the amplitude of (a) the curve $y = 1.5\cos(x)$, and (b) the curve $y = 0.3\sin(x)$.

Solution: (a) We know $-1 \le \cos(x) \le 1$, so $-1.5 \le 1.5\cos(x) \le 1.5$. Since $1.5\cos(x)$ varies at most 1.5 unit away from its average value of 0, we say $1.5\cos(x)$ has an amplitude of 1.5. (b) Similarly, the amplitude of $0.3\sin(x)$ is 0.3 since $-0.3 \le 0.3\sin(x) \le 0.3$ (see Figure 5.7). ∎

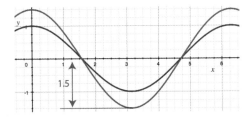
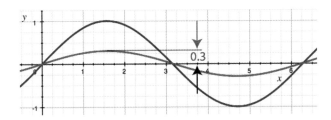

Figure 5.7: (left) The graphs of $y = \cos(x)$ and $y = 1.5\cos(x)$, and (right) the graphs of $y = \sin(x)$ and $y = 0.3\sin(x)$.

What you see in Example 5.3 is that scaling a function stretches (or compresses) its graph vertically. This happens for *every* function, not just for the sine and cosine. In Figure 5.7 you can see that the location of roots is unaffected by scaling (since multiplying zero by any number is still zero).

▷ Scaling the argument affects the period

A moment ago we said that *tone* is affected by the length of a high-low cycle, which is the period of the cosine wave. To describe different tones, we use cosines with different periods, which is achieved by scaling the argument.

Scaling the argument can shorten the period

Example 5.4. Find the period of $\cos(3x)$ when x is in meters.

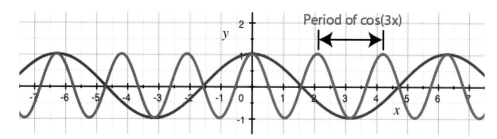

Figure 5.8: The graphs of $y = \cos(x)$ and $y = \cos(3x)$.

Solution: The cosine has a period of 2π. Scaling the argument of the cosine by 3 allows us to reach that number with a smaller value of x. Specifically, the argument of the cosine is 2π when

$$3x = 2\pi \;\Rightarrow\; x = \frac{2\pi}{3}.$$

So this function has a period of $\lambda = \frac{2\pi}{3} \approx 2.09$ meters (see Figure 5.8). ∎

Try It Yourself: Scaling the argument can lengthen the period

Example 5.5. Find the period of $\sin(x/3)$ when x is in meters.

Answer: $\lambda = 6\pi \approx 18.85$ meters (see Figure 5.9). ∎

Full Solution On-line

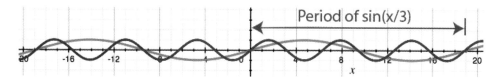

Figure 5.9: The graphs of $y = \sin(x)$ and $y = \sin(x/3)$.

Scaling the argument to affect the period

Example 5.6. Find the period of $\tan(\sqrt{13}\, x)$.

Solution: The function $\tan(\sqrt{13}\, x)$ will complete 1 period when the tangent "sees" a π in its argument. That happens when $\sqrt{13}\, x = \pi$, so $x = \pi/\sqrt{13} \approx 0.8713$. ∎

> Notice that we're not using units in this example. That's because the tangent doesn't typically represent a pressure wave. However, scaling its argument has the same effect.

Examples 5.4–5.6 lead us to the following (correct) conjecture:

> When the period of $f(x)$ is p, the period of $f(\nu x)$ is p/ν. (5.2)

Graphically speaking, this means the graph of f is compressed (horizontally) when $\nu > 1$, and stretched (horizontally) when $\nu \in (0,1)$. If we vary ν between 0.5 and 2, back and forth, the graph of $f(\nu x)$ will compress and expand like a squeezebox.

Note: This happens with *all* functions, not just periodic functions. We've focused on the trigonometric functions because the effect of scaling their argument is easy to see.

Designing period

Example 5.7. Design a cosine wave whose period is (a) 1 meter, and then (b) $\sqrt{3}$ meters.

Solution: Since $\cos(x)$ has a period of 2π meters, the period of $\cos(\nu x)$ is $2\pi/\nu$ meters. (a) To have $2\pi/\nu = 1$ we need $2\pi = \nu$. That is, the curve $y = \cos(2\pi x)$ oscillates with a period of 1 meter. (b) To have $2\pi/\nu = \sqrt{3}$ we need $\nu = 2\pi/\sqrt{3}$. So our answer is that $\cos\left(\frac{2\pi}{\sqrt{3}}x\right)$ has a period of $\sqrt{3}$ meters.   ∎

Try It Yourself: Designing period

Example 5.8. Design a sine wave whose period is (a) 3 meters, and then (b) 1/3 meters.

Answer: See on-line solution.   ∎

Full Solution
On-line

▷ **Negative scalars**

Scaling the output of a function by a negative number exchanges positive and negative values of y. Graphically, the result is that the graph of f is reflected over the x-axis.

If $f(2) > 0$, $-3f(2) < 0$.

Scaling the output by a negative number

Example 5.9. Suppose $f(x) = x(x-1)(x-2)$. Graph $y = f(x)$ and $y = -0.5f(x)$.

Solution: To sketch $y = -0.5f(x)$, we scale the graph of f vertically by a factor of 0.5, and then exchange positive and negative outputs by reflecting the graph over the y-axis (because of the negative). The graphs are shown in Figure 5.10.   ∎

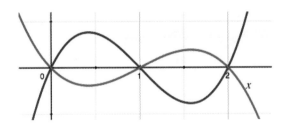

Figure 5.10: The graphs of $y = f(x)$ and $y = -0.5f(x)$ from Example 5.9.

Note: Since $-0 = 0$, the x-intercepts of the graph remain in place (as seen in Figure 5.10).

In similar fashion, scaling the *argument* of a function by a negative number exchanges positive and negative values of x. This has the effect of reflecting the graph of f over the y-axis.

Scaling the argument by a negative number

Example 5.10. Suppose $f(x) = \sqrt{x}$. Graph the curve $y = f(-x)$.

Solution: Points such as $(1, 1)$, $(4, 2)$ and $(9, 3)$ are on the graph of f. By contrast, no point on the curve $y = f(-x)$ has a positive x-coordinate because, for instance, a point over $x = 4$ would have a y-coordinate of $y = \sqrt{-4}$, which isn't a real number. Instead, at

$$x = -1 \quad \text{we see} \quad y = \sqrt{-(-1)} = \sqrt{1} = 1,$$
$$x = -4 \quad \text{we see} \quad y = \sqrt{-(-4)} = \sqrt{4} = 2,$$
$$x = -9 \quad \text{we see} \quad y = \sqrt{-(-9)} = \sqrt{9} = 3,$$

and so on. In short, points on the curve $y = \sqrt{x}$ have been reflected over the y-axis by negating the argument of f. The result is shown in Figure 5.11.   ■

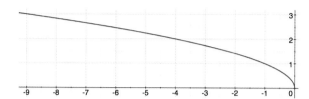

Figure 5.11: The curve $y = \sqrt{-x}$ from Example 5.10.

Scaling the argument by a negative number

Example 5.11. Suppose $f(x) = x(x-1)(x-2)$. Graph $y = f(x)$ and $y = f(-0.5x)$ on the same axes.

Solution: To sketch $y = f(-0.5x)$, we stretch the graph of f horizontally by a factor of 2 (because of the 0.5, just like what happens with the graphs of the trigonometric functions), and then exchange positive and negative values of x by reflecting it over the y-axis (because of the negative). The graphs are shown in Figure 5.12.   ■

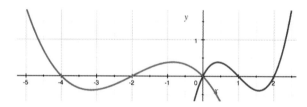

Figure 5.12: The graphs of $y = f(x)$ and $y = f(-0.5x)$ from Example 5.11.

▷ Reciprocals

Reciprocals exchange big numbers with small. For example, 100 is big but $1/100$ is small, and -0.001 is small but $1/(-0.001) = -1000$ is big. Only ± 1 are left unaffected by this exchange.

Graphs of reciprocals of functions

Example 5.12. Sketch the graph of $y = 1/(2t - 4)$.

Solution: The number $2t - 4$ is big and positive when $t > 0$ is large, so its reciprocal is small and positive. Similarly, the number $2t - 4$ is small and positive when t is larger than but very near 2. For such values of t, the y-coordinate on our graph will be big and positive. Similar analysis applies when $t < 2$. The graph is shown in Figure 5.13.   ■

Graphs of reciprocals of functions

Example 5.13. Sketch the graph of $y = \sec(t)$.

Solution: Since $\sec(t) = 1/\cos(t)$, let's begin by graphing the cosine. When $\cos(t)$ is near zero, the number $\sec(t)$ will be large (with the same sign). And $\sec(t) = \pm 1$ when $\cos(t) = \pm 1$ (see Figure 5.13).   ■

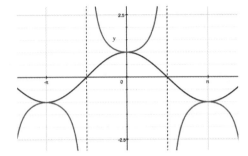

Figure 5.13: (left) The curves y = 2t-4 and y = 1/(2t-4) from Example 5.12 intersect when $y = 1$; (right) the curves $y = \cos(t)$ and $y = \sec(t)$ from Example 5.13 intersect when $y = \pm 1$.

Try It Yourself: Graphs of reciprocals of functions

Example 5.14. Suppose $f(x) = (x-2)(x-5)$. Sketch the curves $y = f(x)$ and $y = 1/f(x)$ on the same axes.

Answer: See on-line solution.   ■

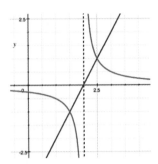

Full Solution On-line

▷ Translating the graph vertically by adding onto the function

Adding onto a cosine or a sine changes the average value. For example, $2 + \cos(x)$ has an average value of 2 instead of 0. This can be helpful when, for example, the oscillation we're describing is removed from our reference point (as in Example 5.15) or when measurements are made in absolute rather than relative units.

In a mathematical context, the verb *translate* means "to move rigidly." When you slide your coffee cup from the left side of your desk to the right, leaving the handle in exactly the same orientation, you're translating it.

Changing the average value

Example 5.15. Suppose the distance from a piston to the top of its cylinder is $D(t) = 1.5\cos(500t) + 6$ centimeters, where time is measured in seconds. Discuss the motion, including (a) the average distance from the piston to the top of the cylinder, (b) the period, and (c) the amplitude of oscillation.

Solution: (a) This function has an average value of 6 cm (see Figure 5.14). Because the oscillatory nature of the motion is described by $1.5\cos(500t)$, we know (b) the period is $2\pi/500 = \pi/250$ seconds, and (c) the amplitude is 1.5 cm.   ■

In Figure 5.14, you see that the graph of $1.5\cos(500t) + 6$ is six units higher than the graph of $1.5\cos(500t)$. In general, the graph of $f(x) + k$ is found by translating the graph of $f(x)$ vertically by k units.

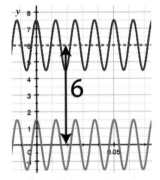

Figure 5.14: The graphs $y = 1.5\cos(500t)$ and $y = 1.5\cos(500t) + 6$ shown with the average value of D.

▷ Translating the graph horizontally by adding into the argument

Let's say that the period of $\cos(x)$ "starts" at $x = 0$. Then the the period of $\cos(x - h)$ will "start" when $x = h$, because that's when the cosine "sees" a zero as its argument. The net effect is that you start drawing the graph at $x = h$ instead of at $x = 0$. In short, the curve $y = \cos(x - h)$ is the same as $y = \cos(x)$, but translated to the right by h units. When working with trigonometric functions, this horizontal translation is called a **phase shift**. Figure 5.15 depicts the graph of $y = \cos(x)$ and $y = \cos(x - \frac{\pi}{4})$.

Translating a graph horizontally by adding into the argument

Example 5.16. Sketch a graph of $y = \sin(x + 2\pi/3)$.

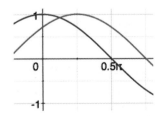

Figure 5.15: The graphs of $y = \cos(x)$ and $y = \cos(x - \pi/4)$.

Solution: We can do this in two steps. (1) Sketch a graph of $y = \sin(x)$, and then (2) take the phase shift into account. In this case, we're trying to graph $y = \sin(x - h)$ where $h = -2\pi/3$. So the graph will be translated horizontally by $-2\pi/3$ units. That is, the graph is shifted to the left. This is shown in Figure 5.16. ∎

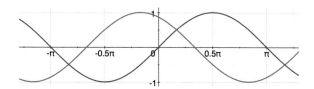

Figure 5.16: The graphs of $y = \sin(x)$ and $y = \sin(x + 2\pi/3)$.

Of course, horizontal translation happens with functions other than sine and cosine.

Translating a graph horizontally by adding into the argument

Example 5.17. Suppose $f(x) = x^3 - x$. Graph both $y = f(x)$ and $y = f(x - 1.5)$ on the same axes.

Solution: The function f has roots at $x = 0$ and $x = \pm 1$, and the x^3 term dominates the behavior when $|x|$ is large (so if you were to zoom out, the graph of f would look like $y = x^3$). From this information, we can sketch the graph of f. To sketch the curve $y = f(x - 1.5)$, we shift the graph of f to the right by 1.5 units (see Figure 5.17). ∎

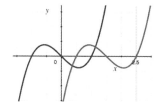

Figure 5.17: The graphs $y = f(x)$ and $y = f(x - 1.5)$ from Example 5.17.

Try It Yourself: Translating a graph horizontally by adding into the argument

Example 5.18. Suppose $f(x) = x(x - 4)$. Graph both $y = f(x)$ and $y = f(x - 2)$ on the same axes.

Answer: See on-line solution. ∎

Full Solution On-line

Let's return to the discussion of sound and pressure waves. We've been speaking of a pressure wave as though it were standing still, as in a photo. Now we want to make it move to the right, as it would in a video. For the sake of discussion, let's assume the pressure wave is proceeding to the right at 3 meters per second (see Figure 5.18). Then

at $t = 0$, the graph is what we see in the "photograph," $y = f(x)$;
at $t = 1$, the wave has moved right by 3 m, so we see $y = f(x - 3)$;
at $t = 2$, the wave has moved right by 6 m, so we see $y = f(x - 6)$;
at $t = 3$, the wave has moved right by 9 m, so we see $y = f(x - 9)$.

$3 = (3)(1)$	$=$ (speed)(time)
$6 = (3)(2)$	$=$ (speed)(time)
$9 = (3)(3)$	$=$ (speed)(time)

In general, when the wave is moving to the right at a speed of $s\frac{\text{m}}{\text{sec}}$, the distance it travels in t seconds is $st =$ (speed)(time). So if our original "photograph" of the pressure shows $y = f(x)$, another photograph after t seconds will show $y = f(x - st)$.

Students sometimes have trouble finding the formula for $f(x - st)$, but the "mapping" notation that was introduced on p. 2 can help. Let's rewrite the function $f(x) = \cos(\nu x)$ as

The skill of finding the formula for $f(x - st)$ will come in handy later, when we calculate something called a *difference quotient* in Chapter 3.

$$f : x \longmapsto \cos(\nu x). \tag{5.3}$$

Equation (5.3) tells us what the function does when we give it an input of x; but remember that a function is an *action*, and that same action is performed no matter

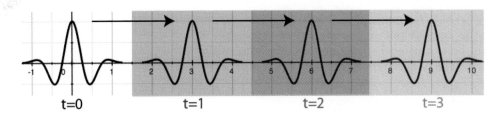

Figure 5.18: One cycle of a traveling wave: $t = 0$, $t = 1$, $t = 2$, and $t = 3$.

what we use as the argument. So we could just as well describe f by

$$f : z \longmapsto \cos(\nu z) \tag{5.4}$$

If we want to use $(x - st)$ as the argument our function, we simply find every occurrence of z in (5.4) and replace it with $(x - st)$. So

| This is a lot like using the find/replace command on a word processor. |

$$f : (x - st) \longmapsto \underbrace{\cos(\nu(x - st))}_{} = \cos(\nu x - \nu s t). \tag{5.5}$$

It might look as if there's more than one variable here, but remember that νx tells us about the physical structure of the wave—that's not changing, and s tells us about the speed—that's not changing. The only thing that's changing here is time, t.

It's common to tighten up the notation by defining $\omega = \nu s$ so that our traveling wave has the form $\cos(\nu x - \omega t)$. The number ω is called the **angular frequency** of the wave, which makes sense when you look at its units (discussed after Example 5.23).

| ω is the lower-case Greek letter "omega." |

Detecting ω and s

Example 5.19. Find (a) the angular frequency, and (b) the speed of the traveling wave $\cos\left(\frac{4}{25}x - \sqrt{7}\,t\right)$ when length is measured in meters and time in seconds.

Solution: (a) The angular frequency is $\omega = \sqrt{7}$ radians per second, which can be read directly from the formula. (b) We can also read $\nu = \frac{4}{25}$ directly from the formula. And since we know that $\omega = \nu s = (4/25)s$, we conclude that $s = 25\sqrt{7}/4$ meters per second. ∎

Try It Yourself: Detecting ω and s

Example 5.20. Find (a) the angular frequency, and (b) the speed of the traveling wave $\cos\left(13x - \sqrt{21}\,t\right)$ when length is measured in miles and time in hours.

Answer: (a) $\sqrt{21}$ radians per hour (b) $\sqrt{21}/13$ miles per hour. ∎

Full Solution On-line

❖ Period, Wavelength, and Avoiding Confusion

We've been using the word "period" to mean the distance between consecutive crests of a cosine wave; but now that we're talking about moving waves, it could also mean the length of time between the arrival of consecutive wave crests at your ear. In order to avoid confusion, physicists and engineers use the word **wavelength** to talk about the *distance* between crests of a wave, and the word **period** to mean the *time interval* between the arrival of the crests at your eardrum.

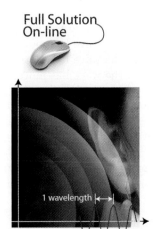

Figure 5.19: Compare to Figure 5.6.

Detecting wavelength

Example 5.21. Determine the wavelength of the traveling wave in Example 5.19.

Solution: Since wavelength doesn't have anything to do with time, let's set $t = 0$. This "photograph" of the wave is just the curve $y = \cos\left(\frac{4}{25}x\right)$, which completes one oscillation when the cosine "sees" 2π as its argument. That happens when $\frac{4}{25}x = 2\pi \Rightarrow x = 25\pi/2$ meters. So the wavelength is $\lambda = 12.5\pi$ meters. ∎

We're using equation (5.2) with $\nu = 4/25$.

Try It Yourself: Detecting wavelength

Example 5.22. Determine the wavelength of the traveling wave in Example 5.20.

Answer: $\lambda = 2\pi/13$. ∎

Full Solution On-line

Designing a traveling wave

Example 5.23. Design a traveling wave with a wavelength of 6 meters and a speed of 5 meters per second.

Solution: We know from equation (5.2) that the wavelength of the curve $y = \cos(\nu x)$ is $2\pi/\nu$. So let's set $6 = 2\pi/\nu$, from which we conclude that $\nu = \pi/3$. To have a speed of 5 m/sec we have $\omega = \nu s = (\pi/3)(5) = 5\pi/3$, so our answer is $\cos\left(\frac{\pi}{3}x - \frac{5\pi}{3}t\right)$. ∎

Try It Yourself: Designing a traveling wave

Example 5.24. Design a traveling wave with a wavelength of 3 centimeters and a speed of 51000 centimeters per second.

Answer: $\cos\left(\frac{2\pi}{3}x - 34000t\right)$. ∎

Full Solution On-line

✥ Dimensional Analysis

In order to talk about units in the present context, you need to know that we always use *radians* to measure angles in calculus. You'll see why in Chapter 3, but for now, take it as a convention: *the argument of the cosine should have units of radians.*

For a refresher on radians, see Appendix A.

With that idea in mind, let's look at the units involved in the argument of $\cos(\nu x - \omega t)$. If length (and so wavelength) is measured in meters,

$$\nu = \frac{2\pi}{\lambda} \quad \begin{matrix} \leftarrow & \text{radians} \\ \leftarrow & \text{m} \end{matrix}$$

Since x has units of meters, the product νx has units of $\left(\frac{\text{radians}}{\text{m}}\right)(\text{m}) = \text{radians}$. Similarly, when time is measured in seconds,

$$\omega = \nu s \quad \text{has units of} \quad \left(\frac{\text{radians}}{\text{m}}\right)\left(\frac{\text{m}}{\text{sec}}\right) = \frac{\text{radians}}{\text{sec}}$$

This is why we say that ω is an angular frequency.

Since both νx and ωt have units of radians, the argument of the cosine has the right dimensions.

✥ Common Difficulties with Transformations

Students sometimes have trouble remembering how the different parameters affect the graph, but this can be largely avoided by thinking in terms of the "mapping" notation that was introduced earlier. For example,

$$\cos : x \longmapsto y$$

reminds us that the cosine takes an input, x, *does something with it*, and arrives at an output, y. So the action of the cosine separates two different states: $x =$ before, and $y =$ after. Keeping this in mind: if we add $+h$ *before* the cosine acts, we shift the graph in the x-direction; and if we add $+k$ *after* the cosine has acted, we shift the graph in the y-direction.

> This is true for all functions, not just the cosine.

$$\cos(x + h) \qquad\qquad \cos(x) + k$$

$$\uparrow \qquad\qquad\qquad \uparrow$$

Adding $+h$ *before* the cosine acts   Adding $+k$ *after* the cosine acts
shifts the graph horizontally   shifts the graph vertically

Similarly, if we scale *before* the cosine acts, we dilate (or contract) the graph in the x-direction; and if we scale *after* the cosine has acted, we dilate (or contract) the the graph in the y-direction.

> This is true for all functions, not just the cosine.

$$\cos(6x) \qquad\qquad 8\cos(x)$$

$$\uparrow \qquad\qquad\qquad \uparrow$$

Scaling *before* the cosine acts   Scaling *after* the cosine acts
dilates (or contracts) the horizontal   dilates (or contracts) the vertical

---

You should know

- the terms *periodic, period, circular functions, even, odd, amplitude, scale, translate, angular frequency, phase shift,* and *wavelength;*

- how wavelength, speed, and angular frequency are related;

- the argument of a trigonometric function should be in radians;

- adding translates the graph (adding in the argument $\Rightarrow$ horizontal shift, adding to the function $\Rightarrow$ vertical shift);

- scaling dilates (or contracts) the graph of a function (scaling the argument $\Rightarrow$ horizontally, scaling the function $\Rightarrow$ vertically).

You should be able to

- calculate the wavelength and frequency of a traveling wave;

- find a formula for $f(x - t)$ when given a formula for $f(x)$;

- sketch $y = \alpha f(\beta(x - h)) + k$ from the graph of f, for given $\alpha, \beta, h, k \in \mathbb{R}$.

❖ 1.5 Skill Exercises

For each of the functions in #1–12, use the technique from p. 40 to find formulas for $f(x + 2), f(x - 5), f(x + 4t)$, and $f(x + h)$.

1. $f(x) = \cos(5x)$

2. $f(x) = \sin(3x)$

3. $f(x) = \tan(7x)$

4. $f(x) = \sec(7x)$

5. $f(x) = 3x + 7$

6. $f(x) = 8x - 9$

7. $f(x) = x^2 + 3x + 8$

8. $f(x) = x^3 + x - 3$

9. $f(x) = \sqrt{x}$

10. $f(x) = x^{1/5}$

11. $f(x) = \frac{x-3}{2x+5}$

12. $f(x) = \frac{x^2}{x+7}$

For each of the functions in #13–18, sketch the specified graph.

13. $y = \cos(2(x - 8))$

15. $y = 3\cos(5x + 35) - 4$

14. $y = \sin(3(x - 9))$

16. $y = 9\sin(\pi x - 0.25\pi^2) + 2$

17. $y = 2f(2x - 1) + 1$ when $f(x) = x^2 + 3x + 8$

18. $y = 3f(4x - 11) - 8$ when $f(x) = x^3 + x - 3$

For #19–20 find numbers α, β, h and k so that the graph depicts $y = \alpha f(\beta(x-h))+k$ when $f(x) = x^2$.

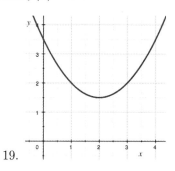

19.

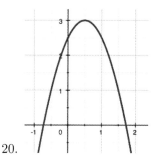

20.

In #21 and 22 $f(x) = \cos(x)$. Find numbers α, β, h and k so that the graph depicts $y = \alpha f(\beta(x - h)) + k$.

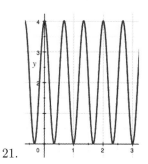

21.

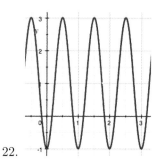

22.

In #23–34, sketch $y = f(t)$ and $y = 1/f(t)$ on the same axes.

23. $f(t) = 3t + 9$

27. $f(t) = t^2 + 1$

31. $f(x) = \sin(t)$

24. $f(t) = 5t - 10$

28. $f(t) = t^2 - 1$

32. $f(x) = \cos(t)$

25. $f(t) = 6t + 8$

29. $f(t) = t^2 + 4t$

33. $f(t) = 2\cos(t) + 3$

26. $f(t) = 2t - 7$

30. $f(t) = t^2 + t - 5$

34. $3\sin(t/2) - 3$.

Determine the angular frequencies of each function in #35-40 (where t is measured in seconds).

35. $f(t) = \sin(4t)$

37. $f(t) = \sin(\pi t)$

39. $f(t) = \cos\left(\frac{4\pi t}{5}\right)$

36. $f(t) = \cos(9t)$

38. $f(t) = \cos(3\pi t)$

40. $f(t) = \cos\left(\sqrt{6}t\right)$

Find the wavelength of each function in #41-46, assuming that x is measured in meters.

41. $f(x) = \sin(8x)$ 43. $f(x) = \cos(4000\pi x)$ 45. $f(x) = \cos\left(\frac{5\pi x}{8}\right)$

42. $f(x) = \cos(400x)$ 44. $f(x) = \sin(19\pi x)$ 46. $f(x) = \cos\left(\sqrt{10}x\right)$

47. Determine the value of ω for which the wavelength of $\cos(\omega x)$ is (a) 6π meters, (b) $\pi/4$ meters, and (c) $2/7$ meters.

48. Determine the value of ω for which the period of $\sin(\omega t)$ is (a) 3π seconds, (b) 10 seconds, and (c) 0.01 seconds.

In exercises #49–54 determine the wavelength, angular frequency, and speed of the traveling wave.

49. $\cos(4x - 19t)$.   51. $\cos(\pi x - 3t)$ 53. $\cos(\sqrt{8}\,x - \sqrt{21}\,t)$.

50. $\cos(3x - 20t)$ 52. $\cos(5x - \pi t)$ 54. $\cos(\sqrt{6}\,x - \sqrt{19}\,t)$.

❖ 1.5 Concept and Application Exercises

Exercises #55–60 concern the application of the *unit step function* (see p. 30) as a tool that allows us to select a window of time in which another function is allowed to be "switched on."

55. Here we see the unit step function used to "switch" another function "on."

    (a) On separate axes, draw the graphs of $y = u(t)$, $y = u(t-1)$, and $y = u(t-2)$. Then describe the graph of the function $u(t-a)$, where a can be an real number.

    (b) Suppose $f(t) = t-3$. On separate axes, draw the graphs of $y = u(t)f(t)$, $y = u(t-1)f(t)$, and $y = u(t-2)f(t)$. Then describe the graph of the function $u(t-a)f(t)$, where a can be an real number.

56. Here we see the unit step function used to "switch" another function "off."

    (a) On separate axes, draw the graphs of $y = 1 - u(t)$, $y = 1 - u(t-1)$, and $y = 1 - u(t-2)$. Then describe the graph of the function $1 - u(t-b)$, where b can be an real number.

    (b) Suppose $f(t) = t-3$. On separate axes, draw the graphs of $y = f(t)(1 - u(t))$, $y = f(t)(1 - u(t-1))$, and $y = f(t)(1 - u(t-2))$. Then describe the graph of the function $f(t)(1 - u(t-b))$, where b can be an real number.

57. Graph the function $u(t)(3t+1)(1 - u(t-1))$.

58. Graph the function $u(t-1)(t^2 - 2)(1 - u(t-3))$.

59. Graph the function $u(t-\pi)\sin(t)(1 - u(t-2\pi))$.

60. Graph the function $u\left(t - \frac{\pi}{4}\right)\cos(2t)\left(1 - u\left(t - \frac{5\pi}{4}\right)\right)$.

61. The greatest integer function, $\lfloor t \rfloor$, was discussed on p. 30. (a) Graph the function $f(t) = t - \lfloor t \rfloor$, and (b) determine its period.

62. Modern signal and image analyses often rely on functions called *wavelets*. The first set of wavelets were the *Haar* functions, which are indexed by the integers m and n as follows:

$$H_{m,n}(t) = 2^{-m/2}H(2^{-m}t - n) \quad \text{where} \quad H(t) = \begin{cases} 1 & \text{if } t \in [0, 0.5) \\ -1 & \text{if } t \in [0.5, 1) \\ 0 & \text{otherwise} \end{cases}$$

   (a) Sketch the graphs of $H_{0,-1}$, $H_{0,1}$, $H_{0,2}$ and $H_{0,3}$ on separate axes.

   (b) Sketch the graphs of $H_{-2,0}$, $H_{-1,0}$, $H_{1,0}$ and $H_{2,0}$ on separate axes.

   (c) Explain how the integers m and n affect the graph.

   (d) Sketch the graphs of $H_{1,0}$, $H_{1,1}$, $H_{1,2}$ and $H_{1,3}$ on the same axes.

   (e) Based on your graphs from part (d), how are the graphs of $H_{m,n}$ related when m is fixed but n is allowed to change?

   (f) Sketch the graphs of $H_{0,1}$, $H_{1,1}$, $H_{2,1}$, and $H_{3,1}$ on the same axes.

   (g) Based on your graphs from part (f), how are the graphs of $H_{m,n}$ related when n is fixed but m is allowed to change?

63. Suppose that $f(0) = -3$. Could f be an even function? If so, draw such a graph. If not, why not?

64. Suppose that $f(0) = 3$. Could f be an even function? If so, draw such a graph. If not, why not?

65. Draw the graph of a function that is neither even nor odd.

66. Draw the graph of a function that is both even and odd.

67. Suppose f is an odd function, and g is an even function. Is the function $h(t) = f(t)/g(t)$ even, odd, or neither? Why?

68. Suppose f and g are both odd functions. Is the function $h(t) = f(t)/g(t)$ even, odd, or neither? Why?

69. Suppose $h(t)$ is the altitude of an object above sea level (measured in feet) after t seconds. Write the following equations as English sentences: $h(8) = 10, h(2) = -17, h(32) = 0, h(t) = 0$.

70. Which of the following might be described quantitatively by a sine or cosine?

   (a) the amount you pay in taxes (as a function of year)

   (b) your height (as a function of days)

   (c) alternating current (as a function of seconds)

   (d) the Dow stock index (as a function of days)

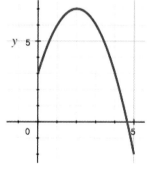

Figure 5.20: Graph for #71.

71. A single period of the function h is shown in Figure 5.20. Use it to determine $h(14)$.

72. Suppose an aircraft flies from New York to Chicago. When it arrives, it is put into a holding pattern because of heavy air traffic, and it circles the city at a radius of 50 miles for an hour before receiving clearance for landing.

   (a) Suppose $H(t)$ is the height of the aircraft at time t. Draw a graph of H.

   (b) Suppose $C(t)$ is the plane's distance from Chicago. Draw a graph of C.

   (c) Suppose $N(t)$ is the plane's distance from New York. Draw a graph of N as a function of time.

   (d) Using your graph from part (c), how would you calculate the length of time it takes the plane to orbit Chicago once?

   (e) Suppose it takes 0.5 hours to circle Chicago once. Design a formula for $N(t)$ that is roughly accurate when the aircraft is circling Chicago.

73. Think of the variable x as representing position, and suppose $f(x)$ is the air pressure at x, where we understand a function value of zero to be the reference (normal, day-to-day) pressure. That is, $f(3) = 0$ says that the air pressure is normal at position $x = 3$.

(a) Draw the graph of f with high pressure near $x = 0$ and normal pressure outside $[-1, 1]$.

(b) If it's not already, make your graph continuous (why is continuity a reasonable assumption)?

(c) Draw the graphs of $y = f(x - 1)$, $y = f(x - 2)$, and $y = f(x - 3)$ on the same axes.

(d) Draw the graphs of $y = f(x + 1)$, $y = f(x + 2)$, and $y = f(x + 3)$ on the same axes.

(e) Draw the graphs of $y = 0.5f(x - 1) + 0.5f(x + 1)$, $y = 0.5f(x - 2) + 0.5f(x + 2)$, and $y = 0.5f(x - 3) + 0.5f(x + 3)$ on the same axes (use different colors).

(f) As t increases, what happens to the graph of $y = 0.5f(x-t)+0.5f(x+t)$?

(g) Clap your hands one time. Really, make it loud.

(h) At the instant your hands came together, the pressure was very high near your *palms,* but was normal at your *ears.* So, at that instant, if the pressure was normal at your ears what did you hear? When *did* you hear something?

(i) What is the physical, "real-world" interpretation of your answer to part (f)?

74. Suppose that h is periodic and $h(x + 6) = h(x)$ for all $x \in \mathbb{R}$.

(a) Could it also be true that $h(x + 2) = h(x)$? If so, draw such a graph. If not, why not?

(b) Could it also be true that $h(x + 1) = h(x)$? If so, draw such a graph. If not, why not?

(c) Could it also be true that $h(x + \pi) = h(x)$? If so, draw such a graph. If not, why not?

75. The functions $\cos(6t)$ and $\sin(5t)$ are both periodic.

(a) Determine the respective periods of $\cos(6t)$ and $\sin(5t)$.

(b) Determine the period of $\cos(6t) + \sin(5t)$.

(c) How many full cycles does $\cos(6t)$ complete in that time? What about $\sin(5t)$?

76. The functions $\cos(pt)$ and $\sin(qt)$ are both periodic.

(a) Determine the respective periods of $\cos(pt)$ and $\sin(qt)$.

(b) What must be true about the numbers p and q in order for the function $f(t) = \cos(pt) + \sin(qt)$ to be periodic?

(c) When the function $f(t) = \cos(pt)+\sin(qt)$ is periodic, what is its period?

77. Suppose $g(t) = 7$ (i.e., a constant function). Is g periodic? If so, what is its period?

78. The function $f(t) = 6\cos(4\pi t)$ is periodic, but $g(t) = \frac{1}{1+t^2}\cos(4\pi t)$ is not. Why not?

79. While simplifying an equation, a teacher writes the line below. Justify the step based on your understanding of the sine and cosine functions.

$$\ldots = t^3\left(2\cos(-6t) + \frac{4}{3}\sin(-6t)\right) = t^3\left(2\cos(6t) - \frac{4}{3}\sin(6t)\right) = \ldots$$

80. You know that the sine function is odd, and the cosine function is even. Determine whether the other four trigonometric functions are *even, odd,* or *neither,* and justify your answer.

81. Suppose $f(x) = 3x$.

    (a) Find a number k so that $f(x+2) = f(x) + k$.

    (b) You know that $f(x+h) \neq f(x) + k$ when $f(x) = \cos(x)$, but vertical and horizontal translation seem to have the same net effect when $f(x) = 3x$. Explain why, in complete sentences (focus on aspects of the graph, not arithmetic).

82. Suppose $f(x) = 2x$.

    (a) Find a number β so that $f(\beta x) = \beta f(x)$.

    (b) You know that $\beta f(x) \neq \beta f(x)$ when $f(x) = \cos(x)$, but it seems to work when $f(x) = 2x$. Explain why, in complete sentences (focus on aspects of the graph, not arithmetic).

83. What does it mean to say that a cosine function has a wavelength of 7 meters? What does *wavelength* have to do with *period?*

84. What does it mean to say that a cosine function has an angular frequency of 7 radians/second? What does *frequency* have to do with *period?*

85. We often talk about light as a traveling wave, just like sound. The speed of light (in a vacuum) is $299,792,458\frac{\text{m}}{\text{sec}}$. The wavelength of blue light is 475 nanometers. What is its frequency?

> Physics

86. Suppose we use an electromagnetic trap to confine an electron in the interval $[0,5]$. The *time-independent Schrödinger equation* tells us that the likelihood of finding the electron at a particular $x \in [0,5]$ is related to the function

> Modern Physics

$$\psi(x) = \sin\left(\sqrt{\frac{4\pi mE}{h}}\, x\right),$$

> The symbol ψ is the lower-case Greek letter called "p-see," or "sigh."

where E is the energy of the electron, m is its mass, and h is Planck's constant. Because the likelihood of finding the electron outside of $[0,5]$ is zero, we require that $\psi(0) = 0$ and $\psi(5) = 0$. What are the possible values of E?

1.6 Exponential Functions

It's difficult to diagnose a problem when you can't see it, so internal medicine is a tricky business. One way to get images of internal organs is to use radioactive tracers, such as the molecule *pertechnetate*. Pertechnetate is a particularly useful molecule in medical imaging because it exposes the patient to radiation for a relatively short period of time. For example, if a patient is injected with 1 milligram of pertechnetate in suspension ...

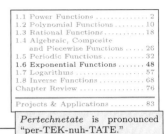

Pertechnetate is pronounced "per-TEK-nuh-TATE."

after	the amount that's still radioactive is	
6.02 hours	1/2 of a mg	
12.04 hours	1/4 of a mg	$(1/4) = (1/2)^2$
18.06 hours	1/8 of a mg	$(1/8) = (1/2)^3$
24.08 hours	1/16 of a mg	$(1/16) = (1/2)^4$
$\vdots$	$\vdots$	
$6.02n$ hours	$(1/2)^n$ of a mg	

Table 3: Radioactive decay of the pertechnetate molecule.

We say the pertechnetate molecule has a radioactive **half-life** of 6.02 hours, which means the amount that is still radioactive after t hours is

$$A(t) = A_0 \left(\frac{1}{2}\right)^{t/6.02},\tag{6.1}$$

where A_0 is the initial mass of the sample. Some people prefer to write this as

$$A(t) = A_0 2^{-t/6.02} = A_0 \left(2^{-1/6.02}\right)^t = A_0(0.01541)^t.\tag{6.2}$$

Since the *exponent* is what varies in this case, we say that $A(t)$ is an **exponential function**.

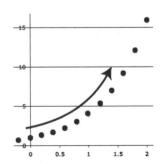

Check this formula against Table 3 so that you understand what the 6.02 is doing in the exponent!

Recall $b^{xy} = (b^x)^y$.

Radioactive half-life

Example 6.1. Thallium-201 has a half-life of 73 hours. (a) Write a function that will tell us the amount of a 4-gram sample that's still radioactive after t hours, and then (b) use it to determine the remaining radioactive mass after 90 hours.

Solution: (a) Using (6.1) as a template, we write $A(t) = 4\,(0.5)^{t/73} = 4(0.9905498)^t$. (b) The amount of the sample that's still radioactive after 90 hours is $A(90) = 1.7018712$ grams. ∎

▷ A Mathematical Pickle

Remember that $4^n = \underbrace{4 \cdot 4 \cdot 4 \cdots 4}_{n \text{ times}}$ when n is an integer, and

$$4^{p/q} = (4^p)^{1/q} = \sqrt[q]{4^p}$$

when p and q are integers, so we understand the meaning of 4^t when t is rational. Several values of 4^t for rational t are shown in Figure 6.1, where you can see an increasing trend.

Understanding numbers like 4^π and $4^{\sqrt{2}}$ is more difficult because the exponent is irrational, but every irrational number can be approximated by rational numbers.

Figure 6.1: $y = 4^t$ for some rational values of t.

Based on the increasing trend we see with rational values of t,

$$3.1 < \pi < 3.15 \quad \Rightarrow \quad 4^{3.1} < 4^\pi < 4^{3.15}$$
$$3.14 < \pi < 3.142 \quad \Rightarrow \quad 4^{3.14} < 4^\pi < 4^{3.142}$$
$$3.141 < \pi < 3.1416 \quad \Rightarrow \quad 4^{3.141} < 4^\pi < 4^{3.142}$$
$$3.1415 < \pi < 3.1416 \quad \Rightarrow \quad 4^{3.1415} < 4^\pi < 4^{3.1416}$$

$$\searrow \qquad \nearrow \qquad\qquad \searrow \qquad \nearrow$$
$$\pi \qquad\qquad\qquad\qquad 4^\pi$$

In baseball, a *pickle* is when the runner is caught between a pair of fielders who toss the ball back and forth as they move closer together, trapping the runner. Defining 4^π is lot like that.

This trick of squeezing down to a specific, well-defined value of 4^π by using rational numbers that get closer and closer to π is important for two reasons. First, it's an example of something called the *limit* that we'll define precisely in Chapter 2. Second, it demonstrates an important mathematical idea: when we don't understand something (e.g., irrational exponents), we often "run back" to things that we *do* understand (e.g. rational exponents).

▷ Running the Bases

When $f(t) = b^t$, we say the number b is the **base** of the exponential function. We typically restrict our attention to positive values of b because (1) that's what we encounter most often in practice, and (2) exponential functions with negative bases have severely restricted domains that make them unsuitable for elementary calculus. Accordingly, we'll say that a number b is an **admissible** base if $b > 0$ and $b \neq 1$. Some graphs of exponential functions are shown in Figure 6.2, and you can see that $f(t) = b^t$ behaves roughly the same for all admissible bases. There are two important things to note:

We don't say that $f(t) = 1^t$ is an exponential function, even though $1 > 0$, because 1^t is constant.

• The number b^t grows with t when $b > 1$, and it eventually surpasses each finite number. Functions with *larger* bases grow faster as t increases.

• The number b^t diminishes as t grows when $b < 1$, and it eventually falls below each $\varepsilon > 0$, no matter how small. Functions with *smaller* bases diminish faster as t increases.

Values of b that are "farther" from 1 result in functions that do things "faster." The number 1 is like a numerical fulcrum for exponential functions.

t	2^t	3^t
0	1	1
1	2	3
2	4	9
3	8	27
4	16	81
5	32	243
⋮	⋮	⋮

Note: Remember that $(1/2)^t$ is the same as 2^{-t}, so its graph is just the reflections $y = 2^t$ across the y-axis (see p. 36).

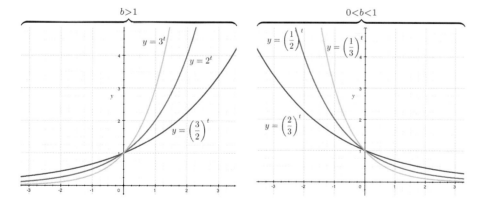

Figure 6.2: Graphs of various exponential functions.

▷ The Amazingly Special Base: e

In Chapter 3 we'll see that tangent lines are central to our study of calculus, and we'll define them in a strict, technical sense. For now, when we say that a line is tangent to the graph of f at the point $(t_0, f(t_0))$, we mean that the line touches the graph, but that it doesn't *cut through* the graph there. We don't yet have the mathematical tools to determine the slopes of tangent lines, but we can infer some important facts without making exact calculations.

- Suppose $f(t) = 2^t$. The left-hand image of Figure 6.3 shows the graph of f along with a line through $(0, 1)$ and $(1, 2)$, whose slope is 1. That line is called a **secant** line because it cuts through the graph of f. Since the secant line is steeper than the tangent line at $(0, 1)$, we know that the slope of the tangent line is less than 1.

- Suppose $f(t) = 4^t$. The right-hand image of Figure 6.3 shows the graph of f along with a secant line through $(0, 1)$ and $(-0.5, 0.5)$, whose slope is 1. In this case, you can see that the slope of the tangent line at $(0, 1)$ is greater than 1.

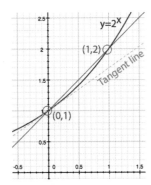 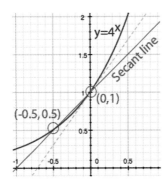

Figure 6.3: (left) Secant and tangent lines on the graphs of $y = 2^x$; (right) Secant and tangent lines on the graph of $y = 4^x$.

In summary, here's what we know about tangent lines and $f(t) = b^t$:

    ⋆ When $b = 2$ the slope of the tangent line at $(0, 1)$ is *less than* 1.

    ⋆⋆ When $b = 4$ the slope of the tangent line at $(0, 1)$ is *greater than* 1.

Part of being a good scientist, mathematician, or engineer is having the capacity to guess at what's right, so let's do it: based on (⋆) and (⋆⋆), it seems reasonable to think that there's some number $b \in (2, 4)$ for which the tangent line to the graph of $f(t) = b^t$ at $(0, 1)$ has a slope of *exactly* 1. In fact, there *is* such a number, and we call it e. It's approximately 2.71828182846, but is irrational.

The function e^t is known as the **Euler** exponential function (named after Leonhard Euler), and is also called the **natural exponential** function. It's *so* important in quantitative fields of study that it has its own notation: **exp**. That is,

$$\exp(t) = e^t, \quad \text{or in the mapping notation} \quad \exp: t \longmapsto e^t.$$

The exp notation is particularly handy when the argument of the exponential function is notationally cumbersome because it allows us to avoid superscripts upon superscripts, for example, and complicated fractions in the exponent.

> The name *Euler* is pronounced "OIL-er."

The number e in vibrations

Example 6.2. A shock absorber combines two basic technologies: a spring and a dashpot, the latter of which draws energy out of the system (imagine that the spring has to push and pull a tire pump). In simple models of this scenario, the up-down vibration of the system is described by

$$y(t) = e^{-at} \cos(bt).$$

Suppose the numbers a and b are positive. How does this function behave as time evolves?

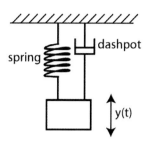

Figure 6.4: A schematic of the shock absorber in Example 6.2.

Solution: The key to understanding the behavior of $y(t)$ is considering the factors e^{-at} and $\cos(bt)$ separately. Because the cosine has an argument of bt, it completes a period in $2\pi/b$ seconds. When we discussed transformations in Section 1.5, we mentioned that the number preceding the cosine (sometimes called a *prefactor*) affects the amplitude of oscillation. Specifically, the function $f(t) = A\cos(bt)$ has an amplitude of A. In our case, the prefactor is not constant, but it affects the amplitude of the vibration in the same way.

$$\text{When} \begin{cases} t = 1 & \text{the vibration has a amplitude of } e^{-a} \\ t = 2 & \text{the vibration has a amplitude of } e^{-2a} \\ t = 3 & \text{the vibration has a amplitude of } e^{-3a} \end{cases}$$

Because the amplitude of the vibration is controlled by the prefactor, we say that the graphs of $y = e^{-at}$ and $y = -e^{-at}$ form an **envelope** for the vibration. You can see this manifested in Figure 6.5, in which $a = 0.25$ and $b = 4\pi$. ∎

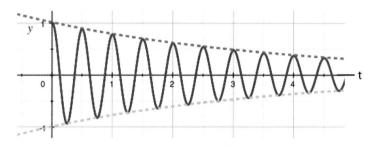

Figure 6.5: The curves $y = e^{-0.25t}$, $y = -e^{-0.25t}$, and $y = e^{-0.25t}\cos(4\pi t)$.

▷ Common Difficulties with Exponential Functions

Students sometimes confuse power functions with exponential functions, but they're really very different. For example, let's briefly contrast the power function $f(t) = t^3$ and the exponential function $g(t) = 3^t$ (whose graphs are shown in Figure 6.6).

- The function f has a root at $t = 0$, but $g(t)$ is *never zero*.

- The function f is odd, but g is neither even nor odd.

At an algebraic level, students sometimes find it difficult to remember that

$$e^{x+y} \neq e^x + e^y.$$

The base e is not special in this sense—it's true that $b^{x+y} \neq b^x + b^y$ for every admissible base. The correct statement is that $b^{x+y} = b^x b^y$. For example,

$$4^5 = 4 \cdot 4 \cdot 4 \cdot 4 \cdot 4 = (4 \cdot 4 \cdot 4)(4 \cdot 4) = 4^3 4^2.$$

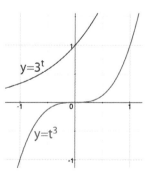

Figure 6.6: Comparing the curves $y = 3^t$ and $y = t^3$.

You should know

- the terms *half-life, exponential, admissible base, prefactor, envelope* and the notation *exp(t);*

- where the number e comes from, and know a decimal approximation for it to at least 6 digits;

- how the relationship between an admissible base and the number 1 dictates the behavior of an exponential function.

You should be able to

- distinguish between power functions and exponential functions (both by formula and by graph);

- use a prefactor to determine an envelope for other functions.

❖ 1.6 Skill Exercises

Use the properties of exponents to rewrite each function in #1–4 as an exponential function whose exponent is just t.

1. $A(t) = (2)^{-5t}$ 2. $A(t) = (0.1)^{2t}$ 3. $A(t) = 14(4)^{3t}$ 4. $A(t) = 13(10)^{-4t}$

Use the properties of exponents to rewrite each function in #5–8 as an exponential function whose base is larger than one.

5. $A(t) = (0.2)^{-5t}$ 6. $A(t) = (0.1)^{-2t}$ 7. $A(t) = 14(0.3)^{t}$ 8. $A(t) = 13(0.4)^{t}$

Without using graphing software, sketch the graphs of the functions in #9–14, using the idea that the first factor affects the magnitude of the second (see Example 6.2).

9. $f(t) = e^{-t}\cos(4\pi t)$ 11. $f(t) = 6e^{0.5t}\sin(t)$ 13. $f(t) = (4t-1)\sin(3t)$

10. $f(t) = 4^{t}\cos(0.25\pi t)$ 12. $f(t) = 5e^{-2t}\sin(2\pi t)$ 14. $f(t) = (3-5t)\cos(4t)$

In #15–20, find formulas for $f(t+2), f(t-5), f(t+\sqrt{2})$, and $f(t+h)$.

15. $f(x) = e^{5t}$ 18. $f(t) = 4\exp(12t^{3} - 7t + 1)$

16. $f(t) = 6(4)^{3t}$ 19. $f(t) = e^{7t}\cos(8t)$

17. $f(t) = \exp(12t^{2} - 7t + 1)$ 20. $f(t) = e^{-3t}\sin(4\pi t)$

❖ 1.6 Concept and Application Exercises

21. Suppose that $f(t) = (0.5)^{t}$ and $g(t) = (0.4)^{t}$. Which function grows faster as t becomes more and more *negative,* and why?

22. Suppose that $f(t) = (0.72)^{t}$ and $g(t) = (0.67)^{t}$. Which function decreases faster as t becomes more and more *positive,* and why?

23. Suppose that $f(t) = 4^t$. Show that $f(a+b) = f(a)f(b)$.

24. Suppose that $f(t) = \exp(t)$. Show that $f(t+h) = f(t)f(h)$.

25. Why do we say that a base of 1 is inadmissible?

26. Suppose $f(t) = (-8)^t$. Then $f(2/3) = 4$, but $f(1/2)$ doesn't make any sense. So the domain of f does not include all the positive numbers. Characterize those numbers that *are* in the domain of f.

The graphs shown in #27 and #28 are of exponential functions. For each exercise, determine (a) which function has the larger base, and (b) whether the bases are larger than or less than 1. In complete sentence, explain how you can tell.

27. 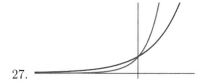   28.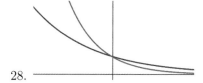

29. Suppose $f(x) = 5^x$.

    (a) Show that there is *not* a number α such that $f(2x) = \alpha f(x)$.

    (b) Show that $f(\beta x) = \alpha f(x)$ if and only if $\alpha = 1$ and $\beta = 1$.

30. Suppose that t months after the national debut of a product line, a company's customer service center processes $0.5 + 2.5t/(1 + 0.25\exp(0.01t))$ calls per month (in hundreds).   —| Business |

    (a) Interpret the following equation as a statement or question in complete sentences.

    $$0.5 + \frac{2.5t}{1 + 0.25\exp(0.01t)} = 2.5$$

    (b) Write an inequality that says, "The number of calls per month t_1 months after the debut is larger than the number of calls per month t_2 months after the debut.

31. Gas molecules are always moving around, but they don't all move at the same speed. Using some clever reasoning, James Clerk Maxwell was able to figure out that when Δv is small, the probability that a particular gas molecule has a velocity between v_0 and $v_0 + \Delta v$ is approximately $p(v_0^2)\Delta v$, where   —| Physical Chemistry |

    $$p(x) = 4\pi \left(\frac{a}{\pi}\right)^{3/2} xe^{-ax},$$

    in which the number a depends on the temperature (see [28, p. 6–9], or [11, p. 242]. As a first analysis, let's suppose that $a = 1$.

    (a) Use a graphing utility to render graphs of $y = x$ and $y = e^x$ on the same axes.

    (b) In the graph from part (a) you can see that $\exp(x)$ grows much faster that the linear function. What does this tell you about the fraction x/e^x?

    (c) Based on your answer from part (b), sketch the graph of p without the use of a graphing utility. (Note that p has a root at $x = 0$.)

32. Beryllium-11, which is denoted by ^{11}Be, has a half-life of 13.81 seconds. Write a function that we could use to calculate the amount of ^{11}Be that remains after t seconds. Assume the original sample has a mass of 2 grams.

33. The horizontal axis in the drawing below is a diagram of one row of a salt crystal, in which the sodium ions (Na$^+$) and chlorine ions (Cl$^-$) alternate.

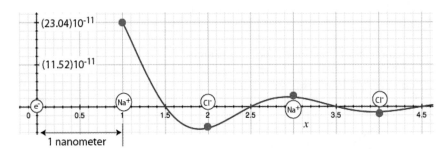

Electrostatic force, like gravity, obeys an inverse square law. On the graph above, the red dot at $x = 1$ indicates the amount of electrostatic force (experienced by an electron at the origin) due to the ion at $x = 1$; similarly for other red dots at $x = 2$, $x = 3$, and $x = 4$. Determine/design a function whose graph passes through the red dots (such as the blue curve shown in the figure).

34. The first microprocessor (called the 4004 chip) was invented in 1971 by Tedd Hoff and had 2,300 transistors. Gordon Moore made a prediction in 1975 that the integrated circuitry on a computer chip would double every 24 months.

   (a) Write a formula describing Moore's law. That is, write the *number* of transistors on a microprocessor as an exponential function (base 2) whose argument is *months since 1971*, $N(m) = \ldots$ You'll need to scale it so that $N(0) = 2300$.

   (b) Use your model from part (a) to predict the number of transistors on a microprocessor made in 2004.

   (c) Check your answer from part (b) against Intel's Itanium 2 processor, produced in 2004.

35. At the end of 2006, TIAA-CREF reported that its Large-Cap Value mutual fund had a *nominal* return of 15.27% (on average, over the 3-year period beginning in 2003). Presuming that the fund grow at the same rate during the entire 3-year period, this means that an investment of $100 into the fund would be worth

$$\underbrace{100}_{\text{initial investment}} + \underbrace{100}_{}\,(0.1527) = 100\,(1 + 0.1527) = 100\,(1.1527) \qquad (6.3)$$

after one year. We can get more return on our investment by leaving it in the account for another year. To calculate the value of the investment after that second year, we repeat the calculation in (6.3) and see

$$\underbrace{100(1.1527)}_{\substack{\text{value at the begin-}\\ \text{ning of year 2}}} + \underbrace{100(1.1527)}_{}\,(0.1527) = \underbrace{100(1.1527)}_{}\,(1.1527) = 100(1.1527)^2.$$

   (a) Find the value of the investment after three years.

(b) Calculate the value of the investment after 10 years.

(c) Find a formula for the value of the investment after n years.

36. In #35 you saw that the value of a $100 investment grows to

Finance

$$100(1.1527) \quad (1.1527)$$

This is the value of the investment after 1
year, including the interest earned

after two years, so you earn interest on your interest! Each time that happens, we say that the interest is *compounded*. In #35, the interest was compounded once per year, but if two investment opportunities are competing for your money, they can "sweeten the deal" by offering to compound the interest more than once per year. For example, suppose that two opportunities are both offering a nominal interest rate of 12% per year, but Company X offers to compound the interest semi-annually (i.e., you get a 6% = 12%/2 return of interest *twice* per year), and Company Y offers to compound the interest quarterly (i.e., you get a 3% = 12%/4 return of interest *four* times per year).

(a) For Company X:

   i. Determine the value of our $100 investment after 0.5, 1.5 and 2 years.

   ii. Find a formula for the value of the investment after n years.

   iii. Use your formula to calculate the value of the investment after 10 years.

(b) For Company Y:

   i. Determine the value of our $100 investment after 0.25, 0.5, and 0.75 years.

   ii. Find a formula the value of the investment after n years.

   iii. Use your formula to calculate the value of the investment after 10 years.

37. Suppose a single computer processor gives equal attention to all jobs. Jobs arrive at a rate of r jobs per second, and on average, each requires d seconds to complete. When we define $p = rd$, and $p < 1$, the fraction of the time that n packets will be in the queue, waiting to be processed, is $f(n) = p(1-p)^n$. For any fixed value of p, this is an exponential function of n.

Computer Science

When $p \geq 1$, the queue keeps getting longer and longer, and jobs do not get processed in a reasonable amount of time.

(a) As n increases, does $f(n)$ increase or decrease?

(b) Suppose $p = 0.75$. What fraction of time will there be *no* packets in line, waiting to be served?

(c) Suppose $p = 0.75$. What fraction of the time will there be 10 packets in line, waiting to be served?

(d) Is it better to have p larger or smaller?

38. Modeling population growth allows us to estimate the future demands on medical facilities, schools, power grids, water-treatment facilities, etc. The table in the margin itemizes the population of Mexico (in millions) in the early 1980s. The third column indicates how much the population grew (in millions) from one year to the next.

City planning

Year	Pop.	Change in pop.
1980	67.38	
1981	69.13	1.75
1982	70.93	1.80
1983	72.77	1.84
1984	74.66	1.89
1985	76.60	1.94
1986	78.59	1.99

(a) Not only is the population increasing, but the *change* in population is increasing. Explain why this makes sense.

(b) The change in population is not constant, so the population is not progressing *arithmetically*. The next simplest model is *geometric*, meaning a sequence of numbers that looks like

$$p_0, \quad \underbrace{p_1 = p_0 r,}_{\substack{\text{population af-}\\\text{ter 1 year}}} \quad \underbrace{p_2 = p_0 r^2,}_{\substack{\text{population af-}\\\text{ter 2 years}}} \quad p_3 = p_0 r^3, \quad p_4 = p_0 r^4, \ldots$$

where p_0 is the population when we begin tracking it. If the progression of population numbers were geometric, what would be the value of the ratio p_2/p_1? What about the value of the ratio p_3/p_2? the value of the ratio p_{k+1}/p_k?

(c) Check the population data in the table. Does it follow the correct pattern? That is, does a geometric model fit the data? If so, what is the number r?

(d) Write down an exponential function of the form $P(t) = P_0 r^t$ and use it to predict the population of Mexico in the year 1998.

(e) According to the Encyclopedia Britannica, there were 95,830,000 people in Mexico in 1998. Was your answer to part (d) close? If not, what mitigating factors could have made you miss?

1.7 Logarithms

In Section 1.6 we talked about using radioactive tracers in medical imaging. Let's return to that discussion for just a moment. Suppose a patient is injected with a tracer but then, for one reason or another, has to wait before getting access to the imaging device. During this time, the tracer is decaying, so there's less and less of it as time goes on. If the patient is forced to wait too long, the drug will become depleted to the point of being useless, and another dose will have to be administered. What constitutes "too long" depends on the particular tracer that's employed, but the basic calculation is always the same.

For instance, the tracer called *pertechnetate* has a half-life of 6.02 hours. So based on our discussion from p. 48, the amount of a 1-gram sample that remains after t hours is

$$A(t) = \left(\frac{1}{2}\right)^{t/6.02} \text{ grams.}$$

If we want to know, for example, how long it takes until only 0.01 grams of the sample remains, we need to solve the equation

$$\frac{1}{100} = \left(\frac{1}{2}\right)^{t/6.02}.$$

To solve this equation for t, we use a mathematical tool called the *logarithm*, which is introduced below.

> The word *logarithm* is pronounced "LOG-er-ith-um."

✦ Basic Ideas and Notation

Said simply, a logarithm is an exponent. Consider the equations

$$3^2 = 9 \quad \text{and} \quad 3^4 = 81.$$

> In Chapter 5 you'll see that there's a connection between logarithms and area.

In these equations, 2 is the exponent that relates 3 to 9, and 4 is the exponent that relates 3 to 81. We encode these facts notationally as follows:

$$2 = \log_3(9) \qquad\qquad 4 = \log_3(81)$$

2 is the exponent that relates 3 to 9   4 is the exponent that relates 3 to 81

Just as we say that 3 is the base of the exponential function 3^t, we say that 3 is the **base** of these logarithms. The examples below address other bases, but as we did with exponential functions, our discussion will be restricted to bases that are positive (and not 1).

Practice with logarithm notation

Example 7.1. Use the notation of logarithms to express the following relationships:

$$25^{1/2} = 5 \quad \text{and} \quad 10^{-1} = 0.1.$$

Solution: Written in logarithm notation, these relationships appear as follows:

$$\tfrac{1}{2} = \log_{25}(5) \qquad \text{and} \qquad -1 = \log_{10}(0.1). \qquad ■$$

1/2 is the exponent that relates 25 to 5   -1 is the exponent that relates 10 to 1/10

Try It Yourself: Practice with logarithm notation

Example 7.2. Use the notation of logarithms to express the facts that (a) $7^4 = 2401$ and (b) $8^{-1/3} = \frac{1}{2}$.

Answer: (a) $4 = \log_7(2401)$, and (b) $-\frac{1}{3} = \log_8(1/2)$. ■

Full Solution On-line

Said a little differently,

The number $\log_3(9)$ solves the equation $3^t = 9$.
The number $\log_{25}(5)$ solves the equation $25^t = 5$.
The number $\log_{10}(0.1)$ solves the equation $10^t = 0.1$.
The number $\log_8(0.5)$ solves the equation $8^t = 0.5$.

Said in generality,

the number $\log_b(c)$ solves the equation $b^t = c$.

> The notation $\log_b(c)$ is read "log base b of c."

Written as an equation, this last sentence is

$$b^{\log_b(c)} = c.$$

Using logarithm notation

Example 7.3. Write the solutions of the equations (a) $4^t = 64$ and (b) $5^t = 12$ in logarithmic notation.

Solution: (a) Based on our discussion above, the equation $4^t = 64$ is solved by the number $t = \log_4(64)$. It's easy to check that this is $t = 3$.

(b) Similarly, the equation $5^t = 12$ is solved by $t = \log_5(12)$. Unlike our answer from part (a), there's not a more familiar way to write this number. ∎

Try It Yourself: Using logarithm notation

Example 7.4. Write the solutions of the equations (a) $2^t = 16$ and (b) $3^t = 15$ in logarithmic notation.

Answer: (a) $\log_2(16)$, which is the number 4, and (b) $\log_3(15)$, which doesn't have a more familiar name. ∎

Full Solution On-line

Do logarithms always make sense?

Example 7.5. What number is $\log_5(-25)$?

Solution: Since 5^t is never negative, there's not a number that solves $5^t = -25$. Therefore, $\log_5(-25)$ isn't a real number. ∎

> We can understand $\log_5(-25)$ if we include complex numbers in our discussion. Complex numbers are introduced in Chapter 9.

When logarithm and exponential functions have the same base

Example 7.6. What number is $\log_7(7^8)$?

Solution: The number $\log_7(7^8)$ solves $7^t = 7^8$, so it must be 8. ∎

When is the logarithm zero?

Example 7.7. When is $\log_b(c) = 0$?

Solution: The number $\log_b(c)$ solves the equation $b^{\log_b(c)} = c$. When $\log_b(c) = 0$, this becomes $b^0 = c$, so $c = 1$. ∎

> Notice that the base was irrelevant to the final answer. This result is true for all admissible b.

▷ An important article ("the")

In our discussion so far, we've said things like, "$\log_{1.8} 10$ is *the* number which solves the equation $1.8^t = 10$." We say "*the* number" instead of "*a* number," because there's only one. As an illustrative example, the left-hand image of Figure 7.1 shows that any particular output of $f(t) = (1.8)^t$ is associated with *exactly one* value of t. Functions such as this, whose domain and range are in a one-to-one

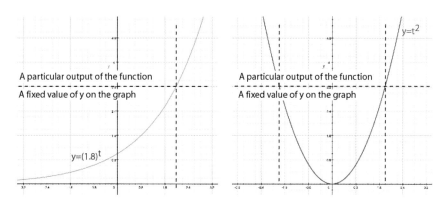

Figure 7.1: The function $f(t) = (1.8)^t$ is one-to-one, but the function $g(t) = t^2$ is not.

correspondence, are called **one-to-one** functions.

Note: Whenever you have *any* function, each input is associated with a single output. The important thing about a one-to-one function is that it works the other way, too: each output is associated with a single input.

▷ Logarithms take products to sums

The number $\log_4(32)$ has a more familiar name. To find it, we're going to write the number 32 in two different ways:

$$32 = 4^{\log_4(32)}. \tag{7.1}$$

$$32 = (16)(2) = \left(4^{\log_4(16)}\right)\left(4^{\log_4(2)}\right) = 4^{\{\log_4(16)+\log_4(2)\}}. \tag{7.2}$$

$$\underbrace{\hspace{7cm}}_{\text{remember } b^x b^y = b^{x+y}}$$

Equations (7.1) and (7.2) tell us that $\log_4(32)$ and $\{\log_4(16) + \log_4(2)\}$ both solve the equation $4^t = 32$, but 4^t is a one-to-one function so there is only one such number. The only logical conclusion is that we must be seeing *two different ways of writing the same number*. That is,

$$\log_4(16 \cdot 2) = \underbrace{\log_4(16)}_{=2} + \underbrace{\log_4(2)}_{=1/2}. \tag{7.3}$$

This isn't unusual: $\cos(\pi/3)$ and $\sqrt{0.25}$ are two different ways of writing 1/2.

So the more familiar name of $\log_4(32)$ is 2.5. Equation (7.3) is an example of a larger fact: for any admissible base b and positive numbers x and y,

$$\log_b(xy) = \log_b(x) + \log_b(y). \tag{7.4}$$

The hypothesis that x and y are positive numbers is an important one. Otherwise, the right-hand side of equation (7.4) does not exist!

Try It Yourself: Practice with the rules of logarithms

Example 7.8. Use equation (7.4) to calculate $\log_8(64 \cdot 2)$.

Answer: 7/3. ∎

Full Solution
On-line

▷ Logarithms interchange exponents and coefficients

Similarly, the number $\log_9(\sqrt{3})$ has a more familiar form. To find it, we're going to write the number $\sqrt{3}$ in two different ways:

$$\sqrt{3} = 9^{\log_9(\sqrt{3})} \tag{7.5}$$

$$\sqrt{3} = 3^{1/2} = (3)^{1/2} = \left(9^{\log_9(3)}\right)^{1/2} = 9^{(1/2)\log_9(3)}. \tag{7.6}$$

Equations (7.5) and (7.6) tell us that $\frac{1}{2}\log_9(3)$ and $\log_9(\,3^{1/2}\,)$ both solve the equation $9^t = \sqrt{3}$. As we did above, since $f(t) = 9^t$ is a one-to-one function we conclude that $\frac{1}{2}\log_9(3)$ and $\log_9(\,3^{1/2}\,)$ must be two different ways of writing the same number. That is,

$$\log_9(3^{1/2}) = \frac{1}{2}\log_9(3). \tag{7.7}$$

So the more familiar name of $\log_9(\sqrt{3})$ is $(\frac{1}{2})(\frac{1}{2}) = \frac{1}{4}$. Equation (7.7) is an example of a larger fact: for any admissible base b, positive x, and any r

$$\log_b(x^r) = r\log_b(x). \tag{7.8}$$

If you don't see that $\log_9(1/2) = 1/2$, consider the following:

The number $\log_9(3)$ solves the equation $9^t = 3$. The number $1/2$ does too. Since 9^t is a one-to-one function, these must be two different ways of writing down the same number. That is,

$$\frac{1}{2} = \log_9(3).$$

Try It Yourself: Practice with the rules of logarithms

Example 7.9. Use equation (7.8) to calculate $\log_5(\sqrt[4]{125})$.

Answer: 3/4. ∎

Full Solution On-line

▷ Logarithms take quotients to differences

When we combine the ideas in (7.4) and (7.8), we get another important algebraic property of the logarithm. Consider the following:

$$\log_9\left(\frac{\sqrt{3}}{81}\right) = \overbrace{\log_9\left(\sqrt{3}\cdot 81^{-1}\right) = \log_9(\sqrt{3}) + \log_9(81^{-1})}^{\text{as in (7.4)}} = \underbrace{\log_9(\sqrt{3}) + (-1)\log_9 81}_{\text{as in (7.8)}}.$$

Since $\log_9(\sqrt{3}) = 1/4$ and $\log_9(81) = 2$, the properties of logarithms have led us to the fact that

$$\log_9\left(\frac{\sqrt{3}}{81}\right) = \left(\frac{1}{4}\right) - (2) = -1.75.$$

Note: The number $\log_9\left(\sqrt{3}/81\right)$ is negative because the base is larger than one, but $\sqrt{3}/81$ is less than one. That is, $\log_9\left(\sqrt{3}/81\right)$ solves the equation

$$9^t = \frac{\sqrt{3}}{81} < 1,$$

which can only happen when $t < 0$.

As above, we've used this example to illustrate a larger fact: when b is an admissible base, and x and y are positive numbers,

$$\log_b\left(\frac{x}{y}\right) = \log_b(x) - \log_b(y). \tag{7.9}$$

Try It Yourself: Practice with the rules of logarithms

Example 7.10. Use equation (7.9) to calculate $\log_2\left(\frac{16}{\sqrt[3]{4}}\right)$.

Answer: 10/3. ∎

Full Solution On-line

1. $\log (1000000)$ 5. $\ln \left(e^3\right)$ 9. $\log_9 (27)$ 13. $\log_3 (26)$

2. $\log (10000)$ 6. $\log_7 \left(7^9\right)$ 10. $\log_8 (4)$ 14. $\log_5 (27)$

3. $\log_2 (1024)$ 7. $\log_4 \left(\frac{1}{64}\right)$ 11. $\log_{64}(16)$ 15. $\log_8 (2)$

4. $\log_3 (81)$ 8. $\log_5 \left(\frac{1}{125}\right)$ 12. $\log_{16}(32)$ 16. $\log_{64} (16)$

Use a calculator to determine the numbers in #17–20.

17. $\log_7 (18)$ 18. $\log_8 (6)$ 19. $\log_3 (10)$ 20. $\log_4 (e)$

Solve the given equations in #21–28.

21. $3^{6x} = 9^{2x}$ 23. $6^{3x} = 5(2^{9x})$ 25. $4^{x^2} = 3^{9x}$ 27. $4^{x^2} = 5(3)^{9x+1}$

22. $3^{6x} = 9^{2x-3}$ 24. $3^{6x} = 5(9^{2x})$ 26. $5^{x^2} = 6^{8x}$ 28. $7^{x^2} = 2(5)^{10x+1}$

29. Suppose $A(t) = 3\,(5)^t$ and $B(t) = 4(\frac{1}{3})^t$. Use the properties of logarithms to rewrite each function as an exponential function (a) whose base is 10; and (b) whose base is e.

30. Suppose $A(t) = 9\,(2)^t$. Rewrite A as an exponential function (a) whose base is 10; and (b) whose base is 0.7.

❖ 1.7 Concept and Application Exercises

31. What's wrong with the following calculation?

$$1 = \log_3(3) = \log_3(1 + 1 + 1) = \log_3(1) + \log_3(1) + \log_3(1) = 0 + 0 + 0 = 0$$

32. What's wrong with the following calculation?

$$\frac{5}{2} = 1 + \frac{3}{2} = \log_4(4) + \log_4(8) = \log_4(4 \cdot 8)$$

$$= \log_4(2 \cdot 4^2) = 2\log_4(2 \cdot 4) = 2\log_4(8) = 2\left(\frac{3}{2}\right) = 3$$

33. Find the error in the following argument:

We know that $3 > 2$, so $3\ln(0.5) > 2\ln(0.5)$. Using the properties of logarithms, we can rewrite this inequality as $\ln(0.5^3) > \ln(0.5^2)$. That is, $\ln(1/8) > \ln(1/4)$. Since exp is an increasing function, larger inputs produce larger outputs, so

$$\frac{1}{8} = \exp\left(\ln\left(\frac{1}{8}\right)\right) > \exp\left(\ln\left(\frac{1}{4}\right)\right) = \frac{1}{4}.$$

34. Without a calculator, determine which of $\log_\pi e$ and $\ln \pi$ is larger.

35. Just as $1968 = 1.968 \times 10^3$, every positive integer can be written in scientific notation as $a \times 10^b$, where a is a number between 1 and 9, and $b \geq 0$ is an integer. Explain why, when N is any positive integer, the number of digits in N is $1 + [\![\log(N)]\!]$, where $[\![x]\!]$ means to truncate x by removing the decimal portion.

36. Suppose $f(t) = \log(t)$ and $g(t) = \ln(t)$. Which grows faster as t gets larger and larger, $f(t)$ or $g(t)$?

37. Suppose $f(t) = \log(t)$ and $g(t) = \ln(t)$. Which grows faster as t moves from 0 toward 1?

38. The *Richter scale* defines the magnitude of an earthquake, M, by the energy it releases, E (measured in joules) according to $M = (-4.4 + \log E)/1.5$.

     Geology

    (a) Find the energy released by the 1906 San Francisco earthquake, which registered $M = 8.2$ on the Richter scale and devastated the city.

    (b) The *Boxing Day Tsunami* of 2004 was caused by an earthquake in the Indian Ocean that registered $M = 9.1$ on the Richter scale. Find the energy released by this earthquake.

    (c) If the energy of one earthquake is 10 times that of another, how much greater is its magnitude on the Richter scale?

39. On page 55 we said that the fraction of the time that n packets will be in the queue, waiting to be served by a single processor, is $f(n) = p(1-p)^n$, where $p < 1$ is a number that depends on the rate at which packets arrive to be served and the length of time that each job requires (both on average).

     Computer Science

    (a) Suppose that $p = 0.1$, and $f(n) = 0.04782969$. Use a logarithm to determine the value of n.

    (b) Interpret the result from part (a). What does it mean?

40. The acidity of a substance is measured by its *pH-factor*, which is defined to be $\text{pH} = -\log[H^+]$, where $[H^+]$ is the concentration of hydrogen ions (in moles per liter). For example, distilled water has 1×10^{-7} moles of hydrogen ions per liter, so it has a pH of 7.

     Chemistry

     Just as a "dozen" means 12 of something, a "mol" means 6.0221415×10^{23} of something. The number 6.0221415×10^{23} is called *Avogadro's number,* and is the ratio of a gram to an atomic mass unit.

    (a) If a solution has a pH of 6, does it have more or fewer moles of hydrogen ions per liter than distilled water? By what factor?

    (b) We say a substance is *acidic* when its pH < 7, and *alkaline* when its pH > 7. Calculate the pH of each of the substances below and classify them as acidic or alkaline.

Substance	$[H^+]$
Arterial blood	3.9×10^{-8} mol/L
Tomatoes	6.3×10^{-5} mol/L
Milk	4.0×10^{-7} mol/L
Coffee	1.2×10^{-6} mol/L

41. Suppose an *annual* rate of interest for an account is r percent is compounded twice per year. In #36 on page 54, you showed that the value of a $100 investment after n years would be

     Finance

$$V(n) = 100 \left(1 + \frac{r}{2}\right)^{2n}.$$

"V" for <u>v</u>alue

    (a) How long will it take for the investment to reach a value of $100,000 if the account earns 12% interest?

    (b) How long will it take for the investment to reach a value of $100,000 if the account earns 12% interest, but compounded 4 times per year instead of twice?

42. It's true that compounding interest more frequently is better (see #41), but there's an upper bound on how good it can get. Later in the book you'll show that the value of the account after t years would be $V(t) = Pe^{rt}$ if the interest were compounded *continually,* where P is the initial investment (called the *principal*).

——| Finance |

   (a) Suppose you want to have \$300,000 after 25 years. Assuming you invest in an account with an annual percentage rate of 7.25%, compounded continuously, how much do you have to invest today? (This is the *present value* of the investment.)

——| $r = 0.0725$ for part (a). |

   (b) Suppose that you could invest \$1000 today in an account that paid interest continuously. What annual percentage rate would you need in order to have \$300,000 after 25 years?

43. When an ideal gas undergoes an adiabatic change (see #24, p. 9), the initial volume and temperature, V_0 and T_0, are related to the final volume and temperature, V_1 and T_1, according to

——| Physical Chemistry |

$$\frac{d}{2} \ln \left(\frac{T_1}{T_0} \right) = -\ln \left(\frac{V_1}{V_0} \right),$$

where d is the number of so-called "degrees of freedom" for the gas. (Near room temperature, the number d is 3 for a monatomic gas and 5 for a diatomic gas.)

   (a) Use the properties of logarithms to show that this equation can also be written as $V_1 T_1^{d/2} = V_0 T_0^{d/2}$.

   (b) Use the Ideal Gas Law as it's expressed in exercise #24 on p. 9 to rewrite this equation as

$$V^\gamma P = (\text{some constant that depends on } V_0 \text{ and } T_0),$$

where $\gamma = (d+2)/d$, which is called the *adiabatic exponent.*

1.8 Inverse Functions

Though we haven't used special vocabulary to describe it, you've already seen cases when the action of one function undoes the work of another. For example, when $k(t) = t^3$ and $h(t) = \sqrt[3]{t}$,

$$h\left(k(t)\right) = \sqrt[3]{k(t)} = \sqrt[3]{t^3} = t \quad \text{and} \quad k\left(h(t)\right) = \left(h(t)\right)^3 = \left(\sqrt[3]{t}\right)^3 = t,$$

so the actions of the functions h and k undo each other. Similarly, when $f(t) = \log_9(t)$ and $g(t) = 9^t$,

$$f\left(g(t)\right) = \log_9\left(g(t)\right) = \log_9(9^t) = t \quad \text{and} \quad g\left(f(t)\right) = 9^{g(t)} = 9^{\log_9(t)} = t.$$

When $f(g(t)) = t$ and $g(f(t)) = t$, we say that f and g are **inverse functions**. Our attention is usually focused on one function or the other, so we might say that g is the inverse of f, but the relationship is symmetric, so f is the inverse of g also. We typically denote the inverse of f by f^{-1} (read as "f inverse"). So in the context of power functions we have

$$
\begin{aligned}
h(t) &= \sqrt[3]{t} \quad \text{and} \quad h^{-1}(t) = t^3, \quad \text{or equivalently} \\
k(t) &= t^3 \quad \text{and} \quad k^{-1}(t) = \sqrt[3]{t}.
\end{aligned}
$$

Similarly, in the case of exponential and logarithmic functions

$$
\begin{aligned}
f(t) &= \log_9(t) \quad \text{and} \quad f^{-1}(t) = 9^t, \quad \text{or equivalently} \\
g(t) &= 9^t \quad \text{and} \quad g^{-1}(t) = \log_9(t).
\end{aligned}
$$

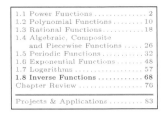

> We want the name of the function to help us remember its special relationship to f, and the name f^{-1} does that.
>
> **WARNING:** the "-1" in f^{-1} is part of the function's name. It's not an algebraic instruction.

Not all functions have inverse functions

Example 8.1. Show that $f(t) = t^2$ does *not* have an inverse function.

Solution: Figure 8.1 shows the graph of $f(t) = t^2$. If you know that the output of f is 9, the input could have been $t = 3$ *or* $t = -3$, and you can't tell which it was. The best you can do is say that $f^{-1}(9) = $ both 3 and -3, but that's not a *function* because it produces more than one output. For this reason, the function $f(t) = t^2$ does not have an inverse function (we say it is not **invertible**). ∎

In Figure 8.1 you see that the graph of $f(t) = t^2$ includes both the point $(3, 9)$ and the point $(-3, 9)$, so the horizontal line $y = 9$ intersects the graph twice. It's this double-intersection that indicates a lack of invertibility.

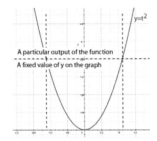

Figure 8.1: Graph of $y = t^2$.

> **Horizontal Line Test:** Suppose that some horizontal line intersects the graph of f more than once. Then the function f is not invertible. On the other hand, if every horizontal line intersects the graph *at most* once, the function f *is* invertible.

Using the Horizontal Line Test

Example 8.2. Use the Horizontal Line Test to show that $f(t) = t^3$ is invertible, but $g(t) = 3(t-1)(t-2)(t-3)$ is not.

Solution: The graphs of $y = f(t)$ and $y = g(t)$ are shown in Figure 8.2. Note that each and every horizontal line intersects the curve $y = t^3$ at most one time, so the function f is invertible; but the line $y = 0.7$ intersects the graph of g three times, so g is not invertible. (*Most* horizontal lines, such as $y = 1.7$, intersect the graph of g only once, but the Horizontal Line Test doesn't allow "partial credit." If a graph fails for *any* line, it fails entirely.) ∎

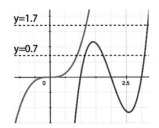

Figure 8.2: The graphs of $y = t^3$ and $y = 3(t-1)(t-2)(t-3)$, along with the horizontal lines $y = 7/10$ and $y = 17/10$.

Finding inverse functions

Example 8.3. The function $f(t) = 9t + 4$ is invertible. Find a formula for f^{-1}.

Solution: The function f generates an output, y, by performing two actions on its argument: multiply by 9, then add 4. To *undo* that procedure, we subtract 4 and then divide by 9. So $f^{-1}: y \longmapsto (y-4)/9$. We can check that

$$f^{-1}(f(t)) = \frac{f(t) - 4}{9} = \frac{9t + 4 - 4}{9} = \frac{9t}{9} = t.$$ ■

Of course, the *action* of f^{-1} doesn't depend on the letter we choose to represent its argument, so we could have written the inverse function in Example 8.3 as $f^{-1}: b \longmapsto (b-4)/9$, or as $f^{-1}: s \longmapsto (s-4)/9$, or even as

$$f^{-1}: t \longmapsto \frac{t-4}{9}. \tag{8.1}$$

The description of f^{-1} we see in (8.1) is nice because (1) we're accustomed to thinking of t as the input variable, and (2) it allows us to graph both f and f^{-1} together, using the horizontal axis to represent the domains of both f and f^{-1} simultaneously (see Figure 8.3).

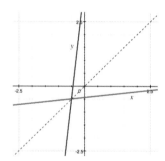 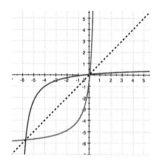

Figure 8.3: (left) The graphs of $f(t)$ and $f^{-1}(t)$ from Example 8.3; (right) the graphs of $f(t)$ and $f^{-1}(t)$ from Example 8.4.

Finding inverse functions algebraically

Example 8.4. Suppose $f(t) = (2t + 1)/(3t + 19)$. (a) Find a formula for $f^{-1}(t)$, and (b) graph f and f^{-1} on the same set of axes.

Solution: (a) Suppose that y is a number in the range of f. The inverse function tells us the value of t that was used to produce y. We can find it algebraically by solving the equation

$$y = \frac{2t + 1}{3t + 19}$$

> You should think of y as a particular, fixed number for our discussion.

for t. After multiplying both sides of this equation by $3t + 19$ we see

$$3yt + 19y = 2t + 1 \tag{8.2}$$

$$t = \frac{1 - 19y}{3y - 2}. \tag{8.3}$$

So $f^{-1}: y \longmapsto (1 - 19y)/(3y - 2)$. Since the letter we use to represent the argument is irrelevant to the action of the function, we can also write this as $f^{-1}: t \longmapsto (1 - 19t)/(3t - 2)$. That is $f^{-1}(t) = (1 - 19t)/(3t - 2)$. (b) The graphs of f and f^{-1} are shown in Figure 8.3.   ■

Don't just read it. Work it out! How do you pass from (8.2) to (8.3)?

Try It Yourself: Finding inverse functions algebraically

Full Solution On-line

Example 8.5. Find a formula for $f^{-1}(t)$ when $f(t) = (3t - 4)/(2 + 8t)$.

Answer: $f^{-1}(t) = (4 + 2t)/(3 - 8t)$. ∎

▷ Graphs of inverse functions

If $f : 1 \to 3$, the inverse action is $f^{-1} : 3 \to 1$, so the point $(1, 3)$ is on the graph of f, and $(3, 1)$ is on the graph of f^{-1}. In short, graphing f^{-1} amounts to swapping the coordinates of every point on the graph of f, which can be done quickly by reflecting the graph of f over the line $y = t$ (e.g., see Figures 8.3 and 8.4).

❖ Restricted Domains and Invertibility

The domain of a function is sometimes restricted for reasons other than its formula. For example, in Section 1.1 we saw that gravitational attraction obeys an inverse-square law, such as $F(r) = 1/r^2$. On the one hand, the graph of F does not formally pass the horizontal line test (because F is an even function), so F is not invertible. On the other, in the context of gravitational force we know that r cannot be negative because it's a distance, and the graph of F *does* pass the Horizontal Line Test when we allow only $r > 0$. The idea of restricting a function's domain is handy.

Restricting domain and invertibility

Example 8.6. Restrict the domain of $f(t) = t^2$ so that the function is invertible.

Solution: The graph of f fails the Horizontal Line Test when the domain of f is unrestricted, but it passes the test when the domain is restricted to $[0, \infty)$. That is, the function $f(t) = t^2$ *is* invertible when its domain is $[0, \infty)$, and we call the inverse the *square-root function* (see Figure 8.4). ∎

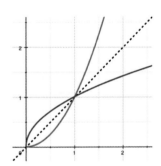

Figure 8.4: $y = f(t)$ over $[0, \infty)$ and $y = f^{-1}(t)$.

▷ Inverse trigonometric functions

Like $g(t) = t^2$, the sine and cosine are not invertible when their domains are unrestricted. The graph of the cosine is shown in Figure 8.5, and you can see that there are many ways to restrict its domain so that there's a one-to-one correspondence between numbers in its domain and range. By convention, we agree to use the restriction $0 \leq \theta \leq \pi$, which corresponds to angles in the 1st and 2nd quadrants.

Notice that the white band, $0 \leq \theta \leq \pi$, are angles that correspond to the 1st and 2nd quadrants on the unit circle.

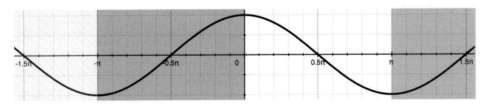

Figure 8.5: There are many ways to restrict the domain of $\cos(\theta)$ so that its invertible. Our convention is to use $0 \leq \theta \leq \pi$.

We can't use the same restriction for the sine (the graph still wouldn't pass the Horizontal Line Test), but the sine is invertible when we restrict its domain to $-\frac{\pi}{2} \leq \theta \leq \frac{\pi}{2}$, which is the convention (see Figure 8.6).

Everything we've discussed about inverse functions applies to the *restricted* sine and cosine functions, but let's pause for a moment to add some language that might

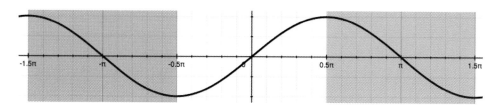

Figure 8.6: There are many ways to restrict the domain of $\sin(\theta)$ so that it's invertible. Our convention is to use $-\pi/2 \le \theta \le \pi/2$, which corresponds to points on the unit circle in the 4th and 1st quadrants.

> The white band in Figure 8.6 is $-\pi/2 \le \theta \le \pi/2$, which are angles that correspond to points on the units circle in the 4th and 1st quadrants.

help you understand these things better. If the sine "takes an angle and tells us a number," its inverse function "takes a number and tells us an angle."

$$\sin : \angle \longmapsto \#   \quad\text{so}\quad   \sin^{-1} : \# \longmapsto \angle$$
$$\cos : \angle \longmapsto \#   \quad\text{so}\quad   \cos^{-1} : \# \longmapsto \angle$$

Calculating with $\sin^{-1}$

Example 8.7. Determine (a) the value of $\sin^{-1}(0.5)$, and (b) the length of arc on the unit circle that subtends that angle.

Solution: (a) We know that $\sin(\pi/6) = 0.5$, so $\sin^{-1}(0.5) = \pi/6$. (b) The length of arc on the circle of radius 1 that subtends $\pi/6$ radians (see Figure 8.7) is $(1)(\pi/6) = \pi/6$ units long.   ∎

> It's also true that
>
> $$\sin(5\pi/6) = 0.5,$$
>
> but
>
> $$\sin^{-1}(0.5) \ne 5\pi/6$$
>
> because $\sin^{-1}$ only selects angles in $[-\pi/2, \pi/2]$, which is the domain restriction that gave rise to $\sin^{-1}$.

In Example 8.7 we saw that $\sin^{-1}(0.5) = \pi/6$ is a length of arc. Accordingly, we refer to $\sin^{-1}(x)$ as the **arcsine** of x, and sometimes write it as $\arcsin(x)$. Similarly, we refer to $\cos^{-1}(x)$ as the **arccosine** of x and sometimes write it as $\arccos(x)$.

Combining trig and inverse-trig functions

Example 8.8. Determine (a) $\cos(\sin^{-1}(0.5))$, and (b) $\sin^{-1}(\sin(3\pi/4))$.

Solution: (a) In Example 8.7 we showed $\sin^{-1}(0.5) = \pi/6$, so

$$\cos\left(\sin^{-1}(0.5)\right) = \cos\left(\frac{\pi}{6}\right) = \frac{\sqrt{3}}{2}.$$

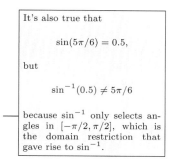

Figure 8.7: The blue arc subtends an angle of $\pi/6$ radians on the unit circle.

(b) Many beginners start by saying, "Sine and arcsine are inverse functions, so they undo one another ..." That's the right spirit but can lead to the wrong answer because, to be more specific, the arcsine is the inverse of the *sine function whose domain is restricted to* $[-\pi/2, \pi/2]$. That is,

$$\sin : \left[-\frac{\pi}{2}, \frac{\pi}{2}\right] \longrightarrow [-1, 1] \quad\text{so}\quad \sin^{-1} : [-1, 1] \longrightarrow \left[-\frac{\pi}{2}, \frac{\pi}{2}\right].$$

So the arcsine will tell us an angle in the interval $[-\pi/2, \pi/2]$. What angle in $[-\pi/2, \pi/2]$ has a sine of $\frac{\sqrt{2}}{2}$? Answer: the angle $\theta = \pi/4$. So

$$\sin^{-1}\left(\sin\left(\frac{3\pi}{4}\right)\right) = \sin^{-1}\left(\frac{\sqrt{2}}{2}\right) = \frac{\pi}{4}   \quad ∎$$

Another important inverse-trigonometric function is the **arctangent**, denoted by $\tan^{-1}(x)$ or $\arctan(x)$, which is defined as the inverse of *the tangent whose domain is restricted to* $(-\pi/2, \pi/2)$. Its basic function is similar to that of the arcsine and arccosine:

$$\tan : \left(-\frac{\pi}{2}, \frac{\pi}{2}\right) \longrightarrow (-\infty, \infty) \quad\text{so}\quad \tan^{-1} : (-\infty, \infty) \longrightarrow \left(-\frac{\pi}{2}, \frac{\pi}{2}\right).$$

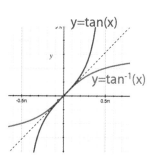

Figure 8.8: Graph of the *restricted* tangent function and the arctangent.

Combining trig and inverse-trig functions

Example 8.9. Calculate (a) $\sin(\tan^{-1}(7/2))$, and (b) find an algebraic formula for $\tan(\sin^{-1}(x))$.

Solution: (a) Since $7/2$ is not associated with any of our standard reference angles, you might be tempted to use your calculator, but this can be done easily by hand. The arctangent tells us the angle θ in $(-\pi/2, \pi/2)$ whose tangent is $7/2$:

$$\theta = \tan^{-1}\left(\frac{7}{2}\right) \quad\Longrightarrow\quad \tan(\theta) = \tan\left(\tan^{-1}\left(\frac{7}{2}\right)\right) = \frac{7}{2}.$$

Let's draw a right triangle with an angle θ whose tangent is $7/2$ (see Figure 8.9). We know from the Pythagorean Theorem that the hypotenuse of this triangle has a length of $\sqrt{53}$, so

$$\sin\left(\tan^{-1}\left(\frac{7}{2}\right)\right) = \sin(\theta) = \frac{7}{\sqrt{53}}.$$

(b) As we did above, let's construct a right triangle with an angle θ whose sine is x (see Figure 8.9). We know from the Pythagorean Theorem that the remaining leg has a length of $\sqrt{1-x^2}$, so

$$\tan\left(\sin^{-1}(x)\right) = \tan(\theta) = \frac{x}{\sqrt{1-x^2}}.$$ ∎

> This skill becomes very important when performing a technique called *trigonometric substitution* in Chapter 6.

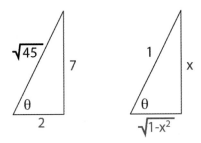

Figure 8.9: (left) A right triangle with $\tan(\theta) = 7/2$; (right) a right triangle with $\sin(\theta) = x$.

Try It Yourself: Combining trig and inverse-trig functions

Example 8.10. Calculate (a) $\cos\left(\sin^{-1}\left(-\frac{\sqrt{3}}{2}\right)\right)$, and then (b) find an algebraic formula for $\cos(\sin^{-1}(x)) = \frac{1}{2}$.

Answer: (a) $\cos\left(\sin^{-1}\left(-\frac{\sqrt{3}}{2}\right)\right) = 0.5$, (b) $\cos(\sin^{-1}(x)) = \sqrt{1-x^2}$. ∎

Full Solution On-line

✧ Common Difficulties with Inverse-trig Functions

The most common, most devastating mistake that beginners make with the inverse-trigonometric functions is mistaking the -1 in $\sin^{-1}$, $\cos^{-1}$, and $\tan^{-1}$ as an algebraic instruction. It's not. That -1 is part of the name of the function. It never, *never, Never, NEVER* means reciprocal.

$$\cos^{-1}(x) \neq \frac{1}{\cos(x)}, \quad \sin^{-1}(x) \neq \frac{1}{\sin(x)}, \quad \text{and} \quad \tan^{-1}(x) \neq \frac{1}{\tan(x)}.$$

> If we mean to communicate the reciprocal of a function value, we make it unmistakably clear by writing $[f(t)]^{-1}$.

We already have a name and notation for $1/\cos(x)$: we call it the secant, and denote it by $\sec(x)$. Similarly, the function $1/\sin(x)$ is the cosecant and is denoted by

$\csc(x)$, and the function $1/\tan(x)$ is called the cotangent and is denoted by $\cot(x)$.

You should know

- the terms *inverse function, invertible, Horizontal Line Test, arcsine, arccosine,* and *arctangent;*

- that if $f : x \to y$, then $f^{-1} : y \to x$;

- that $f(f^{-1}(y)) = y$ and $f^{-1}(f(x)) = x$;

- restrictions on the trigonometric functions that make them invertible.

You should be able to

- use the Horizontal Line Test for injectivity;

- use the graph of f to draw the graph of f^{-1};

- derive a formula for $f^{-1}(t)$ from the formula for $f(t)$;

- calculate numbers such as $\tan^{-1}(1)$ and $\sin^{-1}(\sqrt{2}/2)$ by hand.

❖ 1.8 Skill Exercises

1. Suppose that f is the inverse of g, and $f(7) = 3$. What's $g(3)$?

2. Suppose that f is a one-to-one function whose domain is $[1,3]$ and whose range is $[10,17]$. What are the domain and range of f^{-1}?

3. Find a formula for $f^{-1}(t)$ when $f(t) = t^7 - 15$.

4. Suppose $f(t) = t^2 - 5t + 2$. Find a domain over which f is one-to-one, and find a formula for its inverse.

5. What's wrong with the calculation $\frac{2\pi}{3} = \sin^{-1}\left(\sin\left(\frac{2\pi}{3}\right)\right)$?

6. Draw the graph of

   (a) a function that's not injective, and explain how you can tell;

   (b) a function that *is* injective, and use your graph to determine the domain and range of the inverse function.

7. Figure 8.10 shows the graph of f.

   (a) Find numbers $a < b$ so the function is invertible when restricted to $[a,b]$.

   (b) Find a number x_0 so that, regardless of how close together you choose a and b (but not equal), the function will not have an inverse when we restrict its domain to $[a,b]$.

8. Figure 8.11 shows the graph of f in its entirety.

   (a) Without graphing f^{-1}, determine its domain.

   (b) Without graphing f^{-1}, determine $f^{-1}(2)$ and $f^{-1}(-2)$.

   (c) Without graphing f^{-1}, where is the graph steepest?

   (d) Use the reflection technique to draw the graph of f^{-1}, and see if your answers to parts (a)-(c) were correct.

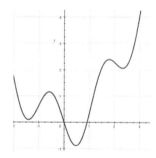
Figure 8.10: Graph for #7.

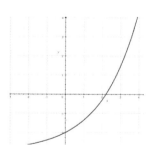
Figure 8.11: Graph for #8.

9. Without the aid of a graphing utility, sketch the graph of $y = \arcsin(\sin(x))$.
   (Hint: the graph over $[0, \pi/2]$ is simple, the rest requires some thought.)

Evaluate the numbers in #10–24 *without a calculator.*

10. $\cos(\sin^{-1}(1))$

11. $\arccos\left(\sin\left(\frac{2\pi}{3}\right)\right)$

12. $\sin(\arctan(-1))$

13. $\tan\left(\cos^{-1}\left(\frac{\sqrt{3}}{2}\right)\right)$

14. $\cos^{-1}(0)$

15. $\arctan(-1)$

16. $\arcsin\left(\frac{-\sqrt{2}}{2}\right)$

17. $\sin^{-1}\left(-\tan\left(\frac{\pi}{4}\right)\right)$

18. $\arcsin\left(\tan\left(\frac{7\pi}{4}\right)\right)$

19. $\arccos\left(\cos\left(-\frac{2\pi}{3}\right)\right)$

20. $\cot(\arccos(0))$

21. $\cos\left(\sin^{-1}\left(-\frac{1}{2}\right)\right)$

22. $\sin\left(\cot^{-1}\left(\sqrt{3}\right)\right)$

23. $\sin(\cos^{-1}(3/4))$

24. $\cos(\sin^{-1}(7/8))$

Rewrite each expression in #25–36 without trigonometric functions.

25. $\sin\left(\tan^{-1}(x)\right)$

26. $\cos\left(\cot^{-1}(x)\right)$

27. $\tan\left(\sin^{-1}(5x)\right)$

28. $\cot\left(\cos^{-1}(5x)\right)$

29. $\sec\left(\tan^{-1}(\sqrt{x})\right)$

30. $\cos\left(\sin^{-1}(\sqrt{x})\right)$

31. $\sin\left(\cos^{-1}(x)\right)$

32. $\sin\left(\cos^{-1}(2x)\right)$

33. $\cot\left(\sin^{-1}\left(\frac{5x}{3}\right)\right)$

34. $\cos\left(\sin^{-1}\left(\frac{\sqrt{3}}{x^2}\right)\right)$

35. $\sec\left(\cot^{-1}\left(\frac{x}{6}\right)\right)$

36. $\csc\left(\tan^{-1}(3\ln x)\right)$

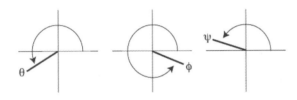

Figure 8.12: For exercises #37–40.

Use the graphs in Figure 8.12 for exercises #37–40.

37. Find a formula for $\sin(\theta)$ that involves NO trig functions if the first-quadrant reference angle for theta is $\theta_{ref} = \cos^{-1}\left(\frac{x}{2}\right)$.

38. Find a formula for $\sec(\phi)$ that involves NO trig functions if the first-quadrant reference angle for phi is $\phi_{ref} = \tan^{-1}\left(e^{3x}\right)$.

> The symbol ϕ is the lower-case Greek letter *phi*, which is pronounced "fee."

39. Find a formula for $\csc(\psi)$ that involves NO trig functions if the first-quadrant reference angle for psi is $\psi_{ref} = \cos^{-1}\left(\frac{1}{2}\sin(x)\right)$.

> The symbol ψ is the lower-case Greek letter *psi*, which is pronounced "sigh."

40. Find a formula for $\cos(\psi)$ that involves NO trig functions if the first-quadrant reference angle for psi is $\psi_{ref} = \tan^{-1}\left(\frac{5x^2}{8}\right)$.

Find a formula for f^{-1} for each of the functions in #41–54. The functions in #53 and #54 have restricted domains.

41. $f(t) = 17t + 8$

42. $f(t) = 9t + 6$

43. $f(t) = \frac{3t+1}{9t-20}$

44. $f(t) = \frac{8t-7}{t+5}$

45. $f(t) = \sqrt[3]{\frac{t-3}{t+1}}$

46. $f(t) = \sqrt[5]{\frac{2t+7}{t-2}}$

47. $f(t) = 37t^3 + 8$

48. $f(t) = 19t^7 + 8$

49. $f(t) = 8 + 2^{3^t}$

50. $f(t) = 9 + 3^{6^t}$

51. $f(t) = e^{5t^3 + 84}$.

53. $f(t) = \ln\left(\frac{1}{t} + 7\right)$, over $(0, \infty)$

52. $f(t) = 7^{t^5 - 19}$

54. $f(t) = \log_5(3t - 9)$, over $(3, \infty)$

Exercises #55–59 address the **hyperbolic sine** (denoted by $sinh$) and **hyperbolic cosine** (denoted by $cosh$), which are defined as follows:

$$\sinh(t) = \frac{1}{2}(e^t - e^{-t}) \quad \text{and} \quad \cosh(t) = \frac{1}{2}(e^t + e^{-t}).$$

55. Show that $\cosh^2(t) - \sinh^2(t) = 1$.

56. The function $\sinh(t)$ is invertible. Find a formula for $\sinh{-1}(t)$. *(Hint: multiply the equation $y = 0.5\sinh(t)$ by e^t. You'll see a $(e^t)^2$ and e^t, so the equation is quadratic in e^t.)*

57. The graph of the hyperbolic sine function, $f(t) = \sinh(t)$, is shown in Figure 8.13. Draw the graph of f^{-1}.

58. Look at the graph of $\cosh(t)$. How would you restrict the domain in order to make it invertible? Make such a restriction and determine a formula for its inverse, $\cosh^{-1}(t)$.

59. The hyperbolic tangent is $\tanh(t) = \sinh(t)/\cosh(t)$. Find a formula for its inverse function, $\tanh^{-1}(t)$.

❖ 1.8 Concept and Application Exercises

60. Find a function f that is its own inverse function. That is, $f^{-1}(t) = f(t)$.

61. Suppose f is an even function. Do we know whether f is invertible? If so, how? If not, provide a pair of examples that show why not.

62. Suppose f is an odd function. Do we know whether f is invertible? If so, how? If not, provide a pair of examples that show why not.

63. Suppose f is neither even nor odd. Do we know whether f is invertible? If so, how? If not, provide a pair of examples that show why not.

64. The function $f(t) = 4t^3 + t - 9$ is invertible. Find $f^{-1}(-9)$. *(Hint: you don't need a formula for $f^{-1}(t)$ in order to do this.)*

65. The function $f(t) = e^{18t^3 - 12}$ is invertible. Find $f^{-1}(1)$.

66. Explain why $f(t) = e^{18t^3 - 12}$ is invertible but $g(t) = e^{18t^4 - 12}$ is not.

67. Suppose $f(t) = 4t^2 + 3t - 10$. Find a domain over which f is one-to-one, and find a formula for its inverse.

68. Suppose $f(t) = -3t^2 + t + 9$. Find a domain over which f is one-to-one, and find a formula for its inverse.

69. Draw the graph of $y = \tan(x)$, restricted to $-\pi/2 \leq x \leq \pi/2$. Then the tangent is invertible.

    (a) Why do we have to restrict the domain of the tangent in order for it to be invertible?

    (b) Use the reflection technique to draw the graph of $y = \tan^{-1}(x)$.

    (c) What are the domain and range of $y = \tan^{-1}(x)$?

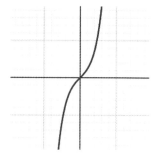

Figure 8.13: Graph for #57.

Chapter 1 Review

❖ True or False?

1. A power function with an even exponent is an even function.

2. A polynomial function whose leading term has an even exponent is an even function.

3. The cosine is an even function.

4. The sine is an even function.

5. All functions are either even or odd.

6. A power function with an odd exponent is invertible.

7. A polynomial with an odd degree is invertible.

8. All odd functions are invertible.

9. Suppose the horizontal line $y = 10$ intersects the graph of f exactly one time. Then f is invertible.

10. Cosine tells us the x-coordinate of a point on the unit circle.

11. The x-coordinate of a point on the unit circle is given by the sine.

12. The wavelengths of $\cos(10x)$ and $\sin(10x)$ are the same.

13. Suppose $f(t) = t^2$. Then $f(t + h) = t^2 + h$.

14. Suppose that $f(t) = 6^t$ and $g(t) = \left(\frac{1}{6}\right)^t$. Then $f(t) = g(-t)$.

15. The exponential function $f(t) = 7^t$ has no y-intercept.

16. The only *admissible* bases are e and 10 because those are the bases that are available on calculators.

17. Suppose $f(t) = 8^t$. Then $f(t + h) = f(t)f(h)$.

18. The change of base Theorem allows us to rewrite the number $\log_4(9)$ as $\log_b(9)/\log_b(4)$ using any number b we want.

19. The number $\log_7(56)$ is equal to 8.

20. $\log_5(x + y) = \log_5(x) + \log_5(y)$.

21. $\log_7(0.75) = (\ln(3) - \ln(4))/\ln 7$.

22. Suppose that $f(t) = 3t - 7$. Then $f^{-1}(t) = \frac{1}{3t-7}$.

23. Suppose the graph of f has a vertical asymptote at $x = 6$. Then the graph of f^{-1} has a horizontal asymptote at $y = 6$.

24. Suppose f is an odd function that's invertible. Then f^{-1} is odd.

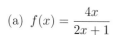 Multiple Choice

25. A clipping from the graph of $f(t) = t^p$, where p is an integer, is shown in Figure 9.1. Based on this clipping,

   (a) the exponent p is even.

   (b) the exponent p is odd.

   (c) we cannot determine the parity of p.

26. Consider the function from #25. Based on its graph,

   (a) the exponent p is positive.

   (b) the exponent p is negative.

   (c) we cannot determine the polarity of p.

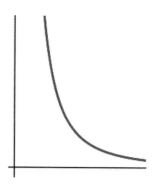

Figure 9.1: For #25.

27. A clipping from the graph of $g(t) = t^{p/q}$, where p and q are integers and the ratio p/q is in lowest terms, is shown in Figure 9.2. Based on this graph, select the best of the options below.

   (a) The number q is even.

   (b) The number q is odd.

   (c) We cannot determine the parity of q from this small clipping.

28. Consider the clipping of the graph of g from #27. Then choose the best of the options below.

   (a) The exponent $p/q > 1$.

   (b) The exponent $p/q < 1$.

   (c) We cannot determine the magnitude of p/q from this small clipping.

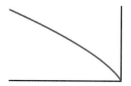

Figure 9.2: For #27.

29. Suppose that $f(t) = \frac{10t^4 + 9t^3 + 10t^2 - 18t + 14}{3t^4 + 8t^2 + 2}$. Select the most accurate of the statements below.

   (a) The graph of f has 4 vertical asymptotes.

   (b) The graph of f has a horizontal asymptote.

   (c) The graph of f has no asymptotes.

   (d) None of the above.

30. Classify each of the following functions as either a power, polynomial, rational, or simply as algebraic.

   (a) $f(x) = \dfrac{4x}{2x + 1}$

   (b) $f(x) = 2x^{1/5}$

   (c) $f(t) = t^3 - 7t + 4t^{-1}$

   (d) $f(t) = 2t^5 + 6t^2 - \dfrac{3}{4}t^9$

   (e) $f(\nu) = \sqrt[4]{\nu}$

   (f) $f(\nu) = \sqrt{2 - 7\nu}$

   (g) $f(\omega) = \dfrac{3\omega - 1}{\sqrt{2\omega + 8}}$

   (h) $f(\omega) = \dfrac{4\omega^2 - 3\omega + 6}{2}$

31. The wavelength of $\cos(10x)$ is ...
       (a) 10   (b) 0.1   (c) 5   (d) None of the above

32. Suppose a traveling wave is modeled by $\cos(3x - 18t)$, where distance is measured in meters and time in seconds. Choose the best option from those below.

(a) The wave speed is 18 meters per second.

(b) The wave's frequency is 18 hertz.

(c) The wave's wavelength is $2\pi/3$ meters.

(d) None of the above.

33. Choose the best of the following statements about logarithms:

   (a) Logarithms take products to sums.

   (b) Logarithms take quotients to differences.

   (c) Logarithmic and exponential functions are inverses.

   (d) All of the above.

34. Determine which of the following equations are correct.

   (a) $\ln(6x + 8y^2) = 6\ln x + 16\ln y$

   (b) $\ln(8xy^4) = 4\ln(8xy)$

   (c) $2\ln(x^2 + 1) = \ln(x^4 + 2x^2 + 1)$

   (d) None of the above.

35. The value of $\sin(\cos^{-1}(-7/8))$ is ...

   (a) $\sqrt{15}/8$.

   (b) nonsense, since $\cos^{-1}(-7/8)$ isn't a number.

   (c) nonsense, since $\cos^{-1}(-7/8)$ is larger than one.

   (d) none of the above.

36. The function $f(t) = \sin^{-1}(t)$ is really

   (a) the cosecant.

   (b) the secant.

   (c) the opposite of the sine function.

   (d) none of the above.

❖ Exercises

37. Suppose one microbeam is connected to another by a sponge-like material. The "lower" microbeam is fixed in place, but the "upper" beam can move up and down. We denote by $y(t)$ the directed distance (in meters) of the upper beam from its mechanical equilibrium after t seconds, and by F the force offered by the connective material. Write an equation that says, "The force experienced by the upper microbeam is proportional to its distance from mechanical equilibrium."

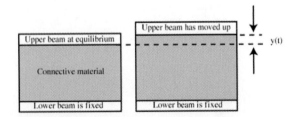

38. Write an equation that says, "The air resistance encountered by an aircraft is proportional to the square of its speed."

39. In #35 on p. 54, you showed that a $100 investment grows to $100(1.1527)^n$ after n years when the annual percentage rate is 15.27% (assuming interest is compounded once each year).

   (a) Suppose that, instead, the investment has a return of r%. Show that the value of a $100 investment is

$$V(t) = 100 \left(1 + \frac{r}{100}\right)^t \quad \text{after } t \text{ years.}$$

   (b) What value of r will double the investment in 7 years?

40. Suppose that one account offers an annual return of 10%, compounded semi-annually, and another offers an annual return of 8%, compounded quarterly. If you invest $100 in each, how long will it take for the accounts to have the same value?

41. Because different investments offer such a wide variety of options, we need a standard by which to compare them. This standard is called the *annual percentage rate* (APR), which is the rate of interest that would be earned each year if it were compounded only once.

   (a) Show that the value of a $100 investment is

$$V(t) = 100 \left(1 + \frac{r}{100m}\right)^{mt}$$

   after t years, when the nominal interest rate is r% and interest is compounded m times per year.

   (b) Find a formula for APR in terms of r and m

   (c) If interest is compounded continuously, the value of a $100 investment is $V(t) = 100e^{\frac{r}{100}t}$ when the nominal interest rate is r%. Find a formula for calculating APR in terms of r.

42. The basic idea of *reducing-balance depreciation* is that instead of *increasing* in value by some percent each year, an asset *decreases* in value each year by some percentage.

   (a) Show that the value of a $100 asset is

$$V(t) = V_0 \left(1 - \frac{r}{100}\right)^t$$

   after t years, where V_0 is the initial value of the asset, and the rate of depreciation is r% per year.

   (b) Suppose that an asset is purchased for $30,000, and that its value depreciates by 8.2% each year. How long will it be until the asset is worth only $1000?

43. Explain the relationship between air velocity and a wing based on the graph in Figure 9.3 on p. 80.

44. The graph of the polynomial $h(t)$ is shown in Figure 9.4.

   (a) What do you know about the leading term?

   (b) What do you know about the constant term?

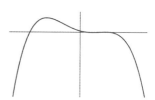

Figure 9.4: For #44.

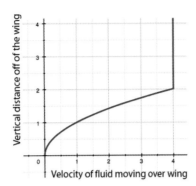

Figure 9.3: Graph for #43.

45. What do you know about the leading coefficients in the numerator and denominator of the rational function that's depicted below?

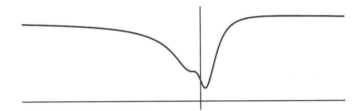

46. Suppose that instead of using a circle for the stereographic projection, we use the ellipse

$$\frac{x^2}{a^2} + \frac{y^2}{b^2} = 1.$$

   (a) What should be the maximum value of $x(t)$, and what value of t would achieve that value?

   (b) What should be the equation of the horizontal asymptote for the graph of $y(t)$?

   (c) Derive the equations for $x(t)$ and $y(t)$ and use them to verify your conjectures from parts (a) and (b).

While the transcendental functions are extremely useful in describing the real world, their function values are notoriously difficult to calculate. For example, while determining the value of $\sin(0)$ is easy, determining the value of $\sin(1)$ is difficult because one radian is not a standard reference angle. An *extremely* important tool in approximating the values of the transcendental functions are **Taylor polynomials**, which you'll learn to design in Chapter 8.

47. The figure below shows Taylor polynomials of degrees 2-28 (even) that have been designed to approximate a specific transcendental function. The last one,

$$T_{28}(x) = 1 - \frac{1}{2}x^2 + \frac{1}{4!}x^4 - \frac{1}{6!}x^6 + \frac{1}{8!}x^8 - \frac{1}{10!}x^{10} + \cdots + \frac{1}{28!}x^{28}$$

is shown in blue. What transcendental function is being approximated by

The exclamation point is not used to add emphasis in mathematics (e.g., 4! does not mean you should say *"four!"* with gusto), but rather indicates a special kind of multiplication called the **factorial**. In short, it means to multiply all of the integers between 1 and the integer that's cited.

$$3! = (3)(2)(1)$$
$$4! = (4)(3)(2)(1)$$
$$5! = (5)(4)(3)(2)(1)$$

The value of 5! is 120, but we often read it as "five factorial."

these polynomials?

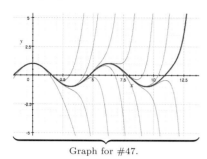

Graph for #47.

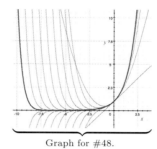

Graph for #48.

48. The graph above shows Taylor polynomials of degrees 1-20 that have been designed to approximate a specific transcendental function. The last one,

$$T_{20}(x) = 1 + x + \frac{1}{2}x^2 + \frac{1}{3!}x^3 + \frac{1}{4!}x^4 + \frac{1}{5!}x^5 + \frac{1}{6!}x^6 + \cdots + \frac{1}{20!}x^{20}$$

is shown in blue. What transcendental function is being approximated by these polynomials?

49. Design a formula that describes a traveling wave whose wavelength is 4 meters, and which travels at 238 meters per second.

50. Suppose that $f(x) = \cos(13x)$. Determine (a) a formula for $f(x - 7t)$, and (b) a formula for $f(x + h)$.

51. Suppose that $f(t) = t^2 + 17t + 1$. Find a formula for $f(t + h)$.

52. Suppose that $A(t) = 2^{kt}$. Find a value of k so that $A(5) = 17$.

53. Solve the equation $\log_3(x^2 + 1) = 2$.

54. For what values of k is the equation $\log_5(6x^2 + 3) = k$ not solvable?

55. Suppose f is a function whose domain includes $t = 0$.

   (a) Show that $\phi(t) = 0.5f(t) + 0.5f(-t)$ is an even function. The function ϕ is called the **even part** of f.

   (b) Show that $\psi(t) = 0.5f(t) - 0.5f(-t)$ is an odd function. The function ψ is called the **odd part** of f.

   (c) Show that $f(t) = \phi(t) + \psi(t)$.

   (d) Show that the even part of e^t is the hyperbolic cosine, $\cosh(t)$, and the odd part is the hyperbolic sine, $\sinh(t)$ (see p. 75).

56. What's the domain of the function $f(t) = \ln(t^2 - 6)$?

57. What is the domain of $f(t) = \sin^{-1}(\tan^{-1}(t))$?

58. Rewrite the following numbers in a simpler form, if there is one.

   (a) $\log_3(81)$ (c) $\log_5(7)$ (e) $\log_{16}(0.25)$

   (b) $\log_4(8)$ (d) $\log_9(3)$ (f) $\log_2(1024)$

59. Use the algebraic rules of logarithms to ...

   (a) expand $\log_4\left(\frac{x^6 y^5}{(x+y)^8}\right)$ into several, simpler logarithms.

   (b) compress $3\log(x) - \log(y) - 7\log(z)$ into a single logarithm.

   (c) compress $3\ln(x) - \log_3(y) - 7\log_2(z)$ into a single logarithm.

60. Write an equation that's solved by the number $\log_7(92)$.

61. How old is an object if the unstable-to-stable ratio of its carbon, R_{now}, is
    0.05% of what we find in living things today?

                                               | Archeology |

62. Find an algebraic formula for $\cos(\tan^{-1}(t))$.

Chapter 1 Projects and Applications

❖ Richardson's Arms Race

On January 23, 2007, the New York Times published an article by Joseph Kahn in which he wrote:

> The Chinese government confirmed today that it had conducted a successful test of a new anti-satellite weapon ... the first successful destruction of a satellite in orbit for more than 20 years.

Stephanie C. Lieggi of the Center for Nonproliferation Studies added:

> Despite Beijing's assurance that the test was not "aimed" at anyone, China's use of an ASAT weapon to destroy a satellite raises a number of questions about Beijing's intentions in space. Beijing has appeared in the past to take the moral high ground with regards to space arms control. Chinese diplomats have called for controls on the weaponization of space for over a decade despite a lack of reciprocity from the United States. Skeptics, particularly in Washington, argued that Beijing was cynically using arms control as a ruse to hide its own military space program. Last week's test appears to lend some credibility to that view. Others, however, argue that it is the billions of dollars the United States has spent on research and development in the military use of space that ultimately triggered China to move forward with ASAT development.

When military spending in one country causes another to increase its own military spending, we call it an *arms race*. In an effort to predict whether a small incident could potentially start a large conflict, English physicist Lewis Fry Richardson (1881–1953) developed the now-famous *Richardson's Arms Race* model.

In its simplest form, Richardson's model describes a situation in which country X adjusts its military spending in response to that of a perceived adversary, country Y, and vice versa. However, increased military spending leaves less money available for road construction, hospitals, education, research, farm subsidies, and social programs, so the model incorporates an internal pressure to reduce military spending.

▷ Part 1: Interpreting the model

In this section we're going to examine a representative example of Richardson's model in order to accomplish two goals. First, we want to develop a sense of what each part of the model tells us. Second, we want to understand how the different parts combine to predict changes in levels of military spending from year to year.

We're going to use x to denote the military spending (in billions of dollars) in country X, and Δx to denote the yearly *change* in that spending. Similarly, we'll use y and Δy to denote the same quantities for country Y. The particular model that we're going to examine describes the change in the two countries' military spending as follows:

$$\Delta x = 0.2y - 0.1x + 10 \qquad (10.1)$$
$$\Delta y = 0.1x - 0.7y + 49. \qquad (10.2)$$

> There are other ways to measure military spending. For example, the dollar amount might not be as important as the percentage of the Gross Domestic Product.

1. Suppose $x = 100$ and $y = 30$ this year.

   (a) Use equations (10.1) and (10.2) to calculate Δx and Δy.

   (b) Determine how much each country will spend on its military next year.

In any given year, the numbers x and y have particular, known values, so we can think of (x, y) as a particular, known point in the Cartesian plane. However, the numbers x and y change from year to year (as you saw in #1), so the point moves.

2. Explain why the point (x, y) must remain in the first quadrant (including the origin and the positive-x and positive-y axes).

3. Would negative Δx (or Δy) make sense? If so, what would it mean?

Now that we have practiced calculating and using Δx and Δy, the next few questions will investigate the role played by each of the constants in the model.

4. Suppose that neither country spent money on its military last year.

   (a) Use equations (10.1) and (10.2) to determine how much each country is spending on its military this year.

   (b) When X and Y decided on last year's budget neither country felt the need to spend money on its military, and yet in part (a) you found that the point (x, y) moved into the first quadrant. Something must have happened in the political landscape during the course of the year to warrant this increase in spending. Determine a plausible explanation for this sudden change in spending levels. (Hint: The numbers 10 and 49 in equations (10.1) and (10.2) are called the **grievance coefficients** for X and Y, respectively.)

The next two questions will focus on country X.

5. Suppose $x = 50$.

   (a) Calculate Δx when $y = 50, 60,$ and 70.

   (b) Describe how Δx depends on y, and based on the meaning of y, explain why that makes sense politically. (Hint: The number 0.2 in equation (10.1) is called the **defense coefficient** for country X.)

6. Suppose $y = 50$.

   (a) Calculate Δx when $x = 50, 60,$ and 70.

   (b) In part (a) you showed that higher levels of military spending in country X result in smaller yearly increments in that spending. Explain why that might happen from a political point of view. (Hint: The number 0.1 in equation (10.1) is called the **fatigue coefficient** for country X.)

7. In this question we'll derive the condition under which country X feels no need to change its military spending.

   (a) By setting $\Delta x = 0$ in equation (10.1) show that country X makes no change in its military spending when (x, y) sits on the line $y = \frac{1}{2}x - 50$, which we'll call ℓ_x.

   (b) Pick a point in the first quadrant that's to the right of ℓ_x and use equation (10.1) to calculate Δx. Based on the sign of Δx, will the point move to the left or to the right in the next year?

(c) Pick a point in the first quadrant that's to the left of ℓ_x and calculate the corresponding value of Δx. Based on the sign of Δx, will the point move to the left or to the right in the next year?

(d) Based on parts (b) and (c), describe how the point (x, y) moves in relation to ℓ_x from one year to the next (if it's not already on ℓ_x).

Now we turn our attention to country Y.

8. Identify the defense and fatigue coefficients for country Y in equation (10.2).

9. In this question we'll derive the condition under which country Y feels no need to change its military spending.

(a) By setting $\Delta y = 0$ in equation (10.2) show that country Y makes no change in its military spending when (x, y) sits on the line $y = \frac{1}{7}x + 70$, which we'll call ℓ_y.

(b) Pick a point in the first quadrant that's above ℓ_y and use equation (10.2) to calculate Δy. Based on the sign of Δy, will the point move up or down in the next year?

(c) Pick a point in the first quadrant that's below ℓ_y and calculate the corresponding value of Δy. Based on the sign of Δy, will the point move up or down in the next year?

(d) Based on parts (b) and (c), describe how the point (x, y) moves in relation to ℓ_y from one year to the next (if it's not already on ℓ_y).

In #7 you saw that country X will not change its military spending if (x, y) sits on ℓ_x, and in #9 you saw that country Y will not change its military spending if (x, y) sits on ℓ_y. For this reason, we refer to the lines ℓ_x and ℓ_y as **nullclines** (meaning "no change").

10. Suppose (x, y) sits on both nullclines, simultaneously. What happens to spending levels from year to year?

11. Find the intersection of ℓ_x and ℓ_y.

▷ Part 2: Computing with the model

The following exercises will lead you through a computational investigation of the example described by equations (10.1) and (10.2). We'll be working with a spreadsheet program, and in the directions below you'll see that we usually refer to numbers by their *location* (in the spreadsheet) rather than by their value.

12. Put the coefficients from (10.1) into cells **A1, B1,** and **C1** respectively.

13. Put the coefficients from (10.2) into cells **A2, B2,** and **C2** respectively.

	A	B	C	D	E
1	0.2	0.1	10		
2	0.1	0.7	49		
3					

14. Enter the number 30 into cell **A4.** (This will be the initial value of x. All subsequent values of x will be below this one.)

15. Enter the number 15 into cell **B4.** (This will be the initial value of y. All subsequent values of y will be below this one.)

16. We're going to type the formula for calculating Δx in cell **C4**. The formula will involve the defense, fatigue, and grievance coefficients for X, but we'll refer to them by their location in the spreadsheet rather than their value. In cell **C4** type the following (including the equals sign):

$$= \text{A\$1*B4 - B\$1*A4 + C\$1} \tag{10.3}$$

you'll understand the role of $ in just a moment, and
why it's on some terms but not others

> Make sure that you know which terms of this formula refer to the defense, fatigue, and grievance coefficients for X, respectively.

If all has gone well, you should see the value 10 appear in cell **C4**.

17. In cell **D4**, type the formula for calculating Δy. It should be very similar to formula (10.3). Remember to use \$ when you cite the locations of the defense, fatigue, and grievance coefficients for Y. If all has gone well, you should see the value 41.5 appear in the cell.

	A	B	C	D	E
1	0.2	0.1	10		
2	0.1	0.7	49		
3					
4	30	15	10	41.5	
5					

Initial x   Initial y   Initial Δx   Initial Δy

18. Now we tell the computer how to change the value of x. Since

$$x_{\text{next}} = x_{\text{current}} + \Delta x \, , \quad \text{and } x \text{ cannot be negative,}$$

in cell **A4**   in cell **C4**

we type the formula $=\textbf{max}(\textbf{0, A4+C4})$ into cell **A5**.

> Depending on the particular program that you're using, the **max** command might require entries to be separated by a semicolon rather than a comma.

19. Similarly, type the formula $=\textbf{max}(\textbf{0, B4+D4})$ into cell **B5.**

20. Now we tell the computer that we want it to calculate Δx and Δy in row 5 in the same way that it did in row 4. Highlight cells **C4** and **D4**. You will see a small box in the lower-right corner of the highlighted cells. Grab this small box with your mouse and pull down 1 row. Then release.

> This "drag and release" feature is amazingly helpful, because it copies formulas while maintaining the relationships between cells based on their relative location in the spreadsheet.

	A	B	C	D	E
1	0.2	0.1	10		
2	0.1	0.7	49		
3					
4	30	15	10	41.5	
5	40	56.5			
6					

Grab this small box with the mouse and drag down

21. Click on cell **C5** and look at its formula. You should see

$$= \text{A\$1*B5 - B\$1*A5 + C\$1}$$

This is the calculation of Δx in year 2. Note that it depends on spending levels in year 2 (found in cells **A5** and **B5**) in exactly the same way that our first calculation of Δx depended on spending levels in year 1. During the drag-and-release step, the spreadsheet updated the formula by maintaining the reference to data addresses *relative to the current cell.* However, the formula still refers to **A$1** and **B$1** because the $ tells the spreadsheet that we are referring to a fixed rather than a relative position in the document.

22. Now we tell the computer to make *all* of the calculations in the same way. Highlight cells **A5** through **D5**. Grab the box in the lower-right corner, just like before, and pull down to row 150.

Here we use the "drag and release" feature to avoid 150 copy/paste commands!

Grab this small box with the mouse and drag down

23. Compare the spending levels in **A150** and **B150** to the coordinates of the intersection that you determined in #11.

24. Change the initial spending levels (in row 4) to $x = 200$ and $y = 160$, and see how it affects the outcome in row 150.

25. Change the initial spending levels (in row 4) to $x = 400$ and $y = 50$, and see how it affects the outcome in row 150.

26. Change the initial spending levels (in row 4) to $x = 450$ and $y = 150$, and see how it affects the outcome in row 150.

Helpful Idea: Try making a scatter plot of your data, so you can see the year-to-year progression of spending levels. If you want to see all the data simultaneously, use columns A–D for the first run, columns E–H for the second, I–L for the third, and M–P for the last. Then include the four scatter plots in the same figure.

27. Draw the lines ℓ_x and ℓ_y in the Cartesian plane. Then plot each of the starting points we've tried, and use an arrow to indicate where it moves in the plane.

▷ Part 3: A broader look at Richardson's model

Until now we've been working with a particular example. More generally, Richardson's model describes the change in military spending in country X as

$$\Delta x = b\,y - a\,x + r, \qquad (10.4)$$

where the numbers a, b and r are all positive. Similarly, the change in military spending for country Y is predicted to be

$$\Delta y = c\,x - d\,y + s, \qquad (10.5)$$

where the numbers c, d and s are all positive.

28. Which constants in equations (10.4) and (10.5) are the grievance, defense, and fatigue coefficients for countries X and Y, respectively?

29. Use equation (10.4) to show that the general form of ℓ_x, the nullcline for x, is

$$y = \frac{a}{b}x - \frac{r}{b}. \qquad (10.6)$$

Notice that because all the constants are positive, the slope of ℓ_x is positive and its y-intercept is negative.

30. Use equation (10.4) to derive a general form of ℓ_y, the nullcline for y.

Both the slope and the y-intercept of ℓ_y should be positive.

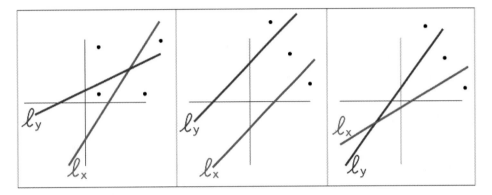

Figure 10.1: (left) Scenario #1; (middle) Scenario #2; (right) Scenario #3.

In the example described by equations (10.1) and (10.2) the lines ℓ_x and ℓ_y intersect in the first quadrant (see Scenario #1 in Figure 10.1), and as you saw in your calculations, this leads to stable, predictable spending levels in the long run. In the next exercises we see that this doesn't always happen.

31. Suppose the lines ℓ_x and ℓ_y are parallel (see Scenario #2 in Figure 10.1). Based on what you know about how the point (x, y) moves from year to year, predict what happens to military spending in the long run.

32. Change the values of a, b, c and d in your spreadsheet (but keep them less than 1) so that the the lines ℓ_x and ℓ_y are parallel. Then initialize the spending levels so that you start from three different positions: (a) above both lines, (b) between the lines, and (c) below both lines. Use complete sentences to describe what your spreadsheet indicates about spending in the long run.

33. Suppose the lines ℓ_x and ℓ_y intersect in the third quadrant (see Scenario #3 in Figure 10.1). Based on what you know about how the point (x, y) moves from year to year, predict what happens to military spending in the long run.

34. Change the values of a, b, c and d in your spreadsheet (but keep them less than 1) so that the the lines ℓ_x and ℓ_y intersect in the third quadrant. Then initialize the spending levels so that you start from three different positions: (a) above both lines, (b) between the lines, and (c) below both lines. Use complete sentences to describe what your spreadsheet indicates about spending in the long run.

35. In the previous exercises, you found that spending levels stabilize if and only if the nullclines intersect in the first quadrant. Show algebraically that this happens when $ad > bc$. (Hint: compare the slopes of ℓ_x and ℓ_y.)

In Part 1 of this project we talked a lot about the political meaning of the coefficients in Richardson's model. Here we close the project with a political interpretation the stability condition that you derived in #35. When we apply the square root to both sides of the stability condition we see

$$\sqrt{ad} > \sqrt{bc}. \tag{10.7}$$

The number $\sqrt{ad}$ is called the **geometric mean** of a and d, and it is a type of average. Since a and d are fatigue coefficients, which quantify social pressure to spend domestically, and b and c are defense coefficients, which quantify the countries' fear of each other, we can understand equation (10.7) as saying the following: Richardson's model predicts that military spending levels will stabilize if the average social pressure for domestic spending is greater than the average fear of the other country.

Geometric Mean: Consider a rectangle whose length and width are ℓ and w, respectively. The perimeter is $2\ell + 2w$. A *square* with the same perimeter has sides of length

$$\frac{2\ell + 2w}{4} = \frac{\ell + w}{2}$$

which we call the **arithmetic mean** (many people just say "average"). Similarly, a rectangle with side lengths ℓ and w has an area of ℓw. A *square* with the same area has sides of length

$$\sqrt{\ell w}$$

which we call the **geometric mean**. In this sense $\sqrt{\ell w}$ is a kind of "average," just as $(\ell + w)/2$ is a kind of average, but it has to do with area instead of perimeter.

The arithmetic mean of two positive numbers is never smaller than their geometric mean. Here's why:

$$(\ell - w)^2 \geq 0$$
$$\ell^2 - 2\ell w + w^2 \geq 0$$

Adding $4\ell w$ to both sides gives us

$$\ell^2 + 2\ell w + w^2 \geq 4\ell w$$
$$(\ell + w)^2 \geq 4\ell w$$

Now square root both sides and divide by 2.

Chapter 2
Limits and Continuity

In this chapter, we develop the concept of and skills for calculating something called a *limit*, which is the genesis of calculus. The *derivative*, the *integral, infinite series*—all of calculus depends on it, so your understanding of the topic is critical.

2.1   Introduction to Limits

Suppose we track the position of Venus. Everything is fine for a while, but then it passes between us and the sun and we're blinded by the glare. While we cannot say where the planet was at that time—not in the sense of having witnessed it—we *can* say where it should have been, and that's the underlying idea of a *limit*.

Consider the graph of $y(t)$ depicted in Figure 1.1. At time $t_0 = 2$ seconds there's a "blinding flash of light" so we don't know $y(2)$, but based on the graph and the data, we could make a pretty convincing argument that $y(2) = 3$.

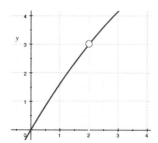

t	$y(t)$
1.875	2.839543269
1.9375	2.920222356
1.998046875	2.997520593
1.999023438	2.998760407
1.999938965	2.999922532
2	?
2.00012207	3.000154934
2.0078125	3.009908823
2.0625	3.078876202
2.125	3.156850962
2.25	3.310096154

Figure 1.1: Based on the graph and the data, we conjecture that $y(2) = 3$.

At a fundamental level, a limit is not about what happens *at* time t_0 but, rather, how the function behaves *near* time t_0. When the number $y(t)$ gets closer and closer to a particular value, say L, as t approaches time t_0 we write

$$\lim_{t \to t_0} y(t) = L, \tag{1.1}$$

and say that L is *the* **limit** *of* $y(t)$ *as* t **approaches** t_0. It's also correct to say that $y(t)$ **converges** *to* L *as* t **tends toward** t_0.

Estimating limits based on graphical evidence

Example 1.1. Use the graph of $y(t)$ shown in Figure 1.2 to estimate $\lim_{t \to 1} y(t)$.

Solution: The graph indicates that $y(1) = 2$, but that's irrelevant to our calculation because the limit is determined by the function values *near* $t = 1$. The number $y(t)$ gets closer and closer to 6 as t approaches 1, so $\lim_{t \to 1} y(t) = 6$. ∎

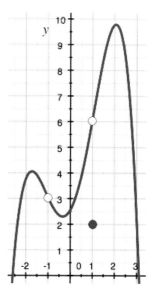

Figure 1.2: Graph for Example 1.1.

Try It Yourself: Graphical limits of sine and cosine

Example 1.2. Use the graphs of $y_1 = \sin(t)$ and $y_2 = \cos(t)$ in Figure 1.3 to estimate (a) $\lim_{t\to 0} \sin(t)$, and (b) $\lim_{t\to 0} \cos(t)$.

Full Solution On-line

Answer: (a) $\lim_{t\to 0} \sin(t) = 0$, and (b) $\lim_{t\to 0} \cos(t) = 1$. ∎

t	$\sin(t)$	$\cos(t)$
0.51950816	0.49645325	0.86806346
0.522576122	0.499114094	0.866536278
0.523470944	0.49988929	0.866089313
0.523582797	0.499986162	0.866033393
0.52359876	0.499999986	0.866025412
$\pi/6$	?	?
0.523598785	0.500000008	0.866025399
0.523599071	0.500000256	0.866025256
0.523603499	0.500004091	0.866023042
0.524203366	0.500523499	0.86572295
0.533272218	0.508353922	0.861148239

Graphs for Example 1.2   Data for Example 1.3

Figure 1.3: (left) Graphs of $y_1 = \sin(t)$ and $y_2 = \cos(t)$; (right) values of $\sin(t)$ and $\cos(t)$ for t near $\pi/6$.

Estimating limits based on numerical evidence

t	$y(t)$
3.875	15.625
3.9375	15.8125
3.9921875	15.9765625
3.999023438	15.99707031
3.999938965	15.99981689
4	?
4.000061035	16.00018311
4.000976563	16.00292969
4.0078125	16.0234375
4.0625	16.1875
4.125	16.375

Figure 1.4: Data for Example 1.4.

Example 1.3. Use the table of data in Figure 1.3 (above) to estimate both (a) $\lim_{t\to \pi/6} \sin(t)$ and (b) $\lim_{t\to \pi/6} \cos(t)$.

Solution: As t gets closer and closer to $\pi/6$, the value of $\sin(t)$ seems to approach the number 0.5, while the value of $\cos(t)$ seems to approach 0.8660254, so we estimate $\lim_{t\to \pi/6} \sin(t) = 0.5$ and $\lim_{t\to \pi/6} \sin(t) = 0.8660254$. ∎

Try It Yourself: Estimating limits based on numerical evidence

Example 1.4. Use the table of data in Figure 1.4 to estimate $\lim_{t\to 4} y(t)$.

Full Solution On-line

Answer: $\lim_{t\to 4} y(t) = 16$. ∎

Interpreting limits

Example 1.5. Suppose $c(t)$ is the concentration of antibiotic in a patient's bloodstream t hours after the first dose is given. Write an equation that says the concentration gets closer and closer to 0.3 mg per liter as time approaches the end of the first hour.

Solution: Since time is approaching the end of the first hour, we use the notation $\lim_{t\to 1}$. The function we monitoring is $c(t)$, and the value it's approaching is 0.3, so our equation is $\lim_{t\to 1} c(t) = 0.3$. ∎

Try It Yourself: Interpreting limits

Example 1.6. Suppose $R(c)$ is the revenue earned (in dollars) from the sale of c tons of corn. In a complete sentence, interpret the equation $\lim_{c\to 200} R(c) = 41,055$.

Full Solution On-line

Answer: See solution on-line. ∎

WARNING: Avoid confusing $\lim_{t\to t_0} y(t)$, which is a number, with our method of finding it—watching the value of $y(t)$ as t approaches t_0.

❖ A Limit Might Not Exist

When no *particular* number is approached by $y(t)$ as $t \to t_0$, we say that $\lim_{t \to t_0} y(t)$ **does not exist**, or that the limit **diverges**. There are three basic ways this might happen, which we discuss in the next three examples.

When the limit doesn't exist because of unbounded growth

Example 1.7. Both gravitational and electrostatic force obey an inverse square law, $F(r) = k/r^2$, where k is a constant that depends on physical qualities of the objects in question. Figure 1.5 shows the graph of $F(r) = 1/r^2$. (a) Use it to explain why $\lim_{r \to 0} 1/r^2$ does not exist, and (b) explain what this means in the context of force.

Solution: (a) As $|r|$ gets smaller and smaller, $F(r)$ grows and grows, eventually surpassing every finite number (e.g., $F(r) > 10,000$ when $|r| < 0.01$). Since $F(r)$ does not settle on any particular value as $r \to 0$, the limit $\lim_{r \to 0} F(r)$ does not exist. (b) Because the number $F(r)$ surpasses every finite number, the force between the two objects grows forever larger as the distance between them shrinks. ∎

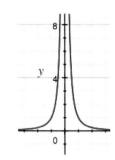

Figure 1.5: Graph for Example 1.7.

When the limit doesn't exist because of oscillation

Example 1.8. Show that $\lim_{t \to 0} \sin(3/t)$ does not exist.

Solution: The data in Table 1 (see Figure 1.6) leads us to believe that $\lim_{t \to 0} \sin(3/t) = 1$, but Table 2 leads us to believe that $\lim_{t \to 0} \sin(3/t) = -1$. In fact, the value of $\sin(3/t)$ oscillates between 1 and -1 infinitely often as $t \to 0$, and never settles on any particular number. Therefore, $\lim_{t \to 0} \sin(3/t)$ does not exist. ∎

t	$\sin(3/t)$
$\frac{6}{\pi}$	1
$\frac{6}{5\pi}$	1
$\frac{6}{9\pi}$	1
$\frac{6}{13\pi}$	1
$\frac{6}{17\pi}$	1
⋮	⋮
0	?

Table 1

t	$\sin(3/t)$
$\frac{6}{3\pi}$	-1
$\frac{6}{7\pi}$	-1
$\frac{6}{11\pi}$	-1
$\frac{6}{15\pi}$	-1
$\frac{6}{19\pi}$	-1
⋮	⋮
0	?

Table 2

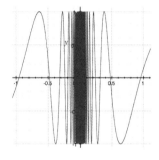

Figure 1.6: (left) Values of $\sin(3/t)$ at various values of t; (right) the graph of $\sin(3/t)$ near zero—oscillations happen more rapidly as t gets closer to $t_0 = 0$, so graphing software eventually loses resolution.

When the limit doesn't exist because of a jump

Example 1.9. Use the graph of $y(t)$ shown in Figure 1.7 to explain why $\lim_{t \to 1} y(t)$ does not exist.

Solution: If we restrict our attention to times *before* $t = 1$ (see Figure 1.8(a)), it appears that $y(t)$ is converging to 2 as t approaches 1. But if we restrict our attention to times *after* $t = 1$, it appears that $y(1)$ was 4. Because $y(t)$ doesn't converge to any *particular* number as t approaches 1, the limit doesn't exist. ∎

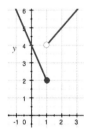

Figure 1.7: Graph for Example 1.9.

❖ One-sided Limits

Even when $\lim_{t \to t_0} f(t)$ doesn't exist, the number $f(t)$ might exhibit some measure of predictability near t_0. Inspired by our analysis of Example 1.9, we use the

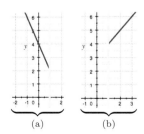

Figure 1.8: Restricting our attention to times (a) before $t = 1$, (b) after $t = 1$.

notation $t \to t_0^-$ to communicate two ideas: (1) that $t \to t_0$, and (2) that we are restricting our attention to times *before* t_0. Similarly, we use $t \to t_0^+$ when restricting ourselves to times *after* t_0. Using this notation, we define **one-sided limits** as follows:

- if the value of $f(t)$ gets closer and closer to the number L_1 as $t \to t_0^-$, we write $\lim_{t \to t_0^-} f(t) = L_1$ and say that L_1 is the **left-hand limit** of f at t_0.

- if the value of $f(t)$ gets closer and closer to the number L_2 as $t \to t_0^+$, we write $\lim_{t \to t_0^+} f(t) = L_2$ and say that L_2 is the **right-hand limit** of f at t_0.

> To help you remember which notation means what, recall that *positive* numbers are found to the *right* of zero on the number line (so + means "right"), and *negative* numbers are found to the *left* of zero (so − means "left").

Practice with one-sided limits

Example 1.10. Figure 1.9 shows the graph of $f(t)$. Use it to determine the one-sided limits at $t_0 = 3$.

Solution: When we restrict ourselves to $t < 3$, the function values converge to 2, so $\lim_{t \to 3^-} f(t) = 2$. Similarly, when we restrict ourselves to $t > 3$, the function values tend toward the number -1, so $\lim_{t \to 3^+} f(t) = -1$. ■

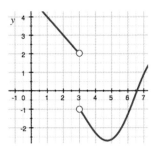

Figure 1.9: Graph for Example 1.10.

Try It Yourself: Practice with one-sided limits

Example 1.11. Determine the one-sided limits at $t_0 = 1$ in (a) Example 1.1, and (b) Example 1.9.

Answer: (a) $\lim_{t \to 1^-} y(t) = 6$ and $\lim_{t \to 1^+} y(t) = 6$, and
(b) $\lim_{t \to 1^-} f(t) = 2$ and $\lim_{t \to 1^+} f(t) = 4$. ■

Example 1.11(a) brings us to the relationship between limits and one-sided limits.

> By saying that two numbers are equal, we mean that they exist in the first place.
>
> This theorem says that, in terms of our example in the chapter opening, the limit exists when the "pre-flash" behavior of the function leads us to the same prediction as the "post-flash" behavior.

Theorem 1.1. The number $\lim_{t \to t_0} y(t)$ exists if and only if $\lim_{t \to t_0^-} y(t)$ and $\lim_{t \to t_0^+} y(t)$ are equal. In that case,

$$\lim_{t \to t_0^-} y(t) = \lim_{t \to t_0} y(t) = \lim_{t \to t_0^+} y(t).$$

As with limits, one-sided limits do not exist when the function values fail to approach a particular number.

One-sided limits might not exist because of oscillation

Example 1.12. Using the function $y(t) = \sin(3/t)$, whose graph is shown in Figure 1.6, explain why $\lim_{t \to 0^+} y(t)$ does not exist.

Solution: As t decreases toward zero, the graph of this function doesn't settle on any *particular* height. That is, the number $y(t)$ doesn't tend toward any specific value as $t \to 0^+$, so $\lim_{t \to 0^+} y(t)$ does not exist. ■

❖ Infinite Limits

When the number $y(t)$ grows and grows as t approaches t_0 from the right, surpassing every finite number (as with gravitational and electrostatic force, as discussed in Example 1.7), we say that $y(t)$ increases **without bound** and write $\lim_{t \to t_0^+} y(t) = \infty$ (or $\lim_{t \to t_0^-} y(t)$ if t approaches t_0 from the left). Similarly, we write $\lim_{t \to t_0^+} y(t) = -\infty$ (or $\lim_{t \to t_0^-} y(t) = -\infty$) when $y(t)$ becomes more and more negative, eventually falling below each and every negative number. In such cases, since no particular number is approached by $y(t)$, the limit *does not exist*. Our use of $\pm\infty$ is intended only to indicate the behavior of $y(t)$ as t approaches t_0.

Infinite limits

Example 1.13. Using the function $y(t) = 1/t$, whose graph is shown in Figure 1.10, explain why neither $\lim_{t\to 0^+} y(t)$ nor $\lim_{t\to 0^-} y(t)$ exists.

Solution: The number $y(t)$ grows and grows as $t \to 0^+$, surpassing all finite values. So $\lim_{t\to 0^+} y(t)$ does not exist, and we write $\lim_{t\to 0^+} y(t) = \infty$. Similarly, when $t \to 0^-$, the number $y(t)$ becomes more and more negative, eventually falling below each and every negative number. So $\lim_{t\to 0^-} y(t)$ does not exist, and we write $\lim_{t\to 0^-} y(t) = -\infty$. ∎

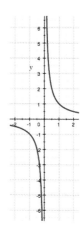

Figure 1.10: Graph for Example 1.13.

Try It Yourself: Infinite limits

Example 1.14. Using the function $f(t) = 4/(t-1)^2$, whose graph is shown in Figure 1.11, determine $\lim_{t\to 1^+} f(t)$ and $\lim_{t\to 1^-} f(t)$.

Answer: $\lim_{t\to 1^-} f(t) = \infty$ and $\lim_{t\to 1^+} f(t) = \infty$. ∎

Full Solution On-line

In Example 1.13 the behavior of $y(t)$ is *different* when $t \to 0^-$ than it is when $t \to 0^+$. By contrast, in Example 1.14 the number $f(t)$ behaves in the *same* way, whether t approaches $t_0 = 1$ from the left or from the right. In the spirit of Theorem 1.1, we write $\lim_{t\to t_0} f(t) = \infty$ when both $\lim_{t\to t_0^-} f(t) = \infty$ and $\lim_{t\to t_0^+} f(t) = \infty$. Similarly, the notation $\lim_{t\to t_0} f(t) = -\infty$ means that both $\lim_{t\to t_0^-} f(t) = -\infty$ and $\lim_{t\to t_0^+} f(t) = -\infty$.

Note: If either (or both) of the one-sided limits $\lim_{t\to t_0^{\pm}} f(t)$ is $\pm\infty$, the graph of f has a vertical asymptote at t_0.

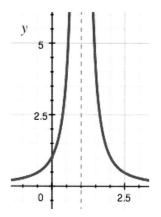

Figure 1.11: Graph for Example 1.14.

❖ In Summary

While there's a lot of vocabulary in this section, there's only *one* idea. Everything in this section revolves around whether $y(t)$ gets closer and closer to a particular number as $t \to t_0$. If so, there's a limit. Otherwise, there's not. For example, while $\lim_{t\to 5}(3t^2 + 4t + 2)$ is a way of writing the number 97 (as we'll see in Example 2.1), $\lim_{t\to 0} \sin(3/t)$ is not a number at all—it's mathematical nonsense. A large part of the beginner's job is to figure out which limits have meaning and which don't.

❖ Common Mistakes with Limits

Beginners often have trouble with the notation of one-sided limits. Think about it this way: we're trying to communicate two pieces of information: (1) Where is t headed? and (2) How is it going there? The notation is built to answer those questions *in order*.

$$t \to -5^+ \qquad\qquad t \to 2^- \qquad\qquad t \to 8$$

t approaches -5 from the right t approaches 2 from the left t approaches 8 from *both* sides

Also, students sometimes think $\lim_{t\to t_0} f(t) = \infty$ means the limit exists, and that its value is infinite, but this is wrong. When we say a limit is infinite we mean, first and foremost, that it *does not exist*. The notation of "$= \infty$" means that $f(t)$ gets larger and larger, surpassing every finite value, so no particular number is approached.

Infinity is not a number. The prefix "in" is used in the sense of "not," as it is in the words *inequality, incomplete, inescapable,* and *inattentive.*

- the terms *limit, one-sided limit, approaches, converges, tends toward, infinite limits,* and the notation $\lim_{t \to t_0} y(t) = L$, $\lim_{t \to t_0^+} y(t) = L$, and $\lim_{t \to t_0^-} y(t) = L$;

- that $\lim_{t \to t_0} y(t)$ has nothing to do with the actual value of $y(t_0)$, but rather tells us what it "should be" based on the behavior of the function *near* t_0;

- how one-sided limits are related to limits (see Theorem 1.1);

- three ways that limits can fail to exist, and two ways that one-sided limits can fail to exist.

- determine when a limit does not exist, based on a graph or table of function values;

- estimate the value of a limit by looking at a graph or a table of function values.

❖ 2.1 Skill Exercises

Use the following graphs for #1–12.

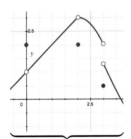

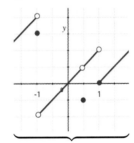

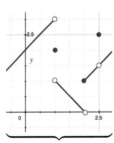

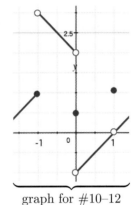

graph for #1–3　　　　graph for #4–6　　　　graph for #7–9　　　　graph for #10–12

1. $f(2)$ and $\lim_{t \to 2} f(t)$

2. $f(3)$ and $\lim_{t \to 3} f(t)$

3. $f(1.5)$ and $\lim_{t \to 1.5} f(t)$

4. $\lim_{t \to 1-} f(t)$ and $\lim_{t \to 1+} f(t)$

5. $\lim_{t \to -1-} f(t)$ and $\lim_{t \to -1+} f(t)$

6. $\lim_{t \to 1.5-} f(t)$ and $\lim_{t \to 1.5+} f(t)$

7. $f(1)$ and $\lim_{t \to 1} f(t)$

8. $f(2)$ and $\lim_{t \to 2-} f(t)$

9. $f(1.5)$ and $\lim_{t \to 1.5} f(t)$

10. $\lim_{t \to 1-} f(t)$ and $\lim_{t \to 1+} f(t)$

11. $\lim_{t \to -1-} f(t)$ and $\lim_{t \to -1+} f(t)$

12. $\lim_{t \to 0-} f(t)$ and $\lim_{t \to 0+} f(t)$

In each of #13–16 use the given table of data to approximate the specified limit.

13. $\lim_{t \to 2+} y(t)$.

t	$y(t)$
2.125	0.5
2.0625	0.25
2.015625	0.0625
2.0078125	0.03125
2.000488281	0.001953125
2.000061035	0.000244141

14. $\lim_{t \to 1+} y(t)$.

t	$y(t)$
1.125	-3.5
1.0625	-3.75
1.0078125	-3.96875
1.00012207	-3.999511719
1.000007629	-3.999969482
1.000000119	-3.999999523

15. $\lim_{t \to 7^-} y(t)$.

t	$y(t)$
6.875	19.5
6.9375	19.75
6.9921875	19.96875
6.99987793	19.99951172
6.999938965	19.99975586
6.999999046	19.99999619

16. $\lim_{t \to 4^-} y(t)$.

t	$y(t)$
3.875	7.5
3.96875	7.875
3.998046875	7.9921875
3.999755859	7.999023438
3.999984741	7.999938965
3.999999046	7.999996185

In each of #17–22 generate a table of function values with which to approximate the specified limit. All angles are measured in radians.

A spreadsheet program can make this very fast: start with $\phi = \pm 1$, and step half of the way toward $\phi = 0$ in each succeeding cell of the spreadsheet.

17. $\lim_{\phi \to 0} \frac{\sin(\phi)}{\phi}$

19. $\lim_{\phi \to 0} \frac{1 - \cos(\phi)}{\phi^2}$

21. $\lim_{\phi \to 0} \frac{3\sin(\phi) - 3\phi + \phi^3}{3\phi^5}$

18. $\lim_{\phi \to 0} \frac{1 - \cos(\phi)}{\phi}$

20. $\lim_{\phi \to 0} \frac{\sin(\phi)}{\phi^2}$

22. $\lim_{\phi \to 0} \frac{\cos(\phi) - 1 + 0.5\phi^2}{\phi^4}$

❖ 2.1 Concept and Application Exercises

23. Suppose that $\lim_{t \to 4} f(t) = 7$. What do you know about $f(4)$?

24. Suppose $f(5) = 6$. What do you know about $\lim_{t \to 5} f(t)$?

25. Use the notation of limits to write an equation that says, "As the amount of fuel dispensed approaches 15 gallons, the purchase price approaches $42."

26. Use the notation of limits to write an equation that says, "As time approaches 8.5 minutes after launch, the speed of the space shuttle approaches 7000 meters per second."

27. Use the notation of limits to write an equation that says, "As speed increases toward c (the speed of light), the inertia of an object grows without bound."

28. Suppose c is the concentration of blood thinner in a patient's bloodstream, and $p(c)$ is the probability that the patient will suffer a stroke due to a blood clot. In complete sentences, explain what the equation $\lim_{c \to 0^+} p(c) = 1$ means in this context.

29. Suppose p is the percentage of pollutants captured by an industrial plant, and $c(p)$ is the cost of production. In complete sentences, explain what the equation $\lim_{p \to 100^-} c(p) = \infty$ means in this context.

30. The notation $[A]$ denotes the concentration of chemical A. Suppose r is the rate at which a reaction produces chemical B. In complete sentences, explain what the equation $\lim_{[A] \to 0^+} r([A]) = 0$ means in this context.

31. Suppose c is the concentration of blood thinner in a patient's bloodstream, and $p(c)$ is the probability that the patient will suffer a stroke due to a blood clot. In complete sentences, explain why $\lim_{c \to 0^-} p(c)$ is a meaningless expression.

32. Suppose p is the percentage of pollutants captured by an industrial plant, and $c(p)$ is the cost of production. In complete sentences, explain why $\lim_{p \to 100^+} p(c)$ is a meaningless expression.

33. The notation $[A]$ denotes the concentration of chemical A. Suppose r is the rate at which a reaction produces chemical B. In complete sentences, explain why $\lim_{[A] \to 0^-} r([A])$ is a meaningless expression.

34. In complete sentences, without using the terms *limit* or *converge*, explain what it means to say that $\lim_{t \to 4} f(t)$ does not exist.

35. In complete sentences, without using the terms *limit* or *converge*, explain why $\lim_{\phi \to 0} \cos(5/\phi)$ is not a number.

Use complete sentences to explain each of the equations in #36–39, without using the terms *limit, converge,* or *infinity.*

36. $\lim_{t \to 4} f(t) = 15$

37. $\lim_{t \to 2+} f(t) = 3$

38. $\lim_{t \to 5-} f(t) = \infty$

39. $\lim_{t \to -3-} f(t) = \lim_{t \to -3+} f(t)$

40. Sketch the graph of a function that satisfies all of the following:

   (a) $\lim_{t \to 1^-} f(t) = 2$

   (b) $\lim_{t \to 1^+} f(t) = 3$

   (c) $\lim_{t \to 0} f(t) = -1$

   (d) $f(0) = 1$

41. Sketch the graph of a function for which $\lim_{t \to 4+} y(t)$ exists, but $\lim_{t \to 4-} y(t)$ does not.

2.2 Calculating Limits

More often than not, limits behave the way you want them to. For example, suppose the value of $y_1(t)$ gets closer and closer to 4 as $t \to t_0$, and the value of $y_2(t)$ gets closer and closer to 5 at the same time. Then—intuitively speaking—the value of the sum $y_1(t) + y_2(t)$ *should* converge to $4 + 5 = 9$ as $t \to t_0$. That's exactly what happens, and all of the limit laws can be understood in this way.

The Limit Laws: Suppose that $\lim_{t \to t_0} f(t) = L$ and $\lim_{t \to t_0} g(t) = M$, that p and q are positive integers, and that β is any real number (constant). Then,

1. $\lim_{t \to t_0} t = t_0$

2. $\lim_{t \to t_0} \beta = \beta$

3. $\lim_{t \to t_0} \left(\beta f(t) \right) = \beta L$

and lastly,

4. $\lim_{t \to t_0} \left(f(t) \pm g(t) \right) = L \pm M$

5. $\lim_{t \to t_0} \left(f(t) g(t) \right) = LM$

6. $\lim_{t \to t_0} \left(\frac{f(t)}{g(t)} \right) = \frac{L}{M}$ when $M \neq 0$

7. $\lim_{t \to t_0} \left(f(t) \right)^{p/q} = L^{p/q}$, provided that $L > 0$ if q is even.

The symbol β is the lower-case Greek letter beta, which many people pronounce as "BAY-tuh."

Said in English:

Limit Law 4 says *the limit of a sum is the sum of the limits.*

Limit Law 5 says *the limit of a product is the product of the limits.*

Limit Law 6 says *the limit of a quotient is the quotient of the limits.*

The limit laws make finding limits of algebraic functions easy.

Limits with polynomials

Example 2.1. Calculate $\lim_{t \to 5} y(t)$ when $y(t) = 3t^2 + 4t + 2$.

Solution: We begin by using Limit Law 4 to separate the terms,

$$\lim_{t \to 5} \left(3t^2 + 4t + 2 \right) = \lim_{t \to 5} \left(3t^2 + 4t \right) + \left(\lim_{t \to 5} 2 \right) = \left(\lim_{t \to 5} 3t^2 \right) + \left(\lim_{t \to 5} 4t \right) + \left(\lim_{t \to 5} 2 \right).$$

Limit Law 2 tells us that $\lim_{t \to 5} 2 = 2$, and Limit Law 3 that

$$\lim_{t \to 5} 4t = 4 \underbrace{\lim_{t \to 5} t}_{\text{Limit Law 1}} = 4(5).$$

Lastly, we use the laws to find out that

$$\overbrace{\lim_{t \to 5} 3t^2}^{\text{Limit Law 3}} = 3 \lim_{t \to 5} t^2 = \underbrace{3 \left(\overbrace{\lim_{t \to 5} t}^{\text{Limit Law 1}} \right)^2}_{\text{Limit Law 7 with } p = 2 \text{ and } q = 1.} = 3(5)^2.$$

Putting all of this together,

$$\lim_{t \to 5} y(t) = \left(\lim_{t \to 5} 3t^2 \right) + \left(\lim_{t \to 5} 4t \right) + \left(\lim_{t \to 5} 2 \right) = 3(5)^2 + 4(5) + 2 = 97. \qquad \blacksquare$$

We saw that $\lim_{t \to 5} y(t) = y(5)$ in Example 2.1. This is a general fact about polynomials, and is important enough that we state it formally:

> **Theorem 2.2.** Suppose $y(t)$ is a polynomial. Then $\lim_{t \to t_0} y(t) = y(t_0)$.

An important limit involving polynomials

Example 2.2. The graph of $y = x^2$ is shown in Figure 2.1. Suppose we connect the point (t, t^2) to the origin by a line segment. Then we find the perpendicular bisector of this segment, and locate its intersection with the y-axis, which we'll call $b(t)$. What is $\lim_{t \to 0} b(t)$?

Solution: First of all, you should make an educated guess about what happens to the y-intercept of the perpendicular bisector as t approaches zero. Do that now, and then read more.

Really, make your guess first (otherwise this is no fun). In order to verify our guess, we begin by noting that the line segment connecting the origin to (t, t^2) has a slope of $\frac{rise}{run} = \frac{t^2 - 0}{t - 0} = t$, so the slope of the perpendicular bisector will be $-1/t$. Further, it will pass through the point $\left(\frac{t+0}{2}, \frac{t^2+0}{2}\right)$, which is the midpoint of the segment connecting (t, t^2) to the origin. Now that we have a slope and a point, we can write down the equation of the perpendicular bisector in point-slope form:

$$y - \frac{t^2}{2} = -\frac{1}{t}\left(x - \frac{t}{2}\right).$$

Rewriting the equation in slope-intercept form, $y = mx + b$, we have

$$y = \left(-\frac{1}{t}\right)x + \left(\frac{1}{2}t^2 + \frac{1}{2}\right),$$

which allows us to see that $b(t) = \frac{1}{2}t^2 + \frac{1}{2}$. Since $b(t)$ is a polynomial,

$$\lim_{t \to 0} b(t) = \lim_{t \to 0} \frac{1}{2}t^2 + \frac{1}{2} = 0.5(0)^2 + \frac{1}{2} = \frac{1}{2}.$$

Many people guess that the limit will be infinite, but it's not, which is why the example has been labeled "important." It demonstrates that intuition can be wrong. The lesson to learn is this: always use your intuition, but always check numerically, too.■

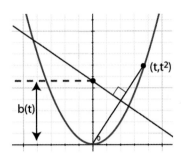

Figure 2.1: Graph for Example 2.2.

> The x-coordinate of the midpoint is just the average x-coordinate, and the y-coordinate of the midpoint is just the average y-coordinate.

> The point-slope form of a line through (x_1, y_1) whose slope is m is
> $$y - y_1 = m(x - x_1).$$

> The function $b(t)$ tells us how the value of the y-intercept depends on our initial choice of t.

A simple limit with a rational function

Example 2.3. Calculate $\lim_{t \to -1} y(t)$ when $y(t) = \frac{t^3 + 2t + 4}{5t + 8}$.

Solution: Let's write the numerator of our function as $f(t) = t^3 + 2t + 4$ and the denominator as $g(t) = 5t + 8$. Both of these are polynomials, so $\lim_{t \to -1} f(t) = f(-1)$ and $\lim_{t \to -1} g(t) = g(-1)$. Since $g(-1) \neq 0$, we can use Limit Law 6 to write

$$\lim_{t \to -1} y(t) = \lim_{t \to -1} \frac{f(t)}{g(t)} = \frac{\lim_{t \to -1} f(t)}{\lim_{t \to -1} g(t)} = \frac{f(-1)}{g(-1)} = \frac{1}{3}. \qquad \blacksquare$$

✧ Introduction to Indeterminate Forms

As we saw in Example 2.3, Limit Law 6 allows us to calculate the limit of a quotient when the limit of the denominator is nonzero. When it *is* zero, the limit of a quotient might exist if the limit of the numerator is also zero. In this case, when both numerator and denominator vanish in the limit, we say that we have a $0/0$ **indeterminate form.**

Factoring a 0/0 indeterminate form

Example 2.4. Determine $\lim_{t \to 4} y(t)$ when $y(t) = \frac{t^2 - 3t - 4}{t^2 + 4t - 32}$.

Solution: We cannot use Limit Law 6 because the limit of the denominator is 0. However, when we factor the polynomials that compose $y(t)$ we see that

$$y(t) = \frac{t^2 - 3t - 4}{t^2 + 4t - 32} = \frac{(t+1)(t-4)}{(t+8)(t-4)} = \frac{t+1}{t+8} \quad \text{when } t \neq 4.$$

When $t \neq 4$, the number $(t-4)$ is not zero, so we can cancel. For example, when $t = 7$ we see $y(7) = \frac{(8)(3)}{(15)(3)} = \frac{8}{15}$.

Since $\lim_{t \to 4} y(t)$ depends on values of t that are *near* 4 but *not equal* to 4,

$$\lim_{t \to 4} y(t) = \lim_{t \to 4} \frac{t+1}{t+8} = \frac{\lim_{t \to 4}(t+1)}{\lim_{t \to 4}(t+8)} = \frac{5}{12}. \qquad \blacksquare$$

Try It Yourself: Factoring a 0/0 indeterminate form

Example 2.5. Determine $\lim_{t \to -1} y(t)$ when $y(t) = \frac{t^2 + 8t + 7}{t^2 + 6t + 5}$.

Answer: $\lim_{t \to -1} y(t) = 3/2.$ $\qquad \blacksquare$

Full Solution On-line

Rationalizing a 0/0 indeterminate form

Example 2.6. Determine $\lim_{t \to 10} y(t)$ when $y(t) = \frac{t - 10}{\sqrt{t+6}-4}$.

Solution: We cannot use Limit Law 6 because the limit of the denominator is 0. However, when we rationalize $y(t)$ we see

$$y(t) = \frac{t-10}{\sqrt{t+6}-4}\left(\frac{\sqrt{t+6}+4}{\sqrt{t+6}+4}\right) = \frac{(t-10)(\sqrt{t+6}+4)}{(t+6)-16} = \frac{(t-10)(\sqrt{t+6}+4)}{t-10} = \sqrt{t+6}+4,$$

so

$$\lim_{t \to 10} y(t) = \underbrace{\lim_{t \to 10}(\sqrt{t+6}+4) = \overbrace{\lim_{t \to 10}\sqrt{t+6} + \lim_{t \to 10} 4}^{\text{Limit Law 2 \& Limit Law 7 with } p=1 \text{ and } q=2.}}_{\text{Limit Law 4}} = \sqrt{16}+4 = 8. \qquad \blacksquare$$

Try It Yourself: Rationalizing a 0/0 indeterminate form

Example 2.7. Determine $\lim_{t \to 5} y(t)$ when $y(t) = \frac{t-5}{\sqrt{t^2+11}-6}$.

Answer: $6/5.$ $\qquad \blacksquare$

Full Solution On-line

Each of the limits in Examples 2.4–2.7 was a 0/0 indeterminate form, but after some algebraic simplification we saw that the limits existed, and all had different values. By contrast, if the limit has the form $n/0$ where $n \neq 0$, we're sure the limit does not exist, but we might be able to say something about the behavior of the function nonetheless.

In essence, this is *why* we say that 0/0 is indeterminate.

Analyzing an n/0 form

Example 2.8. Analyze $\lim_{t \to 1} y(t)$ when $y(t) = \frac{t^2 + 4t + 3}{(t-1)(t^2-1)}$.

Solution: Since this limit results in 8/0, the limit does not exist. However, after factoring we see

$$y(t) = \frac{(t+3)(t+1)}{(t-1)(t-1)(t+1)} = \frac{t+3}{(t-1)^2},$$

which looks more and more like $4/(t-1)^2$ as $t \to 1$. We know from Example 1.14 that $\lim_{t \to 1} 4/(t-1)^2 = \infty$, so ours is too: $\lim_{t \to 1} y(t) = \infty$. $\qquad \blacksquare$

An n/0 form in disguise

Example 2.9. Analyze $\lim_{t\to-3} y(t)$ when $y(t) = \frac{t^2+8t+15}{t^2+6t+9}$.

Solution: This limit is a $0/0$ indeterminate form, which makes us think to try the factor-and-cancel method that worked in previous examples. When we factor the numerator and denominator we see

$$y(t) = \frac{(t+3)(t+5)}{(t+3)(t+3)} = \frac{t+5}{t+3}.$$

So $y(t)$ looks more and more like $2/(t+3)$ as $t \to -3$. Since $\lim_{t\to-3} 2/(t+3)$ results in $2/0$, the limit does not exist. Moreover, unlike Example 2.8, the one-sided limits do not agree since

$$\lim_{t\to-3^-} \frac{2}{t+3} = -\infty \quad \text{and} \quad \lim_{t\to-3^+} \frac{2}{t+3} = \infty. \qquad \blacksquare$$

❖ The Squeeze Theorem

The most important elementary limit is arguably

$$\lim_{\phi\to0} \frac{\sin(\phi)}{\phi}, \tag{2.1}$$

> This limit plays a prominent role in Chapter 3, when calculating rates of change for the sine and cosine.

where ϕ is measured in radians. The limit results in the indeterminate form $0/0$, but since the numerator is not a polynomial we cannot use the factoring technique that's been so handy in previous examples. Instead, we rely on an important fact known as the *Squeeze Theorem* to determine the value of this limit.

In order to develop some intuition before tackling the formal statement of the Squeeze Theorem, draw the graph of a function onto Figure 2.2 that never goes *higher* than the red graph and never goes *lower* than the blue graph. (Really, try it!)

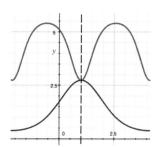

Figure 2.2: Stay between the graphs.

The Squeeze Theorem is a technical way of saying that whenever two graphs squeeze together to a common height, any graph that remains between them is forced to that same height.

Squeeze Theorem: Suppose there is a distance $\delta > 0$ so that

$$\underbrace{f(t) \leq g(t) \leq h(t)}_{\substack{\text{The graph of } g \text{ is caught be-}\\\text{tween the graphs of } f \text{ and } h}} \quad \text{whenever} \quad \underbrace{|t - t_0| < \delta.}_{\substack{\text{This is a precise way of saying}\\\text{that } t \text{ is "close" to } t_0}}$$

Further, suppose that

$$\underbrace{\lim_{t\to t_0} f(t) = L = \lim_{t\to t_0} h(t).}_{\text{This statement means that the graphs squeeze together at time } t_0}$$

Then $\lim_{t\to t_0} g(t)$ exists, and its value is also L.

Using the Squeeze Theorem

Example 2.10. Show that $\lim_{t\to0} t^2 \sin(3/t) = 0$.

Solution: Since $|\sin(3/t)| \leq 1$, we know

$$-t^2 \leq t^2 \sin(3/t) \leq t^2.$$

Since both $\lim_{t \to 0} t^2 = 0$ and $\lim_{t \to 0} -t^2 = 0$, the Squeeze Theorem tells us that $\lim_{t \to 0} t^2 \sin(3/t) = 0$, even though $\lim_{t \to 0} \sin(3/t)$ does not exist. ∎

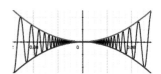

Figure 2.3: The graph of $y = t^2 \sin(3/t)$, along with its upper and lower bounds.

Now let's return to the indeterminate form that began our discussion. Suppose we could show that

$$\cos(\phi) \leq \frac{\sin(\phi)}{\phi} \leq 1 \tag{2.2}$$

when ϕ is near zero. Then, since $\lim_{\phi \to 0} \cos(\phi) = 1$ and $\lim_{\phi \to 0} 1 = 1$, the Squeeze Theorem would allow us to make a conclusion. Specifically,

---

Most Important Elementary Limit: Suppose ϕ is measured in radians. Then

$$\lim_{\phi \to 0} \frac{\sin(\phi)}{\phi} = 1.$$

Proof. We'll use an intuitive argument about distances to establish the first inequality of (2.2) when $\phi > 0$ and near 0. Figure 2.4 shows a clipping of the unit circle, and you can see that the arc from A to C is a more direct (shorter) route than the two-stage path that follows $\overline{AD}$ and then $\overline{DC}$. Further, since ΔDCE is a right triangle, the leg $\overline{DC}$ is shorter than the hypotenuse $\overline{DE}$. This gives us

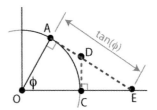

Figure 2.4: Comparing distances.

$$\left(\begin{matrix}\text{length of arc} \\ \text{from } A \text{ to } C\end{matrix}\right) \leq |\overline{AD}| + |\overline{DC}| \leq \underbrace{|\overline{AD}| + |\overline{DE}|}_{\text{See Appendix A for details}} = \tan(\phi). \tag{2.3}$$

> $|\overline{AD}|$ denotes the *length* of the line segment $\overline{AD}$.

Since we're measuring in radians, the length of the arc from A to C is

$$\underbrace{\left(\frac{\phi}{2\pi}\right)}_{\substack{\text{There are } 2\pi \text{ radians (of angle) in the whole circle.} \\ \text{This is the fraction of them we've used by swinging} \\ \text{through } \phi \text{ radians.}}} \overbrace{2\pi}^{\substack{\text{Circumference of} \\ \text{the unit circle}}} = \phi. \tag{2.4}$$

> Here is where measuring angle in radians is important. See exercise #73.

Combining (2.3) with (2.4) gives us $\phi \leq \tan(\phi) = \frac{\sin(\phi)}{\cos(\phi)}$, and cross-multiplying leaves us with

$$\cos(\phi) \leq \frac{\sin(\phi)}{\phi}.$$

> Since $\phi > 0$ and near 0, we know $\cos(\phi) > 0$, so the direction of the inequality is preserved when we cross-multiply.

Next we need to establish the same inequality when $\phi < 0$. Let's write $\phi = -\theta$ so that we can track the negative sign. Then because cosine is an even function (see Chapter 1) and sine is an odd function,

$$\cos(\phi) = \cos(-\theta) = \cos(\theta) \leq \frac{\sin(\theta)}{\theta} = \frac{-\sin(\theta)}{-\theta} = \frac{\sin(-\theta)}{-\theta} = \frac{\sin(\phi)}{\phi}.$$

The second inequality in (2.2) is also a statement about distances. Again, let's begin by considering $\phi > 0$. Remember that $\sin(\phi)$ is the y-coordinate of the point A. So $\sin(\phi) = |\overline{AB}|$. Further, since ΔABC is a right triangle, the leg $\overline{AB}$ is shorter

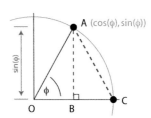

Figure 2.5: $\sin(\phi) = |\overline{AB}|$.

than the hypotenuse $\overline{AC}$, which is shorter than the arc of the circle connecting A to C.

$$\sin(\phi) = |\overline{AB}| \le |\overline{AC}| \le \left(\begin{smallmatrix} \text{length of arc} \\ \text{from } A \text{ to } C \end{smallmatrix} \right) = \phi.$$

Since $\phi > 0$, we can divide and preserve the inequality. That is,

$$\frac{\sin(\phi)}{\phi} \le 1.$$

When $\phi < 0$ we use the substitution from earlier ($\phi = -\theta$) to show that this inequality still holds. Now that (2.2) is fully established, the Squeeze Theorem delivers the result. ∎

The key to using the Most Important Elementary Limit is making sure that the argument of the sine function is the same as the denominator of the fraction.

Using the Most Important Elementary Limit

Example 2.11. Determine the value of $\lim_{\theta \to 0} \frac{\sin(7\theta)}{\theta}$.

Solution: Notice that the argument of this sine function is not the same as the denominator. However, they only differ by a factor of 7, and we can fix that:

$$\lim_{\theta \to 0} \frac{\sin(7\theta)}{\theta} = \lim_{\theta \to 0} \left(7 \, \frac{\sin(7\theta)}{7\theta} \right) = \left(\lim_{\theta \to 0} 7 \right) \left(\lim_{\theta \to 0} \frac{\sin(7\theta)}{7\theta} \right) = (7)(1) = 7. \qquad ∎$$

Try It Yourself: Designing with the Most Important Elementary Limit

Example 2.12. Find numbers a and b for which $\lim_{\phi \to 0} \frac{\sin(a\phi)}{b\phi} = \frac{7}{3}$.

Answer: $a = 7$ and $b = 3$ (other answers are possible). ∎

Full Solution On-line

Using the Most Important Elementary Limit

Example 2.13. Determine $\lim_{\theta \to 0} \frac{\sin(3\theta)}{\sin(5\theta)}$.

Solution: In this case, we rewrite our function as

$$\frac{\sin(3\theta)}{\sin(5\theta)} = \left(\frac{\sin(3\theta)}{1} \right) \left(\frac{1}{\sin(5\theta)} \right) = \frac{3\theta}{5\theta} \left(\frac{\sin(3\theta)}{3\theta} \right) \left(\frac{5\theta}{\sin(5\theta)} \right).$$

Since *the limit of a product is the product of the limits,*

$$\lim_{\theta \to 0} \frac{\sin(3\theta)}{\sin(5\theta)} = \underbrace{\left(\lim_{\theta \to 0} \frac{3}{5} \right)}_{=3/5} \underbrace{\left(\lim_{\theta \to 0} \frac{\sin(3\theta)}{3\theta} \right)}_{=1} \left(\lim_{\theta \to 0} \frac{5\theta}{\sin(5\theta)} \right) = \frac{3}{5} \lim_{\theta \to 0} \frac{5\theta}{\sin(5\theta)}.$$

The remaining limit is also equal to 1 because *the limit of a quotient is the quotient of the limits,* and

$$\lim_{\theta \to 0} \frac{5\theta}{\sin(5\theta)} = \lim_{\theta \to 0} \frac{1}{\frac{\sin(5\theta)}{5\theta}} = \frac{\lim_{\theta \to 0} 1}{\lim_{\theta \to 0} \frac{\sin(5\theta)}{5\theta}} = \frac{1}{1}.$$

So now we have $\lim_{\theta \to 0} \frac{\sin(3\theta)}{\sin(5\theta)} = \frac{3}{5}$. ∎

Next Most Important Elementary Limit: Suppose ϕ is measured in radians. Then

$$\lim_{\phi \to 0} \frac{\cos(\phi) - 1}{\phi} = 0.$$

Proof. Whereas our last proof had a geometric flavor, this one is algebraic. After multiplying by the conjugate of the numerator, we see

$$\frac{\cos(\phi) - 1}{\phi} = \frac{\cos(\phi) - 1}{\phi}\left(\frac{\cos(\phi) + 1}{\cos(\phi) + 1}\right) = \frac{\cos^2(\phi) - 1}{\phi(1 + \cos(\phi))} = \frac{-\sin^2(\phi)}{\phi(1 + \cos(\phi))}.$$

This allows us to calculate

$$\lim_{\phi \to 0} \frac{1 - \cos(\phi)}{\phi} = \lim_{\phi \to 0} \frac{-\sin^2(\phi)}{\phi(1 + \cos(\phi))} = \lim_{\phi \to 0}\left(\frac{\sin(\phi)}{\phi}\frac{-\sin(\phi)}{1 + \cos(\phi)}\right).$$

Using the limit laws for products, quotients, and sums we see that

$$\lim_{\phi \to 0} \frac{1 - \cos(\phi)}{\phi} = \overbrace{\left(\lim_{\phi \to 0} \frac{\sin(\phi)}{\phi}\right)}^{=1}\left(\lim_{\phi \to 0}\frac{-\sin(\phi)}{1 + \cos(\phi)}\right)$$

$$= \frac{-\lim_{\phi \to 0}\sin(\phi)}{\lim_{\phi \to 0}(1 + \cos(\phi))} = \frac{-\lim_{\phi \to 0}\sin(\phi)}{\lim_{\phi \to 0}1 + \lim_{\phi \to 0}\cos(\phi)} = \frac{0}{1 + 1} = 0. \quad \blacksquare$$

Try It Yourself: Using the Next Most Important Elementary Limit

Example 2.14. Repeat the technique in Example 2.11 to determine the value of $\lim_{\theta \to 0} \frac{\cos(7\theta) - 1}{\theta}$.

Answer: Zero. ■

Full Solution
On-line

❖ Common Mistakes with Limits

Beginners often make the mistake of performing a direct substitution, in hopes that $\lim_{t \to t_0} y(t) = y(t_0)$. As often as not, this leaves them with $0/0$, which many people claim is 1, or 0, or meaningless. However we saw several cases in which direct substitution yields $0/0$, but the limit value exists and is neither 1 nor 0.

> The hope that direct substitution will work is not entirely unjustified, as we saw with polynomials, but you should be wary of it for now (we'll talk about when it does or doesn't work in Section 2.5).

> **You should know**
>
> - the limit laws;
> - that $\lim_{\phi \to 0} \frac{\sin(\phi)}{\phi} = 1$ and $\lim_{\phi \to 0} \frac{1 - \cos(\phi)}{\phi} = 0$;
> - the statement and meaning of the *Squeeze Theorem*.

> **You should be able to**
>
> - calculate limits of polynomials;
> - calculate limits of rational functions (using the factoring technique when necessary);
> - design a function of the form $y(t) = \sin(at)/\sin(bt)$ for which $\lim_{t \to 0} y(t) = $ (any specified number).

❖ 2.2 Skill Exercises

Calculate the limits in #1–44. Identify which limit laws you use in your calculation, and at what stage you use them.

1. $\lim_{t \to 0} 3t + 4$

2. $\lim_{t \to 3} 2t - \sqrt{7}$

3. $\lim_{t \to 3} \sqrt{2t - 3}$

4. $\lim_{t \to -1} \sqrt[3]{9t - 1}$

5. $\lim_{t \to 2} t^3 + 4t - 7$

6. $\lim_{t \to 2} t^2 + 7t + 9$

7. $\lim_{t \to 2} 5t^2 - 3t + 1$

8. $\lim_{t \to 0} (3t + 1) \cos(t)$

9. $\lim_{t \to 0} (3t + 1) \sin(t)$

10. $\lim_{t \to \pi/6} \cos(t) \sin(t)$

11. $\lim_{t \to 2} \frac{t^2 + 6t + 2}{t^2 - 3}$

12. $\lim_{t \to 3} \frac{8t^2 - t + 9}{2t^2 + 4t}$

13. $\lim_{t \to 1} \frac{t^2 + 6t - 7}{t^2 + 2t - 5}$

14. $\lim_{t \to 8} \sqrt{t^2 + 4}$

15. $\lim_{t \to 3} \sqrt[5]{5t^2 + 2}$

16. $\lim_{t \to 4} (t^3 + 5t - 1)^{8/3}$

17. $\lim_{t \to 5} \frac{t^2 - t + 9}{t^2 + t + 6}$

18. $\lim_{t \to 3} \frac{t^2 - 2t - 3}{t^2 - t - 6}$

19. $\lim_{t \to -8} \frac{t^2 + t - 56}{t^2 + 7t - 8}$

20. $\lim_{t \to 1} \frac{t^2 + 6t - 7}{t^2 - 3t + 2}$

21. $\lim_{t \to -3} \frac{t^2 + t + 7}{t^2 + 4t + 3}$

22. $\lim_{t \to 2} \frac{t^2 + 5t - 14}{t^2 - 4t + 4}$

23. $\lim_{t \to 3} \left(\frac{t^2 + 6}{t + 7} \right)^{5/3}$

24. $\lim_{t \to 4} \left(\frac{5t^2 - 4t + 1}{3t - 7} \right)^{3/2}$

25. $\lim_{t \to 1} \sqrt[4]{\frac{2t^2 + t - 1}{3t + 17}}$

26. $\lim_{t \to 4} \left(\frac{t^2 - 4t + 1}{3t - 7} \right)^{3/2}$

27. $\lim_{t \to 11} \frac{t - 11}{\sqrt{t + 5} - 4}$

28. $\lim_{t \to 4} \frac{t - 4}{\sqrt{7 - t} - 2}$

29. $\lim_{t \to 1} \frac{\sqrt{10 - t} - 3}{1 - t}$

30. $\lim_{t \to 3} \frac{\sqrt{19 - t} - 4}{t - 3}$

31. $\lim_{t \to 1} \frac{\sqrt[4]{17 - t} - 2}{t - 1}$

32. $\lim_{t \to 2} \frac{t - 2}{\sqrt[4]{83 - t} - 3}$

33. $\lim_{\phi \to 0} \frac{\sin(8\phi)}{7\phi}$

34. $\lim_{\phi \to 0} \frac{8\phi}{\sin(7\phi)}$

35. $\lim_{\phi \to 0} \frac{\sin(8\phi)}{\sin(7\phi)}$

36. $\lim_{\phi \to 0} \frac{\sin(9\phi)}{\sin(24\phi)}$

37. $\lim_{\phi \to 0} \frac{\sin(8\phi)}{\cos(7\phi)}$

38. $\lim_{\phi \to 0} \frac{\cos(8\phi)}{\cos(7\phi)}$

39. $\lim_{\phi \to 0} \frac{1 - \cos(8\phi)}{1 - \cos(7\phi)}$

40. $\lim_{\phi \to 0} \frac{1 - \cos(2\phi)}{\sin(5\phi)}$

41. $\lim_{\phi \to 0} \frac{\sin^2(3\phi)}{\phi \sin(51\phi)}$

42. $\lim_{\phi \to 0} \frac{4\phi^2}{1 - \cos(2\phi)}$

43. $\lim_{\phi \to 2} \frac{3 \sin(4\phi - 8)}{\phi^2 + 12\phi - 28}$

44. $\lim_{\phi \to 4} \frac{\phi^3 - \phi^2 + 12\phi}{\sin^2(\phi - 4)}$

In each of #45–48 you should (a) determine a formula for $m(h)$, and (b) calculate $\lim_{h \to 0} m(h)$.

45. The points $(6, 36)$ and $(6 + h, (6 + h)^2)$ are on the graph of f when $f(t) = t^2$. Denote the slope of the line through these two points by $m(h)$.

46. The points $(5, 25)$ and $(5 + h, (5 + h)^2)$ are on the graph of f when $f(t) = t^2$. Denote the slope of the line through these two points by $m(h)$.

47. The points $(4, 2)$ and $(4 + h, \sqrt{4 + h})$ are on the graph of f when $f(t) = \sqrt{t}$. Denote the slope of the line through these two points by $m(h)$.

48. The points $(1, 1)$ and $(1 + h, \sqrt{1 + h})$ are on the graph of f when $f(t) = \sqrt{t}$. Denote the slope of the line through these two points by $m(h)$.

❖ 2.2 Concept and Application Exercises

49. In Example 2.2 (p. 98) we worked with $f(t) = ax^2$, where $a = 1$. This question asks you to extend the example to $a \neq 1$. Denote the y-intercept of the perpendicular bisector by $b_a(t)$.

     | Geometry |

     (a) Determine a formula for $b_a(t)$, and use it to calculate $\lim_{t \to 0} b_a(t)$, whose value we'll call $B(a)$.

     (b) Determine $\lim_{a \to 0^+} B(a)$.

     (c) Does your answer to part (b) make intuitive sense? If so, explain. If not, why not?

In each of exercises #50-53 design a rational function for which $\lim_{t \to t_0} f(t)$ is a 0/0 indeterminate form, but the actual limit value is L.

50. $t_0 = 3$, $L = 17$ 51. $t_0 = 4$, $L = \sqrt{2}$ 52. $t_0 = 5$, $L = 0$ 53. $t_0 = 0$, $L = 0$

In each of #54–57 use the sine to design a function $f(\phi)$ for which $\lim_{\phi \to 0} f(\phi)$ is a 0/0 indeterminate form, but the actual limit value is L.

54. $L = 2$ 55. $L = -3$ 56. $L = \sqrt{3}$ 57. $L = 0$

58. Explain what's wrong with the following equation (we've misused one of the Limit Laws):

$$\lim_{t \to 0} t \sin\left(\frac{1}{t}\right) = \left(\lim_{t \to 0} t\right)\left(\lim_{t \to 0} \sin\left(\frac{1}{t}\right)\right) = 0 \cdot \left(\lim_{t \to 0} \sin\left(\frac{1}{t}\right)\right) = 0.$$

59. Explain what's wrong with the equation

$$\lim_{\phi \to 0} \frac{\phi \sin(\phi)}{\phi^2} = \frac{\lim_{\phi \to 0} \phi \sin(\phi)}{\lim_{\phi \to 0} \phi^2} = \frac{0}{0} = 1.$$

60. Suppose $f(t) = \sin(3/t)$ and $g(t) = \sin(-3/t)$. Verify that neither $\lim_{t \to 0} f(t)$ nor $\lim_{t \to 0} g(t)$ exists, but $\lim_{t \to 0} \left(f(t) + g(t)\right)$ does. Does this contradict Limit Law 4? If so, how? If not, explain why.

61. Consider the meaning of the notation $\lim_{t \to 1} \sqrt{t - 1}$, and then explain why Limit Law 7 requires $L > 0$ when q is even.

62. Sketch the graphs of functions f and g so that $\lim_{t \to 0} f(t)g(t)$ exists, but $\lim_{t \to 0} f(t)$ does not.

63. Sketch the graphs of functions f and g so that $\lim_{t \to 0} g(t) = 0$, but $\lim_{t \to 0} f(t)/g(t)$ exists.

64. Suppose that $\lim_{t \to 1} f(t)g(t)$ exists, and ...

    (a) $\lim_{t \to 1} g(t) = 0$. Does $\lim_{t \to 1} f(t)$ have to exist also?

    (b) $\lim_{t \to 1} g(t) = 4$. Does $\lim_{t \to 1} f(t)$ have to exist also?

65. Suppose that $\lim_{t \to 1} f(t)/g(t)$ exists, and ...

    (a) $\lim_{t \to 1} f(t) = 0$. Does $\lim_{t \to 1} g(t)$ have to exist also?

    (b) $\lim_{t \to 1} f(t) = 4$. Does $\lim_{t \to 1} g(t)$ have to exist also?

66. Determine the fuel efficiency in miles per gallon if ...
                               | Fuel Efficiency |

    (a) you're driving at 60 miles per hour and using 1 gallon of fuel every 0.9166667 hours.

    (b) your car is parked and off. *(How many miles per hour are you traveling? How many gallons per hour are you using?)*

67. Draw the graph of a function f so that $\cos(t) \le f(t) \le 2 - \cos(t)$. Then determine $\lim_{t \to 2\pi} f(t)$.

68. Use the Squeeze Theorem to determine $\lim_{t \to 0} f(t)$ when

$$f(t) = \begin{cases} t^2 & \text{when} \quad t \text{ is rational} \\ 3t & \text{when} \quad t \text{ is irrational.} \end{cases}$$

69. Suppose f is a function for which $\sin(\phi) \le f(\phi) \le \cos(\phi)$ when ϕ is near zero. Explain why we cannot use the Squeeze Theorem to determine $\lim_{\phi \to 0} f(\phi)$.

70. When the axis of the parabola $y = x^2 - 4$ is tilted ϕ radians clockwise, the x-intercepts shift to

$$x_1(\phi) = \frac{\sin(\phi) - \sqrt{\sin^2(\phi) + 16\cos^2(\phi)}}{2\cos^2(\phi)}$$

$$x_2(\phi) = \frac{\sin(\phi) + \sqrt{\sin^2(\phi) + 16\cos^2(\phi)}}{2\cos^2(\phi)}.$$

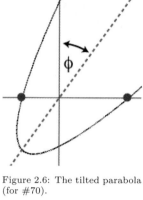

Figure 2.6: The tilted parabola (for #70).

   (a) Notice that $x_1(\phi)$ has the form $\frac{a-b}{2\cos^2(\phi)}$. Multiply it by $(a+b)/(a+b)$ and simplify the expression you get (subtracting and canceling where appropriate).

   (b) Use your expression from part (b) to calculate $\lim_{\phi \to \pi/2} x_1(\phi)$.

   (c) Use the same technique to find $\lim_{\phi \to \pi/2} x_2(\phi)$.

71. Figure 2.7 shows a sector of the unit circle with central angle θ. Denote by $A(\theta)$ the area of the region between the chord PR and the arc PR, and by $B(\theta)$ the area of the triangle PQR.

   (a) Verify that $\lim_{\theta \to 0+} A(\theta) = 0$ and $\lim_{\theta \to 0+} B(\theta) = 0$.

   (b) Calculate $\lim_{\theta \to 0+} \frac{B(\theta)}{A(\theta)}$.

   (c) Based on part (b), which of $A(\theta)$ and $B(\theta)$ tends to zero faster?

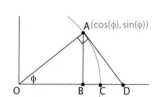

Figure 2.7: For #71.

Checking the details

72. Figure 2.8 depicts the fact that

$$\begin{aligned}(\text{area of } \Delta OAB) &\leq (\text{area of sector } OAC) \\ &\leq (\text{area of } \Delta OAD).\end{aligned} \quad (2.5)$$

   (a) Express the area of ΔOAB, ΔOAD, and the sector OAC in terms of ϕ.

   (b) Substitute these expressions into (2.5) as appropriate, then scale the entire string of inequalities by 2.

   (c) Assuming that $\phi > 0$, divide the inequalities by $\sin \phi$.

   (d) Use the Squeeze Theorem to show $\lim_{\phi \to 0+} \phi/\sin(\phi) = 1$.

   (e) Repeat part (e), assuming that $\phi < 0$. (How does the string of inequalities change?)

   (f) Argue that $\lim_{\phi \to 0-} \phi/\sin(\phi) = 1$.

   (g) Argue that $\lim_{\phi \to 0} \sin(\phi)/\phi = 1$.

Figure 2.8: For #72.

73. Suppose that we measure angle in degrees instead of radians. Then the calculation in (2.4) tells us that ϕ degrees of angle corresponds to $\pi\phi/180$ units of length along the circle. Show that this results in

$$\lim_{\phi \to 0} \frac{\sin(\phi)}{\phi} = \frac{\pi}{180} \quad \text{when } \phi \text{ is measured in degrees.}$$

2.3   Limits at Infinity

Why don't raindrops hurt? Air resistance. (Without it, a raindrop would feel like a baseball that had been dropped on you from 3 feet above. Ouch!) The force experienced by a raindrop due to air resistance depends on its velocity, $v(t)$. More specifically,

$$F_{\text{drag}} = -\gamma \, v(t)^2$$

The symbol γ is the lowercase Greek letter gamma, pronounced "GAM-uh."

where $\gamma > 0$ is constant. When combined with Newton's Second Law of motion, this model of air resistance leads to something called a *differential equation.* You'll learn to solve such equations in Chapter 10, and will see that

$$v(t) = -\sqrt{\frac{mg}{\gamma}} \; \frac{1 - e^{-2\sqrt{mg/\gamma}\,t}}{1 + e^{-2\sqrt{mg/\gamma}\,t}} \;, \qquad (3.1)$$

where m is the mass of the drop, and g is the acceleration per unit mass due to gravity. This formula may appear complicated, but the important information is easy to extract. Since $-2\sqrt{mg/\gamma} < 0$, the exponential terms approach zero as t increases (see Chapter 1). So

$$1 \pm e^{-2\sqrt{mg/\gamma}\,t} \quad \text{gets closer and closer to} \quad 1 \pm 0 = 1$$

as t grows and grows; and consequently the number $v(t)$ approaches $-\sqrt{mg/\gamma}$. This is exactly the idea of a limit, except that instead of examining the behavior of the function as t approaches some finite t_0, we are examining its behavior as t grows without bound (written $t \to \infty$).

Said simply, raindrops don't hurt because the number γ is big enough, relative to the acceleration due to gravity.

In general, when the value of $f(t)$ approaches the number L as $t \to \infty$, we write

$$\lim_{t \to \infty} f(t) = L$$

and say that L is the *limit of $f(t)$* **at infinity**.

Alternatively, we could discuss what happens "as t decreases and decreases (surpassing each negative time) ..." and so talk about $\lim_{t \to -\infty} f(t)$.

In both cases, you should understand that the number $f(t)$ has to get close to L, and *stay close*, as $|t|$ increases.

The ideas and notation are similar when $t \to -\infty$. When $|t| \to \infty$, we sometimes refer to this as the function's **end behavior**, or its behavior **in the far field**. Returning to the idea of the falling raindrop, we write $\lim_{t \to \infty} v(t) = -\sqrt{mg/\gamma}$.

Example of limits at infinity

Example 3.1. The graph of f is shown in Figure 3.1 with its horizontal asymptotes. Use it to make a conjecture about $\lim_{t \to \infty} f(t)$.

Solution: The graph indicates that the number $f(t)$ approaches to 1 as t grows, so $\lim_{t \to \infty} f(t) = 1$. Similarly, $\lim_{t \to -\infty} f(t) = -1$. ∎

Limits of $e^{\beta t}$ and t^{β} as $t \to \infty$, when $\beta < 0$

Example 3.2. Suppose $\beta < 0$. Determine (a) $\lim_{t \to \infty} e^{\beta t}$, and (b) $\lim_{t \to \infty} t^{\beta}$.

Solution: We'll use $\beta = -2$ to illustrate the solutions.

(a) Since $e^{-2t} = \left(\frac{1}{e^2}\right)^t$ is an exponential function whose base is less than 1, its value decreases toward zero as $t \to \infty$ (see Figure 3.2), so $\lim_{t \to \infty} e^{-2t} = 0$.

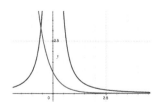

Figure 3.2: The graphs of $y = e^{-2t}$ and $y = 1/t^2$.

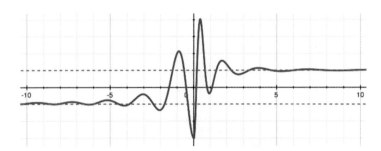

Figure 3.1: The graph of f for Example 3.1.

(b) The function $t \longmapsto 1/t^2$ also decreases toward zero as its argument grows, so $\lim_{t \to \infty} t^{-2} = 0$.

While neither of these functions ever *attains* a value of zero, their values get closer and closer to zero as t grows. The same thing happens with every $\beta < 0$.   ■

❖ Limits at Infinity Might Not Exist

As with limits at finite times, when $f(t)$ fails to approach a particular number as $t \to \infty$, we say that the **limit of $f(t)$ at infinity does not exist** or that it **diverges**. There are two basic ways this could happen.

When limits at infinity fail to exist because of oscillation

Example 3.3. Suppose $f(t) = \sin(t)$. Explain why $\lim_{t \to \infty} f(t)$ does not exist.

Solution: The number $f(t)$ continues to oscillate between -1 and 1 forever, so it never gets closer and closer to a *particular value*.   ■

When a limit at infinity fails to exist because of unbounded growth

Example 3.4. Suppose $g(t) = t + \frac{t}{3}\sin(t)$. Show that $\lim_{t \to \infty} g(t)$ does not exist.

Solution: Since $\sin(t) \geq -1$, we know that

$$g(t) = t + \frac{t}{3}\sin(t) \geq t + \frac{t}{3}(-1) = \frac{2}{3}t.$$

Since $2t/3$ grows arbitrarily large as t increases, and $g(t) \geq 2t/3$ (see Figure 3.3), the value of $g(t)$ will eventually surpasses every finite number. Because it doesn't converge to any *particular value* as t grows, the limit does not exist.   ■

The word "eventually" is important in the solution of Example 3.4. Figure 3.4 shows that $g(t)$ first surpasses 30 when $t = 25.67$ but drops below 30 soon thereafter. It hovers around 30 for a while, but *eventually* rises above 30 and never dips below again. We describe the behavior of surpassing every number and *eventually* never coming back down by saying that the limit at infinity is infinite. That is,

$$\lim_{t \to \infty} g(t) = \infty.$$

WARNING: It's important to remember that this limit *does not exist*, since no particular number is approached.

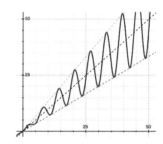

Figure 3.3: The graph of $g(t)$ from Example 3.4, along with the graphs of $y = 2t/3$ (low), $y = t$ (middle), and $y = 4t/3$ (high).

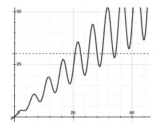

Figure 3.4: The number $g(t)$ hovers around 30 for a while, but eventually it never drops below 30 again.

Unbounded growth does not imply an infinite limit

Example 3.5. Suppose $h(t) = t \sin^2(t)$. Show that $\lim_{t \to \infty} h(t)$ does not exist, but is *not* infinite.

Solution: Though $h(t) > 1000$ when $t = 400.5\pi$, the function value will never *stay* larger than 1000 because $\sin(t) = 0$ whenever t is a multiple of π (see Figure 3.5). Similarly, the value of $h(t)$ will surpass every finite number as $t \to \infty$, but it will always return to zero soon after.　■

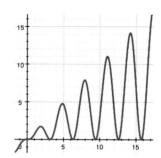

Figure 3.5: Graph for #3.5.

❖ Limit Laws at Infinity

The laws that govern limits at infinity are the same as those that govern limits at finite time. When we rewrite the Limit Laws on p. 97 by replacing t_0 with ∞, we have the limit laws at infinity.

Using the limit laws at infinity

Example 3.6. Find $\lim_{t \to \infty} y_p(t)$ when $y_p(t)$ is the stereographic projection of t onto the circle (see Example 3.1 in Chapter 1).

Solution: We derived the formula $y_p(t) = \frac{2t^2}{t^2+4}$ on p. 18. Because the limits of the numerator and denominator are both infinite (i.e., do not exist), we cannot use the limit law for quotients. However, by factoring a t^2 out of the numerator and denominator, we can rewrite $y_p(t)$ as

$$y_p(t) = \frac{2t^2}{t^2 + 4} = \frac{2t^2}{t^2 \left(1 + \frac{4}{t^2}\right)} = \frac{2}{1 + \frac{4}{t^2}}.$$

Now the limits (at infinity) of the numerator and denominator exist, so

$$\lim_{t \to \infty} y_p(t) = \frac{\lim_{t \to \infty} 2}{\lim_{t \to \infty} \left(1 + \frac{4}{t^2}\right)} = \frac{2}{\left(\lim_{t \to \infty} 1\right) + \left(\lim_{t \to \infty} \frac{4}{t^2}\right)} = \frac{2}{1 + 0} = 2.$$

Since $y_p(t)$ approaches 2, the line $y = 2$ is a horizontal asymptote of the graph (see Figure 3.6).　■

> The importance of horizontal asymptotes is that they tell us the function value is virtually constant when t is very large.

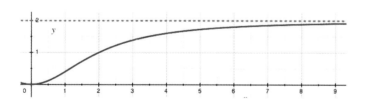

Figure 3.6: The graph of $y = y_p(t)$ and its horizontal asymptote.

Try It Yourself: Using the limit laws at infinity

Example 3.7. Find $\lim_{t \to \infty} x_p(t)$ when $x_p(t)$ is the x-coordinate of the stereographic projection of t onto the circle (see Section 1.3).

Answer: $\lim_{t \to \infty} x_p(t) = 0$.　■

Full Solution On-line

Quotients of infinite limits

Example 3.8. Determine $\lim_{t \to \infty} y(t)$ when $y(t) = \frac{14t + 16 + \sqrt{4t^2 + 11}}{7t + 8}$.

Solution: Because the limits of the numerator and denominator are both infinite (i.e., do not exist), the limit law for quotients cannot be applied. However, some algebraic simplification leads us to the answers:

$$
\begin{aligned}
y(t) &= \frac{14t + 16 \; + \; \sqrt{4t^2 + 11}}{7t + 8} = \frac{14t + 16}{7t + 8} \; + \; \frac{\sqrt{4t^2 + 11}}{7t + 8} \\
&= \frac{2(7t + 8)}{7t + 8} + \frac{\sqrt{4t^2 + 11}}{7t + 8} = 2 + \frac{\sqrt{4t^2 + 11}}{7t + 8}.
\end{aligned}
$$

Next, let's factor out a t from both the numerator and denominator of the remaining fraction so that

$$
y(t) = 2 + \frac{\sqrt{t^2 \left(4 + \frac{11}{t^2}\right)}}{t \left(7 + \frac{8}{t}\right)} = 2 + \frac{|t| \sqrt{4 + \frac{11}{t^2}}}{t \left(7 + \frac{8}{t}\right)}.
$$

 Remember $\sqrt{t^2} = |t|$.

When calculating $\lim_{t \to \infty} y(t)$, we allow t to become more and more positive, so $|t| = t$. In this case,

$$
\begin{aligned}
\lim_{t \to \infty} y(t) &= \lim_{t \to \infty} \left(2 + \frac{t \sqrt{4 + \frac{11}{t^2}}}{t \left(7 + \frac{8}{t}\right)} \right) = \lim_{t \to \infty} \left(2 + \frac{\sqrt{4 + \frac{11}{t^2}}}{7 + \frac{8}{t}} \right) \\
&= \lim_{t \to \infty} (2) + \lim_{t \to \infty} \left(\frac{\sqrt{4 + \frac{11}{t^2}}}{7 + \frac{8}{t}} \right) = 2 + \frac{\lim_{t \to \infty} \sqrt{4 + \frac{11}{t^2}}}{\lim_{t \to \infty} \left(7 + \frac{8}{t}\right)} = 2 + \frac{2}{7}. \quad \blacksquare
\end{aligned}
$$

In this step, we're relying on Limit Law 7 (for powers p/q). Since $\lim_{x \to 0}(4 + x) = 4 > 0$, we know

$$
\lim_{x \to 0} \sqrt{4 + x} = \sqrt{4} = 2.
$$

In our case, $x = 11/t^2$, which converges to zero as $t \to \infty$.

Try It Yourself: Quotients of infinite limits

Example 3.9. Determine $\lim_{t \to -\infty} y(t)$ when $y(t) = \frac{14t + 16 + \sqrt{4t^2 + 11}}{7t + 8}$.

Answer: $\lim_{t \to \infty} y(t) = 2 - \frac{2}{7}$. $\blacksquare$

 Full Solution On-line

Since the answers to Exmples 3.8 and 3.9 are different, the graph of $y(t) = \frac{14t + 16 + \sqrt{4t^2 + 11}}{7t + 8}$ has two horizontal asymptotes: $y = 2 \pm \frac{2}{7}$ (see Figure 3.7).

Figure 3.7: The function graph of $y(t)$ from Example 3.8, and horizontal asymptotes $y = 2 \pm 2/7$.

In the previous examples, both the numerator and denominator of $y(t)$ became infinite as $t \to \infty$. Without algebraic simplification, $\lim_{t \to \infty} y(t)$ had the form ∞/∞ in both cases, but the actual limit values were different. For this reason, we say that ∞/∞ is an **indeterminate form** (like $0/0$). Other indeterminate forms include $\infty - \infty$, 1^∞, $0 \cdot \infty$, and 0^0.

Limit laws at infinity

Example 3.10. Determine $\lim_{t \to \infty} y(t)$ when $y(t) = \sqrt{16t^2 + 3t + 2} - 4t$.

Solution: The limit of each summand is infinite (i.e. does not exist), so this limit results in the $\infty - \infty$ indeterminate form. However, when we multiply by the number 1 (written using the conjugate), we see

$$y(t) = \frac{\sqrt{16t^2 + 3t + 2} - 4t}{1} \left(\frac{\sqrt{16t^2 + 3t + 2} + 4t}{\sqrt{16t^2 + 3t + 2} + 4t} \right) = \frac{16t^2 + 3t + 2 - 16t^2}{\sqrt{16t^2 + 3t + 2} + 4t} = \frac{3t + 2}{\sqrt{t^2 \left(16 + \frac{3}{t} + \frac{2}{t^2}\right)} + 4t} .$$

Since $\sqrt{t^2} = |t|$, we can rewrite this as

$$y(t) = \frac{t\left(3 + \frac{2}{t}\right)}{\sqrt{t^2\left(16 + \frac{3}{t} + \frac{2}{t^2}\right)} + 4t} = \frac{t\left(3 + \frac{2}{t}\right)}{|t|\sqrt{16 + \frac{3}{t} + \frac{2}{t^2}} + 4t},$$

and since $|t| = t$ as $t \to \infty$,

$$\lim_{t \to \infty} y(t) = \lim_{t \to \infty} \frac{t\left(3 + \frac{2}{t}\right)}{4t + t\sqrt{16 + \frac{3}{t} + \frac{2}{t^2}}} = \lim_{t \to \infty} \frac{3 + \frac{2}{t}}{4 + \sqrt{16 + \frac{3}{t} + \frac{2}{t^2}}} = \frac{3}{8}.$$

Consequently, the graph of $y(t)$ has a horizontal asymptote at $3/8$. ∎

> The terms $2/t$, $3/t$ and $2/t^2$ are rational functions whose denominators have a larger degrees than their numerators, so these values converge to zero as t gets larger and larger.

Try It Yourself: Limit laws at infinity

Example 3.11. Determine $\lim_{t \to -\infty} y(t)$ when $y(t) = \sqrt{16t^2 + 3t + 2} - 4t$.

Answer: $\lim_{t \to -\infty} y(t) = \infty$. ∎

Full Solution
On-line

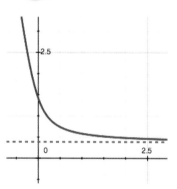

✧ Squeezing at Infinity

As with the limit laws, the statement of the Squeeze Theorem can be adapted to address limits at infinity.

> **Squeeze Theorem:** Suppose that there is a number $M > 0$ for which
> $$\underbrace{f(t) \le g(t) \le h(t)}_{} \qquad \underbrace{\text{whenever} \qquad t > M.}_{}$$
> The graph of g is caught be-   This is a precise way of saying
> tween the graphs of f and h "after t is big enough"
>
> Further, suppose that
> $$\lim_{t \to \infty} f(t) = L = \lim_{t \to \infty} h(t).$$
> Then $\lim_{t \to \infty} g(t)$ exists, and its value is also L.

Figure 3.8: Graph of $y(t)$ from Examples 3.10 and 3.11.

Using the Squeeze Theorem at infinity

Example 3.12. Use the Squeeze Theorem to calculate $\lim_{t \to \infty} \frac{1}{t} \sin(8t)$.

Solution: Since $-1 \le \sin(8t) \le 1$, we know that $-\frac{1}{t} \le \frac{1}{t}\sin(8t) \le \frac{1}{t}$ when $t > 0$. Since both $\lim_{t \to \infty} \frac{1}{t} = 0$ and $\lim_{t \to \infty} -\frac{1}{t} = 0$, we know $\lim_{t \to \infty} \frac{1}{t}\sin(8t) = 0$ (see Figure 3.9). ∎

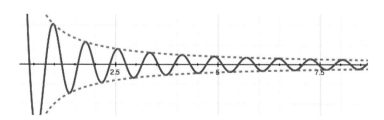

Figure 3.9: The graphs of $y = \pm 1/t$ and $y = \frac{1}{t}\sin(8t)$.

You should know

- the terms *limit at infinity, indeterminate form, end behavior,* and *far field;*

- that the Limit Laws and Squeeze Theorem hold for limits at infinity;

- that $\lim_{t\to\infty} y(t) = \infty$ means the limit does not exist, but that the function values exhibit a specific behavior.

You should be able to

- determine when $\lim_{t\to\infty} y(t) = \infty$ and when it simply doesn't exist;

- calculate limits of rational and/or difference functions.

❖ 2.3 Skill Exercises

Calculate each of the limits in #1–10.

1. $\lim_{t\to\infty} \frac{3t^2-t+7}{8t^2+t+1}$

2. $\lim_{t\to\infty} \frac{\sqrt{t^2+15t+9}}{8t+1}$

3. $\lim_{t\to\infty} \frac{\sqrt{5t^2+9}}{18t+22}$

4. $\lim_{t\to\infty} \frac{\sqrt{2t^6+11}}{4t^3+7}$

5. $\lim_{t\to\infty} \frac{\sqrt{4t^4+t^2+1}}{t^2+3}$

6. $\lim_{t\to\infty} \left(\frac{3t^2+4t+5}{t^2+6}\right)^{3/4}$

7. $\lim_{t\to-\infty} \left(\frac{5t^2-3t+2}{5t^3+6}\right)^{1/3}$

8. $\lim_{t\to-\infty} \left(\frac{6t^3+4t+1}{6t^2+10}\right)^{5/2}$

9. $\lim_{t\to\infty} \frac{4t+\cos(3t)-7}{5t+111}$.

10. $\lim_{t\to\infty} \frac{3t^2-\sin(4t)+7}{8t^2+t+1}$.

11. $\lim_{t\to\infty} \sqrt{t^2+15t+9} - t$

12. $\lim_{t\to\infty} \sqrt{25t^2+9} - (5t+22)$

13. $\lim_{t\to\infty} \sqrt{4t^6+11} + (2t^3+7)$

14. $\lim_{t\to\infty} \sqrt{4t^6+11} - (2t^3+7)$

15. $\lim_{t\to\infty} \sqrt{4t^4+t^2+1} - (t^2+3)$

16. $\lim_{t\to\infty} \sqrt{4t^4+t^2+1} + (t^2+3)$

17. $\lim_{t\to\infty} \sqrt{25t^2+8t-1} - 5t$.

18. $\lim_{t\to\infty} \sqrt{25t^2+8t-1} + 5t$.

19. $\lim_{t\to-\infty} \sqrt{t^2+15t+9} - t$

20. $\lim_{t\to-\infty} \sqrt{25t^2+9} - (5t+22)$

21. $\lim_{t\to-\infty} \sqrt{4t^6+11} + (2t^3+7)$

22. $\lim_{t\to-\infty} \sqrt{4t^6+11} - (2t^3+7)$

23. $\lim_{t\to-\infty} \sqrt{4t^4+t^2+1} - (t^2+3)$

24. $\lim_{t\to-\infty} \sqrt{4t^4+t^2+1} + (t^2+3)$

25. $\lim_{t\to-\infty} \sqrt{25t^2+8t-1} - 5t$.

26. $\lim_{t\to-\infty} \sqrt{25t^2+8t-1} + 5t$

27. $\lim_{x\to\infty} \frac{4x-\sqrt{9x^2+36x}}{8x+1}$

28. $\lim_{x\to\infty} \frac{7x^2-5x+\sqrt{16x^4-7x^3}}{1-x^2}$

29. $\lim_{x\to\infty} \frac{9x-4}{\sqrt{x^2+7x}}$

30. $\lim_{x\to\infty} \frac{\sqrt{x^2+1}-\sqrt{4x^2+5x-12}}{5x-6}$

31. $\lim_{x\to-\infty} \frac{4x-\sqrt{9x^2+36x}}{8x+1}$

32. $\lim_{x\to-\infty} \frac{7x^2-5x+\sqrt{16x^4-7x^3}}{1-x^2}$

33. $\lim_{x\to-\infty} \frac{9x-4}{\sqrt{x^2+7x}}$

34. $\lim_{x\to-\infty} \frac{\sqrt{x^2+1}-\sqrt{4x^2+5x-12}}{5x-6}$

In #35–43 use $\lim_{t\to\pm\infty} y(t)$ and the y-intercept to sketch a rough graph of the following functions.

35. $y = 5 - \dfrac{1}{t^2 + 1}$

36. $y = 5 - \dfrac{t}{t^2 + 1}$

37. $y = 5 - \dfrac{t^2}{t^2 + 1}$

38. $y = 3 + \dfrac{4}{t^2 + 1}$

39. $y = \dfrac{2t}{t^2 + 3} + \dfrac{6}{t^2 - 9}$

40. $y = \dfrac{2t}{t^2 - 4} - \dfrac{4t^2}{3t^2 + 1}$

41. $y = \dfrac{t^3 - 1}{t^2 + 1}$

42. $y = t - \dfrac{t + 4}{t^2 + 5}$

43. $y = \dfrac{5 - 6t^5}{t^4 + t^2 + 1}$

44. Based on the graph of f shown below, make a conjecture about $\lim_{t \to \infty} f(t)$.

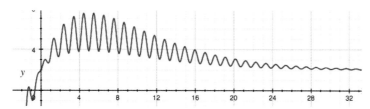

❖ 2.3 Concept and Application Exercises

45. Explain what's wrong with the calculation

$$\lim_{t \to \infty} \frac{2t + 9}{5t + 72} = \frac{\lim_{t \to \infty} 2t + 9}{\lim_{t \to \infty} 5t + 72} = \frac{\infty}{\infty} = 1.$$

46. Suppose $\lim_{t \to \infty} y(t) = 5$. Can $y(t)$ ever be larger than 5? If so, sketch the graph of an example. If not, why not?

47. Suppose $\lim_{t \to \infty} y(t) = \infty$. Can $y(t)$ ever decrease? If so, sketch the graph of an example. If not, why not?

48. Suppose that $f(t) = e^{t+a}$ and $g(t) = e^{t+b}$. Then $\lim_{t \to \infty} f(t) = \infty$ and $\lim_{t \to \infty} g(t) = \infty$.

    (a) By simplifying $f(t)$ and $g(t)$, find a formula for $\lim_{t \to \infty} \frac{f(t)}{g(t)}$ in terms of the numbers a and b.

    (b) Choose values of a and b so that $\lim_{t \to \infty} \frac{f(t)}{g(t)} =$ your age in (i) years, (ii) months, and (iii) days.

    (c) Explain why we say that ∞/∞, like $0/0$, is an *indeterminate form*.

In each of #49–54 use the Squeeze Theorem to calculate the specified limit.

49. $\lim_{t \to -\infty} \dfrac{\cos(t)}{t}$

50. $\lim_{t \to \infty} e^{-0.01t} \sin(t^2)$

51. $\lim_{t \to \infty} \dfrac{1}{t + te^{-t}}$

52. $\lim_{t \to \infty} \dfrac{2 \arctan(t)}{\pi \sqrt{t}}$

53. $\lim_{t \to \infty} \dfrac{3t^4 + \sin^2(6t)}{\sqrt{t^8 + t^6 + 1}}$

54. $\lim_{t \to -\infty} \dfrac{\sin^2(t)e^t}{1 + e^t}$

55. Suppose $\frac{3t-2}{t} \le f(t) \le \frac{3t^3+7}{t^3+1}$ when $t > 30$. Determine $\lim_{t \to \infty} f(t)$.

56. Suppose $\frac{4t-3}{t} \le f(t) \le \frac{4t^5+7}{t^5-1}$ when $t > 92$. Determine $\lim_{t \to \infty} f(t)$.

57. Suppose $f(t)$ is between (or equal to) $g(t) = \frac{\cos(8t)}{t}$ and $h(t) = e^{-t}$ when $t > 6$. What can we say definitively about $\lim_{t \to \infty} f(t)$?

58. Suppose $f(t)$ is between (or equal to) $g(t) = \frac{\sin(3t)}{t^2}$ and $h(t) = \cos(t)$ when $t > 17$. What can we say definitively about $\lim_{t\to\infty} f(t)$?

59. Suppose $a(t)$ is the amount of medication (in mg) in a patient's bloodstream after t days. Use the notation of limits to write an equation that says, "during long-term treatment, the amount of medication in the bloodstream approaches a steady state of 100 mg."

Medical Sciences

60. Suppose $p(t)$ is the percentage of knowledge that someone retains t days after the completion of a course, with no further exposure to the material. Use the notation of limits to write an equation that says, "a person will only retain 5% of a course over time (assuming no further exposure to the course content)."

Cognitive Science

61. Suppose $P(t)$ is the number of individuals in a population, and t is measured in years. Use the notation of limits to write an equation that says, "the population becomes extinct."

Biology

62. Suppose $T(t)$ is the temperature of a circuit board after t seconds. Use the notation of limits to write an equation that says, "as the circuit board is used continually, its temperature levels off at 90°F."

Electrical Engineering

63. Suppose $c(t)$ is the concentration of oxygen in a person's lungs, and c_a is the concentration of oxygen in the atmosphere at sea level. In complete sentences, explain what the equation $\lim_{t\to\infty} c(t) = c_a$ means in this context.

Medical Sciences

64. Suppose $T(t)$ is the temperature in Rochester, NY, where t is measured in seconds from the time since your birth. Make a conjecture about $\lim_{t\to\infty} T(t)$, and support your assertion in complete sentences.

Meteorology

65. Consider a particular molecule in a liquid. A spherical shell of radius r mm and thickness τ mm that surrounds that molecule (depicted in Figure 3.10) has a volume of $4\pi r^2\tau$. If there are ρ molecules/mm^3, on average, the number $4\pi r^2\tau\rho$ is a simple estimate of how many molecules have their centers in this spherical shell (not in the *sphere*, but in the shell). However, each molecule (e.g. the central one) reduces the likelihood of other molecules being nearby, due to electrostatic forces; so the actual number is written as $4\pi r^2\tau\rho\, g(r)$, and the function g is called the *pair correlation function.* What does the equation $\lim_{r\to\infty} g(r) = 1$ mean in this context?

Chemistry

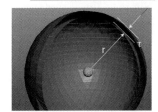

Figure 3.10: A spherical shell of radius r and thickness τ surrounding a central molecule.

66. Suppose a circle of radius r sits atop the x-axis. We'll denote by $(x_r(t), y_r(t))$ the coordinates of the stereographic projection of t onto this circle (see Section 1.3 and Figure 3.11).

(a) What do you think *should* happen to $x_r(t)$ and $y_r(t)$ as $r \to \infty$?

(b) Determine formulas for $x_r(t)$ and $y_r(t)$.

(c) Calculate $\lim_{r\to\infty} x_r(t)$ and $\lim_{r\to\infty} y_r(t)$.

Geometry

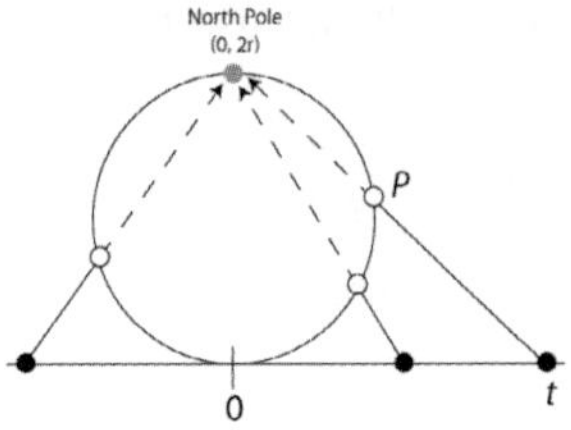

Figure 3.11: For #66.

67. In Example 2.2 (p. 98) we worked with $f(t) = ax^2$, where $a = 1$. This question asks you to extend the example to the case when $a \neq 1$. Denote the y-intercept by $b_a(t)$.

(a) Make an educated guess: will $\lim_{t\to 0} b_a(t)$ be larger than, or less than 1/2 when $a > 1$? What *should* happen as $a \to \infty$?

(b) Determine a formula for $b_a(t)$, and use it to calculate $\lim_{t\to 0} b_a(t)$, whose value we'll call $B(a)$.

(c) Determine $\lim_{a\to\infty} B(a)$.

Geometry

68. The function $P(t) = 10/(1 + 4e^{-0.39t})$ describes the size of a population of microbes, where P is the population size in *thousands* of microbes and t is time in seconds. Determine both the initial population size and the steady-state population size (i.e., the size of $P(t)$ in the long run).

| Biology |

69. Suppose $f(n)$ and $g(n)$ tell us the length of time required by different computer algorithms to complete a given task when the input is "size" n.

| Computer Science |

   (a) Which algorithm (if either) is better for large data sets when $\lim_{n\to\infty} f(n)/g(n) = 0$?

   (b) Which algorithm (if either) is better for large data sets when $\lim_{n\to\infty} f(n)/g(n) = \infty$?

   (c) Which algorithm (if either) is better for large data sets when $\lim_{n\to\infty} f(n)/g(n) = 5$?

   (d) Which algorithm (if either) is better for large data sets when $\lim_{n\to\infty} f(n)/g(n) = 0.2$?

70. The *ozone depletion potential* (ODP) of a chemical is sometimes expressed as the ratio of its *integrated chlorine loading* to that of chlorofluorocarbon CFC-11. The ODP of NH_4ClO_4 (a chemical used in solid rocket fuel) is

| Global Climate |

$$\lim_{t\to\infty} \frac{0.318e^{-t/3}}{2.33(1 + 0.16e^{-t/3} - 1.11e^{-t/46.5} - 0.05e^{-t/1.75})}.$$

Calculate the ODP of NH_4ClO_4 and determine whether it's better or worse for the ozone layer than CFC-11. (See [26] for further reading.)

71. The fraction of fuel energy converted to thrust by a four-stroke engine is described by $\eta(r) = 1 - r^{-\gamma}$, where r is the ratio of initial to final volume (v_i/v_f), and γ is a positive number that depends on a quantity called the *heat capacity*.

| Mechanical Engineering |

   (a) Explain what $r \to \infty$ means in this context.

   (b) Calculate $\lim_{r\to\infty} \eta(r)$, and explain what that tells us about the efficiency of an engine.

72. Suppose a bell tower is set 0.25 miles off the road, and you're driving by at a constant speed of 35 mph. As you pass, the bell tower plays a note whose frequency is ν. Because of the Doppler effect, the frequency you hear depends on time according to

| Physics |

$$f(t) = \frac{761\nu\sqrt{1225t^2 + 0.0625}}{1225t + 761\sqrt{1225t^2 + 0.0625}}.$$

Calculate $\lim_{t\to\infty} f(t)$ and interpret what this means physically.

73. When fluid viscosity plays a greater role than an object's momentum (e.g., a bacterium swimming in blood plasma), the drag force encountered by the object is described by $F_{drag} = -\gamma v$, where $\gamma > 0$ and v is the velocity of the object. When this is combined with Newton's Second Law of motion, this model of fluid resistance results in something called a *linear differential equation*. You'll learn to solve such equations in Chapter 10, and will see that that $v(t) = v_0 e^{-\gamma t/m}$ when no other forces are present, where m is the mass of the object and v_0 is its initial speed. Calculate $\lim_{t\to\infty} v(t)$, and explain the physical meaning of your answer.

| Physics |

2.4   The Technical Definition of a Limit

We have developed a reasonable but rough understanding of limits by answering the question, "Does the number $y(t)$ get closer and closer to a particular value as t approaches t_0 (or ∞), and if so, what value?" In this section, we refine that understanding by making the terms "closer and closer" and "approaches" technically precise.

Consider the graph in Figure 4.1, which depicts $y = f(t)$ and the horizontal line $y = 2$. Based on our discussion in Section 2.1, we conclude that $\lim_{t \to 1} f(t) = 2$. While that conclusion is correct *in spirit*, it's numerically wrong—the value of the limit is actually 1.988. We were fooled into thinking that $\lim_{t \to 1} f(t) = 2$ because the vertical resolution of the figure is 0.05, meaning that we are unable to tell the difference between heights that are less than 0.05 units apart. Since $f(t) \to 1.988$ as $t \to 1$,

$$\underbrace{|f(t) - 2| \to 0.012}_{\substack{\text{the difference between the height of} \\ \text{the graph, and 2, approaches 0.012}}} < \underbrace{0.05}_{\text{resolution}}.$$

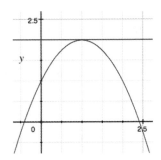

Figure 4.1: How high is the vertex of the parabola?

Zooming in by a factor of 10 changes the resolution to 0.005 (see Figure 4.2), after which it's clear that $\lim_{t \to 1} f(t) \neq 2$.

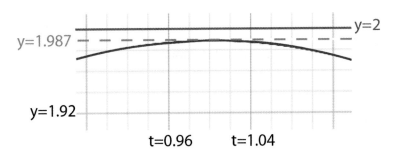

Figure 4.2: Quantitative accuracy is a matter of resolution.

The limit value *appears* to be 1.987 in Figure 4.2, but you already know it's not. As before, detecting the difference between 1.987 and the actual limit value is a matter of resolution. Since $f(t) \to 1.988$ as $t \to 1$,

$$|f(t) - 1.987| < \underbrace{0.005}_{\text{resolution}} \quad \text{once } t \text{ is close enough to } t_0 = 1.$$

However, we can detect the difference with a higher resolution. The basic idea is just this: if we think the limit value is L and we're *wrong*, we'll be able to detect our mistake by using a better resolution; but when the limit value really *is* L, there is no error to detect at *any* resolution. The technical definition of the limit (below) is written to say exactly that:

Technical Definition of the Limit: We say $\lim_{t \to t_0} f(t) = L$ when every $\varepsilon > 0$ is associated with a distance $\delta > 0$ such that

$$\underbrace{|f(t) - L| < \varepsilon}_{\substack{\text{This means we're unable to resolve the} \\ \text{difference between } f(t) \text{ and } L \text{ when the} \\ \text{resolution is } \varepsilon. \text{ They look the same to us.}}} \quad \text{whenever} \quad \underbrace{0 < |t - t_0| < \delta.}_{\substack{\text{This says that } t \text{ is "close} \\ \text{enough" to } t_0.}} \qquad (4.1)$$

The symbols ε and δ are the lower-case Greek letters "epsilon" and "delta," respectively.

The phrase "every $\varepsilon > 0$" means this will work for *all* resolutions.

Recall $|t - t_0|$ is the distance between t and t_0, so we're using δ to quantify the term "close enough."

This does *not* say that the same δ will work for all possible values of $\varepsilon > 0$. Typically, better resolution (i.e., smaller ε) requires us to keep t closer to t_0.

Note: The statement that $0 < |t - t_0| < \delta$ prevents $t = t_0$. The definition is designed this way because, fundamentally, limits are determined by a function's behavior *near* t_0 rather than *at* t_0.

❖ Working Graphically with ε and δ

The inequality $|f(t) - L| < \varepsilon$ is equivalent to $-\varepsilon < f(t) - L < \varepsilon$. Adding L to both sides allows us to write this as

$$L - \varepsilon < f(t) < L + \varepsilon.$$

So the definition of the limit means that the graph of f stays between the heights $L - \varepsilon$ and $L + \varepsilon$ whenever t is "close enough" to t_0 (i.e., within δ units of t_0).

Determining the number δ

Example 4.1. The graph of $y(t)$ in Figure 4.3 shows that $\lim_{t \to 2} y(t) = 3$. If $\varepsilon = 0.5$, what's an appropriate value of δ?

Solution: When $\varepsilon = 0.5$ we have to guarantee that $y(t)$ remains between $3 - 0.5$ and $3 + 0.5$. We begin by marking those heights with dashed lines (see Figure 4.4(left)). Now we can see when the function values get too high or too low. Starting at $t_0 = 2$,

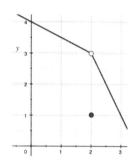

Figure 4.3: Graph for Example 4.1.

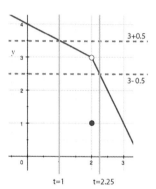

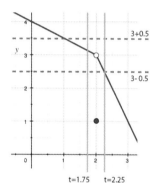

Figure 4.4: (left) Finding where the graph of $y(t)$ leaves the the red-dashed envelope; (right) we must be able to move away from t_0, up to δ units, whether we step left or right.

we can move left all the way to $t = 1$ before the graph gets too high, but only 0.25 units to the right of $t_0 = 2$ before the graph gets too low. Since $|y(t) - 3| < 0.5$ whenever t is within 0.25 units of $t_0 = 2$, we choose $\delta = 0.25$. Any smaller value of $\delta > 0$ would also work.   ∎

Try It Yourself: Determining the number δ

Example 4.2. Suppose f is defined by $f(t) = 0.25t^2$ whenever $t \neq 2$. Then $\lim_{t \to 2} y(t) = 1$. If $\varepsilon = 0.1$, what's an appropriate value of δ?

Answer: Any positive $\delta < \sqrt{4.4} - 2 \approx 0.0976176$ will work.   ∎

Full Solution On-line

When the limit doesn't exist (because of a jump)

Example 4.3. Use the graph of $y(t)$ in Figure 4.5 to explain why $\lim_{t \to 0} y(t) \neq 4$.

Solution: Were it the case that $\lim_{t \to 0} y(t) = 4$, for *every* $\varepsilon > 0$, we could guarantee that $4 - \varepsilon < y(t) < 4 + \varepsilon$ by keeping t close enough to $t_0 = 0$. Figure 4.5 depicts $\varepsilon \approx 0.5$, and you can see that we cannot make any such guarantee. No matter how close we position the vertical lines to $t_0 = 0$, the graph of $y(t)$ strays outside the red-dashed envelope. Numerically speaking, no matter how small we choose δ, we cannot guarantee $|y(t) - 4| < 0.5$. ∎

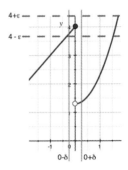

Figure 4.5: Graph for Example 4.3.

When the limit doesn't exist (because of oscillation)

Example 4.4. Use the definition of the limit to explain why $\lim_{t \to 0} \sin(3/t) \neq 0$.

Solution: The graph of $y(t) = \sin(3/t)$, shown on p. 91, continues to oscillate between -1 and 1 as $t \to 0$. So we cannot guarantee that $0 - \varepsilon < y(t) < 0 + \varepsilon$ when ε is small, no matter how near t is to 0 (i.e., *no* positive δ will work). ∎

We've said that $\lim_{t \to t_0} f(t) = L$ when, for every $\varepsilon > 0$, the graph of f remains between the heights $L - \varepsilon$ and $L + \varepsilon$ once t is "near enough" to t_0. Said with a slightly different spin, "the number $f(t)$ gets close to L, and *stays close* as t approaches t_0."

❖ Working Algebraically with ε and δ

While we can use the graph of a function to approximate a value of δ for a given ε, the definition of a limit requires a δ for *every* $\varepsilon > 0$; and when the resolution is changed, all of our graphical work has to be repeated. So instead, we often develop a formula algebraically that calculates an appropriate value of δ for any specified ε.

Finding a formula for δ when $t_0 = 0$

Example 4.5. Suppose $y(t) = 8t^5 + 3t^2 + 9$. Prove that $\lim_{t \to 0} y(t) = 9$.

Solution: Every time someone chooses a resolution $\varepsilon > 0$, no matter how small, we must be able to find a distance $\delta > 0$ so that

$$\underbrace{|y(t) - 9| < \varepsilon}_{\substack{\text{We cannot tell the difference between } y(t) \\ \text{and 9 at the specified resolution}}} \quad \text{whenever} \quad \underbrace{0 < |t - 0| < \delta.}_{t \text{ is "close enough" to } t_0 = 0}$$

> Recall that $|t - 0|$ is the distance between t and zero.
>
> We're writing $|t - 0|$ instead of $|t|$ in order to reinforce the definition, and to visually prepare you for the case when $t_0 \neq 0$.

In just a moment, we'll use an estimation technique to show that

$$|y(t) - 9| < \varepsilon \quad \text{whenever} \quad 0 < |t - 0| < \sqrt{\varepsilon/11}. \tag{4.2}$$

The technique relies on the fact that

$$|t^5| < |t^2| \quad \text{whenever} \quad 0 < |t - 0| < 1.$$

In order to achieve both of these constraints on t, we choose $\delta = \min\{1, \sqrt{\varepsilon/11}\}$. Because this can be done for every $\varepsilon > 0$, we have proven that $\lim_{t \to 0} y(t) = 9$. Now here's the estimation technique:

$$\begin{aligned} |y(t) - 9| &= |8t^5 + 3t^2| & (4.3) \\ &\leq |8t^5| + |3t^2| & (4.4) \\ &\leq 8|t^2| + 3|t^2| = 11|t|^2. \end{aligned}$$

> The transition from (4.3) to (4.4) comes from the **triangle inequality:**
>
> $$|a \pm b| \leq |a| + |b|,$$
>
> with $a = 8t^5$ and $b = 3t^2$.

Since $|y(t) - 9| \leq 11|t|^2$, we know that $|y(t) - 9| < \varepsilon$ whenever $11|t|^2 < \varepsilon$, from which (4.2) follows. ∎

Note: The most important, but most difficult, part of the previous solution was the estimation technique. When working a problem on your own, you should start with it, and record what constraints on t you need as you go.

Finding a formula for δ when $t_0 = 0$

Example 4.6. Suppose $y(t) = 2t^4 - 5t^3 + 13$. Prove that $\lim_{t \to 0} y(t) = 13$.

Solution: Every time someone chooses a resolution $\varepsilon > 0$, no matter how small, we must be able to find a distance $\delta > 0$ so that

$$|y(t) - 13| < \varepsilon \quad \text{whenever} \quad 0 < |t - 0| < \delta. \tag{4.5}$$

When we examine the size of $|y(t) - 13|$ we see

$$|y(t) - 13| = \underbrace{|2t^4 - 5t^3| \leq |2t^4| + |5t^3|}_{\text{by virtue of the triangle inequality}}.$$

When $|t - 0| < 1$ we know that $|t^4| < |t^3|$, which allows us to collect the summands into a single term:

$$|y(t) - 13| \leq |2t^4| + |5t^3| \leq 2|t^3| + 5|t^3| = 7|t|^3. \tag{4.6}$$

Since (4.5) is written in terms of $|t - 0|$, let's rewrite (4.6) as $|y(t) - 13| \leq 7|t - 0|^3$. Then we can say that $|y(t) - 13| < \varepsilon$ whenever $7|t - 0|^3 < \varepsilon$. That is, whenever $|t - 0| < \sqrt[3]{\varepsilon/7}$. The formula $\delta = \min\{1, \sqrt[3]{\varepsilon/7}\}$ incorporates all of our constraints on t, and provides a $\delta > 0$ for every $\varepsilon > 0$, so $\lim_{t \to 0} y(t) = 13$. ∎

Try It Yourself: Finding a formula for δ when $t_0 = 0$

Example 4.7. Suppose $y(t) = 3t^4 + 5t + 11$. Prove that $\lim_{t \to 0} y(t) = 11$.

Answer: $\delta = \min\{1, \varepsilon/8\}$. ∎

Full Solution On-line

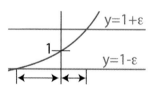

When t_0 is not zero, the techniques that we use are different but the basic ideas are exactly the same.

Finding a formula for δ when $t_0 = 0$ and $y(t)$ is not a polynomial function

Example 4.8. Suppose $y(t) = 3^t$. Prove that $\lim_{t \to 0} y(t) = 1$.

Solution: Every time someone chooses a resolution $\varepsilon > 0$, no matter how small, we must be able to find a distance $\delta > 0$ so that

$$|3^t - 1| < \varepsilon \quad \text{whenever} \quad |t - 0| < \delta.$$

Saying that $|3^t - 1| < \varepsilon$ is the same as saying $-\varepsilon < 3^t - 1 < \varepsilon$. This happens when

$$1 - \varepsilon < 3^t < 1 + \varepsilon.$$

Of course, the left-hand inequality is trivial when ε is big (because 3^t is never negative), so let's assume that we're working with $\varepsilon \in (0, 1)$. Then because $\log_3(3^t) = t$ and the logarithm is an increasing function,

$$\log_3(1 - \varepsilon) < t < \log_3(1 + \varepsilon). \tag{4.7}$$

This is very close to what we want. The inequality $|t - 0| < \delta$ is the same as saying that $-\delta < t < \delta$. This is an interval that's symmetric about $t_0 = 0$, but the interval described by equation (4.7) is not. We *make* a symmetric interval for ourselves by choosing

$$\delta = \min\left\{|\log_3(1 - \varepsilon)|, |\log_3(1 + \varepsilon)|\right\}.$$

Figure 4.6: The curvature of the graph has been exaggerated in order to indicate clearly the effect of the "cupped-up" structure of the graph on our calculation.

As it happens, this reduces to $\delta = \log_3(1+\varepsilon)$ for every $\varepsilon \in (0,1)$. In Figure 4.6 you can see that this happens because the graph of the exponential function leaves the envelope sooner to the right of $t_0 = 0$ than it does to the left. ∎

Finding a formula for δ when $t_0 \neq 0$

Example 4.9. Suppose $y(t) = 3t^2 - 8t + 7$. Prove that $\lim_{t\to 1} y(t) = 2$.

Solution: Every time someone chooses a resolution $\varepsilon > 0$, no matter how small, we must be able to find a distance $\delta > 0$ so that

$$\underbrace{|y(t) - 2| < \varepsilon}_{\substack{\text{We cannot tell the difference between } y(t) \\ \text{and 2 at the specified resolution}}} \quad \text{whenever} \quad \underbrace{0 < |t - 1| < \delta.}_{t \text{ is "close enough" to } t_0 = 1} \quad (4.8)$$

> Recall that $|t-1|$ is the distance between t and 1.

Equation (4.8) says that we can control the size of $y(t) - 2$ by controlling the size $(t-1)$. So when we estimate $|y(t)-2|$, let's rewrite $y(t)$ in terms of $(t-1)$. Following this example, you'll see how to rewrite $y(t)$ as $2 - 2(t-1) + 3(t-1)^2$. Then, when $|t - 1| < 1$,

$$\begin{aligned}
|y(t) - 2| &= |3(t-1)^2 - 2(t-1)| \\
&\leq |3(t-1)^2| + |2(t-1)| \\
&\leq 3|(t-1)| + 2|(t-1)| = 5|t-1|.
\end{aligned}$$

Since $|y(t) - 2| \leq 5|t-1|$ when $|t-1| < 1$, we know that $|y(t) - 2| < \varepsilon$ whenever $5|t-1| < \varepsilon$. Consequently,

$$|y(t) - 2| < \varepsilon \quad \text{whenever} \quad |t - 1| < \varepsilon/5. \quad (4.9)$$

Since we need both $|t-1| < 1$ and $|t-1| < \varepsilon/5$, we choose $\delta = \min\{1, \varepsilon/5\}$. ∎

Don't just read it. Work it out! Where do these inequalities come from?

Rewriting $y(t)$ in terms of $(t - t_0)$

In Example 4.9 we needed to rewrite $y(t) = 3t^2 - 8t + 7$ in terms of $(t-1)$. Here's one way to do it:

1. It's easy to see that $y(1) = 2$, but limits are fundamentally about what happens *near* $t_0 = 1$. So let's get a formula for the value of $y(1 + \Delta t)$:

$$\begin{aligned}
y(1 + \Delta t) &= 3(1 + \Delta t)^2 - 8(1 + \Delta t) + 7 \\
&= 3(1 + 2\Delta t + (\Delta t)^2) - 8 - 8\Delta t + 7.
\end{aligned}$$

Gathering constants together, terms with Δt together, and terms with $(\Delta t)^2$ together, this becomes

$$y(1 + \Delta t) = 2 - 2\Delta t + 3(\Delta t)^2.$$

2. The Δt in this formula is used to "step away from" $t_0 = 1$, so that we can calculate the function value somewhere else. (Notice that when $\Delta t = 0$, meaning that we've not stepped away from $t_0 = 1$ at all, this formula reduces to $y(1) = 2$.) When we want to know the function value at t, our step away from $t_0 = 1$ is $\Delta t = t - 1$. Inserting this Δt into the formula above, we get

$$y(t) = 2 - 2(t - 1) + 3(t - 1)^2.$$

Note: The steps are the same for any t_0, but the coefficients will change.

When we write $y(t)$ using powers of $(t - t_0)$ we refer to t_0 as the **expansion point**, and say that $y(t)$ has been expanded **at** or **about** t_0.

Expansion points

Example 4.10. Determine the expansion points of (a) $y(t) = 5(t+4)^2 + 9(t+4) + 3$, and (b) $y(t) = 3t^2 - 8t + 7$.

Solution: (a) Since $y(t) = 5(t+4)^2 + 9(t+4) + 3$ is written using powers of $(t+4)$, its expansion point is $t_0 = -4$. (b) Since $y(t) = 3t^2 - 8t + 7$ is written using powers of $(t - 0)$, its expansion point is $t_0 = 0$. ∎

Try It Yourself: Rewriting $y(t)$ at a new expansion point

Example 4.11. Rewrite $y(t) = 5t^3 + 3t + 6$ at the expansion point $t_0 = 2$.

Answer: $y(t) = 5(t-2)^3 + 30(t-2)^2 + 63(t-2) + 52$. ∎

Full Solution On-line

Try It Yourself: Practice finding a formula for δ when $t_0 \neq 0$

Example 4.12. Suppose $y(t) = -2t^2 + 20t + 2$. Prove that $\lim_{t \to 3} y(t) = 44$.

Answer: $\delta = \min\{1, \varepsilon/10\}$. ∎

Full Solution On-line

✧ One-sided Limits

We can adapt the technical definition to one-sided limits simply by replacing the phrase "whenever $0 < |t - t_0| < \delta$" with either

$$\text{"whenever } t_0 - \delta < t < t_0\text{"} \quad \text{for} \quad \lim_{t \to t_0^-} y(t)$$

or

$$\text{"whenever } t_0 < t < t_0 + \delta\text{"} \quad \text{for} \quad \lim_{t \to t_0^+} y(t).$$

Working graphically with the one-sided definition

Example 4.13. Figure 4.7 shows the graph of f and $\lim_{t \to 3^-} f(t) = 2$. Find an appropriate value of δ when $\varepsilon = 0.5$.

Solution: As we did with the full limit, we begin by drawing horizontal lines at $y = 2 \pm \varepsilon$. Our job is to constrain the graph to within this envelope by keeping t close enough to $t_0 = 3$, but since we're working with the limit *from the left,* we need only concern ourselves with values of $t < 3$. In Figure 4.7 we've drawn a vertical line at $t_0 = 3$, and another where the graph of f first exits the envelope (as we proceed left from $t_0 = 3$). The distance between these two lines is δ. In this case, it appears that $\delta = 1$. Any smaller value of $\delta > 0$ would also work. ∎

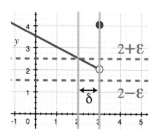

Figure 4.7: Finding δ.

Working algebraically with the one-sided definition

Example 4.14. When $f(t)$ is the function shown below, $\lim_{t \to 10.5^-} f(t) = 3$. Find an appropriate δ when $\varepsilon = 0.25$.

$$f(t) = \begin{cases} \sqrt{t - 1.5} & \text{when } 1.5 < t < 10.5 \\ -2t + 4 & \text{when } t > 10.5. \end{cases}$$

Solution: We need to find a distance δ so that $3 - 0.25 < f(t) < 3 + 0.25$ whenever $10.5 - \delta < t < 10.5$. Since $t < 10.5$, we use the formula $f(t) = \sqrt{t - 1.5}$ and rewrite this as

$$2.75 < \sqrt{t - 1.5} < 3.25 \text{ whenever } 10.5 - \delta < t < 10.5.$$

Since 2.75, $\sqrt{t - 1.5}$ and 3.25 are all positive, we can square them all and maintain the string of inequalities. That is,

$$2.75^2 < t - 1.5 < 3.25^2,$$

from which we conclude that $9.0625 < t < 12.0625$. Of course, we're only considering values of t that are less than 10.5, so we really have the constraint $9.0625 < t < 10.5$. Since $10.5 - 9.0625 = 1.4375$, we can choose any positive $\delta < 1.4375$. Let's choose $\delta = 1.4$. ∎

The red text refers to the horizontal envelope, and green text refers to the vertical lines.

❖ When One-sided Limits are Infinite

We've said $\lim_{t \to t_0} f(t) = L$ when "the number $f(t)$ gets close to L, and *stays close* as t approaches t_0," but what does that mean when L is ∞? Simply put, we say the number $f(t)$ is "close to infinity" when it's positive and "far from zero." Said technically,

---

Infinite One-sided Limits: We write $\lim_{t \to t_0^-} f(t) = \infty$ when each $N > 0$ has an associated distance δ such that

$$f(t) > N \qquad \text{whenever} \qquad t_0 - \delta < t < t_0. \qquad (4.10)$$

$\underbrace{}$ $\underbrace{\phantom{t_0 - \delta < t < t_0}}$

This says $f(t)$ is "far   This forces t to be left of
from zero" t_0, and "close enough"

The phrase "each N" is important because it means that $f(t)$ surpasses every finite value.

You'll adapt this definition in the exercise set to write technical definitions of $\lim_{t \to t_0^+} f(t) = \infty$, $\lim_{t \to t_0^-} f(t) = -\infty$, and $\lim_{t \to t_0^+} f(t) = -\infty$.

Note: Once t is close enough to t_0, the function value can never fall below N again.

Infinite limit at finite time

Example 4.15. Suppose $f(t) = \frac{1}{(t-2)^4}$. Show that $\lim_{t \to 2^-} f(t) = \infty$ by finding a formula that will produce an appropriate δ for any $N > 0$.

Solution: Since everything is positive, $\frac{1}{(t-2)^4} > N$ whenever $\frac{1}{N} > (t - 2)^4$. Taking the fourth root allows us to restate that fact as

This is the distance between t and $t_0 = 2$

$$f(t) > N \quad \text{whenever} \quad \sqrt[4]{\frac{1}{N}} > |t - 2|.$$

So we take $\delta = \sqrt[4]{1/N}$. That is,

$$f(t) > N \quad \text{whenever} \quad 2 - \sqrt[4]{\frac{1}{N}} < t < 2. \qquad ∎$$

Try It Yourself: Infinite limit at finite time

Example 4.16. Suppose $f(t) = \frac{8}{(3t-7)^6}$. Show that $\lim_{t \to 7/3^+} f(t) = \infty$ by finding a formula that will produce an appropriate δ for any $N > 0$.

Answer: $\delta = \frac{1}{3} \sqrt[6]{1/N}$. ∎

Full Solution
On-line

Infinite limit at finite time

Example 4.17. Suppose $f(t) = \frac{1}{9-t^2}$. Show that $\lim_{t \to 3^-} f(t) = \infty$ by finding a formula that will produce an appropriate δ for any $N > 0$.

Solution: Our strategy will be to estimate $f(t)$ with an expression that's easier to manage algebraically. Specifically, since $t < 3$ in this limit,

$$\frac{1}{9-t^2} = \underbrace{\frac{1}{(3-t)(3+t)} > \frac{1}{(3-t)(3+3)}}_{\text{When the denominator is made larger, the value of the fraction is made smaller}} = \frac{1}{6(3-t)}.$$

> The factor of $3 - t$ is what makes $f(t)$ get large as t approaches 3. We want that to happen, so we leave that factor alone and let it do its job. However, the other factor heads toward a value of 6, which doesn't play a role in making the limit infinite.

If we force $\frac{1}{6(3-t)} > N$ by keeping t close to but less than 3,

$$f(t) > \overbrace{\frac{1}{6(3-t)}}^{\text{This is a fact}} > N \ . \tag{4.11}$$
$$\underbrace{\phantom{f(t) > \frac{1}{6(3-t)} > N}}_{\text{We \textit{force} this to happen}}$$

The second inequality in (4.11) is true when $\frac{1}{6N} > 3 - t$, which is the same as $t > 3 - \frac{1}{6N}$. So $\delta = \frac{1}{6N}$, and we have

$$f(t) > N \quad \text{whenever} \quad 3 - \frac{1}{6N} < t < 3. \qquad \blacksquare$$

Try It Yourself: Infinite limit at finite time

Example 4.18. Suppose $f(t) = \frac{1}{t^2+4t-5}$. Show that $\lim_{t \to 1^+} f(t) = \infty$ by finding a formula that will produce an appropriate δ for any $N > 0$.

Answer: $\delta = \frac{1}{7N}$ (other answers are possible). $\qquad \blacksquare$

Full Solution On-line

Infinite oscillations at finite time

Example 4.19. Figure 4.8 depicts the graph of $f(t) = \sin(1/t)/t$. Explain why $\lim_{t \to 0^-} f(t)$ does not exist, but is *not* infinite.

Solution: The limit does not exist because there is no particular number that's approached by $f(t)$. The limit is not infinite because $f(t)$ always returns to zero as t gets closer to 0. $\qquad \blacksquare$

Figure 4.8: Graph of $f(t) = \sin(1/t)/t$.

⁂ Limits at Infinity

The technical definition of a limit includes the idea that t is "close enough" to t_0, but what does that mean when t_0 is ∞? Simply put, we say that t is "close enough" to infinity when it's positive and "far enough" from zero. Stated formally,

> **Limit at Infinity:** Suppose $L \in \mathbb{R}$. We say $\lim_{t \to \infty} y(t) = L$ when each $\varepsilon > 0$ is associated with a number $M > 0$ such that
>
> $$|f(t) - L| < \varepsilon \quad \text{whenever} \quad \underbrace{M < t.}_{\substack{\text{This says that } t \text{ is far enough from} \\ \text{zero (i.e., "close enough" to } \infty)}}$$

> Alternatively, you can think of the statement "whenever $M < t$" as a quantitative way of saying that $|f(t) - L| < \varepsilon$ *eventually*.

> The value of M depends on the specified resolution, ε. Smaller values of ε typically require larger values of M.

> You'll adapt this definition to communicate that $\lim_{t \to -\infty} y(t) = L$ in the exercise set.

Finding a formula for M

Example 4.20. Suppose $y(t) = 2t^2/(t^2+4)$. Show that $\lim_{t\to\infty} y(t) = 2$ by finding a formula that will produce an appropriate M for any resolution, $\varepsilon \in (0,1)$.

Solution: Every time someone picks an $\varepsilon > 0$, no matter how small, we must be able to find a number M so that

$$|y(t) - 2| < \varepsilon \quad \text{whenever} \quad M < t.$$

In this case, a closer look at $y(t)$ will allow us to omit the absolute value. When we write

$$y(t) = \frac{2t^2}{t^2+4} = 2\left(\frac{t^2}{t^2+4}\right)$$

we see that $y(t) < 2$, because $t^2/(t^2+4) < 1$. This allows us to write the inequality $|y(t) - 2| < \varepsilon$ as $2 - y(t) < \varepsilon$. So our goal is to guarantee that

$$2 - \frac{2t^2}{t^2+4} < \varepsilon$$

when t is sufficiently large. The number M is what we use to precisely quantify the meaning of "sufficiently large," and we find it by solving this inequality for t. Multiplying both sides by (t^2+4), which is positive, we see

> The inequality is preserved when we multiply by t^2+4 since $t^2+4 > 0$.

$$
\begin{aligned}
(t^2+4)\left(2 - \frac{2t^2}{t^2+4}\right) &< (t^2+4)\varepsilon \\
2(t^2+4) - \frac{2t^2}{t^2+4}(t^2+4) &< (t^2+4)\varepsilon \\
2t^2 + 8 - 2t^2 &< (t^2+4)\varepsilon \\
8/\varepsilon &< t^2+4 \\
\underbrace{\sqrt{\frac{8}{\varepsilon} - 4}}_{\text{This will be our } M} &< t.
\end{aligned}
$$

> The number $8/\varepsilon > 8$ since $\varepsilon \in (0,1)$, so the number $\frac{8}{\varepsilon} - 4 > 0$.

For example, when $\varepsilon = 0.1$ we know that $|y(t) - 2| < 0.1$ whenever $t > \sqrt{76}$, so $M = \sqrt{76}$. ∎

Finding a formula for M

Example 4.21. Suppose that $x(t) = \sin(t)/e^t$. Show that $\lim_{t\to\infty} x(t) = 0$ by finding a formula that will produce an appropriate M for any resolution, $\varepsilon \in (0,1)$.

Solution: Since $|\sin(t)| < 1$, we know

$$|x(t) - 0| = \left|\frac{\sin t}{e^t} - 0\right| = \frac{|\sin(t)|}{|e^t|} \leq \frac{1}{e^t} = e^{-t}.$$

> Since e^t is always positive, $|e^t| = e^t$.

It follows that $|x(t) - 0| < \varepsilon$ whenever $e^{-t} < \varepsilon$. Since both e^t and ε are positive, we can cross-multiply without affecting the inequality, and write $\varepsilon^{-1} < e^t$. Taking the logarithm of both sides, we see that

$$\underbrace{\ln(\varepsilon^{-1})}_{\text{This will be our } M} < \ln(e^t) = t.$$

> Since the logarithm is an increasing function,
>
> $$a < b \Rightarrow \ln(a) < \ln(b).$$
>
> In this case, we have $a = \varepsilon^{-1}$ and $b = e^t$.

For example, when $\varepsilon = 0.1$ we know that $|x(t) - 0| < 0.1$ whenever $t > \ln(10) \approx 2.3026$, so $M = 2.3026$. ∎

❖ Infinite Limits at Infinity

Earlier, we said that $\lim_{t\to\infty} f(t) = \infty$ if $f(t)$ surpasses each and every finite value and, eventually, never comes back down. Said technically ...

> **Infinite Limits at Infinity:** We say $\lim_{t\to\infty} f(t) = \infty$ if each number $N > 0$ is associated with a number $M > 0$ such that
>
> $$f(t) > N \quad \text{whenever} \quad t > M.$$

You'll adapt this definition to communicate the idea that $\lim_{t\to\infty} f(t) = -\infty$ in the exercise set.

Using the definition of infinite limits at infinity

Example 4.22. Suppose $g(t) = t + \frac{t}{3}\sin(t)$. Show that $\lim_{t\to\infty} g(t) = \infty$ by finding a formula that allows us to determine an appropriate M for each N.

Solution: In Example 3.4 we saw that $g(t) \geq 2t/3$. If t were large enough that $2t/3 > N$, we'd know $g(t) \geq 2t/3 > N$. Saying $2t/3 > N$ is the same as saying that $t > 3N/2$, so this will be our formula for M. For example, $g(t) > 20$ ($N = 20$) whenever $t > M = 3N/2 = 30$. ∎

You should know

- the various technical definitions of limits, such as $\lim_{t\to t_0} y(t) = L$ where t_0 and L can be finite, but might not be;

- that in order for $\lim_{t\to t_0} y(t) = L$, the number $y(t)$ has to get close to L and stay there.

You should be able to

- interpret the $\varepsilon\delta$-definition of the limit in terms of resolution;

- use the graph of a function to estimate an appropriate δ for a given ε;

- derive formulas for choosing δ or for a given ε or N;

- derive formulas for choosing M for a given ε or N.

❖ 2.4 Skill Exercises

For each limit in #1–12 use the appropriate graph below to determine an appropriate $\delta > 0$ for the specified ε.

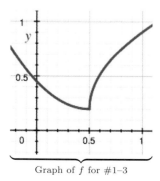

Graph of f for #1–3

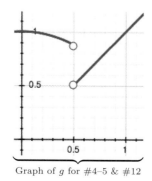

Graph of g for #4–5 & #12

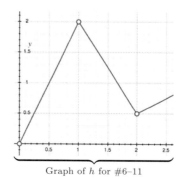

Graph of h for #6–11

1. $\lim_{t \to 0.5+} f(t) = 0.2$; $\varepsilon = 0.3$

2. $\lim_{t \to 0.5-} f(t) = 0.2$; $\varepsilon = 0.3$

3. $\lim_{t \to 0.5} f(t) = 0.2$; $\varepsilon = 0.3$

4. $\lim_{t \to 0.5+} g(t) = 0.5$; $\varepsilon = 0.1$

5. $\lim_{t \to 0.5-} g(t) = 0.9$; $\varepsilon = 0.1$

6. $\lim_{t \to 1-} h(t) = 2$; $\varepsilon = 0.1$

7. $\lim_{t \to 1+} h(t) = 2$; $\varepsilon = 0.1$

8. $\lim_{t \to 1} h(t) = 2$; $\varepsilon = 0.1$

9. $\lim_{t \to 0.5-} h(t) = 1$; $\varepsilon = 0.1$

10. $\lim_{t \to 0.5+} h(t) = 1$; $\varepsilon = 0.1$

11. $\lim_{t \to 0.5} h(t) = 1$; $\varepsilon = 0.1$

12. What value of ε would you choose in order to show that $\lim_{t \to 0.5} f(t) \neq 0.9$?

For each of #13–24, expand the given polynomial about the specified t_0.

13. $t^2 - 3t + 1$, $t_0 = 4$

14. $t^2 - 3t + 1$, $t_0 = 2$

15. $6t^2 + 13t - 7$, $t_0 = -3$

16. $6t^2 + 13t - 7$, $t_0 = -6$

17. $t^3 + 2t + 5$, $t_0 = 1$

18. $t^3 + 2t + 5$, $t_0 = -2$

19. $(t - 4)^2$, $t_0 = 5$

20. $(t + 6)^2$, $t_0 = -1$

21. $-2t^2 - 7t - 6$, $t_0 = 7/6$

22. $-2t^2 - 7t - 6$, $t_0 = 8/3$

23. $-t^2 + t + 17$, $t_0 = \sqrt{6}$

24. $-t^2 + t + 17$, $t_0 = \sqrt{8}$

For each of #25-30, let $y(t) = 3t^2 + 9t + 12$. Then argue that each limit is correct by finding a formula that will produce an appropriate δ for any resolution, $\varepsilon \in (0, 1)$.

25. $\lim_{t \to 0} y(t) = 12$

26. $\lim_{t \to 1} y(t) = 24$

27. $\lim_{t \to 2} y(t) = 42$

28. $\lim_{t \to -1} y(t) = 6$

29. $\lim_{t \to -2} y(t) = 6$

30. $\lim_{t \to -3} y(t) = 12$

For each of #31-36, let $f(t) = 3t - t^2$. Then argue that each limit is correct by finding a formula that will produce an appropriate δ for any resolution, $\varepsilon \in (0, 1)$.

31. $\lim_{t \to 0} f(t) = 0$

32. $\lim_{t \to 1} f(t) = 2$

33. $\lim_{t \to 5} f(t) = -10$

34. $\lim_{t \to -5} f(t) = -40$

35. $\lim_{t \to -1} f(t) = -4$

36. $\lim_{t \to -3} f(t) = -18$

In each of #37–42 verify the specified limits by finding a formula that will produce an appropriate δ for any $\varepsilon \in (0, 1)$.

37. $\lim_{t \to 0} 10^t = 1$

38. $\lim_{t \to 0} 4^t = 1$

39. $\lim_{t \to 1} 10^t = 10$

40. $\lim_{t \to 1} 4^t = 4$

41. $\lim_{t \to 2} 10^t = 100$

42. $\lim_{t \to 2} 4^t = 16$

In each of 43–48 verify the given limit by finding a formula that will produce an appropriate M for any resolution $\varepsilon \in (0, 1)$.

43. $\lim_{t \to \infty} \frac{3}{8t^2 + 1} = 0$

44. $\lim_{t \to -\infty} \frac{7t^3 - 4t + 3}{3t - 9t^3} = \frac{7}{9}$

45. $\lim_{t \to -\infty} e^t \cos(t) = 0$

46. $\lim_{t \to \infty} \frac{3t^2}{8t^2 + 1} = \frac{3}{8}$

47. $\lim_{t \to \infty} \frac{8\sqrt{t^2 + 3t}}{2t} = 4$

48. $\lim_{t \to \infty} \frac{3t - \sqrt{4 + t}}{7 - 6t} = -\frac{1}{2}$

Show that each of the limits in #49-60 is $+\infty$ (or $-\infty$) by finding a formula that will produce an appropriate δ for any $N > 0$ (or $N < 0$).

49. $\lim_{t \to 5+} \frac{1}{t - 5}$

50. $\lim_{t \to 8+} \frac{1}{4t - 32}$

51. $\lim_{t \to 6+} \frac{1}{t^2 - 36}$

52. $\lim_{t \to \sqrt{7}+} \frac{1}{7 - t^2}$

53. $\lim_{t \to 3+} \frac{7}{(t - 3)^4}$

54. $\lim_{t \to 5-} \frac{t^2}{(t - 5)^7}$

55. $\lim_{t \to 2+} \frac{t^2 + 5t + 13}{t^2 + t - 6}$

56. $\lim_{t \to 1+} \frac{t^2 - 1}{t^5 - 2t^4 + t^3}$

57. $\lim_{t \to 9+} \frac{4}{\sqrt{t} - 3}$

58. $\lim_{t \to 25-} \frac{-7}{\sqrt{t} - 5}$

59. $\lim_{t \to 32-} \frac{3}{\sqrt[5]{t} - 2}$

60. $\lim_{t \to 8+} \frac{t}{\sqrt[3]{t} - 2}$

❖ 2.4 Concept and Application Exercises

61. Suppose $y(t) = 4t^2 + 3t + 5$. Without writing it down, imagine the basic form of what you'd see if you wrote an expansion for $y(t)$ about $t_0 = 3$.

    (a) What is the leading coefficient, and why?

    (b) If you calculate $y(3)$ using the formula above, what number do you get?

    (c) If you were to calculate $y(3)$ using the expansion for $y(t)$ about $t_0 = 3$, what would you get?

    (d) Based on what your answers to parts (b) and (c), if we were to expand $y(t)$ about $t_0 = 6$, what would be the constant term?

Write the technical definition for each of the limits in #62-64, and then explain them in complete sentences.

62. $\lim\limits_{t \to -\infty} y(t) = L$ 63. $\lim\limits_{t \to -\infty} y(t) = \infty$ 64. $\lim\limits_{t \to -\infty} e^t = 0$

65. When f is the piecewise-defined function shown below, $\lim_{t \to 3} f(t) = 2$. Suppose $\varepsilon = 0.01$. Find an appropriate δ.

$$f(t) = \begin{cases} -2t + 8 & \text{when } t < 3 \\ \frac{1}{3}t + 1 & \text{when } t > 3 \end{cases}$$

66. Suppose $f(t) = \begin{cases} a_1 t + b & \text{when } t < 0 \\ a_2 t + b & \text{when } t > 0 \end{cases}$ and $\varepsilon > 0$ is a particular, fixed resolution.

    (a) How do we know whether the δ associated with $\lim_{t \to 0^-} f(t)$ or with $\lim_{t \to 0^+} f(t)$ will be larger?

    (b) When will they be the same?

67. When f is the piecewise-defined function shown below, $\lim_{t \to 1} f(t) = 15$. Find a formula that will produce an appropriate δ for any $\varepsilon > 0$.

$$f(t) = \begin{cases} (t-1)^2 + 15 & \text{when } t < 1 \\ (t-1)^4 + 15 & \text{when } t > 1 \end{cases}$$

68. Suppose $f(t) = t^3$, and $\varepsilon > 0$ is fixed. Since $\lim_{t \to 0.5^+} f(t) = 1/8$, there is a maximal distance, δ_+, associated with ε (smaller distances would also work). Similarly, since $\lim_{t \to 0.5^-} f(t) = 1/8$, there is a maximal distance, δ_-, associated with ε. Which of these numbers is smaller? *(You don't need to make any calculations in order to answer!)*

69. Suppose that $f(t) = t^2$ and $\varepsilon > 0$ is fixed. At each t_0, there is a maximal value of δ associated with ε (smaller choices of δ would also work). When we change t_0, the value of δ changes, so we can think of δ as a function of t (for this fixed ε).

    (a) Calculate $\delta(1)$.

    (b) If we increase t away from $t_0 = 1$, do you think $\delta(t)$ will increase, decrease, or stay the same?

    (c) Find a formula for $\delta(t)$.

(d) Plot a graph of δ as a function of t. Was your prediction at $t_0 = 1$ correct?

70. Suppose $|f(t) - 6| < 10^{-47}$ whenever $0 < |t - 8| < 0.001$. What do you know about $\lim_{t \to 8} f(t)$?

71. Suppose $|f(t) - 7| < 10^{-3}$ whenever $0 < |t - 5| < 0.0001$. What do you know about $\lim_{t \to 5} f(t)$?

72. Suppose $\lim_{t \to 3} f(t) = 14$, and the formula for finding δ is $\delta = \sqrt{5\varepsilon}$ whenever $\varepsilon \in (0, 1)$.

    (a) Find δ when $\varepsilon = 0.9999$.

    (b) Based on your answer to part (a), what do you know about $f(2)$?

73. Suppose $\lim_{t \to 7} f(t) = 93$, and the formula for finding δ is $\delta = \varepsilon^2$ whenever $\varepsilon \in (0, 1)$. What do you know about $f(7.24)$?

74. Suppose $\lim_{t \to 7} f(t) = 93$, and the formula for finding δ is $\delta = \varepsilon^4$ whenever $\varepsilon \in (0, 1)$. What do you know about $f(8.01)$?

75. Write the $\varepsilon\delta$-definition of $\lim_{t \to t_0^+} f(t) = L$, when both t_0 and L are finite. Then interpret the definition in terms of resolution and distance, identifying the role played by each part of the formal definition.

76. Write the $N\delta$-definition of $\lim_{t \to t_0^+} f(t) = \infty$, when t_0 is finite. Then identify the role played by each part of the formal definition.

2.5 Continuity

On p. 98 you saw that $\lim_{t \to t_0} y(t) = y(t_0)$ whenever $y(t)$ is a polynomial. In general, any function that exhibits this behavior (limit value = function value) is said to be **continuous** at t_0. The idea of continuity is pervasive in mathematics, and plays a subtle but important role in our models of physical phenomena, such as waves and diffusion. In this section, we'll develop a brief catalog of continuous functions, and discuss the ways in which a function can fail to be continuous. We begin with the formal definition:

Continuity: We say the function f is *continuous* at t_0 when

$$\lim_{t \to t_0} f(t) = f(t_0).$$

Equivalently, we can write this equation as

$$\lim_{\Delta t \to 0} f(t_0 + \Delta t) = f(t_0).$$

We say that f is continuous *on (over) the open interval* (a, b) when it's continuous at each $t \in (a, b)$.

Note: This is only true when both sides exist. When one or both sides fail to exist, the equation is meaningless.

Remember that $t \in (a, b)$ means that t is any number *in* the interval (a, b).

The interval in this definition has to be open so that we can approach any t_0 from *both* sides (which is necessary to define the limit).

These limit equations are equivalent because the argument of f is approaching t_0 in both statements. The only difference is the way we've chosen to communicate that fact.

✧ Some Continuous Functions

Most of the functions you know are continuous on their domains: *power functions, polynomials, trigonometric functions, exponential functions, inverse trigonometric functions,* and *logarithms.* Moreover, we can compose them or combine them algebraically to get other continuous functions.

Algebraic combinations of continuous functions

Example 5.1. Suppose $f(t) = t + 9$ and $g(t) = e^t$. (a) Determine the domain over which g/f is continuous, and (b) do the same for f/g.

Solution: (a) Since $(g/f)(t) = e^t/(t + 9)$ is built with continuous functions, it's continuous on its domain, which is all $t \neq -9$. (b) Similarly, $(f/g)(t) = (t + 9)/e^t$ is continuous over all of $\mathbb{R}$ (its denominator is never zero). ∎

Composing continuous functions

Example 5.2. Suppose $f(t) = t^2 - 4$ and $g(t) = \ln(t)$. (a) Determine the domain over which $g \circ f$ is continuous, and (b) do the same for $f \circ g$.

Solution: (a) The function f has no domain restrictions, but the logarithm only makes sense with *positive* arguments. So $(g \circ f)(t) = \ln(t^2 - 4)$ can only accept values of t for which $t^2 - 4 > 0$. That is, $t \in (-\infty, -2) \cup (2, \infty)$.

(b) Since $g \circ f$ is composed of continuous functions, it will be continuous at all points in its domain. Similarly, $(f \circ g)(t) = (\ln(t))^2 - 4$ is continuous on its domain, $t > 0$. ∎

Using continuity to our advantage

Example 5.3. Suppose $f(t) = (\sqrt{t} - 3)/(t - 9)$. Determine $\lim_{t \to 9} f(t)$.

Solution: We can get rid of the radical by using the conjugate:

$$f(t) = \frac{\sqrt{t}-3}{t-9}\left(\frac{\sqrt{t}+3}{\sqrt{t}+3}\right) = \frac{(\sqrt{t})^2 - 3^2}{(t-9)(\sqrt{t}+3)} = \underbrace{\frac{t-9}{(t-9)(\sqrt{t}+3)} = \frac{1}{\sqrt{t}+3}}_{\text{As long as } t \neq 9}.$$

When we rewrite $f(t)$ as $1/(\sqrt{t}+3)$, we can calculate

$$\lim_{t\to 9} f(t) = \lim_{t\to 9} \frac{1}{\sqrt{t}+3} = \frac{\lim_{t\to 9} 1}{\lim_{t\to 9} \sqrt{t} + \lim_{t\to 9} 3} = \frac{1}{6},$$

where we've relied on the continuity of the square-root function to guarantee that $\lim_{t\to 9}\sqrt{t} = 3$. ∎

From an operational point of view, a function is continuous at t_0 when we can pass the limit "through" the function notation:

$$\lim_{t\to t_0} f(t) = f(t_0) = f\left(\lim_{t\to t_0} t\right).$$

This is true even when the argument of f is something other than t, as long as the limit exists.

Theorem 5.3. Suppose $\lim_{t\to t_0} g(t) = L$, and f is continuous at L. Then

$$\lim_{t\to t_0} f(g(t)) = f\left(\lim_{t\to t_0} g(t)\right) = f(L).$$

Limits and continuous functions

Example 5.4. Calculate $\lim_{\phi\to 0} \ln\left(\frac{\sin(\phi)}{\phi}\right)$, where ϕ is measured in radians.

Solution: Since $\lim_{\phi\to 0} \frac{\sin(\phi)}{\phi} = 1$, and the natural logarithm is continuous at 1,

$$\lim_{\phi\to 0} \ln\left(\frac{\sin(\phi)}{\phi}\right) = \ln\left(\lim_{\phi\to 0} \frac{\sin(\phi)}{\phi}\right) = \ln(1) = 0. \qquad ∎$$

Try It Yourself: Limits and continuous functions

Example 5.5. Calculate $\lim_{t\to 1} \sin\left(\frac{t^2+8t-8}{t-1+6/\pi}\right)$.

Answer: $1/2$. ∎

Full Solution
On-line

❖ Discontinuity

When a function fails to satisfy the definition of continuity at t_0, we say that it is **discontinuous** at t_0. This can happen for three reasons:

(i) both $\lim_{t\to t_0} f(t)$ and $f(t_0)$ exist, but are different;

(ii) the number $f(t_0)$ does not exist;

(iii) the number $\lim_{t\to t_0} f(t)$ does not exist.

▷ Types (i) and (ii): removable discontinuities

Both cases (i) and (ii) are relatively benign, and we can "fix" the function to make it continuous at t_0.

> **Removable Discontinuity:** We say the function f has a *removable discontinuity* at t_0 if
>
> 1. f is discontinuous at t_0, but
>
> 2. we can make f continuous at t_0 by defining (or <u>re</u>defining) its value at t_0.

Removable discontinuities—Type (i)

Example 5.6. Determine whether f is continuous at $t = 3$ when

$$f(t) = \begin{cases} t^2 & \text{when } t \neq 3 \\ 0 & \text{when } t = 3 \end{cases}$$

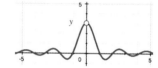

If the function in Example 5.6 seems contrived, that's because it is. Type (i) discontinuities are extremely rare, and are typically not seen in practice. Type (ii) and type (iii) discontinuities are much more common.

Solution: Since $f(t) = t^2$ whenever $t \neq 3$, and t^2 is a polynomial, we know that $\lim_{t \to 3} f(t) = \lim_{t \to 3} t^2 = 9$, but $f(3) = 0$. Since the limit value and the function value are different, the function is not continuous at $t = 3$. We can remove the discontinuity by redefining the function value at $t = 3$ to be 9. ∎

Removable discontinuities—Type (ii)

Example 5.7. Determine if the function $f(\phi) = \frac{\sin(3\phi)}{\phi}$ is continuous at $\phi = 0$.

Solution: The function is *not* continuous at $\phi = 0$ because f is not defined there. However, we know from our work with indeterminate forms that $\lim_{\phi \to 0} \frac{\sin(3\phi)}{\phi} = 3$ (see Figure 5.1), so we can remove the discontinuity by *defining* $f(0) = 3$. ∎

Figure 5.1:   The graph of $\sin(3\phi)/\phi$.

▷ Type (iii): the limit doesn't exist

When $\lim_{t \to t_0} f(t)$ doesn't exist, there's no simple way to "fix" the discontinuity, but one-sided limits can provide us with important quantitative ways of describing the function nonetheless. For example, suppose that

$$g(t) = \begin{cases} t + 4 & \text{if } t < 0 \\ 0.5t^2 + 1.3 & \text{if } t \geq 0. \end{cases} \tag{5.1}$$

In this case, the function is discontinuous at $t = 0$ because $\lim_{t \to 0} g(t)$ does not exist, and we cannot fix that problem by adjusting the function value at $t_0 = 0$. So this is *not* a removable discontinuity.

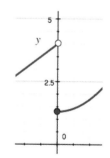

Figure 5.2: The graph of $g(t)$, defined in equation (5.1).

> **Jump Discontinuity:** We say the function f has a *jump discontinuity* at t_0 if
>
> 1. $\lim_{t \to t_0^-} f(t)$ and $\lim_{t \to t_0^+} f(t)$ both exist, but
>
> 2. they're not equal.

Jump discontinuities are often used to describe a sudden change, such as ionic concentration across a cell wall, pressure across a shock wave, or the voltage on the heart's atrioventricular node immediately before and after a heartbeat. In that sense, they are an important part of describing our world. In other situations, we want to avoid them.

Jump discontinuities and artifacts in images

Example 5.8. Suppose we apply the following function to the pixel intensities of an image, as in Example 2.5 in Chapter 1. (a) Calculate the one-sided limits of $f(x)$ at $x = 0.2$, and (b) explain what happens to the image.

$$f(x) = \begin{cases} x & \text{if } x \leq 0.2 \\ 0.8x + 0.2 & \text{if } x > 0.2 \end{cases}$$

Solution: (a) Since $f(x)$ is a polynomial when $x \neq 0.2$, the one-sided limits are easy to calculate:

$$\lim_{x \to 0.2^-} f(x) = \lim_{x \to 0.2^-} x = 0.2 \text{ and}$$
$$\lim_{x \to 0.2^+} f(x) = \lim_{x \to 0.2^+} 0.8x + 0.2 = 0.36.$$

Since these limits are different, the function f has a jump discontinuity at $x = 0.2$.

(b) Pixels that are near each other in an image often have similar intensities. Suppose an image has a large swath of pixels whose intensities are near $x = 0.2$. This $f(x)$ does nothing to the darker pixels ($x \leq 0.2$) but brightens the lighter ones. Visually speaking, the two sets of pixels that merged seamlessly in the original image are now clearly separate. Figure 5.3 shows the graph of f, along with its effect on a test image. ∎

> A sudden jump in intensity might indicate the edge of an object in the scene.

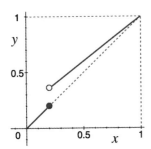  

Figure 5.3: (left) The graphs of $y = x$ and $y = f(x)$ for Example 5.8; (middle) original image; (right) image filtered by the function $f(x)$. Artifacts of the jump discontinuity are apparent on the shirt, face, hair, and next to the circular window in the background.

You might also have a situation in which the limit doesn't exist because one or both of the one-sided limits don't exit. In the special case when $\lim_{t \to t_0} f(t) = \infty$, we say that f has an **infinite discontinuity** at t_0 (we also say that f has a **singularity** at t_0, or that t_0 is a **singular point** of f). Other kinds of discontinuities are not named.

❖ One-sided Continuity

In our discussion of jump discontinuities we saw

$$\lim_{t \to 0^+} g(t) = g(0) \text{ but } \lim_{t \to 0^-} g(t) \neq g(0) \text{ in Figure 5.2, and}$$

$$\lim_{t \to 0.2^+} f(t) \neq f(0.2) \text{ but } \lim_{t \to 0.2^-} f(t) = f(0.2) \text{ in Figure 5.3.}$$

> In both cases, the limit from one side is equal to the function value, but not the limit from the other side.

This doesn't have to happen, but when it does we talk about *one-sided* continuity:

> **One-sided Continuity:** We say the function f is
>
> $$\text{left-continuous at } t_0 \quad \text{when} \quad \lim_{t \to t_0^-} f(t_0) = f(t_0),$$
>
> $$\text{and } \textit{right-continuous at } t_0 \quad \text{when} \quad \lim_{t \to t_0^+} f(t_0) = f(t_0).$$

Instead of saying "left-continuous," sometimes people say "continuous from the left." Similarly for "from the right."

This idea allows us to extend our ideas of continuity to closed intervals.

> **Continuity on [a,b]:** We say the function f is *continuous on (over)* $[a, b]$ when it is continuous at each $t \in (a, b)$, and
>
> 1. it is right-continuous at $t = a$, and
>
> 2. it is left-continuous at $t = b$.

When a function is continuous over a closed interval, it exhibits other important properties that are described by the *Intermediate Value Theorem* and the *Extreme Value Theorem* (below).

✢ The Intermediate Value Theorem (IVT)

In order to develop some intuition before tackling the formal statement of the Intermediate Value Theorem, draw the graph of a continuous function in Figure 5.4 for which $f(-5) = 6$ and $f(4) = 2$. (Really, try it!)

Could you draw such a graph without achieving a height of 3, or a height of 2.001? No, because 3 and 2.001 are between the numbers $f(-5)$ and $f(4)$. In fact, every number between $f(-5)$ and $f(4)$ must be achieved by the function at some time between $t = -5$ and $t = 4$, and this is the central idea expressed below.

Figure 5.4: Draw the graph of a continuous function for which $f(-5) = 6$ and $f(4) = 2$.

> **Intermediate Value Theorem (IVT)** Suppose f is continuous over $[a, b]$, and $f(a) \neq f(b)$. Then every value between $f(a)$ and $f(b)$ is achieved by f at some point $t \in (a, b)$.

You've probably relied on the IVT without knowing it. That can make it seem trivial, but it takes some effort to actually *prove* the IVT, and its consequences are far-reaching. For example, when f is a continuous function with $f(a) < 0$ and $f(b) > 0$ (as in Figure 5.5), we know that $f(x) = 0$ for some $a < x < b$. That is, f must have a root somewhere in (a, b).

Detecting roots

Example 5.9. Show that $f(x) = \cos(x) - x$ has a root somewhere in $[0, 1]$.

Solution: We're not asked to actually *find* the root, but simply to determine whether it's there. Evaluating the function at the endpoints of the interval, we see $f(0) > 0$ and $f(1) < 0$. Since f *is continuous*, and 0 is between $f(0)$ and $f(1)$, the IVT tells us that there is some $t \in (0, 1)$ at which $f(t) = 0$. ∎

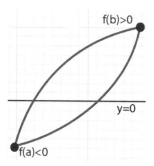

Figure 5.5: We don't know what the graph of f looks like exactly, but we know that there must be a root *somewhere*.

Using the IVT to solve an equation

Example 5.10. Determine whether $t^2 - t = 2$ has a solution in $[-10, 10]$.

Solution: When we rewrite this equation as $t^2 - t - 2 = 0$, the problem becomes finding a root of $f(t) = t^2 - t - 2$. In this case $f(-10)$ and $f(10)$ are both positive, so the IVT does *not* guarantee a root. However, $f(0) < 0$, so the IVT guarantees roots in both $[-10, 0]$ and $[0, 10]$.   ∎

> If the function values at the endpoints are either both positive or both negative, but you believe there *is* a root to be found, check $f(t)$ in the interior of the interval.

▷ The bisection algorithm

Not only can we use the IVT to *detect* roots, but we can *locate* them. Suppose $f(a)$ and $f(b)$ have different signs, so we know there's a root somewhere in $[a, b]$.

1. Evaluate the function at the midpoint of the interval, $t_1 = (a + b)/2$.

> The point t_1 *bisects* the interval $[a, b]$.

2. If $f(t_1) = 0$, you're done! Otherwise ...

   (a) If $f(a)$ and $f(t_1)$ have opposite signs, a root lies between a and t_1. Restart the algorithm, but this time work with the interval $[a, t_1]$.

   (b) If $f(b)$ and $f(t_1)$ have opposite signs, a root lies between t_1 and b. Restart the algorithm, but this time work with the interval $[t_1, b]$.

In practice, it's unlikely that you'll ever find the root exactly. However, each time we restart the algorithm, the interval in which we're looking for the root gets cut in half. So the root's location is known to within $(b - a)(0.5)^k$ after the kth step, which we can make as small as we want by taking sufficiently large values of k.

Using the bisection algorithm to locate the root of a function

Example 5.11. Suppose $f(t) = t^3 + 2$. Determine whether f has a root in $[-2, 0]$ and, if so, use the bisection method to estimate its value.

Solution: It's easy to check that $f(-2) = -6 < 0$ and $f(0) = 2 > 0$. Since f is a polynomial, we know it's continuous, so there must be a root in $[-2, 0]$. To find it, we check the function at the midpoint of the interval, $t_1 = -1$: $f(-1) = 1 > 0$. Then ...

1. Because $f(-2)$ and $f(-1)$ have different polarities, a root must lie in $[-2, -1]$. So we check the function at the midpoint of the interval, $t_2 = -1.5$: $f(-1.5) = -11/8 < 0$.

2. Because $f(-1.5)$ and $f(-1)$ have different polarities, a root must lie in $[-1.5, -1]$. So we check the function at $t_3 = -1.25$ (the midpoint of the interval): $f(-1.25) = 3/64 > 0$.

3. Because $f(-1.5)$ and $f(-1.25)$ have different polarities, a root must lie in $[-1.5, -1.25]$.

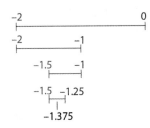

Figure 5.6: The bisection method narrows down on the root.

Let's stop the process and estimate the location of the root as $t_4 = -1.375$ (the midpoint of $[-1.5, -1.25]$). This is certainly not correct, but since -1.375 is 0.0125 units away from each endpoint, we know that we are *at most* 0.0125 units from the actual value.   ∎

✛ Extreme Value Theorem (EVT)

Look at the graph you drew on Figure 5.4 (p. 133). What are the maximum and minimum values of your function on the interval $[-5, 4]$?

The particular values are irrelevant here. The point is that there *are* maximum and minimum values that occur *somewhere* in the interval. The Extreme Value Theorem (EVT) says that this will always happen—a continuous function on a closed interval will always return a maximum and a minimum output, somewhere in the closed interval (at the endpoints or at some point in the interior).

> **Extreme Value Theorem (EVT):** When f is continuous over $[a,b]$, there are numbers $m, M \in [a,b]$ such that
>
> (a) $f(m) \leq f(t)$ for all other $t \in [a,b]$, and
>
> (b) $f(t) \leq f(M)$ for all other $t \in [a,b]$.

> This sounds like a simple idea, but the proof is relatively sophisticated.

Note: Part (a) of the Extreme Value Theorem says that no function value is smaller than the number $f(m)$, and part (b) says that no function value is larger than the number $f(M)$. We'll develop mathematical tools in Chapter 4 that will help us determine the numbers m and M precisely. For now, we'll get rough approximations by working with graphs.

> Little m for *minimum* and capital M for MAXIMUM.

Example of the EVT

Example 5.12. Suppose $f(t) = \sin(t) + 2$. Find the maximum and minimum values of f over $[0, 6\pi]$.

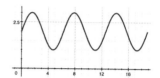

Figure 5.7: The graph of $f(t)$ in Example 5.12.

Solution: The function f achieves its minimum value of 1 at $t = 3\pi/2$, $7\pi/2$, and $11\pi/2$. Any of these could serve as m in the EVT. Similarly, the function f achieves its maximum value of 3 at $t = \pi/2$, $5\pi/2$, and $9\pi/2$. Any of these could serve as M in the EVT. ∎

> A function can achieve its minimum and maximum several times. The EVT simply guarantees that it will happen *at least* once.

You should know

- the terms *continuous* (either at a point or over an interval), *removable discontinuity, jump discontinuity, infinite discontinuity, singularity, singular point, left-continuous, right-continuous,* and *bisection algorithm;*

- the statements and meanings of the Intermediate Value Theorem and Extreme Value Theorem.

You should be able to

- determine the points over which a function is continuous;

- identify the different kinds of discontinuity;

- "fix" a removable discontinuity;

- apply the bisection method to estimate the location of a root.

❖ 2.5 Skill Exercises

For each of the functions graphed in #1–4 you should

(a) Find all points where the function is not continuous and determine the values of the left and right limits at those points.

(b) Find all times when the left and right limits are the same but the function is not continuous. Then determine how we could change the graph in order to make the function continuous there.

(c) Find a value of t (if there are any) at which f is discontinuous, but is left-continuous.

(d) Find a value of t (if there are any) at which f is discontinuous, but is right-continuous.

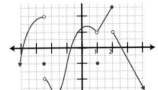

1.

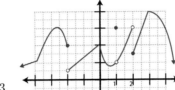

3.

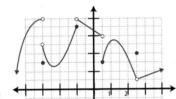

2.

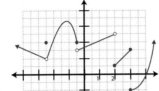

4.

In each of #5–10, determine the value of $f(t_0)$ that would make the function continuous as t_0.

5. $f(t) = \frac{\sin(8t)}{\sin(5t)}$; $t_0 = 0$

7. $f(t) = \frac{t^2 - 2t - 8}{\sqrt{t} - 2}$; $t_0 = 4$

9. $f(t) = \frac{\sin(t - 49)}{\sqrt{t} - 7}$; $t_0 = 49$

6. $f(t) = \frac{t - 1}{\sqrt{t} - 1}$; $t_0 = 1$

8. $f(t) = \frac{t^3 - 1}{\sqrt{t} - 1}$; $t_0 = 1$

10. $f(t) = \frac{1 - \cos(t - 49)}{t + 2\sqrt{7t} - 7}$; $t_0 = 49$

In each of #12–16 (a) calculate the one-sided limits where the formula for $f(t)$ changes; (b) determine any/all values of k for which the given piecewise function is continuous.

11. $f(t) = \begin{cases} k^2 & \text{when } t = 0 \\ 8t + k & \text{when } t \neq 0 \end{cases}$

13. $f(t) = \begin{cases} k - 4 & \text{when } t = 0 \\ \frac{\sin(kt)}{k^2 t} & \text{when } t \neq 0 \end{cases}$

12. $f(t) = \begin{cases} -t^2 + k^2 & \text{when } t < 1 \\ 3t + 8k & \text{when } t \geq 1 \end{cases}$

14. $f(t) = \begin{cases} \sin(kt) & \text{when } t < 0 \\ \cos(kt) & \text{when } t \geq 0 \end{cases}$

15. $f(t) = \begin{cases} k^4 t + 15k - 2 & \text{when } t < 0 \\ (3k^2 + 10k - 10) + (3k + 15 - k^2)t^2 & \text{when } 0 \leq t \leq 1 \\ kt - 5 & \text{when } 1 < t \end{cases}$

16. $f(t) = \begin{cases} -t^2 + k^2 & \text{when } t < 1 \\ (2k^2 - 5k - 6) + (5 + 5k - k^2)t & \text{when } 1 \leq t \leq 2 \\ (t - 4)^2 + 5k & \text{when } 2 < t \end{cases}$

Determine where each of the functions in #17–28 are continuous.

17. $f(t) = t^2 + 8t + 11$

21. $f(t) = \frac{2t + 6}{t^2 + 7t + 12}$

25. $f(t) = \sqrt{t^2 + 5t - 12}$

18. $f(t) = t^6 - 9t^5 + 13$

22. $f(t) = \frac{4t^2 + 7}{t^2 + 8t - 15}$

26. $f(t) = \sqrt[4]{13 + 6t - 8t^2}$

19. $f(t) = \frac{4}{t^2 + 7t + 12}$

23. $f(t) = \cos(t^2 + 1)$

27. $f(t) = \ln(\cos(t) + 0.5)$

20. $f(t) = \frac{t - 3}{t^2 + 8t + 15}$

24. $f(t) = \sin(t^2 - t - 8)$

28. $f(t) = \ln(\sin(t) + 1/\sqrt{2})$

In each of #29–34 you're asked to determine the limit of a function at a value of t_0 where it's continuous.

29. $\lim_{t\to 2} t^4 - 3t^3 + 1$

30. $\lim_{t\to 0} t^3 + 6t^2 - 7t$

31. $\lim_{t\to 1} \cos(t^7 + 6t^3 - 7)$

32. $\lim_{t\to 1} \tan^{-1}(4t^5 - 2t^2 - 1)$

33. $\lim_{t\to 3} \log(2t^4 - 2t^3 - 8)$

34. $\lim_{t\to 4} 10^{-t^4 + 3t^3 - 63}$

In each of #35–38, use the bisection method to locate a root of f in the given interval to within the specified tolerance, ε.

35. $f(t) = t^3 - 7$, $[1, 3]$, $\varepsilon = 0.3$

36. $f(t) = 2^t - 7$, $[1, 3]$, $\varepsilon = 0.3$

37. $f(t) = \sin(t) - 0.25$, $[0, 1]$, $\varepsilon = 0.1$

38. $f(t) = \cos(t) - \frac{1}{3}$, $[0, \pi]$, $\varepsilon = 0.1$

✥ 2.5 Concept and Application Exercises

39. Which of the following are continuous functions of time?

(a) The speed of an arrow, from the moment it's loosed until (gulp) it comes to a full stop

(b) The current flowing through a light bulb throughout a day

(c) The temperature of a light bulb over the course of a day

(d) The height of a tennis ball during a tennis game

(e) The cost of a taxi ride

(f) The speed of light that passes through air, and then water

40. The traffic density along a road at 4:03 p.m. is graphed below (traffic is traveling from left to right). Explain the jump discontinuity in the graph at $x = 0$.

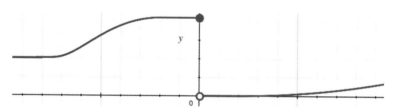

41. Modeled as an ideal pump, the heart's left ventricle instantaneously injects a specified quantity of blood into the systemic arteries, which causes a jump discontinuity in pressure. Use one-sided limits to write equations that say the arterial pressure is 80 mmHg immediately before a beat, and 100 mmHg immediately after.

42. Draw the graph of a function f that has a discontinuity at $t = 3$ even though $\lim_{t\to 3} f(t)$ exists.

43. Suppose that f is a continuous function, and $f(0) = 3$. Could f be an odd function? If so, draw such a graph. If not, why not?

44. Write down a formula for a function that is discontinuous at $t = 2$, but is continuous everywhere else.

45. Find a function that's continuous at $t = 0$, but discontinuous everywhere else.

46. Find functions h and g so that g is continuous at $t = 2$ but $h \circ g$ is not.

47. Find continuous functions h and g so that h/g is continuous at $t = 2$ but g/h is not.

48. Figure 5.8 shows the graph of f in its entirety.

    (a) Find the number m at which f returns its minimum value.

    (b) Explain why there is *no* number M at which f returns a maximum value.

    (c) Why doesn't this violate the EVT?

49. Draw the graph of a function $f(x)$ that achieves a maximum over $[1, 5]$ but not a minimum.

50. Draw the graph of a function $g(x)$ that achieves a minimum over $[1, 5]$ but not a maximum.

51. Draw the graph of a function $h(x)$ that has neither a minimum nor a maximum over $[1, 5]$.

52. Draw the graph of a function that is decreasing, never zero, for which $f(-1) = 2$ and $f(1) = -2$. (Hint: the IVT has something to say about this.)

53. Use the IVT to prove that you weighed 10 pounds at some time in your life. Write your proof using complete sentences, including what assumptions you're making.

54. Use the IVT to prove that you were three feet tall at some point in your life. Write your proof using complete sentences, including what assumptions you're making (there's one in particular that's very important).

55. Use the IVT to prove that, after you throw an object upwards, there must be a moment at which it is neither moving up nor down. Write your proof using complete English sentences, including what assumptions you're making (there's one in particular that's very important).

56. On p. 50 we briefly examined the tangent line to the graph of $f(t) = b^t$ at the point $(0, 1)$. The slope of that line is < 1 when $b = 2$, and > 1 when $b = 4$. We'd *like* to use the IVT to assert that there is some number between 2 and 4 at which the slope is *equal* to 1. What would we have to know about the relationship between the base and the slope in order to use the IVT in that way?

57. Suppose f is a continuous function with $f(0) = 1$ and $f(3) = 64$. Then the IVT guarantees that there is a solution to $f(t) = 58$ somewhere in $[0, 3]$. How many steps of the bisection method are required in order to approximate the value of this solution to within 0.00001?

58. The unit step function (see p. 30) returns values of 0 or 1, but never 0.5. Why doesn't that violate the conclusion of the IVT?

59. Suppose $f(t) = 1/t$. Then $f(-1) = -1$ and $f(1) = 1$, yet $f(t)$ has no root in $[-1, 1]$. Why doesn't that violate the conclusion of the IVT?

60. Suppose a mountain climber ascends a mountain path, starting at 7 AM and arriving at the first way station at 4 PM, where he stays the night. In the morning, he realizes the battery of his digital camera is drained and, more importantly, that he forgot the spare batteries at the base camp. So he leaves the way station at 7 AM and hikes back to the base camp, arriving at 4 PM. Use the IVT to show that there is some point along the path that he crossed at the same time on each day. *(Hint: consider the function $a_1(t) - a_2(t)$ where $a_1(t)$ and $a_2(t)$ tell us the climber's altitude after t hours of hiking on the 1st and 2nd days, respectively.)*

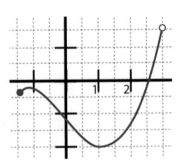

Figure 5.8: Graph for #48.

61. Suppose f is the function $f(t) = \begin{cases} k^3 & \text{when } t = 0 \\ \cos(t+k) & \text{when } t \neq 0 \end{cases}$.

   (a) Use the IVT to show that there *is* a value of k for which $f(t)$ is continuous at zero. (This will involve calculating and comparing limit values.)

   (b) Use the bisection algorithm to approximate that value of k to with 0.001 units.

We denote by $\lfloor t \rfloor$ the **greatest integer function**, or **floor function**, which is defined as $\lfloor t \rfloor = $ "the greatest of all the integers that are less than or equal to t." In each of #62–64, use $f(t) = \lfloor t \rfloor$ and $g(t) = 2\cos(t)$.

62. Draw a graph of $f(t) = \lfloor t \rfloor$. Then find all points of discontinuity and classify each as removable, jump, or infinite.

63. Draw the graph of the composition $f \circ g$, and locate any/all jump discontinuities that occur in $[0, 4]$.

64. Draw the graph of the composition $g \circ f$, and locate any/all jump discontinuities that occur in $[0, 4]$.

We denote by $\lceil t \rceil$ the **least integer function**, also called the **ceiling function**, which is defined as $\lceil t \rceil = $ "the least of all the integers that are greater than or equal to t." In each of #65–68, use $f(t) = \lceil t \rceil$ and $g(t) = t^2$.

65. Draw a graph of $f(t) = \lceil t \rceil$. Then find all points of discontinuity and classify each as removable, jump, or infinite.

66. Draw the graph of the composition $f \circ g$, and locate any/all jump discontinuities that occur in $[0, 4]$.

67. Draw the graph of the composition $g \circ f$, and locate any/all jump discontinuities that occur in $[0, 4]$.

68. Use $f(t)$ to design a function for calculating the amount of money needed to send a first-class letter through the US postal system.

69. The stereographic projection was introduced in Section 1.3. Suppose that instead of using a unit circle, we use a square of side length 1 that has been rotated by $\pi/4$ (as seen in Figure 5.9). As before, a point $(t, 0)$ on the number line follows a linear path toward the "north pole" of the square at $(0, \sqrt{2})$, and stops at its first encounter with the square. This stopping point is the *projection* of t onto the square, and has coordinates $(x(t), y(t))$.

   (a) Sketch graphs of $x(t)$ and $y(t)$, find and classify their discontinuities (removable? jump?). Then explain *why* these functions have discontinuities. (*Hint: it's in the geometry.*)

   (b) Do $x(t)$ and $y(t)$ have discontinuities when the square is rotated by angles other than $\pi/4$? Why, or why not?

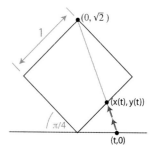

Figure 5.9: For #69.

Chapter 2 Review

❖ True or False?

In #1–3, you may assume that all limits exist and are nonzero.

1. The limit of a sum is the sum of the limits.

2. The limit of a product is the product of the limits.

3. The limit of a quotient is the quotient of the limits.

4. Suppose $g(6) = 19$. Then $\lim_{t \to 6} g(t) = 19$.

5. The expression $0/0$ means one.

6. Suppose f is both left-continuous and right-continuous at $t = 8$. Then f is continuous at $t = 8$.

7. Every function achieves a maximum value on the interval $[1, 2]$.

8. Suppose $\lim_{x \to 9} f(x) = 80$ and $\lim_{x \to 9} h(x) = 82$. If $f(x) \leq g(x) \leq h(x)$, $\lim_{x \to 9} g(x)$ is a number between 80 and 82.

9. The function $f(x) = \frac{x^2 - 4x + 3}{x^2 + 9x - 36}$ is continuous at $x = 3$.

10. The function $f(x) = \frac{x^2 - 4x + 3}{x^2 + 9x - 36}$ is continuous at $x = 1$.

11. Suppose $f(x) = x^2$. Then $\lim_{x \to \infty} f(x) = \infty$.

12. Suppose $f(x) = x^2 \cos^2(x)$. Then $\lim_{x \to \infty} f(x) = \infty$.

13. Suppose $f(x) = \frac{\cos^2(x)}{x^2 + 1}$. Then $\lim_{x \to \infty} f(x) = 0$.

14. Suppose $P(t)$ is the population of the United States t years after 1800. At $t = 20$, the Census Bureau reports that $P(20) = 9638453$. At $t = 207$, the Census Bureau reports that $P(207) > 300000000$. The IVT guarantees some time, say t^*, at which $P(t^*) = 10000000.5$.

15. Suppose f is a continuous function with a root in $[1, 2]$ that we need to approximate to with 0.0001 of it's actual value. If the root is irrational, the bisection method will take the same number of steps, regardless of the function f.

16. A jump discontinuity can be "fixed" by defining/redefining the value of f at a single point.

❖ Multiple Choice

Each of #17–20 refers to the graph of f shown below.

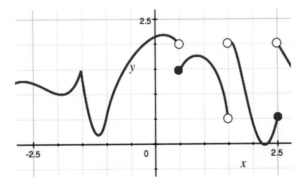

17. At $x = -2$ the function $f(x)$ is
    (a) left-continuous
    (b) right-continuous
    (c) continuous
    (d) all of the above
    (e) none of the above

18. At $x = 0.5$ the function $f(x)$ is
    (a) left-continuous
    (b) right-continuous
    (c) continuous
    (d) all of the above
    (e) none of the above

19. At $x = 1.5$ the function $f(x)$ is
    (a) left-continuous
    (b) right-continuous
    (c) continuous
    (d) all of the above
    (e) none of the above

20. At $x = 2.5$ the function $f(x)$ is
    (a) left-continuous
    (b) right-continuous
    (c) continuous
    (d) all of the above
    (e) none of the above

21. The expression $\lim_{t \to -8+} f(t)$ means, "the limit of $f(t)$ as ...
    (a) t approaches $+8$ from the left."
    (b) t approaches $+8$ from the right."
    (c) t approaches -8 from the left."
    (d) t approaches -8 from the right."
    (e) none of the above

22. Suppose $\lim_{t \to 8+} g(t) = 2$ and $\lim_{t \to 8-} g(t) = 2$. Then,
    (a) $\lim_{t \to 8} g(t)$ exists
    (b) g continuous at $t = 8$
    (c) $g(8) = 2$
    (d) none of the above

23. Suppose $C(r)$ is the area of a circle whose radius is r, and $S(r)$ is the area of the circumscribed square. Then $\lim_{r \to 0+} C(r)/S(r)$ is (a) 1, (b) $2/\pi$, (c) $\pi/2$, (d) $\pi/4$, or (e) none of the above.

24. The value of $\lim_{\phi \to 0} \frac{\sin(8\phi)}{\sin(4\phi)}$ is
    (a) 0, (b) 0.5, (c) 1, (d) 2, or
    (e) none of the above.

25. The value of $\lim_{\phi \to 0} \frac{1 - \cos(8\phi)}{4\phi}$ is
    (a) 0, (b) 0.5, (c) 1, (d) 2, or
    (e) none of the above.

26. When we rewrite $y(t) = 4t^2 + 7t + 5$ using factors of $(t - 2)$ instead of t, the formula is
    (a) $y(t) = 4(t - 2)^2 + 23(t - 2) - 72$
    (b) $y(t) = 4(t - 2)^2 + 33(t - 2) + 7$
    (c) $y(t) = 4(t - 2)^2 + 19(t - 2) - 108$
    (d) none of the above

❖ Exercises

In each of #27-41, calculate the specified limit.

27. $\lim\limits_{t \to 2} t^2 + 3t + 8$

28. $\lim\limits_{t \to \pi/4} \cos(3t) - \sin(4t)$

29. $\lim\limits_{t \to 1} \log_2(t^2 + 9t + 6)$

30. $\lim\limits_{x \to 2} \dfrac{3x^2 - 5x - 2}{x - 2}$

31. $\lim\limits_{x \to -1} \dfrac{x + 1}{x^2 + 5x + 4}$

32. $\lim\limits_{x \to 4} \dfrac{2x^2 + 5x + 3}{2x^2 - 5x - 12}$

33. $\lim\limits_{x \to 0} \left(6 - \dfrac{3}{x^2} \right)$

34. $\lim\limits_{x \to \infty} \left(6 - \dfrac{3}{x^2} \right)$

35. $\lim\limits_{x \to \infty} e^{-0.02x} \sin(9x)$

36. $\lim\limits_{x \to \infty} e^{-0.104x} \left(5 \sin(2x) - 8 \cos(2x) \right)$

37. $\lim\limits_{x \to \infty} \sqrt{9x^2 + 25x + 8} - 3x$

38. $\lim\limits_{\phi \to 0} \dfrac{\tan^2(\beta\phi)}{\phi \sin(\alpha\phi)}$, where $\alpha \neq 0$ and $\beta \neq 0$

39. $\lim\limits_{x \to 3} f(x)$ if $f(x) = \begin{cases} x^2, & x < 3 \\ 12 - x, & x \geq 3 \end{cases}$

40. $\lim\limits_{x \to 16} f(x)$ if $f(x) = \begin{cases} x^2 - 32x + 260, & x < 16 \\ \sqrt{x}, & x > 16 \end{cases}$

41. $\lim\limits_{x \to 4} f(x)$ if $f(x) = \begin{cases} \frac{1}{x}, & x \leq 4 \\ \frac{4}{x} - \frac{5}{8}, & x > 4 \end{cases}$

42. Find $\lim_{t \to 0} f(t)$ when $f(t) = \begin{cases} 0 & \text{when} \quad t \text{ is rational} \\ 1 & \text{when} \quad t \text{ is irrational.} \end{cases}$

43. Design a function $y(t)$ for which $\lim_{t \to 7} y(t) = 22$, but direct substitution of $t = 7$ leaves you with $0/0$.

44. Find a value of k for which the function $f(x) = \begin{cases} kx + k^2 & \text{if } x < 1 \\ -kx - 9 & \text{if } x \geq 1 \end{cases}$ is continuous at $x = 1$.

45. Suppose $f(x) = 1/x$ over $(0, 1]$. Then f does not achieve a maximum. Does this violate the EVT? If so, why do we call it a theorem? If not, explain why not.

46. Suppose $f(x) = x^5$. Then $f(1) = 1$ and $f(2) = 32$. Starting with the interval $[1, 2]$, how many steps will it take the bisection method to estimate the solution of $f(x) = 20$ to within 0.000001?

47. A small sample of bacteria is grown in a laboratory. The size of the sample over time can be described by the function

$$P(t) = \dfrac{500}{2 + 3e^{-0.12t}}$$

where P is the number of bacteria at time t (measured in days).

(a) What was the initial size of the sample?

(b) Is there a maximum size the sample will attain?

48. The graph of f is shown in Figure 6.1.

    (a) Suppose $\varepsilon = 0.2$. What value of δ would you choose in order to argue that $\lim_{t \to 2^-} f(t) = 1$?

    (b) Suppose $\varepsilon = 0.2$. What value of δ would you choose in order to argue that $\lim_{t \to 2^+} f(t) = 1$?

    (c) Suppose $\varepsilon = 0.2$. What value of δ would you choose in order to argue that $\lim_{t \to 2} f(t) = 1$?

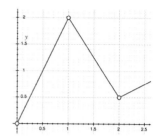

Figure 6.1: Graph for #48.

49. Suppose $f(t) = t^4 + 2t^3 + t + 1$. Prove that $\lim_{t \to 2} f(t) = 35$ by finding a formula that will produce an appropriate δ for any $\varepsilon \in (0, 1)$.

50. In #35 on p. 54 you found that an initial investment of $100 grew to $414.16 over 10 years when the annual percentage rate of interest was 15.27%. In the language of finance, we say that $414.16 is the *future value* of $100 after 10 years. Equivalently, we could also say that $100 is the *present value* of $414.16 (to be paid in 10 years), and in this case we refer to 15.27% as the *discount rate*.

The words "interest" and "future value" both refer to investing *now* and watching an investment grow, while "present value" and "discount rate" refer to today's valuation of a promised sum to be paid in the future.

    (a) Suppose the predicted cash flow of a project is shown in the margin, based on an initial investment of $50,000. (The dollar amounts are annual, *not* cumulative.) Find the present value of each return: 10,000, 12,000 ... if the prevailing discount rate is 10%. Add these numbers together to get the *total present value* (TPV) of the project.

end of year	return paid
1	$ 10,000
2	$ 12,000
3	$ 14,000
4	$ 16,000
5	$ 18,000

    (b) The *net present value* (NPV) of the project is $TPV - 50000$. Calculate the NPV.

    (c) Find the TPV of the project, assuming that the prevailing discount rate is 13%. Then find the NPV in this scenario.

    (d) From a practical standpoint, we should invest in a project if the NPV is positive, and we should avoid investing money if the NPV is negative. You saw in parts (a) and (b) that the NPV depends on the prevailing discount rate (we should invest when it's 10%, but we'd do better to invest our money elsewhere if the prevailing rate is 13%). Show that the NPV for this project is determined by the prevailing discount rate, x, according to

$$NPV(x) = \frac{10000}{1 + 0.01x} + \frac{12000}{(1 + 0.01x)^2} + \frac{14000}{(1 + 0.01x)^3} + \frac{16000}{(1 + 0.01x)^4} + \frac{18000}{(1 + 0.01x)^5}$$

    (e) The discount rate at which $NPV(x) = 0$ is called the *internal rate of return* (IRR) of the investment. You know that $NPV(10) > 0$ and $NPV(13) < 0$. Since $NPV(x)$ is a continuous function, the IVT guarantees that a root exists somewhere between 10 and 13. Use a bisection algorithm to determine the IRR to within an error of 0.001.

Chapter 2 Projects and Applications

❖ The Apothem and Limits at Infinity

A **regular** n-**gon (polygon)** has n sides of equal length. The perpendicular distance from an edge to the center of such a shape, as seen in Figure 7.1a, is called the **apothem**. In the steps below, you'll write the area and perimeter of a regular n-gon in terms of its apothem. In the present context, we're interested in what happens to these quantities in the limit. In Chapter 3, we'll be interested in their *derivatives*.

> Some familiar regular n-gons are . . .
>
> | # sides | name |
> | --- | --- |
> | 3 | equilateral triangle |
> | 4 | square |
> | 5 | pentagon |
> | 6 | hexagon |
> | : | : |
>
> The word *apothem* is pronounced "a-POTH-em."

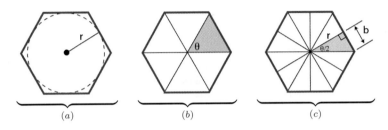

Figure 7.1: A regular hexagon with an apothem of r.

1. Suppose we partition a regular n-gon into a collection of triangles by drawing lines that pass through its vertices (see Figure 7.1b). Explain why the central angle of each triangle is $\theta = 2\pi/n$.

When the area of each triangle is $a(n)$, the area of the whole n-gon is $A(n) = n\,a(n)$. To find a formula for $a(n)$, use an altitude of the triangle to cut it in half (see Figure 7.1c). The "height" of the triangle is the apothem, r, and we'll denote the length of the "base" by b.

> By saying that the area of the n-gon is $n\,a(n)$, We're implicitly asserting that all of the triangles are congruent. You probably believe it, but can you prove it?

2. Use the tangent to relate b, r, and the angle $\theta/2 = \pi/n$.

3. Find the area of the right triangle in Figure 7.1c. (Your formula for $a(n)$ will include both n and r.)

4. Find the area of the entire n-gon (again, your formula will include both n and r).

5. Intuitively speaking, what *should* be true about the areas of n-gons with more and more sides? (We consider only those whose apothems are the same, r.)

6. Calculate the limit $\lim_{n \to \infty} A(n)$. (*Try writing the prefactor of*

$$n \quad as \quad \frac{\pi}{\frac{\pi}{n}} \quad along\ the\ way,\ and\ write\ \tan(\phi)\ as\ \frac{\sin(\phi)}{\cos(\phi)}.)$$

> Since the number n has to be a positive integer, we say this is a *discrete* limit (as opposed to the *continuous* limits that we've seen heretofore). In Chapter 8, we'll see that the discrete limit *must* converge whenever its continuous analog does. That's the case we have here, so go ahead and use what you know of continuous limits.

7. Intuitively speaking, what *should* be true about the perimeters of n-gons with more and more sides? (We consider only those whose apothems are the same, r.)

8. Since the side length of this regular n-gon is $2b$, where b depends on r and n according to your formula from #2, the total perimeter is $P(n) = 2nb$. Calculate $\lim_{n \to \infty} P(n)$.

The **circumradius** of a regular n-gon is the distance from the center to any vertex (see the figure, right). We'll denote this distance by R.

9. Find formulas for $A(n)$ and $P(n)$ in terms of R.

10. Calculate $\lim_{n \to \infty} A(n)$ and $\lim_{n \to \infty} P(n)$ in terms of R.

❖ Implementing the Bisection Algorithm

The following exercises will show you how to approximate $\sqrt{2}$ by using a spreadsheet program to find a root of the equation $x^2 - 2 = 0$ using the bisection algorithm.

1. The first row of cells will act as headers, identifying what each column means to us. In the first six cells of Row 1, type

   left   **right**   **midpoint**   **f(left)**   **f(right)**   **f(midpoint)**
   in cell **A1**   in cell **B1**   in cell **C1**   in cell **D1**   in cell **E1**   in cell **F1**

2. In the second row, we begin by making an educated guess at an interval in which we'll find the root. We know that $\sqrt{2} \in [0, 2]$, so in cell **A2** type **0**, and in cell **B2** type **2**.

3. The midpoint of $[0, 2]$ is just the average of the endpoints, $(0 + 2)/2 = 1$, but *don't* type a 1 into cell **C2**. At this stage we want to switch from typing numbers to typing formulas, so that we can tell the computer to "continue in the same manner." Type **= (A2+B2)/2** into cell **C2** (including the equals sign).

   > Notice how we refer to the initial endpoints by their locations in the spreadsheet rather than their values.

4. Next, we calculate the value of $f(x) = x^2 - 2$ at each of these three points. As we did in the previous step, we're going to reference the numbers by their location, not their value.

   (a) Type **= A2^2 - 2** into cell **D2**.

   (b) Type **= B2^2 - 2** into cell **E2**.

   (c) Type **= C2^2 - 2** into cell **F2**.

Our next job is to tell the computer what to do with all this information, and we have to think about it in general, as if we're in the 10th or 100th step of the algorithm. Suppose that after 10 steps of the algorithm, we know the root lies in the interval $[a, b]$. We continue by calculating the midpoint, $m = (a + b)/2$, and the function values $f(a)$, $f(b)$, and $f(m)$. How do we use these numbers to determine whether the interval in the next iteration is $[a, m]$ or $[m, b]$? One of three things could happen:

(i) The product $f(a)f(m) < 0$, which means that $f(a)$ and $f(m)$ have different polarities. So in our search for a root, we can discard b and make the next interval $[a, m]$.

(ii) The product $f(a)f(m) = 0$, which means that either $f(a) = 0$ or $f(m) = 0$. We don't know which, but we know there's a root in $[a, m]$, so we'll make that our next interval.

(iii) The product $f(b)f(m) < 0$, which means that $f(b)$ and $f(m)$ have different polarities. So in our search for the root, we can discard a and make the next interval $[m, b]$.

There's only one case in which the next interval will be $[m, b]$, so let's test for it. We want to use a command that tells the computer to check the polarity of $f(b)f(m)$.

- If that number is negative, the left endpoint of our next interval is the number m (which is currently stored in cell **C2**).

- Otherwise, the left endpoint of our next interval is the number a (which is currently stored in cell **A2**).

5. The numbers $f(b)$ and $f(m)$ are in cells **E2** and **F2**, respectively, so in cell **A3** type:

$$= \text{if(} \mathbf{E2*F2} < 0, \mathbf{C2}, \mathbf{A2})$$

this expression tells the computer to grab the number in **C2** when **E2*F2**< 0, and to grab the number in **A2** otherwise

This says, "If $f(m)f(b) < 0$ (which indicates a change in polarity, and so a root), use m as the new left end of the interval. Otherwise, keep the left endpoint the same."

6. We type a similar command into cell **B3** to set the right endpoint of the interval: $= \text{if(} \mathbf{E2*F2} < 0, \mathbf{B2}, \mathbf{C2})$

7. Now we're ready to use the power of the spreadsheet! First highlight cells **C2** through **F2**, and put your mouse in the lower right corner of the highlighted area, where there's a small box (the mouse will become a +).

left	right	midpoint	f(left)	f(right)	f(midpoint)
0	2	1	-2	2	-1
1	2				

Figure 7.2: Figure for Step 7.

Click on that box, drag down 1 row, and release (see Figure 7.3). The program will fill in the new cells by using the formulas that you entered in cells **C2** through **F2** (preserving the relative locations of data).

left	right	midpoint	f(left)	f(right)	f(midpoint)
0	2	1	-2	2	-1
1	2				

left	right	midpoint	f(left)	f(right)	f(midpoint)
0	2	1	-2	2	-1
1	2	1.5	-1	2	0.25

Figure 7.3: (top) Dragging down; (bottom) the program copies the formulas into the lower cells for you.

8. We'll do that again, but this time highlight the entire 3rd row. Grab the box in the lower right corner and drag it down 25 or 30 rows (see Figure 7.4).

The farther down you drag, the more iterations you do, and the better your approximation of $\sqrt{2}$ will be.

left	right	midpoint	f(left)	f(right)	f(midpoint)
0	2	1	-2	2	-1
1	2	1.5	-1	2	0.25

Figure 7.4: Figure for Step 8.

9. The value of *f(midpoint)* in the last row should be very close to zero (e.g., 5×10^{-11}), so we'll take the current midpoint to be our approximation of $\sqrt{2}$.

10. Adapt the algorithm to approximate the solution of $\cos(x) = x$.

11. When trying to determine the most intense frequency of light that's emitted by a body whose temperature is known, we have to solve the equation $3e^x - 3 - xe^x = 0$ (see [11, p.292]). Use the bisection algorithm to approximate the solution of this equation.

Since x appears both *in* and *out* of a transcendental function, we have no hope of solving the equation analytically. Numerics to the rescue!

Chapter 3
The Derivative

Instead of knowing only that a microchip is heating up, or that blood pressure is rising, or that chemicals are reacting, it is important in today's quantitative world to also know how quickly such changes happen. We can already discuss this in broad, qualitative terms. For example, Figure 0.1 depicts the speed of a rain drop as a function of time. The speed changes quickly at first, but slowly later on. Similarly, Figure 0.2 depicts the radius of an artery as a function of time. (The

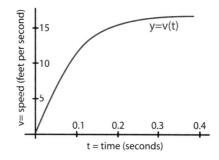

Figure 0.1: Relationship between the speed of a falling raindrop, v (in ft/sec), and time (in seconds).

radius increases when the blood pressure rises due to the heart's contraction.) You can see that the radius is not changing much at first, but the artery dilates quickly between times $t = 0.2$ and $t = 0.4$ seconds.

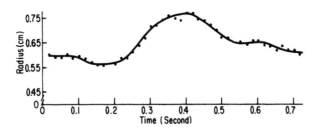

Figure 0.2: Radius of the right pulmonary artery as a function of time. (Experimental data observed by J. M. M. Jarmakani et al., reported in the article "In Vivo Pressure-Radius Relationships of the Pulmonary Artery in Children with Congenital Heart Disease," which was published in the journal *Circulation*, Vol. 43, pp. 585-592.)

The important idea here is that a function is changing quickly where its graph is steep. In this chapter we use slopes of lines to develop something called the *derivative* that allows us to quantify steepness and, in precise language, discuss the rate at which change occurs.

3.1   The Derivative

In the chapter opener, we began a discussion about change and our need to quantify the rate at which it happens. We also said that a function is changing rapidly where its graph is steep, which leads us directly to the question, "Can we quantify steepness?" We do that with *slope* when talking about lines (e.g., $y = 5t$ is steeper than $y = 2t$), and although most functions are nonlinear, the basic idea of a rise-to-run ratio still allows us to discuss the steepness of graphs and the rate at which quantities change.

For example, the graph in Figure 1.1 depicts the radius of an artery as a function of time. You can see that the radius is $r = 0.565$ centimeters at time $t = 0.2$ seconds, and the radius is beginning to increase. In order to gauge the rate at which it increases, we compare that radius to one a moment later. For example, the radius is $r = 0.745$ at time $t = 0.33$, so the radius of the artery increased by 0.18 centimeters in 0.13 seconds.

> The artery expands in response to increased blood pressure, which happens when the heart beats.

For brevity, we say that the *average rate of change* is $\frac{0.18}{0.13} = 1.385$ cm/sec during that time interval. Note that the line connecting the points $(0.2, 0.565)$ and $(0.33, 0.745)$, called a **secant line** has a slope of

> The word *secant* is used here in the geometric rather than the trigonometric sense, meaning, "a line that cuts through a curve."

$$\frac{\text{rise}}{\text{run}} = \frac{0.745 - 0.565}{0.33 - 0.2} = 1.385,$$

which is exactly the average rate of change we calculated a moment ago.

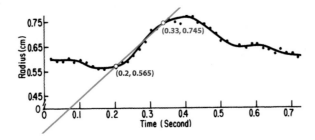

Figure 1.1: Calculating the average rate of change over a time interval is the same as calculating the slope of a line.

With that said, you can see that the secant line connecting $(0.2, 0.565)$ and $(0.33, 0.745)$ is steeper than the graph near time $t = 0.2$. That can happen when the time between measurements is large (0.13 seconds is almost 20% of the t-axis shown on the graph), so we try to improve our estimate of the graph's steepness by using a shorter time frame. Figure 1.2 shows an average rate of change of

$$\frac{0.613 - 0.565}{0.25 - 0.2} = 0.96\frac{\text{cm}}{\text{sec}} \quad \text{between times } t = 0.2 \text{ and } t = 0.25,$$

and

$$\frac{0.575 - 0.565}{0.22 - 0.2} = 0.5\frac{\text{cm}}{\text{sec}} \quad \text{between times } t = 0.2 \text{ and } t = 0.22.$$

Notice that the secant line corresponding to the shorter span of time more accurately approximates the steepness of the curve at time $t_0 = 0.2$ seconds.

More generally, suppose $f(t)$ is the radius of the artery after t seconds. To understand how quickly the radius is changing at time t_0, we compare $f(t_0)$ to the radius a moment later, $f(t_0 + \Delta t)$. The **average rate of change** between times

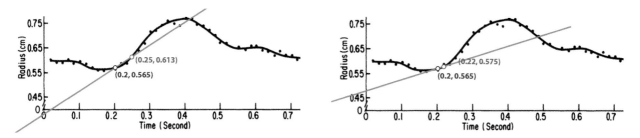

Figure 1.2: (both) Radius of the right pulmonary artery as a function of time. Calculating the average rate of change over a time interval is the same as calculating the slope of a line.

t_0 and $t_0 + \Delta t$ is

$$\frac{f(t_0 + \Delta t) - f(t_0)}{(t_0 + \Delta t) - t_0} = \frac{f(t_0 + \Delta t) - f(t_0)}{\Delta t}.$$

> The symbol Δ is the capital Greek letter delta, and is used to denote a *change* in some quantity. For example, Δt means a change in t.

We often denote this fraction by $\Delta f / \Delta t$, and refer to it as the **difference quotient**.

Calculating an average rate of change

Example 1.1. Suppose a mass is hung from the ceiling on a spring. When it's pushed upwards and then released, its displacement from equilibrium is $f(t) = 4e^{-0.1t}\cos(3t)$, where distance is measured in millimeters and time in seconds. Determine the average rate at which its position changes between $t = 0.75$ and $t = 1.4$ seconds.

> *Equilibrium* is where the mass would come to rest.

Solution: The change in position between these times is $f(1.4) - f(0.75) = 0.6263$ mm, and $\Delta t = 0.65$ seconds, so the average rate of change is

$$\begin{array}{c}\text{Measured in milliimeters} \rightarrow \\ \text{Measured in seconds} \rightarrow\end{array} \frac{\Delta f}{\Delta t} = \frac{0.6263}{0.65} = 0.9635 \frac{\text{mm}}{\text{sec}}. \qquad \blacksquare$$

The number we calculated in Example 1.1 is positive because the object is higher at time $t = 1.4$ than it is at time $t = 0.75$ (resulting in $\Delta f > 0$), but in Figure 1.3 you can see from the graph that the object is descending at time $t = 0.75$, not rising. If we want our calculation to better reflect what's happening when $t = 0.75$, we need to use a smaller span of time.

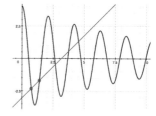

Figure 1.3: The graph of $f(t)$ in Example 1.1

Try It Yourself: Calculating an average rate of change

Example 1.2. Determine the average rate at which the mass in Example 1.1 changes position between $t = 0.75$ and $t = 1.075$ seconds.

Answer: -3.8421 mm/sec $\blacksquare$

Full Solution On-line

The number we calculated in Example 1.2 is negative because the object is lower at time $t = 1.075$ than it is at time $t = 0.75$. That's an improvement over our calculation in Example 1.1 in that it reflects the fact that the object is descending at $t = 0.75$. We can continue to improve our estimate of the rate at which the mass is moving by using smaller and smaller time spans in our calculation.

Calculating average rates of change over progressively smaller Δt

Example 1.3. Determine the average rate at which the mass in Example 1.1 changes position between $t = 0.75$ and $t = 0.75 + \Delta t$ seconds for smaller and smaller values of Δt.

Solution: In the following table (generated with a spreadsheet program), each Δt is 75% smaller than the one preceding it.

Δt (sec)	Δf (mm)	$\frac{\Delta f}{\Delta t} \frac{mm}{sec}$	Δt (sec)	Δf (mm)	$\frac{\Delta f}{\Delta t} \frac{mm}{sec}$
0.1625	-1.025956017	-6.313575492	0.002539063	-0.021328717	-8.400233063
0.040625	-0.322937031	-7.949219224	0.000634766	-0.005345938	-8.421908484
0.01015625	-0.084425626	-8.31267707	0.000158691	-0.001337342	-8.427313886

The progressively smaller values of Δt in Example 1.3 improve our approximation of the rate at which the mass is moving, but we can do better still.

We say that f is **differentiable** at a time, say t_0, when the limit of $\Delta f / \Delta t$ exists as $\Delta t \to 0$. The value of the limit is denoted by $f'(t_0)$, and is called the *first derivative of f at t_0.*

> The word *differentiable* is pronounced "diff-er-EN-chee-uhbul."

Definition of the Derivative: The **first derivative** of f at t_0 is

$$f'(t_0) = \lim_{\Delta t \to 0} \frac{f(t_0 + \Delta t) - f(t_0)}{\Delta t}, \tag{1.1}$$

when the limit exists. We say that f is **differentiable over (on) an open interval** (a, b) when $f'(t_0)$ exists at each $t_0 \in (a, b)$.

> We'll discuss times when this limit *doesn't* exist in Section 3.4.

Because the difference quotient calculates an *average* rate of change over a time span of Δt, and $f'(t_0)$ is determined by limiting $\Delta t \to 0$, we often refer to the derivative as an **instantaneous rate of change**. In the case when $f(t)$ tells us the position of an object at time t, as in Examples 1.1 and 1.2, the number $f'(t_0)$ is called the object's **(instantaneous) velocity** at time t_0.

Calculating velocity

Example 1.4. Suppose the motion of an object along the x-axis is described by $x(t) = 3t^2 + 7t - 4$, where x is measured in feet and time in minutes. Determine the object's velocity at time $t_0 = 2$.

Solution: The velocity at time $t_0 = 2$ is the derivative,

$$
\begin{aligned}
x'(2) &= \lim_{\Delta t \to 0} \frac{x(2 + \Delta t) - x(2)}{\Delta t} \quad \leftarrow \text{Measured in feet} \\
&\phantom{= \lim_{\Delta t \to 0} \frac{x(2 + \Delta t) - x(2)}{\Delta t}} \leftarrow \text{Measured in minutes} \\
&= \lim_{\Delta t \to 0} \frac{(22 + 19\Delta t + 3(\Delta t)^2) - (22)}{\Delta t} \\
&= \lim_{\Delta t \to 0} \frac{19\Delta t + 3(\Delta t)^2}{\Delta t}.
\end{aligned}
$$

Notice that this limit results in a 0/0 indeterminate form, but we can factor the numerator and avoid the 0/0 problem by canceling a factor of Δt.

$$x'(2) = \lim_{\Delta t \to 0} \frac{(19 + 3\Delta t)\Delta t}{\Delta t} = \lim_{\Delta t \to 0} (19 + 3\Delta t) = 19.$$

Because the difference quotient has units of ft/min, so does $x'(2)$. That is, the object is traveling at 19 ft/min.

> The derivative also tells us about the direction of travel (as you might infer from Examples 1.1 and 1.2), but we're going to postpone that discussion until p. 182, and focus on calculation for the moment.

Try It Yourself: Calculating velocity

Example 1.5. Suppose the motion of an object along the y-axis is described by $y(t) = 5t^2 - 12t + 1$, where y is measured in cm and time in hours. Determine the object's velocity at time $t_0 = 1$.

Answer: $y'(1) = -2 \frac{\text{cm}}{\text{hr}}$ ∎

Full Solution On-line

When the independent variable isn't time, or the function doesn't tell us about position, the units associated with the derivative (and so what it means to us) can be very different, but we always understand it as an instantaneous rate of change.

Other instantaneous rates of change (length and temperature)

Example 1.6. Suppose $L(t)$ is the length of a thin metal rod (in millimeters) when its temperature is $t \, °C$. (a) What are the units associated with $L'(35)$, and (b) what does it mean physically?

Solution: (a) Since the difference quotient has units of millimeters per $°C$,

$$\text{Measured in mm} \rightarrow \frac{L(35 + \Delta t) - L(35)}{\Delta t} = \frac{\text{mm}}{°C}.$$
$$\text{Measured in } °C \rightarrow$$

those are the units we assign to its limit, $L'(35)$. (b) For the sake of discussion, suppose $L'(35) = 10^{-8}$. That is,

$$\lim_{\Delta t \to 0} \frac{\Delta L}{\Delta t} = 10^{-8}, \quad \text{from which we conclude that} \quad \frac{\Delta L}{\Delta t} \approx 10^{-8} \quad \text{when}$$

Δt is very near 0. Said differently,

$$\underbrace{\Delta L}_{\text{mm}} \approx \underbrace{\underbrace{10^{-8}}_{\frac{\text{mm}}{°C}} \underbrace{\Delta t}_{°C}}_{\text{mm}} \tag{1.2}$$

when the change in temperature, Δt, is very small. For example, when the temperature of the bar deviates from $35°C$ by $\Delta t = 0.1 \, °C$, the length of the bar changes by about $\Delta L = 10^{-8}(0.1) = 10^{-9}$ mm. In essence, the number $L'(35)$ tells us (approximately) how much the rod will lengthen for each degree that we raise its temperature beyond $35°C$ (provided that the change in temperature is small). ∎

> Just as
> $$\left(\frac{\text{m}}{\text{sec}} \right) (\text{sec}) = \text{m},$$
> our dimensional analysis of (1.2) relies on the fact that
> $$\left(\frac{\text{mm}}{°C} \right) (°C) = \text{mm}.$$
> You should always check that the units are correct when you're working on a problem that has a physical interpretation. Correct units don't guarantee that you're right, but an equation that says sec=kg is certainly wrong.

Other instantaneous rates of change (price and demand)

Example 1.7. Suppose the price of a commodity is $p(x)$ dollars when x people want it. What are the units associated with $p'(250,000)$ and what does the number mean to us?

Solution: The number $p'(250,000)$ has units of $\frac{\text{dollars}}{\text{person}}$, and tells us how the purchase price changes in response to demand. ∎

The derivative in the absence of change

Example 1.8. Suppose $f(t) = c$ is a constant function. Show that $f'(t_0) = 0$ for every t_0.

Solution: The derivative at each time t_0 is the limit of the difference quotient

$$f'(t_0) = \lim_{\Delta t \to 0} \frac{f(t_0 + \Delta t) - f(t_0)}{\Delta t} = \lim_{\Delta t \to 0} \frac{c - c}{\Delta t} = \lim_{\Delta t \to 0} \frac{0}{\Delta t} = \lim_{\Delta t \to 0} 0 = 0. \quad ∎$$

> This makes sense when you think of $f'(t)$ as a rate of change: when f is constant, there *isn't* any change!

> Recall that Δt approaches 0 but never equals 0 when using limits, so we never have 0/0.

❖ Tangent Lines

We've seen that the difference quotient is the slope of a secant line that passes through $(t_0, f(t_0))$. As $\Delta t \to 0$ the secant line "rotates into" a **tangent line** which "glances off" rather than "cuts through" the graph of the function (see Figure 1.4). Since the difference quotient is the *slope* of the secant line, its limit (the derivative) will be the *slope* of the tangent line.

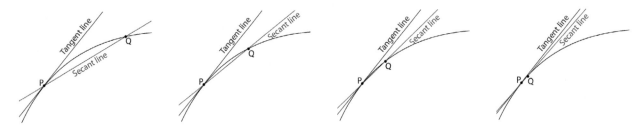

Figure 1.4: The points $P = (t_0, f(t_0))$ and $Q = (t_0 + \Delta t, f(t_0 + \Delta t))$. As $\Delta t \to 0$, the point Q gets closer and closer to P, and the slope of the secant line gets closer and closer to the slope of the tangent line.

The slope of a tangent line

Example 1.9. Determine the slope of the tangent line to the graph of $f(t) = e^t$ at $t = 0$.

Solution: The slope of the tangent line is the limit of the difference quotient,

$$f'(0) = \lim_{\Delta t \to 0} \frac{f(0 + \Delta t) - f(0)}{\Delta t} = \lim_{\Delta t \to 0} \frac{e^{\Delta t} - 1}{\Delta t}.$$

> Notice that the exponent and the denominator are the same. This seemingly minor fact becomes important later.

We don't yet have the analytic tools to determine the value of this limit, but a table of values is enlightening.

Δt	Δf	$\frac{\Delta f}{\Delta t}$	Δt	Δf	$\frac{\Delta f}{\Delta t}$
0.015625	0.015747709	1.00785335	1.52588×10^{-5}	1.52589×10^{-5}	1.000007629
0.001953125	0.001955034	1.000977199	9.53674×10^{-7}	9.53675×10^{-7}	1.000000477
0.00012207	0.000122078	1.000061038	$2.98023223 \times 10^{-8}$	$2.98023228 \times 10^{-8}$	1.000000015

Based on these data we infer that $f'(0) = 1$, which is exactly what we said in Chapter 1 (p. 50). ∎

The slope of a tangent line

Example 1.10. Determine the slope of the tangent line to the graph of $f(t) = \sin(t)$ at $t = 0$.

Solution: The slope of the tangent line is the limit of the difference quotient,

$$f'(0) = \lim_{\Delta t \to 0} \frac{\sin(0 + \Delta t) - \sin(0)}{\Delta t} = \lim_{\Delta t \to 0} \frac{\sin(\Delta t)}{\Delta t}.$$

This limit results in a 0/0 indeterminate form, but recall that in Chapter 2 we saw $\lim_{\phi \to 0} \frac{\sin(\phi)}{\phi} = 1$ (see p. 101). This limit has exactly the same form (except that we're using the symbol Δt instead of ϕ), so its value is 1. ∎

Try It Yourself: The slope of a tangent line

Example 1.11. Determine the slope of the tangent line to the graph of $f(t) = \cos(t)$ at $t = 0$.

Answer: 0 (so this tangent line is horizontal). ■

Though our discussion has focussed on the *slope* of the tangent line, the line itself plays an important role in techniques such as *linear approximation* and algorithms such as *Newton's method,* both of which are discussed in Chapter 4.

Find the equation of a tangent line

Example 1.12. Suppose $f(t) = t^2$. Determine the equation of the tangent line to the graph of f at $t_0 = 0.5$.

Solution: We need a point and a slope in order to write the equation of a line. Since $f(0.5) = 0.25$, the point is $(0.5, 0.25)$. The slope is found by calculating the derivative,

$$
\begin{aligned}
f'(0.5) = \lim_{\Delta t \to 0} \frac{\Delta f}{\Delta t} &= \lim_{\Delta t \to 0} \frac{f(0.5 + \Delta t) - f(0.5)}{\Delta t} = \lim_{\Delta t \to 0} \frac{(0.5 + \Delta t)^2 - 0.25}{\Delta t} \\
&= \lim_{\Delta t \to 0} \frac{0.25 + \Delta t + (\Delta t)^2 - 0.25}{\Delta t} = \lim_{\Delta t \to 0} \frac{\Delta t + (\Delta t)^2}{\Delta t} \\
&= \lim_{\Delta t \to 0} \frac{(1 + \Delta t)\Delta t}{\Delta t} = \lim_{\Delta t \to 0} (1 + \Delta t) = 1.
\end{aligned}
$$

> As we did in Example 1.4 on p. 150, now we factor out a Δt so that we can cancel the Δt in the numerator and avoid the $0/0$ indeterminate form.

So the point-slope equation of the line is $y - 0.25 = 1(t - 0.5)$. In preparation for later ideas, let's rewrite this as $y = 0.25 + 1(t - 0.5)$. ■

> Specifically, we're writing the equation of the line as $y = \cdots$ in preparation for our discussion of *linear approximations* in Chapter 4, and *Taylor polynomials* in Chapter 8

Try It Yourself: Find the equation of a tangent line

Example 1.13. Suppose $f(t) = t^3$. Determine the equation of the tangent line to the graph of f at $t_0 = \frac{1}{2}$.

Answer: $y = \frac{1}{8} + \frac{3}{4}\left(t - \frac{1}{2}\right)$ ■

In general, the tangent line to the graph of f at $(t_0, f(t_0))$ has a slope of $f'(t_0)$, so …

> **Equation of the Tangent Line:** The tangent line to the graph of f at $(t_0, f(t_0))$ is described by the equation $y = f(t_0) + f'(t_0)(t - t_0)$.

❖ A Comment on Language

A cylinder whose radius is r and whose height is h (see Figure 1.5) has a volume of $v = (area\ of\ base) \times (height) = \pi r^2 h$. That's simple enough, but what's the instantaneous rate of change in volume? Before we can answer that question, we need to know whether we're treating r as an independent variable and h as fixed, or vice versa. We communicate that information by saying "with respect to …" For example,

1. Suppose we're talking about a cylinder of a particular (fixed) height, say $h = 4$, so that volume depends only on radius: $v(r) = 4\pi r^2$. In the exercise set you'll show that $v'(3) = 24\pi$, which has units of

$$
\begin{array}{c}
\text{Measured in cm}^3 \to \\
\text{Measured in cm} \to
\end{array}
\quad \frac{v(3 + \Delta r) - v(3)}{\Delta r} = \frac{\text{cm}^3 \text{ of volume}}{\text{cm of radius}}.
$$

So we say that the instantaneous rate of change in volume *with respect to* radius is 24π cubic centimeters (of volume) per centimeter (of radius).

> People sometimes write
> $$\frac{\text{cm}^3}{\text{cm}} = \text{cm}^2.$$
> This dimensional shorthand *does* keep track of units correctly, but it also makes it more difficult to understand what they mean. In this context, units of cm^2 obscure the idea that we're looking at a rate of change in *volume*.

2. Suppose we're talking about a cylinder of a particular (fixed) radius, say $r = 3$, so that volume depends only on height: $v(h) = 9\pi h$. In the exercise set you'll show that $v'(4) = 9\pi$, which has units of

$$\text{Measured in cm}^3 \rightarrow \quad \frac{v(4 + \Delta r) - v(4)}{\Delta r} = \frac{\text{cm}^3 \text{ of volume}}{\text{cm of height}}.$$
$$\text{Measured in cm} \rightarrow$$

So we say that the instantaneous rate of change in volume *with respect to* height is 9π cubic centimeters (of volume) per centimeter (of height).

Figure 1.5: A cylinder of radius r and height h

In cases where no units are specified, we refer to the variables by name, rather than by unit—so $f'(x)$ is the instantaneous rate of change in f *with respect to x*.

❖ Common Difficulties with the Derivative

Many students have difficulty calculating $f'(t_0)$ because they write down $f(t_0 + \Delta t)$ incorrectly. Some mistakenly write it as $f(t_0) + \Delta t$, but that's almost never right. If you're having trouble, try replacing each instance of t with (t) in the formula for f. For example, rewrite

$$f(t) = 5t^2 + 8t + 9 \quad \text{as} \quad f(t) = 5(t)^2 + 8(t) + 9.$$

Then to calculate $f(t_0 + \Delta t)$, you simply replace each occurrence of (t) with $(t + \Delta t)$. In this example,

$$f(t_0 + \Delta t) = 5(t_0 + \Delta t)^2 + 8(t_0 + \Delta t) + 9.$$

You should know

- the terms *average rate of change*, *instantaneous rate of change*, *secant line*, *tangent line*, *difference quotient*, *differentiable*, and *first derivative*;

- that the derivative is the limit of the difference quotient;

- that the *difference quotient* can be interpreted as either an average rate of change or the slope of a secant line;

- that the *derivative* can be interpreted as either an instantaneous rate of change or the slope of a tangent line;

- the meaning of the phrase "with respect to."

You should be able to

- find the equation of the tangent line to the graph of f at $(t_0, f(t_0))$;

- determine the units associated with the derivative and explain what they mean to us.

❖ 3.1 Skill Exercises

In each of #1–4 you are given a function and a specified point on its graph. (a) Graph the function, and (b) on the same axes, graph three secant lines through the given point using $\Delta t = 1, 0.5, 0.01$. Then (c) based on those secant lines, approximate the slope of the tangent line to the graph of f at the given point.

1. $f(t) = e^{-t}$; $(0, 1)$

2. $f(t) = \sin(t)$; $(\frac{\pi}{2}, 1)$

3. $f(t) = \ln(t)$; $(1, 0)$

4. $f(t) = \frac{3t^2}{1+t^2}$; $(2, \frac{12}{5})$

In each of #5–10 use the limit of difference quotients to determine $f'(t_0)$.

5. $f(t) = 3 + t^2$, $t_0 = 1$

6. $f(t) = 4 + t^3$, $t_0 = 2$

7. $f(t) = 1 + 8t + 9t^2$, $t_0 = -4$

8. $f(t) = 2 + 8t - 3t^2$, $t_0 = -2$

9. $f(t) = 5t^2 + t + 8$, $t_0 = 5/4$

10. $f(t) = t^3 + t^2$, $t_0 = -2/3$

In each of #11–14 you should (a) use the limit of the difference quotient to determine $f'(t_0)$, then (b) write the equation of the tangent line to the graph of the function in the form $y = f(t_0) + f'(t_0)(t - t_0)$, and (c) graph the tangent line and the function together on the same axes.

11. $f(t) = 5t^3 - 2t - 3$, $t_0 = 1$

12. $f(t) = -t^4 + t^3 + 1$, $t_0 = -1$

13. $f(t) = 5t^3 - 60t$, $t_0 = 2$

14. $f(t) = 4t^3 - t^2 + 1$, $t_0 = 1/6$

In each of #15–20 you should (a) calculate $f(0)$; (b) use the limit of difference quotients to determine $f'(0)$; and (c) identify where you see your answers from parts (a) and (b) in the formula for $f(t)$.

15. $f(t) = 32 + 8t$

16. $f(t) = 2 - 5t$

17. $f(t) = 2 + 3t + 7t^2$

18. $f(t) = 12 + 9t + t^2$

19. $f(t) = 21 - 4t + 2t^2$

20. $f(t) = -13 - 8t + 5t^2$

In each of #21–26 you should (a) calculate $f(t_0)$; (b) determine the formula for $f(t_0 + \Delta t)$; (c) use the limit of difference quotients to determine $f'(t_0)$; and (d) locate your answers from parts (a) and (c) in your formula for $f(t_0 + \Delta t)$.

21. $f(t) = t^2$, $t_0 = 3$

22. $f(t) = t^2$, $t_0 = 4$

23. $f(t) = t^3$, $t_0 = -2$

24. $f(t) = t^3$, $t_0 = -4$

25. $f(t) = 1 + 2t + 6t^2$, $t_0 = 1$

26. $f(t) = 2 - 7t + 3t^2$, $t_0 = 4$

In each of #27–30 use the limit of the difference quotient to determine the slope of the tangent line at $t = 0$.

27. $f(t) = \sin(7t)$

28. $f(t) = \sin(5.6t)$

29. $f(t) = \cos(2.3t)$

30. $f(t) = \cos(8.7t)$

❖ 3.1 Concept and Application Exercises

31. Using a graphing utility, can you determine which line is tangent to $y = \frac{x}{x^2+1}$?

    (a) $y = -0.11x + 0.62$

    (b) $y = -0.12x + 0.64$

32. Using a graphing utility, can you determine which line is tangent to the curve $y = x^2 - 3x + \frac{1}{x-1}$?

    (a) $y = 2.74x - 7.72$

    (b) $y = 2.75x - 7.75$

33. Suppose $f(t) = 0.1t^2$

    (a) Use a graphing utility to render a graph of f, along with the secant lines that pass through $(1, f(1))$ and each of $(-0.5, f(-0.5))$, $(0, f(0))$, and $(0.5, f(0.5))$.

    (b) Does the tangent line to the graph at $(1, f(1))$ have a slope that is greater or less than the slopes of these secant lines?

    (c) Does the tangent line to the graph at $(1, f(1))$ lie above or below the curve? What about the secant lines?

34. Suppose $f(t) = 0.1t^2$

    (a) Use a graphing utility to render a graph of f, along with the secant lines that pass through $(1, f(1))$ and each of $(1.5, f(1.5))$, $(2, f(2))$, and $(2.5, f(2.5))$.

    (b) Does the tangent line to the graph at $(1, f(1))$ have a slope that is greater or less than the slopes of these secant lines?

    (c) Does the tangent line to the graph at $(1, f(1))$ lie above or below the curve? What about the secant lines?

35. Table 1 (below) shows the slope of the secant line connecting $(2, f(2))$ to $(2 + \Delta x, f(2 + \Delta x))$. Supposing that f is differentiable at $t_0 = 2$, use this data to estimate $f'(2)$.

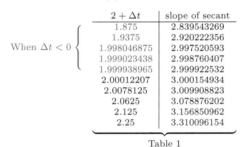

	$2 + \Delta t$	slope of secant	$4 + \Delta t$	slope of secant
	1.875	2.839543269	3.875	15.625
	1.9375	2.920222356	3.9375	15.8125
When $\Delta t < 0$	1.998046875	2.997520593	3.9921875	15.9765625
	1.999023438	2.998760407	3.999023438	15.99707031
	1.999938965	2.999922532	3.999938965	15.99981689
	2.00012207	3.000154934	4.000061035	16.00018311
	2.0078125	3.009908823	4.000976563	16.00292969
	2.0625	3.078876202	4.0078125	16.0234375
	2.125	3.156850962	4.0625	16.1875
	2.25	3.310096154	4.125	16.375

Table 1   Table 2

36. Table 2 (above) shows the slope of the secant line connecting $(4, f(4))$ to $(4 + \Delta t, f(4 + \Delta t))$. Supposing that f is differentiable at $t_0 = 4$, use this data to estimate $f'(4)$.

In each of #37–40 estimate $f'(t_0)$ by making a table such as Table 1 (shown above). *Note: this can be done very quickly using a spreadsheet program.*

37. $f(t) = e^{3t}$, $t_0 = 0$ 39. $f(t) = \ln(t)$, $t_0 = 2$

38. $f(t) = e^{5t}$, $t_0 = 0$ 40. $f(t) = \ln(t)$, $t_0 = 4$

In each of #41–44, in which the position of an object is described by $y(t)$, you should find the object's average velocity over the given interval of time.

41. $y(t) = 2t^2 + 3t - 7$; $[1, 4]$ 43. $y(t) = te^{-t}$; $[1, 2]$

42. $y(t) = t + \sqrt{t}$; $[4, 9]$ 44. $y(t) = t\sin(7.5\pi t)$; $[1, 3]$

45. In 1990, R. J. Rowley published a paper in which he compared two different populations of the same species of sea urchin. One group lived in kelp beds and the other in barrens. The graphs below (adapted from Rowley's original paper) show the size of the average sea urchin from each population over the course of 100 days.

(a) Which species' diameter has a larger instantaneous rate of change initially?

(b) Estimate the average rate of change in the size of the sea urchins for each population during the first 50 days, and during the entire 100 days. Include appropriate units in your answer.

(c) Estimate the instantaneous rate of change at time $t = 50$ for each population. Include appropriate units.

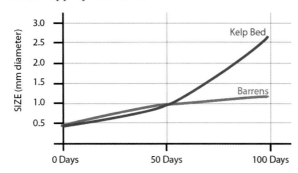

46. The graph below shows the velocity of a motor boat in still water. The acceleration at time t_0 is defined to be the instantaneous rate of change in velocity. That is, $a(t_0) = v'(t_0)$.

   (a) Explain what's happening to the boat during the first 16 seconds, and what might cause the sudden change at $t = 16$.

   (b) Estimate when the boat experiences the greatest acceleration.

   (c) Estimate the average rate of change in $v(t)$ during the first 16 seconds.

   (d) According to the graph, what appears to be the value of $\lim_{t \to \infty} v(t)$?

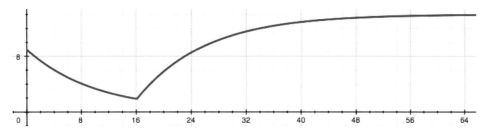

47. Use the graph of f shown below to estimate $f'(0)$, $f'(1)$, $f'(1.6)$, and $f'(2)$.

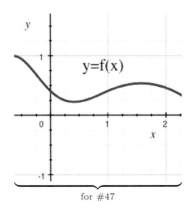

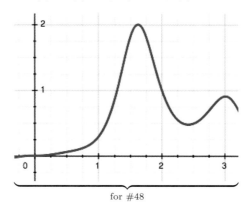

for #47   for #48

48. Based on the graph of $f(t)$ shown above, put the numbers $f'(0.5)$, $f'(1)$, $f'(2)$, and $f'(3)$ in order from smallest to largest.

49. Suppose $f(2) = 4$ and $f'(2) = -3$. Write the equation for the tangent line to the graph of f at $x = 2$.

50. Suppose $f(6) = 9$ and $f'(6) = 17$. Write the equation for the tangent line to the graph of f at $x = 6$.

51. Sketch the graph of a function for which $f'(1)$ and $f'(2)$ are opposites.

52. Sketch the graph of a function for which $f'(1)$ and $f'(3)$ are equal, but not the same as $f'(2)$.

53. Suppose the height of a cylinder is 4 cm.   Solid Geometry

   (a) Write the volume of the cylinder as a function of its radius r, which is also measured in cm.

   (b) Show that $v'(3) < v'(5)$.

   (c) Explain what $v'(3)$ and $v'(5)$ tell us, based on the units.

   (d) Based on the geometry of the cylinder, explain why part (b) makes sense.

54. Suppose the height of a cylinder is 3 cm.   Solid Geometry

   (a) Write the volume of the cylinder as a function of height, h, which is also measured in cm.

   (b) Show that $v'(4)$, $v'(5.1)$ and $v'(\sqrt{37})$ are all the same.

   (c) Explain what $v'(4)$, $v'(5.1)$, and $v'(\sqrt{37})$ tell us, based on the units.

   (d) Based on the geometry of the cylinder, explain why part (b) makes sense.

55. The height (in meters) of an object that's rolling down an inclined plane is given by $s(t) = -3t^2 + 18$, where $t \in (0, 2)$ is measured in seconds.   Physics

   (a) What is the average rate of change in the height with respect to time between 0.5 seconds and 1.5 seconds? (Include proper units.)

   (b) What is the instantaneous rate of change of the height with respect to time when $t = 1/2$? (Include proper units.)

   (c) Use complete sentences to compare and contrast the meaning of these numbers in this context.

56. Suppose the pressure experienced by a gas depends on temperature according to $P = \frac{1}{3}T$, where P is the pressure (in pounds per square inch) and T is the temperature (in degree Fahrenheit). Find the average rate of change of the pressure (including units) when the temperature decreases from $40°$ to $38°$.   Physical Chemistry

57. Suppose two chemicals, A and B, combine to produce a third, C. Using lower-case letters to denote the respective concentrations of these chemicals (in mols per liter), the Law of Mass Action (LMA) says that the rate of production with respect to time is $c'(t) = \gamma\, a(t)\, b(t)$, where $\gamma > 0$ is a constant that depends on physical parameters such as temperature.   Physical Chemistry

   (a) If t is measured in seconds, what are the units of $c'(t)$, $a(t)$, $b(t)$ and γ?

   (b) Explain what the equation $\lim_{t\to\infty} a(t) = 0$ means, and why it makes sense in this context.

   (c) What does the LMA predict about $c'(t)$ when $a(t) = 0$, and why does that make sense in this context?

58. The monthly profit for a particular company can be described by the function $P(x) = -x^2 + 11x - 18$ where P is the profit (in *thousands* of dollars) and x is the number of millions of gallons of oil it produces during the month.

    Business

    (a) If the monthly production of oil increases from 4 million to 6 million gallons, what is the average rate of change of the profit?
    (b) Calculate $P'(12)$ and explain what it means in this context.

59. The Ideal Gas Law can be written as $P = \frac{Nk_B}{V}T$, where N is the number of molecules and k_B is Boltzmann's constant.

    Physical Chemistry

    (a) Suppose that V is held constant (perhaps the gas is in a container) so that pressure depends only on temperature. What are the units of $P'(T)$?
    (b) Show that $P'(300)$, $P'(304)$ and $P'(321)$ are all equal.
    (c) In complete sentences, explain what your answer from part (b) means to us physically.

60. Suppose $y(d)$ is the number of Chinese yen that you can buy for d dollars. What are the units of y' and what does $y'(800)$ tell us?

    Currency Markets

61. Suppose $T(t)$ is the temperature of a cup of coffee, measured in °C, t minutes after it's been poured. What are the units of T', and what is the physical interpretation of $T'(2)$?

    Thermodynamics

62. Suppose $c(h)$ is the number of people using a chat client, h hours after midnight. What are the units of c' and what does $c'(1.5)$ tells us?

    Internet Load

63. Suppose $r(v)$ is the resistance offered by a particular fluid, measured in newtons, when an object moves through it at a speed of $v \frac{m}{sec}$. What are the units of r', and what does $r'(0.01)$ mean to us?

    Fluid Resistance

64. Use the graph of f shown below to determine times t_1 and t_2 at which the tangent line is not horizontal but $f'(t_1) = f'(t_2)$.

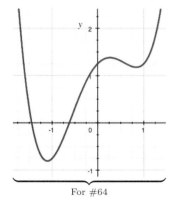

For #64

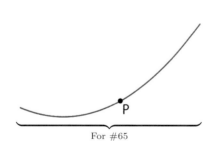

For #65

65. A portion of the graph of f is shown below, and P is at the point $(t_0, f(t_0))$. For parts (a)–(c), take Δt to be a positive number.

    (a) Draw the secant line whose slope is $\frac{f(t_0 + \Delta t) - f(t_0)}{\Delta t}$.
    (b) Draw the secant line whose slope is $\frac{f(t_0) - f(t_0 - \Delta t)}{\Delta t}$.
    (c) Explain the meaning of the number

    $$M(\Delta t) = \frac{1}{2}\left(\frac{f(t_0 + \Delta t) - f(t_0)}{\Delta t} + \frac{f(t_0) - f(t_0 - \Delta t)}{\Delta t} \right).$$

    (d) Suppose that f is differentiable at t_0. Verify that $f'(t_0) = \lim_{\Delta t \to 0} M(\Delta t)$.

3.2   Derivatives of Power Functions

Suppose you're flying over the northern Atlantic in a small plane, hoping to glimpse a whale. What's the probability that you'll see one? Toward the end of World War II a mathematician named Bernard Osgood Koopman (1900–1981) pieced together the first quantitative theory of search and detection, which led  him to the inverse-cube law of sighting. Said simply, the probability of glimpsing an object from a plane is

$$p(s) = \alpha \, s^{-3},$$

This formula is derived at the close of Appendix A, using some elementary ideas about radian measure.

where s is the line-of-sight distance between you and the object, and α is a constant that depends on weather conditions, among other things.

Koopman's inverse-cube law, Halley's inverse-square law (see p. 2), Kepler's Third Law of planetary motion (see p. 4), and many other descriptions of our world rely on power functions. So in this section we turn our attention to power functions and the rate at which they change.

In the examples and exercises of Section 3.1 we practiced calculating the limit of the difference quotient at particular times, t_0. Now, instead of calculating $f'(t_0)$ one point at a time, let's develop general formulas by working with arbitrary t_0.

Determine a general formula for the derivative

Example 2.1. Derive a general formula for $f'(t_0)$ when $f(t) = t^2$.

Solution: We know that

$$
f'(t_0) \;=\; \lim_{\Delta t \to 0} \frac{\Delta f}{\Delta t} = \lim_{\Delta t \to 0} \frac{f(t_0 + \Delta t) - f(t_0)}{\Delta t} = \lim_{\Delta t \to 0} \frac{(t_0 + \Delta t)^2 - (t_0)^2}{\Delta t}
$$

$$
\;=\; \lim_{\Delta t \to 0} \frac{(t_0)^2 + 2t_0\Delta t + (\Delta t)^2 - (t_0)^2}{\Delta t} = \lim_{\Delta t \to 0} \frac{2t_0\Delta t + (\Delta t)^2}{\Delta t} = \lim_{\Delta t \to 0} 2t_0 + \Delta t = 2t_0. \quad \blacksquare
$$

Try It Yourself: Determine a general formula for the derivative

Example 2.2. Derive a general formula for $g'(t_0)$ when $g(t) = t^3$.

Answer: $g'(t_0) = 3(t_0)^2$ $\blacksquare$

Full Solution On-line

It's common practice to omit the subscript, writing $f'(t) = 2t$ and $g'(t) = 3t^2$. This convention also makes the algebraic work of deriving a formula for the derivative much less cumbersome.

Determine a general formula for the derivative

Example 2.3. Derive a general formula for $h'(t)$ when $h(t) = t^{-2}$.

Solution: Let's begin by calculating the numerator of the difference quotient:

$$
\Delta h = h(t + \Delta t) - h(t) = \frac{1}{(t + \Delta t)^2} - \frac{1}{t^2} = \frac{-2t\Delta t - (\Delta t)^2}{(t^2 + 2t\Delta t + (\Delta t)^2)t^2}.
$$

To find $h'(t)$ we need to divide this by Δt and take a limit. A small algebraic maneuver will make the limit-taking process easier.

$$h'(t) = \lim_{\Delta t \to 0} \frac{\Delta h}{\Delta t} = \underbrace{\lim_{\Delta t \to 0} \frac{1}{\Delta t}(\Delta h)}_{\text{Here's the algebraic maneuver}} = \lim_{\Delta t \to 0} \frac{1}{\Delta t}\left(\frac{-2t\Delta t - (\Delta t)^2}{(t^2 + 2t\Delta t + (\Delta t)^2)t^2}\right).$$

Factoring Δt out of the numerator leaves us with

$$h'(t) = \lim_{\Delta t \to 0} \frac{1}{\Delta t}\left(\frac{\Delta t\Big(-2t - (\Delta t)\Big)}{(t^2 + 2t\Delta t + (\Delta t)^2)t^2}\right) = \lim_{\Delta t \to 0}\left(\frac{-2t - (\Delta t)}{(t^2 + 2t\Delta t + (\Delta t)^2)t^2}\right).$$

After applying the quotient, sum, and product laws for limits, we see that

$$h'(t) = \lim_{\Delta t \to 0}\left(\frac{-2t - (\Delta t)}{(t^2 + 2t\Delta t + (\Delta t)^2)t^2}\right) = \frac{-2t}{t^4} = -\frac{2}{t^3}. \qquad \blacksquare$$

Differentiating the square-root function

Example 2.4. Derive a general formula for $k'(t)$ when $k(t) = t^{1/2}$ and $t_0 > 0$.

Solution: As before, we begin by calculating

$$\Delta k = k(t + \Delta t) - k(t) = \sqrt{t + \Delta t} - \sqrt{t} .$$

Then divide by Δt and take the limit as $\Delta t \to 0$.

$$k'(t) = \lim_{\Delta t \to 0} \frac{\Delta g}{\Delta t} = \lim_{\Delta t \to 0} \frac{\sqrt{t + \Delta t} - \sqrt{t}}{\Delta t}.$$

In previous examples, a factor of Δt in the numerator cancelled the Δt in the denominator, thereby avoiding $0/0$ in the limit. We can distill such a Δt from the numerator by using the conjugate. Specifically,

$$k'(t) = \lim_{\Delta t \to 0}\left(\frac{\sqrt{t + \Delta t} - \sqrt{t}}{\Delta t} \overbrace{\frac{\sqrt{t + \Delta t} + \sqrt{t}}{\sqrt{t + \Delta t} + \sqrt{t}}}^{=1}\right)$$

$$= \lim_{\Delta t \to 0} \frac{(t + \Delta t) - t}{\Delta t(\sqrt{t + \Delta t} + \sqrt{t})} = \lim_{\Delta t \to 0} \frac{\Delta t}{\Delta t(\sqrt{t + \Delta t} + \sqrt{t})}.$$

Now we cancel the factor of Δt, and see

$$k'(t) = \lim_{\Delta t \to 0} \frac{1}{\sqrt{t + \Delta t} + \sqrt{t}}.$$

The continuity of the square-root function guarantees

$$\lim_{\Delta t \to 0} \sqrt{t + \Delta t} = \sqrt{t + 0} = \sqrt{t} \quad \text{when } t > 0,$$

so that when we apply the quotient and sum laws to calculate the above limit,

$$k'(t) = \lim_{\Delta t \to 0} \frac{1}{\sqrt{t + \Delta t} + \sqrt{t}} = \frac{1}{\sqrt{t} + \sqrt{t}} = \frac{1}{2\sqrt{t}}. \qquad \blacksquare$$

WARNING   WARNING

Recall that exponents do not distribute over sums:

$$\sqrt{t + \Delta t} \neq \sqrt{t} + \sqrt{\Delta t}.$$

WARNING   WARNING

See Example 2.1 and Example 2.2.

The act of determining a formula for the derivative by calculating the limit of the difference quotient, as in the last several examples, is called **differentiating**. Said differently, when you **differentiate** $f(t)$ you get a formula for $f'(t)$.

When we compile the results of Examples 2.1–2.4 we see

$$f(t) = t^2 \quad \Rightarrow \quad f'(t) = 2t^{2-1} = 2t^1$$

$$g(t) = t^3 \quad \Rightarrow \quad g'(t) = 3t^{3-1} = 3t^2$$

$$h(t) = t^{-2} \quad \Rightarrow \quad h'(t) = -2t^{-2-1} = -2t^{-3}$$

$$k(t) = t^{1/2} \quad \Rightarrow \quad k'(t) = \frac{1}{2}t^{(1/2)-1} = \frac{1}{2}t^{-1/2}.$$

This pattern leads us to the following (correct) conjecture:

Power Rule: Suppose η is any real number and

$$f(t) = t^\eta. \quad \text{Then} \quad f'(t) = \eta t^{\eta-1}. \tag{2.1}$$

Note: We're going to postpone a proof of the Power Rule until we have discussed an idea called *logarithmic differentiation* (p. 223).

Using the Power Rule

Example 2.5. Determine the derivatives of (a) $f(t) = t^6$, (b) $g(t) = t^{9/8}$, and (c) $\ell(t) = t^\pi$.

Solution: In order to use the Power Rule, we need to identify the number η. (a) The number $\eta = 6$, so $f'(t) = 6t^5$. (b) The number $\eta = 9/8$, so $g'(t) = \frac{9}{8}t^{1/8}$. (c) The number $\eta = \pi$, so $\ell'(t) = \pi t^{\pi-1}$.   ■

Try It Yourself: Using the Power Rule

Example 2.6. Determine the derivatives of (a) $h(t) = t^{-5/3}$, (b) $k(t) = t^{-1.3}$, and (c) $m(t) = t^{\sqrt{2}}$.

Answer: (a) $h'(t) = -\frac{5}{3}t^{-8/3}$, (b) $k'(t) = -1.3t^{-2.3}$, (c) $m'(t) = \sqrt{2}\, t^{\sqrt{2}-1}$ ■

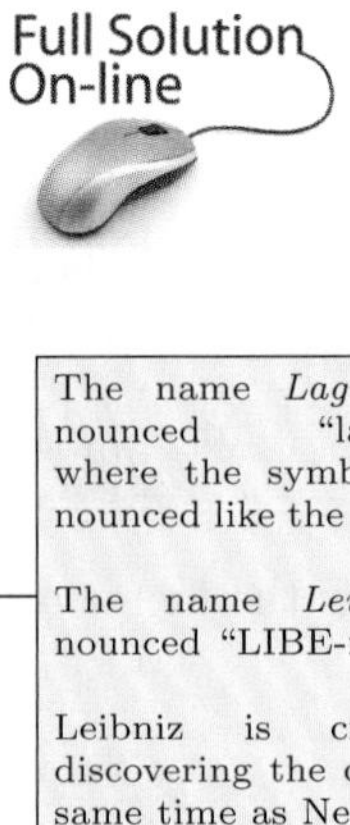

Full Solution On-line

❖ Other Notation

The $f'(t)$ notation that we've been using for the derivative was first developed by **Joseph Louis Lagrange** There are other notations for the derivative that are commonly used in science and engineering, and each plays a role in helping us communicate ideas and facts. **Isaac Newton** used $\dot{f}(t)$, **Gottfried Wilhelm von Leibniz** and used the notation $\frac{df}{dt}$ when referring to the derivative. Sometimes, it's helpful to modify the Liebniz notation by pulling the name of the function off to the right. That is,

$$\frac{d}{dt}f \quad \text{means the same thing as} \quad \frac{df}{dt}.$$

Whereas the fraction that's used in Leibnitz notation reminds us of the difference quotient, $\Delta f/\Delta t$, the notations of Newton and Lagrange remind us that the slope of the tangent line changes from point to point, from moment to moment. When

---

Sidebar notes:

The word *differentiate* is pronounced "diff-er-EN-chee-ate."

Many students talk about *deriving* f when what they mean is *differentiating* f. In common usage, one thing is derived *from* another. For example, we say that "strong acids are derived from the combustion of fossil fuels." It would sound strange to say "we derived fossil fuels to get strong acids." In the same way, it is not correct to say "I derived f to get f'." Instead, we say either "I derived f' from f," or "I differentiated f to get f'."

We have seen examples that lead us to believe the Power Rule, but that's a far cry from an actual proof. We haven't even seen power functions with mildly exotic exponents, such as $1/5$, or more difficult examples in which the exponent is irrational.

The name *Lagrange* is pronounced "lah-GRON-zh," where the symbol zh is pronounced like the "g" in *mirage*.

The name *Leibniz* is pronounced "LIBE-nitz."

Leibniz is credited with discovering the calculus at the same time as Newton.

Newton's notation is commonly used in science and engineering when the independent variable is *time*.

we need to communicate location dependence using Leibniz notation, we employ a vertical bar that is understood to mean "evaluate at ..." So

$$\left.\frac{df}{dt}\right|_{t_0} \quad \text{means the same thing as} \quad f'(t_0).$$

❖ The Sum and Scaling Rules

When used together, the next pair of differentiation facts tell us how to differentiate polynomials, which are built by scaling and adding power functions. In later sections, we'll use them to differentiate sums of scaled functions of any kind (exponential, trigonometric, logarithmic functions, etc.)

> **Sum Rule:** Suppose f and g are differentiable. Then their sum is also differentiable, and its derivative is $f'(t) + g'(t)$.

Written in Leibniz notation, the Sum Rule is

$$\frac{d}{dt}(f+g) = \frac{df}{dt} + \frac{dg}{dt}$$

Proof. Let's begin by writing the sum of our functions as $h(t) = f(t) + g(t)$. Then

$$h'(t) = \lim_{\Delta t \to 0} \frac{h(t+\Delta t) - h(t)}{\Delta t} = \lim_{\Delta t \to 0} \frac{f(t+\Delta t) + g(t+\Delta t) - \left(f(t) + g(t)\right)}{\Delta t}.$$

Gathering like terms, this becomes

$$h'(t) = \lim_{\Delta t \to 0} \frac{f(t+\Delta t) - f(t) + g(t+\Delta t) - g(t)}{\Delta t} = \lim_{\Delta t \to 0} \left(\frac{f(t+\Delta t) - f(t)}{\Delta t} + \frac{g(t+\Delta t) - g(t)}{\Delta t} \right).$$

Because the limit of a sum is the sum of the limits, it follows that

$$h'(t) = \lim_{\Delta t \to 0} \frac{f(t+\Delta t) - f(t)}{\Delta t} + \lim_{\Delta t \to 0} \frac{g(t+\Delta t) - g(t)}{\Delta t} = f'(t) + g'(t). \qquad \blacksquare$$

Using the Sum Rule

Example 2.7. Determine a formula for $f'(t)$ when $f(t) = t^3 + t^2$.

Solution: The Sum Rule tells us that we can differentiate a sum of functions one at a time and then add the results together. So

$$\frac{df}{dt} = \frac{d}{dt}\left(t^3 + t^2\right) = \frac{d}{dt}\left(t^3\right) + \frac{d}{dt}\left(t^2\right) = 3t^2 + 2t. \qquad \blacksquare$$

> **Scaling Rule:** Suppose that f is differentiable at t_0 and $\beta \in \mathbb{R}$. Then $(\beta f)(t)$ is also differentiable at t_0, and its derivative is $\beta f'(t_0)$.

Written in Leibniz notation, the Scaling Rule is

$$\frac{d}{dt}(\beta f) = \beta \frac{df}{dt}$$

Note: You'll prove this in the exercise set by factoring the constant β out of the numerator of the difference quotient and using the fact that the limit of a product is the product of the limits.

Finding the tangent line to the graph of a polynomial

Example 2.8. Suppose $f(t) = 9t^2 + 3t^7$. Determine the equation of the tangent line to the graph of f at $t_0 = -1$.

Solution: In order to write the equation of a line, we need a point and a slope. Since $f(-1) = 6$, the point on the graph is $(-1, 6)$. The slope of the line is $f'(-1)$, so we need to know the derivative of f. The Sum Rule tells us that we can differentiate a sum term by term, so

$$f'(t) = \frac{d}{dt}(9t^2) + \frac{d}{dt}(3t^7)$$

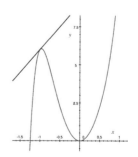

Figure 2.1: The graph of $f(x)$ and the tangent line in Example 2.8

And the Scaling Rule tells us that we can pass the derivative over the scale factors of 9 and 3, so

$$f'(t) = 9\left(\frac{d}{dt}\ t^2\right) + 3\left(\frac{d}{dt}\ t^7\right) = 9(2t^1) + 3(7t^6) = 18t + 21t^6.$$

Therefore, $f'(-1) = 18(-1) + 21(-1)^6 = 3$, and the point-slope equation of the line is $y - 6 = 3(t + 1)$, which we rewrite as $y = 6 + 3(t + 1)$. The graph of f and the tangent line at $t_0 = -1$ are shown in Figure 2.1.   ∎

Try It Yourself: Finding the tangent line to the graph of a polynomial

Example 2.9. Suppose $f(t) = 4t^5 - 2t^3$. Determine the equation of the tangent line to the graph of f at $t_0 = 3$.

Answer: $y = 918 + 1566(t - 3)$ ∎

Full Solution
On-line

> **You should know**
>
> - the terms *differentiate, Leibniz notation, Lagrange notation,* and *Newton notation;*
>
> - the Power Rule, Sum Rule, and Scaling Rule.

> **You should be able to**
>
> - find $f'(t)$ when $f(t)$ is a sum of scaled power functions (e.g. a polynomial).

❖ 3.2 Skill Exercises

In each of #1–12, use the limit of the difference quotient to determine a formula for $f'(x)$. In #10–12, $\alpha, \beta, \gamma \in \mathbb{R}$ are nonzero numbers.

1. $f(x) = 3x^2 - 8x - 2$
2. $f(x) = -7x^2 + 3x + 1$
3. $f(x) = x^3 + 1$
4. $f(x) = \frac{1}{-2x+7}$

5. $f(x) = \frac{1}{3x+1}$
6. $f(x) = \frac{3}{8x-4}$
7. $f(x) = \sqrt{3x + 1}$
8. $f(x) = \sqrt{-2x + 7}$

9. $f(x) = 1/\sqrt{x}$
10. $f(x) = \alpha x^2 + \beta x + \gamma$
11. $f(x) = \frac{1}{\beta x + \gamma}$
12. $f(x) = \sqrt{\beta x + \gamma}$

Recall that $(f + g)(x)$ means $f(x) + g(x)$. Use the graphs of f and g shown below to estimate the derivatives in each of #13–16.

13. $(f + g)'(1)$ 14. $(f - g)'(1.2)$ 15. $(2f + 3g)'(1.4)$ 16. $(6f - 5g)'(1.6)$

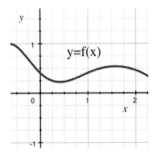

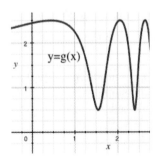

In each of #17–24, use the Power Rule, Sum Rule, and Scaling Rule to find $f'(x)$.

17. $f(x) = 8x + 6x^7 - x^{-3}$

18. $f(x) = 2x^{34.1} + 8x^{-5}$

19. $f(x) = 9 + 5.3x - 6x^4$

20. $f(x) = 8\,x^{7/4} - \frac{1}{2}x^{6/5}$

21. $f(x) = \sqrt{x} + \sqrt[3]{x}$

22. $f(x) = x^{1.4} - \frac{8}{\sqrt{x}}$

23. $f(x) = 3x^\pi + 9x$

24. $f(x) = 2 + x^2\sqrt{x}$

In each of #25–30 determine the specified number.

25. $\left.\frac{df}{dz}\right|_{z=27}$ when $f(z) = z^{5/3}$

26. $\left.\frac{dg}{dc}\right|_{c=9}$ when $g(c) = c^{5/2}$

27. $\left.\frac{dh}{dv}\right|_{v=16}$ when $h(v) = v^{-7/4}$

28. $\left.\frac{dk}{ds}\right|_{s=32}$ when $k(s) = z^{-2/5}$

29. $\dot{g}(3)$ when $g(r) = 4r + 2r^{-3}$

30. $\dot{z}(w)$ when $z(64) = \sqrt[6]{w} + \frac{1}{\sqrt[6]{w}}$

In each of #31–36 you should verify that the given formula for $f'(t)$ is correct by expanding the $f(t)$ into a polynomial and using the Power Rule, Sum Rule, and Scaling Rule to calculate its derivative. Expand the given formula for $f'(t)$ as needed for comparison.

31. $f(t) - (t-6)^2$, $f'(t) = 2(t-6)^1$

32. $f(t) = (5t-6)^2$, $f'(t) = (2)(5)(5t-6)^1$

33. $f(t) = (t-1)^3$, $f'(t) = 2(t-1)^2$

34. $f(t) = (5t-2)^3$, $f'(t) = (3)(5)(5t-1)^2$

35. $f(t) = (t+2)^4$, $f'(t) = 4(t+2)^3$

36. $f(t) = (3t+1)^4$, $f'(t) = (4)(3)(3t+1)^3$

❖ 3.2 Concept and Application Exercises

37. Suppose $f(x) = x^2 + 1$ and $g(x) = x^2 - 5$.

    (a) Sketch the graphs of these two functions on the same axes.

    (b) How do the constant terms ($+1$ and -5) affect the graphs?

    (c) Draw the tangent line to the graph of f at the point $(1, 2)$, and the tangent line to the graph of g at the point $(1, -4)$.

    (d) What's the same about these two lines?

    (e) Based on parts (a)–(d) explain why, intuitively speaking, the constant term of a polynomial *should* vanish when we calculate its derivative.

38. Suppose $f(x) = x^{2/5}$ and $g(x) = x^{8/5}$.

    (a) Determine the equation of the tangent line to the graph of f at $x = 1$.

    (b) Graph f and its tangent line.

    (c) Repeat parts (a) and (b) for $g(x)$ at $x = 1$.

    (d) Suppose $h(x) = x^p$. Based on parts (a)–(c), how can we tell whether the tangent line to the graph of h at $x = 1$ will be above or below the graph of h? Does it matter whether $x = 1$, or could it be some other number? Does the sign of x matter?

39. Suppose $f(x) = x^3 - 3x + 4$ and $g(x) = 3x^2 - 3x$.

(a) Determine where both $f'(x) = g'(x)$ and $f(x) = g(x)$. (i.e., the graphs share a common point, and have the same tangent line there).

(b) Graph both curves and the common tangent line.

For each of the polynomials in #40–43, you should (a) graph the polynomial, then (b) multiply all of the factors together and calculate $f'(t_0)$ for the resulting cubic polynomial, (c) graph the tangent line to the graph at t_0, and lastly (d) compare $f'(t_0)$ to $\lim_{t \to t_0}$ (red factors).

40. $f(t) = (t-3)(t-4)(t-7)$, $t_0 = 7$ 42. $f(t) = (t-1)(t-5)(t-3)$, $t_0 = 3$

41. $f(t) = (t-3)(t-7)(t-4)$, $t_0 = 4$ 43. $f(t) = (t-5)(t-3)(t-1)$, $t_0 = 1$

44. Consider the following alleged method for constructing the tangent line to the graph of $f(x) = x^2$ at $x = b$, when $b \neq 0$: *locate the point $(b/2, 0)$, and draw the straight line connecting it to (b, b^2).* Determine algebraically (i.e., not from a diagram) whether this construction works. If so, explain why it works. If not, find an example to show where it fails.

| Geometry |

45. Suppose $f(x) = x^2 + 6x + 2$.

| Geometry |

(a) Pick two distinct numbers, x_1 and x_2, and find the equation of the tangent line to the graph of f at each.

(b) Determine where lines from part (a) intersect.

(c) What is the relationship between the points you chose in part (a) and the x-coordinate of the intersection that you found in part (b)?

(d) Try this again with another quadratic function, and see if you reach the same result.

46. Suppose the position of an object is described by $y(t)$. Use the derivative to write an equation that says, "as t grows indefinitely, the object's velocity tends to zero."

| Physics |

47. Suppose we rest a circle of radius R into the cup of the parabola $y = ax^2$, where $a > 0$, as seen in the figure below.

| Geometry |

(a) Suppose the circle comes to rest at the points $(\pm t, at^2)$. The center of the circle will be at $(0, b)$. Find a formula for b in terms of a and t.

(b) Now that you know the location of the center of the circle, and of a point on its edge, find a formula for R in terms of a and t.

(c) Solve your formula in part (b) for t, expressed in terms of R and a.

(d) Thinking about the number a as fixed for the moment, you'll notice that there are values of R which are inadmissible according to your formula for t. Why? What does this mean geometrically?

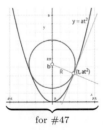

for #47

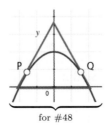

for #48

48. Find points P and Q on the parabola $y = 1 - x^2$ so that the triangle formed by the x-axis and the tangent lines to the parabola at P and Q is equilateral (see figure above).

49. In 1643, the Italian scientist Evangelista Torricelli discovered that fluid leaking from a tank exits at a speed of $v = \sqrt{19.62h}\,\frac{\text{m}}{\text{sec}}$, where h is the distance (in meters) between the center of the hole and the top of the fluid.

    Torricelli's Law

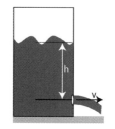

   (a) The area of a hole is πr^2 when its radius is r. Combine this with v to get a *volumetric flow* (i.e., a flow in m^3/sec).

   (b) When the tank is cylindrical, and has a radius of R, the volume of fluid above the hole is approximately $V = \pi R^2 h$. Use this fact to rewrite the volumetric flow from part (a) in terms of V.

   (c) What are the units of dV/dt?

   (d) Write an equation that describes the rate at which the volume of fluid in the tank is changing (your answer will involve both V' and V).

50. Suppose you hold a laser pointer 4 inches above a reflective surface, pointed downward at an angle of $\theta = \pi/6$ radians, as shown in Figure 2.2.

    Geometry

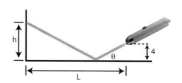

Figure 2.2: For #50.

   (a) Use the fact that the angle of incidence is the angle of reflection to find a formula for the height of the beam on the vertical wall, h, in terms of the distance you hold it from the vertical surface, L.

   (b) Determine a formula for dh/dL.

   (c) Determine the units of dh/dL and explain what it means to us.

51. On p. 16 you saw the *virial expansion* of the Ideal Gas Law, $P = RT\rho + RTB\rho^2$, where R is the *Ideal Gas Constant*, ρ is density in mols per cubic meter, and B is a constant that depends on the gas. Thinking of T as fixed, (a) find a formula for $P'(\rho)$ and determine its units, and (b) explain what $P'(0.35)$ means to us, physically.

    Physical Chemistry

52. The force (in newtons) experienced by the space shuttle due to the Earth's gravity is $F_g(r) = \frac{4.16 \times 10^{19}}{r^2}$, where r is the distance (in meters) between the shuttle and the center of the earth.

    Aerospace

   (a) What is the average rate of change of the gravitational force as the shuttle descends from $r = 6.4 \times 10^6$ to $r = 6.396 \times 10^6$? (Include units.)

   (b) When the shuttle is descending and $r = 6.4 \times 10^6$, what is the instantaneous rate of change of the force due to gravity? (Include units.)

   (c) Use complete sentences to compare and contrast the meaning of these numbers in this context.

53. In a particular electric circuit, the current (measured in amps) is given by $I(R) = 80/R$ where R is the resistance (measured in ohms).

    Electrical Engineering

   (a) If the resistance on the circuit increases from 2 ohms to 10 ohms, find the average rate of change of the current with respect to the resistance.

   (b) Calculate $I'(R)$ and explain what it means to us.

In each of #54–57, in which the position of an object is described by $y(t)$, you should (a) determine a formula for the velocity of the object, (b) determine the time interval(s) during which the velocity is negative, (c) sketch a graph of $y(t)$, and shade the region of your graph in which the the velocity is negative. *(Hint: remember that a continuous function can only change sign at a root.)*

54. $y(t) = t^2 - 3t + 1$

55. $y(t) = 4t^2 + 3t - 11$

56. $y(t) = -t^2 + 5t + 8$

57. $y(t) = -3t^2 + 9t + 11$

In each of #58–61, the position of an object is described by $y(t)$. (a) Determine times t_1 and t_2 when the velocity is zero, and (b) use a graphing utility to graph $y(t)$ and the vertical lines $t = t_1$ and $t = t_2$.

58. $y(t) = 2t^3 + 3t^2 - 72t$

59. $y(t) = 3t^3 - 3t^2 + 4$

60. $y(t) = -5t^3 + 7t^2 + t$

61. $y(t) = -t^3 - 3t^2 + 9t$

62. Kinetic energy depends on mass and velocity according to $K = 0.5mv^2$, where v is the object's speed and m is its mass. Show that the derivative of kinetic energy with respect to velocity is momentum.

          Physics

63. Vector graphics programs are often asked to produce a smooth curve that passes through user-defined points at specified slopes. For historical reasons, the points are called *knots*, and the curve is a called a *spline*. Suppose a user designates two points, P and Q, where Q is 100 points to the right of and 50 points below P. Thinking of P as the origin of a Cartesian plane, find numbers a, b, c, and d so that the graph of $f(x) = ax^3 + bx^2 + cx + d$ has a slope of 0.4 at P and passes through Q with a slope of -2.

          Vector Graphics

64. How does the answer from #63 change if the slope at P is m_1 and the slope at Q is m_2? (That is, develop a more general formula for the spline.)

          Vector Graphics

65. Koopman's formula says that when you're looking for an object, the probability of sighting it when it's s meters away is $p(s) = \beta s^{-3}$, where $\beta > 0$ is a constant. (a) Determine a formula for $p'(s)$, (b) if we take the units of p to be "likelihood", determine the units associated with $p'(s)$.

          Search and Sighting

66. Kepler's Third Law says that the duration of a planet's year, T ($T = 1$ for the Earth), is related to the semimajor axis of its orbit, a (measured in astronomical units, $a = 1$ for the Earth), according to $T = a^{3/2}$. (a) Determine a formula for $T'(a)$. (b) Determine the units of $T'(a)$.

          Astronomy

67. Suppose that $A(r)$ and $C(r)$ are the area and circumference of a circle whose radius is r. Show that $A'(r) = C(r)$.

          Geometry

68. In the apothem project at the end of Chapter 2 (p. 144) you developed a formula for the area and perimeter of a regular n-gon whose apothem is r, $A(r)$ and $P(r)$ respectively. You also developed formulas for the area and perimeter of a regular n-gon whose circumradius is $\tilde{r}$, $\mathcal{A}(\tilde{r})$ and $\mathcal{P}(\tilde{r})$.

          Geometry

   (a) Show that $A'(r) = P(r)$.

   (b) Show that $\mathcal{A}'(\tilde{r}) \neq \mathcal{P}(\tilde{r})$.

Checking the details

69. Suppose that f is differentiable at t_0. Use the limit of the difference quotient to show that, for any number β, the function $\beta f(t)$ is also differentiable at t_0 and its derivative is $\beta f'(t_0)$.

3.3   Derivatives of Trigonometric and Exponential Functions

 A guitar makes sound when its strings vibrate. We can describe this motion mathematically by saying that a point on the string moves up and down according to

$$f(t) = A\sin(\omega t),$$

where the amplitude A depends on how hard the string was plucked, and ω is a constant that depends on physical qualities such as the length, tension, and mass of the string. Since the derivative tells us the instantaneous rate of change, we can determine the velocity of this point on the vibrating string by calculating $f'(t)$.

Differentiating a sine function

Example 3.1. Suppose the position of a point after t seconds is $f(t) = \sin(7t)$, measured in millimeters. Derive a formula for calculating the instantaneous velocity of the point.

Solution: The instantaneous velocity of the point is $f'(t)$, which we calculate as the limit of difference quotients. The difference quotient for f is

$$\frac{\Delta f}{\Delta t} = \frac{\sin(7(t + \Delta t)) - \sin(7t)}{\Delta t} = \frac{\sin(7t + 7\Delta t) - \sin(7t)}{\Delta t}.$$

In order to calculate the limit as $\Delta t \to 0$, we're going to do three things:

1. Use a trigonometric identity to rewrite $\sin(7t + 7\Delta t)$.

2. Gather like terms.

3. Use limit facts we learned in Chapter 2.

Step 1: We can rewrite $\sin(7t + 7\Delta t)$ using the addition formula for the sine:

$$\sin(a + b) = \sin(a)\cos(b) + \sin(b)\cos(a)$$
$$\sin(7t + 7\Delta t) = \sin(7t)\cos(7\Delta t) + \sin(7\Delta t)\cos(7t)$$

so that

$$\frac{\Delta f}{\Delta t} = \frac{\sin(7t)\cos(7\Delta t) + \sin(7\Delta t)\cos(7t) - \sin(7t)}{\Delta t}.$$

Step 2: Now we gather like terms. In this case, "like terms" means those terms that have $\sin(7t)$, which are written in green below.

$$\frac{\Delta f}{\Delta t} = \frac{\sin(7t)\cos(7\Delta t) + \sin(7\Delta t)\cos(7t) - \sin(7t)}{\Delta t}$$

$$= \frac{\sin(7t)\cos(7\Delta t) - \sin(7t)}{\Delta t} + \frac{\sin(7\Delta t)\cos(7t)}{\Delta t}.$$

After factoring a $\sin(7t)$ out of the first term, this becomes

$$\frac{\Delta f}{\Delta t} = \left(\frac{\cos(7\Delta t) - 1}{\Delta t}\right)\sin(7t) + \left(\frac{\sin(7\Delta t)}{\Delta t}\right)\cos(7t).$$

Step 3: In Chapter 2 we established that

$$\lim_{\Delta t \to 0} \left(\frac{\sin(7\Delta t)}{\Delta t} \right) = 7 \qquad \text{(see Example 2.11, p. 102), and}$$

$$\lim_{\Delta t \to 0} \left(\frac{\cos(7\Delta t) - 1}{\Delta t} \right) = 0 \qquad \text{(see Example 2.14, p. 103).}$$

So the limit of the difference quotient is

$$f'(t) = \lim_{\Delta t \to 0} \frac{\Delta f}{\Delta t} = \left(0\right) \sin(7t) + \left(7\right) \cos(7t).$$

Written in Leibniz notation,

$$\frac{d}{dt} \sin(7t) = 7 \cos(7t). \qquad\qquad \blacksquare$$

Try It Yourself: Differentiating a sine function

Example 3.2. Suppose the position of a point after t seconds is $f(t) = \sin(-3t)$, measured in millimeters. Derive a formula for calculating the instantaneous velocity of the point.

Answer: $f'(t) = -3\cos(-3t) \; \frac{\text{mm}}{\text{sec}}$ $\blacksquare$

Full Solution
On-line

Based on the results of Example 3.1 and Example 3.2, you might make the following conjecture (which is correct):

$$\frac{d}{dt} \sin(\omega t) = \omega \cos(\omega t) \quad \text{for each } \omega \in \mathbb{R}. \tag{3.1}$$

> When $\omega = 1$, this tells us that the derivative of the sine is the cosine.

Tangent line to the graph of a sine function

Example 3.3. Suppose $f(t) = 5\sin(2t)$. Determine the equation of the tangent line to the graph of f at $t_0 = \pi/12$.

Solution: To write the equation of a line we need a point and slope. Since $f(\pi/12) = 5/2$, the point is $(\pi/12, 5/2)$. The slope of the line will be $f'(\pi/12)$, so we next determine a formula for $f'(t)$. The Scaling Rule (p. 163) tells us that we can skip over the coefficient of 5 and differentiate $\sin(2t)$, and $\omega = 2$ in this example, so equation (3.1) tells us that

$$f'(t) = 5 \left(\frac{d}{dt} \sin(2t) \right) = 5 \left(2\cos(2t) \right) = 10\cos(2t).$$

This formula allows us to determine that $f'(\pi/12) = 5\sqrt{3}$. Therefore, the equation describing the tangent line is $y - \frac{5}{2} = 5\sqrt{3}\left(t - \frac{\pi}{12}\right)$, which we rewrite as $y = \frac{5}{2} + 5\sqrt{3}\left(t - \frac{\pi}{12}\right)$. $\blacksquare$

In the exercise set, you'll use the addition formula for the cosine to show that

$$\frac{d}{dt} \cos(\omega t) = -\omega \sin(\omega t) \quad \text{for each } \omega \in \mathbb{R}. \tag{3.2}$$

> When $\omega = 1$, this tells us that the derivative of the cosine is the opposite of the sine.

WARNING: You see a negative sign appear in the derivative of the cosine, but *not* in the derivative of the sine.

Differentiating a cosine function

Example 3.4. Suppose the position of a point after t seconds is $f(t) = \cos(17t)$, measured in millimeters. Determine the point's velocity when $t = \pi/2$.

Solution: The instantaneous velocity of the point is $f'(t)$. Using (3.2) with $\omega = 17$, we see that $f'(t) = -17\sin(17t)$, so $f'(\pi/2) = -17\,\frac{\text{mm}}{\text{sec}}$. ∎

Try It Yourself: Tangent line to the graph of a sine function

Example 3.5. Suppose $f(t) = 10\cos(5t)$. Determine the equation of the tangent line to the graph of f at $t_0 = \pi/2$.

Answer: $y = 0 - 50\left(t - \frac{\pi}{2}\right)$. ∎

Full Solution
On-line

❖ Derivatives of Exponential Functions

Exponential functions play a central role in our understanding of the natural world, from population growth to thermodynamics to blood transfusion, so understanding their rates of change is very important. For example, the atmospheric pressure (measured in pascals) is described by

$$P(z) = 101325\left(\frac{1}{2}\right)^{z/5.6},$$

where z is the altitude above sea level (measured in kilometers). Its derivative tells us the rate at which pressure changes with altitude. We'll calculate $P'(z)$ in Example 3.14, after we've worked with some simpler exponential functions. Let's begin by calculating the derivative of the most important one.

Deriving a formula for $f'(t)$ when $f(t) = e^t$

Example 3.6. Derive a formula for $f'(t)$ when $f(t) = e^t$.

Solution: Since $e^{a+b} = e^a e^b$, we can write the difference quotient as

$$\frac{\Delta f}{\Delta t} = \frac{f(t + \Delta t) - f(t)}{\Delta t} = \frac{e^{t+\Delta t} - e^t}{\Delta t} = \frac{e^t e^{\Delta t} - e^t}{\Delta t} = e^t\left(\frac{e^{\Delta t} - 1}{\Delta t}\right).$$

In the limit as $\Delta t \to 0$, this becomes

$$f'(t) = \lim_{\Delta t \to 0}\frac{\Delta f}{\Delta t} = \lim_{\Delta t \to 0} e^t\left(\frac{e^{\Delta t} - 1}{\Delta t}\right) = e^t\underbrace{\left(\lim_{\Delta t \to 0}\frac{e^{\Delta t} - 1}{\Delta t}\right)}_{\text{Since } e^t \text{ does not change as } \Delta t \to 0}. \qquad (3.3)$$

The remaining limit was discussed in Example 1.9 (p. 152), where we said that its value is 1. So equation (3.3) is really

$$f'(t) = e^t.$$

That is, the derivative of the Euler exponential function is itself! ∎

$$\frac{d}{dt}e^t = e^t.$$

> Written in Lagrange notation, the Euler exponential function is described by
>
> $$f'(t) = f(t).$$
>
> The zero function, $f(x) \equiv 0$, is also its own derivative; but if $f(t)$ is the zero function, you have … in a certain sense … nothing, so we call it the *trivial function*. Of *all* the other functions, only e^t and scaled versions of it are described by the equation $f' = f$.

Deriving a formula for $f'(t)$ when $f(t) = e^{3t}$

Example 3.7. Derive a formula for $f'(t)$ when $f(t) = e^{3t}$. Then find an equation for the line that's tangent to the graph of f at $(0, 1)$.

Solution: The difference quotient is

$$\frac{f(t+\Delta t)-f(t)}{\Delta t}=\frac{e^{3(t+\Delta t)}-e^{3t}}{\Delta t}=\frac{e^{3t}e^{3\Delta t}-e^{3t}}{\Delta t}=e^{3t}\left(\frac{e^{3\Delta t}-1}{\Delta t}\right).$$

In the limit as $\Delta t \to 0$, this becomes

$$f'(t)=\lim_{\Delta t\to 0}\frac{\Delta f}{\Delta t}=\lim_{\Delta t\to 0}e^{3t}\left(\frac{e^{3\Delta t}-1}{\Delta t}\right)=e^{3t}\left(\lim_{\Delta t\to 0}\frac{e^{3\Delta t}-1}{\Delta t}\right). \qquad (3.4)$$

Notice that unlike Example 3.6, the denominator of the fraction (Δt) and the exponent in its numerator ($3\Delta t$) are not the same. However,

> Multiplying by 1, written in a suitably helpful form, is an extremely important technique.

$$\lim_{\Delta t\to 0}\frac{e^{3\Delta t}-1}{\Delta t}=\lim_{\Delta t\to 0}3\left(\frac{e^{3\Delta t}-1}{3\Delta t}\right)=3\left(\lim_{\Delta t\to 0}\frac{e^{3\Delta t}-1}{3\Delta t}\right).$$

Since $(3\Delta t)\to 0$ as $\Delta t \to 0$, we can rewrite this as

$$3\lim_{\Delta t\to 0}\frac{e^{3\Delta t}-1}{3\Delta t}=3\left(\lim_{3\Delta t\to 0}\frac{e^{3\Delta t}-1}{3\Delta t}\right)=3(1)=3.$$

This limit has the same *form* as the limit in equation (3.3)—the denominator and the exponent are the same, and go to zero—so its value is 1.

> The equation
> $$f'(t)=3f(t)$$
> describes $f(t)=e^{3t}$ (and scaled versions of it).

Substituting this limit value into equation (3.4), we get $f'(t)=3e^{3t}$. So the slope of the tangent line to the graph of f at $t=0$ is $f'(0)=3$. Therefore, the equation of the tangent line is

$$y=f(0)+f'(0)(t-0)=1+3t. \qquad \blacksquare$$

Try It Yourself: Use the definition to differentiate $f(t)=e^{-4t}$

Example 3.8. Use the definition to differentiate $f(t)=e^{-4t}$. Then find an equation for the tangent line to the graph of f at $(0,1)$.

Answer: $f'(t)=-4e^{-4t}$, and the tangent line is $y=1-4(t-0)$. $\qquad \blacksquare$

Full Solution On-line

Based on the results of Example 3.7 and Example 3.8, you might make the following conjecture (which is correct):

$$\frac{d}{dt}e^{\omega t}=\omega e^{\omega t} \quad \text{for each } \omega \in \mathbb{R}. \qquad (3.5)$$

> The equation
> $$f'(t)=\omega f(t)$$
> describes $f(t)=e^{\omega t}$ (and scaled versions of it).

Differentiating exponential functions with other bases

Example 3.9. Determine a formula for $f'(t)$ when $f(t)=4^t$.

Solution: Our strategy is to rewrite the formula for $f(t)$ so that it looks like $e^{\omega t}$, after which we can determine a formula for $f'(t)$ using equation (3.5). Specifically, we know that $u=e^{\ln(u)}$ whenever $u>0$. Our "u" will be 4^t. That is,

> The restriction that $u>0$ keeps u in the domain of the logarithm.

$$f(t)=4^t=\underbrace{e^{\ln(4^t)}=e^{t\ln(4)}}=e^{\omega t} \quad \text{where } \omega=\ln(4).$$

Using properties of logarithms, see p. 61

Now equation (3.5) tells us that $f'(t)=\omega e^{\omega t}=\ln(4)\,4^t$. $\qquad \blacksquare$

Try It Yourself: Differentiating exponential functions with other bases

Example 3.10. Determine a formula for $f'(t)$ when $f(t) = 7^t$.

Answer: $f'(t) = \ln(7)\, 7^t$ ∎

Full Solution
On-line

Based on the results of Example 3.9 and Example 3.10, you might make the conjecture (which is correct) that

$$\frac{d}{dt}b^t = \ln(b)\, b^t \quad \text{for each } b > 0. \tag{3.6}$$

Differentiating exponential functions with other bases

Example 3.11. Suppose $f(t) = 7 + 2(5^t)$. Determine the equation of the tangent line at $t = 3$.

Solution: We know that $f(3) = 257$, so the tangent line passes through the point $(3, 257)$. It's slope is $f'(3)$, so we need a formula for $f'(t)$. Using the Sum and Scaling Rules (see p. 163), we see that

$$f'(t) = \frac{d}{dt}(7) + \frac{d}{dt}\left(2(5^t)\right) = 0 + 2\frac{d}{dt}5^t.$$

> Recall that the derivative of a constant is zero (see Example 1.8 on p. 151)

In this case, the base of the exponential function is $b = 5$, so equation (3.6) allows us to finish the calculation by writing

$$f'(t) = 2\frac{d}{dt}5^t = 2\left(\ln(5)\, 5^t\right) = 2\ln(5)\, 5^t.$$

Now we can determine $f'(3) = 250\ln(5)$, so the equation of the tangent line is $y - 257 = 250\ln(5)(t-3)$, which we rewrite as $y = 257 + 250\ln(5)(t-3)$. ∎

The following examples address the situation, like that of atmospheric pressure on p. 171, in which the exponent is scaled by a coefficient.

Differentiate an exponential function whose variable is scaled

Example 3.12. Derive a formula for $f'(x)$ when $f(x) = 6^{1.7x}$.

Solution: Since $6^{1.7x} = (6^{1.7})^x$, we can treat this as an exponential function whose base is $6^{1.7}$. Then from equation (3.6) it follows that

$$f'(x) = \ln(6^{1.7})\,(6^{1.7})^x = 1.7\ln(6)\, 6^{1.7x}.$$ ∎

Try It Yourself: Differentiate an exponential function whose variable is scaled

Example 3.13. Derive a formula for $f'(z)$ when $f(z) = 3^{z/9.2}$.

Answer: $f'(z) = \frac{1}{9.2}\ln(3)\, 3^{z/9.2}$. ∎

Full Solution
On-line

Based on the results of Example 3.12 and Example 3.13, you might make the following conjecture (which is correct):

$$\frac{d}{dt}b^{\omega t} = \omega\ln(b)\, b^{\omega t} \quad \text{for any } b > 0 \text{ and any number } \omega. \tag{3.7}$$

Differentiating exponential functions with other bases (Atmospheric Pressure)

Example 3.14. When z is the altitude above sea level (in kilometers), the pressure in the Earth's troposphere (in pascals, Pa) is approximated by

$$P(z) = 101325 \left(\frac{1}{2}\right)^{z/5.6}.$$

(a) Determine a formula for $P'(z)$, and (b) determine its units.

Solution: Using $b = \frac{1}{2}$ and $\omega = \frac{1}{5.6}$, we know from equation (3.7) that

$$\frac{d}{dz}\left(\frac{1}{2}\right)^{z/5.6} = \frac{1}{5.6}\ln\left(\frac{1}{2}\right)\left(\frac{1}{2}\right)^{z/5.6}.$$

Therefore, the Scaling Rule (p. 163) tells us that $P'(z) = \frac{101325\ln(0.5)}{5.6}\left(\frac{1}{2}\right)^{z/5.6}$.

(b) Though we didn't explicitly use the difference quotient, we know that

$$P'(z) = \lim_{\Delta z \to 0} \frac{P(z + \Delta z) - P(z)}{\Delta z} \quad \begin{array}{l} \leftarrow\text{Measured in Pa} \\ \leftarrow\text{Measured in km} \end{array},$$

so $P'(z)$ has units of pascals per kilometer. To understand what this means physically, consider $P'(0) = -12541.6318$. That is,

$$\lim_{\Delta z \to 0} \frac{\Delta P}{\Delta z} = -12541.6318, \text{ from which we conclude that } \frac{\Delta P}{\Delta z} \approx -12541.6318 \text{ when}$$

Δz is very near zero. Said differently, $\Delta P \approx -12541.6318\Delta z$, so $P'(0)$ tells us approximately how much the pressure will change for each kilometer (or fraction thereof) that we rise above sea level. ∎

You should be able to

- determine $f'(t)$ when $f(t)$ is either $\sin(\omega t)$, $\cos(\omega t)$, $e^{\omega t}$, or $b^{\omega t}$, for any $\omega \in \mathbb{R}$ and any $b > 0$.

✧ 3.3 Skill Exercises

For each of #1-12, determine the equation of the tangent line to the graph of f at the specified value of x_0.

1. $f(x) = e^{3x}$, $x_0 = 0$
2. $f(x) = e^{-3x}$, $x_0 = 0$
3. $f(x) = 2^x$, $x_0 = 4$
4. $f(x) = \left(\frac{1}{2}\right)^x$, $x_0 = 4$
5. $f(x) = 7\cos(x)$, $x_0 = \pi/3$
6. $f(x) = 3\sin(x)$, $x_0 = \pi/4$

7. $f(x) = \cos(2x)$, $x_0 = \pi/3$
8. $f(x) = \sin(5x)$, $x_0 = \pi/4$
9. $f(x) = 7\cos(2x)$, $x_0 = \pi/3$
10. $f(x) = 3\sin(5x)$, $x_0 = \pi/4$
11. $f(x) = \beta\cos(\omega x)$, $x_0 = t$
12. $f(x) = \beta\sin(\omega x)$, $x_0 = t$

13. Suppose $f(x) = e^{-x}$. Find $f'(x)$ by (a) using equation (3.5, p. 172) with $\omega = -1$, and (b) using equation (3.6, p. 173) with $b = 1/e$.

14. The hyperbolic trigonometric functions, $\cosh(t) = 0.5(e^t + e^{-t})$ and $\sinh(t) = 0.5(e^t - e^{-t})$, were introduced on p. 75.

(a) By substituting the formulas for $\cosh(t)$ and $\sinh(t)$, show that

$$(\cosh(t))^2 - (\sinh(t))^2 = 1.$$

(b) Show $\frac{d}{dt}\sinh(t) = \cosh(t)$.

(c) Show $\frac{d}{dt}\cosh(t) = \sinh(t)$.

Notice that the sign doesn't change, as it does with the circular functions, sine and cosine!

15. Show that $\frac{d}{dt}\sinh(\omega t) = \omega\cosh(\omega t)$

16. Show that $\frac{d}{dt}\cosh(\omega t) = \omega\sinh(\omega t)$.

17. At what points does the graph of $\cosh(t)$ have a slope of 1? *(Hint: To solve the relevant equation, multiply by e^t. The result will be quadratic in e^t.)*

18. At what point does the graph of $\sinh(t)$ have a slope of 1? *(Hint: To solve the relevant equation, multiply by e^t. The result will be quadratic in e^t.)*

The equation $y' = y$ says that the derivative of the function y is just itself. On p. 171, we saw that this equation describes the function $y(x) = e^x$. In #19–22 determine a function that's described by the given equation.

19. $y' = 7y$ 20. $y' = -9y$ 21. $\frac{dy}{dx} = 3y$ 22. $\frac{dy}{dx} = -8y$

❖ 3.3 Concept and Application Exercises

23. Suppose $f(x) = \sin(x) + \cos(x)$.

    (a) Use a graphing utility to render a graph of f, and determine the value(s) of x where f increasing the fastest.

    (b) Overlay a graph of $f'(x)$ on the same screen. Describe what happens to $f'(x)$ where f is increasing most rapidly.

24. Suppose $f(x) = 2x + \sin(x)$.

    (a) Determine a formula for $f'(x)$

    (b) What is the smallest possible slope of a tangent line to the graph of f, and at what value(s) of x does it occur?

    (c) Graph $f(x)$ and $f'(x)$ on the same screen, along with the vertical line(s) at the x-coordinate(s) you found in part (b).

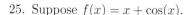

25. Suppose $f(x) = x + \cos(x)$.

    (a) Determine a formula for $f'(x)$

    (b) What is the smallest possible slope of a tangent line to the graph of f, and at what value(s) of x does it occur?

    (c) Graph $f(x)$ and $f'(x)$ on the same screen, along with the vertical line(s) at the x-coordinate(s) you found in part (b).

    (d) Based on the graph from part (c), describe the behavior of f near the x-values you found in part (b).

26. On the right triangle pictured below, the length of the hypotenuse and the length of the horizontal leg are related by $x = h\cot(\theta)$. When h remains

Geometry

constant, the number x depends only on θ. If length is measured in millimeters and angle in radians, explain the meaning of $dx/d\theta$.

for #26   for #27

27. Suppose a piston is attached to a crank shaft as in the image above. (a) Use the Law of Cosines to write the relationship between x and θ. (b) If length is measured in millimeters and angle in radians, explain what $\dfrac{dx}{d\theta}$ tells us. *(You don't need to calculate $dx/d\theta$ in order to answer part (b).)*

28. Suppose the panels of a bifold door are 16 inches wide.

(a) Determine the relationship between the distance from the fixed pin to the sliding pin, x, and the angle between the first panel and the track, θ, as shown in the schematic below.

(b) Determine a formula for $dx/d\theta$ and explain what it means to us.

(c) If you could determine a formula for $d\theta/dx$, what would it tell us?

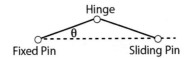

Hinge

Fixed Pin   Sliding Pin

29. The Gateway Arch in St. Louis is described by

$$y = 693.8597 - 68.7672\cosh(0.0100333x),$$

where y is the height of the arch at its midline (in feet), and x measures the distance (in feet) from the point immediately below the summit of the arch.

(a) How high is the arch at its summit?

(b) Use a graphing utility to graph both the curve described above and the parabola $y = 625.0925\left(1 - \left(\frac{x}{299.2261}\right)^2\right)$ over the interval $[-300, 300]$.

(c) Based on your graphs from part (b), speculate about why the cosh is preferred over the parabola in this context.

30. According to weather.com, the daily high temperature in Rochester, NY is modeled by $T(t) = 56 + 25\cos(\frac{\pi t}{6})$ degrees Fahrenheit, where t is measured in months, and $t = 0$ is the beginning of August.

(a) Determine a formula for $T'(t)$, including proper units.

(b) Explain what $T'(t)$ means to us.

(c) Graph $T(t)$ and its tangent line at $t = 10$.

31. According to absak.com, Anchorage, Alaska gets an average of $D(t) = 12.415 + 6.765\sin(\frac{\pi t}{6})$ hours of daylight, where t is measured in months and $t = 0$ is March 15.

(a) Determine a formula for $D'(t)$, including proper units.

(b) Explain what $D'(t)$ means to us.

(c) Graph $D(t)$ and the tangent line to its graph on June 15.

32. The lit fraction of the visible moon can be modeled by $M(t) = \frac{1}{2} + \frac{1}{2}\cos\left(\frac{2\pi}{29}t\right)$, where t is the number of days elapsed since the last new moon.

Astronomy

(a) Determine the values of $M(0)$, $M(14.5)$ and $M(29)$.

(b) Determine a formula for $M'(t)$.

(c) Determine $M'(0)$, and $M'(14.5)$ and explain what they mean to us.

33. The ocean tides are affected by various forces, the strongest of which is called the *semidiurnal main force*. The height of the tide (in feet) caused by this force is given by $H(t) = 1.99306\cos\left(\frac{2\pi}{12.421}t\right)$, where t is measured in hours and $t = 0$ represents 6 AM.

Oceanography

(a) Determine a formula for $H'(t)$ and use it to determine $H'(5)$ and $H'(6.2105)$, including proper units.

(b) Explain the meaning of $H'(t)$.

(c) Graph the function H along with its tangent lines at time $t = 5$ and $t = 6.2105$.

34. When a person has a systolic blood pressure of 100 mmHg (millimeters of mercury), a diastolic blood pressure of 70 mmHg, and a heart rate of 75 beats per minute, we can roughly model the blood pressure as $P(t) = 85 + 15\sin(150\pi t)$ where t is measured in minutes.

Medical Science

(a) Determine a formula for $P'(t)$, including proper units, and explain what it means to us.

(b) Graph the function $P(t)$, and its tangent line at $t = 1/150$

35. The function $q(t) = 5e^{-0.2t}$ describes the charge q (in coulombs) on a capacitor as time t (in seconds) passes.

Electrical Engineering

(a) On average, how is the charge changing per second between 0.5 seconds and 0.8 seconds? (Include proper units.)

(b) Determine how quickly the charge is changing with respect to time after 0.8 seconds. (Include proper units.)

(c) Use complete sentences to compare and contrast the meaning of these numbers in this context.

36. The current in a particular circuit is described by $I(t) = 2\cos\left(\frac{\pi}{6}t\right)$ amperes, where t is measured in seconds.

Electrical Engineering

(a) Find the current's average rate of change between $t = 1$ and $t = 12$ seconds.

(b) Calculate $I'\left(\frac{5\pi}{3}\right)$, including proper units, and explain what it means in this context.

37. During the first year of a regional advertisement campaign to introduce a new product, the number of people who are aware of the product after t days is $P(t) = 80,000(2^{0.01t})$. On average, how many people per day are becoming aware of the product between the 100th and 200th day of the campaign? Is the marketing campaign particularly successful?

Advertising

38. Suppose the altitude of a falling object is $y(t) = 50 - 2t + 2e^{-0.4t}$ meters after t seconds. Determine (a) a formula for its velocity, including proper units, and (b) the *terminal velocity* of the object, $\lim_{t\to\infty} y'(t)$.

Newtonian Free Fall

39. Suppose that when an object is dropped from rest its velocity is $v(t) = -2 + 2e^{-0.4t}$ $\frac{m}{sec}$ at time t seconds. Determine a formula for $v'(t)$, and based on its units, explain what $v'(t)$ tells us. | Newtonian Free Fall |

40. Suppose that $T(t) = 72 + 40e^{-0.2t}$ is the temperature of cup of hot chocolate in degrees Fahrenheit, and time is measured in seconds. | Newtonian Cooling |

    (a) Determine the initial temperature of the object.

    (b) Determine a formula for $T'(t)$, including proper units, and explain what it tells us.

    (c) Determine whether the object's temperature is changing more quickly at time $t = 0$ or $t = 10$.

    (d) What is the temperature of the room in which the cup is sitting?

41. Inductors resist changes in current. Suppose the current passing through an inductor is $c(t)$. Write an equation that says, "the rate of change in current with respect to time is proportional to the current." | Electrical Engineering |

42. On p. 48 we said that if we start with A_0 milligrams of the molecule pertechnetate, there will be $A(t) = A_0 \left(\frac{1}{2}\right)^{t/6.02}$ milligrams t hours later. Show that the ratio $A'(t)/A(t)$ is a negative constant. | Radioactive Decay |

43. The e-life of a radioactive substance is the length of time it takes to reduce the amount of it by a factor of e (i.e., from A_0 to A_0/e). Calculate the e-life of pertechnetate using the formula cited in #42, and compare it to the ratio you calculated in that exercise. | Radioactive Decay |

44. We could describe population growth in a simple way by asserting that more babies are born in larger cities than in smaller towns (if only because there are more people in a big city to *have* babies). Suppose $P(t)$ is a population at time t. Write an equation that says, "The rate at which the population is growing (in people per year) is directly proportional to the current population." | Biology |

This is called the Malthusian population model.

45. Skeletal muscles contract when *thick filaments* and *thin filaments* are pulled past each other by *crossbridges* between them (see [16]). When a crossbridge is extended a distance of x meters beyond its equilibrium (x is very small!), one model asserts that it has a potential energy of | Medical Science |

$$\mathcal{E} = \frac{\alpha}{\beta}e^{\beta x} - \alpha x \text{ joules,}$$

    where α and β are constants pertaining to physical aspects of the muscle.

    (a) Verify that the units of $\frac{d\mathcal{E}}{dx}$ are newtons (force).

    (b) Find a formula for $\mathcal{E}'(x)$.

46. Let us denote by A the maximum distance that an inter-filament crossbridge can be extended beyond its equilibrium (see #45). If v is the velocity at which the crossbridges contract, the fraction of crossbridges that are extended between x and $x + \Delta x$ meters beyond their equilibrium at any given moment is approximately $u(x)\Delta x$, where | Medical Science |

$$u(x) = \frac{\gamma}{v}e^{(x-A)/v},$$

    in which γ is a constant that depends on properties of the muscle. For the purposes of this exercise, suppose that both $A = 1$ and $\gamma = 1$.

(a) Suppose $v = 0.5$. Use a graphing utility to render a graph of u, and zoom in on the graph at $x = 0.5$ until it looks linear. Use that graph to estimate $u'(0.5)$.

(b) Determine a formula for $u'(x)$ and use it to calculate $u'(0.5)$ exactly. *(Hint: rewrite $(x - A)/v$ as $x/v - A/v$ and use properties of exponents to simplify $u(x)$ before differentiating.)*

47. Repeat #46b when $v = 0.6, 0.7, 0.8, 0.9$, and $v = 1$, and denote the values by $f(v)$. Then plot your data points for f, and determine whether $f'(0.7)$ is positive or negative.

Checking the details

After calculating the derivatives of $\sin(7t)$ and $\sin(-3t)$ in the examples, we inferred a general formula. In #48 you're asked to verify that our claim is correct. Exercises #49–50 are similar, but deal with different functions.

48. Show that $\frac{d}{dt}\sin(\omega t) = \omega \cos(\omega t)$ when ω is any constant by using the sum identity for the sine to calculate the limit of the difference quotient.

49. Show that $\frac{d}{dt}\cos(\omega t) = -\omega \sin(\omega t)$ when ω is any constant by using the sum identity for the cosine to calculate the limit of the difference quotient.

50. Show that $\frac{d}{dt}e^{\omega t} = \omega e^{\omega t}$ when ω is any constant by calculating the limit of the difference quotient.

3.4   The Derivative as a Function

In the previous sections we emphasized that $f'(t_0)$ tells us the instantaneous rate of change in $f(t)$ at time t_0. We also emphasized that $f'(t_0)$ tells us the slope of the tangent line to the graph of f at time t_0. We have not emphasized, but it is an important fact, that $f'(t)$ is a function unto itself. For example,

> when $f(t) = \sin(t)$ its derivative is the function $f'(t) = \cos(t)$,　and
>
> when $g(t) = t^2 - 4t + 3$ its derivative is the function $g'(t) = 2t - 4$.

The two functions $f(t)$ and $f'(t)$ are closely related, and in this section we explore that relationship in greater detail.

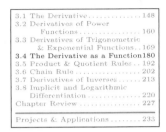
✛ The Domain of the Derivative

Note that t_0 has to be in the domain of f in order for $f'(t_0)$ to exist because $f(t_0)$ appears explicitly in our definition of the derivative:

$$f'(t_0) = \lim_{\Delta t \to 0} \frac{f(t_0 + \Delta t) - f(t_0)}{\Delta t}.$$

So the domain of f' is never larger than the domain of f. In fact, it's often smaller. For example, consider the function

$$f(t) = \text{"the largest integer} \leq t\text{,"}$$

which is often denoted by $\lfloor t \rfloor$ and is called the **greatest integer function** or **floor function**. As you see in Figure 4.1, the secant lines that connect $(1, f(1))$ and $(1 + \Delta t, f(1 + \Delta t))$ get steeper and steeper as $\Delta t \to 0^-$. Their slopes surpass every finite value, so

$$\lim_{\Delta t \to 0^-} \frac{f(1 + \Delta t) - f(1)}{\Delta t} = \infty \quad \text{(i.e., the limit does not exist)}.$$

Because the limit that defines $f'(1)$ does not exist, the domain of f' does not include $t_0 = 1$. Based on this example, you might think that a function must be continuous where it's differentiable. That's exactly right.

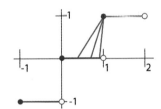

Figure 4.1: The graph of $y = \lfloor t \rfloor$ along with several secant lines.

> **Theorem 4.4.** Suppose $f'(t_0)$ exists. Then f is continuous at t_0.

Proof. This follows from the definition of the derivative:

$$f'(t_0) = \lim_{\Delta t \to 0} \frac{f(t_0 + \Delta t) - f(t_0)}{\Delta t}.$$

To say $f'(t_0)$ exists means that this quotient converges to a particular number in the limit, even though its denominator tends to 0. That can only happen if its numerator also converges to 0 (giving us an indeterminate form). That is, when $f'(t_0)$ exists, we know

$$\lim_{\Delta t \to 0} \Big(f(t_0 + \Delta t) - f(t_0) \Big) = 0$$

$$\lim_{\Delta t \to 0} f(t_0 + \Delta t) - \lim_{\Delta t \to 0} f(t_0) = 0$$

$$\lim_{\Delta t \to 0} f(t_0 + \Delta t) = f(t_0),$$

which is the definition of continuity we saw in Chapter 2 (see p. 129).　∎

However, continuity alone is insufficient to guarantee that $f'(t_0)$ exists.

> This is the heart of the matter.

> Using the sum law for limits and the fact that $f(t_0)$ is constant.

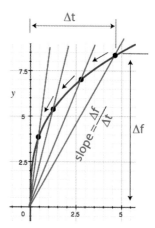

Continuity is not enough to guarantee differentiability

Example 4.1. Show that $f(t) = 5t^{1/3}$ is not differentiable at $t_0 = 0$.

Solution: Using the Scaling Rule and Power Rule, we arrive at

$$\frac{d}{dt}\left(5t^{1/3}\right) = \frac{5}{3t^{2/3}} \quad \begin{cases} \text{which doesn't exist when } t = 0, \text{ even} \\ \text{though the cube-root function is contin-} \\ \text{uous at } t = 0. \end{cases}$$

The difference quotient tells us the slope of the secant line connecting $(0, f(0))$ and $(0 + \Delta t, f(0 + \Delta t))$. That line gets steeper and steeper as $\Delta t \to 0$, and becomes vertical in the limit. Consequently, $\lim_{\Delta t \to 0} \Delta f / \Delta t = \infty$ (i.e., does not exist). Figure 4.2 tells the story visually in the case of $\Delta t \to 0^+$. ∎

Figure 4.2: The curve $y = 5t^{1/3}$ crosses the y-axis at $(0, 0)$. The y-axis is the tangent line to the graph at $t = 0$.

Continuity is not enough to guarantee differentiability

Example 4.2. Use the graph of f shown in Figure 4.4 to explain why f is not differentiable at $t_0 = 1$.

Solution: In the left-hand graph of Figure 4.4, you can see that secant lines corresponding to $\Delta t > 0$ all have a positive slope, so $\lim_{\Delta t \to 0^+} \frac{\Delta f}{\Delta t} \geq 0$. Similarly, secant lines corresponding to $\Delta t < 0$ all have negative slope, so $\lim_{\Delta t \to 0^-} \frac{\Delta f}{\Delta t} \leq 0$. These one-sided limits could only be equal if they were both zero, but they're not, so the limit doesn't exist. ∎

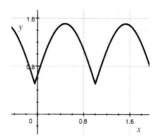

Figure 4.3: The graph of f for Example 4.2.

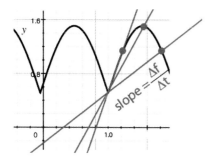
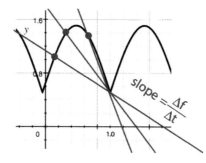

Figure 4.4: (left) Secant lines have negative slope when $\Delta t < 0$; (right) secant lines have positive slope when $\Delta t > 0$.

We say that the graph of a function has a **corner** at t_0 when the one-sided limits of the difference quotient are different (as in Example 4.2). In the special case when both one-sided limits are infinite but opposite in sign, we say the graph of f has a **cusp** at t_0. In summary, then ...

The Domain of f': The domain of f' includes t_0 only when the function f is continuous at t_0, and the graph of f does not have a cusp, corner, or vertical tangent line at t_0.

Here you're seeing that f can be used to tell us about f'. In a moment you'll see that f' also tells us about f.

❖ The Range of f'

The range of f' is the set of all possible outputs the function can produce. In the current context, we are only interested in whether a given output is positive or negative. When $f'(t_0)$ is positive, the tangent line to the graph of f at t_0 is headed up, so you expect to see the graph of f increasing; and when $f'(t_0) < 0$ you expect the graph to follow the tangent line downward. This is borne out in the following examples ...

Increase or decrease is determined by the sign of f'

Example 4.3. Graph the functions $g(t) = t^2 - 4t + 3$ and $g'(t) = 2t - 4$ on the same axes, and verify that $g(t)$ is increasing at each value of t where $g'(t) > 0$.

Solution: The graphs $y = g(t)$ and $y = g'(t)$ are shown in Figure 4.5, where you can see that $g'(t) > 0$ when $t > 2$. The vertex of the parabola $y = g(t)$ is also at $t = 2$, after which $g(t)$ is increasing.   ∎

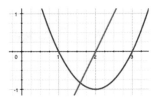

Figure 4.5: The graphs $y = g(t)$ and $y = g'(t)$.

Increase or decrease is determined by the sign of f'

Example 4.4. Graph the functions $f(t) = \sin(t)$ and $f'(t) = \cos(t)$ on the same axes, and verify that $f(t)$ is increasing at each value of t where $f'(t) > 0$.

Solution: The graphs $y = f(t)$ and $y = f'(t)$ are shown in Figure 4.6. In the time frame that's shown, you can see that $f'(t) > 0$ when $t < \pi/2$, which is the interval over which $f(t)$ is increasing.   ∎

Thus we are led to the following fact:

Increase-Decrease Theorem: Suppose f is differentiable at each $t \in (a, b)$.

- If $f'(t) > 0$ throughout (a, b), the function f is increasing on the interval.

- If $f'(t) < 0$ throughout (a, b), the function f is decreasing on the interval.

Figure 4.6: The graphs $y = f(t)$ and $y = f'(t)$.

Proof. This result relies on something called the Mean Value Theorem, which is discussed in Chapter 4, so we're going to postpone a proof until then.

Note: When t_0 lies in an open interval over which $f'(t)$ is continuous, the fact that $f'(t_0) > 0$ implies $f'(t) > 0$ in an interval around t_0. This allows us to make a conclusion about the behavior of f by checking f' at a single point.

Physical interpretation and increase/decrease

Example 4.5. Interpret the answer to Example 1.4 (p. 150) in its physical context.

Solution: In Example 1.4 an object is moving along the x-axis, and we found that $x'(2) = 19$ ft/min. Because $x'(2) > 0$, the Increase-Decrease Theorem tells us that the number $x(t)$ is increasing, which means that the object is moving to the right. Since the object is already to the right of the origin (i.e., $x(2) > 0$) we can say that it's moving *away* from the origin at 19 ft/min.   ∎

> We didn't have a formula for $x'(t)$ on p. 150, but now we're in a position to say that $x'(t) = 6t + 7$, which is continuous everywhere.

Physical interpretation and increase/decrease

Example 4.6. Suppose that C is the steady-state concentration of oxygen in the lung, and α is the fraction of oxygen that's absorbed during each respiratory cycle. What does it mean to say that $C'(\alpha) < 0$?

Solution: Simply put, larger values of α correspond to smaller values of C. In this context, it means that better absorption of O_2 across the blood-gas membrane in the lung results in a lower concentration of resident O_2 in the lung.   ∎

Increase & invertibility

Example 4.7. Suppose $y(x) = x + 15x^{15}$. Use the derivative to prove that y is an invertible function of x.

Solution: Since $y'(x) = 1 + 225x^{14}$, which is always positive, the value of $y(x)$ is always increasing. No output is achieved twice, so the function is invertible. Alternatively, we could say the height of the graph is always increasing because $y'(x) > 0$, so it passes the Horizontal Line Test. ∎

❖ The Roots of f'

In Example 4.3 you see that $g(t)$ changes from decreasing to increasing at $t = 2$, which is where g' has a root (i.e., $g'(2) = 0$). Similarly, in Example 4.4 you see that $f(t)$ changes from increasing to decreasing at $t = \pi/2$, which is where f' has a root. Lest we overgeneralize, note the following example, which illustrates that a function doesn't *have* to change from increasing to decreasing (or vice versa) where its derivative has a root.

Graphs don't have to change direction where $f'(t) = 0$

Example 4.8. Suppose $f(t) = t^3$. Show that $f'(t)$ has a root at $t = 0$ but the graph of f does not change the direction it's headed.

Solution: The derivative of f is $f'(t) = 3t^2$, which has a root at $t = 0$. However, larger values of t *always* result in larger $f(t)$, so f is increasing *everywhere*. In visual terms, the graph of $f'(t)$ is above the t-axis both before and after $t = 0$, so the graph of f is rising both before and after—it hasn't changed the direction it's headed (see Figure 4.7). ∎

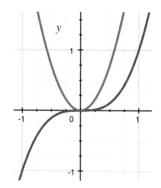

Figure 4.7: The graphs $y = t^3$ and $y = 3t^2$.

Similarly, in Example 4.2 you see that the graph of f changes from decreasing to increasing at $t = 1$, and $f'(1)$ *does not exist*. By contrast, in Example 4.1 you saw that $f'(0)$ does not exist, but the function $f(t) = t^{1/3}$ is always increasing. So like the roots of f', the function f can—*but doesn't have to*—change from increasing to decreasing (or vice versa) at times when $f'(t)$ does not exist. These ideas lead us to make two definitions:

A number t_0 in the domain of f is said to be a **first-order critical number** of f when either $f'(t_0) = 0$ or $f'(t_0)$ does not exist.

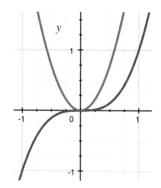 It's common to omit the term "first-order."

People sometimes refer to a critical number as a **critical point** and in the context of velocity we say that t_0 is a **stationary point** when $f'(t_0) = 0$.

We say that t_0 is a **turning point** when t_0 is in the domain of f and f changes from increasing to decreasing (or vice versa) at t_0.

❖ The Graph of f'

Sometimes we need only a rough understanding of f' rather than the precision offered by a formula for $f'(t)$. Other times, a formula for $f(t)$ is unavailable, so deriving a formula for $f'(t)$ is impossible. In such cases, we can get a rough sketch of $y = f'(t)$ from the graph of f.

Sketching a graph of f' from the the graph of f

Example 4.9. Suppose $f(t)$ is the arterial radius at time t, as depicted in Figure 4.8. Sketch a graph of $f'(t)$.

Solution: We do this in three steps:

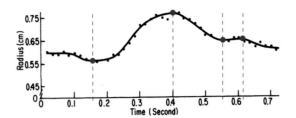
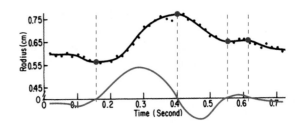

Figure 4.8: (left) The graph of f with critical points marked; (right) the graph of f' superimposed on the graph of f.

Step 1: Begin by locating the critical points of f. Since the graph of f is well behaved (no corners, cusps, discontinuities, etc.), these happen where $f'(t) = 0$ (i.e., where the tangent line is horizontal). Such times have been marked by blue dots • in the left-hand graph of Figure 4.8.

Step 2: Now that we know the times at which $f'(t) = 0$, we know the t-intercepts of the graph of f'. Between these times, the graph is either above or below the t-axis. Recall that the graph of f is increasing when $f'(t) > 0$, and decreasing when $f'(t) < 0$; so the graph of f' should be above the t-axis when the graph of f is rising, and below the t-axis when the graph of f is falling.

Step 3: Since $f'(t)$ tells us the slope of the tangent line, the graph of f' should be far away from the axis when the graph of f is steep, and close to the axis when the graph of f is nearly horizontal.

The graph on the right-hand side of Figure 4.8 shows steps 2 and 3.   ∎

Try It Yourself: Sketching a graph of f' from the graph of f

Full Solution
On-line

Example 4.10. Sketch a graph of f' based on the graph of f in Figure 4.9.

Answer: See solution on-line.   ∎

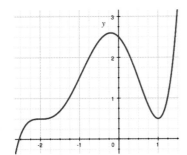
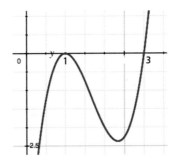

Figure 4.9: (left) The graph of f for Example 4.10; (right) the graph of $y = g'(x)$ for Example 4.11.

On the other hand, sometimes we understand f' and want to recover information about f. If we have a formula for $f'(t)$, there are ways of recovering the formula for $f(t)$ that you'll see in Chapter 5. If we have only a graph of f', we can't calculate exact values of $f(t)$ but we can extract important qualitative information and make approximations.

> This often happens when we understand the forces involved in a given scenario, and need to predict how a physical system will respond. We'll discuss this in more detail in Chapter 4.

Sketching a graph based on the sign of the derivative

Example 4.11. Figure 4.9 depicts the graph of g'. Use it to determine where $g(x)$ is decreasing, and sketch the graph of g.

Solution: It's extremely important to keep in mind that Figure 4.9 depicts the graph of g', not the function g itself, so it tells us whether the graph of g is increasing or decreasing but it does *not* tell us the height of the graph g. As in previous examples, we're going to do this in three steps:

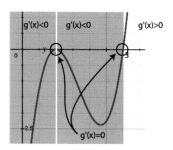

Step 1: Locate the critical points. Since g' exists everywhere, these happen where $g'(t) = 0$—that is, where the graph of g' touches the t-axis. The graph of g will have a horizontal tangent line at these points, so we'll have to make it "flatten out" when we sketch it.

Figure 4.10: Steps 1 & 2 of Example 4.11 combined.

Step 2: Between the critical points, use the graph of g' to determine whether $g'(t)$ is positive or negative. We know from the Increase-Decrease Theorem that the graph of g is increasing over intervals where $g'(t) > 0$, and decreasing over intervals where $g'(t) < 0$.

Step 3: Sketch the graph of g so that it's decreasing in where $g'(t) < 0$, "flattens out" at the critical points, and increases where $g'(t) > 0$.

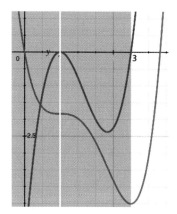

Steps 1 and 2 have been combined in Figure 4.10 (the plane is shaded wherever $g'(t)$ is negative). The graph of g is shown in Figure 4.11. ∎

Figure 4.11: Step 3 of Example 4.11.

Try It Yourself: Sketching a graph based on the sign of the derivative

Example 4.12. Use the graph of $g'(x)$, shown below, to sketch a graph of $g(x)$.

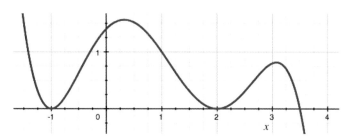

Answer: Solution on-line. ∎

Full Solution On-line

❖ Common Difficulties

Students often confuse *turning points* with *critical points*, so always keep the example of $f(t) = t^3$ in mind. It has a critical point at $t = 0$ but the graph continues to increase.

It's also common for students to infer the converse of the Increase-Decrease Theorem, but the assertion that $f'(t_0)$ *must be positive if f is increasing* at t_0 is wrong. The function $f(t) = t^3$ serves as a prime counterexample, because it's always increasing but $f'(0) = 0$. What you can say with confidence is that, if f is increasing in a neighborhood of t_0, we know $f'(t_0)$ is *not negative*.

You should know

- the terms *cusp*, *corner*, *first-order critical number (point)*, and *stationary point*;

- that differentiability implies continuity, but not vice versa;

- conditions under which f' does not exist;

- the Increase-Decrease Theorem;

- the function f' tells us about the function f, and vice versa.

You should be able to

- determine when $f(t)$ is increasing or decreasing based on the sign of $f'(t)$;

- sketch the graph of f' based on the graph of f, and vice versa;

- interpret increase or decrease in a physical context.

❖ 3.4 Skill Exercises

In #1–2 the graph of f is shown. Use it to determine all numbers at which f is not differentiable.

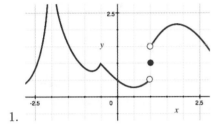   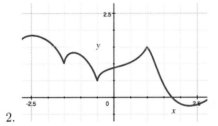

1.   2.

In #3–12 you're given formula for $f'(x)$. You should (a) use it to locate any/all critical points of f, (b) determine the intervals over which f is decreasing, and (c) sketch a graph of f based on your answers from parts (a) and (b).

3. $f'(x) = 3x + 7$

4. $f'(x) = -6x + 13$

5. $f'(x) = x^2 - 7$

6. $f'(x) = 2x^2 + x - 12$

7. $f'(x) = x^3 + 7x$

8. $f'(x) = x^3 + 2x^2 - 6x$

9. $f'(x) = \frac{2x+1}{4x-3}$

10. $f'(x) = \frac{3x+1}{x^2+9x+18}$

11. $f'(x) = \tan(x)$

12. $f'(x) = \sec(x)$

In #13–18 you should (a) graph $y = f(x)$, (b) find a formula for $f'(x)$ when $x < 1$, (c) find a formula for $f'(x)$ when $x > 1$, (d) determine whether the function is differentiable at $x = 1$, and (e) sketch a graph of $f'(x)$.

13. $f(x) = \begin{cases} 3x - 4 & \text{if } x \leq 1 \\ 8 - 7x & \text{if } x > 1 \end{cases}$

14. $f(x) = \begin{cases} 4x - 2 & \text{if } x < 1 \\ 5x - 1 & \text{if } x \geq 1 \end{cases}$

15. $f(x) = \begin{cases} 6x - 2 & \text{if } x \leq 1 \\ 5 - x & \text{if } x > 1 \end{cases}$

16. $f(x) = \begin{cases} 8 - 2x & \text{if } x < 1 \\ 7x - 1 & \text{if } x \geq 1 \end{cases}$

17. $f(x) = \begin{cases} 1 - x^2 & \text{if } x < 1 \\ x - x^3 & \text{if } x \geq 1 \end{cases}$

18. $f(x) = \begin{cases} 2x - x^2 & \text{if } x \leq 1 \\ \cos(2\pi x) & \text{if } x > 1 \end{cases}$

In #19–22 you're given a graph of $f(x)$. Sketch a graph of $f'(x)$.

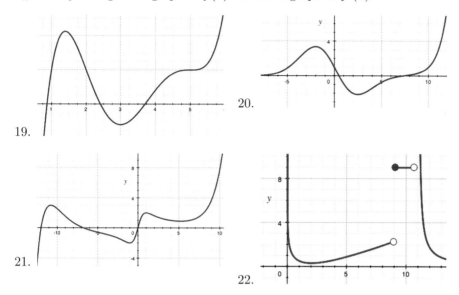

19.   20.

21.   22.

In #23–26 you're seeing the graph of *the derivative, $f'(x)$*. You should (a) use it to locate any/all critical points of f, (b) determine the intervals over which f is decreasing, and (c) sketch a graph of f based on your answers to parts (a) and (b).

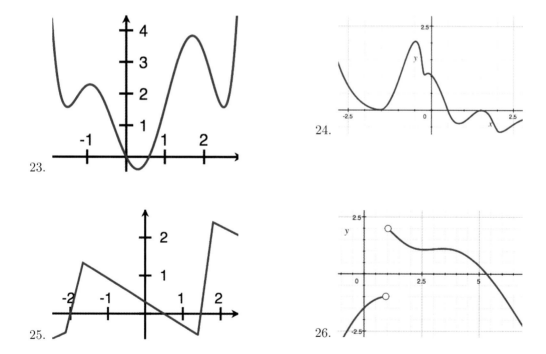

23.   24.

25.   26.

❖ 3.4 Concept and Application Exercises

In each of #27–30 the function $x(t)$ describes the displacement of an object (in meters) from $x = 0$, after t seconds. Suppose that you are standing at $x = 0$, facing the positive x-axis. Use $x(t)$ and $x'(t)$ to write equations and/or inequalities that describe the motion of the object when it's ...

27. 4 m in front of you at time $t = 3$, traveling away from you at 5 m/sec.

28. 7 m in front of you at time $t = 4$ seconds, traveling toward you at 1 m/sec.

29. 8 m behind you at time $t = 10$ seconds, traveling away from you at 6 m/sec.

30. 9 m behind you at time $t = 12$ seconds, traveling toward you at 16 m/sec.

In each of #31–34 the function $y(t)$ describes the displacement of an object (in meters) from $y = 0$, after t seconds. Suppose that you are at $y = 0$. Based on $y(t)$ and $y'(t)$, determine the time intervals (if any) over which the object is (a) below you and moving away from you, (b) below you and moving toward you, (c) above you and moving away from you, and (d) above you and moving toward you.

31. $y(t) = t^2 - 7t + 12$

32. $y(t) = -t^2 + 7t + 10$

33. $y(t) = t^3 + 7t^2 - t$

34. $y(t) = t^3 + 8t^2 - 4t$

35. Suppose $T(t)$ is a the temperature of a microchip (in °C) after t seconds of operation. What does the equation $T'(t) = 0$ tell you about the temperature of the chip? | Computer Engineering |

36. Suppose $T(t)$ is a the temperature of a microchip (in °C) after t seconds of operation. Use the derivative to write an equation that says, "the microchip is cooling off when $t = 100$ seconds." | Computer Engineering |

37. Suppose V is the volume of the heart's left ventricle. Explain what the equation $V'(t) < 0$ tells us about the heart. | Medical Science |

38. Suppose P is the probability that an ultrasound technician can detect the gender of a fetus. Explain what the equation $P'(t) > 0$ means in this context. | Medical Science |

39. Suppose D is the drag force on a baseball due to air resistance, and r is a measure of the ball's roughness. What does the equation $D'(r) < 0$ mean in this context? | Baseball |

The populations of two plant species, A and B, are shown in Figure 5.3. Use this graph to answer #40–42.

40. Determine a time when $A'(t) > 0$ but $B'(t)$ is not.

41. Find the critical points of $A(t)$.

42. Are there critical points of $A(t)$ that are not turning points?

43. Explain why $t = 0$ is a turning point for $f(t) = -t^2$, but not for $g(t) = t^{-2}$, even though both function are increasing when $t < 0$ and decreasing when $t > 0$.

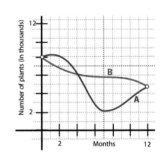

Figure 4.12: For #40–42.

44. Are there values of a and b for which $f(x) = ax + \sin(bx)$ is (a) always increasing, and (b) always decreasing? If so, find them. If not, explain why not.

45. Are there values of a and b for which $f(x) = ax + e^{bx}$ is (a) always increasing, and (b) always decreasing? If so, find them. If not, explain why not.

46. Suppose $f(x) = x^3 + 9x^2 + 14x - 7$. Use $f'(x)$ to determine whether f is invertible.

47. Suppose $f(x) = 6x - 7e^{-x}$. Use $f'(x)$ to determine whether f is invertible.

48. Suppose $f(x) = x^3 + bx^2 + 14x - 7$. Use $f'(x)$ to determine the values of b (if any) for which f is invertible.

49. Suppose $f(x) = ax + be^{-x}$. Determine whether there are values of a and b for which f is invertible.

50. Sketch the graph of a function $f(x)$ for which $\lim_{x \to 3^-} f'(x)$ and $\lim_{x \to 3^+} f'(x)$ are both positive, but not the same.

51. Sketch the graph of a function $f(x)$ for which $\lim_{x \to 3^-} f'(x)$ and $\lim_{x \to 3^+} f'(x)$ are both negative, but not the same.

52. The graph of f is shown in Figure 4.13. Use secant lines (see p. 181) to explain why

$$\lim_{\Delta x \to 0^-} \frac{f(2 + \Delta x) - f(2)}{\Delta x} = -\infty \quad \text{and} \quad \lim_{\Delta x \to 0^+} \frac{f(2 + \Delta x) - f(2)}{\Delta x} = \infty \ .$$

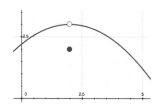

Figure 4.13: Graph for #52.

53. Suppose $f(x) = |x|$.

    (a) Show that $f'(0)$ does not exist because the limit from the left (of the difference quotient) and the limit from the right are not equal.

    (b) Draw a graph of $f'(x)$.

54. Determine any/all points at which $f(x) = |5x - 7|$ fails to be differentiable.

Exercises #55–60 explore how a technique called *centered differencing* can be used to extend our understanding of derivatives by accommodating corners, and some cusps. These exercises should be done in order.

55. The graph of f is shown in Figure 4.14, and the point P is at $(t_0, f(t_0))$.

    (a) Draw the line segment whose slope is $M(\Delta t) = \frac{f(t_0 + \Delta t) - f(t_0 - \Delta t)}{2\Delta t}$.

    (b) Suppose that f is differentiable at t_0. Verify (analytically, not graphically) that $\lim_{\Delta t \to 0} M(\Delta t) = f'(t_0)$. *(Hint: revisit #65 on p. 159.)*

    (c) Explain why this method of calculating $f'(t_0)$ is called *centered differencing*.

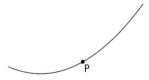

Figure 4.14: Graph for #55.

56. Suppose $f(t) = \begin{cases} 3t + 7 & \text{when } t < 1 \\ 4t + 6 & \text{when } t \geq 1 \end{cases}$.

    (a) Draw a graph of f and verify that f is not differentiable at $t = 1$.

    (b) Determine $f'(t)$ for each $t < 1$.

    (c) Determine $f'(t)$ for each $t > 1$.

    (d) Calculate $\lim_{\Delta t \to 0} M(\Delta t)$ at $t = 1$.

57. Suppose the one-sided limits of the difference quotient exist at t_0. We'll denote $\lim_{\Delta t \to 0^-} \frac{f(t_0 + \Delta t) - f(t_0)}{\Delta t}$ by $f'_-(t_0)$, and $\lim_{\Delta t \to 0^+} \frac{f(t_0 + \Delta t) - f(t_0)}{\Delta t}$ by $f'_+(t_0)$. Show that $\lim_{\Delta t \to 0} M(\Delta t)$ is the average of $f'_-(t_0)$ and $f'_+(t_0)$. *(Hint: revisit #65 on p. 159.)*

58. Suppose $f(t) = |t|$. Calculate $\lim_{\Delta t \to 0} M(\Delta t)$ at $t = 0$.

59. Suppose $f(t) = |t^3 - 8|$. Calculate $\lim_{\Delta t \to 0} M(\Delta t)$ at $t = 2$.

60. Suppose $f(t) = t^{2/3}$. Calculate $\lim_{\Delta t \to 0} M(\Delta t)$ at $t = 0$.

61. Suppose $f(t) = |t^4 - 81|$.

    (a) Graph $f(t)$ and explain why f is not differentiable at $t = 3$.

    (b) Determine whether the structure you see at $t = 3$ is a corner or a cusp.

In exercises #62–63 you should use a graphing utility to display the curve $y = xe^{-\beta x^2}$ for several values of β near $\beta = 1$. Notice that the highest point on the graph depends on β, so let's write its coordinates as $(x(\beta), y(\beta))$.

62. Do the graphs lead you to conclude that $x'(1)$ positive or negative?

63. Do the graphs lead you to conclude that $y'(1)$ positive or negative?

In exercises #64–65 you should use a graphing utility to display the curves $y = \beta x^2$ and $y = \sin(x)$ for several values of β that are very near $\beta = 0.2$. Note that the point of intersection depends on β, so let's write its coordinates as $(x(\beta), y(\beta))$.

64. Do the graphs lead you to conclude that $x'(0.2)$ is positive or negative?

65. Do the graphs lead you to conclude that $y'(0.2)$ is positive or negative?

66. This exercise extends #47 from p. 166, in which you located the points of tangency where a circle of radius R would rest in the cup of the parabola $y = ax^2$ (when $a > 0$). Denote by y_c the y-coordinate of the point of tangency.

    (a) Use your answer to part (c) of #47 (p. 166) to develop a formula for y_c in terms of a and R.

    (b) Thinking of a as known and fixed, show that $\frac{dy_c}{dR} > 0$.

    (c) Interpret part (b) as a rate of change, and explain its meaning in this geometric context.

67. This exercise extends #47 from p. 166, in which you located the points of tangency where a circle of radius R would rest in the cup of the parabola $y = ax^2$ (when $a > 0$). Denote by y_c the y-coordinate of the point of tangency.

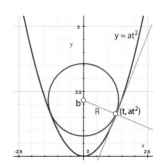

    (a) Use your answer to part (c) of #47 (p. 166) to develop a formula for y_c in terms of a and R.

    (b) Thinking of R as known and fixed, show that $\frac{dy_c}{da} > 0$.

    (c) Interpret part (b) as a rate of change, and explain its meaning in this geometric context.

Figure 4.15: For #66–67.

68. The efficiency of a four-stroke engine is described by $\eta(r) = 1 - r^{-\gamma}$, where r is the compression ratio (v_i/v_f) and γ is a positive number that depends on a quantity called the *heat capacity*. Show that $\eta'(r) > 0$, and explain what that tells us about the relationship between the compression ratio and efficiency.

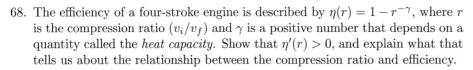

Mechanical Engineering

69. The **logistic** population model describes the change in population as $P'(t) = P(t)(K - P(t))$, where K is a constant called the **carrying capacity** that depends on the environment.

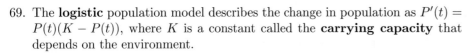

Biology

    (a) If t is measured in years, what are the units of P'?

    (b) Suppose $0 < P(t) < K$. Based on the sign of $P(t)(K - P(t))$, what does the logistic model predict will happen to the population?

    (c) Suppose $K < P(t)$. Based on the sign of $P(t)(K - P(t))$, what does the logistic model predict will happen to the population?

70. The *effort force, E,* required to life a weight of w is $E(L) = w\ell L^{-1}$ when you use a lever whose *effort arm* has length of L and whose *resistance arm* has a length of ℓ (see the schematic below). Show that $E'(L) < 0$ and explain what

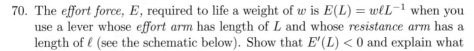

Mechanical Engineering

that means about the relationship between the length of the effort arm and the amount of force needed to lift the weight.

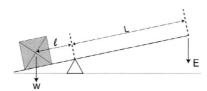

71. Suppose two chemicals, say A and B, combine to form a third, C. We'll denote the concentration of the product after t seconds by $c(t)$. Figure 4.16 depicts the relationship between $c'(t)$ and $c(t)$. The numbers in the graph are not realistic but its qualitative aspects are. Note that the reactants have not had time to produce any C at time $t = 0$, so $c(0) = 0$.

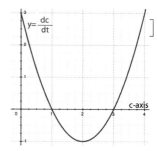

Figure 4.16: For #71

   (a) Does the graph below indicate that $c'(0)$ is positive or negative?

   (b) Based on part (a), what happens to the value of c when t is small?

   (c) Check the sign of c', given the new value of c.

   (d) How large will c grow, according to this graph?

   (e) What stops c from growing, physically speaking?

3.5   The Product Rule & Quotient Rule

In this section we develop quick ways of differentiating products and quotients, which are ubiquitous in science and engineering.

❖ Product Rule

Objects generally expand when you heat them, but not necessarily at the same rate in all directions. Suppose we heat a rectangular metal plate whose length and width are $f(t)$ and $g(t)$, respectively. Then the area of the plate is $a(t) = f(t)g(t)$ at time t, and the rate at which it's changing is given by the derivative

$$a'(t) = \lim_{\Delta t \to 0} \frac{a(t + \Delta t) - a(t)}{\Delta t}.$$

When we write $f(t + \Delta t)$ as $f(t) + \Delta f$, and $g(t + \Delta t)$ as $g(t) + \Delta g$ as indicated in Figure 5.1, we have

$$a(t + \Delta t) = \Big(f(t) + \Delta f \Big) \Big(g(t) + \Delta g \Big),$$

so the numerator of the difference quotient is

$$
\begin{aligned}
\Delta a &= a(t + \Delta t) - a(t) \\
&= \Big(f(t) + \Delta f \Big) \Big(g(t) + \Delta g \Big) - f(t)g(t) \\
&= f(t)g(t) + f(t)\Delta g + g(t)\Delta f + \Delta g \Delta f - f(t)g(t) \\
&= f(t)\Delta g + g(t)\Delta f + \Delta f \Delta g.
\end{aligned}
$$

When we divide this by Δt, we see that the difference quotient is

$$\frac{\Delta a}{\Delta t} = f(t)\frac{\Delta g}{\Delta t} + g(t)\frac{\Delta f}{\Delta t} + \frac{\Delta f}{\Delta t}\Delta g,$$

so the derivative is

$$a'(t) = \lim_{\Delta t \to 0} \frac{\Delta a}{\Delta t} = \lim_{\Delta t \to 0} \left(f(t)\frac{\Delta g}{\Delta t} + g(t)\frac{\Delta f}{\Delta t} + \frac{\Delta f}{\Delta t}\Delta g \right). \tag{5.1}$$

When f and g are both differentiable at time t we know that

$$\lim_{\Delta t \to 0} \frac{\Delta g}{\Delta t} = g'(t) \quad , \text{ and } \quad \lim_{\Delta t \to 0} \frac{\Delta f}{\Delta t} = f'(t).$$

> The first and last summands have the same magnitude but opposite signs.

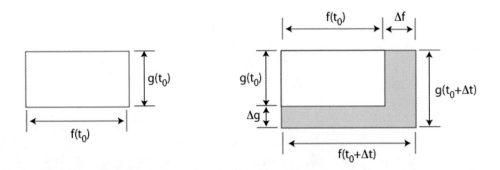

Figure 5.1: A rectangle whose area changes over time (perhaps due to heating).

And since differentiability implies continuity at t (see p. 180),

$$\lim_{\Delta t \to 0} g(t + \Delta t) = g(t)$$

$$\lim_{\Delta t \to 0} g(t + \Delta t) - g(t) = 0$$

$$\lim_{\Delta t \to 0} (g(t + \Delta t) - g(t)) = 0$$

$$\lim_{\Delta t \to 0} \Delta g = 0.$$

Now using the limit laws to evaluate (5.1), we see

$$a'(t) = f(t)g'(t) + g(t)f'(t) + (0)g'(t).$$

This derivation started with the idea that $f(t)$ and $g(t)$ are lengths, but the mathematics is the same when any or all of quantities is (are) negative, and it doesn't matter whether they're increasing or decreasing. Discarding the story of area, we have a general rule:

Product Rule: Suppose f and g are differentiable at t. Then their product is also differentiable at t and

$$(fg)'(t) = f(t)g'(t) + g(t)f'(t).$$

> Written in Leibniz notation, the Product Rule is
>
> $$\frac{d}{dt}(fg) = f\frac{dg}{dt} + g\frac{df}{dt}$$
>
> A lot of people remember the formula by reciting a mantra to themselves:
>
> *The first times the derivative of the second, plus the second times the derivative of the first.*
>
> Other people find that a more colloquial phase is helpful: *each factor "takes a turn" with the derivative.*

Applying the Product Rule

Example 5.1. Find the critical point(s) of $y(t) = te^{-t}$.

Solution: The function $y(t)$ is a product of $f(t) = t$, and $g(t) = e^{-t}$. Since $f'(t) = 1$ and $g'(t) = -1e^{-t}$, the Product Rule tells us that

$$y'(t) = f(t)g'(t) + g(t)f'(t) = t\left(- e^{-t} \right) + e^{-t}(1) = (1 - t)e^{-t}.$$

Since the exponential term is never zero, $y'(t)$ is only zero at $t = 1$. ∎

The Product Rule in context

Example 5.2. Suppose the vertical displacement of a vibrating mass from its initial position is described by $y(t) = 1.2e^{-2t}\cos(3t)$, where distance is measured in centimeters, and time in seconds. Find the velocity of the mass at time $t = 1.25$ seconds.

Solution: Whenever you read or hear the word *velocity* (without the adjective "average") it means *instantaneous velocity,* so we need to calculate $y'(1.25)$. When we write $y(t)$ as the product of $f(t) = 1.2e^{-2t}$ and $g(t) = \cos(3t)$, the Product Rule tells us that

$$y'(t) = f(t)g'(t) + g(t)f'(t) = \left(1.2e^{-2t}\right)\left(- 3\sin(3t)\right) + \cos(3t)\left(1.2(-2)e^{-2t}\right)$$

$$= -3.6e^{-2t}\sin(3t) - 2.4e^{-2t}\cos(3t).$$

Therefore, $y'(1.25) \approx 0.3305533 \; \frac{\text{cm}}{\text{sec}}$. More specifically,

> The units of y' come from the difference quotient.

1. Since $y(1.25) = -0.0808 < 0$, the mass is below its initial position at time $t = 1.25$.

2. Since $y'(1.25) > 0$, the Increase-Decrease Theorem (p. 182) tells us that the number $y(t)$ is increasing at/near $t = 1.25$, so the mass is rising.

> It's important to notice the way that we're using information from y' to understand and describe the behavior of $y(t)$.

Incorporating these two pieces of information, we can say that the mass is 0.0808 centimeters below its initial position, and is rising at 0.3305533 centimeters per second (see Figure 5.2).   ■

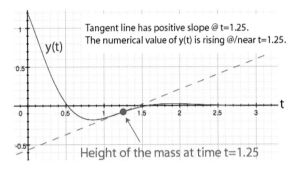

Figure 5.2: The graph of $y(t) = 1.2e^{-2t}\cos(3t)$, and its tangent line at time $t = 1.25$ seconds.

Try It Yourself: The Product Rule in context

Example 5.3. Suppose the horizontal displacement of a mass from its initial position is described by $x(t) = \sin^2(t)$, where distance is measured in meters and time in minutes. Describe the motion of the mass at time $t = 3\pi/4$.

Answer: The mass is 0.5 meters to the right of $x = 0$, and is headed toward the left at 1 meter per minute.   ■

Full Solution
On-line

❖ When There Are More Than Two Factors

When there are more than two factors, we separate them into two groups: (1) the first factor, and (2) all the others. For example, when $f(x) = h(x)g(x)k(x)$ we write $f(x) = h(x)\boxed{g(x)k(x)}$, and use the Product Rule to calculate

$$f'(x) = \frac{d}{dx}\left(h(x)\boxed{g(x)k(x)}\right) = h(x)\frac{d\boxed{g(x)k(x)}}{dx} + \frac{dh}{dx}\boxed{g(x)k(x)}.$$

Since

$$\frac{d\boxed{g(x)k(x)}}{dx} = \frac{d}{dx}\left(g(x)k(x)\right) = g(x)k'(x) + g'(x)k(x)$$

we can rewrite $f'(x)$ as

$$\begin{aligned} f'(x) &= h(x)\left(g(x)k'(x) + g'(x)k(x)\right) + h'(x)g(x)k(x) \\ &= h(x)g(x)k'(x) + h(x)g'(x)k(x) + h'(x)g(x)k(x). \end{aligned}$$

Notice that this is just an extension of the pattern we saw a moment ago.

Said colloquially, each factor in the original product "takes a turn" with the derivative.

Applying the Product Rule when there are more than two factors

Example 5.4. Suppose $f(x) = x^2 e^{8x}\sin(7x)$ Find $f'(x)$.

Solution: The function has three factors, so $f'(x)$ will be a sum of three terms:

$$\begin{aligned} f'(x) &= x^2 e^{8x}(\sin(7x))' + x^2(e^{8x})'\sin(7x) + (x^2)'e^{8x}\sin(7x) \\ &= x^2 e^{8x}7\cos(7x) + x^2 8e^{8x}\sin(7x) + 2xe^{8x}\sin(7x) \end{aligned}$$
    ■

Try It Yourself: Applying the Product Rule when there are more than two factors

Example 5.5. Find a formula for $z'(t)$ when $z(t) = \sin^3(t)$.

Answer: $z'(t) = 3\sin^2(t)\cos(t)$　　　　　　　　■

Full Solution On-line

❖ Quotient Rule

Suppose we have a gas in a container. We could increase the pressure of the gas by squeezing the container, or we could decrease the pressure by cooling it. What happens if we do both? A simple but unhelpful answer is that the change in pressure depends on how much squeezing and how much cooling we do. We can make that answer more precise (and useful) by relating pressure to temperature and volume via the Ideal Gas Law:

$$P = \frac{N k_B T}{V}. \tag{5.2}$$

P	=	Pressure
T	=	Temperature
V	=	Volume
N	=	# of atoms
k_B	=	Boltzmann's constant

The point of this discussion is that even a simple situation can lead us to differentiate a quotient (in order to find $P'(t)$). This is accomplished with ...

Quotient Rule: Suppose $f(t)$ and $g(t)$ are both differentiable at t, and $g(t) \neq 0$. Then the quotient of $f(t)/g(t)$ also differentiable at t and

$$\frac{d}{dt}\left(\frac{f}{g}\right) = \frac{g(t)f'(t) - f(t)g'(t)}{(\,g(t)\,)^2}. \tag{5.3}$$

Many people have trouble remembering whether the numerator begins with $g(t)f'(t)$ or $f(t)g'(t)$. So over the years, a "mantra" has been handed down from generation to generation: the derivative of the function "high"/"low" is

low d high minus high d low, all over low low.

Notice that we start and end with the same word, "low."

Proof. In the following computation we will suppress the dependence on t for readability's sake. So the notation for $f(t + \Delta t)$ used at the beginning of the section becomes $f + \Delta f$, and $g(t + \Delta t)$ is written as $g + \Delta g$. Similarly, $f(t)$ and $g(t)$ will be written simply as f and g. Using this notation,

$$\frac{d}{dt}\left(\frac{f}{g}\right) = \lim_{\Delta t \to 0} \frac{\frac{f + \Delta f}{g + \Delta g} - \frac{f}{g}}{\Delta t} = \lim_{\Delta t \to 0} \frac{1}{\Delta t}\left(\frac{f + \Delta f}{g + \Delta g} - \frac{f}{g}\right).$$

When we rewrite these fractions with a common denominator, this becomes

$$\lim_{\Delta t \to 0} \frac{1}{\Delta t}\left(\frac{(f + \Delta f)g}{(g + \Delta g)g} - \frac{(g + \Delta g)f}{(g + \Delta g)g}\right) = \lim_{\Delta t \to 0} \frac{1}{\Delta t}\left(\frac{fg + g\Delta f - (gf + f\Delta g)}{(g + \Delta g)g}\right).$$

Because fg is the same as gf, the difference in the numerator leaves us with

$$\lim_{\Delta t \to 0} \frac{g\Delta f - f\Delta g}{(g + \Delta g)g\Delta t}.$$

Note that the numerator of the difference quotient has begun to resemble the formula in equation (5.3). Since

$$f' = \lim_{\Delta t \to 0} \frac{\Delta f}{\Delta t} \quad \text{and} \quad g' = \lim_{\Delta t \to 0} \frac{\Delta g}{\Delta t},$$

let's rewrite our expression as

$$\lim_{\Delta t \to 0}\left(g\frac{\Delta f}{\Delta t} - f\frac{\Delta g}{\Delta t}\right)\frac{1}{(g + \Delta g)g} = (g f' - f g')\frac{1}{g^2}.　■$$

The fact that

$$\lim_{\Delta t \to 0} \Delta g = 0$$

was discussed on p. 193.

Applying the Quotient Rule

Example 5.6. Use the Quotient Rule to find $h'(t)$ when $h(t) = \frac{8t+3}{5t+1}$.

Solution: In this case, $f(t) = 8t + 3$ and $g(t) = 5t + 1$, so the derivative of $h(t)$ is

$$h'(t) = \frac{g(t)f'(t) - f(t)g'(t)}{(\,g(t)\,)^2} = \frac{(5t+1)(8) - (8t+3)(5)}{(5t+1)^2} = -\frac{7}{(5t+1)^2}. \qquad \blacksquare$$

Try It Yourself: Applying the Quotient Rule

Example 5.7. Use the Quotient Rule to find $g'(t)$ when $h(t) = \frac{9t}{3t+4}$.

Answer: $h'(t) = \frac{36}{(3t+4)^2}.$ $\blacksquare$

Full Solution On-line

The Quotient Rule in context

Example 5.8. Suppose that pressure, temperature, and volume are related according to the Ideal Gas Law. Find a formula for $P'(t)$, and determine its units.

Solution: The function $P(t)$ is shown in equation (5.2). Since N and k_B are constants, the Quotient Rule tells us

$$P'(t) = \frac{V(t)(Nk_BT(t))' - V'(t)Nk_BT(t)}{(\,V(t)\,)^2} = Nk_B\frac{V(t)T'(t) - V'(t)T(t)}{(\,V(t)\,)^2}.$$

The units of pressure are pascals, so $P'(t)$ has units of pascals per second. $\blacksquare$

> The units of $P'(t)$ are the same as those of the difference quotient, $\Delta P/\Delta t$.

Using the Quotient Rule to differentiate tan(θ)

Example 5.9. Use the Quotient Rule to show that $\frac{d}{d\theta}\tan(\theta) = \sec^2(\theta)$.

Solution: Since $\tan(\theta) = \frac{\sin(\theta)}{\cos(\theta)}$, the Quotient Rule tells us that

$$\frac{d}{d\theta}\tan(\theta) = \frac{d}{d\theta}\left(\frac{\sin(\theta)}{\cos(\theta)}\right) = \frac{1}{\cos^2(\theta)}\left(\cos(\theta)\frac{d}{d\theta}\sin(\theta) - \sin(\theta)\frac{d}{d\theta}\cos(\theta)\right)$$

$$= \frac{1}{\cos^2(\theta)}\left(\cos(\theta)\cos(\theta) - \sin(\theta)\left(-\sin(\theta)\right)\right)$$

$$= \frac{1}{\cos^2(\theta)}\left(\cos^2(\theta) + \sin^2(\theta)\right) = \frac{1}{\cos^2(\theta)} = \sec^2(\theta). \qquad \blacksquare$$

Try It Yourself: Using the Quotient Rule to differentiate cot(θ)

Example 5.10. Use the quotient rule to find a formula for $\frac{d}{d\theta}\cot(\theta)$.

Answer: $\frac{d}{d\theta}\cot(\theta) = -\csc^2(\theta).$ $\blacksquare$

Full Solution On-line

     The Quotient Rule can also be used to find the derivatives of the secant and cosecant. In the exercise set, you'll complete the following table:

Derivatives of Trigonometric Functions:			
$\frac{d}{d\theta}\sin(\theta)$	$=\cos(\theta)$	$\frac{d}{d\theta}\cos(\theta)$	$=-\sin(\theta)$
$\frac{d}{d\theta}\tan(\theta)$	$=\sec^2(\theta)$	$\frac{d}{d\theta}\cot(\theta)$	$=-\csc^2(\theta)$
$\frac{d}{d\theta}\sec(\theta)$	$=\sec(\theta)\tan(\theta)$	$\frac{d}{d\theta}\csc(\theta)$	$=-\csc(\theta)\cot(\theta)$

> **Helpful Tip#1:**
>
> Whenever you take the derivative of a "co"-function, the result begins with a negative sign.
>
> **Helpful Tip#2:**
>
> The patterns in the 2nd column of the table are the same as in the 1st column, except that we've added the sound "co" to those functions that lacked it, and removed the sound from those that had it.

❖ Connection to Percent Rate of Change

The Product Rule and Quotient Rule are nicely and simply related by an idea called **percent rate of change**. Simply put, the percent rate of change in f is the ratio $f'(t)/f(t)$. Provided that both $f(t)$ and $g(t)$ are nonzero:

$$\frac{d}{dt}(fg) = \underbrace{(fg)\left(\frac{f'}{f} + \frac{g'}{g}\right)}_{} \quad \text{and} \quad \frac{d}{dt}\left(\frac{f}{g}\right) = \underbrace{\left(\frac{f}{g}\right)\left(\frac{f'}{f} - \frac{g'}{g}\right)}_{}.$$

(current value)×(sum of percent rates of change)   (current value)×(difference of percent rates of change)

Note: The derivative of a product is a sum, and the derivative of a quotient is a difference. This relationship (products to sums, and quotients to differences) has a lot to do with the rules of logarithms. You'll be able to see why later, when we talk about a technique called *logarithmic differentiation*.

> To understand percent rate of change at an intuitive level, consider this: spending an extra \$1000 per month would be a hardship for many college students, but its impact on a multinational company is much less significant because it makes so much more money.
>
> Similarly, whether $f'(t) = 1000$ is considered "large" depends on the value of $f(t)$. We make this quantitatively precise by calculating the ratio $f'(t)/f(t)$, which is the percent rate of change.

❖ Common Difficulties with the Product Rule and Quotient Rule

People sometimes forget that the derivative of a quotient is *not* the quotient of the derivatives, but it's easy to check. For example, suppose that $g(t) = t$ and $f(t) = t$. Then as long as $t \neq 0$, we have $1 = \frac{f(t)}{g(t)}$. Since this quotient is constant, its derivative is zero, but the quotient of the derivatives is

$$\frac{f'(t)}{g'(t)} = \frac{1}{1} = 1 \neq 0.$$

Similarly, the derivative of a product is *not* the product of the derivatives. Using f and g from above, we have $f(t)g(t) = t^2$. So the derivative of the product is $2t$, but multiplying the derivatives of f and g together gives us

$$f'(t)g'(t) = (1)(1) = 1 \neq 2t.$$

You should know

- the terms *Product Rule* and *Quotient Rule;*
- the derivatives of all the trigonometric functions.

You should be able to

- find the derivative of a product of two or more functions;
- find the derivative of a quotient function.

❖ 3.5 Skill Exercises

In each of #1–18 use the Product Rule to determine a formula for $f'(x)$.

1. $f(x) = (2x + 4)(6x - 1)$

2. $f(x) = (3x - 5)(7 - 8x)$

3. $f(x) = (13x + 7)^2$

4. $f(x) = (8x^2 - 2x + 1)^2$

5. $f(x) = x\sin(6x)$

6. $f(x) = x^2\cos(7x)$

7. $f(x) = 7x^3 e^{5x}$

8. $f(x) = 4x^{11} e^{5x}$

9. $f(x) = e^{0.1x}\cos(5x)$

10. $f(x) = e^{2.7x}\sin(3x)$

11. $f(x) = \sin(7x)\cos(5x)$

12. $f(x) = \tan(2x)\tan(7x)$

13. $f(x) = \cot^2(18x)$

14. $f(x) = x\sin(3x)\tan(9x)$

15. $f(x) = x^2 9^x \cos(10x)$

16. $f(x) = x^4 e^{5x}\sin(3x)$

17. $f(x) = \sin^3(5x)$

18. $f(x) = \cos^3(8x)$

In each of #19–36 use the Quotient Rule to determine a formula for $f'(x)$.

19. $f(x) = \frac{4x+8}{9x+3}$

20. $f(x) = \frac{9x+3}{4x+8}$

21. $f(x) = \frac{3x^2+9x+1}{2x^2-8x-10}$

22. $f(x) = \frac{x^2+x+3}{3x^2-7x+2}$

23. $f(x) = \frac{e^x}{x^2+1}$

24. $f(x) = \frac{3x+2}{e^{6x}}$

25. $f(x) = \sec(x)$

26. $f(x) = \csc(x)$

27. $f(x) = \frac{\sin(8x)}{x^2+x+1}$

28. $f(x) = \frac{\cos(2x)}{x^3-5x+2}$

29. $f(x) = \frac{\sqrt{x}}{4x-1}$

30. $f(x) = \frac{3x+2}{x^{1/3}}$

31. $f(x) = \frac{\tan(x)}{x^2-4}$

32. $f(x) = \frac{\sqrt{x}}{\cot(x)}$

33. $f(x) = \frac{\cos(9x)}{1+\sqrt{x}}$

34. $f(x) = \frac{x^2+\sqrt{x}}{e^x\sec(9x)}$

35. $f(x) = \frac{e^{6x}\cos(2x)}{x^2+2}$

36. $f(x) = \frac{x^3\sin(4x)}{e^{3x}+2}$

In each of #37–48 determine the equation of the tangent line to the graph of f at $(t_0, f(t_0))$.

37. $f(t) = te^{-t}$; $t_0 = 3$

38. $f(t) = t^2 e^t$; $t_0 = -2$

39. $f(t) = t\cos(2t)$; $t_0 = 0$

40. $f(t) = t^6\sin(7t)$; $t_0 = 0$

41. $f(t) = \cos(0.5\pi t)e^{2t}$; $t_0 = 1$

42. $f(t) = \sin(0.25\pi t)e^{3t}$; $t_0 = 1$

43. $f(t) = \tan(4t)$; $t_0 = \pi/16$

44. $f(t) = \cot(9t)$; $t_0 = \pi/54$

45. $f(t) = \frac{t-4}{t+8}$; $t_0 = 2$

46. $f(t) = \frac{3t-\sqrt{7}}{t\sqrt{6}+18}$; $t_0 = -1$

47. $f(t) = \frac{\cos(5t)}{9t+61}$; $t_0 = \pi/2$

48. $f(t) = \frac{\sin(2t)}{t^2+1}$; $t_0 = \pi$

In each of #49–54, you should (a) use the Product Rule to find the derivative (see Example 5.3), and (b) expand the product algebraically, and then calculate the derivative using the Sum Rule and Scaling Rule.

49. $f(x) = (x+5)^2$

50. $f(x) = (3x+7)^2$

51. $f(x) = (1+e^x)^2$

52. $f(x) = (5+3e^x)^2$

53. $f(x) = (e^x + e^{-x})^2$

54. $f(x) = (e^x - e^{-x})^2$

55. The hyperbolic tangent is $\tanh(\theta) = \frac{\sinh(\theta)}{\cosh(\theta)}$. Find a formula for $\frac{d}{d\theta}\tanh(\theta)$.

56. Suppose $f(x) = \frac{3x+2}{\sin(x)}$. Find a formula for $f'(x)$ by (a) using the Quotient Rule, and (b) writing $1/\sin(x)$ as $\csc(x)$ and using the Product Rule.

57. Suppose $f(x) = \frac{\cos(x)}{e^x}$. Find a formula for $f'(x)$ by (a) using the Quotient Rule, and (b) writing $1/e^x$ as e^{-x} and using the Product Rule.

In #58–61 use the graphs of f and g in Figure 5.3 to estimate the given derivative.

58. $(fg)'(1)$ 59. $(f^2)'(1)$ 60. $(f/g)'(1)$ 61. $(g/f)'(1)$

Use the data in the table below to calculate the specified derivatives in #62–65.

t	$f(t)$	$f'(t)$	$g(t)$	$g'(t)$
0	3	20	1/2	-5
2	12	8	-1	-1/3

62. $(fg)'(0)$ 63. $(fg)'(2)$ 64. $(f/g)'(0)$ 65. $(f/g)'(2)$

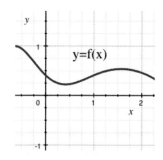

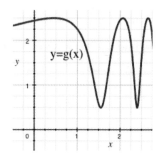

Figure 5.3: Graphs for # 58–61.

❖ 3.5 Concept and Application Exercises

66. When the volume of a vessel depends on the pressure of its contents according to $V = V_0 + CP$, where V_0 is a constant, we say that it is *compliant*. In the case of the human heart, pressure in the left ventricle, P, increases during the beginning of systole, and the compliance parameter, C, decreases. Find a formula for $V'(t)$.

Medical Sciences

67. Suppose $f(v)$ is the fuel consumption of a commercial aircraft (in liters per kilometer) when it travels at v kilometers per hour. You know that $f(200) = 0.8$, and that $f'(200) = 0.007$.

Fuel Efficiency

   (a) Based on its units, what information does the function $vf(v)$ give us?

   (b) Calculate $(vf(v))'$ when $v = 200$, and explain what it tells you about fuel consumption.

68. The Ideal Gas Law is typically written as $PV = Nk_B T$.

Physical Chemistry

   (a) Write temperature as a function of pressure and volume.

   (b) If pressure and volume are both changing in time, find a formula for the instantaneous rate of change in temperature with respect to time.

69. The Ideal Gas Law can be written as $P = \frac{Nk_B T}{V}$.

Physical Chemistry

   (a) Suppose that T is held constant. What are the units of $\frac{dP}{dV}$?

   (b) Show that $\frac{dP}{dV}$ is negative, and explain what that means, physically.

In each of #70–75 a mass is moving along the y-axis according to $y(t)$, where length is measured in meters and time in seconds. Describe the motion of the mass at time $t = 1$, both quantitatively by determining the numbers $y(1)$ and $y'(1)$, and qualitatively using words like "above, below, toward," and "away."

70. $y(t) = e^{3t} \sin(5t)$

71. $y(t) = e^{t\sqrt{3}} \cos(t\sqrt{7})$

72. $y(t) = t^3 \sin(8t)$

73. $y(t) = (t^7 - 6t + 1)\cos(t\sqrt{8})$

74. $y(t) = (t^8 - 3t - 9)e^{-9t}$

75. $y(t) = (t^2 - \sqrt{3}t + \pi)e^{-t\sqrt{7}}$

76. Sketch the graphs of functions f and g so that $f'(4) \neq 0$ and $g'(4) \neq 0$, but $(fg)'(4) = 0$.

77. Sketch the graphs of functions f and g so that $f'(4) \neq 0$ and $g'(4) \neq 0$, but $(f/g)'(4) = 0$.

78. Suppose $f'(x_0) = 0$ and $g'(x_0) = 0$. Show that x_0 is *also* a critical point of the product function fg.

79. Suppose that x_0 is a critical point of f, but *not* of g. How could x_0 be a critical point of the product function, fg?

80. Find functions f and g that are always increasing, but for which the product fg is not (i.e., not always). *(Hint: make it happen at a particular point, say $x = 1$.)*

81. Consider the ratio x/y, where $x, y \in \mathbb{R}$ are positive but not necessarily rational numbers. If we're allowed to change the values of x and y *slightly*, to $x + \Delta x$ and $y + \Delta y$, what relationship between Δx and Δy ensures that the value of the fraction *increases*?

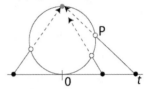

Figure 5.4: Diagram to accompany #82–83.

On p. 114 (exercise #66), you derived formulas for the x and y coordinates of the stereographic projection of the number t onto a circle of radius r that sits atop the x-axis, at the origin (see Figure 5.4). In #82–83, think of $t > 0$ as being fixed, and radius of the circle as variable.

82. Consider the x-coordinate of P:

    (a) Intuitively speaking, should the x-coordinate of the projection become larger or smaller when r increases?

    (b) Calculate the rate of change in x with respect to r, and check your answer to part (a) quantitatively.

___ Geometry

83. Consider the y-coordinate of P:

    (a) Intuitively speaking, should the y-coordinate of the projection become larger or smaller when r increases?

    (b) Calculate the rate of change in y with respect to r, and check your answer to part (a) quantitatively.

___ Geometry

84. Suppose you hold a laser pointer 4 inches above a reflective surface, 10 inches away from a vertical surface, as shown in the figure below ($L = 10$).

    (a) Use the fact that the angle of incidence is the angle of reflection to find a formula for the height of the beam on the vertical wall, h, in terms of the angle at which the beam hits the reflective surface, θ.

    (b) Determine a formula for $dh/d\theta$.

    (c) Determine the units of $dh/d\theta$ and explain what it means to us.

___ Geometry

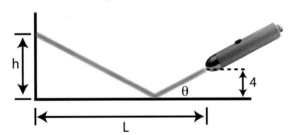

85. Ohm's Law tells us that the voltage drop across a resistor is determined by the product of the current and the resistance. That is, $V(t) = I(t)R(t)$ where $V(t)$ measures the voltage (in volts), $I(t)$ is the current (in amperes), and $R(t)$ is the resistance (in ohms).

    (a) Find a formula for $V'(t)$.

    (b) Show that $I'(t)/I(t) = -R'(t)/R(t)$ when voltage is constant, and confirm that the units on this equation are consistent.

___ Electrical Engineering

86. Objects emit light when they're hot. When the temperature of a so-called *black-body* (such as a kiln, or star) is T kelvins, the frequency of light that is the most intense is determined by the critical point of

$$u(x) = \frac{x^3}{e^x - 1}$$

(called the Planck distribution, see [11]). Determine a formula for $u'(x)$.

___ Physics

87. Suppose $f(p)$ is the *number* of webcams sold when the price is p.

___ Economics

(a) Explain why $f(p)$ is never negative.

(b) In complete sentences, explain what $f'(p)$ means to us and why it's never positive. *(Hint: Consider the units of f'.)*

(c) Explain why the revenue from sales is $R(p) = pf(p)$.

(d) Determine a formula for $R'(p)$, and explain what it means to us.

(e) Show that your formula from part (b) it can be rewritten as $R'(p) = f(p)(1 - E(p))$, where $E(p) = -p\frac{f'(p)}{f(p)}$ is called the *elasticity of demand starting at price level p.*

(f) Economists say that demand is *elastic* when $E(p) > 1$, and that it's *inelastic* when $E(p) < 1$. Based on the formulation of R' in part (e), explain what *elastic* and *inelastic* demand means in terms of *revenue growth*.

Exercises #88–89 are based on the discussion of elasticity in #87. You should (a) determine whether the demand is *elastic* or *inelastic,* and (b) determine whether to increase the price based on your answer to part (a).

88. Suppose $f(p) = \frac{4,500}{p+50} - 30$, and the current price of webcam is \$30.

89. Suppose $f(p) = \frac{4,500}{p+50} - 30$, and the current price of webcam is \$37.

90. The *velocity of money,* denoted by ν, is easiest to understand if you think in terms of a cash economy. In that case, it's the number of times that an average unit of currency changes hands due to purchasing (in a specified time frame). For example, suppose you and and a friend constitute a very small economy in which there are only 100 dollars in circulation. You buy a pair of red socks from her for \$10, she buys an iPod from you for \$80, and you buy her dog for \$50, all in the same month. A total of \$140 changed hands during the month. Since there is \$100 in economy, each unit of currency must have changed hands $\nu = 140/100 = 1.4$ times in that month. In general, we total all of the purchases made in a month, P, and divide that number by the total amount of currency in circulation, M. So the velocity of money is $\nu = P/M$.

Economics

(a) Find a formula for ν' in terms of P, P', M and M'.

(b) Suppose the velocity of money remains constant over several months. Show that $\frac{P'}{P} = \frac{M'}{M}$.

(c) The fractional rate of change in P, which you see above as P'/P, is called *inflation* (inflation is measured in *percent* per year, not dollars). What does the fractional rate of change in M mean to us?

(d) What does the equation $\frac{P'}{P} = \frac{M'}{M}$ say about the relationship between inflation and the amount of money in circulation when the velocity of money is constant?

3.6 The Chain Rule

Both $f(t) = \sin(t)$ and $g(t) = \sin(5t)$ are sine functions, but g completes a cycle five times faster than f does. Intuitively then, the value of g must *change* five times faster than f does. At time $t = 0$ (when both functions are at the beginning of a cycle), it's easy to see that this works:

$$\begin{array}{ccc} f'(t) & = & \cos(t) \\ f'(0) & = & 1 \end{array} \quad \text{and} \quad \begin{array}{ccc} g'(t) & = & 5\cos(5t) \\ g'(0) & = & 5. \end{array}$$

The *Chain Rule* is way of extending this idea to cases where the argument of a function is even more complicated, such as $h(t) = \sin(5t + \sqrt{t^2 + e^t})$. Before citing the general formula, let's develop some intuition about what *should* be true . . .

✾ Developing Intuition: A Hill with Constant Slope

As depicted in Figure 6.1, suppose you're climbing a hill with constant slope, say 0.3 meters (vertical) per meter (horizontal), and the x-coordinate of your position is described by $x(t) = 0.5t + 1$. It's easy to calculate the rate of change in your altitude with respect to time by putting together two pieces of information:

1. For each meter you travel horizontally, you rise 0.3 meters vertically.

2. You travel horizontally at a rate of $x'(t) = 0.5$ meters per second.

Therefore, if your altitude at time t is denoted by $A(t)$,

This answer has the units we expect for a
change in altitude per unit time

$$\frac{dA}{dt} = \underbrace{\left(0.3 \; \frac{\text{m vertical}}{\text{m horizontal}}\right)}_{\text{Slope of the hill}} \underbrace{\left(0.5 \; \frac{\text{m horizontal}}{\text{sec}}\right)}_{\text{Horizontal rate of travel}} = \overbrace{0.15 \; \frac{\text{m vertical}}{\text{sec}}}.$$

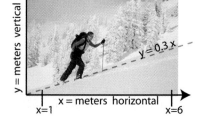

Figure 6.1: Trekking uphill.

✾ Developing Intuition: A Hill with Non-constant Slope

The same ideas apply when the slope of the hill or the horizontal rate of travel is not constant. For example, the sand dune shown in Figure 6.2 coincides nicely with the graph of $f(x) = 50 - 0.01x^2$ when $0 \le x \le 25$ (where x and y are both measured in feet). Suppose that you walk, slip, fall, roll, tumble down the sand dune so that your horizontal position is described by $x(t) = 1 + 1.5t^2$ for the first few seconds of your descent. In order to determine the rate of change in your altitude at a specific moment, we need to know how fast you're moving to the right, and how steep the hill is.

Let's talk about what's happening at time $t = 2$. The horizontal coordinate of your position is $x(2) = 7$, where the slope of the hill is $f'(7) = -0.14$ feet (vertical) per foot (horizontal), and you're moving to the right at $x'(2) = 6$ ft/sec. So the rate of change in your altitude is

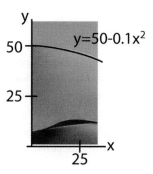

Figure 6.2: Hill with non-constant slope.

$$A'(2) = \underbrace{\left(-0.14 \; \frac{\text{ft vertical}}{\text{ft horizontal}}\right)}_{\substack{\text{Slope of the hill} \\ \text{when } x = 7}} \underbrace{\left(6 \; \frac{\text{ft horizontal}}{\text{sec}}\right)}_{\text{Horizontal speed, } x'(2)} = -0.84 \; \frac{\text{ft vertical}}{\text{sec}}. \qquad (6.1)$$

❖ Writing our Calculations in More General Terms

Since your altitude after t seconds is

$$A(t) = \text{Height}(x(t)) = f(x(t)),$$

we can rewrite equation (6.1) as

$$\frac{d}{dt}f(x(t))\bigg|_{t=2} = \frac{dA}{dt}\bigg|_{t=2} = \overbrace{\left(\frac{\text{ft vertical}}{\text{ft horizontal}}\right)}^{\substack{\text{Slope of hill (de-}\\\text{pends on position)}}} \overbrace{\left(\frac{\text{ft horizontal}}{\text{sec}}\right)}^{\text{Horizontal speed}}$$

$$= f'(7)\ x'(2)\ \frac{\text{ft vertical}}{\text{sec}} = f'(x(2))\ x'(2)\ \frac{\text{ft vertical}}{\text{sec}}.$$

This pattern for calculating the derivative of $f(x(t))$ works in general, and is called the **Chain Rule**:

> **Chain Rule:** Suppose $f(x)$ and $x(t)$ are differentiable functions. Then $f(x(t))$ is differentiable and
> $$\frac{d}{dt}f(\,x(t)\,) = f'(\,x(t)\,)\ x'(t).$$

Proof. While the stories of hills and altitude were helpful in developing intuition, they were just stories. Now it's time to prove that we're right. According to our definition, the derivative of $f(x(t))$ is

$$\frac{d}{dt}f(x(t)) = \lim_{\Delta t \to 0} \frac{f(x(t + \Delta t)) - f(x(t))}{\Delta t}.$$

Guided by the form of the Chain Rule that we've asserted above, let's multiply and divide this expression by $\Delta x = x(t + \Delta t) - x(t)$ so that we see a difference quotient for $x(t)$:

$$\frac{d}{dt}f(x(t)) = \lim_{\Delta t \to 0} \frac{f(x(t + \Delta t)) - f(x(t))}{\Delta x}\frac{\Delta x}{\Delta t}$$

$$= \left(\lim_{\Delta t \to 0} \frac{f(x(t + \Delta t)) - f(x(t))}{\Delta x}\right)\left(\lim_{\Delta t \to 0}\frac{\Delta x}{\Delta t}\right).$$

The second factor is exactly $x'(t)$, by definition. And since $x(t + \Delta t) = x(t) + \Delta x$, we can rewrite the first factor as

$$\lim_{\Delta t \to 0} \frac{f(x(t) + \Delta x) - f(x(t))}{\Delta x},$$

which is *almost* the definition of $f'(\,x(t)\,)$ except that we see $\Delta t \to 0$ instead of $\Delta x \to 0$. Luckily, we know that $x(t)$ is continuous since it's differentiable (see p. 180), from which it follows that $\Delta x \to 0$ as $\Delta t \to 0$ (see p. 193). This allows us to rewrite

$$\lim_{\Delta t \to 0} \frac{f(x(t) + \Delta x) - f(x(t))}{\Delta x} \quad \text{as} \quad \lim_{\Delta x \to 0} \frac{f(x(t) + \Delta x) - f(x(t))}{\Delta x},$$

which is exactly $f'(x(t))$, by definition. ∎

Note: Many people prefer to write the Chain Rule in Leibniz notation,

$$\frac{df}{dt} = \frac{df}{dx}\frac{dx}{dt},$$

The vertical bar means "evaluate at ..."

Notice that the Chain Rule addresses the *composition* of functions $f(x(t))$, so we could also call it the *Composition Rule*, but we don't.

Recall that the limit of a product is the product of the limits.

because the dx terms on the right-hand side appear to cancel, thereby arriving at df/dt algebraically. We'll come back to this notation later.

This is the same basic idea as "canceling" the "horizontal units" in our story of changing altitude.

In the following examples, we will write each $h(t)$ as the composition of an "outer function," $f(x)$, with an "inner function," $x(t)$. It's often true that the inner function is grouped for you by parentheses or some other delimiter.

Verifying the Chain Rule with $h(t) = \sin(7t)$

Example 6.1. Use the Chain Rule to differentiate $h(t) = \sin(7t)$.

Solution: By defining $f(x) = \sin(x)$ and $x(t) = 7t$, we can write this function as $h(t) = f(x(t))$. Since $f'(x) = \cos(x)$ and $x'(t) = 7$, the Chain Rule tells us that

We found the derivative of $\sin(7t)$ on p. 169, so here we can test the new method on a case with which we're already familiar.

$$\frac{dh}{dt} = \underbrace{f'(x(t))}_{\substack{\text{When calculating the deriva-}\\ \text{tive of the outer function, leave}\\ \text{the inner function alone.}}} x'(t) = \underbrace{\cos(x(t))}\ 7 = 7\cos(7t),$$

which agrees with the formula on p. 169.   ∎

Verifying the Chain Rule with $h(t) = e^{3t}$

Example 6.2. Use the Chain Rule to differentiate $h(t) = e^{3t}$.

Solution: By defining $f(x) = e^x$ and $x(t) = 3t$, we can write this function as $h(t) = f(x(t))$. Since $f'(x) = e^x$ and $x'(t) = 3$, the Chain Rule tells us that

We found the derivative of e^{3t} on p. 171, so here we can test the new method on a case with which we're already familiar.

$$\frac{dh}{dt} = f'(x(t))\ x'(t) = e^{x(t)}\ 3 = 3e^{3t},$$

which agrees with the formula on p. 171.   ∎

Practice with the Chain Rule

Example 6.3. Suppose $h(t) = (t^6 + 5t - 1)^{4.6}$. Find a formula for $h'(t)$.

Solution: In this case, we write $h(t) = f(x(t))$ where $f(x) = x^{4.6}$ and $x(t) = t^6 + 5t - 1$. Since $f'(x) = 4.6x^{3.6}$ and $x'(t) = 6t^5 + 5$, the Chain Rule tells us that

$$\frac{dh}{dt} = f'(x(t))\ x'(t) = 4.6(x(t))^{3.6}\ x'(t) = 4.6(t^6 + 5t - 1)^{3.6}(6t^5 + 5).$$ ∎

Practice with the Chain Rule

Example 6.4. Suppose $h(t) = e^{\sin(t)}$. Calculate $h'(t)$.

Solution: Let's write $h(t) = f(x(t))$ where $f(x) = e^x$ and $x(t) = \sin(t)$. Since $f'(x) = e^x$ and $x'(t) = \cos(t)$, the Chain Rule tells us that

$$\frac{dh}{dt} = f'(x(t))\ x'(t) = e^{x(t)}\cos(t) = e^{\sin(t)}\cos(t).$$ ∎

Practice with the Chain Rule

Example 6.5. Suppose $h(t) = \sin(t^2 + 3t - 7)$. Calculate $h'(4)$.

Solution: Let's write $h(t) = f(x(t))$ where $f(x) = \sin(x)$ and $x(t) = t^2 + 3t - 7$. Since $f'(x) = \cos(x)$ and $x'(t) = 2t + 3$, the Chain Rule tells us that

> Recall that the derivative of $\sin(x)$ is $\cos(x)$ provided that x is in radians.

$$\frac{dh}{dt} = f'\big(x(t)\big)\, x'(t) = \cos\big(x(t)\big)\, x'(t) = \cos\big(t^2 + 3t - 7\big)(2t + 3).$$

Therefore, $h'(4) = 11\cos(21)$. ∎

Try It Yourself: Practice with the Chain Rule

Full Solution On-line

Example 6.6. Suppose $h(t) = (9t^4 - 3t + \cos(t))^4$. Determine $h'(0)$.

Answer: $h'(0) = -12$. ∎

✣ Repeated Applications of the Chain Rule

When deciding how to separate an expression into an inner and an outer function so that you can use the Chain Rule to determine its derivative, it's best to group as much as possible as the inner function, leaving only a single operation as the outer.

Repeated applications of the Chain Rule

Example 6.7. Determine a formula for $h'(t)$ when $h(t) = \sqrt{t^2 + 3 + e^{\sin(t)}}$.

Solution: The square root provides a natural grouping for us, so we set

$$f(x) = \sqrt{x} = x^{1/2} \quad \text{and} \quad x(t) = t^2 + 3 + e^{\sin(t)}.$$

We know from the Power Rule that $f'(x) = \frac{1}{2}x^{-1/2} = \frac{1}{2\sqrt{x}}$, so the Chain Rule tells us that

$$h'(t) = f'\big(x(t)\big)\, x'(t) = \frac{1}{2\sqrt{t^2 + 3 + e^{\sin(t)}}}\, x'(t).$$

Now we have to find a formula for $x'(t)$.

$$x'(t) = \frac{d}{dt}\left(t^2 + 3 + e^{\sin(t)}\right) = 2t + 0 + \frac{d}{dt}e^{\sin(t)}.$$

In Example 6.4, we used the Chain Rule to show that

$$\frac{d}{dt}e^{\sin(t)} = e^{\sin(t)}\cos(t),$$

so $x'(t) = 2t + e^{\sin(t)}\cos(t)$. Substituting this back into the formula for $h'(t)$ gives

$$h'(t) = \frac{1}{2\sqrt{t^2 + 3 + e^{\sin(t)}}}\left(2t + e^{\sin(t)}\cos(t)\right).$$ ∎

Try It Yourself: Repeated applications of the Chain Rule

Example 6.8. Use the technique of Example 6.7 to determine a formula for $h'(t)$ when $h(t) = \cos\big(8t + 9 - 5^{\sin(6t)}\big)$.

Full Solution On-line

Answer: $h'(t) = -\sin\big(8t + 9 - 5^{\sin(6t)}\big)\big(8 - 6\ln(5)\,5^{\sin(6t)}\cos(6t)\big)$. ∎

❖ Longer Chains and Leibniz Notation

When more than two functions are composed together (as in the examples below), Leibniz notation is often helpful for keeping calculations well organized and understandable. In Leibniz notation, the derivative of $f((x(t))$ with respect to t is

$$\frac{df}{dt} = \frac{df}{dx}\frac{dx}{dt}.$$

Now suppose that instead of just $x(t)$ we have $x(h(t))$. Then

$$\frac{dx}{dt} = \frac{dx}{dh}\frac{dh}{dt}.$$

Combining this with our first equation, we have

$$\frac{df}{dt} = \frac{df}{dx}\frac{dx}{dh}\frac{dh}{dt}.$$

Differentiating longer chains

Example 6.9. Suppose $f(x) = 3^x$, $x(h) = h^2 + 4$, and $h(t) = \sin(5t)$. Determine the derivative of $f\big(x(h(t))\big)$ with respect to t.

Solution: The value of $f\big(x(h(t))\big)$ depends on x, which depends on h, which depends on t. We can think of this chain of functions one link at a time by using the Chain Rule to write

$$\frac{df}{dt} = \frac{df}{dx}\frac{dx}{dh}\frac{dh}{dt}.$$

We know that $\frac{df}{dx} = \ln(3)\,3^x$, that $\frac{dx}{dh} = 2h$, and $\frac{dh}{dt} = 5\cos(5t)$, so

$$\frac{df}{dt} = \ln(3)\,3^x\,2h\,5\cos(5t).$$

We're not done yet because the dependence on t is still hidden in x and h rather than occurring explicitly in our formula. Since x depends on h according to $x(h) = h^2+4$, we rewrite the formula as

$$\frac{df}{dt} = \ln(3)\,3^{h^2+4}\,2h\,5\cos(5t).$$

The result of this step is that our formula no longer depends on x. Our final step in the process is to use our formula for the function h, after which our expression becomes

$$\frac{df}{dt} = \ln(3)\,3^{\sin^2(5t)+4}\,2\sin(5t)\,5\cos(5t).$$

We can simplify this expression by using the fact that $2\sin(5t)\cos(5t) = \sin(10t)$ and $5\ln(3) = \ln(3^5) = \ln(243)$, so that

$$\frac{df}{dt} = \ln(243)\,3^{\sin^2(5t)+4}\,\sin(10t). \qquad\blacksquare$$

Differentiating longer chains

Example 6.10. Suppose $g(t) = t^2 + 7t + 1$, $k(t) = \tan(6t)$, and $\ell(t) = e^{9t}$. Determine the derivative of $g(k(\ell(t)))$ with respect to t.

Solution: This is a case where the notation can accidentally obscure what's really happening. The instructions above declared each individual function in terms of t, but we're talking about a composite function whose output is produced by g, which depends on the value of k, which depends on the value of ℓ, which depends on t. In order to better reflect this composition with our notation, let's write $g(k) = k^2 + 7k + 1$ and $k(\ell) = \tan(6\ell)$. Then the Chain Rule tells us

$$\frac{dg}{dt} = \frac{dg}{dk} \frac{dk}{d\ell} \frac{d\ell}{dt}.$$

Since $\frac{dg}{dk} = 2k + 7$, and $\frac{dk}{d\ell} = 6\sec^2(6\ell)$, and $\frac{d\ell}{dt} = 9e^{9t}$, this expression becomes

$$\frac{dg}{dt} = \left(2k + 7\right) 6\sec^2(6\ell)\, 9e^{9t} = 54\left(2k + 7\right)\, \sec^2(6\ell)\, e^{9t}.$$

As before, the dependence on t is not yet explicit so we have work to do. We begin by using the formula for $k(\ell)$ to write

$$\frac{dg}{dt} = 54\left(2\tan(6\ell) + 7\right)\sec^2(6\ell)e^{9t}.$$

Lastly, we use the formula for $\ell(t)$ to write

$$\frac{dg}{dt} = 54\left(2\tan(6e^{9t}) + 7\right)\sec^2(6e^{9t})e^{9t}. \qquad \blacksquare$$

Try It Yourself: Differentiating longer chains

Example 6.11. Using the functions from Example 6.9, determine a formula for the derivative of $h\big(f(x(t))\big)$ with respect to t.

Answer: $\frac{d}{dt}h\big(f(x(t))\big) = 10\ln(3)t\,\cos\left(5\left(3^{t^2+4}\right)\right)\, 3^{t^2+4}. \qquad \blacksquare$

Full Solution On-line

❖ Changing Units with the Chain Rule

The Chain Rule is often used to help us change *scale*. It helps us zoom in and zoom out, mathematically.

> Later, this will be an extremely helpful idea when we learn about a technique of *integration* called *u-substitution*.

Zooming in using the Chain Rule

Example 6.12. Suppose $T(m)$ is your temperature (in °F) after m minutes of exercise, and your temperature is increasing 0.5 °F per minute at $t = 4$ minutes (i.e., $T'(4) = 0.5$). Determine the rate of change in your temperature with respect to seconds.

Solution: The number of minutes that have passed is related to the number of seconds, s, according to

$$m = \left(\frac{1}{60}\right)s, \quad \text{so} \quad \frac{dm}{ds} = \frac{1}{60}\,\frac{\min}{\sec}.$$

> When verbalizing the relationship between minutes and seconds, we often say "a minute is 60 seconds." That leads some people to equate m and $60s$, but then $m = 60$ when $s = 1$, which is wrong.
>
> When changing scale, it's often best to begin with the smaller quantity, and to use the word "every." For example, "every 60 seconds is a minute." This leads you partition your seconds into groups of 60, of which there are $s/60$.

This fact allows us to change the time scale with the Chain rule:

$$\underbrace{\frac{dT}{ds}}_{\frac{°F}{\sec}} = \underbrace{\frac{dT}{dm}}_{\frac{°F}{\min}}\; \underbrace{\frac{dm}{ds}}_{\frac{\min}{\sec}} = \frac{1}{2}\frac{1}{60} = \frac{1}{120}.$$

This might seem much smaller than our original rate of change (which was 0.5 °F/min), but it's not. If you get data every second, but your friend only gets data every minute, the same change that's gradual for you seems sudden to her. ∎

Try It Yourself: Changing scale with the chain rule

Example 6.13. Suppose the length of a metal rod is increasing with temperature. At 35°C, the instantaneous rate of change in length is 1.2×10^{-6} meters per degree celsius. What is the instantaneous rate of change in length per degree Fahrenheit at this temperature?

Answer: $dL/df = 6.67 \times 10^{-7}$ when $dL/dc = 1.2 \times 10^{-6}$. ∎

> You'll need to look up the formula that relates °C to °F.

Full Solution On-line

Changing units isn't done on a whim in science or engineering, but because a problem has a "natural" unit of measurement. For example, on p. 48 we said the pertechnetate molecule has a radioactive half-life of 6.02 hours, and wrote the amount remaining after t hours as

$$A(t) = A_0 \left(\frac{1}{2}\right)^{t/6.02}, \tag{6.2}$$

where A_0 is the initial mass of the sample. If we had said …

the half-life is	the exponent in (6.2) would be
361.2 minutes	$t/361.2$, where t is measured in minutes
21672 seconds	$t/21672$, where t is measured in seconds
0.2508333 days	$t/0.2508333$, where t is measured in days

What's important to notice is that the exponent in each and every example is expressed as a fraction of the half-life. So it seems natural to use the half-life as our unit of time: 1 unit of time should mean 1 half-life. In these units, which we'll denote by τ, equation (6.2) is just

> The symbol τ is the lower-case Greek letter *tau*, whose pronunciation rhymes with the word "now."

$$A(\tau) = A_0 \left(\frac{1}{2}\right)^{\tau}.$$

In those units of time, the amount of A is changing at a rate of $\frac{dA}{d\tau} = A_0 \ln(0.5) \left(\frac{1}{2}\right)^{\tau}$. If we want to know the rate of change with respect to seconds, minutes, or hours, we use the Chain Rule to calculate:

$$\frac{dA}{ds} = \frac{dA}{d\tau}\frac{d\tau}{ds} \qquad \frac{dA}{dm} = \frac{dA}{d\tau}\frac{d\tau}{dm} \qquad \frac{dA}{dh} = \frac{dA}{d\tau}\frac{d\tau}{dh} \ .$$

$$\underbrace{\qquad\qquad}_{\substack{d\tau/ds \;=\; 1/21672 \\ \text{since } \tau = s/21672}} \qquad \underbrace{\qquad\qquad}_{\substack{d\tau/dm \;=\; 1/361.2 \\ \text{since } \tau = m/361.2}} \qquad \underbrace{\qquad\qquad}_{\substack{d\tau/dh \;=\; 1/6.02 \\ \text{since } \tau = h/6.02}}$$

> The fraction $h/6.02$ calculates the number of half-lives that have passed in h hours. For example, two half-lives have passed after 12.04 hours since $12.04/6.02 = 2$.

You should know

- the term *Chain Rule*.

You should be able to

- find the derivative of a composite function (i.e., $f(g(t))$);
- use the Chain Rule to change scale (units).

✢ 3.6 Skill Exercises

In each of #1–36 combine the Chain Rule with other differentiation techniques as appropriate, to determine a formula for $f'(x)$.

1. $f(x) = \sqrt{13x - 4}$

2. $f(x) = \sqrt[3]{4 - 13x}$

3. $f(x) = \cos(14x^2 - 7)$

4. $f(x) = \sin(7 - 14x^2)$

5. $f(x) = e^{\tan(x)}$

6. $f(x) = 3^{\cos(x)}$

7. $f(x) = \cos^3(3x + 9)$

8. $f(x) = \sin^3(x^6)$

9. $f(x) = 8\tan^3(2x) - 4\sec^4(7x)$

10. $f(x) = \tan^2(7x) - \sec^2(7x)$

11. $f(x) = \cos(\sin(5x))$

12. $f(x) = \sin(\cos(5x))$

13. $f(x) = (3x^2 + 7)^6(8x^3 - 4)^2$

14. $f(x) = (4x + 10)^{12}(5 - 3x)^9$

15. $f(x) = x^2\sqrt{4x + 9}$

16. $f(x) = \cos(x)\sqrt{10x - 1}$

17. $f(x) = x^3 7^{\cos(x)}$

18. $f(x) = \cos(7x)e^{\sin(6x)}$

19. $f(x) = \csc\left(\frac{12x^2 - 5}{19x^3 - 3x^2}\right)$

20. $f(x) = \cot\left(\frac{3 - 14x^3}{9 + x^4}\right)$

21. $f(t) = \left(\frac{2x + 5}{9x - 2}\right)^5$

22. $f(x) = \left(\frac{4x + 1}{8x - 7}\right)^2$

23. $f(x) = \left(\frac{1 + e^{3x}}{1 - e^{3x}}\right)^4$

24. $f(x) = \left(\frac{1 + \cos(3x)}{2 - \sin(4x)}\right)^8$

25. $f(x) = \frac{(3x - 15)^3}{(19x - 2)^7}$

26. $f(x) = \frac{(4 - 9x)^4}{(10x + 2)^5}$

27. $f(x) = \frac{2 + e^x}{\sqrt{x^2 + 9x + 1}}$

28. $f(x) = \frac{8x - 7}{x + \sqrt[4]{5\sin(x) + 9}}$

29. $f(x) = \sec\left(\sqrt{\frac{4x}{3x + 1}}\right)$

30. $f(x) = \tan\left(\sqrt[5]{\frac{5 - 4x}{x^2 + 7}}\right)$

31. $f(x) = \tan(\cos(\sec(2x)))$

32. $f(x) = \cot(\sin(\csc(3x))$

33. $f(x) = \left(\left((x^2 + 1)^2 + 1\right)^2 + 1\right)^2$

34. $f(x) = \sqrt{1 + \sqrt{1 + \sqrt{1 + \sqrt{x}}}}$

35. $f(x) = \cosh(7 + x^2\sin(7x))$

36. $f(x) = 8^{\cos(7x + \sin(8x))}$

In each of #37–40, determine a formula for $f'(x)$.

37. $f(x) = \left(\frac{\alpha x + \beta}{\gamma x + \delta}\right)^n$ 38. $f(x) = \sec^n(x)$ 39. $f(x) = e^{(x/\sigma)^2}$ 40. $f(x) = e^{(\sigma x)^2}$

In each of #41–44, find a formula for $f'(x)$ by writing $f(x) = g(x)/h(x)$ as $f(x) = g(x)[h(x)]^{-1}$ and using a combination of the Product and Chain rules rather than the Quotient Rule.

41. $f(x) = \frac{x^3 - 2x + 1}{x^2 + 9}$

42. $f(x) = \frac{e^x}{x^2 + 1}$

43. $f(x) = \frac{7x + 1}{\cos(x) + 3}$

44. $f(x) = \frac{8^x}{x + \sqrt{x}}$

In each of #45–52, suppose $f(x) = \sqrt[3]{x}$, $g(x) = \sin(x)$, and $h(x) = 3x^8 - 2x + 1$. Find a formula for the derivative of the indicated function.

45. $f(g(x))$ 47. $g(g(x))$ 49. $f(h(x))$ 51. $f(g(g(x)))$

46. $g(f(x))$ 48. $h(g(x))$ 50. $f(g(h(x)))$ 52. $h(g(f(x)))$

Use the graphs of f and g in Figure 6.3 to estimate the derivatives in #53–54.

53. $\left.\frac{d}{dx}f(g(x))\right|_{x=1.8}$

54. $\left.\frac{d}{dx}f(g(x))\right|_{x=1.2}$

55. Suppose $f(t) = (5t + 3)^2$. Determine $f'(t)$ by (a) writing $x(t) = 5t + 3$ and using the Chain Rule, (b) using the Product Rule, and (c) multiplying the factors of $f(t)$ together and differentiating the resulting polynomial.

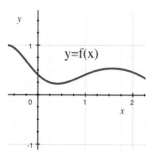

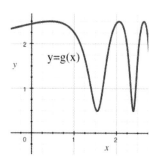

Figure 6.3: Graphs for #53–54.

56. Suppose $f(t) = \sin(6t)$. By writing $x(t) = 6t$, use the Chain Rule to determine $f'(t)$ and verify that it agrees with (3.1), p. 170.

57. Suppose $f(t) = e^{4t}$. By writing $x(t) = 4t$, use the Chain Rule to determine $f'(t)$ and verify that it agrees with (3.5), p. 172.

58. Use the data for the functions f and g, shown in the tables below, to determine $\frac{d}{dt} f(g(t))$ at times $t_0 = 1$ and $t_0 = 2$.

t	$g(t)$	$g'(t)$
1	10	-3
2	8	-7
3	-1	-2

x	$f(x)$	$f'(x)$
8	1	5
9	3	-4
10	7	-3

59. Use the data for the functions f and g, shown in the tables below, to determine $\frac{d}{dt} f(g(t))$ at times $t_0 = 3$ and $t_0 = 4$.

t	$g(t)$	$g'(t)$
2	6	-3
3	4	-7
4	1	0

x	$f(x)$	$f'(x)$
1	8	-5
4	2	4
6	9	-3

Determine the general formula for $f'(x)$ in each of #60–63.

60. $f(x) = \sqrt{g(x)}$

61. $f(x) = \tan\left(g(x) - h(x)\right)$

62. $f(x) = e^{g(x)}$

63. $f(x) = \left[h(x)g(x)\right]^{-4}$

64. Suppose the rate of change in the length of a metal bar with respect to temperature is $d\ell/dT = 4 \frac{\text{feet}}{°\text{F}}$. Use the Chain Rule to express this rate of change in terms of feet per $°\text{C}$.

65. Suppose the mass of a radioactive sample is changing (due to radioactive decay) at $0.1 \frac{\text{kg}}{\text{day}}$. Use the Chain Rule to express this rate of change in terms of kilograms per hour.

❖ 3.6 Concept and Application Exercises

66. Find nonlinear functions f and g that are both always increasing, but for which $\frac{d}{dt} f(g(t))$ is zero at some time.

67. Suppose $\frac{d}{dt} f(x(t)) \Big|_{t=3} = 2$, and $x'(3) < 0$. Is the graph of f sloped up or down at $x(3)$, in the sense of its tangent line?

68. Suppose $\frac{d}{dt} f(x(t)) \Big|_{t=4} = -13$, and $x'(4) > 0$. Is the graph of f sloped up or down at $x(4)$, in the sense of its tangent line?

69. Suppose your position on the x-axis is $x(t)$ after t seconds. If $\frac{d}{dt} f(x(t)) \Big|_{t=3} = 2$ and $f'(x(3)) < 0$, are you headed to the right or to the left?

70. Suppose your position on the x-axis is $x(t)$ after t seconds. If $\frac{d}{dt} f(x(t)) \Big|_{t=5} = 12$ and $f'(x(5)) > 0$, are you headed to the right or to the left?

71. Suppose $f(x) = 1 + \sin^2(x) + \cos(2x) - \cos^2(x)$. (a) Show $f(x)$ is constant by confirming that $f'(x) = 0$. (b) Determine the constant value of $f(x)$.

72. Suppose $f(x) = \csc^2(5x) - \cot^2(5x)$. (a) Show $f(x)$ is constant (on its domain) by confirming that $f'(x) = 0$. (b) Determine the constant value of $f(x)$.

73. The *Lennard-Jones* model of the potential energy between atoms of argon gas (see p. 25) says

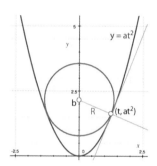

$$u(x) = K\left(1759.37x^{-12} - x^{-6}\right),$$

where x is the distance between the centers of two atoms, measured in angstroms, and K has a numerical value of 70.3748752.

(a) Find a formula for $\frac{du}{dx}$.

(b) The force experienced because of potential is $-\frac{du}{dm}$ when distance is measured in *meters,* and force is measured in *newtons.* Find a formula for $\frac{du}{dm}$.

(c) Use your answer from part (b) to determine the distance at which the argon atoms experience no force due to the potential energy between them.

In each of #74–77 the points $(\pm x_c, y_c)$ are where a circle of radius R rests in the cup of the parabola $y = ax^2$ when $a > 0$ (see #47, p. 166); and one of the numbers, R or a, is fixed. (a) Based on the geometry, explain whether you expect the specified derivative to be positive, negative, or zero, and (b) check numerically to see whether your prediction was correct by determining a formula for it.

74. $a = 3$, $\frac{dx_c}{dR}$ 75. $a = 4$, $\frac{dy_c}{dR}$ 76. $R = 5$, $\frac{dy_c}{da}$ 77. $R = 6$, $\frac{dx_c}{da}$

78. When filling a cup with water, the height of liquid depends on the volume according to

$$h(v) = \sqrt[3]{\frac{27}{\pi}}(v + \pi)^{1/3} - 3.$$

(a) Find a formula for $h'(v)$, including the proper units.

(b) What does it mean, physically, that $h'(v) > 0$?

79. When a traveling wave on the open ocean has a wavelength of λ feet and is traveling over water of depth d feet, its velocity is

$$v = \sqrt{\frac{g\lambda}{2\pi}\tanh\left(\frac{2\pi d}{\lambda}\right)} \; \frac{\text{ft}}{\text{sec}}$$

where g is the force per unit mass due to gravity (32 ft/sec^2) and tanh is the hyperbolic tangent (see #55 on p. 198).

(a) A tsunami wave can have $\lambda = 328,083$ feet (see [34]). The ocean has an average depth of 12,200 feet (see [35]). How fast does such a wave travel across the ocean? (Convert your answer to miles per hour.)

(b) Thinking of λ as a particular, fixed number, find a formula for $v'(d)$.

(c) What are the units of $v'(d)$, and what does it mean to us, physically?

80. The *crossbridge density* function, $u(x)$, was introduced in #46 on p. 178:

$$u(x) = \frac{\gamma}{v} e^{(x-A)/v}.$$

For the purposes of this exercise, suppose that both $A = 1$ and $\gamma = 1$. Suppose that $m = u'(0.5)$, which depends on v. Determine a formula for $\frac{dm}{dv}$ and use it to calculate $m'(0.7)$. What does the number $m'(0.7)$ tell you about the dependence of $u'(0.5)$ on v?

Medical Science

Thermal Physics

81. The *multiplicity* of an object, denoted by Ω, is the number of ways that "packets" of energy can be distributed among its atoms. Suppose two objects are in a room, say A and B.

(a) Suppose the multiplicity of object A is $\Omega_A = 3$ and the multiplicity of object B is $\Omega_B = 4$. If we consider the objects together, how many different ways are there to arrange the energy packets (assuming that the objects cannot exchange energy packets)?

(b) Suppose the multiplicity of object A is $\Omega_A = 300$ and the multiplicity of object B is $\Omega_B = 40,000$. If we consider the objects together, how many different ways are there to arrange the energy packets (assume that the objects cannot exchange energy packets)?

(c) Suppose the two objects have a combined total of Q energy packets, q of which are in object A. Then we write the multiplicity of object A as $\Omega_A(q)$ and the multiplicity of object B as $\Omega_B(Q - q)$. What is the combined multiplicity of the objects, $\Omega = \ldots$?

(d) When the two objects are touching, they can exchange energy packets. The Second Law of Thermodynamics says that such exchanges almost always happen in such a way that the total multiplicity increases. Rewrite the inequality $\frac{d\Omega}{dq} > 0$ in terms of Ω_A, Ω_B and their derivatives.

(e) Does $d\Omega/dq$ tell us what happens when a packet of energy is transferred from A to B, or vice versa?

82. Adding thermal energy to an object typically increases its temperature, but the *amount* of energy you need to add depends on how hot the object is already. The ratio

Physical Chemistry

$$\frac{\text{thermal energy added}}{\text{resulting increase in temperature}}$$

is called the *heat capacity* of the object; it has units of joules/kelvin and we denote it by C_v. The relationship between the heat capacity of a solid crystal and its temperature is

C_v is called the heat capacity at constant volume.

Recall that $\exp(x)$ is another notation for e^x.

Equation (6.3) is derived from the Einstein model of a solid crystal, which treats each atom of the crystal as a so-called *harmonic oscillator* (see [11, p. 307]).

$$C_v(T) = 3Nk_B \left(\frac{\varepsilon}{k_B T} \right)^2 \frac{\exp\left(\frac{\varepsilon}{k_B T} \right)}{\left(\exp\left(\frac{\varepsilon}{k_B T} \right) - 1 \right)^2} \tag{6.3}$$

where N is the number of atoms, k_B is Boltzmann's constant, and ε measures the energy of the oscillators (which we take to be constant). Determine a formula for $C_v'(T)$, including the proper units, and explain what it tell us about a crystal.

3.7 Derivatives of Inverse Functions

Due to its wave-like nature, when light is shot at a screen through a small, single slit, we see a diffraction pattern such as the one shown in Figure 7.1 (see [22] or [9]).

Figure 7.1: The diffraction pattern from a single-slit experiment.

Some simple right-triangle trigonometry (see Figure 7.2) predicts that the angle off the vertical at which we'll find the first dark spot, called a *fringe*, is

$$\theta = \sin^{-1}\left(\frac{\lambda}{a}\right),$$

where λ is the wavelength of the light, and a is the width of the slit. We can adjust the diffraction pattern by changing the width of the slit, and the rate of change in θ with respect to a is determined by the derivative of the arcsine. This formula, like the derivatives of all inverse functions, can be derived by using the Chain Rule.

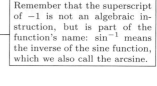

Figure 7.2: Diagram of the single-slit diffraction experiment. The height of the curve indicates the intensity of the light along the screen.

✧ Derivatives of Inverse Trigonometric Functions

The basic strategy for calculating derivatives of inverse functions is based on the idea that $f(f^{-1}(x)) = x$. On the left-hand side of that identity, we see a composition of functions whose derivative calls for the Chain Rule; but on the right-hand side we see only x, whose derivative is 1. Of course, these must be the same.

The derivative of the arcsine

Example 7.1. Derive a formula for $\frac{d}{dx}\sin^{-1}(x)$.

Solution: Because $\sin(\sin^{-1}(x))$ is another way of writing x, when $-1 \le x \le 1$, we know that its derivative is just 1. However, we could use the Chain Rule to write

$$\frac{d}{dx}\sin\left(\sin^{-1}(x)\right) = \underbrace{\cos\left(\sin^{-1}(x)\right)}_{\text{Derivative of the outside function, sine}}\frac{d}{dx}\sin^{-1}(x).$$

As we said a moment ago, though it looks unfamiliar, this must be a 1. That is,

$$\cos\left(\sin^{-1}(x)\right)\frac{d}{dx}\sin^{-1}(x) = 1.$$

That means

$$\frac{d}{dx}\sin^{-1}(x) = \frac{1}{\cos\left(\sin^{-1}(x)\right)},$$

which becomes a useful formula once we simplify the denominator of the fraction. Since $\sin^{-1}(x)$ is an angle whose sine is x, as seen in Figure 7.3, we know that $\cos(\sin^{-1}(x)) = \sqrt{1-x^2}$. Therefore, $\frac{d}{dx}\sin^{-1}(x) = \frac{1}{\sqrt{1-x^2}}$. ∎

Remember that the superscript of -1 is not an algebraic instruction, but is part of the function's name: $\sin^{-1}$ means the inverse of the sine function, which we also call the arcsine.

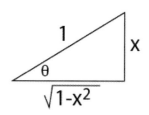

Figure 7.3: the angle θ is $\sin^{-1}(x)$.

Try It Yourself: The derivative of the arccosine

Example 7.2. Derive a formula for $\frac{d}{dx}\cos^{-1}(x)$.

Answer: $\frac{d}{dx}\cos^{-1}(x) = -\frac{1}{\sqrt{1-x^2}}$. ∎

Full Solution
On-line

In the exercise set, you'll apply the same technique to complete the following table of derivatives.

Helpful Tip: Whenever you take the derivative of a "co"-function, the result begins with a negative sign, just like the derivatives of the trigonometric functions (see the table on p. 196).

Derivatives of Inverse Trigonometric Functions:

$$\frac{d}{dx}\sin^{-1}(x) \;=\; \frac{1}{\sqrt{1-x^2}} \qquad\qquad \frac{d}{dx}\cos^{-1}(x) \;=\; -\frac{1}{\sqrt{1-x^2}}$$

$$\frac{d}{dx}\sec^{-1}(x) \;=\; \frac{1}{x\sqrt{x^2-1}} \qquad\qquad \frac{d}{dx}\csc^{-1}(x) \;=\; -\frac{1}{x\sqrt{x^2-1}}$$

$$\frac{d}{dx}\tan^{-1}(x) \;=\; \frac{1}{1+x^2} \qquad\qquad \frac{d}{dx}\cot^{-1}(x) \;=\; -\frac{1}{1+x^2}$$

Practice with derivatives of the inverse trigonometric functions

Example 7.3. Suppose $f(a) = \sin^{-1}(13/a)$. Determine $f'(a)$ when $a > 13$.

Solution: By defining $x = \frac{13}{a}$, we can use the Chain Rule to see

$$f'(a) = \frac{d}{da}\sin^{-1}(x) = \underbrace{\frac{1}{\sqrt{1-x^2}}}_{\text{Derivative of the outside function, } \sin^{-1}}\frac{dx}{da} = \frac{1}{\sqrt{1-\left(\frac{13}{a}\right)^2}}\left(-\frac{13}{a^2}\right) = -\frac{13}{a\sqrt{a^2-169}}.$$ ∎

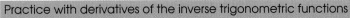

Try It Yourself: Practice with derivatives of the inverse trigonometric functions

Example 7.4. Determine $f'(b)$ when $f(b) = \cos^{-1}\left(\sqrt{5b+7}\right)$.

Answer: $f'(b) = -\frac{5}{2\sqrt{-42-65b-25b^2}}$. ∎

Full Solution
On-line

❖ Derivatives of Logarithms

The same basic technique of composing a function with its inverse allows us to calculate the derivative of a logarithm.

The derivative of the natural logarithm

Example 7.5. Determine $f'(x)$ when $f(x) = \ln(x)$.

Solution: When $x > 0$, we know that $e^{\ln(x)} = x$, so its derivative is 1. However, we could use the Chain Rule to write

Recall that only $x > 0$ are in the domain of the natural logarithm.

$$\frac{d}{dx}e^{\ln(x)} = e^{\ln(x)}\frac{d}{dx}\ln(x).$$

As we said a moment ago, though it looks unfamiliar, this must be a 1. That is,

$$e^{\ln(x)}\frac{d}{dx}\ln(x) = 1.$$

Since $e^{\ln(x)} = x$, this is the same as saying that

$$x\frac{d}{dx}\ln(x) = 1,$$

from which we conclude that $\frac{d}{dx}\ln(x) = \frac{1}{x}$.   ■

The derivative of other logarithms

Example 7.6. Determine $f'(x)$ when $f(x) = \log_4(x)$.

Solution: Using the Change of Base Theorem, we write $f(x) = \frac{\ln(x)}{\ln(4)} = \frac{1}{\ln(4)}\ln(x)$, so $f'(x) = \frac{1}{\ln(4)x}$.   ■

Of course, the number 4 in Example 7.6 could be replaced by any admissible base, b, and the calculation would work out. More generally:

$$\frac{d}{dx}\log_b(x) = \frac{1}{\ln(b)x}.$$

As you saw in Example 7.5, when the base of the logarithm is the number e, this simplifies:

$$\frac{d}{dx}\ln(x) = \frac{1}{x}.$$

Practice with derivatives of logarithms

Example 7.7. Determine $h'(t)$ when $h(t) = \ln(t^2 + e^{5t})$.

Solution: Let's write $h(t)$ as $\ln(x(t))$, where $x(t) = t^2 + e^{5t}$. Then according to the Chain Rule,

$$h'(t) = \frac{d}{dt}\ln\left(x(t)\right) = \underbrace{\frac{1}{x(t)}}_{\text{derivative of the "outside" function, ln}} x'(t) = \frac{1}{t^2 + e^{5t}}(2t + 5e^{5t}) = \frac{2t + 5e^{5t}}{t^2 + e^{5t}}. \quad ■$$

Try It Yourself: Practice with derivatives of logarithms

Example 7.8. Determine $h'(t)$ when $h(t) = \log_3(t^4 + \sin(8t))$.

Answer: $h'(t) = \frac{4t^3 + 8\cos(8t)}{\ln(3)(t^4 + \sin(8t))}$.   ■

Full Solution
On-line

Using rules of logarithms to simplify differentiation

Example 7.9. Determine $h'(x)$ when $h(x) = \ln\left(\frac{7x^3\sqrt{6x-1}}{(4x+3)^5}\right)$.

Solution: We can make this a relatively easy task by using the rules of logarithms to simplify $h(x)$ before differentiating. Because logarithms take products and quotients to sums and differences,

$$h(x) = \ln\left(\frac{7x^3\sqrt{6x-1}}{(4x+3)^5}\right) = \ln(7x^3\sqrt{6x-1}) - \ln(4x+3)^5 = \ln(7) + \ln(x^3) + \ln(6x-1)^{1/2} - \ln(4x+3)^5.$$

And because logarithms take powers to products, we can simplify this further to

$$h(x) = \ln(7) + 3\ln(x) + \frac{1}{2}\ln(6x-1) - 5\ln(4x+3).$$

Now using the Chain Rule gives us

$$h'(x) = \frac{3}{x} + \frac{1}{2}\frac{1}{6x-1}(6) - 5\frac{1}{4x+3}(4) = \frac{3}{x} + \frac{3}{6x-1} - \frac{20}{4x+3}. \quad ■$$

Since $\ln(7)$ is a number (constant), its derivative is zero.

Try It Yourself: Using rules of logarithms to simplify differentiation

Example 7.10. Determine $h'(t)$ when $h(t) = \log_2\left(\frac{1+\sin^2(3t)}{5t^4\sqrt[3]{18t-1}}\right)$.

Answer: $h'(t) = \frac{1}{\ln(2)}\left(\frac{3\sin(6t)}{1+\sin^2(3t)} - \frac{4}{t} - \frac{6}{18t-1}\right)$.   ■

❖ General Facts About Change & Inverse Functions

Figure 7.4 shows the curve $y = f(x)$. Notice that when $x \approx 1$, a small variation in x corresponds to a large change in y. This happens because the graph of f is steep—i.e., because $f'(x)$ is large when $x \approx 1$.

Now turn the book sideways. Notice that when $y \approx 3$, a small variation in y corresponds to an every smaller change in x. Said in the language of calculus, the derivative of the *inverse function* is tiny. In brief, the derivative of f^{-1} is small when f' is large, and vice versa. More specifically:

Theorem 7.5. Suppose f is an invertible function for which $f(x_0) = y_0$, and x_0 is not a critical point of f. Then f^{-1} is differentiable at y_0 and

$$(f^{-1})'(y_0) = \frac{1}{f'(x_0)}.$$

Proof. This fact can be established by composing the functions f and f^{-1} and differentiating, as we've done above. Treating y as the independent variable, we can write

$$f\left(f^{-1}(y)\right) = y.$$

Differentiating with respect to our independent variable, y, we use the Chain Rule on the left-hand side and arrive at

$$f'\left(f^{-1}(y)\right)\frac{d}{dy}f^{-1}(y) = 1 \quad \Rightarrow \quad \frac{d}{dy}f^{-1}(y) = \frac{1}{f'\left(f^{-1}(y)\right)}.$$

When (x_0, y_0) is a point on the graph, $f^{-1}(y_0) = x_0$, which allows us to rewrite this equation as

$$(f^{-1})'(y_0) = \frac{1}{f'(x_0)}. \qquad ■$$

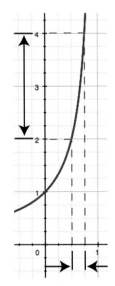

Figure 7.4: The graph of an invertible function f whose derivative near $x = 1$ is large.

Calculating the derivative of an inverse function

Example 7.11. When the temperature in degrees Celsius is c, the temperature in degrees Fahrenheit is $f = \frac{9}{5}c + 32$. Determine (a) the number $f'(c)$, and (b) the derivative of the inverse function.

Solution: (a) Since $f(c)$ is linear, its derivative is constant: $f'(c) = \frac{9}{5}$. (b) Solving the given formula for c as a function of f yields $c = \frac{5}{9}f - \frac{160}{9}$, whose derivative is $c'(f) = \frac{5}{9}$. Notice that these are reciprocals, which is exactly what Theorem 7.5 promises.   ■

> The number $f'(c) = 9/5$ because f increases by 9 when c increases by 5. Said differently, c increases by 5 when f increases by 9, which is why $c'(f) = 5/9$. The point is that f and c increase together, and because the relationship between them is invertible, we can think of either quantity as depending on the other.

Calculating the derivative of an inverse function

Example 7.12. Suppose $f(x) = \sin(x)$. Calculate $(f^{-1})'(0.5)$ by (a) using Theorem 7.5, and (b) using the formula for the derivative of the arcsine.

Solution: (a) Since $f(\pi/6) = 0.5$, the point $\left(\frac{\pi}{6}, 0.5\right)$ is on the graph of f, so

$$(f^{-1})'(0.5) = \frac{1}{f'(\pi/6)} = \frac{1}{\cos(\pi/6)} = \frac{2}{\sqrt{3}}.$$

(b) We know $f^{-1}(y) = \sin^{-1}(y)$, whose derivative is

$$(f^{-1})'(y) = \frac{1}{\sqrt{1-y^2}} \quad \Rightarrow \quad (f^{-1})'(0.5) = \frac{1}{\sqrt{1-\frac{1}{4}}} = \frac{1}{\sqrt{3/4}} = \frac{2}{\sqrt{3}}. \qquad \blacksquare$$

Try It Yourself: Calculating the derivative of an inverse function

Example 7.13. Suppose $f(x) = x^3$. Calculate $(f^{-1})'(8)$ by (a) using Theorem 7.5, and (b) using the formula $f^{-1}(y) = y^{1/3}$.

Answer: $(f^{-1})'(8) = 1/12.$ $\qquad\blacksquare$

Full Solution
On-line

You should know

- that $f'(x_0)$ and $(f^{-1})'(y_0)$ are reciprocals when (x_0, y_0) is a point on the graph of f.

You should be able to

- differentiate inverse functions, including logarithms and inverse trigonometric functions.

❖ 3.7 Skill Exercises

Find the derivative of each function in #1–26.

1. $f(t) = \tan^{-1}(4t)$

2. $f(t) = \cos^{-1}(1-t)$

3. $f(t) = \sin^{-1}(3t^2)$

4. $f(t) = \sin^{-1}(\sqrt{t})$

5. $f(t) = \sec^{-1}(\sin(t))$

6. $f(t) = \tan^{-1}(e^{-3x})$

7. $f(t) = t\csc^{-1}(7t)$

8. $f(t) = t^2 \sin^{-1}(2t-1)$

9. $f(t) = \frac{2t}{\cos^{-1}(t)}$

10. $f(t) = \frac{\sin^{-1}(3t^2)}{3t^2}$

11. $f(t) = 3t^2 \sin^{-1}(3t^2)$

12. $f(t) = \ln\left(\tan^{-1}(t)\right)$

13. $f(t) = \ln\left(\cot^{-1}(4t)\right)$

14. $f(t) = \sin^{-1}\left(\ln(t)\right)$

15. $f(t) = \tan^{-1}\left(\ln(t^2+1)\right)$

16. $f(t) = \sin\left(\cos^{-1}(8t)\right)$

17. $f(t) = \tan\left(\sin^{-1}(2t)\right)$

18. $f(t) = \frac{\cos^{-1}(t)}{\sin^{-1}(t)}$

19. $f(t) = \sqrt{\tan^{-1}(t)}$

20. $f(t) = \left(\sec^{-1}(t)\right)^{-2/3}$

21. $f(t) = e^{-t}\sin^{-1}(2t)$

22. $f(t) = e^{\cot^{-1}(t)}$

23. $f(t) = \ln\left(\cos(t)\right)$

24. $f(t) = -\ln\left(\sin(t)\right)$

25. $f(t) = \ln\left(\sec(t) + \tan(t)\right)$

26. $f(t) = \ln\left(\csc(t) - \cot(t)\right)$

For each function below find the derivative in two ways: (a) using the Chain Rule immediately, without first using the rules of logarithms to simplify the expression and (b) by first simplifying with the properties of logarithms. Show that both methods yield equivalent derivatives.

27. $f(t) = \ln(5t^4)$

28. $f(t) = \ln(12t^9)$

29. $f(t) = \ln\left(\frac{3}{t^2}\right)$

30. $f(t) = \ln\left(\frac{9}{\sqrt{t}}\right)$

31. $f(t) = \ln\left(\frac{2t+1}{4t-3}\right)$

32. $f(t) = \ln\left(\frac{9t^2}{7t+2}\right)$

33. $f(t) = \ln(t^2 \sin(t))$

34. $f(t) = \ln(te^{-t})$

35. $f(t) = \log_2(7t^2)$

36. $f(t) = \log_4(t\cos(t))$

37. $f(t) = \ln(5t\sec(t))$

38. $f(t) = \ln\left(\frac{e^t}{t}\right)$

Find the derivative of each function below by first using the properties of logarithms to simplify the function.

39. $f(t) = \ln\left(\frac{e^{0.5t}}{12t+7}\right)$

40. $f(t) = \ln\left((3t^2 + 4t)^9 \sqrt[4]{t-1}\right)$

41. $f(t) = \ln\left(\frac{t^{3/2}}{\sqrt{t-1}}\right)$

42. $f(t) = \ln\left(\frac{4(2t-1)^2}{1-9t}\right)$

43. $f(t) = \log_5\left(\frac{t^2}{t^2-1}\right)$

44. $f(t) = \log_8\left(\frac{\sqrt{t}}{4t+3}\right)$

❖ 3.7 Concept and Application Exercises

45. Suppose $f(x) = 3x + 4$, so that $f(1) = 7$. (a) Use Theorem 7.5 to calculate $(f^{-1})'(7)$, and (b) derive a formula for f^{-1} and use it to calculate $(f^{-1})'(7)$.

46. Suppose $f(x) = -2x + 5$, so that $f(2) = 1$. (a) Use Theorem 7.5 to calculate $(f^{-1})'(1)$, and (b) derive a formula for f^{-1} and use it to calculate $(f^{-1})'(1)$.

47. Suppose that f is always increasing. Show that f^{-1} is too.

48. Suppose that f is always decreasing. Show that f^{-1} is too.

49. Suppose f is an invertible function for which $f(2) = 3$ and $f'(2) = 7$. Determine $(f^{-1})'(3)$, and explain what it tells us about the inverse function.

50. Suppose f is an invertible function for which $f(1) = 4$ and $f'(1) = -9$. Determine $(f^{-1})'(4)$, and explain what it tells us about the inverse function.

51. The length of time (in hours) that it takes 1 gram of pertechnetate to decay to A grams is $t(A) = -6.02 \log_2(A)$. Calculate $t'(0.5)$ and explain what it mean in this context.

52. When a satellite is positioned h kilometers above the Earth's surface, the angle to the Earth's horizon is θ (see Figure 7.5). Determine a formula for $d\theta/dh$, and its units.

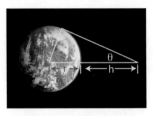

Figure 7.5: For #52.

53. Sound is often measured in *decibels, D*, which are calculated by $D = 10\log(10^{12}W)$ where W is the *sound power level* in watts. Determine $\frac{dD}{dW}$ for $W = 6 \times 10^{-11}$ watts and interpret its meaning.

54. The acidity of a substance is measured by its pH-factor, which is defined to be pH$= -\log[H^+]$, where $[H^+]$ denotes the concentration of hydrogen ions, measured in moles per liter. Find the first derivative of pH with respect to $[H^+]$ and give an interpretation of it in context.

55. Suppose $t = -60,000\ln(1 - 1/P)$ is the time (in days) between births in a population, of P individuals. Determine dt/dp when there are 2000 members of the population, and explain what it means to us in this context.

56. When a 6-foot-tall adult speaks to an h-foot-tall child who's 1 foot away, he looks down at an angle of $\theta = \tan^{-1}(6 - h)$. Calculate $\theta'(4)$ and explain what it means in this context.

57. The *Richter scale* defines the magnitude of an earthquake, M, by the energy it releases, E, according to $M(E) = \frac{-4.4 + \log(E)}{1.5}$. Determine $M'(E)$ and explain what it means when $E = 30{,}000$.

58. The pressure, P, in an isothermal segment of an atmospheric column is related to altitude, a, according to $a = -\frac{k_B T}{mg} \ln\left(\frac{P}{P_0}\right)$, where P_0 is the pressure at sea level, k_B is Boltzmann's constant, T is the temperature of the air (in kelvins), g is the acceleration per unit mass due to gravity, and m is the average mass of a molecule of air in the column. (a) Determine a formula for $a'(P)$, and (b) explain what it means to us that $a'(P) < 0$.

59. A camera operator standing on the ground is pointing his steadicam at an eagle that's flying 50 feet above and x feet in front of him (see Figure 7.6). If the camera's angle of inclination is θ, calculate $\theta'(40)$, and interpret its meaning in this context.

Figure 7.6: For #59

60. The *multiplicity* of an object, denoted by Ω, is the number of ways that "packets" of energy can be distributed among its atoms. That depends on the number of packets, so let's write it as $\Omega(q)$. The entropy of an object is $S = k_B \ln(\Omega)$, where k_B is Boltzmann's constant and has units of joules/kelvin. Verify that $\frac{dS}{dq} = \frac{k_B \Omega'}{\Omega(q)}$, and based on its units explain why we define an object's temperature using $\frac{dS}{dq}$ (the units of q are joules).

61. Suppose a satellite is positioned as in #52, and you are on the surface of the Earth in the same plane as the satellite. If θ is the angle from the satellite to your position (see Figure 7.7) and ϕ describes your position, (a) determine a formula for θ in terms of ϕ, and (b) determine a formula for $d\theta/d\phi$.

Figure 7.7: For #61

3.8   Implicit and Logarithmic Differentiation

We now have a variety of techniques for calculating rates of change when we have an explicit formula for calculating y at each value of x. However, there are times when the relationship between variables is complicated enough that writing it in the form $y = \cdots$ is prohibitively difficult, or even impossible. A famous example is the **folium of Descartes**, which is the curve described by $x^3 + y^3 = 3xy$ and shown in Figure 8.1 (left).

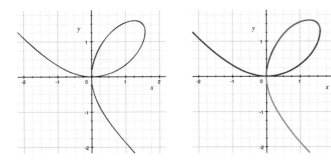

Figure 8.1: (left) The folium of Descartes; (right) partitioning the folium into graphs of functions: $y = f(x)$, $y = g(x)$, and $y = h(x)$.

The folium fails the vertical line test when taken as a whole, so it's not the graph of a function, but Figure 8.1 (right) shows how we can partition the curve into graphs of functions. We say that these functions, say $f(x)$, $g(x)$, and $h(x)$, are defined **implicitly** by the equation $x^3 + y^3 = 3xy$. Though we can't solve this equation for an explicit formula by which to calculate a unique y for each x, the fact that y *is* a function of x ($y = f(x)$, $y = g(x)$, or $y = h(x)$) allows us to use the rules of differentiation that we developed in previous sections.

> The word "implicit" is used in contrast to the word "explicit" from the first paragraph, because we can't solve the folium's equation for an explicit formula by which to calculate y for a given x.

Differentiation of an implicit function

Example 8.1. Suppose $x^3 + y^3 = 3xy$ and $y = f(x)$, whose graph is depicted in Figure 8.1 (right). Determine a formula for dy/dx.

Solution: When we substitute $y = f(x)$ into the equation of the folium, we see $x^3 + (f(x))^3 = 3xf(x)$. Since $x^3 + (f(x))^3$ is the same as $3xf(x)$, their derivatives must be equal:

$$\frac{d}{dx}\left(x^3 + \left(f(x)\right)^3\right) = \frac{d}{dx}\left(3xf(x)\right).$$

Using the Chain Rule on the left-hand side and the Product Rule on the right, we calculate

$$3x^2 + 3\left(f(x)\right)^2 f'(x) = 3f(x) + 3xf'(x),$$

which is easily solved for

$$f'(x) = \frac{f(x) - x^2}{\left(f(x)\right)^2 - x}.$$

Since we don't have an explicit formula for $f(x)$ at hand, the inclusion of $f(x)$ in our equation is not practical. So we switch back to using y, and write

$$\frac{dy}{dx} = \frac{y - x^2}{y^2 - x}. \qquad \blacksquare$$

Try It Yourself: Differentiation of an implicit function

Example 8.2. Suppose $x^3 + y^3 = 3xy$ and $y = g(x)$, whose graph is the red curve show in the right-hand image of Figure 8.1. Determine a formula for dy/dx.

Answer: $\frac{dy}{dx} = \frac{y-x^2}{y^2-x}$.   ■

Full Solution On-line

The calculations in Example 8.2 are the same as those in Example 8.1, so we typically omit the step of writing $y = f(x)$ or $y = g(x)$, and work directly with y. This technique is called **implicit differentiation**.

Implicit differentiation

Example 8.3. The equation $1 = x\sin(x) + y\sin(y)$ defines a nested family of curves, three of which are depicted in Figure 8.2. Determine the x-coordinates of points on the blue curve where the tangent line is horizontal.

Solution: The tangent line is horizontal when its slope is zero, so we need a formula for y'. When we differentiate implicitly we see

$$\frac{d}{dx}(1) = \frac{d}{dx}\Big(x\sin(x) + y\sin(y)\Big)$$
$$0 = x\cos(x) + \sin(x) + y'\sin(y) + y\cos(y)y',$$

which we solve for

$$y' = -\frac{x\cos(x) + \sin(x)}{\sin(y) + y\cos(y)}.$$

The value of this fraction is zero when $x\cos(x) + \sin(x) = 0$. This happens at $x = 0$, clearly, and the graph of $f(x) = x\cos(x) + \sin(x)$ shows two others that correspond to points on the blue curve—one in $[1.5, 2.5]$, and another in $[-2.5, -1.5]$ (see Figure 8.3). Since $f(x)$ is a continuous function, we apply the bisection method and find that $x = \pm 2.02876$.   ■

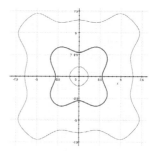

Figure 8.2: Three of the curves defined by $1 = x\sin(x) + y\sin(y)$.

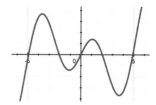

Figure 8.3: The graph of $f(x) = x\cos(x) + \sin(x)$ shows that there's a root in $[1.5, 2.5]$.

WARNING: One of the most common errors in implicit differentiation is forgetting to use the Chain Rule.

Try It Yourself: Implicit differentiation

Example 8.4. The equation $\sin(y) + \sin(x) = x^2 + y^2$ describes a loop, which is depicted in the left-hand graph of Figure 8.4. Determine the x-coordinates of points on the loop where the tangent line is horizontal.

Answer: $x = 0.450183$ (using the bisection algorithm).   ■

Full Solution On-line

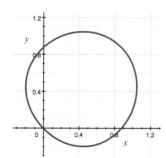
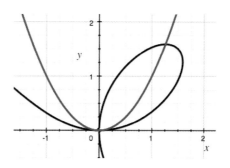

Figure 8.4: (left) For Example 8.4—the curve described by $\sin(x) + \sin(y) = x^2 + y^2$; (right) for Example 8.5—the folium of Descartes and the parabola $y = x^2$.

In Example 8.3 and Example 8.4 the equation $y' = 0$ reduced to an equation in only one variable. That doesn't always happen.

Implicit differentiation

Example 8.5. Find the x-coordinate of the point in the first quadrant that's on the folium of Descartes where the tangent line is horizontal.

Solution: We determined that $y' = \frac{y-x^2}{x-y^2}$ in Example 8.2. This fraction can only be 0 when $y = x^2$ (i.e., when the numerator is 0). The right-hand graph in Figure 8.4 shows both the folium and the parabola $y = x^2$. To find the point of intersection, we substitute $y = x^2$ into the equation of the folium, and arrive at the equation $x^3 + x^6 = 3x^3$. When we rewrite this as

$$x^3(x^3 - 2) = 0,$$

we see that either $x = 0$ or $x = \sqrt[3]{2}$, the latter of which is the point we want.   ∎

❖ Logarithmic Differentiation

The function $y = x^x$ is neither a power function (since the exponents varies) nor an exponential function (since the base varies), so none of our current techniques of differentiation apply. However, we can use the logarithm to rewrite this relationship between x and y in an implicit form. When $x > 0$,

$$\ln(y) = \ln(x^x)$$
$$\ln(y) = x\ln(x).$$

Now differentiating the left-hand side with the Chain Rule and the right-hand side with the Product Rule, we see

$$\frac{1}{y}y' = \ln(x) + 1 \quad \Rightarrow \quad y' = (\ln(x) + 1)y.$$

We return to an explicit formulation by substituting the definition of y into the formula for y'.

$$y' = (\ln(x) + 1)x^x.$$

This technique of applying the logarithm and using the rules of logarithms to simplify before differentiating is called **logarithmic differentiation**.

Try It Yourself: Using logarithmic differentiation

Example 8.6. Determine an explicit formula for $y'(x)$ when $y = x^{\sin(x)}$, and $x > 0$.

Answer: $y' = \left(\cos(x)\ln(x) + \frac{\sin(x)}{x} \right)x^{\sin(x)}.$ ∎

Full Solution
On-line

Sometimes we use logarithmic differentiation to simplify the differentiation process, even though we *could* calculate the derivative directly.

Using logarithmic differentiation

Example 8.7. Determine an explicit formula for $y'(x)$ when $y = \left(\frac{(8x+4)(2x+5)}{\sqrt{x+10}} \right)^3.$

Solution: Instead of using the Chain Rule and Quotient Rule, we apply the logarithm and get the implicit form

$$\ln(y) = \ln\left(\frac{(8x+4)(2x+5)}{\sqrt{x+10}} \right)^3.$$

Using the rules of logarithms on the right-hand side, we can rewrite this as

$$\ln(y) = 3\ln(8x+4) + 3\ln(2x+5) - \frac{3}{2}\ln(x+10),$$

which has *more* terms, but each one is simpler than what we started with. Now differentiating with the Chain Rule,

$$\frac{1}{y}y' = \frac{24}{8x+4} + \frac{6}{2x+5} - \frac{3}{2(x+10)}.$$

By multiplying both sides by y, and substituting the explicit formula for y in terms of x, we arrive at

$$y' = \left(\frac{24}{8x+4} + \frac{6}{2x+5} - \frac{3}{2(x+10)}\right)\left(\frac{(8x+4)(2x+5)}{\sqrt{x+10}}\right)^3. \qquad \blacksquare$$

Try It Yourself: Using logarithmic differentiation

Example 8.8. Determine an explicit formula for $y'(x)$ when $y = \frac{(12x+1)^6\sqrt[5]{x+7}}{(3x+11)^4}$.

Answer: $y' = \left(\frac{72}{12x+1} + \frac{1}{5(x+7)} - \frac{12}{3x+11}\right)\frac{(12x+1)^6\sqrt[5]{x+7}}{(3x+11)^4}$. $\qquad \blacksquare$

Full Solution
On-line

❖ Proof of the Power Rule

Now that we are equipped with logarithmic differentiation, we can easily prove the Power Rule. Consider $y = x^\eta$ when η is any real number. When $x > 0$,

$$\ln(y) = \ln(x^\eta) = \eta\ln(x).$$

Differentiating both sides of this equation, we get

$$\frac{1}{y}y' = \frac{\eta}{x} \quad\Rightarrow\quad y' = \frac{\eta}{x}y = \frac{\eta}{x}x^\eta = \eta x^{\eta-1}.$$

If x^η is defined when $x < 0$, we set $u = -x > 0$ and write $y = (-u)^\eta = (-1)^\eta u^\eta$. Since u is a function of x, we use the Chain Rule to calculate

$$\frac{dy}{dx} = \frac{dy}{du}\frac{du}{dx} = (-1)^\eta \eta u^{\eta-1}(-1).$$

> For example, $x^{1/2}$ does not include negative x in its domain, but $x^{1/3}$ does.

Substituting $u = -x$ and keeping track of the factors of -1, this becomes

$$\frac{dy}{dx} = (-1)^{\eta+1}\eta(-x)^{\eta-1} = (-1)^{2\eta}\eta x^{\eta-1} = \left((-1)^2\right)^\eta \eta x^{\eta-1} = \eta x^{\eta-1}.$$

❖ Looking Back to the Product Rule and Quotient Rule

On p. 197 we wrote the Product Rule with the sum of percent rates of change, and the Quotient Rule with their difference, and said that the formulas were closely related to the rules of logarithms. Now with logarithmic differentiation you can see why. In the work that follows, we'll assume that f and g are positive-valued functions, and we'll suppress the dependence on t by writing f and g instead of $f(t)$ and $g(t)$. With that said, you know that logarithms take products to sums, so

$$\ln(fg) = \ln(f) + \ln(g).$$

Now using the Chain Rule to differentiate both sides with respect to time we have

$$\frac{1}{fg}(fg)' = \frac{1}{f}f' + \frac{1}{g}g'.$$

Don't just read it. Work it out!

After multiplying both sides by fg we see the formula cited on p. 197,

$$(fg)' = (fg)\left(\frac{f'}{f} + \frac{g'}{g}\right).$$

Similarly, because logarithms take quotients to differences,

$$\ln\left(\frac{f}{g}\right) = \ln(f) - \ln(g).$$

Differentiating both sides of this equation with respect to t yields

$$\frac{1}{(f/g)}\left(\frac{f}{g}\right)' = \frac{1}{f}f' - \frac{1}{g}g' \quad \Rightarrow \quad \left(\frac{f}{g}\right)' = \left(\frac{f}{g}\right)\left(\frac{f'}{f} - \frac{g'}{g}\right).$$

You should know

- the terms *implicit function*, *implicit differentiation*, and *logarithmic differentiation*.

You should be able to

- calculate derivatives of implicitly defined functions;

- use the rules of logarithms to simplify the differentiation process.

❖ 3.8 Skill Exercises

Use implicit differentiation to find a formula for y' in each of #1–12.

1. $x^2 + y^2 = 4$

2. $(2x - 3)^2 + (5y + 2)^2 = 10$

3. $x^2 + 2x + y^3 = 2$

4. $4x^3 - 2y^2 = 9$

5. $4x^2 + 9xy - 2y^2 = 6$

6. $xe^y = 4y + 1$

7. $e^x \sin(y) = 7$

8. $\tan(2x + y) = 4$

9. $y^2 \ln x = x^2$

10. $x = (y - x)^4$

11. $\sin^{-1}(2x) = y^2$

12. $\tan^{-1}(x^2 y) = y^3 + x^4$

In #13–24 find the equation of the tangent line to the curve at the given point.

13. $x^2 + y^2 e^{-0.1y} = 4$; $(2, 0)$

14. $e^{-y/5}x^2 + y = 25$; $(5, 0)$

15. $7x^{1/5} + 2y^{1/7} = 9$; $(1, 1)$

16. $2^x + 3^y + 4^{x+y} = 1043$; $(4, 1)$

17. $e^{\cos(x)} + e^{\cos(y)} = 2\sqrt{e}$; $(\pi/3, \pi/3)$

18. $3e^{\cos(x)} + 7e^{\sin(y)} = 10\sqrt{e}$; $(\pi/2, 0)$

19. $4(x + y) = e^x - e^y$; $(0, 0)$

20. $y^2 - x = 2^y + 5^x$; $(1, 3)$

21. $\sin(x + y) = \frac{1}{2}$; $(\pi/9, \pi/18)$

22. $e^{x+y} = 6$; $(\ln(\sqrt[4]{6}), \ln(\sqrt[4]{216}))$

23. $\cos(3x + 4y) = 0$; $(\pi/2, \pi/2)$

24. $\log_3(8x - 2y) = 2$; $(1, 0.5)$

25. Determine where the tangent line is vertical when $y = (y - x)^4$.

26. Determine where the tangent line is vertical when $y = (2y + 4x)^4$.

27. Determine the point on $4(x + y) = e^x - e^y$ where the tangent line is horizontal. *(Hint: You may need the bisection method for part of your solution.)*

28. Determine the point on $y^3 = (y - x)^2$ where the tangent line is vertical.

Use logarithmic differentiation to find a formula for y' in each of #29–48.

29. $y = x^{2x}$

30. $y = 4x^{x^2 - 3}$

31. $y = x^{\tan(x)}$

32. $y = x^{\cot(x)}$

33. $y = x^{e^{x^2}}$

34. $y = x^{5^{3x}}$

35. $y = x^{5x} \sin(3x)$

36. $y = x^{-6x} \cos(4x)$

37. $y = \dfrac{x^{\sin(x)}}{\cos(x)}$

38. $y = \dfrac{x^{\csc(x)}}{e^{8x}}$

39. $y = x^2 + x^x$

40. $y = \sqrt{x^x} - 3x^{-7x}$

41. $y = \sqrt{(x^4 - 3x)^{45}(x^7 + 6x^6)^{13}}$

42. $y = \left((2x - 11)^{-21}\left(\frac{1}{x}\right)^{0.005}\right)^8$

43. $y = 9x^{12} \sin(\pi x)\sqrt{x^2 - 1}$

44. $y = 15x^8 \sin^6(3x) \tan\left(\frac{\pi}{3}x\right) e^{35x}$

45. $y = \dfrac{30x^4 \sqrt[4]{8x^3 - 5}}{\tan(3x)}$

46. $y = \dfrac{\sin^3(4x)(2x^{11} + 14x)^{10}}{\sqrt{x^6 + x^7}}$

47. $y = \sqrt{\dfrac{(8x - 7)e^{-0.3x}}{x^4 \sqrt[3]{11x^3 - 2x^4}}}$

48. $y = \sqrt{\dfrac{2x}{\sqrt{\frac{2x}{\sqrt{x}}}}}$

❖ 3.8 Concept and Application Exercises

49. Use implicit differentiation to show that the x-intercept and y-intercept of any tangent line to the curve $\sqrt{x} + \sqrt{y} = \sqrt{c}$ sum to c.

50. The equation $2y\ln(y) = x^2 - 4$ describes a curve in the upper half-plane that has two *branches* (see Figure 8.5). Determine how close these two branches come together by finding the values of x at which the tangent line is vertical.

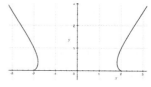

Figure 8.5: For #50.

51. Find a formula for y' when $\frac{x^2}{a^2} + \frac{y^2}{b^2} = 1$, where b and a are positive numbers.

52. Find a formula for y' when $\frac{x^2}{a^2} - \frac{y^2}{b^2} = 1$, where b and a are positive numbers.

53. Find all points on the curve described by $x^{2/3} + y^{2/3} = 1$ where the tangent line has a slope of -1.

54. The equation $x^2 + xy + y^2 = 1$ describes an ellipse that's been rotated away from the x-axis. Find the rightmost and leftmost point of this ellipse by locating the points at which the tangent line is vertical.

55. The equation $x^2 - xy + y^2 = 3$ describes an ellipse that's been rotated away from the x-axis. Find the highest and lowest points of this ellipse by locating the points at which the tangent line is horizontal.

Exercises #56 and #57 refer to the curve described by $(x^2 + y^2)^2 = 2x^2 - 2y^2$, shown in Figure 8.6.

Figure 8.6: For #56 and #57.

56. Find all points on the curve where the tangent line is horizontal.

57. Find all points on the curve where the tangent line is vertical.

The equation $x\cos(x) + y\cos(y) = 1$ describes a family of curves, two of which are shown in Figure 8.7. Exercises #58 and #59 refer to this equation.

58. Find the rightmost and leftmost points of the circle-like curve in the first quadrant by determining where the tangent line is vertical.

59. Find the highest and lowest points of the circle-like curve in the first quadrant by determining where the tangent line is horizontal.

60. As the number α decreases from 0.6 to 0.5, the curve described by $x^2 + y^2 e^{-\alpha y} = 2$ pinches together and separates, as shown in Figure 8.8.

   (a) Find a formula for y' when $x^2 + y^2 e^{-\alpha y} = 2$.

   (b) Show that when $x = 0$, the only way for $y' \neq 0$ is for $y = 2/\alpha$.

   (c) The point $(0, 2/\alpha)$ only sits on the curve for one specific value of α, which is where the curve pinches together. Find this value of α by substituting $y = 2/\alpha$ into the original equation, and choosing α so that the only solution is $x = 0$.

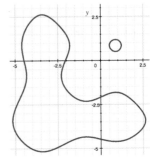

Figure 8.7: For #59 and #58.

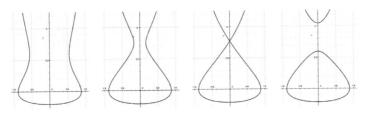

Figure 8.8: Graphs of $x^2 + y^2 e^{-\alpha y} = 2$ for #60, where α decreases from 0.6 to 0.5 from left to right.

61. In the initial test phase of the tracking system, four satellites will track an object as it passes over the North Pole at an altitude of 379 kilometers. Initially, the satellites (A, B, C and D) are in equatorial orbit so that units A and B are diametrically opposed (i.e., on opposite sides of the planet), and each unit can see two others. They hold minimal altitude to fulfill this requirement so, for example, if the satellites drop one kilometer in altitude, they will lose sight of one another.

   (a) Look up the equatorial radius of the Earth and determine how high the satellites are, initially.

   (b) How much does the altitude of the satellites need to be increased so that they can track the object as it passes over the North Pole? (Take into account that the Earth's polar and equatorial radii are different.)

62. Suppose $f(x)$ is differentiable, and $f(x + y) = k$ where k is constant. Find a formula for dy/dx.

63. Suppose $f(x)$ is differentiable, and $f(ax + by) = k$ where a, b and k are constants. Find a formula for dy/dx.

64. The **normal line** to a curve at the point P is the line through P that's orthogonal to the tangent line. Suppose that an ellipse is centered at the origin. Show that if the normal line at each point of the ellipse passes through the origin, the ellipse must really be a circle.

Chapter 3 Review

❖ True or False?

1. When differentiating a power function, $f(x) = x^n$, you should decrease the exponent if $n > 0$, and increase it if $n < 0$.

2. When f is increasing near x_0, and $f'(x_0)$ exists, we know that $f'(x_0) > 0$.

3. When $f'(x_0) > 0$ and f' is continuous at x_0, we know $f(x)$ is increasing near x_0

4. When $f'(x_0) > 0$, we know $f(x)$ is increasing near x_0.

5. If the tangent line at x_0 has a slope of zero, there is a first-order critical point at x_0.

6. All first-order critical points happen where the slope of the tangent line is zero.

7. If $f'(x_0)$ exists, the function f is continuous at x_0.

8. If f is continuous at x_0, we know that $f'(x_0)$ exists.

9. All first-order critical points are turning points.

10. The tangent line touches the graph of f at exactly one point.

Figure 9.1 depicts the relationship between pressure, x (measured in megapascals), and flow rate, y (measured in liters per minute). Data points (from [36]) are indicated by blue circles, and the red curve is the graph of an approximating polynomial. Each of #11–12 refers to this approximating function.

11. $y'(0.029) > y'(0.29)$

12. The units of y' are MPa per L/min.

13. The derivative of a sum is the sum of the derivatives.

14. The derivative of a difference is the difference of the derivatives.

15. The derivative of a product is the product of the derivatives.

16. The derivative of a quotient is the quotient of the derivatives.

17. Suppose that $f'(x) > 0$ over (a, b). Then $f(a) < f(b)$.

18. $\frac{d}{dt} f(g(t)) = f'(g'(t))$

19. The derivative of f has a jump discontinuity at a corner.

20. The derivative of $f(x) = \cos(x)$ is $f'(x) = \cos(x)$.

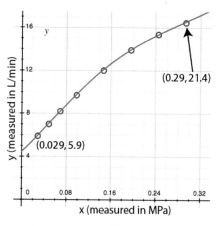

Figure 9.1: For #11–12.

❖ Multiple Choice

21. The derivative is defined as ...

    (a) the slope of the tangent line   (d) all of the above

    (b) the instantaneous rate of change   (e) none of the above

    (c) the limit of difference quotients

22. A *critical point* is a value of x where ...

    (a) f achieves its largest value   (d) all of the above

    (b) f' changes polarity   (e) none of the above

    (c) $f'(x) = 0$ or does not exist

23. The number $\left.\frac{df}{dx}\right|_{x_0}$ is ...

    (a) a slope   (d) all of the above

    (b) the same as $f'(x_0)$ (e) none of the above

    (c) a rate of change

24. Suppose $q = f/g$. When we use the Quotient Rule to find q', the numerator is ...

    (a) $fg' - gf'$ (d) $gf' + fg'$

    (b) $fg' + gf'$ (e) none of the above

    (c) $gf' - fg'$

25. Suppose $y'(2) > 0$. Then the object whose position is $y(t)$ is ...

    (a) speeding up   (d) moving toward the origin

    (b) slowing down   (e) not enough information

    (c) moving away from the origin

26. Suppose $y = \sin^{-1}(x)$. Then ...

    (a) $y' = -\sin^{-2}(x)$ (d) $y' = \frac{1}{\sqrt{1-x^2}}$

    (b) $y' = -\csc(x)\cot(x)$ (e) none of the above

    (c) $y' = \frac{1}{1+x^2}$

27. Suppose (x_0, y_0) is on the graph of the invertible function f. Then $(f^{-1})'(y_0) = \dots$

    (a) $-f^{-2}(y_0)f'(x_0)$ (d) $1/f'(y_0)$

    (b) $-f^{-2}(y_0)$ (e) none of the above

    (c) $1/f'(x_0)$

28. Suppose $y = \ln(3x + 1)$. Then ...

    (a) $y' = \ln(3)$

    (b) $y' = \frac{1}{3} + \frac{1}{x} + 1$

    (c) $y' = \frac{1}{3x+1}$

    (d) $y' = \frac{3}{\ln(3x+1)}$

    (e) none of the above

29. Suppose $f(x) = \cos(\sin(x))$. Then ...

    (a) $f'(x) = -\sin(\cos(x))$

    (b) $f'(x) = -\sin(\cos(x))\cos(x)$

    (c) $f'(x) = -\sin(\cos(x))\sin(x)$

    (d) $f'(x) = \cos(\cos(x))\cos(x)$

    (e) none of the above

30. Suppose $f(x) = e^{\cos(x)}$. Then $f'(x)$ is ...

    (a) $e^{-\sin(x)}$

    (b) $-e^{\cos(x)}$

    (c) $-\cos(x)e^{\sin(x)}$

    (d) $-\sin(x)e^{\cos(x)}$

    (e) None of the above.

❖ Exercises

For each of #31–38 find the equation of the tangent line to the graph of f at the specified point.

31. $f(x) = 3x^2 + 4x - 7$, $x = 2$

32. $f(x) = -2x^5 + 4x^3 + x + 2$, $x = -1$

33. $f(x) = \sin(4x) + x + 1$, $x = \pi/6$

34. $f(x) = \exp(-8x) - \sin(x)$, $x = 0$

35. $f(x) = (x^2 - 7)e^{2x}$, $x = 3$

36. $f(x) = \frac{2x+6}{12x^2+3x-1}$, $x = -3$

37. $f(x) = \sqrt{3 + x^6}$, $x = 1$

38. $f(x) = e^{\cos(x)}$, $x = 0$

Use implicit differentiation to find y' in each of #39–46.

39. $x\sin(y) = y\cos(x)$

40. $x + y^2 - x^3 + y^4 = 0.5$

41. $e^{x+y} = e^{xy}$

42. $\ln\left(\frac{x}{y^2}\right) = 4x + y$

43. $\tan(y) - y = x^2$

44. $\sqrt{x - y} = y + x$

45. $x^2 + y^3 = xe^y$

46. $\frac{1+e^x}{1-e^y} = 2$

47. The volume of a cube is given by $V(t) = [\ell(t)]^3$, where V is measured in m^3 and the side length, $\ell(t)$, which depends on time (e.g. a melting ice cube), is measured in meters.

    (a) Determine a general formula for $V'(t)$.

    (b) Identify the units of each factor in of $V'(t)$, and verify that they combine to yield the proper units for $V'(t)$.

48. The electric power of a circuit is given by $P = RI^2$ where P is the electric power (in watts), R is the resistance (in ohms) and I is the current (in amps). If the current and the resistance are changing with time, find the general form for dP/dt and then perform a dimensional analysis to show that the units are consistent on both sides of the equation.

49. Figure 9.2 depicts the graph of $e^{0.5x} = \sin(2x)\cos(2y)$.

(a) Explain why this equation allows only negative values of x.

(b) Derive a formula for dy/dx.

(c) Notice that the left-hand side of the defining equation approaches zero as $x \to -\infty$. Since $\sin(2x)$ does *not* approach zero as $x \to -\infty$ (the limit does not exist), what value(s) must y approach as $x \to -\infty$?

(d) For the values of y you found in part (c), what happens to dy/dx as $x \to -\infty$?

(e) For what values of x will our formula for dy/dx be undefined? As x approaches one of these values, what happens to the behavior of y?

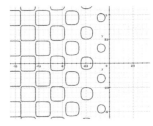

Figure 9.2: For #49.

Use logarithmic differentiation to find y' in each of #50–53.

50. $y = \frac{(9x-15)^7(1+\sin(3x))^5}{x^2+1}$

52. $y = (x^6+9)^{\arctan(x)}$

51. $y = \frac{x^4(2-x)^6}{(8x^4+1)^3}$

53. $y = (x^4+3)^{\sin(8x+1)}$

54. Suppose $f'(x) = (x-2)(x-3)^2$.

    (a) Locate the critical point(s) of f, and determine whether each is a turning point.

    (b) Determine the interval(s) on which f is increasing.

    (c) Sketch a graph of f.

55. Suppose $f(x) = (x+1)(x-3)(x-5)$.

    (a) Locate the critical point(s) of f, and determine whether each is a turning point.

    (b) Determine the interval(s) on which f is increasing.

    (c) Sketch a graph of f.

56. Two competitors in an egg-rolling competition push their eggs across a straight track that's 7.5 feet long, and then back. Their distance from the start/finish mark is shown in Figure 9.3.

    (a) Does one player stay in the lead during the entire race? If so, who?

    (b) At what time(s) is A farthest ahead of B? At what time(s) is B farthest ahead of A?

    (c) Estimate how fast A and B are moving after one second.

    (d) When is A is moving at -5 ft/sec?

    (e) What is happening to the numbers $A(t)$ when $\frac{dA}{dt} = -5$?

    (f) What does it mean, physically, to say that $\frac{dA}{dt} = -5$?

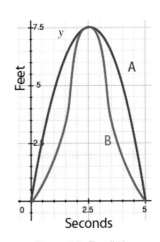

Figure 9.3: For #56.

57. Graph $f(x) = x^4 - 2x^2 - x$ and the tangent line at $x = -1$. Based on that graph, the tangent line appears to be "doubly tangent" to the graph, meaning that it's tangent at two different places. Verify that this is indeed the case.

58. Without first drawing the graph of f, draw the graph of f' so that both (a) and (b) are true:

    (a) The function f has critical points at $x = 0$, $x = 2$ and $x = 5$.

    (b) The graph of f is increasing when $x < 2$.

59. The graph of f is shown in Figure 9.4. Use it to determine those points at which $f'(x)$ is (a) positive, (b) negative, and (c) zero.

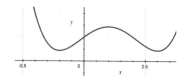

Figure 9.4: For #59.

60. Suppose $f(x) = \sqrt{x}$, and $b(t)$ is the y-intercept of the tangent line to the graph of f at $(t, f(t))$. Show that $b(t) = \frac{1}{2}f(t)$.

61. Suppose $f(x) = \frac{1}{\sqrt{x}}$, and $b(t)$ is the y-intercept of the tangent line to the graph of f at $(t, f(t))$. Show that $b(t) = \frac{3}{2}f(t)$.

62. Momentum is defined as the product of mass and speed, $p = mv$. Thinking of m as a particular, fixed amount of mass (a) show that $\frac{dp}{dv}$ is constant, and (b) explain what that means to us, physically.

63. Suppose we are casting a steel rod. Its mass will be $m(x)$ grams when its length is x cm. What are the units associated with $m'(3)$, and what does it mean in this context?

64. Suppose T is the amount of thrust required for an aircraft to maintain its airspeed, and s is the airspeed. Is the equation $\frac{dT}{ds} = 2342$ sensible? If so, what does it mean? If not, why not?

65. Suppose f is the fuel economy of a vehicle (in mpg), and v is its speed. Does the equation $\frac{df}{dv} = -2$ sensible? If so, what does it mean? If not, why not?

66. Wind results from pressure differences in the Earth's atmosphere. Suppose the velocity of wind (in miles per hour) is $v(s) = s^{-2} + (1.01)^{-s}$, where $1 \le s < 10$ is the distance (in miles) from the center of a region of high pressure.

   (a) What are the units of the difference quotient for v?

   (b) What is the *average* rate of change in wind speed between 1 and 3 miles from the center of the high-pressure zone?

   (c) What is the *average* rate of change in wind speed between 6 and 7 miles from the center of the high-pressure zone?

   (d) Is the wind growing stronger or weaker as we move away from the center of the high-pressure zone?

67. Suppose $E(T)$ is the amount of energy stored in a substance, measured in calories per gram, when its temperature is T kelvin. What are the units of $E'(T)$, and what is the physical meaning of $E'(T)$?

68. Suppose $m(t)$ is the amount of un-metabolized anesthetic in your body, measured in milligrams, t hours after it's administered. What are the units of m', and what is the physical meaning of $m'(t)$?

69. Suppose $V(P)$ is the volume of an artery when the blood pressure is P. Use the first derivative to write an inequality that says, "increasing the pressure causes the artery to expand."

70. When the volume of a vessel depends on the pressure of its contents according to $V = V_0 + CP$, where V_0 is a constant, we say that the vessel is *compliant*. The number C, called the *compliance parameter*, can depend on volume. Use the first derivative to write an inequality that says, "the vessel becomes less compliant as it swells."

71. In any particular region of the body, the resistance to blood flow is $R(f) = \tilde{\eta} + \tilde{\rho}f$ when the flow rate is f cubic meters per second (called a *volumetric flow rate*, based on its units), where $\tilde{\eta}$ and $\tilde{\rho}$ are positive constants that depend on the blood's viscosity and density (respectively).

    (a) Show that $R'(f)$ is constant.

    (b) What is the physical meaning of the fact in part (a)?

72. In a resistor-capacitor circuit, the charge on the capacitor during discharge varies according to $q(t) = q_0 e^{-t/RC}$, where t is measured in seconds, q_0 is the initial charge on the capacitor, and RC is a number called the *capacitive time constant* (CTC), which has dimensions of time.   ─┤ Electrical Engineering

    (a) If the CTC is 1 second, what is the rate of change of charge on the capacitor at times $t = 1$ and $t = 10$?

    (b) If q is the charge on the capacitor, what does $q'(t)$ represent physically?

    (c) Based on the formula for $q'(t)$, is it better to have a larger or a smaller CTC if we want a large current?

73. A study of tuberculosis patients estimated the probability that an infected person in the age bracket $30 - 50$ years will die *after* age τ is $L(\tau) = e^{-0.06(\tau - 30))}$.   ─┤ Public Health

    (a) What does it mean that $L(30) = 1$?

    (b) Find a formula for $L'(\tau)$, and verify that $L'(\tau) < 0$.

    (c) Does it make *sense* that $L'(\tau) < 0$? If not, what do you think is wrong with the model? If so, why?

74. Exercise #66 on p. 168 asked you about Kepler's Third Law, which says that the length of time it takes a planet to orbit the sun (in Earth-years) is related to the shape of the its orbit by $T = a^{3/2}$, where a denotes the the length of the semi-major axis of the planet's (elliptical) orbit, measured in astronomical units. You showed $T'(a) = 1.5 a^{1/2}$. What does it mean, physically speaking, that $T'(a) > 0$?

75. In Chapter 1 you saw a famous equation in modern physics that relates an object's speed to its inertia (which is quantified by mass):   ─┤ Modern Physics

$$m = \frac{m_0}{\sqrt{1 - \left(\frac{v}{c}\right)^2}},$$

where m_0 denotes the object's mass at rest, c is the speed of light, and v is the object's speed. Suppose we measure mass in kilograms, and speed in meters per second. Determine a formula for $m'(v)$, including appropriate units, and explain what it means about the relationship between speed and inertia.

76. Louis de Broglie was awarded the Nobel Prize for Physics in 1929 for his work on the wave-like nature of matter. The *de Broglie wavelength* of an object is   ─┤ Modern Physics

$$\lambda = \frac{h}{mv},$$

─┤ The name *Broglie* is pronounced "broy-lee"

where m is the object's mass as discussed in #75, and h is Planck's constant. Show that $\lambda'(v) < 0$ and interpret this fact in context.

77. Quantum effects begin to manifest in an ideal gas when the average distance between its molecules is smaller than its *thermal de Broglie wavelength,* which is roughly the average de Broglie wavelength of all its molecules (see #76). The thermal de Broglie wavelength is   ─┤ Modern Physics

$$\Lambda = \sqrt{\frac{h^2}{2mk_B T}},$$

where m is the average mass of a molecule of the gas, h is Planck's constant, k_B is Boltzmann's constant, and T is the temperature. Show that $\Lambda'(T) < 0$.

Chapter 3 Projects and Applications

✤ Parabolic Reflectors

Whether used for satellite TV or detecting the structure of the universe, dishes that send and receive signals are built as *paraboloids* (i.e., parabolas in 3D). This is because parabolas have a unique and important feature: rays (signals) that strike the parabola perpendicular to its directrix are all directed to its focus (see the left-hand image in Figure 10.1). Your job in this exercise is to verify this property

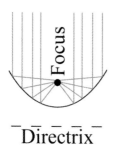
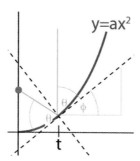

Figure 10.1: Reflecting incoming rays to the focus.

analytically for the parabola $y = ax^2$, where $a > 0$ is constant. To do this, you'll use the fact that the angle of incidence equals the angle of reflection.

1. Verify that the tangent line to the parabola $x = 1/2a$ has a slope of 1.

2. Show that if a light ray hits the parabola at $x = 1/2a$, it will reflect to the point $(0, 1/4a)$.

3. Show that if a light ray hits the parabola at any spot, say $x = t$, it will reflect to the point $(0, 1/4a)$. (*You should derive an equation for the line along which the reflected ray travels, and find a formula for the y-intercept. Your job is to verify that it depends on a (which characterizes the parabola) but not on t.*)

Note: In order to determine the angle at which the incident ray reflects off the dish, you first need to know the slope of the tangent line to the parabola at (t, at^2). The rest involves manipulating the trigonometric functions to get the information you want (e.g., the slope of the line along which the reflected ray travels). Try to work in terms of ϕ, as shown in the right-hand image of Figure 10.1, because you can easily relate it to the slope of the tangent line by using the tangent function. You should consider reviewing the techniques demonstrated on p. 71.

✤ The Cumulative Area Function

Sometimes we face a situation in which we know more about a function's derivative than we do about the function itself. The following exercises lead you through a geometric (i.e., rather than physical) example in which this is true.

Suppose that $f(x)$ is a continuous function whose values are non-negative. The **cumulative area function** is

$$A(t) = \begin{cases} \text{the area between the the graph of } f \text{ and the } x\text{-axis} \\ \text{from } x = 0 \text{ to } x = t \end{cases}$$

The assumption of non-negativity is not actually necessary, but it makes this introduction easier to understand.

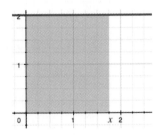

Figure 10.2: For #1.

▷ Simple Cases

The following pair of simple examples help us to establish a pattern that will be important later.

1. Figure 10.2 depicts the graph of the constant function $f(x) = 2$.

    (a) Based on the geometry you see in the graph, find a formula for $A(t)$.

    (b) Use the rules of differentiation from this chapter to determine $A'(t)$.

    (c) What do you notice about $A'(t)$ and the height of the graph, $f(t)$?

2. The graph of $f(x) = 0.2x + 1$ is shown in Figure 10.3.

    (a) Based on the geometry you see in the graph, find a formula for $A(t)$.

    (b) Use the rules of differentiation from this chapter to determine $A'(t)$.

    (c) What do you notice about $A'(t)$ and the height of the graph, $f(t)$?

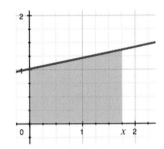

Figure 10.3: For #2.

▷ More Complicated Cases

It was easy to find a formula for $A(t)$ in #1 and #2 because $f(x)$ was a linear function, so the shaded region was bounded by lines (see Figures 10.2 and 10.3). When $f(x)$ is a nonlinear function, the upper boundary of the shaded region is a curve instead of a line, so finding a simple formula for $A(t)$ is much more difficult (and sometimes impossible). However, it's often true that we can derive a formula for $A'(t)$ by using the definition of the derivative as a limit of difference quotients:

$$A'(t) = \lim_{\Delta t \to 0} \frac{\Delta A}{\Delta t} = \lim_{\Delta t \to 0} \frac{A(t + \Delta t) - A(t)}{\Delta t}.$$

3. Suppose $f(x) = x^2$. Figure 10.4 shows the region whose area is $A(t)$. When $\Delta t > 0$, the number $A(t + \Delta t)$ is the area of a larger region. With only the calculus that we've covered up to this point, we can't find the difference $\Delta A = A(t + \Delta t) - A(t)$ exactly, but we can estimate it in a useful way. Specifically, based on the geometry shown in Figure 10.4,

    (a) explain why $t^2 \Delta t \leq A(t + \Delta t) - A(t)$, and

    (b) explain why $A(t + \Delta t) - A(t) \leq (t + \Delta t)^2 \Delta t$.

    Putting these inequalities together, we see that

    $$t^2 \Delta t \leq \Delta A \leq (t + \Delta t)^2 \Delta t.$$

    (c) Divide this string of inequalities by Δt, the result of which will be that you've trapped the difference quotient between two known quantities.

    (d) Use the Squeeze Theorem to show that

    $$\lim_{\Delta t \to 0^+} \frac{A(t + \Delta t) - A(t)}{\Delta t} = t^2.$$

    We began by using the geometry shown in Figure 10.4, which depicts $\Delta t > 0$.

    (e) How does the figure change when $\Delta t < 0$?

    (f) How does our calculation change when $\Delta t < 0$?

4. Suppose $f(x) = x^3$. Use the trapping technique you learned in #3 and the Squeeze Theorem to verify that $A'(t) = t^3$.

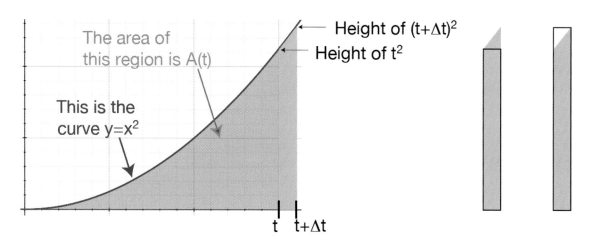

Figure 10.4: (left) A depiction of $A(t)$ and the difference ΔA in the case when $\Delta t > 0$; (right) estimating the size of ΔA.

5. Suppose $f(x) = \frac{1}{x}$. Use the trapping technique you learned in #3 and the Squeeze Theorem to verify that $A'(t) = \frac{1}{t}$.

6. Suppose $f(x) = e^x$. Use the trapping technique you learned in #3 and the Squeeze Theorem to verify that $A'(t) = e^t$.

> You'll need to use the fact that exponential functions are continuous.

7. Suppose $f(x) = e^{-x}$. Use the trapping technique you learned in #3 and the Squeeze Theorem to verify that $A'(t) = e^{-t}$.

In parts #3–7 the function $f(x)$ was positive when $x \geq 0$, and either always increasing or always decreasing, but neither of those characteristics is needed for $A'(t) = f(t)$. In fact, we only need $f(x)$ to be continuous for this to work! We'll prove it in Chapter 5.

Chapter 4
Applications of the Derivative

Whereas the previous chapter focused on developing an understanding of and ability to calculate with the derivative, this chapter is devoted to its applications.

4.1   Linear Approximation

Suppose that 2 seconds after launch, a model rocket is 500 feet above the ground and rising at 250 feet per second. How high will it be 1 second later, at time $t = 3$? (Really, take a moment to think about it.)

Let's use $h(t)$ to denote the altitude of the rocket (in feet) t seconds after launch. Then $h(2) = 500$ and $h'(2) = 250$, so

$$h(3) \approx \underbrace{500}_{\text{height at time } t = 2} + \underbrace{250}_{\text{rate of ascent (ft/sec)}} \underbrace{(1)}_{\text{length of time between } t = 3 \text{ and } t = 2 \text{ seconds}} = 750 \text{ feet.}$$

This is an approximation because it's unlikely that the rocket's velocity *remains* 250 feet per second, but if its change in velocity is small, the approximation is pretty good. Similarly, we could approximate

$$h(4) \approx 500 + 250 \, (2) = 1000 \text{ feet}$$
$$h(5) \approx 500 + 250 \, (3) = 1250 \text{ feet}$$
$$\vdots \qquad \vdots \qquad \vdots \quad \vdots$$
$$h(t) \approx 500 + \underbrace{250(t - 2)}_{\text{seconds between time } t \text{ and time 2}}.$$

Figure 1.1:   Model rocket in flight.

The linear function that you see on the right-hand side of this approximation is called the **linear approximation** of, or **linearization** of $h(t)$, and its graph is exactly the tangent line to the graph of h at time $t_0 = 2$. More generally ...

Linearization: Suppose $h(t)$ is differentiable at t_0. The linearization of $h(t)$ at t_0 is the function
$$\ell(t) = h(t_0) + h'(t_0)(t - t_0).$$

Think of t_0 as the moment at which we measure data (such as height, $h(t_0)$, and velocity, $h'(t_0)$). Then $(t - t_0)$ is how long it's been since we measured.

Using a linear approximation to estimate altitude

Example 1.1. Suppose that 7 seconds after launch, a model rocket is 950 feet in the air, and is rising at 100 feet per second. If $h(t)$ is the altitude of the rocket

t seconds after launch, (a) write down the linear approximation of $h(t)$ near time $t_0 = 7$, and (b) use it estimate the altitude of the rocket at time $t = 7.5$.

Solution: (a) Since $h(7) = 900$ and $h'(7) = 100$, the linear approximation is $\ell(t) = 950 + 100(t - 7)$. (b) Using this formula we estimate $h(7.5) \approx \ell(7.5) = 1000$. ∎

Just as we used the linear approximation of h to estimate altitude, we can use the linear approximation to estimate any function value. We just need to know the function value at a specific time, and how quickly the function is changing.

Linear approximation of the sine

Example 1.2. Suppose $f(t) = \sin(t)$. (a) Determine the linear approximation of $f(t)$ at $t_0 = 0$, and (b) use it to estimate $\sin(0.25)$.

Solution: (a) Since $f'(t) = \cos(t)$, the linear approximation of f about $t_0 = 0$ is

$$\ell(t) = \underbrace{f(0)}_{=0} + \underbrace{f'(0)}_{=1}(t - 0) = 0 + t = t.$$

(b) The linear approximation tells us that $f(0.25) \approx \ell(0.25) = 0.25$. ∎

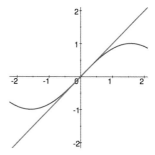

Figure 1.2: Graphs of $y = \sin(t)$ and $y = t$ near $t_0 = 0$.

Try It Yourself: Linear approximation of the logarithm

Example 1.3. Suppose $f(t) = \ln(t)$. (a) Determine the linear approximation of $f(t)$ at $t_0 = 1$. (b) Use it to estimate $f(0.97)$.

Answer: (a) $\ell(t) = t - 1$, and (b) $f(0.97) \approx -0.03$. ∎

Full Solution On-line

In the previous examples, we used linear approximation to estimate function values that are difficult to calculate precisely. The next pair of examples use the same technique, but the particular function is not immediately apparent. The basic strategy of our estimation will be this: find a number that, when changed slightly, makes the whole computation much easier. Treat that number as a variable, and make a linear approximation of the resulting function.

Linear approximation of a power function

Example 1.4. Use linear approximation to estimate $\sqrt{1.25}$.

Solution: It's a lot easier to compute $\sqrt{1}$ than $\sqrt{1.25}$, so as we suggested above, let's replace the number 1.25 with a variable and consider the function $f(x) = \sqrt{x}$. Because calculating $f(1)$ is easy, we can estimate

$$f(1.25) \approx \underbrace{f(1) + f'(1)(1.25 - 1)}_{\text{This is the linear approximation of } f \text{ at } x_0 = 1} = 1 + (0.5)(0.25) = 1.125.$$ ∎

Linear approximation of a power function

Example 1.5. You overhear your neighbor say that 10 years ago he replaced his furnace for $3000. If inflation has hovered about 3% per year, use linear approximation to estimate how much would it cost to replace the furnace today.

Solution: Let's use c to denote the cost of replacing the furnace today, and write the cost of replacing the furnace n years ago by c_n. Then

$$\underbrace{c}_{\text{cost this year}} = \overbrace{c_1}^{\text{cost last year}} + \overbrace{0.03c_1}^{\text{3\% of last year's cost}} = 1.03c_1.$$

For the same reason, $c_1 = 1.03c_2$, which allows us to write the equation $c = 1.03c_1$ as $c = (1.03)^2 c_2$. Similarly,

$$c_2 = 1.03c_3 \quad \Rightarrow \quad c = (1.03)^3 c_3$$
$$c_3 = 1.03c_4 \quad \Rightarrow \quad c = (1.03)^4 c_4$$
$$\vdots \qquad\qquad \vdots$$
$$c_9 = 1.03c_{10} \quad \Rightarrow \quad c = (1.03)^{10} c_{10}.$$

We know that the cost 10 years ago was $c_{10} = 3000$, so the cost today is

$$c = 3000(1.03)^{10}.$$

The value of c would be a lot easier to compute if the base of that 10th power were 1 instead of 1.03, which reminds us of the technique that we used in Example 1.4. So let's write today's cost of replacing the furnace as

$$c(1.03), \quad \text{where} \quad c(x) = 3000x^{10}.$$

Because calculating $c(1)$ is easy, we can approximate $c(1.03)$ as

$$c(1.03) \approx \underbrace{c(1) + c'(1)(1.03 - 1)}_{\text{This is the linear approximation of } c \text{ at } x_0 = 1} = 3000 + (30,000)(0.03) = 3900. \qquad \blacksquare$$

❖ Newton's Method

We began this section with a story of a model rocket headed skyward. Now consider the end of that flight, when it comes back down.

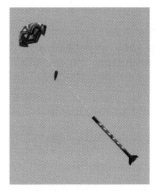

Figure 1.3: A model rocket falling back to Earth, with its parachute deployed.

Using linearization to estimate the landing time

Example 1.6. Suppose that 120 seconds after launch, the model rocket is 30 feet above the ground, and is falling at 5 feet per second. Use linear approximation to estimate the time at which it lands.

Solution: Let's continue to denote the altitude of the model rocket by $h(t)$. Then the information we have about the descent is $h(120) = 30$ and $h'(120) = -5$ (negative because $h(t)$ is decreasing), so the linearization of h at $t_0 = 120$ is

$$\ell(t) = 30 - 5(t - 120).$$

The rocket arrives at the ground when $h(t) = 0$, which we approximate by finding out when $\ell(t) = 0$. This is an easy equation to solve:

$$0 = 30 - 5(t - 120) \quad \Rightarrow \quad t = 126.$$

So our estimate is that the rocket will reach the ground at $t = 126$ seconds after launch. $\qquad \blacksquare$

   In the left-hand image of Figure 1.4, which depicts the graphs of $h(t)$ and $\ell(t)$, you can see that $t = 126$ is *not* a root of h—our approximation of the landing time is too early. However, $t = 126$ is closer to the landing time than $t = 120$. If we repeat the procedure but start at $t = 126$, we can get an even better estimate of the landing time.

Try It Yourself: Approximating the location of a root with a tangent line

Example 1.7. Suppose that 126 seconds after launch, the model rocket is 10.5 feet above the ground and falling at 1.95 feet per second. Use linear approximation to estimate the time at which it lands.

Answer: $\ell(t) = 10.5 - 1.95(t - 126)$, which is zero when $t = 131.38$ seconds. ∎

Full Solution On-line

> While it's unlikely that the linear approximations will ever bring us to the exact root, they'll bring us closer and closer, and eventually "close enough." Your definition of "close enough" will depend on what you're doing—i.e., how much error your calculations can allow.

In the right-hand image of Figure 1.4 you can see that $t = 131.38$ is not a root of h either, but we came *closer* to the root of $h(t)$ each time we used a linear approximation. We can get better and better estimates of the landing time (i.e., the root of $h(t)$) by using linear approximations over and over again.

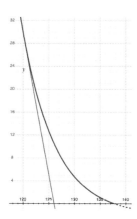
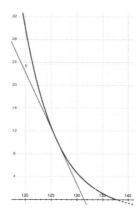

Figure 1.4: (both) The graph of $h(t)$ and a tangent line; (left) linear approximation at $t = 120$ predicts that the rocket will land at $t = 126$; (right) starting at $t = 126$, linear approximation predicts that the rocket will land at $t = 131.38$.

Since the process of estimating a root using linear approximation is always the same, we can program a computer to do this for us. Toward writing down the algorithm, let's denote our initial guess at the time of the root by t_0, and the successive estimates that we make based on linear approximations by $t_1, t_2, t_3, \ldots$ named in order so that t_n is our estimate after the nth iteration. The linearization of h at time t_n is $\ell(t) = h(t_n) + h'(t_n)(t - t_n)$, which is zero when

> Beginners often ask how to find t_0, the starting place for this method. The short story is that choosing t_0 is as much an art as a science. Typically, you have some idea about the root's location (perhaps from looking at a graph, or based on a conclusion of the Intermediate Value Theorem), and you use that to make the first guess.

$$0 = h(t_n) + h'(t_n)(t - t_n). \quad \text{We solve this for} \quad t = t_n - \frac{h(t_n)}{h'(t_n)},$$

which is our next estimate of the root. This algorithm is called **Newton's method.**

Newton's method:
$$t_{n+1} = t_n - \frac{h(t_n)}{h'(t_n)} \tag{1.1}$$

Using Newton's method to search for a root

Example 1.8. Suppose $h(t) = -t^3 + t + 1$. Approximate the root of h by performing two steps of Newton's method, starting from $t_0 = 1$.

Solution: Since $h'(t) = -3t^2 + 1$ we know that

$$t_1 = t_0 - \frac{h(t_0)}{h'(t_0)} = 1 - \frac{1}{-2} = 1.5.$$

Similarly,

$$t_2 = t_1 - \frac{h(t_1)}{h'(t_1)} = 1.5 - \frac{-0.875}{-5.7501} = 1.3478.$$

These linear approximations are depicted in Figure 1.5. Notice that each linear approximation brings us closer to the root. ∎

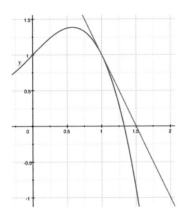 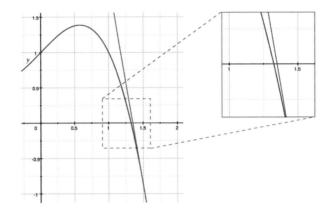

Figure 1.5: (left) The first step of Newton's method in Example 1.8; (right) the second step of Newton's method in the same example.

Using Newton's method to approximate $\sqrt[3]{2}$

Example 1.9. Estimate $\sqrt[3]{2}$ by performing three steps of Newton's method.

Solution: Newton's method is an algorithm for finding roots of functions, so the first thing we need is a function whose root is at $\sqrt[3]{2}$. Many beginners try $h(t) = t - \sqrt[3]{2}$, but this function requires us to use $\sqrt[3]{2}$, which is the number whose value we're trying to approximate in the first place.

Instead, let's begin by writing an equation that specifies the number we want: $t = \sqrt[3]{2}$. The cube root is what makes calculation difficult, so let's cube both sides of our equation and rewrite it as $t^3 = 2$, which is the same as saying $t^3 - 2 = 0$. That is, calculating $\sqrt[3]{2}$ is the same as finding the root of $h(t) = t^3 - 2$.

Now that we have a function, h, we need an initial guess at its root. When we start with $t_0 = 1.5$, Newton's method tells us that

$$
\begin{aligned}
t_1 &= t_0 - \frac{h(t_0)}{h'(t_0)} = 1.5 - \frac{h(1.5)}{h'(1.5)} = 1.296296296 \\[2mm]
t_2 &= t_1 - \frac{h(t_1)}{h'(t_1)} = 1.296296296 - \frac{h(1.296296296)}{h'(1.296296296)} = 1.260932225 \\[2mm]
t_3 &= t_2 - \frac{h(t_2)}{h'(t_2)} = 1.260932225 - \frac{h(1.260932225)}{h'(1.260932225)} = 1.259921861.
\end{aligned}
$$

The actual root of h occurs at the irrational number $\sqrt[3]{2} = 1.25992104989\ldots$ so Newton's method has led to us to an approximation that's correct to six decimal places in only three steps. ∎

Try It Yourself: Using Newton's method to approximate $\sqrt[4]{7}$

Example 1.10. Estimate $\sqrt[4]{7}$ by performing three steps of Newton's method, starting at $t_0 = 2$.

> It's easy to implement Newton's method with a spreadsheet program, which makes it very fast. Step-by-step instructions can be found on p. 318.

Full Solution
On-line

Answer: $t_3 = 1.626623403$, using $f(t) = t^4 - 7$. ∎

It's important to understand that Newton's method is an algorithm for finding the root of a function—*any* function, as long as we can calculate its derivative.

Using Newton's method to find critical points

Example 1.11. Use Newton's method to find the critical point of $g(t) = t^2 - \sin(t)$.

Solution: The critical point of g happens where $g'(t) = 0$, which is near $t_0 = 0.5$ (see Figure 1.6). Since we're looking for a root of g', this is the function that plays the role of h in equation (1.1), and Newton's method takes the form

$$t_{n+1} = t_n - \frac{g'(t_n)}{g''(t_n)}. \quad \begin{array}{l} \leftarrow \quad \text{the function whose root we're finding} \\ \leftarrow \quad \text{the derivative of that function} \end{array}$$

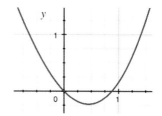

Figure 1.6: $y = t^2 - \sin(t)$.

Since $g'(t) = 2t - \cos(t)$ and $g''(t) = 2 + \sin(t)$, starting with $t_0 = 0.5$ yields ...

t	$g'(t)$	$g''(t)$
$t_0 = 0.500000000$	0.1224174381096270000000	2.479425538604200
$t_1 = 0.450626693$	0.0010790504957215500000	2.435529752624670
$t_2 = 0.450183648$	0.0000000883536128704066	2.435130891704640
$t_3 = 0.450183611$	0.0000000000000000000000	2.435130859036710

Based on this table, we'd like to say the critical point of g occurs at t_3, but it's always best to allow for round-off error when using a computer, so let's just say that we've found the critical point to within machine precision. ∎

A spreadsheet was used to generate this table of data. On p. 318 you can see how to do it for yourself.

Try It Yourself: Using Newton's method to find critical points

Example 1.12. Use Newton's method to find the critical point of $g(t) = \cos(3t) - t\sin(t)$ nearest to $t_0 = 2$.

Answer: $t^* = 2.1238433714$. ∎

Full Solution
On-line

When the numbers t_0, t_1, t_2 ... that are generated by Newton's method get closer and closer to a particular value, as you've seen in all the examples so far, we say that Newton's method **converges**. That doesn't always happen ...

Newton's method can get caught in a cycle

Example 1.13. The graph of $f(t) = t^3 - t + 0.5$ is shown in Figure 1.7. Starting at $t_0 = 1$, use the graph of f to locate the third iterate of Newton's method, t_3.

Solution: We find t_1 by drawing the tangent line to the graph of f at $(t_0, f(t_0))$ and following it to the t-axis. We continue by repeating the process at $(t_1, f(t_1))$ ...

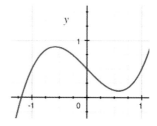

Figure 1.7: Graph of $f(t) = t^3 - t + 0.5$.

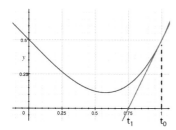

Draw the tangent line to the graph of f at $\left(t_0, f(t_0)\right)$ and follow it to the t-axis to find t_1.

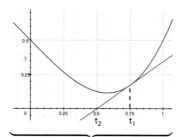

Draw the tangent line to the graph of f at $\left(t_1, f(t_1)\right)$ and follow it to the t-axis to find t_2.

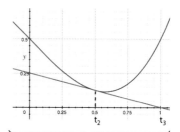

Draw the tangent line to the graph of f at $\left(t_2, f(t_2)\right)$ and follow it to the t-axis to find t_3.

Starting at $t_0 = 1$ leads to $t_1 = 0.75$, $t_2 = 0.5$, and $t_3 = 1$, which is where we started! If we continue, we'll revisit these values over and over again. ∎

Unlike previous examples, the successive iterations of Newton's method in Example 1.13 do *not* lead us closer and closer to a particular value of t. When this happens, we say that the sequence of approximations **diverges**.

We'll talk about sequences, convergence, and divergence in greater depth in Chapter 8.

Try It Yourself: Newton's method can fail to converge

Example 1.14. Suppose $f(t) = t^{1/3}$ and we try to solve the equation $f(t) = 0$ by starting Newton's method at $t_0 = 1$. Calculate t_4.

Answer: $t_1 = -2$, $t_2 = 4$, $t_3 = -8$, $t_4 = 16$, $t_5 = -32$. ∎

Full Solution On-line

Note: Newton's method is a powerful and typically reliable method for finding the root of a function. The previous pair of examples are intended to introduce you to its limitations, and the idea of divergence.

▷ **Common Errors with Newton's Method**

Newton's method says $t_{n+1} = t_n - \frac{f(t_n)}{f'(t_n)}$, but many people have trouble remembering whether the denominator of the fraction is $f(t)$ or $f'(t)$. Here are two mnemonics that might help you.

1. (Graphical) In order for Newton's method to "work" we need tangent lines to intersect the t-axis. That won't happen if the tangent line is horizontal (i.e., when $f'(t) = 0$), so our formula should break down at such a point. Putting $f'(t_n)$ in the denominator of the fraction makes that happen.

2. (Dimensional) We started our discussion of Newton's method by looking for the root of $h(t)$, which was the time at which the model rocket landed. In this context, the variable t was measured in seconds, so

$$\underbrace{t_{n+1}}_{\text{seconds}} = \underbrace{t_n}_{\text{seconds}} - \left(\begin{array}{c}\text{this} \quad \text{fraction} \\ \text{must have units} \\ \text{of seconds}\end{array}\right).$$

Since $h(t)$ was measured in feet and the units of $h'(t)$ were feet per second, the fraction cannot be $h'(t_n)/h(t_n)$ because that fraction carries units of $\frac{\text{ft/sec}}{\text{ft}} = \frac{1}{\text{sec}}$. Rather, it must be $h(t_n)/h'(t_n)$, whose units are $\frac{\text{ft}}{\text{ft/sec}} = \text{seconds}$.

Units can help you reason your way to or through a formula.

✧ **Differentials**

At the beginning of this section we talked about a model rocket whose altitude was changing at $h'(2) = 250$ feet per second. We estimated that

$$h(t) \approx \underbrace{500}_{\substack{\text{altitude at} \\ \text{time } t_0 = 2}} + \underbrace{250(t-2)}_{\substack{\text{our approximation of the } change \text{ in} \\ \text{altitude since time } t_0 = 2}}.$$

The part of this linear approximation that estimates *change* is called the **differential** of h and is denoted by dh. So in our example of the model rocket, $dh = 250(t-2)$. More generally,

$$h(t) \approx h(t_0) + \underbrace{h'(t_0)(t-t_0)}_{\text{our approximation of the change in } h \text{ since time } t_0} \quad \text{so} \quad dh = h'(t_0)(t-t_0),$$

Think of t_0 as the moment at which we measure data (such as height and velocity). Then $\Delta t = t - t_0$ is how long it's been since we measured.

and we say that the differential, dh, is a **linear approximation of change** in h. If we subtract $h(t_0)$ from both sides of this approximation, we see

$$h(t) - h(t_0) \quad \approx \quad h'(t_0)(t - t_0),$$

which is often written as

$$\underbrace{\Delta h}_{\text{the actual change in } h} \quad \approx \quad \underbrace{dh.}_{\text{the linear approximation of change in } h}$$

To remind ourselves that, generally speaking, this approximation is only good when t is very close to t_0, we often write

$$dh = h'(t_0)\, dt, \tag{1.2}$$

where dt, called the **differential** of t, is understood to be some very small increment of t.

Note: The notation of equation (1.2) is new, but it expresses exactly the same idea about estimating change that you saw at the beginning of the section.

▷ A graphical way to understand differentials

In Example 1.4 we estimated $\sqrt{1.25}$ using the linear approximation

$$\sqrt{x} \approx 1 + \frac{1}{2}(x - 1).$$

The graph of the right-hand side is just the tangent line to the graph of $f(x) = \sqrt{x}$ at $x = 1$. As indicated in Figure 1.8, we use the symbols Δx and Δy when talking about actual changes in the function value, and we use the symbols dx and dy to discuss the linear approximation of change.

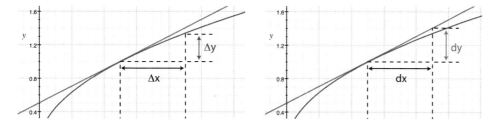

Figure 1.8: (both) The graph of $y = \sqrt{x}$ and the line $y = 1 + 0.5(x - 1)$; (left) the term Δy refers to a change in y; (right) the term dy refers to the linear approximation of the change in y.

▷ A connection to Leibniz notation

We've tended to use Lagrange notation for the derivative (i.e., h'), but many people find Leibniz notation helpful when learning about differentials. By writing h' as dh/dt, the idea that

$$\text{(linear approximation of change in height)} = \text{(rate)} \times \text{(time)}$$

is written as

$$dh = \left(\frac{dh}{dt}\right) dt, \tag{1.3}$$

which has a certain simple elegance to it. With that said, it's important to remember that $\frac{dh}{dt}$ (velocity) is a number that changes from one moment to the next. For example, when $h(t) = 3t^2 + 7t$ we know that $h'(t_0) = 6t_0 + 7$, so we could write

$$dh = (6t_0 + 7) \; dt.$$

At time $t_0 = 0$ this says that the linear approximation of change in h is 7 times larger than a corresponding change in t. At time $t_0 = 1$, it says that the linear approximation of change in h is 13 times larger than a corresponding change in t.

> If you divide this equation by dt, you see
>
> $$\frac{dh}{dt} = 6t_0 + 7.$$

Interpreting differentials

Example 1.15. Interpret and discuss the equations (a) $dA = 2 \; dP$, where A and P are the area and perimeter of a circle, and (b) $dh = -0.002 \; da$, where a is the age of a child (in years) and h is his or her height (in feet).

Solution: (a) This equation says that the (linear approximation of) increase in a circle's area is twice as large as the increase in its perimeter. Though it's not obvious, this is true when the perimeter of a circle is 4π units.

(b) Think about the differential da as a small increment in the child's age, such as one day, hour, or minute. This equation says that the corresponding change in height is *negative,* so the child is *shrinking!* While some medical condition might be at work here, at first glance this equation doesn't make sense.

▷ **Differentials in science and engineering**

In engineering and the physical sciences, the Product Rule is often written as

$$d(fg) = f \; dg + g \; df, \tag{1.4}$$

which emphasizes the relationship between the *change in the product* and the *change in the factors*. Similarly, the Quotient Rule is often written as

$$d\left(\frac{f}{g}\right) = \frac{g \; df - f \; dg}{g^2}. \tag{1.5}$$

> More precisely, equation (1.4) relates the *linear approximation* of change in the product to the *linear approximation* of change in the factors.

Note that the independent variable is irrelevant in these equations. They express the linear approximation of change, regardless of why or how things are changing.

Differentials and momentum

Example 1.16. Momentum is the product of mass and velocity, $p = mv$. Use differentials to write the formula for calculating the linear approximation of change in momentum.

Solution: Both mass and velocity can change (e.g., a rocket changes its velocity by throwing fuel out the back, which reduces the mass of the rocket). So we use the Product Rule to write $dp = m \; dv + v \; dm$. ∎

Try It Yourself: Differentials and ideal gas

Example 1.17. The Ideal Gas Law says that $P = Nk_BT/V$. Use differentials to write the formula for calculating the linear approximation of change in pressure when both temperature and volume are changing.

Answer: $dP = \frac{1}{V^2}\left(Nk_BV \; dT - Nk_BT \; dV\right)$. ∎

Full Solution
On-line

▷ Differentials and error propagation

If we want to know the height of a mountain or building, we can measure the length of its shadow and use some basic trigonometry:

$$h = x \tan(\theta),$$

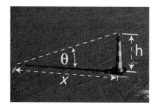

Figure 1.9: Calculating the height of a tower.

where θ is the angle shown in Figure 1.9 and x is the length of the shadow. Our measurements of x and θ will contain error (either due to the finite precision of the measuring device, or human error), so our calculation of h has some uncertainty in it. Differentials are often used to approximate how much uncertainty results from measurement error.

Estimating error with differentials

Example 1.18. Suppose you measure the shadow of a building as 100 feet long, and $\theta = \pi/3$. Approximate (a) the height of the building, and (b) the uncertainty in your answer from part (a) if you can only measure the angle θ to within 0.01 radians.

Solution: (a) Based on these measurements, the height of the building is $h = 100\tan(\pi/3) = 100\sqrt{3}$ feet. (b) Since there's some uncertainty in θ, let's treat it as a variable and write $h = 100\tan(\theta)$. Then the change in h due to a small variation in θ is approximately

> We're ignoring the uncertainty in the length of the shadow, x. We'll allow for uncertainty in both x and θ when we study the *total derivative* of a bivariate function in Chapter 13.

$$dh = \frac{dh}{d\theta}\, d\theta = 100\sec^2(\theta)\, d\theta.$$

When $\theta = \pi/3$ this becomes $dh = 100(0.5)^2\, d\theta = 400\, d\theta$, so if the *actual* angle differs from our measurement by as much as $d\theta = 0.01$ radians, we have $dh = \pm 400(0.01) = \pm 4$ feet. So we approximate that the building is $100\sqrt{3} \pm 4$ feet tall. ∎

Estimating error with differentials

Example 1.19. The kinetic energy of an object is $K = 0.5mv^2$ when its mass is m kilograms and it's traveling at a speed of v meters per second. Suppose a mass of 8 kg is measured to be traveling at 3 meters per second. Approximate (a) its kinetic energy, and (b) the possible error in your calculation from part (a) if there is an uncertainty of 0.1 meter per second in the measurement of speed.

Solution: (a) We use the measurement of speed to calculate the object's kinetic energy as $K = 0.5(8)(3)^2 = 36$ joules. (b) Since we're not entirely sure about v, let's leave it as a variable and write $K = 4v^2$. Then $K'(3) = 24$, so

> Recall that energy is measured in joules (see Appendix E).

$$dK = 24\, dv.$$

Based on this equation, an error of $dv = 0.1$ in our measurement of velocity changes our calculation of kinetic energy by approximately $dK = 2.4$ joules. So we approximate the actual kinetic energy as being somewhere between $36 - 2.4 = 33.6$ and $36 + 2.4 = 38.4$ joules. ∎

You may wonder why we use differentials rather than something that seems simpler. Specifically, why didn't we just calculate $K(3.1)$ and $K(2.9)$ in Example 1.19, then see how much they differ from $K(3)$? This makes a lot of sense when there's only one variable, but extending the technique into the multivariable setting is much more difficult than extending the technique of differentials.

❖ Common Difficulties with Differentials

Some students mistake dh for h', but they are different things. Whereas h' tells us the instantaneous *rate* of change, the differential is a linear approximation of *how much change* happens. In the story of the model rocket, for example, the derivative is h' and has units of ft/sec, while the differential dh has units of feet.

Units can help you reason your way to or through a formula.

You should know

- the terms *linear approximation, linearization, Newton's method, converge, diverge,* and *differential;*

- that Newton's method is an algorithm for finding roots of functions;

- why Newton's method "fails" when $h'(t) = 0$;

- that differentials give us a linear approximation of change;

- that, graphically speaking, the differentials dy tells us about change along the tangent line, whereas Δy is the actual change in the function value.

You should be able to

- write the linear approximation of $f(t)$ about t_0;

- execute several steps of Newton's method by hand;

- use a linear approximation of f about t_0 to estimate $f(t_0 + \Delta t)$;

- write the Product Rule and Quotient Rule in terms of differentials;

- use differentials to estimate error.

❖ 4.1 Skill Exercises

1. Suppose $f(2) = 3$ and $f'(2) = 7$. Estimate $f(2.001)$.

2. Suppose $f(5) = 12$ and $f'(5) = 3$. Estimate $f(4.99)$.

3. Suppose $f'(8) = 11$. Approximate the change in $f(t)$ when t is increased from 8 to 8.0001.

4. Suppose $f'(-2) = 6$. Approximate the change in $f(t)$ when t is decreased from -2 to -2.17.

Use linear approximation at $t = 1$ to estimate the numbers in #5–8.

5. $f(1.01)$ when $f(t) = 1 + t^9$ 7. $f(0.9)$ when $f(t) = \ln(t)$

6. $f(0.9)$ when $f(t) = 13t^8$ 8. $f(1.1)$ when $f(t) = \sin(\pi t/6)$

In #9–12 use linearization at the specified t_0 to approximate the following numbers.

9. $\sqrt{24}$, $t_0 = 25$ 11. $\cos(3)$, $t_0 = \pi$

10. $\sqrt[3]{65}$, $t_0 = 64$ 12. $e^{0.1}$, $t_0 = 0$

In #13–18 you should execute four steps of Newton's method using the given function, starting at the specified point.

13. $f(t) = t^2 - 17$; $t_0 = 3$

14. $f(t) = t^3 - 17$; $t_0 = 2$

15. $f(t) = \tan(t) - t - 1$; $t_0 = 1$

16. $f(t) = \sin(t) - 0.8t$; $t_0 = 1$

17. $f(t) = 4t + 3$, $t_0 = 17$

18. $f(t) = -8t + 5$, $t_0 = \pi/73$

For each of #19–22 you should (a) write a nonlinear equation that's solved by the number α, (b) pick a starting point for Newton's method, and (c) use Newton's method to approximate the value of α correct to three decimal places (use your calculator to make the comparison).

19. $\alpha = \sqrt{17}$ 20. $\alpha = \sqrt{40}$ 21. $\alpha = \sqrt[4]{9}$ 22. $\alpha = \sqrt[3]{19}$

In #23–26 use Newton's method to locate a critical point of the given function (to within 10^{-5} the actual value) by starting at the specified t_0.

23. $f(t) = t^3 - \sin(t)$; $t_0 = 1$

24. $f(t) = t^3 - \sin(t)$; $t_0 = -1$

25. $f(t) = t^4 - 2t^3 + 0.5t - 1$; $t_0 = -0.2$

26. $f(t) = t^4 - 2t^3 + 0.5t - 1$; $t_0 = 2$

In #27–30 you should (a) evaluate $f(t_0)$, and (b) calculate df when $dt = 0.01$.

27. $f(t) = \sin(t)$, $t_0 = \pi/3$

28. $f(t) = \ln(t)$, $t_0 = 4$

29. $f(t) = \sqrt{4t + 1}$, $t_0 = 2$

30. $f(t) = \frac{7t}{t^2+2}$, $t_0 = -3$

In #31–34 you should (a) evaluate $f(t_0)$, and (b) use linearization to calculate a range of possible function values when the uncertainty in t_0 is $dt = 0.1$.

31. $f(t) = t^2$, $t_0 = 1$

32. $f(t) = e - t$, $t_0 = \ln(5)$

33. $f(t) = \cot(t)$, $t_0 = \pi/6$

34. $f(t) = \sinh(t)$, $t_0 = 0$

❖ 4.1 Concept and Application Exercises

In #35–38 you should (a) write the formula for $f(t_0 + \Delta t)$ at the specified t_0, just as you did when learning to calculate $f'(t_0)$ as the limit of difference quotients, then (b) discard all $(\Delta t)^2$ and $(\Delta t)^3$ terms in your formula from part (a), and (c) by checking the value of the remaining coefficients, verify that what remains is the linearization of f at t_0 when $\Delta t = t - t_0$.

35. $f(t) = 4 + 3t + t^2$, $t_0 = 1$

36. $f(t) = 2 - 5t^2$, $t_0 = 3$

37. $f(t) = 1 + t + t^3$, $t_0 = 1$

38. $f(t) = 4 + 5t^2 - t^3$, $t_0 = 2$

39. Suppose $g(t) = 9t + 1$. Show that the linear approximation of g about $t_0 = 0$ is, in fact, $g(t)$. Why does that make sense?

40. Show that Newton's method finds the root of $f(t) = 4t + 5$ in a single step, no matter what t_0 is used to start the algorithm.

41. Show that Newton's method finds the root of $f(t) = mt + b$ in a single step, no matter what t_0 is used to start the algorithm.

42. In order to use Newton's method to generate a decimal representation of $\sqrt{15}$ we need to write down a function for which $\sqrt{15}$ is a root. In complete sentences, explain why $f(x) = x - \sqrt{15}$ is not a good choice but $g(x) = x^2 - 15$ is.

43. Show that Newton's method can be written as $t_{n+1} = 2t_n - at_n^2$ when $f(t) = \frac{1}{t} - a$. Note: this is useful because it allows computers to calculate $1/a$ using Newton's method (provided that $a \neq 0$) without actually dividing (division can amplify round-off error).

44. Show that Newton's method can be written as $t_{n+1} = \frac{1}{2}t_n + \frac{a}{2t_n}$ when $f(t) = t^2 - a$. Note: this is useful because it allows computers to calculate $\sqrt{a}$ using Newton's method (when $a > 0$).

45. Suppose $f(t) = t^2 + t - 8$. Find the value(s) of t_0 at which Newton's method cannot start. *(Hint: see the discussion on p. 242.)*

46. Suppose $f(t) = t^3 + 2t^2 - 7t + 1$. Find the value(s) of t_0 at which Newton's method cannot start. *(Hint: see the discussion on p. 242.)*

47. Suppose $f(t) = \sqrt{|t|}$, whose only root is at $t = 0$. Perform six iterations of Newton's method, starting at $t_0 = 1$, and compare the function values at t_6 and t_0.

48. Suppose $f(t) = t^{1/3}$, whose only root is at $t = 0$.

    (a) Perform six iterations of Newton's method, starting at $t_0 = 1$, and compare the function values at t_6 and t_0.

    (b) Suppose $f(t) = t^{1/3}$, whose only root is at $t = 0$. Show that no matter where we start, each iterate of Newton's method is farther away from the root than the one before it.

49. Suppose that f is a differentiable function, and after starting Newton's method at some t_0, we find that $f(t_{13}) = 0$. Show that all further iterations of Newton's method leave us at t_{13}.

50. Suppose $f(\phi) = \begin{cases} \pi - 2\phi \sin\left(\frac{\pi}{\phi}\right) & \text{when } \phi \neq 0 \\ \pi & \text{when } \phi = 0. \end{cases}$

    (a) Verify that f is continuous everywhere.

    (b) Use Newton's method to calculate $\phi_1, \phi_2,$ and ϕ_3 when $\phi_0 = 1/2$.

    (c) Verify that $\phi_{n+1} = 1/2^{n+1}$ when $\phi_n = 1/2^n$, so that Newton's method converges to $\phi = 0$.

    (d) Verify that $\phi = 0$ is not a root of f, and explain what (if anything) went wrong with Newton's method.

51. The function $f(t) = t^2 e^{-t^2}$ has a single root at $t = 0$, and critical points at $t = \pm 1$. Figure 1.10 shows that Newton's method will pull us away from the root if we start it at any $t_1 > 1$. Similarly, if we start at t_2, which is less than but very close to 1, Newton's method will shoot us far enough to the left that we land beyond the critical point at $t = -1$, and the method pushes us farther away in successive steps. Yet if we start Newton's method at some $t \approx 0$, Newton's method brings us to the root at $t = 0$. How far away from $t = 0$ can we start Newton's algorithm and get it to converge to the root?

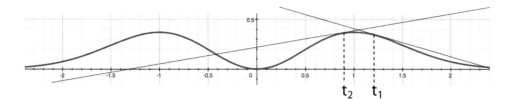

Figure 1.10: The graph of f in #51.

52. In #51 you found an interval around $t = 0$, called a *basin of attraction*, inside which Newton's method converges to the root of $f(t)$. We can vary the width of the basin by introducing a parameter into the exponential term. Specifically, let's consider $f(t) = t^2 e^{-\beta t^2}$, where $\beta > 0$.

   (a) Find a formula for the width of the basin. This depends on β so we'll denote it by $w(\beta)$.

   (b) Show that $w'(\beta) < 0$ and explain what that tells us about the way the basin changes with β.

53. In pure water, the pressure at a depth of $y = 1$ meter is 1.11135×10^5 pascals, and $p'(1) = 9810$.

   (a) Write the linearization of pressure as a function of depth, and identify the units associated with each part of the formula.

   (b) Use your answer from part (a) to approximate the change in pressure when we increase our depth from 1 to $1 + \Delta y$ meters.

54. In complete sentences, compare and contrast the meaning of (a) Δf and df, and then (b) Δx and dx.

In each of #55–59 determine whether the given equation makes sense. Use complete sentences to explain your reasoning.

55. $dv = -6\,dh$, where v is the volume of a rectangular box, and h is its height.

56. $da = 5\,dw$, where a is the area of a rectangle, and w is its width.

57. $df = -32214\,dx$, where f is the gravitational attraction between two asteroids and x is the distance between them.

58. $dp = 14\,da$, where $p(a)$ is the air pressure at an altitude of a meters (above sea level).

59. $dp = -1.25\,dt$, where p is the amount that you pay for a call on your GoPhone, and t is the length of time that you talk.

60. Suppose $f(v)$ is the drag force experienced by an object traveling at $v\frac{\text{m}}{\text{sec}}$. How do the situations described by $df = 3\,dv$ and $df = 3v\,dv$ differ? *(Hint: for the latter, begin by comparing the cases $v = 2$ and $v = 10$.)*

In each of #62–63 write an equation with differentials (such as those in #55–59) that communicates the specified idea.

61. When you increase your distance from the Earth, the change in the force due to the Earth's gravity decreases by 1% as much. (Use the variable f to mean the force due to the Earth's gravity, and the variable r to represent distance from the Earth's center.)

62. When you increase the length of a rectangle by a little bit, the change in area is five times larger than the change in length. (Use a to denote area and ℓ to represent length.)

63. The change in momentum is m times larger than the change in speed. (Use p to denote momentum and v to represent speed.)

64. When you stretch a spring by a little bit, the force required to hold it increases by three times as much. (Use the variable x to mean the amount that you've stretched a spring beyond its natural length, and the variable f to represent the force that's required to hold it there.)

65. In complete sentences, explain what the equation $df = 7\,dx$ means to us numerically (avoid using the words "differential" and "derivative").

66. In complete sentences, explain what the equation $df = k\,dx$ means to us numerically (avoid using the words "differential" and "derivative"), and what the sign of k tells us about the relationship between f and x.

67. *Aa* is a Hawiian word (pronounced "ah-ah") for a type of flow lava. When aa lava is ejected from a volcanic vent at an average rate of q m^3/sec, the length of the flow is $L(q) = \beta\sqrt{q}$ meters, where β is a constant that depends on the physical properties of the lava at the front of the flow. Suppose that a particular flow of aa lava on Mt. Etna has $\beta = 830$.

    | Volcanology |

    (a) Determine the units of dL/dq.

    (b) Suppose q is measured to be 95.3 m^3/sec. How long do we project the flow to be?

    (c) Suppose there's an uncertainty of 2 m^3/sec in the measurement. Use differentials to estimate the possible error in the projected flow length, L.

68. When we talked briefly about a single-slit diffraction experiment on p. 213, we said the line connecting the slit to the first *fringe* is separated from vertical by an angle of $\theta = \sin^{-1}(\lambda/a)$, where λ is the wavelength of the light being used, and a is the width of the slit through which is passes. Suppose the slit has a width of $a = 1 \times 10^{-6}$ meters, and we use light whose wavelength is $\lambda = 460 \times 10^{-9}$ meters. (a) Determine the angular separation of the first fringe from the vertical. (b) Estimate the error in θ if the machine that's used to produce the light can only guarantee the wavelength to within 1 nanometer $(1 \times 10^{-9}$ meter).

69. A tank in the shape of a lower hemisphere (see Figure 1.11) is 10 feet in diameter. When the fluid in the tank is b feet deep, its volume is

$$V = \frac{\pi b}{6}(3a^2 + b^2),$$

where a is the radius of the circular surface of the fluid.

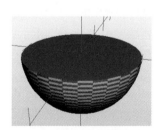

Figure 1.11: For #69.

    (a) Find a formula for a in terms of b.

    (b) How much fluid is in the tank when the depth is measured to be 8 feet?

    (c) Suppose there's an uncertainty of 0.0625 feet in our measurement of the depth. Use differentials to approximate the range of volumes due to this uncertainty.

4.2 Related Rates

Suppose a batter stands two feet from home plate, and watches as a fastball sails straight over the plate into the catcher's mitt (*strike!*). If the pitch was thrown at 130 feet per second (nearly 90 mph), how fast were the batter's eyes moving (in radians per second) when the ball was 10 feet in front of the plate? We'll answer that particular question in a moment, but first let's stop to talk about this scenario in some generality.

There are two quantities that are changing together: (1) the *distance* between the ball and home plate, and (2) the *angle* between the batter and the ball. This situation—i.e., two quantities changing together—is very common. For example, the *extension* of an elastic material and the *restoring force* it exerts are quantities that change together. The *concentration* of a reactant and the *temperature* of a solution are quantities that change together. The list goes on and on . . .

Typically you know the rate at which one of the quantities is changing, and since the quantities are related, you can use that information to determine the rate at which the other is changing. Here is some advice for dealing with such problems:

1. Sketch a diagram of the scenario and label it—including quantities that remain fixed, and quantities that change.

2. Use the geometry that you see in your diagram (triangles, circles, etc.) to relate the changing quantities.

3. In these kinds of problems, you're often told the moment in which we're interested. You should only use that information *after* you have a formula for calculating the rate of change in which you're interested.

4. In problems that address physical situations, continually check units as you work. If you ever see an equation that says something like length = mass, you've made a mistake.

> In the baseball story above, for example, you need to derive a formula for $d\theta/dt$ before using the fact that the ball is 10 feet in front of the plate.

> When an equation describes a physical situation, the left-hand and right-hand sides must describe the same *kind* of quantity. When they do, we say that the equation is *dimensionally consistent*. When they don't, the equation is certainly wrong.

Relating lengths

Example 2.1. Suppose a camera is filming the launch of a rocket from 3 miles away. How fast is the distance from the camera to the rocket changing when the rocket is 4 miles high and traveling at 0.2 miles per second?

Solution: Step 1 is to sketch a diagram of the scenario. This has been done in Figure 2.1, where we've labeled the height of the rocket as y. The distance from the camera to the rocket, which we'll call ℓ, can be related to the height of the rocket using the Pythagorean Theorem:

$$\ell^2 = 9 + y^2. \tag{2.1}$$

Keep in mind that both ℓ and y depend on time, but the horizontal distance from the camera to the launch pad does not. Since we're interested in the *rate at which the distance from the camera to the rocket is changing*, we differentiate both sides of equation (2.1) with respect to time. Using the Chain Rule on both sides, we see

$$2\ell\,\ell' = 2y\,y' \quad \Rightarrow \quad \ell' = \frac{y\,y'}{\ell}.$$

Let's pause to check the units of our equation. Since ℓ is measured in miles, and t in seconds, we know that ℓ' has units of miles per second (as does y'). When we

Figure 2.1: Diagram for Example 2.1.

check the units in our formula, we see

$$\frac{y\,y'}{\ell} \quad \substack{\leftarrow \\ \leftarrow} \quad \frac{(\text{miles})(\text{miles per second})}{\text{miles}}$$

which reduces to miles per second, so the equation is dimensionally correct. Since $y' = 0.2$ and $\ell = 5$ when $y = 4$, this formula tells us that

$$\ell' = \frac{(4)(0.2)}{5} = \frac{4}{25} \text{ miles per second.}$$

Note that $\ell' > 0$ so the distance from the camera to the rocket is increasing, which makes sense in this scenario. If you were to see $\ell' < 0$, you'd know there was a mistake somewhere in the calculations. ■

The batter and the pitch problem

Example 2.2. Answer the question about the batter's eyes that led off this section.

Solution: Step 1 is to produce a labeled diagram of the situation, which you see in Figure 2.2. We've denoted the distance from the ball to home plate as y, and labeled the angle from the batter to the ball as θ, both of which depend on time. Step 2: Since we have a right triangle, we can relate these quantities by way of the tangent:

$$\tan(\theta) = \frac{y}{2}, \tag{2.2}$$

where θ is measured in radians. Keep in mind that both θ and y depend on time, but the horizontal distance from the batter to the plate does not. Since we're interested in the *rate at which θ is changing,* we differentiate both sides of (2.2) with respect to time. Using the Chain Rule on the left-hand side:

$$\sec^2(\theta)\,\theta' = \frac{y'}{2} \quad \Rightarrow \quad \theta' = \cos^2(\theta)\,\frac{y'}{2}.$$

Recall that radians are a dimensionless quantity. With this in mind, we can check whether our equation is dimensionally consistent:

θ' has dimensions of $\frac{1}{\text{second}}$ because radians are a dimensionless quantity,

$\cos^2(\theta)$ has units of $\left(\frac{\text{length}}{\text{length}}\right)^2$, so it's a dimensionless quantity, and

$\frac{y'}{2}$ has units of $\frac{\text{feet per second}}{\text{feet}}$, which is the same as $\frac{1}{\text{second}}$.

Since both sides of our equation have dimensions of 1/second, our equation for $d\theta/dt$ is dimensionally consistent.

We know that y is decreasing at a rate of 130 feet per second, so $y' = -130$; and the hypotenuse of the right triangle has a length of $\sqrt{104}$ when $y = 10$, so $\cos(\theta) = 2/\sqrt{104}$. Substituting these numbers into our formula for θ', we see that

$$\theta' = 0.5 \left(\frac{2}{\sqrt{104}}\right)^2 \left(-130\right) = -\frac{260}{104} = -2.5 \text{ radians per second.}$$

Notice that our work resulted in $d\theta/dt < 0$. That means the angle between the batter and the ball is decreasing, which makes sense in this scenario. If you were to see $\theta' > 0$ as your final result, you'd know there was a mistake somewhere in the calculations. ■

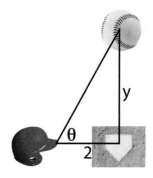

Figure 2.2: A diagram of the baseball scenario for Example 2.2.

A radian is the ratio "length of arc per length of radius,"

$$\frac{\text{length of arc}}{\text{length of radius}},$$

which is why a circular arc that subtends θ radians has a length of θR. With units of length/length, a radian is dimensionless. For more detail, see Appendices A and E.

Warning: beginners commonly make the mistake of writing 130 instead of -130 for y'.

Example 2.1 focused on related lengths, and Example 2.2 dealt with a changing angle. Our next example deals with both.

Multiple Changing Quantities

Example 2.3. Suppose that, rather than being fixed in place, the camera in Figure 2.3 is on a moving truck. If the camera's angle of inclination is increasing at 47 radians per hour when it's 1.5 miles from the launch pad, and the rocket is 4 miles high and rising at 0.2 miles per second, is the truck moving toward or away from the launch pad?

Solution: Step 1 is typically to a diagram of the scenario, but we already have Figure 2.3, in which the height of the rocket is y, the distance between it and the camera is x, and the camera's angle of inclination is θ. As we did in Example 2.2, we can relate these quantities by using the tangent:

$$\tan(\theta) = \frac{y}{x}.$$

Since we're interested in whether x is increasing or decreasing, let's isolate x by rewriting our equation as

$$x = y \cot(\theta).$$

Then when we differentiate with respect to time, applying the Product Rule and Chain Rule to the right-hand side, we see

$$\frac{dx}{dt} = \frac{dy}{dt} \cot(\theta) - y \csc^2(\theta) \frac{d\theta}{dt}.$$

Our last step is to specify the particular moment in which we're interested by inserting appropriate numbers on the right-hand side of the equation. In this case, the word "appropriate" means that we have to be careful with our units. While θ' is given to us radians per *hour,* the rate of the rocket's ascent is specified in miles per *second.* Let's convert that to miles per hour, so that all the terms on the right-hand side of our equation have the same units:

$$\text{rate of ascent} = 0.2 \frac{\text{mi}}{\text{sec}} \times 3600 \frac{\text{sec}}{\text{hr}} = 720 \frac{\text{mi}}{\text{hr}}.$$

When $x = 1.5$ and $y = 4$, right-triangle trigonometry reveals that $\cot(\theta) = 3/8$ and $\csc^2(\theta) = 73/64$, so our equation for x' becomes

$$\frac{dx}{dt} = 720 \left(\frac{3}{8}\right) - 4 \left(\frac{73}{64}\right)(47) = 55.5625.$$

Since $\frac{dx}{dt} > 0$ the number $x(t)$ is increasing, so the truck is driving away from the launch pad (at roughly 55 miles per hour). ∎

Relating volume and height

Example 2.4. An inverted conical tank is being drained at a rate of 3 cubic feet per minute from its tip. The height of the tank is 8 feet, and its radius is 4 feet. At what rate is the height of the fluid in the tank changing when there are 1.5 cubic feet of fluid in the tank?

Solution: Figure 2.4 depicts this scenario as seen from the side. We can relate the volume of fluid to the variables r and h by way of the volume formula for a cone:

$$V = \frac{1}{3}\pi r^2 h,$$

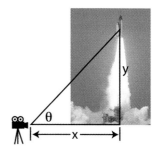

Figure 2.3: Diagram for Example 2.3.

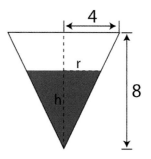

Figure 2.4: Schematic for Example 2.4.

where V, r and h all change with time. At this point it appears that we have three variables instead of two, much as we saw in Example 2.3, but we don't have data for all of them. Specifically, no information about r was provided in the description of this scenario, so let's use some geometry to remove it from our formula for volume. Due to the similar triangles that you see in Figure 2.4, we know that

$$\frac{4}{8} = \frac{r}{h} \quad \Rightarrow \quad r = \frac{1}{2}h.$$

> A common mistake in this kind of problem is focusing on the volume of the *container*, but we're interested in the volume of *fluid*.

This allows us to reduce the volume formula to a relationship between two variables:

$$V = \frac{1}{3}\pi r^2 h = \frac{1}{3}\pi \left(\frac{h}{2}\right)^2 h = \frac{\pi}{12}h^3. \tag{2.3}$$

Since we're interested in the *rate at which the height is changing,* we differentiate both sides of this equation with respect to time. Using the Chain Rule on the left-hand side, we have

$$\frac{dV}{dt} = \frac{\pi}{4}h^2\frac{dh}{dt}.$$

Let's pause to check the dimensions of our formula. Since volume is measured in cubic feet, and time in seconds, the left-hand side has units of $\frac{ft^3}{sec}$. On the right-hand side, the number $\pi/4$ is dimensionless, but h is measured in feet. So h^2 has units of ft^2 and dh/dt has units of $\frac{ft}{sec}$. Together, the product of these factors has units of $\frac{ft^3}{sec}$. So our equation for dV/dt is dimensionally correct.

> Note that $\pi/4$ is dimensionless in this example, but the previous examples include constant terms that carry units (the 9 in equation (2.1) and the 2 in equation (2.2)). Whether a quantity carries units has to do with the role it plays, not whether it's constant.

We know that V is *decreasing* at 3 cubic feet per minute, so $V' = -3$, and we can use equation (2.3) to determine that $h = \sqrt[3]{18/\pi}$ when $V = 1.5$. Substituting these numbers into our equation, we see

$$-3 = \frac{\pi}{4}\left(\frac{18}{\pi}\right)^{2/3}\frac{dh}{dt} \quad \Rightarrow \quad \frac{dh}{dt} = -\frac{4}{3\pi}\left(\frac{18}{\pi}\right)^{-2/3} \approx -0.132548\frac{ft}{min}.$$

Note that $h' < 0$, so the height is decreasing, which make sense in this scenario. If you were to see $h' > 0$ as the final result, you'd know there was a mistake somewhere in the calculations. ∎

Try It Yourself: Relating volume and height

Example 2.5. A conical tank (point up) is being drained at a rate of 4 cubic feet per minute from the center of its circular base. The height of the tank is 4 feet, and its radius is 5 feet. At what rate is the radius of the fluid in the tank changing when there are 2 cubic feet of fluid in the tank?

Answer: $dr/dt = 0.06448567$ feet per minute. ∎

Full Solution
On-line

❖ Common Difficulties with Related Rates

Perhaps the most common problem has to do with substituting values into equations too early. For example, if we set $y = 4$ in equation (2.1), we see that $\ell^2 = 25$. Now differentiating leads us to $\ell' = 0$, which is clearly wrong. The problem is that substituting $y = 4$ picks out a particular moment—much as taking a photograph captures a particular instant, and there is *no change* in a photograph. Instead, specify the particular moment in which you're interested only after you have an equation that describes change.

You should know

- to identify all quantities in your problem as constant or as variable, and to differentiate all variables simultaneously;

- to substitute values of variables in your equations *only at the very last step of your work.*

You should be able to

- draw a diagram and use the geometry it exhibits to write down relationships among quantities (e.g. using Pythagorean Theorem, similar triangles, Law of Cosines, the equation of a circle, etc.);

- use those relationships in order to relate rates of change.

❖ 4.2 Concept and Application Exercises

1. Suppose a particle is traveling along the parabola $y = x^2$, and its horizontal rate of change is $\frac{dx}{dt} = 2$ when $x = 3$ What is $\frac{dy}{dt}$ at that moment?

2. Suppose a particle is traveling along the ellipse $x^2 + 4y^2 = 1$, and its horizontal rate of change is $\frac{dx}{dt} = 12$ when $x = 0.5$ and $y = 0.25\sqrt{3}$. What is $\frac{dy}{dt}$ at that moment?

3. A 12-foot ladder is leaning against a wall. The ladder slips and begins to fall. If the foot of the ladder moves away from the wall at a constant rate of 2 feet per second, how fast is the top of the ladder approaching the ground when the base is 7 feet away from the wall?

4. Suppose the wall in #3 makes a 80° angle with the ground, sloped away from the base of the ladder. How fast is the top of the ladder approaching the ground when the base is 7 feet away from the base of the wall?

5. Suppose that ship A is 50 miles directly south of ship B at time $t = 0$ hours. Ship A is sailing east at 10 miles per hour, and ship B is sailing due south at 15 miles per hour.

   (a) Is the distance between the ships increasing or decreasing at the moment when ship A has traveled 10 miles eastward?

   (b) Is the distance between the ships increasing or decreasing at the moment when ship A has traveled 30 miles eastward?

6. A lighthouse sits 1 km off the shore and its light completes 3 revolutions per minute (see Figure 2.5). How fast is the light moving along the shore when the angle between the light and shore is $\frac{\pi}{6}$?

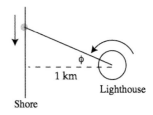

Figure 2.5: For #6.

7. The kinetic energy of an object is $K = v^2$ when its mass is 2 kg, where v is its speed (in m/sec). Suppose such an object is traveling at 4 meters per second, and is experiencing an acceleration of $3 \frac{m}{sec}$ per second. Determine the rate of change in its kinetic energy.

8. The kinetic energy of an object is $K = v^2$ when its mass is 2 kg, where v is its speed (in m/sec). Suppose such an object is traveling at 4 meters per second, and its kinetic energy is decreasing at a rate of $3 \frac{J}{sec}$. Determine the rate of change in its speed.

9. Suppose air is blown into a spherical balloon at the rate of 10 cubic centimeters per second. How fast is the radius changing when the volume is 20 cubic cm?

10. Suppose air is blown into a spherical bubble of liquid glass at 5 cubic inches per second. How fast is the surface area changing when the volume is 64 cubic inches?

11. The Aegis cruiser is one of the Navy's most powerful warships, and one of its most important jobs is protecting aircraft carriers. Suppose that a Silkworm anti-ship missile has been launched and is heading due west at 350 miles per hour, exactly half a mile north of the Aegis cruiser (see Figure 2.6). At what rate (radians per second) do the Aegis cruiser's phalanx cannons have to rotate in order to successfully track the hostile missile when it is 3/4 mile east of the cruiser (i.e., when $x = 3/4$ in the diagram)?

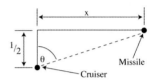

Figure 2.6: For #11.

12. Recalculate the answer to #11 under the assumption that the cruiser is traveling northward at 20 miles per hour.

13. Suppose P is 1 foot from the taller (5-foot) pole in Figure 2.7 and moving to the right at 4 feet per second. What's the rate of change in θ?

14. Suppose P is 1 foot from the taller (5-foot) pole in Figure 2.7 and $\frac{d\theta}{dt} = 2.7$ radians per second. How fast is P moving, and in what direction?

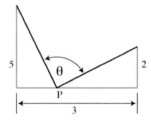

Figure 2.7: For #13 and #14.

15. A horizontal trough is 16 meters long, and its ends are isosceles trapezoids with an altitude of 4 meters, a lower base of 4 meters and an upper base of 6 meters. Water is being poured into the trough at the rate of 5 cubic meters per minute. How fast is the water level rising when the water is 2 meters deep?

16. A horizontal tank is 2 meters long, and its ends are circles with a radius of 0.5 meters. Water is entering the tank at a rate of 1.2 cubic meters per minute. How fast is the water level rising when the water is 0.25 meters deep?

Exercises #17–21 involve a plane at an altitude of 2 kilometers that flies directly over a radar tracking station at 360 km/hr, as depicted in Figure 2.8.

17. How fast is the distance between the tracking station and the plane changing after 1 minute?

18. How fast is the angle of inclination (i.e., the angle θ in Figure 2.8) changing after 1 minute?

19. Suppose that the plane is climbing at an angle of 30°. How fast is the angle of inclination changing after 1 minute?

20. Suppose that the plane is climbing at an angle of 30°. How fast is the distance between the tracking station and the plane changing after 1 minute?

21. Suppose that the plane in #17 is climbing at an angle of ϕ radians.

    (a) How fast is the angle of inclination changing after 1 minute? (Your answer will depend on ϕ.)

    (b) We'll refer to your answer from part (a) as r. Make a reasoned conjecture about whether $\frac{dr}{d\phi}$ is positive or negative. Explain your reasoning in complete sentences.

    (c) Determine a formula for r as a function of ϕ, and then calculate $\frac{dr}{d\phi}$ and check your answer from part (a).

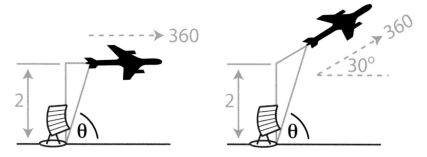

Figure 2.8: (left) Diagram for #17 and #18; (right) diagram for #19–21.

22. A billboard that's 54 feet wide is perpendicular to a straight road, and is 18 feet east of the nearest point on the road (see Figure 2.9). A car approaches from the south at 35 miles per hour. What's the rate of change in the angle subtended by the billboard from the passenger's point of view when the car is 0.25 miles south of the billboard?

23. Suppose an 80-foot building is on level ground, and its shadow is 60 feet long. If the angle of inclination between the sun and ground is increasing at 0.25 radians per minute, at what rate is the shadow's length changing? (Assume that the sun will pass directly over the building.)

24. Suppose the building from #23 sits on a hill whose angle of inclination is $32°$, as seen in Figure 2.10. At what rate is the length of the shadow changing?

25. Suppose a tanker accident has spilled oil off the California coast. The oil forms a very thin cylinder, whose height is the thickness of the slick. When the radius of the cylinder is 500 feet, the thickness of the slick is 0.01 feet and is decreasing at 0.001 feet/hour. At what rate is the area of the slick increasing? *(Hint: the shape of the slick is changing, but not its volume.)*

26. Suppose that oil-eating bacteria have been deposited on the slick from #25 and are consuming 5 cubic feet per hour. When the radius of the slick is 500 feet, the thickness of the slick is .01 feet and is decreasing at .001 feet/hour. At what rate is the area of the slick changing?

27. A door that is 0.9 m wide is swinging shut at the rate of 0.25 radians per second. Find the rate at which the distance between the jamb and the end of the door's free edge is changing when the door has $\pi/6$ radians remaining through which to close.

28. A door is 90 cm wide and its hinges are 10 cm from the wall. The door stop is 8 cm long, and sits 80 cm from the hinges. If the door swings open at the rate of 0.25 radians per second, determine the rate at which the distance between the end of the door stop and the free edge of the door is changing when the door is $\pi/3$ radians open.

29. A boat is being pulled toward the dock with a rope that attaches to the bow and passes through a loop on the end of the dock that's 1 meter above the bow. If the rope is being pulled at a rate of 1 meter per second, how fast is the boat approaching the dock when it is 8 meters away?

30. A bucket of cement mixture is placed on level ground at point A, 13.5 meters below a pulley. A rope is attached to the bucket, threaded through the pulley and released. The free end hangs 1 meter above point A. A worker grasps the

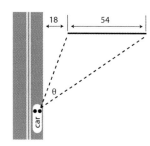

Figure 2.9: For #22.

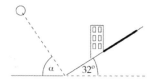

Figure 2.10: For #24.

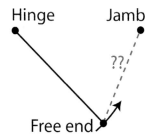

Figure 2.11: Diagram for #27.

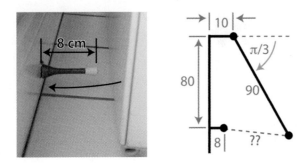

Figure 2.12: Diagrams for #28.

free end and, keeping it at a height of 1 meter, walks away from point A at a rate of 1.6 meters per second. How fast is the bucket ascending when the worker is exactly 9 meters from point A? (Assume that the pulley and the bucket are "points.")

31. If the crankshaft of the piston mechanism shown in Figure 2.13 is rotating at 2000 rpm, find the velocity of the piston when $\phi = \frac{\pi}{2}$.

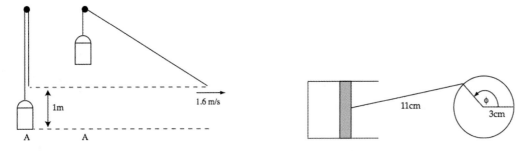

Figure 2.13: (left) Diagram for #30; (right) diagram for #31.

32. Suppose the hour hand of a clock is 1 foot long, and the minute hand is 1.5 feet long. Determine the rate at which the distance between the tips of the hands is changing at 3 PM.

33. A cylindrical tank 10 feet in diameter is being drained at a rate of 4 cubic feet per minute. At what rate is the height of fluid in the tank changing when there are 2 cubic feet of water in the tank?

34. An inverted conical tank (i.e., "point-down," like an ice-cream cone) is 10 feet high and, at its top, has a diameter of 12 feet. The tank is being filled with purified water at a rate of 4 cubic feet per second. At what rate is the height of the water changing when it is 7 feet deep?

35. A tank in the shape of a lower hemisphere (see Figure 2.14) that is 7 feet in diameter is leaking at a rate of 2 cubic feet per hour from its lowest point. At what rate is the height of fluid in the tank changing when the water in the tank is 1 foot deep? *(Hint: see #69 on p. 250.)*

36. A tank in the shape of an upper-hemisphere (see Figure 2.14) that is 10 feet in diameter is being drained at a rate of 5 cubic feet per minute from an outlet in its circular base. At what rate is the height of fluid in the tank changing when there are 12 cubic feet of water in the tank? *(Hint: see #69 on p. 250.)*

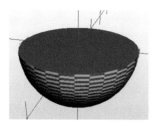
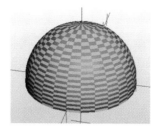

Figure 2.14: (left) Figure for #35; (right) figure for #36.

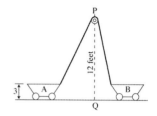

Figure 2.15: For #37.

37. Carts A and B, which are 3 feet high, are connected by a 39-foot rope that passes over a pulley P (see Figure 2.15). The point Q is on the floor, 12 feet directly beneath P, between the carts. Cart A is being pulled away from Q at a rate of 2 feet per second. How fast is cart B moving toward Q at the instant when cart A is 5 feet from Q? (You should assume the rope stays taut.)

38. Suppose a bifold door is closing, and the sliding pin is moving at 0.1 meters per second (see Figure 2.16). If the panels are each 1/3 of a meter wide, how fast is the distance from the sliding pin to the midpoint of opposite panel changing when the sliding pin is 5/6 of a meter from the fixed hinge?

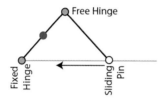
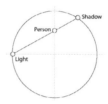

Figure 2.16: (left) Diagram for #38; (right) diagram for #39.

39. Suppose a person walks at 3 feet per second, from north to south across a circular room of radius 30 feet (see Figure 2.16). If a light sits at the westernmost point of the room, how fast is this person's shadow moving when he is 25% of the way across the room? *(Hint: If the origin of the Cartesian plane is at the center of the room, and the angle between the x-axis and the shadow is θ, you can determine the speed of the shadow by knowing $d\theta/dt$.)*

40. Suppose a person walks at 3 feet per second, from north to south, across a circular room whose radius is 60 feet. If a light sits at the westernmost point of the room, how fast is this person's shadow moving when she is 25% of the way across the room? *(See the hint for #39.)*

41. By writing the radius of the room as r feet, show that the answer to #39 doesn't actually depend on the radius.

42. Suppose the person in the circular room of #39 is standing still, 15 feet north of the room's center. However, the light source is moving along the edge of the room in a counterclockwise direction at 2 radians per second. How fast is the person's shadow moving on the wall when the light source is at $(-15\sqrt{2}, 15\sqrt{2})$?

4.3   Higher-Order Derivatives

Newton's Second Law of motion says that an object's momentum changes *instantaneously* in response to force. In fact, we define force to be $F = p'(t)$, where p is the object's momentum. An object's momentum is the product of its mass and velocity, so we can write the Second Law as

$$F = (mv)' = mv' + vm'.$$

In the common case when $m' = 0$ (i.e., mass is constant), this reduces to $F = mv'$. Since v' is the instantaneous rate of change in v, which we call **acceleration**, we often denote it by a. Thus we have arrived at the famous $F = ma$.

Since acceleration is $a = \frac{d}{dt}(v)$, and velocity is the derivative of position, $v = \frac{d}{dt}y(t)$, we have

$$a = \frac{d}{dt}v = \frac{d}{dt}\left(\frac{d}{dt}y(t)\right). \tag{3.1}$$

That is, acceleration is the derivative of *the derivative of* position. That's hard to say, so instead we refer to acceleration as the **second derivative** of position with respect to time, and we write

$$\underbrace{a = y''(t) \quad \text{or} \quad a = \frac{d^2y}{dt^2}, \quad \text{or} \quad a = \frac{d^2}{dt^2}y(t), \quad \text{or} \quad a = \left(\frac{d}{dt}\right)^2 y(t).}_{\text{These are all different notations for the second derivative}}$$

It's extremely important to understand that $y''(t) = \frac{d}{dt}y'(t)$. The *second* derivative tells us how the *first* derivative is changing. Specifically, according to the Increase-Decrease Theorem (p. 182):

Suppose y' is differentiable at each $t \in (a, b)$.

- If $y''(t) > 0$ throughout (a, b), the function y' is increasing on the interval.

- If $y''(t) < 0$ throughout (a, b), the function y' is decreasing on the interval.

❖ Calculating Second Derivatives

Since $y''(t)$ is the derivative of $y'(t)$, we find it by differentiating $y(t)$ twice.

Calculating second derivatives

Example 3.1. Suppose $y(t) = \frac{1}{12}t^4 - \frac{1}{6}t^3 - 3t^2$. Determine formulas for $y'(t)$ and $y''(t)$, and discuss their graphs.

Solution: Because y'' is just the derivative of y',

$$y'(t) = \frac{d}{dt}\left(\frac{1}{12}t^4 - \frac{1}{6}t^3 - 3t^2\right) = \frac{1}{3}t^3 - \frac{1}{2}t^2 - 6t$$

$$y''(t) = \frac{d}{dt}\left(y'(t)\right) = \frac{d}{dt}\left(\frac{1}{3}t^3 - \frac{1}{2}t^2 - 6t\right) = t^2 - t - 6.$$

In the left-hand graph of Figure 3.1 you can see that $y(t)$ is increasing when $y'(t) > 0$, and that $y(t)$ is decreasing when $y'(t) < 0$. Similarly, the right-hand image of Figure 3.1 shows that $y'(t)$ is increasing when $y''(t) > 0$, and $y'(t)$ is decreasing when $y''(t) < 0$. ∎

Mass *might* not be constant. For example, a rocket ejects fuel as it flies, so $m' < 0$; and molecules of hemoglobin bind with oxygen in the lungs, so $m' > 0$.

In everyday language, the word *accelerate* means "to speed up." In the technical lexicon *acceleration* means only that velocity is *changing*. The object could either speed up or slow down because of an acceleration.

When you are traveling in a car, the first derivative has to do with your speed and the direction you're headed. The second derivative is what you *feel* (the changes in velocity).

Important!

Important!

Important!

Important!

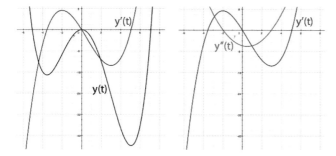

Figure 3.1: (left) The graphs $y(t)$ and $y'(t)$, and (right) the graphs $y'(t)$ and $y''(t)$.

Calculating second derivatives

Example 3.2. Find the second derivatives of $f(t) = e^t$ and $g(t) = e^{-t}$.

Solution: Because y'' is just the derivative of y',

$$f''(t) = \frac{d}{dt}\left(f'(t)\right) = \frac{d}{dt}\left(e^t\right) = e^t = f(t)$$

$$g''(t) = \frac{d}{dt}\left(g'(t)\right) = \frac{d}{dt}\left(-e^{-t}\right) = (-1)^2 e^{-t} = e^{-t} = g(t). \qquad \blacksquare$$

> The equation $y'' = y$ describes e^t and e^{-t}, as well as sums and/or multiples of them.

Try It Yourself: Calculating second derivatives

Example 3.3. Find the second derivatives of $f(t) = e^{3t}$ and $g(t) = e^{4t}$.

Answer: $f''(t) = 9e^{3t}$ and $g''(t) = 16e^{4t}$. $\qquad \blacksquare$

Full Solution On-line

Calculating second derivatives

Example 3.4. Find the second derivatives of $f(t) = \sin(t)$ and $g(t) = \cos(t)$.

Solution: Because y'' is just the derivative of y',

$$f''(t) = \frac{d}{dt}\left(f'(t)\right) = \frac{d}{dt}\left(\cos(t)\right) = -\sin(t) = -f(t)$$

$$g''(t) = \frac{d}{dt}\left(g'(t)\right) = \frac{d}{dt}\left(-\sin(t)\right) = -\cos(t) = -g(t). \qquad \blacksquare$$

> The equation $y'' = -y$ describes $\sin(t)$ and $\cos(t)$, as well as sums and/or multiples of them.

Try It Yourself: Calculating second derivatives

Example 3.5. Determine formulas for the second derivatives of $f(t) = \sin(5t)$ and $g(t) = \cos(6t)$.

Answer: $f''(t) = -25\sin(5t)$ and $g''(t) = -36\cos(6t)$. $\qquad \blacksquare$

Full Solution On-line

When calculating the second derivative of an implicitly defined function, we differentiate twice and then replace any occurrence of y' with its formula, so that we finish with an expression in terms of only x and y.

Finding second derivatives of implicit functions

Example 3.6. Find an expression for y'' when $x^3 + y^3 = 3xy$.

Solution: In Example 8.1 (p. 220) we determined that $y' = \frac{y - x^2}{y^2 - x}$. Differentiating this again, using the Quotient Rule on the right-hand side, gives us

$$y'' = \frac{d}{dx}\left(\frac{y - x^2}{y^2 - x}\right) = \frac{(y^2 - x)(y' - 2x) - (y - x^2)(2yy' - 1)}{(y^2 - x)^2}$$

When we gather the terms in the numerator with factors of y', we see

$$y'' = \frac{y + x^2 - 2xy^2 + (2yx^2 - x - y^2)y'}{(y^2 - x)^2} = \frac{y + x^2 - 2xy^2}{(y^2 - x)^2} + \frac{2yx^2 - x - y^2}{(y^2 - x)^2}y'.$$

Now replacing y' with $\frac{y - x^2}{y^2 - x}$ yields

$$y'' = \frac{y + x^2 - 2xy^2}{(y^2 - x)^2} + \frac{(2yx^2 - x - y^2)(y - x^2)}{(y^2 - x)^3}. \qquad \blacksquare$$

Try It Yourself: Finding second derivatives of implicit functions

Example 3.7. Use implicit differentiation to find y'' when $x^2 + y^2 = 1$.

Answer: $y'' = -\frac{1}{y^3}$. $\qquad \blacksquare$

Full Solution On-line

✧ Interpretation of Second Derivatives in Physical Context

Now that we can calculate second derivatives, let's interpret them in the context of physical examples. The most familiar one is, of course, motion. Suppose an object's position on the x-axis is $x(t)$ after t seconds, and $x'(t) > 0$ (so the object is moving to the right).

- During intervals of time in which $\frac{d}{dt}x'(t) > 0$, the number $x'(t)$ becomes larger—meaning more positive than it already is—so the object will *speed up*.

- During intervals in which $\frac{d}{dt}x'(t) < 0$, the number $x'(t)$ becomes smaller— meaning less positive than it is—so the object *slows down*.

The details are slightly different when $x'(t) < 0$, but the larger facts are the same. In short ...

> You should try doing the same thought experiment in the case when the object is moving to the left.

Suppose an object's position is described by $x(t)$. Then the object ...

- speeds up during intervals in which x' and x'' are both positive or both negative;

- slows down during intervals in which x' and x'' have opposite signs.

Comparing acceleration and velocity

Example 3.8. Suppose the position of an object on the x-axis is described by $x(t) = \frac{1}{3}t^3 - 6t^2 + 20t + 1$. Determine when the object is speeding up and when it's slowing down.

Solution: Let's look at the formulas for $x'(t)$ and $x''(t)$:

$$\underbrace{x'(t) = t^2 - 12t + 20,}_{v(t)} \quad \text{and} \quad \underbrace{x''(t) = 2t - 12}_{a(t)}.$$

In order to answer this question, we need to determine when $x'(t)$ and $x''(t)$ have the same sign. According to the Intermediate Value Theorem (p. 133) the sign of a continuous function can only change at a root, so we solve the equations

$$0 = x'(t) = t^2 - 12t + 20 = (t - 10)(t - 2) \quad \Rightarrow \quad t = 10 \text{ or } t = 2.$$

and

$$0 = x''(t) = 2t - 12 = 2(t - 6) \quad \Rightarrow \quad t = 6.$$

The graph of $x'(t)$ is a parabola that's cupped upwards (see Figure 3.2), so $x'(t)$ will be positive outside the interval $[2, 10]$. Similarly, the graph of $x''(t)$ is a line with positive slope (see Figure 3.2), so $x''(t)$ is positive on $(6, \infty)$.

The graphs of $x'(t)$ and $x''(t)$ are shown together in Figure 3.2, which makes it easy to see when they have the same sign. We conclude that the object is ...

speeding up $\begin{cases} \text{when } x'(t) \text{ and } x''(t) \text{ have the same sign.} \\ \text{This happens when } t \in (2, 6) \cup (10, \infty). \end{cases}$

slowing down $\begin{cases} \text{when } x'(t) \text{ and } x''(t) \text{ have different signs.} \\ \text{This happens when } t \in (-\infty, 2) \cup (6, 10). \end{cases}$ ∎

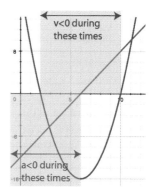

Figure 3.2: The graphs $y = x'(t)$ and $y = x''(t)$.

Try It Yourself: Comparing acceleration and velocity

Example 3.9. Suppose the position of an object is described by $x(t) = -t^3 - 4t^2 + 10t + 2$. Determine when the object is speeding up.

Answer: Speeding up when $t \in \left(-\frac{8+\sqrt{184}}{6}, -\frac{4}{3}\right) \cup \left(-\frac{8-\sqrt{184}}{6}, \infty\right)$. ∎

Full Solution On-line

One of the most important things we do with derivatives is *describe* the physical world. The second derivative appears in our description of force, as you saw earlier, but also in our description of phenomena such as wave propagation, and diffusion.

Newton's Second Law and the second derivative

Example 3.10. Suppose a baseball with a mass of m kilograms is rising after being hit by a bat, and its height is $y(t)$ meters after t seconds. The force it initially experiences due to air resistance is proportional to the square of its speed, and the force on the ball due to gravity is constant. Use derivatives to write an equation that describes the force experienced by the baseball.

Solution: The force due to air resistance is proportional to the *square* of the speed, so $F_{\text{drag}} = -\beta \, v^2$, where $\beta > 0$ is some constant that depends on physical characteristics of the ball's surface; and since the ball has a mass of m kilograms, the force it experiences due to gravity is $F_g = -9.8m$ (negative, because F_g pulls "down"). So we can write the net force experienced by the mass as

$$F_{\text{net}} = F_g + F_{\text{drag}} = -9.8m - \beta v^2.$$

On the other hand, Newton's Second Law tells us that the net force experienced by the mass is $F_{\text{net}} = ma = m \, y''(t)$. When we put these facts together, we see

$$\underbrace{m}_{\text{mass}} \quad \underbrace{y''(t)}_{\text{acceleration}} \quad \underbrace{=}_{\text{is}} \quad \underbrace{(-9.8m)}_{F_g} + \underbrace{(-\beta v^2)}_{F_{\text{drag}}}.$$

Since the velocity of the ball is $v = y'$, we can rewrite this as

$$m \, y''(t) = -9.8m - \beta \left(y'(t)\right)^2.$$ ∎

> Air resistance is a force that always acts in the direction opposite to the velocity. So when $v > 0$ (because the ball is rising) our formula guarantees that $F_{\text{drag}} < 0$.

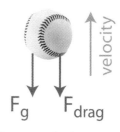

Figure 3.3: The force diagram for Example 3.10.

> This is an example of something called a *separable differential equation*. In Chapter 10 you'll learn how it's used to find a formula for $y(t)$.

Newton's Second Law and the second derivative

Example 3.11. If the nuclei in a diatomic molecule get too far apart, they are pulled back together due to electrical forces and other quantum mechanical phenomena; and if the nuclei come too close, they are pushed apart. Between "too far" and "too close," there's an equilibrium distance at which the net force on the nuclei is zero. Use the second derivative to write an equation that says, "The force experienced by the nuclei is directly proportional to their displacement from equilibrium."

Solution: We'll use y to denote the displacement of the nuclei from the equilibrium distance, where $y > 0$ means the nuclei are too far apart and $y < 0$ means they are too close together. The formula $F = -ky$ (where $k > 0$) says that force is proportional to y (we write $-ky$ so that the force is negative (pulling) when the nuclei are too far apart, and positive (pushing) when they are too close).

Additionally, Newton's Second Law tells us that $F = ma = my''$, where m is the mass of a nucleus. Equating these two descriptions of net force, we see

$$\underbrace{my''}_{\text{(mass)} \times \text{(acceleration)}} = \underbrace{-ky}_{\text{net force}}.$$

■

Too far (y>0): pulled together

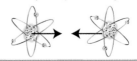

Just right (y=0): no force

Too close (y<0): pushed apart

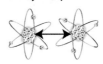

Figure 3.4: The force experienced by nuclei in a diatomic molecule depends on their distance apart.

Other interpretations of the second derivative

Example 3.12. Suppose $s(t)$ denotes the area of Arctic sea ice t days into the calendar year (e.g., $t = 23/24$ at 11 PM on January 1). Suppose $s'(t) < 0$ and $s''(t) > 0$ when $t \in (239, 242)$. What do those inequalities mean in this context?

Solution: Before trying to interpret these inequalities in context, let's begin by considering what they tells us about the curve $y = s'(t)$. The first tells us that the graph of $s'(t)$ is below the horizontal axis when $239 < t < 242$. However, $\frac{d}{dt} s'(t) = s''(t) > 0$ during that interval, so the graph of $s'(t)$ is headed upward. Consequently, we know that $0 > s'(241) > s'(240)$.

It's late August when $t = 240$, so $s'(241) < 0$ tells us that the area of sea ice is decreasing in late summer. However, $s'(t)$ is not *as* negative on day 241 as it was the day before. So while the area of Arctic sea ice is decreasing, we're not losing sea ice is as quickly on day 241 as we were on day 240.

■

> The graph of $s'(t)$ is actually below the horizontal axis for a much longer interval of time, but that's not indicated by the data here. The graph of $s'(t)$ typically rises above the horizontal axis in September.

▷ **Dimensional analysis**

Like the *first* derivative, the units associated with the *second* derivative will depend on what is being described, but the difference quotient will always determine what those units are.

Dimensional analysis of the second derivative

Example 3.13. Suppose $x(t)$ is the position of an object (measured in meters from $x = 0$) after t seconds. (a) Determine the units of $x''(t)$, and (b) interpret their meaning in this context.

Solution: The difference quotient tells us that $x'(t)$ has units of m/sec, and also

$$x''(t) = \lim_{\Delta t \to 0} \frac{x'(t + \Delta t) - x'(t)}{\Delta t} \quad \begin{matrix} \leftarrow & \text{m/sec} \\ \leftarrow & \text{sec} \end{matrix},$$

which we often write as $\frac{m}{\text{sec}^2}$. (b) The units of x'' ($\frac{\text{m/sec}}{\text{sec}}$) indicate that it measures the rate at which velocity is changing.

■

Dimensional analysis of the second derivative

Example 3.14. Suppose the cost of a carbon nanotube is $c(x)$ dollars when its length is x centimeters. (a) Determine the units of $c''(x)$, and (b) interpret their meaning in this context.

Solution: The difference quotient tells us that $c'(x)$ has units of \$/cm, and also

$$c''(x) = \lim_{\Delta x \to 0} \frac{c'(x + \Delta x) - c'(x)}{\Delta x} \quad \begin{matrix} \leftarrow \\ \leftarrow \end{matrix} \quad \begin{matrix} \$/\text{cm} \\ \text{cm} \end{matrix},$$

which is sometimes written as $\frac{\$}{\text{cm}^2}$. (b) Suppose that making a *long* carbon nanotube is more difficult than making a *short* one, so that increasing the length from 35 to 36 cm is more costly than increasing it from 5 to 6 cm. Said numerically, $c'(35) > c'(5)$. The number $c''(x)$ measures the rate at which $c'(x)$ changes. ∎

> The function $c'(x)$ is analogous to velocity, and $c''(x)$ is analogous to acceleration. They tell you the same kind of information as the familiar functions of velocity and acceleration.

❖ Second Derivatives and Graphs: Concavity

Because $f''(t)$ tells us the rate of change in $f'(t)$, and $f'(t)$ tells us the slope of the tangent line to the graph of f, the second derivative tells us whether the graph of f has a cupped-up or cupped-down structure. Specifically:

- When $f''(t) > 0$ in (a, b), the number $f'(t)$—i.e., the slope of the tangent line—is increasing. For example, in the left-hand image of Figure 3.5 the slope starts negative, become less negative, and then becomes positive as we look from left to right.

- When $f''(t) < 0$ in (a, b), the number $f'(t)$—i.e., the slope of the tangent line—is decreasing. For example, in the right-hand image of Figure 3.5 the slope starts positive, become less positive, and then becomes negative as we look from left to right.

Whether the graph is headed up or down, we see a cupped-up structure when $f''(t) > 0$ and say that the graph is **concave up**. Similarly, we see a cupped-down structure when $f''(t) < 0$ and say that the graph is **concave down**.

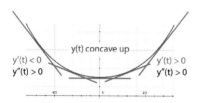
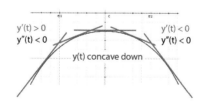

Figure 3.5: (left) The graph of $y(t)$ is concave up when $y''(t) > 0$ over an interval; (right) the graph of $y(t)$ is concave down when $y''(t) < 0$ over an interval.

Calculating concavity

Example 3.15. Suppose $f(t) = t^3$. Use the second derivative to determine when the graph of f is concave up and when it's concave down.

Solution: Since $f'(t) = 3t^2$, the second derivative is $f''(t) = \frac{d}{dt}\left(3t^2\right) = 6t$, which is positive when $t > 0$ and negative when $t < 0$. So the graph of f is concave up on the interval $(0, \infty)$ and concave down on the interval $(-\infty, 0)$. The shaded portion in Figure 3.6 shows those times when $f''(t) < 0$, and you can see the cupped-down structure there. ∎

Calculating concavity

Example 3.16. Suppose $f(t) = \frac{1}{3}t^3 - 6t^2 + 20t + 1$. Use the second derivative to determine the intervals over which the graph of f is concave up, and the intervals over which it's concave down.

Solution: The second derivative of f is $f''(t) = 2t - 12$, which is positive when $t > 6$. Therefore, the graph of f will be concave up when $t > 6$, as seen in Figure 3.6. The shaded portion of the graph shows the times when $f''(t) < 0$, where you can see that the graph exhibits the cupped-down structure. ∎

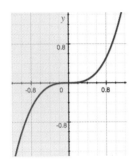
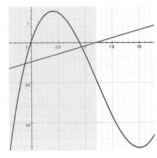

Figure 3.6: (left) Graph of $y = t^3$ for Example 3.15; (right) the graphs $y = f(t)$ and $y = f''(t)$ for Example 3.16.

In Examples 3.15 and 3.16, we saw the concavity of the graph change from "down" to "up" where the second derivative changed sign, i.e., when $f''(t_0) = 0$. This leads us to some new vocabulary.

> We say that t_0 is a **second-order critical number** of f when either $f''(t_0) = 0$ or $f''(t_0)$ does not exist.

The term "second-order" refers to the fact that we're looking at the *second* derivative.

> We say the graph of f has an **inflection point** at t_0 when it changes concavity at t_0 (i.e., it's concave up on one side, and concave down on the other).

WARNING: These are *not* two different terms for the same event. Concavity might not change at a second-order critical point!

Second-order critical points are not always inflection points

Example 3.17. Suppose $f(t) = t^4$. Find all second-order critical points and determine whether they are inflection points.

Solution: The second derivative is $f''(t) = 12t^2$, which is zero only when $t = 0$ and positive everywhere else. Since the concavity never changes, $t = 0$ is a second-order critical point, but *not* an inflection point. ∎

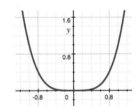

Figure 3.7: Graph of $y = t^4$ for Example 3.17.

Identifying inflection points

Example 3.18. Suppose $g(x) = e^{-x^2}$, which is called a **Gaussian**. Locate all second-order critical points and determine which (if any) are inflection points.

Solution: We begin by calculating the derivatives of g.

$$g'(x) = -2xe^{-x^2}$$
$$g''(x) = 4x^2e^{-x^2} - 2e^{-x^2} = (4x^2 - 2)e^{-x^2}$$

The word *Gaussian* is pronounced "Gow-shun," where "gow" rhymes with "now."

This function is named after the German mathematician and scientist Carl Friedrich Gauss (1777–1855).

Since $g''(x)$ is defined everywhere, the only second-order critical points are where $g''(x) = 0$. Since the exponential function is never zero, this happens only at $x = \pm 1/\sqrt{2}$. To determine whether these are inflection points, we sample g'' at selected points around $x = \pm 1/\sqrt{2}$ and look for a change in its sign:

$$g''(-10) > 0 \qquad g''(0) < 0 \qquad g''(10) > 0.$$

> Since g'' is continuous, it can only change sign at a root, so we only need to sample it at 3 points.

We conclude that g'' changes from positive to negative at $x = -1/\sqrt{2}$, and from negative to positive at $x = 1/\sqrt{2}$. That is, both of them are inflection points. ■

Try It Yourself: Identifying inflection points

Example 3.19. Suppose $f(x) = 3x^5 - 5x^4$. Locate all second-order critical points and determine which (if any) are inflection points.

Answer: $x = 0$ and $x = 1$, only the latter of which is an inflection point. ■

❖ Yet Higher Derivatives

We find the first derivative by differentiating once, and the second by differentiating twice. If we differentiate a third time, the function we find is called—you guessed it—the **third derivative**, which is commonly denoted by

$$f'''(x) \quad \text{or} \quad \frac{d^3 f}{dx^3} \quad \text{or} \quad \frac{d^3}{dx^3} f.$$

> In the fall of 1972 President Nixon announced that the rate of increase of inflation was decreasing. This was the first time a sitting president used the third derivative to advance his case for reelection. (see [37])

The third derivative tells you, for example, the rate at which acceleration (force) is changing. In this context, $f'''(t)$ is called the *jerk*.

Finding the third derivative

Example 3.20. Suppose $f(x) = 5x^6$ and $g(x) = 7x$. Find formulas for $f'''(x)$ and $g'''(x)$.

Solution: Applying our rule for power functions repeatedly, we see

$$
\begin{aligned}
f'(x) &= 30x^5 & g'(x) &= 7x^0 = 7 \\
f''(x) &= \tfrac{d}{dx} f'(x) = 150x^4, & g''(x) &= \tfrac{d}{dx} g'(x) = 0 \\
f'''(x) &= \tfrac{d}{dx} f''(x) = 600x^3 & g'''(x) &= \tfrac{d}{dx} g''(x) = 0. \quad \blacksquare
\end{aligned}
$$

> The derivative of a constant function is zero.

When we work with higher derivatives, the Lagrange (prime) notation is modified, replacing the prime with a superscript that indicates the number of derivatives. In general, the nth derivative of the function f with respect to x is written

$$f^{(n)}(x) \quad \text{or} \quad \frac{d^n f}{dx^n} \quad \text{or} \quad \frac{d^n}{dx^n} f \quad \text{or} \quad \left(\frac{d}{dx} \right)^n f.$$

> As there are both physical and geometric interpretations of $f'(x)$ and $f''(x)$, there are both physical and geometric interpretations of higher derivatives. The fourth derivative can be understood in terms of shear force, and appears in the *beam equation*, which describes the deflection of a beam when it's subjected to a load.

❖ Common Difficulties with the Second Derivative

Students often confuse *inflection points* with *second-order critical points,* so keep the example of $f(t) = t^4$ in mind. It has a second-order critical point at $t = 0$, but never changes concavity.

You should know

- the terms *acceleration, second derivative, concavity, second-order critical number, inflection point, Gaussian,* and n^{th} *derivative;*

Full Solution On-line

- the second derivative tells us about acceleration;

- that concavity *can* change at second-order critical points, but doesn't have to.

- read and write equations that describe Newton's Second Law in terms of derivatives;

- find inflection points;

- determine higher-order derivatives, $\frac{d^k f}{dx^k}$ for $k \geq 2$;

- use the sign of the second derivative to determine where the graph of f is concave up and where it's concave down.

✥ 4.3 Skill Exercises

Find the second derivative of each function in #1–12.

1. $f(x) = 3x + 4$

2. $f(x) = -10x + 2$

3. $f(x) = 4x^2 + x^5 - 8$

4. $f(x) = 3x^2 - 2x - 5$

5. $f(x) = \frac{1}{3}x^4 + \frac{1}{5}x^{-3} + 7$

6. $f(x) = 2x^{17} - 32x^{-1} + 8x$

7. $f(x) = 2.3x^{1.4} + 0.15x^{-0.3}$

8. $f(x) = 5x^{-3/2} - 17x^{1/10}$

9. $f(x) = x^{1/3} - \sqrt{x^2 + 4}$

10. $f(x) = 6\sqrt[3]{4x + 1}$

11. $f(x) = \frac{5x+2}{2x-1}$

12. $f(x) = \frac{5-8x}{4x+3}$

In each of #13–24 you should (a) calculate $f''(x_0)$, and (b) determine whether the graph of $f(x)$ is concave up or concave down at x_0.

13. $f(x) = \cos(x)$; $x_0 = 7\pi/6$

14. $f(x) = \sin(x)$; $x_0 = -\pi/6$

15. $f(x) = \cos(7x)$; $x_0 = 0$

16. $f(x) = \sin(2x)$; $x_0 = \pi/4$

17. $f(x) = e^{2x}$; $x_0 = 1$

18. $f(x) = e^{-5x}$; $x_0 = 1$

19. $f(x) = \ln(x)$; $x_0 = 1$

20. $f(x) = \ln(5x)$; $x_0 = 12$

21. $f(x) = \tan(x)$; $x_0 = \pi/4$

22. $f(x) = \sec(x)$; $x_0 = 0$

23. $f(x) = 8\sin(3x^2)$; $x_0 = 0$

24. $f(x) = 2\csc(3 - 2x)$; $x_0 = 1.5$

In each of #25–34 you should (a) render a graph of $f''(x)$ using a graphing utility, (b) locate the second-order critical points (using Newton's method if necessary), and (c) based on the graph of f'', determine which of the second-order critical points (if any) are inflection points.

25. $f(x) = e^x \sin(x)$

26. $f(x) = x^2 \cos(3x)$

27. $f(x) = x \ln(x)$

28. $f(x) = \frac{\ln(x)}{x}$

29. $f(x) = e^{14x}$

30. $f(x) = x^2 e^{-0.1x}$

31. $f(x) = \cosh(x)$

32. $f(x) = \sinh(x)$

33. $f(x) = \frac{1-e^{2x}}{e^{2x}+1}$

34. $f(x) = \frac{1}{1+e^{-x}}$

35. Suppose $f(x) = x^3$. Find formulas for $f'(x)$ and $f''(x)$, and then draw the graphs of f, f' and f'' on the same axes.

36. Suppose $f(x) = x^4$. Find formulas for $f'(x)$ and $f''(x)$, and then draw the graphs of f, f' and f'' on the same axes.

In each of #37–42, determine $\frac{d^n f}{dx^n}$ for the specified n.

37. $f(x) = x^6 - x^5 + 3x + 9$; $n = 10$

38. $f(x) = x^7 + 8x^3 - 3x^2 + 2$; $n = 54$

39. $f(x) = \cos(x)$; $n = 34$

40. $f(x) = \sin(x)$; $n = 297$

41. $f(x) = e^x$; $n = 12$

42. $f(x) = xe^x$; $n = 4$

43. Suppose $f(x) = xe^x$. Determine a formula for $\frac{d^k f}{dx^k}$.

44. Are there numbers b and c for which the graph of $f(x) = bx + \sin(cx)$ is concave up all the time? If so, find some. If not, why not?

Each of #45–46 shows the graphs of $f(x)$, $f'(x)$, and $f''(x)$. Your job is to determine which is which!

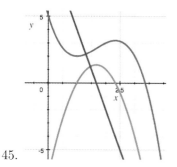

45.

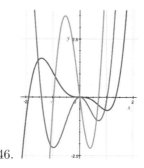

46.

Each of #47-48 shows the graph of $f''(x)$. Use the information it contains to sketch a graph of $f'(x)$.

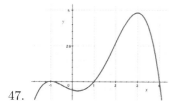

47.

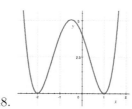

48.

In each of #49–56, use implicit differentiation to find a formula for y''

49. $x^2 - y^2 = 1$

50. $x^2 + y^3 = y$

51. $y^4 + x^4 = 4xy$

52. $y^6 + x^6 = 6xy$

53. $x^4 + 6\sin(y) = 9$

54. $\tan(x) + \sin(y) = 4$

55. $e^x - y^2 = x$

56. $e^x - e^y = 7x + y$

In Example 3.5 you saw that the equation $y'' = -36y$ describes the function $\cos(6t)$. For each of #57–64 find two different (i.e., not multiples of each other) functions that are described by the equation.

57. $y'' = -9y$ 59. $y'' = -7y$ 61. $y'' = 49y$ 63. $y'' = 2y$

58. $y'' = -16y$ 60. $y'' = -8y$ 62. $y'' = 64y$ 64. $y'' = 11y$

❖ 4.3 Concept and Application Exercises

Each function in #65–80 describes the movement of a mass on the y-axis. Determine (a) when the mass is headed toward (and away from) the origin, and (b) when the mass is speeding up (and slowing down).

65. $y(t) = -6t + 10$ 69. $y(t) = -t^2 + 12t + 5$ 73. $y(t) = t^{-2/3}$ 77. $y(t) = e^{-t^3}$

66. $y(t) = 5t + 11$ 70. $y(t) = -4t^2 - 9t$ 74. $y(t) = t^{3/2} - t$ 78. $y(t) = e^{-t^2}$

67. $y(t) = t^2 + 2t - 2$ 71. $y(t) = t^3 - 9t^2 + t$ 75. $y(t) = \cos^2(t)$ 79. $y(t) = e^t(t^2 + 1)$

68. $y(t) = 3t^2 - 4t - 8$ 72. $y(t) = t^3 + 3t^2$ 76. $y(t) = \tan^2(t)$ 80. $y(t) = \frac{e^t}{t^2+1}$

81. Draw the graph of a function for which $x = 2$ is an inflection point, but *not* a first-order critical point.

82. Draw the graph of a function for which $x = 2$ is both an inflection point *and* a first-order critical point, but not a root.

83. Without first drawing the graph of f, draw the graph of f' so that both (a) and (b) are true:

     (a) The function f is differentiable everywhere, and has first-order critical points at $x = 2$ and $x = 10$.

     (b) The graph of f is concave up on *only* $(2, 6)$, and is decreasing on *only* $(10, \infty)$.

84. Suppose an object has a positive velocity (that stays positive). Can it have an acceleration that's *always* negative? Explain in terms of both derivatives and intuition.

85. Suppose an object has a positive acceleration (that stays positive). Can it have a velocity that's *always* negative? Explain in terms of both derivatives and intuition.

86. When filling a cup with water, the height of liquid depends on the volume according to

$$h(v) = \sqrt[3]{\frac{27}{\pi}}(v + \pi)^{1/3} - 3.$$

     (a) Find a formula for $h''(v)$, and show that $h''(v) < 0$.

     (b) What does it mean to us, physically, that $h''(v)$ is negative?

     (c) What does part (a) tell you about the relationship between the graph of h and the tangent line at any particular point? *(Hint: see Figure 3.5(right).)*

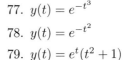

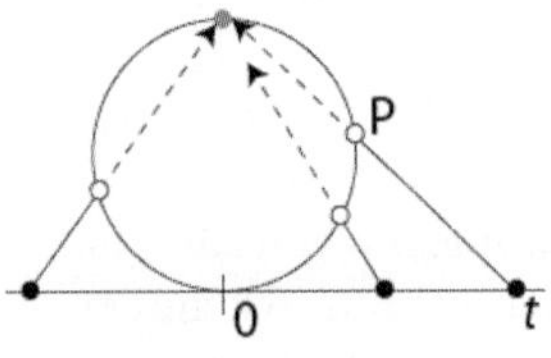

Figure 3.8: An example of the cup in #86.

87. On p. 200 you revisited the stereographic projection of the positive number t onto the circle of radius r that sits atop the x-axis at the origin (see Figure 3.9). Specifically, you saw that the rate of change in x with respect to r (the radius of the circle) is always positive. Continuing to think of $t > 0$ as a particular, fixed number, answer the following:

     (a) Intuitively, what *should* be true about $x''(r)$?

(0,2r) is the "North Pole"

Figure 3.9: For #87.

(b) Determine the formula for $x''(r)$, and check your answer to part (a) quantitatively.

On p. 166 you found the coordinates of the points at which a circle of radius R would rest in the cup of the parabola $y = ax^2$ when $a > 0$ (see Figure 3.10), which we'll denote by $(\pm x_c, y_c)$. Use those formulas for #88–89.

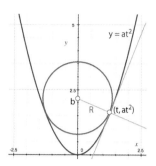

Figure 3.10: For #88–89.

88. Think of R as a particular, fixed number so that y_c depends only on a.

   (a) Show $y_c''(a) < 0$

   (b) What does this fact mean to us geometrically?

89. Think of a as a particular, fixed number, so that y_c depends only on R.

   (a) Show $y_c''(R) < 0$

   (b) What does this fact mean to us geometrically?

90. Suppose $y(x)$ is a function for which $y''(x) = -y(x)$. When the height on the graph gets larger, what happens to the concavity? What does this mean graphically?

91. In #52 on p. 249 you derived a formula $w(\beta)$, for the width of the basin inside which Newton's method converges to the root of $f(t) = t^2 e^{-\beta t^2}$, where $\beta > 0$. Show that $w''(\beta) > 0$ and explain what that means about the way the basin changes with β.

92. Suppose $y(t) = 3t^2 - 8t + 7$.

   (a) Calculate $y'(1)$ and $y''(1)$.

   (b) In Example 4.9 (p. 120) you saw how to rewrite the function $y(t) = 3t^2 - 8t + 7$ as $y(t) = 3(t - 1)^2 - 2(t - 1) + 2$. Where do your answers from part (a) arise in this rewritten version of $y(t)$?

93. Suppose $z(t) = -4t^2 + 20t + 9$.

   (a) Calculate $z'(2)$ and $z''(2)$.

   (b) Use the technique from p. 120 to rewrite $z(t)$ with powers of $(t - 2)$.

   (c) Where do your answers from part (a) arise in the rewritten version of $z(t)$ that you found in part (b)?

94. The graph of $f(x) = \sqrt{4 - x^2}$ is a semicircle. In Chapter 11 you'll see that a circle of radius R has a constant curvature of $1/R$.

   (a) Determine a formula for $f''(x)$.

   (b) Show that $f''(x)$ is always negative, but is not constant.

   (c) Are the words "concavity" and "curvature" synonyms?

In each of #95–98 use derivatives to write equations and/or inequalities that describe the given scenario.

95. The tangent line to the graph of f is horizontal at $x = 4$, and its graph is concave down there.

96. The graph of f is concave down and increasing at $x = 3$.

97. The graph of f is always concave down.

98. The graph of f is always concave up.

In each of #99–102 an object's horizontal position is described by $x(t)$. Use x, x' and x'' to write equations and/or inequalities that express each of the scenarios.

99. Acceleration is proportional to velocity.

100. Acceleration is inversely proportional to velocity.

101. The object is moving to the right, but is slowing down.

102. The object is moving to the left, and is speeding up.

103. Suppose $f(x)$ is the restoring force (measured in newtons) offered by a spring that has been stretched x meters beyond its natural length.

    | Mechanical Engineering |

  (a) Use the first derivative to write an equation that says, "increasing the extension of the spring causes the restoring force to increase."

  (b) Use the second derivative to write an equation that says, "the force offered by the spring increases by smaller and smaller amounts as the spring is extended more and more."

104. Suppose a mass of m kilograms is attached to a spring. Its horizontal position is described by $x(t)$, where $x > 0$ means the spring is extended beyond its natural length, and $x < 0$ means it's compressed (see Figure 3.11). The force exerted by the spring is proportional to the amount that it's extended or compressed (such a spring is said to be *linear*). As in Example 3.10, write an equation that describes the force experienced by the mass due to the spring.

    | Mechanical Engineering |

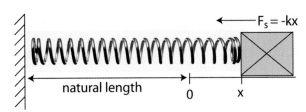

Figure 3.11: Diagram for #104.

105. Suppose the mass in #104 is moving in a fluid (e.g., air) and experiences a drag force that's proportional to its velocity. Adjust the equation from #104 to include the drag force.

106. In order to understand something about a population we often select some of its members (called a sample of the population) to question or measure. It's easy to calculate the mean value measured in your sample, $\hat{\mu}$, but different samples will result in different values of $\hat{\mu}$. However, the *Central Limit Theorem* tells us that values of $\hat{\mu}$ from various samples will cluster around the population mean, μ, in a so-called *normal distribution* when the sample size is large enough. When $\mu = 0$, the normal distribution is described by the function

    | Statistics |

$$p(x) = \frac{1}{\sigma\sqrt{2\pi}} \exp\left(-\frac{x^2}{2\sigma^2}\right).$$

  (a) Use a graphing utility to plot a graph of p for several values of $\sigma > 0$.

  (b) Find the inflection points of the graph of p.

  (c) Without doing any calculation, find the inflection points of $f(x) = \frac{1}{\sigma\sqrt{2\pi}} \exp\left(-\frac{(x-\mu)^2}{2\sigma^2}\right)$ (*Hint: how is $f(x)$ related to $p(x)$?*)

4.4 Local and Global Extrema

Imagine that you walk up a hill in your home town, and find its highest point. While you're standing there, no point near you has a higher altitude, so we say that you've found an *extremum*. But unless your home town is in Tibet, you've not found the highest point on the planet, so we say that you've found a *local* extremum. The highest point on the planet is on Mt. Everest, and we say that it's a *global* extremum. The graph shown in Figure 4.1 exhibits these same kinds of extrema.

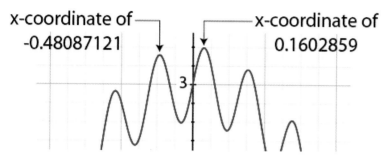

Figure 4.1: The function $f(x) = 3 - 0.5x^2 + 0.5\sin(10x)$ has several critical points, including $x = -0.48087121$ and $x = 0.1602859$.

The function whose graph you're seeing is $f(x) = 3 - 0.5x^2 + 0.5\sin(10x)$. Its largest possible value is produced at the critical number $x = 0.1602859$, so we say that f has a *global maximum* there. This function has another critical point at $x = -0.48087121$; and while the function value there is not the *largest possible* number that f can produce, it's larger than any other function value in the immediate vicinity (see Figure 4.2). So we say that f has a *local maximum* at $x = -0.48087121$. More generally ...

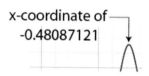

Figure 4.2: Zooming in on the graph of f in the vicinity of $x = -0.48087121$.

> - The function f has a **global (absolute) maximum** at x_0 if $f(x_0) \geq f(x)$ for all other x in the domain.
>
> - The function f has a **local (relative) maximum** at x_0 if there is an open interval containing x_0 throughout which $f(x_0) \geq f(x)$.

Similarly, when discussing minima ...

> - The function f has a **global (absolute) minimum** at x_0 if $f(x_0) \leq f(x)$ for all other x in the domain.
>
> - The function f has a **local (relative) minimum** at x_0 if there is an open interval containing x_0 throughout which $f(x_0) \leq f(x)$.

Further, we say that f has an **extremum** at x_0 if f achieves either a maximum or minimum there (whether local or global).

While these definitions formally articulate the idea of local extrema that we developed at the beginning of this section, they don't allow us to refer to the endpoint of a closed interval as a local extremum, though we might like to. For example, consider the function that's defined as $f(x) = |x|$ over $[-2, 1]$, but is undefined everywhere else. As you see in Figure 4.3, the function value at $x_0 = 1$ is the largest

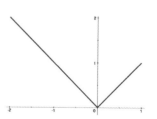

Figure 4.3: The graph of $f(x) = |x|$ restricted to $[-2, 1]$.

of all those nearby, so in the spirit of our discussion we'd like to say that $x_0 = 1$ is a local maximum; but the definition says that we have to consider the function values in an open interval that includes x_0, and our function isn't even defined when $x > 1$, so we can't check the function values there! For this reason, we extend the previous definitions as follows:

> When working on the interval $[a, b]$ ...
>
> - We say that f has a **local (relative) maximum** at b if there is an interval of the form $(c, b]$ throughout which $f(x) \leq f(b)$.
>
> - We say that f has a **local (relative) minimum** at b if there is an interval of the form $(c, b]$ throughout which $f(x) \geq f(b)$.
>
> Similarly, we say that f has a local maximum or minimum at $x = a$ if the relevant inequality is satisfied throughout an interval of the form $[a, c)$.

> A consequence of this extension is that every global minimum is also a local minimum (but not vice versa), and every global maximum is also a local maximum (but not vice versa).

Classifying extrema

Example 4.1. Suppose $f(\theta) = \sin(4\theta)$ when $0 \leq \theta \leq \pi$. Locate all local minima of f, and determine which is (are) global.

Solution: The graph of f is shown in Figure 4.4. Note that when θ is near but not equal to zero, $\sin(\theta) > \sin(0)$, so this function has a local minimum at $\theta = 0$. There are also local minima at $\theta = 3\pi/16$ and $\theta = 7\pi/16$, where the value of the function is -1. Because this function cannot produce a value less than -1, we know that the local minima at $\theta = 3\pi/16$ and $\theta = 7\pi/16$ are also global minima. ∎

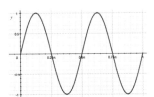

Figure 4.4: The graph of $f(\theta)$ for Example 4.1.

Note: In the previous example, the function f was only defined on $[0, \pi]$. It's often the case that our discussion is constrained to a given interval because of physical or financial aspects of the problem we're trying to solve (there is only *this* much material, or *that* much money).

Try It Yourself: Classifying extrema

Example 4.2. The graph of $f(\theta) = \sin(4\theta)$, restricted to $0 \leq \theta \leq \pi$, is shown in Figure 4.4. Locate all local maxima of f, and determine which is (are) global.

Answer: $\theta = \frac{\pi}{8}, \frac{7\pi}{8}$ and $\theta = \pi$, the first two of which are global maxima. ∎

Full Solution On-line

Extrema and location

Example 4.3. The graph of $f(t) = t^3 - 6t^2 + 11t - 5.5$ is shown in Figure 4.5. (a) Find an interval $[\alpha, \beta]$ so that by restricting f to $[\alpha, \beta]$ the global maximum of f occurs at $t = \beta$. Then (b) find different α and β for which the global maximum of f occurs inside (α, β), a local minimum at $t = \alpha$, and the global minimum at $t = \beta$,.

Solution: (a) When f is restricted to $[0, 1]$, there are no critical points on the interval. Since the function is always increasing when $t < 1$, the global maximum occurs at $t = 1$.

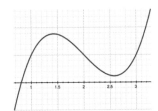

Figure 4.5: Graph of f for Example 4.3.

(b) Looking at the graph, we see that this happens when f is restricted to $[1, 2.25]$. In contrast to part (a), restricting f to $[1, 2.25]$ makes $t = 1$ a local minimum. ∎

Note: An important element of Example 4.3 is the idea that you cannot say whether f has an extremum at t_0 unless you know what the domain is.

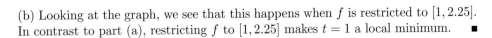

In the last few examples we've seen local extrema occur at functions' critical points, and at the endpoints of their domains. The following theorem says that those are the *only* places that we'll find local extrema.

> **Fermat's Theorem:** Suppose f has a local extremum at x^*. Then x^* is either a critical point of f or an endpoint of the function's domain.

> The name *Fermat* is pronounced "Fair-MAH."

Proof. We need to show that local extrema in the *interior* of a function's domain occur only at critical points. The graph of f might have a corner or cusp at x^*, in which case $f'(x^*)$ does not exist and—by definition—x^* is a critical point. If $f'(x^*)$ *does* exist, we need to establish that it *must* be zero. Exercise #41 (p. 280) will lead you through a simple proof of this fact. ∎

Fermat's Theorem and parabolas

Example 4.4. Show that the vertex of $y = ax^2 + bx + c$ is at $x = -b/2a$.

Solution: The vertex is where a parabola achieves its minimum (or maximum) height, so we can treat this as a problem of finding extrema. Fermat's Theorem tells us that they occur only at critical points, so we calculate $y' = 2ax + b$, which is zero only when $x = -b/2a$. So the vertex must be (and is) at $x = -b/2a$. ∎

> The point of this example is to apply Fermat's Theorem in a familiar setting. Here we've arrived at the formula for the location of a parabola's vertex that you learned in precalculus.

The following example is important because it shows that the converse of Fermat's Theorem is not true.

The converse of Fermat's Theorem is not true

Example 4.5. Suppose $f(x) = x^3$ when $x \in (-1, 1)$. Show that $x_0 = 0$ is a critical point, but that $f(0)$ is not a local extremum.

Solution: We know that $f'(x) = 3x^2$, which is 0 when $x = 0$. So $x_0 = 0$ is a critical point. However, $f(x) > f(0)$ when $x > 0$, and $f(x) < f(0)$ when $x < 0$, so the function has neither a local minimum nor a local maximum at $x = 0$. ∎

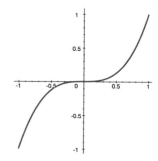

Figure 4.6: The graph of $y = x^3$ restricted to $[-1, 1]$.

❖ First Derivative Test

In Example 4.5 we saw that a critical point did *not* necessarily correspond to a local extremum. We now establish a test that uses the first derivative to decide whether f has an extremum at a critical point.

> **First Derivative Test:** Suppose f is differentiable throughout (a, b), except possibly at $x_0 \in (a, b)$, and x_0 is a critical point of f.
>
> - If $f'(x) > 0$ when $x < x_0$, and $f'(x) < 0$ when $x > x_0$, the function f has a local maximum at x_0 (it might be a *global* maximum).
>
> - If $f'(x) < 0$ when $x < x_0$, and $f'(x) > 0$ when $x > x_0$, the function f has a local minimum at x_0 (it might be a *global* minimum).
>
> - If $f'(x)$ has the same sign before and after x_0, the function has neither a local maximum nor a local minimum at x_0.

> The hypothesis of differentiability is meant to avoid ill-behaved functions. The basic idea is that the function has changed from increasing to decreasing, or vice versa, and we can tell because f' has changed sign.

Note: The particular values of f' before and after x_0 are irrelevant. All that matters is whether f' changes from positive to negative, or vice versa.

Classifying extrema using the First Derivative Test

Example 4.6. Exercise #35 on p. 25 introduced the *Lennard-Jones potential energy* between atoms in an argon gas:

$$u(x) = K \frac{1759.37 - x^6}{x^{12}},$$

where $x > 0$ is the distance between the centers of two atoms, measured in angstroms, and K has a numerical value of 70.3748752. (a) Find and classify all critical points of this function, and (b) interpret their meaning.

Solution: (a) After rewriting $u(x) = K \left(\frac{1759.37}{x^{12}} - \frac{1}{x^6} \right)$ it's easier to see that

$$u'(x) = K \left(\frac{-12(1759.37)}{x^{13}} + \frac{6}{x^7} \right).$$

At first glance it appears that $x = 0$ is a critical point, since $u'(0)$ is not defined, but $x = 0$ is not in the domain of this function. So we solve the equation $u'(x) = 0$:

$$0 = \frac{K}{x^{13}} \left(-12(1759.37) + 6x^6 \right)$$

when $x = \sqrt[6]{2(1759.37)} \approx 3.899$ angstroms. To perform the First Derivative Test, we check the sign of $u'(x)$ at points before and after $x = \sqrt[6]{2(1759.37)}$:

$$u'(3) = \frac{K}{3^{13}} \left(-12(1759.37) + 6(3^6) \right) = \frac{K}{3^{13}} \left(-12(1759.37) + 4374 \right) < 0,$$

$$u'(10) = \frac{K}{10^{13}} \left(-12(1759.37) + 6(10^6) \right) = \frac{K}{10^{13}} \left(-12(1759.37) + 6000000 \right) > 0.$$

> Notice that we're only continuing the calculation until we can tell whether u' is positive or negative. The particular value is irrelevant.
>
> If you graph $y = u(x)$ near $x = 3.899$, you see what appears to be a horizontal line. So the graph is unhelpful in this case, but the derivative tells us what's happening.

Since the graph of u is decreasing before $x = \sqrt[6]{2(1759.37)}$ and increasing after, we conclude that $u(x)$ has a local minimum at $x = \sqrt[6]{2(1759.37)}$. In fact, since $u(x)$ is decreasing at each $x < \sqrt[6]{2(1759.37)}$, and increasing ever after, the function has a *global* minimum at $x = \sqrt[6]{2(1759.37)}$. (b) If we accept that atoms move to minimize potential energy, this model predicts that atoms of this gas will tend to stay 3.899 angstroms away from one another (in the absence of other forces). ∎

Try It Yourself: Classifying extrema using the First Derivative Test

Example 4.7. Suppose that $f(x) = x^5 - 2x^3 + x + 0.5$. Find all critical points of the function and use the First Derivative Test to classify each as the location of a local minimum or a local maximum.

Answer: There are local maxima at $x = -1$ and $x = \sqrt{1/5}$, and local minima at $x = -\sqrt{1/5}$ and $x = 1$. ∎

Full Solution On-line

❖ Second Derivative Test

Instead of checking the sign of the first derivative on the left and right sides of a critical point, we can sometimes determine whether we've found a local maximum or minimum by checking the second derivative *at* the critical point. Simply said, a first-order critical point is a local minimum if the graph is concave up there, and a local maximum if the graph is concave down there (e.g., see Figure 4.5 on p. 274.)

Second Derivative Test: Suppose x_0 is a first-order critical point of f, and $f''(x_0)$ exists.

- If $f''(x_0) < 0$, the function f has a local maximum at x_0 (it might be a *global* maximum).

- If $f''(x_0) > 0$, the function f has a local minimum at x_0 (it might be a *global* minimum).

Note: The Second Derivative Test makes no conclusion when $f''(x_0) = 0$. You'll see why in Example 4.9.

Classifying extrema using the Second Derivative Test

Example 4.8. Suppose $f(x) = x^3 - 6x^2 + 8x - 3$. Locate and classify all critical points of f.

Solution: Since $f'(x) = 3x^2 - 12x + 8$, which exists everywhere, the critical numbers are those for which $f'(x) = 0$. We solve this equation with the quadratic formula, and find two solutions:

$$x_1 = \frac{6 + 2\sqrt{3}}{3} \approx 3.15470 \quad \text{and} \quad x_2 = \frac{6 - 2\sqrt{3}}{3} \approx 0.8453.$$

Since $f''(x) = 6x - 12 = 6(x - 2)$, which is positive at x_1 and negative at x_2, we conclude that f has a local minimum at x_1 and a local maximum at x_2. ∎

No conclusion from the Second Derivative Test

Example 4.9. Suppose that $f(t) = t^4$ and $g(t) = -t^6$. (a) Show that $t = 0$ is a critical point of both functions and (b) that $f''(0) = 0$ and $g''(0) = 0$, and then (c) classify the critical point for each function.

Solution: (a) The first derivatives are $f'(t) = 4t^3$ and $g'(t) = -6t^5$, both of which are zero when $t = 0$. (b) The second derivatives are $f''(t) = 12t^2$ and $g''(t) = -30t^4$, both of which are zero at $t = 0$. (c) The function f achieves a global *minimum* at $t = 0$, and the function g achieves a global *maximum* at $t = 0$. ∎

Note: When $f''(x)$ is easy to derive and work with, as in Example 4.8, the Second Derivative Test is a good method for classifying extrema. However, as seen in Example 4.9, the Second Derivative Test *can* fail to provide you with useful information.

You should know

- the terms *local (relative) maximum, local (relative) minimum, global (absolute) maximum, global (absolute) minimum, extremum,* and *extrema;*

- that the Second Derivative Test is inconclusive when $f''(x_0) = 0$.

You should be able to

- use the First Derivative Test to classify critical points as corresponding to local maxima, local minima, or neither;

- use the Second Derivative Test to classify critical points as corresponding to local maxima or local minima.

❖ 4.4 Skill Exercises

In #1–6 you should (a) locate all critical points and (b) use the First Derivative Test to determine whether each is the location of a local minimum, maximum, or neither.

1. $f(x) = -x^2 + 3x + 7$

2. $f(x) = 3x^2 + 6x - 4$

3. $f(x) = x^3 - x^2 + x \; 2$

4. $f(x) = -x^3 + 7x^2 + x + 1$

5. $f(x) = |8x - 5|$

6. $f(x) = |10x^2 + x - 5|$

7. $f(x) = x^2 e^{-x}$

8. $f(x) = \sqrt[3]{x}\, e^x$

In #9–14 you should (a) locate all critical points of f and (b) use the Second Derivative Test to determine whether each is the location of a local minimum, maximum, or neither.

9. $f(x) = x^3 - 4x^2 - 3x + 5$

10. $f(x) = -x^3 + x^2 + 8x - 2$

11. $f(x) = 0.25x^4 - 2x^2 - 1$

12. $f(x) = x^8 - 2x^4 - 3$

13. $f(x) = e^{\cos(x)}$

14. $f(x) = \arctan(x) - \frac{1}{4}x$

In #15–16 (a) starting at $t_0 = 0$, use Newton's method to locate a critical point of f, and (b) use the Second Derivative Test to determine whether each is the location of a local minimum, maximum, or neither. (A spreadsheet might be useful.)

15. $f(t) = (t+1)^{-2/3} \cos(t+1)$

16. $f(t) = (t-1) \arctan(t)$

In #17–22 you should (a) use a graphing utility to render a graph of the specified function over $[1,5]$, and (b) locate the global extrema of f on that interval.

17. $f(t) = 6 - 7t$

18. $f(t) = 4 + 3t$

19. $f(t) = -t^2 - t + 9$

20. $f(t) = t^2 - 6t + 1$

21. $f(t) = t^3 + t^2 - 7t + 1$

22. $f(t) = 8t^3 - 27t^2 + 29t + 8$

In #23 and #24 you should (a) locate all critical points of the function whose graph is shown, and (b) classify each as a local minimum, local maximum, or neither.

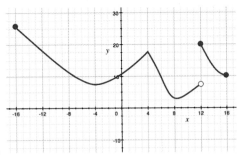

23.

24.

In #25–30 you should (a) use a graphing utility to render a graph of $f(t) = t^3 + 2t^2 - 7t + 1$, and (b) choose an interval $[a,b]$ that contains at least one critical point of f, with $a \neq b$, so that the stated condition is met when we restrict f to $[a,b]$.

25. The global maximum of f occurs at $t = a$, and the global minimum at $t = b$.

26. The global maximum of f occurs at $t = b$, and the global minimum at $t = a$.

27. The global maximum of f occurs at $t = a$, but the global minimum is *not* at $t = b$.

28. The global minimum of f occurs at $t = b$, but the global maximum is *not* at $t = a$.

29. The global maximum of f occurs at both $t = a$ and $t = b$.

30. The global extrema of f occur at neither $t = a$ nor $t = b$.

31. Suppose $f(x) = \alpha e^x + \beta e^{-x}$, where α and β can be any nonzero constants. (a) Use f' to show that f has a single critical point when the product $\alpha\beta > 0$, and none when $\alpha\beta < 0$. Then (b) assuming that $\alpha\beta > 0$ use the Second Derivative Test to determine an additional relationship between α and β that guarantees the critical point a local maximum.

32. Suppose $f(x) = \frac{\alpha}{3}x^3 + \frac{\beta}{2}x^2 - (\alpha + \beta)x + 7$, where α and β can be any nonzero constants. (a) Show that $x = 1$ is critical point of f, and then (b) use the Second Derivative Test to determine the relationship between α and β that guarantees $x = 1$ is a local minimum.

❖ 4.4 Concept and Application Exercises

33. Verify that $f(t) = (t - 2)^5$ has a critical point at $t = 2$, but that it's neither a local minimum nor a local maximum.

34. Verify that $f(t) = t^{1/3}$ has a critical point at $t = 0$, but that it's neither a local minimum nor a local maximum.

35. Suppose $f(t) = t^4 + 4t^3 + 6t^2 + 4t + 1$.

    (a) Verify that $t = -1$ is a both a first-order critical point *and* a second-order critical point for f.

    (b) What can we conclude about the critical point at $t = -1$ based on the fact that $f'(-1) = 0$ and $f''(-1) = 0$?

    (c) Use the First Derivative Test to classify $t = -1$ as a local minimum or local maximum.

    (d) Use a graphing utility to render a graph of $f(t)$ and visually verify your answer from part (c).

36. Explain why you cannot use the Second Derivative Test to classify the critical point in #5.

37. Design a function f for which $t = 1$ is a local maximum, but $t = 0$ is neither a local minimum nor a local maximum.

38. Design a function f with a domain that includes $[0, 1]$, but that does not have a global maximum or a global minimum when restricted to that interval.

39. A friend is trying to classify the local extrema of $f(x) = x^3 - 3x + 1$. He found a critical point at $x = 1$, and decided to use the First Derivative Test to classify it. He calculated $f'(-2) > 0$ and $f'(2) > 0$, and concluded that $x = 1$ is not a local extremum. (a) Use a graphing utility to plot the graph of f, and (b) explain what your friend did wrong.

40. A friend is trying to classify the local extrema of $f(x) = x^{12} - 3x^4$. She found a critical point at $x = 0$, and decided to use the Second Derivative Test to classify it. She calculated that $f''(0) = 0$ and, since it's neither positive nor negative, concluded that $x = 0$ is not a local extremum. (a) Use a graphing utility to plot the graph of f, and (b) explain what your friend did wrong.

Checking the details

41. Suppose that f achieves a maximum at $t^* \in (a, b)$, and that f is differentiable at t^*. In this exercise, you'll prove that $f'(t^*)$ must be zero, without relying on the continuity of f'.

   (a) Explain why $f(t^* + \Delta t) - f(t^*) \le 0$ as long as $t^* + \Delta t$ is in (a, b), which we assume for this whole exercise.

   (b) Explain why $\frac{f(t^* + \Delta t) - f(t^*)}{\Delta t} \le 0$ when $\Delta t > 0$.

   (c) Use part (b) to explain why $f'(t^*) \le 0$.

   (d) Explain why $\frac{f(t^* + \Delta t) - f(t^*)}{\Delta t} \ge 0$ when $\Delta t < 0$.

   (e) Use part (d) to explain why $f'(t^*) \ge 0$.

   (f) Use parts (c) and (e) to explain why $f'(t^*)$ must be zero.

4.5   Mean Value Theorem and L'Hôpital's Rule

Suppose you run a race with a friend. Your friend promises to run at a steady, constant rate from start to finish, and though you may speed up and slow down at will, by some fluke you tie. Could you have been running faster than your friend at each and every moment, or slower the whole time? No, because one of you would have won the race uncontested. If you start and finish together, there must have been some time at which you were running at the same speed.

Let's say this in the language of functions. Suppose your position is described by $g(t)$. If you start the race at time $t = \alpha$ and finish at time $t = \beta$, the total distance you covered is $g(\beta) - g(\alpha)$, and you did it in $(\beta - \alpha)$ seconds. Your friend covered the same distance in the same time, at a constant speed of

$$\frac{\text{distance}}{\text{time}} = \frac{g(\beta) - g(\alpha)}{\beta - \alpha}.$$

The assertion that there is some time, say t^*, when both of you were running at the same speed can be written as

$$\underbrace{g'(t^*)}_{\text{Your speed at time } t^*} = \underbrace{\frac{g(\beta) - g(\alpha)}{\beta - \alpha}}_{\text{Your friend's speed (constant) at time } t^*}. \qquad (5.1)$$

This equation can also be understood from a graphical point of view. Since the secant line connecting $(\alpha, g(\alpha))$ and $(\beta, g(\beta))$ has a slope of $\frac{\text{"rise"}}{\text{"run"}} = \frac{g(\beta)-g(\alpha)}{\beta-\alpha}$, equation (5.1) says there is a time (and maybe more than one) when the tangent line to the graph of g has the same slope as that secant line.

We arrived at equation (5.1), which is known as the Mean Value Theorem, by way of an intuitive argument. Now we need to establish it as a mathematical fact, and we do that by using something called Rolle's Theorem.

✥ Rolle's Theorem

In the special case when $g(\alpha) = g(\beta)$, equation (5.1) says there is *some* point at which the tangent line is horizontal. This fact is known as Rolle's Theorem.

> **Rolle's Theorem:** Suppose g is continuous over $[\alpha, \beta]$, differentiable over (α, β), and $g(\alpha) = g(\beta)$. Then there is a number $t^* \in (\alpha, \beta)$ at which $g'(t^*) = 0$.

Proof. It's our job to establish this assertion mathematically, regardless of stories of running races. To do it, we consider two cases:

Case 1: The function g is constant, which implies that $g'(t) = 0$ at each and every $t \in (\alpha, \beta)$. In this case, Rolle's Theorem is trivial.

Case 2: The function g is *not* constant (but $g(\alpha) = g(\beta)$). Because g is continuous, the Extreme Value Theorem (p. 134) guarantees that it achieves a maximum and a minimum value on $[\alpha, \beta]$. Since g is not constant, either the minimum or the maximum must be in (α, β). Suppose it occurs at t^*. Then Fermat's Theorem (p. 275) tells us that $g'(t^*) = 0$.

In either case, there's at least one $t^* \in (\alpha, \beta)$ at which $g'(t^*) = 0$. There might be more. ∎

This intuitive argument relies on the idea that your velocity is continuous so, for example, it cannot *jump* from 4 m/sec to 6 m/sec.

Figure 5.1: An example graph of g and its secant line.

We're using the word *mean* in the sense of "average"

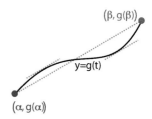

Figure 5.2: Try connecting the dots with the graph of a continuous function whose derivative always exists but is never zero.

Many students wonder why the intervals of continuity and differentiability in the statement of Rolle's Theorem aren't both open, or both closed. In our proof, we established that g must achieve either its maximum or its minimum value in (α, β). Since we can only conclude that $g'(t^*) = 0$ if $g'(t^*)$ exists, and we know that $t^* \in (\alpha, \beta)$ *but we don't know exactly where*, we can only guarantee that $g'(t^*) = 0$ by requiring that g be differentiable *everywhere* in (α, β). Now why isn't an open interval of continuity good enough? You'll explore that question in the exercise set by way of some examples.

> On a technical note, our proof relies on the Extreme Value Theorem, which cannot be applied to an open interval.

❖ Mean Value Theorem (MVT)

Rolle's Theorem is easily extended to the case when $g(\alpha) \neq g(\beta)$. This extension is called the Mean Value Theorem.

Mean Value Theorem (MVT): Suppose g is continuous over $[\alpha, \beta]$, and differentiable over (α, β). Then there is a number $t^* \in (\alpha, \beta)$ at which

$$g'(t^*) = \frac{g(\beta) - g(\alpha)}{\beta - \alpha} . \tag{5.2}$$

> In the context of the running story above, $g(t) =$ your position, $c(t) =$ your friend's position (constant rate of change), and $h(t) =$ the distance between you.

Proof. Suppose $c(t) = mt + b$ is a linear function for which $c(\alpha) = g(\alpha)$ and $c(\beta) = g(\beta)$, as seen in Figure 5.1. Then the function

$$h(t) = g(t) - c(t)$$

is differentiable over (α, β) and continuous on $[\alpha, \beta]$ because it's the difference of functions that are. Further, $h(\alpha) = 0$ and $h(\beta) = 0$, so Rolle's theorem guarantees that there is some $t^* \in (\alpha, \beta)$ for which $h'(t^*) = 0$. That is,

$$g'(t^*) = c'(t^*).$$

Since $c'(t^*) = m$, which is the slope of its graph,

$$g'(t^*) = c'(t^*) = m = \frac{c(\beta) - c(\alpha)}{\beta - \alpha} = \frac{g(\beta) - g(\alpha)}{\beta - \alpha}. \qquad \blacksquare$$

Don't just read it. Work it out! Where does this equation come from?

    Unlike many other things in this book, the Mean Value Theorem is not used to solve problems directly, but rather to facilitate our understanding of other facts and phenomena.

❖ Consequences of the Mean Value Theorem

In [14], Purcell and Varberg write "The Mean Value Theorem is the midwife of calculus—not very important or glamorous by itself, but often helping to deliver other theorems that are of major significance."

▷ The Increase-Decrease Theorem

On p. 182 we stated the Increase-Decrease Theorem, which claims that if $f'(t)$ remains positive throughout (α, β), the function f is increasing on (α, β). Now that we know the Mean Value Theorem, we can prove it.

    Suppose t_1 and t_2 are any two times in (α, β), and for convenience let's assume they've been indexed so that $t_1 < t_2$. Since t_1 and t_2 are inside an interval where f

is differentiable, the function is continuous on $[t_1, t_2]$, so the Mean Value Theorem tells us that there is a time $t^* \in (t_1, t_2)$ at which

$$\frac{f(t_2) - f(t_1)}{t_2 - t_1} = f'(t^*).$$

Since f' is always positive (that was our hypothesis), this allows us to conclude that

$$f(t_2) - f(t_1) = \underbrace{f'(t^*)}_{>0} \underbrace{(t_2 - t_1)}_{>0} > 0.$$

Equivalently, we could write $f(t_2) > f(t_1)$, which says that later times correspond to larger function values. That's exactly what it means for f to be increasing. ■

▷ Lack of change

Intuitively, if the rate of change in $g(t)$ is always zero, there won't *be* any change. Mathematically, to prove that it's true we need the Mean Value Theorem. Stated formally:

Theorem 5.6. Suppose $g'(t) = 0$ at each $t \in (\alpha, \beta)$. Then g is constant over the interval (α, β).

Proof. Suppose $t_0, t_1 \in (\alpha, \beta)$ are distinct. Since f is differentiable on (α, β), it's continuous on the closed interval $[t_0, t_1]$. This allows us to use the MVT to say that there is some t^* at which

$$\frac{g(t_1) - g(t_0)}{t_1 - t_0} = \underbrace{g'(t^*) = 0.}_{\text{Since } g'(t) \text{ is zero at every } t \in (\alpha, \beta)}$$

That can only happen when $g(t_1) = g(t_0)$. Since t_0 and t_1 are arbitrary, the function g is constant over the entire interval. ■

▷ The uniqueness of the Euler exponential function

On p. 171 we said that any nontrivial function for which $g' = g$ (i.e., whose derivative is just itself) must be a scaled version of $f(t) = e^t$. We can prove this fact by showing that the ratio $f(t)/g(t)$ is constant for any such g, which we do with the Quotient Rule. Since $f' = f$ and $g' = g$,

> There's a world of difference between saying that *we don't know* any others, and saying that there *aren't* any.

$$\frac{d}{dt}\left(\frac{g}{f}\right) = \frac{g'f - f'g}{f^2} = \frac{gf - fg}{f^2} = 0.$$

Therefore, the quotient g/f is constant (see Theorem 5.6). That is, there is some number C for which

$$\frac{g(t)}{f(t)} = C \quad \Rightarrow \quad g(t) = Cf(t) = Ce^t. \qquad ■$$

❖ Cauchy's Extension of the Mean Value Theorem

Let's return to the story of running a race, but with two changes. First, your friend is no longer bound by the promise to run at a constant rate; and second, instead of running a particular distance, you decide to sprint for a specified length of time.

When your trusted time-keeper calls "go!" you run, and when the time-keeper stops the race, you find that you've traveled twice as far as your friend. In order for that to happen, there must have been at least one moment during the race when you ran exactly twice as fast as your friend.

Let's say this in the language of functions. Suppose your position is described by $f(t)$, your friend's position is described by $g(t)$, and the race was run from time $t = \alpha$ to $t = \beta$. Then you traveled a distance of $f(\beta) - f(\alpha)$ and he traveled a distance of $g(\beta) - g(\alpha)$. So the factor by which you outdistanced him is

$$\frac{f(\beta) - f(\alpha)}{g(\beta) - g(\alpha)}.$$

We want to say that there was at least one moment, say t^*, at which you were running exactly *that* many times faster than him. That is,

$$f'(t^*) = \left(\frac{f(\beta) - f(\alpha)}{g(\beta) - g(\alpha)} \right) g'(t^*).$$

The proof of this claim is attributed to Augustin Louis Cauchy (1789–1857), and the formal statement below articulates the conditions under which it's true.

> **Cauchy's Mean Value Theorem:** Suppose f and g are both continuous on $[\alpha, \beta]$, differentiable on (α, β), and $f(\alpha) \neq f(\beta)$ and $g(\alpha) \neq g(\beta)$. Then there is a number $t^* \in (\alpha, \beta)$ at which
>
> $$f'(t^*) = \left(\frac{f(\beta) - f(\alpha)}{g(\beta) - g(\alpha)} \right) g'(t^*).$$

Proof. In the context of the racing story above, you start at $f(\alpha)$, and at time $t \in (\alpha, \beta)$ you are at position $f(t)$, which is a distance of $f(t) - f(\alpha)$ from your starting place. So the fraction of your total distance that you've run at time t is

$$\frac{f(t) - f(\alpha)}{f(\beta) - f(\alpha)}.$$

Similarly, the fraction of your friend's total distance completed at time t is

$$\frac{g(t) - g(\alpha)}{g(\beta) - g(\alpha)}.$$

The difference between these fractions is

$$h(t) = \frac{f(t) - f(\alpha)}{f(\beta) - f(\alpha)} - \frac{g(t) - g(\alpha)}{g(\beta) - g(\alpha)}.$$

In the exercise set, you'll verify that $h(t)$ satisfies the hypotheses of Rolle's Theorem (see p. 281), so we know there is some time $t^* \in (\alpha, \beta)$ at which $h'(t^*) = 0$. Since the only non-constant terms in $h(t)$ are $f(t)$ and $g(t)$, its derivative is simply

$$h'(t^*) = \frac{f'(t^*)}{f(\beta) - f(\alpha)} - \frac{g'(t^*)}{g(\beta) - g(\alpha)}.$$

After setting this to zero, we arrive at the result with some simple algebra. ∎

Note: In the exercise set, you'll remove the hypothesis that $f(\alpha) \neq f(\beta)$ and $g(\alpha) \neq g(\beta)$ by restating the relationship between $f'(t^*)$ and $g'(t^*)$ as

$$f'(t^*)(g(\beta) - g(\alpha)) = g'(t^*)(f(\beta) - f(\alpha)).$$

Suppose your friend's speed at time t is $s(t)$. If your speed was always at most $1.9s(t)$, you would only have gone (at most) 1.9 times as far. If your speed was always at least $2.1s(t)$, you would have run (at least) 2.1 times as far.

Of course, instead of "twice," this story could have read "three times as far (fast)," or "1.4 times as far (fast)." The particular factor is irrelevant.

This relationship comes from applying Rolle's Theorem to the helper function

$$\tilde{h}(t) = (f(t) - f(\alpha))(g(\beta) - g(\alpha)) - (f(\beta) - f(\alpha))(g(t) - g(\alpha)),$$

which is just a scaled version of the $h(t)$ that we used in the proof.

Like the Mean Value Theorem itself, Cauchy's extension of the MVT has value in its ability to help us understand other things. On p. 719 we'll interpret the result geometrically in the context of *parametric curves,* but for now we can use it to understand why something called *L'Hôpital's Rule* works.

> The name L'Hôpital is pronounced "lope-ee-TAHL."

✥ L'Hôpital's Rule

In Chapter 2 we saw several limits that resulted in indeterminate forms. These were mostly limits of rational functions, in which case we can resolve the limit value with a factor-and-cancel technique. We also saw $\lim_{\phi \to 0} \frac{\sin(\phi)}{\phi} = 1$, which we calculated using a geometric argument and the Squeeze Theorem. When a function is complicated enough the factor-and-cancel technique doesn't work, and relevant geometry is elusive, as in

$$\lim_{t \to 0} \frac{e^t - 1 - t}{t^2} = ?,$$

we can use the derivative to help determine the value of the limit. Specifically,

---

L'Hôpital's Rule: Suppose f and g are differentiable in some open interval that contains t_0 (except possibly <u>at</u> t_0), and neither $g(t)$ nor $g'(t)$ is zero in that interval (except possibly <u>at</u> t_0). Further, suppose $\lim_{t \to t_0} f(t)/g(t)$ has the form $0/0$ or ∞/∞. Then

$$\lim_{t \to t_0} \frac{f(t)}{g(t)} = \lim_{t \to t_0} \frac{f'(t)}{g'(t)},$$

provided that the right-hand side exists, or is $\pm\infty$.

Note: The limit in L'Hôpital's rule can be one-sided, or at infinity; and the words "open interval" can be replaced by "non-trivial interval," by which we mean to prohibit the interval $[t_0, t_0]$.

Proof. We will address the $0/0$ indeterminate form. In this case $\lim_{t \to t_0} f(t) = 0$, so either $f(t_0) = 0$ or f has a removable discontinuity at t_0. Since the limit we're trying to establish has nothing to do with the actual value of f at t_0, let's remove the discontinuity (if there is one) by defining $f(t_0) = 0$. Similarly we set $g(t_0) = 0$, so

$$\frac{f(t)}{g(t)} = \frac{f(t) - f(t_0)}{g(t) - g(t_0)}.$$

At each t, Cauchy's Mean Value Theorem guarantees a t^* for which

$$\frac{f(t)}{g(t)} = \frac{f(t) - f(t_0)}{g(t) - g(t_0)} = \frac{f'(t^*)}{g'(t^*)}.$$

Since t^* is between t and t_0, we know $t^* \to t_0$ as $t \to t_0$, which allows us to say that

$$\lim_{t \to t_0} \frac{f(t)}{g(t)} = \lim_{t \to t_0} \frac{f'(t^*)}{g'(t^*)} = \lim_{t^* \to t_0} \frac{f'(t^*)}{g'(t^*)}. \qquad \blacksquare$$

Using L'Hôpital's Rule in a familiar example

Example 5.1. Use L'Hôpital's Rule to calculate $\lim_{t \to 3} \frac{t^2 - 7t + 12}{t^2 - 2t - 3}$.

Solution: Since this limit results in the indeterminate form $0/0$, we can apply L'Hôpital's Rule with $f(t) = t^2 - 7t + 12$ and $g(t) = t^2 - 2t - 3$:

$$\lim_{t \to 3} \frac{f(t)}{g(t)} = \lim_{t \to 3} \frac{f'(t)}{g'(t)} = \lim_{t \to 3} \frac{2t - 7}{2t - 2} = -\frac{1}{4}.$$

You should check that the factor-and-cancel method gives us the same answer. ∎

Try It Yourself: Using L'Hôpital's Rule in a familiar example

Example 5.2. Use L'Hôpital's Rule to calculate $\lim_{\phi \to 0} \frac{\sin(\phi)}{\phi}$.

Answer: $\lim_{\phi \to 0} \frac{\sin(\phi)}{\phi} = 1$, just as we saw in Chapter 2. ∎

Full Solution On-line

Multiple applications of L'Hôpital's rule

Example 5.3. Use L'Hôpital's Rule to calculate $\lim_{t \to 0} \frac{e^t - 1 - t}{t^2}$.

Solution: Since the limit results in a $0/0$ indeterminate form, L'Hôpital's rule tells us that its value is the same as

$$\lim_{t \to 0} \frac{e^t - 1}{2t}.$$

This limit *also* results in a $0/0$ indeterminate form, so we can use L'Hôpital's rule to calculate *its* value. Differentiating numerator and denominator, we know its value is the same as

$$\lim_{t \to 0} \frac{e^t}{2} = \frac{1}{2}.$$ ∎

Try It Yourself: Multiple applications of L'Hôpital's rule

Example 5.4. Determine the value of $\lim_{\phi \to 0} \frac{\phi - \sin(\phi)}{\phi^3}$.

Answer: $1/6$. ∎

Full Solution On-line

Sometimes the limit just doesn't exist

Example 5.5. Calculate $\lim_{t \to 0.5^-} \tan(\pi t)/\ln(1 - 2t)$.

Solution: As $t \to 0.5^-$ the argument of the tangent approaches $\pi/2$ from the left, so the tangent blows up. Similarly, $\lim_{t \to 0.5^-} 1 - 2t = 0$, so the logarithm blows up too (but negative). This leaves us with

$$\lim_{t \to 0.5^-} \frac{\tan(\pi t)}{\ln(1 - 2t)} = \frac{\infty}{-\infty}.$$

Since this is an indeterminate form, we can apply L'Hôpital's Rule. After differentiating the numerator and denominator with the Chain Rule, we calculate

$$\lim_{t \to 0.5^-} \frac{\pi \sec^2(\pi t)}{-\frac{2}{1 - 2t}} = \lim_{t \to 0.5^-} \frac{\pi}{2} \frac{2t - 1}{\cos^2(\pi t)} = \frac{0}{0}.$$

Since this limit resulted in a $0/0$ indeterminate form, we can apply L'Hôpital's Rule a second time and calculate

$$\lim_{t \to 0.5^-} \frac{\pi}{2} \frac{2}{-2\pi \cos(\pi t) \sin(\pi t)}.$$

Since the denominator tends to zero in the limit but the numerator does not, the limit is (negative) infinite, which means that the graph of $\tan(\pi t)/\ln(1 - 2t)$ has a vertical asymptote at $t = 1/2$. ∎

There are situations in which L'Hôpital's Rule does not apply

Example 5.6. Explain why L'Hôpital's Rule does not apply to $\lim_{t \to 0} \frac{t+4}{t+3}$.

Solution: L'Hôpital's Rule does not apply in this situation because the limit does *not* result in a $0/0$ or ∞/∞ indeterminate form. In fact, since the limits of the numerator and denominator both exist and are nonzero, the Limit Laws tell us that

$$\lim_{t \to 0} \frac{t+4}{t+3} = \frac{\lim_{t \to 0}(t+4)}{\lim_{t \to 0}(t+3)} = \frac{4}{3}. \qquad \blacksquare$$

> Since the hypotheses of L'Hôpital's Rule are not met in this case, it makes no guarantee about $\lim_{t \to 0} f'(t)/g'(t)$. In fact, you can easily check that
> $$\lim_{t \to 0} \frac{f'(t)}{g'(t)} = 1.$$

❖ Using L'Hôpital's Rule with Limits at Infinity

In the Note on p. 285, we said that L'Hôpital's Rule can be applied to limits at infinity (i.e., $t \to \pm\infty$). Here's why: if we define a temporary helper variable $\tau = 1/t$, saying that $t \to \infty$ is the same as saying that $\tau \to 0^+$. So

$$\lim_{t \to \infty} \frac{f(t)}{g(t)} = \lim_{\tau \to 0^+} \frac{f(1/\tau)}{g(1/\tau)}.$$

Provided that this limit results in an indeterminate form, we can apply L'Hôpital's Rule. After using the Chain Rule to perform the requisite differentiation, we see

$$\lim_{\tau \to 0^+} \frac{f'(1/\tau)\left(\frac{-1}{\tau^2}\right)}{g'(1/\tau)\left(\frac{-1}{\tau^2}\right)} = \lim_{\tau \to 0^+} \frac{f'(1/\tau)}{g'(1/\tau)} = \lim_{t \to \infty} \frac{f'(t)}{g'(t)}.$$

Using L'Hôpital's Rule to compare power functions to logarithms

Example 5.7. Calculate $\lim_{t \to \infty} \ln(t)/t^2$.

Solution: As mentioned above, this limit results in a ∞/∞ indeterminate form. So we use L'Hôpital's Rule to write

$$\lim_{t \to \infty} \frac{\ln(t)}{t^2} = \lim_{t \to \infty} \frac{1/t}{2t} = \lim_{t \to \infty} \frac{1}{2t^2} = 0. \qquad \blacksquare$$

Try It Yourself: Using L'Hôpital's Rule to compare power functions to logarithms

Example 5.8. Calculate $\lim_{t \to \infty} \ln(t)/\sqrt{t}$.

Answer: $\lim_{t \to \infty} t^{-1/2} \ln(t) = 0.$ $\qquad \blacksquare$

Full Solution
On-line

In the exercise set you'll generalize Examples 5.7 and 5.8 to prove the following.

> Suppose $p > 0$. Then t^p grows faster than $\ln(t)$ as $t \to \infty$. Specifically,
> $$\lim_{t \to \infty} \frac{\ln(t)}{t^p} = 0.$$

Using L'Hôpital's Rule to compare power functions to exponentials

Example 5.9. Calculate $\lim_{t \to \infty} t^2/e^t$.

Solution: This limit leads us to the indeterminate form

$$\lim_{t \to \infty} \frac{t^2}{e^t} = \frac{\infty}{\infty}.$$

When we apply L'Hôpital's Rule, we get the same indeterminate form

$$\lim_{t \to \infty} \frac{t^2}{e^t} = \lim_{t \to \infty} \frac{2t}{e^t} = \frac{\infty}{\infty}.$$

Since $\lim_{t \to \infty} 2t/e^t$ results in an indeterminate form, we can apply L'Hôpital's Rule to *it:*

$$\lim_{t \to \infty} \frac{t^2}{e^t} = \lim_{t \to \infty} \frac{2t}{e^t} = \lim_{t \to \infty} \frac{2}{e^t} = 0.$$ ■

Try It Yourself: Using L'Hôpital's Rule to compare power functions to exponentials

Example 5.10. Calculate $\lim_{t \to \infty} t^4/e^t$.

Answer: $\lim_{t \to \infty} t^4/e^t = 0$. ■

Full Solution
On-line

In the exercise set you'll generalize Examples 5.9 and 5.10 to prove the following.

Suppose $p > 0$ and $b > 1$. Then b^t grows faster than t^p as $t \to \infty$. Specifically,

$$\lim_{t \to \infty} \frac{t^p}{b^t} = 0.$$

❖ Other Indeterminate Forms

When limits result in $0 \cdot \infty$, 1^∞ or ∞^0 we can often rewrite the relevant function as a quotient and apply L'Hôpital's rule.

Working with the $0 \cdot \infty$ indeterminate form

Example 5.11. Calculate $\lim_{t \to 0^+} t \ln(t)$.

Solution: When we rewrite

$$\lim_{t \to 0^+} t \ln(t) = \lim_{t \to 0^+} \frac{\ln(t)}{t^{-1}}$$

we see that it results in an ∞/∞ indeterminate form. So we apply L'Hôpital's rule, and see that

$$\lim_{t \to 0^+} \frac{\ln(t)}{t^{-1}} = \lim_{t \to 0^+} \frac{1/t}{-t^{-2}} = \lim_{t \to 0^+} \left(\frac{1}{t}\right)(-t^2) = \lim_{t \to 0^+} -t = 0.$$ ■

Try It Yourself: Working with the $0 \cdot \infty$ indeterminate form

Example 5.12. Calculate $\lim_{t \to 0^+} \sqrt{t} e^{1/t}$.

Answer: ∞. ■

Full Solution
On-line

Working with the 1^∞ indeterminate form

Example 5.13. Calculate $\lim_{t \to 0^+} (1+t)^{1/t}$.

Solution: When $f(t)$ is a positive function, we know $f(t) = e^{\ln(f(t))}$, so

$$\lim_{t \to 0^+} f(t) = \lim_{t \to 0^+} e^{\ln(f(t))}.$$

And since the exponential function is continuous,

$$\lim_{t \to 0^+} \exp\left(\ln\Big(f(t)\Big)\right) = \exp\left(\lim_{t \to 0^+} \ln\Big(f(t)\Big)\right),$$

so we can determine $\lim_{t \to 0^+} f(t)$ by calculating $\lim_{t \to 0^+} \ln(f(t))$. In our case, $f(t) = (1+t)^{1/t}$, so the logarithm allows us to rewrite the exponent as a coefficient:

> Many students are tempted to say that $0 \cdot \infty$ is zero, but remember that we're talking about a *limit*. The notation $0 \cdot \infty$ means that one factor of a product gets smaller and smaller while the other gets larger and larger. The value of the limit depends on how quickly each factor changes.

> We're using $t \to 0^+$ in this example because logarithm requires a positive argument.

> Remember that $\exp(u)$ is another way of writing e^u. It's especially useful in avoiding superscripts on superscripts, which make the page (even more) difficult to read.

> This is what leads us to think about the logarithm in the first place.

$$\lim_{t \to 0^+} \ln\left((1+t)^{1/t}\right) = \lim_{t \to 0^+} \frac{1}{t}\ln(1+t).$$

This limit results in a $0/0$ indeterminate form, so we can use L'Hôpital's rule to calculate its value as

$$\lim_{t \to 0^+} \ln\left(f(t)\right) = \lim_{t \to 0^+} \frac{\ln(1+t)}{t} = \lim_{t \to 0^+} \frac{1/(1+t)}{1} = 1.$$

Many students mistakenly stop here, and say that the limit value is 1, but we're not done!

Therefore,

$$\lim_{t \to 0^+} (1+t)^{1/t} = \exp\left(\lim_{t \to 0^+} \ln\left(f(t)\right)\right) = \exp(1) = e.$$ ∎

Try It Yourself: Working with the 1^∞ indeterminate form

Example 5.14. Calculate $\lim_{t \to 0^+}(1+5t)^{7/t}$.

Answer: e^{35}. ∎

Full Solution
On-line

Working with the ∞^0 indeterminate form

Example 5.15. Calculate $\lim_{t \to \infty}(1+3t)^{1/\ln(t^2)}$.

Solution: We use the same technique here that we did in Examples 5.13 and 5.14. Consider the logarithm of our quantity:

$$\ln\left((1+3t)^{1/\ln(t^2)}\right) = \frac{1}{\ln(t^2)}\ln(1+3t) = \frac{\ln(1+3t)}{2\ln(t)},$$

which results in an $\frac{\infty}{\infty}$ indeterminate form when we let $t \to \infty$. So we apply L'Hôpital's Rule and see that

$$\lim_{t \to \infty} \frac{\ln(1+3t)}{2\ln(t)} = \lim_{t \to \infty} \frac{\frac{3}{1+3t}}{\frac{2}{t}} = \lim_{t \to \infty} \frac{3t}{2+6t} = \frac{1}{2}.$$

Since the logarithm of our function converges to $1/2$ in the limit, our function converges to $e^{1/2} = \sqrt{e}$ as $t \to \infty$. ∎

Note: The technique of using the logarithm is often useful when a limit is difficult to evaluate because of an exponent.

✵ Common Difficulties with L'Hôpital's Rule

Students' most common problem with L'Hôpital's Rule comes from jumping to the quotient of derivatives, f'/g', without first checking that the original limit is an indeterminate form. This leads to the wrong answer (see Example 5.6, p. 287).

At the conceptual level, many students assert that the indeterminate form 1^∞ is really just 1, because 1 to any power is 1. But remember that L'Hôpital's rule deals with *limits*, so the base gets closer and closer to 1 but never really *is* 1, and the exponent gets larger and larger but never really *is* infinite at any time.

The same kind of conceptual problem often happens with $0 \cdot \infty$, where it might seem that the limit value should be zero. But we're talking about the *limit* of two factors, one tending to zero and the other growing without bound. The value of the limit depends on which factor "wins" that race.

|You should know|

- the statement and meaning of Rolle's Theorem and the Mean Value Theorem;

- Cauchy's extension of the Mean Value Theorem;

- the statement and meaning of L'Hôpital's rule;

- when L'Hôpital's Rule does *not* apply.

You should be able to

- Determine the value of t^* that satisfies the Mean Value Theorem;

- Use L'Hôpital's rule to calculate limits of the form $\frac{0}{0}, \frac{\infty}{\infty}, 0 \cdot \infty$, and 1^∞ and other similar indeterminate forms.

✤ 4.5 Skill Exercises

In each of #1–4 (a) verify that the average rate of change in $f(t)$ over the specified interval is zero, and (b) determine the value of t^* guaranteed by Rolle's Theorem.

1. $f(t) = t^2 - 4t + 5; [1,3]$

2. $f(t) = t^2 - 9t + 8; [2,7]$

3. $f(t) = \sin(t); [1.25\pi, 1.75\pi]$

4. $\sec(t); [-0.25\pi, 0.25\pi]$

In each of #5–8 (a) calculate the average rate of change in $f(t)$ over the specified interval, and (b) determine the value of t^*, guaranteed by the MVT.

5. $f(t) = t^3 + 2t + 7, [2,5]$

6. $f(t) = t^3 - 8t^2 - t + 7, [0,3]$

7. $f(t) = e^{2t}, [0,4]$

8. $f(t) = \cos(2t), [0,\pi]$

In #9–14 determine (a) whether L'Hôpital's Rule is applicable, and (b) the value of the limit.

9. $\displaystyle\lim_{t \to 6} \frac{5t+9}{8t+1}$

10. $\displaystyle\lim_{t \to 5} \frac{3t^2 - 16t + 5}{12t^2 + 4t + 9}$

11. $\displaystyle\lim_{t \to -1} \frac{t^4 - 3t^3 - 7t^2 + 15t + 18}{t^2 - t - 2}$

12. $\displaystyle\lim_{t \to 2} \frac{4t^2 - 3t + 14}{t^4 - 8t + 10}$

13. $\displaystyle\lim_{t \to 1} \frac{e^{2(t-1)} - 2t + 1}{t^4 - 2t^3 - 8t^2 + 18t - 9}$

14. $\displaystyle\lim_{t \to 1} \frac{\ln(t^2)}{t - 1}$

In each of #15 and #16 (a) verify that the given limit results in an indeterminate form, (b) use L'Hopital's rule to rewrite the limit, (c) verify that the resulting limit also results in an indeterminate form, (d) use L'Hopital's rule again, (e) explain why L'Hopital's rule is not helpful in this case, and (f) find another way to determine the limit value.

15. $\displaystyle\lim_{x \to \infty} \frac{(x^2 + 3)^{5/9}}{x^{10/9}}$

16. $\displaystyle\lim_{x \to \infty} \frac{x^{14/3}}{(2x^2 + 1)^{7/3}}$

Each limit in #17–22 is the limit of a difference quotient (i.e., a derivative). (a) Use L'Hopital's rule to determine the value of the limit, (b) figure out what function is being differentiated, and at what value of t_0, and (c) determine $f'(t_0)$ using what you know about derivatives.

17. $\lim\limits_{h \to 0} \dfrac{\sin(3(0+h)) - 0}{h}$

20. $\lim\limits_{h \to 0} \dfrac{\cos\left(\frac{\pi}{6} + h\right) - \frac{1}{2}}{h}$

18. $\lim\limits_{h \to 0} \dfrac{\tan(4(0+h)) - 0}{h}$

21. $\lim\limits_{h \to 0} \dfrac{e^h - 1}{h}$

19. $\lim\limits_{h \to 0} \dfrac{\sin\left(\frac{\pi}{4} + h\right) - \frac{1}{\sqrt{2}}}{h}$

22. $\lim\limits_{h \to 0} \dfrac{\ln(1+h) - 0}{h}$

In #23–30 (a) identify the form of the indeterminate limits (e.g. 1^∞ or 0^0, etc.), and (b) use L'Hôpital's Rule to determine the value of the limit.

23. $\lim\limits_{t \to 0^+} \left(1 + \frac{7}{8t}\right)^{3t}$

27. $\lim\limits_{\phi \to 0^+} \left(\sin(3\phi)\right)^{\sin(8\phi)}$

24. $\lim\limits_{t \to 0^+} \left(1 - \frac{3}{5t}\right)^{8t}$

28. $\lim\limits_{\phi \to 0^+} \left(\sin(4\phi)\right)^{\sin(5\phi)}$

25. $\lim\limits_{t \to \infty} \left(1 + 3t\right)^{1/t}$

29. $\lim\limits_{\phi \to 0^+} \left(\sec(3\phi)\right)^{\csc(2\phi)}$

26. $\lim\limits_{t \to \infty} \left(1 + 4t\right)^{1/(7t)}$

30. $\lim\limits_{\phi \to 0^+} \left(\cos(3\phi)\right)^{\cot(2\phi)}$

31. Design a function $f(t)$ so that $\lim_{t \to 0} f(t)$ has the indeterminate form 1^∞, but the limit value is actually 9.

32. Design a function $f(t)$ so that $\lim_{t \to 0} f(t)$ has the indeterminate form 1^∞, but the limit value is actually 1/5.

33. How should we define $f(1)$ so that f is continuous at $t = 1$, when $f(t) = \frac{\ln(t)}{t^2 - 1}$?

34. Show that $\lim_{t \to 0^+} f(t)$ exists when $f(t) = t \ln(t)$, but $\lim_{t \to 0^+} f'(t)$ does not. In complete sentences, explain what this tells you about the graph of f near $t = 0$, and include a graph of f to help make your point.

35. Suppose $g(t) = \sqrt{t}$, and $f(t) = t \ln(t)$. Based on $\lim_{t \to 0^+} \frac{f'(t)}{g'(t)}$, would you say that the graph of f or the graph of g is steeper near $t = 0$?

❖ 4.5 Concept and Application Exercises

In a certain sense the hypotheses of Rolle's Theorem prohibit "non-physical" functions. In #36–37 (a) verify that $f(0) = f(8)$, (b) verify that $f'(t)$ is never zero on $[0, 8]$, (c) explain why this *not* a counterexample that disproves Rolle's Theorem, and (d) explain the story that would be told by the graph if $f(t)$ were the position of a person at time t. Pay special attention to the position and velocity at times $t = 0, 4, 8$.

36.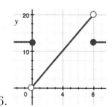

37.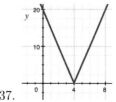

38. The MVT guarantees at *least* one value of t^* at which $f'(t^*)$ equals the average rate of change over an interval. Draw the graph of a function f over $[1, 5]$ for which there are two values of t^*.

39. Suppose $f(x) = \sin(\pi/x)$. Find a number α so that both (a) Rolle's Theorem guarantees a point $t^* \in (\alpha, 0.5)$ at which $f'(t^*) = 0$ and (b) there are exactly 20 critical points of f in $(\alpha, 0.5)$.

40. A boy was 36 inches tall on one day, and 38 inches tall 281 days later. What does the MVT tell us?

41. Suppose $f(x) = 1 + \sqrt{x}$.

    (a) The MVT guarantees a $t^* \in (1, 9)$ for which $f'(t^*) = $ what number?

    (b) Verify that this happens by finding such a value of t^*.

42. Consider the function $f(t) = \begin{cases} 2t + 1 & \text{when } t < 3 \\ 4t - 5 & \text{when } t \geq 3. \end{cases}$

    (a) What are the possible values of $f'(t)$?

    (b) Calculate the average rate of change in f over $[1, 7]$.

    (c) Compare the numbers in parts (b) and (c), and explain why this is not an exception to the MVT.

43. The populations of two plant species, A and B, are shown in Figure 5.3.

    (a) Verify that $A(t)$ and $B(t)$ have the same average rate of change over $[0, 12]$.

    (b) If we denote by m the average rate of change from part (a), find the times at which $B'(t) = m$, and those at which $A'(t) = m$.

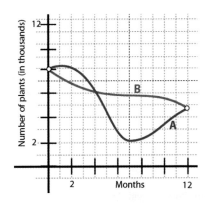

Figure 5.3: For exercises #43 and #44.

44. The populations of two plant species, A and B, are shown in Figure 5.3.

    (a) Verify that $A(0) = B(0)$ and $A(12) = B(12)$.

    (b) In light of your answer from part (a), what does Cauchy's extension of the MVT tell us must happen at some time, $t^* \in (0, 12)$?

    (c) Use the graphs of $A(t)$ and $B(t)$ to estimate such a time.

45. Two of your friends arrive at different answers when calculating $\lim_{t \to \infty} \frac{t + \sin(t)}{t + 1}$, and come to you for help. Based on their arguments below, decide which of them is right, and explain why the other is wrong.

Friend #1: Because the sine remains between -1 and 1, we know that

$$\frac{t-1}{t+1} \le \frac{t+\sin(t)}{t+1} \le 1.$$

Since $\lim_{t\to\infty} \frac{t-1}{t+1} = 1$, the Squeeze Theorem tells us that $\lim_{t\to\infty} \frac{t+\sin(t)}{t+1} = 1$.

Friend #2: Since $\lim_{t\to\infty} t + \sin(t) = \infty$ (the oscillation of the sine is overpowered by the growth of t) and $\lim_{t\to\infty} t + 1 = \infty$, this limit results in an ∞/∞ indeterminate form, so we can apply L'Hôpital's Rule, which tells us that

$$\lim_{t\to\infty} \frac{t+\sin(t)}{t+1} = \lim_{t\to\infty} \frac{1+\cos(t)}{1}.$$

Since $\lim_{t\to\infty} 1 + \cos(t)$ does not exist (due to the oscillation of the cosine), neither does our limit.

46. Figure 5.4 shows a sector of the unit circle with central angle θ. Denote by $A(\theta)$ the area between the line segment PR and the arc PR, and by $B(\theta)$ the area of the right triangle $\triangle PQR$. Determine $\lim_{\theta\to 0^+} A(\theta)/B(\theta)$.

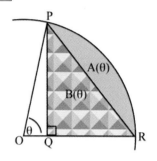

Figure 5.4: For #46.

Checking the details

47. Suppose $p > 0$, and $b > 1$. Show that $\lim_{t\to\infty} \log_b(t)/t^p = 0$.

48. Suppose that $p > 0$, and $1 > b > 0$. Use L'Hôpital's Rule to show that $\lim_{t\to\infty} \log_b(t)/t^p = 0$.

49. Suppose that $p > 0$ is an integer, and $b > 1$. Use L'Hôpital's Rule to show that $\lim_{t\to\infty} t^p/b^t = 0$.

50. Suppose that $p > 0$ is *not* an integer, and $b > 1$. Use the result from #49 to argue that $\lim_{t\to\infty} t^p/b^t = 0$.

51. In the proof of the MVT we used $c(t) = mt + b$ to build the function $h(t)$. Determine the value of b so that $c(\alpha) = g(\alpha)$.

52. Check that $h(t)$, from the proof of Cauchy's Mean Value Theorem on p. 284, satisfies the hypotheses of Rolle's Theorem (see p. 281).

53. Consider functions h and $\tilde{h}$, from the proof of Cauchy's Mean Value Theorem and the note below it. Verify that

$$\tilde{h}(t) = \underbrace{\Big(f(\beta) - f(\alpha)\Big)\Big(g(\beta) - g(\alpha)\Big)}_{\text{notice that this part is constant}} h(t).$$

54. In this exercise, we remove the hypotheses that $g(\alpha) \ne g(\beta)$ and $f(\alpha) \ne f(\beta)$ from the statement of Cauchy's MVT.

   (a) Verify that $\tilde{h}$, as seen in #53, satisfies the hypotheses of Rolle's Theorem, even if $f(\alpha) = f(\beta)$ or $g(\alpha) = g(\beta)$.

   (b) Find a relationship between f' and g' at some time $t \in (\alpha, \beta)$ by applying Rolle's Theorem to $\tilde{h}$.

4.6   Curve Sketching

The ubiquity of graphing software today lets us quickly and easily plot curves and surfaces. So while this section will focus on sketching graphs of functions, you should understand that the graphs are a means to an end. The actual goal of this section is to solidify and synthesize your understanding of how the first and second derivatives govern the behavior of functions.

Sketching the graph of a function is typically done in two steps (more if you need greater accuracy):

---

1. Determine where the graph is increasing and where it's decreasing;

2. Determine where the graph is concave up and where it's concave down.

Sketch the graph of a cubic function

Example 6.1. Sketch the graph of $f(x) = x^3 + 2x^2 - x + 1$.

Solution: Notice that $f'(x) = 3x^2 + 4x - 1$ is continuous everywhere, so it can only change from positive to negative (or vice versa) at a root:

$$3x^2 + 4x - 1 = 0 \quad \text{at} \quad x_1 = \underbrace{\frac{-2 - \sqrt{7}}{3}}_{\approx -1.549} \quad \text{and} \quad x_2 = \underbrace{\frac{-2 + \sqrt{7}}{3}}_{\approx 0.215}.$$

Because the graph of f' is a parabola that's cupped upward (sketched in Figure 6.1), we know that ...

- $f'(x) < 0$ when $x \in (x_1, x_2)$, so f is decreasing on that interval;

- $f'(x) > 0$ outside of $[x_1, x_2]$, so f is increasing there.

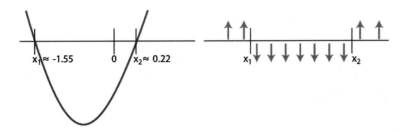

Figure 6.1: (left) A sketch of $y = f'(x)$; (right) arrows indicate whether the graph of f is increasing or decreasing.

     Using this information, we sketch our first approximation of the graph (see Figure 6.2). Based on that sketch, you can infer that the graph of f will be concave down at x_1 (which is a local maximum), and concave up at x_2 (which is a local minimum). In step 2 of the sketching process, we find out where the change in concavity occurs.

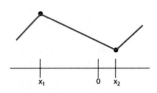

Figure 6.2: Our first approximation of $y = x^3 + 2x^2 - x + 1$.

     The second derivative is $f''(x) = 6x + 4$, which is continuous everywhere, so it can only change from positive to negative (or vice versa) at a root:

$$6x + 4 = 0 \quad \Longleftrightarrow \quad x = -2/3.$$

Since the graph of f'' is a line with positive slope (as depicted in Figure 6.3) we conclude that

- $f''(x) < 0$ when $x < -2/3$, meaning that the graph of f is concave down;

- $f''(x) > 0$ when $x > -2/3$, meaning the graph of f is concave up.

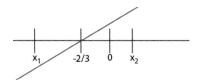 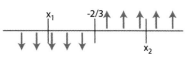

Figure 6.3: (left) A sketch of $y = f''(x)$; (right) arrows indicate whether the graph of f is concave up or concave down.

When we impose this concavity information on our sketch, we get our second approximation (the red graph shown in Figure 6.4).

If we need greater accuracy, we can determine the y-intercept ($f(0) = 1$), calculate the function values at the critical points ($f(x_1) \approx 3.6311$ and $f(x_2) \approx 0.8874$), and find the root of this function ($x \approx -2.5469$ from Newton's Method). ∎

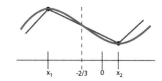

Figure 6.4: Our second approximation of $y = x^3 + 2x^2 - x + 1$, shown in red. Notice that the second approximation increases and decreases with the first approximation. The dashed vertical guideline indicates where the graph changes concavity.

Try It Yourself: Sketch the graph of a cubic polynomial

Example 6.2. Sketch the graph of $f(x) = 5x^3 - x^2 + x + 2$.

Answer: See on-line solution. ∎

Full Solution On-line

Concavity and graphing

Example 6.3. Suppose $g(x) = x^3 e^{-x}$. Sketch the graph of g.

Solution: Before applying the calculus, let's note that this function has a single root at $x = 0$. Using the Product Rule, we determine that $g'(x) = x^2 e^{-x}(3 - x)$, which has roots at $x = 0$ and $x = 3$. Since $g'(x)$ is continuous everywhere, these critical values are the only places where it can change from positive to negative (or vice versa), but does it?

- When $x \approx 0$ we have $e^{-x} \approx 1$ and $(3 - x) \approx 3$, so $g'(x) \approx 3x^2$ when $x \approx 0$. For this reason the graph of g' "bounces off" of the x-axis at zero, just like the parabola $y = 3x^2$ (see Figure 6.5).

- When $x \approx 3$ we have $x^2 \approx 9$ and $e^{-x} \approx e^{-3}$, so $g'(x) \approx 9e^{-3}(3 - x)$ when $x \approx 3$. For this reason the graph of g' "crashes through" the x-axis at $x = 3$, just like the line $y = 9e^{-3}(3 - x)$.

The presentation here is a graphical version of the First Derivative Test. Alternatively, we could classify the critical points by checking g' numerically:

$$g'(-1) = 4e > 0$$
$$g'(1) = 2e^{-1} > 0$$
$$g'(4) = -16e^{-4} < 0$$

Note that the sign of $g'(x)$ doesn't change across $x = 0$. However, $g'(x)$ changes from positive to negative across $x = 3$, indicating a local maximum.

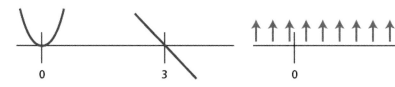

Figure 6.5: (left) A rough sketch of the graph of g' near its roots; (right) arrows show whether g is increasing or decreasing.

Based on the graph of g' sketched in Figure 6.5, we know that $g'(x) > 0$ when $x \in (-\infty, 0) \cup (0, 3)$. It follows that $g(x)$ is increasing on $(-\infty, 3)$, and is decreasing

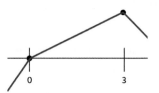

on $(3, \infty)$. This basic behavior is depicted in Figure 6.6.

Since the only root of g occurs at $x = 0$, its graph cannot intersect the axis again. So while the graph is concave down at $x = 3$, it must change its concavity sometime thereafter. In the next step of our sketching process, we find out where.

Figure 6.6: Our first approximation of $y = g(x)$. The only root of g occurs at $x = 0$.

The second derivative of g is $g''(x) = (x^2 - 6x + 6)xe^{-x}$. Since $g''(x)$ is continuous everywhere, it can only change sign at its roots. The exponential term is never zero, so we have roots only at $x = 0$ and $x = 3 \pm \sqrt{3}$ (which is where $x^2 - 6x + 6 = 0$).

Since we know where $x^2 - 6x + 6$ is zero, we can rewrite $g''(x)$ as

$$g''(x) = \underbrace{\left(x - (3 + \sqrt{3})\right)\left(x - (3 - \sqrt{3})\right)}_{\text{This is the factored form of } x^2 - 6x + 6} xe^{-x}.$$

The roots of $g''(x)$ arise from the first three factors, each of which is linear. So we know the graph of $g''(x)$ crashes through the x-axis at $x = 0$, $3 - \sqrt{3}$, and $3 + \sqrt{3}$, changing sign each time (see Figure 6.7).

In case you want to see the calculations of $g''(x)$ between and around the critical points:

$g''(-1) = -13e < 0$
$g''(1) = \exp(-1) > 0$
$g''(3) = -9\exp(-3) < 0$
$g''(5) = 5\exp(-5) > 0$

Notice that the sign of $g''(x)$ changes across each of the second-order critical points.

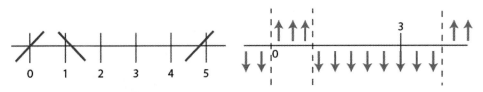

Figure 6.7: (left) A rough sketch of the graph of g'' near its roots; (right) arrows show whether the graph of g is concave up or concave down.

Since $g''(-1) = -13e < 0$,

$$g''(x) < 0 \text{ when } \underbrace{x \in (-\infty, 0) \cup (3 - \sqrt{3}, 3 + \sqrt{3})}_{\text{The graph of } g \text{ is concave down here}} \quad \text{and} \quad g''(x) > 0 \text{ when } \underbrace{x \in (0, 3 - \sqrt{3}) \cup (3 + \sqrt{3}, \infty)}_{\text{The graph of } g \text{ is concave up here}}.$$

This concavity information is shown in Figure 6.8. Lastly, we mention that the exponential decay dominates the polynomial growth when x is large (this was demonstrated using L'Hôpital's Rule in Section 4.5), so $\lim_{x \to \infty} g(x) = 0$. ∎

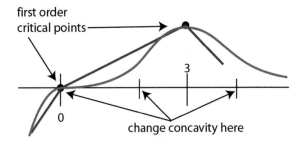

first order critical points

change concavity here

Figure 6.8: The graph $y = g(x)$ from Example 6.3. Note that $x = 0$ is both a first- and second-order critical point.

Try It Yourself: Derivatives and graphing

Example 6.4. Suppose $g(x) = (x^2 - x)e^{-x}$. Sketch the graph of g.

Answer: See solution on-line. ∎

Full Solution On-line

❖ Working with the Graph of the Derivative

In order to solidify and synthesize the relationship between a function and its derivatives yet further, let's work with graphs instead of formulas. In Example 6.5 we will use the graph of $y'(x)$ to determine when $y(x)$ increases and decreases, and in Example 6.6 we'll use the same y' to discuss intervals of concavity.

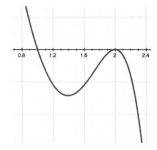

Figure 6.9: Graph for Examples 6.5, 6.6, and 6.7.

Extracting increase and decrease from the graph of $y'(x)$

Example 6.5. The graph of $y'(x)$ is shown in Figure 6.9. Use it to determine (a) intervals over which $y(x)$ is increasing, (b) intervals over which $y(x)$ is decreasing, (c) all first-order critical points, (d) all turning points.

Solution: In the left-hand image of Figure 6.10, the plane has been shaded at values of x where $y'(x)$ is negative. (a) Since $y'(x) > 0$ when $x \in (-\infty, 1)$, the graph of $y(x)$ continues to climb until $x = 1$. So $y(x)$ is increasing on $(-\infty, 1]$.

> You might wonder why $x = 1$ is included in both parts (a) and (b). Remember that the terms "increasing" and "decreasing" describe a relationship between function values over intervals. In this case, $x = 1$ is at the end of one interval and at the beginning of the next.

(b) Similarly, $y(x)$ is decreasing when $y'(x) < 0$. This happens when $x \in (1, 2)$, so $y(x)$ is decreasing on $(1, 2)$. The derivative is also negative on $(2, \infty)$, so we say that $y(x)$ is decreasing on $(2, \infty)$. All told, the function is decreasing on the union of these intervals $[1, 2] \cup [2, \infty) = [1, \infty)$.

(c) Notice that $y'(1) = 0$ and $y'(2) = 0$, so $x = 1$ and $x = 2$ are first-order critical points.

(d) The graph of $y(x)$ exhibits different behavior at $x = 1$ than it does at $x = 2$:

$$\text{at } x = 1 \cdots \begin{cases} \text{the derivative changes from positive to negative, so the graph} \\ \text{of } y(x) \text{ changes from increasing to decreasing;} \end{cases}$$

$$\text{at } x = 2 \cdots \begin{cases} \text{the derivative is negative on both the left and the right of} \\ x = 2, \text{ so the graph of } y(x) \text{ decreases, flattens out, and then} \\ \text{decreases some more.} \end{cases}$$

So we say that $x = 1$ is a turning point, but $x = 2$ is not.   ■

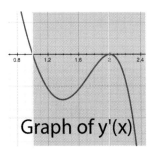
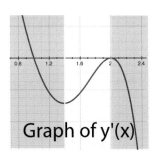

Figure 6.10: (left) The function $y(x)$ is increasing when $x \in (-\infty, 1]$, and is decreasing when $x \in [1, \infty)$; (right) the function $y(x)$ is concave down when $x \in (-\infty, 1.4) \cup (2, \infty)$, and is concave up when $x \in (1.4, 2)$.

Extracting concavity information from the graph of $y'(x)$

Example 6.6. The graph of $y'(x)$ is shown in Figure 6.9. Use it to determine (a) intervals over which the graph of $y(x)$ is concave down, (b) intervals over which the graph of $y(x)$ is concave up, and (c) inflection points.

Solution: (a) We know that the graph of $y(x)$ is concave down when $y''(x) < 0$. Since we have the graph of $y'(x)$ at our disposal, let's rewrite this concavity condition in terms of y':

$$\frac{d}{dx}y'(x) < 0,$$

which happens exactly *when the graph of $y'(x)$ is decreasing.* In Figure 6.10, the plane has been shaded at those values of x where $y'(x)$ is decreasing. From this we conclude that the graph of y is concave down when $x \in (-\infty, 1.4) \cup (2, \infty)$.

(b) Similarly, the graph of $y(x)$ will be concave up where $\frac{d}{dx}y'(x) > 0$, which happens when $x \in (1.4, 2)$.

(c) Inflection points are the places where $y''(x) = \frac{d}{dx}y'(x)$ changes from positive to negative, or vice versa—that is, when $y'(x)$ changes from increasing to decreasing, or vice versa. From our graph, we see this at $x = 1.4$ and $x = 2$.   ■

Our next task is to combine the increase-decrease and concavity information acquired in the previous examples.

Graphing $y(x)$ based on the graph of $y'(x)$

Example 6.7. The graph of $y'(x)$ is shown in Figure 6.9. Use it to sketch a graph of $y(x)$.

Solution: Based on the information about the graph that we acquired in Examples 6.5 and 6.6, we can describe the curve as follows:

over $(-\infty, 1)$ the graph of $y(x)$ is increasing, and is concave down;
over $(1, 1.4)$ the graph of $y(x)$ is decreasing, and is concave down;
over $(1.4, 2)$ the graph of $y(x)$ is decreasing, and is concave up;
over $(2, \infty)$ the graph of $y(x)$ is decreasing, and is concave down.

The left-hand graph in Figure 6.11 depicts such a curve.   ■

> If you and a friend were to answer this question independently, you might start with $y(1) = 2$ while she starts at $y(1) = 3$, because information about particular function values is unavailable. From there, your graphs both increase or decrease according to the same y' information, so they will remain a constant (vertical) distance apart.

Try It Yourself: Working with the graph of g'

Example 6.8. The right-hand image in Figure 6.11 depicts the graph of $g'(t)$. Use it to (a) find all first-order critical points, (b) find all turning points, (c) determine the interval(s) over which $g(x)$ is increasing, (d) find the interval(s) of over which the graph of g is concave up, (d) find all inflection points, and (e) draw the graph of g.

Answer: The graph of g is concave up over $(-\infty, -3) \cup (-3, -1) \cup (1.5, 3)$, and increasing over $(3, 0)$. Full solutions, along with a graph of g, are on-line.   ■

Full Solution
On-line

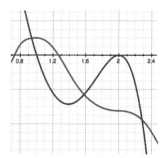 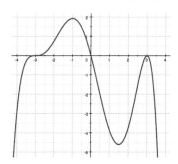

Figure 6.11: (left) A sketch of $y(x)$ superimposed on the graph of $y'(x)$ from Example 6.5; (right) the graph for Example 6.8.

You should know

- sketching the graph of a function is typically a two-step process;

- the graph of f *can* change from increasing to decreasing at first-order critical points, but doesn't have to;

- the graph of f *can* change concavity at second-order critical points, but doesn't have to.

You should be able to

- use the roots of $f(x)$ to "anchor" your graph of f;

- sketch the graph of f based on a formula for $f(x)$;

- sketch the graph of f based on a formula for $f'(x)$;

- sketch the graph of f based on a graph of $f'(x)$.

❖ 4.6 Skill Exercises

In #1–22 you should (a) determine the interval(s) over which $f'(x) > 0$, (b) determine the intervals over which $f''(x) > 0$, and (c) use that information to sketch a graph of $f(x)$.

1. $f(x) = x^2 + 2x - 8$

2. $f(x) = 7 + 6x - x^2$

3. $f(x) = x^3 + 3x^2 - 24x - 5$

4. $f(x) = 2x^3 + x^2 - 3x + 1$

5. $f(x) = \frac{1}{4}x^4 + \frac{4}{3}x^3 + 2x^2 + 1$

6. $f(x) = x^4 + x^3 - 12x^2 - 1$

7. $y = \frac{3}{10}x^{10/3} - \frac{6}{7}x^{7/3}$

8. $y = \frac{15}{7}x^{5/7} - \frac{5}{9}x^{9/7}$

9. $f(x) = -\frac{1}{3}(1 - x^2)^{3/2}$

10. $f(x) = \sqrt{4 - x^2}$

11. $y = \sqrt[3]{27 - x^3}$

12. $f(x) = (1 + x^2)^{-1/5}$

13. $f(x) = \frac{\sqrt{x}}{x^2 + 1}$

14. $f(x) = \frac{x}{\sqrt{x^2 + 1}}$

15. $f(x) = (x^2 - x)e^{-x}$

16. $f(x) = \frac{1}{2}x + e^{-3x}$

17. $y = x - 3\ln(x + 4)$

18. $y = \ln(x^2 + 1) - \ln(x)$

19. $y = e^{x/3}\left(3\sin(2x) - 18\cos(2x)\right)$

20. $y = e^{-5x}\left(2\sin\left(\frac{1}{2}x\right) - 20\cos\left(\frac{1}{2}x\right)\right)$

21. $y = \tan(x) - 2x$ for $-\frac{\pi}{2} \leq x \leq \frac{\pi}{2}$

22. $y = 2x + 3\cos(x) - \frac{1}{2}\sin(2x)$ for $0 \leq x \leq \pi$

In each of #23–30 the function $f'(x)$ is given. You should (a) render a graph of f' over the specified interval using graphing software (and then print it), and (b) sketch the curve $y = f(x)$ on the same graph, with $f(0) = 0$.

23. $f'(x) = e^{\cos(x)} - 1$; $[-8, 8]$

24. $f'(x) = e^{1-x^2} - 1$; $[-5, 5]$

25. $f'(x) = \left((x^2 - x + 1)^{2/3} - 1 \right)^2$; $[-1, 2]$

26. $f'(x) = \cos(4\sqrt{x})$; $[0, 10]$

27. $f'(x) = \sqrt{5x^6 - 3x^2 + 1} - 1$; $[-2, 2]$

28. $f'(x) = \sqrt{5x^6 - 3x^2 + 1} - x$; $[-1, 1]$

29. $f'(x) = e^{x/5} \sin(3x)$; $[0, 8]$

30. $f'(x) = e^{-x/2} \cos(4x)$; $[-3, 3]$

In each of #31–36 the graph of $f'(x)$ is given. Based on the given graph of $f'(x)$ you should (a) sketch the graph of f with $f(0) = 0$, and (b) sketch the graph of $f''(x)$.

31.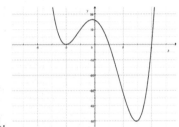

32.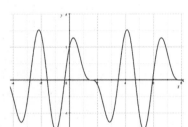

33.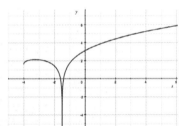

34.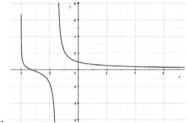

35.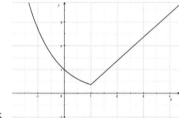

36.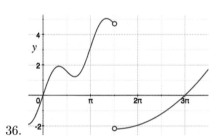

4.7 Optimization

Imagine you're on an African photo-safari, standing in a field of tall grasses, enjoying the sights. When you lower your camera to take a drink of water, you notice a lioness, crouched in a stalking stance and looking right at you. Uh-oh ...

You are 15 meters from your truck (direct line of sight) which is parked on the dirt road, 6 meters behind you. If you run straight toward the truck, you'll have to stay in the tall grass the whole time, and that will slow you down. You can run faster on the road, but if you head straight for the road you'll actually have to run a longer distance, which takes more time. If you had a moment to think about it, you might realize that you could minimize your time by way of a compromise, running neither directly at the truck, nor directly at the road.

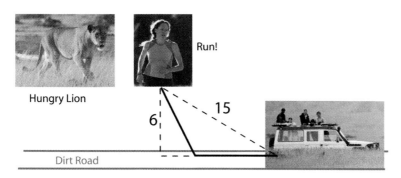

Figure 7.1: If you get home, you'll have a great story to tell!

Whether we are trying to *minimize* time, *maximize* efficiency, *minimize* cost, *maximize* blood flow, *minimize* potential energy, or *maximize* height ... we refer to the process of finding our answer as **optimization**, and the function whose minimum or maximum value we're finding is called the **objective function**.

In this section, we're going to demonstrate (and then you're going to practice) optimizing objective functions. Typically, the first steps of the process are:

1. sketch and label a diagram of the scenario (when possible);

2. look for relationships in the geometry of that diagram (e.g., rectangles, trapezoids, triangles, circles, etc.);

3. formulate the objective function using the variables from the diagram and the observed relationships.

Minimizing time by way of a compromise

Example 7.1. Bundled in your safari gear, you can run at 3 m/sec on the road, but at only 1.5 m/sec in the grasses. (a) What piecewise-linear route should you use in order to minimize the length of time that it will take you to get to the car, given the distances shown in Figure 7.1? (b) How quickly can you make it to the truck?

Solution: The Pythagorean Theorem tells us that the remaining leg of the large dashed right triangle has a length of $\sqrt{189}$. Let's separate this length at the place where you emerge from the grass, onto the road. We'll denote the length of one segment by x, and the other by $\sqrt{189} - x$, as shown in Figure 7.2. With this notation, the distance you run in the grass is $\sqrt{36 + x^2}$ meters. Since you run at

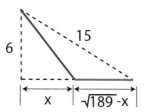

Figure 7.2: Relevant geometry for Example 7.1.

1.5 m/sec in the grass, that takes you $(\sqrt{36 + x^2})/1.5$ seconds. Similarly, you run a distance of $\sqrt{189} - x$ meters on the road, at 3 m/sec, so that takes you $(\sqrt{189} - x)/3$ seconds. All together, your total time is

$$T(x) = \left(\begin{smallmatrix}\text{time} & \text{in} \\ \text{the grass}\end{smallmatrix}\right) + \left(\begin{smallmatrix}\text{time} & \text{on} \\ \text{the road}\end{smallmatrix}\right) = \frac{\sqrt{36 + x^2}}{1.5} + \frac{\sqrt{189} - x}{3} \quad \text{seconds.}$$

A graph of T is shown in Figure 7.3. The number $T(x)$ is smallest at the critical point, so we calculate

$$T'(x) = \frac{2x}{3\sqrt{36 + x^2}} - \frac{1}{3},$$

which is zero when

$$\frac{2x}{3\sqrt{36 + x^2}} = \frac{1}{3}.$$

After multiplying both sides of this equation by $3\sqrt{36 + x^2}$, we have

$$2x = \sqrt{36 + x^2}.$$

Squaring both sides results in

$$4x^2 = 36 + x^2 \quad \Rightarrow \quad x = \pm\sqrt{12} = \pm 2\sqrt{3}.$$

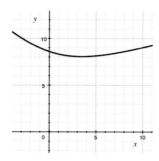

Figure 7.3: A graph of $T(x)$.

The negative solution is an artifact of the squaring maneuver that we performed during the calculation, and you can verify that $T'(-2\sqrt{3}) \neq 0$. However, a quick check shows that $T''(2\sqrt{3}) > 0$, from which we conclude that $x = 2\sqrt{3}$ is a minimum.

So the answers to our questions are these: (a) you can minimize your time by running directly toward the spot on the road that's $\sqrt{189} - \sqrt{4.5} \approx 10.28$ meters behind the truck, and (b) it will take you $T(2\sqrt{3}) \approx 8.0467$ seconds. Let's hope the lioness has a lame paw. ∎

> Moreover, a negative value of x would mean that we come out of the grass *farther away* from the truck than if we had run directly toward the road ($x = 0$)! That wouldn't make any sense, so you should be wary of the negative solution in this context.

Minimizing distance

Example 7.2. We want to stabilize two radio transmitters, the heights of which are 200 ft and 172 ft, by running guide wires from the towers to the ground. If the towers stand 100 ft apart, and we intend to connect the guide wires to an anchor in the ground that's somewhere between them, where should we place the anchor in order to minimize the amount of steel cable that we need? (Assume that the guide wires will connect to the top of each tower.)

Solution: The location of the anchor separates the 100-foot distance into two segments, one of length x and the other of length $100 - x$, as shown in Figure 7.4. Since each tower stands at a right angle to the ground, we can use the Pythagorean Theorem to express the lengths of cable in terms of x:

$$\sqrt{(200)^2 + x^2} \;=\; \text{length of cable needed for the taller tower}$$
$$\sqrt{(172)^2 + (100 - x)^2} \;=\; \text{length of cable needed for the shorter tower.}$$

So the total length of cable required is

$$\ell(x) = \sqrt{(200)^2 + x^2} + \sqrt{(172)^2 + (100 - x)^2}.$$

We can find the minimum value of this function by locating a critical point. So we use the Chain Rule to write

$$\ell'(x) = \frac{x}{\sqrt{(200)^2 + x^2}} - \frac{100 - x}{\sqrt{(172)^2 + (100 - x)^2}},$$

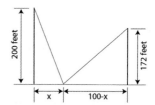

Figure 7.4: A schematic of the two towers and their guide wires for Example 7.2.

which is zero when

$$\frac{x}{\sqrt{(200)^2 + x^2}} = \frac{100 - x}{\sqrt{(172)^2 + (100 - x)^2}}.$$

After squaring both sides of this equation, and cross-multiplying, we see

$$x^2\Big((172)^2 + (100 - x)^2\Big) = (100 - x)^2\Big((200)^2 + x^2\Big)$$
$$(172)^2 x^2 + x^2(100 - x)^2 = (200)^2(100 - x)^2 + x^2(100 - x)^2.$$

Subtracting $x^2(100 - x)^2$ from both sides leaves us with

$$(172)^2 x^2 = (200)^2(100 - x)^2$$
$$\frac{(172)^2}{(200)^2} x^2 = 10000 - 200x + x^2.$$

After collecting our terms on one side of the equation, we use the quadratic formula to find that

$$x = \frac{5000}{93} \approx 53.763 \quad \text{or} \quad x = \frac{5000}{7} \approx 714.288.$$

The larger of these values would mean that we should put the anchor far off to the side, rather than in between the towers. That's clearly not what we want, but a quick check shows that ℓ' changes from negative to positive at $5000/93$, indicating a minimum value. So our solution is to place the anchor 53.76 feet from the taller tower, which will result in a total length of $\ell(53.76) = 385.21$ feet of cable. ∎

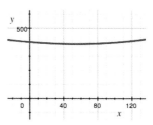

Figure 7.5: The graph of $\ell(x)$ for Example 7.2.

> The larger value of x is an artifact of the squaring maneuver that we used to eliminate radicals from the equation.

Maximizing revenue

Example 7.3. Suppose the selling price of your company's product, p dollars, is related to the demand for it according to $p = 100(5x + 6)e^{-x}$, where x is measured in thousands. (a) What level of demand maximizes revenue, and (b) what price is associated with that level of demand?

Solution: (a) If x units are sold for \$$p$ each, the revenue generated is $r = xp$, so our objective function is
$$r(x) = 100(5x^2 + 6x)e^{-x}.$$

In the graph of $r(x)$ that's shown in Figure 7.6 you can see that a global maximum is achieved somewhere near $x = 2.5$. We can locate it precisely by calculating the critical points of $r(x)$. When we differentiate with the Product Rule, we find that

$$r'(x) = 100(10x + 6)e^{-x} - 100(5x^2 + 6x)e^{-x}$$
$$= 100(6 + 4x - 5x^2)e^{-x}.$$

The exponential factor cannot be zero, so we locate the critical points of r by using the quadratic formula to determine that $6 + 4x - 5x^2 = 0$ when

$$x = \frac{2 + \sqrt{34}}{5} \approx 1.566 \quad \text{or} \quad x = \frac{2 - \sqrt{34}}{5} \approx -0.766.$$

The negative value of x is meaningless in this context, but a quick check shows that r' changes from positive to negative at $x = (2 + \sqrt{34})/5$, indicating a maximum, so we conclude that the company should produce 1566 units. (b) Using the formula for p that was given, we determine that the selling price at this level of production will be $p = \$288.88$. ∎

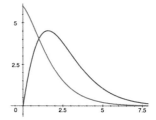

Figure 7.6: The graphs of price as a function of demand and revenue as a function of demand, where the y-axis is measured in hundreds and the x-axis in thousands.

Maximizing probability

Example 7.4. Gas molecules are always moving around, but they don't all move at the same speed. Using some clever reasoning, James Clerk Maxwell was able to figure out that when Δv is small, the probability that any particular gas molecule has a speed between v and $v + \Delta v$ is approximately $p(v)\Delta v$, where

$$p(v) = \sqrt{\frac{\alpha}{\pi}}\, v^2 e^{-\alpha v^2},$$

in which the number α depends on the temperature [28, p. 6–9]. If we assume that α is a particular fixed number, what is the most likely speed for a particle of gas according to Maxwell's formula?

Solution: For any particular Δv, the maximum value of $p(v)\Delta v$ happens where $p(v)$ achieves its maximum value. (A typical graph of p is shown in Figure 7.7.) We locate the v where $p(v)$ achieves its maximum value by determining where $p'(v) = 0$. Using the Product and Chain Rules, we find

$$p'(v) = \sqrt{\frac{\alpha}{\pi}}\left(2ve^{-\alpha v^2} + v^2 e^{-\alpha v^2}(-2\alpha v)\right) = 2\sqrt{\frac{\alpha}{\pi}}(1 - \alpha v^2)ve^{-\alpha v^2},$$

which is zero when $v = 0$ and when $v = 1/\sqrt{\alpha}$. A quick check shows that p' changes from negative to positive at $v = 0$, and vice versa at $p = 1/\sqrt{\alpha}$, so we conclude that p has a minimum value at $v = 0$ and a maximum at $v = 1/\sqrt{\alpha}$. That is, the most likely speed of a particle of the gas is $v = 1/\sqrt{\alpha}$. ∎

> Roughly said, the *probability* of an event is a number in $[0, 1]$ that quantifies how likely it is to occur. A larger number indicates that an event is more likely to happen.
>
> The function $p(v)$ is an example of something called a *probability density function* (pdf), which we'll study later using something called the *definite integral* (see p. 563).

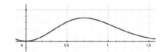

Figure 7.7: A typical graph of $y = p(v)$.

❖ Common Difficulties with Optimization

In Example 7.4 the objective function was provided by James Clerk Maxwell. By contrast, Examples 7.1 and 7.2 required us to determine the objective function on our own, and *that's* often the hardest part of the process. So in order to provide practice designing objective functions, text books (this one included) often provide examples and exercises that rely on very simple geometry.

Using geometry to design the objective function

Example 7.5. Suppose a dog owner is laying cable for an invisible fence at the side of her house. She has exactly 140 feet of cable, and plans to lay it so that the dog's portion of the yard is rectangular. No cable is necessary on the side of the yard where the house serves as a boundary. What width maximizes the dog's area of the yard?

Solution: We begin by sketching and labeling a diagram of this situation. If the dog's yard will be x feet wide and y feet long, its area is $A = xy$, and

$$\underbrace{140}_{\text{length of cable available}} = \underbrace{2x + y.}_{\text{two widths and one length}}$$

After solving this equation for y, we can rewrite the area as $A = x(140 - 2x)$. To find the maximum area, we simply calculate $A' = -4x + 140$, which is zero when $x = 35$ feet (we know that this is a maximum because the graph of $A(x)$ is a parabola that's cupped down). ∎

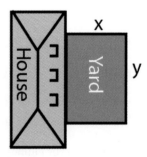

Figure 7.8: Schematic for Example 7.5.

Minimizing distance

Example 7.6. How close does the line $y = 3x - 4$ come to the point $(1, 10)$?

Solution: The distance between the points (x, y) and $(1, 10)$ is $\sqrt{(x-1)^2 + (y-10)^2}$. Since we're concerned only with points on the line $y = 3x - 4$, we can write the relevant distance as a function of a single variable (substituting $3x - 4$ for y):

$$r(x) = \sqrt{(x-1)^2 + (3x-14)^2}.$$

Our goal is to determine the minimum value of $r(x)$, so r is our objective function. The graph of r is shown in Figure 7.9, and you can see that a minimum occurs near $x = 4$. We locate it precisely by finding the critical point of r. Upon differentiating with the Chain Rule, we see

$$r'(x) = \frac{10x - 43}{\sqrt{(x-1)^2 + (3x-14)^2}},$$

which can only be zero when the numerator is. So the global minimum value of r occurs when $x = 4.3$. That is, the point on the line that's closest to $(1, 10)$ is $(4.3, 8.9)$, and the distance between them is $r(4.3) = 3.4785$. ∎

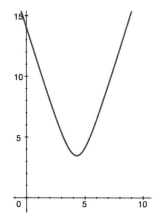

Figure 7.9: The graph of $r(x)$ for Example 7.6.

You should know

- the terms *local, relative, global,* and *absolute extrema;*

- the First Derivative Test;

- the Second Derivative Test;

- that you cannot classify a critical point at x_0 as a max or a min based on $f''(x_0) = 0$;

- that extrema can occur either inside a closed interval, or at its end points.

You should be able to

- use the First Derivative Test to classify first-order critical points;

- use the Second Derivative Test to classify first-order critical points;

- compare and contrast the First and Second Derivative tests;

- locate and classify all critical points of a continuous function on a closed interval.

❖ 4.7 Skill Exercises

1. Determine the maximum product of two numbers that sum to 10.

2. Determine the minimum product of two numbers that differ by 7.

3. Suppose x and y satisfy the equation of the ellipse $2x^2 + 0.2y^2 = 1$.

   (a) Use a graphing utility to render a graph of the ellipse.

   (b) Use a graphing utility to render a graph of the line $x + y = 1$, and visually locate points on the ellipse whose coordinates sum to 1.

   (c) Use a graphing utility to render a graph of the line $x + y = 1.5$, and visually locate points on the ellipse whose coordinates sum to 1.5.

   (d) Use a graphing utility to render a graph of the line $x + y = 2$, and visually locate points on the ellipse whose coordinates sum to 2.

(e) What happens to the line $x + y = c$ when we increase c? Does its slope change?

(f) Determine the point(s) on the ellipse whose coordinates have the largest sum.

4. Repeat #3 with the ellipse $4x^2 + 0.3y^2 = 1$.

5. Determine the point(s) on the ellipse $3x^2 + 0.1y^2 = 1$ with coordinates that have the most negative sum.

6. Determine the point(s) on the ellipse $4x^2 + 0.3y^2 = 1$ with coordinates that have the most negative difference.

In #7–10, determine how close the specified curve comes to the point $(10, 1)$.

7. $y = 3x + 7$ 9. $y = x^3 + x + 1$

8. $y = x^2 + 2x + 8$ 10. $y = \sin(x)$

Exercises #11–14 concern a 10-cm length of wire that we cut into two pieces, each of which is then bent into a particular shape. Where should we clip the wire if our goal is to minimize the total area of these shapes, when the pieces are bent into ...

11. a circle and a square?   13. a square and an equilateral triangle?

12. a circle and an equilateral triangle?   14. a pentagon and a square?

15. We are given a rectangular sheet of material that is 10 cm long and 8 cm wide, and are told to construct an open-topped box (see Figure 7.10). So we make square cuts from each corner and fold in order to make the box. How much do we clip from each corner in order to maximize the volume?

Figure 7.10: For #15.

16. Where should P be placed in Figure 7.11 in order to maximize the angle θ?

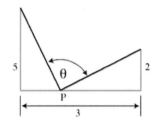

17. A closed cylinder (i.e., with both top and bottom) is to be made so that its top and bottom are of stronger material than its sides. The material for the top and bottom costs $10 per square foot, and the material for the sides costs $4 per square foot. If the volume of the cylinder is to be 1 cubic foot, what dimensions minimize the cost of building this container?

Figure 7.11: For #16.

18. A closed cylinder (i.e., with both top and bottom) is to be made so that its top and bottom are of stronger material than its sides. The material for the top and bottom costs $10 per square foot, and the material for the sides costs $4 per square foot. If the cylinder can cost at most $40, what is the maximum volume of this container?

19. A fence that is 8 feet tall runs parallel to the face of a building 4 feet away. What is the length of the shortest ladder that will reach from the ground, over the fence, to the side of the building?

20. Figure 7.12 is a schematic of the last leg of a "ropes course" that you're designing, in which participants will travel down the hill on a zip line. The tallest tree in your way is 34 feet tall, and is 100 feet from the tower where participants will start their descent. The base of the tower is 48 feet above the field on which people will land. (a) If you need the zip line to clear the tree by 10 feet (and can't cut it down), what's the minimum amount of cable you'll need? (b) How high does the tower have to be built?

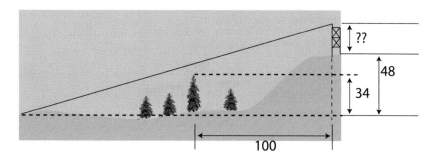

Figure 7.12: Diagram of the zip line for #20.

21. Suppose you want to keep the public out of a wetland area that borders a lake, and federal funding will pay to erect 5,000 meters of fence. What's the largest rectangular area that you can enclose? (You don't have to fence the lakeside.)

22. Suppose you're writing a federal grant to pay for the enclosure of a rectangular portion of a large wetland area that borders a lake. You need to enclose 4 square miles near the lake, and appropriate fencing will cost $10 per foot (but you do not need to fence the lakeside). What's the smallest level of federal funding that you need?

Exercises #23–24 return to the story of the dog owner in Example 7.5 (p. 304). Suppose she owns both male and female dogs, and wants to separate them by stretching a line of cable through the yard. The cable cannot be cut into pieces, but must remain one long contiguous piece.

23. Assuming both dogs will have an equally sized rectangular portion of the yard. What's the largest area she can enclose with her 140 feet of cable if the dividing line is parallel to the side of the yard whose length is marked as x in the diagram? Two possible configurations for cable placement are shown in the images below.

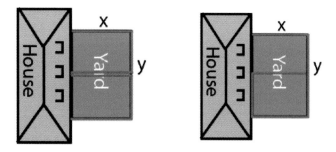

Figure 7.13: Diagram of cable configurations for #23.

24. Suppose the dividing line stretches diagonally across the field, from corner to corner.

    (a) What's the largest area she can enclose with her 140 feet of cable? We'll call this number A^* in the remaining parts of this exercise.

    (b) Use a graphing utility to render a graph of the perimeter equation that you used in part (a), along with the hyperbolas $xy = A$ for values of

$A > A^*$ (don't make A *too* big.) Your graphing window should include at least $0 \le x \le 50$ and $0 \le y \le 20$.

  (c) Repeat part (b) with $0 < A < A^*$.

  (d) Based on what you saw in parts (b) and (c) above, how are the scenarios of $A > A^*$ and $A < A^*$ different, graphically speaking?

  (e) If we were to solve this problem by looking at the curves and trying different values of A, what would we see when $A = A^*$?

25. A soccer goal is 24 feet wide. Suppose a soccer player is dribbling the ball toward the end line, 36 feet outside the left post, as seen in Figure 7.14. At what distance from the end line is her shooting angle largest?

26. A soccer goal is 24 feet wide. Suppose a soccer player is dribbling the ball toward the end line, x feet outside the left post (e.g., $x = 36$ in exercise #25). (a) Determine a formula for the distance from the end line, y, at which her shooting angle is largest when $x > 0$, (b) show that $dy/dx > 0$ and explain what that means in this context, and (c) graph the relationship between y and x. What familiar curve is this?

27. A soccer goal is 24 feet wide. Suppose a soccer player is dribbling the ball along the edge of the penalty box, 54 feet from the goal, as seen in Figure 7.14. Use calculus to show that his shooting angle is maximized when he is halfway across the field.

28. A soccer goal is 24 feet wide. Suppose a soccer player is dribbling the ball along a path that's parallel to the edge of the penalty box, y feet from the goal ($y = 54$ in exercise #28). (a) Use calculus to show that his shooting angle is maximized when he is halfway across the field, regardless of y, and (b) graph the shooting angle as a function of y.

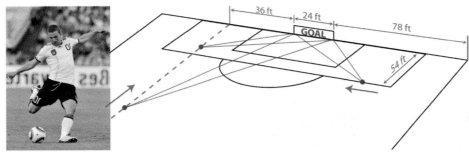

Figure 7.14: Diagram for #25–28.

Exercises #29–30 are inspired by the *World Championship Punkin Chunkin* competition, held in Delaware every fall since 1986, in which teams use catapults, trebuchets, and compressed air canons to see who can hurl a pumpkin the farthest.

29. Suppose that a pumpkin is hurled eastward across flat ground. It's launched at an initial velocity of 100 ft/sec, with an angle of inclination of β radians, where $\beta \in (0, \pi/2)$. Then neglecting air resistance, after t seconds of flight it has traveled $x(t) = 100 \cos(\beta)t$ feet horizontally and has an altitude of $y(t) = -16t^2 + 100 \sin(\beta)t$ feet. Use calculus to prove that the range is maximized by $\beta = \pi/4$.

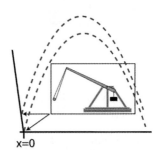

Figure 7.15: For #30.

30. Suppose that immediately behind the launcher is a hill that makes a $\pi/3$ angle with the horizontal. By pushing the launcher up the hill, we raise the starting altitude of the pumpkin, but we also back it up (see Figure 7.15). If the effective range is measured from the bottom of the hill, where the launcher was originally placed, how far should we push the launcher up the hill in order to maximize the effective range? *(You'll need to adjust the equations for $y(t)$ and $x(t)$ from #29 in order to account for the new starting position.)*

31. The strength of a rectangular beam is proportional to the product of its width and the square of its height. Consider a rectangular beam of height h and width w that's cut from a circular log of radius r, as shown in Figure 7.16. What are the dimensions (in terms of r) of the beam that maximize its strength?

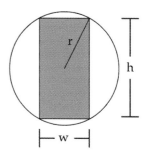

Figure 7.16: For #31.

32. Suppose a trough is made with a trapezoidal face. The base has a length of 5 centimeters, and the side flaps have a length of 2 centimeters each, as shown in Figure 7.17. What angle of inclination maximizes the cross-sectional area of the trough (i.e., the area of the trapezoid)?

Figure 7.17: For #32.

In 2003, Tim Pennings of Hope College published a short article about the strategy used by his dog, Elvis, to fetch balls out of Lake Michigan (see [33]). When Pennings tossed a ball into the water, down the beach a little way from Elvis, the dog faced the same kind of problem that led off this section: he can run at 6.4 m/sec but can only swim at 0.911 m/sec. Exercises #33–37 have to do with this scenario.

33. Suppose that when Pennings throws the ball, it lands 6 meters out into the water, and 5 meters south of his position. Where should Elvis plunge into the lake (and then head directly toward the ball) in order to minimize the time it takes to reach the ball? (Answer in terms of y, as seen in Figure 7.18.)

34. Suppose that when Pennings throws the ball, it lands 6 meters out into the water, and 7 meters south of his position. Where should Elvis plunge into the lake (and heading directly toward the ball) in order to minimize the time it takes to reach the ball? (Answer in terms of y, as seen in Figure 7.18.)

35. Suppose that when Pennings throws the ball, it lands 6 meters out into the water. Verify that the number y, which is shown in Figure 7.18, is always the same (it doesn't matter how far down the shore Pennings throws the ball).

36. Suppose that when Pennings throws the ball, it lands x meters out into the water ($x =$ the perpendicular distance from the ball to the shore). Assuming that Elvis always wants to reach the ball in the minimum amount of time, find a formula for y in terms of x.

37. Different dogs run and swim at different rates. Suppose Fido can run at r m/sec and can swim at s m/sec, where $r > s$. Find a formula for y in terms of x, r, and s (where $x =$ the perpendicular distance from the ball to shore).

38. Suppose that light is emitted from a source, and has to travel through two different mediums (e.g., air and glass, or water) in order to arrive at a detector (e.g., your eye). The source and detector are separated by a horizontal distance of L, the source is a distance of A from the interface, and the detector is a distance of B, as seen in Figure 7.19. Fermat's Principle is that light always travels along the path of *least time*, which we can use to explain refraction, as follows: If the light were to cross the interface at a horizontal distance of x from the detector ...

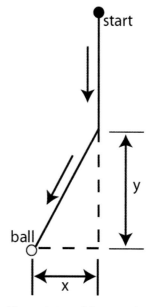

Figure 7.18: Schematic for #33–37.

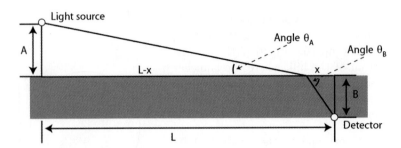

Figure 7.19: Diagram for exercise #38.

(a) Find an expression in terms of x and B for the distance it would travel through the "lower" medium, d_B.

(b) Suppose the speed of light in the lower medium is v_B. Find a formula for the time that the light would spend in the "lower" medium, T_B.

(c) Show that $T'_B(x) = \cos(\theta_B)/v_B$, where θ_B is the angle that the light's path makes with the horizontal on the "lower" side of the interface.

(d) Find an expression in terms of x, L and A for the distance it would travel through the "upper" medium, d_A.

(e) Suppose the speed of light in the lower medium is v_A. Find a formula for the time that the light would spend in the "upper" medium, T_A.

(f) Show that $T'_A(x) = -\cos(\theta_A)/v_A$, where θ_A is the angle that the light's path makes with the horizontal on the "upper" side of the interface.

(g) The time it takes the light to travel from source to detector is $T = T_A + T_B$. Show that $T(x)$ has a critical point only where $\cos(\theta_A)/v_A = \cos(\theta_B)/v_B$.

(h) Use this fact to explain why, whenever light crosses the interface between two mediums, it always bends toward the medium in which it travels more slowly.

Exercises #39–40 deal with a circular lake whose radius is 1 kilometer (see Figure 7.20). Imagine you are on the east side of the lake, and you need to get to the west side as soon as possible. You can paddle through the water at 1.5 m/sec.

39. Suppose you can run with your kayak at 2 m/sec. What route should you take in order to arrive at the west side of the lake (with your kayak) in the least amount of time?

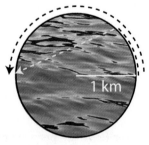

Figure 7.20: For #39 and #40.

40. Suppose that you can run with your kayak at r m/sec. What's the smallest value of r for which it's better to run all the way around the lake than to do any paddling?

41. Suppose you're beginning an origami project by folding your paper as shown in Figure 7.21. If the paper is 6 inches wide, (a) where should you make the fold in order to maximize the area of triangle A, and (b) what is that maximum area?

42. Suppose you're beginning an origami project by folding your paper as shown in Figure 7.21. If the paper is 6 inches wide, what is the minimum area of triangle B?

43. Suppose a hallway that's 12 feet wide meets a hallway that's 8 feet wide at a right angle, as seen in Figure 7.22 when $\phi = 0$. What's the longest rigid pipe that can be taken around the corner (without tipping it upwards)?

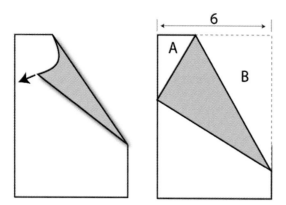

Figure 7.21: Grab the upper right-hand corner of a sheet of paper, and pull it to the left side of the sheet.

44. Suppose a hallway that's 10 feet wide meets a hallway that's 7 feet wide at a right angle, as seen in Figure 7.22 when $\phi = 0$. Determine the longest pipe that can be taken around the corner, provided that you are allowed to tip it. The ceiling is 11 feet above the floor.

45. Suppose a hallway that's 12 feet wide meets hallway that's 8 feet wide, as seen in Figure 7.22 with $\phi = \pi/4$. What's the longest rigid pipe that can be taken around the corner (without tipping it upwards)?

46. Suppose a hallway that's 12 feet wide meets another that's 8 feet wide, as seen in Figure 7.22.

   (a) Suppose $\phi = 0.24\pi$. Determine the longest length of pipe, ℓ, that can be taken around the corner (without tipping).

   (b) Suppose $\phi = 0.26\pi$. Determine the longest length of pipe, ℓ, that can be taken around the corner (without tipping).

   (c) Based on your answers to parts (a) and (b), approximate $\ell'(0.25\pi)$.

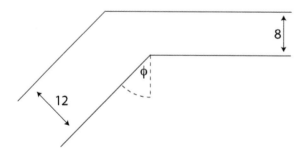

Figure 7.22: Diagram for #43–46.

47. In Example 2.5 of Chapter 1 we changed pixel intensities using the function $f(x) = 3x - 5x^2 + 4x^3 - x^4$. What pixels experience the greatest change in intensity? (Hint: find extrema of $f(x) - x$ with Newton's method.)

48. Suppose we adjust pixel intensities of an image as in Example 2.5 of Chapter 1 using the function $f(x) = 4(x - 0.5)^3 + 0.5$. What pixels experience the greatest change in intensity?

Chapter 4 Review

❖ True or False

In exercises #1–3 suppose the linear approximation of f at $t = 5$ is $y = 7 + 3(t-5)$.

1. Then $f(5) = 7$.

2. Then $f'(5) > 0$.

3. Then $f(6) = 10$.

4. Newton's method always converges.

5. Iterations of Newton's method use the formula $t_{n+1} = t_n - \frac{f(t_n)}{f'(t_n)}$.

6. Suppose $h'(6) > 0$. Then $dh > 0$ when $dt > 0$.

7. Suppose $f''(t) > 0$ when $t \in (8, 10)$. Then $f'(t)$ is increasing in $(8, 10)$.

8. Acceleration is the second derivative of position.

9. Suppose the position of an object is $x(t)$. The object speeds up when $x'(t)$ and $x''(t)$ are both negative.

10. Suppose t_0 is a second-order critical number of f. Then $(t_0, f(t_0))$ is an inflection point on the graph of f.

11. Suppose t_0 is a second-order critical number of f. Then t_0 is a first-order critical point also.

12. Suppose $f'(t_0) > 0$. Then t_0 cannot be a second order critical point of f.

13. Suppose f is a polynomial and $f''(t)$ is always positive. Then $f(5) > f(2)$.

14. Suppose the graph of f is concave up when $t \in (2, 6)$. Then $f''(4) > 0$.

15. Suppose $f''(t) < 0$ when $t \in (-10, -8)$. Then the graph of f is concave down on that interval.

16. Suppose f is differentiable on $(1, 7)$, and achieves its maximum value at $t = 6$. Then $t = 6$ is a critical number for f.

17. Suppose $t = 9$ is a first order critical point of f. Then $t = 9$ is either a local min or a local max.

18. Suppose f is differentiable everywhere, and achieves its global maximum at $t = 8$. Then $t = 8$ is a critical point of f.

19. Suppose f achieves its global maximum at $t = 8$. Then $f(8) > f(t)$ for all $t \neq 8$.

20. Rolle's Theorem is a special case of the Mean Value Theorem.

❖ Multiple Choice

21. L'Hôpital's Rule tells us that ...

    (a) $\lim_{t \to 2} \frac{2t^2 + 7}{8t - 1} = \lim_{t \to 2} \frac{4t}{8} = 1$
    (d) $\lim_{z \to 1} \frac{14z + 7}{z^2 - 1} = \lim_{z \to 1} \frac{14}{2z} = 7$

    (b) $\lim_{x \to 3} \frac{x^2 - 9}{3x - 5} = \lim_{x \to 3} \frac{2x}{3} = 2$
    (e) all of the above

    (c) $\lim_{\theta \to \pi} \frac{\sin^2(\theta)}{\theta - \pi} = \lim_{\theta \to \pi} \frac{2\sin(\theta)\cos(\theta)}{1} = 0$
    (f) none of the above

22. The linear approximation of $f(t) = -2t^2 + 3t + 2$ at $t = 3$ is ...

    (a) $y = -7 - 3(t - 9)$ (c) $y = -9 - 7(t - 3)$ (e) $y = -3 - 9(t - 7)$

    (b) $y = -7 - 9(t - 3)$ (d) $y = -9 - 3(t - 7)$ (f) none of the above

23. Suppose $y(t)$ is the linear approximation of $f(t)$ at time $t = 7$, and $f'(7) > 0$. Which (if any) of the following are guaranteed to be true?

    (a) $f(8) > y(8)$ (c) $y(8) = f(8)$ (e) $y(8) > y(7)$

    (b) $y(8) > f(8)$ (d) $f(8) > f(7)$ (f) none of the above

24. Suppose a model rocket is $y(t)$ feet off the ground t seconds after launch, and $y'(9) = 10$. Then linear approximation suggests the rocket will ...

    (a) rise 6 feet in the next minute   (d) fall 10 feet in the next second

    (b) rise 60 feet in the next minute   (e) fall 100 feet in the next second

    (c) rise 600 feet in the next minute   (f) none of the above

25. Suppose a model rocket is $y(t)$ feet off the ground t seconds after launch. If $y(9) = 200$ and $y'(9) = -10$. Then linear approximation suggests the rocket will land ...

    (a) in 10 seconds   (c) in 200 seconds

    (b) in 20 seconds   (d) none of the above

26. Suppose that we use Newton's method to find a root of f, and start at $t_0 = 1$. If t_6 is a root, but t_5 is not. Then we're certain that ...

    (a) $t_6 = 1$ (c) $t_5 < t_6$ (e) $t_7 = t_6$

    (b) $t_6 > 1$ (d) $t_6 < t_5$ (f) none of the above

27. Suppose Newton's method diverges when $t_0 = 1$, and $f'(t_n)$ is never zero. Then ...

    (a) $f(t_n) \neq 0$ for all n (c) at each step, $t_{n+1} < t_n$ (e) there is some $t_n < 0$

    (b) at each step, $t_{n+1} > t_n$ (d) there is some $t_n > 10$ (f) none of the above

28. The differential dh ...

    (a) denotes the linear approximation of change in h (c) denotes the actual change in h

    (b) is the same as $h'(t)$ (d) none of the above

29. Suppose h is a polynomial function for which $h'(2) > 0$ and $h''(2) > 0$. Then by taking very small $\Delta t > 0$ we can guarantee ...

    (a) $dh < \Delta h$ (c) $dh = \Delta h$

    (b) $dh > \Delta h$ (d) none of the above

30. Suppose f is a polynomial function for which $f''(t) > 0$ in $(1, 2)$, and $f''(t) < 0$ in $(2, 3)$. Then ...

   (a) $t = 2$ is a second-order critical point of f

   (b) $t = 2$ is an inflection point on the graph of f

   (c) $f''(2) = 0$

   (d) all of the above

   (e) none of the above

31. Saying that f has a *relative maximum* at t_0 is the same as saying that t_0 is the location of a ...

   (a) relative minimum

   (b) local minimum

   (c) absolute maximum

   (d) global maximum

   (e) local maximum

   (f) none of the above

32. Suppose f is a polynomial function for which $f'(t) \neq 0$ when $t \in [6, 12]$, and f'' has a single root at in $[6, 12]$ at $t = 7$. Then on $[6, 12]$, the graph of f is steepest ...

   (a) at $t = 7$

   (b) at $t = 6$

   (c) at $t = 12$

   (d) none of the above is guaranteed

❖ Exercises

33. Suppose $f(t) = t^2 + t - 6$. (a) Show that $t = 2$ is a root, and (b) use a spreadsheet or calculator to determine the first three iterations of Newton's method when $t_0 = 1$ (i.e., determine t_3).

34. Suppose that $f(7) = 12$ and $f'(7) = 13$. Use linear approximation to estimate $f(6.9)$.

35. Suppose $f(t) = 4t^5 - 5t^4$. Write the linear approximation of f at $t = 1$.

36. Suppose $f(t) = e^{\cos(t)}$. Write the linear approximation of f at $t = \pi/2$.

37. Suppose ρ is the average density of a foam "stress ball," measured in $\frac{\text{g}}{\text{cm}^3}$, and F is the amount of force (in newtons) applied by your squeezing hand. Does the equation $d\rho = 0.2\, dF$ make any sense? If so, explain what it tells us. If not, why not?

38. Suppose P_1 and P_2 are the points where the line $y = mx$ intersects the ellipse $4x^2 + y^2 = 9$, and D is the distance between them. If m is changing over time at a constant rate of $m'(t) = 1/2$, determine dD/dt when $m = 1$.

39. When granular materials are poured onto a flat surface, they form a conical pile. The internal angle between the surface of the pile and the horizontal is called the *angle of repose* (see Figure 8.1). Suppose that table sugar, whose angle of repose is $35°$, is being poured onto a table at a rate of 2 cubic centimeters per second. How fast is the radius of the pile changing when there are 17 cubic centimeters of sugar on the table?

40. In Figure 8.2 the angle subtended by the blades is ϕ. Note that the dashed lines passing through the pivot point are parallel to the cutting edges of the scissors. The tapering of the blades won't matter in this problem, so assume the blades are rectangular, with a width of 1 cm.

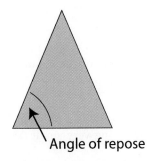

Angle of repose

Figure 8.1: The angle of repose depends on material characteristics such as density and the aspect ratio of the grains.

(a) Suppose the blades are closing at a rate of $d\phi/dt = -\pi/10$ radians per second. How fast is the cut point moving away from the pivot point when $\phi = \pi/6$? (*Hint: draw a line that passes through both the cut point and the pivot point.*)

(b) Suppose the distance from the pivot point to end of the blade (i.e., end of the rectangle) is 2 meters, and the tip of each blade is moving at 37,000 meters per second as the scissors close. Show that the cut point moves faster than the speed of light as the scissors are finishing the cut.

Figure 8.2: (both) Images for #40; (right) the angle $\phi > 0$ when the cut is complete.

41. Suppose $f(t) = 0.05t^5 + 0.5t^4 + 2t^3 + 4t^2 + t - 1$. (a) Use f'' to show that the graph of f has an inflection point at $t = -2$; (b) verify that $f''(t) = (t+2)^3$; and (c) use a graphing utility to plot the graph of f near $t = -2$.

42. Suppose $g(t) = \frac{1}{6}t^3 + t^2 - t - \frac{19}{15}$. (a) Use g'' to show that the graph of g has an inflection point at $t = -2$; (b) show that the tangent line to the graph of g at $t = -2$ is the same as the tangent line to the graph of f (from #41) at $t = -2$; (c) use a graphing utility to plot the graphs of f, g, and the common tangent line in the window $-2.4 \le x \le 0.4$ and $-1.6 \le y \le 1.2$; (d) determine which graph is pulling away from the tangent line faster when $|t - 2|$ is small; and (e) based on the formulas for f'' and g'', explain why that makes sense.

43. Suppose the position of an object on the x axis is given by $x(t) = (t+1)^2(t-4)$. Determine when the object is speeding up.

44. Suppose the position of an object on the y axis is given by $y(t) = \sin(t)$. Determine when the object is slowing down.

45. Suppose $p(c)$ is the profit a company earns by making c netPC computers. If both p' and p'' are positive when $c \in (0, 30000)$, is the change in profit greater when (a) we increase production from $c = 200$ to $c = 201$, or (b) we increase it from $c = 300$ to $c = 301$?

46. Suppose $d(c)$ is the demand for netPC computers when they're sold at c dollars. If $d'(c) < 0$ and $d''(c) > 0$ when $c \in (0, 400)$, is the drop in demand greater when we increase price from $c = 200$ to $c = 201$, or when we increase it from $c = 300$ to $c = 301$?

47. Suppose $g(t) = te^{-t}$. (a) Determine where the graph of g is increasing, decreasing, concave up, concave down; (b) determine the intercepts of the graph; (c) determine the asymptotic behavior of $g(t)$ as $t \to \pm\infty$, and (d) sketch a graph of g.

48. Suppose $h(t) = (t+1)(t-7)(t-15)$. (a) Determine where the graph of h is increasing, decreasing, concave up, concave down; (b) determine the intercepts of the graph; (c) determine the asymptotic behavior of $h(t)$ as $t \to \pm\infty$, and (d) sketch a graph of h.

49. Suppose $f(t) = t^2 e^t$. Determine a formula for $\frac{d^4 f}{dt^4}$.

50. Suppose $k(t) = e^{2t}/(4t + 5)$. Determine a formula for $\frac{d^3 k}{dt^3}$.

51. Suppose $f'(t) = (t-4)^2 \sin^2(t^6 + 1)$ What does the First Derivative Test tell you about the critical point at $t = 4$?

52. Suppose $f''(t) = \ln(t)\sqrt{t^3 + 2}$, and $t = 3$ is a first-order critical point. What does the Second Derivative Test tell you about $t = 3$?

53. Suppose f is the constant function $f(t) = 7$.

    (a) Verify that $t = 2$ is a first-order critical point.

    (b) Does f have a local minimum at $t = 2$?

    (c) What does the First Derivative Test tell you about the critical point at $t = 2$?

54. Suppose f is the constant function $f(t) = 13$.

    (a) Verify that $t = 5$ is a first-order critical point.

    (b) Does f have a global maximum at $t = 5$?

    (c) What does the First Derivative Test tell you about the critical point at $t = 5$?

55. Figure 8.3 (left) shows the graph of f', the graph of f'', and a third curve that was included for decoration. Which is which?

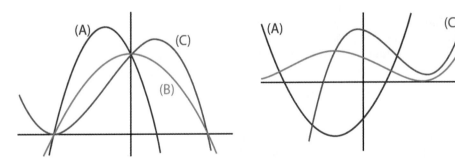

Figure 8.3: (left) Graphs for #55; (right) graphs for #56.

56. Figure 8.3 (right) shows the graph of f, the graph of f'', and a third curve that was included for decoration. Which is which?

57. Suppose $f(t) = 1 + 2t - t^2$ and $g(t) = 1 - 2t - 4t^2 + t^3$. (a) Show that $f(0) = g(0)$ and $f(4) = g(4)$; and (b) what does Rolle's Theorem tell us must happen somewhere in $[0, 4]$? (Hint: consider the function $f(x) - g(x)$.)

58. Suppose $f(t) = 1 + 2t - t^2$. (a) Calculate $f(0)$ and $f(5)$; (b) what does the Mean Value Theorem tell us must happen at some time in $[0, 5]$? and (c) find such a time in $[0, 5]$.

59. Suppose $f(t) = 1 + 2t - t^2$ and $g(t) = 1 - 2t - 4t^2 + t^3$. (a) Show that $f(0) \neq f(5)$ and $g(0) \neq g(5)$; (b) what does Cauchy's extension of the Mean Value Theorem tell us must happen somewhere in $[0, 5]$? and (c) find such a time t.

In #60 and #61 (a) use a graphing utility to render a graph of $f(t) = t^3 + 2t^2 - 7t + 1$, and (b) choose an interval $[a, b]$ containing at least one critical point of f, with $a \neq b$, over which the specified conditions are true.

60. The global minimum of f over $[a, b]$ occurs at $t = a$, but the global maximum is *not* at $t = b$.

61. The global minimum of f over $[a, b]$ occurs at both $t = a$ and $t = b$.

62. In Example 4.4, we verified that the vertex of the parabola $y = ax^2 + bx + c$ is at $x_v = -b/2a$.

    (a) Find the y-coordinate of the vertex, and denote it by y_v.

    (b) Suppose that a, c are particular fixed numbers (if you'd like, use $a = 3$ and $c = 5$). Then y_v is a function of b. Show that it's an *increasing* function of b when the vertex lies in the right half-plane (i.e., $x_v > 0$), and is a *decreasing* function when the vertex lies in the left half-plane.

63. Suppose the dog owner in Example 7.5 (p. 304) owns both male and female dogs, and wants to separate them by stretching a line of cable through the yard. What's the largest area she can enclose with her 140 feet of cable if the dividing line is parallel to the side of the yard whose length is marked as y in the diagram?

64. Suppose $f(x) = |x|$. Then $f(-1) = f(1)$, so average change in f over $[-1, 1]$ is zero. However, there is no point in that interval at which $f'(x) = 0$. Explain why this is not an exception to Rolle's Theorem.

65. Consider the function $f(t) = -1/t$.

    (a) Determine $f(-2)$ and $f(2)$.
    (b) Calculate the average rate of change in f over $[-2, 2]$.
    (c) Find a formula for $f'(t)$ and show that it's always positive.
    (d) Explain why parts (b) and (c) don't contradict the MVT.

66. Suppose $f(x) = \begin{cases} 0 & \text{when } 0 \leq x < 1 \\ 2 & \text{when } x = 1. \end{cases}$

    (a) Show that the average rate of change in f over $[0, 1]$ is 2, but there is no point at which $f'(x^*) = 2$.
    (b) Explain why this is not a contradiction of the MVT.

Calculate each of limits in #67–72, using L'Hôpital's Rule where necessary.

67. $\lim_{t \to 7} \frac{t^2 - 8t + 7}{15t^2 + 82}$

68. $\lim_{\phi \to 0} \frac{\tan^{-1}(\phi) - \phi + \frac{1}{3}\phi^2}{\phi^5 + \phi^7}$

69. $\lim_{r \to 0} \frac{\sin(r^2)}{6r^2 + r^3}$

70. $\lim_{r \to 0} \frac{\cos(r) - 1 + r^2}{3r^4 + 9r^5}$

71. $\lim_{x \to \infty} \frac{4^x + x^2}{3^x + x}$

72. $\lim_{x \to \infty} \frac{x \ln(x)}{x^2 + 1}$

73. Suppose $f(t) = (t - 4)^2$. At each t_0 we can write the equation of the tangent line to the graph of f, and we'll denote by $b(t_0)$ the y-intercept of that line. (a) Determine a formula for b as a function of t_0, and (b) show that its graph is concave down.

74. Suppose f is a function for which $f''(t)$ is always positive, and $b(t_0)$ is defined as in #73. Show that the graph of b is increasing when $t_0 < 0$ and decreasing when $t_0 > 0$.

Chapter 4 Projects and Applications

❖ Newton's Method with a Spreadsheet

There are subtleties to Newton's method, but once they're understood, what's left is an algorithm. Luckily, in the modern era we have computers to do much of the grunt work for us. In the following steps, you'll see how to find the roots of

$$f(t) = t^3 - 19t^2 + 83t - 60, \quad \text{whose derivative is} \quad f'(t) = 3t^2 - 38t + 83,$$

by using a spreadsheet program to implement Newton's method.

1. Use a graphing utility to render a graph of f, and get a rough estimate of the roots.

We'll begin by naming the columns of the spreadsheet based on what data they'll hold for us.

2. After opening the spreadsheet program, type the letter **t** in cell **A1**.

3. In cell **B1** type **f(t)**.

4. In cell **C1** type **fp(t)** (*fp* for "f prime of t").

Now let's look for the root that's near $t = 0$.

5. In cell **A2** type the number **0**.

In a spreadsheet program, we reference numbers by their location rather than by their value. So in the next step you'll see that calculating f and f' is done by referencing cell A2.

6. In cell **B2** type **=A2^3-19*A2^2+83*A2-60**

7. In cell **C2** type **=3*A2^2-38*A2+83**

Now that cells **B2** and **C2** hold the value of the function and the derivative, respectively, we can implement Newton's method. Our initial guess at the root was $t = 0$, and our next guess will go into the next row of the spreadsheet. Remember that we will reference the parts of the formula based on where they are in the spreadsheet.

$$t_1 = \underbrace{t_0}_{\text{in cell } \mathbf{A2}} - \frac{f(t_0)}{f'(t_0)} \begin{matrix} \leftarrow \text{in cell } \mathbf{B2} \\ \leftarrow \text{in cell } \mathbf{C2} \end{matrix}$$

8. In cell **A3** type **=A2-B2/C2**

If everything has gone well, you should now see the number 0.722891566 in cell A3, and the organizational work of using the spreadsheet is now done. The only thing we need to do now is tell the computer to perform these calculations over and over again. That's done with a "select-and-drag" technique.

9. Select cells **B2** and **C2**. In the lower right-hand corner of this pair of selected cells, you'll see a small box (see the left-hand image in Figure 9.1)

10. Grab that box with your mouse and pull down (see the right-hand image in Figure 9.1).

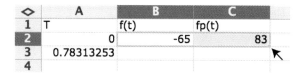

◇	A	B	C
1	T	f(t)	fp(t)
2	0	-65	83
3	0.78313253		
4			

◇	A	B	C
1	T	f(t)	fp(t)
2	0	-65	83
3	0.78313253	-11.1723421	55.0808535
4			

Figure 9.1: (left) Selecting cells B2 and C2 in step 9; (right) grabbing the corner and pulling down in step 10.

Now the formulas for calculating the value of the function and the derivative have been copied into row 3. When you use this select-and-drag technique, the *relative* position of numbers is preserved. For example, cell **C2** refers to the number in **A2** (two cells to its left). If you click on cell **C3** you'll see that it refers to **A3** (two cells to *its* left). Lastly, we use the select-and-drag technique to repeat the Newton's method step, and recalculate the value of the function and its derivative many times over.

11. Select cells **A3** through **C3**. In the lower right-hand corner, you'll see that small box again.

12. Grab that box with the mouse and pull it down 20 rows or so.

13. If all has gone well, you should see that the function value is zero at $t = 0.899238912$. Find the other two roots by adjusting our initial guess in cell **A2**. (You won't need to change anything else!)

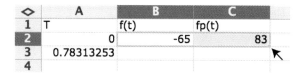

Figure 9.2: Steps #11 and 12.

✢ Optimization and Apothem

Exercises #11–14 on p. 306 focused on a wire of length 10 cm that was cut into two pieces, each of which is bent into a regular polygon.

1. Verify that the circle in #11 (on p. 306) can be inscribed in the square.

2. Verify that the circle in #12 (on p. 306) can be inscribed in the equilateral triangle.

3. Verify that the square and the equilateral triangle in #13 (on p. 306) have the same inscribed circle.

The radius of the inscribed circle is called the *apothem,* and it was introduced first in Chapter 2 (see p. 144). In short, the apothem is the minimal distance from the center of a regular polygon to one of its edges. Exercises #1–3 seem to suggest that the sum of enclosed areas is minimized when the regular polygons have the same apothem. In the following steps, you'll verify that as being true.

4. Derive formulas for the area and perimeter of a regular n-sided polygon that has an apothem of r. *(See p. 144 if you need help.)*

Now suppose that we bend one piece of wire into a regular n-sided polygon whose apothem is x, and the other into a regular m-sided polygon whose apothem is y.

5. Explain why $0 \leq x \leq \frac{5}{n} \cot\left(\frac{\pi}{n}\right)$.

6. Using the formula for perimeter that you developed in #4, write an equation that says the sum of the perimeters is 10 cm. This equation will be called the *perimeter constraint.*

7. Get a formula for $\frac{dy}{dx}$ by differentiating the perimeter constraint with respect to x.

8. Using x and y as in #6, write a formula for the combined area of these regular polygons, A_c.

9. If we think of y as a function of x, as we did in #7, the area A_c becomes a function of only x. Determine a formula for $\frac{dA_c}{dx}$.

10. Show that there is only one critical point of A_c in $\left[0, \frac{5}{n} \cot\left(\frac{\pi}{n}\right)\right]$.

11. Verify that the critical point from #10 is at a global minimum of A_c.

12. Based on the value of the critical point, what can you conclude about the apothems of these polygons?

✥ Energy and Force

Though enumerated separately, the following exercises are all driving toward a single idea: the relationship between energy and force.

1. The metric unit of energy is a *joule,* denoted by J.

    (a) An object with a mass of m kilograms has a kinetic energy of $K = \frac{1}{2}mv^2$ when traveling at v meters per second. Write down the relationship among joules, meters, seconds, and kilograms.

    (b) An object with a mass of m kilograms has a potential energy of $U = mgh$ when it's h meters above the ground, where g is the acceleration per kilogram due to gravity. Verify that the units associated with U are joules by examining the units of its factors.

    > Here we're setting $U = 0$ at ground level.

2. The metric unit of force is the *newton,* denoted by N.

    (a) Suppose an object is moving on the x-axis, so x is a function of time, $x(t)$. Its *linear momentum* is defined as the product of its mass and velocity: $p = m\frac{dx}{dt}$. What are the units of momentum in terms of meters, kilograms, and seconds?

    (b) Newton's Second Law can be written as $F = \frac{dp}{dt}$. How are newtons related to meters, kilograms, and seconds?

    > Newton's Second Law says that momentum changes because of force. In fact, it states that momentum changes *only* because of force—see Appendix E.

    (c) Assuming that mass is constant, use the definition of momentum from part (a) to rewrite Newton's Second Law as a relationship between F and $\frac{d^2x}{dt^2}$.

3. Suppose a small mass is sliding freely along the curve shown in Figure 9.3.

    > By "freely" we mean that it's not driven forward by a motor, engine, rocket, or other propulsion.

    (a) If the object's horizontal position is $x(t)$, what does $\frac{dx}{dt}$ tell us?

    (b) Suppose that $\frac{dx}{dt} > 0$ when the object is at $x = -3/4$.

    > You might prefer to think about a ball rolling along the hill, which is fine except that we mean to avoid issues of rotational kinetic energy in this discussion.

        i. Is the number $x(t)$ increasing or decreasing?

        ii. Is the object moving toward the right or toward the left?

    (c) Based on your experience with hills, and the slope of this one, is $\frac{dx}{dt}$ increasing or decreasing at $x = -3/4$?

    (d) Write down your answer from part (b) as an inequality involving $\frac{d^2x}{dt^2}$.

4. This question is a slight variation on the last, and is intended to drive home a particular point about the physical scenario.

    (a) Suppose the object is at $x = -3/4$ and that $\frac{dx}{dt} < 0$. Is the object moving toward the right or toward the left?

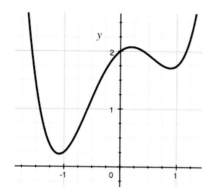

Figure 9.3: This is curve along which the mass is moving in #3 and #4. You're keeping track of the object's x-coordinate as it moves.

(b) Based on your experience with hills, and the slope of this one, explain why the number $\frac{dx}{dt}$ is decreasing at $x = -3/4$.

(c) Write down your answer from part (b) as an inequality involving $\frac{d^2x}{dt^2}$.

In #3 and #4, you saw that $\frac{d^2x}{dt^2}$ is nonzero. Based on Newton's Second Law as you wrote it down in #2(b), that means the mass is experiencing a force. This force is inexorably tied to the potential energy of the mass. Said colloquially, Nature changes spontaneously in order to reduce potential energy. We experience that change in terms of force.

Note that if we take the horizontal line at $y = 0$ to be the ground, the potential energy of the sliding mass depends on the height of the curve (see #1(b)). But potential energy might be present for reasons other than altitude (e.g., a spring that's stretched beyond its natural length carries potential energy, as do ions in the vicinity of other charges). So in the following questions, we'll not restrict ourselves to a particular context. Rather, we'll just say that the potential energy of an object changes from point to point on the x-axis (without worrying about why).

5. Suppose the potential energy at x is $U(x) = (x-1)^2(x+1)^2 + 0.75x + 1$ joules. The graph of $U(x)$ is shown in Figure 9.3.

    (a) Verify that $\frac{dU}{dx}$ has units of newtons (i.e., force).

    (b) Calculate the number $\frac{dU}{dx}$ at $x = -3/4$.

In #3 and #4 you said that $\frac{d^2x}{dt^2}$ is negative, which, based on Newton's Second Law as you wrote it down in #2b, means that the *force is negative* (meaning, "toward the left"). By contrast, in #5 you determined that $\frac{dU}{dx}$ has units of force, but that it is *positive* at $x = -3/4$. This is because

$$F = -\frac{dU}{dx}.$$

6. Find locations x_1, x_2, and x_3 (labeled from smallest to largest) where the mass in #5 experiences no force. These locations are called **equilibria**.

    Consider using Newton's method.

7. On the x-axis, just to the left and right of x_1 (which you found in #6), draw arrows that point in the direction of the force that's experienced by the mass (these will point left or right, depending on the sign of $\frac{dU}{dx}$). Then do the same for x_2 and x_3.

We refer to the x-axis with superimposed arrows (as drawn in #7) as a **phase line**. Phase lines are used to classify equilibria (we'll learn more about them in Chapter 10). Specifically ...

8. When the arrows on the left and right sides of an equilibrium point toward it, we say the equilibrium is **stable**. If we wanted to put the the mass at a stable equilibrium but made a small error in our placement, the force experienced by the mass would push it *toward* the equilibrium. Determine which of x_1, x_2, and x_3 are stable.

9. When the arrows on the left and right sides of an equilibrium point away from it, we say the equilibrium is **unstable**. If we wanted to put the the mass at a such an equilibrium but made a small error in our placement, the force experienced by the mass would push it *away* the equilibrium. Determine which of x_1, x_2, and x_3 are unstable.

10. Suppose the potential energy at x is $U(x) = x^4 - 32x^3 + 35x^2 - 373.75x + 2305$. Locate the equilibria, and classify each as stable or unstable.

❖ Quantum Mechanics

A whole new understanding of the physical world was born in the first half of the 20th century when quantum mechanics was discovered. Some of the most important work in the field was initially spurred by Erwin Schrödinger (who won the Nobel Prize in Physics in 1933 with Paul Dirac). His famous theory says that the probability of finding an electron in any particular interval is determined by the square of the function that's described by

$$-\frac{h}{4\pi m}y'' + V(x)y = Ey \tag{9.1}$$

in which h is Planck's constant, m is the mass of an electron, $V(x)$ is a function called the *potential*, and E is the *energy* of the electron. Today, we refer to (9.1) as the *time-independent Schrödinger equation* (TISE), and you know enough that you can use it to find a formula for $y(x)$ in some simple cases.

Figure 9.4: Erwin Schrödinger

▷ The infinite square well

The "infinite square well" is so named because of the graph of the potential function

$$V(x) = \begin{cases} 0 & \text{when } x \in [0, L] \\ \infty & \text{otherwise,} \end{cases}$$

(see Figure 9.6), which we understand to mean that it's impossible for the electron to leave the interval $[0, L]$. That means we're looking for a function that vanishes

Figure 9.5: Paul Dirac

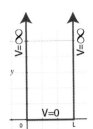

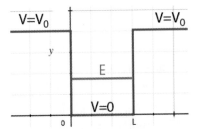

Figure 9.6: (left) The infinite square well; (right) the finite square well.

(i.e. is always zero) outside $[0, L]$. If we assume that the function is continuous at $x = 0$ and $x = L$, we know the following three things about it:

$$-\frac{h}{4\pi m}y'' = Ey \text{ in } [0, L] \tag{9.2}$$
$$y(0) = 0$$
$$y(L) = 0.$$

Equation (9.2) is just (9.1) with $V = 0$. It looks a lot more familiar if we rewrite it as

$$y'' = -\omega^2 y, \quad \text{where} \quad \omega = \sqrt{\frac{4\pi m E}{h}}. \tag{9.3}$$

1. By finding its second derivative, show that $y = A\sin(\omega x) + B\cos(\omega x)$ is described by (9.3) when A and B are any constants.

2. Based on the data at $x = 0$, show that $B = 0$.

3. If A is *also* zero, there's no electron at all. The only other way to make $y(L) = 0$ is for $\sin(\omega L) = 0$, which can only happen when ωL is an integer multiple of π. That is, $\omega L = k\pi$, where k is an integer. Solve this equation for E (which is bundled up in ω).

Since k is an integer, the formula you found for E prohibits most values—only specific energy levels are allowed. We say that the energy is *quantized*.

▷ The finite square well

In this case, we suppose there is a positive number $V_0 > E$ so that

$$V(x) = \begin{cases} 0 & \text{when } x \in [0, L] \\ V_0 & \text{otherwise.} \end{cases}$$

Since we have two different conditions (one inside $[0, L]$ and the other outside), we write (9.1) twice:

$$y'' = -\omega^2 y \quad \text{when } x \in (0, L) \tag{9.4}$$

$$y'' = -\frac{4\pi m(E - V_0)}{h} y \quad \text{otherwise,} \tag{9.5}$$

where $\omega = \sqrt{4\pi m E/h}$, like before. Our strategy for solving the TISE in this case will be to do four things:

(a) find an "inside" function, $y_i(x)$, described by (9.4);

(b) find a "left" function, $y_\ell(x)$, described by (9.5) on $(-\infty, 0)$;

(b) find a "right" function, $y_r(x)$, described by (9.5) on (L, ∞);

(d) link them together.

You already know that (9.4) is solved by the function $y_i(x) = A_i\sin(\omega x) + B_i\cos(\omega x)$, so part (a) is done. For parts (b) and (c), notice that

$$-\frac{4\pi m(E - V_0)}{h} > 0 \quad \text{since} \quad V_0 > E.$$

So we can make (9.5) a little easier on the eyes by writing it as

$$y'' = \lambda^2 y \quad \text{where} \quad \lambda = \sqrt{\frac{4\pi m(V_0 - E)}{h}}. \tag{9.6}$$

> We no longer have the conditions that $y_i(0) = 0$ and $y_i(L) = 0$, which resulted from the *infinite* part of the infinite square well (the electron could not exist outside of $[0, L]$).

4. By finding its second derivative, show that $y_r(x) = A_r e^{\lambda x} + B_r e^{-\lambda x}$ is described by (9.6) when A_r and B_r are any constants.

5. Since the probability of finding the electron at infinity is zero, we require that $\lim_{x \to \infty} y_r(x) = 0$. What does this tell you about the number A_r?

6. By finding its second derivative, show that $y_\ell(x) = A_\ell e^{\lambda x} + B_\ell e^{-\lambda x}$ is described by (9.6) when A_ℓ and B_ℓ are any constants.

7. Since the probability of finding the electron at infinity is zero, we require that $\lim_{x \to -\infty} y_\ell(x) = 0$. What does this tell you about the number B_ℓ?

Now we link the pieces together: $y(x) = \begin{cases} y_\ell(x) & \text{when } x < 0 \\ y_i(x) & \text{when } x \in [0, L] \\ y_r(x) & \text{when } x > L \end{cases}$ (9.7)

Based on the definition of $y(x)$ given in (9.7), write equations that say

8. $y(x)$ is continuous at $x = 0$,

9. $y'(x)$ is continuous at $x = 0$,

10. $y(x)$ is continuous at $x = L$, and

11. $y'(x)$ is continuous at $x = L$.

12. Use these equations to show that, if $A_\ell = 0$, *all* the coefficients must be zero.

From the standpoint of classical mechanics, the condition $V_0 > E$ means the electron doesn't have enough energy to leave the potential well, $[0, L]$. Accordingly, we call the exterior of the well, $(-\infty, 0) \cup (L, \infty)$, the *classically forbidden region* (CFR).

What you showed in #12 is that Schrödinger's equation predicts something different: if the electron is anywhere at all (i.e., at least one of our coefficients is nonzero), the number A_ℓ cannot be zero. Consequently, there is a nonzero probability that the electron will be found outside the well! This fact is at the heart of an amazing behavior called *quantum tunneling*.

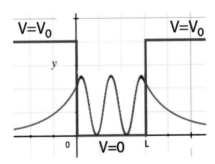

Figure 9.7: The probability of finding an electron is related to the square of $y(x)$.

Chapter 5
Integration

It's easy to calculate the distance that an object travels if its velocity is constant, but more complicated if the object is speeding up. For example, suppose we use a radar gun to determine the velocity of a car as it speeds up, and we see this data:

t (sec)	0	3	6	9	12
v (ft/sec)	1	10	25	45	55

We can use this data to approximate the distance that the car traveled. Because it was speeding up, we know that the car was moving *at most* 10 ft/sec during the first three seconds, so it traveled *at most*

$$\left(10 \ \frac{\text{ft}}{\text{sec}}\right)(3 \ \text{sec}) = 30 \text{ feet during that time.}$$

Similarly, the velocity of the car was *at most* 25 ft/sec between $t = 3$ and $t = 6$, which is a 3-second interval, so it traveled *at most*

$$\left(25 \ \frac{\text{ft}}{\text{sec}}\right)(3 \ \text{sec}) = 75 \text{ feet during that time.}$$

If we make similar estimates of the car's motion in the remaining intervals of time ($6 \leq t \leq 9$ and then $9 \leq t \leq 12$), we can approximate the distance it traveled:

$$\Delta x \approx \left(10 \ \frac{\text{ft}}{\text{sec}}\right)(3 \ \text{sec}) + \left(25 \ \frac{\text{ft}}{\text{sec}}\right)(3 \ \text{sec}) + \left(45 \ \frac{\text{ft}}{\text{sec}}\right)(3 \ \text{sec}) + \left(55 \ \frac{\text{ft}}{\text{sec}}\right)(3 \ \text{sec}) = 405 \text{ feet.}$$

Our calculation has overestimated of the car's actual change in position (we kept saying "at most"), but we have *an* approximation, and that's the point. This chapter focuses on developing an understanding of such approximations—how good they are, how to make them better, how to make them quickly, and most importantly, how to pass from approximation to exactitude.

5.1   Estimating Net Change

In the chapter opener we estimated that a car traveled 405 feet in 12 seconds, but recognized that our approximation was an overestimation of the vehicle's actual change in position because, for example, we assumed the velocity of the car was 45 ft/sec during the *entire* 3 seconds between $t = 6$ and $t = 9$. In fact, the car was moving much slower at $t = 6$, and sped up to 45 mph during that time.

We can improve our approximation by using more data. Suppose we use our radar gun nine times instead of five and see

t (sec)	0	1.5	3	4.5	6	7.5	9	10.5	12
v (ft/sec)	1	7	10	13	25	32	45	48	55

Because the car is speeding up, we know that it was moving *at most* 7 ft/sec during the first 1.5 seconds, so it traveled *at most*

$$\left(7 \ \frac{\text{ft}}{\text{sec}}\right)(1.5 \ \text{sec}) = 10.5 \text{ feet in that span of time.}$$

Similarly, the velocity of the car was *at most* 10 ft/sec between $t = 1.5$ and $t = 3$, which is a span of 1.5 seconds, so it traveled *at most*

$$\left(10 \ \frac{\text{ft}}{\text{sec}}\right)(1.5 \ \text{sec}) = 15 \text{ feet during that span of time.}$$

As in the chapter opener, when we approximate the distance traveled by the car during these 12 seconds by making such estimates in each interval of 1.5 seconds, we see

$$\Delta x \ \approx \ \left(7 \ \frac{\text{ft}}{\text{sec}}\right)(1.5 \ \text{sec}) + \left(10 \ \frac{\text{ft}}{\text{sec}}\right)(1.5 \ \text{sec}) + \left(13 \ \frac{\text{ft}}{\text{sec}}\right)(1.5 \ \text{sec}) + \left(25 \ \frac{\text{ft}}{\text{sec}}\right)(1.5 \ \text{sec})$$

$$+ \underbrace{\left(32 \ \frac{\text{ft}}{\text{sec}}\right)(1.5 \ \text{sec})}_{\text{for } t \in [6,7.5]} + \underbrace{\left(45 \ \frac{\text{ft}}{\text{sec}}\right)(1.5 \ \text{sec})}_{\text{for } t \in [7.5,9]} + \underbrace{\left(48 \ \frac{\text{ft}}{\text{sec}}\right)(1.5 \ \text{sec})}_{\text{for } t \in [9,10.5]} + \underbrace{\left(55 \ \frac{\text{ft}}{\text{sec}}\right)(1.5 \ \text{sec})}_{\text{for } t \in [10.5,12]} = 352.5 \ \text{feet.}$$

By increasing the number of times that we use the radar gun from five to nine, we reduced our overestimate from 405 to 352.5 feet. There's still error, but there's less of it because our assumption of constant velocity was made over smaller intervals of time. Imagine how good the estimate would be if we knew the car's velocity at a hundred, a thousand, or a million moments during this twelve-second interval! Of course, the thought of writing a million summands is daunting, so a shorthand has been developed to make such calculations manageable.

❖ Sigma Notation

Sigma notation is a shorthand that uses the capital Greek letter sigma, Σ, to compress and keep track of sums. Any time you see Σ you should ...

> Σ is the upper-case Greek letter sigma. When reading Greek, it makes the "s" sound, as in the word *sum*.

1. realize that you're looking at a summation;

2. figure out what kind of thing is being added;

3. figure out how many of them are being added.

Sigma notation is best learned by demonstration rather than explanation, so ...

Sigma notation as an indication of summation

Example 1.1. Suppose we index the U.S. states in alphabetical order, and append the District of Columbia at the end of the list:

index	state
1	Alabama
2	Alaska
3	Arizona
⋮	⋮
50	Wyoming
51	Washington D.C.

If p_k denotes the population of location k, what is the value of $\sum_{k=1}^{51} p_k$?

Solution: The first thing to notice is Σ, which indicates that this is a summation. More specifically, we're adding the numbers p_k, each of which is the population of some region. The index is denoted by k, and it ranges from 1 to 51, so

$$\overset{51}{\underset{k=1}{\sum}} p_k \quad \text{means} \quad \overbrace{p_1}^{\text{the population of Alabama}} + p_2 + p_3 + \cdots + p_{50} + \underbrace{p_{51}}_{\text{the population of Washington D.C.}}$$

Now we see that $\sum_{k=1}^{51} p_k$ is the population of the United States. On July 20, 2010, the "population clock" of the US Census Bureau (`http://www.census.gov/main/www/popclock.html`) estimated that number to be 309,784,902. ∎

Sigma notation as an indication of summation

Example 1.2. The atmosphere comprises many kinds of molecules, including O_2, N_2, H_2O, and CO_2. Suppose we index the types of atmospheric molecules as follows:

index	species
1	oxygen
2	nitrogen
3	water

The index 4 will refer to the collection of all other types of molecule in the atmosphere. When we talk about the pressure from a specific kind of molecule, as if all the others were absent, we use the term *partial pressure*. Based on our indexing scheme, we'll denote the partial pressure due to O_2 by P_1, the partial pressure due to N_2 by P_2, and so on. Now interpret the meaning of $\sum_{k=1}^{4} P_k$.

Solution: The first part of $\sum_{k=1}^{4} P_k$. is the letter sigma, which indicates that we're adding. Next to the Σ you see the kind of thing that's being added together: partial pressures. Below the Σ you see $k=1$ and above it you see the number 4. This tells you that the letter k represents the index, that it starts at 1, and that it increments all the way up to 4. In short,

$$\overset{4}{\underset{k=1}{\sum}} P_k = \overbrace{P_1}^{\text{pressure due only to } O_2} + \underbrace{P_2}_{\text{pressure due only to } N_2} + P_3 + P_4.$$

Because we've accounted for the pressure contributions of all the different kinds of molecules in the atmosphere, this sum is the atmospheric pressure. ∎

When using sigma notation to do calculus, the summands often depend explicitly on the index k, as seen in the next example.

When the summands depend explicitly on the index of summation

Example 1.3. Determine the value of $\sum_{k=1}^{5}(3+2k)$.

Solution: The Σ tells us that this is a summation, but this time there's no context to provide an interpretation—this is just addition. The index starts at 1, ends at 5, and the summand changes every time k increments. Specifically,

$$\overset{5}{\underset{k=1}{\sum}}(3+2k) = \underbrace{\left(3+2(1)\right)}_{k=1} + \underbrace{\left(3+2(2)\right)}_{k=2} + \underbrace{\left(3+2(3)\right)}_{k=3} + \underbrace{\left(3+2(4)\right)}_{k=4} + \underbrace{\left(3+2(5)\right)}_{k=5}, \tag{1.1}$$

the value of which is 45.   ■

Try It Yourself: When the summands depend explicitly on the index of summation

Example 1.4. Determine the value of $\sum_{k=1}^{4}(3k^2 - 7)$.

Answer: 62.   ■

▷ Addition facts and sigma notation

Sigma notation becomes a powerful tool once we express some simple facts in terms of it. For example, adding a constant to itself over and over again is really just multiplication.

Summing a constant

Example 1.5. Determine the value of $\sum_{k=1}^{6} 2$.

Solution: The Σ tells us that we're adding, but unlike Examples 1.3 and 1.4, this time the summand doesn't change with k (it's always the number 2).

$$\sum_{k=1}^{6} 2 = \underbrace{2}_{k=1} + \underbrace{2}_{k=2} + \underbrace{2}_{k=3} + \underbrace{2}_{k=4} + \underbrace{2}_{k=5} + \underbrace{2}_{k=6} = 6(2) = 12. \qquad ■$$

Try It Yourself: Summing a constant

Example 1.6. Determine the value of $\sum_{k=1}^{4} 15$.

Answer: 60.   ■

> When β is a constant, $\sum_{k=1}^{n} \beta = n\beta$.

Another simple but important fact about addition is that the order in which we add numbers is irrelevant.

Calculating with sigma notation

Example 1.7. Determine $\sum_{k=1}^{8} k$.

Solution: When we write out this summation, we see

$$\sum_{k=1}^{8} k = 1 + 2 + 3 + 4 + 5 + 6 + 7 + 8. \qquad (1.2)$$

Of course, we can add it any order, so this is the same as

$$\sum_{k=1}^{8} k = 8 + 7 + 6 + 5 + 4 + 3 + 2 + 1. \qquad (1.3)$$

which we mention because it leads to an important fact. Specifically, when we stack equations (1.2) and (1.3), we see that the numbers in each column add to 9. So if we add them together, we get

$$
\begin{array}{rcl}
\sum_{k=1}^{8} k & = & 1 + 2 + 3 + 4 + 5 + 6 + 7 + 8 \\
+ \sum_{k=1}^{8} k & = & 8 + 7 + 6 + 5 + 4 + 3 + 2 + 1 \\
\hline
2\sum_{k=1}^{8} k & = & 9 + 9 + 9 + 9 + 9 + 9 + 9 + 9 = 8 \times 9
\end{array}
$$

from which it follows that $\sum_{k=1}^{8} k = \frac{(8)(9)}{2} = 36$. ∎

Example 1.8. Use the stacking technique demonstrated in Example 1.7 to determine $\sum_{k=1}^{16} k$.

Answer: 136. ∎

Full Solution
On-line

The stacking technique used in Examples 1.7 and 1.8 works in general, so let's write it down in generality:

$$\sum_{k=1}^{n} k = \frac{n(n+1)}{2}. \qquad (1.4)$$

There are other summation formulas that are often handy, two of which are cited below (though we won't derive them here):

On-line sources list formulas for summing k^m for many different values of m.

$$\sum_{k=1}^{n} k^2 = \frac{n(n+1)(2n+1)}{6}. \qquad (1.5)$$

$$\sum_{k=1}^{n} k^3 = \left(\frac{n(n+1)}{2}\right)^2. \qquad (1.6)$$

Another simple but important fact about addition is that we can factor out terms that are common to all the summands.

Example 1.9. Determine the value of $\sum_{k=1}^{5} 7k$.

Solution: The point of this example is that we can write

$$\sum_{k=1}^{5} 7k = 7(1) + 7(2) + 7(3) + 7(4) + 7(5) = 7(1 + 2 + 3 + 4 + 5) = 7\sum_{k=1}^{5} k.$$

Equation (1.4) tells us that $\sum_{k=1}^{5} k = \frac{5(6)}{2} = 15$, so this sum is $7(15) = 105$. ∎

Example 1.10. Determine the value of $\sum_{k=1}^{6} 4k^2$ by factoring out the 4.

Answer: 364. ∎

Full Solution
On-line

As you saw in Examples 1.9 and 1.10,

When β is any constant, and each a_k is a number (possibly changing with k)

$$\sum_{k=1}^{n} \beta a_k = \beta \sum_{k=1}^{n} a_k.$$

Now let's begin to use these ideas in concert.

Quick calculation using sigma notation

Example 1.11. Determine the value of $\sum_{k=1}^{5}(3 + 2k)$.

Solution: This is the sum from Example 1.3, but now we have some tools to expedite the calculation. Look back to the addition that's laid out in equation (1.1) on p. 327. Because the order in which we add numbers is irrelevant, we could group all the 3s together, and group all the terms that have the form $2k$ ...

$$\sum_{k=1}^{5}(3 + 2k) = \Big(3 + 3 + 3 + 3 + 3\Big) + \Big(2(1) + 2(2) + 2(3) + 2(4) + 2(5)\Big).$$

That is to say,

$$\sum_{k=1}^{5}(3 + 2k) = \sum_{k=1}^{5}3 + \sum_{k=1}^{5}2k = 5(3) + 2\sum_{k=1}^{5}k.$$

We use equation (1.4) to quickly determine that $\sum_{k=1}^{5} k = 15$, so that

$$\sum_{k=1}^{5}(3 + 2k) = 15 + 2(15) = 45.$$

This is exactly the answer that we calculated on p. 327.   ■

Try It Yourself: Quick calculation using Σ notation

Example 1.12. Use the regrouping technique demonstrated in Example 1.11 to calculate $\sum_{k=1}^{7}(4 + 8k^2)$.

Answer: 1148.   ■

Full Solution
On-line

In Example 1.11 we grouped all the 3s together, and in Example 1.12 we grouped all the 4s. In general, because the order of addition is irrelevant:

$$\sum_{k=1}^{n}(a_k + b_k) = \sum_{k=1}^{n}a_k + \sum_{k=1}^{n}b_k.$$

Similarly,

$$\sum_{k=1}^{n}(a_k - b_k) = \sum_{k=1}^{n}a_k - \sum_{k=1}^{n}b_k.$$

With these facts in hand, we can return to the motivating problem: using information about the rate of change (e.g., velocity) to estimate a net change in position.

❖ Riemann Sums

Let's return to the story of approximating the change in a car's position based on knowledge of its velocity. Imagine you're at the drag strip where velocity is being measured, and you get a turn to drive. So you hop in the car, buckle up, and push the gas pedal to the floor. After a moment, you look down at the speedometer ...

Estimating change in position when $v(t)$ is known

Example 1.13. Suppose the car's velocity is $v(t) = 40 + 22t$ ft/sec (where $t = 0$ is the moment you first looked at the speedometer). Estimate the vehicle's change in position between $t = 0$ and $t = 5$ by dividing that interval of time into fifty subintervals and calculating the car's velocity in each one.

Solution: If we divide $[0, 5]$ into fifty subintervals of equal length, each will span $\frac{5}{50} = 0.1$ seconds, which we denote by Δt (see the section of $[0, 5]$ depicted below).

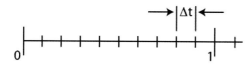

> This is a lot like the estimation that we did at the beginning of the section, but in this case we have a function instead of a table of data.

Let's denote by v_k the velocity that's measured in the kth subinterval. Then the distance traveled by the car during the first span of Δt seconds is about (rate)×(time)= $v_1 \Delta t$, the distance traveled during the second span of Δt seconds is about (rate)×(time)= $v_2 \Delta t$, and so on. Adding these up, we estimate the car's change in position to be

$$\Delta x \approx v_1 \Delta t + v_2 \Delta t + v_3 \Delta t + \cdots + v_{50} \Delta t = \sum_{k=1}^{50} v_k \Delta t.$$

> Remember that if the units don't make sense, your calculation is surely wrong. So let's check the dimensions here: v_k has units of ft/sec, and Δt has units of seconds, so each summand has units of
> $$\left(\frac{\text{ft}}{\text{sec}}\right) \text{sec} = \text{ft}.$$
> So our equation says that Δx has units of length, which is correct.

Before we can finish our calculation, we need to know the numbers v_k. As before, let's measure the car's velocity at the last moment of each subinterval—so our first velocity measurement happens at time $0.1 = 0 + \Delta t$. It is helpful to think of this as one "step" away from our starting time . . .

The right endpoint of the second subinterval of time is at $0.2 = 0 + 2\Delta t$ . . .

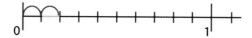

The right endpoint of the third subinterval is at $0.3 = 0 + 3\Delta t$ . . .

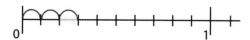

In general, the right endpoint of the kth subinterval is at $0 + k\Delta t$. Using our velocity function at these times, we see that $v_k = 40 + 22(k\Delta t)$, so our approximation of the car's change in position is

$$\Delta x \approx \sum_{k=1}^{50} \Big(40 + 22(k\Delta t)\Big)\Delta t = \sum_{k=1}^{50} \Big(40\Delta t + 22(\Delta t)^2 k\Big) = \sum_{k=1}^{50} 40\Delta t + \sum_{k=1}^{50} 22(\Delta t)^2 k.$$

Notice that $40\Delta t$ and $22(\Delta t)^2$ are constants (they don't involve k), so

$$\Delta x \approx 50(40\Delta t) + 22(\Delta t)^2 \sum_{k=1}^{50} k.$$

Lastly, we evaluate the remaining sum using equation (1.4) on p. 329, and use the fact that $\Delta t = 0.1$ to approximate

$$\Delta x \approx 50(40\Delta t) + 22(\Delta t)^2 \left(\frac{50(50+1)}{2} \right) = 480.5 \text{ feet} \qquad \blacksquare$$

Try It Yourself: Estimating change in position when $v(t)$ is known

Example 1.14. Repeat Example 1.13 but using five hundred subintervals.

Answer: 475.55 feet. $\qquad \blacksquare$

Full Solution
On-line

In Example 1.13 and Example 1.14 we checked the car's velocity at the last moment of each subinterval, but mathematically speaking, there's no reason we have to use the *last* moment. Whether at the beginning of a subinterval, the end, or somewhere in between, the time at which we decide to calculate the car's velocity in the kth subinterval is called the **sample time** or **sample point**, and is often denoted by t_k^*. Using this notation, the car's velocity at that time is $v(t_k^*)$ and our approximation of its change in position is

> The "k" in the notation t_k^* keeps track of which subinterval we're talking about (the first? the second?), and the asterisk indicates that this is the moment in that subinterval when we've chosen to sample the velocity.

$$\Delta x \approx \sum_{k=1}^{n} v(t_k^*)\Delta t,$$

where n is the number of subintervals that we use. We call this a **Riemann sum**, after the German mathematician Bernhard Riemann (1826-1866). More specifically, we say that it's a **left-sampled** Riemann sum when the first moment of each subinterval is the sample time, a **right-sampled** Riemann sum when the last moment of each subinterval is the sample time, and a **midpoint-sampled** Riemann sum when the sample time of each subinterval is at its midpoint (see Figure 1.1).

> The name *Riemann* is pronounced "REE-mahn."

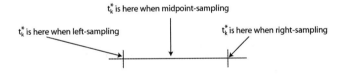

t_k^* is here when midpoint-sampling

t_k^* is here when left-sampling

t_k^* is here when right-sampling

Figure 1.1: The sample time, t_k^*, can be anywhere in the kth interval.

Approximating change in altitude with a left-sampled Riemann sum

Example 1.15. Suppose the altitude of a radio-controlled plane is changing at a rate of $y'(t) = 10 - 5t$ ft/sec. Estimate its change in altitude between $t = 1$ and $t = 2$ seconds by using a left-sampled Riemann sum with twelve subintervals.

Solution: Notice that $y'(t) > 0$ between $t = 1$ and $t = 2$ seconds, so the plane is rising during this example. If we divide our one-second interval into twelve subintervals, each will have a duration of $1/12$ of a second, which we'll call Δt. As before, our approximation of the plane's change in altitude is made by a Riemann sum:

$$\Delta y \approx \sum_{k=1}^{12} \underbrace{y'(t_k^*)}_{\text{(rate)}} \underbrace{\Delta t}_{\text{(time)}} = \sum_{k=1}^{12} \underbrace{\left(10 - 5t_k^* \right)}_{\text{ft/sec}} \underbrace{\Delta t}_{\text{sec}} .$$

To continue our calculation, we need a formula for t_k^*. If we were right-sampling, we'd find t_k^* by starting at $t = 1$ and stepping Δt units to the right k times, so t_k^* would be $1 + k\Delta t$ (as seen in Figure 1.2). But we're sampling at the *left* endpoint

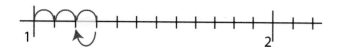

Figure 1.2: The right endpoint of the 3rd subinterval is at $1 + 3\Delta t$. The left endpoint of that subinterval is one step backwards from there—at $1 + 3\Delta t - \Delta t = 1 + 2\Delta t$.

of each subinterval instead, so we take a step backwards from the right endpoint and have $t_k^* = 1 + k\Delta t - \Delta t = 1 + (k-1)\Delta t$.

Substituting this into our Riemann sum, we see

$$\Delta y \approx \sum_{k=1}^{12} \left(10 - 5t_k^*\right)\Delta t = \sum_{k=1}^{12} \left(10 - 5(1 + (k-1)\Delta t)\right)\Delta t.$$

We're going to use equation (1.4) from p. 329 to evaluate this sum; but before we can do that, we need to gather terms according to how many factors of k they have (some have a single factor, others don't have a factor of k at all):

$$\Delta y \approx \sum_{k=1}^{12} 5\Delta t + 5(\Delta t)^2 - 5(\Delta t)^2 k.$$

In preparation for using the equation (1.4) from p. 329, we separate this into two sums:

$$\Delta y \approx \sum_{k=1}^{12} 5\Delta t + 5(\Delta t)^2 - \sum_{k=1}^{12} 5(\Delta t)^2 k.$$

> The first of these sums says to add $5 + 5(\Delta t)^2$ to itself 12 times.

Since $5(\Delta t)^2$ and $5\Delta t + 5(\Delta t)^2$ are constant, this is really

$$\Delta y \approx 12\left(5\Delta t + 5(\Delta t)^2\right) - 5(\Delta t)^2 \sum_{k=1}^{12} k.$$

Lastly, we use equation (1.4) to evaluate the remaining sum, and use the fact that $\Delta t = \frac{1}{12}$ to approximate

$$\Delta y \approx 12\left(5\Delta t + 5(\Delta t)^2\right) - 5(\Delta t)^2 \left(\frac{12(12 + 1)}{2}\right) = \frac{65}{24} \text{ feet.} \qquad \blacksquare$$

Try It Yourself: Approximating change with a left-sampled Riemann sum

Example 1.16. Repeat Example 1.15, but this time use twenty-one subintervals.

Answer: The plane rises by $\Delta y \approx \frac{55}{21}$ feet. $\qquad \blacksquare$

Full Solution
On-line

Approximating change in altitude with a midpoint-sampled Riemann sum

Example 1.17. The radio-controlled plane from Example 1.15 is descending when $t > 2$ because $y'(t) < 0$. Approximate its change in altitude between times $t = 2$ and $t = 4$ seconds by using a midpoint-sampled Riemann sum with ten subintervals.

Solution: If we divide the two-second interval between $t = 2$ and $t = 4$ into ten subintervals of equal length, each will have a duration of $\Delta t = \frac{2}{10} = \frac{1}{5}$ seconds. Our approximation of the plane's change in altitude is

$$\Delta y \approx \sum_{k=1}^{10} y'(t_k^*)\Delta t = \sum_{k=1}^{10} \left(10 - 5(t_k^*)\right)\Delta t.$$

To continue our calculation, we need a formula for t_k^*. If we were right-sampling, we'd find t_k^* by starting at $t = 2$ and stepping Δt units to the right k times, so t_k^* would be $2 + k\Delta t$ (as seen in Figure 1.3). But we're sampling at the midpoint of each subinterval instead, so let's take half a step back from the right endpoint, so $t_k^* = 2 + k\Delta t - \frac{1}{2}\Delta t = 2 + (k - \frac{1}{2})\Delta t$.

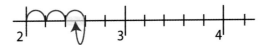

Figure 1.3: The right endpoint of the 3rd subinterval is at $2 + 3\Delta t$, so the midpoint is at $2 + 3\Delta t - \frac{1}{2}\Delta t$.

Substituting this formula for t_k^* into our Riemann sum, we see

$$\Delta y \approx \sum_{k=1}^{10} \left(10 - 5 \left(2 + k\Delta t - \frac{1}{2}\Delta t \right) \right) \Delta t.$$

We're going to use equation (1.4) from p. 329 to evaluate this sum; but before we can do that, we need to gather terms according to how many factors of k they have. So we rewrite the Riemann sum as

$$\Delta y \approx \sum_{k=1}^{10} \frac{5}{2}(\Delta t)^2 - 5(\Delta t)^2 k.$$

Because the order of addition is irrelevant, we can begin by adding up all those terms without a factor of k, then add those terms with a single factor of k:

$$\Delta y \approx \sum_{k=1}^{10} \frac{5}{2}(\Delta t)^2 - \sum_{k=1}^{10} 5(\Delta t)^2 k.$$

> The first sum adds $2.5(\Delta t)^2$ to itself 10 times. In the second sum, every summand has a factor of $5(\Delta t)^2$, which we can factor out.

Since $2.5(\Delta t)^2$ and $5(\Delta t)^2$ are constant, this is really

$$\Delta y \approx 25(\Delta t)^2 - 5(\Delta t)^2 \sum_{k=1}^{10} k.$$

Now using equation (1.4) from p. 329 to perform the remaining summations with $n = 10$, and using the fact that $\Delta t = 1/5$,

$$\Delta y \approx 25(\Delta t)^2 - 5(\Delta t)^2 \frac{10(10 + 1)}{2} = -10 \text{ feet.} \qquad \blacksquare$$

Try It Yourself: Approximating change with a midpoint-sampled Riemann sum

Example 1.18. Use a midpoint-sampled Riemann sum with thirty subintervals to approximate the change in the altitude of the radio-controlled plane from Example 1.15 between times $t = 1$ and $t = 4$ seconds.

Full Solution
On-line

Answer: $\Delta y \approx -7.5$ feet. $\qquad \blacksquare$

Among other things, these examples have illustrated the following formulas:

When partitioning $[a, b]$ into n subintervals of equal length, $\Delta t = \frac{b-a}{n}$, the sample point in the kth subinterval is ...

$$
\begin{aligned}
t_k^* &= a + k\Delta t && \text{when right-sampling} \\
t_k^* &= a + \left(k - \tfrac{1}{2}\right)\Delta t && \text{when midpoint-sampling} \\
t_k^* &= \underbrace{a}_{\text{start time}} + \underbrace{(k-1)\Delta t}_{\text{some number of seconds}} && \text{when left-sampling}
\end{aligned}
$$

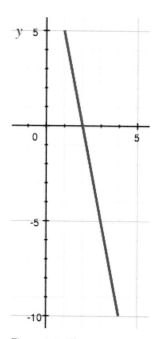

❖ Net Change v. Total Change

Suppose you're practicing the 100-meter sprint. You set up on the starting blocks, sprint to the finish line, then walk back and set up again. On the one hand, you've traveled 200 meters. On the other, you're right back where you started. To communicate these ideas concisely, we say that you traveled a **total distance** of 200 meters, and that your **net change** in position is zero. Net change in position is often called **displacement**.

A similar thing happened with the airplane in Example 1.18—the airplane's vertical velocity was positive and then negative (see Figure 1.4), so it rose and then fell. In contrast to the running example above, in which the amount of forward and backward motion is the same, the airplane in Example 1.18 rises by about $\frac{55}{21}$ feet (see Example 1.16) and then descends about 10 feet (see Example 1.17). So the *net change* in altitude between time $t = 1$ and $t = 4$ seconds is approximately

$$
\overbrace{\frac{55}{21}}^{\Delta y \text{ during the climb}} + \underbrace{(-10)}_{\Delta y \text{ during the descent}} \approx -7.5 \text{ feet,}
$$

which is fundamentally why we find that $\Delta y \approx -7.5$ in Example 1.18.

Interpreting Riemann sums as net change

Example 1.19. Suppose the airplane in Example 1.18 had an altitude of twelve feet at time $t = 1$. Approximate its altitude at time $t = 4$.

Solution: Since we calculated the net change in altitude to be $\Delta y \approx -7.5$ feet,

$$
y(4) = \overbrace{y(1)}^{\text{starting altitude}} + \underbrace{\Delta y}_{\text{change in altitude}} \approx 12 + (-7.5) = 4.5 \text{ feet.} \qquad \blacksquare
$$

Try It Yourself: Interpreting Riemann sums as net change

Example 1.20. Suppose the airplane in Example 1.18 had an altitude of forty-one feet at time $t = 1$. Approximate its altitude at time $t = 4$.

Answer: $y(4) \approx 33.5$ feet. $\qquad \blacksquare$

Figure 1.4: The graph of $v(t)$ from Example 1.18. The plane was rising for 1 second and descending for 2 seconds.

Full Solution On-line

In summary, we've been using *rate of change* (velocity) to estimate *net change* (in position). Let's say this in the language of differential calculus, writing $y'(t)$ instead of $v(t)$ for velocity:

$$\sum_{k=1}^{n} y'(t_k^*)\Delta t \approx \Delta y = \text{net change in } y.$$

This relationship holds even when $y(t)$ represents something other than position, and even when the independent variable is something other than time.

Net change in temperature over time

Example 1.21. Suppose the temperature of a chemical solution is changing at a rate of $y'(t) = -3t + 14$ °F per hour. (a) Use a right-sampled Riemann sum with thirty subintervals to approximate the net change in the temperature of the solution between times $t = 4$ and $t = 6$, and (b) estimate the solution's temperature at time $t = 6$, assuming its temperature was 104°F at time $t = 4$.

Solution: A quick dimensional analysis confirms that the Riemann sum in this example calculates a net change in temperature (it has units of °F):

$$\Delta y \approx \sum_{k=1}^{30} \underbrace{y'(t_k^*)}_{\text{°F/sec}} \underbrace{\Delta t}_{\text{sec}} = \sum_{k=1}^{30}(-3t_k^* + 14)\Delta t.$$

Since we're right-sampling, starting at time $t = 4$, the sample point in the kth subinterval is at $t_k^* = 4 + k\Delta t$. Substituting this into our Riemann sum yields

$$\sum_{k=1}^{30}\left(-3(4 + k\Delta t) + 14\right)\Delta t = \underbrace{60\Delta t - 3(\Delta t)^2 \frac{30(31)}{2}}_{\text{After taking the simplifying steps seen in previous examples.}}.$$

If we divide the two-hour span of time between $t = 4$ and $t = 6$ into thirty subintervals of equal length, each has a duration of $\Delta t = \frac{6-4}{30} = \frac{1}{15}$ hours. Substituting this value of Δt into our formula yields the approximation

$$\text{net change in temperature} = \Delta y \approx -\frac{33}{15} \text{ °F.}$$

(b) If we assume that $y(4) = 104$, the temperature at time $t = 6$ is

$$y(6) = \overbrace{y(4)}^{\text{starting temp}} + \underbrace{\Delta y}_{\text{change in temp}} \approx 104 + \left(-\frac{33}{15}\right) = 101.8 \text{ °F.} \qquad \blacksquare$$

Net change in revenue due to price increase

Example 1.22. Suppose that customer demand for a certain commodity, denoted by $D(p)$ changes at rate of $D'(p) = -40 - \frac{5000}{(p+1)^2} \frac{\text{units}}{\$}$, where p is the cost of the commodity (in dollars). Write down a left-sampled Riemann sum with forty-one subintervals that estimates the net change in demand when the producer raises the price of the commodity from \$50 to \$53.

Solution: In this case, the independent variable is price, not time, but the structure of the Riemann sum is the same. A quick dimensional analysis verifies that it calculates the right kind of quantity:

$$\Delta D \approx \sum_{k=1}^{41} \underbrace{D'(p_k^*)}_{\text{units/\$}} \underbrace{\Delta p}_{\$} = \sum_{k=1}^{41} \left(-40 - \frac{5000}{(p_k^*+1)^2} \right) \Delta p.$$

If we think about the price increase from \$50 to \$53 as the result of 41 very small price increases, each price increase is $\Delta p = \frac{3}{41}$ dollars (i.e., about 7 cents). Then, since we're left-sampling and starting at a price of $p = 50$, we have

> We're just subdividing the interval between $p = 50$ and $p = 53$ into 41 subintervals.

$$p_k^* = 50 + (k-1)\Delta p = 50 + (k-1)\frac{3}{41}.$$

Substituting this into our Riemann sum yields

$$\Delta D \approx \sum_{k=1}^{41} \left(-40 - \frac{5000}{(50 + (k-1)\frac{3}{41} + 1)^2} \right) \frac{3}{41}. \qquad \blacksquare$$

We were not asked to calculate in Example 1.22, but only to write down the appropriate Riemann sum. You'll notice that none of the summation formulas from earlier in the section have the index k in a denominator. Had we been asked for an approximation, we would either have to calculate forty-one numbers and add them up, or use a computer to do it for us.

> Spreadsheet programs make calculating Riemann sums very fast and easy. Step-by-step instructions are provided on p. 398.

Using graphs to estimate change

Example 1.23. Figure 1.5 provides information about the rate at which posts regarding President Obama appeared on twitter.com during several days in January, 2010. If P denotes the cumulative number of posts regarding President Obama, use a right-sampled Riemann sum with six subintervals to estimate the increase in P during the twelve-day period January 2 through 13.

Solution: We estimate the increase in P during these twelve days by calculating

$$\Delta P \approx \sum_{k=1}^{6} \underbrace{P'(t_k^*)}_{\frac{\text{posts}}{\text{day}}} \underbrace{\Delta t}_{\text{days}} \quad \text{where} \quad \Delta t = \frac{12}{6} = 2 \text{ days}$$

and t_k^* is the right-endpoint of the kth subinterval. That is t_1^* is January 4, t_2^* is January 6, and so on. From the graph, we can see that

$$
\begin{aligned}
\Delta P &= P'(t_1^*)\Delta t + P'(t_2^*)\Delta t + P'(t_3^*)\Delta t + P'(t_4^*)\Delta t + P'(t_5^*)\Delta t + P'(t_6^*)\Delta t \\
&= 23000(2) + 30000(2) + 30000(2) + 23000(2) + 26000(2) + 40000(2).
\end{aligned}
$$

So the cumulative number of posts regarding President Obama increased by approximately 344,000 during those 12 days. $\qquad \blacksquare$

Try It Yourself: Using graphs to estimate change

Example 1.24. Using the graph in Figure 1.5, estimate the increase in P during the twelve-day period January 2 through 13 by using a left-sampled Riemann sum with three subintervals.

Answer: $\Delta P \approx 300,000$. $\qquad \blacksquare$

Full Solution On-line

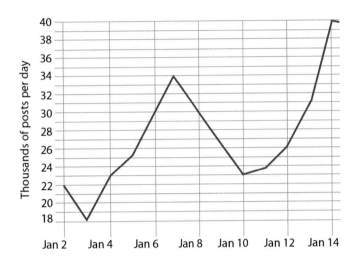

Figure 1.5: The graph for Examples 1.23 and 1.24 reflects interpolated data from trendrr.com (a digital media tracking company) regarding the rate at which posts appeared on twitter.com.

❖ Common Difficulties with Riemann Sums

Some students try to memorize the formulas for t_k^* on p. 335, but it's easy to get them mixed up—especially when there are so many other things to remember (such as the summation formulas on p. 329). Instead, consider memorizing *only* how to find the right endpoint of a subinterval (equation (1.4) on p. 329), and step backward from there as necessary to locate the midpoint or the left endpoint.

Also, some students get confused by the fact that k always ranges from 1 to n in a Riemann sum, regardless of the times at which the integration starts and stops. It happens that way because k tells us which subinterval we're talking about, not what time it is. Whether we're calculating the net change during $1 \leq t \leq 4$ or during $5 \leq t \leq 17$, the index $k = 1$ refers to the first subinterval and $k = n$ refers to the last. Since we always include all subintervals, from first to last, the number k always ranges from 1 to n in a Riemann sum.

> The formula for t_k^* holds information about the start time and the length of each subinterval, from which you can determine the stop time.

You should know

- the terms *sample time (point)*, *Riemann sum*, *left-sampled*, *right-sampled*, *midpoint-sampled*, *total change*, *net change*, and the notation t_k^*;

- that Riemann sums determine net change;

- the summation formulas on p. 329, and the rules of Σ notation on p. 328–330.

You should be able to

- expand Σ notation to demonstrate understanding of what's being added, how many things are being added, and how they depend (or not) on the index of summation;

- write down a formula for t_k^* when left-, right-, or midpoint-sampling;

- use the rules of Σ notation on p. 328–330 and the summation formulas on p. 329 to determine the value of a Riemann sum for a specified number of subintervals;

- perform a dimensional analysis to establish the units associated with a Riemann sum;

- approximate the value of $y(t)$, given $y(t_0)$ and an approximation of Δy.

✧ 5.1 Skill Exercises

Use the summation formulas from p. 329 to calculate each of #1–12.

1. $\displaystyle\sum_{k=1}^{15} 3$

2. $\displaystyle\sum_{k=1}^{8} 4$

3. $\displaystyle\sum_{k=1}^{6} (12k - 7)$

4. $\displaystyle\sum_{k=1}^{135} (3k + 2)$

5. $\displaystyle\sum_{k=1}^{225} \frac{2k + 1}{3}$

6. $\displaystyle\sum_{k=1}^{400} \frac{7k - 3}{5}$

7. $\displaystyle\sum_{k=1}^{90} (4 - k^2)$

8. $\displaystyle\sum_{k=1}^{70} (2 + 3k^2)$

9. $\displaystyle\sum_{k=1}^{60} (k^2 + 30k)$

10. $\displaystyle\sum_{k=1}^{87} (3k^2 - 2k)$

11. $\displaystyle\sum_{k=1}^{12} (3k^2 - 4k + 7)$

12. $\displaystyle\sum_{k=1}^{16} (2 - 8k - 11k^2)$

In each of #13–18 use the appropriate formula from p. 335 to write down t_k^*, given the particular interval, sampling scheme, and number of subintervals.

13. $[0, 4]$, right-sampling, $n = 300$

14. $[0, 5]$, left-sampling, $n = 271$

15. $[3, 7]$, left-sampling, $n = 1000$

16. $[3, 9]$, right-sampling, $n = 5000$

17. $[-5, 2]$, midpoint-sampling, $n = 30$

18. $[-4, 17]$, midpoint-sampling, $n = 50$

In each of #19–22 you should approximate the net change in $y(t)$ over the stated interval using a right-sampled Riemann sum with the indicated number of subintervals.

19. $y'(t) = 4t + 3$; $n = 10$ over $[1, 6]$

20. $y'(t) = 5 - 2t$; $n = 7$ over $[-1, 9]$

21. $y'(t) = 5t^2 + 8$; $n = 6$ over $[0, 3]$

22. $y'(t) = 4t^3$; $n = 8$ over $[2, 5]$

In each of #23–26 you should approximate the net change in $y(t)$ over the stated interval using a left-sampled Riemann sum with the indicated number of subintervals.

23. $y'(t) = 8t + 1$; $n = 6$ over $[2, 3]$

24. $y'(t) = 3 - 5t$; $n = 4$ over $[3, 7]$

25. $y'(t) = 9 - t^2$; $n = 5$ over $[1, 3]$

26. $y'(t) = 6t - t^2$; $n = 3$ over $[0, 2]$

In each of #27–30 you should approximate the net change in $y(t)$ over the stated interval using a midpoint-sampled Riemann sum with the indicated number of subintervals.

27. $y'(t) = 7t - 6$; $n = 4$ over $[10, 18]$ 29. $y'(t) = t^2 - 4$; $n = 8$ over $[-1, 1]$

28. $y'(t) = 2t + 9$; $n = 12$ over $[2, 15]$ 30. $y'(t) = t^3$; $n = 6$ over $[0, 3]$

❖ 5.1 Concept and Application Exercises

31. If r_1, $r_2...r_{50}$ represent the revenues earned from a company's sales force in each state, interpret the expression $\sum_{n=1}^{50} r_n$.

Business

32. If P_1, $P_2...P_{120}$ represent the populations in each of Kentucky's 120 counties, interpret the expression $\sum_{n=1}^{120} P_n$.

Demographics

33. Suppose that t_k is the time it took Lance Armstrong to ride the kth stage in the 2009 Tour de France (there are 21 stages). What does $\sum_{k=1}^{21} t_k$ tell us?

Athletics

34. Suppose that m_k is the mass (in atomic mass units) of the kth atom in sucrose. Determine the value of $\sum_{k=1}^{45} m_k$.

Chemistry

35. Suppose d_k is the distance that racers traveled in the kth stage of the 2009 Tour de France. Use Σ notation to write an expression that describes the total distance traveled by racers who finished all 21 stages.

Athletics

36. Suppose $v_1, v_2...v_{17}$ are the capacities (in gallons) of your company's oil storage facilities. Use Σ-notation to write an expression that describes the total volume of oil that your company can hold at any given time.

Business

37. The rate at which a tank leaks depends on how much is in it (more fluid $\Rightarrow$ more pressure $\Rightarrow$ faster loss). Suppose the leak in a specific tank is described by the following measured data:

Engineering

Time (sec)	0	1.3	4.1	6	7.2	8
Rate (L/sec)	20	14	12	9	7	6

(a) Use right-sampling to estimate the amount of fluid leaked in the first 8 seconds.

(b) Is your answer from part (a) an overestimate or an underestimate? Explain how you can tell.

38. Use the same data as in #37 for this question.

Engineering

(a) Use left-sampling to estimate the amount of fluid leaked in the first 8 seconds.

(b) Is your answer from part (a) an overestimate or an underestimate? Explain how you can tell.

39. The rate of blood flow rises and falls as the heart beats. Consider the following flow data for a given artery:

Medical Science

Time (sec)	0	0.2	0.4	0.6	0.8	1
Rate (cm^3/sec)	0	80	140	55	6	2

(a) Use right-sampling to estimate the volume of fluid that passes through the artery in 1 second.

(b) Use left-sampling to estimate the volume of fluid that passes through the artery in 1 second.

40. The US Census (see [38]) indicates that the average rate of change of population in Monroe County, NY, is as follows:

Demographics

July 1 of ...	2001	2002	2004	2005	2006
Rate (people/year)	357	-99	-2235	-2247	-681.5

(a) Use right-sampling to estimate the net change in population between July 1 of 2001 and 2006.

(b) Use left-sampling to estimate the net change in population between July 1 of 2001 and 2006.

41. Suppose a particular vehicle consumes $f(m)$ $\frac{\text{gallons}}{\text{mi}}$ after traveling m miles in a given trip (f is not constant because it depends on engine temperature, among other things). For each of the cases below calculate Δm, and determine a formula for m_k^*. Then use Σ-notation to write a formula for approximating the amount of fuel used during the first 53 miles of a trip.

     Engineering

  (a) 4000 samples of f, and right-sampling

  (b) 3000 samples of f, and left-sampling

  (c) 2000 samples of f, and midpoint-sampling

42. Suppose a computer performs $f(t)$ floating-point operations per second (flops) t seconds after it starts. For each of the cases below calculate Δt, and determine a formula for t_k^*. Then use Σ-notation to write a formula for approximating the total number of flops between $t = 1$ and $t = 7$.

     Computer Science

  (a) 100 samples of f, and right-sampling

  (b) 200 samples of f, and left-sampling

  (c) 300 samples of f, and midpoint-sampling

43. Suppose a car traveling at 90 kph slows to a full stop in 9 seconds. During that time, its velocity is $v(t) = 90 - 36000t$ $\frac{\text{km}}{\text{hr}}$, where t is measured in hours. Estimate the distance the car traveled while stopping by using a left-sampled Riemann sum with 10 intervals.

     Physics

44. Suppose a car speeds up as it leaves an intersection, and its velocity is $v(t) = 0.05t^2$ miles per hour t seconds after the traffic light turns green. Use a right-sampled Riemann sum with 100 subintervals to estimate the distance traveled by the car in the first 30 seconds after the traffic light turns green.

     Physics

45. Suppose a scuba diver's depth changes at a rate of $v(t) = 1 - (4 - t)^2$ $\frac{\text{ft}}{\text{sec}}$. (a) Estimate her net change in depth when $t \in [0, 4]$ by using a midpoint-sampled Riemann sum with 15 subintervals. (b) If she was 10 ft under water at time $t = 0$, estimate her depth at time $t = 4$.

     Recreation

46. In this exercise we'll denote by $T(x)$ the length of time it takes an ant to dig a horizontal tunnel of length x into a wall of sand, and we'll make the following assumptions: (1) The ant walks into the tunnel, grabs a clump of sand, and walks back to the entrance where he drops it out of the tunnel. Then he walks back in for more. (2) The ant will walk at the same speed whether carrying a clump of sand or not. (3) The cross-sectional area of the tunnel is constant.

     Entomology

  (a) Construct a formula for T in terms of x (it will be a summation, and will depend on both x and various other parameters that you introduce while framing the problem).

  (b) Use your formula to answer the question, "If we double the length of the tunnel, what happens to the time required to dig it?"

Exercises #47–50 refer to the left-hand image in Figure 1.6, which depicts the heart rate of a person during a workout.

47. Use a right-sampled Riemann sum with 4 subintervals to estimate the number of heartbeats that occurred between 10 and 18 minutes into the workout.

48. Use a right-sampled Riemann sum with 4 subintervals to estimate the number of heartbeats that occurred between 12 and 20 minutes into the workout.

49. Use a left-sampled Riemann sum with 10 subintervals to estimate the number of heartbeats that occurred between 10 and 20 minutes into the workout.

50. Use a left-sampled Riemann sum with 3 subintervals to estimate the number of heartbeats that occurred between 11 and 20 minutes into the workout.

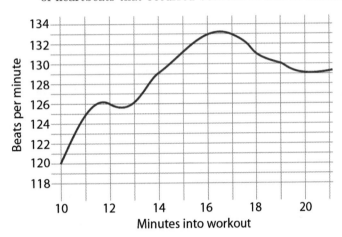 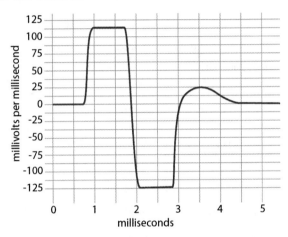

Figure 1.6: (left) For exercises #47–50; (right) for exercises #51–54.

Exercises #51–54 refer to the right-hand image in Figure 1.6, which depicts the rate at which the voltage across a neuron's cell wall changes during an *action potential* (i.e., as it fires).

51. Use a right-sampled Riemann sum with 4 subintervals to estimate the change in voltage that occurs between times $t = 1$ and $t = 3$.

52. Use a right-sampled Riemann sum with 10 subintervals to estimate the change in voltage that occurs between times $t = 0$ and $t = 5$.

53. Use a left-sampled Riemann sum with 6 subintervals to estimate the change in voltage that occurs between times $t = 1$ and $t = 4$.

54. Use a left-sampled Riemann sum with 10 subintervals to estimate the change in voltage that occurs between times $t = 0$ and $t = 5$.

5.2　Area and the Definite Integral

In Section 5.1 we used Riemann sums to approximate the net change in the position of a car, the altitude of a plane, the temperature of a chemical solution, and the demand for a commodity. We can also understand the Riemann sum

$$\sum_{k=1}^{n} f(t_k^*)\Delta t$$

in a geometric sense by looking at the graph of f. The height of each point on the graph is determined by the function value, and Δt is a width on the horizontal axis. For the moment, let's assume that $f(t)$ is never negative, so

$$f(t_k^*)\Delta t = (\text{height}) \times (\text{width}) = \text{area of a rectangle}.$$

Figure 2.1 shows that adding the areas of such rectangles—i.e., a Riemann sum—allows us to approximate the area between the graph of f and the horizontal axis, and *more* subintervals gives us a *better* approximation.

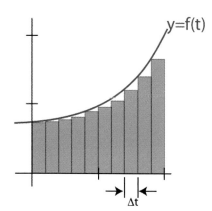
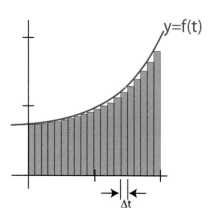

Figure 2.1: (left) A left-sampled Riemann sum with $n = 10$ subintervals approximates the area between the graph of f and the horizontal axis; (right) a left-sampled Riemann sum with $n = 20$ subintervals does an even better job approximating that area.

As we continue to use more and more subintervals, there is less and less error in our approximation (with Riemann sums) of the area between the graph of f and the horizontal axis. In the limit, the error vanishes altogether.

Suppose $f(t)$ is continuous and non-negative over $[a, b]$. Then the area between the horizontal axis and the graph of f over $[a, b]$ is

$$\lim_{n\to\infty} \sum_{k=1}^{n} f(t_k^*)\Delta t,$$

where $\Delta t = (b - a)/n$ and t_k^* denotes a sample point in the kth subinterval of a given Riemann sum.

On p. 350 we'll cite a theorem that guarantees the existence of this limit (because f is continuous). For now, we're focusing on the basic idea.

(Note: this limit can exist even when f has discontinuities.)

Let's begin by verifying this claim with some familiar examples.

Calculating area as the limit of Riemann sums

Example 2.1. Suppose $f(t) = -2 + 4t$. Determine the area between the graph of f and the horizontal axis when $1 \leq t \leq 3$ by using (a) geometry, and (b) the limit of Riemann sums.

Solution: (a) The region between the graph and the horizontal axis can be seen as a triangle that sits atop a rectangle (see Figure 2.2). The height of the rectangle is $f(1) = 2$, and the height of the triangle is $f(3) - f(1) = 11 - 3 = 8$. Since the base has a length of $3 - 1 = 2$, the area of this region is

$$(\text{Total area}) = (\text{area of rectangle}) + (\text{area of triangle}) = (2)(2) + \frac{1}{2}(2)(8) = 12.$$

(b) A right-sampled Riemann sum for this function has the form

$$\sum_{k=1}^{n} f(t_k^*)\Delta t = \sum_{k=1}^{n}(-2 + 4t_k^*)\Delta t = \sum_{k=1}^{n}(-2 + 4(1 + k\Delta t))\Delta t,$$

After distributing factors and gathering like terms, as we did in the previous section, this becomes

$$\sum_{k=1}^{n} 2\Delta t + 4(\Delta t)^2 k \;=\; \sum_{k=1}^{n} 2\Delta t + \sum_{k=1}^{n} 4(\Delta t)^2 k$$

$$=\; \sum_{k=1}^{n} 2\Delta t + 4(\Delta t)^2 \sum_{k=1}^{n} k$$

$$=\; 2n\Delta t + 4(\Delta t)^2 \frac{n(n+1)}{2}.$$

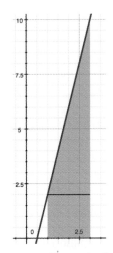

Figure 2.2: The graph of $y = -2 + 4t$. The area we're calculating in Example 2.1 is shaded green.

We're using the summation formula on p. 329:

$$\sum_{k=1}^{n} k = \frac{n(n+1)}{2}.$$

Since we're dividing $[1,3]$ into n subintervals, each has a length of $\Delta t = \frac{3-1}{n} = \frac{2}{n}$. So our Riemann sum becomes

$$\sum_{k=1}^{n} f(t_k^*)\Delta t = 2n\left(\frac{2}{n}\right) + 4\left(\frac{2}{n}\right)^2 \frac{n(n+1)}{2} = 4 + 8\left(\frac{n+1}{n}\right).$$

Therefore, the area between the graph of f and the horizontal axis is

$$\lim_{n\to\infty} \sum_{k=1}^{n} f(t_k^*)\Delta t = \lim_{n\to\infty}\left(4 + 8\left(\frac{n+1}{n}\right)\right) = 4 + 8 = 12,$$

which is what we found in part (a). ■

Try It Yourself: Calculating area as the limit of Riemann sums

Example 2.2. Suppose $f(t) = 1 + 7t$. Determine the area between the graph of f and the horizontal axis when $2 \leq t \leq 13$ by using (a) geometry, and (b) the limit of Riemann sums.

Answer: (a) 588.5 square units, (b) 588.5 square units. ■

Full Solution On-line

The real power of this new method is its ability to compute in more complicated cases (i.e., when simple geometry is insufficient).

Calculating area as the limit of Riemann sums

Example 2.3. Suppose $f(t) = t^2$. Determine the area between the graph of f and the horizontal axis when $0 \leq t \leq 1$ by using the limit of Riemann sums.

Solution: A right-sampled Riemann sum for this function has the form

$$\sum_{k=1}^{n} f(t_k^*)\Delta t = \sum_{k=1}^{n} (t_k^*)^2 \Delta t = \sum_{k=1}^{n} (0 + k\Delta t)^2 \Delta t = \sum_{k=1}^{n} (\Delta t)^3 k^2 = (\Delta t)^3 \sum_{k=1}^{n} k^2.$$

After using equation (1.5) from p. 329 to perform the sum, we're left with

$$\sum_{k=1}^{n} f(t_k^*)\Delta t = (\Delta t)^3 \left(\frac{n(n+1)(2n+1)}{6} \right).$$

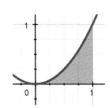

Since we're dividing $[0,1]$ into n subintervals, each has a length of $\Delta t = \frac{1}{n}$. Substituting this into our formula, and taking the limit, we see

Figure 2.3: The graph of $y = t^2$. The area we're calculating in Example 2.3 is shaded green.

$$\lim_{n \to \infty} \sum_{k=1}^{n} f(t_k^*)\Delta t = \lim_{n \to \infty} \left(\frac{1}{n} \right)^3 \left(\frac{n(n+1)(2n+1)}{6} \right) = \frac{1}{3}. \qquad \blacksquare$$

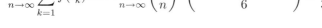

Try It Yourself: Calculating area as the limit of Riemann sums

Example 2.4. Suppose $f(t) = 13 + 2t - t^2$. Determine the area between the graph of f and the horizontal axis when $1 \le t \le 2$ by using the limit of Riemann sums.

Full Solution On-line

Answer: $41/3$. $\qquad\qquad\qquad\qquad\qquad\qquad\qquad\qquad\blacksquare$

We can generalize this technique to determine the area between curves. Suppose that $f(t) > g(t)$ throughout $[a,b]$. Then a rectangle of width Δt that reaches from the graph of f down to the graph of g at t_k^* has a height of

$$(\text{high}) - (\text{low}) = f(t_k^*) - g(t_k^*),$$

so the area between the graphs is

$$A \approx \sum_{k=1}^{n} \Big(f(t_k^*) - g(t_k^*) \Big) \Delta t,$$

as shown in Figure 2.4. As before, we calculate the exact area with a limit.

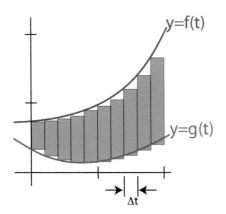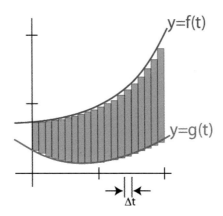

Figure 2.4: (left) A left-sampled Riemann sum with $n = 10$ subintervals approximates the area between the curves $y = f(t)$ and $y = g(t)$; (right) a left-sampled Riemann sum with $n = 20$ subintervals does an even better job approximating that area.

Suppose f and g are continuous functions, and $f(t) > g(t)$ over the interval $[a, b]$. Then the area between the curves $y = f(t)$ and $y = g(t)$ is

$$\lim_{n \to \infty} \sum_{k=1}^{n} \Big(f(t_k^*) - g(t_k^*) \Big) \Delta t,$$

where $\Delta t = (b - a)/n$ and t_k^* denotes a sample point in the kth subinterval of a given Riemann sum.

On p. 350 we'll cite a theorem that guarantees the existence of this limit (because f and g are continuous). For now, we're focusing on the basic idea.

Area between curves

Example 2.5. Determine the area between the graphs of $f(t) = 3 + t^2$ and $g(t) = 2t$ over $[0, 2]$.

Solution: Because the graph of $f(t) > g(t)$ when $t \in [0, 2]$, as seen in Figure 2.5, the Riemann sum

$$\sum_{k=1}^{n} \Big(f(t_k^*) - g(t_k^*) \Big) \Delta t = \sum_{k=1}^{n} \Big(3 + (t_k^*)^2 - 2t_k^* \Big) \Delta t$$

approximates the area, where $\Delta t = \frac{2-0}{n}$. When using a right-sampled Riemann sum, we have $t_k^* = 0 + k\Delta t$, so our approximation is

$$
\begin{aligned}
\sum_{k=1}^{n} \Big(3 + (t_k^*)^2 - 2t_k^* \Big) \Delta t &= \sum_{k=1}^{n} \Big(3 + (k\Delta t)^2 - 2k\Delta t \Big) \Delta t \\
&= \sum_{k=1}^{n} 3\Delta t + k^2 (\Delta t)^3 - 2(\Delta t)^2 k \\
&= \sum_{k=1}^{n} 3\Delta t + (\Delta t)^3 \sum_{k=1}^{n} k^2 - 2(\Delta t)^2 \sum_{k=1}^{n} k \\
&= (3\Delta t)n + (\Delta t)^3 \frac{n(n+1)(2n+1)}{6} - 2(\Delta t)^2 \frac{n(n+1)}{2}.
\end{aligned}
$$

When we rewrite $\Delta t = 2/n$, our approximation of the area becomes

$$6 + \frac{4(n+1)(2n+1)}{3n^2} - \frac{4(n+1)}{n}.$$

The actual area is found in the limit:

$$\text{Area} = \lim_{n \to \infty} \left(6 + \frac{4(n+1)(2n+1)}{3n^2} - \frac{4(n+1)}{n} \right) = \frac{14}{3}. \qquad \blacksquare$$

The word **lamina** means "a thin layer, or plate," of something. For example, the region of the plane in Figure 2.5 might be a lamina. You can see its length and width, but not its thickness because you're looking directly down on it. Figure 2.6 shows the same lamina from another angle, and if we take its thickness and material composition into account we can determine its mass.

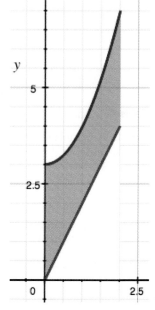

Figure 2.5: The graphs of $f(t) = 3 + t^2$ and $g(t) = 2t$ for Example 2.5.

Calculating mass

Example 2.6. Suppose an aluminum bracing is 0.5 cm thick, and its footprint is shown in Figure 2.5 (where x and y are measured in cm). Determine the mass of the bracing. (Aluminum has a density of $2700 \, \frac{\text{kg}}{\text{m}^3}$.)

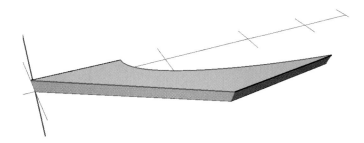

Figure 2.6: The lamina from Figure 2.5 as seen from another angle.

Solution: We can determine the mass of the bracing once we know its volume. We found the area of the footprint to be $14/3$ cm^2 in Example 2.5, so the volume is

$$V = (\text{thickness}) \times (\text{area of the footprint}) = \left(\frac{1}{2}\right) \times \left(\frac{14}{3}\right) = \frac{7}{3} \text{ cm}^3,$$

which is the same as $\frac{7}{3} \times 10^{-6}$ m^3. Therefore, the mass is

$$m = (\text{density}) \times (\text{volume}) = \left(2700\frac{\text{kg}}{\text{m}^3}\right) \times \left(\frac{7}{3} \times 10^{-6} \text{ m}^3\right) = 0.0063 \text{ kg} = 6.3 \text{ g.} \quad \blacksquare$$

Try It Yourself: Area between curves

Example 2.7. Determine the area between the graphs of $f(t) = 8 + t - t^2$ and $g(t) = -3 + 2t^2$ over $[1, 2]$.

Answer: $11/2$. ■

Full Solution
On-line

❖ Net Area

Our discussion so far has been restricted to functions that remain non-negative over the interval in which we're interested, but now let's talk about other cases.

The limit of Riemann sums when $f(t)$ **changes sign**

Example 2.8. Calculate the limit of right-sampled Riemann sums over the interval $[1, 4]$ when $f(t) = 10 - 5t$.

Solution: A right-sampled Riemann sum for this function has the form

$$\sum_{k=1}^{n} f(t_k^*)\Delta t = \sum_{k=1}^{n} \left(10 - 5(1 + k\Delta t)\right)\Delta t = \sum_{k=1}^{n} 5\Delta t - 5(\Delta t)^2 k.$$

When we separate this into two sums and use our summation formulas, we see

$$\sum_{k=1}^{n} f(t_k^*)\Delta t = \sum_{k=1}^{n} 5\Delta t - \sum_{k=1}^{n} 5(\Delta t)^2 k = 5n\Delta t - 5(\Delta t)^2 \left(\frac{n(n+1)}{2}\right).$$

Since we're dividing $[1, 4]$ into n subintervals of equal length, $\Delta t = \frac{3}{n}$. Substituting this into our Riemann sum and calculating the limit as $n \to \infty$ yields

$$\lim_{n \to \infty} \sum_{k=1}^{n} f(t_k^*)\Delta t = \lim_{n \to \infty} \left(5n\left(\frac{3}{n}\right) - 5\left(\frac{3}{n}\right)^2 \left(\frac{n(n+1)}{2}\right)\right)$$

$$= \lim_{n \to \infty} \left(15 - \frac{45}{2}\left(\frac{n+1}{n}\right)\right) = -\frac{15}{2}. \quad \blacksquare$$

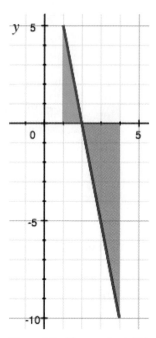

The limit of Riemann sums in Example 2.8 was negative, so it cannot be an area. But then what does it mean? To understand it, look at the graph provided in Figure 2.7. There you see that the line $y = 10 - 5t$ defines two triangles. Using basic geometry, we can determine that the upper triangle has an area of 2.5 square units, and the lower one has an area of 10 square units. The answer in Example 2.8 is simply the difference of these two numbers:

$$-7.5 = 2.5 - 10.$$

We worked with this very function, over this very interval, in Example 1.18 on p. 334. In that physical context, a remote controlled plane rose and then fell, and we said that the *net change* in its altitude was -7.5 feet. Extending that language, we say that -7.5 is the **net area** of $f(t) = 10 - 2t$ over $[1, 4]$. Said loosely,

$$\text{net area} = \begin{pmatrix} \text{the } \textbf{area above} \text{ the hor-} \\ \text{izontal axis bounded by} \\ \text{the graph of } f \end{pmatrix} - \begin{pmatrix} \text{the } \textbf{area below} \text{ the hor-} \\ \text{izontal axis bounded by} \\ \text{the graph of } f \end{pmatrix}.$$

Figure 2.7: The graph of $f(t)$ from Example 2.8.

> **Net Area:** Suppose that $f(t)$ is continuous over the interval $[a, b]$. Then the **net area** bounded by the graph of f over $[a, b]$ is
>
> $$\lim_{n \to \infty} \sum_{k=1}^{n} f(t_k^*)\Delta t,$$
>
> where $\Delta t = (b - a)/n$ and t_k^* denotes a sample point in the kth subinterval of a given Riemann sum.

On p. 350 we'll cite a theorem that guarantees the existence of this limit (because f is continuous). For now, we're focusing on the basic idea.

Try It Yourself: Finding net area

Example 2.9. Determine the net area between the horizontal axis and the graph of $f(t) = -4 + 3t^2$ over $[0, 5]$.

Answer: 105. ■

Full Solution On-line

❖ Net Change

In Section 5.1 we estimated net change by using Riemann sums. For example, when the velocity of an object is $v(t)$, the object's

$$\text{net change in position is} \approx \sum_{k=1}^{n} v(t_k^*)\Delta t.$$

Riemann sums *approximate* net change because they assume a constant rate of change (e.g., velocity) over each span of Δt seconds. But $\Delta t \to 0$ as $n \to \infty$, so in a sense the limit allows us to account for the rate of change at every moment.

Recall that when talking about limits, n never really *is* infinity, so Δt never really *is* zero.

> **Net Change:** Suppose $y(t)$ is a function whose rate of change, $y'(t)$, is continuous over the interval $[a, b]$. Then the **net change** in $y(t)$ over $[a, b]$ is
>
> $$\lim_{n \to \infty} \sum_{k=1}^{n} y'(t_k^*)\Delta t,$$
>
> where $\Delta t = (b - a)/n$ and t_k^* denotes a sample point in the kth subinterval of a given Riemann sum.

Net change in temperature over time

Example 2.10. Suppose the temperature of a chemical solution is changing at a rate of $y'(t) = -3t + 14\ °$F per hour. (a) Calculate the net change in the temperature of the solution between times $t = 4$ and $t = 6$, and (b) determine the solution's temperature at time $t = 6$, assuming its temperature was $104°$F at time $t = 4$.

Solution: This is the same scenario as in Example 1.21 (on p. 336), except that now we want an exact figure rather than an approximation of net change in part (a). As before, we set up a right-sampled Riemann sum, but this time with n subintervals (planning to limit $n \to \infty$):

$$\Delta y \approx \sum_{k=1}^{n} y'(t_k^*)\Delta t = \sum_{k=1}^{n} \left(2\Delta t - 3(\Delta t)^2 k\right) = 2n\Delta t - 3(\Delta t)^2\,\frac{n(n+1)}{2}.$$

Since we're dividing $[4,6]$ into n subintervals of equal size, each has a length of $\Delta t = \frac{2}{n}$. When we substitute this into our Riemann sum we have

$$\Delta y \approx \left(2n\left(\frac{2}{n}\right) - 3\left(\frac{2}{n}\right)^2 \frac{n(n+1)}{2}\right) = 4 - 6\frac{n+1}{n}\ .$$

Now taking the limit as $n \to \infty$,

$$\Delta y = \lim_{n\to\infty} \sum_{k=1}^{n} y'(t_k^*)\Delta t = \lim_{n\to\infty}\left(4 - 6\frac{n+1}{n}\right) = -2.$$

(b) Since the net change in temperature is $\Delta y = -2\ °$F, we have

$$y(6) = y(4) + \overbrace{\Delta y}^{\text{starting temp}} \approx 104 + \underbrace{(-2)}_{\text{change in temp}} = 102\ °\text{F}. \qquad \blacksquare$$

Try It Yourself: Net change in altitude over time

Example 2.11. Suppose the altitude of a radio-controlled airplane is changing at a rate of $y'(t) = 2t - 4$ feet per second. (a) Calculate the net change in the plane's altitude between times $t = 1$ and $t = 5$, and (b) determine the plane's altitude at time $t = 5$, assuming its altitude was 37 feet at time $t = 1$.

Answer: (a) 8 feet, (b) 45 feet. $\blacksquare$

Full Solution
On-line

❖ The Definite Integral

The quantities that we call *net change* and *net area* are calculated using a limit, which leads to some important questions . . .

1. Are we sure the limit converges?

2. If so, does the limit value depend on whether we left-sample, right-sample, midpoint-sample, or choose a different sampling scheme altogether?

When the limit converges and its value is independent of the sampling method, we say that f is **integrable** over $[a, b]$. In the larger scheme of things, integrability depends on both the function $f(t)$ and the interval $[a, b]$. However, the following theorem states a simple condition that guarantees integrability.

> This definition will be superseded by the one on p. 355, in which we also allow for other partitioning schemes.

Theorem 2.7. Suppose that f is a continuous function over $[a, b]$, or has a finite number of jump discontinuities. Then f is integrable over $[a, b]$.

The proof of this theorem is beyond the scope of this course, but can be found in [42].

Note: Functions that have a finite number of jump and/or removable discontinuities are also integrable.

When f is integrable over $[a, b]$, we refer to the number

$$\lim_{n \to \infty} \sum_{k=1}^{n} f(t_k^*) \Delta t$$

as the **definite integral** of f over $[a, b]$. Because the notation of Riemann sums is cumbersome, a shorthand notation has been developed for the definite integral. It begins with the symbol $\int$, called the **integral sign**, that looks like an elongated "S" for "sum:"

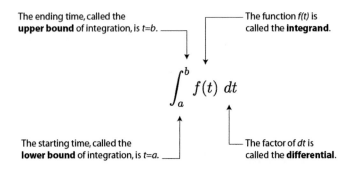

The ending time, called the **upper bound** of integration, is $t=b$.

The function $f(t)$ is called the **integrand**.

$$\int_a^b f(t)\ dt$$

The starting time, called the **lower bound** of integration, is $t=a$.

The factor of dt is called the **differential**.

When we read this notation aloud, we say "the integral of f from a to b." If there's room for confusion we add the phrase, "with respect to time," or "with respect to t."

The term "differential" is more commonly used in the context of differential calculus, but in Section 6.1 you'll see a technique called *substitution* in which dt changes in accord with the ideas we developed in Section 4.1. Beyond that, it's useful to have language for the "dt" so that we can talk about it.

The differential is sometimes "absorbed" into the integrand. For example,

$$\int \frac{1}{t}\ dt$$

is sometimes written as

$$\int \frac{dt}{t}.$$

Note: It's important to recognize that this is only new notation, not a new idea. However, it's *standard* notation, so you need to become well acquainted with it.

Net change in temperature over time

Example 2.12. Suppose the temperature of a chemical solution is changing at a rate of $y'(t) = -3t + 14$ °F per hour. (a) Write the net change in the temperature of the solution between times $t = 4$ and $t = 6$ using integral notation, and (b) use your answer from part (a) to write an equation that relates $y(6)$ to $y(4)$.

Solution: (a) Notice that we're not asked to calculate, but just to write (this is solely practice with the notation). The net change is determined by the limit of Riemann sums:

$$\Delta y = \int_4^6 (-3t + 14)\ dt.$$

In Example 2.10 we determined the value of this integral to be $\Delta y = -2$. (b) Using this notation, the equation we used in part (b) of Example 2.10 can be written as

temperature at time $t = 4$

$$y(6) = \quad y(4) \quad + \int_4^6 (-3t + 14)\ dt. \qquad \blacksquare$$

net change in temperature between times $t = 4$ and $t = 6$

Net change in revenue due to price increase

Example 2.13. Suppose that customer demand for a certain commodity, denoted by $D(p)$, changes at rate of $D'(p) = -40 - \frac{5000}{(p+1)^2} \frac{\text{units}}{\$}$, where p is the cost of the commodity (in dollars). (a) Use integral notation to write down the net change in demand when the producer raises the price of the commodity from \$50 to \$53, and (b) use your answer from part (a) to write an equation that relates $D(53)$ to $D(50)$.

Solution: (a) The net change in demand is determined by the limit of Riemann sums, so we write

$$\Delta D = \int_{50}^{53} \left(-40 - \frac{5000}{(p+1)^2} \right) dp.$$

(b) With this notation in hand, the demand at a price of \$53 is related to the demand at a price of \$50 by the equation

$$D(53) = \overbrace{D(50)}^{\text{demand at a price of \$50}} + \underbrace{\int_{50}^{53} \left(-40 - \frac{5000}{(p+1)^2} \right) dp.}_{\text{net change in demand between price \$50 and \$53}}$$

■

It's common practice to omit the parentheses that delimit the integrand, so that

$$\int_{4}^{6} (-3t + 14) \, dt \quad \text{is written} \quad \int_{4}^{6} -3t + 14 \, dt$$

and

$$\int_{50}^{53} \left(-40 - \frac{5000}{(p+1)^2} \right) dp \quad \text{is written} \quad \int_{50}^{53} -40 - \frac{5000}{(p+1)^2} \, dp.$$

Transitioning between integral and Riemann sum notation

Example 2.14. Write the definite integral $\int_{3}^{7} \sin(8t + 1) \, dt$ as the limit of right-sampled Riemann sums.

Solution: The integral notation tells us that we're working with the function $f(t) = \sin(8t + 1)$, so

$$\int_{3}^{7} \sin(8t + 1) \, dt = \lim_{n \to \infty} \sum_{k=1}^{n} \sin(8t_k^* + 1) \Delta t.$$

The integral notation also tells us that t starts at 3 and ends at 7, so $\Delta t = \frac{4}{n}$. A right-sampling technique would use $t_k^* = 3 + k\Delta t = 3 + \frac{4k}{n}$, so

$$\int_{3}^{7} \sin(8t + 1) \, dt = \lim_{n \to \infty} \sum_{k=1}^{n} \sin\left(8\left(3 + \frac{4k}{n} \right) + 1 \right) \frac{4k}{n}.$$

■

❖ Properties of the Definite Integral

Suppose that f and g are integrable over $[a, b]$, and that m, M, and β are constants. Then ...

1. The region over (or under) the interval $[a, b]$ that lies between the line $y = m$ and the horizontal axis is a rectangle, so the net area calculated by the definite integral is:

$$\int_{a}^{b} m \, dt = m(b - a). \tag{2.1}$$

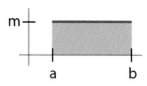

Figure 2.8: Depiction of equation (2.1) when $m > 0$.

> Notice that the lower and upper bounds of integration tell us where the price starts and stops, respectively.

2. Suppose $f(t)$ is never smaller than $m > 0$ over $[a, b]$. Then the region bounded by the graph of f contains the rectangle whose height is m, so

$$\int_a^b f(t)\ dt \geq m(b-a). \tag{2.2}$$

In fact, the assumption that $m > 0$ is not necessary (it just makes the scenario easier to visualize). Equation (2.2) is true whenever $f(t) \geq m$ over $[a, b]$ because each Riemann sum satisfies the inequality

$$\sum_{k=1}^n f(t_k^*)\Delta t \geq \sum_{k=1}^n m\Delta t = nm\Delta t = nm\left(\frac{b-a}{n}\right) = m(b-a).$$

Therefore, the limit of the Riemann sums also obeys this inequality.

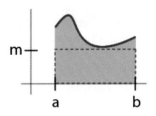

Figure 2.9: Depiction of equation (2.2) when $m > 0$.

3. Suppose $f(t)$ is never larger than $M > 0$ over $[a, b]$. Then

$$\int_a^b f(t)\ dt \leq M(b-a). \tag{2.3}$$

In fact, the assumption that $M > 0$ is not needed (it just makes the scenario easier to visualize). Equation (2.3) is true whenever $f(t) \leq M$ over $[a, b]$ because each Riemann sum satisfies the inequality

$$\sum_{k=1}^n f(t_k^*)\Delta t \leq \sum_{k=1}^n M\Delta t = nM\Delta t = nM\left(\frac{b-a}{n}\right) = M(b-a).$$

Therefore, the limit of the Riemann sums also obeys this inequality.

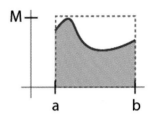

Figure 2.10: Depiction of equation (2.3) when $M > 0$ and $f(t) > 0$.

4. Scaling $f(t)$ by β stretches the graph vertically (and if $\beta < 0$, it reflects the graph over the horizontal axis). The effect is to scale the net area by β, so

$$\int_a^b \beta f(t)\ dt = \beta \int_a^b f(t)\ dt. \tag{2.4}$$

5. Similarly, if $g(t) \leq f(t)$ over $[a, b]$,

$$\int_a^b g(t)\ dt \leq \int_a^b f(t)\ dt. \tag{2.5}$$

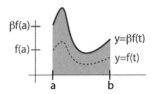

Figure 2.11: Depiction of equation (2.4) when $\beta > 1$ and $f(t) > 0$ over $[a, b]$.

6. If $f(t)$ changes sign somewhere in $[a, b]$, the net area is smaller than it would be if $f(t)$ were forced to remain positive the whole time, so

$$\left| \int_a^b f(t)\ dt \right| \leq \int_a^b |f(t)|\ dt \tag{2.6}$$

7. Because the order of addition doesn't matter, we can rewrite

$$\sum_{k=1}^n \Big(f(t_k^*) \pm g(t_k^*) \Big)\Delta t \quad \text{as} \quad \sum_{k=1}^n f(t_k^*)\Delta t \pm \sum_{k=1}^n g(t_k^*)\Delta t$$

(i.e., first adding all the terms involving f, and then all those involving g). Then using the fact that the limit of a sum (as $n \to \infty$) is the sum of the limits, we get

$$\int_a^b f(t) \pm g(t)\ dt = \int_a^b f(t)\ dt \pm \int_a^b g(t)\ dt. \tag{2.7}$$

We need both limits to exist in order to use the fact that "the limit of a sum is the sum of the limits." We know that both of our limits do exist because f and g are both integrable.

8. Suppose $f(t)$ is a velocity. When we add the net change in position that occurs during $[1, 2]$ to the net change that occurs during $[2, 4]$, we get the net change in position between times $t = 1$ and $t = 4$. Written in integral notation,

> You saw a geometric discussion of this same fact on p. 348 during the discussion of Figure 2.7.

$$\int_1^2 f(t)\ dt + \int_2^4 f(t)\ dt = \int_1^4 f(t)\ dt.$$

Said more generally, when f is integrable over an interval that contains times $a, b,$ and c

$$\int_a^b f(t)\ dt + \int_b^c f(t)\ dt = \int_a^c f(t)\ dt. \tag{2.8}$$

Though we arrived at equation (2.8) by considering the case when $a < b < c$, it's important to note that the equation is true even when those times are ordered differently.

9. Imagine that you have a slider that controls the position of an object on the y axis. The object starts at a height of $y = 2$ and ends up at $y = 8$, as you push the slider left-to-right from $t = a$ to $t = b$, so the net change in position is $8 - 2 = 6$ units.

t=a   t=b

Then when you move move the slider backwards, from $t = b$ to $t = a$, the object moves "backwards" along its trajectory, from a height of 8 to a height of 2, which is a net change of -6 units on the y-axis. Said in integral notation,

> You can also understand this fact algebraically by way of the Riemann sums. If time is increasing from $t = a$ to $t = b$ in n steps of equal size, each has a length of
>
> $$\Delta t = \frac{b - a}{n} > 0$$
>
> If we're stepping backwards from $t = b$ to $t = a$ with n steps of equal length, each step is
>
> $$\Delta t = \frac{a - b}{n} < 0.$$
>
> So the Riemann sums that we write down, stepping to the right and stepping to the left, will be the same except for a negative sign in the Δt.

$$\underbrace{\int_b^a y'(t)\ dt}_{\substack{\text{The net change that happens when} \\ \text{we pull the slider backwards}}} = -\underbrace{\int_a^b y'(t)\ dt}_{\substack{\text{The net change that happens when} \\ \text{the slider is pushed forwards}}}$$

More generally, whenever $f(t)$ is integrable over $[a, b]$,

$$\int_b^a f(t)\ dt = -\int_a^b f(t)\ dt.$$

At a purely algebraic level, swapping the bounds of integration negates the value of a definite integral.

❖ Partitioning

Throughout the entire chapter we've talked about subintervals of *equal length*, but it's not mathematically necessary that all subintervals have the same length. In fact, when working with Riemann sums, it's often advantageous to determine the length of a subinterval based on how steep the graph of the integrand is. For example, in Figure 2.12 you see a depiction of two Riemann sums that are being used to approximate

$$\int_0^5 e^{\cos(t)}\ dt.$$

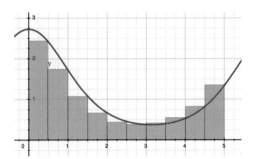 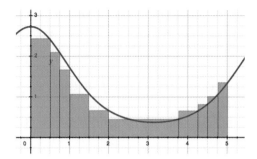

Figure 2.12: (left) A right-sampled Riemann sum with a regular partition; (right) a right-sampled Riemann sum in using a partition that is not regular.

Both have the same number of subintervals, but the Riemann sum depicted on the right is closer to the actual value of the integral because it samples the function more frequently where it's changing quickly.

Whenever we divide $[a, b]$ into subintervals, the endpoints of those subintervals are called **partition points**, and we label them in increasing order from 0 to n,

$$a = t_0 < t_1 < t_2 < \cdots < t_{n-1} < t_n = b.$$

We say the set $\{t_1, t_2, \ldots, t_n\}$ is a **partition** of $[a, b]$, and if the length of all the subintervals is the same, we say it's a **regular** partition of $[a, b]$. We can extend our notation for Riemann sums to allow for subintervals of different lengths by writing

$$\sum_{k=1}^{n} f(t_k^*) \underbrace{(t_k - t_{k-1})}_{\text{the length of the } k^{\text{th}} \text{ subinterval}}.$$

We're using the word *regular* in the sense of "recurring at uniform intervals," not in the sense of "usual" or "customary."

While non-uniform subintervals allow us to make better approximations with Riemann sums, the idea necessitates a slight change the way we discuss and define the definite integral. Instead of talking about the *number* of subintervals in a partition, we talk about their *size*. A partition's **norm**, which is denoted by $\|\Delta t\|$, is defined to be

$$\|\Delta t\| = \text{the maximum value of } \underbrace{|t_k - t_{k-1}|}_{\text{length of the } k^{\text{th}} \text{ subinterval}}, \quad k = 1 \ldots n.$$

If the subintervals of a partition are allowed to have different sizes, we can achieve the goal of having more and more subintervals by picking one and subdividing it over and over again, but that doesn't make our approximation across the rest of $[a, b]$ any better.

The norm of a partition

Example 2.15. Determine the value of $\|\Delta t\|$ if $\{t_0, t_1, \ldots, t_{30}\}$ is a regular partition of the interval $[2, 5]$.

Solution: The norm of a partition is simply the length of the longest subinterval. Since this is a regular partition, all of the subintervals have a length of $\Delta t = \frac{3}{30} = 0.1$. Therefore, $\|\Delta t\| = 0.1$. ∎

The norm of a partition

Example 2.16. Determine $\|\Delta t\|$ for both partitions shown in Figure 2.13.

Solution: Although the partition used in the right-hand Riemann sum has *more* subintervals, the largest subinterval in both partitions is the one that stretches from $t = 2$ to $t = 4$, so $\|\Delta t\| = 2$ in both cases. ∎

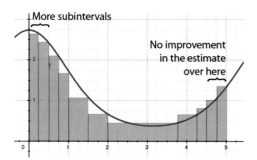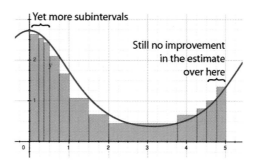

Figure 2.13: When we use irregular partitions, increasing the number subintervals no longer implies that all of the subintervals get smaller—we can make *more* subintervals by mincing one that we already have, without touching any of the others.

In order for Riemann sums to approach the net area in the limit, the widths of subintervals should vanish in the limit. That's the same as saying that $\|\Delta t\| \to 0$, so we write the definite integral as

$$\underbrace{\lim_{\|\Delta t\| \to 0} \sum_{k=1}^{n} f(t_k^*)\,(t_k - t_{k-1}).}_{n \to \infty \text{ when } \|\Delta t\| \to 0}$$

When this limit converges to the same number, regardless of the partitioning that's used and regardless of the sampling scheme, we say that f is **integrable** over $[a, b]$, and refer to the limit as the **definite integral** of f over $[a, b]$. As before, we use integral notation to denote the limit, so

$$\int_a^b f(t)\,dt = \lim_{\|\Delta t\| \to 0} \sum_{k=1}^{n} f(t_k^*)\,(t_k - t_{k-1}).$$

You might be wondering, "Will I get a different number doing *this* instead of using regular partitions?" Since neither the partitioning nor the sampling scheme affect this limit when f is integrable, the limit value will be the same. We close the section by noting that all continuous functions over $[a, b]$ are integrable in this new sense.

> In the exercise set, you'll show that limiting $\|\Delta t\| \to 0$ is equivalent to limiting $n \to \infty$ when dealing with regular partitions.

> This definition is meant to supersede the introduction on p. 349.

You should know

- the terms *lamina, net area, integrable, definite integral, integral sign, upper (and lower) bound of integration, integrand, differential, partition, partition points, regular partition,* and *norm (of a partition);*

- the definite integral is the limit of Riemann sums;

- the definite integral can be interpreted as a net area, or as a net change;

- the basic properties of the definite integral.

You should be able to

- calculate net area (change) as the limit of Riemann sums;

- calculate the area between the graphs of two functions;

- rewrite integral notation in as the limit of a Riemann sum, and vice versa;

- determine the norm of a given partition.

❖ 5.2 Skill Exercises

Write each of the limits in #1–4 using integral notation.

1. $\lim_{n\to\infty} \sum_{k=1}^{n} \sin\left(3 + \frac{5k}{n}\right) \frac{5}{n}$

2. $\lim_{n\to\infty} \sum_{k=1}^{n} \cos\left(\frac{\pi}{2} + \frac{\pi k}{n}\right) \frac{\pi}{n}$

3. $\lim_{n\to\infty} \sum_{k=1}^{n} \ln\left(2 + \frac{10k}{n}\right) \frac{5}{n}$

4. $\lim_{n\to\infty} \sum_{k=1}^{n} \exp\left(3 + \frac{9k}{n}\right) \frac{27}{n}$

Write each of the integrals in #5–10 as the limit of Riemann sums.

5. $\int_{1}^{6} \ln(t)\, dt$, right-sampled

6. $\int_{0}^{\pi} \sin(3t)\, dt$, right-sampled

7. $\int_{1}^{7} \ln(4 + \cos(t))\, dt$, left-sampled

8. $\int_{2}^{6} \cos(3 + \ln(t))\, dt$, left-sampled

9. $\int_{-1/2}^{1/2} \tan(4t + 1)\, dt$, midpoint-sampled

10. $\int_{-3}^{\pi} \cos(\sin(t))\, dt$, midpoint-sampled

11. Calculate $\int_{0}^{4} t^2$ as the limit of Riemann sums using (a) right-sampling, and (b) left-sampling.

12. Calculate $\int_{7}^{10} t$ as the limit of Riemann sums using (a) right-sampling, and (b) midpoint-sampling.

13. Use the definition of the definite integral as the limit of Riemann sums to calculate (a) $\int_{1}^{2} 5t\, dt$, (b) $\int_{1}^{2} t^2\, dt$, and (c) $\int_{1}^{2} 5t + t^2\, dt$.

14. Use the definition of the definite integral as the limit of Riemann sums to calculate (a) $\int_{2}^{4} 5t^2\, dt$, and (b) $\int_{2}^{4} t^2\, dt$. Then compare them.

❖ 5.2 Concept and Application Exercises

15. Suppose $q(t)$ is the number of calls received per hour, t hours after a company's customer service center opens at 8:00 AM each day. Use the definite integral to write an equation that says, "The customer service center received 205 calls between 9 AM and noon." | Business |

16. Suppose $q(t)$ is the number of calls received per hour, t hours after a company's customer service center opens at 8:00 AM each day. Use the definite integral to write an equation that we would solve to answer the question, "How many hours did it take before we fielded 312 calls?" | Business |

17. It's often helpful to stratify a population based on age—newborns, toddlers, ..., senior citizens, centenarians—because diseases (among other things) affect them differently. Suppose the population of a city is $\int_{0}^{150} f(a)\, da$ people, where a is measured in years (and $f(a)$ is zero when a is large). What are the units associated with $f(a)$ and what does it mean to us? | Sociology |

18. Suppose the net electric charge on a rod is $\int_{0}^{1} \sigma(x)\, dx$ coulombs, where x is measured in meters. What are the units associated with $\sigma(x)$ and what does it mean to us? | Physics |

19. Suppose that a brown and a gray horse are running, and the gray is faster. We'll denote the speed of the brown horse by $b(t)$, and the speed of the gray horse by $g(t)$, both measured in meters per second. We know from the properties of integrals that

$$b(t) \leq g(t) \text{ on } [0, 60] \quad \Rightarrow \quad \int_0^{60} b(t) \, dt \leq \int_0^{60} g(t) \, dt.$$

Explain what this means in the context of the running horses.

20. Suppose that a sales representative earns at least $\$m$ per day. Denote the rate of her sales (in dollars) by $g(t)$. We know from the properties of integrals that

$$m \leq g(t) \text{ on } [0, 30] \quad \Rightarrow \quad 30m \leq \int_0^{30} g(t) \, dt.$$

Explain what this fact means in the context of the sales representative.

Each of #21–24 refers to the graph of f shown in the left-hand image of Figure 2.14.

21. Estimate $\int_0^8 f(t) \, dt$ using a right-sampled Rieman sum with $\Delta t = 2$.

22. Estimate $\int_2^3 f(t) \, dt$ by interpreting the number as net area.

23. Estimate $\int_3^4 f(t) \, dt$ by interpreting the number as net area.

24. Estimate $\int_4^6 f(t) \, dt$ by interpreting the number as a net area.

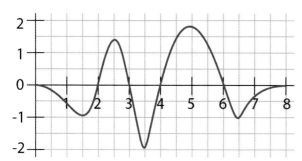 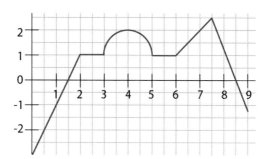

Figure 2.14: (left) Graph for #21–24; (right) graph for #25–30.

Each of #25–30 refers to the graph of f shown in the right-hand image of Figure 2.14.

25. Calculate $\int_1^2 f(t) \, dt$

26. Calculate $\int_8^9 f(t) \, dt$

27. Calculate $\int_2^3 f(t) \, dt$

28. Calculate $\int_5^6 f(t) \, dt$

29. Calculate $\int_6^8 f(t) \, dt$

30. Calculate $\int_0^5 f(t) \, dt$

In each of #31–34 find the total area enclosed by the graph of f and the horizontal axis over the given interval.

31. $f(t) = t^2 - 4$, $[3, 7]$

32. $f(t) = t^3 + 1$, $[1, 2]$

33. $f(t) = t^2 - 4$, $[0, 5]$

34. $f(t) = t^3 + 1$, $[-2, 0]$

In each of #35–40 (a) sketch the graphs of the given functions over the specified interval, and (b) use the definition of the definite integral as the limit of Riemann sums to determine the area between them.

35. $f(t) = 3t + 4$, $g(t) = \frac{t}{2}$, $[0, 5]$

36. $f(t) = -2t + 12$, $g(t) = -t + 1$, $[0, 3]$

37. $f(t) = 3t + 12$, $g(t) = -2t^2 + 10$, $[1, 4]$

38. $f(t) = 13t^2 - 1$, $g(t) = 2t - 8$, $[2, 7]$

39. $f(t) = -t^2 + 9$, $g(t) = -2t^2 + 8$, $[-2, 2]$

40. $f(t) = 2t^2 + 1$, $g(t) = -2t^2 - 1$, $[-3, 3]$

41. Determine the value of $\int_{-3}^{3} \sqrt{9 - t^2}\, dt$. (*Hint: draw a graph of the integrand.*)

42. Consider the function $f(t) = 0.5 + \sin(t)$.

    (a) Use a graphing utility to render a graph of f.

    (b) Find numbers a and b so that $\int_a^b f(t)\, dt$ is negative.

    (c) Find numbers a and b so that $\int_a^b f(t)\, dt$ is positive.

    (d) Find numbers a and b so that $\int_a^b f(t)\, dt$ is zero.

43. Suppose the velocity of a flock of migrating birds is $f(t)$, the graph of which is shown in the left-hand image of Figure 2.15. (a) Approximate the distance traveled by the flock during the seven-hour time period shown, and (b) explain in complete sentences how you arrived at your answer.

Biology

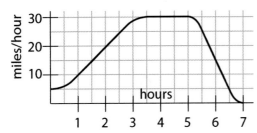
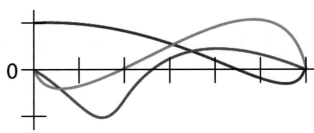

Figure 2.15: (left) Graph for #43; (right) graph for #44.

44. Suppose a security guard enters a hallway through a door. He walks the hallway, and returns to the door whence he entered. Determine which of the graphs in the right-hand image of Figure 2.15 most likely depicts his velocity (ft/min) during that tour of the hallway, and explain your reasoning.

45. Calculate $\int_0^5 \lceil 2t \rceil\, dt$, where $\lceil x \rceil$ denotes the least integer that's greater than or equal to x (e.g., $\lceil \pi \rceil = 4$ and $\lceil 7 \rceil = 7$). (*Hint: sketch the graph of the integrand.*)

46. Calculate $\int_0^2 \lfloor t^2 \rfloor\, dt$, where $\lfloor x \rfloor$ denotes the greatest integer that's less than or equal to x (e.g., $\lfloor \pi \rfloor = 3$ and $\lfloor 7 \rfloor = 7$).

47. Suppose f is an odd function. Explain why $\int_{-a}^{a} f(t)\, dt = 0$.

48. Suppose f is an even function. Explain why $\int_{-a}^{a} f(t)\, dt = 2 \int_0^a f(t)\, dt$.

Checking the details

49. Explain why, when working with regular partitions, calculating the limit as $n \to \infty$ is the same as calculating as $\|\Delta t\| \to 0$.

50. Explain why, when working with regular partitions, $\|\Delta t\| \to 0$ can approach zero *only if* $n \to \infty$.

5.3   Numerical Methods of Integration

In Section 5.2 we said that all continuous functions are integrable, meaning that the limit of Riemann sums exists and that it's independent of both the sampling scheme and partitioning method we use. However, knowing that a limit *exists* is different than knowing what it is! For example, if we wanted to know the area that's bounded between the graph of $f(t) = \ln(2 + \sin(t))$ and the horizontal axis over $[1, 5]$, we would write

$$\int_1^5 \ln(2 + \sin(t)) \, dt = \lim_{n \to \infty} \sum_{k=1}^{n} \ln\left(2 + \sin\left(1 + \frac{4k}{n}\right)\right) \frac{4}{n},$$

but none of our summation formulas address the case when k is "caught inside" a sine, a logarithm, or both. Since we're unable to determine this limit analytically, we can only approximate it, and that leads to an important question. How many subintervals do we need in order to make our approximation good? In general, the answer to that questions depends on what function you're using, and what you mean by "good."

Throughout our discussion, we'll use the term *error* to mean the discrepancy between our approximation and the actual value of the integral, and the error tolerance will always be specified for you. The relevant question will be how many subintervals you need in order to approximate a definite integral to that given accuracy.

❖ Riemann Sums: Endpoint Sampling

A Riemann sum typically *approximates* $\int_a^b f(t) \, dt$ because it treats $f(t)$ as constant across subintervals, even though the function might increase (or decrease). For example, in the middle image of Figure 3.2 you see a case in which a left-sampled Riemann sum underestimates the actual area that's bounded between the graph of a linear function and the horizontal axis.

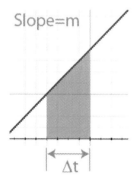 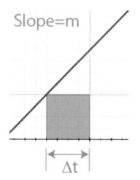 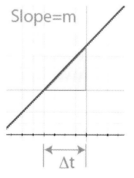

Figure 3.2: (left) The actual area between the graph of a linear function and the horizontal axis, over a single subinterval; (middle) a rectangle from a left-sampled Riemann sum; (right) the amount by which the Riemann sum underestimates the actual area is the area of a triangle; (note) a right-sampled Riemann sum *overestimates* the area bounded by a decreasing function.

In this example of a linear function, the Riemann sum neglects the area of a triangle in each subinterval. The base of that triangle has a length of Δt, and if the graph of f has a slope of m, the height of the triangle is $|m|\Delta t$, so its area is

$$\frac{1}{2}(\text{base}) \times (\text{height}) = \frac{1}{2}|m|(\Delta t)^2.$$

Figure 3.1: The area that's bounded between the graph of $f(t) = \ln(2 + \sin(t))$ and the horizontal axis.

Depending on what you're approximating, an estimation that's within 0.01 units of the actual value might be good enough, but your error tolerance might be only 0.00001 units, or less . . .

By *endpoint-sampling* we mean to address left-sampled and right-sampled Riemann sums.

Recall that slope means "rise over run," so $m\Delta t$ is

$$\left(\frac{\text{rise}}{\text{run}}\right)(\text{run}) = \text{rise}.$$

To determine the overall error in an endpoint-sampled (i.e., left or right-sampled) Riemann sum, we simply add the areas of the triangles from all the subintervals:

$$\sum_{k=1}^{n} \frac{1}{2}|m|(\Delta t)^2 = n\frac{1}{2}|m|(\Delta t)^2 = n\frac{1}{2}|m|\left(\frac{b-a}{n}\right)^2 = \frac{|m|(b-a)^2}{2n}. \tag{3.1}$$

We use $|m|$ instead of m because the error depends on how *steep* the line is, not on whether it has a positive or negative slope.

Note that the error gets worse when the line is steeper (i.e., larger m), but you could compensate by using more subintervals. You should convince yourself of this geometrically by graphing several examples with steeper and steeper lines.

We use the same basic technique to estimate error when f is nonlinear, but we have to account for the fact that $f'(t)$ is not constant. Since the error in our approximation depends on how steep the graph is, when we use an endpoint-sampled Riemann sum, we ask the question, "How big can $|f'(t)|$ get?" (We use absolute value because we don't care whether the graph is rising or falling, but only how quickly.)

Suppose we find that $|f'(t)|$ never exceeds some number, which we'll call M_1, throughout $[a, b]$. From the standpoint of accumulating error, the worst-case scenario would be that $|f'(t)|$ is *always* M_1. By assuming that worst-case scenario, we overestimate the actual amount of error in our approximation using the same "triangle" argument that led us to equation (3.1).

The notation used here is intended to remind you of its meaning: "M" for <u>m</u>aximum <u>m</u>agnitude, and 1 for "first derivative."

> **Error estimation for endpoint-sampled Riemann sums:** Denote by $E(n)$ the value of an endpoint-sampled Riemann sum with n subintervals that approximates $\int_a^b f(t)\, dt$, and suppose that $|f'(t)| \leq M_1$ for every $t \in [a, b]$. Then $\int_a^b f(t)\, dt$ differs from $E(n)$ by no more than $\frac{M_1(b-a)^2}{2n}$. Written technically,
>
> $$E(n) - \frac{M_1(b-a)^2}{2n} \leq \int_a^b f(t)\, dt \leq E(n) + \frac{M_1(b-a)^2}{2n}.$$

We know that $\int_a^b f(t)\, dt$ is within $M_1(b-a)^2/2n$ of our estimate, $E(n)$. So it could be as large as $E(n)+$ that amount, or as small as $E(n)-$ that amount.

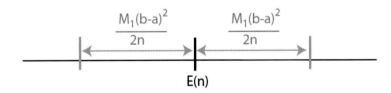

Figure 3.3: The value of $E(n)$ shown on the number line. The error formula cited above guarantees that the number $\int_a^b f(t)\, dt$ is within $M_1(b-a)^2/2n$ of $E(n)$, but doesn't tell us whether it's larger than or less than our approximation.

Guaranteeing accuracy using error estimation

Example 3.1. How many subintervals are needed to guarantee that a right-sampled Riemann sum is within 10^{-4} of $\int_1^5 \ln(2 + \sin(t))\, dt$?

Solution: A moment ago we said that $E(n)$ differs from $\int_a^b f(t)\, dt$ by no more than $\frac{M_1(b-a)^2}{2n}$. So our basic strategy will be to choose n large enough that

$$\frac{M_1(b-a)^2}{2n} \leq \frac{1}{10000}.$$

We're using $M_1(b-a)^2/2n$ as a proxy for the actual discrepancy between the integral and the Riemann sum.

In this case, $a = 1$, and $b = 5$. To determine M_1 we examine the derivative of $f(t) = \ln(2 + \sin(t))$. Since $f'(t) = \cos(t)/(2 + \sin(t))$ never exceeds 1, let's use

$M_1 = 1$. Then to achieve

$$\frac{1 \cdot (5-1)^2}{2n} \leq \frac{1}{10000} \quad \text{we need} \quad n \geq 80000.$$ ∎

Try It Yourself: Guaranteeing accuracy using error estimation

Example 3.2. How many subintervals are needed to guarantee that a left-sampled Riemann sum is within 10^{-3} of $\int_0^5 e^{\cos(t)} \, dt$?

Answer: $n = 18750$. ∎

Full Solution On-line

✥ Trapezoid Rule

When f is an increasing function, as you saw in Figure 3.2, left-sampled Riemann sums underestimate the net area, and right-sampled Riemann sums overestimate it. With one approximation too high and the other too low, we might get a better estimate of the actual value by averaging the results.

right-sampling: $\displaystyle\int_a^b f(t) \, dt \approx \overbrace{f(t_1)\Delta t}^{\text{1st subinterval}} + \overbrace{f(t_2)\Delta t}^{\text{2nd subinterval}} + \cdots\cdots\cdots + \overbrace{f(t_n)\Delta t}^{n^{\text{th}} \text{ subinterval}}$

left-sampling: $\displaystyle\int_a^b f(t) \, dt \approx \overbrace{f(t_0)\Delta t}^{\text{1st subinterval}} + \overbrace{f(t_1)\Delta t}^{\text{2nd subinterval}} + \cdots\cdots\cdots + \overbrace{f(t_{n-1})\Delta t}^{n^{\text{th}} \text{ subinterval}}$

average: $\displaystyle\int_a^b f(t) \, dt \approx \underbrace{\frac{f(t_0)+f(t_1)}{2}\Delta t}_{\text{1st subinterval}} + \underbrace{\frac{f(t_1)+f(t_2)}{2}\Delta t}_{\text{2nd subinterval}} + \cdots + \underbrace{\frac{f(t_{n-1})+f(t_n)}{2}\Delta t}_{n^{\text{th}} \text{ subinterval}}$

Because $\frac{f(t_{n-1})+f(t_n)}{2}\Delta t$ is the area of a trapezoid (as you'll show in the exercise set), we refer to this average or the right and left-sampled Riemann sums as the **trapezoid rule**. If we denote by $T(n)$ the trapezoid-rule approximation of $\int_a^b f(t) \, dt$ that uses n subintervals,

$$T(n) = \frac{\Delta t}{2}\left(f(t_0) + 2f(t_1) + 2f(t_2) + \cdots + 2f(t_{n-1}) + f(t_n)\right).$$

> Factor a $\Delta t/2$ out of the "average" approximation above.

Using the trapezoid rule

Example 3.3. Use the trapezoid rule with $n = 6$ to estimate $\int_0^\pi \sin(t) \, dt$.

Solution: Since $n = 6$ we have $\Delta t = \frac{\pi - 0}{6} = \frac{\pi}{6}$, and the partition points are

$$t_0 = 0, \quad t_1 = \frac{\pi}{6}, \quad t_2 = \frac{\pi}{3}, \quad t_3 = \frac{\pi}{2}, \quad t_4 = \frac{2\pi}{3}, \quad t_5 = \frac{5\pi}{6}, \quad \text{and} \quad t_6 = \pi.$$

So the trapezoid rule tells us that

$$\int_0^\pi \sin(t) \, dt \approx \frac{\pi/6}{2}\left(\sin(0) + 2\sin\left(\frac{\pi}{6}\right) + 2\sin\left(\frac{\pi}{3}\right) + 2\sin\left(\frac{\pi}{2}\right) + 2\sin\left(\frac{2\pi}{3}\right) + 2\sin\left(\frac{5\pi}{6}\right) + \sin(\pi)\right),$$

which is $\frac{\pi}{6}(4 + 2\sqrt{3}) \approx 3.908194$. ∎

Full Solution
On-line

Try It Yourself: Using the trapezoid rule

Example 3.4. Use the trapezoid rule with $n = 4$ to estimate $\int_3^5 2t + 3 \, dt$.

Answer: 22. ∎

Because the region between the graph of a linear function and the horizontal axis *is* a trapezoid (see the left-hand image in Figure 3.2), the trapezoid rule tells us the *exact* value of the definite integral when $f(t)$ is a linear function. So the error in a trapezoid-rule approximation depends on the extent to which f is not linear—i.e., the extent to which f' is not constant. We quantify the rate at which f' changes by using $\frac{d}{dt} f' = f''$, so you might expect the error formula for the trapezoid rule to depend on the second derivative. It does.

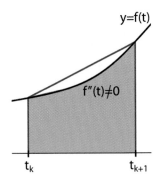

y=f(t)

$f''(t) \neq 0$

t_k t_{k+1}

Figure 3.4: The area bounded by the graph of f is not the area of a trapezoid when $f''(t) \neq 0$.

Error estimation for the trapezoid rule: Denote by $T(n)$ the value of a trapezoid rule approximation of $\int_a^b f(t) \, dt$ with n subintervals, and suppose that $|f''(t)|$ never exceeds M_2 when $t \in [a, b]$. Then $\int_a^b f(t) \, dt$ differs from $T(n)$ by no more than $\frac{M_2(b-a)^3}{12n^2}$. Written technically,

$$T(n) - \frac{M_2(b-a)^3}{12n^2} \;\leq\; \int_a^b f(t) \, dt \;\leq\; T(n) + \frac{M_2(b-a)^3}{12n^2}.$$

Guaranteeing accuracy using error estimation

Example 3.5. How many subintervals are needed to guarantee that a trapezoid rule approximation is within 10^{-4} of $\int_1^5 \ln(2 + \sin(t)) \, dt$?

Solution: A moment ago we said that $T(n)$ differs from the integral value by no more than $\frac{M_2(b-a)^3}{12n^2}$. So our basic strategy is to choose n large enough that

$$\frac{M_2(b-a)^3}{12n^2} \leq \frac{1}{10000},$$

> We're using $M_2(b-a)^3/12n$ as a proxy for the actual discrepancy between the integral and the Riemann sum.

where $a = 1$ and $b = 5$. To determine M_2 we examine the second derivative of $f(t) = \ln(2 + \sin(t))$. We know from Example 3.1 that $f'(t) = \cos(t)/(2 + \sin(t))$, so using the Quotient Rule we see that

$$f''(t) = \frac{-2\sin(t) - 1}{(2 + \sin(t))^2}.$$

Figure 3.5 shows the graph of f'', and you can see that $|f''(t)|$ never exceeds 1, so let's use $M_2 = 1$. Then our inequality is

> You can check by finding the critical points of f''.

$$\frac{1 \cdot (5-1)^3}{12n^2} \leq \frac{1}{10000} \quad \Rightarrow \quad n \geq \sqrt{\frac{160000}{3}} = 230.9401.$$

By choosing $n = 231$ we guarantee that $\frac{M_2(b-a)^3}{12n^2}$ is smaller than 10^{-4}. Since the actual discrepancy between the integral and a right-sampled Riemann sum is smaller than that fraction, it will also be less than 10^{-4}. ∎

Try It Yourself: Guaranteeing accuracy using error estimation

Example 3.6. How many subintervals are needed to guarantee that an approximation of $\int_0^5 e^{\cos(t)} \, dt$ using the trapezoid rule is within 10^{-3} of the acutal value?

Answer: $n = 170$, using $M_2 = 2.75$. ∎

Full Solution
On-line

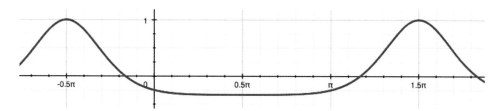

Figure 3.5: The graph of f'' for Example 3.7.

❖ Riemann Sums: Midpoint Sampling

When f is a linear function, each rectangle from a midpoint-sampled Riemann sum underestimates area in one half of the subinterval, and overestimates area in the other half. Because f is linear, these errors sum to *zero* (see Figure 3.6).

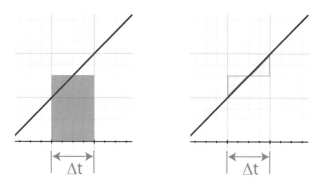

Figure 3.6: (left) A single rectangle from a midpoint-sampled Riemann sum; (right) triangles show how the rectangle overestimated in one half of the subinterval, and underestimated in the other half—by exactly the same amount!

Consequently, a midpoint-sampled Riemann sum tells the exact value of a definite integral when $f(t)$ is linear. That is, it tells us the exact value of a definite integral when $f'(t)$ is constant. In general, the error in a midpoint-sampled Riemann sum depends on the extent to which $f'(t)$ is *not* constant, which is why the formula below relies on f''.

> **Error estimation for midpoint-sampled Riemann sums:** Denote by $M(n)$ the value of a midpoint-sampled Riemann sum with n subintervals that approximates $\int_a^b f(t)\, dt$, and suppose that $|f''(t)|$ never exceeds M_2 when $t \in [a, b]$. Then $\int_a^b f(t)\, dt$ differs from $M(n)$ by no more than $\frac{M_2(b-a)^3}{24n^2}$. Written technically,
>
> $$M(n) - \frac{M_2(b - a)^3}{24n^2} \;\leq\; \int_a^b f(t)\, dt \;\leq\; M(n) + \frac{M_2(b - a)^3}{24n^2}\;.$$

Guaranteeing accuracy using error estimation

Example 3.7. How many subintervals are needed to guarantee that a midpoint-sampled Riemann sum is within 10^{-4} of $\int_1^5 \ln(2 + \sin(t))\, dt$?

Solution: As before, our basic strategy is to solve the inequality

$$\frac{M_2(b - a)^3}{24n^2} \leq \frac{1}{10000}$$

You've already seen this happen, but we didn't point it out because we were focused on other things at the time. In Example 1.18 (p. 334) we used a midpoint-sampled Riemann sum to approximate the net change in a linear function, and arrived at a value of -7.5. We calculated the definite integral of the same linear function over the same span of time in Example 2.8 (p. 347), and determined the value to be -7.5. So midpoint sampling told us the exact value of the definite integral.

We're using $M_2(b - a)^3/24n$ as a proxy for the actual discrepancy between the integral and the Riemann sum.

for n. We still have $a = 1$, $b = 5$, and we determined that $M_2 = 1$ in Example 3.5. So our inequality is

$$\frac{1 \cdot (5-1)^3}{24n^2} \leq \frac{1}{10000} \quad \Rightarrow \quad n \geq \sqrt{\frac{80000}{3}} = 163.2993.$$

By choosing $n = 164$ we guarantee that $\frac{M_2(b-a)^3}{24n^2}$ is smaller than 10^{-4}. Since the actual discrepancy between the integral and a midpoint-rule approximation is smaller than that fraction, it will also be less than 10^{-4}.   ∎

Try It Yourself: Guaranteeing accuracy using error estimation

Example 3.8. How many subintervals are needed to guarantee that a midpoint-sampled Riemann sum is within 10^{-3} of $\int_0^5 e^{\cos(t)} \, dt$?

Full Solution On-line

Answer: $n = 86$ using $M_2 = 2.75$.   ∎

Whereas left- and right-endpoint sampling takes only the height of a graph into account, midpoint sampling also accounts for its average rate of change between the partition points. After seeing the dramatic benefit of accounting for the average rate of change ($n = 80000$ in Example 3.1 and $n = 164$ in Example 3.7), it seems worthwhile to investigate the inclusion of even more information, such as concavity, in our approximation scheme.

❖ Simpson's Rule

The simplest curve that has nonzero concavity is arguably a parabola, so let's talk about using parabolas to approximate a curve, as seen in Figure 3.7.

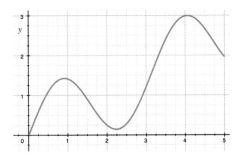 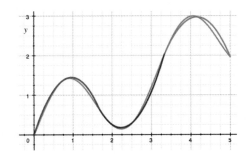

Figure 3.7: (left) The graph of $f(t) = 0.5t + \sin(2t)$; (right) the graph of the same function, overlaid with approximating parabolas that pass through the points $(t_k, f(t_k))$.

Since parabolas have *three* parameters, $y = at^2 + bt + c$, we can design one that passes through the *three* points $(t_{k-1}, f(t_{k-1}))$, $(t_k, f(t_k))$, and $(t_{k+1}, f(t_{k+1}))$, as seen in Figure 3.8. In Section 5.4, after we know a little more about the definite integral, we'll show that the net area between such a parabola and the horizontal axis over $[t_{k-1}, t_{k+1}]$ is

$$\int_{t_k - \Delta t}^{t_k + \Delta t} at^2 + bt + c \, dt = \frac{\Delta t}{3} \Big(f(t_{k-1}) + 4f(t_k) + f(t_{k+1}) \Big).$$

So when we approximate the graph of f with parabolas over the points $\{t_0, t_1, t_2\}$,

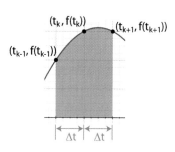

Figure 3.8: A parabola that passes through $(t_{k-1}, f(t_{k-1}))$, $(t_k f(t_k))$, and $(t_{k+1}, f(t_{k+1}))$, and the net area between it and the horizontal axis.

and $\{t_2, t_3, t_4\}$, and $\{t_4, t_5, t_6\}$, ... and $\{t_{n-2}, t_{n-1}, t_n\}$, the net area is

$$\int_a^b f(t)\, dt \;\approx\; \frac{\Delta t}{3}\Big(f(t_0) + 4f(t_1) + f(t_2)\Big)$$

$$+ \frac{\Delta t}{3}\Big(f(t_2) + 4f(t_3) + f(t_4)\Big)$$

$$\vdots$$

$$+ \frac{\Delta t}{3}\Big(f(t_{n-2}) + 4f(t_{n-1}) + f(t_n)\Big).$$

Notice that the partition point t_2 is used to calculate the area bounded by both the first parabola *and* the second, so it occurs twice in this sum. Similarly, the number $f(t_k)$ occurs twice in this approximation when k is any even index (except for $k = 0$ and $k = n$, which occur only once).

After factoring the common $\Delta t/3$ out of this sum, we can rewrite it as

$$\int_a^b f(t)\, dt \approx \frac{\Delta t}{3}\Big(f(t_0) + 4f(t_1) + 2f(t_2) + 4f(t_3) + 2f(t_4) + \cdots + 2f(t_{n-2}) + 4f(t_{n-1}) + f(t_n)\Big),$$

which is called **Simpson's method** or **Simpson's rule**. Notice that, since the parabolas begin and end at partition points with even indices, we can use this approximation method only when n is even.

> Since t_0 is an endpoint of the interval of integration, the number $f(t_0)$ is used to calculated the area bounded by a single parabola. The same is true of t_n.

Using Simpson's algorithm

Example 3.9. Use Simpson's method to approximate $\int_1^2 1/t\, dt$ with four subintervals.

Solution: Since we're using four subintervals, each has a length of $\Delta t = \frac{2-1}{4} = \frac{1}{4}$. That means our partition points are

$$t_0 = 1, \qquad t_1 = \tfrac{5}{4}, \qquad t_2 = \tfrac{3}{2}, \qquad t_3 = \tfrac{7}{4}, \qquad \text{and} \qquad t_4 = 2.$$

Therefore, we approximate

$$\int_1^2 \frac{1}{t}\, dt \;\approx\; \frac{\Delta t}{3}\left(f(1) + 4f\left(\frac{5}{4}\right) + 2f\left(\frac{3}{2}\right) + 4f\left(\frac{7}{4}\right) + f(2)\right)$$

$$= \frac{1/4}{3}\left(1 + 4\left(\frac{4}{5}\right) + 2\left(\frac{2}{3}\right) + 4\left(\frac{4}{7}\right) + \frac{1}{2}\right) = \frac{1747}{2520}. \qquad \blacksquare$$

Try It Yourself: Using Simpson's method of approximation

Example 3.10. Use Simpson's method to approximate $\int_0^\pi \sin(t)\, dt$ with six subintervals.

Answer: $\int_0^\pi \sin(t)\, dt \approx \frac{\pi}{18}(8 + 2\sqrt{3})$. $\qquad \blacksquare$

Full Solution On-line

The error formula for Simpson's method is much like the others:

Error estimation for Simpson's method: Denote by $S(n)$ the number we calculate using Simpson's method with n subintervals to approximate $\int_a^b f(t)\, dt$, and suppose that $|f''''(t)|$ never exceeds M_4 when $t \in [a, b]$. Then $\int_a^b f(t)\, dt$ differs from $S(n)$ by no more than $\frac{M_4(b-a)^5}{180n^4}$. Written technically,

$$S(n) - \frac{M_4(b-a)^5}{180n^4} \;\leq\; \int_a^b f(t)\, dt \;\leq\; S(n) + \frac{M_4(b-a)^5}{180n^4}.$$

> By making use of parabolas, Simpson's method takes height, slope, and concavity (i.e., f, f', and f'') into account, so you might wonder why this error estimate relies on the *fourth* derivative instead of the magnitude of f'''. We'll discuss that on p. 379, once we have a few more mathematical tools at the ready.

Guaranteeing accuracy using error estimation

Example 3.11. How many subintervals are needed to guarantee that a Simpson's method approximates $\int_1^5 \ln(2 + \sin(t))\, dt$ to within 10^{-4} of its actual value?

Solution: As before, our strategy is to solve the inequality

$$\frac{M_4(b-a)^5}{180n^4} \le \frac{1}{10000}$$

for n. We know that $a = 1$ and $b = 5$, but to determine M_4 we need to know about the *fourth* derivative of $f(t) = \ln(2 + \sin(t))$. Figure 3.9 shows the graph of $f''''(t)$, and you can see that its *magnitude* never exceeds 4, so we'll use $M_4 = 4$. This makes our inequality

> Instead of using the graph, you could verify that $|f''''(t)|$ never exceeds 4 by finding its critical points and calculating the value of $f''''(t)$ at each.

$$\frac{4(5-1)^5}{180n^4} \le \frac{1}{10000} \quad \Rightarrow \quad n \ge 21.84.$$

from which we conclude that $n = 22$ (recall that n has to be an even integer).   ■

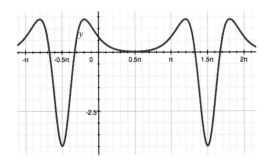

Figure 3.9: A graph of $f''''(t)$ for Example 3.11.

Try It Yourself: Guaranteeing accuracy using error estimation

Example 3.12. How many subintervals are needed to guarantee that a Simpson's method approximates $\int_0^5 e^{\cos(t)}\, dt$ to within 10^{-3} of its actual value?

Answer: $n = 22$.   ■

Full Solution On-line

✧ Remembering the Error Formulas

We've used M_1, M_2 and M_4 in the error formulas. The letter "M" reminds us of the words "<u>m</u>aximum <u>m</u>agnitude," and the subscript tells us the derivative whose magnitude we're maximizing. If you can remember which derivative controls the error for each kind of approximation scheme, the rest of the error formula is easy to remember because they all have the form

$$\frac{M_k(b-a)^{k+1}}{\# \, n^k} \text{ , where the symbol \# is 2, 12, 24, or 180.}$$

You should know

- the terms *error*, *Trapezoid Rule*, and *Simpson's method;*

- the error formulas for endpoint-sampled Riemann sums, midpoint-sampled Riemann sums, and Simpson's method;

- the coefficient pattern for the trapezoid rule: 1, 2, 2, 2, 2, 2, 2 ... 2, 2, 1;

- the coefficient pattern for Simpson's method: 1, 4, 2, 4, 2, 4, 2 ... 2, 4, 1.

You should be able to

- implement an endpoint-sampled or midpoint-sampled Riemann sum (from previous sections);

- implement the trapezoid rule and Simpson's method;

- use the error formulas to determine how many subintervals are needed to achieve a desired accuracy, using any of the methods discussed in this section.

✣ 5.3 Skill Exercises

In each of #1–6 approximate the definite integral using the trapezoid rule with the specified number of subintervals.

1. $\int_0^5 3t + 1 \, dt$, $n = 3$

2. $\int_0^3 5t - 2 \, dt$, $n = 5$

3. $\int_1^4 t^5 + 3t - 1 \, dt$, $n = 6$

4. $\int_2^6 2t^4 - 8t + 3 \, dt$, $n = 7$

5. $\int_0^{2\pi} \sin(x) \, dx$, $n = 4$

6. $\int_0^{2\pi} \cos(x) \, dx$, $n = 8$

In each of #7–12 approximate the definite integral using Simpson's rule with the specified number of subintervals.

7. $\int_0^{12} 4x + 1 \, dx$, $n = 4$

8. $\int_1^9 3x - 8 \, dx$, $n = 8$

9. $\int_2^{12} 8x^2 - 3x \, dx$, $n = 10$

10. $\int_3^{21} 9x^2 + x \, dx$, $n = 6$

11. $\int_0^{2\pi} \sin(x) \, dx$, $n = 4$

12. $\int_0^{2\pi} \cos(x) \, dx$, $n = 8$

In each of #13–18 use the error formulas from this section to determine how much error could be in the specified approximation, where n is the number of subintervals, and the numerical method used is indicated by the letter E (endpoint sampling), M (midpoint sampling), T (trapezoid rule), or S (Simpson's rule).

13. $\int_0^6 3t + 8 \, dt$, $n = 10$, E

14. $\int_0^6 7t - 12 \, dt$, $n = 6$, E

15. $\int_1^6 13t^2 + 1 \, dt$, $n = 5$, M

16. $\int_1^6 2t^3 + 11 \, dt$, $n = 15$, M

17. $\int_5^7 13t^6 + 1 \, dt$, $n = 4$, S

18. $\int_5^7 -7t^8 + 9t^4 \, dt$, $n = 8$, S

In each of #19–26 determine the smallest number of subintervals required to approximate the definite integral to within ε of its actual value when using the given method of integration (indicated by the letter E,M,T, or S, as in the previous set of exercises). You might find it helpful to render a graph of the appropriate derivative of the integrand using a graphing utility.

19. $\int_0^4 \frac{1}{24}x^4 - \frac{1}{2}x^2 \, dx$, $\varepsilon = 0.01$, E

20. $\int_1^5 x^2 + 3x - 4 \, dx$, $\varepsilon = 0.001$, E

21. $\int_2^5 2x^4 + 3x \, dx$, $\varepsilon = 0.01$, T

22. $\int_1^5 x^2 + 3x - 4 \, dx$, $\varepsilon = 0.001$, T

23. $\int_3^6 7t + 4$, $\varepsilon = 10^{-61}$, M

24. $\int_2^5 \sin(t) \, dt$, $\varepsilon = 10^{-5}$, M

25. $\int_0^1 t^7 + 14t$, $\varepsilon = 10^{-15}$, S

26. $\int_2^5 \cos(t) \, dt$, $\varepsilon = 10^{-4}$, S

❖ 5.3 Concept and Application Exercises

In Section 5.4 you'll see that

$$\ln(9) = \int_1^9 \frac{1}{t}\, dt, \quad \text{and more generally} \quad \ln(x) = \int_1^x \frac{1}{t}\, dt \quad \text{when} \quad x > 0.$$

This is an important fact because we can use numerical methods to calculate these integrals to any degree of accuracy that we desire. In each of #27–30 (a) determine the number of subintervals that are needed to approximate the specified logarithm to within ε of its actual value by using the given method, and (b) calculate that approximation (using a spreadsheet program or calculator as necessary).

27. $\ln(2)$, $\varepsilon = 0.1$, left-sampling

28. $\ln(3)$, $\varepsilon = 0.01$, right-sampling

29. $\ln(4)$, $\varepsilon = 0.1$, trapezoid rule

30. $\ln(9)$, $\varepsilon = 0.01$, Simpson's rule

31. In Section 5.4 you'll see that $\int_0^1 \frac{4}{1+x^2}\, dx = \pi$. This is an important fact because we can use numerical methods to calculate this integral, and so the value of π, to any degree of accuracy we desire. Use the trapezoid rule to calculate the value of the integral to within 10^{-3} of its actual value.

32. Use Simpson's method to approximate the integral in #31 to within 10^{-4} of its actual value, and then use that approximation to estimate the value of π.

33. Suppose we approximate $\int_a^b f(t)\, dt$ using a right-sampled Riemann sum. What condition on f would guarantee that our approximation is an underestimate?

34. Suppose we approximate $\int_a^b f(t)\, dt$ using the trapezoid rule. What condition on f would guarantee that our approximation is an overestimate?

35. Suppose $|f''(t)| < 2$. How many subintervals are needed to approximate $\int_2^7 f(t)\, dt$ to within 10^{-6} of its actual value using the midpoint rule?

36. Suppose that $f(t)$ is a cubic polynomial. Show that Simpson's rule calculates $\int_a^b f(t)\, dt$ *exactly* using only two subintervals.

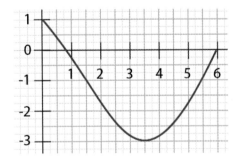
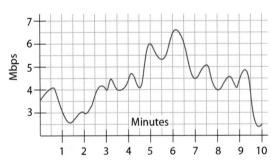

Figure 3.10: (left) For #37 and #38; (right) chart for exercises #39 and #40.

37. Suppose that the left-hand image in Figure 3.10 depicts the graph of $f''(t)$. How many subintervals are required to approximate $\int_0^6 f(t)\, dt$ to within 0.001 of its actual value using the trapezoid rule?

38. Suppose that the left-hand image of Figure 3.10 depicts the graph of $f''''(t)$. How many subintervals are required to approximate $\int_0^6 f(t)\, dt$ to within 0.001 of its actual value using Simpson's rule?

Exercises #39–40 refer to the right-hand image of Figure 3.10, which depicts the transfer of data across a home wireless network in megabits per second. Note that time is measured differently on the vertical and horizontal axes.

39. Use the trapezoid rule with $\Delta t = 1$ to estimate the total bit traffic over the network in the 10 minutes shown in Figure 3.10.

40. Use Simpson's method with $\Delta t = 1$ to estimate the total bit traffic over the network in the 10 minutes shown in Figure 3.10.

41. Use Simpson's method to estimate the volume enclosed by a guitar that's 4.25 inches deep, shown in Figure 3.11 (width measurements in inches, taken every 2 inches).

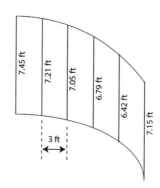

Figure 3.11: (left) For #41, with width measurements shown 2 inches apart; (right) for #42.

42. Data regarding the width of a contoured swimming pool is shown in Figure 3.11. If the pool is built with a level base, determine the volume of water needed to make a uniform depth of 4 feet throughout the pool.

43. Use the trapezoid rule to approximate the net area between the graphs of f and g over $[0, 1.25]$, based on the data below.

t	0	0.25	0.5	0.75	1	1.25
$f(t)$	20	14	12	9	7	6
$g(t)$	10	13	11	5	2	0

44. Figure 3.12 depicts the thrust (in newtons) provided by a model rocket engine. Newton's Second Law of motion is often written as $F_{net} = p'(t)$, where p denotes momentum, so

$$\int_{t_1}^{t_2} F_{net}(t)\, dt = \int_{t_1}^{t_2} p'(t)\, dt = p(t_2) - p(t_1)$$

is the net change in an object's momentum, called the *impulse*. In this context, $F_{net}(t) = F_{rocket}(t) + F_{gravity}$.

(a) Determine the (negative) gravitational force on a model rocket whose mass is 0.1 kg.

(b) Use the trapezoid rule to approximate the impulse on that model rocket during the first second of its launch.

(c) Use Simpson's method to approximate the impulse on that model rocket during the first second of its launch.

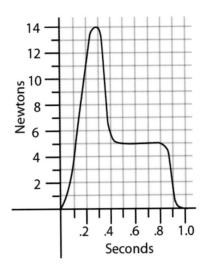

Figure 3.12: Thrust (in newtons) provided by the model rocket engine in #44.

45. The graph below (drawn to reflect data from Figure 1 in [32]) depicts trades per minute in the foreign exchange market. Use a trapezoid rule with $\Delta t = 2$ hours to approximate the trade volume over the 24 hours shown.

Business

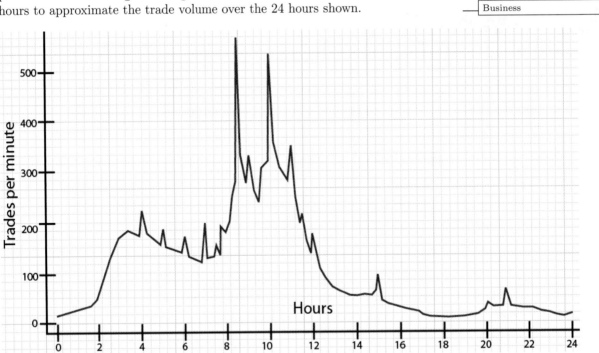

Checking the details

46. Sketch the graph of a continuos function $f(t) > 0$, and explain why the number $\frac{f(a)+f(b)}{2}(b-a)$ is the area of a trapezoid.

5.4   Antiderivatives and the Fundamental Threorem

In Example 2.12 (see p. 350) we worked with the rate of change in the temperature of a solution, $y'(t)$. We wrote the net change in the temperature between times $t = 4$ and $t = 6$ as a definite integral, and concluded that

$$\overbrace{y(6) = \quad y(4)}^{\text{temperature at time } t=4} + \underbrace{\int_4^6 y'(t)\, dt.}_{\substack{\text{net change in temperature be-}\\ \text{tween times } t=4 \text{ and } t=6}} \tag{4.1}$$

In general, when *any* quantity is changing during $[a, b]$ we expect to see

$$f(b) = \overbrace{f(a)}^{\text{value at time } t=a} + \overbrace{\int_a^b f'(t)\, dt,}^{\text{net change between times } t=a \text{ and } t=b}$$

which is often written as

$$f(b) - f(a) = \int_a^b f'(t)\, dt.$$

Fundamental Theorem of Calculus (FTC), Part 1: Suppose f is differentiable on an interval that contains $[a, b]$, and $f'(t)$ is continuous on $[a, b]$. Then

$$f(b) - f(a) = \int_a^b f'(t)\, dt.$$

Part 2 is found on p. 384.

This fact is sometimes called the *Evaluation Theorem*.

Proof. Though we have introduced this formula in the context of net change, the numerical fact is actually a consequence of the Mean Value Theorem. The basic idea behind this proof is that a net change, $f(b) - f(a)$, can be accomplished by many smaller changes.

Suppose $a = t_0 < t_1 < t_2 < \cdots < t_{n-1} < t_n = b$ is a regular partition of $[a, b]$. Then

$$\overbrace{f(t_n) - f(t_0)}^{\substack{\text{net change between} \\ t=a \text{ and } t=b}} = \underbrace{\overbrace{f(t_1) - f(t_0)}^{\substack{\text{net change in the} \\ \text{1st subinterval}}} + \overbrace{f(t_2) - f(t_1)}^{\substack{\text{net change in the} \\ \text{2nd subinterval}}} + \overbrace{f(t_3) - f(t_2)}^{\substack{\text{net change in the} \\ \text{3rd subinterval}}} + \cdots + \overbrace{f(t_n) - f(t_{n-1})}^{\substack{\text{net change in the} \\ n^{\text{th}} \text{ subinterval}}}}$$

Notice that both $f(t_1)$ and $-f(t_1)$ appear in the sum, as do $f(t_2)$ and $-f(t_2)$, etc. The only terms that don't have opposites in the sum are $-f(t_0)$ and $f(t_n)$.

Using Σ notation, we can write this equation as

$$f(b) - f(a) = \sum_{k=1}^n \Big(f(t_k) - f(t_{k-1}) \Big).$$

Now we look to the formula we're trying to establish for guidance. We know that the differential in the definite integral is the infinitesimal analog of Δt, which the right-hand side of our current expression lacks. We can insert it by writing

$$f(b) - f(a) = \sum_{k=1}^n \frac{f(t_k) - f(t_{k-1})}{\Delta t} \Delta t\,.$$

Further, since f satisfies the hypotheses of the Mean Value Theorem over each subinterval, we know there is a time $t_k^* \in [t_{k-1}, t_k]$ at which the instantaneous rate of change equals the average rate of change over the subinterval:

$$\frac{f(t_k) - f(t_{k-1})}{\Delta t} = f'(t_k^*).$$

Substituting this into our equation, we see

$$f(b) - f(a) = \sum_{k=1}^{n} f'(t_k^*)\, \Delta t.$$

Since the definite integral is the limit of Riemann sums, we apply the limit:

$$\lim_{n \to \infty} \left(f(b) - f(a) \right) = \lim_{n \to \infty} \sum_{k=1}^{n} f'(t_k^*)\, \Delta t.$$

> This Riemann sum was chosen in a very special way. However, the function f' is integrable because it's continuous, so the limit of Riemann sums exists, and is independent of the sampling scheme that's used.

The left-hand side doesn't depend on n, so its limit is just $f(b) - f(a)$; and f' is integrable since it's continuous, so the limit on the right-hand side is the definite integral of f' from $t = a$ to $t = b$. ∎

In addition to validating equation (4.1), the Fundamental Theorem of Calculus provides us with an entirely new tool for calculating definite integrals.

Using the Fundamental Theorem of Calculus to calculate a definite integral

Example 4.1. Calculate $\int_4^6 -3t + 14\ dt$.

Solution: In Example 1.21 (see p. 336) we used a limit of Riemann sums to determine that this definite integral is the number -2. However, the Fundamental Theorem of Calculus gives us another (simpler) way to make the calculation. Since our integrand is the derivative of $y(t) = -\frac{3}{2}t^2 + 14t + 1$, the Fundamental Theorem of Calculus tells us that our definite integral is

$$\int_4^6 y'(t)\, dt = y(6) - y(4) = (31) - (33) = -2.$$

It's an important fact that we could have used $y(t) = -\frac{3}{2}t^2 + 14t + 2$ in this calculation, or

$$y(t) = -\frac{3}{2}t^2 + 14t + 3, \quad \text{or} \quad y(t) = -\frac{3}{2}t^2 + 14t + 4 \ldots$$

because the constant term affects the height of the graph but not its net change (as seen in Figure 4.1). For example, using the general $y(t) = -\frac{3}{2}t^2 + 14t + C$, we have

$$\int_4^6 y'(t)\, dt = y(6) - y(4) = (30 + C) - (32 + C) = -2.$$

Notice that the $+C$ is subtracted away in the calculation. ∎

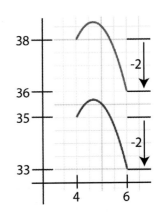

Figure 4.1: The graphs of $y = -1.5t^2 + 14t + 5$ and $y = -1.5t^2 + 14t + 3$ between $t = 4$ and $t = 6$.

The real importance of the Fundamental Theorem of Calculus is that it provides a way to calculate definite integrals that were impossible before.

Using the Fundamental Theorem of Calculus to calculate a definite integral

Example 4.2. Use the Fundamental Theorem of Calculus to determine $\int_0^{\pi/2} \cos(t)\, dt$.

Solution: Like all definite integrals, this one is the limit of Reimann sums. But when we write it down as such (using right-sampling) we see

$$\int_0^{\pi/2} \cos(t)\,dt = \lim_{n\to\infty} \sum_{k=1}^n \cos(t_k^*)\,\Delta t = \lim_{n\to\infty} \sum_{k=1}^n \cos\left(\frac{k\pi}{2n}\right)\frac{\pi}{2n},$$

in which the index of summation occurs in the argument of the cosine. That prevents us from using our summation formulas. However, since $\cos(t)$ is the derivative of the function $y(t) = \sin(t)$, we can use the Fundamental Theorem of Calculus to rewrite our definite integral as

$$\int_0^{\pi/2} y'(t)\,dt = y(\pi/2) - y(0) = (1) - (0) = 1.$$

As in the previous example, it's important to note that we could have used $y(t) = \sin(t) + C$ in our calculation, where C is any constant, because

$$\int_0^{\pi/2} y'(t)\,dt = y(\pi/2) - y(0) = (1 + C) - (0 + C) = 1.$$

Notice that the $+C$ is subtracted away in the calculation.   ■

It's common to use a vertical bar to mean "evaluate at," so the central equation of the Fundamental Theorem of Calculus looks like

$$\int_a^b f'(t)\,dt = \left. f(t) \right|_{t=a}^{t=b}$$

$$\underbrace{\qquad\qquad}_{\text{This notation means } f(b) - f(a)}$$

> Note that the ending time, $t = b$, appears above the starting time, $t = a$, on both sides.

More commonly, the "$t =$" is omitted from the vertical bar, and we write

$$\int_a^b f'(t)\,dt = \left. f(t) \right|_a^b$$

Let's take a moment to be explicitly clear about the fact that, while the letter t appears in the equation above, it's not acting as the argument of a function (because $f(b) - f(a)$ is a number, not a function). We often refer to the variable of integration as a **dummy variable**, because it amounts to a placeholder.

> When you're done calculating a definite integral, you have a number. No longer do you see the dummy variable.

Practice with a new notation

Example 4.3. Determine (a) $\int_1^9 1/t\,dt$, and (b) $\int_0^1 \frac{1}{1+t^2}\,dt$.

Solution: (a) Since $f'(t) = 1/t$ is the derivative of $f(t) = \ln(t)$, the Fundamental Theorem of Calculus tells us that

$$\int_1^9 \frac{1}{t}\,dt = \left. \ln(t) \right|_1^9 \quad,\text{ which is the same as } \quad \ln(9) - \ln(1) = \ln(9).$$

> More generally, if we change the 9 to x (which we take to be positive), the same calculation shows us that
> $$\int_1^x \frac{1}{t}\,dt = \ln(x).$$

(b) Since $f'(t) = 4/(1+t^2)$ is the derivative of $f(t) = 4\arctan(t)$, the Fundamental Theorem of Calculus tells us that

$$\int_0^1 \frac{4}{1+t^2}\,dt = \left. 4\tan^{-1}(t) \right|_0^1 \quad,$$

which is the same as saying

$$\int_0^1 \frac{4}{1+t^2}\,dt = 4\tan^{-1}(1) - 4\tan^{-1}(0) = 4\left(\frac{\pi}{4}\right) - 0 = \pi \qquad ■$$

❖ Antiderivatives

In Examples 4.1–4.3 we calculated the definite integral by finding a function whose derivative was the integrand. We say that such a function is an *antiderivative* of the integrand. More formally …

> We say the function $F(t)$ is an **antiderivative** of $f(t)$ on the interval (a, b) when $F'(t) = f(t)$ at every $t \in (a, b)$.

This is not a new idea, but only new language applied to a relationship that you already know. For example, both of the following sentences communicate the same relationship between functions:

- $2t$ is the derivative of t^2.

- t^2 is an antiderivative of $2t$.

Notice we say that t^2 is *an* antiderivative, not *the* antiderivative, because the derivative of $t^2 + C$ is $2t$ when C is *any* constant. In fact, this additive constant is the extent to which any two antiderivatives on an interval can differ.

> **Theorem 4.8.** Suppose that $F(t)$ and $G(t)$ are both antiderivatives of $f(t)$ over the interval (a, b). Then the difference between them is constant on that interval.

Proof. We know that $F'(t) = f(t)$ and $G'(t) = f(t)$ because F and G are both antiderivatives of f. Therefore,

$$\frac{d}{dt}\Big(F(t) - G(t)\Big) = F'(t) - G'(t) = f(t) - f(t) = 0.$$

Since its derivative is always zero on (a, b), the function $F(t) - G(t)$ is constant on that interval (see Theorem 5.6 on p. 283). ∎

We can rephrase Theorem 4.8 by saying that $F(t) = G(t) + C$ on (a, b) when F and G are both antiderivatives of f over that interval. Graphically speaking, this means the graph of F is found by translating the graph of G vertically by C units. For example, the graphs shown in Figure 4.2 are all antiderivatives of $f(t) = 3t^2$.

Figure 4.3 emphasizes the role played by the interval (a, b) in the statement of the theorem. The curves you see there are the graphs of

$$F(t) = \frac{1}{t^2} \quad \text{and} \quad G(t) = \begin{cases} \frac{1}{t^2} - 1 & \text{when } t < 0 \\ \frac{1}{t^2} + \frac{1}{4} & \text{when } t > 0 \end{cases}.$$

It's easy to show that $F'(t) = G'(t)$ at all t in the domain, but the difference between these functions changes. The number $F(t) - G(t)$ is always 1 on the interval $(-\infty, 0)$, but is always $-1/4$ on the interval $(0, \infty)$. In general, when the domain of an antiderivative is disconnected, we can vertically translate the different pieces of its graph by different amounts without changing the derivative anywhere.

❖ Indefinite Integrals

The collection of antiderivatives of a particular $f(t)$, any one of which could be used in the Fundamental Theorem of Calculus, is called the **indefinite integral** of f and is denoted by

$$\int f(t)\, dt.$$

Consider the following two sentences:

- George W. Bush is the son of George Herbert Walker Bush.

- George Herbert Walker Bush is the father of George W. Bush.

The relationship between these two Presidents is the same in both sentences, but whether we use the term *son* or *father* depends on who is the subject of the sentence. Similarly, whether we use the word *derivative* or *antiderivative* to describe the relationship between functions depends on which function is the subject of the sentence.

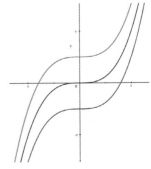

Figure 4.2: The graphs of antiderivatives can differ only by a vertical shift.

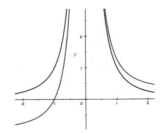

Figure 4.3: The graphs of antiderivatives can differ only by a vertical shift.

Because all antiderivatives of $f(t)$ on a given interval differ by a constant, we typically write

$$\int f(t)\, dt = (\text{an antiderivative that you know}) + C.$$

Note: Pursuant to the discussion accompanying Figure 4.3, *whenever a formula for an indefinite integral is given, you should understand it to be valid on an interval.*

> ▷ Indefinite integrals of power functions

The indefinite integral of a constant function

Example 4.4. Determine (a) $\int 5\, dt$, and (b) $\int \beta\, dt$, where β is any constant.

Solution: (a) Since $\frac{d}{dt}(5t) = 5$, and antiderivatives can only differ by a constant, we conclude that

$$\int 5\, dt = 5t + C.$$

All of the functions with a derivative of 5 ...   have this form.

(b) Similarly, since $\frac{d}{dt}(\beta t) = \beta$, we conclude that

$$\int \beta\, dt = \beta t + C.$$

All of the functions with a derivative of β ...   have this form.

Determining the indefinite integral of a power function

Example 4.5. Determine (a) $\int 2t\, dt$, and (b) $\int 3t^2\, dt$.

Solution: (a) Since $\frac{d}{dt}(t^2) = 2t$, and antiderivatives can only differ by a constant, we know

$$\int 2t\, dt = t^2 + C.$$

All of the functions with a derivative of $2t$ ...   have this form.

(b) Since $\frac{d}{dt}(t^3) = 3t^2$, we know

$$\int 3t^2\, dt = t^3 + C.$$

All of the functions with a derivative of $3t^2$ ...   have this form.

Try It Yourself: Determining the indefinite integral of a power function

Example 4.6. Determine (a) $\int -t^{-2}\, dt$, and (b) $\int -5t^{-6}\, dt$.

Answer: (a) $t^{-1} + C$, and (b) $t^{-5} + C$.

We're unlikely to find unscaled power functions in practice, so the next examples address the presence of coefficients.

Determining the indefinite integral of a scaled power function

Example 4.7. Determine (a) $\int t\, dt$, and (b) $\int 5t\, dt$.

If this were a definite integral you would see bounds of integration,

$$\int_a^b f(t)\, dt,$$

and we could calculate it by evaluating any antiderivative of f at $t = b$ and $t = a$. But the bounds of integration are not specified in the indefinite integral, so you just identify what antiderivative you *would* use. Since all other antiderivatives differ from yours by additive constants, we include a "$+C$" whenever we mean to talk about the whole collection of antiderivatives.

Full Solution
On-line

Solution: (a) We know that differentiating t^2 would give us $2t$. That differs from our integrand by a factor of 2, so let's ask, "What would cause a factor of 2 to disappear from our formula for the derivative?" The answer is, "a coefficient of $1/2$."

$$\frac{d}{dt}\left(\frac{1}{2}t^2\right) = t.$$

Since antiderivatives can only differ by an additive constant,

$$\frac{d}{dt}\left(\frac{1}{2}t^2\right) = t \quad \Longrightarrow \quad \int t\ dt = \frac{1}{2}t^2 + C.$$

(b) In this case we see a coefficient of 5 instead of the expected 2, so let's ask, "What would cause a factor of 2 to become a 5 in our formula for the derivative?" The answer is, "a coefficient of $5/2$."

$$\frac{d}{dt}\left(\frac{5}{2}t^2\right) = 5t \quad \Longrightarrow \quad \int 5t\ dt = \frac{5}{2}t^2 + C. \qquad \blacksquare$$

Try It Yourself: Determining the indefinite integral of a scaled power function

Example 4.8. Determine (a) $\int -5t^8\ dt$, and (b) $\int -3t^{-6}\ dt$.

Answer: (a) $-\frac{5}{9}t^9 + C$, and (b) $\frac{3}{5}t^{-5} + C$. $\qquad \blacksquare$

Full Solution On-line

Based on Examples 4.4–4.8, we make the following conjecture (which is correct):

> Suppose that β and η are constants, and $\eta \neq -1$. Then
> $$\int \beta\, t^\eta\ dt = \frac{\beta}{\eta+1}t^{\eta+1} + C.$$

Notice that attempting to use $\eta = -1$ in this formula would result in a division by zero.

The formula above does not apply to $\eta = -1$, so we list that case separately. Since

$$\frac{d}{dt}\Big(\beta \ln(t)\Big) = \frac{\beta}{t} \quad \text{when } t > 0 \text{ and } \beta \text{ is any constant,}$$

we know that

$$\int \beta t^{-1}\ dt = \beta \ln(t) + C \quad \text{when} \quad t > 0.$$

But $1/t$ exists when $t < 0$, while $\ln(t)$ does not. This apparent problem is resolved by the absolute value.

> $$\int \frac{\beta}{t}\ dt = \beta \ln|t| + C \quad \text{when } \beta \text{ is any constant.}$$

This formula works over either interval, $t > 0$ or $t < 0$.

▷ Indefinite integrals of some trigonometric functions

The indefinite integral of the cosine function

Example 4.9. Determine (a) $\int \cos(t)\ dt$, (b) $\int \cos(4t)$, and (c) $\int 7\cos(4t)\ dt$.

Solution: (a) Since $\cos(t)$ is the derivative of $\sin(t)$, we know that

$$\int \cos(t)\ dt = \sin(t) + C.$$

(b) Similarly, $4\cos(4t)$ is the derivative of $\sin(4t)$. We don't see the coefficient of 4 in our integrand, so as we did with the power functions in previous examples, we infer that $\sin(4t)$ has a coefficient of $1/4$. That is,

$$\int \cos(4t)\ dt = \frac{1}{4}\sin(4t) + C.$$

> You should check by verifying that the derivative of $\frac{1}{4}\sin(4t) + C$ is $\cos(4t)$.

(c) In this case we see a 7 instead of the expected coefficient of 4, so we infer that $\sin(4t)$ has a coefficient of $7/4$. That is,

$$\int 7\cos(4t)\ dt = \frac{7}{4}\sin(4t) + C. \qquad \blacksquare$$

> You should check by verifying that the derivative of $\frac{7}{4}\sin(4t) + C$ is $7\cos(4t)$.

Try It Yourself: The indefinite integral of the sine function

Example 4.10. Determine (a) $\int \sin(t)\ dt$, (b) $\int \sin(9t)$, and (c) $\int 13\sin(9t)\ dt$.

Answer: (a) $-\cos(t) + C$, (b) $-\frac{1}{9}\cos(9t) + C$, and (c) $-\frac{13}{9}\cos(9t) + C$. $\qquad \blacksquare$

Full Solution On-line

Based on the results of Examples 4.9 and Example 4.10, you might make the following conjecture (which is correct):

Suppose that β and ω are constants, and $\omega \neq 0$. Then

$$\int \beta\ \cos(\omega t)\ dt = \frac{\beta}{\omega}\sin(\omega t) + C \qquad \text{and} \qquad \int \beta\ \sin(\omega t)\ dt = -\frac{\beta}{\omega}\cos(\omega t) + C.$$

For similar reasons, as you'll show in the exercise set,

Suppose that β and ω are constants, and $\omega \neq 0$. Then

$$\int \beta\ \sec^2(\omega t)\ dt = \frac{\beta}{\omega}\tan(\omega t) + C$$

$$\int \beta\ \csc^2(\omega t)\ dt = -\frac{\beta}{\omega}\cot(\omega t) + C$$

and

$$\int \beta\ \sec(\omega t)\tan(\omega t)\ dt = \frac{\beta}{\omega}\sec(\omega t) + C$$

$$\int \beta\ \csc(\omega t)\cot(\omega t)\ dt = -\frac{\beta}{\omega}\csc(\omega t) + C.$$

▷ **Indefinite integrals of exponential functions**

The indefinite integral of an exponential function

Example 4.11. Determine (a) $\int e^{7t}\ dt$, (b) $\int 4^t \ln(4)$, and (c) $\int 5(4^t)\ dt$.

Solution: (a) We know the derivative of e^{7t} is $7e^{7t}$. We don't see the factor of 7 in our integrand, so we infer the presence of a coefficient of $1/7$. That is,

$$\int e^{7t}\ dt = \frac{1}{7}e^{7t} + C$$

> You should check by verifying that the derivative of $\frac{1}{7}e^{7t} + C$ is e^{7t}.

(b) The derivative of 4^t is $4^t \ln(4)$, so

$$\int 4^t \ln(4)\ dt = 4^t + C.$$

(c) We see a coefficient of 5 instead of the expected $\ln(4)$, so we infer the presence of a coefficient of $5/\ln(4)$. That is,

$$\int 5(4^t)\, dt = \frac{5}{\ln(4)}\, 4^t + C \qquad \blacksquare$$

You should check by verifying that the derivative of $\frac{5}{\ln(4)}\, 4^t + C$ is $5(4^t)$.

Full Solution On-line

Try It Yourself: The indefinite integral of an exponential function

Example 4.12. Determine (a) $\int 3e^{-8t}\, dt$, and (b) $\int 6(5^t)$

Answer: (a) $-\frac{3}{8}e^{-8t} + C$, and (b) $\frac{6}{\ln(5)}\, 5^t + C$ $\qquad \blacksquare$

Based on the results of Example 4.11 and Example 4.12 you might make the following conjecture (which is correct):

When β is any constant, and b is a positive number that's not 1,

$$\int \beta\, b^t\, dt = \frac{\beta}{\ln(b)}\, b^t + C.$$

▷ **Properties of the indefinite integral**

By now, you understand that finding an antiderivative (and so the indefinite integral) amounts to "undoing" the differentiation process. Since differentiation obeys the Scaling Rule and the Sum Rule, so does the indefinite integral.

Suppose F and G are antiderivatives of f and g respectively, and β is any constant. Then

$$\int \beta\, f(t)\, dt = \beta F(t) + C \quad \text{and} \quad \int f(t) + g(t)\, dt = F(t) + G(t) + C.$$

Written in integral notation,

$$\underbrace{\int \beta\, f(t)\, dt = \beta \int f(t)}_{\text{this property is called } homogeneity} \quad \text{and} \quad \underbrace{\int f(t) + g(t)\, dt = \int f(t)\, dt + \int g(t)\, dt}_{\text{this property is called } additivity}.$$

You might wonder why the second of these facts doesn't display a "$+C$" with F and a "$+C$" with G. We only need to write $+C$ once because adding a constant to F and another to G amounts to adding a single constant to the sum. For example,

$$(F+1) + (G+3) = F + G + 4.$$

The indefinite integral of the hyperbolic cosine function

Example 4.13. Determine $\int \cosh(t)\, dt$.

Solution: Recall that $\cosh(t) = \frac{1}{2}e^t + \frac{1}{2}e^{-t}$. The additivity of the indefinite integral allows us to write

$$\int \cosh(t)\, dt = \int \frac{1}{2}e^t + \frac{1}{2}e^{-t}\, dt = \int \frac{1}{2}e^t + \int \frac{1}{2}e^{-t}\, dt.$$

Then because the indefinite integral is homogeneous ($\beta = 1/2$ here), we know

$$\int \frac{1}{2}e^t + \int \frac{1}{2}e^{-t}\, dt = \frac{1}{2}\int e^t + \frac{1}{2}\int e^{-t} = \frac{1}{2}e^t - \frac{1}{2}e^{-t} + C = \sinh(t) + C. \qquad \blacksquare$$

Try It Yourself: The indefinite integral of the hyperbolic sine function

Example 4.14. Determine $\int \sinh(t)\, dt$.

Answer: $\cosh(t) + C$. $\qquad \blacksquare$

Full Solution On-line

✥ Addendum: Returning to Simpson's Rule

During our discussion of Simpson's method on p. 364, we asserted that the net area between the horizontal axis and a parabola that passes through the points

$$(t_{k-1}, f(t_{k-1})), \quad (t_k, f(t_k)), \quad \text{and} \quad (t_{k+1}, f(t_{k+1}))$$

is

$$\frac{\Delta t}{3} \Big(f(t_{k-1}) + 4f(t_k) + f(t_{k+1}) \Big)$$

> Because we are working with a regular partition in this discussion of Simpson's Rule, the partition points are separated by a distance of Δt. Therefore
>
> $$t_{k+1} = t_k + \Delta t$$
> $$t_{k-1} = t_k - \Delta t$$
>
> So when we shift the graph so that t_k to ends up at zero, the neighboring partition points end up at $\pm\Delta t$.

This formula is easy to derive with the Fundamental Theorem of Calculus. In order to simplify our calculations, let's begin by shifting the parabola sideways (which doesn't change the net area), so that passes through

$$(-\Delta t, f(t_{k-1})), \quad (0, f(t_k)), \quad \text{and} \quad (\Delta t, f(t_{k+1})). \tag{4.2}$$

The fact that our parabola passes through these particular points will be useful in a moment. For now, let's focus on finding the net area between the horizontal axis and a parabola of the form $y = at^2 + bt + c$, over $[-\Delta t, \Delta t]$. This requires an integration that's easy to do with an antiderivative …

$$\int_{-\Delta t}^{\Delta t} at^2 + bt + c \, dt = \frac{a}{3}t^3 + \frac{b}{2}t^2 + ct \Big|_{-\Delta t}^{\Delta t} = \frac{\Delta t}{3} \Big(2a(\Delta t)^2 + 6c \Big). \tag{4.3}$$

Don't just read it. Use the Fundamental Theorem of Calculus to work out equation (4.3) for yourself!

Now let's rewrite this expression in terms of our function values. Because the parabola $y = at^2 + bt + c$ passes through the points listed in (4.2), we know that

$$
\begin{aligned}
f(t_{k-1}) &= y(-\Delta t) &= a(\Delta t)^2 - b\Delta t + c \tag{4.4}\\
f(t_k) &= y(0) &= c \\
f(t_{k+1}) &= y(\Delta t) &= a(\Delta t)^2 + b\Delta t + c. \tag{4.5}
\end{aligned}
$$

Adding equations (4.4) and (4.5) gives us

$$f(t_{k-1}) + f(t_{k+1}) = 2a(\Delta x)^2 + 2c,$$

which allows us to rewrite the net area in (4.3) as

$$\frac{\Delta t}{3}\Big(2a(\Delta t)^2 + 6c \Big) = \frac{\Delta t}{3}\Big(2a(\Delta t)^2 + 2c + 4c \Big) = \frac{\Delta t}{3}\Big(f(t_{k-1}) + f(t_{k+1}) + 4c \Big).$$

And because $c = f(t_k)$, we can replace the remaining $4c$ with $4f(t_k)$. That is,

$$\int_{-\Delta t}^{\Delta t} at^2 + bt + c \, dt = \frac{\Delta t}{3}\Big(f(t_{k-1}) + 4f(t_k) + f(t_{k+1}) \Big).$$

▷ Revisiting the error approximation formula for Simpson's method

The three coefficients of $y = at^2 + bt + c$ allow us to approximate the *height, slope,* and *concavity* of a function's graph, which is why Simpson's method is so good at approximating the net area between a graph and the horizontal axis. It seems reasonable to think that we can do even better if we approximate the height, slope, concavity, and *rate of change in concavity* (f, f', f'' and f''' respectively) by using a third-degree polynomial instead of a quadratic—i.e., if we approximate the graph of f using $y = \beta t^3 + at^2 + bt + c$. Surprisingly, that doesn't improve our approximation of net area.

In the exercise set, you'll use the Fundamental Theorem of Calculus to show that the net area between the curve $y = \beta t^3 + at^2 + bt + c$ and the horizontal axis over $[-\Delta t, \Delta t]$ is

$$\int_{-\Delta t}^{\Delta t} \beta t^3 + at^2 + bt + c \ dt = \frac{\Delta t}{3}\left(2a(\Delta t)^2 + 6c\right),$$

which is exactly what we found using a parabola. This means that the parabolic approximation is better than we thought. It approximates net area as though it included information about $f, f', f'', and f'''$. This is one way to understand why the error formula for Simpson's method includes a bound on the *fourth* derivative (we've accounted for the first three). Another way uses *Taylor Series*, which we'll see in Chapter 8.

You should know

- the terms *dummy variable, antiderivative,* and *indefinite integral;*

- the relationship between a function and its antiderivatives.

You should be able to

- determine antiderivatives (indefinite integrals) of scaled power functions, trigonometric functions, and exponential functions;

- determine antiderivatives (indefinite integrals) of sums of the above-specified functions;

- use the Fundamental Theorem of Calculus to calculate the value of a definite integral.

❖ 5.4 Skill Exercises

In #1–24 (a) determine the indefinite integral, (b) check your answer to part (a) by differentiating it.

1. $\int 8t + 7 \ dt$

2. $\int 7t + 5 \ dt$

3. $\int 8t^6 + 7t^2 \ dt$

4. $\int 2t^3 - 18t \ dt$

5. $\int e^t - 8 \ dt$

6. $\int 7e^t + t \ dt$

7. $\int 7e^t \ dt$

8. $\int 8e^t \ dt$

9. $\int 6^t \ dt$

10. $\int 3^t \ dt$

11. $\int 7(4^t) \ dt$

12. $\int 9(2^t) \ dt$

13. $\int \frac{3}{t} + 6t \ dt$

14. $\int \frac{4}{t} + 9^t \ dt$

15. $\int 2\sin(t) \ dt$

16. $\int 9\cos(t) \ dt$

17. $\int \sin(2t) \ dt$

18. $\int \cos(9t) \ dt$

19. $\int 3\sin(8t) \ dt$

20. $\int 4\sin(7t) \ dt$

21. $\int_0^\pi 9\csc^2(13t) \ dt$

22. $\int_0^\pi 2\csc^2(14t) \ dt$

23. $\int 5\sec^2(7t) + 8t^2 \ dt$

24. $\int 3\sec^2(9t) - 3(7)^t \ dt$

In #25–28 write the integrand as a sum of functions before integrating.

25. $\int \dfrac{t^2 - 1}{t}\, dt$ 26. $\int \dfrac{t - \sqrt{t}}{t^2}\, dt$ 27. $\int \dfrac{1 + t^3}{\sqrt{t}}\, dt$ 28. $\int \dfrac{t + 3}{\sqrt[5]{t}}\, dt$

In #29–40 evaluate the definite integral using the Fundamental Theorem of Calculus.

29. $\displaystyle\int_1^3 4t - 7\, dt$ 33. $\displaystyle\int_{-1}^1 3t^5 + 1\, dt$ 37. $\displaystyle\int_0^1 t^{4/7}\, dt$

30. $\displaystyle\int_{-1}^6 -2t + 8\, dt$ 34. $\displaystyle\int_{-1}^1 2t^{17} + 8t^{23}\, dt$ 38. $\displaystyle\int_1^{32} t^{2/5}\, dt$

31. $\displaystyle\int_0^1 4t^2 + 2t - 3\, dt$ 35. $\displaystyle\int_0^9 \sqrt{t}$ 39. $\displaystyle\int_1^8 4t^{1/3} + 2t^{-3}\, dt$

32. $\displaystyle\int_1^2 9t^5 + 7t^3 + 2t\, dt$ 36. $\displaystyle\int_8^{27} \sqrt[3]{t}\, dt$ 40. $\displaystyle\int_1^{16} 8t^{5/2} + 11t^{-7/4}\, dt$

41. Differentiate to show that $\ln(1 + t)$ is an antiderivative of $\frac{1}{1+t}$, but $\ln(1 + t^2)$ is *not* an antiderivative of $\frac{1}{1+t^2}$

42. Suppose $f(t) = t^2$. Use the Fundamental Theorem of Calculus to show that $\int_a^c f(t)\, dt + \int_c^b f(t)\, dt = \int_a^b f(t)\, dt$, no matter the values of a, b, and c.

❖ 5.4 Concept and Application Exercises

In #43–46 (a) sketch the graph of the integrand over the interval of integration, (b) determine the value of the integral based on the geometry of your graph, and (c) evaluate the definite integral using the Fundamental Theorem of Calculus.

43. $\displaystyle\int_0^8 4\, dt$ 44. $\displaystyle\int_0^8 -2\, dt$ 45. $\displaystyle\int_0^4 t\, dt$ 46. $\displaystyle\int_{-3}^3 5t\, dt$

In each of #47–52, determine the net area between the horizontal axis and the graph of $f(t)$ over the specified interval.

47. $f(t) = 3t^2 + 1$, $[0, 5]$ 50. $f(t) = 7\sin(13t) + \frac{4}{t}$, $[1, 8]$

48. $f(t) = 4t^7 - 8t$, $[-3, 5]$ 51. $f(t) = \sec^2(t) - t^2$, $[-\pi/4, \pi/4]$

49. $f(t) = 3\cos(8t) + t$, $[-\pi, 1.5\pi]$ 52. $f(t) = \csc^2(t) + \sqrt{t}$, $[\pi/4, \pi/2]$

53. Find a value of T so that the net area between the graph of $f(t) = 3t^2 - 4$ and the horizontal axis over $[0, T]$ is zero.

54. Near the surface of the Earth, the acceleration due to gravity is virtually constant at $-9.8\ \frac{\text{m}}{\text{sec}}$ per second. | Kinetics |

   (a) Since $v'(t) = -9.8$, use the indefinite integral to write a formula for $v(t)$.

   (b) What does the "$+C$" in your formula from part (a) mean in this context?

   (c) Since $y'(t) = v(t)$, use the indefinite integral to write a formula for $y(t)$.

   (d) You saw another constant of integration in part (c), say "$+D$." What does this constant mean in this context?

55. Suppose a rocket is ascending at a rate of $v(t) = 10t^2 + 14t \ \frac{\text{ft}}{\text{sec}}$ during the first 6 seconds of its flight. If the rocket starts on the ground, how long is it before it reaches an altitude of 100 feet?

Physics

56. Suppose a rocket is ascending at a rate of $v(t) = 10t^2 + 14t \ \frac{\text{ft}}{\text{sec}}$ during the first 6 seconds of its flight, at which time it runs out of fuel.

Physics

　　(a) What's the rocket's altitude when the engine shuts down?

　　(b) After the engine shuts down, the rocket coasts with a velocity of $v(t) = -32(t-6) + v_0$, where t is the time since launch and v_0 is your answer from part (a). What time does it hit the ground, and how fast is it going?

57. Consider the bracket that stretches across $[0, 4]$ whose upper and lower boundaries are the curves $y = 4 + 0.5x^2$ and $y = 0.25x$, where distance is measured in inches. If the bracket will be 0.1 inch thick and made out of aluminum, how much will it weigh?

Engineering

58. Due to the permeability of a cell wall to certain ions, we often model the electric properties of organic cells with RC-circuits. In one such model, the voltage across the cell wall changes at a rate of $V'(t) = -20e^{-t/10}$ mV/second. What is the net change in the voltage across the cell wall between $t = 0$ and $t = 2$ seconds?

Biology

59. Suppose a computer chip cools at a rate of $T'(t) = -0.5e^{-0.1t} \ \frac{\text{°F}}{\text{sec}}$ after the machine is turned off.

Computer Engineering

　　(a) What is the net change in the chip's temperature during the first minute after the machine is turned off?

　　(b) If the chip's operating temperature is 85°F, how long does it take to cool down to 81°F?

60. Suppose that $f''(t) = 2t - 4$, $f'(1) = 3$ and $f(2) = 8$. Find a formula for $f(t)$.

61. Suppose $f''(t) = (1-t)^2$.

　　(a) Verify that $t = 1$ is a second-order critical point of f, but not an inflection point.

　　(b) Find an antiderivative of f'' by integrating, and make sure that $t = 1$ is not a first-order critical point of f.

　　(c) Find a formula for $f(t)$ by integrating again.

　　(d) Use a graphing utility to render a graph of f near $t = 1$, and describe what you see.

62. You know that $f(t) = t^{-4}$ is always positive, so the equation

$$\int_{-3}^{1} t^{-4} \ dt = -\frac{1}{3}t^{-3}\Big|_{-3}^{1} = -\frac{28}{81}$$

must be wrong. Find the error.

Checking the details

63. Use the Fundamental Theorem of Calculus to show that

$$\int_{-\Delta t}^{\Delta t} \beta t^3 + at^2 + bt + c \ dt = \frac{\Delta t}{3}\Big(2a(\Delta t)^2 + 6c\Big).$$

Verify the equalities in #64–67 by differentiating the proposed antiderivataive (assume that β and ω are constants, and $\omega \neq 0$).

64. $\int \beta \, \sec^2(\omega t) \, dt = \frac{\beta}{\omega} \tan(\omega t) + C$

65. $\int \beta \, \csc^2(\omega t) \, dt = -\frac{\beta}{\omega} \cot(\omega t) + C$

66. $\int \beta \, \sec(\omega t) \tan(\omega t) \, dt = \frac{\beta}{\omega} \sec(\omega t) + C$

67. $\int \beta \, \csc(\omega t) \cot(\omega t) \, dt = -\frac{\beta}{\omega} \csc(\omega t) + C$

68. Suppose $f(t) = \beta \ln|t| + C$ where $\beta \neq 0$ and C is any constant.

    (a) Sketch a representative graph of f over $(-2, 2)$, using $\beta = 1$ and $C = 0$.

    (b) Write $f(t)$ as $\beta \ln(x) + C$ where $x = |t|$.

    (c) Determine dx/dt when $t < 0$.

    (d) Use the fact that $\frac{df}{dt} = \frac{df}{dx}\frac{dx}{dt}$ to show that $f'(t) = \frac{\beta}{t}$ when $t < 0$.

5.5   More About the Fundamental Theorem

Part 1 of the Fundamental Theorem of Calculus was introduced on p. 371. It tells us that

$$y(t) - y(a) = \int_a^t y'(\tau)\, d\tau.$$

Since $y(a)$ is a number, differentiating both sides of this equation with respect to time yields

$$y'(t) = \frac{d}{dt} \int_a^t y'(\tau)\, d\tau, \tag{5.1}$$

which expresses the fundamental relationship between integration and differentiation. In short, it says that

$$\int_a^t y'(\tau)\, d\tau \ \text{ is an antiderivative of } \ y'(t), \tag{5.2}$$

regardless of whether we can find a familiar formula for the integral. For example, consider the following ...

Finding an antiderivative

Example 5.1. Use equation (5.2) to find an antiderivative for (a) $y'(t) = 2t + 1$, and (b) $y'(t) = e^{\cos(t)}$.

Solution: (a) Equation (5.2) tells us that an integral of the form $\int_3^t y'(\tau)\, d\tau$ is an antiderivative. In this case we can find a familiar form of the integral by applying our usual integration procedure:

$$\int_3^t y'(\tau)\, d\tau = \int_3^t 2\tau + 1 \ d\tau = \tau^2 + \tau \Big|_3^t = t^2 + t - 12,$$

> Different lower bounds result in different constants, so instead of "-12" we might see "$+5.1$" or "-7," none of which affect the derivative.

whose derivative is exactly $2t + 1$. (b) We are unable to find a familiar expression for an antiderivative of $e^{\cos(t)}$. However the function is continuous everywhere, so it's integrable. That is, the integral

$$\int_0^t e^{\cos(\tau)}\, d\tau$$

exists, and its value depends on the upper bound of integration, t. So even though we cannot find a familiar expression for it, this integral *is* a function of t. Moreover, equation (5.1) tells us that

$$e^{\cos(t)} = \frac{d}{dt} \int_0^t e^{\cos(\tau)}\, d\tau,$$

so $\int_0^t e^{\cos(\tau)}\, d\tau$ is an antiderivative of $e^{\cos(t)}$. ∎

Fundamental Theorem of Calculus (FTC), Part 2: Suppose that $f(t)$ is continuous over $[a, b]$. Then at each $t \in (a, b)$

$$\frac{d}{dt} \int_a^t f(\tau)\, d\tau = f(t).$$

Part 1 is found on p. 371

Proof. For ease of notation, let's define $g(t) = \int_a^t f(\tau)\, d\tau$. We have to prove that

$$\lim_{\Delta t \to 0} \frac{g(t + \Delta t) - g(t)}{\Delta t} = f(t). \tag{5.3}$$

Our strategy will be to establish (5.3) by finding m and M for which

$$m \le \frac{g(t + \Delta t) - g(t)}{\Delta t} \le M.$$

It will be true that both m and M converge to $f(t)$ as $\Delta t \to 0$, so equation (5.3) will follow from the Squeeze Theorem.

We'll begin by assuming that $\Delta t > 0$ and developing a formula for the difference quotient of g. The numerator is

$$g(t + \Delta t) - g(t) = \int_a^{t+\Delta t} f(\tau)\, d\tau - \int_a^t f(\tau)\, d\tau.$$

Equation (2.8) on p. 353 tells us that we can separate the first of these integrals into two parts:

$$\int_a^{t+\Delta t} f(\tau)\, d\tau = \int_a^t f(\tau)\, d\tau + \int_t^{t+\Delta t} f(\tau)\, d\tau,$$

which allows us to rewrite the numerator of the difference quotient as

$$g(t + \Delta t) - g(t) = \int_a^t f(\tau)\, d\tau + \int_t^{t+\Delta t} f(\tau)\, d\tau - \int_a^t f(\tau)\, d\tau.$$

Notice that the first and last terms on the right-hand side are the same, but one is added and the other subtracted, so their net contribution is zero. This leaves us with

$$g(t + \Delta t) - g(t) = \int_t^{t+\Delta t} f(\tau)\, d\tau.$$

Next, we use this expression for $g(t+\Delta t) - g(t)$ to establish the string of inequalities that was mentioned in our initial discussion of the proof's structure. When the closed interval $[t, t + \Delta t]$ is contained in (a, b), the Extreme Value Theorem guarantees that the continuous function f achieves a maximum value, M, so

$$\int_t^{t+\Delta t} f(\tau)\, d\tau \le \int_t^{t+\Delta t} M\, dt = M\Delta t. \tag{5.4}$$

The continuous function f also achieves a minimum value, m, somewhere in the closed interval $[t, t + \Delta t]$. So

$$\int_t^{t+\Delta t} f(\tau)\, d\tau \ge \int_t^{t+\Delta t} m\, dt = m\Delta t. \tag{5.5}$$

Writing these inequalities together, we have

$$m\Delta t \le g(t + \Delta t) - g(t) \le M\Delta t,$$

from which it follows that

$$m \le \frac{g(t + \Delta t) - g(t)}{\Delta t} \le M. \tag{5.6}$$

Don't just read it. Work it out using the properties of definite integrals (see p. 352)!

Remember that $\Delta t > 0$, so dividing by Δt preserves the order of inequalities.

Recall that m is a particular value of f, say $m = f(t^*)$ where $t^* \in [t, t + \Delta t]$. Limiting $\Delta t \to 0^+$ forces $t^* \to t$ so, because f is continuous,

$$\lim_{\Delta t \to 0^+} m = \lim_{t^* \to t} m = \lim_{t^* \to t} f(t^*) = f(t).$$

The same reasoning explains why $M \to f(t)$ as $\Delta t \to 0^+$. Now we can apply the Squeeze Theorem to equation (5.6) and conclude that

$$\lim_{\Delta t \to 0^+} \frac{g(t + \Delta t) - g(t)}{\Delta t} = f(t)$$

The proof when $\Delta t < 0$ is virtually the same, but requires careful attention to inequalities.   ∎

> The number $t + \Delta t$ is smaller than t when $\Delta t < 0$, which reverses the inequalities in equations (5.4) and (5.5); and dividing by Δt changes inequalities when $\Delta t < 0$.

❖ Understanding in Terms of Graphs and Net Area

We can cast the Fundamental Theorem in graphical terms by understanding $\int_a^t f(\tau)\, d\tau$ as the net area bounded by the graph of f and the horizontal axis between times a and t. In this context, we refer to $\int_a^t f(\tau)\, d\tau$ as the **net area function**, or as the **cumulative area function** if $f(t)$ is never negative. For example, suppose Figure 5.1 shows the graph of f, and $a = 1$.

- In graphs (a) and (b) you see the net area increase as t grows from 1 to 2, because $f(t)$ is positive in that interval.

- In graphs (c) and (d) you see the net area decrease as t grows from 2 to 3, because $f(t)$ is negative in that interval (so there's more and more area beneath the horizontal axis as t grows).

- In graphs (e) and (f) you see the net area increase as t grows from 3 to 3.5 and beyond, because $f(t)$ is positive.

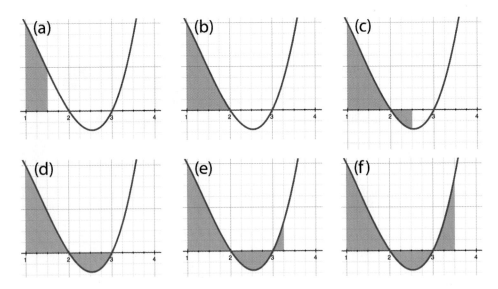

Figure 5.1: Progressive depictions of net area as the upper bound of integration grows—(a) and (b) net area is increasing; (c) and (d) net area is decreasing, (e) and (f) net area is increasing.

In summary,

$$\int_a^t f(\tau)\, d\tau \quad \text{is} \quad \begin{cases} \text{increasing where } f(t) > 0 \\ \text{decreasing where } f(t) < 0 \end{cases},$$

which makes sense since $f(t)$ is the derivative of $\int_1^t f(\tau)\,d\tau$.

Graphing the net area function

Example 5.2. Graph $f(t) = 0.5t^3 - 2t^2 + 0.5t + 3$ and the associated net area function (starting at $a = 1$) on the same axes.

Solution: The net area function associated with f (starting at $a = 1$) is

$$g(t) = \int_1^t f(\tau)\,d\tau = \int_1^t \frac{1}{2}\tau^3 - 2\tau^2 + \frac{1}{2}\tau + 3\,d\tau.$$

We can simplify the formula for the net area function by using the Fundamental Theorem of Calculus, but we need an antiderivative to do that. Since this integrand is a polynomial, finding one is easy:

$$\int \frac{1}{2}\tau^3 - 2\tau^2 + \frac{1}{2}\tau + 3\,d\tau = \int \frac{1}{2}\tau^3\,d\tau - \int 2\tau^2\,d\tau + \int \frac{1}{2}\tau\,d\tau + \int 3\,d\tau$$

$$= \frac{1}{8}\tau^4 - \frac{2}{3}\tau^3 + \frac{1}{4}\tau^2 + 3\tau + C.$$

Now choosing $C = 0$, we can use the Fundamental Theorem of Calculus to write

$$g(t) = \int_1^t f(\tau)\,d\tau = \int_1^t \frac{1}{2}\tau^3 - 2\tau^2 + \frac{1}{2}\tau + 3\,d\tau$$

$$= \frac{1}{8}\tau^4 - \frac{2}{3}\tau^3 + \frac{1}{4}\tau^2 + 3\tau \Big|_1^t$$

$$= \underbrace{\frac{1}{8}t^4 - \frac{2}{3}t^3 + \frac{1}{4}t^2 + 3t}_{\text{our antiderivative evaluated at } \tau = t \ldots} - \underbrace{\frac{65}{24}}_{\ldots \text{and at } \tau = 1}.$$

> Recall that the value of C is irrelevant when using the Fundamental Theorem of Calculus. We take $C = 0$ to make the calculation simple.

The graphs of f and g are shown together in Figure 5.2. Note that the graph of f is the very one that was used in Figure 5.1. Note also that g is increasing (decreasing) wherever $f(t)$ is positive (negative), as we said it would. ∎

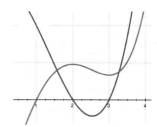

Figure 5.2: The graphs of f and g for Example 5.2.

Let's pause to remark that $g(1) = 0$ in Example 5.2, which makes sense because the region beneath the graph of f from 1 to 1 has no width, so its area is zero. Let's also note that $g(t) < 0$ when $t < 1$, even though $f(t)$ is positive. At an intuitive level, this happens because our interpretation of

$$\sum_{k=1}^n f(\tau_k^*)\Delta\tau \quad \text{as an area relies on } \Delta\tau = \frac{b-a}{n} > 0,$$

so that we can understand $\Delta\tau$ as a width; but the number $\frac{b-a}{n}$ is *negative* when $b < a$, so that $f(\tau_k^*)\Delta\tau < 0$ when $f(\tau_k^*) > 0$. That is, starting at time 1 and moving *left* to time t accumulates "negative" area, so $g(t)$ will be negative.

> The number $\Delta\tau < 0$ when $b < a$, meaning that $\Delta\tau$ is a step backwards along the horizontal axis, or a step backwards in time.

From an algebraic standpoint, it happens because

$$\int_1^t f(\tau)\,d\tau = -\underbrace{\int_t^1 f(\tau)\,d\tau}.$$

> Since $t < 1$, the lower bound of this integral is smaller than the upper bound, which is the way we've been thinking about definite integrals.

❖ Practice Using the Fundamental Theorem of Calculus

Understanding the Fundamental Theorem of Calculus and *using* the Fundamental Theorem of Calculus are two different things, so let's get some practice . . .

Using the Fundamental Theorem of Calculus

Example 5.3. Suppose $g(t) = \int_0^t e^{\cos(\tau)} \, d\tau$. Determine the slope of the tangent line to the graph of g at $t = \pi$.

Solution: A more familiar expression for $g(t)$ is unavailable, but we don't need one to answer this question. The Fundamental Theorem of Calculus tells us that differentiating g leaves us with the integrand. So

$$g'(t) = e^{\cos(t)} \quad \Rightarrow \quad g'(\pi) = e^{-1}. \qquad \blacksquare$$

Try It Yourself: Using the Fundamental Theorem of Calculus

Example 5.4. Suppose $g(t) = \int_3^t \cos\left(\ln(\tau + 4)\right) d\tau$. (a) Determine a formula for $g'(t)$, and (b) calculate the slope of the tangent line to the graph of g at $t = 6$.

Answer: (a) $g'(t) = \cos\left(\ln(t + 4)\right)$, and (b) $g'(6) = \cos(\ln(10))$. $\qquad \blacksquare$

Full Solution
On-line

Combining the Fundamental Theorem of Calculus with the Chain Rule

Example 5.5. Determine $g'(t)$ when $g(t) = \int_3^{8t} \cos\left(\ln(\tau + 4)\right) d\tau$.

Solution: The Fundamental Theorem of Calculus doesn't address integrals in which the upper bound of integration is itself a function of t, so let's think of the function g as

$$g(x) = \int_3^x \cos\left(\ln(\tau + 4)\right) d\tau, \quad \text{where } x = 8t.$$

> The upper bound of integration is moving to the right at 8 units per second.

Now the Chain Rule tells us that

$$\frac{dg}{dt} = \frac{dg}{dx}\frac{dx}{dt} = g'(x)\, x'(t).$$

We know that $x'(t) = 8$, and the Fundamental Theorem of Calculus tells us

$$g'(x) = \frac{d}{dx}\int_3^x \cos\left(\ln(\tau + 4)\right) d\tau = \cos\left(\ln(x + 4)\right).$$

Substituting this into our Chain Rule equation, and using the fact that $x = 8t$, we have

$$\frac{d}{dt}\int_3^{8t} \cos(\ln(\tau + 4)) \, d\tau = \overbrace{\cos\left(\ln(8t + 4)\right)}^{g'(x) \text{ with } x=8t} \underbrace{8}_{x'(t)}. \qquad \blacksquare$$

> The term $8t$ tells us where the upper bound of integration is at any given time, and the factor of 8 that you see at the end of this equation tells us how quickly the upper bound of integration is moving.

Try It Yourself: The Fundamental Theorem of Calculus with the Chain Rule

Example 5.6. Find the derivative of $\int_8^{13t} \tan\left(3^\tau + 4\right) d\tau$ with respect to t.

Answer: $13 \tan(3^{13t} + 4)$. $\qquad \blacksquare$

Full Solution
On-line

Examples 5.5 and 5.6 hint at more general fact: the rate at which the net area function changes depends on how quickly the upper bound of integration is moving.

Suppose f is continuous on an open interval containing $[a, x(t)]$, and $x(t)$ is differentiable. Then

$$\frac{d}{dt} \int_a^{x(t)} f(\tau)\, d\tau = f(x(t))\, x'(t).$$

> If $x(t) < a$, we have to change the interval notation, so the hypotheses are, "Suppose f is continuous on an open interval containing $[x(t), a]$, and $x(t)$ is differentiable."

Combining the Fundamental Theorem of Calculus with the Chain Rule

Example 5.7. Determine $g'(t)$ when $g(t) = \int_0^{\cos(t)} \ln\left(7^\tau + \tau\right)\, d\tau$.

Solution: Unlike Examples 5.5 and 5.6, this upper bound of integration is not always moving to the right. Nonetheless, the formula we've developed is still correct, so

$$g'(t) = \ln\left(7^{\cos(t)} + \cos(t)\right)\left(-\sin(t)\right).$$ ∎

Using the Fundamental Theorem of Calculus with properties of the definite integral

Example 5.8. Find the derivative of $\int_t^6 \sec\left(\tau^2 + 1\right)\, d\tau$ with respect to t.

Solution: The Fundamental Theorem of Calculus addresses net area functions in which the variable appears as the upper bound of integration, so let's begin by rewriting our function as

$$g(t) = -\int_6^t \sec\left(\tau^2 + 1\right)\, d\tau.$$

Now the Fundamental Theorem of Calculus allows us to differentiate, and we see

$$g'(t) = -\sec\left(t^2 + 1\right).$$ ∎

Try It Yourself: Using FTC with properties of the definite integral

Example 5.9. Find the derivative of $\int_{\tan(t)}^9 e^\tau\, d\tau$ with respect to t.

Answer: $-e^{\tan(t)} \sec^2(t)$. ∎

Full Solution On-line

Using Fundamental Theorem of Calculus with properties of the definite integral

Example 5.10. Find the derivative of $\int_{\tan(t)}^t e^\tau\, d\tau$ with respect to t.

Solution: We begin by writing this function as a pair of integrals (since the integrand is continuous everywhere, it doesn't matter where we do it):

$$\int_{\tan(t)}^t e^\tau\, d\tau = \int_{\tan(t)}^9 e^\tau\, d\tau + \int_9^t e^\tau\, d\tau.$$

This allows us to calculate

$$\frac{d}{dt} \int_{\tan(t)}^t e^\tau\, d\tau = \frac{d}{dt} \int_{\tan(t)}^9 e^\tau\, d\tau + \frac{d}{dt} \int_9^t e^\tau\, d\tau.$$

We found the first of these derivatives in Example 5.9, and the second is addressed directly by the Fundamental Theorem of Calculus:

$$-e^{\tan(t)} \sec^2(t) + e^t.$$ ∎

Try It Yourself: Using FTC with properties of the definite integral

Example 5.11. Find the derivative of $\int_t^{\sin(8t)} \sec\left(\tau^2 + 1\right) \, d\tau$ with respect to t.

Full Solution On-line

Answer: $-\sec\left(t^2 + 1\right) + 8\cos(8t)\sec\left(\sin^2(8t) + 1\right)$. ∎

You should know

- the terms *net area function*, and *cumulative area function;*

- the statement and meaning of the Fundamental Theorem of Calculus;

- that the net (cumulative) area function is increasing where $f(t) > 0$ and decreasing where $f(t) < 0$.

You should be able to

- determine a formula for position of an object as a function of time, when given a formula for its velocity;

- determine where the net (cumulative) area function is increasing based on a graph of the integrand;

- use the Fundamental Theorem of Calculus to differentiate the net (cumulative) area function and variations of it.

❖ 5.5 Skill Exercises

In #1–8 determine a formula for $A'(t)$.

1. $A(t) = \int_0^t x^3 - 4x \, dx$

2. $A(t) = \int_\pi^t \cos(4x) \, dx$

3. $A(t) = \int_{10}^{t^2} 3x^5 - 4x \, dx$

4. $A(t) = \int_8^{\sqrt{t}} 5x^3 + 8 \, dx$

5. $A(t) = \int_{\cos(t)}^3 \sin(x) \, dx$

6. $A(t) = \int_{\tan(t)}^8 \frac{1}{1+x^3} \, dx$

7. $A(t) = \int_{t^2}^t \sin(x) \, dx$

8. $A(t) = \int_{\cos(t^3)}^{t^3} \frac{1}{1+x^3} \, dx$

9. Suppose that f is an even function. Determine a formula for $\frac{d}{dt} \int_{-t}^t f(x) \, dx$.

10. Suppose that f is an odd function. Determine a formula for $\frac{d}{dt} \int_{-t}^t f(x) \, dx$.

11. Suppose $2x^6 - 7x = \int_0^x f(t) \, dt$. Determine a formula for $f(x)$.

12. Suppose $2^x + \cos(\pi x) - 7 = \int_x^3 f(t) \, dt$. Determine a formula for $f(x)$.

13. Suppose $3x + 7 = \int_a^x f(t) \, dt$. What is the value of a?

14. Suppose $\ln(x) + x^2 = \int_a^x f(t) \, dt$. What is the value of a?

15. Locate the first-order critical points of $A(t) = \int_2^t 8x^3 + 5x^2 - 7x \, dx$.

16. Suppose $A(t) = \int_{-2}^t |x| \, dx$. Show that $A'(0)$ exists, but $A''(0)$ does not.

❖ 5.5 Concept and Application Exercises

In each of #17–24 draw the graph of a function f so that $A(t) = \int_0^t f(x)\,dx$ has the given characteristics.

17. $A(2) > 0$, and $A(t)$ is increasing at $t = 2$.

18. $A(3) < 0$, and $A(t)$ is decreasing at $t = 3$.

19. $A(1) > 0$, and $A(t)$ is decreasing at $t = 1$.

20. $A(4) < 0$, and $A(t)$ is increasing at $t = 4$.

21. increasing and concave up at $t = -1$.

22. increasing and concave down at $t = -2$.

23. decreasing and concave up at $t = -3$.

24. decreasing and concave down at $t = -4$.

25. Suppose $F(t) = \int_0^t f(x)\,dx$ and $G(t) = \int_2^t f(x)\,dx$. Compare and contrast the graphs of F and G.

26. Suppose $A(t) = \int_3^t f(x)\,dx$. Without knowing anything about f (other than the fact that it's integrable), locate a root of A.

Exercises #27–32 ask about $A(t) = \int_0^t f(x)\,dx$, where f is the function whose graph is shown in the left-hand image of Figure 5.3.

27. Is $A(t)$ increasing or decreasing at $t = 1$?

28. Is $A(t)$ increasing or decreasing at $t = 2.5$?

29. Determine the critical points of $A(t)$.

30. At what value(s) of t is the graph of A steepest?

31. On what interval(s) is the graph of A linear?

32. On what interval(s) is the graph of A parabolic?

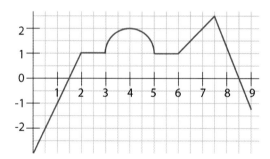 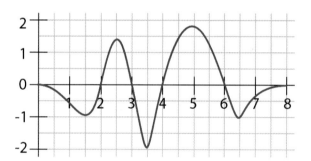

Figure 5.3: (left) Graph for exercises #27–32; (right) graph for exercises #33–37.

Exercises #33–37 ask about $A(t) = \int_0^t f(x)\,dx$, where f is the function whose graph is shown in the right-hand image of Figure 5.3.

33. Find all (any) local maxima of $A(t)$ on $[0, 8]$.

34. Find all (any) local minima of $A(t)$ on $[0, 8]$.

35. On what interval(s) is the graph of A concave down?

36. On what interval(s) is the graph of A concave up?

37. Sketch a graph of $A(t)$.

38. Suppose $A(t) = \int_{-1}^t f(x)\,dx$ where the graph of f is shown in the left-hand image of Figure 5.4. Sketch a graph of A.

39. Suppose $f(x) = \sin(x^2)$, and $A(t) = \int_0^t f(x)\,dx$. We'll denote by $B(t)$ the area of the right triangle shown in the right-hand image of Figure 5.4, whose hypotenuse stretches from $(0,0)$ to $(t, f(t))$. Calculate $\lim_{t \to 0^+} A(t)/B(t)$.

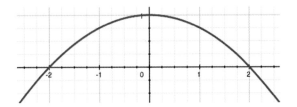
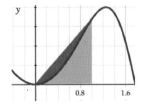

Figure 5.4: (left) Graph for exercise #38; (right) graph for exercise #39.

40. Suppose $A(t) = \int_t^6 f(x)\, dx$ where f is the function whose graph is shown in the right-hand image of Figure 5.3. Determine (a) the value of $A'(5)$, and (b) the value of $A'(3)$

41. Suppose $A(t) = \int_t^7 f(x)\, dx$ where f is the function whose graph is shown in the left-hand image Figure 5.3. Put the numbers $A'(0.5), A'(4), A'(8.5)$ in increasing order.

42. A circular pipe has a radius of 1 inch, but flow through the pipe is blocked by a valve. When the valve is open (either fully or partially, as shown in Figure 5.5) the flow is proportional to the opened area.

    Engineering

    (a) Verify that, when the left-edge of the valve is at x, the flow is

    $$f(x) = 2k \int_0^x \sqrt{1 - (w-1)^2}\, dw.$$

    where k is a constant of proportionality.

    (b) Suppose the valve is sliding to the right at $0.25\ \frac{\text{in}}{\text{sec}}$. What's the rate of change in the flow through the pipe when the pipe is 75% open?

    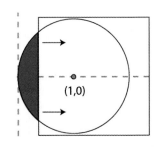

    Figure 5.5: For #42.

43. The area between the horizontal axis and the semicircle $y = \sqrt{16 - x^2}$ over $[-4, x]$ is $A(x) = \int_{-4}^x \sqrt{16 - p^2}\, dp$. Determine (a) where the graph of A is increasing, and (b) where the graph of A is concave up.

44. The area enclosed by the ellipse $3x^2 + 2y^2 = 27$ over $[-3, t]$ is $A(t) = \int_{-3}^t 0.5\sqrt{27 - 3x^2}\, dx$. Determine (a) where the graph of A is increasing, and (b) where the graph of A is concave up.

45. Suppose $A(t) = \int_0^t f(x)\, dx$, where $f(x)$ is a differentiable function that's always positive and has a unique global minimum at $x = 5$ (meaning that $f(x) > f(5)$ when $x \neq 5$). What do you know about $A'(5)$ and $A''(5)$?

46. Suppose $A(t) = \int_0^t f(x)\, dx$, where $f(x)$ is a function that's always decreasing. What do you know about $A'(5)$ and $A''(5)$?

47. Suppose $A(7) < A(5)$ when $A(t) = \int_0^t f(x)\, dx$, where f is a non-constant, continuous function. What do you know about f in the interval $[5, 7]$?

48. Suppose $A(7) = A(5)$ when $A(t) = \int_t^0 f(x)\, dx$, where f is a non-constant, continuous function. What do you know about f in the interval $[5, 7]$?

49. Suppose that a is a fixed, positive number, and consider the following two functions:

    Mathematics

    $$f(t) = \int_1^{at} \frac{1}{x}\, dx \quad \text{and} \quad g(t) = \int_1^a \frac{1}{x}\, dx + \int_1^t \frac{1}{x}\, dx.$$

(a) Verify that $f'(t) = g'(t)$.

(b) Verify that $f(1) = g(1)$.

(c) Based on parts (a) and (b), what must be true about f and g?

(d) What does your answer from part (c) tell us about the natural logarithm?

50. In light of the fact that $g'(t) = \cos(t)$ when $g(t) = 5 + \sin(t)$, consider the equation:

$$\sin(t) + 5 = \int_a^t \cos(x) \; dx = \sin(x)\Big|_a^t = \sin(t) - \sin(a),$$

which seems to indicate that there's some real number a for which $\sin(a) = -5$, but the range of the sine function is $[-1, 1]$. Find the problem in the calculation.

Chapter 5 Review

❖ True or False

1. $\sum_{k=1}^{n} k^2$ is the same as $\left(\sum_{k=1}^{n} k\right)^2$.

2. $\sum_{k=1}^{n} 9k^2$ is the same as $9 \sum_{k=1}^{n} k^2$.

3. If $v(t) > 0$ is the speed of a car that's speeding up, a right-sampled Riemann sum overestimates the distance the car travels.

4. The subintervals of a regular partition have different lengths.

5. If f is a continuous, non-constant function for which $\int_1^2 f(t) \, dt = 0$, the function value must be negative somewhere in $[1, 2]$.

6. If we approximate $\int_3^4 2t^3 + 7 \, dt$ using Simpson's method, our approximation will be *exact* (i.e., no error).

7. The number $\int_0^1 3t^2 - 4 \, dt$ is the area bounded between the curve $y = 3t^2 - 4$ and the horizontal axis when $0 \le t \le 1$.

8. All continuous functions are integrable over $[4, 5]$.

9. The numbers $\int_3^4 2^x \, dx$ and $\int_3^4 2^t \, dt$ are the same.

10. The terms *antiderivative* and *indefinite integral* are synonyms.

11. Suppose that F and G are two antiderivatives of f. Then $F(0) = G(0)$.

12. $\int t^3 \, dt = 3t^2 + C$.

13. Suppose f is continuous. Then $\int_5^t f(x) \, dx$ is differentiable.

14. If $\rho(x) > 0$ has units of g/cm, and x is measured in cm, the integral $\int_1^6 \rho(x) \, dx$ is a mass.

Suppose $a(t)$ and $b(t)$ are the number of plants of species A and B, respectively, after t years. The rates of change in these populations are graphed together in Figure 6.1. Assume that $a(0) = b(0)$.

15. $A(2) > B(2)$

16. $A(3.5) > B(3.5)$

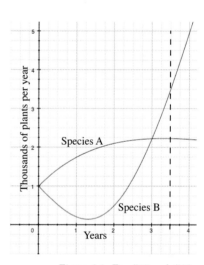

Figure 6.1: For #15 and #16.

❖ Multiple Choice

17. Suppose $f(0) = 1$ and $f'(t) \le 3$. Then we're certain that

    (a) $f(4)$ cannot be negative.   (d) $f(4)$ cannot exceed 15.
    (b) $f(4)$ must be larger than 1.
    (c) $f(4)$ cannot be below -11.   (e) none of the above

18. $\int \frac{1}{t} \, dt =$

    (a) $\ln |t| + C$ (c) $\ln(t)$ (e) none of the above
    (b) $\ln(t) + C$ (d) $\ln |t|$

13. Grab the knob in the lower right corner of cell **E2** and drag downward to cell **E52** to calculate the function value at all of your partition points.

Now that we know the function value at all of our partition points, we have to decide whether we want to use right- or left-sampling. It doesn't matter, mathematically speaking, but our choice will affect what we tell the computer to do in the remaining steps. For the sake of discussion, let's use a right-sampled Riemann sum. In that case, we don't use the function value at $t_0 = 3$, but make our first calculation at time $t_1 = 3.02$.

14. Calculate (function value at $t = 3.02$)×(step size) by entering the formula =**E3*B$4** in cell **F3**.

15. Calculate the cumulative total of the first two terms by entering the formula =**F3+E4*B$4** in cell **F4**.

16. Grab the knob on the lower-right corner of cell **F4** with your mouse, and pull downward to cell **F52**.

After doing this, click on cells **F5**, **F6**, and **F7**. You'll see that the formula has been copied in a way that preserves the *relative* position of data in the spreadsheet. The formulas in rows 5, 6 and 7 use the function values in rows 5, 6 and 7, respectively. However, all of them call on cell **B4** for the step size. This is due to the $ that precedes the 4 in the formula. It tells the computer that this is an *absolute* rather than a *relative* reference.

The value of the Riemann sum is the number you see in cell **F52**. Since we'll always use the same number of summands when $n = 50$, no matter the upper and lower bounds, we can calculate values of related integrals by simply changing the upper and lower bounds. The spreadsheet makes all the necessary adjustments for us.

17. Estimate the value of $\int_0^\pi \sin(t)\, dt$ by changing the upper and lower bounds in your spreadsheet to 0 and 3.14159 respectively.

18. It's easy to use *fewer* subintervals—simply by disregarding or deleting cells that are too low in the spreadsheet. For example, change the lower and upper bounds of integration to 0 and 1, and set $n = 10$ (by entering 10 in cell **B3**). You should see that $t = 1$ appears in cell **D12**. All cells below that are irrelevant to the calculation.

19. It's only a little more difficult to use *more* subintervals. For example, leave the lower and upper bounds of integration as 0 and 1, but set $n = 60$. You should find that $t = 1$ doesn't appear in column D at all.

By making the step size smaller, we need more steps—i.e., more rows. Instead of dragging column D, then column E, and then column F in three separate steps, you can do it all at once.

20. Highlight cells **D52**, **E52**, and **F52**. There will be a knob in the lower-right corner, just like before. Grab it and drag down to row 62.

Of course, we can adapt the spreadsheet in a variety of ways ...

21. Use your spreadsheet to estimate $\int_3^{10} t^2\, dt$ using a right-sampled Riemann sum with 70 summands.

22. How would you change the spreadsheet so that you have a left-sampled Riemann sum instead?

23. How would you change the spreadsheet so that you're calculating with the trapezoid rule instead?

24. How would you change the spreadsheet so that you're calculating with Simpson's rule instead?

❖ An Introduction to Stirling's Approximation

We refer to the product ...

$$3 \cdot 2 \cdot 1 \quad \text{as} \quad \text{"3 factorial," and denote it by 3!}$$
$$4 \cdot 3 \cdot 2 \cdot 1 \quad \text{as} \quad \text{"4 factorial," and denote it by 4!}$$
$$5 \cdot 4 \cdot 3 \cdot 2 \cdot 1 \quad \text{as} \quad \text{"5 factorial," and denote it by 5!}$$
$$\vdots \qquad \vdots$$
$$n \cdot (n-1) \cdots 3 \cdot 2 \cdot 1 \quad \text{as} \quad \text{"n factorial," and denote it by } n!$$

The factorial arises when calculating the number of possible *combinations,* and *permutations* of a set (see Appendix C), and due to the fact that

$$\frac{d^n}{dt^n}(t^n) = n!$$

we'll see them in our study of *power series* in Chapter 8. Because the factorial comes up so often in science and mathematics, many people have worked on formulas for approximating it easily and accurately, especially when n is large, and in this project we're going to use the ideas of this chapter to do just that. Our approximation involves three facts:

- Since $x = e^{\ln(x)}$ when $x > 0$, we can write $n! = e^{\ln(n!)}$.

- We know that $\ln(n!) = \sum_{k=1}^{n} \ln(k)$, and $n \ln(n) = \ln(n^n)$.

- We can approximate $\sum_{k=1}^{n} \ln(k)$ using an integral.

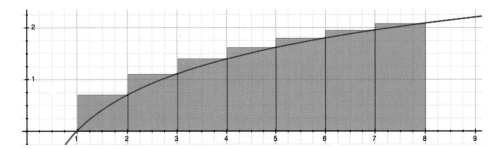

Figure 7.2: The curve $y = \ln(x)$ and rectangles whose areas are $\ln(2)$, $\ln(3)$, $\ln(4)$, $\ldots$, $\ln(8)$.

In all of the steps below, we assume that $n \geq 2$ is an integer.

1. Use Figure 7.2 to explain why $\int_1^n \ln(x)\ dx < \ln(n!)$.

   > We'll use this kind of graphical thought process again when we develop something called the *Integral Test* in Chapter 8.

2. Verify that $x \ln(x) - x$ is an antiderivative of $\ln(x)$ by differentiating it. (You'll see where this comes from in Chapter 6.)

3. Use the Fundamental Theorem of Calculus to calculate $\int_1^n \ln(x)\ dx$.

4. Use the fact that $n \ln(n) = \ln(n^n)$ to verify that $\left(\frac{n}{e}\right)^n \leq \exp\left(\int_1^n \ln(x)\ dx\right)$.

   > Recall that $\exp(u)$ is another way of writing e^u.

5. Use Figure 7.3 to explain why $\ln(n!) < \int_2^{n+1} \ln(x)\ dx$.

6. Calculate $\int_2^{n+1} \ln(x)\ dx$.

7. Verify that $\exp\left(\int_2^{n+1} \ln(x)\ dx\right) = \frac{e^2}{4}\left(\frac{n+1}{e}\right)^{n+1}$.

8. Use your work from #1–7 to show that

$$\left(\frac{n}{e}\right)^n \le n! \le \frac{e^2}{4}\left(\frac{n+1}{e}\right)^{n+1}.$$

The factorial will play an important role in Chapter 8, and this string of inequalities gives us some hint of the rate at which it grows. The famous approximation formula derived by James Stirling (1692–1770) describes the factorial with greater precision. Stirling's approximation formula is often written as

$$n! \approx \sqrt{2\pi n}\left(\frac{n}{e}\right)^n,$$

in which you see the n^n growth that we established with our inequalities. We'll derive Stirling's formula in a project at the end of Chapter 8.

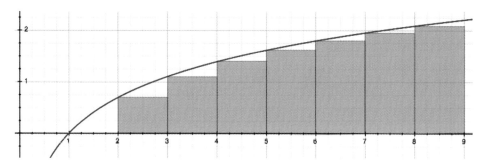

Figure 7.3: The curve $y = \ln(x)$ and rectangles whose areas are $\ln(2)$, $\ln(3)$, $\ln(4)$, $\ldots$, $\ln(8)$.

✧ Pollution and Pulmonary Distress

In order to study the effects of pollutants on individuals whose cardiopulmonary system is already distressed, Elder et al. (see [5]) exposed rats to atmospheres that comprised combinations of ultrafine carbon particles (UFC), ozone, and a toxin called *lipopolysaccharide* (LPS) that is often found in patients with bacterial lung infections. In response to these toxicants, cells in the rats' lungs produced a burst of so-called *reactive oxygen species* (ROS). This typically triggers the production of antioxidants that neutralize unused ROS, preventing them from damaging healthy tissue.

> ROS is also a byproduct of normal cellular respiration.

The research group measured the amount of ROS produced by using a technique called *luminol-enhanced chemiluminescence* in which cells from the rats' lungs are incubated in a chemical called *luminol* whose reaction with ROS results in the emission of light. Both the level and the duration of the ROS response are important, so the cumulative emission of light over time is what we measure, and that's done with integration. However, continuous data about the light emission is not available (because measurements take time, and can only be done so often) so a numerical technique is required.

1. Use the trapezoid rule to determine the cumulative ROS (CROS) response for each rat, based on the chemiluminescence data reported (as a dimensionless quantity) in the tables below.

2. Sum the CROS measurements for all young rats and, separately, sum the CROS measurements for all old rats. Based on these two numbers, which group produced more ROS?

More information is needed before conclusions about causal relationships can be drawn, but this data suggests that toxicants either do more damage to old lungs, or older animals are less able to produce antioxidants to neutralize excess ROS, which can further contribute to the damaging effects of toxicants.

3. Determine the average CROS for the three rats in each category (young and old), and record the the number as a signed deviation from the average CROS of rats (young and old, respectively) exposed to normal atmosphere.

4. Verify that the average CROS for the old rats exposed to the three-toxicant mixture is an order of magnitude larger than sum of the average CROS detected in old rats that were exposed to ozone, to LPS, or to UFC individually; and check that this does *not* happen with the young rats.

This last point may indicate that older animals exceed a threshold beyond which they cannot produce antioxidants.[1]

Old rats exposed to normal air			
Min	Rat #1	Rat #2	Rat #3
0	0.152	0.174	0.152
4	0.412	0.225	0.315
8	0.371	0.162	0.277
12	0.311	0.146	0.200
16	0.287	0.135	0.204
20	0.243	0.101	0.157

Young rats exposed to normal air			
Min	Rat #16	Rat #17	Rat #18
0	0.010	0.027	0.016
4	0.009	0.035	0.000
8	0.026	0.023	0.024
12	0.015	0.026	0.012
16	0.017	0.034	0.020
20	0.013	0.022	0.026

Old rats exposed to UFC			
Min	Rat #4	Rat #5	Rat #6
0	0.043	0.027	0.045
4	0.143	0.114	0.121
8	0.150	0.124	0.158
12	0.095	0.113	0.107
16	0.099	0.072	0.129
20	0.108	0.059	0.093

Young rats exposed to UFC			
Min	Rat #19	Rat #20	Rat #21
0	0.123	0.325	0.129
4	0.144	0.340	0.114
8	0.085	0.216	0.075
12	0.080	0.161	0.056
16	0.075	0.156	0.050
20	0.051	0.127	0.060

Old rats exposed to ozone			
Min	Rat #7	Rat #8	Rat #9
0	0.163	0.135	0.273
4	0.542	0.897	0.621
8	0.424	0.795	0.472
12	0.405	0.777	0.405
16	0.327	0.679	0.333
20	0.249	0.547	0.281

Young rats exposed to ozone			
Min	Rat #22	Rat #23	Rat #24
0	0.104	0.040	0.062
4	0.358	0.324	0.533
8	0.385	0.483	0.512
12	0.385	0.519	0.525
16	0.286	0.447	0.399
20	0.249	0.350	0.319

Old rats exposed to LPS			
Min	Rat #10	Rat #11	Rat #12
0	0.151	0.190	0.125
4	1.367	2.045	1.242
8	1.371	1.755	1.056
12	1.391	1.724	1.056
16	1.237	1.525	0.922
20	1.055	1.320	0.782

Young rats exposed to LPS			
Min	Rat #25	Rat #26	Rat #27
0	0.181	0.151	0.165
4	0.503	0.816	0.292
8	0.639	0.971	0.545
12	0.661	0.965	0.559
16	0.617	0.772	0.501
20	0.530	0.706	0.498

Old rats exposed to LPS & UFC & ozone			
Min	Rat #13	Rat #14	Rat #15
0	0.297	0.232	0.264
4	10.690	7.100	4.620
8	9.390	6.400	4.132
12	8.478	5.942	3.754
16	7.050	4.807	3.309
20	5.787	4.033	2.724

Young rats exposed to LPS & UFC & ozone			
Min	Rat #28	Rat #29	Rat #30
0	0.140	0.054	0.071
4	0.367	0.214	0.250
8	0.681	0.534	0.473
12	0.945	0.652	0.697
16	0.925	0.715	0.799
20	0.822	0.654	0.729

[1]Thanks to Dr. Alison Elder, University of Rochester School of Medicine.

Chapter 6
Techniques of Integration

In Chapter 5 we used the definite integral to calculate the net change in, or accumulation of, a quantity, and we saw that it's easy to do with the Fundamental Theorem of Calculus *provided that we can find an antiderivative of the integrand.* In this chapter, we extend our ability to find antiderivatives using our knowledge of the Chain Rule and Product Rule.

6.1 Substitution

In this section we talk about a technique of integrating called *substitution* (or *change of variable*) in which we use the Chain Rule "in reverse."

❖ The Basic Idea

You know from the Chain Rule that

$$\frac{d}{dt}\sin(3t^2) = \cos(3t^2)\,6t. \tag{1.1}$$

That is, the function $\sin(3t^2)$ is an antiderivative of $\cos(3t^2)\,6t$, so when we integrate with respect to time we see

$$\int \cos(3t^2)\,6t\,dt = \sin(3t^2) + C. \tag{1.2}$$

But how would you integrate $\cos(3t^2)\,6t$ if you didn't already know equation (1.1)? The key is noticing that $6t$ is the derivative of $3t^2$. Since you're familiar with differentiation, that might make you think $\cos(3t^2)\,6t$ is the result of having differentiated some function using the Chain Rule, and you could find it by reversing the process. Let's try. Since $6t$ is the derivative of $3t^2$,

$$\cos(3t^2)\,6t \quad \text{has the basic form} \quad \cos(u)\,u',$$

where $u = 3t^2$. We know that u' came from the Chain Rule, but what did we differentiate to get $\cos(u)$? The answer is $\sin(u)$, of course. So we guess that $\sin(u)$ is an antiderivative, and replacing u with $3t^2$ leads us to write our answer as $\sin(3t^2)+C$, which is exactly what we said in equation (1.2).

The central idea for you to understand is that grouping terms appropriately allows us to see the integrand as the result of a differentiation with the Chain Rule. Once we know that, we can find an antiderivative by working our way through the differentiation process in reverse.

✤ Working with the Differential

The basic idea presented above is streamlined by employing the identity

$$du = u'(t) \, dt, \qquad\qquad (1.3)$$

which was introduced on p. 243.

Working with the differential

Example 1.1. Determine $\int \cos(3t^2) \, 6t \, dt$.

Solution: As we did above, let's set $u = 3t^2$. Then $u' = 6t$ so that

$$du = 6t \, dt,$$

and we can write the integral as

$$\int \cos(u) \, du \quad \text{which is} \quad \sin(u) + C.$$

You might be tempted to stop here, but we're not quite done. The original indefinite integral refers to a family of functions whose first derivative is $\cos(3t^2) \, 6t$, which is a function of t, so our answer must be in terms of t. The last step of the substitution technique is returning to the original variable, replacing u with $3t^2$, after which we see $\sin(3t^2) + C$. ∎

> You can think of the substitution variable as a way of masking the dependence on t so that we can see the Chain-Rule structure of the integrand. When we're done reverse-engineering the differentiation process, we need to remove the mask.

Working with the differential

Example 1.2. Use substitution to determine $\int (t^2 + 4)^9 \, 3t \, dt$.

Solution: In this integrand we see a quadratic term, $t^2 + 4$, and a linear term, $3t$. Since the derivative of a quadratic function is linear, let's set $u = t^2 + 4$. Then $u' = 2t$ so that

$$du = (2t) \, dt.$$

Our integral involves the term $(3t) \, dt$ rather than $(2t) \, dt$, but if we multiply this equation by $\frac{3}{2}$ we see

$$\frac{3}{2} du = (3t) \, dt.$$

Now we can see that the structure of our original integral is

$$\int u^9 \left(\frac{3}{2} \, du \right) \quad \text{which we write as} \quad \int \frac{3}{2} u^9 \, du$$

because it looks more familiar. Now the integration is easy:

$$\int \frac{3}{2} u^9 \, du = \frac{3}{20} u^{10} + C,$$

after which returning to the original variable yields $\frac{3}{20}(t^2 + 4)^{10} + C$. ∎

> Instead of using substitution, you could expand this integrand into a 19th-degree polynomial, but that would be tedious. Substitution expedites the integration process.
>
> In later examples, substitution will be necessary rather than just convenient.

Try It Yourself: Working with the differential

Example 1.3. Use substitution to determine $\int (t^3 - 2t + 11)^{17}(12t^2 - 8) \, dt$.

Answer: $\frac{1}{2}(t^3 - 2t + 11)^{18} + C$. ∎

Full Solution
On-line

A more complicated differential

Example 1.4. Use substitution to determine $\int (\sqrt{t} - 1)^{17}\, dt$.

Solution: Unlike previous examples, we don't see a factor that seems ready to play the role of u'. However, we know that integrating power functions is easy, so let's try $u = \sqrt{t} - 1$. With that definition of u, we see $u' = \frac{1}{2\sqrt{t}}$, so

$$du = \frac{1}{2\sqrt{t}}\, dt.$$

We don't have a factor of $\frac{1}{2\sqrt{t}}$ in our integrand, but since $\sqrt{t} = u + 1$ we can rewrite the relationship between dt and du as

$$dt = 2\sqrt{t}\, du, \quad \text{which is the same as saying} \quad dt = 2(u+1)\, du.$$

Now substituting into our integral leaves us with

$$\int u^{17}\, 2(u+1)\, du = \int 2u^{18} + 2u^{17}\, du = \frac{2}{19}u^{19} + \frac{1}{9}u^{18} + C.$$

Returning to the original variable, we have $\frac{2}{19}(\sqrt{t} - 1)^{19} + \frac{1}{9}(\sqrt{t} - 1)^{18} + C$. ■

> Alternatively, you could insert a factor of $1/(2\sqrt{t})$ by multiplying $(2\sqrt{t})/(2\sqrt{t})$ onto the integrand.

Try It Yourself: A more complicated differential

Example 1.5. Use the substitution $u = \sqrt[3]{t} - 4$ to determine $\int (\sqrt[3]{t} - 4)^9\, dt$.

Answer: $\frac{1}{4}(\sqrt[3]{t} - 4)^{12} + \frac{24}{11}(\sqrt[3]{t} - 4)^{11} + \frac{24}{5}(\sqrt[3]{t} - 4)^{10} + C$. ■

Full Solution On-line

✥ Applying Substitution to Integrals of Rational Functions

Recall that a rational function is ratio of polynomials. The examples below address cases when the denominator is **irreducible**, meaning that it cannot be factored. The case of a factorable denominator is discussed in Section 6.5.

> It's an important but subtle fact that only quadratic and linear polynomials can be irreducible.

Using a substitution to determine an indefinite integral

Example 1.6. Determine $\int \frac{1}{8t+3}\, dt$.

Solution: When we set $u = 8t + 3$, we see that $du = 8\, dt$, which is the same as saying $dt = \frac{1}{8}\, dx$. This allows us to rewrite the integral as

$$\int \frac{1}{u}\left(\frac{1}{8}\, du\right) = \int \frac{1}{8}\frac{1}{u}\, du = \frac{1}{8}\ln|u| + C.$$

When we return to the original variable, we have that

$$\int \frac{1}{8t + 3}\, dt = \frac{1}{8}\ln|8t + 3| + C.$$ ■

Try It Yourself: Using substitution to determine an indefinite integral

Example 1.7. Use substitution to determine $\int \frac{2t}{3t^2 + 5}\, dt$.

Answer: $\frac{1}{3}\ln|3t^2 + 5| + C$. ■

Full Solution On-line

Using the arctangent

Example 1.8. Determine $\int \frac{1}{3t^2+5}\,dt$.

Solution: Based on the pattern exhibited in Examples 1.6 and 1.7, you might think that the substitution $u = 3t^2 + 5$ would lead us to an answer of $\ln(3t^2 + 5) + C$, but when we check by differentiating with the Chain Rule, we see that

$$\frac{d}{dt}\ln(3t^2+5) + C = \frac{1}{3t^2+5}\,(6t)$$

which is not our integrand, so $\ln(3t^2 + 5) + C$ is incorrect. We can see why our substitution didn't work if we pay attention to the differential. When $u = 3t^2 + 5$ we have $du = 6t\,dt$. That's the same as saying that $dt = \frac{1}{6t}\,du$, and we don't see a factor of $1/t$ in our integrand. When this kind of thing happened in Examples 1.4 and 1.5 we wrote t in terms of u, so you might try to rewrite $u = 3t^2 + 5$ as $t = \cdots$ but we see a $\pm$ occur in the solution and don't know which to choose. In short, we're stuck with writing our integral as

$$\int \frac{1}{3t^2+5}\,dt = \int \underbrace{\frac{1}{u}\,\frac{1}{6t}\,du}_{!!!!!!}$$

where we see both t and u in the same integral! In all of the previous examples, the original variable of integration vanished when we made our substitution. Never were t and u in the same integral, and that turns out to be a good guiding principle: if you ever seen the original and new variables of integration in the same integral, your attempt at substitution has been unsuccessful.

So now what? You can often determine the underlying form of an integrand by temporarily setting all the coefficients to 1. Doing that here yields $\frac{1}{t^2+1}$, which we first saw on p. 214 as the derivative of the arctangent. This leads us to expect an arctangent as our answer.

With that in mind, our strategy is to manipulate our integrand algebraically so that it has the form $\frac{1}{u^2+1}$. Let's begin by factoring out a 5 so that we see a "+1" in the denominator:

$$\int \frac{1}{3t^2+5}\,dt = \int \frac{1}{5}\,\frac{1}{\frac{3}{5}t^2+1}\,dt = \frac{1}{5}\int \frac{1}{\frac{3}{5}t^2+1}\,dt.$$

In order for this integrand to have the form $\frac{1}{u^2+1}$, we need

$$u^2 = \frac{3}{5}t^2, \quad \text{which happens when} \quad u = \frac{\sqrt{3}}{\sqrt{5}}\,t.$$

Taking this to be our new variable, the differential is

$$du = \frac{\sqrt{3}}{\sqrt{5}}\,dt \quad \Rightarrow \quad \frac{\sqrt{5}}{\sqrt{3}}\,du = dt.$$

Now we can write our integral as

$$\frac{1}{5}\int \frac{1}{u^2+1}\,\frac{\sqrt{5}}{\sqrt{3}}\,du = \frac{\sqrt{5}}{5\sqrt{3}}\int \frac{1}{u^2+1}\,du = \frac{\sqrt{5}}{5\sqrt{3}}\tan^{-1}(u) + C.$$

Now substituting our definition of u into this formula, we have

$$\int \frac{1}{3t^2+5}\,dt = \frac{\sqrt{5}}{5\sqrt{3}}\tan^{-1}\left(\frac{\sqrt{3}}{\sqrt{5}}\,t\right) + C. \qquad \blacksquare$$

The fundamental difference between the integrands in Examples 1.7 and 1.8 is that the former has a factor of t in the numerator (which plays a part in du) but the latter does not.

You might be tempted to treat t as constant when looking at the integral on the right-hand side, but it's not. Since $u = 3t^2 + 6$, the variable u changes when t does, and vice versa.

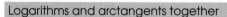

Try It Yourself: Using the arctangent

Example 1.9. Use the substitution technique to determine $\int \frac{5}{19+4t^2}\, dt$.

Answer: $\frac{5}{2\sqrt{19}} \tan^{-1}\left(\frac{2}{\sqrt{19}}\, t\right) + C.$ ∎

Full Solution On-line

Logarithms and arctangents together

Example 1.10. Determine $\int \frac{2t+1}{3t^2+5}\, dt$.

Solution: As in previous examples, we'd like to use the substitution $u = 3t^2 + 5$, but then $du = 6t\, dt$ and our numerator is $2t + 1$. The "+1" in the numerator prevents this substitution from working—at least immediately—but we can separate the integrand into two fractions, and write the integral as

$$\int \frac{2t}{3t^2+5} + \frac{1}{3t^2+5}\, dt = \int \frac{2t}{3t^2+5}\, dt + \int \frac{1}{3t^2+5}\, dt.$$

The first of these integrals was done in Example 1.7, and the second in Example 1.8, so our answer is that

$$\int \frac{2t+1}{3t^2+5}\, dt = \frac{1}{3} \ln|3t^2+5| + \frac{\sqrt{5}}{5\sqrt{3}} \tan^{-1}\left(\frac{\sqrt{3}}{\sqrt{5}} t\right) + C.$$ ∎

Try It Yourself: Using the arctangent

Example 1.11. Use the substitution technique to determine $\int \frac{3t-5}{19+4t^2}\, dt$.

Answer: $\frac{3}{8} \ln|19+4t^2| - \frac{5}{2\sqrt{19}} \tan^{-1}\left(\frac{2}{\sqrt{19}} t\right) + C.$ ∎

Full Solution On-line

Completing the square in order to use the arctangent

Example 1.12. Determine $\int \frac{7}{t^2+6t+13}\, dt$.

Solution: Unlike previous examples, this denominator has a linear term. We get around this by using a technique called *completing the square* to write the denominator as

$$t^2 + 6t + 13 = t^2 + 6t + 9 - 9 + 13 = (t+3)^2 + 4.$$

> The technique of completing the square is very handy when dealing with quadratic expressions. If you need a review, see Appendix B.

This allows us to write our integral as

$$\int \frac{7}{t^2+6t+13}\, dt = \int \frac{7}{(t+3)^2+4}\, dt$$

in which the grouping of $t + 3$ leads us to make the substitution $u = t + 3$. Then $du = dt$, so our integral becomes

$$\int \frac{7}{(t+3)^2+4}\, dt = \int \frac{7}{u^2+4}\, du$$

in which the integrand looks a lot like the derivative of the arctangent—if only the 7 and the 4 were both ones, instead … With that in mind, let's factor out the 7 and the 4 so that we see a "+1" in the denominator.

$$\int \frac{7}{u^2+4}\, du = \int \frac{7}{4} \frac{1}{\frac{1}{4}u^2+1}\, du = \frac{7}{4} \int \frac{1}{\frac{1}{4}u^2+1}\, du$$

> Avoid using the same letter to mean different things in the same calculation. We've already used the letters t and u in our work, so we have to choose something else.

This denominator will have the form $\frac{1}{v^2+1}$ if

$$v^2 = \frac{1}{4}u^2, \quad \text{which happens when} \quad v = \frac{1}{2}u.$$

Taking this to be our new variable, we have

$$dv = \frac{1}{2}\, du \quad \Rightarrow \quad du = 2\, dv.$$

Substituting this into our integral, we see that

$$\frac{7}{4} \int \frac{1}{v^2 + 1}\, 2\, dv = \frac{7}{2} \int \frac{1}{v^2 + 1}\, dv = \frac{7}{2}\tan^{-1}(v) + C.$$

In our final step, we return to the original variable. Since $v = \frac{1}{2}u$ and $u = t + 3$, we have

$$\int \frac{7}{t^2 + 6t + 13}\, dt = \frac{7}{2}\tan^{-1}\left(\frac{t+3}{2}\right) + C. \qquad \blacksquare$$

> We could have arrived at this answer using a single substitution by rewriting the integral as
> $$\frac{7}{4} \int \frac{1}{\frac{(t+3)^2}{4} + 1}\, dt$$
> and then setting
> $$u = \frac{t+3}{2}.$$

Completing the square in order to use the arctangent

Example 1.13. Determine $\int \frac{1}{3t^2 - 24t + 53}\, dt$.

Solution: Seeing the linear term in the denominator leads us to complete the square, and rewrite our integral as

$$\int \frac{1}{3t^2 - 24t + 53}\, dt = \int \frac{1}{3(t - 4)^2 + 5}\, dt.$$

> If you need to review this technique, see Appendix B.

The grouping of $(t - 4)$ leads us to make the substitution $u = t - 4$. Then $du = dt$ so our integral becomes

$$\int \frac{1}{3u^2 + 5}\, du = \frac{\sqrt{5}}{5\sqrt{3}}\tan^{-1}\left(\frac{\sqrt{3}}{\sqrt{5}}\, u\right) + C.$$

> We determined this integral in Example 1.8.

We finish by returning to the original variable. Since $u = t - 4$, we have found that

$$\int \frac{1}{3t^2 - 24t + 53}\, dt = \frac{\sqrt{5}}{5\sqrt{3}}\tan^{-1}\left(\frac{\sqrt{3}}{\sqrt{5}}\,(t - 4)\right) + C. \qquad \blacksquare$$

Try It Yourself: Completing the square in order to use the arctangent

Example 1.14. Determine $\int \frac{1}{2t^2 - 10t + 71}\, dt$.

Answer: $\frac{1}{\sqrt{42}}\tan^{-1}\left(\sqrt{\frac{2}{21}}\,(t - 5)\right) + C.$ $\qquad \blacksquare$

Full Solution On-line

✦ Applying Substitution to Some Trigonometric Integrals

The term *trigonometric integrals* typically refers to those with integrands that are products of trigonometric functions. In Examples 1.15–1.18 we use substitution to complete a table of elementary trigonometric integrals, and then we proceed to more complicated cases that can also be done with substitution.

Integrating the tangent function

Example 1.15. Use substitution to determine $\int \tan(t)\, dt$.

Solution: Since tangent is the ratio of sine to cosine,

$$\int \tan(t)\, dt = \int \frac{\sin(t)}{\cos(t)}\, dt = \int \frac{1}{\cos(t)}\sin(t)\, dt.$$

If we choose $u = \cos(t)$ we see $u' = -\sin(t)$, so the differential is

$$du = -\sin(t)\, dt \quad \Rightarrow \quad \sin(t)\, dt = -du.$$

Substituting this into our integral, we see

$$\int \frac{1}{\cos(t)}\ \sin(t)\ dt = \int \frac{1}{u}\ (-du) = -\int \frac{1}{u}\ du = -\ln|u| + C.$$

Returning to our original variable, we see

$$\int \tan(t)\ dt = -\ln|\cos(t)| + C,$$

which is often written as $\ln|\sec(t)| + C$.   ■

> We can use the laws of exponents to move the -1 coefficient into the exponent of the argument.

Try It Yourself: Integrating the cotangent function

Example 1.16. Use substitution to determine $\int \cot(t)\ dt$.

Answer: $\ln|\sin(t)| + C$, which is sometimes written as $-\ln|\csc(t)| + C$.   ■

Full Solution
On-line

Integrating the secant

Example 1.17. Use the substitution technique to determine $\int \sec(t)\ dt$.

Solution: We know that $\sec(t) = 1/\cos(t)$, but unlike Example 1.15 there's no factor of sine in the integrand to help us build dx. Though it's not obvious, the right thing to do is multiply and divide the integrand by $\sec(t) + \tan(t)$.

$$\int \sec(t)\ dt = \int \sec(t)\ \underbrace{\frac{\sec(t) + \tan(t)}{\sec(t) + \tan(t)}}_{=1}\ dt = \int \frac{\sec^2(t) + \sec(t)\tan(t)}{\sec(t) + \tan(t)}\ dt.$$

This might look more complicated to you, but notice that the numerator of this integrand is exactly the derivative of the denominator! Setting

$$u = \sec(t) + \tan(t) \quad \Rightarrow \quad du = \Big(\sec(t)\tan(t) + \sec^2(t)\Big)\ dt,$$

which allows us to rewrite the integral as

$$\int \frac{1}{u}\ du = \ln|u| + C.$$

Returning to the original variable yields $\ln|\sec(t) + \tan(t)| + C$.   ■

Try It Yourself: Integrating the cosecant

Example 1.18. Determine $\int \csc(t)\ dt$ by multiplying and dividing the integrand by $\csc(t) + \cot(t)$, and then using the substitution technique.

Answer: $-\ln|\csc(t) + \cos(t)| + C$.   ■

Full Solution
On-line

Let's collect these integrals of the trigonometric function (and those we knew previously) together:

$$\int \sin(t)\ dt = -\cos(t) + C \qquad\qquad \int \cos(t)\ dt = \sin(t) + C$$

$$\int \tan(t)\ dt = \ln|\sec(t)| + C \qquad\qquad \int \cot(t)\ dt = -\ln|\csc(t)| + C$$

$$\int \sec(t)\ dt = \ln|\sec(t) + \tan(t)| + C \qquad \int \csc(t)\ dt = -\ln|\csc(t) + \cot(t)| + C$$

> **Helpful Tip#1:**
>
> Whenever the result of an integration (in this table) involves a "co"-function, it begins with a negative sign.
>
> **Helpful Tip#2:**
>
> The 2nd column of the table is largely the same as in the 1st column, except that we've added the sound "co" to those functions that lacked it, and removed the sound from those that had it.

Products of sine and cosine

Example 1.19. Use substitution to determine $\int \cos^5(t) \sin^4(t) \, dt$

Solution: The strategy for an integrand such as this is to remove one factor to act as u', and then to rewrite everything in terms of u. Specifically, since cosine is raised to an *odd power*, we choose $u' = \cos(t)$, and move a factor of cosine next to the differential in preparation for substitution:

$$\int \cos^5(t) \sin^4(t) \, dt = \int \cos^4(t) \sin^4(t) \, \underbrace{\cos(t) \, dt}_{\text{this will be } du} \ .$$

> More so than in previous examples, we have to separate the integrand in order to "find" a du in this integral.

Then the Pythagorean identity allows us to rewrite the remaining factors of cosine (of which there are now an even number) in terms of the sine:

$$\cos^2(t) + \sin^2(t) = 1 \quad \Rightarrow \quad \cos^2(t) = 1 - \sin^2(t).$$

> Here's why we chose use a cosine to act as u' rather than a sine: it left us with an *even* number of powers, which we can express in terms of sine.

So our integrand is really

$$\int \left(1 - \sin^2(t)\right)^2 \sin^4(t) \, \cos(t) \, dt.$$

Now the substitution $u = \sin(t)$, $du = \cos(t) \, dt$ allows us to recast our integrand as a polynomial rather than a product of transcendental functions:

$$\int (1 - u^2)^2 u^4 \, du = \int u^4 - 2u^6 + u^8 \, du = \frac{1}{5}u^5 - \frac{2}{7}u^7 + \frac{1}{9}u^9 + C.$$

Now returning to our original variable, we have that

$$\int \cos^5(t) \sin^4(t) \, dt = \frac{1}{5}\sin^5(t) - \frac{2}{7}\sin^7(t) + \frac{1}{9}\sin^9(t) + C. \qquad \blacksquare$$

Try It Yourself: Products of sine and cosine

Example 1.20. Use substitution to determine $\int \cos^4(8t) \sin^3(8t) \, dt$. *(Hint: there are an odd number of factors of sine.)*

Answer: $\frac{1}{56}\cos^7(8t) - \frac{1}{40}\cos^5(8t) + C.$ $\blacksquare$

Full Solution On-line

Products of tangent and secant

Example 1.21. Use substitution to determine $\int \sec^4(8t) \tan^6(8t) \, dt$.

Solution: Like Examples 1.19 and 1.20, we have to separate this integrand in order to "make" du for ourselves. We know that $8\sec^2(8t)$ is the derivative of $\tan(8t)$, so let's move two factors of secant next to the differential in preparation for a substitution:

$$\int \sec^4(8t) \tan^6(8t) \, dt = \int \sec^2(8t) \tan^6(8t) \, \underbrace{\sec^2(8t) \, dt}_{\text{this will be } du} \ .$$

> The important thing to understand about this technique is that the Pythagorean Theorem allows us to exchange even powers of one trigonometric function for even powers of another.
>
> You might wonder why we didn't move $\sec(8t) \tan(8t)$ next to the differential of the original integral in preparation for the substitution $u = \sec(8t)$. That would have left us with tangent to an *odd* power, and we cannot rewrite it in terms of the secant using the Pythagorean identity.

The Pythagorean identity, $1 + \tan^2(t) = \sec^2(t)$, allows us to rewrite the remaining factors of secant in terms of the tangent:

$$\int \sec^2(8t) \tan^6(8t) \, \sec^2(8t) \, dt = \int \left(1 + \tan^2(8t)\right) \tan^6(8t) \, \sec^2(8t) \, dt.$$

Now setting $u = \tan(8t)$, we have $du = 8\sec^2(8t)\ dt$, so that $\sec^2(8t)\ dt = \frac{1}{8}\ du$. Substituting this into our integral, we see

$$\int \left(1 + u^2\right) u^6\ \frac{1}{8} du = \frac{1}{8}\int u^6 + u^8\ du = \frac{1}{56}u^7 + \frac{1}{72}u^9 + C.$$

We finish by returning to our original variable, writing

$$\int \sec^4(8t)\tan^6(8t)\ dt = \frac{1}{56}\tan^7(8t) + \frac{1}{72}\tan^9(8t) + C. \qquad \blacksquare$$

Try It Yourself: Products of tangent and secant

Example 1.22. The derivative of $\sec(9t)$ is $9\sec(9t)\tan(9t)$. Use this fact to determine $\int \sec^3(9t)\tan^5(9t)\ dt$ using a substitution.

Answer: $\frac{1}{63}\sec^7(9t) - \frac{2}{45}\sec^5(9t) + \frac{1}{27}\sec^3(9t) + C.$ $\qquad \blacksquare$

Full Solution
On-line

❖ Definite Integrals

Some people calculate definite integrals by using substitution as a kind of "stepping stone" that allows them to find an antiderivative of the integrand, after which they can evaluate it at the upper and lower bounds of integration as instructed by the Fundamental Theorem of Calculus. Alternatively, you could calculate a definite integral by changing the variable, changing the bounds of integration accordingly, and never returning to the original variable or bounds. This latter technique is often called a *change of variable,* and our goal in the following discussion to show you how it's done and why it works.

> Remember that a definite integral is a number, not a family of functions.

The Chain Rule was introduced on p. 202 in the context of walking on a hill. We imagined that you were walking along the curve $y = f(x)$, and since your horizontal position changed with time as you walked, we called it $x(t)$. Using that notation, your altitude is $f(x(t))$ at time t, and it's changing at a rate of

$$\underbrace{\frac{d}{dt} f(x(t))}_{\frac{\text{ft-vertical}}{\text{sec}}} = f'(x)\ \underbrace{x'(t)}_{\substack{\text{slope of the hill:} \\ \text{ft-vertical/ft-horizontal}}}.$$

(with labels: $\frac{\text{ft-vertical}}{\text{sec}}$ over $\frac{d}{dt}f(x(t))$, and ft-horizontal/sec over $x'(t)$)

> Written in Leibniz notation, your altitude is changing at a rate of
> $$\frac{df}{dt} = \frac{df}{dx}\frac{dx}{dt}.$$

So the net change in altitude that occurs between times $t = a$ and $t = b$ is

$$\int_a^b \underbrace{\frac{d}{dt} f(x(t))}_{\frac{\text{ft-vertical}}{\text{sec}}}\ \underbrace{dt}_{\text{sec}}\ = \int_a^b f'(x(t))\ x'(t)\ dt.$$

The next example uses this idea to lay the conceptual framework for the change of variable technique.

Developing intuition for changing varibles

Example 1.23. Suppose a hill is described by $y = f(x)$, where $f(x) = 10x - x^2$ and distance is measured in feet. As you climb it, your horizontal position after t seconds is $x(t) = 1 + 2t$. (a) Write down a definite integral that calculates the net change in altitude between times $t = 0$ and $t = 2$, and (b) calculate that net change by using the function f.

Solution: (a) The net change in altitude between times $t = 0$ and $t = 2$ is given by the definite integral

$$\int_0^2 \underbrace{\frac{df}{dt}}_{\text{ft/sec}} \underbrace{dt}_{\text{sec}} \ .$$

The Chain Rule tells us that $\frac{df}{dt} = \frac{df}{dx}\frac{dx}{dt}$, which allows us to write this integral more explicitly. Since $f'(x) = 10 - 2x$ and $x'(t) = 2$, our integral is

$$\int_0^2 f'(x(t)) \ x'(t) \ dt = \int_0^2 \Big(10 - 2(1 + 2t)\Big)(2) \ dt.$$

(b) We start walking at $x(0) = 1$, so our initial altitude is $f(1) = 9$. We finish walking at $x(2) = 5$, where the altitude is $f(5) = 25$. So the net change in altitude is $f(5) - f(1) = 16$ feet. ∎

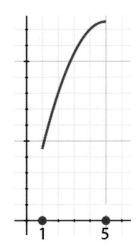

Figure 1.1: The graph of $f(x) = 10x - x^2$ is shown above. The motion described by $x(t) = 1 + 2t$ takes you from $x = 1$ to $x = 5$.

In Example 1.23 you saw two ways of writing down the same number:

$$\int_0^2 \Big(10 - 2(1 + 2t)\Big)(2) \ dt \ = \ \text{net change in altitutde} \ = \ f(5) - f(1)$$

We know from the Fundamental Theorem of Calculus that $f(5) - f(1) = \int_1^5 f'(x) \ dx$, so we can rewrite this equality as

$$\int_0^2 \Big(10 - 2(1 + 2t)\Big)(2) \ dt = \int_1^5 10 - 2x \ dx.$$

> Note that the integrand has changed according to the substitution $x = 1 + 2t$.

The left-hand side of this equation calculates the net change in altitude based on your progress up the hill as a function of time, so the bounds of integration tell us *when* to start and stop calculating. By contrast, the right-hand side of this equation calculates the same number using only the slope of the hill, in which case what matters is not when but *where* you start and stop. (Note the bounds of integration!)

> **Substitution (Change of Variable):** Suppose that $x'(t)$ is continuous on $[a, b]$, and f' is continuous on the range of $x(t)$. Then
>
> $$\int_a^b f'(x(t)) \ x'(t) \ dt = \int_{x(a)}^{x(b)} f'(x) \ dx.$$

> The hypotheses of continuity are meant to ensure that the definite integral exists in the first place.

Using the Change of Variable Formula

Example 1.24. Calculate the number $\int_0^2 \sqrt{2t^3 + 9} \ (6t^2) \ dt$.

Solution: The key to calculating this integral is noticing that $6t^2$ is the derivative of $2t^3 + 9$. That leads us to define $x(t) = 2t^3 + 9$ so that $x'(t) = 6t^2$ and $dx = 6t^2 \ dt$. Since our initial position is $x(0) = 9$ and our final position is $x(2) = 25$, we can rewrite our definite integral as

$$\underbrace{\int_0^2 \sqrt{2t^3 + 9} \ (6t^2) \ dt}_{\text{bounds have changed from \textit{when} to \textit{where}}} = \int_9^{25} \sqrt{x} \ dx = \frac{2}{3}x^{3/2}\Big|_{x=9}^{x=25} = \frac{2}{3}(125 - 27) = \frac{196}{3}. \quad ∎$$

Changing variable and changing scale

Example 1.25. Calculate $\int_0^3 \sqrt{1+60t}\, dt$.

Solution: When we set $x(t) = 1 + 60t$, we start at $x(0) = 1$ and end at $x(3) = 181$. Further, $dx = 60\, dt$ so that $dt = \frac{1}{60}\, dx$. Therefore,

$$\int_0^3 \sqrt{1+60t}\, dt = \int_1^{181} \sqrt{x}\, \frac{1}{60} dx = \frac{1}{90} x^{3/2} \Big|_1^{181} = \frac{1}{90}\left(181^{3/2} - 1\right). \qquad \blacksquare$$

Note: You can think of the change of variable in Example 1.25 as a change of *scale*. Suppose you and a friend both measure time, but with different watches. She measures in minutes, and you in seconds. Then whenever she says that 1 minute has passed, you say that 60 seconds have passed. More generally, if she says that t minutes have passed, you say that $60t$ seconds have passed. If you started your stop watch 1 second before she did, the moment she calls $t = 0$, you say is $x = 1$. This is exactly the change of variable $x = 1 + 60t$.

Try It Yourself: Using the Change of Variable Formula

Full Solution
On-line

Example 1.26. Calculate the number $\int_0^1 e^{t^2+t}\, (2t + 1)\, dt$.

Answer: $e^2 - 1$. *(Note that $2t + 1$ is the derivative of $t^2 + t$.)* $\blacksquare$

⁜ Common Difficulties

Many students think of the differential as doing nothing more than sitting at the end of the integrand, so that we know where it stops. While true, it's also true that the rudder sits at the end of the boat, and your chances of getting where you want to go are slim if you can't use it correctly. In a nutshell, *handling the differential correctly is essential to successful substitutions.*

If you ever see two variables in the same integral, something has gone wrong (see Example 1.8, p. 406). While we mentioned this idea earlier as a guiding principle, it's important enough that we should repeat it:

> **Guiding Principal for Substitution:** You should never see the original variable of integration and the substitution variable in the same integral.

Lastly, avoid using same letter to mean two or more different things during a calculation (e.g., using the letter t as the substitution variable, even though it's the original variable of integration). That makes the bookkeeping of changing variables much more difficult, and almost invariably leads to problems, whether in the bounds of integration or in handling the differential.

You should know

- the terms *substitution* and *change of variables;*

- that the substitution technique is just the Chain Rule "in reverse;"

- that after a successful substitution you will not see the original variable occur anywhere.

You should be able to

- use substitution to determine indefinite integrals;

- use a change of variables to calculate a definite integral.

❖ 6.1 Skill Exercises

In each of #1–8 (a) define u to be the argument of the integrand, (b) determine the relationship between du and dt, (c) use the substitution to calculate the integral, and (d) return to the original variable.

1. $\displaystyle\int \tan(4t)\, dt$

2. $\displaystyle\int \tan\left(\frac{2t}{3}\right) dt$

3. $\displaystyle\int \sec(5t)\, dt$

4. $\displaystyle\int \sec(\pi t)\, dt$

5. $\displaystyle\int 6\cot\left(\frac{t}{4}\right) dt$

6. $\displaystyle\int 5\cot(3t)\, dt$

7. $\displaystyle\int 4\csc\left(\frac{\pi}{6}t\right) dt$

8. $\displaystyle\int 3\csc(2\pi t)\, dt$

In each of #9–14 use the specified change of variable to rewrite the definite integral in terms of x and dx, including new bounds of integration. (You are not being asked to calculate the integral, just to practice rewriting it in terms of a new variable.)

9. $\displaystyle\int_0^4 \frac{1}{2t+1}\, dt; \; x = 2t+1$

10. $\displaystyle\int_1^2 \frac{1}{(3t+7)^2}\, dt; \; x = 3t+7$

11. $\displaystyle\int_1^9 \frac{4t}{t^2+3}\, dt; \; x = t^2+3$

12. $\displaystyle\int_{-1}^1 \frac{t^2}{8t^3-3}\, dt; \; x = 8t^3-3$

13. $\displaystyle\int_e^{e^2} \frac{1}{t\ln(t)}\, dt; \; x = \ln(t)$

14. $\displaystyle\int_6^7 \frac{\ln(2t+1)}{4t+2}\, dt; \; x = 2t+1$

In each of #15–30 (a) find a factor of the integrand that is the derivative of something else (up to a constant scale factor), (b) use that factor to identify the terms that you'll group together as u, and (c) rewrite the integral in terms of u and du. (You are not being asked to calculate the integral, just to practice rewriting it in terms of a new variable.)

15. $\displaystyle\int 2t(t^2+7)^{11}\, dt$

16. $\displaystyle\int 3t^2(t^3-8)^9\, dt$

17. $\displaystyle\int 3t\sqrt{t^2+1}\, dt$

18. $\displaystyle\int t^4 \sqrt[3]{7-2t^5}\, dt$

19. $\displaystyle\int te^{t^2}\, dt$

20. $\displaystyle\int t^3 e^{8t^4-2}\, dt$

21. $\displaystyle\int \sec^2(t)e^{\tan(t)}\, dt$

22. $\displaystyle\int \cos(3t)2^{\sin(3t)}\, dt$

23. $\displaystyle\int \frac{10t}{(t^2-8)^2}\, dt$

24. $\displaystyle\int \frac{12t+18}{\sqrt{3t^2+9t-11}}\, dt$

25. $\displaystyle\int \sin^{12}(t)\cos(t)\, dt$

26. $\displaystyle\int \cos^5(6t)\sin(6t)\, dt$

27. $\displaystyle\int \cot^9(7t)\csc^2(7t)\, dt$

28. $\displaystyle\int \sec^4(t)\tan(t)\, dt$

29. $\displaystyle\int \cos^{4/3}(3t)\sin(3t)\, dt$

30. $\displaystyle\int \csc^{3/2}(t)\cot(t)\, dt$

In #31–34 use a substitution to rewrite the integrand in the form $\frac{a}{u^2+1}$, for some constant a, and then determine the indefinite integral.

31. $\displaystyle\int \frac{3}{4t^2+1}\, dt$

32. $\displaystyle\int \frac{5}{9t^2+1}\, dt$

33. $\displaystyle\int \frac{7}{t^2+5}\, dt$

34. $\displaystyle\int \frac{13}{8t^2+5}\, dt$

In #35–40 complete the square in the denominator and then use a substitution, as demonstrated in Example 1.12 (p. 407).

35. $\displaystyle\int \frac{1}{t^2 + 2t + 8}\, dt$ 37. $\displaystyle\int \frac{2}{t^2 + 3t + 7}\, dt$ 39. $\displaystyle\int \frac{6}{2t^2 + 5t + 8}\, dt$

36. $\displaystyle\int \frac{1}{t^2 + 4t + 37}\, dt$ 38. $\displaystyle\int \frac{3}{t^2 + 5t + 8}\, dt$ 40. $\displaystyle\int \frac{5}{3t^2 + 4t + 16}\, dt$

Determine the integrals in #41–48 by separating each into a sum of integrals in such a way that designating u as the denominator will work as a substitution for at least one of them.

41. $\displaystyle\int \frac{2t + 6}{t^2 + 1}\, dt$ 45. $\displaystyle\int \frac{2t + 6}{t^2 + 4t + 10}\, dt$

42. $\displaystyle\int \frac{2t - 3}{t^2 + 1}\, dt$ 46. $\displaystyle\int \frac{2t - 3}{t^2 + 4t + 10}\, dt$

43. $\displaystyle\int \frac{4t + 12}{t^2 + 11}\, dt$ 47. $\displaystyle\int \frac{4t + 12}{t^2 + 8t + 101}\, dt$

44. $\displaystyle\int \frac{6t - 15}{t^2 + 11}\, dt$ 48. $\displaystyle\int \frac{6t - 15}{t^2 + 8t + 101}\, dt$

In #49–66 use the substitution technique to evaluate the indefinite integral.

49. $\displaystyle\int (3t + 4)^{29}\, dt$ 55. $\displaystyle\int (5\sin(t) + 16)^{80} \cos(t)\, dt$ 61. $\displaystyle\int \frac{\ln(t)}{t}\, dt$

50. $\displaystyle\int (20t - 2)(5t^2 - t + 7)^{17}$ 56. $\displaystyle\int \sec^2(5t)\sqrt{7\tan(5t) + 9}\, dt$ 62. $\displaystyle\int \ln(t^{1/t})\, dt$

51. $\displaystyle\int \sin(3t + 7)\, dt$ 57. $\displaystyle\int \frac{2e^t}{\sqrt{e^t + 5}}\, dt$ 63. $\displaystyle\int \frac{\sin(\ln(t))}{t}\, dt$

52. $\displaystyle\int \cos(7t + 3)\, dt$ 58. $\displaystyle\int e^{3t} \csc(e^{3t}) \cot(e^{3t})\, dt$ 64. $\displaystyle\int \sec(\ln(t)) \tan(\ln(t)) \frac{1}{t}\, dt$

53. $\displaystyle\int \sec^2(9t - 5)\, dt$ 59. $\displaystyle\int \frac{\sin(\sqrt{t})}{\sqrt{t}}\, dt$ 65. $\displaystyle\int \frac{(6t^{5/2} + 4\sqrt{t}) \cos(t^3 + 2t)}{2\sqrt{t}}\, dt$

54. $\displaystyle\int \tan(8t + 1)\, dt$ 60. $\displaystyle\int \frac{\cos^7(\sqrt{t})}{5\sqrt{t}} \sin(\sqrt{t})\, dt$ 66. $\displaystyle\int \frac{12\ln(t)}{t} \sqrt{(\ln(t))^2 + 1}\, dt$

Calculate the indefinite integrals in #67–72 by designating the grouped quantity as u. (The remaining factors comprise du, but also have some terms that you'll need to rewrite in terms of u.)

67. $\displaystyle\int 9t^5(t^3 + 8)^{1/4}\, dt$ 69. $\displaystyle\int t^9(t^5 + 13)^{1/2}\, dt$ 71. $\displaystyle\int \frac{t}{\sqrt{2t - 1}}\, dt$

68. $\displaystyle\int 7t(t + 6)^{1/5}\, dt$ 70. $\displaystyle\int t^3(t^2 - 1)^{1/7}\, dt$ 72. $\displaystyle\int t^2\sqrt{t^2 - 5}\, dt$

In each of #73–84 use substitution and trigonometric identities to determine the integrals.

73. $\displaystyle\int \sin^3(t)\cos^2(t)\,dt$

74. $\displaystyle\int \sin^6(t)\cos^5(t)\,dt$

75. $\displaystyle\int \sin^5(t)\cos^7(t)\,dt$

76. $\displaystyle\int \sin^3(t)\cos^3(t)\,dt$

77. $\displaystyle\int \tan^7(t)\sec^4(t)\,dt$

78. $\displaystyle\int \tan^8(t)\sec^6(t)\,dt$

79. $\displaystyle\int \tan^3(t)\sec(t)\,dt$

80. $\displaystyle\int \tan^5(t)\sec^3(t)\,dt$

81. $\displaystyle\int \cot^3(t)\csc^5(t)\,dt$

82. $\displaystyle\int \cot^4(t)\csc^6(t)\,dt$

83. $\displaystyle\int \cos^2(4t)\,dt$

84. $\displaystyle\int \sin^4(3t)\,dt$

Use a substitution to calculate the definite integral in each of #85–92.

85. $\displaystyle\int_0^7 3t(4t^2+8)^6\,dt$

86. $\displaystyle\int_0^{\sqrt{\pi}} 13t\sin(6t^2)\,dt$

87. $\displaystyle\int_{-1}^3 (2t+1)(5t^2+5t-1)^6\,dt$

88. $\displaystyle\int_{-2}^1 (6t^2-10t)(5t^2-2t^3)^6\,dt$

89. $\displaystyle\int_0^{\pi/2} \sin(\phi)\cos^{10}(\phi)\,d\phi$

90. $\displaystyle\int_0^{\pi/4} \sec^4(\phi)\tan^8(\phi)\,d\phi$

91. $\displaystyle\int_0^3 7t\sqrt{5-\sqrt{9-t^2}}\,dt$

92. $\displaystyle\int_0^9 \frac{4}{(8+\sqrt{t})^5}\,dt$

Show that each integral in #93–96 can be written as $\int_0^1 \sqrt{x}\,dx$ using an appropriate substitution.

93. $\displaystyle\int_{-13}^{-12} \sqrt{t+13}\,dt$

94. $\displaystyle\int_{\pi/2}^{\pi} \sin(t)\sqrt{\cos(t)+1}\,dt$

95. $\displaystyle\int_1^e \frac{1}{t}\sqrt{\ln(t)}\,dt$

96. $\displaystyle\int_2^4 (0.5t-1)\sqrt{0.25t^2-t+1}\,dt$

❖ 6.1 Concept and Application Exercises

97. Suppose that $f(x)=1$ when $x\in[0,2]$ and is zero otherwise. On three separate sets of axes, draw $y=f(x)$, $y=f(x/3)$ and $y=3f(x)$.

    (a) Calculate $\int_0^2 f(x)\,dx$.

    (b) What's the relationship between $\int_0^2 f(x)\,dx$ and $\int_0^2 3f(x)\,dx$?

    (c) What's the relationship between $\int_0^2 f(x)\,dx$ and $\int_0^6 f(x/3)\,dx$?

    (d) Given your answers from parts (b) and (c), what's the relationship between $\int_0^2 3f(x)\,dx$ and $\int_0^6 f(x/3)\,dx$?

    (e) Confirm your answer from part (d) by using the change of variable $u=x/3$ to calculate $\int_0^6 f(x/3)\,dx$.

98. Suppose $f(x)=3+\cos(\pi x/4)$ when $x\in[0,2]$ and is zero otherwise. On three separate sets of axes, draw $y=f(x)$, $y=f(x/3)$ and $y=3f(x)$. Then answer parts (a)–(e) from #97.

99. Suppose that f is an odd function. Use the substitution $x=-t$ to show that $\int_0^a f(t)\,dt=-\int_{-a}^0 f(t)\,dt$. What can we conclude about $\int_{-a}^a f(t)\,dt$?

100. Use a change of variable to show that $\int_0^{2^n} \frac{1}{2^n+t}\, dt$ is the area between the curve $y = 1/t$ and the horizontal axis over $[2^n, 2^{n+1}]$. Then calculate the value of this integral and show that it does not depend on n.

101. You know that $\int_{-1}^{1} 2\sqrt{1-x^2}\, dx = \pi$ because the area of the unit circle is π. Similarly, to calculate the area of the ellipse $\frac{x^2}{a^2} + \frac{y^2}{b^2} = 1$ we calculate

$$\int_{-a}^{a} 2b\sqrt{1 - \left(\frac{x}{a}\right)^2}\, dx.$$

Use a change of variable to show the value of this integral is πab.

Geometry

102. Pat and Chris decide to head to the Louisiana coast to help clean up oil. They clean at rates of

$$p'(t) = \frac{1.5}{1+0.1t}\ \frac{\text{ft}^2}{\text{min}} \quad \text{and} \quad c'(t) = \frac{2.5}{1+0.01t^2}\ \frac{\text{ft}^2}{\text{min}} \quad \text{respectively.}$$

(a) How much can Pat clean in a half hour?

(b) How long does it take Pat to clean 60 square feet, by himself?

(c) How much can Chris clean in an hour?

(d) How long does it take Chris to clean 20 square feet, by himself?

(e) How long does it take Pat and Chris to clean 60 square feet together? (Newton's Method will be helpful.)

In #103–106 (a) use a graphing utility to render a graph of $f(t)$, and (b) find the net area between the graph of f and the horizontal axis over the specified interval.

103. $f(t) = t\sin(t^2);\ [0, \sqrt{\pi}]$

105. $f(t) = \frac{5}{\sqrt{t}}e^{\sqrt{t}};\ [1, 9]$

104. $f(t) = t^4\sqrt{t^2+4};\ [-1, 1]$

106. $f(t) = \frac{3t}{8+t^2};\ [1, 4]$

In #107–110, the upper and lower boundaries of a lamina are specified over a given interval. (a) Calculate the area of the lamina (all distance is measured in cm), (b) determine the mass of the resulting plate, supposing that it will be 0.5 cm thick and made of aluminum. (Aluminum has a density of 2700 $\frac{\text{kg}}{\text{m}^3}$.)

Engineering

107. $f(x) = \frac{3x}{1+x^2},\ g(x) = 2x - x^2;\ [0, 2]$

108. $f(x) = 0.4x\sqrt{x^2+1},\ g(x) = 1 - 0.2\sqrt{x^2+3};\ [-1, 1]$

109. $f(x) = \cos(x)e^{\sin(x)},\ g(x) = -2x + 0.5x^2;\ [0, \pi]$

110. $f(x) = xe^{-x^2},\ g(x) = 0.5x(3-x);\ [0, 1]$

111. Suppose $f(x) = \cos(x)e^{\sin(x)}$ and $g(x) = \sin(x)e^{\cos(x)}$. (a) Graph f and g on the same axes, (b) use Newton's method to locate the first negative and first positive value of x where the graphs intersect, and (c) determine the area bounded by the graphs between those values of x.

112. Suppose $f(x) = x\cos\left(\frac{\pi}{2}x^2\right)$. (a) Use Simpson's method with four subintervals to estimate the area bounded between the graph of f and the x-axis over $[0, 1]$, and (b) determine the exact value of the area by integrating.

113. Suppose an object moves vertically with a velocity of $y'(t) = t\sqrt{4 - t^2}\ \frac{\text{cm}}{\text{sec}}$ when $t \in [0, 2]$, and its initial altitude is $y(0) = 3$ cm. What is its altitude at time $t = 2$?

Altitude

114. Suppose a person exhales $\frac{1}{6}(t - 1.5)^{-2/3}$ $\frac{\text{L}}{\text{sec}}$ for 1 second. How much air has exited his lungs?

115. Suppose a patient receives plasma at a rate of $5(10 - t)^{2.1}$ $\frac{\text{ml}}{\text{min}}$ for 10 minutes.

   (a) How long is it before the patient finishes the first liter of plasma?

   (b) How much plasma has the patient received at the end of 10 minutes?

116. Suppose an aircraft is flying over the ocean, and the pilot is searching visually for a target. Koopman's sighting law says that, if we denote the distance across the water between the pilot and the target by r, the probability of detecting the target decreases with distance at a rate of

$$p'(r) = -3khr(h^2 + r^2)^{-5/2},$$

where h is the (constant) altitude of the plane, and k is a positive constant that accounts for weather conditions among other things.

   (a) Note that $p'(r) < 0$. What does that mean in this context?

   (b) Integrate $p'(r)$ to find a formula for $p(r)$.

   (c) What must be true about $\lim_{r \to \infty} p(r)$?

   (d) In light of your answer from part (a), determine the constant of integration (the "$+C$") from part (b).

117. Suppose an ion is y meters above the center of a charged disk of radius R meters, as depicted in Figure 1.2. If the disk has a uniform charge density of σ coulombs per m^2, a ring of radius r and width Δr accommodates $2\pi r(\Delta r)\sigma$ coulombs of charge. So if the ion has a charge of q coulombs, that ring attracts (repels) the ion toward (from) the disk with a vertical force of approximately

$$\frac{1}{4\pi\varepsilon_0} \frac{2\pi\sigma qy \; r}{(y^2 + r^2)^{3/2}} \Delta r,$$

where ε_0 is a number called the *permittivity constant*. (We say *approximately* because not all of the charge is exactly r units away from the center.)

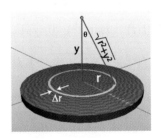

Figure 1.2: For #117. Note that side-to-side force is balance by symmetry.

   (a) Write a Riemann sum with n subintervals that approximates the cumulative vertical force on the ion due to the entire disk.

   (b) Limiting $n \to 0$ in part (a) leads us to a definite integral. Write the integral that calculates the cumulative force experience by the ion due to the charged disk.

   (c) Calculate the value of the integral from part (b). Your answer will include the constants $R, y, \sigma, \pi, \varepsilon_0$ and q.

118. The **convolution** of functions f and g, which arises in the study of *differential equations* and in the theory of probability, is denoted by $f * g$ and defined by

$$(f * g)(t) = \int_0^t f(x)g(t - x) \; dx.$$

Students often ask whether the order of these functions is important. Use substitution to prove that $(f * g)(t) = (g * f)(t)$.

6.2 Integration by Parts

In Section 6.1 we used our understanding of the Chain Rule to help us determine antiderivatives. In this section, we use our understanding of the Product Rule to extend our abilities even further.

❖ The Basic Idea

When $u(t)$ and $v(t)$ are differentiable functions, the Product Rule tells us that

$$\frac{d}{dt}\Big(u(t)v(t)\Big) = u'(t)v(t) + u(t)v'(t).$$

When we integrate this equation with respect to time we see

$$u(t)v(t) + C = \int u'(t)v(t)\ dt + \int u(t)v'(t)\ dt,$$

which is often rewritten as

$$\int u(t)v'(t)\ dt = u(t)v(t) - \int u'(t)v(t)\ dt + C$$

The derivative is "on v" on the left-hand side of this equation, but it's "on u" on the right-hand side, so you can think of this technique as a way of moving the derivative from one factor to the other.

This equation, called the **integration by parts** formula, is useful when we need to integrate $u(t)v'(t)$, but integrating $u'(t)v(t)$ is easier.

Integrating by Parts

Example 2.1. Determine $\int t\cos(t)\ dt$.

Solution: Suppose $u(t) = t$ and $v'(t) = \cos(t)$. Then $v(t) = \sin(t)$, and the Product Rule tells us that

$$\frac{d}{dt}\Big(t\sin(t)\Big) = 1\sin(t) + \overbrace{t\cos(t)}^{\text{here's our integrand}}.$$

When we rewrite this equation in order to isolate our integrand, we see

$$t\cos(t) = \frac{d}{dt}\Big(t\sin(t)\Big) - \sin(t),$$

from which we conclude that

$$\int t\cos(t)\ dt = \int \underbrace{\frac{d}{dt}\Big(t\sin(t)\Big)}_{\text{This part is easy!}} - \sin(t)\ dt = t\sin(t) - \int \sin(t)\ dt = t\sin(t) + \cos(t) + C. \quad\blacksquare$$

Try It Yourself: Integrating by Parts

Example 2.2. Determine $\int t\sin(t)\ dt$.

Answer: $-t\cos(t) + \sin(t) + C$. $\quad\blacksquare$

Full Solution On-line

The choice of which factor to designate as $u(t)$ and which to call $v'(t)$ is important because it often happens that one choice makes the integral easier (or at least no worse) while others make it more difficult. For instance, we could have set $u(t) = \cos(t)$ and $v'(t) = t$ in Example 2.1, but then $v(t) = 0.5t^2$ and

$$\frac{d}{dt}\left(\frac{t^2}{2}\cos(t)\right) = t\cos(t) - \frac{t^2}{2}\sin(t).$$

Isolating our integrand in this equation, we see

$$t\cos(t) = \frac{d}{dt}\left(\frac{t^2}{2}\cos(t)\right) + \frac{t^2}{2}\sin(t)$$

so that

$$\int t\cos(t)\ dt = \int \frac{d}{dt}\left(\frac{t^2}{2}\cos(t)\right) + \frac{t^2}{2}\sin(t)\ dt = \frac{t^2}{2}\cos(t) + \int \frac{t^2}{2}\sin(t)\ dt.$$

We started with a trigonometric function and *one* factor of t, but now have to integrate a trigonometric function and *two* factors of t. Instead of making the integral easier, this choice of u and v' made it more complicated!

Rule of Thumb #1 for Integration by Parts: Designate factors of the integrand as u and v' so that

(a) the function $u(t)$ is a factor whose derivative is simpler (or at least no worse to deal with) than itself, and

(b) the factor designated as $v'(t)$ is something that you can integrate (because you need to know v).

Note: The word "simpler" is admittedly vague. Power functions become simpler when you differentiate them in the sense that the power is reduced. Logarithms become simpler when you differentiate them in the sense of becoming "more familiar," since the derivative of a logarithm is a power function (e.g., $(\ln(t))' = 1/t$).

Integration by parts was successful in the previous examples because we chose $u(t)$ according to part (a) of Rule of Thumb #1, although we had not yet articulated it. In the next pair of examples, part (b) drives our choices.

Using Rule of Thumb #1

Example 2.3. Determine $\int t\tan^{-1}(t)\ dt$.

Recall that $\tan^{-1}(t)$ means the arctangent, not the reciprocal of the tangent.

Solution: In contrast to previous examples, we cannot set $u(t) = t$ because that would leave us with $v'(t) = \tan^{-1}(t)$ and finding $v(t)$ is (at least for now) beyond us. So we have to set $u(t) = \tan^{-1}(t)$ and take $v'(t) = t$. Then $v(t) = 0.5t^2$ and the Product Rule tells us that

$$\frac{d}{dt}\left(\frac{t^2}{2}\tan^{-1}(t)\right) = \underbrace{t\tan^{-1}(t)}_{\text{here's our integrand}} + \frac{t^2}{2}\frac{1}{1+t^2}.$$

Isolating our integrand in this equation, we see

$$t\tan^{-1}(t) = \frac{d}{dt}\left(\frac{t^2}{2}\tan^{-1}(t)\right) - \frac{1}{2}\left(\frac{t^2}{1+t^2}\right),$$

from which we conclude that

$$\int t\tan^{-1}(t)\ dt = \int \frac{d}{dt}\left(\frac{t^2}{2}\tan^{-1}(t)\right) - \frac{1}{2}\left(\frac{t^2}{1+t^2}\right)\ dt = \frac{t^2}{2}\tan^{-1}(t) - \frac{1}{2}\int \frac{t^2}{1+t^2}\ dt.$$

Alternatively, you could perform a polynomial division (see Appendix B for a review) to show that $$\frac{t^2}{1+t^2} = 1 - \frac{1}{1+t^2}.$$

The numerator and denominator of the remaining integrand differ by only a "+1," so we can simplify it easily by adding zero (in the form of $+1-1$):

$$\frac{1}{2}\int \frac{t^2+1-1}{1+t^2}\ dt = \frac{1}{2}\int \frac{t^2+1}{1+t^2} - \frac{1}{1+t^2}\ dt = \frac{1}{2}\int 1 - \frac{1}{1+t^2}\ dt = \frac{1}{2}\left(t - \tan^{-1}(t)\right) + C.$$

Substituting this into the equation above, we have our answer:

$$\int t \tan^{-1}(t)\, dt = \frac{t^2}{2} \tan^{-1}(t) - \frac{t}{2} + \frac{1}{2} \tan^{-1}(t) + C. \qquad \blacksquare$$

Try It Yourself: Using Rule of Thumb #1

Full Solution
On-line

Example 2.4. Determine $\int t \ln(t)\, dt$.

Answer: $\frac{1}{2}t^2 \ln(t) - \frac{1}{4}t^2 + C$. $\qquad \blacksquare$

Repeated integration by parts

Example 2.5. Determine $\int t^2 e^t\, dt$.

Solution: The factor of t^2 becomes simpler if we differentiate it (the number of factors of t is reduced), so let's set $u(t) = t^2$ and $v'(t) = e^t$. Then $v(t) = e^t$ and

$$\frac{d}{dt}\left(t^2 e^t\right) = \underbrace{t^2 e^t + 2t e^t}_{\text{here's our integrand}}.$$

After isolating our integrand in this equation, we see that

$$\int t^2 e^t\, dt = \int \frac{d}{dt}\left(t^2 e^t\right) - 2t e^t\, dt = t^2 e^t - 2\int t e^t\, dt.$$

The remaining integral is also computed with an integration by parts. Because the factor of t becomes simpler if we differentiate it, let's think of the integrand as the product of $u(t) = t$ and $v'(t) = e^t$. Then $v(t) = e^t$, and

$$\frac{d}{dt}\left(t e^t\right) = t e^t + e^t.$$

After isolating our integrand in this equation, we see that

$$\int t e^t\, dt = \int \frac{d}{dt}\left(t e^t\right) - e^t\, dt = t e^t - e^t + C.$$

Substituting this into our equation above, we have our answer:

$$\int t^2 e^t\, dt = t^2 e^t - 2t e^t + 2e^t + C. \qquad \blacksquare$$

Note that we differentiated the power function in the first step of Example 2.5, and differentiated the power function in the second step too. This points to another rule of thumb concerning integration by parts.

> **Rule of Thumb #2 for Integration by Parts:** If you have to integrate by parts more than once, let the factors play the same roles in each step.

Try It Yourself: Repeated integration by parts

Full Solution
On-line

Example 2.6. Determine $\int t^2 \cos(t)\, dt$.

Answer: $t^2 \sin(t) + 2t \cos(t) - 2\sin(t) + C$. $\qquad \blacksquare$

Cyclic integration by parts

Example 2.7. Determine $\int e^t \cos(t)\, dt$.

Solution: Neither of these factors becomes simpler if you differentiate it, and neither gets worse, so this integrand offers no direction about choosing u and v'. It turns out that it doesn't matter in cases like this, so let's choose $u(t) = \cos(t)$ and $v'(t) = e^t$. Then $v(t) = e^t$, and

$$\frac{d}{dt}\Big(\cos(t)e^t\Big) = -\sin(t)e^t + \cos(t)e^t.$$

After isolating our integrand in this equation, we see that

$$\int \cos(t)e^t \ dt = \int \frac{d}{dt}\Big(\cos(t)e^t\Big) + \sin(t)e^t \ dt = \cos(t)e^t + \int \sin(t)e^t \ dt. \quad (2.1)$$

The remaining integral also requires an integration by parts. Our rule of thumb tells us that, since the trigonometric function played the role of u in the first step, it should also play the role of u in the second step of the process. So let's set $u(t) = \sin(t)$ and $v'(t) = e^t$. Then $v(t) = e^t$ and

$$\frac{d}{dt}\Big(\sin(t)e^t\Big) = \cos(t)e^t + \sin(t)e^t.$$

After isolating our integrand in this equation, we see that

$$\int \sin(t)e^t \ dt = \int \frac{d}{dt}\Big(\sin(t)e^t\Big) - \cos(t)e^t \ dt = \sin(t)e^t - \int \cos(t)e^t \ dt$$

The remaining integral is exactly what we started with, but all is not lost. When we substitute this expression for $\int \sin(t)e^t \ dt$ into equation (2.1), we see that

$$\int \cos(t)e^t \ dt = \cos(t)e^t + \sin(t)e^t - \int \cos(t)e^t \ dt + C.$$

Note that $\int \cos(t)e^t \ dt$ occurs on both sides of the equation, but with different signs. This allows us to isolate $\int \cos(t)e^t \ dt$ by adding it to both sides of the equation. When we do, we see that

$$2 \int \cos(t)e^t \ dt = \cos(t)e^t + \sin(t)e^t + C.$$

We arrive at our answer by dividing both sides by 2:

$$\int \cos(t)e^t \ dt = \frac{1}{2}e^t\Big(\cos(t) + \sin(t)\Big) + C. \qquad \blacksquare$$

> You can write "$+C/2$" if you'd like, but it's not necessary. Recall that "$+C$" means we can add *any* constant to our formula and still have an antiderivative. Dividing by two doesn't change that.

Try It Yourself: Cyclic integration by parts

Example 2.8. Determine $\int e^{2t} \sin(3t) \ dt$.

Answer: $\left(\frac{2}{13}\sin(3t) - \frac{3}{13}\cos(3t)\right)e^{2t}$. $\qquad \blacksquare$

Full Solution
On-line

❖ When There's Only One (or One *Kind*) of Factor

Integration by parts is sometimes helpful when the derivative of an integrand is simpler than the integrand itself. Since we need two factors in order to integrate by parts, we set $v'(t) = 1$. Then

$$\int u(t) \ dt = \int u(t) \ (1) \ dt = \int u(t)v'(t) \ dt.$$

When there's only one factor

Example 2.9. Determine $\int \ln(t)\, dt$.

Solution: Because the derivative of the logarithm is $1/t$, which is a simpler function than $\ln(t)$, let's set $u(t) = \ln(t)$ and $v'(t) = 1$. Then $v(t) = t$ and

$$\frac{d}{dt}(t\ln(t)) = \ln(t) + 1.$$

After isolating our integrand in this equation, we see that

$$\int \ln(t)\, dt = \int \frac{d}{dt}\Big(t\ln(t)\Big) - 1 \, dt = t\ln(t) - t + C. \qquad \blacksquare$$

When there's only one factor

Example 2.10. Determine $\int \tan^{-1}(t)\, dt$.

Solution: Because the derivative of the arctangent is $1/(t^2+1)$, which is a simpler function than $\tan^{-1}(t)$, let's set $u(t) = \tan^{-1}(t)$ and $v'(t) = 1$. Then $v(t) = t$ and

$$\frac{d}{dt}\Big(t\tan^{-1}(t)\Big) = \tan^{-1}(t) + (t)\left(\frac{1}{1+t^2}\right).$$

After isolating our integrand in this equation, we see that

$$\int \tan^{-1}(t)\, dt = \int \frac{d}{dt}\Big(t\tan^{-1}(t)\Big) - \frac{t}{t^2+1} \, dt = t\tan^{-1}(t) - \int \frac{t}{t^2+1}\, dt.$$

In the remaining integral, the numerator looks like the derivative of the denominator (up to a factor of 2) so we can use the substitution technique. We set $x = 1 + t^2$, and have $dx = 2t\, dt$, which we rewrite as $dt = \frac{1}{2}\, dx$. This substitution results in

> We're using x as the substitution variable because the letter u has already been used in the context of this problem.

$$\int \frac{t}{1+t^2}\, dt = \int \frac{1}{x}\frac{1}{2}\, dx = \frac{1}{2}\ln|x| + C.$$

> The absolute value is included as a reminder, but isn't actually necessary since $x = 1 + t^2$ is never negative.

That is, after returning to our original variable,

$$\int \tan^{-1}(t)\, dt = t\tan^{-1}(t) - \frac{1}{2}\ln(1+t^2) + C. \qquad \blacksquare$$

Integrating by parts when there's a single repeated factor

Example 2.11. Determine $\int \sec^3(t)\, dt$.

Solution: This integrand has three factors, but they're all the same! If we choose one of them as $u(t) = \sec(t)$, the other two must constitute $v'(t) = \sec^2(t)$. So $v(t) = \tan(t)$ and

$$\frac{d}{dt}\Big(\sec(t)\tan(t)\Big) = \sec(t)\tan^2(t) + \sec^3(t).$$

After isolating our integrand we see

$$\int \sec^3(t)\, dt = \int \frac{d}{dt}\Big(\sec(t)\tan(t)\Big) - \sec(t)\tan^2(t)\, dt = \sec(t)\tan(t) - \int \sec(t)\tan^2(t)\, dt$$

At this point, you might be tempted to integrate by parts again, but the $\tan^2(t)$ in the integrand provides another option. We can use the Pythagorean relationship between the tangent and the secant: $\tan^2(t) = \sec^2(t) - 1$. Substituting this into

> This identity comes from the Pythagorean relationship between sine and cosine:
>
> $$\sin^2(t) + \cos^2(t) = 1.$$
>
> Dividing this equation by $\cos^2(t)$ leaves us with
>
> $$\tan^2(t) + 1 = \sec^2(t).$$

the equation above, we see

$$\int \sec^3(t)\ dt\ =\ \sec(t)\tan(t) - \int \sec(t)(\sec^2(t) - 1)\ dt$$

$$=\ \sec(t)\tan(t) - \int \sec^3(t)\ dt + \int \sec(t)\ dt.$$

As in Example 2.8, we seem to have come full circle, but the occurrences of $\int \sec^3(t)\ dt$ on the left and right sides of the equation have different signs. When we add $\int \sec^3(t)\ dt$ to both sides and then divide by two, we find that

$$\int \sec^3(t)\ dt = \frac{1}{2}\sec(t)\tan(t) + \frac{1}{2}\int \sec(t)\ dt.$$

The remaining integral was done in Example 1.17 (see p. 409). Using that result, we have

$$\int \sec^3(t)\ dt = \frac{1}{2}\sec(t)\tan(t) + \frac{1}{2}\ln|\sec(t) + \tan(t)| + C. \qquad \blacksquare$$

✦ Definite Integrals

The technique for definite integrals is exactly the same. Because

$$u(t)v'(t) = \frac{d}{dt}\Big(u(t)v(t)\Big) - u'(t)v(t)$$

we can write

$$\int_a^b u(t)v'(t)\ dt = u(t)v(t)\Big|_a^b - \int_a^b v(t)u'(t)\ dt.$$

Integration by parts with a definite integral

Example 2.12. Calculate $\int_0^{\sqrt{\pi}} \phi^3 \cos(\phi^2)\ d\phi$.

Solution: In order to simplify the argument of the cosine, let's begin by using the substitution $x = \phi^2$. Then

$$x(0) = 0, \quad x(\sqrt{\pi}) = \pi \ \text{ and } \ dx = 2\phi\ d\phi \ \Rightarrow \phi\ d\phi = \frac{1}{2}\ dx.$$

Substituting these into our integral, we see

$$\int_0^{\sqrt{\pi}} \phi^3 \cos(\phi^2)\ d\phi = \int_0^{\sqrt{\pi}} \phi^2 \cos(\phi^2)\ \phi\ d\phi = \frac{1}{2}\int_0^\pi x\cos(x)\ dx.$$

This is exactly the integrand that we saw in Example 2.1. Using that integration by parts, we see that

$$\frac{1}{2}\int_0^\pi x\cos(x)\ dx = \frac{1}{2}\Big(x\sin(x)\Big|_0^\pi - \int_0^\pi \sin(x)\ dx\Big) = \frac{1}{2}\Big(0 + \cos(x)\Big|_0^\pi\Big) = -1. \qquad \blacksquare$$

Try It Yourself: Integration by parts with a definite integral

Example 2.13. Calculate $\int_0^{\pi/4} \theta \sec^2(\theta)\ d\theta$.

Answer: $\frac{\pi}{4} - \ln\left(\frac{\sqrt{2}}{2}\right)$. $\qquad \blacksquare$

Full Solution On-line

❖ Working with Differentials

Integration by parts is often written, discussed, and done in the language of differentials. Specifically, since $du = u'(t)\ dt$ and $dv = v'(t)\ dt$, we can write the integration by parts formula as follows:

$$\int u\ dv = uv - \int v\ du$$

Integrating by parts with the differential formula

Example 2.14. Determine $\int t^2 \ln(t)\ dt$.

Solution: Since $\ln(t)$ becomes simpler when we differentiate it, we choose $u(t) = \ln(t)$ and $v'(t) = t^2$. Then our integral has the form

$$\int t^2 \ln(t)\ dt = \int u\ dv.$$

When working with the differential formula, many people set up a table to organize the relevant information:

$$u = \ln(t) \qquad dv = t^2\ dt$$
$$du = \frac{1}{t}\ dt \qquad v = \frac{1}{3}t^3$$

Now the differential formula tells us that

$$\int \underbrace{t^2 \ln(t)\ dt}_{u\ dv} = \underbrace{\frac{1}{3}t^3 \ln(t)}_{uv} - \int \underbrace{\frac{1}{3}t^2\ dt}_{v\ du}.$$

Upon calculating the remaining integral we see that

$$\int t^2 \ln(t)\ dt = \frac{1}{3}t^3 \ln(t) - \frac{1}{9}t^3 + C \qquad \blacksquare$$

Try It Yourself: Integrating by parts with the differential formula

Example 2.15. Determine $\int t\ \tan^{-1}(t)\ dt$ by using the differential formula for integration by parts. *(Hint: use $v(t) = 0.5t^2 + 0.5$.)*

Answer: $\frac{1}{2}(t^2 + 1)\arctan(t) - \frac{1}{2}t + C.$ $\qquad \blacksquare$

Full Solution
On-line

You should know

- the phrase *integration by parts;*

- how the technique of integration by parts is derived from the Product Rule;

- rules of thumb (for determining one factor as g' and the other as f, and for repeated applications of the technique).

You should be able to

- use integration by parts to determine definite and indefinite integrals;

- use a table of $f, df, g,$ and dg to perform integration by parts using the differential formula.

✦ 6.2 Skill Exercises

Use integration by parts to determine the indefinite integrals in #1–20.

1. $\int t\sin(2t)\,dt$

2. $\int 8t\cos\left(\frac{\pi}{3}t\right)dt$

3. $\int t\sec^2(t)\,dt$

4. $\int t\sec(t)\tan(t)\,dt$

5. $\int 9te^{6t}\,dt$

6. $\int 5te^{-t/10}\,dt$

7. $\int 3t\left(5^{2t}\right)dt$

8. $\int 7t\left(3^{-7t}\right)dt$

9. $\int 9t^2\cos(4t)\,dt$

10. $\int 5t^2\cos(3t)\,dt$

11. $\int t^2 e^{-0.2t}\,dt$

12. $\int t^2 e^{-0.1t}\,dt$

13. $\int t^3 e^{t/2}\,dt$

14. $\int t^3\cos(t/5)\,dt$

15. $\int \sin(t)e^{3t}\,dt$

16. $\int \cos(3t)e^{2t}\,dt$

17. $\int t\sin(\ln(t))\,dt$

18. $\int t\ln(t)\,dt$

19. $\int te^{at}\,dt$

20. $\int t^2\sin(bt)\,dt$

Calculate the definite integral in #21–26.

21. $\int_1^2 t^2\ln(2t)\,dt$

22. $\int_0^{0.125} \sin^{-1}(4t)\,dt$

23. $\int_0^{0.125} \cos^{-1}(8t)\,dt$

24. $\int_{-\pi/16}^{\pi/16} t^3\cos(4t)\,dt$

25. $\int_2^3 t^4\ln(t^2)\,dt$

26. $\int_1^4 t^{1/3}\ln(t^{-1/2})\,dt$

Determine the integrals in #27–30 by using the substitution $x = \sqrt{t}$ and then integrating by parts.

27. $\int e^{\sqrt{t}}\,dt$

28. $\int \ln(\sqrt{t})\,dt$

29. $\int \frac{\ln(\sqrt{t})}{\sqrt{t}}\,dt$

30. $\int \sin^{-1}(\sqrt{t})\,dt$

Calculate the definite integrals in #31–34 by using a substitution and then integrating by parts.

31. $\int_0^{\sqrt[3]{\pi/2}} t^8\sin(t^3)\,dt$

32. $\int_0^{\pi/4} \tan^3(t)\sec^2(t)e^{\tan^2(t)}\,dt$

33. $\int_{\pi/4}^{\pi/3} \frac{\ln(\cot(t))}{\sin^2(t)}\,dt$

34. $\int_0^{\ln(\sqrt{2})} \frac{\sin^{-1}(e^{-t})}{e^t}\,dt$

35. Determine a general formula for $\int e^{at}\sin(bt)\,dt$ when a and b are nonzero.

36. Determine a general formula for $\int t^a\ln(t^b)\,dt$ when a and b are nonzero.

In #37–44 use integration by parts to determine an antiderivative for the given function.

37. $f(\theta) = \csc(\theta)\sec^2(\theta)$

38. $f(\theta) = \csc^2(\theta)\sec(\theta)$

39. $f(\theta) = \sin(\theta)\sin(\theta)$

40. $f(\theta) = \cos(\theta)\cos(\theta)$

41. $f(\theta) = \cos(2\theta)\cos(7\theta)$

42. $f(\theta) = \sin(5\theta)\sin(8\theta)$

43. $f(\theta) = \sin(3\theta)\cos(5\theta)$

44. $f(\theta) = \sin(6\theta)\cos(7\theta)$

Mixed Practice: This set of exercises is intended to provide you with practice in determining which integration technique is appropriate for a given integral. Exercises 45–56 may require either the techniques from this section **or** from the previous section.

45. $\int 5t \cos(t)\, dt$

46. $\int 7t \sin(13t^2)\, dt$

47. $\int te^{-2t}\, dt$

48. $\int \frac{2t+4}{t^2+9}\, dt$

49. $\int \frac{t}{e^{4t^2}}\, dt$

50. $\int \frac{6t}{\sqrt{3t^2-4}}\, dt$

51. $\int \frac{3}{t^2-8t+17}\, dt$

52. $\int \frac{\ln(2t-4)}{t-2}\, dt$

53. $\int e^{3t} \cos(5t)\, dt$

54. $\int \frac{2t+5}{t^2+3t+8}\, dt$

55. $\int t \arctan(t)\, dt$

56. $\int \frac{t^2}{t^3+10t^2}\, dt$

✤ 6.2 Concept and Application Exercises

In #57–62 (a) use a graphing utility to render a graph of f, and (b) calculate the net area between the graph and the horizontal axis over the given interval.

57. $f(t) = t \sin(3t);\ [0, \pi]$

58. $f(t) = t^2 2^t;\ [0, 1]$

59. $f(t) = t^3 e^{-t^2};\ [0, 1]$

60. $f(t) = \tan^{-1}(t-5);\ [0, 5]$

61. $f(t) = (6t - 3)\ln(t^2 - t + 2);\ [0, 2]$

62. $f(t) = (8t + 4)\ln(t^2 + t + 5);\ [-1, 1]$

In #63–66 (a) use a graphing utility to render plots of f and g,(b) use Newton's method to determine the values of $t > 0$ where the curves intersect, and (c) determine the net area bounded by the curves between those points.

63. $f(t) = \ln(t),\ g(t) = (t-2)^2$

64. $f(t) = 2te^{-t};\ g(t) = t^2 e^{-t}$

65. $f(t) = \sin(t);\ g(t) = \tan^{-1}(t^2)$

66. $f(t) = 30te^{-t},\ g(t) = \frac{20}{t^2+3t+3}$

67. Suppose a runner's velocity is $v(t) = 4te^{-0.1t}\ \frac{m}{sec}$ after t seconds.

    (a) How far does the runner travel during the first 10 seconds?

    (b) How long would it take the runner to travel 100 meters? (Newton's Method will be helpful.)

68. A mass on a spring oscillates with a velocity of $y'(t) = e^{-t/5}\sin(\pi t)\ \frac{cm}{sec}$. Determine a formula for $y(t)$, assuming the mass is initially at $y(0) = 0$.

69. The charge on a capacitor in an LRC circuit varies according to $q'(t) = e^{-0.2t}\cos(3t)$ coulombs per second. Determine a formula for $q(t)$, assuming the charge on the capacitor is initially $q(0) = 1$.

70. Use integration by parts to show that $\int_0^1 f''(x)f(x)\, dx < 0$ whenever f is a non-constant function whose second derivative is continuous, and $f(0) = f(1) = 0$.

71. Figure 2.1 depicts the graph of f over $[a, b]$. Integrate by parts to show that $\int_a^b (x-a)f'(x)\, dx$ is the area of the shaded region (interpret both parts of the resulting formula in terms of the figure).

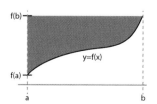

Figure 2.1: For #71.

72. On p. 418 (#118) you were introduced to the convolution

Mathematics

$$(f * g)(t) = \int_0^t f(x)g(t-x) \, dt,$$

which arises in *differential equations* and in probability theory. Use integration by parts to show that $(f*g)(t) = (g*f)(t)$ when $f(t) = \sin(t)$ and $g(t) = \cos(t)$.

Signal Analysis

A fact important in modern signal analysis is that continuous functions with roots at $t = 0$ and $t = 1$ can be "built" by scaling and adding the functions $\sqrt{2}\sin(k\pi t)$, where $k > 0$ is an integer (note that $\sqrt{2}\sin(k\pi t)$ has roots at $t = 0$ and $t = 1$ when k is an integer). That is, we can "build" $f(t)$ as

$$f(t) = A_1\sqrt{2}\sin(\pi t) + A_2\sqrt{2}\sin(2\pi t) + A_3\sqrt{2}\sin(3\pi t) + \cdots$$

where, as it turns out,

$$A_k = \int_0^1 \sqrt{2}\sin(k\pi t)f(t) \, dt.$$

In exercises #73–74 (a) determine a general formula for A_k, and (b) use a graphing utility to render graphs of both f and $y = \sum_{k=1}^n A_k\sqrt{2}\sin(k\pi t)$ on the same axes for each of $n = 2, 3, 4, 5$ (i.e., four graphs per exercise).

73. $f(t) = 2t - 2t^2$

74. $f(t) = t^3 - \frac{4}{3}t^2 + \frac{1}{3}t$

6.3   Overview of Trigonometric Integrals

In Section 6.1 you saw that some products of trigonometric functions can be integrated with the substitution technique, and in Section 6.2 you saw that others can be done with integration by parts. In this section we provide an overview of the topic, so that you get a sense of when to use one technique or the other, and we introduce a new technique as well: the use of half-angle formulas to reduce the complexity of an integrand.

❖ Products of Sine and Cosine

When you see a product that involves only sines and/or cosines, the proper technique of integration is determined by the answer to a single question: **Is there an odd number of factors of cosine (or sine)?** If so, substitution will work. If not, the half-angle identities

$$\cos^2(\theta) = \frac{1}{2} + \frac{1}{2}\cos(2\theta) \qquad \text{and} \qquad \sin^2(\theta) = \frac{1}{2} - \frac{1}{2}\cos(2\theta),$$

can be used to reduce the complexity of the integrand.

Even and odd powers

Example 3.1. Determine $\int \cos^7(\theta)\sin^4(\theta)\,d\theta$.

Solution: There are seven factors of cosine (an odd number) so our strategy will be to use substitution, and one factor of cosine will be used in the differential of the new variable. Specifically, we take $u' = \cos(\theta)$ so that $du = \cos(\theta)\,d\theta$ and $u = \sin(\theta)$. Toward completing this substitution, let's write the integral as

$$\int \cos^6(\theta)\sin^4(\theta)\,\cos(\theta)d\theta.$$

We can write the remaining four factors of cosine in terms of u using the Pythagorean identity:

$$\cos^6(\theta) = \left(\cos^2(\theta)\right)^3 = \left(1 - \sin^2(\theta)\right)^3 = (1 - u^2)^3.$$

This allows us to rewrite the integral as

$$\int (1-u^2)^3 u^4\,du = \int u^4 - 3u^6 + 3u^8 - u^{10}\,du = \frac{1}{5}u^5 - \frac{3}{7}u^7 + \frac{1}{3}u^9 - \frac{1}{11}u^{11} + C.$$

Returning to the original variable gives us

$$\int \cos^7(\theta)\sin^4(\theta)\,d\theta = \frac{1}{5}\sin^5(\theta) - \frac{3}{7}\sin^7(\theta) + \frac{1}{3}\sin^9(\theta) - \frac{1}{11}\sin^{11}(\theta) + C. \quad \blacksquare$$

> This technique worked because there were an *odd* number of cosines. Using one of them to form *du* left an even number, which we were able to convert to sines using the Pythagorean identity. The same kind of technique works if there are an odd number of sines.

Even powers

Example 3.2. Determine (a) $\int \cos^2(\theta)\,d\theta$, and (b) $\int \cos^4(\phi)\,d\phi$.

Solution: (a) The half-angle identity is useful because it allows us to exchange two factors of cosine for a single cosine with twice the argument:

$$\int \cos^2(\theta)\,d\theta = \int \frac{1}{2} + \frac{1}{2}\cos(2\theta)\,d\theta = \frac{\theta}{2} + \frac{1}{4}\sin(2\theta) + C.$$

> The half-angle identity effectively cuts the number of factors in half (in this case from 2 to 1).

(b) We have a pair of $\cos^2(\theta)$ in this integrand. By applying the half-angle identity to each we can rewrite the integral as

$$\int \cos^2(\theta)\cos^2(\theta) = \int \left(\frac{1}{2} + \frac{1}{2}\cos(2\phi)\right)\left(\frac{1}{2} + \frac{1}{2}\cos(2\phi)\right) d\phi$$

$$= \int \frac{1}{4} + \frac{1}{2}\cos(2\phi) + \frac{1}{4}\cos^2(2\phi)\, d\phi$$

$$= \frac{\phi}{4} + \frac{1}{4}\sin(2\phi) + \frac{1}{4}\int \cos^2(2\phi)\, d\phi. \tag{3.1}$$

> The half-angle identity effectively cuts the number of factors in half. We started with four factors, but now have only two.

The substitution $\theta = 2\phi$, $d\theta = 2\,d\phi$ reduces the remaining integral to what we saw in part (a), so

$$\frac{1}{4}\int \cos^2(2\phi)\, d\phi = \frac{\phi}{8} + \frac{1}{32}\sin(4\phi) + C.$$

Inserting this into equation (3.1) leads us to our final answer.

$$\int \cos^4(\phi) = \frac{3\phi}{8} + \frac{1}{4}\sin(2\phi) + \frac{1}{32}\sin(4\phi) + C \qquad \blacksquare$$

> You could also calculate the integrals in Example 3.2 using an integration by parts with $v' = \cos(\theta)$ and $u =$ all other factors.

Try It Yourself: Products of sine and cosine

Example 3.3. Determine (a) $\int \cos^2(\theta)\sin^3(\theta)\, d\theta$, and (b) $\int \cos^2(\phi)\sin^2(\theta)\, d\phi$.

Answer: (a) $\frac{1}{3}\cos^3(\theta) - \frac{1}{5}\cos^5(\theta) + C$ using substitution, and (b) $\frac{\theta}{8} - \frac{1}{32}\sin(4\theta) + C$ using the half-angle identities. $\blacksquare$

Full Solution On-line

Note: If there are odd powers of *both* sine and cosine, it doesn't matter which you use as u.

❖ Products of Secant and Tangent

Our strategy for integrating products of secant and tangent is similar to what we used in the above examples. But since neither is secant the derivative of tangent nor tangent the derivative of secant, we end up asking two questions instead of only one.

- **Is there an even number of secants?** If so, we can use two of them to make $du = \sec^2(\theta)\, d\theta$, and use the substitution $u = \tan(\theta)$.

> We mean a *positive* even number of factors.

- **Presuming that there are factors of secant in the integrand, is there an odd number of tangents?** If so, we can use a factor of tangent to make $du = \sec(\theta)\tan(\theta)\, d\theta$, and use the substitution $u = \sec(\theta)$.

> We mean a *positive* odd number of factors.

In both cases, we are looking for a substitution by checking whether we can make the needed differential.

Even number of secants

Example 3.4. Determine $\int \sec^6(\theta)\tan^2(\theta)\, d\theta$.

Solution: Since there is an even number of factors of secant in this integrand, we will use two of them in the differential, $du = \sec^2(\theta)\, d\theta$. In preparation for that substitution, let's rewrite the integral as

$$\int \sec^4(\theta)\tan^2(\theta)\ \sec^2(\theta)\, d\theta.$$

In order to use the substitution $u = \tan(\theta)$, we need to rewrite the remaining factors of secant in terms of tangent. Since there is an even number remaining, we can do that with the Pythagorean identity, $\sec^2(\theta) = 1 + \tan^2(\theta)$. That is, our integral has the form

$$\int \left(\sec^2(\theta)\right)^2 \tan^2(\theta) \ \sec^2(\theta) \, d\theta = \int \left(1 + \tan^2(\theta)\right)^2 \tan^2(\theta) \ \sec^2(\theta) \, d\theta.$$

Now we can make the substitution, and write our integral as

$$\int (1 + u^2)^2 \, u^2 \, du = \frac{1}{3} u^3 + \frac{2}{5} u^5 + \frac{1}{7} u^7 + C.$$

> Expand the integrand so that you see a polynomial, then integrate.

After returning to our original variable, we have

$$\int \sec^6(\theta) \tan^2(\theta) \, d\theta = \frac{1}{3} \tan^3(\theta) + \frac{2}{5} \tan^5(\theta) + \frac{1}{7} \tan^7(\theta) + C. \qquad \blacksquare$$

Try It Yourself: Using substitution to integrate products of secant and tangent

Example 3.5. Determine (a) $\int \sec^7(\theta) \tan^3(\theta) \, d\theta$, and (b) $\int \sec^4(\theta) \, d\theta$.

Answer: (a) $\frac{1}{9} \sec^9(\theta) - \frac{1}{7} \sec^7(\theta) + C$, and (b) $\tan(\theta) + \frac{1}{3} \tan^3(\theta) + C$. $\qquad \blacksquare$

Full Solution On-line

There are cases in which the answer to both of the above questions is, "Not yet," but we can use the Pythagorean relationship between tangent and secant before attempting to integrate.

Products of tangents

Example 3.6. Determine (a) $\int \tan^4(\theta) \, d\theta$, and (b) $\int \tan^5(\theta) \, d\theta$.

Solution: (a) When we have the product of an even number of tangents, we can rewrite the product in terms of secant by using the Pythagorean identity:

$$\int \left(\tan^2(\theta)\right)^2 d\theta = \int \left(\sec^2(\theta) - 1\right)^2 d\theta = \int \sec^4(\theta) - 2\sec^2(\theta) + 1 \, d\theta.$$

The last two summands are easy to integrate, and the first has an even number of secants, so we can use the substitution $u = \tan(\theta)$. In this case, that yields

$$\int \tan^4(\theta) \, d\theta = \frac{1}{3} \tan^3(\theta) - \tan(\theta) + \theta + C.$$

(b) When we have the product of an odd number of tangents, we can move one factor to the side and rewrite the others in terms of the secant.

$$
\begin{aligned}
\int \tan^4(\theta) \tan(\theta) \, d\theta &= \int (\sec^2(\theta) - 1)^2 \tan(\theta) \, d\theta \\
&= \int \left(\sec^4(\theta) - 2\sec^2(\theta) + 1\right) \tan(\theta) \, d\theta \\
&= \int \sec^4(\theta) \tan(\theta) \, d\theta - 2 \int \sec^2(\theta) \tan(\theta) \, d\theta + \int \tan(\theta) \, d\theta.
\end{aligned}
$$

The last of these integrals was determined to be $\ln|\sec(\theta)| + C$ on p. 408 using the substitution technique, and the others have an odd number of tangents so we can integrate using the substitution $u = \sec(\theta)$. This yields

> Alternatively, we can also use $u = \tan(\theta)$ since these integrands have an even number of secants. Doing it that way, our answer looks different, but in fact differs from this one by only a constant. Try it!

$$\int \tan^5(\theta) \, d\theta = \frac{1}{4} \sec^4(\theta) - \sec^2(\theta) + \ln|\sec(\theta)| + C. \qquad \blacksquare$$

Lastly we come to cases in which the answer to both of the above questions is simply "No," so we are unable to use the substitution technique. When that happens, we employ integration by parts.

Odd powers of secant

Example 3.7. Determine $\int \sec^5(\theta)\, d\theta$.

Solution: In this case, we have an odd number of factors of secant, but no factors of tangent, so the substitution techniques from previous examples are inapplicable. However, we can write our integrand as $\sec^3(\theta)\sec^2(\theta)$, and the factor of $\sec^2(\theta)$ is easy to integrate on its own. This leads us to try integrating by parts with

$$
\begin{aligned}
dv &= \sec^2(\theta)\, d\theta & u &= \sec^3(\theta) \\
v &= \tan(\theta) & du &= 3\sec^3(\theta)\tan(\theta)\, d\theta.
\end{aligned}
$$

Then

$$
\int \underbrace{\sec^3(\theta)}_{u}\, \underbrace{\sec^2(\theta)\, d\theta}_{dv} = \underbrace{\sec^3(\theta)\tan(\theta)}_{uv} - \int \underbrace{\tan(\theta)}_{v}\, \underbrace{3\sec^3(\theta)\tan(\theta)\, d\theta}_{du}\ .
$$

The two factors of tangent in the remaining integral allow us to use the Pythagorean identity to rewrite this equation as

$$
\begin{aligned}
\int \sec^5(\theta)\, d\theta &= \sec^3(\theta)\tan(\theta) - \int 3(\sec^2(\theta) - 1)\sec^3(\theta)\, d\theta \\
&= \sec^3(\theta)\tan(\theta) - 3\int \sec^5(\theta)\, d\theta + 3\int \sec^3(\theta)\, d\theta
\end{aligned}
$$

where we see the integral of $\sec^5(\theta)$ appear on both sides. As we did in Section 6.2, when we saw this kind of "cycling behavior," we proceed by adding $3\int \sec^5(\theta)\, d\theta$ to both sides:

$$
4\int \sec^5(\theta)\, d\theta = \sec^3(\theta)\tan(\theta) + 3\int \sec^3(\theta)\, d\theta
$$

Now dividing by 4 yields

$$
\int \sec^5(\theta)\, d\theta = \frac{1}{4}\sec^3(\theta)\tan(\theta) + \frac{3}{4}\int \sec^3(\theta)\, d\theta. \tag{3.2}
$$

We're not done, but note our accomplishment: we had *five* factors of secant, and now have only *three*. So integrating by parts made the problem simpler, and integrating by parts again finishes the problem. That was done on p. 423, and we found

$$
\int \sec^3(\theta)\, d\theta = \frac{1}{2}\sec(\theta)\tan(\theta) + \frac{1}{2}\ln|\sec(\theta) + \tan(\theta)| + C. \tag{3.3}
$$

Substituting that fact into equation (3.2) yields the answer

$$
\int \sec^5(\theta)\, d\theta = \frac{1}{4}\sec^3(\theta)\tan(\theta) + \frac{3}{8}\sec(\theta)\tan(\theta) + \frac{3}{8}\ln|\sec(\theta) + \tan(\theta)| + C. \ \blacksquare
$$

The other case in which the answers to the above questions is simply "No" can be reduced to the kind of integral you saw in Example 3.7.

$\sec^n(\theta)\tan^k(\theta)$ where n is odd, and k is even

Example 3.8. Determine $\int \sec^3(\theta)\tan^2(\theta)\, d\theta$.

Solution: In this case, we have an odd number of factors of secant and an even number of factors of tangent, so the substitution techniques from previous examples are inapplicable. However, we can exchange an even number of tangents for an even

We could try $du = \sec^2(\theta)\, d\theta$ in a substitution, but then we cannot write the rest of the integrand in terms of $u = \tan(\theta)$ using the Pythagorean identity because there are an odd number of factors of secant remaining. Similarly, we could try $du = \sec(\theta)\tan(\theta)$, but then we cannot write the rest of the integrand in terms of $u = \sec(\theta)$ using the Pythagorean identity because there are an odd number of factors of tangent remaining. This is why the parity (even or oddness) of the exponents matters.

number of secants using the Pythagorean identity. When we do that, we see

$$\int \sec^3(\theta) \tan^2(\theta)\, d\theta = \int \sec^3(\theta) \Big(\sec^3(\theta) - 1 \Big)\, d\theta = \int \sec^5(\theta) - \sec^3(\theta)\, d\theta.$$

The integral of $\sec^3(\theta)$ was cited in equation (3.3) and used to determine the integral of $\sec^5(\theta)$ in Example 3.7. Using the results determined there, we can write

$$\int \sec^3(\theta) \tan^2(\theta)\, d\theta = \frac{1}{4} \sec^3(\theta) \tan(\theta) - \frac{1}{8} \sec(\theta) \tan(\theta) - \frac{1}{8} \ln|\sec(\theta) + \tan(\theta)| + C. \qquad \blacksquare$$

❖ Products of Cosecant and Cotangent

Because the differential and Pythagorean relationships between cosecant and cotangent parallel those between secant and tangent, the same techniques work in the same situations for the same reasons.

Powers of cosecant and cotangent

Example 3.9. Determine $\int \csc^4(\theta) \cot^2(\theta)\, d\theta$.

Solution: Since we have an even number of cosecants, and $-\csc^2(\theta)$ is the derivative of $\cot(\theta)$, plan to make the substitution $u = \cot(\theta)$. In order for that to work, we have to move two factors of $\csc(\theta)$ to the differential, and write the others in terms of the cotangent using the Pythagorean identity $\cot^2(\theta) + 1 = \csc^2(\theta)$. That is, we write our integral as

$$
\begin{aligned}
\int \csc^2(\theta) \cot^2(\theta)\, \csc^2(\theta)\, d\theta &= \int \left(\cot^2(\theta) + 1 \right) \cot^2(\theta)\, \csc^2(\theta)\, d\theta \\
&= \int \left(\cos^4(\theta) + \cot^2(\theta) \right) \csc^2(\theta)\, d\theta.
\end{aligned}
$$

Now the substitution $u = \cot(\theta)$, $du = -\csc(\theta)\, d\theta$ allows us rewrite the integral as

$$\int (u^4 + u^2)\, (-du) = -\frac{1}{5} u^5 - \frac{1}{3} u^3 + C.$$

Returning to our original variable, we have our answer:

$$\int \csc^4(\theta) \cot^2(\theta)\, d\theta = -\frac{1}{5} \cot^5(\theta) - \frac{1}{3} \cot^3(\theta) + C. \qquad \blacksquare$$

Try It Yourself: Powers of cosecant and cotangent

Example 3.10. Determine $\int \csc(\theta) \cot^3(\theta)\, d\theta$.

Answer: $\csc(\theta) - \frac{1}{3} \csc^3(\theta) + C.$ $\qquad \blacksquare$

Full Solution
On-line

❖ Summary

Whether dealing with products of sine and cosine, products of secant and tangent, or products of cosecant and cotangent, try substitution first. It's often easiest to make the differential you want, and then try to rewrite the remaining factors in the integrand using the Pythagorean identity.

$du = \cos(\theta)\,d\theta$　　　　　$\Rightarrow$　rewrite all other factors in terms of $u = \sin(\theta)$

$du = \sin(\theta)\,d\theta$　　　　　$\Rightarrow$　rewrite all other factors in terms of $u = -\cos(\theta)$

$du = \sec^2(\theta)\,d\theta$　　　　$\Rightarrow$　rewrite all other factors in terms of $u = \tan(\theta)$

$du = \sec(\theta)\tan(\theta)\,d\theta$　$\Rightarrow$　rewrite all other factors in terms of $u = \sec(\theta)$

$du = \csc^2(\theta)\,d\theta$　　　　$\Rightarrow$　rewrite all other factors in terms of $u = -\cot(\theta)$

$du = \csc(\theta)\cot(\theta)\,d\theta$　$\Rightarrow$　rewrite all other factors in terms of $u = -\csc(\theta)$

When substitution doesn't work out, you can reduce the complexity of the integrand by using either the half-angle identities for sine and cosine (see Example 3.2 on p. 429) or an integration by parts (see Example 3.7 on p. 432).

You should know

- what questions to ask in order to determine whether a substitution will work (see p. 429 and p. 430);

- to use the half-angle identities (for sine and cosine) or integration by parts (for secant and tangent) if substitution doesn't work.

You should be able to

- determine indefinite integrals of products of sine and cosine, products of secant and tangent, or products of cosecant and cotangent.

❖ 6.3 Skill Exercises

Determine the indefinite integrals in #1–6.

1. $\displaystyle\int \cos^7(\theta)\,d\theta$

2. $\displaystyle\int \sin^5(12\theta)\,d\theta$

3. $\displaystyle\int \sin^3(\theta)\cos^2(\theta)\,d\theta$

4. $\displaystyle\int \sin^4(16\theta)\cos^3(16\theta)\,d\theta$

5. $\displaystyle\int \sin^3(5\theta)\cos^3(5\theta)\,d\theta$

6. $\displaystyle\int \sin^7(\theta)\cos^5(\theta)\,d\theta$

In #7–12 find an antiderivative of the specified function.

7. $f(\theta) = \sin^4(8\theta)$

8. $f(\theta) = \cos^6(3\theta)$

9. $f(\theta) = \sin^2(\theta)\cos^2(\theta)$

10. $f(\theta) = \sin^2(3\theta)\cos^4(3\theta)$

11. $f(\theta) = \sin^8(4\theta)\cos^6(4\theta)$

12. $f(\theta) = \sin^4(7\theta)\cos^8(7\theta)$

In #13–18 find the indefinite integral by using $du = \sec^2(\theta)\,d\theta$ (or $du = \csc^2(\theta)\,d\theta$ as appropriate).

13. $\displaystyle\int \tan^4(\theta)\sec^2(\theta)\,d\theta$

14. $\displaystyle\int \tan^3(\theta)\sec^2(\theta)\,d\theta$

15. $\displaystyle\int \cot^2(2\theta)\csc^4(2\theta)\,d\theta$

16. $\displaystyle\int \cot^5(2\theta)\csc^4(2\theta)\,d\theta$

17. $\displaystyle\int \sec^6(\theta)\,d\theta$

18. $\displaystyle\int \csc^{10}(5\theta)\,d\theta$

In #19–22 determine the indefinite integral by writing $du = \sec(\theta)\tan(\theta)\,d\theta$ (or $du = \csc(\theta)\cot(\theta)\,d\theta$ as appropriate).

19. $\displaystyle\int \tan^3(\theta)\sec(\theta)\,d\theta$

21. $\displaystyle\int \cot^5(\theta)\csc^4(\theta)\,d\theta$

20. $\displaystyle\int \tan^5(\theta)\sec^3(\theta)\,d\theta$

22. $\displaystyle\int \tan^9(\theta)\sec^5(\theta)\,d\theta$

In #23–28 determine the indefinite integral.

23. $\displaystyle\int \tan^4(\theta)\,d\theta$

25. $\displaystyle\int \csc^5(2\theta)\,d\theta$

27. $\displaystyle\int \tan^4(\theta)\sec(\theta)\,d\theta$

24. $\displaystyle\int \cot^5(\theta)\,d\theta$

26. $\displaystyle\int \tan^2(\theta)\sec^3(\theta)\,d\theta$

28. $\displaystyle\int \cot^7(\theta)\,d\theta$

In #29–32 first convert the integrands into products of tangents and secants (or cotangents and cosecants) and then use the techniques discussed in this section to find the antiderivative.

29. $\displaystyle\int \frac{\sin^2(\theta)}{\cos^3(\theta)}\,d\theta$

30. $\displaystyle\int \frac{\sin^3(\theta)}{\cos^8(\theta)}\,d\theta$

31. $\displaystyle\int \frac{\cos^4(\theta)}{\sin^7(\theta)}\,d\theta$

32. $\displaystyle\int \frac{\cos^3(\theta)}{\sin^9(\theta)}\,d\theta$

In #33–36 rewrite the integrand in terms of sine and cosine, and then determine the definite integral.

33. $\displaystyle\int \tan^3(\theta)\csc^2(\theta)\,d\theta$

35. $\displaystyle\int \frac{\sec^4(\theta)}{\csc^2(\theta)}\,d\theta$

34. $\displaystyle\int \sec(\theta)\cos^3(\theta)\,d\theta$

36. $\displaystyle\int \cot^8(\theta)\sec^5(\theta)\,d\theta$

Mixed Practice: This set of exercises is intended to provide you with practice in determining which integration technique is appropriate for a given integral. Exercises #37–48 may require techniques from previous sections.

37. $\displaystyle\int t^2\sin(t)\,dt$

41. $\displaystyle\int \sin^2(t)\cos^3(t)\,dt$

45. $\displaystyle\int \cos^2(t) - \sin^2(t)\,dt$

38. $\displaystyle\int \tan^6(t)\,dt$

42. $\displaystyle\int \frac{\sec^2(t)}{\tan^{19}(t)}\,dt$

46. $\displaystyle\int e^{-t}\Big(\sin(t) + \cos(t)\Big)\,dt$

39. $\displaystyle\int \sec^2(t)e^{\tan(t)}\,dt$

43. $\displaystyle\int \sec^6(t)\,dt$

47. $\displaystyle\int \frac{\cos^3(t)}{\sin^7(t)}\,dt$

40. $\displaystyle\int \frac{\ln(\cot(t))}{\sin(t)\cos(t)}\,dt$

44. $\displaystyle\int \csc^3(t)\cot(t)\,dt$

48. $\displaystyle\int \tan^2(t) + \cot^2(t)\,dt$

❖ 6.3 Concept and Application Exercises

49. Suppose $f(\theta) = \sec^2(\theta)\tan^3(\theta)d\theta$.

    (a) Determine an antiderivative of f by using the substitution $u = \tan(\theta)$.

    (b) Determine an antiderivative of f by using the substitution $u = \sec(\theta)$.

(c) Use a graphing utility to render plots of the functions from parts (a) and (b) on the same screen. In complete sentences, explain what you see and why it makes sense.

50. Suppose $f(\theta) = \sec^4(\theta) \tan^5(\theta)$. Answer parts (a)–(c) of #49.

51. Suppose $f(\theta) = \cos(\theta) \sin^3(\theta)$.

   (a) Determine an antiderivative of f by using the substitution $u = \sin(\theta)$.

   (b) Determine an antiderivative of f by using the substitution $u = \cos(\theta)$.

   (c) Use a graphing utility to render plots of the functions from parts (a) and (b) on the same screen. In complete sentences, explain what you see and why it makes sense.

52. Suppose $f(\theta) = \cos^5(\theta) \sin^3(\theta)$. Answer parts (a)–(c) of #51.

53. Working as we did in Example 3.7, use integration by parts to show that for each positive integer k,

$$\int \sec^{2k+1}(\theta)\, d\theta = \frac{1}{2k} \tan(\theta) \sec^{2k-1}(\theta) + \frac{2k-1}{2k} \int \sec^{2k-1}(\theta)\, d\theta.$$

54. Use the formula in #53 and the result of Example 3.7 to determine $\int \sec^7(\theta)\, d\theta$.

55. In signal analysis we use functions of the form $A_k \cos(k\pi t)$, where k is a positive integer. Determine a value of A_k so that

$$\int_0^1 A_k^2 \cos^2(k\pi t)\, dt = 1,$$

and verify that the value does not depend on k.

56. In signal analysis we use functions of the form $A_k \cos(k\pi t)$, where k is a positive integer.

   (a) Determine a value of A_k so that $\int_0^L A_k^2 \cos^2(k\pi t)\, dt = 1$ when L is a positive integer, and verify that the value does not depend on k.

   (b) In complete sentences, explain the way in which A_k changes as L gets larger, and why that make sense (think geometrically, about area).

57. Suppose m and n are positive integers. What conditions on m and n allow us to use substitution when determining $\int \sin^n(\theta) \cos^m(\theta)\, d\theta$?

58. Suppose m and n are positive integers. What conditions on m and n allow us to use substitution when determining $\int \sec^n(\theta) \tan^m(\theta)\, d\theta$?

6.4   Trigonometric Substitution

The technique of integration called trigonometric substitution, which is demonstrated below, is also sometimes called *reverse substitution* or *inverse substitution* because it amounts to thinking of the given integrand as the result of an unwise substitution—one that we try to undo. As you might suspect from the name of this section, we use trigonometric identities in the process. Specifically:

$$\cos^2(\theta) + \sin^2(\theta) = 1 \quad \text{which is the same as} \quad \cos^2(\theta) = 1 - \sin^2(\theta)$$
$$1 + \tan^2(\theta) = \sec^2(\theta) \quad \text{which is the same as} \quad \tan^2(\theta) = \sec^2(\theta) - 1.$$

As you'll see in Examples 4.1–4.6, these identities are particularly helpful when the integrand includes terms such as $\sqrt{t^2 \pm 1}$ or $\sqrt{1 - t^2}$ because "undoing the substitution" leaves us with a perfect square under the radical.

Using the tangent in a trigonometric substitution

Example 4.1. Determine $\int \sqrt{1 + t^2} \, dx$.

Solution: The $1 + t^2$ that we see under the radical is reminiscent of the identity $1 + \tan^2(\theta) = \sec^2(\theta)$. As mentioned above, let's imagine that our integral comes from a substitution that someone else made. Specifically, that person used $t = \tan(\theta)$, which implies

$$\frac{dt}{d\theta} = \sec^2(\theta) \quad \Rightarrow \quad dt = \sec^2(\theta) \, d\theta.$$

> More formally, what we're actually doing is defining a new variable, θ, that's related to t by the equation
> $$t = \tan(\theta).$$
> Said differently,
> $$\theta = \tan^{-1}(t).$$

So we can think of the integral as

$$\int \sqrt{1 + \tan^2(\theta)} \, \sec^2(\theta) \, d\theta = \int \sqrt{\sec^2(\theta)} \, \sec^2(\theta) \, d\theta = \int |\sec(\theta)| \, \sec^2(\theta) \, d\theta.$$

And since $\theta = \arctan(t)$, the angle $\theta \in (-\pi/2, \pi/2)$. That's important because the secant is positive at all such angles, so we can omit the absolute values and write our integral as we did back on p. 423 …

$$\int \sec^3(\theta) \, d\theta = \frac{1}{2} \sec(\theta) \tan(\theta) + \frac{1}{2} \ln|\sec(\theta) + \tan(\theta)| + C.$$

See Example 2.11 on p. 423.

Now that the integration is complete, we need to return to the original variable. It's often easiest to do this by drawing the relationship between t and θ on a right triangle. For example, the left-hand image in Figure 4.1 depicts the relationship $\tan(\theta) = t/1$. In the right-hand image, we've completed the triangle using the Pythagorean Theorem, after which it's easy to see that $\sec(\theta) = \sqrt{1 + t^2}$. This allows us to write our answer as

$$\frac{1}{2} t \sqrt{1 + t^2} + \frac{1}{2} \ln|\sqrt{1 + t^2} + t| + C. \qquad \blacksquare$$

Note: Whether you think of the technique from Example 4.1 as "undoing an unwise substitution" or as "judiciously defining a new variable θ," what matters is that it works because the Pythagorean identity allows us to rewrite the term under the radical as a perfect square.

Using the tangent in a trigonometric substitution

Example 4.2. Determine $\int \frac{1}{\sqrt{7 + 5t^2}} \, dt$.

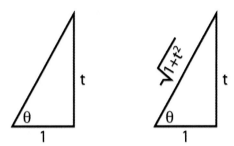

Figure 4.1: (left) The relationship between t and θ is $\tan(\theta) = t/1$; (right) finishing the triangle with the Pythagorean Theorem.

Solution: On p. 406 we introduced the idea of temporarily setting all the coefficients of an integrand to 1 in order to see its underlying structure. Doing that here, the denominator becomes $\sqrt{1 + t^2}$, which leads to expect that we'll use the same basic technique here that was successful in Example 4.1. In order to use the identity $1 + \tan^2(\theta) = \sec^2(\theta)$, we need to see "1+" under the radical, so let's begin by factoring out the 7 and writing our integrand as

$$\frac{1}{\sqrt{7 + 5t^2}} = \frac{1}{\sqrt{7\left(1 + \frac{5}{7}t^2\right)}} = \frac{1}{\sqrt{7}} \frac{1}{\sqrt{1 + \frac{5}{7}t^2}}.$$

In order for the term under the radical to have the form $1 + \tan^2(\theta)$, we need

$$\frac{5}{7}t^2 = \tan^2(\theta), \quad \text{which happens when } \quad t = \frac{\sqrt{7}}{\sqrt{5}}\tan(\theta).$$

That relationship between t and θ implies

$$\frac{dt}{d\theta} = \frac{\sqrt{7}}{\sqrt{5}}\sec^2(\theta) \quad \Rightarrow \quad dt = \frac{\sqrt{7}}{\sqrt{5}}\sec^2(\theta)\, d\theta.$$

This allows us rewrite

$$\int \frac{1}{\sqrt{7}} \frac{1}{\sqrt{1 + \frac{5}{7}t^2}}\, dt \quad \text{as the integral} \quad \int \frac{1}{\sqrt{7}} \frac{1}{\sqrt{1 + \tan^2(\theta)}} \frac{\sqrt{7}}{\sqrt{5}}\sec^2(\theta)\, d\theta.$$

Since $1 + \tan^2(\theta) = \sec^2(\theta)$, this integral is really

$$\frac{1}{\sqrt{5}} \int \frac{1}{\sqrt{\sec^2(\theta)}}\sec^2(\theta)\, d\theta = \frac{1}{\sqrt{5}} \int \frac{1}{|\sec(\theta)|}\sec^2(\theta)\, d\theta.$$

As in Example 4.1, we can omit the absolute values because

$$t = \frac{\sqrt{7}}{\sqrt{5}}\tan(\theta) \quad \Rightarrow \quad \theta = \arctan\left(\frac{\sqrt{5}}{\sqrt{7}}t\right) \in \left(-\frac{\pi}{2}, \frac{\pi}{2}\right),$$

where the cosine (and so the secant) is positive. Then we can cancel a factor of $\sec(\theta)$ and reduce our integral to

$$\frac{1}{\sqrt{5}} \int \sec(\theta)\, d\theta = \frac{1}{\sqrt{5}} \ln|\sec(\theta) + \tan(\theta)| + C.$$

Lastly, we return to the original variable using the relationship between t and θ, rewritten here as $\tan(\theta) = \sqrt{5}\, t/\sqrt{7}$. The associated right triangle is depicted in

Figure 4.2, from which we conclude that $\sec(\theta) = \sqrt{7 + 5x^2}/\sqrt{7}$. Therefore, our integral is

$$\int \frac{1}{\sqrt{7 + 5t^2}} \, dt = \frac{1}{\sqrt{5}} \ln \left| \frac{\sqrt{7 + 5t^2}}{\sqrt{7}} + \frac{\sqrt{5}\, t}{\sqrt{7}} \right| + C.$$

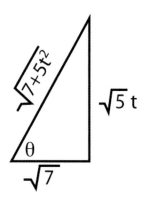

We can make this answer a little simpler by using the properties of logarithms. Specifically, logarithms take quotients to differences so

$$\frac{1}{\sqrt{5}} \ln \left| \frac{\sqrt{7 + 5t^2} + \sqrt{5}\, t}{\sqrt{7}} \right| + C \;=\; \frac{1}{\sqrt{5}} \left(\ln \left| \sqrt{7 + 5t^2} + \sqrt{5}\, t \right| - \ln(\sqrt{7}) \right) + C$$

$$= \frac{1}{\sqrt{5}} \ln \left| \sqrt{7 + 5t^2} + \sqrt{5}\, t \right| - \frac{\ln(\sqrt{7})}{\sqrt{5}} + C.$$

Figure 4.2: A right triangle in which $\tan(\theta) = \sqrt{5}\, t/\sqrt{7}$.

In the same way that $\int 2t \, dt = t^2 + C$ and $\int 2t \, dt = t^2 - 4 + C$ describe the same family of functions because C can be *any* constant, we typically omit the $-\ln(\sqrt{7})/\sqrt{5}$ from our answer, and describe this family of functions as

$$\int \frac{1}{\sqrt{7 + 5t^2}} \, dt = \frac{1}{\sqrt{5}} \ln \left| \sqrt{7 + 5t^2} + \sqrt{5}\, t \right| + C.$$

Alternatively, if you're uncomfortable with that idea, you can define a new constant, $C_1 = -\ln(\sqrt{7})/\sqrt{5} + C$ and write

$$\int \frac{1}{\sqrt{7 + 5t^2}} \, dt = \frac{1}{\sqrt{5}} \ln \left| \sqrt{7 + 5t^2} + \sqrt{5}\, t \right| + C_1. \qquad \blacksquare$$

Try It Yourself: Using the tangent in a trigonometric substitution

Example 4.3. Determine $\int \frac{5}{\sqrt{9 + 16t^2}} \, dt$.

Answer: $\frac{5}{4} \ln \left| \sqrt{9 + 16t^2} + 4t \right| + C$. $\qquad \blacksquare$

Full Solution
On-line

Using the secant in a trigonometric substitution

Example 4.4. Determine $\int \sqrt{t^2 - 1} \, dt$ when $t \geq 1$.

Solution: We've been using the identity $\tan^2(\theta) + 1 = \sec^2(\theta)$ to rewrite radicands as perfect squares, but this time we see $t^2 - 1$ instead of $t^2 + 1$, and that requires a minor change in our technique. Let's rewrite the Pythagorean relationship between secant and tangent as $\sec^2(\theta) - 1 = \tan^2(\theta)$. Then the term under the radical would be a perfect square if $t^2 = \sec^2(\theta)$. That happens when

$$t = \sec(\theta), \quad \text{where } \theta \in \left[0, \frac{\pi}{2} \right) \text{ since } t \geq 1.$$

The secant ranges from 1 to ∞ over $[0, \pi/2)$. This allows us to reach all $t \geq 1$.

Consequently, $dt = \sec(\theta) \tan(\theta) \, d\theta$, and we can rewrite our integral as

$$\int \sqrt{\sec^2(\theta) - 1} \, \sec(\theta) \tan(\theta) \, d\theta = \int \sqrt{\tan^2(\theta)} \, \sec(\theta) \tan(\theta) \, d\theta = \int |\tan(\theta)| \, \sec(\theta) \tan(\theta) \, d\theta.$$

Since $\theta \in [0, \pi/2)$ the tangent is not negative, which means that the absolute value in this integrand is irrelevant. So let's omit it and write our integral as

$$\int \sec(\theta) \tan^2(\theta) \, d\theta = \int \sec(\theta)(\sec^2(\theta) - 1) \, d\theta = \int \sec^3(\theta) - \sec(\theta), \, d\theta.$$

The integral of $\sec^3(\theta)$ was an example of integration by parts on p. 423, and the integral of $\sec(\theta)$ was done with substitution on p. 409. Upon recalling these results, we see that our integral simplifies to

$$\frac{1}{2} \sec(\theta) \tan(\theta) - \frac{1}{2} \ln|\sec(\theta) + \tan(\theta)| + C.$$

The final step of our process is returning to the original variable, which we do by drawing a right triangle in which $\sec(\theta) = t/1$. This has been done in Figure 4.3. Based on this triangle, our answer is

$$\frac{1}{2}t\sqrt{t^2-1} - \frac{1}{2}\ln|t + \sqrt{t^2-1}| + C \qquad \blacksquare$$

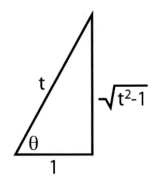

Figure 4.3: A right triangle in which $\sec(\theta) = t/1$.

Using the secant in a trigonometric substitution

Example 4.5. Determine $\int \frac{1}{\sqrt{3t^2-2}}\,dt$ when $t > \sqrt{2/3}$.

Solution: When we determine the structure of this integrand by temporarily setting the coefficients to 1, we see $1/\sqrt{t^2-1}$, which is similar in nature to the integral that we did in Example 4.4. That leads us to try the same basic technique, but we'll need the constant term to be a 1 to do it (as it is in the Pythagorean identity), so we begin by factoring the 2 out of the radical:

$$\frac{1}{\sqrt{3t^2-2}} = \frac{1}{\sqrt{2\left(\frac{3}{2}t^2-1\right)}} = \frac{1}{\sqrt{2}}\frac{1}{\sqrt{\frac{3}{2}t^2-1}}.$$

In order for the term under the radical to have the form $\sec^2(\theta)-1$, we need $\sec^2(\theta) = \frac{3}{2}t^2$. This happens when

$$t = \sqrt{\frac{2}{3}}\ \sec(\theta), \qquad \text{where } \theta \in \left(0, \frac{\pi}{2}\right) \text{ since } t > \frac{\sqrt{2}}{\sqrt{3}}.$$

> The secant ranges from 1 to ∞ over $[0, \pi/2)$. This allows us to reach all $t > \sqrt{2/3}$.

Consequently, $dt = \frac{\sqrt{2}}{\sqrt{3}}\ \sec(\theta)\tan(\theta)\,d\theta$, and we can rewrite our integral as

$$\int \frac{1}{\sqrt{2}}\frac{1}{\sqrt{\sec^2(\theta)-1}}\frac{\sqrt{2}}{\sqrt{3}}\sec(\theta)\tan(\theta)\,d\theta = \frac{1}{\sqrt{3}}\int \frac{1}{\sqrt{\tan^2(\theta)}}\sec(\theta)\tan(\theta)\,d\theta$$

$$= \frac{1}{\sqrt{3}}\int \frac{1}{|\tan(\theta)|}\sec(\theta)\tan(\theta)\,d\theta.$$

Since $\tan(\theta) > 0$ when $\theta \in (0, \pi/2)$, we can omit the absolute values and cancel the factors of $\tan(\theta)$, thereby arriving at

$$\frac{1}{\sqrt{3}}\int \sec(\theta)\,d\theta = \frac{1}{\sqrt{3}}\ln|\sec(\theta) + \tan(\theta)| + C.$$

The triangle in Figure 4.4 depicts the relationship between θ and x, and you can see that $\tan(\theta) = \sqrt{3t^2-2}/\sqrt{2}$. Inserting this into our formula we arrive at

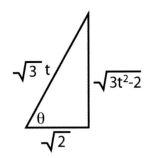

Figure 4.4: A right triangle in which $\sec(\theta) = \sqrt{3}\,t/\sqrt{2}$.

$$\int \frac{1}{\sqrt{3t^2-2}}\,dt = \frac{1}{\sqrt{3}}\ln\left|\frac{\sqrt{3}\,t}{\sqrt{2}} + \frac{\sqrt{3t^2-2}}{\sqrt{2}}\right| + C$$

which, using the properties of logarithms (as in Example 4.2), we rewrite as

$$\int \frac{1}{\sqrt{3t^2-2}}\,dt = \frac{1}{\sqrt{3}}\ln\left|\sqrt{3}\,t + \sqrt{3t^2-2}\right| + C \text{ when } t > \sqrt{\frac{2}{3}}. \qquad \blacksquare$$

Try It Yourself: Using the secant in a trigonometric substitution

Example 4.6. Determine $\int \sqrt{8t^2-7}\,dx$ when $t > \sqrt{7/8}$.

Answer: $\frac{\sqrt{7}}{2}t\sqrt{8t^2 - 7} - \frac{7}{2\sqrt{8}}\ln\left|\sqrt{8}t + \sqrt{8t^2 - 7}\right| + C$ when $t > \sqrt{7/8}$.   ∎

In the examples so far, we've used the Pythagorean relationship between the tangent and secant to address radicands of the form $(\;)^2 \pm 1$. When the integrand has the form $1 - (\;)^2$ we need to use the Pythagorean relationship between the sine and cosine instead.

Using the sine in a trigonometric substitution

Example 4.7. Determine $\int \frac{1}{\sqrt{7 - 3t^2}}\,dt$.

Solution: As before, we begin by factoring the constant term out of the radical,

$$\int \frac{1}{\sqrt{7 - 3t^2}}\,dt = \int \frac{1}{\sqrt{7\left(1 - \frac{3}{7}t^2\right)}}\,dt = \int \frac{1}{\sqrt{7}}\frac{1}{\sqrt{1 - \frac{3}{7}t^2}}\,dt.$$

Now the term under the radical has the form $1 - (\;)^2$, which reminds us of the Pythagorean identity

$$\cos^2(\theta) + \sin^2(\theta) = 1 \quad \text{written in the form} \quad \underbrace{1 - \sin^2(\theta)}_{1-(\;)^2} = \cos^2(\theta).$$

In order for the term under the radical to have the form $1 - \sin^2(\theta)$, we need $\sin^2(\theta) = \frac{3}{7}t^2$. This happens when

$$t = \frac{\sqrt{7}}{\sqrt{3}}\,\sin(\theta), \quad \text{where } \theta \in (-\pi/2, \pi/2).$$

> The domain of the integrand is $\left(-\sqrt{7/3}, \sqrt{7/3}\right)$. Our new characterization of t as a function of θ has to reach all those values of t. We do this by letting the sine range from -1 to 1, as it does over the interval $(-\pi/2, \pi/2)$.

It follows that $dt = \frac{\sqrt{7}}{\sqrt{3}}\cos(\theta)\,d\theta$, and substituting this into our integral we see

$$\int \frac{1}{\sqrt{7}}\frac{1}{\sqrt{1 - \sin^2(\theta)}}\frac{\sqrt{7}}{\sqrt{3}}\cos(\theta)\,d\theta = \frac{1}{\sqrt{3}}\int \frac{1}{\sqrt{\cos^2(\theta)}}\cos(\theta)\,d\theta = \frac{1}{\sqrt{3}}\int \frac{\cos(\theta)}{|\cos(\theta)|}\,d\theta.$$

Since $\cos(\theta) > 0$ when $\theta \in (-\pi/2, \pi/2)$, we can omit the absolute value sign, and cancel a factor of $\cos(\theta)$. This leaves us with

$$\frac{1}{\sqrt{3}}\int 1\,d\theta = \frac{1}{\sqrt{3}}\theta + C.$$

Lastly, we return to the original variable by rewriting the equation

$$t = \sqrt{\frac{7}{3}}\,\sin(\theta) \quad \text{as} \quad \theta = \sin^{-1}\left(\sqrt{\frac{3}{7}}\,t\right).$$

So our answer is that

$$\int \frac{1}{\sqrt{7 - 3t^2}}\,dt = \frac{1}{\sqrt{3}}\sin^{-1}\left(\frac{\sqrt{3}}{\sqrt{7}}\,t\right) + C.$$ ∎

Using the sine in a trigonometric substitution

Example 4.8. Determine $\int \sqrt{3 - 8t^2}\,dt$.

Solution: As in Example 4.7, our solution will make use of the trigonometric identity $\cos^2(\theta) = 1 - \sin^2(\theta)$. However, the last steps of this solution are somewhat more complicated than what we saw in Example 4.7, and will require us to use the double-angle identity for the sine:

$$\sin(2\theta) = 2\sin(\theta)\cos(\theta).$$

As before, we begin by factoring the constant term out of the radical,

$$\int \sqrt{3 - 8t^2}\, dt = \int \sqrt{3\left(1 - \frac{8}{3}t^2\right)}\, dt = \int \sqrt{3}\,\sqrt{1 - \frac{8}{3}t^2}\, dt.$$

In order for the term under the radical to have the form $1 - \sin^2(\theta)$, we need $\sin^2(\theta) = \frac{8}{3}t^2$. This happens when

$$t = \sqrt{\frac{3}{8}}\sin(\theta), \quad \text{where } \theta \in [-\pi/2, \pi/2]$$

so that all t in the domain of the integrand are achieved. It follows that $dt = \frac{\sqrt{3}}{\sqrt{8}}\cos(\theta)\, d\theta$, and we rewrite our integral as

$$\int \sqrt{3}\,\sqrt{1 - \sin^2(\theta)}\,\frac{\sqrt{3}}{\sqrt{8}}\cos(\theta)\, d\theta = \frac{3}{\sqrt{8}}\int \sqrt{\cos^2(\theta)}\,\cos(\theta)\, d\theta = \frac{3}{\sqrt{8}}\int |\cos(\theta)|\,\cos(\theta)\, d\theta.$$

Since $\cos(\theta)$ remains non-negative when $\theta \in [-\pi/2, \pi/2]$, we can omit the absolute values and write our integral as

$$\frac{3}{\sqrt{8}}\int \cos^2(\theta)\, d\theta.$$

We integrated cosine-squared in Example 3.2 (p. 429). Based on that work, we have

$$\frac{3}{\sqrt{8}}\int \cos^2(\theta)\, d\theta = \frac{3}{\sqrt{8}}\left(\frac{1}{2}\theta + \frac{\sin(2\theta)}{4}\right) + C.$$

Figure 4.5 depicts the triangle that we use to return to the original variable. You can see that it's easy to determine $\sin(\theta)$ in terms of t, but not $\sin(2\theta)$. That prompts us to continue simplifying our answer by using the double-angle identity for the sine:

$$\frac{3}{\sqrt{8}}\left(\frac{1}{2}\theta + \frac{\sin(2\theta)}{4}\right) + C = \frac{3}{\sqrt{8}}\left(\frac{1}{2}\theta + \frac{2\sin(\theta)\cos(\theta)}{4}\right) + C.$$

Now we can return to the original variable. The triangle in Figure 4.5, which depicts the relationship between t and θ, shows

$$\sin(\theta) = \frac{\sqrt{8}t}{\sqrt{3}} \quad \text{and} \quad \cos(\theta) = \frac{\sqrt{3 - 8t^2}}{\sqrt{3}}.$$

Substituting these expressions into our answer, we see that

$$\begin{aligned}
\int \sqrt{3 - 8t^2}\, dt &= \frac{3}{\sqrt{8}}\left(\frac{1}{2}\theta + \frac{1}{2}\sin(\theta)\cos(\theta)\right) + C \\
&= \frac{3}{\sqrt{8}}\left(\frac{1}{2}\sin^{-1}\left(\sqrt{\frac{8}{3}}\,t\right) + \frac{1}{2}\left(\frac{\sqrt{8}t}{\sqrt{3}}\right)\left(\frac{\sqrt{3 - 8t^2}}{\sqrt{3}}\right)\right) + C \\
&= \frac{3}{2\sqrt{8}}\sin^{-1}\left(\sqrt{\frac{8}{3}}\,t\right) + \frac{t}{2}\sqrt{3 - 8t^2} + C.
\end{aligned}$$

∎

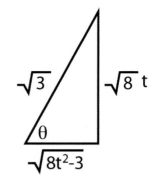

Figure 4.5: A right triangle in which $\sin(\theta) = \sqrt{8}\,t/\sqrt{3}$.

Try It Yourself: Using the sine in a trigonometric substitution

Example 4.9. Consider the disk of radius r, centered at the origin. Verify that its area is $\int_{-r}^{r} 2\sqrt{r^2 - x^2}\ dx$, and use trigonometric substitution $x = r\sin(\theta)$ to determine the value of that integral.

Answer: πr^2.   ∎

Full Solution On-line

> **Summary of trigonometric substitution:** Suppose $a > 0$ and $b > 0$ are constants, and t is the variable of integration. When your integrand includes
>
> $\sqrt{a^2 t^2 + b^2}$ consider using the substitution $t = \dfrac{b}{a}\tan(\theta)$
>
> $\sqrt{a^2 t^2 - b^2}$ consider using the substitution $t = \dfrac{b}{a}\sec(\theta)$
>
> $\sqrt{b^2 - a^2 t^2}$ consider using the substitution $t = \dfrac{b}{a}\sin(\theta)$

The technique of trigonometric substitution is sometimes helpful when there is not a square root in the integrand, or when the radicand doesn't have one of the forms listed above.

Using trigonometric substitution in the absence of a radical

Example 4.10. Determine $\int \frac{1}{(1+t^2)^3}\ dt$.

Solution: Seeing the factor of $1 + t^2$ in the denominator leads us to try

$$t = \tan(\theta) \quad \Rightarrow \quad dt = \sec^2(\theta)\ d\theta.$$

After substituting this into the integral, we see

$$\int \frac{1}{(1+\tan^2(\theta))^3}\ \sec^2(\theta)\ d\theta = \int \frac{1}{\sec^6(\theta)}\ \sec^2(\theta)\ d\theta = \int \frac{1}{\sec^4(\theta)}\ d\theta = \int \cos^4(\theta)\ d\theta.$$

In Example 3.2 (p. 429) we found that

$$\int \cos^4(\theta)\ d\theta = \frac{3}{8}\theta + \frac{1}{4}\sin(2\theta) + \frac{1}{32}\sin(4\theta) + C.$$

The triangle that helps us return to the original variable is shown in Figure 4.6, and you can see that it's easy to determine $\sin(\theta)$ in terms of t, but not $\sin(2\theta)$ or $\sin(4\theta)$. Using the double-angle formula for the sine allows us to make the necessary change in our answer:

$$\frac{1}{4}\sin(2\theta) = \frac{1}{4}\Big(2\sin(\theta)\cos(\theta)\Big) = \frac{1}{2}\sin(\theta)\cos(\theta) = \frac{1}{2}\cdot\frac{t}{1+t^2},$$

and similarly

$$\frac{1}{32}\sin(4\theta) = \frac{1}{32}\sin(2(2\theta)) = \frac{1}{32}\Big(2\sin(2\theta)\cos(2\theta)\Big) = \frac{1}{16}\sin(\theta)\cos(\theta)\ \cos(2\theta).$$

Finally, we use the double-angle formula for the cosine to write this as

$$\frac{1}{16}\sin(\theta)\cos(\theta)\ \cos(2\theta) = \frac{1}{16}\sin(\theta)\cos(\theta)\Big(\cos^2(\theta) - \sin^2(\theta)\Big) = \frac{1}{16}\frac{t}{1+t^2}\left(\frac{1}{1+t^2} - \frac{t^2}{1+t^2}\right).$$

Putting this all together, we have

$$\int \frac{1}{(1+t^2)^2}\ dt = \frac{3}{8}\tan^{-1}(t) + \frac{1}{2}\frac{t}{1+t^2} + \frac{1}{16}\frac{t - t^3}{(1+t^2)^2} + C\ .$$ ∎

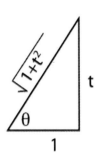

Figure 4.6: Using a triangle to relate θ to t.

Try It Yourself: Using trigonometric substitution with a different kind of radicand

Example 4.11. Use $t = \tan^2(\theta)$ to determine $\displaystyle\int \sqrt{\dfrac{t+1}{t}}\, dt$ when $t > 0$.

Answer: $\sqrt{t+t^2} + \ln\left|\sqrt{t+1} + \sqrt{t}\right| + C$. ∎

Full Solution On-line

❖ Disconnected Domains

In Example 4.5 we established that

$$f(t) = \frac{1}{\sqrt{3}} \ln\left|\sqrt{3}\,t + \sqrt{3t^2 - 2}\right| \tag{4.1}$$

is an antiderivative of $1/\sqrt{3t^2 - 2}$ over the interval $(\sqrt{2/3}, \infty)$, but the domain of $1/\sqrt{3t^2 - 2}$ also includes $(-\infty, -\sqrt{2/3})$, which we did not discuss. While we could rework the problem over that interval, it's easier to make use of the fact that $1/\sqrt{3t^2 - 2}$ is an even function.

Our experience tells us that when $f'(t)$ is an even function, it has an antiderivative that's odd (see Table 1). Recall that a function is odd when $f(t)$ and $f(-t)$ are opposites. That is, when $f(t) = -f(-t)$. We can use this idea to extend the formula in (4.1) to values of $t < -\sqrt{2/3}$. Specifically, suppose that t is some number in $(-\infty, -\sqrt{2/3})$. Then $-t$ is a positive number that's larger than $\sqrt{2/3}$, so we can use equation (4.1) to calculate $f(-t)$. If we negate that function value, we see $-f(-t)$, which is the same as $f(t)$ when f is odd. That is,

$$
\begin{aligned}
f(t) = -f(-t) \;\; &= \;\; -\frac{1}{\sqrt{3}} \ln\left|\sqrt{3}\,(-t) + \sqrt{3(-t)^2 - 2}\right| \\
&= -\frac{1}{\sqrt{3}} \ln\left|-\sqrt{3}\,t + \sqrt{3t^2 - 2}\right| \\
&= -\frac{1}{\sqrt{3}} \ln\left|\sqrt{3}\,t - \sqrt{3t^2 - 2}\right|.
\end{aligned}
$$

Table 1	
$f'(t)$	an antiderivative that's an odd function
$3t^2$	t^3
$\cos(t)$	$\sin(t)$
$\sec^2(t)$	$\tan(t)$
$\cosh(t)$	$\sinh(t)$
⋮	⋮

The last of these equalities is true because $|a - b| = |b - a|$. We've made this last step so that the final formula exhibits a simple symmetry.

Merging this with the formula in (4.1), we obtain an odd antiderivative

$$
f(t) = \begin{cases}
-\frac{1}{\sqrt{3}} \ln\left|\sqrt{3}\,t - \sqrt{3t^2 - 2}\right| & \text{when } t < -\sqrt{\frac{2}{3}} \\[2mm]
\frac{1}{\sqrt{3}} \ln\left|\sqrt{3}\,t + \sqrt{3t^2 - 2}\right| & \text{when } t > \sqrt{\frac{2}{3}}.
\end{cases}
$$

Of course, we could add any constant to this function and still have an antiderivative, because such a constant changes the height of the graph but not its slope. At this point it's important to note that, unlike the functions that you see in Table 1, the domain of $1/\sqrt{3t^2 - 2}$ is disconnected. This allows us to raise or lower the branches of the graph independently, so we write

$$
\int \frac{1}{\sqrt{3t^2 - 2}}\, dt = \begin{cases}
-\frac{1}{\sqrt{3}} \ln\left|\sqrt{3}\,t - \sqrt{3t^2 - 2}\right| + C_\ell & \text{when } t < -\sqrt{\frac{2}{3}} \\[2mm]
\frac{1}{\sqrt{3}} \ln\left|\sqrt{3}\,t + \sqrt{3t^2 - 2}\right| + C_r & \text{when } t > \sqrt{\frac{2}{3}}
\end{cases}
$$

where C_ℓ and C_r *can* be the same number but don't have to be. The graphs of two antiderivatives are shown in Figure 4.7.

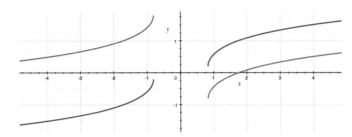

Figure 4.7: The graphs of two antiderivatives of $1/\sqrt{3t^2 - 2}$. Notice that the domain is separated into two branches. The blue curve depicts the case when C_ℓ and C_r are both zero. The red curve depicts a case when $C_\ell > 0$ and $C_r < 0$.

Try It Yourself: Using odd reflection

Example 4.12. Determine $\int \sqrt{8t^2 - 7}\, dt$. *(Hint: see Example 4.6, p. 440.)*

$Answer:$
$$\begin{cases} -\frac{\sqrt{7}}{2}t\sqrt{8t^2 - 7} - \frac{7}{2\sqrt{8}}\ln\left|\sqrt{8}t - \sqrt{8t^2 - 7}\right| + C_\ell & \text{when } t < -\sqrt{7/8} \\[2mm] \frac{\sqrt{7}}{2}t\sqrt{8t^2 - 7} - \frac{7}{2\sqrt{8}}\ln\left|\sqrt{8}t + \sqrt{8t^2 - 7}\right| + C_r & \text{when } t > \sqrt{7/8}\,. \end{cases}$$
∎

Full Solution On-line

✤ Common Difficulties

The most common problem associated with the application of this technique has to do with tracking/using/manipulating the differential. Some students don't write it down at all, so they can't track how it changes. Other students use the same letter for their new variable (instead of introducing θ), which leads them to neglect the change in the differential.

```
━━━━━━ You should know ━━━━━━

• the Pythagorean trigonometric identities that relate sine and cosine, tangent
  and secant.
```

```
━━━━━━ You should be able to ━━━━━━

• use trigonometric substitution to calculate integrals when the integrand
  includes terms of the form $\sqrt{a^2x^2 \pm b^2}$, or $\sqrt{b^2 - a^2x^2}$.
```

✤ 6.4 Skill Exercises

Calculate the indefinite integral in #1–20.

1. $\displaystyle\int \sqrt{9 - t^2}\, dt$

2. $\displaystyle\int \sqrt{16 - 25t^2}\, dt$

3. $\displaystyle\int \sqrt{5 - 4t^2}\, dt$

4. $\displaystyle\int \sqrt{3 - 7t^2}\, dt$

5. $\displaystyle\int \frac{dt}{\sqrt{12 - 5t^2}}$

6. $\displaystyle\int \frac{dt}{\sqrt{10 - 2t^2}}$

7. $\displaystyle\int \frac{dt}{t\sqrt{25 - 49t^2}}$

8. $\displaystyle\int \frac{dt}{t\sqrt{14 - 36t^2}}$

9. $\displaystyle\int \frac{t^2}{\sqrt{64 - 9t^2}}\, dt$

10. $\displaystyle\int \frac{t^2}{\sqrt{3 - 11t^2}}\, dt$

11. $\displaystyle\int \sqrt{t^2 - 81}\, dt$

12. $\displaystyle\int \sqrt{4t^2 - 25}\, dt$

13. $\int \frac{1}{t^2}\sqrt{5t^2-36}\,dt$ 15. $\int \frac{dt}{t\sqrt{4t^2-3}}$ 17. $\int \sqrt{t^2+12}\,dt$ 19. $\int \frac{dt}{\sqrt{6t^2+1}}$

14. $\int \frac{1}{t^2}\sqrt{t^2-16}\,dt$ 16. $\int \frac{dt}{t\sqrt{100t^2-1}}$ 18. $\int \sqrt{4t^2+15}\,dt$ 20. $\int \frac{dt}{\sqrt{3t^2+11}}$

In each of #21–26 begin by completing the square under the radical.

21. $\int \sqrt{t^2+8t}\,dt$ 23. $\int \sqrt{6t-t^2}\,dt$ 25. $\int \sqrt{2t^2+3t}\,dt$

22. $\int \sqrt{t^2-10t}\,dt$ 24. $\int \sqrt{4t-t^2}\,dt$

26. $\int \sqrt{2t-3t^2}\,dt$

Mixed Practice: This set of exercises is intended to provide you with practice in determining which integration technique is appropriate for a given integral. Exercises 27–42 may require techniques from previous sections.

27. $\int \frac{t}{\sqrt{9t^2-4}}\,dt$ 31. $\int \frac{8t+1}{t^2+4}\,dt$ 35. $\int t^2\cos\left(\frac{\pi}{8}t\right)dt$ 39. $\int \frac{3t}{(t^2+1)^2}\,dt$

28. $\int \frac{dt}{t\sqrt{9t^2-4}}$ 32. $\int \sqrt{6t-5t^2}\,dt$ 36. $\int t^2\ln(t^2)\,dt$ 40. $\int \tan^7(t)\sec^2(t)\,dt$

29. $\int \left(\ln(t)\right)^2 dt$ 33. $\int \frac{t-1}{\sqrt{6t+5}}\,dt$ 37. $\int \frac{2}{(t^2+1)^2}\,dt$ 41. $\int t\sin^{-1}(t)\,dt$

30. $\int t^3 e^{6t}\,dt$ 34. $\int \frac{t-1}{6t^2+5}\,dt$ 38. $\int \frac{t^3}{(t^2+1)^2}\,dt$ 42. $\int \sqrt{4t^2+8t}\,dt$

❖ 6.4 Concept and Application Exercises

43. A circular pipe has a radius of 1 inch, but flow through the pipe is blocked by a valve. When the valve is open (either fully or partially, as shown in Figure 4.8) the flow is proportional to the opened area. On p. 392 (exercise #42) you verified that, when the left-edge of the valve is at x, the flow is

$$f(x) = 2k\int_0^x \sqrt{1-(w-1)^2}\,dw.$$

Perform the integration and write a formula for $f(x)$. *(Hint: begin with a substitution.)*

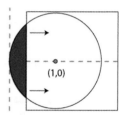

Figure 4.8: For #43.

44. Suppose an ion is y meters above the center of a charged line segment of length $2L$ meters, as depicted in Figure 4.9. If the line has a uniform charge density of σ coulombs per meter, a subinterval of length Δx accommodates $\sigma\Delta x$ coulombs of charge. So if the ion has a charge of q coulombs, that subinterval attracts (repels) the ion toward (from) the line segment with a vertical force of approximately

$$\frac{1}{4\pi\varepsilon_0}\frac{q\sigma y}{(y^2+x^2)^{3/2}}\Delta x,$$

where ε_0 is a number called the *permittivity constant*. (We say *approximately* because not all of the charge is exactly x units away from the center.)

(a) Write a Riemann sum with n subintervals that approximates the cumulative vertical force on the ion due to the entire line segment.

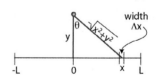

Figure 4.9: For #44. Note that side-to-side force is balanced by symmetry.

❖ Rational Functions that Are Not in Lowest Terms

A rational function is not in lowest terms when the degree of its numerator is larger than or equal to the degree of the denominator.

A rational function that's not in lowest terms

Example 5.11. Determine $\int \frac{t+5}{t+7}\, dt$.

Solution: Since the numerator and denominator have the same degree, this rational integrand is *not* in lowest terms. In simple cases like this, where the numerator and denominator are *almost* the same linear expression, adding zero is very helpful . . .

$$\int \frac{t+5}{t+7}\, dt = \int \frac{t+5+2-2}{t+7}\, dt = \int \frac{t+7}{t+7} - \frac{2}{t+7}\, dt = \int 1 - \frac{2}{t+7}\, dt,$$

which is $t - 2\ln|t+7| + C$.   ■

In more complicated cases, polynomial division can be used to rewrite the integrand as the sum of a polynomial and a rational function that *is* in lowest terms.

> For a review of polynomial division, see Appendix B.

A rational function that's not in lowest terms

Example 5.12. Determine $\int \frac{2t^3 + 21t^2 + 80t + 124}{t^2 + 6t + 13}\, dt$.

Solution: Since the this rational function is not in lowest terms, we begin with a polynomial division, which yields

$$\frac{2t^3 + 21t^2 + 80t + 124}{t^2 + 6t + 13} = 2t + 9 + \frac{7}{t^2 + 6t + 13}.$$

Therefore,

$$\int \frac{2t^3 + 21t^2 + 80t + 124}{t^2 + 6t + 13}\, dt = \int 2t + 9 + \frac{7}{t^2 + 6t + 13}\, dt.$$

> Note that the denominator is irreducible, so integrating this rational function does not involve a PFD.

The first of these integrals results in a polynomial, and the other was done in Example 1.12 on p. 407. So we conclude

$$\int \frac{2t^3 + 21t^2 + 80t + 124}{t^2 + 6t + 13}\, dt = t^2 + 9t + \frac{7}{2}\tan^{-1}\left(\frac{t+3}{7}\right) + C. \qquad ■$$

❖ In Summary

We typically integrate rational functions by using the substitution technique of Section 6.1, but that often requires us to rewrite them using algebraic techniques such as completing the square and partial fractions decomposition. Most of the work in this section was devoted to the (sometimes extensive) algebra of the latter.

Whenever you have to integrate a rational function in lowest terms, you should determine whether the denominator can be factored into a product of polynomials.

- If so, use a partial fractions decomposition.

- If not, consider using a substitution (especially if the numerator is linear and the denominator is quadratic) and/or completing the square in hopes of rewriting the integrand in the form $1/(u^2 + 1)$.

You should know

- the terms *lowest terms* and *partial fractions decomposition (PFD)*.

You should be able to

- perform polynomial division to reduce a rational function to lowest terms;

- use the techniques of completing the square and PFD to integrate rational functions.

❖ 6.5 Skill Exercises

Determine the indefinite integrals in #1–12, each of which involves an integrand whose denominator can be written as a product of distinct linear factors.

1. $\displaystyle \int \frac{2}{t^2 - 4}\, dt$

5. $\displaystyle \int \frac{3t - 1}{5t^2 + 25t + 30}\, dt$

9. $\displaystyle \int \frac{4t^2 + 1}{t^3 - 4t^2 + 3t}\, dt$

2. $\displaystyle \int \frac{3}{t^2 - 4t}\, dt$

6. $\displaystyle \int \frac{5t + 9}{2t^2 + 10t - 28}\, dt$

10. $\displaystyle \int \frac{5 - t}{t^3 - 16t}\, dt$

3. $\displaystyle \int \frac{1}{t^2 - 5t + 4}\, dt$

7. $\displaystyle \int \frac{t + 5}{8t^2 - 6t + 1}\, dt$

11. $\displaystyle \int \frac{5t}{4t^2 - 28t - 32}\, dt$

4. $\displaystyle \int \frac{4}{t^2 + t - 6}\, dt$

8. $\displaystyle \int \frac{27t - 2}{3t^2 - 10t - 8}\, dt$

12. $\displaystyle \int \frac{t^2 + 6t}{10t^3 + 13t^2 + 4t}\, dt$

Determine the indefinite integrals in #13–20, each of which involves an integrand whose denominator is irreducible.

13. $\displaystyle \int \frac{6}{7t^2 + 9}\, dt$

15. $\displaystyle \int \frac{5 - 4t}{7t^2 + 3}\, dt$

17. $\displaystyle \int \frac{8}{t^2 + 4t + 5}\, dt$

19. $\displaystyle \int \frac{t - 4}{5t^2 + 6t + 10}\, dt$

14. $\displaystyle \int \frac{2t + 1}{3t^2 + 4}\, dt$

16. $\displaystyle \int \frac{7 - 6t}{9t^2 + 2}\, dt$

18. $\displaystyle \int \frac{3t + 1}{t^2 + 2t + 9}\, dt$

20. $\displaystyle \int \frac{2t}{3t^2 + 8t + 1}\, dt$

Determine the indefinite integrals in #21–28, each of which involves an integrand whose denominator has a repeated linear factor.

21. $\displaystyle \int \frac{2}{4t^3 + t^2}\, dt$

23. $\displaystyle \int \frac{3}{t(t + 4)^2}\, dt$

25. $\displaystyle \int \frac{6t + 1}{t(t + 2)^3}\, dt$

27. $\displaystyle \int \frac{4t}{(t - 2)^2(t^2 + 1)}\, dt$

22. $\displaystyle \int \frac{5t - 1}{t^3 + 6t^2 + 9t}\, dt$

24. $\displaystyle \int \frac{2}{t(3t - 5)^2}\, dt$

26. $\displaystyle \int \frac{3t + 1}{t(2t - 1)^3}\, dt$

28. $\displaystyle \int \frac{5}{t^4 + 3t^3 + 4t^2}\, dt$

Determine the indefinite integrals in #29–30, each of which has an integrand whose denominator includes a repeated quadratic factor.

29. $\displaystyle \int \frac{t^4 + (t^2 + 1)^2}{t(t^2 + 1)^2}\, dt$

30. $\displaystyle \int \frac{9t^2 - 9}{t(4t^2 + 9)^2}\, dt$

In #31–36 rewrite the integrand in lowest terms before using a partial fractions decomposition to integrate.

31. $\displaystyle \int \frac{t^2 + 4t + 9}{t - 5}\, dt$

33. $\displaystyle \int \frac{6t^3 + 2}{t^2 - 2t + 1}\, dt$

35. $\displaystyle \int \frac{9t^3 - 5t}{t^2 + 2t + 8}\, dt$

32. $\displaystyle \int \frac{8t^2 - 5t + 2}{7t + 1}\, dt$

34. $\displaystyle \int \frac{9t^4 - 15t}{t^2 + 5t + 6}\, dt$

36. $\displaystyle \int \frac{10t^4 + 3}{4t^2 + t + 18}\, dt$

In #37–42 (a) adjust the integrand algebraically by adding zero (e.g. $+1-1$) and/or multiplying by 1 (e.g., $2/2$) so that it takes the form $A + B/(Ct + D)$, and then (b) determine the indefinite integral using the substitution technique.

37. $\displaystyle \int \frac{t + 1}{t + 2}\, dt$

39. $\displaystyle \int \frac{2t - 1}{t + 5}\, dt$

41. $\displaystyle \int \frac{15 - t}{4t + 1}\, dt$

38. $\displaystyle \int \frac{t - 3}{t + 4}\, dt$

40. $\displaystyle \int \frac{5t + 3}{t - 2}\, dt$

42. $\displaystyle \int \frac{7t + 8}{6t - 2}\, dt$

In #43–46 determine the indefinite integral assuming that a, b, c, and d are nonzero constants.

43. $\displaystyle \int \frac{t + a}{t + 1}\, dt$

44. $\displaystyle \int \frac{t + a}{t + b}\, dt$

45. $\displaystyle \int \frac{t + 3}{ct + d}\, dt$

46. $\displaystyle \int \frac{at + b}{ct + d}\, dt$

Use a partial fractions decomposition to evaluate each of #47–58.

47. $\displaystyle \int \frac{3t}{2t^2 + 26t - 28}\, dt$

50. $\displaystyle \int \frac{7}{t^3 + 8t}\, dt$

53. $\displaystyle \int \frac{1 - t}{13t - 16}\, dt$

56. $\displaystyle \int \frac{17}{t^3 + 4t^2 - 21t}\, dt$

48. $\displaystyle \int \frac{t - 7}{3t - 1}\, dt$

51. $\displaystyle \int \frac{t^3 + 1}{t^2 - 3t + 2}\, dt$

54. $\displaystyle \int \frac{8t - 1}{5t^2 + 1}\, dt$

57. $\displaystyle \int \frac{9t - 1}{t^3 - 5t^2}\, dt$

49. $\displaystyle \int \frac{13}{8t^2 + 12}\, dt$

52. $\displaystyle \int \frac{5t}{t^2 + 8t + 16}\, dt$

55. $\displaystyle \int \frac{16t^2}{t^4 - 1}\, dt$

58. $\displaystyle \int \frac{t^2 + 5t - 8}{6t + 1}\, dt$

Determine the indefinite integrals in #59–64, each of which is reduced to a rational function by a change of variable (remember to return to t in the end).

59. $\displaystyle \int \frac{3}{t\sqrt{t + 9}}\, dt$

61. $\displaystyle \int \frac{t}{t^2 - \sqrt{t}}\, dt$

63. $\displaystyle \int \frac{8}{t + 5\sqrt{t + 6}}\, dt$

60. $\displaystyle \int \frac{\sqrt{t + 16}}{t}\, dt$

62. $\displaystyle \int \frac{t^2}{t + \sqrt{t}}\, dt$

64. $\displaystyle \int \frac{7}{8 + \sqrt[3]{t - 5}}\, dt$

Mixed Practice: This set of exercises is intended to provide you with practice in determining which integration technique is appropriate for a given integral. Exercises 65–76 may require techniques from previous sections.

65. $\displaystyle \int \frac{t + 5}{t - 6}\, dt$

68. $\displaystyle \int \frac{\ln(3t)}{2t}\, dt$

71. $\displaystyle \int \frac{t^2 + 1}{t^3 + 4t^2}\, dt$

74. $\displaystyle \int \ln(t^2 - 3t + 2)\, dt$

66. $\displaystyle \int \frac{6t + 5}{3t^2 + 5t - 4}\, dt$

69. $\displaystyle \int \frac{7t - 6}{t^2 + 8}\, dt$

72. $\displaystyle \int \frac{4}{t + 12}\, dt$

75. $\displaystyle \int \frac{3}{(7t^2 + 1)^2}\, dt$

67. $\displaystyle \int \frac{t - 4}{t^2 - 3t + 4}\, dt$

70. $\displaystyle \int (4 - 7t^2)^{3/2}\, dt$

73. $\displaystyle \int \ln(t^2 - 3t + 17)\, dt$

76. $\displaystyle \int \frac{\cos(2t)}{7 - \sin(2t)}\, dt$

✢ 6.5 Concept and Application Exercises

77. Shade the region of the ab-plane in which $\int \frac{1}{at^2+bt+1}\, dt$ is composed of logarithms.

78. Shade the region of the ac-plane in which $\int \frac{1}{at^2+t+c}\, dt$ is an arctangent $(+C)$.

79. Shade the region of the bc-plane in which $\int \frac{1}{t^2+bt+c}\, dt$ comprises neither a logarithm nor an arctangent.

80. Explain why the partial fractions decomposition of $\frac{1}{(x-a)(x-b)}$ is a multiple of $\frac{1}{x-a} + \frac{1}{x-b}$, but the partial fractions decomposition of $\frac{x}{(x-a)(x-b)}$ is not.

81. Suppose a runner sprints with velocity $v(t) = \frac{t^2}{2+0.1t^2}\ \frac{\text{m}}{\text{sec}}$.

    (a) How far does the runner go in 10 seconds?

    (b) How long does it take the runner to travel 100 m? (Newton's Method will be helpful.)

82. The fraction of oxygen gas absorbed by the lung, A, depends on the length of time a breath of air spends in the lung (called the *residency time*), T. Suppose $A'(T) = \frac{2\alpha\beta T}{(\beta+T^2)^2}$.

    (a) Physically speaking, what must be true about $A(0)$?

    (b) Determine a formula for $A(T)$.

    (c) Graph $A(T)$ for various $\alpha > 1$ when $\beta = 1$.

    (d) Explain the effect of α on the graph of A, and why α must be in $(0,1)$ for this model to be physically realistic.

    (e) Graph $A(T)$ for various values of positive β when $\alpha = 1$.

    (f) Explain the effect of β on the graph of A, and what that means about the physical system.

83. Consider the expression $\frac{A_1}{x-a_1} + \frac{A_2}{x-a_2} + \frac{A_3}{x-a_3}$.

    (a) If we add these summands, what's the degree of the denominator?

    (b) If we add these summands, what's the degree of the numerator (at most)?

    (c) Answer parts (a) and (b) regarding $\frac{A_1}{x-a_1} + \frac{A_2}{x-a_2} + \frac{A_3x+B_3}{x^2+a_3}$, where $a_3 > 0$.

    (d) Explain why partial fractions decomposition only works when the rational function is in lowest terms.

84. Suppose $Q(x) = (x-a_1)(x-a_2)\cdots(x-a_n)$, where all the a_k are distinct and $P(x)$ is a polynomial of degree $< n$. Then there are numbers $A_1, A_2, \ldots A_n$ for which

$$\frac{P(x)}{Q(x)} = \frac{A_1}{x-a_1} + \frac{A_2}{x-a_2} + \cdots + \frac{A_n}{x-a_n}. \tag{5.8}$$

    (a) Prove that $A_k = P(a_k)/Q'(a_k)$. (*Hint: begin by multiplying both sides of equation (5.8) by $Q(x)$ and differentiating. Consider focusing on A_1, in particular, rather than trying to establish the pattern in full generality at first.*)

    (b) Use the result of part (a) to determine the partial fractions decomposition of $(x^2 - 3x + 1)/(x^3 - 4x^2 - 5x)$.

Athletics

Medical Science

On p. 477 you'll see something called the *Laplace Transform* that is commonly applied to problems in mechanical and electrical engineering. This transformation takes functions of t and, using something called an *improper integral* (see Section 6.6), creates a new function of a new variable, s. Here are some examples:

Engineering

$$\sin(bt) \longleftrightarrow \frac{b}{s^2 + b^2}, \quad \cos(bt) \longleftrightarrow \frac{s}{s^2 + b^2}, \quad \text{and} \quad t^n \longleftrightarrow \frac{n!}{s^{n+1}}$$

where $n!$, called n *factorial,* denotes the product $n \cdot (n-1) \cdot (n-2) \cdots (3)(2)(1)$. In #85–88 (a) use a partial fractions decomposition to write $f(s)$ in terms of the rational functions cited above, and (b) determine the associated function of t.

85. $f(s) = \dfrac{2s^2 + 1}{s^4 + s^2}$

87. $f(s) = \dfrac{s^3 + 4s^2 + 20}{s^4 + 5s^2}$

86. $f(s) = \dfrac{18 + 2s^2 - 3s^3}{s^5 + 9s^3}$

88. $f(s) = \dfrac{2s^4 - 7s^2 - 49}{s^5 + 7s^3}$

6.6 Improper Integrals

An integral is said to be **improper** when the interval of integration continues forever or the integrand has a singularity. We use limits to define and work with both of these cases.

❖ Integrals Over Semi-infinite Intervals

> The word "semi-infinite" literally means "half-infinite," and is used to describe intervals that are infinite in one direction but not the other.

Suppose $y(t)$ is the altitude of an object at time t. Then you know that

$$y(T) = y(1) + \underbrace{\int_1^T y'(t)\,dt}_{\substack{\text{net change in altitude be-} \\ \text{tween time } t = 1 \text{ and } t = T}}$$

so we can determine the object's altitude at time $T = 10$, $T = 1000$, $T = 1000000$, and so on. And if we want to know what happens in the long run, we can calculate the limit

$$\lim_{T \to \infty} y(T) = y(1) + \lim_{T \to \infty} \int_1^T y'(t)\,dt.$$

Calculating long-term behavior with an integral

Example 6.1. Suppose an object begins at an altitude of $y(1) = 2$ meters, and $y'(t) = 1/t^2$. Determine the long-term behavior of this object.

Solution: Since $y'(t) > 0$ when $t \geq 1$ this object is *always* rising, but

$$\int_1^T \frac{1}{t^2}\,dt = -\frac{1}{t}\Big|_1^T = 1 - \frac{1}{T},$$

so in the long run it rises to an altitude of

$$\lim_{T \to \infty} y(T) = y(1) + \lim_{T \to \infty} \int_1^T \frac{1}{t^2}\,dt = 2 + \lim_{T \to \infty}\left(1 - \frac{1}{T}\right) = 2 + 1 = 3 \text{ m.} \quad\blacksquare$$

Calculating long-term behavior with an integral

Example 6.2. Suppose an object begins at an altitude of $y(1) = 5$ meters, and $y'(t) = 1/t$. Determine the long-term behavior of this object.

Solution: As in Example 6.1, this object is rising because $y'(t) > 0$ when $t \geq 1$. And like before, we begin by calculating

$$\int_1^T \frac{1}{t}\,dt = \ln|t|\Big|_1^T = \ln(T) - \ln(1) = \ln(T).$$

Now we see something different from Example 6.1. In the long run,

$$\lim_{T \to \infty} y(T) = y(1) + \lim_{T \to \infty} \int_1^T \frac{1}{T}\,dt = 5 + \lim_{T \to \infty} \ln(T) = \infty.$$

So this object is forever rising, surpassing every finite altitude! $\quad\blacksquare$

In order to avoid writing the limit in calculation after calculation, a short-hand notation has been developed.

> **Integrals Over Semi-infinite Intervals:** Suppose that $f(t)$ is a continuous function and a is any real number. Then
>
> $$\text{by } \int_{a}^{\infty} f(t) \, dt \quad \text{we mean} \quad \lim_{T \to \infty} \int_{a}^{T} f(t) \, dt,$$
>
> $$\text{and by } \int_{-\infty}^{a} f(t) \, dt \quad \text{we mean} \quad \lim_{T \to \infty} \int_{-T}^{a} f(t) \, dt.$$
>
> The improper integrals are said to be **convergent** when the corresponding limit exists, and **divergent** otherwise.

People often express this idea by saying, "An improper integral is the limit of definite integrals."

Note: When the upper bound of integration is ∞, the n subintervals of any Riemann sum would have to span the semi-infinite interval $[a, \infty)$, so each would have to be infinitely long, which doesn't make any sense. So don't take the notation at face value, but instead remember that we are using a limit.

Calculating an improper integral

Example 6.3. Show that $\int_{0}^{\infty} \sin(t) \, dt$ diverges.

Solution: Some people mistakenly assert that this integral has a value of zero, but remember that an improper integral is really the limit of definite integrals:

$$\int_{0}^{\infty} \sin(t) \, dt = \lim_{T \to \infty} \int_{0}^{T} \sin(t) \, dt = \lim_{T \to \infty} 1 - \cos(T) \, dt.$$

Since $1 - \cos(T)$ does not approach a particular number as $T \to \infty$, this limit does not exist. Said differently, the improper integral diverges. ∎

Comparing integrals over semi-infinite intervals

Example 6.4. Suppose an object's initial altitude is $y(0) = 1$ meter, and it rises at a rate of $y'(t)$ that is always $\geq \frac{1}{2} \frac{m}{\text{sec}}$. Determine $\lim_{t \to \infty} y(t)$.

Solution: Since $y'(t)$ is always $0.5 \frac{m}{\text{sec}}$ or more, this object rises by at least 1 meter every 2 seconds. Because that never stops, this object surpasses every finite altitude over time. Written mathematically, $\lim_{T \to \infty} y(T) = \infty$. We can arrive at the same conclusion by using the equation that we derived at the beginning of this section:

$$\lim_{T \to \infty} y(T) = y(0) + \lim_{T \to \infty} \int_{0}^{T} y'(t) \, dt.$$

Because $y'(t) \geq 0.5$, we know that

$$\int_{0}^{T} y'(t) \, dt \geq \int_{0}^{T} \frac{1}{2} \, dt = \frac{T}{2}.$$

Since the value of the integral is always larger than $T/2$, it surpasses every finite value in the limit. Written mathematically,

$$\lim_{T \to \infty} y(T) = y(0) + \lim_{T \to \infty} \int_{0}^{T} y'(t) \, dt = 1 + \infty = \infty. \qquad ∎$$

Calculating an improper integral

Example 6.5. Suppose $a > 0$. Show $\int_{a}^{\infty} 1/t^p \, dt$ converges if and only if $p > 1$.

Solution: We addressed the case of $p = 1$ in Example 6.2, with $a = 1$ (the integral diverged, though we didn't have that vocabulary yet), so here we restrict our discussion to other values of p. We know that

$$\int_a^T t^{-p} \, dt = \frac{1}{1-p} t^{1-p} \Big|_a^T = \frac{1}{1-p}(T^{1-p} - a^{1-p}).$$

So

$$\int_a^\infty t^{-p} \, dt = \lim_{T \to \infty} \int_a^T t^{-p} \, dt = \lim_{T \to \infty} \frac{1}{1-p}(T^{1-p} - a^{1-p}),$$

which is $a^{1-p}/(p-1)$ when $p > 1$ and diverges otherwise. ∎

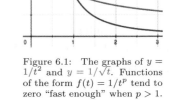

Figure 6.1: The graphs of $y = 1/t^2$ and $y = 1/\sqrt{t}$. Functions of the form $f(t) = 1/t^p$ tend to zero "fast enough" when $p > 1$.

Note: Examples 6.3—6.5 allude to the fact that the convergence of $\int_1^\infty f(t) \, dt$ requires not only that $f(t) \to 0$ as $t \to \infty$, but also that it happens "fast enough."

Try It Yourself: Calculating an improper integral

Example 6.6. Determine whether $\int_0^\infty e^{-t} \, dt$ converges and, if so, determine its value.

Answer: The integral converges and has a value of 1. ∎

Full Solution On-line

▷ **Comparison theorems for integrals over semi-infinite intervals**

In Examples 6.1–6.6 you've seen that some improper integrals represent numbers (i.e., the integral converges) and others don't (i.e., the integral diverges). When the integrand remains non-negative, we can often make a quick comparison to determine whether an improper integral converges. In fact, we did just that in Example 6.4. Here's another example.

Comparing integrals over semi-infinite intervals

Example 6.7. Make a conjecture about whether $\int_1^\infty 1/(t^2 + \sqrt{3 + \sin(t)}) \, dt$ converges.

Solution: Since $\sqrt{3 + \sin(t)}$ is a positive number,

$$\frac{1}{t^2 + \sqrt{3 + \sin(t)}} \leq \frac{1}{t^2}.$$

> The fraction on the left has a larger denominator, so its value is smaller.

It follows that

$$\int_1^T \frac{1}{t^2 + \sqrt{3 + \sin(t)}} \, dt \leq \int_1^T \frac{1}{t^2} \, dt$$

for each $T > 1$ (see Figure 6.2). After evaluating the integral on the right-hand side

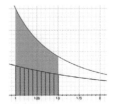 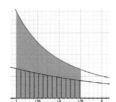 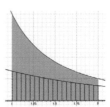 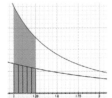

Figure 6.2: (all) The curves $y = 1/(t^2 + \sqrt{3 + \sin(t)})$ and $y = 1/t^2$; (left to right) the area $\int_1^T 1/t^2 \, dt$ for $T = 1.25, 1.5, 1.75$, and 2. Note that the (green) area between the horizontal axis and the curve $y = 1/t^2$ is larger than the (striped) area between the curve $y = 1/(t^2 + \sqrt{3 + \sin(t)})$ and the horizontal axis.

of this inequality, we see that

$$\int_1^T \frac{1}{t^2 + \sqrt{3 + \sin(t)}} \, dt \le 1 - \frac{1}{T} < 1.$$

Since $1/(t^2 + \sqrt{3 + \sin(t)})$ is never negative, the remaining integral continually increases as T grows larger and larger, but our last inequality guarantees that its value will never exceed 1. This is reminiscent of a function (of T) whose graph is increasing toward a horizontal asymptote, so we believe that the limit exits. That is, our conjecture is that this integral converges. ∎

The following theorem guarantees that our conjecture from Example 6.7 is correct.

Direct Comparison Theorem for Improper Integrals (Part 1): Suppose that $f(t)$ and $g(t)$ are continuous functions whose value is non-negative when $t \in [a, \infty)$, and that $f(t) \le g(t)$.

   (a) If $\int_a^\infty g(t) \, dt$ converges, so does $\int_a^\infty f(t) \, dt$.

   (b) If $\int_a^\infty f(t) \, dt$ diverges, so does $\int_a^\infty g(t) \, dt$.

We used part (a) of this theorem in Example 6.7, where $g(t) = 1/t^2$.

Since f and g are non-negative functions, the line $y = 0$ and the curve $y = g(t)$ constitute an envelope that contains the graph of f, much in the sense of the Squeeze Theorem, which is one way to understand why part (a) is true. The non-negativity also prevents divergence due to oscillation, as you saw in Example 6.3.

Proof. The technically rigorous proof of part (a) relies on an idea called the *supremum* that is beyond the scope of this book (though see [42]), so it's omitted here. However, the ideas in Example 6.7, including the notion of $\int_a^T f(t) \, dt$ as a function of T whose graph has a horizontal asymptote, are fundamentally correct.

Part (b) of the theorem is easier to prove: since $f(t) \le g(t)$, we know that

$$\int_a^T f(t) \, dt \le \int_a^T g(t) \, dt \le \int_a^\infty g(t) \, dt$$

for each $T \ge a$. Since $\lim_{T \to \infty} \int_a^T f(t) \, dt$ exceeds every finite number, the improper integral $\int_a^\infty g(t) \, dt$ cannot be one of them. ∎

Comparing integrals over semi-infinite intervals

Example 6.8. Determine whether $\int_2^\infty \frac{t \sin^2(t)}{7t^4 - 5} \, dt$ converges.

Solution: Since the integrand is never negative over our interval of integration, we can apply the Direct Comparison Theorem. The first step is deciding whether we think that the integral converges or diverges, which is often done with a roughly hewn approximation. Because $\sin^2(t)$ never gets larger than 1, this integrand looks a lot like $t/(7t^4) = 1/(7t^3)$ when t gets big, so we expect the integral to converge (see Example 6.5).

We prove that we're right by finding a larger function $g(t)$ for which $\int_2^\infty g(t) \, dt$ converges. We typically find such a $g(t)$ by starting with a copy of our integrand, and making algebraic adjustments that increase its value. For example, since $\sin^2(t)$ never exceeds 1, we know that

$$\frac{t \sin^2(t)}{7t^4 - 5} \le \frac{t}{7t^4 - 5}.$$

And since $t^4 > 5$ when $t \geq 2$ (i.e., over our interval of integration), we can make the denominator smaller by subtracting t^4 instead of subtracting 5. That further increases the size of the fraction, so

$$\frac{t}{7t^4 - 5} \leq \frac{t}{7t^4 - t^4} = \frac{t}{6t^4} = \frac{1}{6t^3}.$$

Therefore,

$$\int_2^T \frac{t \sin^2(t)}{7t^4 - 5}\, dt \leq \int_2^T \frac{1}{6t^3},$$

and the right-hand side converges in the limit (see Example 6.5). It follows that our integral also converges in the limit. ∎

Comparing integrals over semi-infinite intervals

Example 6.9. Determine whether $\int_3^\infty \frac{4\sqrt[3]{t}}{5t + \sqrt{t}}\, dt$ converges.

Solution: Because the integrand is never negative when $t \geq 3$, we can use the Direct Comparison Theorem to make this determination. Since the largest exponent in the numerator is $1/3$, and the largest in the denominator is 1, we think of this integrand as behaving roughly like $t^{1/3}/t = 1/t^{2/3}$. So we expect this integral to diverge (see Example 6.5).

We prove that we're right by finding a smaller function $f(t)$ for which $\int_3^\infty f(t)\, dt$ diverges. As before, this is typically done by starting with a copy of our integrand, and then making changes that reduce its value. For example, since $\sqrt{t} < t$ over our interval of integration, we can increase the size the denominator (and so reduce the value of the function) by replacing $\sqrt{t}$ with t:

$$\frac{4\sqrt[3]{t}}{5t + t} \leq \frac{4\sqrt[3]{t}}{5t + \sqrt{t}}.$$

The left-hand side of this inequality is really

$$\frac{4\sqrt[3]{t}}{5t + t} = \frac{4\sqrt[3]{t}}{6t} = \frac{2}{3t^{2/3}}.$$

Consequently,

$$\int_3^T \frac{2}{3t^{2/3}}\, dt \leq \int_3^T \frac{4\sqrt[3]{t}}{5t + \sqrt{t}}\, dt.$$

The left-hand side diverges as $T \to \infty$ (see Example 6.5), so the right-hand side does also. That is, our integral diverges. ∎

Finding a comparable function and establishing that $f(t) \leq g(t)$ in order to use the Direct Comparison Theorem can be difficult. Instead, an equivalent test can be done that relies on limits instead of inequalities.

Limit Comparison Theorem for Improper Integrals (Part 1): Suppose $f(t)$ and $g(t)$ are continuous and non-negative when $t \in [a, \infty)$, and that $\lim_{t \to \infty} \frac{f(t)}{g(t)} = L$ is a positive number. Then either

   (a) both $\int_a^\infty g(t)\, dt$ and $\int_a^\infty f(t)\, dt$ converge, or

   (b) both $\int_a^\infty g(t)\, dt$ and $\int_a^\infty f(t)\, dt$ diverge.

> This limit says that $f(t) \approx Lg(t)$ when t is large, and positive scale factors don't affect whether an improper integral remains finite in the limit.

Proof. If you've studied the technical definition of limits at infinity (in Chapter 2) you'll be able to prove this fact by using the Direct Comparison Theorem, as outlined in the exercise set. If not, here's an intuitive version of the proof: the fundamental idea is that limit $\lim_{t\to\infty} f(t)/g(t) = L$ tells us that $f(t) \approx L\,g(t)$ when t is extremely large. So intuitively speaking,

$$\int_N^T f(t)\,dt \approx \int_N^T L\,g(t)\,dt$$

when $N > 0$ is immense and T is even larger. Since these integrals are roughly equal for every such T, it cannot be the case that one of them converges to a particular number as as $T \to \infty$ while the other exceeds every finite value. ∎

> We know that
> $$\int_a^\infty f(t)\,dt$$
> is the same as
> $$\int_a^N f(t)\,dt + \int_N^\infty f(t)\,dt$$
> for every $N > a$. The first of these integrals certainly exists because f is continuous. So we need only worry about whether
> $$\int_N^\infty f(t)\,dt$$
> converges. The same is true of g, of course.

Using the Limit Comparison Theorem over semi-infinite intervals

Example 6.10. Use the Limit Comparison Theorem to determine whether the integral $\int_3^\infty \frac{4\sqrt[3]{t}}{5t+\sqrt{t}}\,dt$ converges.

Solution: In Example 6.9 we said that $f(t) = \frac{4\sqrt[3]{t}}{5t+\sqrt{t}}$ is roughly like $g(t) = 1/t^{2/3}$, so we calculate

$$\lim_{t\to\infty} \frac{f(t)}{g(t)} = \lim_{t\to\infty} \frac{\frac{4\sqrt[3]{t}}{5t+\sqrt{t}}}{\frac{1}{t^{2/3}}} = \lim_{t\to\infty} \frac{4t}{5t+\sqrt{t}}.$$

This limit leaves us with an ∞/∞ indeterminate form, so we apply L'Hôpital's Rule and calculate

$$\lim_{t\to\infty} \frac{4t}{5t+\sqrt{t}} = \lim_{t\to\infty} \frac{4}{5+\frac{1}{2\sqrt{t}}} = \frac{4}{5}.$$

Since this limit exists and is nonzero, and $\int_3^\infty g(t)\,dt$ diverges, so does $\int_3^\infty f(t)\,dt$ (which is what we found in Example 6.9). ∎

> We've said that our integrand behaves as $t^{-2/3}$ when t is large. The positive limit value that you see here validates that assertion.

Using the Limit Comparison Theorem over semi-infinite intervals

Example 6.11. Use the Limit Comparison Theorem to determine whether the integral $\int_2^\infty \frac{t\sin^2(t)}{7t^4-5}\,dt$ converges.

Solution: In Example 6.8 we said that this integrand is roughly like $1/(7t^3)$ when t is extremely large. So it seems natural to check

$$\lim_{t\to\infty} \frac{\frac{t\sin^2(t)}{7t^4-5}}{\frac{1}{7t^3}} = \lim_{t\to\infty} \frac{7t^4\sin^2(t)}{7t^4-5}.$$

While $\frac{7t^4}{7t^4-5}$ tends to 1 in the limit, the factor of $\sin^2(t)$ continues to oscillate forever, so this limit does not exist. That prevents us from using the Limit Comparison Theorem directly. However, our first inequality of Example 6.8 was

$$\frac{t\sin^2(t)}{7t^4-5} \le \frac{t}{7t^4-5},$$

so if we can prove that $\int_2^\infty \frac{t}{7t^4-5}\,dt$ converges, the Direct Comparison Theorem guarantees that our integral does too. We do this by comparing $\frac{t}{7t^4-5}$ to $\frac{1}{7t^3}$ in the limit:

$$\lim_{t\to\infty} \frac{\frac{t}{7t^4-5}}{\frac{1}{7t^3}} = \lim_{t\to\infty} \frac{7t^4}{7t^4-5} = 1.$$

> We've said that $\frac{t}{7t^4-5}$ behaves as $\frac{1}{7t^3}$ when t is large. The positive limit value that you see here validates that assertion.

Since this limit exists and is nonzero, and $\int_2^\infty \frac{1}{7t^3}\,dt$ converges, so does $\int_2^\infty \frac{t}{7t^4-5}$. It follows from the Direct Comparison Theorem that $\int_2^\infty \frac{t\sin^2(t)}{7t^4-5}\,dt$ does too. ∎

Try It Yourself: Using the Limit Comparison Theorem over semi-infinite intervals

Example 6.12. Use the Limit Comparison Theorem to determine whether the integral $\int_5^\infty \frac{5+\sqrt{t}}{t^2+6t+19}\, dt$ converges.

Answer: It does. ■

Full Solution
On-line

❖ Integrals Over Infinite Intervals

As we saw with semi-infinite intervals, Riemann sums don't make sense when the notation $\int_{-\infty}^\infty f(t)\, dt$ is taken at face value, so we say that such integrals are improper. However, as we did with integrals over semi-infinite intervals, we can understand them in terms of limits.

Integrals Over Infinite Intervals: Suppose that $f(t)$ is a continuous function and $a \in \mathbb{R}$. Then

$$\text{by } \int_{-\infty}^\infty f(t)\, dt \text{ we mean } \int_{-\infty}^a f(t)\, dt + \int_a^\infty f(t)\, dt,$$

so that $\int_{-\infty}^\infty f(t)\, dt$ is a number only when both $\int_{-\infty}^a f(t)\, dt$ and $\int_a^\infty f(t)\, dt$ are.

People commonly have two questions about this definition:

1. Does the number a matter?

   Answer: No. If a_1 and a_2 are two real numbers,

   $$\int_{-\infty}^\infty f(t)\, dt = \int_{-\infty}^{a_1} f(t)\, dt + \int_{a_1}^\infty f(t)\, dt$$

   $$= \lim_{T \to \infty} \int_{-T}^{a_1} f(t)\, dt + \lim_{T \to \infty} \int_{a_1}^T f(t)\, dt$$

   $$= \lim_{T \to \infty} \left(\int_{-T}^{a_2} f(t)\, dt + \int_{a_2}^{a_1} f(t)\, dt \right) + \lim_{T \to \infty} \left(\int_{a_1}^{a_2} f(t)\, dt + \int_{a_2}^T f(t)\, dt \right).$$

   Since f is continuous, the integral $\int_{a_1}^{a_2} f(t)\, dt$ is a real number—and note that it does not depend on T, so the limits above do not affect it. Therefore,

   $$\overbrace{\phantom{\int_{a_2}^{a_1} f(t)\, dt + \int_{a_1}^{a_2} f(t)\, dt}}^{\text{these are opposites}}$$

   $$\int_{-\infty}^\infty f(t)\, dt = \lim_{T \to \infty} \left(\int_{-T}^{a_2} f(t)\, dt \right) + \int_{a_2}^{a_1} f(t)\, dt + \int_{a_1}^{a_2} f(t)\, dt + \lim_{T \to \infty} \left(\int_{a_2}^T f(t)\, dt \right)$$

   $$= \int_{-\infty}^{a_2} f(t)\, dt + \int_{a_2}^\infty f(t)\, dt.$$

2. How does this definition differ from $\lim_{T \to \infty} \int_{-T}^T f(t)\, dt$?

   Answer: There are cases where $\lim_{T \to \infty} \int_{-T}^T f(t)\, dt$ exists but $\int_{-\infty}^\infty f(t)\, dt$ does not, according to our definition. For example, when $f(t) = t$, we have

   $$\lim_{T \to \infty} \int_{-T}^T t\, dt = \lim_{T \to \infty} 0 = 0,$$

   but

   $$\lim_{T \to \infty} \int_0^T t\, dt = \lim_{T \to \infty} \frac{1}{2}T^2 = \infty.$$

Since $\lim_{T\to\infty}\int_0^T t\,dt$ is not a number, neither is $\int_{-\infty}^{\infty} t\,dt$ according to our definition.

However, if the improper integral $\int_{-\infty}^{\infty} f(t)\,dt$ exists according to our definition, so does $\lim_{T\to\infty}\int_{-T}^{T} f(t)\,dt$, and they have the same value. This happens because the limit of a sum is the sum of the limits (when both limits exist):

$$\int_{-\infty}^{\infty} f(t)\,dt = \lim_{T\to\infty}\int_{-T}^{a} f(t)\,dt + \lim_{T\to\infty}\int_{a}^{T} f(t)\,dt$$

$$= \lim_{T\to\infty}\left(\int_{-T}^{a} f(t)\,dt + \int_{a}^{T} f(t)\,dt\right) = \lim_{T\to\infty}\int_{-T}^{T} f(t)\,dt.$$

In this sense, the definition that we've stated is the more restrictive of the two.

✥ Integrands with Discontinuities or Singularities

Integrals are also said to be improper when the interval of integration includes a discontinuity or singularity of the integrand. For example, consider

$$\int_0^1 \frac{1}{\sqrt{t}}\,dt$$

in which a left-sampled Riemann sum is impossible because the height of the first rectangle would be determined by $1/\sqrt{t}$ at $t = 0$. Since we're talking about calculus, it should come as no surprise that we get around this problem with a limit. Specifically, we define

$$\int_0^1 \frac{1}{\sqrt{t}}\,dt \quad \text{to be} \quad \lim_{\ell\to 0^+}\int_\ell^1 \frac{1}{\sqrt{t}}\,dt,$$

which is depicted in Figure 6.3.

> We're using ℓ for ℓimit of integration.

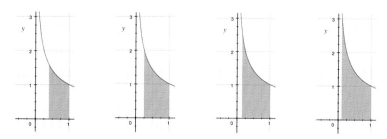

Figure 6.3: (left to right) The area $\int_\ell^1 1/\sqrt{t}\,dt$ for $\ell = 0.4, 0.2, 0.15,$ and 0.1.

Calculating an improper integral

Example 6.13. Determine the value of $\int_0^1 1/\sqrt{t}\,dt$.

Solution: The area between the curve $y = 1/\sqrt{t}$ and the t-axis over $[\ell, 1]$ is

$$\int_\ell^1 t^{-1/2}\,dt = 2t^{1/2}\Big|_\ell^1 = 2 - 2\sqrt{\ell},$$

so

$$\int_0^1 \frac{1}{\sqrt{t}}\,dt = \lim_{\ell\to 0^+}\int_\ell^1 t^{-1/2}\,dt = \lim_{\ell\to 0^+} 2 - 2\sqrt{\ell} = 2. \qquad \blacksquare$$

Of course, you know from experience that limits don't always exist ...

Calculating an improper integral

Example 6.14. Show that $\int_0^1 1/t \, dt$ is not a number.

Solution: As before, we begin by calculating

$$\int_\ell^1 \frac{1}{t} \, dx = \ln|t|\Big|_\ell^1 = \ln(1) - \ln(\ell) = 0 - \ln(\ell) = \ln\left(\frac{1}{\ell}\right).$$

So the definition of our integral is

$$\int_0^1 \frac{1}{t} \, dt = \lim_{\ell \to 0^+} \int_\ell^1 \frac{1}{t} \, dt = \lim_{\ell \to 0^+} \ln\left(\frac{1}{\ell}\right) = \infty. \qquad \blacksquare$$

More generally,

Integrals of Discontinuous Functions: Suppose f is a continuous function on the interval $(a, b]$, but is discontinuous at $t = a$. Then by

$$\int_a^b f(t) \, dt \quad \text{we mean} \quad \lim_{\ell \to a^+} \int_\ell^b f(t) \, dt.$$

Similarly, when f is continuous on the interval $[a, b)$ but is discontinuous at $t = b$, by

$$\int_a^b f(t) \, dt \quad \text{we mean} \quad \lim_{\ell \to b^-} \int_a^\ell f(t) \, dt.$$

The improper integral $\int_a^b f(t) \, dt$ is said to be **convergent** when the corresponding limit exists, and **divergent** otherwise.

Calculating an improper integral

Example 6.15. Show that $\int_0^1 1/t^p \, dt$ converges if and only if $p < 1$.

Solution: The case of $p = 1$ was addressed in Example 6.14 (the integral diverged), so here we address the case of all other values of p. We know that

$$\int_\ell^1 \frac{1}{t^p} \, dt = \int_\ell^1 t^{-p} \, dt = \frac{1}{1-p} t^{1-p}\Big|_\ell^1 = \frac{1}{1-p}(1 - \ell^{1-p}).$$

So

$$\int_0^1 \frac{1}{t^p} \, dt = \lim_{\ell \to 0^+} \int_\ell^1 \frac{1}{t^p} \, dt = \lim_{\ell \to 0^+} \frac{1}{1-p}\left(1 - \ell^{1-p}\right),$$

which is $1/(1-p)$ when $p < 1$ (such as $p = 0.5$ in Example 6.13), and diverges when $p > 1$ (because the exponent on ℓ is negative when $p > 1$). $\qquad \blacksquare$

Calculating an improper integral

Example 6.16. Determine whether $\int_0^2 1/\sqrt[3]{2-t} \, dt$ converges and, if so, find its value.

Solution: Since this integrand is continuous over $[0, 2)$, by

$$\int_0^2 \frac{1}{\sqrt[3]{2-t}} \, dt \quad \text{we mean} \quad \lim_{\ell \to 2^-} \int_0^\ell \frac{1}{\sqrt[3]{2-t}} \, dt,$$

In Example 6.5 on p. 461 we showed that $\int_1^\infty 1/t^p \, dt$ converges if and only if $p > 1$. By contrast, in this example we see that $\int_0^1 1/t^p \, dt$ converges if and only if $p < 1$. So as long as $p \neq 1$, exactly one of

$$\int_0^a \frac{1}{t^p} \, dt \quad \text{and} \quad \int_a^\infty \frac{1}{t^p} \, dt$$

converges. By remembering when $\int_a^\infty 1/t^p \, dt$ converges, you know when $\int_a^\infty 1/t^p \, dt$ does.

When we perform the substitution

$$x = 2 - t \Rightarrow \begin{cases} x &=& 2 \text{ when } t = 0 \\ x &=& 2 - \ell \text{ when } t = \ell \\ dx &=& -dt, \end{cases}$$

we see that

$$\int_0^\ell \frac{1}{\sqrt[3]{2-t}} \, dt = \int_2^{2-\ell} \frac{1}{\sqrt[3]{x}} \, (-dx) = \int_{2-\ell}^2 x^{-1/3} \, dx = \frac{3}{2} x^{2/3} \Big|_{2-\ell}^2 = \frac{3}{2} \Big(2^{2/3} - (2-\ell)^{2/3} \Big).$$

Therefore,

$$\int_0^2 \frac{1}{\sqrt[3]{2-t}} \, dt = \lim_{\ell \to 2^-} \int_0^\ell \frac{1}{\sqrt[3]{2-t}} \, dt = \lim_{\ell \to 2^-} \frac{3}{2} \Big(2^{2/3} - (2-\ell)^{2/3} \Big) = 1.5 \sqrt[3]{4},$$

which has clearly converged.   ∎

▷ More comparison theorems

As we did with integrals over infinite intervals, when $f(t) \geq 0$ we can often make a quick comparison to determine whether an improper integral converges.

Comparing improper integrals

Example 6.17. Make a conjecture about whether $\int_1^2 \frac{1}{(1+te^{-t})\sqrt[3]{2-t}} \, dt$ converges.

Solution: Since $1 + te^{-t}$ is always larger than 1 when $t \in [1, 2]$, we know that

$$\frac{1}{(1+te^{-t})\sqrt[3]{2-t}} \leq \frac{1}{\sqrt[3]{2-t}}.$$

> The fraction on the left has a larger denominator, so it's the smaller number.

It follows that

$$\int_1^\ell \frac{1}{(1+te^{-t})\sqrt[3]{2-t}} \, dt \leq \int_1^\ell \frac{1}{\sqrt[3]{2-t}} \, dt$$

for each $\ell \in (1, 2)$, as shown in Figure 6.4.

Since the integrands are always positive, larger values of ℓ result in larger integral values, so

$$\int_1^\ell \frac{1}{(1+te^{-t})\sqrt[3]{2-t}} \, dt \leq \underbrace{\lim_{\ell \to 2^-} \int_1^\ell \frac{1}{\sqrt[3]{2-t}} \, dt = 1.5 \sqrt[3]{4}}_{\text{see Example 6.16}}.$$

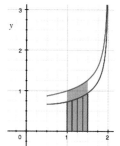 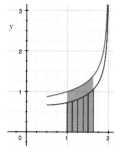 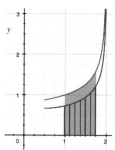 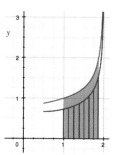

Figure 6.4: (left to right) The curves $y = 1/\sqrt{2-x}$ and $y = 1/((1 + xe^{-x})\sqrt{2-x})$, and the area $\int_1^\ell \frac{1}{\sqrt{2-x}} \, dx$ for $\ell = 1.5, 1.625, 1.75,$ and 1.9. Note that the (blue-striped) area between the horizontal axis and the curve $y = 1/((1+xe^{-x})\sqrt{2-x})$ is always smaller than the (green) area between the curve $y = 1/\sqrt{2-x}$ and the horizontal axis.

So the area

$$\int_1^\ell \frac{1}{(1+te^{-t})\sqrt[3]{2-t}}\,dt$$

continually increases as $\ell \to 2^-$, but it can never grow larger than $1.5\sqrt[3]{4}$. Based on these facts, we make the conjecture that the limit exists, and the value of this improper integral is no larger than $1.5\sqrt[3]{4}$. ∎

The following theorem guarantees that our conjecture from Example 6.17 is right.

Direct Comparison Theorem for Improper Integrals (Part 2): Suppose that $f(t)$ and $g(t)$ are continuous functions whose value is non-negative when $t \in (a, b)$, and that $f(t) \le g(t)$.

  (a) If $\int_a^b g(t)\,dt$ converges, so does $\int_a^b f(t)\,dt$.

  (b) If $\int_a^b f(t)\,dt$ diverges, so does $\int_a^b g(t)\,dt$.

Note: The proof of this theorem is closely related to that of the Direct Comparison Theorem for Improper Integrals (Part 1), so we omit it here. However, you can glean some intuition about why it's true by thinking of these integrals as areas, as done in Example 6.17.

Comparing improper integrals

Example 6.18. Use the Direct Comparison Theorem to determine whether the integral $\int_0^1 \csc(t)\,dt$ converges.

Solution: Because $\sin(t) < t$ when $t \in (0, 1)$, we know that $\csc(t) > \frac{1}{t}$ when $t \in (0, 1)$. Since we know that $\int_0^1 \frac{1}{t}\,dt$ diverges (see Example 6.14), so does $\int_0^1 \csc(t)\,dt$. ∎

> The functions t and $\sin(t)$ have the same value at $t = 0$, but the derivative of t is always larger than the derivative of $\sin(t)$, so t grows faster than $\sin(t)$.

Try It Yourself: Comparing improper integrals

Example 6.19. Use the Direct Comparison Theorem to determine whether the integral $\int_0^1 \frac{1}{t+\sqrt{t}}\,dt$ converges.

Answer: It converges. ∎

Full Solution On-line

    As we saw with integrals over infinite intervals, we can can use limits instead of direct comparison in this setting.

Limit Comparison Theorem for Improper Integrals (Part 2): Suppose $f(t)$ and $g(t)$ are continuous and non-negative when $t \in (a, b]$, and that $\lim_{t \to a^+} \frac{f(t)}{g(t)} = A$ is a nonzero number. Then either

  (a) both $\int_a^b g(t)\,dt$ and $\int_a^b f(t)\,dt$ converge, or

  (b) both $\int_a^b g(t)\,dt$ and $\int_a^b f(t)\,dt$ diverge.

The same is true when $f(t)$ and $g(t)$ are continuous and non-negative over $[a, b)$, and $\lim_{t \to b^-} \frac{f(t)}{g(t)} = B$ is a nonzero number.

Note: The proof of this theorem is almost identical to that of the Limit Comparison Theorem for Improper Integrals (Part 1), so we omit it here.

Comparing improper integrals

Example 6.20. Use the Limit Comparison Theorem to determine whether $\int_0^1 \frac{1}{t^2+\sqrt[4]{t}}\, dt$ converges.

Solution: When t is near zero, t^2 is minuscule, so the denominator is roughly $\sqrt[4]{t}$. This leads us to compare our integrand to $1/\sqrt[4]{t}$ in the limit:

$$\lim_{t\to 0} \frac{\frac{1}{\sqrt[4]{t}}}{\frac{1}{t^2+\sqrt[4]{t}}} = \lim_{t\to 0} \frac{t^2+\sqrt[4]{t}}{\sqrt[4]{t}} = \lim_{t\to 0}(t^{7/4}+1) = 1.$$

Since the limit is a nonzero number, and $\int_0^1 \frac{1}{t^{1/4}}\, dt$ converges (see Example 6.15), our integral does too.   ■

Try It Yourself: Comparing improper integrals

Example 6.21. Use the Limit Comparison Theorem to determine whether the integral $\int_0^1 \frac{1}{t+\sin(t)}\, dt$ converges by comparing the integrand to $1/t$.

Answer: It diverges.   ■

Full Solution On-line

❖ Calculating the Value of an Improper Integral

In general, there are two questions to ask about an improper integral.

1. Is it a number? (i.e., does the limit converge?)

2. If so, what number is it? (i.e., what does the limit converge to?)

A good deal of this section has been devoted to answering the first question, and we conclude by addressing the second. When the integrand is simple enough that we can find an antiderivative, as in $\int_1^\infty t^{-p}\, dt$, we can calculate the value of an improper integral exactly; but when no antiderivative is available, we have to approximate. Of course, as soon as the word "approximation" enters a technical conversation, we have to ask, "How accurate is the approximation?" In many cases, we can approximate the value of the integral to within any positive error tolerance, no matter how small. This is best illustrated by example.

> The improper integral
> $$\int_1^\infty \frac{1}{t^2}\, dt$$
> converges, so it's a number. Specifically, it's the number 2. You should think of the improper integral as just another way of writing the number 2, much as $\log(100)$ and $\sqrt[3]{8}$ are ways of writing the number 2.

Approximating the value of an improper integral

Example 6.22. Approximate $\int_5^\infty \frac{1}{t^2+\sqrt{3+\sin(t)}}\, dt$ to within 0.01 of its actual value.

Solution: Our strategy for making the desired estimate will rely on the fact that

$$\int_5^\infty \frac{1}{t^2+\sqrt{3+\sin(t)}}\, dt = \int_5^\ell \frac{1}{t^2+\sqrt{3+\sin(t)}}\, dt + \int_\ell^\infty \frac{1}{t^2+\sqrt{3+\sin(t)}}\, dt$$

when the number $\ell > 5$. In brief, we're going to do two things: (1) approximate the second integral as having a value of zero, and (2) approximate the first integral to a high degree of accuracy using a numerical method. Of course, this is only a good approximation scheme if the second integral is *near* zero, which means that ℓ must be chosen "large enough." More specifically, since an error of 0.01 is tolerable in this example, we'll choose ℓ large enough that the second integral is less than 0.005, and then approximate the first integral to within 0.005 of its actual value using a Riemann sum.

> Consider the second integral: intuitively speaking, when ℓ is larger we're integrating over "less" of the real line, so larger numbers ℓ correspond to smaller integral values.

> The number 0.005 is half of the specified error tolerance.

Step 1 (finding ℓ): In Example 6.7 on p. 462 we showed that this improper integral converged because the value of the integrand is always less than $1/t^2$. Consequently,

$$\int_{\ell}^{T} \frac{1}{t^2 + \sqrt{3 + \sin(t)}}\, dt \leq \int_{\ell}^{T} \frac{1}{t^2}\, dt = \frac{1}{\ell} - \frac{1}{T},$$

whence

$$\lim_{T \to \infty} \int_{\ell}^{T} \frac{1}{t^2 + \sqrt{3 + \sin(t)}}\, dt \leq \frac{1}{\ell}.$$

It follows that

$$\int_{\ell}^{\infty} \frac{1}{t^2 + \sqrt{3 + \sin(t)}}\, dt \leq 0.005 \quad \text{when} \quad 200 \leq \ell.$$

Step 2 (using Riemann sums): Since the magnitude of our integrand's second derivative never exceeds 0.01 on $[5, 200]$, our error formula from p. 363 tells us that a midpoint-sampled Riemann sum with n subintervals approximates the actual value of the integral to within

$$\frac{M_2(b-a)^3}{24n^2} = \frac{(0.01)(195)^3}{24n^2} = \frac{3089.53125}{n^2}$$

of its actual value. So we want a value of n such that

$$\frac{3089.53125}{n^2} \leq 0.005 \quad \Rightarrow \quad 786.07014 \leq n.$$

It might seem as if $n = 787$ is a lot of subintervals, but a modern spreadsheet program makes short work of the calculation, and we find that

$$\int_{5}^{200} \frac{1}{t^2 + \sqrt{3 + \sin(t)}}\, dt \approx 0.190648757.$$

Step 3 (putting the pieces together): Now we have

$$\int_{5}^{\infty} \frac{1}{t^2 + \sqrt{3 + \sin(t)}}\, dt = \int_{5}^{200} \frac{1}{t^2 + \sqrt{3 + \sin(t)}} + \int_{200}^{\infty} \frac{1}{t^2 + \sqrt{3 + \sin(t)}}$$

$$\approx \underbrace{0.190648757}_{\text{accurate to within } 0.005} + \underbrace{0.00.}_{\text{accurate to within } 0.005}$$

So the actual value of $\int_{5}^{\infty} \frac{1}{t^2 + \sqrt{3 + \sin(t)}}\, dt$ is within 0.01 units of 0.190648757. ∎

> It is not plainly obvious that $|f''(t)| \leq 0.01$ across $[5, 200]$, but it's easy to check by performing the following steps:
>
> 1. Derive a formula for $f''(t)$ and check that $f''(5) \leq 0.01$.
>
> 2. Show that $f''(t)$ remains positive, but is always decreasing over $[5, 200]$.
>
> Therefore, the maximum magnitude of f'' occurs at $t = 5$.

❖ Common Difficulties with Convergence

When $f(t)$ is a decreasing function, it's true that $\int_{a}^{\infty} f(t)\, dt$ cannot converge unless $\lim_{t \to \infty} f(t) = 0$, but people often mistakenly think that $\lim_{t \to \infty} f(t) = 0$ *guarantees* the convergence of $\int_{a}^{\infty} f(t)\, dt$. It does not. The function $f(t) = 1/\sqrt{t}$ is a great example:

$$\lim_{t \to \infty} \frac{1}{\sqrt{t}} = 0 \quad \text{but} \quad \int_{1}^{\infty} \frac{1}{\sqrt{t}}\, dt \text{ diverges.}$$

Similarly, people sometimes think that an integral cannot possibly converge if the integrand has a singularity. That's not true either, and the function $f(t) = 1/\sqrt{t}$ is a great example here, too:

$$\lim_{t \to 0^+} \frac{1}{\sqrt{t}} = \infty \quad \text{but} \quad \int_{0}^{1} \frac{1}{\sqrt{t}}\, dt \text{ converges.}$$

You should know

- the term *improper integral, converge,* and *diverge;*

- that improper integrals are understood as, and defined by limits;

- that integrals over semi-infinite intervals can only converge if the integrand tends to zero "fast enough;"

- that integrating a function up to a singularity can only converge if the function blows up "slowly enough;"

- the Direct Comparison Theorem for Improper Integrals (parts 1 and 2);

- the Limit Comparison Theorem for Improper Integrals (parts 1 and 2).

You should be able to

- calculate simple improper integrals;

- use the Direct and Limit Comparison Theorems to determine whether improper integrals converge;

- determine the value of a convergent improper integral to within any positive error tolerance.

✥ 6.6 Skill Exercises

Calculate the value of each integral in #1–18 as the limit of definite integrals.

1. $\displaystyle\int_1^\infty \frac{1}{t^8}\,dt$

2. $\displaystyle\int_1^\infty \frac{1}{t^{13}}\,dt$

3. $\displaystyle\int_0^\infty \frac{2t}{(3t^2+1)^2}\,dt$

4. $\displaystyle\int_0^\infty \frac{5t}{(\sqrt{t}+1)^9}\,dt$

5. $\displaystyle\int_2^\infty e^{-t/4}\,dt$

6. $\displaystyle\int_1^\infty 3^{-5t}\,dt$

7. $\displaystyle\int_3^\infty t^2 e^{-t}\,dt$

8. $\displaystyle\int_1^\infty t^{-4}\ln(t)\,dt$

9. $\displaystyle\int_1^\infty \frac{6}{t\sqrt{t+1}}\,dt$

10. $\displaystyle\int_{19}^\infty \frac{6}{t\sqrt{t+8}}\,dt$

11. $\displaystyle\int_0^\infty e^{-t}\cos(t)\,dt$

12. $\displaystyle\int_0^\infty e^{-2t}\sin(8t)\,dt$

13. $\displaystyle\int_{-\infty}^0 te^t\,dt$

14. $\displaystyle\int_{-\infty}^0 (t^2-1)e^t\,dt$

15. $\displaystyle\int_{-\infty}^1 \frac{10}{t^2+8t+25}\,dt$

16. $\displaystyle\int_{-\infty}^6 \frac{5}{t^2+4t+13}\,dt$

17. $\displaystyle\int_{-\infty}^3 \frac{2}{t^2-12t+32}\,dt$

18. $\displaystyle\int_{-\infty}^1 \frac{2}{t^2-16t+28}\,dt$

In #19–22 (a) verify that the integrand has a singularity at $t=0$, and (b) determine whether the improper integral converges.

19. $\displaystyle\int_0^\infty \frac{1}{(t+1)\sqrt{t}}\,dt$

20. $\displaystyle\int_0^\infty \frac{e^{-t}}{t^{2/3}}\,dt$

21. $\displaystyle\int_0^\infty \frac{1}{t^{1/3}\sqrt{2t+11}}\,dt$

22. $\displaystyle\int_0^\infty \frac{\ln(t)}{(t+7)^2}\,dt$

In #23–40 (a) explain why the integral is improper, (b) determine whether it converges, and (c) explain your reasoning (e.g., Direct Comparison Theorem, Limit Comparison Theorem, calculation).

23. $\displaystyle\int_0^\infty \frac{2}{4t^2+9}\, dt$

24. $\displaystyle\int_5^\infty \frac{t+1}{t^2-4}\, dt$

25. $\displaystyle\int_0^\infty \frac{1}{2t^2-3t-2}\, dt$

26. $\displaystyle\int_0^\infty \frac{1}{2t^2-3t+2}\, dt$

27. $\displaystyle\int_4^\infty \frac{10t}{t^2+8}\, dt$

28. $\displaystyle\int_2^\infty \frac{te^{-2t}}{\ln(t)}\, dt$

29. $\displaystyle\int_0^\infty \frac{1}{\sqrt{t+1}}\, dt$

30. $\displaystyle\int_2^\infty \frac{dt}{(t^2-1)^{3/2}}\, dt$

31. $\displaystyle\int_0^\infty \frac{t^2}{t^4-1}\, dt$

32. $\displaystyle\int_4^\infty \frac{\sin^2(t)}{e^t-3}\, dt$

33. $\displaystyle\int_1^\infty \frac{2+e^{-x}}{x-e^{-x}}\, dx$

34. $\displaystyle\int_0^1 \frac{\ln x}{\sqrt[3]{x}-x}\, dx$

35. $\displaystyle\int_0^{\pi/2} \frac{1}{(t-\frac{\pi}{2})\cos(t)}\, dt$

36. $\displaystyle\int_1^2 \frac{t}{e^t\sqrt{t^2-1}}\, dt$

37. $\displaystyle\int_{-\infty}^{-2} e^t\, dt$

38. $\displaystyle\int_{-\infty}^0 \frac{2}{t-3}\, dt$

39. $\displaystyle\int_{-\infty}^{-2} \frac{t^2}{t^2+5}\, dt$

40. $\displaystyle\int_{-\infty}^0 \frac{5}{\sqrt[4]{2t+3}}\, dt$

In exercises #41–58 (a) determine whether the integral converges, and if so (b) calculate its value.

41. $\displaystyle\int_{-\infty}^\infty \frac{4}{t^2+1}\, dt$

42. $\displaystyle\int_{-\infty}^\infty \frac{t}{t^2+4}\, dt$

43. $\displaystyle\int_{-\infty}^\infty \frac{e^{\arctan(t)}}{t^2+1}\, dt$

44. $\displaystyle\int_{-\infty}^\infty te^{-t^2}\, dt$

45. $\displaystyle\int_0^1 \frac{1}{3t-2}\, dt$

46. $\displaystyle\int_1^3 \frac{1}{t^2-4}\, dt$

47. $\displaystyle\int_{-2}^0 \frac{1}{(t+1)^{2/3}}\, dt$

48. $\displaystyle\int_0^4 \frac{t-2}{(t^2-4t+3)^{4/5}}\, dt$

49. $\displaystyle\int_0^{\pi/2} \tan(t)\, dt$

50. $\displaystyle\int_2^3 \frac{1}{\sqrt{3-t}}\, dt$

51. $\displaystyle\int_4^8 \frac{7}{(t-4)^3}\, dt$

52. $\displaystyle\int_{-4}^4 \frac{3}{\sqrt{16-t^2}}\, dt$

53. $\displaystyle\int_0^{\pi/4} \csc^2(t)\, dt$

54. $\displaystyle\int_0^{\pi/2} \tan^2(t)\, dt$

55. $\displaystyle\int_1^5 \frac{6}{(5-t)^{4/5}}\, dt$

56. $\displaystyle\int_{-2}^2 \frac{6}{(4-t^2)^{0.9}}\, dt$

57. $\displaystyle\int_0^\infty \frac{1}{t^2+7t+12}\, dt$

58. $\displaystyle\int_1^\infty \frac{1}{t^2+11t+30}\, dt$

Using the technique demonstrated in Step 1 of Example 6.22 (p. 471), determine a value of ℓ for which $\int_1^\ell f(t)\, dt$ is within 0.01 of $\int_1^\infty f(t)\, dt$.

59. $f(t) = \frac{1}{1+\sqrt{t}}e^{-t}$

60. $f(t) = \frac{1}{2t^6-1}e^{-t/2}$

61. $f(t) = \frac{\sqrt{t}}{\sqrt{t}+t^6}$

62. $f(t) = \frac{\sqrt{t}}{t^6-0.5\sqrt{t}}$

Mixed Practice: This set of exercises is intended to provide you with practice in determining which integration technique is appropriate for a given integral. Exercises 63–70 may require techniques from previous sections.

63. $\displaystyle\int_3^5 \frac{t}{\sqrt{t^2-4}}\, dt$

64. $\displaystyle\int_0^3 \frac{1}{\sqrt{9-t^2}}\, dt$

65. $\displaystyle\int_0^2 \frac{2}{t^2-1}\,dt$

68. $\displaystyle\int_0^{\pi/6} \tan(t)\,dt$

66. $\displaystyle\int_1^4 \frac{4}{t^2+3t}\,dt$

69. $\displaystyle\int_1^2 \frac{1}{t\sqrt{t^2+2}}\,dt$

67. $\displaystyle\int_0^{\pi/6} \cot(t)\,dt$

70. $\displaystyle\int_1^5 \frac{t}{\sqrt{t-1}}\,dt$

❖ 6.6 Concept and Application Exercises

71. Suppose you hear the following conversation in a study center.

Student #1: I'm supposed to integrate a function from 1 to infinity. How is that possible?
Student #2: What do you know about the function?
Student #1: It's continuous, always positive, and its graph approaches a horizontal asymptote at $y = 0$.
Student #2: Oh, then you'll get a finite number when you integrate, guaranteed.
Student #1: Really? Are you sure.
Student #2: I think so, yeah. Let's ask somebody to make sure.

Suppose they ask you whether they're right. Give the students your answer, and support your assertions with illuminating examples.

72. Since $f(t) = 1/(t^2 + 5t + 6)$ is positive at each $t \in [1, \infty)$, its integral over this interval cannot be zero. So the calculation

$$\int_1^\infty \frac{1}{t^2+5t+6}\,dt = \int_1^\infty \frac{1}{t+2} - \frac{1}{t+3}\,dt$$
$$= \int_1^\infty \frac{1}{t+2}\,dt - \int_1^\infty \frac{1}{t+3}\,dt = \infty - \infty = 0$$

must have an error in it. Find the problem.

73. Explain why $\displaystyle\int_0^\infty e^{-t}\sin(2t)\,dt$ converges, but $\displaystyle\int_0^\infty e^t\sin(2t)\,dt$ does not.

74. Determine all values of p, if any, for which $\displaystyle\int_0^\infty \frac{1}{x^p(2+x)^{1-p}}\,dx$ converges.

75. Determine whether we can use the Direct Comparison Theorem to make a conclusion about the convergence of $\int_1^\infty \frac{1}{t^2+\cos(t)}\,dt$ by considering $\int_1^\infty \frac{1}{t^2}\,dt$. Explain your answer in complete sentences.

76. Determine whether we can use the Limit Comparison Theorem to make a conclusion about the convergence of $\int_1^\infty \frac{1}{t^2+\cos(t)}\,dt$ by considering $\int_1^\infty \frac{1}{t^2}\,dt$. Explain your answer in complete sentences.

77. In exercise #44 (p. 446) you calculated the force experienced by an ion that's y meters above a charged line segment. Now determine the force it experiences due to a whole line by calculating $\int_{-\infty}^\infty \frac{q\sigma y}{(y^2+x^2)^{3/2}}\,dx$. (Note that the force is inversely proportional to y.)

      Physics

78. In exercise #117 (p. 418) you calculated the force experienced by an ion that's y meters above a charged line disk. Now determine the force it experiences due to a whole plane by calculating $\int_0^\infty \frac{2\pi\sigma y\ r}{(y^2+r^2)^{3/2}}\,dr$. (Note the surprising fact that the force does not depend on y.)

      Physics

79. In #46 on p. 447, you saw that

Search Theory

$$P = 1 - \exp\left(\int_{-\infty}^{\infty} \frac{kh}{(\, h^2 + \chi^2 + (vt)^2 \,)^{3/2}} \, dt\right)$$

is the probability of detecting a ship that's traveling with velocity v along the line $x = \chi$ during the interval $[t_1, t_2]$. Now find the probability that the ship is detected at all, at any time, by calculating a formula for

$$P(\chi) = 1 - \exp\left(\int_{-\infty}^{\infty} \frac{kh}{(\, h^2 + \chi^2 + (vt)^2 \,)^{3/2}} \, dt\right).$$

80. In #79 you saw that $P(\chi) = 1 - e^{-2m/(h^2+\chi^2)}$ is the probability of detecting a ship that's traveling along the line $x = \chi$, where h is the altitude of the pilot above the Cartesian plane, and m is a constant that accounts for the ship's speed among other things. For the purposes of this exercise, set $h = 1$ and $m = 1$.

Search Theory

(a) Use a graphing utility to render a graph of $P(\chi)$.

(b) Show that $f(t) = t$ and $g(t) = 1 - e^{-t}$ have the same value at $t = 0$, and $f'(t) > g'(t)$ for all $t > 0$.

(c) Use part (b) to show that $\frac{1}{1+\chi^2} > 1 - e^{-1/(1+\chi^2)}$ for all χ.

(d) Use the result of part (c) to find ℓ so that $\int_{-\ell}^{\ell} P(\chi) \, d\chi$ is within 0.005 of $\int_{-\infty}^{\infty} P(\chi) \, d\chi$, much as we did in Example 6.22 (p. 471).

(e) Use Simpson's method to determine $\int_{-\ell}^{\ell} P(\chi) \, d\chi$ to within 0.005 of its actual value.

(f) The number $W = \int_{-\infty}^{\infty} P(\chi) \, d\chi$ is called the *effective sweep width*. Determine the effective sweep width to within 0.01 of its actual value when $P(\chi) = 1 - e^{-1/(1+\chi^2)}$. *(Hint: a spreadsheet program will be helpful.)*

(g) To see why W is called a sweep width, consider an idealized machine, called a *definite range detector,* that certainly detects a target when it's within a specific range, $W/2$, and certainly fails to detect it otherwise. In this case, the function $p(\chi) = 1$ if $|\chi| \le W/2$ and $p(\chi) = 0$ otherwise. Graph $P(\chi)$ and calculate $\int_{-\infty}^{\infty} P(\chi) \, d\chi$ in this case.

In #80 you were introduced the idea of a sweep width. Different electronic detectors result in different functions $P(x)$, due to their engineering and the conditions of the environment in which they're deployed. In #81–84 (a) use a graphing utility to render a graph of P, and (b) calculate the effective sweep width to within 0.01 of its actual value.

Search Theory

81. $P(x) = \frac{15+5x^2}{20+2x^4}$

83. $P(x) = x^2 e^{-|x|}$

82. $P(x) = \frac{1+x^2}{2+x^6}$

84. $P(x) = 0.7|x|e^{-0.01x^2}$

The *Laplace transform* is a mathematical tool used by engineers to handle models in which an object experiences a discontinuous force—i.e., a jolt. This transform uses an improper integral to take one function as an "input," $f(t)$, and it produces another function as its "output," $F(s)$. More specifically, the Laplace transform of the function $f(t)$ is:

Engineering

$$F(s) = \int_0^{\infty} f(t)e^{-st} \, dt.$$

In each of #85–92 calculate the Laplace transform of the given function. Treat b as an arbitrary constant.

85. $f(t) = t$ 87. $f(t) = e^{3t}$ 89. $f(t) = \sin(t)$ 91. $f(t) = \cos(t)$

86. $f(t) = t^2$ 88. $f(t) = e^{bt}$ 90. $f(t) = \sin(bt)$ 92. $f(t) = \cos(bt)$

93. Suppose $F(s)$ is the Laplace transform of $f(t)$. Verify that $4F(s)$ is the Laplace transform of $4f(t)$. (*Be careful to write the improper integrals as the limit of definite integrals.*)

94. Suppose $F(s)$ and $G(t)$ are the Laplace transforms of $f(t)$ and $g(t)$, respectively. Verify that $5F(s) + 7G(s)$ is the Laplace transform of $5f(t) + 7g(t)$. (*Be careful to write the improper integrals as the limit of definite integrals, and use the limit laws.*)

95. On p. 482 you'll study the properties of the Gamma function, which is defined as $\Gamma(n) = \int_0^\infty t^{n-1} e^{-t}\, dt$. Explain why the domain of Γ is $n > 0$.

96. A text on thermal physics ([11, p. 297]) claims that the free energy of a photon gas is proportional to $\int_0^\infty x^2 \ln(1 - e^{-x})\, dx$. Determine the value of this integral to within 0.001 of its actual value. (*Hint: integrate by parts over $[\varepsilon, L]$ and limit $L \to \infty$ and $\varepsilon \to 0^+$, then use a Riemann sum.*)

Physics

Checking the details

97. In this exercise you'll use the technical definition of the the limit to prove the Limit Comparison Theorem for Improper Integrals (Part 1) from p. 464. A subtle but important part of this argument is the fact that

$$\int_\tau^T f(t)\, dt \quad \text{and} \quad \int_\tau^T g(t)\, dt$$

grow with T because $f(t)$ and $g(t)$ are non-negative.

(a) Use the technical definition of limits at infinity to explain why there is some time τ after which $\left| \frac{f(t)}{g(t)} - L \right| < \frac{L}{2}$.

(b) Use part (a) to explain why $\frac{L}{2} g(t) < f(t)$ when $\tau < t$.

(c) Part (b) guarantees that $\int_\tau^T \frac{L}{2} g(t)\, dt < \int_\tau^T f(t)\, dt$ for every $T > \tau$. Use that fact to explain why, if $\lim_{T \to \infty} \int_\tau^T g(t)\, dt$ diverges, so does $\lim_{T \to \infty} \int_\tau^T f(t)\, dt$.

(d) Use part (a) to explain why $f(t) < \frac{3L}{2} g(t)$ when $\tau < t$.

(e) Part (d) guarantees that $\int_\tau^T f(t)\, dt < \int_\tau^T \frac{3L}{2} g(t)\, dt$ for every $T > \tau$. Use that fact to explain why, if $\lim_{T \to \infty} \int_\tau^T f(t)\, dt$ diverges, so does $\lim_{T \to \infty} \int_\tau^T g(t)\, dt$.

(f) In parts (c) and (e) you showed that if either of $\int_\tau^\infty f(t)\, dt$ and $\int_\tau^\infty g(t)\, dt$ diverges, the other does too. Use this fact to explain why, if $\int_\tau^\infty f(t)\, dt$ converges, the integral $\int_\tau^\infty g(t)\, dt$ must also converge, and vice versa.

Chapter 6 Review

❖ True or False

1. The substitution $x = t$, $dx = dt$ is useful.

2. Integrating with a substitution amounts to "undoing" a Chain Rule differentiation.

3. Integrating by parts relies on the Product Rule.

4. The polynomial $t^2 + 3t - 8$ is irreducible.

5. The polynomial $t^2 + 3t + 8$ is irreducible.

6. The rational function $\frac{t^3+t-3}{t^2+7}$ is written in lowest terms.

7. The integral $\int_1^2 \frac{1}{t}\, dt$ is improper.

8. The integral $\int_1^2 \frac{1}{t-2}\, dt$ is improper.

9. The integral $\int_0^7 \frac{1}{t^6}\, dt$ converges.

10. The integral $\int_4^\infty \frac{1}{t^6}\, dt$ converges.

11. $\int_5^\infty \frac{1}{t^4+1}\, dt > \int_4^\infty \frac{1}{t^4+1}\, dt$.

12. $\int_2^\infty \frac{1}{t^7-1}\, dt > \int_2^\infty \frac{1}{t^7+1}\, dt$.

13. Since $\int_1^\infty \frac{1}{t^2+1}\, dt$ converges, the Direct Comparison Theorem guarantees that $\int_1^\infty \frac{1}{t^2+\sin^2(t)}\, dt$ does too.

14. The number $\int_1^{12} \frac{1}{t^3+1}\, dt$ is within 0.01 of $\int_1^\infty \frac{1}{t^3+1}\, dt$.

❖ Multiple choice

15. The partial fractions decomposition of $\frac{3}{t^2+5t}$ is

    (a) $\frac{1/5}{t} - \frac{1/5}{t+5}$

    (b) $\frac{1/6}{t} - \frac{1/4}{t+5}$

    (c) $\frac{1/4}{t} - \frac{1/6}{t+5}$

    (d) Any of these will work

    (e) None of the above

16. The partial fractions decomposition of $\frac{3}{t(t^2+4)^2}$ has the form

    (a) $\frac{A}{t} + \frac{B}{t^2+4} + \frac{C}{(t^2+4)}$

    (b) $\frac{A}{t} + \frac{Bt+D}{t^2+4} + \frac{Ct+E}{(t^2+4)}$

    (c) $\frac{A}{t} + \frac{Bt+D}{t^2+4} + \frac{Ct+E}{(t^2+4)^2}$

    (d) any of (a), (b) or (c) will work

    (e) none of the above

17. One step of integration by parts allows us to rewrite $\int e^{4t} \sin(9t)\, dt$ as

    (a) $\frac{1}{4}e^{4t} \sin(9t) - \frac{9}{4} \int e^{4t} \cos(9t)\, dt$

(b) $\frac{1}{4}e^{4t}\cos(9t) - \frac{9}{4}\int e^{4t}\cos(9t)\ dt$

(c) $\frac{1}{9}e^{4t}\cos(9t) - \frac{9}{4}\int e^{4t}\sin(9t)\ dt$

(d) $\frac{1}{9}e^{4t}\cos(9t) - \frac{9}{4}\int e^{4t}\cos(9t)\ dt$

(e) none of the above

18. The integral $\int \frac{1}{1-x^2}\ dx$ can be done with

(a) the trigonometric substitution $x = \tan(\phi)$

(b) the trigonometric substitution $x = \cot(\phi)$

(c) a partial fractions expansion

(d) all of the above

(e) none of the above

19. The integral $\int \frac{3}{4-x^2}\ dx$ can be done with

(a) the trigonometric substitution $x = 2\sin(\phi)$

(b) the trigonometric substitution $x = 2\cos(\phi)$

(c) a partial fractions expansion

(d) all of the above

(e) none of the above

20. The integral $\int \frac{2x}{9-x^2}\ dx$ can be done with

(a) the trigonometric substitution $x = 3\sin(\phi)$

(b) the change of variable $u = 9 - x^2$

(c) a partial fractions expansion

(d) all of the above

(e) none of the above

21. The integral $\int t\sqrt{8+t}\ dt$ can be done with

(a) the trigonometric substitution $t = 8\tan(\phi)$

(b) integration by parts

(c) a partial fractions expansion

(d) all of the above

(e) none of the above

22. To determine $\int \sin^2(t)\ dt$, begin by ...

(a) using the substitution $x = \sin(t)$

(b) using the identity $\sin^2(t) = 1 - \cos^2(t)$

(c) using the identity $\sin^2(t) = \frac{1}{2}(1 - \cos(2t))$

(d) any of the above will work

(e) none of the above

23. To determine $\int \sin^2(t)\cos^3(t)\ dt$, begin by ...

(a) using the substitution $x = \cos^3(t)$

(b) rewriting $\cos^3(t)$ as $\cos^2(t)\cos(t)$

(c) rewriting $\sin^2(t)$ as $1 - \cos^2(t)$

(d) integrate $\sin^2(t)$ and $\cos^3(t)$ separately using the power rule

(e) none of the above

24. The integral $\int_1^2 \frac{1}{(t^2-4)^{1/3}}\, dt$

(a) does not exist because the integrand is undefined at $t = 2$

(b) is calculated using a trigonometric substitution

(c) is an improper integral that converges

(d) is an improper integral that diverges

(e) none of the above

❖ Exercises

25. Determine $\displaystyle\int \frac{1}{1-e^t}\, dt$ by using the substitution $x = e^{-t}$.

26. Complete the square in the denominator in order to determine $\displaystyle\int \frac{1}{x^2 - 2x + 7}\, dx$.

27. Determine $\int e^{\sqrt{t}}\, dt$ by using the substitution $x = \sqrt{t}$ and then integrating by parts.

28. Determine $\int \frac{1}{(t^4+1)t}\, dt$ by rewriting the integrand in the form $\frac{A}{t} + \frac{Bt^3 + Ct^2 + Dt + E}{t^4 + 1}$.

Determine the integrals in #29–40 using substitution.

29. $\displaystyle\int t^2 \sqrt[3]{t-1}\, dt$

30. $\displaystyle\int (t-4)\sqrt[3]{t+2}\, dt$

31. $\displaystyle\int \frac{5t}{(t^2-6)^2}\, dt$

32. $\displaystyle\int \frac{2}{\sqrt{9t-11}}\, dt$

33. $\displaystyle\int \frac{t^2}{1-81t^4}\, dt$

34. $\displaystyle\int t\sqrt{t+1}\, dt$

35. $\displaystyle\int \cot\left(\frac{t}{4}\right)\, dt$

36. $\displaystyle\int \sec(2t)\, dt$

37. $\displaystyle\int t^3 e^{t^4}\, dt$

38. $\displaystyle\int \sin^2(3t)\cos(3t)\, dt$

39. $\displaystyle\int \frac{1}{8-5t}\, dt$

40. $\displaystyle\int \frac{7t}{4t^2+11}\, dt$

Determine the integrals in #41–46 using partial fractions decompositions or other techniques discussed in Section 6.5.

41. $\displaystyle\int \frac{t-4}{t+12}\, dt$

42. $\displaystyle\int \frac{5t+4}{3t^2+7}\, dt$

43. $\displaystyle\int \frac{1}{t^2 - 3t - 4}\, dt$

44. $\displaystyle\int \frac{10}{(t-3)^2}\, dt$

45. $\displaystyle\int \frac{2t^3}{(t^2+5)^2}\, dt$

46. $\displaystyle\int \frac{9t+5}{2t^2 + t - 6}\, dt$

Determine the integrals in #47–52 using integration by parts.

47. $\displaystyle\int t\sin(10t)\, dt$

48. $\displaystyle\int_1^4 t^5 \ln(t)\, dt$

49. $\displaystyle\int t^3 \sin(4t)\, dt$

50. $\displaystyle\int_0^1 e^{4t}\sin(10t)\, dt$

51. $\displaystyle\int \sin^{-1}\left(\frac{t}{4}\right)\, dt$

52. $\displaystyle\int \ln(t^2)\, dt$

Determine the integrals in #53–56 using a trigonometric substitution.

53. $\int \sqrt{t^2 + 4}\, dt$

54. $\int \dfrac{t^2}{\sqrt{16 - 6t - t^2}}\, dt$

55. $\int \dfrac{\sqrt{9t^2 - 1}}{t}\, dt$

56. $\int \dfrac{6}{t^2 \sqrt{8t^2 - 3}}\, dt$

Determine whether the integrals in #57–64 converge.

57. $\int_{4}^{\infty} \dfrac{1}{t^3 - 4t}\, dt$

58. $\int_{-\infty}^{0} e^t \sin(8t)\, dt$

59. $\int_{0}^{4} \dfrac{7}{\sqrt[3]{t - 4}}\, dt$

60. $\int_{0}^{7} \dfrac{5t}{\sqrt{49 - t^2}}\, dt$

61. $\int_{0}^{\infty} \dfrac{t - \sin^2(t)}{t^2 + 15}\, dt$

62. $\int_{1}^{\infty} \dfrac{1}{t^5 + 4t}\, dt$

63. $\int_{0}^{\pi/4} \csc(t)\, dt$

64. $\int_{-1}^{0} \dfrac{1}{(x + 1)^{2/3}}\, dx$

All of the techniques studied in the chapter are represented in the following set of integrals, which serves as the final "mixed practice" exercise set. Where the integral is definite, you should (a) determine whether it's improper, and if so (b) determine whether it converges, and (c) calculate its value.

65. $\int_{3}^{12} t\sqrt{3t - 9}\, dt$

66. $\int 3\csc(12t)\, dt$

67. $\int \dfrac{2 - 7t}{4t^2 + 9}\, dt$

68. $\int \dfrac{t - 6}{t - 9}\, dt$

69. $\int_{1}^{3} t^2 e^{-0.01t}\, dt$

70. $\int_{0}^{6\pi} t\sin(4t)\, dt$

71. $\int \dfrac{e^{2t}}{e^{2t} - 3}\, dt$

72. $\int \dfrac{\sec^2(t)}{\tan(t) + 4}\, dt$

73. $\int_{0}^{1} \dfrac{1}{\sqrt{7 - t^2}}\, dt$

74. $\int \dfrac{1}{(t^2 + 1)^2}\, dt$

75. $\int_{0}^{\infty} \dfrac{1}{t^2 + 4}\, dt$

76. $\int_{0}^{\infty} \dfrac{1}{t^2 - 4}\, dt$

77. $\int \tan\left(\dfrac{\pi}{3}t\right) dt$

78. $\int (e^t + e^{-t})^2\, dt$

79. $\int e^{0.2t} \cos(3t)\, dt$

80. $\int \cos(\ln(t))\, dt$

81. $\int_{0}^{1} \dfrac{1}{\sqrt{t}}\, dt$

82. $\int_{0}^{\infty} \dfrac{1}{\sqrt{t}}\, dt$

83. $\int \sin^3(t)\, dt$

84. $\int \cos^4(t)\, dt$

85. $\int_{0}^{\pi} t^2 \cos(2t)\, dt$

86. $\int_{-1}^{1} te^{3t}\, dt$

87. $\int \dfrac{8t - 1}{4t^3 + 4t^2 + t}\, dt$

88. $\int \dfrac{t^2 + 11t}{(t^2 + 6)^2}\, dt$

89. $\int \dfrac{t}{\sqrt{4t^2 + 4t + 10}}\, dt$

90. $\int (t^2 - 1)^{3/2}\, dt$

91. $\int \dfrac{t^3 + t - 7}{t + 2}\, dt$

92. $\int \dfrac{t^2 + t - 7}{t^2 + 5}\, dt$

93. $\int \sec^4(t)\tan(t)\, dt$

94. $\int \sin(8t)\sin(7t)\, dt$

95. $\int t^3 e^{t^2}\, dt$

Chapter 6 Projects and Applications

❖ The Gamma Function

The **Gamma function**, denoted by $\Gamma(x)$, is often used as an extension of the factorial (see p. 400) to non-integer numbers. In short,

$$\Gamma(x) = \int_0^\infty t^{x-1} e^{-t} \, dt,$$

which is a continuous function of $n \in (0, \infty)$.

▷ Basic facts about $\Gamma(x)$

The exercises below will familiarize you with the basic properties of this function.

1. Verify that $\Gamma(2) = 1$.

2. Integrate by parts to show that $\Gamma(n+1) = n\Gamma(n)$, being careful to write the improper integral as the limit of definite integrals.

3. Use #1 and #2 to show that $\Gamma(3) = 2!$, and then show that $\Gamma(4) = 3!$.

4. If $\Gamma(n) = (n-1)!$ when $n > 0$ integer, use #2 to show that $\Gamma(n+1) = n!$.

Based on the results of #1–4 we know that $\Gamma(n+1) = n!$ when $n > 0$ is an integer. Henceforth, let's *define* the factorial as $n! = \Gamma(n+1)$.

5. Use the Gamma function to calculate $0!$.

As mentioned above, the Gamma function is defined for all $x \in (0, \infty)$, not just the integers.

6. Explain why $\lim_{x \to \infty} \Gamma(x)$ does not exist. *(Hint: see #4.)*

7. Show that $\lim_{x \to 0^+} \frac{\Gamma(x)}{1/x}$ is finite by calculating its value. *(Hint: the result from #2 and the fact that $\Gamma(x)$ is continuous will be handy.)*

Since $\lim_{x \to 0^+} \frac{\Gamma(x)}{1/x}$ is finite, we know that $\lim_{x \to 0^+} \Gamma(x) = \infty$.

▷ Extending the domain of Γ

In #2 you showed that $\Gamma(n+1) = n\Gamma(n)$, which we can use to extend the domain of Γ to negative numbers. For example, rewriting the identity as

$$\Gamma(x) = \frac{1}{x}\Gamma(x+1). \tag{8.1}$$

allows us to define $\Gamma(x)$ when $x \in (-1, 0)$.

8. It's true, but not obvious, that $\Gamma(0.5) = \sqrt{\pi}$. Use this fact in conjunction with equation (8.1) to calculate $\Gamma(-0.5)$.

9. Use equation (8.1) and the continuity of Γ to explain why $\lim_{x \to 0^-} \Gamma(x) = -\infty$.

10. Use #9 and equation (8.1) to explain why $\lim_{x \to -1^+} \Gamma(x) = -\infty$.

11. Use your result from #8 to calculate $\Gamma(-1.5)$.

12. Use equation (8.1) and your results from #9 and #10 to determine $\lim_{x \to -1^-} \Gamma(x)$ and $\lim_{x \to -2^+} \Gamma(x)$.

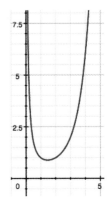

Figure 8.1: The graph of the curve $y = \Gamma(n)$ when $n \in (0, 5)$. Note the critical point in $(0, 1)$. How do you think we determine its location precisely?

13. Use equation (8.1) and your result from #8 to calculate $\Gamma(-2.5)$.

14. Use equation (8.1) and your results from #12 to determine $\lim_{x \to -2^-} \Gamma(x)$ and $\lim_{x \to -3^+} \Gamma(x)$.

15. Explain why we cannot use equation (8.1) to define $\Gamma(n)$ when $n < 0$ is an integer.

16. Based on the results of #8–15, sketch a rough graph of $\Gamma(x)$ when $x < 0$.

❖ The Seasons

The Earth's axis of rotation is not perpendicular to the plane in which our planet moves around the sun, and consequently some days have more hours of sunlight than others. Said simply, it's summer (longer days) in America when the northern hemisphere is tilted toward the sun; and the day that gets the most sunlight, called the *summer solstice*, happens when the Earth's axis (which doesn't change direction over the course of a year) is tilted *directly* toward our star.

> Actually, the Earth's axis *does* change direction each year, but only slightly.

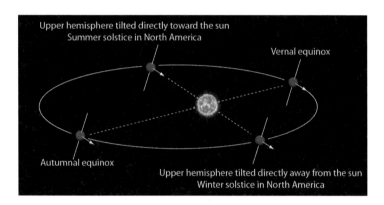

Figure 8.2: The axis of Earth's rotation is always tipped 23° in the direction of the (parallel) arrows.

By contrast, it's winter in America when the northern hemisphere is tipped away from the sun, and the day that gets the least sunlight, called the *winter solstice*, happens when the Earth's axis is tipped *directly* away. Twice a year, the Earth's axis of rotation is tipped neither toward nor away from sun, so day and night have equal length. These days are called the *vernal (spring)* and *autumnal equinoxes*. The important thing for you to keep in mind is that seasons have to do with the Earth's tilt with respect to the plane of its orbit, not with the orbit itself.

Since the Earth's orbit is roughly circular, let's talk about the planet's position in its orbit using an angle. We'll designate perigee (the point at which Earth is closest to the sun) to be at $\theta = 0$, and measure θ in radians (see Figure 8.3).

In the project called *Planetary Orbits* at the end of Chapter 9 we'll discuss the fact that the distance from the sun to the Earth, designated by r (which changes with time), is related to the angle θ (which also changes with time) by the equation

$$H = r^2 \frac{d\theta}{dt}, \tag{8.2}$$

where $H > 0$ is constant. Moreover, you'll use this fact to help establish that

$$r = \frac{a\varepsilon}{1 + \varepsilon \cos(\theta)}, \tag{8.3}$$

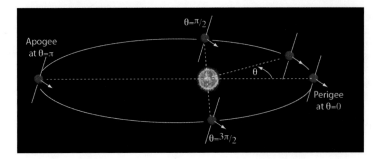

Figure 8.3: The oblong character of the Earth's orbit has been exaggerated. The actual orbit is closer to being circular.

where $\varepsilon \in (0,1)$ is a number called the *eccentricity* of the orbit, and $a > 0$ is a constant that depends (inconsequentially) on H, the mass of the sun, and the gravitational constant.

> Smaller ε correspond to more circular orbits, and larger ε indicate more oblong orbits.

▷ Time, angle, and a new technique of integration

Instead of using the Gregorian calendar to define the beginning of a year, let's say that a year begins when Earth is at perigee (see Figure 8.3). The Earth is at apogee when one-half of a year has passed, and we say that one year has passed when the planet returns to perigee. Said differently, half a year has passed when $\theta = \pi$, and $\theta = 2\pi$ when a whole year has passed. In this sense, the angle θ can be understood as a different way of measuring time, and in the following steps (based on an exposition by Simoson [1]) you'll derive an explicit formula that relates θ to t.

If we cross-multiply in equation (8.2) and use the relationship expressed by equation (8.3), we see

$$\frac{dt}{d\theta} = \frac{(a\varepsilon)^2}{H} \frac{1}{(1 + \varepsilon \cos(\theta))^2},$$

which tells us about the relationship between changes in t and changes in θ. In fact, if we integrate both sides with respect to θ we see

$$\int t'(\theta)\, d\theta = \int \frac{(a\varepsilon)^2}{H} \frac{1}{(1 + \varepsilon \cos(\theta))^2}\, d\theta.$$

The left-hand side is just $t(\theta)$, but the right-hand side is more complicated. The standard (though not obvious) approach to integrating so-called *rational trigonometric functions* like this is to use the substitution

$$x = \tan\left(\frac{\theta}{2}\right) \quad \Rightarrow \quad dx = \sec^2\left(\frac{\theta}{2}\right) \frac{1}{2}\, d\theta. \tag{8.4}$$

> This particular substitution is often called the *Weierstrass substitution*, after the German mathematician Karl Weierstrass. It's closely related to the stereographic projection (see formulas on p. 18).

1. Draw a right-triangle with an interior angle of $\theta/2$ whose tangent is $x/1$, as we did with trigonometric substitutions (e.g., see p. 439).

2. Use the triangle to determine a formula for $\sec^2\left(\frac{\theta}{2}\right)$ in terms of x, and differentiate it with respect to x in order to verify that $\frac{d\theta}{dx} = \frac{2}{1+x^2}$. Consequently,

$$d\theta = \frac{2}{1 + x^2}\, dx.$$

3. Use the half-angle identity for the cosine, $2\cos^2\left(\frac{\theta}{2}\right) = 1 + \cos(\theta)$, to verify that

$$\cos(\theta) = \frac{1 - x^2}{1 + x^2}.$$

4. Though it's not relevant to the integration that we're trying to perform at the moment, use the double-angle identity for the sine $\sin(\theta) = 2\sin\left(\frac{\theta}{2}\right)\cos\left(\frac{\theta}{2}\right)$ to verify that

$$\sin(\theta) = \frac{2x}{1+x^2}.$$

In #3 and #4 above, you see why this method works: rational functions of sine and cosine become rational functions of x after we make this substitution, and though rational functions of x can be difficult to integrate, they're more familiar than rational functions of sine and cosine.

5. By using the facts you established in #2 and #3, show that

$$t(\theta) = \frac{2(a\varepsilon)^2}{H} \int \frac{1+x^2}{\left((1+\varepsilon)+(1-\varepsilon)x^2\right)^2}\, dx.$$

6. Factor $(1+\varepsilon)$ out of the denominator so that our formula becomes

$$t(\theta) = \frac{2(a\varepsilon)^2}{H(1+\varepsilon)^2} \int \frac{1+x^2}{(1+\alpha x^2)^2}\, dx, \quad \text{where } \alpha = \frac{1-\varepsilon}{1+\varepsilon}.$$

Note that the numerator and denominator *almost* cancel, but a factor of α is missing from the numerator. This leads us to multiply and divide by α, so that we see

$$t(\theta) = \frac{2(a\varepsilon)^2}{\alpha H(1+\varepsilon)^2} \int \frac{\alpha+\alpha x^2}{(1+\alpha x^2)^2}\, dx.$$

This integrand simplifies nicely when we add zero to the numerator. Specifically, when we add and subtract 1 from the numerator we see

$$\int \frac{\alpha+\alpha x^2}{(1+\alpha x^2)^2}\, dx = \int \frac{\alpha-1}{(1+\alpha x^2)^2} + \frac{1+\alpha x^2}{(1+\alpha x^2)^2}\, dx$$

$$= \int \frac{\alpha-1}{(1+\alpha x^2)^2}\, dx + \int \frac{1}{1+\alpha x^2}\, dx.$$

7. Use a trigonometric substitution to show that

$$\int \frac{1}{1+\alpha x^2}\, dx = \frac{1}{\sqrt{\alpha}}\tan^{-1}(\sqrt{\alpha}\, x) + C_1.$$

8. Use the same trigonometric substitution to show that

$$\int \frac{\alpha-1}{(1+\alpha x^2)^2}\, dx = \frac{\alpha-1}{2\sqrt{\alpha}}\tan^{-1}(\sqrt{\alpha}\, x) + \frac{\alpha-1}{2}\frac{x}{1+\alpha x^2} + C_2.$$

9. Use your answers from #7 and #8 to verify that

$$t(\theta) = \frac{2(a\varepsilon)^2}{\alpha H(1+\varepsilon)^2}\left(\frac{\alpha+1}{2\sqrt{\alpha}}\tan^{-1}\left(\sqrt{\alpha}\tan\left(\frac{\theta}{2}\right)\right) + \frac{\alpha-1}{2}\frac{\tan\left(\frac{\theta}{2}\right)}{1+\alpha\tan^2\left(\frac{\theta}{2}\right)} + C\right).$$

10. Based on the meaning of θ and t, explain why $C = 0$.

11. Verify that

$$\frac{\tan\left(\frac{\theta}{2}\right)}{1+\alpha\tan^2\left(\frac{\theta}{2}\right)} = \frac{\sin\left(\frac{\theta}{2}\right)\cos\left(\frac{\theta}{2}\right)}{\cos^2\left(\frac{\theta}{2}\right)+\alpha\sin^2\left(\frac{\theta}{2}\right)}.$$

12. Verify that factoring $\frac{\alpha+1}{2\sqrt{\alpha}}$ out of the expression in parentheses results in

$$t(\theta) = \gamma\left(\tan^{-1}\left(\sqrt{\alpha}\tan\left(\frac{\theta}{2}\right)\right) + \frac{(\alpha-1)\sqrt{\alpha}}{\alpha+1}\frac{\sin\left(\frac{\theta}{2}\right)\cos\left(\frac{\theta}{2}\right)}{\cos^2\left(\frac{\theta}{2}\right)+\alpha\sin^2\left(\frac{\theta}{2}\right)}\right) \qquad (8.5)$$

where $\gamma = \frac{(\alpha+1)(a\varepsilon)^2}{\alpha^{3/2}H(1+\varepsilon)^2}$.

▷ Reformulating to avoid discontinuity

When you graph $t(\theta)$ as it's formulated in equation (8.5) you see something like the image in Figure 8.4.

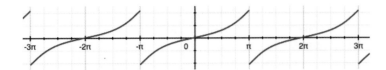

Figure 8.4: An example graph of $t(\theta)$ as formulated in equation (8.5) in which $\varepsilon = 0.37$ and $\gamma = 1$.

Of course, there's not really a jump discontinuity when $\theta = \pi$. (We don't jump back in time when the Earth reaches apogee.) This discontinuity is an artifact of the tangent function, and we can get around it by using trigonometric identities to rewrite $t(\theta)$ so that it agrees with the formula in (8.5) when $-\pi < \theta < \pi$ and extends continuously beyond that interval. As you'll show in #13–15, those trigonometric identities lead us to

$$t(\theta) = \gamma \left(\frac{\theta}{2} + \tan^{-1} \left(-\frac{\sin(\theta)}{R + \cos(\theta)} \right) \right) + \frac{\sqrt{\alpha}}{1 + \alpha} \cdot \frac{\sin(\theta)}{Q - \cos(\theta)} \right), \qquad (8.6)$$

where $R = \frac{\sqrt{\alpha}+1}{\sqrt{\alpha}-1}$ and $Q = \frac{\alpha+1}{\alpha-1}$, a graph of which is shown in Figure 8.5.

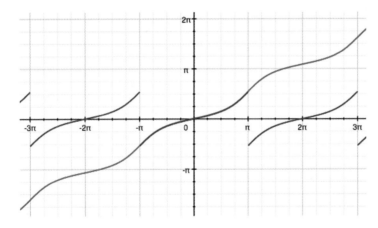

Figure 8.5: The graph of our reformulated $t(\theta)$ superimposed on Figure 8.4, with $\varepsilon = 0.37$ and $\gamma = 1$.

Changing from the formula in (8.5) to the one written in (8.6) is accomplished in two steps. First we will use the sum identity for the tangent to show that

$$\tan^{-1} \left(\sqrt{\alpha} \tan \left(\frac{\theta}{2} \right) \right) = \frac{\theta}{2} + \tan^{-1} \left(\frac{\left(\sqrt{\alpha} - 1 \right) \sin \left(\frac{\theta}{2} \right) \cos \left(\frac{\theta}{2} \right)}{\cos^2 \left(\frac{\theta}{2} \right) + \sqrt{\alpha} \sin^2 \left(\frac{\theta}{2} \right)} \right) \qquad (8.7)$$

Then we will use the addition and half-angle identities for sine and cosine to arrive at equation (8.6).

Step 1 (using the addition identity for the tangent): If we could devise a β for which $\sqrt{\alpha} \tan \left(\frac{\theta}{2} \right) = \tan \left(\frac{\theta}{2} + \beta \right)$, we'd have

$$\tan^{-1} \left(\sqrt{\alpha} \tan \left(\frac{\theta}{2} \right) \right) = \tan^{-1} \left(\tan \left(\frac{\theta}{2} + \beta \right) \right) = \frac{\theta}{2} + \beta,$$

so let's try to find such a β. The sum identity for the tangent is

$$\tan\left(\frac{\theta}{2} + \beta\right) = \frac{\tan\left(\frac{\theta}{2}\right) + \tan(\beta)}{1 - \tan\left(\frac{\theta}{2}\right)\tan(\beta)}.$$

13. Use the sum identity for the tangent to solve the equation

$$\sqrt{\alpha}\tan\left(\frac{\theta}{2}\right) = \tan\left(\frac{\theta}{2} + \beta\right)$$

for β, and use your result to validate equation (8.7).

Step 2 (using other identities to complete the transformation): Now we make use of the following identities:

Pythagorean Identity: $\sin^2\left(\frac{\theta}{2}\right) + \cos^2\left(\frac{\theta}{2}\right) = 1$

Double-angle Identity: $\sin(\theta) = 2\sin\left(\frac{\theta}{2}\right)\cos\left(\frac{\theta}{2}\right)$

Half-angle Identity: $\sin^2(\theta) = \frac{1}{2}(1 - \cos(\theta))$

14. Use the trigonometric identities shown above to rewrite

$$\frac{(\sqrt{\alpha} - 1)\sin\left(\frac{\theta}{2}\right)\cos\left(\frac{\theta}{2}\right)}{\cos^2\left(\frac{\theta}{2}\right) + \sqrt{\alpha}\sin^2\left(\frac{\theta}{2}\right)} \qquad \text{as} \qquad -\frac{\sin(\theta)}{R + \cos(\theta)}.$$

15. Use the trigonometric identities shown above to rewrite

$$\frac{(\alpha - 1)\sqrt{\alpha}}{\alpha + 1}\frac{\sin\left(\frac{\theta}{2}\right)\cos\left(\frac{\theta}{2}\right)}{\cos^2\left(\frac{\theta}{2}\right) + \alpha\sin^2\left(\frac{\theta}{2}\right)} \qquad \text{as} \qquad \frac{\sqrt{\alpha}}{1 + \alpha}\cdot\frac{\sin(\theta)}{Q - \cos(\theta)}.$$

▷ Analysis in Nondimensional Time

We can "nondimensionalize" our measurement of time by defining

$$T(\theta) = \frac{1}{t(2\pi)}t(\theta),$$

which tells the *fraction of a year* corresponding to θ radians.

16. Verify that $T(\theta)$ is dimensionless, and plot a graph of $T(\theta)$ for $\varepsilon = 0.5$.

17. In physical terms, explain why the graph of $T(\theta)$ is always increasing.

18. Note that the graph of $T(\theta)$ alternates between concave up and concave down. What is happening to the planet's orbital velocity when the graph of $T(\theta)$ is concave up?

Suppose the vernal equinox occurs when the planet is at $\theta = \phi$. Then the summer solstice occurs at $\theta = \phi + \frac{\pi}{2}$, and the fraction of the year between them is

$$k(\phi) = T\left(\phi + \frac{\pi}{2}\right) - T(\phi),$$

whose derivative is

$$k'(\phi) = -\frac{R\sin(\phi) + 1}{1 + R^2 + 2\sin(\phi)} - \frac{R\cos(\phi) - 1}{1 + R^2 - 2\cos(\phi)} - \frac{\sqrt{\alpha}}{\alpha + 1}\left(\frac{Q\sin(\phi) + 1}{\left(Q + \sin(\phi)\right)^2} + \frac{Q\cos(\phi) - 1}{\left(Q - \cos(\phi)\right)^2}\right).$$

19. Verify that k has a critical point when $\cos(\phi)$ and $\sin(\phi)$ are opposites.

20. It turns out that the *only* first-order critical points of k in $[0, 2\pi]$ are at $\phi = 3\pi/4$ and $\phi = 7\pi/4$. Use the first derivative test to classify each as a minimum or maximum.

21. Explain why your answer from #20 makes sense based on the planet's orbital velocity.

22. Use a graphing utility to graph $k(\phi)$ for 15 values of ε (evenly distributed between 0 and 1) all on the same screen.

23. Let's define $K = k(3\pi/4)$ which, as you see in your graphs from #22, depends on $\varepsilon \in (0, 1)$. Based upon your graphs, is $dK/d\varepsilon$ positive or negative?

24. Based upon your graphs from #22, is $K''(\varepsilon)$ ever zero?

25. Based on the meaning of ε (see the comment box on p. 484), explain why $\lim_{\varepsilon \to 0} K(\varepsilon) = 0.25$ and $\lim_{\varepsilon \to 1} K(\varepsilon) = 1$.

Chapter 7
Applications of Integration

When we first discussed the relationship between a car's velocity and the distance it travels (see p. 325), we began with an approximation in which we treated the car's velocity as if it were constant over short periods of time. Then we passed to an exact value in the limit. The applications of the integral are far too numerous for us to provide a comprehensive catalog here, but all of them rely on that basic idea: approximate with a Riemann sum by treating some quantity as if it's constant over small intervals, and then pass to an exact value in the limit. For example, when calculating ...

- the length of a curve, we'll treat its slope as if it's constant over small intervals;

- volume, we'll treat radius as if it's constant over small intervals;

- hydrostatic force, we'll treat pressure as if it's constant over small intervals;

- mass, we'll treat density as if it's constant over small intervals.

As indicated by the list above, we'll begin with geometric questions (because they're easy to visualize) and then move into discussions of other applications areas.

7.1   Arc Length and Surface Area

Consider the task of designing a suspension bridge. One thing you'd need to know is the length of the main cable between the towers, which hangs in the shape of a parabola due to the load it supports (with no load, a cable hangs in a shape called a *catenary*). How do you calculate the length of a parabolic arc, a catenary, or any other curve for that matter?

Figure 1.1: A cable hangs in the shape of a parabola when under a uniform load that far exceeds the weight of the cable itself.

Calculations like this are difficult because the slope of the curve changes from point to point. However, when a curve is the graph of a differentiable function (as is a parabola), it looks linear if we zoom in far enough. That leads us to approximate the length of a curve $y = f(x)$ as the sum of many, many short line segments, as seen in Figure 1.2.

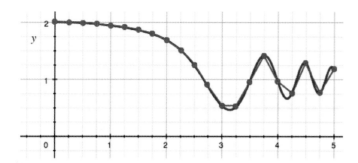

Figure 1.2: The curve $y = f(x)$ is approximated by a piecewise-linear path.

More specifically, to approximate the length of $y = f(x)$ over $[a, b]$, we begin by choosing a regular partition of the interval: $a = x_0 < x_1 < x_2 < \cdots < x_n = b$. The line segments that we'll use to approximate the curve are those connecting consecutive points $(x_k, f(x_k))$. If the length of our subintervals is Δx, and the kth line segment has a slope of m_k (see Figure 1.3), we can use the Pythagorean Theorem to write its length as

$$\sqrt{(\Delta x)^2 + (m_k \Delta x)^2} = \sqrt{(1 + m_k^2)(\Delta x)^2} = \sqrt{1 + m_k^2} \, |\Delta x| \text{ units.}$$

So the collection of line segments has a cumulative length of

$$\underbrace{\sqrt{1 + (m_1)^2} \, |\Delta x|}_{\substack{\text{length of the 1st} \\ \text{line segment}}} + \underbrace{\sqrt{1 + (m_2)^2} \, |\Delta x|}_{\substack{\text{length of the 2nd} \\ \text{line segment}}} + \cdots + \underbrace{\sqrt{1 + (m_n)^2} \, |\Delta x|}_{\substack{\text{length of the last} \\ \text{line segment}}}.$$

> Recall that *regular* means that all subintervals are the same size. This is not technically necessary, but makes writing down the idea easier.

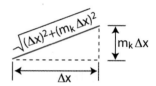

Figure 1.3: We use the Pythagorean Theorem to calculate the length of a line segment.

Now let's write this in terms of the function f. Since the kth linear segment is secant to the graph of f, the Mean Value Theorem tells us that there is a point in the kth subinterval at which the tangent line to the graph of f has a slope of m_k. That is, there is some point x_k^* in the kth subinterval at which $f'(x_k^*) = m_k$. So we can rewrite the cumulative length of our line segments as

$$\sum_{k=1}^{n} \sqrt{1 + \left(f'(x_k^*) \right)^2} \, \Delta x,$$

where we've dropped the absolute value from the Δx since $a \le b$. Of course, more subintervals lead to a better approximation of our curve, and so a better approximation of its length. With this in mind, we define the length of the curve $y = f(x)$ over $[a, b]$ to be the limit,

$$\lim_{n \to \infty} \sum_{k=1}^{n} \sqrt{1 + \left(f'(x_k^*) \right)^2} \, \Delta x.$$

Said formally,

> **Arc length:** Suppose $f'(x)$ is continuous on $[a, b]$. Then the length of the curve $y = f(x)$ over $[a, b]$ is
> $$\int_a^b \sqrt{1 + \left(f'(x) \right)^2} \, dx.$$

Calculating arc length

Example 1.1. Determine the arc length of the curve $y = \frac{1}{2}x^2 - \frac{1}{4}\ln(x)$ between $x = 1$ and $x = 2$.

Solution: This curve is the graph of $f(x) = \frac{1}{2}x^2 - \frac{1}{4}\ln(x)$. Since $f'(x) = x - \frac{1}{4x}$, the arc length integral is

$$\int_1^2 \sqrt{1 + \left(x - \frac{1}{4x}\right)^2}\, dx = \int_1^2 \sqrt{1 + \left(x^2 - \frac{1}{2} + \frac{1}{16x^2}\right)}\, dx$$

$$= \int_1^2 \sqrt{x^2 + \frac{1}{2} + \frac{1}{16x^2}}\, dx = \int_1^2 \sqrt{\left(x + \frac{1}{4x}\right)^2}\, dx = \int_1^2 \left|x + \frac{1}{4x}\right|\, dx.$$

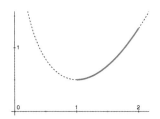

Figure 1.4: The curve $y = \frac{1}{2}x^2 - \frac{1}{4}\ln(x)$ between $x = 1$ and $x = 2$.

Since x is positive in this interval of integration, so is $x + \frac{1}{4x}$, meaning that we can omit the absolute value and write the arc length as

$$\int_1^2 x + \frac{1}{4x}\, dx = \frac{1}{2}x^2 + \frac{1}{4}\ln(x)\Big|_1^2 = \left(2 + \frac{1}{4}\ln(2)\right) - \left(\frac{1}{2} + 0\right) = \frac{3}{2} + \frac{1}{4}\ln(2).\ \blacksquare$$

Try It Yourself: Calculating arc length

Example 1.2. Determine the arc length of the curve $y = \ln(x) - \frac{1}{8}x^2$ between $x = 1$ and $x = 7$.

Answer: $\ln(7) + 6$. $\blacksquare$

Full Solution On-line

The arc length of a parabola

Example 1.3. Determine the arc length of the parabola $y = 0.5x^2$ between $x = 1$ and $x = 3$.

Solution: Since $y' = x$, the arc length integral is

$$\text{arc length} = \int_1^3 \sqrt{1 + x^2}\, dx.$$

In Example 4.1 (p. 437) we used the integration technique called trigonometric substitution to determine that

$$\frac{1}{2}x\sqrt{1 + x^2} + \frac{1}{2}\ln\left|\sqrt{1 + x^2} + x\right|$$

is an antiderivative of $\sqrt{1 + x^2}$. Using that fact in conjunction with the Fundamental Theorem of Calculus, we calculate

$$\int_1^3 \sqrt{1 + x^2}\, dx = \frac{1}{2}x\sqrt{1 + x^2} + \frac{1}{2}\ln\left|\sqrt{1 + x^2} + x\right|\Big|_1^3,$$

which is $\frac{3(\sqrt{10} - \sqrt{2})}{2} + \frac{1}{2}\ln\left|\frac{\sqrt{10} + 3}{\sqrt{2} + 1}\right|$. $\blacksquare$

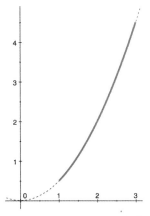

Figure 1.5: The curve $y = \frac{1}{2}x^2$ between $x = 1$ and $x = 3$.

Try It Yourself: Using the tangent in a trigonometric substitution

Example 1.4. Determine the arc length of the parabola $y = 3x^2$ between $x = 2$ and $x = 4$ using the trigonometric substitution $x = \frac{1}{6}\tan(\theta)$.

Answer: $2\sqrt{576} - \sqrt{170} + \frac{1}{2}\ln\left|\frac{\sqrt{576} + 24}{\sqrt{170} + 12}\right|$. $\blacksquare$

Full Solution On-line

❖ Surface Area

If we revolve the graph of a function f about the horizontal axis, we generate a **surface of revolution**, as seen in the top row of Figure 1.6. We can calculate the area of that surface in much the same way as we calculate arc length. Specifically, we begin by approximating the graph of f with line segments, and revolve *them* about the axis (see the bottom row of Figure 1.6). The area of the resulting surface is an approximation of the area we actually want to know, and we pass from approximation to our exact answer in the limit.

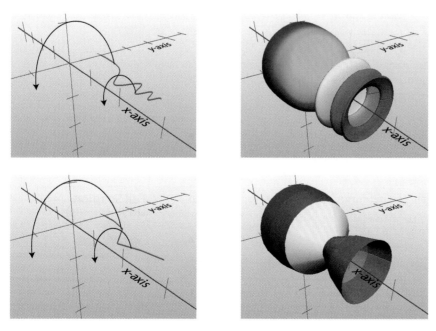

Figure 1.6: (upper left) Imagine that you're looking at the graph of $y = f(x)$ on a friend's paper, from above the 4th quadrant; (upper right) revolving that graph about the x-axis sweeps out a surface; (lower left) imagine that you're looking at the piecewise linear approximation of $y = f(x)$ on a friend's paper, from above the 4th quadrant; (lower right) revolving that graph about the x-axis sweeps out a surface, each segment of which is a frustum.

Each line segment in our approximation of $y = f(x)$ creates a shape called a **frustum** when revolved around the x-axis (see Figure 1.7). Said simply, a frustum is what's left of a cone after you slice off its tip (see Figure 1.7).

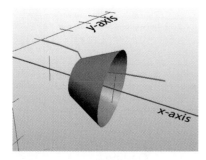 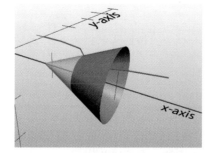

Figure 1.7: (left) As seen from above the 4th quadrant, a non-vertical line segment makes a frustum when it is revolved about the x-axis; (right) a frustum is just part of a cone.

Since these frustums approximate the surface of revolution, we can estimate its area by adding theirs. So let's talk about the area of a frustum. A moment ago we said that a frustum is what's left of a cone after you slice off its tip, so we calculate its area by subtracting the area of the tip from the area of the whole cone. In the exercise set you'll show that the area of a cone is πLR when its base has a radius of R and the length from the tip to its circular edge is L (see Figure 1.8).

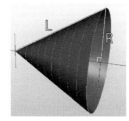

Figure 1.8: The surface area of this open-ended cone is πLR.

In order for the πLR formula to be useful in the present context, we need to restate it in terms of the slope of the line segment that generated the frustum in the first place. Since this line segment is not vertical, it has a finite slope m, and we can extend it until it hits the horizontal axis. Suppose we do that, as shown in Figure 1.9, and the extended line segment has a horizontal extension of H. Since this line segment has a slope of m, its height changes by mH from left to right, and by virtue of the Pythagorean Theorem we know that its length is $\sqrt{1+m^2}\,H$.

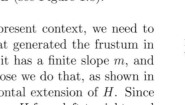

How do the picture and the argument change when $m < 0$?

Figure 1.9: (left) A line segment with slope $m > 0$ that approximates the graph of f over a small interval; (middle) extending the line segment until it hits the x-axis; (right) use the slope, m, and Pythagorean Theorem to finish labeling the diagram.

Revolving the extended line about the x axis makes a cone with $R = mH$ and $L = \sqrt{1+m^2}\,H$, so its area is

$$A_{\text{cone}} = \pi LR = \pi\left(\sqrt{1+m^2}\,H\right)(mH) = \pi mH^2\sqrt{1+m^2}.$$

Similarly, if the cone's tip (i.e., the part not generated by the original line segment, see Figure 1.10) has a horizontal extension of h, the area of the tip is

$$A_{\text{tip}} = \pi mh^2\sqrt{1+m^2}.$$

The difference in these numbers is the area of the frustum that's generated by revolving the original line segment about the x-axis:

$$A_{\text{frustum}} = A_{\text{cone}} - A_{\text{tip}} = \pi m(H^2 - h^2)\sqrt{1+m^2}.$$

By factoring the difference of squares in our formula, it becomes

$$A_{\text{frustum}} = \pi m(H+h)(H-h)\,\sqrt{1+m^2}.$$

The number $(H - h)$, which is the "height" of the frustum, is a length along the x-axis, so let's call it $\triangle x$ and write

$$A_{\text{frustum}} = \pi m(H+h)\triangle x\,\sqrt{1+m^2}\,.$$

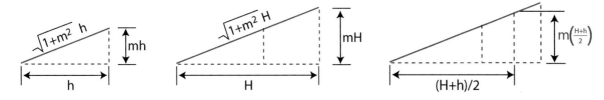

Figure 1.10: (left) A cone made from this line segment would have an area of $\pi mh^2\sqrt{1+m^2}$; (middle) also shown above, repeated here for comparison with the leftmost image; (right) $(H+h)/2$ is the midpoint of $[h, H]$, and $m(h+H)/2$ is the height of the line segment there.

The number $H+h$ doesn't have an obvious geometric interpretation in this context, but $(H+h)/2$ is the midpoint of the interval $[h, H]$, and the height of the line segment there is $m\left(\frac{H+h}{2}\right)$, as shown in Figure 1.10. We can introduce this into our formula by writing

$$A_{\text{frustum}} = 2\pi\, m\left(\frac{H+h}{2}\right)\Delta x \sqrt{1+m^2}\,.$$

Recall that the line segments that generate our frustums come from approximating the curve $y = f(x)$. This fact allows us to make the final step of writing our area formula in terms of f. As we did when developing the arc length integral, we'll denote the slope of the kth line segment by m_k, and use x_k^* to indicate the point in the kth subinterval at which $f'(x_k^*) = m_k$. When the number of subintervals, n, is sufficiently large (so that Δx is sufficiently small), the distance from the x-axis to the midpoint of the kth segment is approximately $|f(x_k^*)|$, so the frustum that's created by revolving the kth line segment about the x-axis has a surface area of

$$A_k \approx 2\pi|f(x_k^*)|\sqrt{1+\left(f'(x_k^*)\right)^2}\,\Delta x. \tag{1.1}$$

> The graphs we've shown have all been above the horizontal axis, so $f(x) > 0$. Of course, that doesn't always happen. The absolute value is used here to account for cases when $f(x) < 0$.

When we add the areas of our frustums together, we have an approximation of the surface area:

$$A \approx \sum_{k=1}^{n} A_k = \sum_{k=1}^{n} 2\pi|f(x_k^*)|\sqrt{1+\left(f'(x_k^*)\right)^2}\,\Delta x,$$

and we pass from approximation to an exact calculation in the limit,

$$A = \lim_{n\to\infty} \sum_{k=1}^{n} A_k = \lim_{n\to\infty} \sum_{k=1}^{n} 2\pi|f(x_k^*)|\sqrt{1+\left(f'(x_k^*)\right)^2}\,\Delta x.$$

> **Surface Area:** Suppose $f'(x)$ is continuous on $[a, b]$, and $\mathcal{S}$ is the surface that is generated by revolving the graph of f over $[a, b]$ about the x-axis. Then the area of $\mathcal{S}$ is
>
> $$\int_a^b 2\pi|f(x)|\sqrt{1+\left(f'(x)\right)^2}\,dx.$$

Before applying this formula in examples, let's pause a moment to note the basic structure of the formula for A_k shown in equation (1.1). Since $|f(x_k^*)|$ is the distance from the axis of revolution (the x-axis) to the edge of the surface (the graph of f), it's a radius. That means $2\pi|f(x_k^*)|$ is the circumference of a circle. You'll recognize the remaining factors as approximating the arc length of a small segment of the graph of f, in essence providing "height" to the circle. That is,

$$A_k = (\text{circumference}) \times (\text{height}) = \text{area}.$$

Using the surface area formula in a simple case

Example 1.5. Suppose $f(x)$ is a constant function whose value is always $r > 0$, and $\mathcal{S}$ is the surface generated by revolving the graph of f about the x-axis, when $0 \le x \le h$. Determine the area of $\mathcal{S}$.

Solution: Since $f'(x) = 0$, the surface area formula tells us that the area is

$$A = \int_0^h 2\pi|f(x)|\sqrt{1+\left(f'(x)\right)^2}\,dx = \int_0^h 2\pi r\,dx = 2\pi r x\Big|_{x=0}^{x=h} = 2\pi rh,$$

which is the standard formula for the surface area of a cylinder.   ∎

Using the surface area formula

Example 1.6. Suppose $f(x) = e^x$, and $\mathcal{S}$ is the surface generated by revolving the graph of f about the x-axis, when $0 \leq x \leq \ln(3)$. Determine the area of $\mathcal{S}$.

Solution: Since $f'(x) = e^x$, the surface area formula tells us that

$$A = \int_0^{\ln(3)} 2\pi e^x \sqrt{1 + (e^x)^2} \; dx.$$

> Because e^x is always positive, we know that $|e^x| = e^x$.

In this case, the substitution $u = e^x$, $du = e^x \, dx$ allows us to rewrite the integral as

$$A = \int_1^3 2\pi \sqrt{1 + u^2} \; du = 2\pi \int_1^3 \sqrt{1 + u^2} \; du.$$

We calculated the value of this integral in Example 1.3 (see p. 491). Multiplying that answer by 2π gives us the area of this surface:

$$2\pi \left(\frac{3(\sqrt{10} - \sqrt{2})}{2} + \frac{1}{2} \ln \left| \frac{\sqrt{10} + 3}{\sqrt{2} + 1} \right| \right) = 3\pi(\sqrt{10} - \sqrt{2}) + \pi \ln \left| \frac{\sqrt{10} + 3}{\sqrt{2} + 1} \right|. \quad \blacksquare$$

Try It Yourself: Using the surface area formula

Example 1.7. Suppose $f(x) = \sqrt{x}$, and $\mathcal{S}$ is the surface that's generated by revolving the graph of f about the x-axis, when $1 \leq x \leq 4$. Determine the area of $\mathcal{S}$.

Answer: $\sqrt{2}(9 - \sqrt{3})\pi$. $\blacksquare$

Full Solution On-line

❖ Revolving About the y-axis

In order to understand the change in the surface area formula that occurs when revolving the graph of a function about the y-axis, let's take a moment to examine the surface area integral in some more detail. The distance from the axis of revolution to the curve—i.e., the radius of the surface at x—is $f(x)$, so $2\pi f(x)$ is the circumference of the circle that's made by revolving the point $(x, f(x))$ about the x-axis. The remaining part of the expression was used earlier in this section to calculate arc length.

$$A = \int_a^b \overbrace{2\pi f(x)}^{\text{circumference}} \underbrace{\sqrt{1 + \Big(f'(x)\Big)^2}}_{\text{this is the arc length element}} dx \quad .$$

So you can think of this integral as adding up (circumference)×(length) = (area). If we revolve about the y-axis instead, the arc length element doesn't change, but the distance to the axis of revolution does. In this case, the point $(x, f(x))$ is $|x|$ units away from the axis of revolution. So we expect the surface area to be calculated with the integral

$$A = \int_a^b \overbrace{2\pi |x|}^{\text{circumference}} \underbrace{\sqrt{1 + \Big(f'(x)\Big)^2}}_{\text{this is the arc length element}} dx \quad .$$

In fact, this intuition is correct.

Surface Area: Suppose $f'(x)$ is continuous on $[a, b]$, where either $0 \leq a$ or $b \leq 0$, and $\mathcal{S}$ is the surface that is generated by revolving the graph of f over $[a, b]$ about the y-axis. Then the area of $\mathcal{S}$ is

$$A = \int_a^b 2\pi |x| \sqrt{1 + \left(f'(x)\right)^2}\, dx.$$

Proof. We will prove that the formula is correct in the case that $0 < a$. The key to proving this formula mathematically is realizing that when a line has a slope of m with respect to the x-axis, its slope with respect to the y-axis is

$$\frac{\Delta x}{\Delta y} = \frac{1}{m}.$$

With this in hand, we can use the same geometric arguments from before to write the area of a frustum about the y-axis as

$$A_{\text{frustum}} = \pi \left(\frac{1}{m}\right) (H + h)(H - h) \sqrt{1 + \left(\frac{1}{m}\right)^2}.$$

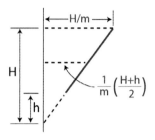

Figure 1.11: Rotating a (non-vertical) line segment about the y-axis makes a frustum.

After factoring the $1/m^2$ out of the radicand, this becomes

$$2\pi \left(\frac{1}{m}\right) \left(\frac{H + h}{2}\right) (H - h) \sqrt{\left(\frac{1}{m}\right)^2 (m^2 + 1)} = 2\pi \left(\frac{1}{m}\right) \left(\frac{H + h}{2}\right) \left(\frac{H - h}{m}\right) \sqrt{m^2 + 1}$$

when $m > 0$. As indicated in Figure 1.11, the number $(H + h)/(2m)$ is the x-coordinate of the line segment's midpoint, and you can see that $(H - h)$ is the segment's vertical extension, which we write as Δy. So

How does the argument change when $m < 0$?

$$A_{\text{frustum}} = 2\pi\, x \left(\frac{\Delta y}{m}\right) \sqrt{m^2 + 1} = 2\pi\, x\, \Delta x\, \sqrt{m^2 + 1},$$

where we've used the fact that $\Delta y/m = \Delta x$. At this point, the rest of the argument is the same: the Mean Value Theorem guarantees that there's a point in the kth subinterval at which the slope of the tangent line is m_k (the slope of the kth line segment). If we call this point x_k^*,

$$A = \lim_{n \to \infty} \sum_{k=1}^n 2\pi\, x_k^* \sqrt{(f'(x_k^*))^2 + 1}\, \Delta x = \int_a^b 2\pi x \sqrt{1 + (f'(x))^2}\, dx. \quad \blacksquare$$

Using the surface area formula

Example 1.8. Suppose $f(x) = x^2$, and $\mathcal{S}$ is the surface generated by revolving the graph of f about the y-axis, when $1 \leq x \leq 2$. Determine the area of $\mathcal{S}$.

Solution: The surface area formula tells us that

$$A = \int_1^2 2\pi |x| \sqrt{1 + (2x)^2}\, dx = \int_1^2 2\pi x \sqrt{1 + 4x^2}\, dx.$$

Because the interval of integration restricts $x \geq 1$, we know that $|x| = x$.

When we make the substitution $u = 1 + 4x^2$, $du = 8x\, dx$, this integral becomes

$$\int_1^2 2\pi x \sqrt{1 + 4x^2}\, dx = \int_5^{17} \frac{\pi}{4} \sqrt{u}\, du = \frac{\pi}{6} u^{3/2} \Big|_5^{17} = \frac{\pi}{6} \left(17^{3/2} - 5^{3/2}\right). \quad \blacksquare$$

Try It Yourself: Using the surface area formula

Example 1.9. Suppose $f(x) = \ln(x)$, and $\mathcal{S}$ is the surface generated by revolving the graph of f about the y-axis, when $4 \leq x \leq 5$. Determine the area of $\mathcal{S}$.

Answer: $\left(5\sqrt{26} - 4\sqrt{17} + \ln\left(\frac{5 + \sqrt{26}}{4 + \sqrt{17}}\right)\right) \pi.$ $\blacksquare$

Full Solution On-line

✥ Combining the Surface Area Formulas

In the discussion above we've indicated that when a surface is generated by revolving the graph of f over $[a, b]$ about a coordinate axis, whether the x-axis or y-axis, the formula we use to calculate its area always has the same structure:

$$A = \int_a^b \overbrace{2\pi r}^{\text{circumference}} \underbrace{ds}_{\text{arc length element}}$$

where $r \geq 0$ is measured from the graph of f to the appropriate axis. When the axis of revolution is vertical, you must be careful not to double-count area. For example, consider the curve $y = x^2$, restricted to $[0, 1]$. When we revolve it about the y-axis, we get a parabolic "bowl" called a paraboloid. If we restrict the same curve to $[-1, 1]$ instead, revolving it about the y-axis gives us the same surface, but inattentive use of the integral formula above will yield *twice* the actual area because the portion of the curve over $[-1, 0]$ sweeps out the same surface as the segment over $[0, 1]$.

> The surface area formula stated on p. 496 includes the hypothesis that "$0 \leq a$ or $b \leq 0$" in order to avoid double-counting area in precisely this fashion.

You should know

- the arc length formula;

- the formulas for calculating the area of a surface of revolution generated by revolving the graph of a function about the x-axis and y-axis;

- how the surface area integrals are understood as adding the areas of cylinders.

You should be able to

- calculate the arc length of the graph of f from $x = a$ to $x = b$;

- calculate the area of a surface of revolution generated by revolving the graph of a function about the x-axis and y-axis.

✥ 7.1 Skill Exercises

1. Suppose $\mathcal{C}$ is the segment of $y = 3x$ in which $0 \leq x \leq 4$. Determine the length of $\mathcal{C}$ using (a) the distance formula, and (b) the integral that we developed in this section.

2. Suppose $\mathcal{C}$ is the segment of $y = 2 - 7x$ in which $1 \leq x \leq 3$. Determine the length of $\mathcal{C}$ using (a) the distance formula, and (b) the integral that we developed in this section.

In #3–12 determine the length of the given curve.

3. $y = 5 + 6x^{3/2}$ over $[1, 2]$

4. $y = \frac{1}{3}(x^2 + 2)^{3/2}$ over $[0, 1]$

5. $y = \frac{1}{96}x^6 + \frac{1}{x^4}$ over $[2, 3]$

6. $y = \frac{2}{3}x^{9/2} + \frac{1}{30x^{5/2}}$ over $[1, 3]$

7. $y = \ln(\sec(x))$ over $[0, \pi/4]$

8. $y = x^{2/3}$ over $[1, 7]$

9. $y = e^{-x}$ over $[0, 5]$

10. $y = \ln(x)$ over $[1, 2]$

11. $y = \frac{1}{2}\ln(\sin(2x))$ over $\left[\frac{\pi}{6}, \frac{\pi}{4}\right]$

12. $y = \frac{1}{2}(\cos(x) - \sec(x))$ over $\left[\frac{\pi}{6}, \frac{\pi}{4}\right]$

A curve $\mathcal{C}$ is defined in each of #13–18. Determine the area of the surface that's generated by revolving $\mathcal{C}$ about the x-axis.

13. $y = 3x$ when $x \in [0, 1]$

14. $y = 6 - 2x$ when $x \in [1, 8]$

15. $y = \sqrt{x}$ when $x \in [0, 5]$

16. $y = x^3$ when $x \in [1, 7]$

17. $y = \frac{1}{2}x^2$ when $x \in [0, 1]$

18. $y = \sin(x)$ when $x \in [0, \pi]$

A curve $\mathcal{C}$ is defined in each of #19–24. Determine the area of the surface that's generated by revolving $\mathcal{C}$ about the y-axis.

19. $y = -5x$ when $x \in [1, 2]$

20. $y = 3 - 7x$ when $x \in [2, 7]$

21. $y = 3x^2$ when $x \in [-1, 2]$

22. $y = 5 + 2x^2$ when $x \in [0, 3]$

23. $y = \ln(x) - 0.125x^2$ when $x \in [1, 4]$

24. $y = 0.5e^x + 0.5e^{-x}$ when $x \in [-1, 0]$

A curve $\mathcal{C}$ is defined in each of #25–30. Approximate the arc length of $\mathcal{C}$ using Simpson's method with the specified number of subintervals. A spreadsheet or programmable calculator might be helpful.

25. $y = x^2 - x^3$, $x \in [1, 2]$, $n = 6$

26. $y = 1 + x + x^8$, $x \in [-1, 1]$, $n = 6$

27. $y = 1/(x^2 - x^3)$, $x \in [2, 7]$, $n = 8$

28. $y = 1/(1 + x + x^8)$, $x \in [-1, 1]$, $n = 8$

29. $y = e^{x/2}$, $x \in [0, 3]$, $n = 10$

30. $y = \ln(x + x^2)$, $x \in [2, 5]$, $n = 10$

A curve $\mathcal{C}$ is defined in each of #31–36. Use the trapezoid rule with n subintervals to approximate the area of the surface that's generated by revolving $\mathcal{C}$ about the x-axis. A spreadsheet or programmable calculator might be helpful.

31. $y = x^6 + 1$, $x \in [1, 2]$, $n = 5$

32. $y = 1 + x + 5x^2$, $x \in [-1, 1]$, $n = 7$

33. $y = 1/(x^3 - 1)$, $x \in [2, 7]$, $n = 9$

34. $y = 1/(1 + x^8)$, $x \in [-1, 1]$, $n = 6$

35. $y = \sqrt[3]{x}$, $x \in [1, 7]$, $n = 10$

36. $y = \sqrt[5]{1 + x^2}$, $x \in [2, 5]$, $n = 4$

A curve $\mathcal{C}$ is defined in each of #37–42. Use a midpoint-sampled Riemann sum with n subintervals to approximate the area of the surface that's generated by revolving $\mathcal{C}$ about the y-axis. A spreadsheet or programmable calculator might be helpful.

37. $y = x^9$, $x \in [1, 2]$, $n = 5$

38. $y = 1 + x^4$, $x \in [-1, 1]$, $n = 7$

39. $y = 1/x$, $x \in [2, 7]$, $n = 9$

40. $y = x^2/(1 + x^8)$, $x \in [-1, 1]$, $n = 6$

41. $y = \arctan(x)$, $x \in [0, 1]$, $n = 10$

42. $y = \arcsin(x)$, $x \in [0, 1]$, $n = 4$

A curve $\mathcal{C}$ is defined in each of #43–48. Use the error formulas from Section 5.3 (p. 359) to determine the number of subintervals that are needed to approximate the length of $\mathcal{C}$ to within 0.001 of its actual value using the specified numerical method: endpoint-sampling (E), midpoint sampling (M), or trapezoid rule (T).

43. $y = 2x - x^2$, $x \in [1, 2]$, E

44. $y = 1 + x^4$, $x \in [-1, 1]$, E

45. $y = 1/x$, $x \in [2, 7]$, T

46. $y = x^2/(1 + x^8)$, $x \in [-1, 1]$, T

47. $y = \arctan(x)$, $x \in [0, 1]$, M

48. $y = \arcsin(x)$, $x \in [0, 1]$, M

❖ 7.1 Concept and Application Exercises

49. The graph of $y = \sqrt{r^2 - x^2}$, $x \in [-r, r]$ is a semicircle. Use the arc length formula on p. 490 to show that its length is πr.

50. The graph of $y = \sqrt{r^2 - x^2}$, $x \in [-r, r]$ is a semicircle, and it sweeps out a sphere when revolved about the x-axis. Use the surface area formula on p. 494 to show that the surface area of that sphere is $4\pi r^2$.

51. Suppose the profile of a gumdrop sitting on the x-axis (and rotationally symmetric about the y-axis) is the curve $y = 1.1 + -x^2$, where $x \in [-1, 1]$ and length is measured in centimeters (see Figure 1.12). Determine the total surface area of the gumdrop.

52. Suppose we thicken the gumdrop from #51 by inserting a column of height 1 and radius r into its middle (see Figure 1.12).

    (a) Determine the total surface are of the gumdrop as a function of r.

    (b) Determine whether the rate of change in surface area with respect to r is greater when r is small or large, and explain why that makes sense in the context of this rotational surface.

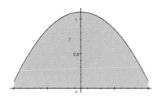 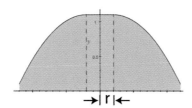

Figure 1.12: (left) The gumdrop from # 51; (right) the gumdrop from #52.

53. Suppose the curve $\mathcal{C}$ is the graph of $f(x) = \frac{1}{30}(2 + \cos(x))$ over $[0, 2\pi]$. Show that the arc length of $\mathcal{C}$ is larger than the area of the surface that's obtained by revolving it about the x-axis. *(Hint: no actual calculation is needed!)*

54. Consider the integrals $\int_a^b f'(t)\,dt$ and $\int_a^b \sqrt{1 + (f'(t))^2}\,dt$, where f' is differentiable and $f''(t)$ is continuous. If we want to approximate these integrals to within 0.001 of their actual value using a left-sampled Riemann sum, which will require fewer subintervals? *(Hint: see Section 5.3 on p. 359.)*

55. Suppose the curve $\mathcal{C}$ is the portion of the graph of $f(x) = \frac{1}{\sqrt{x}} - 1$ that's in the first quadrant.

    (a) Show that the area between $\mathcal{C}$ and the x-axis is finite.

    (b) Write down the integral for calculating the arc length of $\mathcal{C}$.

    (c) Show that the integral in part (b) diverges, and explain why that makes sense in this context.

(d) Write down the integral that calculates the area of the surface that's made by revolving C about the y-axis.

(e) Show that the integral in part (d) converges.

56. Suppose the curve C is the portion of $y = \sin(nx)$ over $[0, \pi]$, where $n \geq 1$ is an integer.

(a) Determine the area of the surface that's generated by revolving C about the x-axis as a function of n.

(b) Determine whether that function of n is increasing or decreasing by using a graphing utility to render a plot of it.

(c) Explain why your answer to part (b) makes sense in the context of this surface of revolution.

57. Suppose $f(x)$ is a member of the function-family $\int g(x) - \frac{1}{g(x)} \, dx$, where g is a positive-valued function and you can integrate both $g(x)$ and $1/g(x)$. Show that the arc length formula simplifies to something that you can integrate exactly.

Checking the details

58. To make a cone, we clip a sector out of a disk and connect the resulting edges, as shown in Figure 1.13.

(a) Suppose the disk has a radius of L, and we remove a sector of it whose arc subtends θ radians. What is the area that remains?

(b) What is the perimeter of the remaining circular arc?

(c) If we were to make a cone by connecting the edges, as indicated in the figure, the circumference of its circular base would be the number you found in part (b). If we denote the radius of that circular base by R, determine a formula for R in terms of L and θ.

(d) Using your formulas from parts (a) and (c), show that the area of the cone is $\pi L R$.

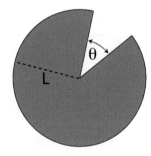

Figure 1.13: Making a cone by removing a sector of a disk (for #58).

7.2 Solids of Revolution and the Method of Slicing

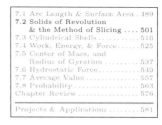
In the previous section, we revolved a curve about an axis to produce a surface. In this section we continue the discussion by considering the volume enclosed by that surface.

❖ Horizontal Axes of Revolution

In the same way that a surface is created by revolving a curve about an axis, a **solid of revolution** is created by revolving an area. For example, suppose $\mathcal{R}$ is the region that extends from $x = 0$ to $x = 8$ whose upper and lower edges are the graph of $f(x) = 8\sqrt{x}e^{-0.5x}$ and the x-axis, respectively. If we revolve $\mathcal{R}$ about the x-axis, as if it were connected to the axis by a hinge, it would sweep out a solid that looks a lot like a Hershey's kiss (see Figure 2.1).

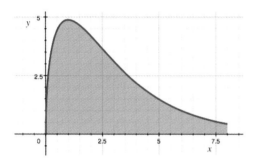
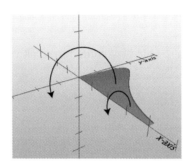
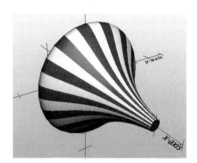

Figure 2.1: (left) $\mathcal{R}$ is the region between the graph of f and the x-axis over $[0, 8]$; (middle) we revolve $\mathcal{R}$ about the x-axis, as if it were connected on a hinge; (right) that sweeps out a volume that (in this case) looks a lot like a Hershey's kiss.

❖ Disks

When we were first talking about area, we began by using an approximation with Riemann sums. Let's do the same here.

Recall that each rectangle associated with our Riemann sum has a width of Δx, and the kth one has a height of $f(x_k^*)$, where x_k^* is some point in the kth subinterval. In Figure 2.2 you see a right-sampled Riemann sum, and one of the rectangles has been revolved around the x-axis, sweeping out a solid disk.

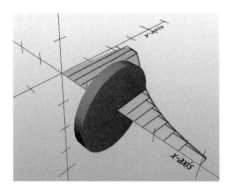
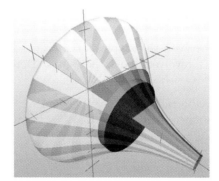

Figure 2.2: (left) Each rectangle sweeps out a disk when its revolved around the x-axis; (right) a single disk shown with the entire solid.

Since the radius of the disk is the height of the rectangle, $f(x_k^*)$, and it has a thickness of Δx, the volume of this solid disk is

$$\underbrace{\pi\Big(f(x_k^*)\Big)^2}_{\text{area}}\ \underbrace{\Delta x}_{\text{thickness}}.$$

By adding the volumes of the disks from each rectangle, we get an estimate of the volume of this solid of revolution:

$$V \approx \sum_{k=1}^{n} \pi\Big(f(x_k^*)\Big)^2\ \Delta x.$$

We pass from approximation to an exact calculation in the limit,

$$V = \lim_{n\to\infty} \sum_{k=1}^{n} \pi\Big(f(x_k^*)\Big)^2\ \Delta x = \int_a^b \pi\Big(f(x)\Big)^2\ dx.$$

Let's begin our set of examples by deriving a pair of volume formulas that might be familiar to you from solid geometry.

Using disks to calculate the volume of a solid of revolution

Example 2.1. The graph of $f(x) = \sqrt{r^2 - x^2}$ is a semi-circle of radius r, which produces a sphere when revolved about the x-axis. If length is measured in centimeters, determine the volume enclosed by that sphere.

Solution: The graph of f extends from $x = -r$ to $x = r$, so the volume of the solid of revolution is

$$V = \int_{-r}^r \pi\underbrace{\Big(\sqrt{r^2-x^2}\Big)^2}_{\text{cm}^2}\ \underbrace{dx}_{\text{cm}} = \int_{-r}^r \underbrace{\pi\Big(r^2-x^2\Big)\ dx}_{\text{cm}^3}.$$

Because the integrand is an even function, we can rewrite this integral as

$$2\int_0^r \pi\Big(r^2-x^2\Big)\ dx = 2\pi\left(r^2 x - \frac{1}{3}x^3\right)\Big|_{x=0}^{x=r} = 2\pi\left(r^3 - \frac{1}{3}r^3\right) = \frac{4}{3}\pi r^3.$$

Notice that r (which is measured in cm) is cubed in our formula. So our expression for volume has units of cm^3, which makes sense.   ∎

Try It Yourself: Using disks to calculate the volume of a solid of revolution

Example 2.2. The triangle of area between the graph of $f(x) = mx$ and the x-axis, $0 \le x \le h$, produces a cone when revolved about the x-axis. Determine the volume of that cone.

Answer: $V = \frac{1}{3}\pi(mh)^2 h$. Note that the radius at the right end of the cone is $f(h) = mh$. If we denote this radius by r, our answer is $V = \frac{1}{3}\pi r^2 h$ (i.e., the volume of a cone is $1/3$ the volume of a cylinder).   ∎

Full Solution
On-line

Using disks to calculate the volume of a solid of revolution

Example 2.3. Supposing that distance is measured in mm, determine the volume of the solid shown in Figure 2.1.

Solution: The bounding curve is the graph of $f(x) = 8\sqrt{x}e^{-0.5x}$, and x ranges from 0 to 8, so the volume of the solid is

$$V = \int_0^8 \underbrace{\pi \left(8\sqrt{x}e^{-0.5x}\right)^2}_{\text{mm}^2} \underbrace{dx}_{\text{mm}} = 64\pi \underbrace{\int_0^8 xe^{-x}\ dx}_{\text{integrate by parts}} = 64\pi(1 - 9e^{-8})\ \text{mm}^3. \quad\blacksquare$$

A horizontal axis of revolution that is not the x-axis

Example 2.4. Suppose that $\mathcal{R}$ is the region bounded between $y = x^2$ and $y = 5$ when $x \in [0, 2]$. If all distances are measured in centimeters, calculate the volume of the solid that's created by revolving $\mathcal{R}$ about the line $y = 5$.

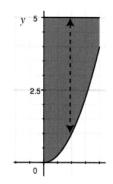

Solution: Unlike previous examples, this axis of revolution is not the x-axis, but the technique for calculating volume is exactly the same. Each vertical line segment that connects $y = x^2$ to the axis of revolution produces a disk when it's revolved, and the radius of that disk is the length of the line segment: (higher)$-$(lower)$= 5 - x^2$, as seen in Figure 2.3. Therefore, the volume of this solid is

Figure 2.3: The region $\mathcal{R}$ in Example 2.4.

$$V = \int_0^2 \underbrace{\pi(5 - x^2)^2}_{\text{cm}^2}\ \underbrace{dx}_{\text{cm}} = \pi \underbrace{\int_0^2 25 - 10x^2 + x^4\ dx}_{\text{cm}^3} = \frac{446\pi}{15}\ \text{cm}^3. \quad\blacksquare$$

In summary:

Disk Method for Calculating Volume: Suppose $\mathcal{R}$ is the region between the graph of f and the line $y = c$ when $x \in [a, b]$. Then the solid that's created by revolving $\mathcal{R}$ about the line $y = c$ has a volume of

$$V = \int_a^b \pi\left(c - f(x)\right)^2\ dx.$$

Note: Due to the squaring in our formula, it doesn't matter whether we use $c - f(x)$ or $f(x) - c$. However, the idea that the length of a vertical line segment is calculated as *(higher)$-$(lower)* will be important in later material.

✥ Washers

When the axis of revolution is not in contact with the region that's being revolved about it, we see a "hole" in our solid. For example, suppose $\mathcal{R}$ is the region that extends from $x = 0$ to $x = 1.5$ whose lower boundary is the line $y = 1$ and whose upper boundary is the graph of a function f, as seen in Figure 2.4. Because the lower boundary is a horizontal line, the solid that results from revolving $\mathcal{R}$ about the x-axis looks as though its middle has been "drilled out."

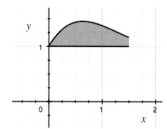 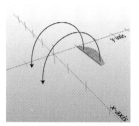 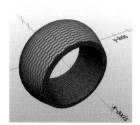

Figure 2.4: (left) The region $\mathcal{R}$ and the line $y = 1$ in the Cartesian plane; (middle) revolving $\mathcal{R}$ about the line $y = 1$; (right) the solid of revolution that's swept out.

We use a regular partition and Riemann sum to estimate the volume of such a solid, just like before, but this time each rectangle sweeps out a "washer" (see Figure 2.5). A washer is just a disk with its center "drilled out" so we can determine its volume by calculating the volume of a disk and then subtracting the volume that has been drilled out.

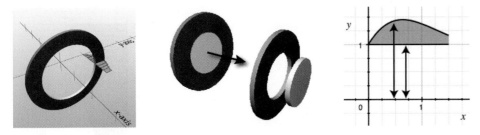

Figure 2.5: (left) Each rectangle sweeps out a washer when revolved about the line $y = 1$; (middle) a washer is just a disk with the middle drilled out; (right) calculating the radii of the larger and smaller disks.

Let's talk about the washer that's swept out by the kth rectangle. Since the kth rectangle has a width of Δx, that will be the thickness of the washer. Thinking of the washer as the difference of two disks, we'll need to know two radii. The radius of the large disk is determined by drawing a line segment perpendicular to the axis of revolution that stretches from the axis to the boundary curve that's *farthest* away (see Figure 2.5). In our case, the length of this segment is $R = f(x_k^*)$, so the volume of the large disk at x_k^* is

$$V_\ell(x_k^*) = \pi R^2 \Delta x = \underbrace{\pi \Big(f(x_k^*) \Big)^2}_{\text{area}} \underbrace{\Delta x}_{\text{thickness}} \ .$$

> We're using subscripts ℓ, s, and w to indicate when V denotes the volume of the <u>l</u>arge disk, the <u>s</u>mall disk, and the <u>w</u>asher.

Similarly, to find the radius of the small disk that we drill out, draw a line segment perpendicular to the axis of revolution that stretches from the axis to the boundary curve that's *closest*. In our current discussion the closer boundary is a line, but it can be the graph of any function in general, $y = g(x)$. The length of this segment is $r = g(x_k^*)$ in the case of our current discussion (where g is the function that's always 1), so the volume of the small disk is

$$V_s(x_k^*) = \pi r^2 \Delta x = \underbrace{\pi \Big(g(x_k^*) \Big)^2}_{\text{area}} \underbrace{\Delta x}_{\text{thickness}} \ .$$

The volume of the washer, V_w, is just the difference of these numbers

$$V_w(x_k^*) = V_\ell(x_k^*) - V_s(x_k^*) = \pi \Big(f(x_k^*) \Big)^2 \Delta x - \pi \Big(g(x_k^*) \Big)^2 \Delta x.$$

We estimate the volume of the solid by adding the volumes of the washers from each rectangle in our Riemann sum,

$$V \approx \sum_{k=1}^{n} V_w(x_k^*) = \sum_{k=1}^{n} \pi \Big(f(x_k^*) \Big)^2 \Delta x - \sum_{k=1}^{n} \pi \Big(g(x_k^*) \Big)^2 \Delta x.$$

We arrive at an exact calculation in the limit,

$$V = \lim_{n \to \infty} \sum_{k=1}^{n} V_w(x_k^*) = \underbrace{\int_0^{1.5} \pi \Big(f(x) \Big)^2 dx}_{\substack{\text{this calculates the volume} \\ \text{as if there were no hole}}} - \underbrace{\int_0^{1.5} \pi \Big(g(x) \Big)^2 dx}_{\substack{\text{this piece amounts to} \\ \text{drilling out the middle}}} .$$

Calculating a volume using washers

Example 2.5. Suppose $\mathcal{R}$ is the region that stretches from $x = 0$ to $x = 1.5$ whose upper boundary is $y = 2 + \sin\left(\frac{\pi x}{2}\right)$ and whose lower boundary is the line $y = 1.25$ (see Figure 2.6). Determine the volume of the solid that's generated by revolving $\mathcal{R}$ about the x-axis.

Solution: If the region were to stretch all the way to the axis, it would have a volume of

$$V_\ell = \int_0^{1.5} \pi\left(2 + \sin\left(\frac{\pi x}{2}\right)\right)^2 dx.$$

However, drilling out the middle reduces the volume by

$$V_s = \int_0^{1.5} \pi\left(1.5\right)^2 dx,$$

so the volume of the solid is

$$V - V_\ell - V_s = \int_0^{1.5} \pi\left(2 + \sin\left(\frac{\pi x}{2}\right)\right)^2 dx - \int_0^{1.5} \pi\left(1.5\right)^2 dx$$

$$= \pi \int_0^{1.5} 1.75 + \sin\left(\frac{\pi x}{2}\right) + \underbrace{\sin^2\left(\frac{\pi x}{2}\right)}_{\text{use the half-angle identity}} dx = \frac{27\pi}{8} + \frac{5 + \sqrt{2}}{2}. \quad \blacksquare$$

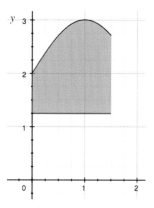

Figure 2.6: The region $\mathcal{R}$ for Example 2.5.

Of course, the same ideas and techniques apply when neither boundary is horizontal, and when the axis of revolution is something other than $y = 0$.

Calculating the volume of a solid of revolution with a hole

Example 2.6. Suppose $\mathcal{R}$ is the region between the curves $y = 5 - x$ and $y = 0.2x^2 - 2x - 1$ that extends from $x = 1$ to $x = 4$, and $\mathcal{S}$ is the solid that's generated by revolving $\mathcal{R}$ about the line $y = 6$. If length is measured in millimeters, determine the volume of $\mathcal{S}$.

Solution: After sketching a graph of the region $\mathcal{R}$, we see that the parabolic piece of the boundary is farthest from the axis of revolution. So the larger radius of the washer at x_k^* is found by stretching a line segment from the axis to the parabola (see Figure 2.7). The length of this segment is

$$R = \underbrace{6}_{\text{high}} - \underbrace{(0.2x^2 - 2x - 1)}_{\text{low}}, = 7 + 2x - 0.2x^2.$$

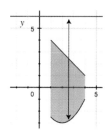

Figure 2.7: Finding the larger radius.

To determine the smaller radius, we stretch a line segment from the axis of revolution to the closer boundary segment (see Figure 2.8). Its length is

$$r = \underbrace{6}_{\text{high}} - \underbrace{(-x + 5)}_{\text{low}} = 1 + x.$$

So the volume of $\mathcal{S}$ is

$$V = \int_1^4 \underbrace{\pi(7 + 2x - 0.2x^2)^2}_{\text{area, in mm}^2} \underbrace{dx}_{\substack{\text{thickness} \\ \text{in mm}}} - \int_1^4 \pi(1 + x)^2 \, dx = \frac{37548\pi}{125} \text{ mm}^3. \quad \blacksquare$$

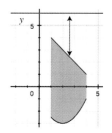

Figure 2.8: Finding the smaller radius.

Try It Yourself: Calculating the volume of a solid of revolution with a hole

Example 2.7. Suppose $\mathcal{R}$ is the region between the curves $y = x - x^2$ and $y = x^2 + x + 3$ that extends from $x = -2$ to $x = 2$, and $\mathcal{S}$ is the solid that's generated by revolving $\mathcal{R}$ about the line $y = -7$. If length is measured in centimeters, determine the volume of $\mathcal{S}$.

Answer: $\frac{1156\pi}{3}$ cm^3.   ∎

Full Solution
On-line

In summary:

> Suppose that $\mathcal{R}$ is the region between the continuous curves $y = f(x)$ and $y = g(x)$ that extends from $x = a$ to $x = b$, and that the line $y = c$ does not intersect $\mathcal{R}$ (except perhaps tangentially, at the boundary). Then the solid that's generated by revolving $\mathcal{R}$ about $y = c$ has a volume of
>
> $$V = \int_a^b \pi \Big(f(x) - c\Big)^2 \, dx - \int_a^b \pi \Big(g(x) - c\Big)^2 \, dx,$$
>
> when $y = g(x)$ is closer to the line $y = c$ than is $y = f(x)$ over $[a, b]$.

✧ Vertical Axes of Revolution

In all of the examples so far, we've revolved a region of the plane about a horizontal line. In that situation, the disks/washers we use to determine volume are stacked horizontally, and each has a thickness of Δx (dx in the limit). By contrast, if we make a solid by revolving a region about a vertical line, the disks/washers that we use will stack vertically, and each will have a thickness of Δy (dy in the limit).

Revolving about a vertical line

Example 2.8. Suppose $\mathcal{R}$ is the region between the parabola $y = x^2$ and the x-axis that extends from $x = 0$ to $x = 2$, and $\mathcal{S}$ is the solid that's generated by revolving $\mathcal{R}$ about the line $x = 2$. If length is measured in centimeters, determine the volume of $\mathcal{S}$.

Solution: We begin by sketching the region and the axis of revolution, as you see in Figure 2.9. Since the axis is the right-hand boundary of $\mathcal{R}$, this solid will not have a hole in it, so we're using disks (not washers) to calculate the volume; and since the axis of revolution is vertical, those disks will stack vertically, as indicated in Figure 2.9.

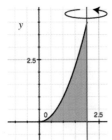
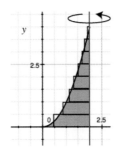
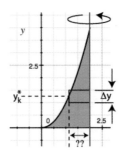

Figure 2.9: (left) The region $\mathcal{R}$ and the axis of revolution; (middle) rectangles are stacked vertically so that they make disks when spun about the axis of revolution; (right) the width of the kth rectangle is the radius of the kth disk.

In order to determine the volume of each disk, we need to know its thickness and its radius. The thickness of each disk will be Δy, as you see in Figure 2.9, and the radius of the kth disk is the width of the kth rectangle (i.e., the horizontal distance from the axis of revolution to the parabola, at the height y_k^*). Since

$$y = x^2 \ \text{ is the same as } \ x = \sqrt{y}$$

on the portion of the parabola that's our boundary, the left edge of the kth rectangle is at $x = \sqrt{y_k^*}$ (see Figure 2.9). Since the axis of revolution is at $x = 2$, the width of the kth rectangle is $2 - \sqrt{y_k^*}$, which means that the volume of the kth disk is

$$V_k = \pi \underbrace{\left(2 - \sqrt{y_k^*}\right)^2}_{\text{area}} \ \underbrace{\Delta y}_{\text{thickness}}.$$

> Just as vertical length is calculated as "high" – "low," horizontal length is calculated as "right" – "left."

We estimate the volume of the entire solid by adding the volumes of the disks,

$$V \approx \sum_{k=1}^{n} V_k = \sum_{k=1}^{n} \pi \left(2 - \sqrt{y_k^*}\right)^2 \Delta y,$$

and pass to an exact calculation in the limit:

$$V = \lim_{n \to \infty} \sum_{k=1}^{n} V_k = \int_0^4 \pi \left(2 - \sqrt{y}\right)^2 \, dy = \frac{8\pi}{3} \ \text{cm}^3.$$

Note that the variable of integration is y (as indicated by the differential), so the bounds of integration are the minimum and maximum heights of the region $\mathcal{R}$. ∎

Revolving about a vertical line

Example 2.9. Suppose $\mathcal{R}$ is the region between the parabola $y = 2 - x^2$ and the line $y = 2 - x$ that extends from $x = 0$ to $x = 1$, and $\mathcal{S}$ is the solid that's generated by revolving $\mathcal{R}$ about the line $x = 1.25$. If length is measured in centimeters, determine the volume of $\mathcal{S}$.

Solution: After drawing the region $\mathcal{R}$ and the vertical line $x = 1.25$, we see that we're using washers to calculate the volume (see Figure 2.10); and since the axis of revolution is vertical, those washers will stack vertically (i.e., in the direction of increasing y). That means each of them will have a height of Δy.

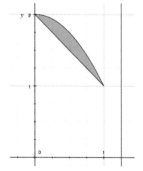
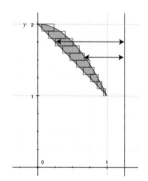
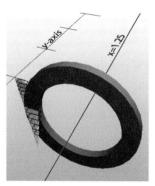

Figure 2.10: (left) The region $\mathcal{R}$ and the axis of revolution for Example 2.9; (middle) stretch an arrow from the axis to the farther boundary curve in order to calculate the "larger" radius, or to the closer boundary curve to calculate the "smaller" radius; (right) each rectangle from sweeps out a washer when $\mathcal{R}$ revolves about $x = 1.25$.

As we did in Examples 2.6 and 2.7, we calculate the volume of the washers in two steps: (1) calculate the volume of a large, solid disk, and (2) subtract away the volume of a smaller disk.

We determine the radius of the large disk by stretching a line segment from the axis of revolution to the boundary curve that's farthest from it—in this case, that's the line $y = 2 - x$ (see the middle image of Figure 2.10). Since

$$y = 2 - x \ \text{ is the same as } \ x = 2 - y,$$

a horizontal line segment at a height of y_k^* intersects the far boundary at $x = 2 - y_k^*$; and since the axis of revolution is at $x = 1.25$, that means that radius of the large disk is $1.25 - (2 - y_k^*) = y_k^* - 0.75$. Therefore, its volume is

$$V_\ell = \underbrace{\pi(y_k^* - 0.75)^2}_{\text{area, in cm}^2} \ \underbrace{\Delta y}_{\substack{\text{thickness} \\ \text{in cm}}} \quad \text{cm}^3.$$

Similarly, we determine the radius of the smaller disk by stretching a line segment from the axis of revolution to the boundary curve that's closest to it—in this case, that's the parabola $y = 2 - x^2$ (see Figure 2.10). Since

$$y = 2 - x^2 \ \text{ is the same as } \ x = \sqrt{2 - y}$$

on the portion of the parabola that's our boundary, a horizontal line segment at a height of y_k^* intersects the parabola at $x = \sqrt{2 - y_k^*}$. Therefore, the radius of the small disk is $1.25 - \sqrt{2 - y_k^*}$, and its volume is

$$V_s = \underbrace{\pi(1.25 - \sqrt{2 - y_k^*})^2}_{\text{area, in cm}^2} \ \underbrace{\Delta y}_{\substack{\text{thickness} \\ \text{in cm}}} \quad \text{cm}^3.$$

The volume of the kth washer is just the difference of these numbers:

$$V_k = \pi(y_k^* - 0.75)^2 \Delta y - \pi(1.25 - \sqrt{2 - y_k^*})^2 \Delta y \ \ \text{cm}^3.$$

Adding the volumes of these washers and applying a limit, we have

$$V = \lim_{n \to \infty} \sum_{k=1}^n V_k = \int_1^2 \underbrace{\pi(y - 0.75)^2}_{\text{area, in cm}^2} \ \underbrace{dy}_{\substack{\text{thickness} \\ \text{in cm}}} - \underbrace{\int_1^2 \pi(1.25 - \sqrt{2 - y})^2 \, dy}_{\text{volume, in cm}^3} = \frac{\pi}{4} \ \ \text{cm}^3.$$

Note that the variable of integration is y (as indicated by the differential), so the bounds of integration are the minimum and maximum heights of the region $\mathcal{R}$. ∎

Try It Yourself: Revolving about a vertical line

Example 2.10. Suppose $\mathcal{R}$ is the region between $y = \sqrt{x}$ and $y = 0.5x$ that extends from $x = 0$ to $x = 4$, and $\mathcal{S}$ is the solid that's generated by revolving $\mathcal{R}$ about the line $x = -2$. If length is measured in microns, determine the volume of $\mathcal{S}$.

Full Solution
On-line

Answer: $\frac{48\pi}{5}$ cubic microns. ∎

❖ Generalizing the Method: Slicing

The basic idea behind all of the calculations that we've made so far has been to approximate the volume of a solid by adding terms of the form $A_k \Delta x$, where A_k is the cross-sectional area of the solid at x_k^* and Δx is a small "thickness." Until now, the cross sections have been either washers or disks, but the basic idea works in more general settings.

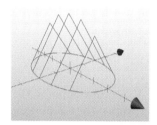

Figure 2.11: The triangles from Example 2.11 as seen from a vantage point above the 4th quadrant.

Calculating volume of a solid that's not rotationally symmetric

Example 2.11. Consider the ellipse $4x^2 + 9y^2 = 36$. At each value of $x \in [-3, 3]$ an isosceles triangle sits perpendicular to the Cartesian plane. The bases of these triangles are parallel to the y-axis and stretch across the ellipse; the height of the triangle at x is $\sqrt{9 - x^2}$ (see Figures 2.11 and 2.12). If all lengths are measured in centimeters, what's the volume enclosed by these triangles?

Solution: In order to use the method of slicing, we must be able to describe the area of a cross section. In this case, the cross section at x is a triangle whose height is $\sqrt{9 - x^2}$ and whose base, which stretches across the ellipse, has a length of $\frac{4}{3}\sqrt{9 - x^2}$. So the area of the cross section at x_k^* is

$$A_k = \frac{1}{2}(\text{base}) \times (\text{height}) = \frac{2}{3}\left(9 - (x_k^*)^2\right) \text{ cm}^2,$$

and the volume enclosed is

$$V \approx \sum_{k=1}^{n} A_k \Delta x = \sum_{k=1}^{n} \frac{2}{3}\left(9 - (x_k^*)^2\right)\Delta x.$$

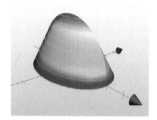

Figure 2.12: The surface created by triangles in Example 2.11, colored by height, as seen from a vantage point above the 4th quadrant.

As usual, we pass from approximation to an exact calculation in the limit:

$$V = \lim_{n \to \infty} \sum_{k=1}^{n} A_k \Delta x = \int_{-3}^{3} \frac{2}{3}(9 - x^2)\, dx = 24 \text{ cm}^3. \qquad \blacksquare$$

Try It Yourself: Calculating volume of a solid that's not rotationally symmetric

Example 2.12. Consider the triangular region in the first quadrant between the x-axis and the line $y = 2x$, when $x \in [0, 5]$. At each value of x a square whose base stretches from $(x, 0)$ to $(x, 2x)$ sits perpendicular to the Cartesian plane. If all lengths are measured in centimeters, what's the volume enclosed by these squares?

Answer: $V = 500/3$ cm^3. $\qquad \blacksquare$

Full Solution On-line

❖ Calculating Mass

In Example 2.4 we calculated the volume of a solid to be $446\pi/15$ cm^3. If it were made of a material with a uniform density of 3 grams per cubic centimeter, its mass would be

$$(\text{density}) \times (\text{volume}) = \left(3\frac{\text{g}}{\text{cm}^3}\right) \times \left(\frac{446\pi}{15} \text{ cm}^3\right) = \frac{446\pi}{5} \text{ g}.$$

Calculating mass is more involved if the object's density varies from point to point. Toward understanding the proper approach to the calculation, think back to when we first discussed the area of a region that's bounded between the graph of a function and the horizontal axis. We first approximated the area by treating the function value as if it were constant over small intervals, and then passed to an exact calculation in the limit. The problem we face with a variable density is fundamentally the same, so we approach it the same way. We begin with an approximation of mass in which we treat the density as if it's constant over small disks:

$$m \approx \sum_{k=1}^{n} \rho_k V_k, \tag{2.1}$$

where V_k is the volume of the kth disk, and ρ_k is the (constant) density of the kth disk. Then we pass to an exact answer in the limit.

Calculating mass

Example 2.13. Suppose $\mathcal{S}$ is the solid described in Example 2.4 (p. 503), and its density depends on x according to $\rho(x) = (4 - x) \frac{\text{g}}{\text{cm}^3}$. Determine the mass of $\mathcal{S}$.

Solution: Incorporating information about our particular region and density function, the approximation in (2.1) becomes

$$m \approx \sum_{k=1}^{n} \underbrace{(4 - x_k^*)}_{\rho_k} \; \underbrace{\pi \left(5 - (x_k^*)^2\right)^2 \Delta x}_{V_k}.$$

We pass from approximation to an exact figure in the limit,

$$m = \lim_{n \to \infty} \sum_{k=1}^{n} \underbrace{(4 - x)}_{\text{g/cm}^3} \underbrace{\pi \left(5 - x^2\right)^2 \Delta x}_{\text{cm}^3} = \int_0^2 \pi (4 - x) \left(5 - x^2\right)^2 dx = \frac{1474\pi}{15} \text{ g.} \quad \blacksquare$$

Try It Yourself: Calculating mass

Example 2.14. Suppose $\mathcal{S}$ is the solid described in Example 2.3 (p. 502), and its density depends on x according to $\rho(x) = 1 + x^2 \frac{\text{g}}{\text{mm}^3}$. Determine the mass of $\mathcal{S}$.

Answer: $\left(448 - 49088e^{-8}\right) \pi$ grams. $\quad \blacksquare$

Full Solution
On-line

❖ Common Difficulties

Students often have difficulty setting up these volume integrals, so here are some pointers for you.

- If the axis of revolution is horizontal, the disks/washers will stack horizontally, so the differential will be dx and the bounds of integration will be the smallest and largest values of x corresponding to the region that's being revolved.

- If the axis of revolution is vertical, the disks/washers will stack vertically, so the differential will be dy and the bounds of integration will be the smallest and largest values of y corresponding to the region that's being revolved.

- The differential tells you the variable of integration, and your integrand has to be expressed entirely in that variable. This is can be tricky when integrating in y, because most familiar curves are expressed in terms of x.

- When calculating with washers, it's tempting to combine the two integrals into one, since they have the same bounds of integration, but many students make errors in the formula when they do this. Keeping the integral separate means you have to write π, $\int_a^b$, and dx (or dy) twice instead of once, but there's less room for making mistakes with the formula.

You should know

- the terms *solid of revolution, disks,* and *washers.*

You should be able to

- use disks or washers to determine the volume of a solid of revolution when the axis of revolution is horizontal (an integral in x);

- use disks or washers to determine the volume of a solid of revolution when the axis of revolution is vertical (an integral in y);

- perform a dimensional analysis of your integral, to make sure that you're calculating volume.

❖ 7.2 Skill Exercises

In each of #1–6 revolving the region between the line and $y = 0$ about the given axis results in a cone. (a) Use the method of disks to calculate the volume of that cone with an integral, and (b) use the formula $V = \frac{1}{3}\pi r^2 h$ from solid geometry to check yourself.

1. $y = 2x$, $0 \le x \le 3$, about $y - 0$
2. $y = 4 - 0.5x$, $0 \le x \le 8$, about $y = 0$
3. $y = 4 - 2x$, $0 \le x \le 2$, about $x = 0$

4. $y = 9 - x/3$, $0 \le x \le 27$, about $x - 0$
5. $y = 7x$, $0 \le x \le 2$, about $x = 2$
6. $y = -14 - 7x$, $-5 \le x \le -2$, about $x = -5$

In #7–18 (a) sketch the indicated region (using a graphing utility if necessary), and (b) determine the volume of the solid obtained by rotating the region about the indicated axis using the disk method.

7. Between $y = 0, y = 6x - x^2$; about $y = 0$.

8. Between $y = (x - 3)^2 + 1$, $y = 0$, $x = 0$, and $x = 4$; about $y = 0$.

9. Between the x-axis and $y = \frac{1}{x-1}$ where $-4 \le x \le 0$; about $y = 0$.

10. Between $y = x + \sin(x)$, $y = 0$ and $x = 2\pi$; about $y = 0$.

11. Between $y = \sqrt{2 - x}$, $y = 0$, and $x = 0$; about $x = 0$.

12. Between $y = 8 - \frac{x}{2}$, $y = 2$, $y = 5$, and $x = 0$; about $x = 0$.

13. Between $y = 1/\sqrt{x}$, $y \ge 1$ and $x = 0$; about $x = 0$.

14. Between $y = \cos(x)$, $x = -\frac{\pi}{2}$ and $x = \frac{\pi}{2}$; about $x = 0$. *(Hint: Use symmetry and integration by parts.)*

15. Between $y = 4x - x^3$, $x = 0$, $x = 2$, and $y = 6$; about $y = 6$.

16. Between $y = \cos\left(\frac{x}{2}\right)$, $y = -5$, $x = -\pi$, and $x - \pi$; about $y = -5$.

17. $y = x^2$, $y = 0$, $y = 4$, and $x = 8$; about $x = 8$.

18. $y = 1/x$, $y = 1$, $y = 7$, and $x = -4$; about $x = -4$.

In #19–30 (a) sketch a plot of the indicated region, then (b) determine the volume of the solid obtained by rotating the region about the indicated axis using the washer method.

19. Between $y = x^2$ and $y = x + 6$; about $y = 0$.

20. Between $y = 4 - x^2$ and $y = 8 - 2x^2$; about $y = 0$.

21. Between $y = 4/\sqrt{x^2 + 1}$ and $y = \sqrt{x}$ for $0 \le x \le 2.3$; about $y = 0$.

22. Between $y = 1/\sqrt[4]{1 - x^2}$ and $y = \frac{x^2}{\sqrt[4]{1-x^2}}$; about $y = 0$.

23. Between $y = \sqrt{x}$, $y = 6 - x$ and $y = 0$; about $x = 0$.

24. For $x \ge 0$, between $y = x^2$, $y = \frac{2}{3}x^2$ and $y = 2$; about $x = 0$.

25. For $x \ge 0$, between $x^2 + y^2 = 1$ and $\frac{1}{4}x^2 + y^2 = 1$; about $x = 0$.

26. Between $y = \frac{x}{x+1}$, $y = \frac{3}{8}(x - 1)$, and $y = 0$; about $x = 0$.

27. Between $y = x^{3/2} - x^{1/2}$, $y = 2 + 2\sqrt{x}$, and $x = 0$; about $y = 6$.

28. Between $y = \frac{x}{x^2+1}$, $y = \sqrt{16 - x/4}$, $x = 0$, and $x = 5$; about $y = 6$.

29. Enclosed by $x^2 + y^2 = 4$; about $x = 4$.

30. Enclosed by $y = x$, $y = 1 - x$, and $y = 4 - 2x$; about $x = 4$.

Mixed Practice For #31–36 (a) sketch the indicated region, then (b) determine the volume of the solid obtained by rotating the region about the indicated axis using either the disk or the washer method.

31. Between $y = \frac{x}{3-x}$, $x = 3$, $y = 0$ and $y = 8$; about $x = 3$.

32. Between $y = \frac{1}{2}x$, $y = \frac{1}{4}x^2$; about $x = 6$.

33. Between $y = \cos(x)$ and $y = \sin(x)$ for $\frac{-3\pi}{4} \le x \le \frac{\pi}{4}$; about $y = 2$.

34. Between $y = \cosh(x)$, $y = 2$; about $y = 2$.

35. Between $y = x\sqrt{x + 2}$, $y = 0$ for $-2 \le x \le 0$; about $y = 0$.

36. Enclosed by $y = \sec(x)$, $y = \tan(x)$, $x = -1$ and $x = \frac{\pi}{4}$; about $y = 2$.

❖ 7.2 Concept and Application Exercises

37. Suppose you wanted to determine the volume of a pyramid with a square base. The base measures 100 m on each side, and the height of the pyramid is 150 m.

   (a) Describe mathematically the volume of a horizontal slice of the pyramid whose altitude is y and whose thickness is Δy.

   (b) Write down an approximation to the volume of the pyramid using sigma notation.

   (c) Write an exact expression for the volume by applying a limit to your answer from part (b).

   (d) Determine the volume of the pyramid.

38. Use the procedure outlined in #37 to determine the volume of a pyramid of height 60 meters with an equilateral triangular base of side-length 100 meters.

39. Consider the circle $x^2 + y^2 = 4$. At each value of $x \in [-2, 2]$ a rectangle sits perpendicular to the Cartesian plane. The bases of these rectangles stretch across the circle, and the height of the rectangle at x is $x^2 + 1$ (see Figure 2.13). If all lengths are measured in centimeters, what's the volume enclosed by these rectangles?

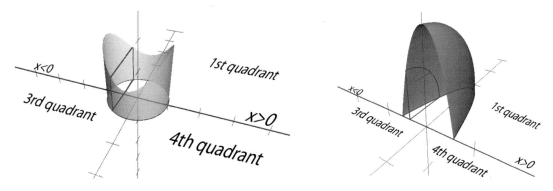

Figure 2.13: (both) Seen from a vantage point over the fourth quadrant; (left) for #39; (right) for #40.

40. Consider the region in the Cartesian plane whose lower boundary is $y = 0$ and whose upper boundary is the parabola $y = 9 - x^2$. At each $x \in [-3, 3]$ a quarter-circle sits perpendicular to the Cartesian plane. Its center is at $(x, 0)$ and its radius, which is perpendicular to $y = 0$, stretches to the parabolic upper boundary (see Figure 2.13). If all lengths are measured in millimeters, what's the volume enclosed between these circles and the Cartesian plane?

41. Consider the half-disk of points in quadrants 1 and 2 for which $x^2 + y^2 \leq 1$. At each $x \in [-1, 1]$ a rectangle sits perpendicular to the Cartesian plane. Its height is $x^{2/3}$ and its base reaches from $y = 0$ to the circular boundary of the half-disk. If all lengths are measured in centimeters, use Simpson's method with 10 subintervals to approximate the volume that's enclosed by these rectangles (see Figure 2.14).

42. Consider the rectangle of points (x, y) for which $x \in [0, 1]$ and $y \in [-2, 2]$. At each $x \in [0, 1]$ two parabolic arcs sit perpendicular to the Cartesian plane. The higher one has a height of $0.25(x + 4)(4 - y^2)$ at the point (x, y), and the lower one has a height of $0.2(x + 4)(4 - y^2)$, as seen in Figure 2.14. If all lengths are measured in millimeters, determine the volume enclosed between the parabolic surfaces.

43. A solid of revolution is formed by revolving the first-quadrant region under $y = 5x - x^2$ around the x-axis, where x and y are measured in meters. If the solid has a density of $\rho = x + 1$ kg/m^3, (a) write a Riemann sum with n subintervals that approximates the mass of the object, (b) limit $n \to \infty$ and calculate the mass exactly.

Engineering

44. A solid of revolution is formed by revolving the first-quadrant region under $y = e^{-x}$ for $0 \leq x \leq 1$ around the x-axis, where x and y are measured in meters. If the solid has a density of $\rho = e^{-x/2}$ kg/m^3, (a) write a Riemann sum with n subintervals that approximates the mass of the object, (b) limit $n \to \infty$ and calculate the mass exactly.

Engineering

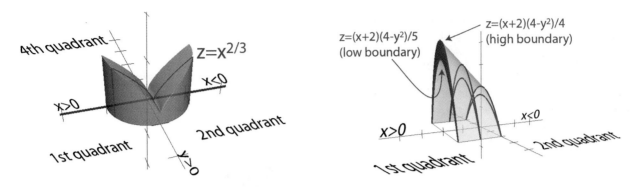

Figure 2.14: (both) Seen from a vantage point above the first quadrant; (left) for #41; (right) for #42.

45. Archimedes discovered that an object, either partially or wholly submerged in water, experiences an upward push (called a buoyant force) equal to the weight of the displaced water (water weighs $62.42796 \ \frac{lbs}{ft^3}$).

   (a) Suppose a ball of radius r feet is dropped into water. Determine the buoyant force on the sphere when the lowest point of the sphere is h feet below the surface. (The ball is not entirely submerged until $h \geq 2r$.)

   (b) Suppose the ball has a uniform density is $20 \ \frac{lbs}{ft^3}$. How deep will it sink into the water? (Your answer will be in terms of r.)

46. Suppose a spherical buoy of radius 3 meters sinks into the water until 1/4 of its volume is submerged. How deep under water is the buoy's lowest point?

47. Suppose the profile of a buoy is the area between the parabola $y = x^2$ and the horizontal line $y = 1$, when $x \in [-1, 1]$ and all lengths are measured in meters. The buoy is rotationally symmetric about the y-axis, and floats vertex-down in the water due to its mass distribution. If the total mass of the buoy is 800 kg, use Archimedes' principle (see #45) to determine how far it sinks when placed in the water.

48. In preparation for a new show, the famous escape artist Rick "The Stick" wants to add a grand finale. His last trick (of the show) will be a race against the clock in which he's chained atop a pillar that's inside a glass tank, and the tank will be filling with water. In order to make the show exciting, he wants to be completely under water for exactly 1 minute before escaping from his chains. As one of Rick's consultants, you should use the following facts to determine (a) the height of the pillar atop which he will stand (to the nearest centimeter), and (b) given your answer to part (a), the rate at which the water is rising at the moment Rick finally escapes his bonds.

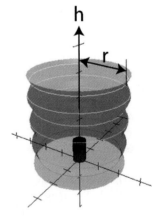

Figure 2.15: Rick's tank in #48, with the cement pillar inside.

   • It will take Rick exactly 3 minutes to escape his bonds.

   • The cement pillar on which he stands will have a radius of 1/3 meter.

   • Rick is exactly 2 meters tall, but because he is *so* slender you may assume his volume is zero.

   • Water will enter the vat at exactly 4 cubic meters per minute.

   • The vat is radially symmetric about its vertical axis, and its profile is described by $r = 2 + 0.1\cos(6h)$, where both r and h are measured in meters, and $h = 0$ at the floor (see Figure 2.15).

You want rain to bead up on your windshield, but companies that make contact lenses want to avoid your tears beading up on the lens. Rather, they want your tears to *wet* the lens. Whether a liquid beads or wets depends on surface chemistry at the solid-liquid interface. When the solid surface is homogeneous, a droplet will form a *spherical cap,* two versions of which are shown in Figure 2.16. The angle θ identified in Figure 2.16 is called the *contact angle,* which is the subject of exercises #49 and #50.

49. Suppose a droplet of 1 microliter sits on a solid surface, and the contact angle is $\pi/3$. What is the area of the solid-liquid interface?

50. Suppose a droplet of 1 microliter sits on a solid surface, and the contact angle is $2\pi/3$. What is the area of the solid-liquid interface?

Figure 2.16: Image for exercise #49–50. We say that the liquid is *beading* when $\theta \geq \pi/2$ and is *wetting* otherwise.

51. A spherical glass marble of radius 3 mm forms with an air bubble (see Figure 2.17). If the gas bubble is symmetric with respect to the x-axis, and can be described by $y = \sqrt{x^3 - 4x}$ where x and y are measured in millimeters, determine the volume of glass used to form the marble (in mm^3).

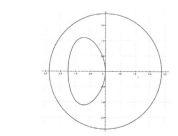

Figure 2.17: (left) The glass marble from #51; (right) looking at the marble from the side.

52. Figure 2.18 shows a surface of revolution known as **Gabriel's horn** (also called **Torricelli's trumpet**) that is formed by revolving the curve $y = 1/x$ about the x-axis for $x \geq 1$.

   (a) If you filled the horn with paint, how much would you need (assuming x and y are measured in meters)?

   (b) If you then decided to paint the outside of the horn with a coat of paint only 1 micron thick, does the horn hold enough paint to get the job done? How much of an excess (or shortage) do you have?

Figure 2.18: Gabriel's horn as seen from a vantage point above the 4th quadrant.

7.3   Volumes by Cylindrical Shells

Suppose $\mathcal{R}$ is the region that extends from $x = 0$ to $x = 4$, whose lower boundary is the x-axis, and whose upper boundary is the graph of $f(x) = 5 - 0.25x^2 + \cos(2\pi x)$, as shown in Figure 3.1. When we revolve $\mathcal{R}$ about the y-axis, we create the solid

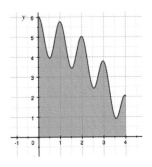
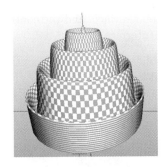

Figure 3.1: (left) The region $\mathcal{R}$, (right) the solid produced by revolving $\mathcal{R}$ about the y-axis.

that's shown in the right-hand image of Figure 3.1. Theoretically speaking, the method of disks and/or washers could be used to calculate this volume, but we would need to use two concentric disks/washers when $y \approx 3$, and three of them when $y \approx 4.5$, and we'd have to keep track of when to switch from one to two to three (see Figure 3.2). Worse yet, instead of stretching from one boundary curve to another, some of these washers stretch from a curve to itself, and that makes it extremely difficult to determine their width. In brief, the method of washers is simply impractical. So instead, we use something called *the method of cylindrical shells* to calculate the volume, which is the subject of this section.

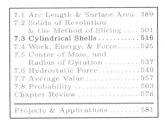

Figure 3.2: The method of washers can be impractical.

❖ Vertical Axes of Revolution

When we were first talking about area, we began by using an approximation with Riemann sums. Let's do the same here. Recall that each rectangle associated with a midpoint-sampled Riemann sum has a width of Δx, and the kth one has a height of $f(x_k^*)$, where x_k^* is the midpoint of kth subinterval. In Figure 3.3 you see a midpoint-sampled Riemann sum, and one of the rectangles has been revolved around the y-axis, sweeping out a *cylindrical shell.*

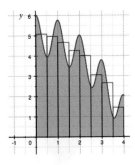
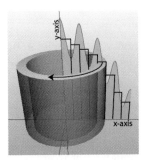

Figure 3.3: (left) A midpoint-sampled Riemann sum approximates the area of $\mathcal{R}$, (right) revolving a rectangle about the y-axis produces a cylindrical shell.

In the same sense that a washer is just a disk with the middle "drilled out," a cylindrical shell is just a cylinder with the middle drilled out, so our calculation of its volume is done in the same way: (1) find the volume of a solid cylinder, and (2) subtract away the volume that is drilled out. A cylinder of height $f(x_k^*)$ that extends from the y-axis to x_k has a volume of

$$(\text{area of the base}) \times (\text{height}) = \pi(x_k)^2 f(x_k^*).$$

Similarly, a cylinder of the same height that extends from the y-axis to x_{k-1} has a volume of

$$(\text{area of the base}) \times (\text{height}) = \pi(x_{k-1})^2 f(x_k^*).$$

The difference in these numbers is the volume of the kth cylindrical shell,

$$V_k = \pi(x_k)^2 f(x_k^*) - \pi(x_{k-1})^2 f(x_k^*) = \pi\Big((x_k)^2 - (x_{k-1})^2\Big) f(x_k^*).$$

By factoring the difference of squares in this expression, we can rewrite it as

$$V_k = \pi(x_k + x_{k-1})(x_k - x_{k-1})f(x_k^*) = \pi(x_k + x_{k-1})\Delta x f(x_k^*),$$

and since $x_k^* = \frac{1}{2}(x_k + x_{k-1})$, we can write this as

$$V_k = 2\pi\left(\frac{x_k + x_{k-1}}{2}\right)f(x_k^*)\Delta x = 2\pi x_k^* f(x_k^*)\Delta x.$$

Note the basic structure of our formula for V_k: $2\pi x_k^*$ is the circumference of a circle whose radius is x_k^*, and $f(x_k^*)$ is a height on the graph of f. Their product is the surface area of a cylinder. Multiplying this by Δx gives it width, and so volume. For the sake of discussion, suppose that length is measured in millimeters. Then our formula gives us

$$V_k = \underbrace{\overbrace{2\pi x_k^*}^{\substack{\text{circum-}\\\text{ference}\\\text{in mm}}}\ \overbrace{f(x_k^*)}^{\substack{\text{height}\\\text{in mm}}}}_{\text{area in mm}^2}\ \underbrace{\Delta x}_{\substack{\text{width}\\\text{in mm}}}\ \text{mm}^3.$$

We approximate the volume of the solid of revolution by adding the volumes of the cylindrical shells,

$$V \approx \sum_{k=1}^{n} V_k = \sum_{k=1}^{n} 2\pi x_k^* f(x_k^*)\Delta x,$$

and pass from approximation to an exact figure in the limit,

$$V = \lim_{n\to\infty}\sum_{k=1}^{n} V_k = \int_0^4 2\pi x f(x)\, dx.$$

Using cylindrical shells to calculate volumes

Example 3.1. Suppose length is measured in millimeters. Determine the volume of the solid shown in Figure 3.1 (on p. 516).

Solution: Our formula for volume tells us that

$$V = \int_0^4 2\pi x f(x)\, dx = \int_0^4 2\pi x\Big(5 - 0.25x^2 + \cos(2\pi x)\Big)\, dx$$

$$= \int_0^4 10\pi x - \frac{\pi}{2}x^3\, dx + \underbrace{\int_0^4 2\pi x \cos(2\pi x)\, dx}_{\text{integrate by parts}} = 48\pi \ \text{mm}^3. \qquad \blacksquare$$

Don't just read it. Take this opportunity to practice the integration by parts technique.

The basic ideas of our derivation are the same even when the lower boundary is not the x-axis.

Using cylindrical shells when the lower boundary is not the x-axis

Example 3.2. Suppose $\mathcal{R}$ is the region that extends from $x = 0$ to $x = \pi$ that is bounded above by the graph of $f(x) = 2x + 1$ and below by the graph of $g(x) = \sin(x)$. If length is measured in centimeters, determine the volume of the solid that's generated by revolving $\mathcal{R}$ about the y-axis.

Solution: When we draw the region $\mathcal{R}$, shown in Figure 3.4, we see that horizontal rectangles are impractical at small values of y, so we use vertical rectangles instead.

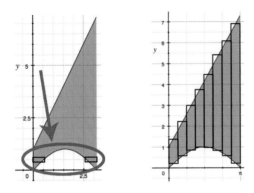

Figure 3.4: (left) Horizontal rectangles are impractical for the region $\mathcal{R}$ from Example 3.2; (right) the region $\mathcal{R}$ from Example 3.2, overlaid with rectangles from a Riemann sum.

The height of such a rectangle is the distance between the upper and lower boundaries (see the right-hand image of Figure 3.4):

$$\text{height of } k^{\text{th}} \text{ rectangle} = f(x_k^*) - g(x_k^*) = (2x_k^* + 1) - \sin(x_k^*).$$

So the volume of the kth cylindrical shell is

$$V_k = \underbrace{\overbrace{2\pi x_k^*}^{\text{circumference}} \overbrace{\Big(f(x_k^*) - g(x_k^*)\Big)}^{\text{height}}}_{\text{area}} \underbrace{\Delta x}_{\text{width}},$$

and the volume of the solid of revolution is

$$V \approx \sum_{k=1}^{n} V_k = \sum_{k=1}^{n} 2\pi x_k^* \Big(f(x_k^*) - g(x_k^*)\Big)\Delta x.$$

In the limit as $n \to \infty$ we see

$$V = \int_0^\pi 2\pi x \Big(f(x) - g(x)\Big)\, dx = \int_0^\pi 2\pi x \Big(2x + 1 - \sin(x)\Big)\, dx$$

$$= \int_0^\pi 4\pi x^2 + 2\pi x \, dx - \underbrace{\int_0^\pi 2\pi x \sin(x)\, dx}_{\text{integrate by parts}} = \frac{4}{3}\pi^4 + \pi^3 - 2\pi^2 \text{ mm}^3. \qquad \blacksquare$$

Don't just read it. Take this opportunity to practice the integration by parts technique.

In the previous examples, the radius of the cylindrical shell made by the kth rectangle has always been the distance between the rectangle's center and the axis of revolution, $x_k^* - 0 = x_k^*$. When the axis of revolution is not the y-axis (the line $x = 0$), the basic idea is the same but the formula looks a little different.

Using cylindrical shells when the axis of revolution is removed from $\mathcal{R}$

Example 3.3. Suppose $\mathcal{R}$ is the region that extends from $x = -1$ to $x = 1$ that is bounded above by the graph of $f(x) = 2 - x^2$ and below by the graph of $g(x) = x - 1$. If length is measured in centimeters, determine the volume of the solid that's generated by revolving $\mathcal{R}$ about the line $x = 1.5$.

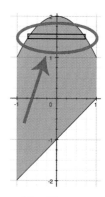

Solution: When we draw the region $\mathcal{R}$, shown in Figure 3.5, we see that horizontal rectangles stretch from a boundary curve to itself when y is large, which indicates that horizontal rectangles are impractical, so we use vertical rectangles instead. One such rectangle is shown in Figure 3.6. Because the distance between the axis of revolution and the dashed line at x_k^* is $1.5 - x_k^*$, and the height of the dashed line is

$$f(x_k^*) - g(x_k^*) = \left(2 - (x_k^*)^2\right) - (x_k^* - 1) = 3 - x_k^* - (x_k^*)^2,$$

Figure 3.5: When rectangles stretch from a boundary curve to itself, deriving a general formula for length is often impractical and sometimes impossible.

revolving it about the axis would produce a cylinder whose surface area is

$$\underbrace{2\pi(1.5 - x_k^*)}_{\text{circumference in cm}} \underbrace{\left(3 - x_k^* - (x_k^*)^2\right)}_{\text{height in cm}} \text{ cm}^2.$$

Multiplying this by Δx gives the cylinder some width, and so volume. The volume of the kth cylindrical shell is

$$V_k = \underbrace{2\pi(1.5 - x_k^*)\left(3 - x_k^* - (x_k^*)^2\right)}_{\text{area in cm}^2} \underbrace{\Delta x}_{\substack{\text{width} \\ \text{in cm}}} \text{ cm}^3,$$

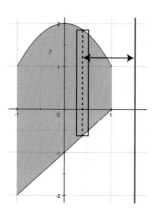

so the volume of the solid is

$$V \approx \sum_{k=1}^{n} V_k = \sum_{k=1}^{n} 2\pi(1.5 - x_k^*)\left(3 - x_k^* - (x_k^*)^2\right)\Delta x.$$

Figure 3.6: The region $\mathcal{R}$ for Example 3.4.

In the limit, we have an exact calculation of the volume,

$$V = \lim_{n \to \infty} \sum_{k=1}^{n} V_k = \int_{-1}^{1} 2\pi(1.5 - x)(3 - x - x^2)\, dx = \frac{52\pi}{3} \text{ cm}^3. \qquad \blacksquare$$

The technique developed in Examples 3.1–3.4 is summarized below.

Suppose $\mathcal{R}$ is the region that extends from $x = a$ to $x = b$, where $a < b$, whose upper and lower boundaries are the graphs of the continuous functions f and g, respectively. If the line $x = c$ does not intersect $\mathcal{R}$ (except perhaps tangentially at the boundary), the solid that's generated by revolving $\mathcal{R}$ about the line $x = c$ has a volume of

$$V = \int_a^b 2\pi|x - c|\left(f(x) - g(x)\right)\, dx.$$

Note: If $x \leq c$ throughout $\mathcal{R}$, we know $|x - c| = x - c$. If $x \geq c$ throughout $\mathcal{R}$, we know $|x - c| = c - x$. So in either case we can drop the absolute value from the integrand.

Try It Yourself: Using cylindrical shells when the axis of revolution is removed from $\mathcal{R}$

Example 3.4. Suppose $\mathcal{R}$ is the region that extends from $x = 1$ to $x = 3$ that is bounded above by the graph of $f(x) = 1 + x^2$ and below by the graph of $g(x) = 2x - 1$. If length is measured in microns, determine the volume of the solid that's generated by revolving $\mathcal{R}$ about the line $x = -2$.

Answer: 40π cubic microns. $\blacksquare$

Full Solution On-line

❖ Horizontal Axes of Revolution

In all of the previous examples, the axis of revolution was vertical, and the rectangles that we used to make cylindrical shells stretched vertically from one boundary curve to another (see Figures 3.3–3.6). Similarly, we see cylindrical shells when both our rectangles and the axis of revolution are horizontal.

Shells with a horizontal axis of revolution

Example 3.5. Suppose $\mathcal{R}$ is the bounded region between the parabola $x = y^2$ and the line $y = x - 2$. Determine the volume of the solid that's generated by revolving $\mathcal{R}$ about the line $y = 2.5$.

Solution: When we draw the region $\mathcal{R}$, as shown in the left-hand image of Figure 3.7, we see that vertical rectangles stretch from a boundary curve to itself when x is small, which indicates that vertical rectangles will be impractical for this calculation, so we use horizontal rectangles instead (see the middle image of Figure 3.7). The kth rectangle, whose midline is at a height of y_k^*, has a width of

$$\begin{pmatrix} x\text{-coordinate on the} \\ \text{right hand boundary} \end{pmatrix} - \begin{pmatrix} x\text{-coordinate on the} \\ \text{left hand boundary} \end{pmatrix} = \underbrace{(y_k^* + 2)}_{\text{rewrite the equation of the line as } x = \cdots} - (y_k^*)^2,$$

and its distance from the axis of revolution is $2.5 - y_k^*$, so revolving it about the axis of revolution would create a cylinder whose surface area is

$$\underbrace{2\pi(2.5 - y_k^*)}_{\text{circumference}} \underbrace{\left(y_k^* + 2 - (y_k^*)^2 \right)}_{\substack{\text{``height'' of} \\ \text{the cylinder}}}.$$

Therefore, the associated cylindrical shell has a volume of

$$2\pi(2.5 - y_k^*)\left(y_k^* + 2 - (y_k^*)^2 \right)\Delta y.$$

We approximate the volume of the solid of revolution by adding the volumes of our shells,

$$V \approx \sum_{k=1}^{n} 2\pi(2.5 - y_k^*)\left(y_k^* + 2 - (y_k^*)^2 \right)\Delta y.$$

Then taking a limit gives us the exact volume,

$$V = \lim_{n \to \infty} \sum_{k=1}^{n} 2\pi(2.5 - y_k^*)(y_k^* + 2 - y_k^*)\Delta y = \int_{-1}^{2} 2\pi(2.5 - y)(y + 2 - y^2)\, dy = 18\pi \text{ units}^3. \quad \blacksquare$$

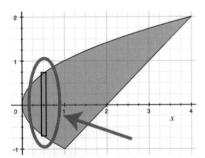
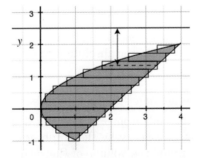

Figure 3.7: (left) When rectangles stretch from a boundary curve to itself, deriving a general formula for length is often impractical and sometimes impossible; (middle) rectangles stretch left to right since the axis of revolution is horizontal; (right) a cylindrical shell generated by one of the rectangles.

Note that the differential is dy in the integral of Example 3.5, because the rectangles are stacked vertically. Consequently, all of the terms in the integrand must be written in terms of y, and the bounds of integration are the least and greatest values of y in $\mathcal{R}$. In general ...

Suppose $\mathcal{R}$ is the region that extends from $y = a$ to $y = b$, where $a < b$, whose left and right boundaries are the continuous curves $x = g(y)$ and $x = f(y)$, respectively. If the line $y = c$ does not intersect $\mathcal{R}$ (except perhaps tangentially at the boundary), the solid that's generated by revolving $\mathcal{R}$ about the line $y = c$ has a volume of

$$V = \int_a^b 2\pi |y - c|\Big(f(y) - g(y)\Big)\ dy.$$

Note: If $y \leq c$ throughout $\mathcal{R}$, we know $|y - c| = y - c$. If $y \geq c$ throughout $\mathcal{R}$, we know $|y - c| = c - y$. So in either case we can drop the absolute value from the integrand.

Try It Yourself: Shells with a horizontal axis of revolution

Example 3.6. Suppose $\mathcal{R}$ is the bounded region between the parabolas $x = y^2 - 2$ and $x = -y^2$. Determine the volume of the solid that's generated by revolving $\mathcal{R}$ about the line $y = -5$.

Answer: $\frac{80\pi}{3}$ units^3.   ■

Full Solution
On-line

✧ Washers or Shells?

To determine whether to use washers or shells when calculating the volume of any particular solid, begin by sketching the region $\mathcal{R}$ and draw line segments that stretch from one boundary curve to another. If those segments are

> As a general rule of thumb, you should avoid stretching line segments from a boundary curve to itself.

- perpendicular to the axis of revolution, you are using washers;

- parallel to the axis of revolution, you are using cylindrical shells.

Deciding whether to try washers or shells

Example 3.7. The three regions below are to be revolved about an axis in order to make a solid. Determine (a) whether a vertical axis of revolution demands washers or shells in order to calculate the volume of that solid, and (b) then do the same for horizontal axes.

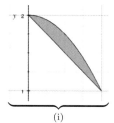
(i)

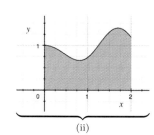
(ii)

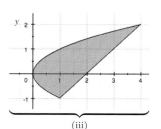
(iii)

Solution: (i) In this case vertical line segments stretch from one boundary curve to the other, and so do horizontal line segments. This means that we will always have a choice of using washers or shells, regardless of whether the axis of revolution is horizontal or vertical, and can make our choice based on personal preference.

(ii) Horizontal line segments sometimes reach from the upper boundary curve to itself (e.g., at heights of $y = 0.75$ and $y = 1.25$), but vertical line segments always reach from one boundary curve to another, so we should be thinking about vertical rectangles (whose width is dx). (a) When the axis of revolution is vertical, these line segments produce cylindrical shells, and (b) when the axis of revolution is horizontal, the same line segments produce washers.

(iii) Vertical line segments sometimes reach from the left-hand boundary curve to itself (e.g., at $x = 0.5$), but horizontal line segments always reach from one boundary curve to another, so we should be thinking about horizontal rectangles (whose width is dy). (a) When the axis of revolution is horizontal, these rectangles produce cylindrical shells, and (b) when the axis is vertical, they produce washers. ∎

Deciding between washers and shells

Example 3.8. Suppose $\mathcal{R}$ is the region bounded above by $y = (x - 1)^2$ and below by $y = -x$ when $x \in [0, 2]$. If all lengths are measured in meters, calculate the volume of the solid generated by revolving $\mathcal{R}$ about (a) the line $y = 3$, and (b) the line $x = -1$.

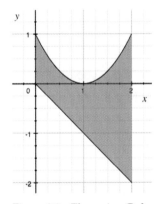

Figure 3.8: The region $\mathcal{R}$ for Example 3.8.

Solution: The horizontal line $y = 0.5$ intersects the parabola twice, which indicates that horizontal rectangles will be impractical. However, vertical lines always stretch from one boundary curve to the other, so we'll be using vertical rectangles in our calculations.

(a) Since the axis of revolution is horizontal, our vertical rectangles will sweep out washers when they're revolved about it. Reaching from the line $y = 3$ to the farther boundary tells us the outer radius of our washer: $3 - (-x) = 3 + x$. Similarly, reaching from the axis of revolution to the closer boundary curve tells us the inner radius of our washer: $3 - (1 - x)^2$. So the volume of the solid is

$$V = \int_0^2 \pi(3 + x)^2 \, dx - \int_0^2 \pi\left(3 - (1 - x)^2\right)^2 \, dx = 57.3864 \text{ m}^3.$$

(b) Since the axis of revolution is vertical, our vertical rectangles will sweep out cylindrical shells when they're revolved about it. The height of the rectangle at x is $(\text{higher}) - (\text{lower}) = (1 - x)^2 - (-x) = 1 - x + x^2$, and its distance to the axis—which is the radius of the cylindrical shell—is simply $(\text{right}) - (\text{left}) = x - (-1) = x + 1$. So the volume produced by revolving $\mathcal{R}$ about the line $x = -1$ is

$$V = \int_0^2 2\pi(x + 1)(1 - x + x^2) \, dx = 12\pi \text{ m}^3. \qquad ∎$$

The important thing for you to see in Example 3.8 is that the orientation of the rectangles depends on $\mathcal{R}$, but whether they produce washers or cylindrical shells depends on their relationship to the axis of revolution.

You should know

- the term *cylindrical shell*;
- the units associated with each part of the volume integral.

You should be able to

- determine when calculating the volume of a given solid of revolution will require washers or shells, based on the orientation of the axis of revolution and the geometry of the region that's being revolved about it;

* use the method of cylindrical shells (when possible) to calculate the volume of a solid of revolution, regardless of whether the axis is vertical or horizontal.

❖ 7.3 Concept and Application Exercises

In #1–12 (a) use a graphing utility to render a plot of the indicated region, and (b) determine the volume of the solid obtained by rotating the region about the indicated axis using the method of cylindrical shells.

1. Between $x = 0$, $y = x^2 - 5$, $y = 1$ for $x \geq 0$; about $x = 0$

2. Between $y = 4x - x^2$ and $y = 16 - x^2$ for $x \geq 0$; about $x = 0$

3. Between $y = \sin(x)$ and $y = 0$ for $0 \leq x \leq \pi$; about $x = 0$

4. Enclosed by $y = x + \sin(x)$, $x = 2\pi$, $y = 2\pi$; about $x = 0$

5. Enclosed by $y = 0$, $y = (x - 1)^3 + 1$, $y = 2 - x$; about $y = 0$

6. Between $y = e^x$ and $y = e^{-x}$ for $0 \leq x \leq 2$; about $y = 0$

7. Between $y = \sqrt{x}$, $y = 6 - x$, and $x = 0$; about $y = 0$

8. Between $4x^2 + y^2 = 1$, $0.25x^2 + y^2 = 1$ for $y \geq 0$; about $y = 0$

9. Enclosed by $y = \frac{1}{x}\arctan(x)$, $y = \arctan(x^2)$, $x = 0$, $x = 1$; about $x = 0$

10. Enclosed by $y = 0$, $y = \sqrt{x}$, $x = 1$; about $x = 2$

11. Enclosed by $y = 0$, $y = \ln(x)$, $x = 2$; about $y = 6$

12. Between $y = \cos(x)$ and $y = \sin(x)$ for $\frac{-3\pi}{4} \leq x \leq \frac{\pi}{4}$; about $x = \frac{\pi}{4}$

Exercises #13–24 refer to the regions in the figure below. For each exercise (a) determine which method(s) could be used to calculate the volume of the solid that's generated by revolving the region about the indicated axis, (b) explain why the methods you rejected won't work, or aren't as efficient, (c) write down the integral for calculating the volume, and (d) calculate the integral from part (c).

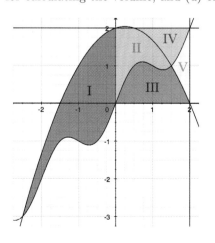

The upper boundary of Region I is the parabola $y = 2 + \frac{1}{3}x - \frac{2}{3}x^2$, and its lower boundary is the curve $y = x + \frac{1}{2}\sin(\pi x)$. The same is true of Region II. The upper boundary of Region IV is the line $y = 2$, and the right-hand boundary of Region V is the line $x = 2$.

13. I about $x = 0$

14. II about $x = 0$

15. III about $x = 0$

16. IV about $x = 0$

17. V about $y = 0$

18. III about $y = 0$

19. III & V about $y = 0$

20. II & IV about $y = 0$

21. I & II about $x = 2$

22. II & IV about $x = 2$

23. IV & V about $y = 3$

24. III & V about $y = -1$

Mixed Practice In #25-30 (a) sketch the indicated region and the specified axis of rotation, and (b) determine the volume of the solid obtained by rotating the region about the indicated axis using either the disk, washer or cylindrical shells method as appropriate.

25. Between $y = 0$; $y = \sqrt{x}$, $x = 1$; about $y = 2$

26. Between $x = 0$, $y = x^3$, $y = 8$; about $x = 0$

27. Between $y = \sin(x)$ and $y = 2$ when $0 \leq x \leq \pi$; about $x = 0$

28. Between $y = \dfrac{e^x}{1 + e^x}$, $y = x^2 - 2x + 2$, $x = 0$ and $x = 3$; about $y = 0$

29. Between $y = 0$, $y = \sqrt{x}$, and $y = 6 - x$; about $x = 8$

30. Bounded by $y = 40$, $y = (x - 2)^2(x + 3)^2$ for $-1 \leq x \leq 3$; about the line $y = 40$

In #31–36 use Simpson's method with the specified number of subintervals to approximate the volume of the solid that's generated by revolving the specified region about the given axis.

31. Between $y = 0$ and $y = \sin(\pi x^3)$ when $x \in [0, 1]$; about $x = 0$; $n = 6$

32. Between $y = 3$ and $y = \sin(\pi x^3)$ when $x \in [0, 1]$; about $x = 2$; $n = 6$

33. Between $y = 0$ and $y = e^{-x^2} \sin(\pi x)$ when $x \in [0, 1]$; about $x = 4$; $n = 8$

34. Between $y = 0$ and $y = e^{-x^2} \sin(\pi x)$ when $x \in [0, 1]$; about $x = -1$; $n = 8$

35. Between $y = 7 - x$ and $y = x + \sin(\pi x)$ when $x \in [1, 2]$; about $x = 3$; $n = 10$

36. Between $y = 7 - 2x$ and $y = x + \cos(\pi x)$ when $x \in [0, 2]$; about $x = -4$; $n = 10$

In #37–42 a volume is generated by revolving a region about the given axis. Determine the number of subintervals needed to approximate the volume of that solid to within 0.01 of its actual value by using the specified numerical method (R = right-sampled Riemann sum, M = midpoint rule, T = trapezoid rule).

37. Between $y = 0$ and $y = e^{\sin(2x)}$ when $x \in [0, \pi]$; about $x = 8$; R

38. Between $y = 0$ and $y = e^{\cos(2x)}$ when $x \in [0, \pi]$; about $x = 4$; R

39. Between $y = 0$ and $y = x + \sin(x)$ when $x \in [0, \pi]$; about $x = 0$; M

40. Between $y = 0$ and $y = x^2 - \cos(3\pi x)$ when $x \in [0, 0.5]$; about $x = 1$; M

41. Between $y = 4$ and $y = 5x - e^x$ when $x \in [0, 1]$; about $x = 5$; T

42. Between $y = -5$ and $y = \sin(\pi x) - 2x$ when $x \in [-1, 1]$; about $x = 1$; T

7.4　Work, Energy, and Force

In our day-to-day language we often treat the words *energy* and *effort* as synonyms, but *energy* also has a technical meaning that is central to our understanding of the physical world. In this section we develop that technical meaning, and then make energy calculations in various simple situations.

❖ Force and Energy

The discussion below shows you how the idea of *energy* arises naturally when working with Newton's Second Law. At its conclusion, we will arrive at a simple formula that can be used in calculation.

Suppose that you pick up this book, and raise it h meters in T seconds. During that time the net force experienced by the book is

$$F_{\text{net}} = F_\ell + F_g,$$

where F_ℓ is the lifting force that you apply and F_g denotes the force due to gravity, both of which we'll think of as functions of time. So Newton's Second Law says

For a review of Newton's Laws of Motion, see Appendix E.

$$\text{net force} \;=\; (\text{mass of the book}) \times (\text{acceleration})$$
$$F_\ell + F_g \;=\; my''.$$

If we multiply this equation by y' we see

$$\left(F_\ell + F_g\right) y' = m\, y'\, y'',$$

which might look like a mere typographic change, but in fact does something important. Since m is constant and y'' is the derivative of y', the right-hand side of this equation is exactly the derivative of $\frac{1}{2}m(y')^2$ with respect to time. This allows us to use the Fundamental Theorem of Calculus to integrate:

Don't just read it. Check it for yourself using the Chain Rule!

$$\int_0^T \left(F_\ell(t) + F_g(t)\right) y'(t)\, dt \;=\; \int_0^T my'(t)y''(t)\, dt$$

$$\int_0^T F_\ell(t)\, y'(t) + F_g(t)\, y'(t)\, dt \;=\; \frac{1}{2}m\left(y'(t)\right)^2 \Bigg|_0^T$$

$$\int_0^T F_\ell(t)y'(t)\, dt + \int_0^T F_g(t)y'(t)\, dt \;=\; \frac{1}{2}m\left(y'(T)\right)^2 - \frac{1}{2}m\left(y'(0)\right)^2. \quad (4.1)$$

Let's pause a moment to note that the expression on the right-hand side of this equation has units of $\text{kg}\left(\frac{\text{m}}{\text{sec}}\right)^2$. Similarly, the integrals on the left-hand side calculate

$$(\text{force}) \times (\text{velocity}) \times (\text{time}) \quad \text{so their units are} \quad \left(\text{kg}\frac{\text{m}}{\text{sec}^2}\right)\left(\frac{\text{m}}{\text{sec}}\right)(\text{sec}) = \text{kg}\left(\frac{\text{m}}{\text{sec}}\right)^2.$$

Both sides of equation (4.1) are calculating the same *kind* of thing. We call it *energy*, and refer to its units as *joules* (denoted by J). In the larger scheme of things, energy is a common way of describing physical phenomena. In this case, because $\frac{1}{2}m(y')^2$ depends on velocity, we call it **kinetic energy** and denote it by K. This leads us to write the change in kinetic energy that you see on the right-hand side of equation (4.1) as ΔK, so it becomes

The word *kinetic* means "relating to or resulting from motion."

$$\int_0^T F_\ell(t)y'(t)\, dt + \int_0^T F_g(t)y'(t)\, dt = \Delta K. \qquad (4.2)$$

The first integral on the left-hand side of this equation calculates your contribution to the overall change in kinetic energy, and we call it the **work** that you do. The other integral calculates the work done by gravity. When we isolate the part of this equation that has to do with you, we see

$$\int_0^T F_\ell(t)y'(t)\ dt = \Delta K - \int_0^T F_g(t)y'(t)\ dt.$$

When an object moves a short distance, as in the scenario of lifting the book, we approximate the force due to gravity as being constant. So instead of writing F_g, let's write the force due to gravity as $-mg$ (negative because gravity pulls *down*) where $g = 9.8\ \frac{m}{sec^2}$. This makes the integral on the right-hand side of our equation very easy to compute:

> The force due to gravity is *not* constant, but depends on the distance between the book and the planet according to the inverse square law. However, when the change in altitude is small, the change in the force due to gravity is negligible.

$$\begin{aligned} \int_0^T F_\ell(t)y'(t)\ dt &= \Delta K - \int_0^T (-mg)y'(t)\ dt \\ &= \Delta K + mg\ y(t)\Big|_0^T \\ &= \Delta K + mg\Big(y(T) - y(0)\Big). \end{aligned}$$

This is commonly expressed as

$$\int_0^T F_\ell(t)y'(t)\ dt = \Delta K + \Delta U \tag{4.3}$$

where the number

> Equation (4.3) is just a rewritten version of equation (4.2) which is just a rewritten version of (4.1). While we've introduced new vocabulary and notation along the way, these equations all say the same thing.

$$\Delta U = mg\Big(y(T) - y(0)\Big), \quad \text{which is also written as} \quad \Delta U = mg\Delta h,$$

is called the change in **gravitational potential energy** Note that when gravity is the only force present, equation (4.3) says that

> E.g., if you drop the book rather than lifting it, so $F_\ell = 0$.

$$0 = \Delta K + \Delta U = \Delta(K + U).$$

That is, the sum of kinetic and potential energy, called **mechanical energy**, does not change over time. We say that it's *conserved*. Equation (4.3) also yields the simple formula that was promised earlier.

Suppose an object has mass of m kilograms, and you lift it a distance of Δh meters. If there has been no change in the object's kinetic energy, and there are no forces other than yours and the force due to gravity, the work you did in lifting was

$$W = \Delta U = mg\Delta h,$$

where $g = 9.8\ \frac{m}{sec^2}$ is the acceleration per unit mass due to gravity.

❖ Examples of Energy Calculation

Now that we have a basic understanding of energy, let's make some calculations.

Calculation of work

Example 4.1. How much work is required to lift a 61 kg, rectangular fish tank (including the fish) from the floor to the top of a display stand that's 2 m tall?

Solution: Since the altitude of the fish tank is increased by 2 m, the change in its gravitational potential energy is $\Delta U = mg\Delta h = (61)(9.8)(2) = 1195.6$ joules. Equation (4.3) tells us that this is exactly the amount of energy you have to supply in order to get the job done (i.e., the work). ∎

Calculation of work in the case of "variable distance"

Example 4.2. Suppose that 20 m of rope is hanging over the side of a building, and the rope has a constant mass density of $\rho = 1.2$ kg/m. Determine the amount of work required to pull the rope to the top of the building.

Solution: The driving ideas in this example are the same as those in the last, but mass near the top of the rope doesn't travel as far as mass near the end of the rope. To account for that, let's think about the rope as n little segments of line, each having a length of Δy. Then we can determine the total work by adding the work required to lift each segment.

> In Example 4.1 the water at the top of the tank rose 2 meters and was still at the top of the tank, and the water at the bottom of the tank rose 2 meters and was still at the bottom of the tank. All of the water had the same change in altitude. In Example 4.2, different segments of rope travel different distances.

Let's index our segments of rope so that the "first" one is nearest to the top of the building, whose altitude we'll call $y = 0$, and the nth segment is at the bottom of the rope, where $y = -20$. As usual, we'll denote the midpoint of the kth segment by y_k^*.

Because the rope has a constant density, each segment has a mass of $\rho\Delta y$, and the mass in the kth segment has to travel a distance of

$$\text{final altitude} - \text{initial altitude} = 0 - y_k^*,$$

so the change in its potential energy as we lift it to the top of the building is

$$\Delta U_k = mg\Delta h = (\rho\Delta y)(9.8)(-y_k^*).$$

> Recall that y_k^* is a negative number, so $-y_k^* > 0$.

By summing the changes in all the segments of rope, we can approximate the total change in potential energy as

$$\Delta U \approx \sum_{k=1}^{n} \Delta U_k = \sum_{k=1}^{n} -9.8 y_k^* \rho \Delta y.$$

We pass from approximation to an exact figure in the limit,

$$\Delta U = \lim_{n\to\infty} \sum_{k=1}^{n} \Delta U_k = \int_{-20}^{0} -9.8 y \rho \, dy = \int_{0}^{20} 9.8 y \rho \, dy = \left. \frac{9.8\rho}{2} y^2 \right|_0^{20} = 1960\rho.$$

> The bounds of integration are the minimum and maximum values of y.

Since $\rho = 1.2 \ \frac{\text{kg}}{\text{m}}$ we have $\Delta U = 2352$ J, and equation (4.3) tells us that this is exactly the amount of work we have to do. ∎

Try It Yourself: Calculation of work in the case of "variable distance"

Example 4.3. Suppose that a rope hangs over the side of a building, from $y = 0$ to $y = -10$ meters, and it has a mass density of $\rho = 1.4 + 0.2\sin(y)$ kg/m. Determine the amount of work required to pull the rope to the top of the building.

Answer: 68.43067 J. ∎

Full Solution
On-line

Calculation of work in the case of "variable distance"

Example 4.4. Suppose a V-shaped trough is filled with water. The trough is 1 meter high, 1.5 meters wide at the top, and 7 meters long (see Figure 4.1). If the water is allowed to exit through a valve in the bottom of the trough, how much work is done by gravity as it empties the trough?

Solution: In the previous example, different segments of rope traveled different distances. In this example, water at the top of the tank has to be moved farther than water at the bottom of the tank, so our approach will be largely the same. What makes this case different from the example of pulling up the rope is the fact that there's more water (i.e., more mass) at the top than at the bottom.

Let's partition the 1 meter height of the trough into n subintervals of equal length, each of which corresponds to a rectangular "slab" of water whose height is Δy and whose length is 7 meters (see the middle image of Figure 4.1). This allows us to estimate the work that's required to empty the trough by calculating the change in the potential energy of each slab, and then adding those numbers together.

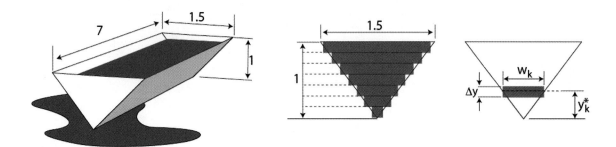

Figure 4.1: (left) The triangular trough, with water spilling out of the bottom; (middle) partitioning the vertical dimension of the trough; (right) similar triangles allow us to write w_k in terms of y_k^*.

Let's denote by y_k^* the midpoint of the kth subinterval, and say that the width of the kth slab is the width of the trough at that height. Then the volume of the kth slab is

$$(\text{length}) \times (\text{width}) \times (\text{height}) = 7w_k\Delta y \text{ m}^3,$$

and since water has a mass density of 1000 $\frac{\text{kg}}{\text{m}^3}$, the mass of of the kth slab is

$$m_k = 1000\underbrace{\,}_{\text{kg/m}^3} \underbrace{(7w_k\Delta y)}_{\text{m}^3} = 7000w_k\Delta y \text{ kg}.$$

On average, all this mass has to travel from an altitude of y_k^* to an altitude of zero, so

$$\Delta h_k = (\text{final altitude}) - (\text{initial altitude}) = 0 - y_k^* = -y_k^*,$$

and its change in potential energy is approximately

$$\Delta U_k = m_k\, g\,\Delta h_k = (7000w_k\Delta y)(9.8)(-y_k^*) = -68600w_k y_k^*\Delta y.$$

We can reduce our formula for ΔU_k to a single variable by using similar triangles to write w_k in terms of y_k^*. Specifically, note that the triangular face of the trough is *similar* to the triangle whose height is y_k^*. Since ratios of corresponding side lengths are equal in similar triangles,

$$\frac{w_k}{y_k^*} = \frac{\text{base}}{\text{altitude}} = \frac{1.5}{1} \quad \Rightarrow \quad w_k = 1.5y_k^*,$$

which allows us to write $\Delta U_k = -68600(1.5y_k^*)y_k^*\Delta y = -102900(y_k^*)^2\Delta y$. Now we approximate the total change in potential energy of the entire mass of water as

$$\Delta U \approx \sum_{k=1}^{n}\Delta U_k = \sum_{k=1}^{n} -102900(y_k^*)^2\Delta y.$$

We get an exact calculation in the limit,

$$\Delta U = \lim_{n\to\infty} \sum_{k=1}^{n} \Delta U_k = \int_0^1 -102900 y^2 \, dy = -34300 \text{ J.}$$

The bounds of integration are the minimum and maximum values of y.

Since the potential energy of the water was reduced by 34300 due to gravity's (downward) pull, gravity did 34300 joules of work. ∎

Calculation of work in the case of "variable distance"

Example 4.5. Suppose an inverted cone (point down) is filled with water. The cone is 0.5 meters high, and has a radius of 0.2 meters at the top. How much work is required to pump all the water to the top of the tank?

Solution: Like the last example, we use a regular partition to separate the vertical distance of 0.5 meters into n subintervals, each of which corresponds to a "disk" of water whose thickness is Δy (see Figure 4.2). Let's say that the altitude of the inverted cone's tip is $y = 0$, and the topmost part of this inverted cone is at $y = 0.5$; and as we did above, let's denote by y_k^* the midpoint of the kth subinterval, so that the diameter of the kth disk is the diameter of the cone at that height. Then the volume of the kth disk is

$$\text{(area of a circle)} \times \text{(thickness)} = \pi \left(\frac{d_k}{2}\right)^2 \Delta y \text{ m}^3,$$

and since water has a mass density of $1000 \frac{\text{kg}}{\text{m}}$, the mass of the kth slab is

$$m_k = \underbrace{1000}_{\text{kg/m}^3} \underbrace{\left(\pi \left(\frac{d_k}{2}\right)^2 \Delta y\right)}_{\text{m}^3} = 250\pi (d_k)^2 \, \Delta y \text{ kg.}$$

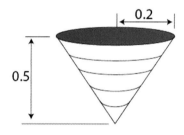 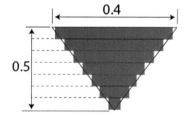 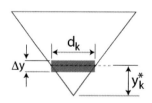

Figure 4.2: (left) The conical tank; (middle) partitioning the vertical dimension of the tank; (right) similar triangles allow us to write d_k in terms of y_k^*.

The mass in the kth disk has an altitude of y_k^*, on average, and it's being lifted to an altitude of $y = 0.5$, so its potential energy changes by approximately

$$\Delta U_k = m_k \, g \, \Delta h = \left(250\pi (d_k)^2 \, \Delta y\right)(9.8)(0.5 - y_k^*).$$

As we did in Example 4.4, we can use similar triangles to rewrite this entirely in terms of y_k^*. Since ratios of corresponding side lengths are equal in similar triangles,

$$\frac{d_k}{y_k^*} = \frac{\text{top diameter}}{\text{altitude}} = \frac{0.4}{0.5} = 0.8 \quad \Rightarrow \quad d_k = 0.8 y_k^*,$$

so the change in the potential energy of the kth slab of water is about

$$\Delta U_k = \left(250\pi (0.8 y_k^*)^2 \Delta y\right)(9.8)(0.5 - y_k^*) = 1568\pi \left(0.5(y_k^*)^2 - (y_k^*)^3\right)\Delta y \text{ J.}$$

We can now approximate the change in potential energy for the entire mass of water by adding the changes in all the slabs:

$$\Delta U \approx \sum_{k=1}^{n} U_k = \sum_{k=1}^{n} 1568\pi \left(0.5(y_k^*)^2 - (y_k^*)^3\right) \Delta y.$$

We pass from approximation to an exact calculation in the limit,

$$\Delta U = \lim_{n \to \infty} \sum_{k=1}^{n} U_k = \int_0^{0.5} 1568\pi \left(0.5y^2 - y^3\right) dy = \frac{49\pi}{6} \text{ J.}$$

> The bounds of integration are the minimum and maximum values of y.

Equation (4.3) tells us that this is exactly the amount of work needed to move the water.　∎

✣ Elasticity and Energy

Be it foam rubber, an arterial wall, or any host of things, elastic materials extend (or compress) when they experience a force, so we often talk about them as though they were springs. The relationship between force applied to a spring, F, and the spring's extension, x, is expressed in its simplest form by Hooke's Law:

$$x = \frac{F}{k},$$

where $k > 0$ quantifies stiffness of the spring and is called the **spring constant**. Springs that are described by Hooke's Law are called **linear springs**.

> In order for this equation to make sense, the number k must carry units of newtons per meter.
>
> Note that for a given force, F, larger values of k result in a smaller extension of the spring. Said differently, larger values of k correspond to stiffer springs.

Alternatively, we can write Hooke's Law as $F = kx$, which is often understood as a description of the amount of force needed to hold the spring steady at an extension of x. Newton's Third Law tells us that objects act *on each other* (rather than one thing acting on another), so if we exert a force of $F = kx$ on the spring, it exerts a force in the opposite direction $F_s = -kx$.

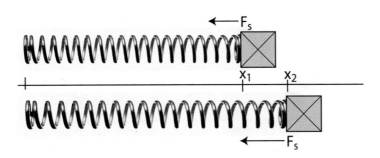

Figure 4.3: The force exerted by a spring is in the direction opposite and proportional to its extension (according to Hooke's Law).

We can reframe our discussion in terms of *energy* by way of Newton's Second Law, as we did before. Suppose a mass of m kg is on the end of a spring, and you pull it from $x = x_1$ to $x = x_2$ over the course of T seconds (see Figure 4.3). The net force on the mass during that motion is

$$F_{\text{net}} = F_p + F_s,$$

where F_p is the force that you apply by pulling and F_s is the force due to the spring, both of which change over time. So Newton's Second Law says

$$\text{(net force)} = \text{(mass)} \times \text{(acceleration)}$$
$$F_p + F_s = mx''.$$

As before, multiplying this equation by x' allows us to integrate the right-hand side:

$$\Big(F_p(t) + F_s(t)\Big)x'(t) = m\,x'(t)\,x''(t)$$

$$\int_0^T F_p(t)x'(t)\,dt + \int_0^T F_s(t)x'(t)\,dt = \int_0^T m\,x'(t)\,x''(t)\,dt. \qquad (4.4)$$

Since m is constant, the right-hand side of this equation is the net change in the kinetic energy of the mass:

$$\int_0^T m\,x'(t)\,x''(t)\,dt = \frac{1}{2}m\Big(x'(t)\Big)^2\Big|_0^T = \frac{1}{2}m\Big(x'(T)\Big)^2 - \frac{1}{2}m\Big(x'(0)\Big)^2 = \Delta K,$$

so our equation says

$$\int_0^T F_p(t)x'(t)\,dt + \int_0^T F_s(t)x'(t)\,dt = \Delta K.$$

The first integral on the left-hand side calculates your contribution to the change in kinetic energy—i.e., the *work* you did. The other calculates the work done by the spring. When we isolate the part of this equation that has to do with you, we see

$$\int_0^T F_p(t)x'(t)\,dt = -\int_0^T F_s(t)x'(t)\,dt + \Delta K. \qquad (4.5)$$

Since $dx = \frac{dx}{d\tau}\,d\tau = x'(\tau)\,d\tau$, we often make a change of variable and rewrite this equation as

$$\int_{x_1}^{x_2} F_p(x)\,dx = -\int_{x_1}^{x_2} F_s(x)\,dx + \Delta K. \qquad (4.6)$$

And because $F_s = -kx$ according to Hooke's Law,

$$-\int_{x_1}^{x_2} F_s(x)\,dx = \int_{x_1}^{x_2} kx\,dx = \frac{k}{2}x^2\Big|_{x_1}^{x_2} = \frac{k}{2}(x_2)^2 - \frac{k}{2}(x_1)^2. \qquad (4.7)$$

We call this number the change in the spring's potential energy, and denote it by ΔU. So equation (4.6) can be summarized as

$$\int_{x_1}^{x_2} F_p(x)\,dx = \Delta U + \Delta K. \qquad (4.8)$$

If the mass is held still at x_1 and at x_2, we know $\Delta K = 0$ so that

$$\int_{x_1}^{x_2} F_p(x)\,dx = \Delta U.$$

The left-hand side of the equation is the amount of energy that you contributed in order to stretch the spring (i.e., the *work* you did). The right-hand side is the change in the spring's potential energy. This equation is important because it allows us to calculate the energy that you need to contribute by simply evaluating ΔU, which depends not on whether or how the force varies, but only on the initial and terminal extensions of the spring.

Elastic potential energy (springs)

Example 4.6. Suppose a linear spring exerts 200 N of force when it's extended 0.01 meters beyond its natural length. How much work is required to extend the spring from 0.01 m to 0.02 m?

Sidebar:

If you are confused by this change of variable, consider the following: when integrating

$$\int_1^2 \sqrt{t^2 + 3}\,(2t)\,dt$$

we use the substitution

$$x = t^2 + 3$$

and

$$dx = x'(t)\,dt$$

to rewrite the integral as

$$\int_4^7 \sqrt{x}\,dx.$$

Note that the bounds of integration have changed from initial and terminal t to initial and terminal x. We're doing the same kind of thing as we pass from equation (4.5) to (4.6), except that the explicit relationship between x and t has not been specified.

You should perform a dimensional analysis to convince yourself that $0.5kx^2$ has units of joules.

Solution: According to equation (4.7), the change in the spring's potential energy is

$$\Delta U = \frac{k}{2}(0.02)^2 - \frac{k}{2}(0.01)^2 = \frac{1.5k}{10000} \text{ J.}$$

We can determine the value of k by using what we know of the force exerted by the spring: when $x = 0.01$ this spring offers a force of 200 newtons, so

$$-200 = -k(0.01) \quad \Rightarrow \quad k = 20000.$$

Therefore, $\Delta U = 3$ joules, which is the amount of energy that we must supply (i.e., the *work* we must do) to stretch the spring. ∎

> Since the spring was pulled to the right, from $x = 0$ to $x = 0.01$, the force exerted by the spring is to the left. That's why we've set $F_s = -200$. If you ever get confused about the signs of the numbers involved, remember that k is supposed to be positive.

❖ Work = Force × Distance

Although Newton's Second Law is written in terms of time, many forces depend on position (e.g., gravitational and electrostatic force) so the ability to determine changes in energy by calculating integrals in terms of position is useful.

> Newton's Second Law is
> $$F = \frac{dp}{dt}$$
> where p =momentum.

You saw such an integral a moment ago. When discussing the change in a spring's potential energy as we stretch it from x_1 to x_2, we saw equation (4.8):

$$\int_{x_1}^{x_2} F_p(x) \, dx = \Delta U + \Delta K$$

where $F_p(x)$ is the force or your pull. Similarly, changing variables from time to position allows us to rewrite equation (4.3) as

$$\int_{y_1}^{y_2} F_\ell(y) \, dy = \Delta U + \Delta K,$$

where y_1 and y_2 are the initial and final altitudes of the book, respectively, and F_ℓ is the force of your lifting.

In both of these integrals, since F is a force and dx is a displacement, we're calculating work by adding (force)×(distance), and the basic idea is true in general. The energy that you contribute to a system is the integral of force over distance:

$$W = \int_{z_1}^{z_2} F(z) \, dz$$

where F is the force you apply (which is often treated as being constant), and the bounds of integration are the initial and terminal positions.

Calculating work by integrating force with respect to position

Example 4.7. Suppose you park your car on a low-grade hill, turn off the engine, and step out to talk to a friend. You realize that you left the car in neutral when it starts to roll down the hill (ack!). You and your friend try to stop the roll by pushing on the car, and though you provide a combined force of 1000 newtons directly up hill, the car continues to roll backwards. How much work do you do as it pushes you back 3 meters (at which time you think to turn the wheel and let the curb stop the car)?

Solution: In this example, you and your friend are providing a constant force but the car rolls from $z_1 = 0$ to $z_2 = -3$, so the amount of work you've done is

$$W = \int_0^{-3} 1000 \, dz = -3000 \text{ J.}$$

Let's take a moment to interpret this negative answer. The central idea is that *the work you do is your contribution to a change in kinetic energy.* Twice in this section we've used Newton's Second Law to derive equations of the form

$$\Delta K = -\Delta U + \int_{x_1}^{x_2} F(x)\, dx,$$

where F is the force that you apply. The car descended the hill in this example, so $\Delta U < 0$. Without your participation, the car's kinetic energy would have been $\Delta K = -\Delta U > 0$. Since you did (negative) work in this example, the car's kinetic energy was 3000 joules less, so it was moving slower. ∎

Note: In general, your contribution to the change in kinetic energy (work) will be negative when an object moves in the direction opposite to the force you apply.

We can calculate changes in potential energy with the same kind of integral (i.e., with respect to position rather than time). For example, in equation (4.7) we wrote the change in the spring's potential energy as

$$\Delta U = -\int_{x_1}^{x_2} F_s(x)\, dx,$$

where F_s is the force offered by the spring. Similarly, when a mass of m kilograms moves from an altitude of y_1 to y_2, its gravitational potential energy changes by

$$\Delta U = mg(y_2 - y_1) = -\int_{y_1}^{y_2} F_g\, dy,$$

where $F_g = -mg$ is the force due to gravity. More generally, the change in potential energy is

$$\Delta U = -\int_{z_1}^{z_2} F_a(z)\, dz$$

where F_a is the force *against which* you're working, and the bounds of integration are the initial and terminal positions.

Escape velocity and potential energy

Example 4.8. Disregarding air resistance, how fast would we need to throw an object in order for it to escape the Earth's gravitational pull?

Solution: We know that the object's mechanical energy is conserved during its trip, since gravity is the only force involved. That is, kinetic energy decreases by the same amount that potential energy increases, so our strategy will be to determine the increase in potential energy for an object that breaks free of the Earth's gravity.

Unlike previous examples, the distances involved in this problem are so large that we cannot treat the force due to gravity as constant. Rather, we must use the fact that gravity obeys an inverse-square law:

$$F_g = -\frac{GMm}{r^2}$$

where M is the mass of the Earth, m is the mass of the object in question, G is the universal gravitational constant, and r is the distance between the center of the Earth and the center of the object.

If we denote the radius of the Earth by R, the change in the object's potential energy as it rises to an altitude of A is

$$\Delta U = -\int_{R}^{R+A} F_g \; dr = -\int_{R}^{R+A} \underbrace{-\frac{GMm}{r^2}}_{\text{force that depends on position, units of N}} \overset{\text{displacement, units of m}}{dr} = \frac{GMm}{R} - \frac{GMm}{R+A}.$$

As the object continues to rise, $A \to \infty$, so the change in potential energy approaches GMm/R. In order for the object's potential energy to increase by GMm/R, its kinetic energy must decrease by the same amount; and in order for its kinetic energy to decrease by GMm/R, the object must have at least that much to begin with. That is,

$$\frac{1}{2}m\Big(v(0)\Big)^2 \ge \frac{GMm}{R} \qquad \Rightarrow \qquad v(0) \ge \sqrt{\frac{2GM}{R}}. \qquad\qquad \blacksquare$$

Try It Yourself: Work against force

Example 4.9. How much work is required to move an object from $x = 3$ to $x = 7$ meters when you must work against a force of $F(x) = -x^2 - \sin(\pi x)$ newtons?

Answer: $\frac{316}{3}$ J. $\qquad\qquad\qquad\qquad\qquad\qquad\qquad\qquad\qquad\qquad\blacksquare$

Full Solution On-line

You should know

- the terms *work, kinetic energy, potential energy, mechanical energy, linear springs, spring constant,* and the relationship among them.

You should be able to

- calculate the work required to lift, pump, or otherwise change the altitude of a mass;

- calculate the work required to move an object through a variable force.

❖ 7.4 Concept and Application Exercises

1. Suppose that 50 m of rope is hanging over the side of a building, and the rope has a constant mass density of $\rho = 0.7$ kg/m. Determine the amount of work required to pull the rope to the top of the building.

2. Suppose that 30 m of rope is hanging over the side of a building, and the rope has a constant mass density of $\rho = 2.4$ kg/m. Determine the amount of work required to pull the rope to the top of the building.

3. Suppose that 12 m of rope is hanging over the side of a building. It extends from $y = 0$ (at the top of the building) to $y = -12$, and has a mass density of $\rho = e^{-0.1y}$ kg/m. Determine the amount of work required to pull the rope to the top of the building.

4. Suppose that 8 m of rope is hanging over the side of a building. It extends from $y = 0$ (at the top of the building) to $y = -8$, and has a mass density of $\rho = \frac{1}{10}y + \frac{3}{2}$ kg/m. Determine the amount of work required to pull the rope to the top of the building.

5. Suppose that 15 m of rope with a mass density of $\rho = 0.4$ kg/m is lying on the ground. Determine the amount of work required to hoist one end to the top of a 10-m tall building.

axis $\omega/2\pi$ times each second. Combining these quantities, we see that the midpoint travels at a speed of

$$v_k = \overbrace{2\pi|x_k^* - c|}^{\text{m per rotation}} \underbrace{\left(\frac{\omega}{2\pi}\right)}_{\text{rotations per second}} = |x_k^* - c|\omega \text{ meters per second.}$$

Furthermore, when n is large the mass in the kth subinterval is $m_k \approx \rho(x_k^*)\Delta x$, where $\rho(x)$ describes density as a function of position. So the mass in the kth subinterval has a kinetic energy of approximately

$$\frac{1}{2}m_k v_k^{\,2} = \frac{1}{2}\rho(x_k^*)\Delta x\Big(\omega(x_k^* - c)\Big)^2.$$

Adding the energy contributions from all the subintervals, we estimate the kinetic energy of the entire object to be

$$K \approx \sum_{k=1}^n \frac{1}{2}\underbrace{\rho(x_k^*)\Delta x}_{\text{kg}}\underbrace{\Big(\omega(x_k^* - c)\Big)^2}_{\text{(m/sec)}^2}\ \text{J.}$$

> We can drop the absolute values because the squaring in this expression forces the number to be non-negative.

Of course, since the order of multiplication is irrelevant, we can shift the factor of $\rho(x_k^*)\Delta x$ to the right-hand side of the equation where we expect it, and then arrive at an exact answer in the limit:

$$K = \int_a^b \frac{1}{2}\Big(\omega(x - c)\Big)^2 \rho(x)\ dx.$$

Calculating kinetic energy of a rotating object

Example 5.4. Suppose an object extends from $x = 1.2$ to $x = 1.4$ meters, and is rotating around the y-axis at 8π radians per second. If the object's density is described by $\rho(x) = 2 + x^2\ \frac{\text{kg}}{\text{m}}$, determine its kinetic energy.

Solution: The number $c = 0$ in this example, and $\omega = 8\pi$. Since $\rho(x)$ is a polynomial, this is an easy calculation:

$$K = \int_{1.2}^{1.4} \frac{1}{2}(8\pi x)^2(2 + x^2)\ dx = \frac{1882976}{46875}\pi^2\ \text{J.} \qquad \blacksquare$$

❖ Gyration of Linear Objects

Let's note that the object in the previous example has a mass of

$$m = \int_{1.2}^{1.4} \underbrace{\rho(x)}_{\text{kg/m}}\ \underbrace{dx}_{\text{m}} = \int_{1.2}^{1.4} 2 + x^2\ dx = \frac{277}{375}\ \text{kg.}$$

If all of it were at a point that lies $r = \frac{19}{5}\sqrt{\frac{163}{1385}}$ meters away from the axis of rotation, and that point orbited the axis at the same angular velocity (8π radians per second), it would have a kinetic energy of

$$\frac{1}{2}mv^2 = \frac{1}{2}\left(\frac{277}{375}\right)\left(8\pi\left(\frac{19}{5}\sqrt{\frac{163}{1385}}\right)\right)^2 = \frac{1882976}{46875}\pi^2\ \text{J—}\textit{exactly the same energy!}$$

We call this special radius the **radius of gyration**, and find it simply by solving the equation

$$K = \frac{1}{2}m(\omega r)^2,$$

once K and m have been determined.

Try It Yourself: Calculating the radius of gyration

Example 5.5. Suppose an object extends from $x = 3$ to $x = 3.25$ meters, and is rotating at 3π radians per second about the axis at $x = 4$. If the object's density is described by $\rho(x) = 1.2x + \cos(4x) \frac{\text{kg}}{\text{m}}$, determine (a) its kinetic energy, and (b) the radius of gyration.

Answer: (a) 40.1345 J, (b) 0.8763345 m. ■

Full Solution On-line

✤ Center of Mass of a Lamina

On p. 346 we defined the word *lamina* to mean "a thin layer or plate" of something. Intuitively speaking, its center of mass is the point $(\overline{x}, \overline{y})$ where you can balance it on the tip of your finger. The basic ideas and techniques for finding this point are exactly the same as we saw with linear objects.

> The center of mass might not be a point on the actual lamina, depending on the geometry. For example, a washer's center of mass it at its center, where there's a hole.

Suppose the upper edge of the lamina is the graph of f, the lower edge is the graph of g, and the lamina extends horizontally from $x = a$ to $x = b$ (see Figure 5.4). Whether because its thickness is not perfectly constant, or because the plate is not made of a uniform material, the density of the lamina might not be constant. At this point in our study of calculus, we will address the cases when density depends on x, but not on both x *and* y.

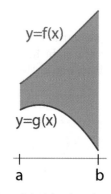

Figure 5.4: A lamina whose upper and lower boundaries are graphs of functions.

▷ Finding $\overline{x}$ when density depends on x

Equation (5.2) on p. 538 tells us that we need to know the moment with respect to the y-axis, and the total mass of the lamina in order to calculate $\overline{x}$. As you see in Figure 5.5, we'll begin by using a regular partition with n subintervals to subdivide $[a, b]$. The rectangle that sits over the kth subinterval has a height of $f(x_k^*) - g(x_k^*)$, a width of Δx, and a roughly uniform density of $\rho(x_k^*)$ when n is large. So the mass of the lamina is

$$m \approx \sum_{k=1}^{n} \underbrace{\rho(x_k^*)}_{\frac{\text{kg}}{\text{m}^2}} \underbrace{\overbrace{\left(f(x_k^*) - g(x_k^*)\right)}^{=\text{height}} \overbrace{\Delta x}^{=\text{width}}}_{\text{area, measured in m}^2}.$$

We pass from approximation to an exact calculation in the limit, and have

$$m = \lim_{n \to \infty} \sum_{k=1}^{n} \rho(x_k^*)\left(f(x_k^*) - g(x_k^*)\right)\Delta x = \int_a^b \rho(x) \overbrace{\left(f(x) - g(x)\right)}^{\text{high}-\text{low}} dx.$$

When x_k^* is the midpoint of the kth subinterval, the average distance between the y-axis and the mass in the kth rectangle is x_k^*, so the lamina's first moment with respect to the y-axis is

$$M_y \approx \sum_{k=1}^{n} \underbrace{x_k^*}_{\text{m}} \underbrace{\rho(x_k^*)\left(f(x_k^*) - g(x_k^*)\right)\Delta x}_{\text{kg}}.$$

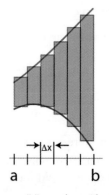

Figure 5.5: A midpoint-sampled Riemann sum approximates the area between the graphs of f and g.

Again, we pass to an exact calculation in the limit:

$$M_y = \lim_{n \to \infty} \sum_{k=1}^{n} x_k^* \, \rho(x_k^*) \Big(f(x_k^*) - g(x_k^*) \Big) \Delta x = \int_a^b x\rho(x) \Big(f(x) - g(x) \Big) \, dx.$$

With these numbers in hand, we can calculate the x-coordinate of the centroid,

$$\overline{x} = \frac{M_y}{m} = \frac{\int_a^b x\rho(x) \Big(f(x) - g(x) \Big) \, dx}{\int_a^b \rho(x) \Big(f(x) - g(x) \Big) \, dx}.$$

Calculating $\overline{x}$ for a lamina

Example 5.6. Suppose $\mathcal{R}$ is the region between $y = 5 - 2x + x^2$ and $y = 1 + \frac{x}{2}$, when $1 \le x \le 2$. If distance is measured in cm, and the density in $\mathcal{R}$ is described by $\rho(x) = 1.5x^3 - 1 \, \frac{g}{cm^2}$, determine $\overline{x}$.

Solution: In situations like this, it's often helpful to sketch a graph of the region in question. Figure 5.6 shows the region for this example, and you can see that the parabola is above the line, so the mass is

$$m = \int_1^2 \underbrace{(1.5x^3 - 1)}_{\frac{g}{cm^2}} \underbrace{\overbrace{\Big((5 - 2x + x^2) - \Big(1 + \frac{x}{2}\Big) \Big)}^{\text{high} - \text{low} = \text{height in cm}} \overbrace{}^{\substack{\text{width} \\ \text{in cm}}}}_{\text{area, in cm}^2} \, dx = \frac{149}{12} \text{ g,}$$

and the first moment with respect to the y-axis is

$$M_y = \int_1^2 \underbrace{x}_{\text{cm}} \underbrace{(1.5x^3 - 1)\overbrace{\Big((5 - 2x + x^2) - \Big(1 + \frac{x}{2}\Big) \Big)}^{\text{high} - \text{low} = \text{height in cm}}}_{\text{(density)} \times \text{(area)} = \text{mass in g}} \, dx = \frac{17743}{840} \text{ g-cm.}$$

So we calculate $\overline{x} = \frac{M_y}{m} = \frac{17743}{10430} \approx 1.7011$ cm.   ■

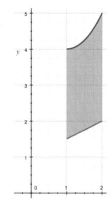

Figure 5.6: The lamina for Example 5.6.

Note: We've consistently kept track of units during this discussion, highlighting what each piece in the integral means. This both acts as a self-checking mechanism and guides the development of the formulas in any given situation.

▷ Finding $\overline{y}$ when density depends on x

The same reasoning that led us to the formula for $\overline{x}$ also leads us to say that $\overline{y} = M_x/m$ where M_x denotes the first moment of the lamina with respect to the x-axis. Since density doesn't change as we move away from the x-axis (ρ depends on only x), we can think about each rectangle in Figure 5.5 (p. 542) as having constant density. In Example 5.1 (p. 539) we saw that the first moment in such a circumstance is

$$\text{(mass of the object)} \times \text{(average distance to the axis)}.$$

The mass of the rectangle over the kth subinterval is

$$m_k = \text{(density)} \times \text{(area)} = \underbrace{\rho(x_k^*)}_{\text{g/cm}^2} \underbrace{\Big(f(x_k^*) - g(x_k^*) \Big) \Delta x}_{\text{cm}^2};$$

and since it stretches from $y = g(x_k^*)$ up to $y = f(x_k^*)$, the average distance between the x-axis and points in the kth rectangle is $(f(x_k^*) + g(x_k^*))/2$. Therefore, the first moment of the kth rectangle with respect to the x-axis is

$$\underbrace{\left(\frac{f(x_k^*) + g(x_k^*)}{2} \right)}_{\text{average distance}} \underbrace{\rho(x_k^*)\Big(f(x_k^*) - g(x_k^*)\Big)\Delta x}_{\text{mass}},$$

and the first moment of the lamina with respect to the x-axis is

$$M_x \approx \sum_{k=1}^{n} \left(\frac{f(x_k^*) + g(x_k^*)}{2} \right) \rho(x_k^*)\Big(f(x_k^*) - g(x_k^*)\Big)\Delta x.$$

As before, we pass from approximation to an exact calculation in the limit,

$$M_x = \lim_{n \to \infty} \sum_{k=1}^{n} \left(\frac{f(x_k^*) + g(x_k^*)}{2} \right) \rho(x_k^*)\Big(f(x_k^*) - g(x_k^*)\Big)\Delta x = \int_a^b \left(\frac{f(x) + g(x)}{2} \right) \rho(x)(f(x) - g(x))\,dx.$$

This formula can be written in a more compact but less intuitive way by multiplying together the factors $(f(x) - g(x))$ and $(f(x) + g(x))$ to arrive at

$$M_x = \int_a^b \frac{\rho(x)}{2}\underbrace{\Big(f(x)^2 - g(x)^2\Big)}_{\text{this still has the basic form of "high"} - \text{"low"}}\,dx.$$

Calculating $\overline{y}$ for a lamina

Example 5.7. Suppose $\mathcal{R}$ is the region in Example 5.6. Determine $\overline{y}$.

Solution: Since the upper boundary is the graph of $f(x) = 5 - 2x + x^2$, the lower boundary is the graph of $g(x) = 1 + \frac{x}{2}$, and the density is described by $\rho(x) = 1.5x^3 - 1$, the lamina's first moment with respect to the x-axis is

$$M_x = \int_1^2 \frac{(1.5x^3 - 1)}{2}\Big((5 - 2x + x^2)^2 - (1 + 0.5x)^2\Big)\,dx = \frac{13345}{336} \text{ g-cm.}$$

We determined the mass of this lamina to be $149/12$ grams in Example 5.6, so the y-coordinate of the centroid is $\overline{y} = \frac{M_x}{m} = \frac{13345}{4172} \approx 3.1987$ cm (see Figure 5.7). ∎

The following box gathers the formulas we've derived so far.

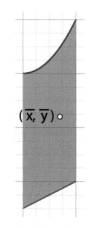

$(\overline{x}, \overline{y})\circ$

Figure 5.7: The region $\mathcal{R}$ from Examples 5.6 and 5.7 with the centroid located.

Center of Mass of a Lamina: Suppose $\mathcal{R}$ is the region in the Cartesian plane that extends from $x = a$ to $x = b$ (where $a < b$) whose upper boundary is the graph of the function f, and whose lower boundary is the graph of the function g. If the density in $\mathcal{R}$ varies according to $\rho(x)$, and the functions f, g and ρ are continuous, the region's center of mass is at the point $(\overline{x}, \overline{y})$, where

$$\overline{x} = \frac{M_y}{m}, \qquad \overline{y} = \frac{M_x}{m},$$

and

$$m = \int_a^b \rho(x)\Big(f(x) - g(x)\Big)\,dx,$$

$$M_y = \int_a^b x\rho(x)\Big(f(x) - g(x)\Big)\,dx,$$

$$M_x = \int_a^b \frac{\rho(x)}{2}\Big(f(x)^2 - g(x)^2\Big)\,dx.$$

Notice that when ρ is constant, it cancels in the fractions M_y/m and M_x/m. So a lamina's center of mass depends solely on its geometry when density is constant. In this special case, the lamina's center of mass is often called its **centroid**.

Try It Yourself: Locating the center of mass of a lamina

Example 5.8. Suppose $\mathcal{R}$ is the region between $y = 9 - 2x$ and $y = x^2$, when $0 \leq x \leq 1$. If distance is measured in cm, and the density in $\mathcal{R}$ is described by $\rho(x) = 1 + x^2 \; \frac{\text{g}}{\text{cm}^2}$, find the center of mass.

Answer: $\overline{x} = \frac{158}{299} \approx 0.5284$ and $\overline{y} = \frac{8693}{2093} \approx 4.1534$.   ∎

Full Solution
On-line

❖ Kinetic Energy and Gyration of Laminae

Suppose $\mathcal{R}$ is a region in the Cartesian plane, such as the lamina described in the box above, and that it's revolving at ω radians per second about the line $x = c$. We can approximate the lamina's kinetic energy by subdividing $[a, b]$ into n subintervals with a regular partition, and treating the distance to the axis of revolution as being constant in each of them.

In light of our discussion so far, we approximate the mass of the kth rectangle as $\rho(x_k^*)(f(x_k^*) - g(x_k^*))\Delta x$, where x_k^* denotes the midpoint of the kth subinterval. The points in this rectangle are all approximately $|x_k^* - c|$ meters from the axis of revolution (see Figure 5.8), so we estimate its kinetic energy to be

$$\frac{1}{2}m_k v_k^2 = \frac{1}{2} \underbrace{\rho(x_k^*)(f(x_k^*) - g(x_k^*))\Delta x}_{\text{kg}} \; \underbrace{\left(\omega(x_k^* - c)\right)^2}_{(\text{m/sec})^2}.$$

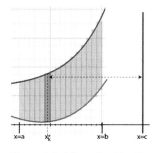

Figure 5.8: Measuring the distance from a rectangle to the axis of revolution.

Upon adding the contributions from each subinterval, we estimate the kinetic energy of the lamina as a whole to be

$$K \approx \sum_{k=1}^{n} \frac{1}{2} \rho(x_k^*)(f(x_k^*) - g(x_k^*))\Delta x \left(\omega(x_k^* - c)\right)^2.$$

After making this expression more familiar by moving $\rho(x_k^*)(f(x_k^*) - g(x_k^*))\Delta x$ to the end, we take a limit to pass from approximation to an exact calculation:

$$K = \int_a^b \frac{1}{2}\left(\omega(x - c)\right)^2 \rho(x)(f(x) - g(x)) \; dx. \tag{5.5}$$

Calculating kinetic energy for a rotating lamina

Example 5.9. Suppose $\mathcal{R}$ is the region that extends from $x = 0$ to $x = 3$ whose upper and lower edges are the lines $y = 2x + 1$ and $y = -2x - 1$, respectively. If the density in $\mathcal{R}$ is described by $\rho(x) = e^{-x} \; \frac{\text{kg}}{\text{m}^2}$, and $\mathcal{R}$ is revolving at 10π radians per second about the line $x = 6$, determine (a) the kinetic energy of the lamina, and (b) the radius of gyration.

Solution: (a) The number $c = 6$ in this example, and $\omega = 10\pi$, so the kinetic energy of the lamina is

$$K = \int_0^3 \frac{1}{2}\left(10\pi(x - 6)\right)^2 e^{-x}\left((2x + 1) - (-2x - 1)\right) \; dx = \left(6200 - 4100e^{-3}\right)\pi^2 \approx 59176.89 \text{ J}.$$

(b) The mass of the lamina is

$$m = \int_0^3 e^{-x}\left((2x + 1) - (-2x - 1)\right) \; dx = 6 - 18e^{-3} \approx 5.1038 \text{ kg}.$$

If the mass were all concentrated r meters away from the axis of revolution and orbited the axis at 10π radians per second, its kinetic energy would be $\frac{1}{2}m(10\pi r)^2$.

The radius of gyration is the value of r for which this is the number K that we determined in part (a).

$$K = \frac{1}{2}m(10\pi r)^2 \quad \Rightarrow \quad r = \frac{1}{10\pi}\sqrt{\frac{2K}{m}} \approx 2.32 \text{ m.}$$ ■

❖ Common Difficulties

This is a section in which students often feel overwhelmed by the sheer number of formulas, so here's a word of advice: don't bother remembering them. Instead, understand how distance, mass, and velocity combine to tell us what we want to know. Then you need only remember that:

- area = (height)×(width);

- mass = (density)×(length) or (density)×(area), depending on the context;

- first moment = (distance)×(mass);

- velocity = (angular velocity)×(radius);

and the definite integral is what we use to add up the contributions from the relevant values of x.

You should know

- the terms *center of mass, first moment, average mass density, mass density at a point, radius of gyration,* and *lamina;*

- that the first moment of an object is defined as (mass)×(distance).

You should be able to

- calculate m, M_x, and M_y;

- explain where length, area, and density occur in the formulas for m, M_x, M_y, and K;

- determine the center of mass of a 1-dimensional object (e.g., we described a baseball bat using only 1 dimension);

- determine the center of mass of a 2-dimensional object (e.g., we described laminae using only 2 dimensions);

- determine the kinetic energy of 1- and 2-dimensional objects that are rotating about a given axis.

❖ 7.5 Concept and Application Exercises

For questions #1–4 determine the center of mass of the linear object with the given mass density and length.

1. $\rho(x) = 2x + 1 \frac{\text{kg}}{\text{m}}$ when $x \in [0, 5]$.

2. $\rho(x) = 4 - \frac{1}{5}x^2 \frac{\text{kg}}{\text{m}}$ when $x \in [0, 2]$.

3. $\rho(x) = 2^{x/3} \frac{\text{kg}}{\text{m}}$ when $x \in [1, 4]$.

4. $\rho(x) = \frac{1}{2}\sin(x) + 1 \frac{\text{kg}}{\text{m}}$ when $x \in [0, 2\pi]$.

5. A linear object is 4 m long and has a mass of 16 kg. If its center of mass is 7/3 meters from its left end, determine its mass density function $\rho = ax + b$ (where x is the distance from the left end).

6. A linear object is 8 m long and has a mass of 80 kg. If its center of mass is 44/15 meters from its left end, determine its mass density function $\rho = ax + b$ (where x is the distance from the left end).

7. Suppose a linear object's mass density is described by $\rho(x) = \frac{1}{2}x + 1$, where x is the distance from the left end of the object. If the center of mass is at $x = 4$, determine the length and mass of the object.

8. Suppose a linear object's mass density is described by $\rho = 3x + 2$, where x is the distance from the left end of the object. If the center of mass is at $x = 12$, determine the length and mass of the object.

In questions #9–12 consider a linear object that extends from $x = 0.4$ to $x = 3.2$ where x is in meters. Determine the object's kinetic energy if its angular velocity about the y-axis is ω and it has the given mass density, $\rho(x)$.

9. $\rho(x) = 2x + 1 \frac{\text{kg}}{\text{m}}$; $\omega = 7\pi \frac{\text{radians}}{\text{sec}}$

10. $\rho(x) = \frac{1}{3}(x^2 + 1) \frac{\text{kg}}{\text{m}}$; $\omega = 10\pi \frac{\text{radians}}{\text{sec}}$

11. $\rho(x) = e^{x/4} \frac{\text{kg}}{\text{m}}$; $\omega = \pi/3 \frac{\text{radians}}{\text{sec}}$

12. $\rho(x) = \sin(x) \frac{\text{kg}}{\text{m}}$; $\omega = 4\pi \frac{\text{radians}}{\text{sec}}$

Exercises #13–16 refer to a linear object that extends from $x = 0$ to $x = 10$ meters, and that is rotating about the y-axis at a rate of 6π radians per second. (a) Graph the density function $\rho(x)$, and (b) determine the object's radius of gyration.

13. $\rho = 3 - \frac{1}{4}x \frac{\text{kg}}{\text{m}}$

14. $\rho = 2 - \cos\left(\frac{\pi}{10}x\right) \frac{\text{kg}}{\text{m}}$

15. $\rho = \sqrt{0.5x + 4} \frac{\text{kg}}{\text{m}}$

16. $\rho = \frac{1}{x^2 - 10x + 26} \frac{\text{kg}}{\text{m}}$

For the regions described in questions #17–20 (a) sketch the region (using a graphing utility if necessary) and indicate where you think its center of mass is, then (b) calculate the center of mass and compare it to your initial estimate. Assume x and y are measured in meters.

17. Enclosed by $y = 1$, $x = 1$ and $y = e^x$ with $\rho = 8 \frac{\text{kg}}{\text{m}^2}$

18. Enclosed by $y = 0$, $x = 4$ and $y - \sqrt{x}$ with $\rho = 15 \frac{\text{kg}}{\text{m}^2}$

19. Enclosed by $y = 0$, $x = \pi$, $x = 2\pi$, and $y = 2 + \cos(x)$ with $\rho = 5 \frac{\text{kg}}{\text{m}^2}$

20. Enclosed by $y = 1 - x^2$, and $y = 2 + x - x^2$ for $y \geq 0$ with $\rho = 2 \frac{\text{kg}}{\text{m}^2}$

In #21–28, all lengths are measured in centimeters. Determine the location of each region's center of mass.

21. Enclosed by $y = 9 - x^2$, and $y = 7 - x$; with $\rho = x + 5 \frac{\text{g}}{\text{cm}^2}$

22. Enclosed by $y = x^2 - 1$ and $3y - 4x = 4$; with $\rho = x^2 + 1 \frac{\text{g}}{\text{cm}^2}$

23. Enclosed by $y = 0$, $y = \sqrt{x + 1}$, $x = 0$ and $x = 8$; with $\rho = 2 + x \frac{\text{g}}{\text{cm}^2}$

24. Enclosed by $y = 0$, $y = \sqrt[3]{x - 2}$, $x = 2$ and $x = 11$; with $\rho = 2x + 1 \frac{\text{g}}{\text{cm}^2}$

25. Enclosed by $y = 0$ and $y = 4 - x^2$; with $\rho = \frac{1}{x+3} \frac{\text{g}}{\text{cm}^2}$

26. Enclosed by $y = 0$ and $y = 6x - x^2$; with $\rho = \frac{1}{8-x} \frac{\text{g}}{\text{cm}^2}$

27. Enclosed by $x = 1$, $x = 3$, $y = 2 + \cos(\pi x)$, and $y = \frac{1}{2}x - 1$; with $\rho = x \ \frac{g}{cm^2}$

28. Enclosed by $x = 0$, $x = 1$, $y = 0$ and $y = e^{-x}$; with $\rho = e^x \ \frac{g}{cm^2}$

For questions #29–32, determine (a) the kinetic energy and (b) the radius of gyration of the rotating lamina described below. Assume x and y are measured in meters.

29. From $x = 1$ to $x = 4$ and between $y = \frac{1}{x}$ and $y = x + 1$; with $\rho = x \ \frac{kg}{m^2}$ and rotating at $\pi/4$ radians per second about the line $x = 5$.

30. From $x = 0$ to $x = 3$ and between $y = \frac{1}{2}x$ and $y = x + 2$; with $\rho = \frac{x+1}{x+3} \ \frac{kg}{m^2}$ and rotating at 3π radians per second about the line $x = 5$

31. From $x = 0$ to $x = \pi$ and between $y = \sin(x)$ and $y = 0$; with $\rho = 6 \ \frac{kg}{m^2}$ and rotating at 9π radians per second about the line $x = 2\pi$.

32. From $x = \frac{\pi}{4}$ to $d = \frac{5\pi}{4}$ and between $y = \sin(x)$ and $y = \cos(x)$; with $\rho = 3 \ \frac{kg}{m^2}$ and rotating at 10π radians per second about the line $x = 8$.

33. Optical filters let certain wavelengths of light pass through to detectors but block others, and each can be described by a graph such as the one in Figure 5.9. The horizontal axis of the graph indicates the wavelength of light in nanometers, λ, and the vertical axis is a dimensionless quantity called the *relative spectral response* (RSR) that quantifies the strength of a detector's response to each wavelength of light (the detector is behind the filter).

Imaging Science

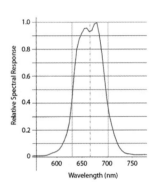

In order to discuss or compare different filters, it's often helpful to characterize them in terms of ideal filters whose RSR is 1 in some interval and is 0 otherwise. Such an idealized filter should have the same cumulative RSR, so we look for

$$\int_0^\infty RSR_a(\lambda) \, d\lambda = \int_0^\infty RSR_i(\lambda) \, d\lambda,$$

where $RSR_a(\lambda)$ is the detector's response when behind the *actual* filter, and RSR_i would be its response with the *idealized* filter in place. Since RSR_i is only 1 over some interval, say $[\lambda_1, \lambda_2]$, this equation becomes

$$\int_0^\infty RSR_a(\lambda) \, d\lambda = \int_{\lambda_1}^{\lambda_2} 1 \, d\lambda = \lambda_2 - \lambda_1.$$

Figure 5.9: A graph of $RSR(\lambda)$ in the red band from IKONOS satellite data.

That is, the number $\int_0^\infty RSR_a(\lambda) \, d\lambda$ is the width of the interval over which the idealized filter allows a response of 1. For this reason, we refer to the number $\int_0^\infty RSR_a(\lambda) \, d\lambda$ as the *effective bandwidth* of the filter. In this exercise, we'll work with a simulated function:

$$RSR(\lambda) = 0.8014 + \left((\lambda - 600)^3(800 - \lambda)^3 + 10(\lambda - 776)^3(876 - \lambda)^3\right) \times 10^{-12}$$

when $\lambda \in [738.1109, 840.4689]$ and $RSR(\lambda) = 0$ otherwise.

(a) Use a graphing utility to render a plot of $y = RSR(\lambda)$ in the window $[738.1109, 840.4689] \times [-0.1, 1.1]$.

(b) Determine the effective bandwidth of this simulated filter.

(c) Thinking of the graph from part (a) as defining a lamina in the plane, calculate the λ-coordinate of the centroid, λ_*.

(d) Draw the graph of $RSR_i(\lambda)$ onto the graph from part (a), centered at $\lambda = \lambda_*$.

7.6 Hydrostatic Force

Divers experience pressure due to the mass of water above them. Suppose that, looking from below, the area of the diver's silhouette is A (see Figure 6.1), and that she's at a depth of D meters. Then the amount of water directly above her is just AD. If ρ is the density of water, which we take to be constant, the mass of water above the diver is ρAD so its weight is $g\rho AD$, where g is the acceleration due to gravity. *Pressure* is defined to be the force per unit area, so the downward pressure of the water is

$$ P = \frac{g\rho AD}{A} = g\rho D = (\text{gravity}) \times (\text{density}) \times (\text{depth}). $$

We said that this is the *downward* pressure, but you know from experience that hydrostatic pressure is the same in all directions. While swimming, or perhaps when watching documentaries, you've seen that very small air bubbles are spherical. That happens precisely because the water pressure on the air is the same from all directions. This fact is important to engineering applications involving water.

Figure 6.1: A diver's silhouette.

> Since y is a *depth* in this scenario, $y = 0$ at the top of the water, and $y = 5$ at the river bed.

Hydrostatic force

Example 6.1. Suppose a vertical dam has a rectangular face that is 25 meters wide, and the water behind it is 5 meters deep. How much hydrostatic force is sustained by the dam?

Solution: Since pressure depends on *depth*, the base of the dam experiences more pressure than the top. We account for that fact by partitioning the face of the dam into rectangles at different depths. Each rectangle has a width of 25 meters and a height of Δy, as seen in Figure 6.2. If y_k^* is the average depth of the kth rectangle, it experiences an average pressure of

$$ (\text{gravity}) \times (\text{density}) \times (\text{depth}) = g\rho y_k^*. $$

And since the kth rectangle has an area of $25\Delta y$, the force it sustains is

$$ F_k = \underbrace{(\text{pressure})}_{\text{N}} \times \underbrace{(\text{area})}_{\text{N/m}^2} = (g\rho y_k^*)\underbrace{(25\Delta y)}_{\text{m}^2} = 25g\rho y_k^*\Delta y. $$

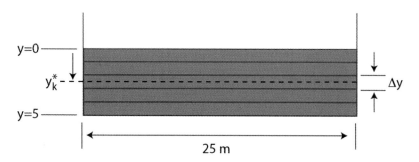

Figure 6.2: The base of the dam experiences more pressure than the part near the top of the water.

We approximate the hydrostatic force sustained by the dam by adding the forces sustained by all the rectangles,

$$ F \approx \sum_{k=1}^{n} F_k = \sum_{k=1}^{n} 25g\rho y_k^*\Delta y, $$

and pass from approximation to an exact calculation in the limit:

$$F = \lim_{n \to \infty} \sum_{k=1}^{n} F_k = \int_0^5 25g\rho y \; dy = \frac{1}{2}625\rho g.$$

The density of pure water is 1000 $\frac{\text{kg}}{\text{m}^3}$, and the acceleration due to gravity is 9.8 $\frac{\text{m}}{\text{sec}^2}$, so $F = 3062500$ newtons. ∎

   You might ask why the total amount of water held back by the dam doesn't play a role in our calculation. Consider holding your hand a foot under water. You feel a certain amount of water pressure, and that pressure is the same whether your hand is in a bathtub, a pond, a swimming pool, or one of the Great Lakes. The *expanse* of water is not relevant to pressure you feel, but only the depth.

Hydrostatic force

Example 6.2. Suppose a vertical dam has a trapezoidal face that is 4 meters tall, 30 meters wide at the top, and 12 meters wide at the bottom. The water behind the dam is 3.5 meters deep. How much hydrostatic force is sustained by the dam?

Solution: Unlike Example 6.1, the width of this dam changes with depth, and understanding the exact relationship between depth and width will be central to our calculation. Let's again define y to mean depth, so $y = 0$ at the top of the water and $y = 3.5$ at the base of the dam. Based on the data we have about this dam (depicted in Figure 6.3), we know that its right-hand edge stretches from the point $(6, 3.5)$ to $(15, -0.5)$.

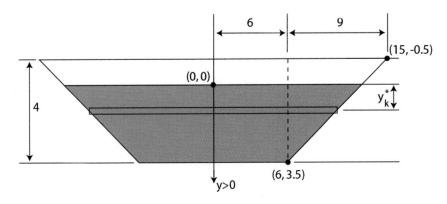

Figure 6.3: The trapezoidal face of a dam from Example 6.2.

   Since the y-coordinate along the right-hand edge drops by 4 as the x-coordinate increases by 9, the edge has a slope of $-4/9$ in this coordinate system. Now writing the point-slope form of a line, we can describe the edge as

$$y - 3.5 = -\frac{4}{9}(x - 6).$$

This equation allows us to determine the width of the dam at a depth of y. Solving it for x, we see $x = 6 - \frac{9}{4}(y - 3.5)$, which is the distance from the center to the right-hand edge. So the width of the dam is twice this number:

$$\text{width at a depth of } y \text{ is } 2x = 12 - \frac{9}{2}(y_k^* - 3.5).$$

As before, we're going to approximate the face of the dam with rectangles. If the kth rectangle has an average depth of y_k^* and a height of Δy, its area is

$$(\text{width}) \times (\text{height}) = \left(12 - \frac{9}{2}(y_k^* - 3.5)\right)\Delta y,$$

so the hydrostatic force it sustains is

$$F_k = \underbrace{(\text{pressure})}_{\text{N}} \times \underbrace{(\text{area})}_{\text{m}^2} = \underbrace{(g\rho y_k^*)}_{\text{N/m}^2}\left(12 - \frac{9}{2}(y_k^* - 3.5)\right)\Delta y.$$

The cumulative force sustained by the face of the dam is $F \approx \sum_{k=1}^{n} F_k$, and we pass from approximation to an exact calculation in the limit:

$$F = \lim_{n\to\infty}\sum_{k=1}^{n} F_k = \int_0^{3.5} g\rho y\left(12 - \frac{9}{2}(y - 3.5)\right)\,dy = 1035431.25 \text{ N},$$

where we've used $g = 9.8\ \frac{\text{m}}{\text{sec}^2}$ and $\rho = 1000\ \frac{\text{kg}}{\text{m}^3}$.   ■

Try It Yourself: Hydrostatic force

Example 6.3. Suppose a dam has a trapezoidal face that is 7 meters tall, 30 meters wide at the top, and 22 meters wide at the bottom. The water behind the dam is 6 meters deep. How much force is sustained by the dam?

Answer: 4284000 N (using $\rho = 1000$ and $g = 9.8$).   ■

Hydrostatic force

Example 6.4. Suppose an aquarium wants to install a circular window of radius 1.5 meters that will look into an open-air tank. The window will be installed so that its lowest point is at a depth of 2.6 meters (see Figure 6.4). How much hydrostatic force must be sustained by the window?

Solution: In previous examples, we've used y to represent depth. In this example, that complicates the relevant equations, so instead we'll set our origin at the center of the window, and orient the axes so that $y > 0$ is above the center (as usual).

As in Examples 6.2 and 6.3, we have a situation in which width changes with depth. Specifically, since the window is circular and has a radius of 1.5, we know that its edge is described by $x^2 + y^2 = 1.5^2$, so a horizontal rectangle whose midline is at y_k^* (see Figure 6.5) has a width of

$$2x_k = 2\sqrt{1.5^2 - (y_k^*)^2}.$$

If its height is Δy, its area is $(\text{width})\times(\text{height}) = 2\sqrt{1.5^2 - (y_k^*)^2}\Delta y$. Because such a rectangle has an average depth of $1.1 - y_k^*$, the hydrostatic force it sustains is

$$F_k = \underbrace{g\rho(1.1 - y_k^*)}_{\text{pressure}} \times \underbrace{2\sqrt{1.5^2 - (y_k^*)^2}\,\Delta y}_{\text{area}}.$$

The cumulative force sustained by the window is $F \approx \sum_{k=1}^{n} F_k$, and we pass from approximation to an exact calculation in the limit:

$$F = \lim_{n\to\infty}\sum_{k=1}^{n} F_k = \int_{-1.5}^{1.1} 2g\rho(1.1 - y)\sqrt{1.5^2 - y^2}\,dy.$$

Full Solution On-line

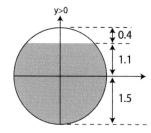
Figure 6.4: The circular, partially submerged window for Example 6.4.

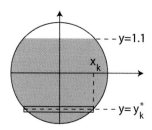
Figure 6.5: A horizontal rectangle whose midline is at y_k^*.

This is approximately 77088.73 N, when we use $g = 9.8\ \frac{m}{sec^2}$ and $\rho = 1000\ \frac{kg}{m^3}$. ∎

Full Solution
On-line

Try It Yourself: Hydrostatic force

Example 6.5. Suppose an aquarium wants to install a circular window of radius 3 meters at the penguin exhibit. So that patrons can watch the birds both on land and in the water, the window will be installed so that the half of it is below the water line. How much hydrostatic force must it sustain?

Answer: 1796580 N.   ∎

Hydrostatic force on an inclined plane

Example 6.6. Suppose a rectangular dam is tilted at an angle of $\pi/3$ from the horizontal (see Figure 6.6). If the dam is 80 meters across, and the water behind it is 30 meters deep, determine the amount of force sustained by the face of the dam.

Solution: As in previous examples, we divide the face of the dam into strips of height Δy. Because the plane where the dam meets the water is inclined at an angle of $\pi/3$, the area of each strip is

$$(\text{length}) \times (\text{width}) = (80) \times \left(\frac{\Delta y}{\sin(\pi/3)}\right) = \frac{160}{\sqrt{3}}\Delta y.$$

If we denote the depth of the kth strip by y_k^*, we can write the force that it sustains as approximately

$$F_k \approx (\text{pressure}) \times (\text{area}) = (g\rho y_k^*)\frac{160}{\sqrt{3}}\Delta y,$$

so the total force sustained by the dam is

$$F \approx \sum_{k=1}^{n} F_k \approx \sum_{k=1}^{n} g\rho y_k^* \frac{160}{\sqrt{3}}\Delta y.$$

When we pass from approximation to an exact calculation with the limit, we see

$$F = \lim_{n \to \infty} \sum_{k=1}^{n} F_k = \int_0^{30} \frac{160 g\rho}{\sqrt{3}} y\ dy = \frac{705600000}{\sqrt{3}}\ \text{N},$$

where we've used $\rho = 1000\ \frac{kg}{m^3}$ and $g = 9.8\ \frac{m}{sec^2}$.   ∎

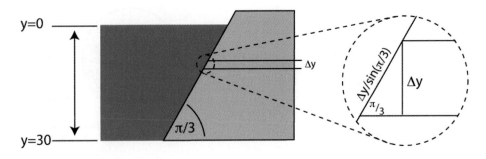

Figure 6.6: The tilted face of the dam in Example 6.6.

You should know

- the term *hydrostatic force;*

- that hydrostatic pressure is $P = (\text{gravity}) \times (\text{density}) \times (\text{depth})$;

- that force $= (\text{pressure}) \times (\text{area})$.

You should be able to

- calculate the hydrostatic force sustained by a dam.

❖ 7.6 Concept and Application Exercises

In all of the exercises below, use $g = 9.8 \, \frac{m}{sec^2}$ for the acceleration per unit mass due to gravity and $\rho = 1000 \frac{kg}{m^3}$ for the mass density of water.

1. A rectangular retaining wall is 120 m across and 23 m high. If it holds back water that is 20 meters deep, how much hydrostatic force must the retaining wall sustain?

2. A rectangular viewing window at an ocean aquarium measures 4 meters wide and 2 meters tall. If the surface of the water is 1 meter above the top of the window, determine the hydrostatic force pressing against the window.

3. A storage tank has a vertical face in the shape of an isosceles triangle (point down) with an upper edge of length 4 meters and a height of 6 meters. Determine the force against the vertical face if the tank is filled to the top with water.

4. Using the tank described in #3, determine the force against the vertical face if the tank is only filled to a depth of 4 meters.

5. An aquarium has a below-water-level viewing window in the shape of a trapezoid which has a 4-meter upper edge, a 1-meter lower edge and a height of 3 meters. Determine the force against the window if the surface of the water is even with the window's upper edge.

6. Using the viewing window described in #5, determine the force against the window if the surface of the water is 1 meter below the window's upper edge.

7. Suppose a dam is 60 m tall and is shaped as a trapezoid with an upper width of 220 m and a lower width of 72 m. The water behind the dam is 51 m deep. How much hydrostatic force does this dam sustain?

8. Using the viewing window described in #5, determine the force against the window if the surface of the water is 2 meters above the window's upper edge.

9. A horizontal cylindrical storage tank with circular ends is filled with water. If the radius of the circular end is 10 meters and the tank is filled exactly halfway, determine the force against the circular end.

10. Using the tank described in #9, determine the force against the circular end if the water is only 2 meters deep.

11. Using the tank described in #9, determine the force against the circular end if the water is 8 meters deep.

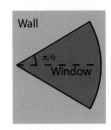

Figure 6.7: The wedge-shaped window in #12.

12. An aquarium has an underwater viewing window in the shape of a circular sector with radius 2 meters (see Figure 6.7). The window extends $\pi/6$ radians above and below the horizontal. If the surface of the water is $1/2$ meter above the highest point on the window, determine the force against the window.

13. The trailer of a tractor-trailer detaches during an accident on a bridge and plummets into a lake that is 11 m deep. Miraculously, the trailer comes to rest on the bottom intact and upright. Its back end is held up above the (flat) lake bed by the rear wheels while the front end rests on bottom, so the bed of the trailer is inclined at an angle of $6.75°$ (see Figure 6.8). Determine the hydrostatic force acting against the long, vertical side of trailer which measures 10 meters $\times$ 3.5 meters.

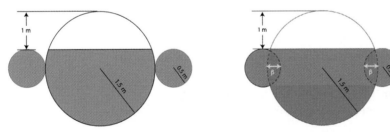

Figure 6.8: The sunken trailer in #13.

14. If the trailer at the bottom of the lake in #13 had dimensions of 17 meters $\times$ 3.5 meters and rested on the lake bed at an inclination of $3.6°$, determine the hydrostatic force acting against the long, vertical side of trailer.

15. A viewing window at an aquarium is shaped from three circles whose centers are aligned, as in Figure 6.9. If the middle window has a diameter of 3 meters and the other two have diameters of 1 meter, determine the force acting against the entire window structure when the surface of the water is 1 meter below the top of the central window.

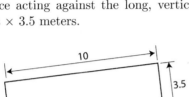

Figure 6.9: (left) For #15; (right) for #16.

16. Suppose the window from #15 is redesigned to use less glass by moving each of the outer circles β meters toward the center of the structure, where $\beta \in (0, 1)$, as depicted in Figure 6.9. (a) Write the hydrostatic force acting against the window structure as a function of β, and (b) determine the hydrostatic force when $\beta = 0.25$.

17. A curved window in the shape of a parabolic arc is installed as a viewing window in an aquarium. If the window can be described as the area between $y = 0$ and $y = 4x - x^2$ where x and y are in meters, and if the surface of the water is 2 meters above the top of the parabola, determine the hydrostatic force on the window.

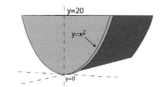

Figure 6.10: Water sits behind a parabolic dam in #18.

18. A dam face is shaped like a parabolic arc and can be described by the region between $y = \frac{2}{90}x^2$ and $y = 20$ where x and y are in meters (see Figure 6.10). If the water behind the dam is 20 meters deep, determine the hydrostatic force against the dam.

19. A dam face in the shape of a trapezoid has an upper base of 60 meters, a lower base of 20 meters, and stands 40 meters high. Near the bottom of the dam is a circular drainage pipe whose radius is 4 meters and whose lowest point is 1 meter from the base of the dam. Determine the hydrostatic force against the (shaded) dam face when the water behind it is 30 m deep.

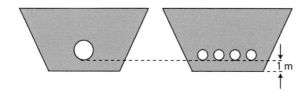

Figure 6.11: (left) The dam face for #19; (right) the dam face for #20.

20. An alternate configuration for the dam in question #19 is to have four smaller pipes, each with a radius of 1 meter, and each with its lowest point 1 meter from the base of the dam (see Figure 6.11). Determine the amount of hydrostatic force sustained by the dam face in this configuration.

21. A rectangular dam has a length of 12 meters, and the water behind the dam is 6 meters deep. If the dam is inclined at $\pi/4$ radians away from the water, determine the hydrostatic force on the dam.

22. A trough with a triangular-shaped cross-section is filled with water. The sides of the trough form an angle of $\pi/3$ radians with the horizontal. If the trough is 2 meters long and 0.5 meters deep, determine the hydrostatic force sustained by one of the long sides of the trough.

23. A rectangular-shaped swimming pool has length of 10 meters and a width of 5 meters. The pool is 1 meter deep at its shallow end, and 4 meters deep at its deep end (see Figure 6.12). Determine the hydrostatic force acting against (a) the 5-meter wide vertical wall at the shallow end, (b) the 5-meter wide vertical wall at the deep end, and (c) the 10-meter wide vertical wall along the side.

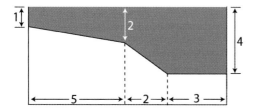

Figure 6.12: The profile of the swimming pool from #23.

24. Consider a 50-meter-long dam behind which the water is 25 meters deep. A cross section of the dam shows that the side next to the water is in the shape of a parabolic arc, as depicted in Figure 6.13. If we place the origin of a coordinate system at the bottom of the dam, we can use $y = x^2$ as a

Figure 6.13: For #24.

description of the height of the dam (x and y in meters). Determine the amount of hydrostatic force acting against the dam face.

25. A rectangular tank is 2 meters long, 1.2 meters wide and 0.5 meters deep, and is filled to the top with water. If the water is leaking out of the tank at a constant rate of 0.2 m^3 per second, formulate the hydrostatic force acting against the smaller vertical face of the tank as a function of time.

26. A trough with a cross-section in the shape of an isosceles triangle (point down) is filled to the top with water. The top of the trough is 0.5 meters wide and the trough is 0.25 meters deep. If the water is leaking out of the trough at a rate of 20 cm^3 per second, find a function that describes the hydrostatic force acting against the end-face of the trough as a function of time.

27. A rectangular tank is 1 m long, 0.3 m wide, and 0.25 m tall. Due to the materials used, each vertical side of the tank can withstand a maximum force of 100 newtons. What volume of water (in m^3) can the tank hold before rupturing?

28. Suppose the tank in #27 is filled with high-fructose corn syrup, which has a mass density of 1394.8 $\frac{kg}{m^3}$ (see [39]). How much corn syrup (in m^3) can the tank hold?

29. A tank that is 2 meters long and 1 meter deep has a trapezoidal end-face with an upper edge that is 0.8 meters long and a lower edge that is 0.6 meters long. The trapezoidal end-face can withstand a total of $\frac{4655}{6}$ newtons of force. What volume of water (in m^3) can the tank hold?

30. Using the tank from #29, determine how much olive oil (density 800$\frac{kg}{m^3}$, see [40]) the tank can hold.

7.7 Average Value

Suppose that you're running down the street, and your velocity after t seconds is $v(t)$ meters per second. Then we can use the definite integral to write your net change in position as

$$\underbrace{\int_a^b v(t)}_{\frac{m}{sec}} \underbrace{dt}_{sec} \ \ m.$$

You accomplished this over a time span of $b - a$ seconds, so you must have been running at

$$\begin{array}{c} m \to \\ sec \to \end{array} \frac{\int_a^b v(t)\,dt}{b - a} \ \ \text{meters per second, on average.}$$

In fact, we define the average value of a continuous function in exactly this way.

> Suppose that f is continuous. The **average value** of f over $[a, b]$ is denoted by $\langle f \rangle$ and defined to be
>
> $$\langle f \rangle = \frac{1}{b - a} \int_a^b f(t)\,dt.$$

Calculating average value

Example 7.1. Suppose $f(t) = 3t$. Determine the average value of f over $[0, 4]$.

Solution: Since the function values range from 0 to 12, our experience with averages leads us to think that the average value of f is 6. *Because f is a linear function,* this ends up being right:

$$\langle f \rangle = \frac{1}{4 - 0} \int_0^4 3t\,dt = \frac{1}{4}\left(\frac{3}{2}t^2\Big|_0^4\right) = \frac{1}{4}\left(24 - 0\right) = 6.$$

The relationship among the function, its integral, and its average value is depicted in Figure 7.1. ∎

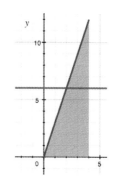

Figure 7.1: The graph of $f(t)$ and the horizontal line $y = \langle f \rangle$ for Example 7.1.

Calculating average value

Example 7.2. Suppose $f(t) = \sqrt{t}$. Determine the average value of f over $[0, 144]$.

Solution: Since the function values range from 0 to 12, you might expect the average value of f to be 6, but it's not:

$$\langle f \rangle = \frac{1}{144 - 0} \int_0^{144} \sqrt{t}\,dt = \frac{1}{144}\left(\frac{2}{3}t^{3/2}\Big|_0^{144}\right) = 8.$$

The relationship among the function, its integral, and its average value is depicted in Figure 7.2. ∎

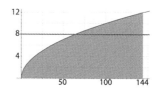

Figure 7.2: The graph of $f(t)$ and the horizontal line $y = \langle f \rangle$ for Example 7.2.

Our guess at the average value was correct in Example 7.1 because $f'(t)$ is constant. By contrast, the nonlinear function in Example 7.2 ascends quickly when t is small, and so leaves the lower function values in only a short span of time. Over the rest of the interval, the function produces relatively large function values that draw the average value above 6.

You know that a continuous function f attains minimum and maximum values on $[a, b]$, say m and M respectively. Based on our everyday usage of the word *average,* it seems reasonable to expect $\langle f \rangle$ to be some number between them.

Theorem: Suppose f is a continuous function whose minimum and maximum values on $[a, b]$ are m and M respectively. Then $m \le \langle f \rangle \le M$.

Proof. This follows directly from basic properties of the definite integral. Since $f(t) \le M$, we know

$$\int_a^b f(t)\,dt \le \int_a^b M\,dt = M(b-a).$$

Therefore $\langle f \rangle \le M$. A similar inequality shows $m \le \langle f \rangle$. ∎

In Figures 7.1 and 7.2 you see that there is a time at which a continuous function actually achieves its average value. While those functions were very simple, this fact is true even when f is more complicated, as long as it's continuous.

Mean Value Theorem for Integrals: Suppose f is a continuous function. Then there is a $t^* \in (a, b)$ at which $f(t^*) = \langle f \rangle$.

Proof. Since f is continuous, it achieves a minimum and a maximum value on $[a, b]$, say m and M respectively. The Intermediate Value Theorem guarantees that f achieves all values between m and M, and since $\langle f \rangle$ is one of those values, there must be some time at which it's achieved. ∎

Calculating average value

Example 7.3. Suppose $f(t) = \sqrt{t}$. Determine the number $t^* \in (0, 144)$ for which $f(t^*) = \langle f \rangle$, as guaranteed by the Mean Value Theorem for Integrals.

Solution: We know from Example 7.2 that $\langle f \rangle = 8$. So we solve the equation $\sqrt{t^*} = 8$ and find that $t^* = 64$. ∎

Try It Yourself: Calculating average value

Example 7.4. Suppose $f(t) = 3t$. Determine the number $t^* \in (0, 4)$ guaranteed by the MVT for Integrals.

Answer: $t^* = 2$. ∎

Full Solution
On-line

Note: Examples 7.3 and 7.4 indicate that the average value of a linear function is achieved at the midpoint of the interval, but the average value of a nonlinear function might be achieved somewhere else.

✧ An Interpretation in Terms of Area

The Mean Value Theorem for Integrals tells us that $\langle f \rangle = f(t^*)$. Consequently,

$$\int_a^b f(t)\,dt = f(t^*)(b-a). \tag{7.1}$$

When $f(t)$ remains non-negative over $[a, b]$, the number on the left-hand side is an area, so the number on the right-hand side should be too. Since $f(t^*)$ is a "height" on the graph of f, and $(b - a)$ is the width of the interval $[a, b]$, we can understand their product as

$$f(t^*)(b - a) = (\text{height}) \times (\text{width}) = \text{the area of a rectangle},$$

which is depicted in Figure 7.3. If $f(t)$ is ever negative in $[a, b]$, the integral on the left-hand side of equation (7.1) tells us *net area*, rather than area, but we can still interpret the right-hand side as (height)×(width) by allowing the rectangle to lie below the horizontal axis if needed (in which case its "height," $f(t^*)$, is negative).

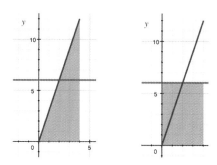

Figure 7.3: (left) The area between the graph of f (from Example 7.1) and the horizontal axis, over $[0, 4]$; (right) the rectangle of height $\langle f \rangle$ and width 4 has the same area.

Geometric interpretation when $\langle f \rangle < 0$

Example 7.5. Suppose $f(t) = t^2 - 9$. Sketch the graph of f and a rectangle with the same net area over $[0, 4]$.

Solution: The net area of f over this interval is $\int_0^4 f(t)\, dt = -44/3$. A rectangle over the same interval with that net area has

$$-\frac{44}{3} = (\text{width}) \times (\text{"height"}) = 4(\text{"height"}) \quad \Rightarrow \quad (\text{"height"}) = -\frac{11}{3},$$

as shown in Figure 7.4. Note that the "height" of the rectangle is exactly $\langle f \rangle$. ■

❖ Average Value and Improper Integrals

Limits allow us to talk about the average value of a function over semi-infinite intervals. For example, when f is a continuous function on $[0, \infty)$ we define its average value over that interval to be

$$\langle f \rangle = \lim_{L \to \infty} \frac{1}{L} \int_0^L f(t)\, dt.$$

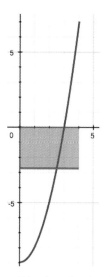

Figure 7.4: A rectangle with the same net area lies below the horizontal axis when $\langle f \rangle < 0$.

Average value over a semi-infinite interval

Example 7.6. Determine the average value of $\sin^2(t)$ over $[0, \infty)$.

Solution: Let's begin by using the half-angle identity to write

$$\int_0^L \sin^2(t)\, dt = \int_0^L \frac{1}{2} - \frac{1}{2}\cos(2t)\, dt = \frac{1}{2}L - \frac{1}{4}\sin(2L).$$

Then we use a limit to calculate the average value,

$$\langle f \rangle = \lim_{L \to \infty} \frac{1}{L}\left(\frac{1}{2}L - \frac{1}{4}\sin(2L)\right) = \frac{1}{2}. \qquad ■$$

> You might think this result is obvious because $\sin^2(t)$ ranges between 0 and 1, but $\sin^4(t)$ has the same range and its average value is not $1/2$.

When the graph of f has a vertical asymptote at $t = a$, we can use limits to calculate an average value over $(a, b]$,

$$\langle f \rangle = \lim_{L \to a} \frac{1}{b - L} \int_L^b f(t)\, dt.$$

Similarly, when the graph of f has a vertical asymptote at $t = b$, we say

$$\langle f \rangle = \lim_{L \to b} \frac{1}{L - a} \int_a^L f(t)\, dt.$$

Average value in the presence of a vertical asymptote

Example 7.7. Determine the average value of $t^{-1/3}$ over $(0, 8]$.

Solution: We begin by calculating $\int_L^8 t^{-1/3}\, dt = \frac{3}{2}(4 - L^{2/3})$. Then

$$\langle f \rangle = \lim_{L \to 0} \frac{1}{8 - L} \int_L^8 t^{-1/3}\, dt = \lim_{L \to 0} \frac{3}{2(8 - L)}(4 - L^{2/3}) = \frac{3}{4}. \qquad \blacksquare$$

You should know

- the term *average value*, and the notation $\langle f \rangle$;
- the Mean Value Theorem for Integrals.

You should be able to

- calculate the average value of a continuous function;
- determine the height of a rectangle (perhaps negative) whose net area is $\int_a^b f(x)\, dx$.

✧ 7.7 Skill Exercises

In #1–10 determine the average value for each function over the given interval.

1. $f(t) = t^2 + 4t$ over $[0, 10]$
2. $f(t) = t\sqrt{t - 4}$ over $[4, 13]$
3. $f(t) = \sin(t)$ over $[0, \pi]$
4. $f(t) = t^{-2}\tan(1/t)$ over $[1, 2]$
5. $f(t) = \sin(2t) + 10\cos(t/4)$ over $[0, 2\pi]$

6. $f(t) = e^t$ over $[0, 1]$
7. $f(t) = (t^2 - 1)e^{-0.5t}$ over $[0, 6]$
8. $f(t) = \sinh(t)$ over $[-1, 2]$
9. $f(t) = 2t - t/(t^2 + 1)$ over $[-1, 3]$
10. $f(t) = 8/(t^3 + 6t)$ over $[1, 5]$

In #11–16 determine the value of t^* in the given interval that satisfies the MVT for Integrals.

11. $f(t) = t^2$; $[-1, 3]$
12. $f(t) = t\sqrt{t^2 - 1}$; $[2, 5]$

13. $f(t) = \cos(t)$; $[\pi/6, \pi/2]$
14. $f(t) = \tan(t)$; $[-\pi/3, \pi/4]$

15. $f(t) = e^t$; $[0, 1]$
16. $f(t) = \ln(t)$; $[1, 4]$

In #17–22 use a limit to calculate the average value over the specified interval.

17. $f(t) = t^{-3}$ over $[1, \infty]$
18. $f(t) = 1/(t^2 + 1)$ over $[0, \infty]$
19. $f(t) = \arctan(t)$ over $[0, \infty)$

20. $f(t) = \sin^4(t)$ over $[0, \infty)$
21. $f(t) = 1/\sqrt{t}$ over $(0, 3]$
22. $f(t) = \csc(\sqrt{t})$ over $(0, 0.25\pi^2]$

❖ 7.7 Concept and Application Exercises

23. Draw the graph of a continuous function over $[0, 10]$ so that both (a) $f(t)$ is negative for a longer period of time (i.e., over more of the interval) than it's positive, and (b) its average value is $\langle f \rangle = 0$.

24. Suppose f is a positive-valued function for which $\int_0^\infty f(t)\,dt$ converges. Show that $\langle f \rangle = 0$ over $[0, \infty)$.

25. Show that the average value of $f(t) = t + \sin(2k\pi t)$ over $[0, 1]$ is the same for all integer values of k.

26. Suppose f and g are continuous over $[a, b]$. Prove that $\langle f + g \rangle = \langle f \rangle + \langle g \rangle$.

27. Suppose the velocity of an object is $v(t) = t\sin(t)$ meters/second. What is its average velocity during the first π seconds? | Kinetics |

28. Suppose an object experiences an acceleration of $a(t) = t^2 - 4t + 1 \, \frac{m}{s^2}$, where t is measured in seconds. Determine the object's average acceleration between $t = 2$ and $t = 6$ seconds. | Kinetics |

29. Suppose an object starts 3 meters to the right of the origin and travels along the x-axis with a velocity of $v(t) = t + \sin(4t) \, \frac{m}{s}$, where t is measured in seconds. Determine the object's average position during the first 5 seconds of travel. | Kinetics |

30. Suppose an object starts at rest 4 meters to the left of the origin and then travels along the x-axis. If the object experiences an acceleration of $a(t) - e^{-t/2}$ $\frac{m}{s^2}$, where t is measured in seconds, determine its average position between times $t = 1$ and $t = 10$ seconds. | Kinetics |

31. During six months of heavy trading, a company's stock price fluctuated according to $P(t) = -t^2 + 4t + 45$ dollars per share, where t is measured in months. | Business |

    (a) Determine the average stock price during the sixth month of trading.

    (b) Calculate the price at the end of month 6 and determine whether you'll sell your shares in the company. Explain your reasoning in complete sentences.

32. The temperature of a cup of hot chocolate is $T(t) = 20 + 75e^{-0.02t}$ degrees Celsius after t minutes. | Thermodynamics |

    (a) What is the average temperature of the hot chocolate over the first 30 minutes?

    (b) What was the temperature of the drink initially, and after 30 minutes had passed?

    (c) At what time was the hot chocolate exactly the average temperature from part (a)?

33. Suppose the size of a population of sea stars is $P(t) = 7620/(5 + e^{-0.236t})$ individuals, where t is measured in years. Determine the average size of the sea star population (a) during the first 10 years of observation, $0 \le t \le 10$, and (b) during the second 10 years of observation, $10 \le t \le 20$. (Hint: the integration can be done with the substitution technique.) | Biology |

34. A rod is 8 meters long and its density is $\rho(x) = \frac{12}{\sqrt{x+1}} \, \frac{kg}{m}$, where x is measured (in meters) from one end. Determine (a) the average density of the rod, and (b) a location on the rod where its density is the same as the average. | Engineering |

35. Find the vertical height of a rectangle with horizontal length 4 that has the same area as the region in the first quadrant between $y = \sqrt{4-x}$ and $y = 0$.

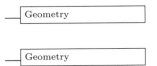

36. Find the vertical height of a rectangle with horizontal length 5 that has the same area as the region between $x = -2$, $x = 3$, $y = 0$, and

$$y = \begin{cases} 2 - \dfrac{1}{2}x^2, & x \le 0 \\[2mm] \dfrac{2}{9}(x-3)^2, & x > 0. \end{cases}$$

Consider a listening buoy that's put in place to detect submarines. When there is a circular area of radius r about the buoy in which the submarine will be detected, we say that the *sweep width* of the detector is $2r$. However, whether due to variations in water temperature and salinity or in the topography of the sea floor, the region in which the buoy will detect the submarine might *not* be circular. In this case we define the effective sweep width to be the *mean diameter* of the detection region, denoted here by $\langle D \rangle$, which is found as an integral average

$$\langle D \rangle = \frac{1}{\pi} \int_0^\pi w(\theta)\, d\theta,$$

where $w(\theta)$ is the apparent width of the region as seen from an angle of θ radians. It may help you to think of the detection region as a solid body. If we rotate it by θ radians and shine parallel light rays on it, the length of the shadow is $w(\theta)$ (see Figure 7.5). Exercises #37–44 address the mean diameter of various regions.

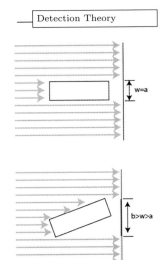

Figure 7.5: (both) Shining parallel light rays onto rectangle whose width and length are a and b, respectively; (top) when the rectangle is not rotated at all, the shadow has a length of $w(0) = a$; (bottom) when the rectangle is rotated by some angle $\theta \in (0, \pi/2)$, the shadow has a length of $w(\theta) \in (a, b)$.

37. Calculate the mean diameter of a line segment of length 2.

38. Derive a formula for the mean diameter of a line segment of length L.

39. Calculate the mean diameter of a rectangle whose width and length are 2 and 6, respectively.

40. Derive a formula for the mean diameter of a rectangle whose width and length are a and b, respectively. (Your final formula should not have an integral sign.)

41. Consider an ellipse whose semi-minor and semi-major axes have lengths of 2 and 5, respectively. Approximate the mean diameter of the ellipse by using a Riemann sum with 20 subintervals.

42. Write a formula for the mean diameter of an ellipse whose semi-minor and semi-major have lengths of a and b, respectively. (Leave your formula as an integral.)

43. In #40 you derived a formula for $\langle D \rangle$ in the case of a rectangle. (a) Intuitively (geometrically) speaking, what should happen to $\langle D \rangle$ as $a \to 0^+$? (b) Verify your answer from part (a) by calculating $\lim_{a \to 0^+} \langle D \rangle$.

44. In #42 you derived a formula for $\langle D \rangle$ in the case of an ellipse. (a) Intuitively (geometrically) speaking, what should happen to $\langle D \rangle$ as $a \to b^-$? (b) Verify your answer from part (a) by calculating $\lim_{a \to b^-} \langle D \rangle$.

7.8 Probability

How likely is it that a fair coin toss will result in *heads?* Since *heads* is one of two possible outcomes, both of which are equally likely (this is the meaning of "fair"), we commonly say that the chances are 1 out of 2. More technically, we say that the *probability* of seeing *heads* is 1/2. In general, the **probability** of an event is a number between 0 and 1 that quantifies how likely it is to happen.

Calculating probabilities

Example 8.1. Suppose you toss a standard six-sided die. (a) What's the probability of seeing a 3? (b) What's the probability of seeing either a 2 or a 3?

Solution: (a) There are six possible events that are equally likely, and rolling a 3 is only one of them, so the probability of seeing a three is 1/6. (b) Allowing for a roll of either 2 or 3 means that two of the six possible events are being counted, so the probability is $2/6 = 1/3$. ∎

Suppose we denote the probability that an event will happen by p, and the probability that it *will not* happen by q. Then $p + q = 1$. Said differently, the probability that the event will *not* happen is $1 - p$. For example, since the probability of rolling a three is 1/6 with a standard six-sided die, the probability of rolling something else is $1 - 1/6 = 5/6$.

Try It Yourself: Calculating probabilities

Example 8.2. Suppose you toss a standard six-sided die. What's the probability of seeing neither a 2 nor a 3?

Answer: 2/3. ∎

Application to search and rescue

Example 8.3. Suppose a swimmer is somewhere in a region of the ocean whose area is A square meters, and the swimmer's suit has a radio frequency identification (RFID) chip in it. A boat is sent in search of the swimmer, and signaling equipment on the boat will detect the RFID chip if the boat comes within r meters of the swimmer. Determine the probability that the boat's detector will fail to locate the swimmer at the moment that it's switched on.

Solution: The detector will sense the swimmer if he's within a disk of r meters that's centered at the detector. The area of this disk is πr^2, and the area of the entire region is A, so the fraction of the whole area that's seen by the detector at the moment it's switched on is $\pi r^2 / A$ (which is less than 1 since A is very large). We take this to be the probability of finding the swimmer at the moment the detector is switched on, so the probability of *failing* to detect the swimmer is $1 - \frac{\pi r^2}{A}$. ∎

Full Solution On-line

Figure 8.1: A depiction of the scenario in Example 8.3.

We have made a number of simplifying assumptions in this example in order to develop the pertinent ideas. For instance, the detector has been idealized in the sense that detection is *certain* within a specified distance, and cannot happen beyond that distance.

Application to search and rescue

Example 8.4. Suppose that, having failed to locate the swimmer in Example 8.3, the boat proceeds through the water at a constant rate of v meters per second. Derive a formula for the probability that the boat will *fail to detect* the swimmer within a small span of Δt seconds.

Solution: Since the boat moves through the water at v meters per second, the new area scanned by the detector during Δt seconds is $2rv\Delta t$ (see Figure 8.2). Said differently, the fraction of the whole area that's swept out by the detector is $2rv\Delta t/A$, so we take this to be the probability of finding the swimmer in a small span of Δt seconds. Therefore, the probability of *not* finding the swimmer in a

In Example 8.5 we will see what happens when $\Delta t \to 0^+$, so here we use the phrase "small span" to mean *really* small.

given span of Δt seconds is $1 - 2rv\Delta t/A$.   ■

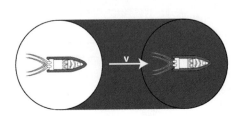
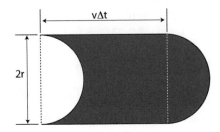

Figure 8.2: The boat travels $v\Delta t$ meters in Δt seconds, and the detection region has a diameter of $2r$, so the new area swept out by the detector in Δt seconds is $2rv\Delta t$.

❖ Independent Events

Suppose that we both toss a coin and roll a die, both of which are fair. What's the probability that we'll see both *heads* <u>and</u> a roll of 3? Let's list the possible outcomes:

Figure 8.3: Flipping a penny and tossing a die are independent events.

Die:	1	1	2	2	3	3	4	4	5	5	6	6
Coin:	H	T	H	T	H	T	H	T	H	T	H	T

Since there are 12 events, all of which are equally likely, the single one that interests us occurs with a probability of 1/12. Note that

$$\frac{1}{12} = \left(\frac{1}{2}\right)\left(\frac{1}{6}\right) = \left(\begin{array}{c}\text{probability of}\\ \text{seeing } heads\end{array}\right) \times \left(\begin{array}{c}\text{probability of}\\ \text{rolling a 3}\end{array}\right),$$

which is indicative of an important fact about probability: when two events are completely unrelated, the probability of *both* happening is just the product of their individual probabilities of occurring.

Application to ecology

Example 8.5. A tiger is wandering randomly through its territory at a steady pace of v meters per second, and the territory has an area of A square meters. The animal has a collar with a radio transmitter that will be detected if carried within r meters of a stationary sensor. Derive a formula for the probability, $p(t)$, that the sensor will detect the tiger within t seconds after it's switched on.

Solution: As in the search-and-rescue examples above, this scenario comes down to a small disk (of radius r about the collar) moving through a large area. A successful detection happens when the disk overlaps a point (the sensor). The difference is that the rescuers from previous examples were actively searching, whereas the tiger is just wandering. We take this to mean that the tiger may revisit some areas before moving on to others, so in any small span of Δt seconds, no matter how long the tiger has been wandering, the probability of being detected is $2rv\Delta t/A$. We'll use this fact in just a moment.

Though our goal is to formulate $p(t)$, many people find it easier to start with the probability that the sensor will *fail* to detect the tiger within t seconds. We'll call that number $f(t)$. Then $f(t + \Delta t)$ is the probability that it will fail to detect the tiger within $t + \Delta t$ seconds. In order for that to happen, the sensor must fail to

> The final step of the solution will be to write $p(t) = 1 - f(t)$.

detect the tiger within t seconds *and* fail to detect the tiger in the Δt seconds after that. That is,

$$f(t + \Delta t) \;=\; \overbrace{f(t)}^{\substack{\text{the probability of failing to detect the}\\ \text{tiger in the first } t \text{ seconds}}} \underbrace{\left(1 - \frac{2rv}{A}\Delta t\right)}_{\substack{\text{the probability of failing to detect the}\\ \text{tiger in the following } \Delta t \text{ seconds}}}.$$

> Recall that when two events are independent, like the coin flip and the die roll above, the probability of *both* happening is calculated by multiplying their individual probabilities.

When we distribute the $f(t)$ on the right-hand side of this equation, it becomes

$$f(t + \Delta t) = f(t) - \frac{2rv}{A} f(t)\ \Delta t.$$

Then subtracting $f(t)$ from both sides and dividing by Δt leads us to

$$\frac{f(t + \Delta t) - f(t)}{\Delta t} = -\frac{2rv}{A} f(t).$$

> Note that the left-hand side of this equation is the difference quotient for f, and the right-hand side does not depend on Δt.

Considering smaller and smaller Δt (i.e., in the limit as $\Delta t \to 0^+$) we have discovered that

$$f'(t) = -\frac{2rv}{A} f(t),$$

which says that $f'(t)$ is just a scaled version of $f(t)$. We know that only exponential functions have this characteristic, so

$$f(t) = C \exp\left(-\frac{2rv}{A}t\right),$$

> We proved that only exponential functions are described by $f' = \lambda f$ (when $\lambda \neq 0$) on p. 283.
>
> Recall that $\exp(\lambda t)$ is another way of writing $e^{\lambda t}$. It's useful when the exponent has fractions, or subscripts, or superscripts, or is complicated in any way.

where C can be any number. The last part of developing our formula is determining the appropriate value of C in this scenario. Notice that $f(0) = C$, so we can understand C as the probability that the sensor will fail to detect the tiger at the instant that it's switched on. If we assume that the tiger is tagged at some random point in its range, this probability is $1 - \frac{\pi r^2}{A}$ (as in Example 8.3), so

$$f(t) = \left(1 - \frac{\pi r^2}{A}\right) \exp\left(-\frac{2rv}{A}t\right).$$

So the probability that the sensor will detect the tiger within t seconds after its engaged is $p(t) = 1 - f(t)$.   ∎

Suppose we want to know the probability that the tiger from Example 8.5 is found at some time $t \in [a, b]$. The number $p(b)$ is the probability that the tiger is found at any time in $[0, b]$, so $p(b)$ is too large; but if we subtract the probability that the tiger is found at time $t \in [0, a]$, we get our answer. The probability of detecting the tiger at some time $t \in [a, b]$ is

$$p(b) - p(a) = \int_a^b p'(t)\ dt. \tag{8.1}$$

This relationship between probability and the definite integral is important to the study of *continuous random variables,* which we discuss in some detail below.

❖ Probability Density

Instead of talking about the likelihood of an event happening within a certain span of time, we might be interested in the likelihood that the amount of active ingredient in a medication is within 1% of the manufacturer's claim, the chances that a stock price will end the day within a certain range of dollar values, or any of a host of other scenarios. In general, when a quantity arises from a random process and can be realized in a continuous range of values (whether a finite or infinite interval), we say that it is a **continuous random variable**.

If X is a continuous random variable, it's common to denote by $P(a \leq X \leq b)$ the probability that a realization of X lies in $[a, b]$. It's a subtle but important fact that every continuous random variable has an associated non-negative function f for which

$$P(a \leq X \leq b) = \int_a^b f(x) \, dx.$$

> By "realization," we mean "the value of any specific sampling of the variable." For example, *weight* is a continuous random variable because it can take any value in $(0, \infty)$. Your friend's weight is a realization of that random variable. Your weight is another. These two realizations might be the same, or might not be.

The function f is called the *probability density function* of X, and it plays a central role in calculating statistical quantities like *expected value* (average), *variance*, and *standard deviation* (the latter two of which describe how tightly realizations of the random variable tend to cluster around the mean).

A **probability density function (pdf)** for a random variable X is a function $f(x)$ whose values are never negative, for which

$$\int_{-\infty}^{\infty} f(x) \, dx = 1 \quad \text{and} \quad P(a \leq X \leq b) = \int_a^b f(x) \, dx.$$

Calculating probability with a pdf

Example 8.6. Suppose that X is a continuous random variable whose probability density function is

$$f(x) = \begin{cases} 1 - |x| & \text{when } |x| \leq 1 \\ 0 & \text{otherwise} \end{cases}$$

Determine the probability that a realization of X will occur in $[0.5, 0.75]$.

Solution: Since $|x| = x$ in $[0.5, 0.75]$, we have

$$P(0.5 \leq X \leq 0.75) = \int_{0.5}^{0.75} 1 - |x| \, dx = \int_{0.5}^{0.75} 1 - x \, dx = \frac{3}{32}. \qquad \blacksquare$$

The probability of realizing any particular value is zero

Example 8.7. Suppose X is a continuous random variable whose probability density function is $f(x)$. Show that the probability of seeing any particular value is zero.

Solution: We can say that $x = a$ by writing $x \in [a, a]$. While that's a little clumsy for common usage, it allows us to use the probability density function to determine

> If you continue to study probability, later you'll find that the idea of a pdf is extended to include something called a *generalized function*. In that case, there *can* be a nonzero probability of observing a particular value.

$$P(X = a) = P(a \leq X \leq a) = \int_a^a f(x) \, dx = 0. \qquad \blacksquare$$

One of the most important probability density functions is

$$f(x) = \frac{1}{\sqrt{2\pi}} \exp\left(-\frac{x^2}{2}\right),$$

> The notation $\exp(u)$ means e^u. It is particularly handy when u is a complicated expression because it makes the exponent easier to read.

which is called the **standard normal distribution**. The proof that

$$\int_{-\infty}^{\infty} \frac{1}{\sqrt{2\pi}} \exp\left(-\frac{x^2}{2}\right) dx = 1$$

is typically presented using a so-called *double integral* in *polar coordinates*, both of which will be developed later.

Transformed normal distributions

Example 8.8. Show that $g(x) = \frac{1}{\sigma\sqrt{2\pi}} \exp\left(-\frac{(x-\mu)^2}{2\sigma^2}\right)$ is a probability density function.

Solution: The values of $g(x)$ are never negative, so we need only check that the total integral is 1. Recall that

$$\int_{-\infty}^{\infty} g(x)\,dx \quad \text{is defined to be} \quad \int_{-\infty}^{0} g(x)\,dx + \int_{0}^{\infty} g(x)\,dx.$$

The change of variable $z = \frac{x-\mu}{\sigma} \Rightarrow dz = \frac{1}{\sigma}dx$ allows us to write

> Books on statistics will call this change of variables a *z-transform*.

$$
\begin{aligned}
\int_{0}^{\infty} g(x)\,dx &= \lim_{T\to\infty} \int_{0}^{T} \frac{1}{\sigma\sqrt{2\pi}} \exp\left(-\frac{(x-\mu)^2}{2\sigma^2}\right) dx \\
&= \lim_{T\to\infty} \int_{-\mu/\sigma}^{(T-\mu)/\sigma} \frac{1}{\sqrt{2\pi}} \exp\left(-\frac{z^2}{2}\right) dz = \int_{-\mu/\sigma}^{\infty} \frac{1}{\sqrt{2\pi}} \exp\left(-\frac{z^2}{2}\right) dz
\end{aligned}
$$

and

$$
\begin{aligned}
\int_{-\infty}^{0} g(x)\,dx &= \lim_{T\to\infty} \int_{-T}^{0} \frac{1}{\sigma\sqrt{2\pi}} \exp\left(-\frac{(x-\mu)^2}{2\sigma^2}\right) dx \\
&= \lim_{T\to\infty} \int_{-(T+\mu)/\sigma}^{-\mu/\sigma} \frac{1}{\sqrt{2\pi}} \exp\left(-\frac{z^2}{2}\right) dz = \int_{-\infty}^{-\mu/\sigma} \frac{1}{\sqrt{2\pi}} \exp\left(-\frac{z^2}{2}\right) dz,
\end{aligned}
$$

both of which are convergent improper integrals. Their sum is

$$\int_{-\infty}^{-\mu/\sigma} \frac{1}{\sqrt{2\pi}} \exp\left(-\frac{z^2}{2}\right) dz + \int_{-\mu/\sigma}^{\infty} \frac{1}{\sqrt{2\pi}} \exp\left(-\frac{z^2}{2}\right) dz = \int_{-\infty}^{\infty} \frac{1}{\sqrt{2\pi}} \exp\left(-\frac{z^2}{2}\right) dz,$$

which is 1, according to the comments above. ∎

The general form of the **normal distribution** is

$$\frac{1}{\sigma\sqrt{2\pi}} \exp\left(-\frac{(x-\mu)^2}{2\sigma^2}\right).$$

Any continuous random variable whose probability density function has this form is said to be **normally distributed**.

❖ Expected Value

Suppose your website makes money by posting advertisements. You get 0.5 cents for displaying an ad, and another 0.25 cents if a user clicks on it. On any given day, your site gets 10,000 hits, and 2000 of them typically result in a click on the banner ad at the top of the page, so the banner ad typically generates

$$(8000 \text{ display only}) \times \left(0.5 \ \frac{\text{cents}}{\text{display}}\right) + (2000 \text{ clicks}) \times \left(0.75 \ \frac{\text{cents}}{\text{click}}\right) = 5500 \text{ cents}$$

each day. So on average, you expect to earn $5500/10000 = 0.55$ cents per hit. Alternatively, we say that the *expected value* of the ad is 0.55. Note that

$$
\begin{aligned}
\frac{5500}{10000} &= \frac{(8000)(0.5) + (2000)(0.75)}{10000} \\
&= \frac{(8000)(0.5)}{10000} + \frac{(2000)(0.75)}{10000} \\
&= (0.5)\left(\frac{4}{5}\right) + (0.75)\left(\frac{1}{5}\right).
\end{aligned}
$$

We've written the expected value like this in order to display its relationship to probability. The probability that the banner ad generates 0.75 cents of profit is

$$
P(0.75) = \frac{2000}{10000} = \frac{1}{5},
$$

and the probability that it generates a profit of 0.5 cents is

$$
P(0.5) = \frac{8000}{10000} = \frac{4}{5},
$$

so the structure of our formula for expected value is

$$
\text{expected value} = \sum_k (\text{value}_k)(\text{probability of earning value}_k). \qquad (8.2)
$$

We can extend this formula to continuous random variables. Suppose the probability density function for X is the continuous function $f(x)$, whose value is zero outside of $[a, b]$. When we use a regular partition to subdivide $[a, b]$ into subintervals of length Δx, we can say the probability that a realization of X occurs in $[x_{k-1}, x_k]$ is $\int_{x_{k-1}}^{x_k} f(x)\, dx$, and the Mean Value Theorem for Integrals (see p. 558) guarantees that there is some point $x_k^* \in [x_{k-1}, x_k]$ at which this probability is

$$
\int_{x_{k-1}}^{x_k} f(x)\, dx = f(x_k^*)\Delta x.
$$

When Δx is small, we can approximation all $x \in [x_{k-1}, x_k]$ as being roughly equal to x_k^*, so when we use equation (8.2) to estimate the expected value of X, which is often denoted by $\mathcal{E}(X)$ or μ, we see

$$
\mathcal{E}(X) \approx \sum_{k=1}^{n} \underbrace{x_k^*}_{} \underbrace{f(x_k^*)\Delta x}_{}. \qquad (8.3)
$$

this is playing the   this is the probability that a re-
role of (value$_k$)   alization of X has that value

We pass from approximation to an exact calculation in the limit

$$
\mathcal{E}(X) = \lim_{n \to \infty} \sum_{k=1}^{n} x_k^* f(x_k^*)\Delta x = \int_a^b x\, f(x)\, dx = \int_{-\infty}^{\infty} x\, f(x)\, dx,
$$

where the last equality is true because $f(x) = 0$ outside of $[a, b]$. However, this integral still tells us the expected value, even when the probability density function of X is never zero.

Expected Value of a Continuous Random Variable: Suppose X is a continuous random variable whose probability density function is $f(x)$. Then the **expected value** of X is

$$
\mathcal{E}(X) = \int_{-\infty}^{\infty} x\, f(x)\, dx.
$$

Calculating expected value

Example 8.9. Determine the expected value of the random variable X whose probability density function is

$$f(x) = \begin{cases} 0 & \text{if } x < 0 \\ 2e^{-2x} & \text{if } x \geq 0. \end{cases}$$

Solution: Since $f(x) = 0$ when $x < 0$, the expected value of X is

$$\mathcal{E}(X) = \int_{-\infty}^{\infty} x\, f(x)\, dx = \int_{-\infty}^{0} x\, f(x)\, dx + \int_{0}^{\infty} x\, f(x)\, dx = \int_{0}^{\infty} x\, f(x)\, dx.$$

When we use the change of variable $u = 2x$, $du = 2\, dx$, we can rewrite the remaining integral as

$$\int_{0}^{\infty} x\, f(x)\, dx = \lim_{T \to \infty} \int_{0}^{T} x\left(2e^{-2x}\right) dx = \lim_{T \to \infty} \int_{0}^{2T} \frac{u}{2}\, e^{-u}\, du,$$

which is calculated using an integration by parts.

$$\lim_{T \to \infty} \int_{0}^{2T} \frac{u}{2}\, e^{-u}\, du = \lim_{T \to \infty} \left(-Te^{-2T} - \frac{1}{2}e^{-2T} + \frac{1}{2}\right) = \frac{1}{2}. \qquad \blacksquare$$

> If you don't understand why $\lim_{T \to \infty} Te^{-2T} = 0$, try writing it as
>
> $$\lim_{T \to \infty} \frac{T}{e^{2T}}.$$
>
> The limit results in ∞/∞, so we can apply L'Hôpital's Rule.

Try It Yourself: Calculating expected value

Example 8.10. Determine the expected value of the normal distribution $g(x) = \frac{1}{\sigma\sqrt{2\pi}} \exp\left(-\frac{(x-\mu)^2}{2\sigma^2}\right)$. *(Hint: change variables as we did in Example 8.8.)*

Answer: $\mathcal{E}(X) = \mu$. $\qquad \blacksquare$

Full Solution On-line

In fact, the expected value of any random variable X is typically denoted by μ, or by μ_x if there's the possibility of confusion with other random variables.

❖ Variance and Standard Deviation

Suppose X is a continuous random variable. We know how to calculate its expected value, μ, but we also know that actually seeing that value occur has a probability of zero (see Example 8.7). That leads us to ask, "how far do realizations of X vary away from μ, on average?" To answer the question, we could calculate the expected value of $|X - \mu|$. In practice however, we calculate $\sqrt{\mathcal{E}(\,(X-\mu)^2\,)}$ instead. Algebraically, this has the benefit of avoiding the absolute value, and later we'll see that it arises naturally from an understanding of vectors. The number $\sqrt{\mathcal{E}(\,(X-\mu)^2\,)}$ is called the *standard deviation* of X and the radicand is called its *variance*.

> **Variance and Standard Deviation:** Suppose X is a continuous random variable whose probability density function is $f(x)$ and whose expected value, μ, is finite. Then the **variance** of X, denoted by $\text{Var}(X)$, is defined to be
>
> $$\text{Var}(X) = \mathcal{E}\left(\,(X-\mu)^2\,\right) = \int_{-\infty}^{\infty} (x-\mu)^2 f(x)\, dx,$$
>
> and the **standard deviation** of X, denoted by $\sigma(X)$ or σ_x, is defined as
>
> $$\sigma(X) = \sqrt{\text{Var}(X)}.$$

The relationship between variance and standard deviation is typically written as

$$\text{Var}(X) = \sigma^2.$$

In Figure 8.4 you see different normal distributions that have the same expected value (zero) but different standard deviations. On p. 272 (exercise #106) you showed that the inflection points of the graph occur at $\pm\sigma$. In this sense you can understand σ as a measure of the width of a normal distribution about its mean.

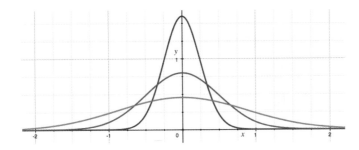

Figure 8.4: Normal distributions with $\sigma = 1/4$, $\sigma = 1/2$, and $\sigma = 7/8$, all with $\mu = 0$.

Calculating variance of a normal distribution

Example 8.11. Suppose X is a normally distributed continuous random variable. Determine (a) its variance and (b) its standard deviation.

Solution: (a) Because X is normally distributed, its probability density function is

$$f(x) = \frac{1}{\sigma\sqrt{2\pi}} \exp\left(-\frac{(x-\mu)^2}{2\sigma^2}\right),$$

and $\mathcal{E}(X) = \mu$ (see Example 8.10), so the variance of X is

$$\text{Var}(X) = \frac{1}{\sigma\sqrt{2\pi}} \int_{-\infty}^{\infty} (x-\mu)^2 \exp\left(-\frac{(x-\mu)^2}{2\sigma^2}\right) dx.$$

Since this is a convergent improper integral, we can write it as

$$\text{Var}(X) = \frac{1}{\sigma\sqrt{2\pi}} \lim_{T\to\infty} \int_{-T}^{T} (x-\mu)^2 \exp\left(-\frac{(x-\mu)^2}{2\sigma^2}\right) dx.$$

Then the change of variable that we've used in all our calculations with the normal distribution, $z = \frac{x-\mu}{\sigma}$ and $dz = \frac{dx}{\sigma}$, allows us to rewrite it as

$$\text{Var}(X) = \frac{1}{\sigma\sqrt{2\pi}} \lim_{T\to\infty} \int_{-(T+\mu)/\sigma}^{(T-\mu)/\sigma} (\sigma z)^2 e^{-z^2/2} (\sigma\, dz) = \frac{\sigma^2}{\sqrt{2\pi}} \int_{-\infty}^{\infty} z^2 e^{-z^2/2}\, dz.$$

Now an integration by parts allows us to calculate. Specifically, let's designate

$$\begin{aligned} u &= z & v' &= z e^{-z^2/2} \\ u' &= 1 & v &= -e^{-z^2/2}. \end{aligned}$$

Then we can write this improper integral as

$$\frac{\sigma^2}{\sqrt{2\pi}} \int_{-\infty}^{\infty} z^2 e^{-z^2/2}\, dz = \lim_{L\to\infty} \frac{\sigma^2}{\sqrt{2\pi}} \int_{-L}^{L} z^2 e^{-z^2/2}\, dz$$

$$= \lim_{L\to\infty} \left(-\frac{\sigma^2}{\sqrt{2\pi}} L e^{-L^2/2} + \frac{\sigma^2}{\sqrt{2\pi}} \int_{-L}^{L} e^{-z^2/2}\, dz \right)$$

$$= \frac{\sigma^2}{\sqrt{2\pi}} \int_{-\infty}^{\infty} e^{-z^2/2}\, dz = \sigma^2.$$

(b) We've done the hard work already, by calculating $\mathrm{Var}(X) = \sigma^2$. The standard deviation is the square root of this number, σ. ∎

Calculating probability with normal distributions

Example 8.12. Suppose X is a normally distributed continuous random variable. Determine the probability that a realization of X will occur within one standard deviation of $\mathcal{E}(X)$.

Solution: Since X is normally distributed, we know its probability density function has the form

$$g(x) = \frac{1}{\sigma\sqrt{2\pi}} \exp\left(-\frac{(x-\mu)^2}{2\sigma^2} \right),$$

where $\mu = \mathcal{E}(X)$ and σ is the standard deviation. The probability that a realization of X is in $[\mu - \sigma, \mu + \sigma]$ is

$$\int_{\mu-\sigma}^{\mu+\sigma} \frac{1}{\sigma\sqrt{2\pi}} \exp\left(-\frac{(x-\mu)^2}{2\sigma^2} \right)\, dx = \int_{-1}^{1} \frac{1}{\sqrt{2\pi}} \exp\left(-\frac{z^2}{2} \right)\, dz,$$

where $z = (x-\mu)/\sigma$. Because we don't know an antiderivative of the integrand, we cannot make this calculation with the Fundamental Theorem of Calculus, and have to use a numerical method. When we employ Simpson's method with 20 subintervals, we calculate 0.682690. Note that the particular values of σ and μ are not relevant to the calculation. Our answer is true for *all* normal distributions. ∎

Try It Yourself:　Calculating standard deviation

Example 8.13. Suppose X is the continuous random variable from Example 8.9. Calculate the standard deviation of X.

Answer: $\sigma = 1/\sqrt{8}$. ∎

Full Solution
On-line

✥ Connections

Before practicing the ideas of this section in the exercise set, let's stop to mention the connection to ideas that we've discussed previously. In both examples below we think of the region between the graph of the probability density function f and the x-axis as a thin plate whose density is constant $\rho\ \frac{\text{unit mass}}{\text{unit area}}$.

▷ Center of mass

On p. 542 we found that the x-coordinate of this plate's center of mass is

$$\bar{x} = \frac{M_y}{m} = \frac{\int_{-\infty}^{\infty} x\rho f(x)\, dx}{\int_{-\infty}^{\infty} \rho f(x)\, dx}, \quad \text{which reduces to} \quad \int_{-\infty}^{\infty} x f(x)\, dx$$

since ρ is constant and f has a total integral of 1 (it's a probability density function). This is the integral for expected value! In this sense, we might say that the center of mass is the x-coordinate where we *expect* to find the mass of the plate.

▷ Kinetic energy

Suppose the plate is loosely attached to the line $x = \mu$ (where μ is the expected value of the random variable), and can revolve about that axis. If we spin it at ω radians per second, equation (5.5) on p. 545 tells us that the plate's kinetic energy is

$$\int_{-\infty}^{\infty} \frac{\omega^2 \rho}{2} (x - \mu)^2 f(x) \, dx = \frac{\omega^2 \rho}{2} \mathrm{Var}(X).$$

This means that a plate whose mass is concentrated near $x = \mu$ (i.e., $\mathrm{Var}(X)$ is small) will have less kinetic energy than the one whose mass is spread out (i.e., $\mathrm{Var}(X)$ is large). For example, Figure 8.5 shows two normal distributions with $\mu = 0$. When we spin these "plates" about $x = 0$, mass located at large x travels faster than mass close to the axis of revolution. When more of the mass is father away from $x = 0$, more of it is traveling fast, so we have a greater kinetic energy.

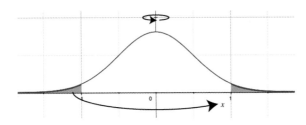 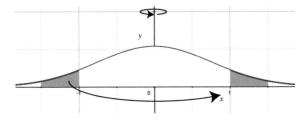

Figure 8.5: (left) when $\sigma = 0.5$, only a little bit of mass travels fast; (right) when $\sigma = 0.75$, more of the mass is far away (and so traveling fast) so the rotational kinetic energy of the plate is larger.

Regardless of their interpretations, the integrals

$$\int_{-\infty}^{\infty} (x - \mu) f(x) \, dx \quad \text{and} \quad \int_{-\infty}^{\infty} (x - \mu)^2 f(x) \, dx$$

are often called the **first moment** and **second moment** of f about μ.

You should know

- the terms *continuous random variable*, *probability density function (pdf)*, *expected value*, *normal distribution*, *variance*, and *standard deviation*;

- the notation $P(a \leq X \leq b), \mathcal{E}(X), \mu$ (or μ_x), $\mathrm{Var}(X)$, and σ (or σ_x);

- that σ and $\mathrm{Var}(X)$ measure the extent to which realizations of X tend to vary from $\mathcal{E}(X)$.

You should be able to

- use a given pdf to calculate $P(a \leq X \leq b)$;

- calculate $\mathcal{E}(X)$, $\mathrm{Var}(X)$, and σ_x for a continuous random variable when given the associated pdf.

✧ 7.8 Skill Exercises

In #1–6 (a) determine the value of k so that $f(x)$ is a probability density function, and then (b) find the expected value of $f(x)$.

1. $f(x) = \dfrac{k}{x^2 + 1}$

2. $f(x) = \begin{cases} k\sin(x) & \text{when } x \in [0, \pi] \\ 0 & \text{otherwise} \end{cases}$

3. $f(x) = \begin{cases} \cos(x) & \text{when } x \in [-k, k] \\ 0 & \text{otherwise} \end{cases}$

4. $f(x) = \begin{cases} \frac{x^2}{x^2+1} & \text{when } x \in [-k, k] \\ 0 & \text{otherwise} \end{cases}$

5. $f(x) = \begin{cases} kx^2 & \text{when } x \in [-1, 2] \\ 0 & \text{otherwise} \end{cases}$

6. $f(x) = \begin{cases} e^{kx} & \text{when } x \in [0, \infty) \\ 0 & \text{otherwise} \end{cases}$

In #7–12 (a) show that $f(x)$ is a probability density function, and (b) find its variance.

7. $f(x) = 2 - 2x$ over $[0, 1]$ and zero otherwise

8. $f(x) = \dfrac{3}{16}\sqrt{x}$ over $[0, 4]$ and zero otherwise

9. $f(x) = \dfrac{3}{2}x - \dfrac{3}{4}x^2$ over $[0, 2]$ and zero otherwise

10. $f(x) = \ln(x)$ over $[1, e]$ and zero otherwise

11. $f(x) = \frac{1}{\pi}\arccos(x)$ over $[-1, 1]$ and zero otherwise

12. $f(x) = \dfrac{4\arctan(x)}{\pi - 2\ln(2)}$ over $[0, 1]$ and zero otherwise

❖ 7.8 Concept and Application Exercises

13. Suppose the random variable X has units of kg. What are the units associated (a) with $\mathcal{E}(X)$, (b) with $\text{Var}(X)$, and (c) with $\sigma(X)$?

14. Based on the graph of f shown in Figure 8.6, explain why f is not a probability density function.

15. Based on the graph of f shown in Figure 8.6, explain why f is not a probability density function.

Figure 8.6: (left) the graph of f for #14; (right) the graph of f for #15

16. Suppose $f(x) = x^{-2}$ when $x \in [1, \infty)$ and $f(x) = 0$ otherwise. (a) Verify that $f(x)$ is a probability density function, and (b) calculate its expected value.

17. The *uniform distribution* is probability distribution defined by the function

$$ f(x) = \begin{cases} \frac{1}{b-a}, & a \leq x \leq b \\ 0, & \text{otherwise} \end{cases} . $$

The uniform distribution is analogous to the probability distribution of rolling a fair die in the sense that all outcomes are equally likely. Specifically, in the case of the uniform distribution, all subintervals of $[a, b]$ *of the same length* are equally likely to include a realization of the random variable.

(a) Show that the uniform distribution meets the criteria for a probability distribution function for any $a, b \in \mathbb{R}$ provided $a < b$.

(b) What is the expected value and standard deviation of the uniform distribution?

(c) For $a = 1$ and $b = 25$ show that any subinterval of length 3 has the same probability of occurring.

18. The exponential distribution

$$f(t) = \begin{cases} 0, & t < 0 \\ \lambda e^{-\lambda t}, & t \geq 0 \end{cases}$$

is often used to model waiting times, where $\lambda > 0$.

    (a) Show that the exponential distribution meets the criteria for a probability distribution function.

    (b) Determine the expected value of the exponential distribution.

    (c) Determine the standard deviation of the exponential distribution.

19. When you visit your doctor, suppose you expect to wait five minutes in the examination room before your doctor comes in for your appointment, and the distribution of wait times is exponentially distributed (see #18).

    (a) Find the value of λ that yields an expected wait time of five minutes.

    (b) What is the probability that you will wait no more than eight minutes for the doctor?

    (c) What is the probability that you will have to wait at least 15 minutes?

    (d) What is the standard deviation for your function?

    (e) What is the probability that your wait will fall within one standard deviation from the expected value?

20. Suppose a company has developed a long-lasting battery to be used in hybrid and all-electric cars. The experimental data collected by the Research & Development Division indicates that the probability density function

$$f(t) = \begin{cases} \frac{1}{94478}(18t^3 - t^4), & \text{for } t \in [0, 18] \text{ years} \\ 0, & \text{otherwise} \end{cases}$$

can be used to predict when the battery will have to be replaced.

    (a) What is the probability that a battery will last ten years?

    (b) What is the probability that a battery will last ten years or less?

    (c) What is the probability that a battery will last 15-20 years?

    (d) The Marketing Division would like to advertise the battery as "guaranteed for 15 years." Is this claim supported by the experimental data?

    (e) What is the probability that a battery will fail within 1 standard deviation of the expected value? Support your answer using the techniques from this section.

Exercises #21–#26 focus on the presence of substances (vitamins, minerals, etc.) in the blood stream of adult humans. Each substance is normally distributed over the population.

21. The amount of hemoglobin in blood plasma has a mean of 24.5 mg/L and a standard deviation of 7.25 mg/L. Write down the integral that determines the probability that a particular measurement of hemoglobin falls between 10 and 17.25 mg/L.

22. The amount of vitamin C is normally distributed with a mean of 17 mg/L and a standard deviation of 12.5. Write down the integral that determines the probability that a particular measurement of vitamin C falls within one standard deviation of the mean.

     — Medical Science

23. The mineral zinc supports the human immune system and plays a role in regulating genetic activity. The following questions refer to the integral

     — Medical Science

$$\int_6^{6.7} \frac{1}{0.525\sqrt{2\pi}} \exp\left(-\frac{(x-5.65)^2}{2(0.525)^2}\right) dx$$

in which the integrand is the probability density function for the distribution of zinc levels (mg/L) across the adult population.

(a) What is the mean of the zinc distribution?

(b) What is the standard deviation of the zinc distribution

(c) What does the value of this integral mean to us?

24. Carbon dioxide (CO_2) is usually thought of as the waste product in respiration, but CO_2 also helps dilate the smooth muscle tissues, regulates the cardiovascular system, regulates the alkaline/acid balance of the body and plays a role in the proper functioning of the digestive system. The following questions refer to the integral

     — Medical Science

$$\int_{0.5}^{1} \frac{1}{0.1\sqrt{2\pi}} \exp\left(-\frac{(x-1.2)^2}{2(0.1)^2}\right)$$

in which the integrand is the probability density function for the distribution of CO_2 levels (g/L) across the adult population.

(a) What is the mean of the CO_2 distribution?

(b) What is the standard deviation of the CO_2 distribution

(c) What does the value of this integral mean to us?

25. Vitamin B12 helps prevent heart disease. If the distribution of B12 has a mean of 415 nanograms per liter (ng/L) and a standard deviation of 142.5 ng/L, use Simpson's Rule (with $n = 20$) to approximate the probability that somebody's vitamin B12 levels fall within two standard deviations of the mean.

     — Medical Science

26. The amount of calcium in the blood (in mg/L) has a mean of 9.265 and a standard deviation of 0.3825. Use Simpson's Rule (with $n = 16$) to approximate the probability of somebody's calcium levels falling between 0 and 8.5 mg/L.

     — Medical Science

27. The American Diabetes Association recommends that blood sugar levels be kept within 70–130 mg/dL (milligrams per decaliter). The American adult population has a mean blood sugar level of 97.3 mg/dL and a standard deviation of 9.3 mg/dL. Use Simpson's Rule with $n = 50$ to approximate the probability that a patient's blood sugar will fall *outside* the recommended range.

     — Medical Science

Chapter 7 Review

❖ True or False

1. Suppose C is the portion of $y = x^2$ for which $x \in [0, 1]$. The surface generated by revolving C about the y-axis has a larger area than the surface generated by revolving C about the x-axis.

2. Suppose f is a function for which $|f(x)| \leq 2\sqrt{1 + (f'(x))^2}$, which is always finite, and C is the portion of its graph over $[a, b]$. Then revolving C about the x-axis produces a surface of revolution whose area is numerically larger than the volume it encloses.

3. Suppose we generate a solid of revolution by revolving a region $\mathcal{R}$ about the line $x = x_0$. If $x = x_0$ is tangent to $\mathcal{R}$, we can always use the method of disks to calculate the volume of the resulting solid.

4. Suppose $f(x) \geq 0$ and C is a segment of the graph of f. If the area between C and the horizontal axis is finite, so is the length of C.

5. To find the volume of a solid of revolution that has been rotated about the y-axis, you must use the cylindrical shells method.

6. Suppose $\mathcal{R}$ is the region between the line $y = 4$ and the curve $y = \sin(x)$, $0 \leq x \leq 3\pi$. If we produce a solid by revolving $\mathcal{R}$ about the y-axis, the method of cylindrical shells is more appropriate for calculating the volume than the method of disks (washers).

7. Suppose $\mathcal{R}$ is the region between the line $y = 4$ and the curve $y = \sin(x)$, $0 \leq x \leq 3\pi$. If we produce a solid by revolving $\mathcal{R}$ about the line $y = 10$, the method of cylindrical shells is more appropriate for calculating the volume than the method of disks (washers).

8. The integral $\int_3^{10} 2\pi(8 - x)e^x \, dx$ is the calculation of a volume using the cylindrical shells method.

9. If you pull against a spring, but the spring is stronger than you and so contracts, you have done *negative* work.

10. A lamina's center of mass must be contained somewhere inside the lamina.

11. A vertical dam face sustains less hydrostatic pressure than one that's tilted, given the same depth of water.

12. Suppose $f(t) = 0$ when $t < 0$, and $f(t) = e^{-3t}$ when $t \geq 0$. Then f is a probability density function.

13. If f is a probability density function, the probability that a realization of a continuous random variable falls within $[a, b]$ is $\int_a^b f(x) \, dx$.

14. Suppose X is a continuous random variable whose probability density function, f is zero outside of $[a, b]$. Then the expected value of X is $\frac{1}{b-a} \int_a^b f(x) \, dx$.

15. Suppose f is a continuous function. Then $\lim_{\Delta x \to 0} \frac{1}{2\Delta x} \int_{a-\Delta x}^{a+\Delta x} f(t) \, dt = f(a)$.

16. Suppose f is a continuous function that's positive over $[a, b)$, zero at $x = b$, and negative over $(b, c]$. If the average value of f is positive, $b - a > c - b$.

✦ Multiple Choice

17. The x-coordinate of a lamina's center of mass is ...

   (a) $\bar{x} = m/M_x$
   (b) $\bar{x} = m/M_y$
   (c) $\bar{x} = M_x/m$

   (d) $\bar{x} = M_y/m$
   (e) none of the above

18. To find the arclength of $y = 4x^7$ from $x = 1$ to $x = 2$ we use ...

   (a) $\int_1^2 \sqrt{1 - (4x^7)^2}\, dx$

   (b) $\int_1^2 \sqrt{1 - (28x^6)^2}\, dx$

   (c) $\int_1^2 \sqrt{1 - 28x^2}\, dx$

   (d) $\int_1^2 \sqrt{1 - 4x^7}\, dx$

   (e) none of the above

19. Suppose S is the surface of revolution generated by revolving the graph of f about the interval $[a, b]$ on the x-axis. Then the area of S is ...

   (a) $\int_a^b 2\pi |f(x)| \sqrt{1 + (f'(x))^2}$
   (b) the integral of (circumference)·(arclength element)
   (c) approximated by the sum of areas of frustums
   (d) all of the above
   (e) none of the above

20. The integral $\int_0^2 2\pi x \left((2x) - (x^2)\right) dx$ can be interpreted as calculating the volume of a solid of revolution that's generated by revolving the region between $y = 2x$ and $y = x^2$ ...

   (a) about the y-axis using the washers method
   (b) about the x-axis using the washers method
   (c) about the x-axis using the shells method
   (d) about the y-axis using the shells method
   (e) none of the above

21. To find the volume of the solid obtained by revolving the first quadrant region under $y = 4 - x^2$ about the x-axis with the disk method we use ...

   (a) $\int_0^2 2\pi(4 - x^2)^2\, dx$

   (b) $\int_0^2 \pi(4 - x^2)^2\, dx$

   (c) $\int_0^4 \pi(4 - y)\, dy$

   (d) $\int_0^2 2\pi x(4 - x^2)\, dx$

   (e) none of the above

22. The integral $\int_0^{0.5927} 2\pi x\sqrt{1 - x^2}\, dx$ can be interpreted as ...

   (a) the first moment of a linear object

    (b) a volume of a solid of revolution

    (c) a surface area

    (d) the probability that the realization of a random variable will occur between $0 \le x \le 0.5927$

    (e) all of the above

23. Suppose X is a continuous random variable whose expected value is μ. The probability that a realization of X will be μ is ...

    (a) zero

    (b) one

    (c) higher than the probability of achieving any other value

    (d) dependent on the variance of X

    (e) none of the above

24. Suppose f is a continuous function that's positive over $[1, 2)$ and negative over $(2, 10]$, and $\langle f \rangle$ is its the average value over $[1, 10]$. Then ...

    (a) $\langle f \rangle$ is positive   (d) $\langle f \rangle$ is non-positive

    (b) $\langle f \rangle$ is non-negative   (e) we need more information in order

    (c) $\langle f \rangle$ is negative   to guarantee anything about $\langle f \rangle$

✤ Exercises

25. Suppose $\mathcal{C}$ is the portion of $y = 4^x$ for which $x \in [0, 1]$. Use the trapezoid rule with $\Delta x = 0.1$ to approximate the length of $\mathcal{C}$.

26. Suppose $\mathcal{C}$ is the portion of $y = \frac{4}{65}x^{13/4} + \frac{1}{x^{5/4}}$ for which $x \in [1, 4]$. Determine the arc length of $\mathcal{C}$.

27. Suppose $\mathcal{C}$ is the portion of $y = \sin(x)$ for which $x \in [0, \pi/4]$. Determine the area of the surface that's generated by revolving $\mathcal{C}$ about the x-axis.

28. Suppose $\mathcal{R}$ is the region between $y = e^{-0.01x}$ and the interval $x \in [0, \infty)$. Show that revolving $\mathcal{R}$ about the y-axis produces a solid of revolution whose volume is finite but whose surface area is infinite.

In #29–32 $\mathcal{R}$ is the region between the curves $y = 2 - \sin(\pi x)$ and $y = x^2 - 4x$ when $x \in [0, 4]$.

29. Determine the volume of the solid that's generated by revolving $\mathcal{R}$ about the line $x = 0$.

30. Determine the volume of the solid that's generated by revolving $\mathcal{R}$ about the line $y = -5$.

31. Determine the x-coordinate of the center of mass of $\mathcal{R}$, assuming it has constant density.

32. Determine the y-coordinate of the center of mass of $\mathcal{R}$, assuming it has constant density.

In #33–36 $\mathcal{R}$ is the region between the curves $y = 9 - x^2$ and the horizontal axis when $x \in [-1, 2]$.

33. Determine the volume of the solid that's generated by revolving $\mathcal{R}$ about the line $x = 6$.

34. Determine the volume of the solid that's generated by revolving $\mathcal{R}$ about the line $y = 0$.

35. Determine the x-coordinate of the center of mass of $\mathcal{R}$, assuming its density is $\rho(x) = 7 - x$.

36. Determine the y-coordinate of the center of mass of $\mathcal{R}$, assuming its density is $\rho(x) = 7 - x$.

37. Consider the region in the Cartesian plane whose lower boundary is $y = 0$ and whose upper boundary is the curve $y = \sin(x)$ for $x \in [0, \pi]$. At each x an isosceles triangle sits perpendicular to the Cartesian plane. Its base stretches from $(x, 0)$ to $(x, \sin(x))$, and its altitude has a length of x. If all lengths are measured in millimeters, what's the volume enclosed by these triangles?

38. Suppose $\mathcal{R}$ is the region between the curves $y_1 = 1.5x^2 - 3.5x + 3$ and $y_2 = x^3 - 4.5x^2 + 5.5x + 3$, where $x \in [0, 3]$. At each $x \in [0, 3]$ a semicircle sits perpendicular to the Cartesian plane, it's diameter stretching from y_1 to y_2. If all lengths are measured in centimeters, determine the volume enclosed by these circles.

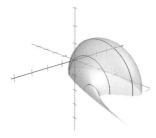

Figure 9.1: The volume described in #38, as seen from a vantage point above the 1st quadrant.

39. Suppose $\mathcal{R}$ is the region from #38. Determine the volume of the solid that's generated by revolving $\mathcal{R}$ about the line $x = 5$.

40. Suppose $\mathcal{R}$ is the region from #38 and has a density of $\rho(x) = 1 + 3x$. If $\mathcal{R}$ revolves about the line $x = 4$ at 2 radians per second, what is its kinetic energy?

41. Suppose $\mathcal{R}$ is the region between the curves $y = \frac{1}{32}x^2$ and $y = \sqrt[3]{x}$ when $y \in [0, 2]$, and that it has a constant density of $\rho \frac{\text{kg}}{\text{m}^2}$. If $\mathcal{R}$ is revolving about the line $y = -3$ at 4 radians per second, and all lengths are measured in meters, determine its radius of gyration.

42. Suppose the vertical face of a dam is constructed in the shape of an irregular trapezoid. The lower edge of the trapezoid is 30 meters across. The upper edge is 60 meters across, but extends only 5 meters farther to the right. The dam is 40 meters high and the water is currently at the top of the dam. What is the hydrostatic force acting against the face of the dam?

Figure 9.2: For #42 (not drawn to scale).

43. A water storage tank is 15 meters long, and its end faces are equilateral triangles (tip down) with sides of length 4 meters. If the tank is being filled at a constant rate of $1.3 \frac{\text{m}^3}{\text{sec}}$, find a formula that describes the hydrostatic force acting against one of the end faces as a function of time.

44. A spherical tank with a radius of 2 meters is filled completely with water. If the water is allowed to drain out of the bottom of the tank, how much work is done by gravity in emptying the tank?

45. If a 40-meter rope with a total mass of 35 kg is hanging from the side of a scaffolding, determine the amount of work required to haul up the entire length of rope.

46. Suppose a 100-meter rope with a total mass of 45 kg is hanging from the side of a ravine. A 4-kg bucket holding 15 kg of water is attached to the end of the rope. The rope is hauled up at a rate of 3 meters per second, but the water is leaking out of the bucket at $0.25 \frac{\text{kg}}{\text{sec}}$. Determine the amount of work done in winding up the entire length of rope.

47. Suppose $f(x) = 4 - 0.5x$ when $x \in [0, a]$ and $f(x) = 0$ otherwise. Find a value of a for which f is a probability density function.

48. Suppose that X is a non-negative continuous random variable whose probability density function is $f(x) = \begin{cases} 4e^{-4x} & \text{when } x \geq 0 \\ 0 & \text{otherwise} \end{cases}$. Determine (a) the expected value, (b) the variance, (c) the standard deviation of X, and the probability that a realization of X will occur in $[0, 2]$.

49. Find the average value of $f(x) = 3x + 14x^2$ over the interval $[1, 3]$.

50. Suppose $f(x) = \frac{25}{12}x^{-3}$ when $x \in [1, 5]$, and $f(x) = 0$ otherwise.

    (a) Determine the average value of f over $[1, 5]$.

    (b) Calculate the first moment of f about $x = 0$, which we'll denote by $\bar{x}$.

    (c) Show that $f(\bar{x})$ is not your answer from part (a).

51. The expected wait time on a pizza is 20 minutes after placing the order. Answer the following questions, assuming an exponential distribution of wait times (i.e., $f(t) = \lambda e^{-\lambda t}$ for some $\lambda > 0$ when $t \geq 0$, and $f(t) = 0$ for $t < 0$).

    (a) What is the likelihood that you will receive your pizza at most 20 minutes after placing the order?

    (b) What is the likelihood that your pizza will take at least 30 minutes to arrive?

    (c) What is the likelihood that your pizza will arrive between 5 and 10 minutes after placing the order?

    (d) Find the standard deviation and interpret it in terms of pizza delivery times.

Chapter 7 Projects and Applications

❖ Ideal Gas, Energy, and the n-dimensional Ball

When a molecule moves in three dimensions, it has a velocity in the x direction, in the y direction, and in the z direction. If we call these velocities x', y' and z' respectively, we say that the molecule's speed is

$$\|v\| = \sqrt{(x')^2 + (y')^2 + (z')^2}$$

and its kinetic energy is

$$K = \frac{1}{2}m\|v\|^2 = \frac{1}{2}m\Big((x')^2 + (y')^2 + (z')^2\Big)$$

where m is the mass of the molecule. Said differently,

$$(x')^2 + (y')^2 + (z')^2 = \frac{2K}{m}. \tag{10.1}$$

The notation $\| \ \|$ denotes the so-called *magnitude* of a vector, which we'll discuss in Chapter 11.

In the same way that $(x')^2 + (y')^2 = R^2$ describes a circle of radius R in the $x'y'$-plane, equation (10.1) describes a sphere of radius $\sqrt{2K/m}$ in 3-dimensional space.

If the kinetic energy is shared between two molecules of the same gas, we have

$$K = \frac{1}{2}m\Big((x_1')^2 + (y_1')^2 + (z_1')^2\Big) + \frac{1}{2}m\Big((x_2')^2 + (y_2')^2 + (z_2')^2\Big)$$

where we've used subscripts to distinguish between the two molecules (which probably have different speeds). Like before, we can multiply this equation by $2/m$ and rewrite it as

$$(x_1')^2 + (y_1')^2 + (z_1')^2 + (x_2')^2 + (y_2')^2 + (z_2')^2 = \frac{2K}{m},$$

which describes a sphere of radius $\sqrt{\frac{2K}{m}}$ in 6-dimensional space. In general, the equation

$$\sum_{k=1}^{n}(x_k')^2 + (y_k')^2 + (z_k')^2 = \frac{2K}{m} \tag{10.2}$$

describes the kinetic energy of a gas with n identical molecules, and can be understood as a sphere in $3n$-dimensional space. When calculating a thermal quantity called the *entropy*, which is fundamental to the behavior of physical systems, what matters is the surface area of such a sphere (see [11, p. 68]). One way to determine that area is by differentiating the volume of the sphere with respect to the radius, so let's find a formula for the volume enclosed by a sphere of radius r in n dimensions.

▷ The volume of an n-dimensional ball

A *unit ball* is the set of points whose distance from a specified location is no larger than 1. In our case, the specified location will always be the origin. For example:

- the unit ball in one dimension is the interval $[-1, 1]$, whose 1-dimensional volume (commonly called *length*) is 2.

- the unit ball in two dimensions is the disk described by $x^2 + y^2 \leq 1$, whose 2-dimensional volume (commonly called *area*) is π.

- the unit ball in three dimensions is the collection of points described by $x^2 + y^2 + z^2 \leq 1$. On p. 502 we showed that its 3-dimensional volume is $\frac{4}{3}\pi$.

Let's denote the volume of the unit ball in n dimensions by C_n, so that $C_1 = 2$, $C_2 = \pi$, and $C_3 = \frac{4}{3}\pi$. Later in this project you'll be able to derive a formula for C_n. For now, let's talk about what happens when we dilate the unit ball by a factor of r.

- Dilating the interval $[-1, 1]$ by a factor of r gives us the interval $[-r, r]$ whose 1-dimensional volume is $2r$.

- In two dimensions we have $x^2 + y^2 \leq r$, whose 2-dimensional volume is πr^2.

- In three dimensions we get the collection of points described by $x^2 + y^2 + z^2 \leq r$. On p. 502 we showed that its 3-dimensional volume is $\frac{4}{3}\pi r^3$.

In general, dilating the unit ball by a factor of r in n-dimensions results in a ball whose volume is $C_n r^n$, which we'll denote by $V_n(r)$. That is,

$$V_n(r) = C_n r^n \tag{10.3}$$

> We get a factor of r for each perpendicular "direction" in n-space, for fundamentally the same reason that we do when $n = 1, 2$ or 3.

Differentiating this formula with respect to r is easy, but we need to know the value of C_n in order to calculate with it. We can derive a formula for C_n by thinking of $V_n(r)$ as the volume of a solid of revolution. We've already done that twice ...

- In Example 4.9 (p. 443) we used integration to calculate the area of the disk enclosed by $x^2 + y^2 = r^2$. By rewriting the equation of the circle as $y^2 = r^2 - x^2$ we see that each value of x corresponds to a line segment of radius $\sqrt{r^2 - x^2}$. This fact allowed us to calculate the area of the disk with the definite integral

$$\text{area} = \int_{-r}^{r} 2\sqrt{r^2 - x^2} \, dx$$

which, said loosely, adds up the "areas" of lines whose width is dx (see Figure 10.1).

Figure 10.1: Using line segments to calculate the area of a two-dimensional ball (a disk).

- If we rewrite the equation of the sphere as $y^2 + z^2 = r^2 - x^2$, we see that each value of x corresponds to a disk of radius $\sqrt{r^2 - x^2}$ that extends in the y and z directions, as seen in Figure 10.2. Such a disk has an area of $\pi(r^2 - x^2)$. If we give these each a thickness of dx and use a definite integral to add them up, we see

$$\int_{-r}^{r} \pi(r^2 - x^2) \, dx,$$

which is exactly the integral we calculated in Example 2.1 (p. 502).

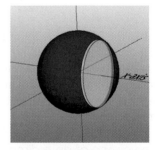

Figure 10.2: Using disks to calculate the volume of a ball in three dimensions.

To calculate the volume of a two-dimensional ball (a disk) we integrated the volumes of one-dimensional balls (line segments). To calculate the volume of a three-dimensional ball we integrated the volumes of two-dimensional balls (disks). The same basic technique works in general. We can calculate the volume of a ball in $(n + 1)$ dimensions by integrating the volumes of n-dimensional balls. Specifically, the sphere of radius r in $(n + 1)$-dimensions is described by the equation

$$\sum_{k=1}^{n+1} x_k^2 = r^2.$$

When we rewrite it as

$$\sum_{k=1}^{n} x_k^2 = r^2 - x_{n+1}^2,$$

we see that each value of x_{n+1} corresponds to an n-dimensional ball of radius $\sqrt{r^2 - x_{n+1}^2}$. According to equation (10.3), the volume of such a ball is

$$V_n\left(\sqrt{r^2 - x_{n+1}^2}\right) = C_n\left(\sqrt{r^2 - x_{n+1}^2}\right)^n.$$

To find the volume of the $(n+1)$-dimensional ball, we simply multiply these n-dimensional volumes by a "thickness" of dx_{n+1}, and use a definite integral add them as x_{n+1} varies from its minimum to its maximum value:

$$V_{n+1}(r) = \int_{-r}^{r} C_n\left(\sqrt{r^2 - x_{n+1}^2}\right)^n dx_{n+1}.$$

1. Rewrite this integral using the trigonometric substitution $x_{n+1} = r\sin(\theta)$.

2. Use your answer from #1 to explain why $C_{n+1} = C_n \int_{-\pi/2}^{\pi/2} \cos^{n+1}(\theta)\, d\theta$.

The next steps of calculation are devoted to getting a handle on the remaining integral, which we'll denote by I_{n+1}. That is,

$$I_{n+1} = \int_{-\pi/2}^{\pi/2} \cos^{n+1}(\theta)\, d\theta \quad \text{so} \quad C_{n+1} = I_{n+1}C_n.$$

3. Calculate I_0 and I_1

4. Assuming that $n > 0$, separate the integrand of I_{n+1} into $\cos(\theta)\cos^n(\theta)$, and integrate by parts.

5. Use the equation that results from step #4 to show that $I_{n+1} = \frac{n}{n+1}I_{n-1}$ when $n > 0$.

6. Use the results of #3 and #4 to calculate values of $I_3, I_4, \ldots, I_{10}$.

7. Use the results of #6 to calculate $C_4, C_5, \ldots C_{10}$.

8. Since $V_n(r) = C_n r^n$, each C_n that you calculated in #7 is the volume of a unit ball in n dimensions. Notice that C_5 is the largest, which happens because—as you should verify now—the number $I_n > 1$ only until $n = 5$.

9. Write down a general formula for C_n in terms of n.

10. The unit ball has a diameter of 2, so it sits in a cube whose volume is 2^n in n dimensions. Show that the ratio $C_n/2^n \to 0$ as n grows large, and explain what this means geometrically.

❖ Asteroid Mining

There are many problems to overcome if we want to establish mining operations that extract minerals from the asteroids in our solar system, not the least of which is the survival of the miners who do it. In this project, you're going to worry about two particular problems that lead to an unhappy outcome:

- Since asteroids are so small (relative to the Earth), they exert much less gravitational force on objects. This means that the escape velocity on such an asteroid is much smaller than it is on Earth, so a miner who was bounding back to his ship could accidently escape the asteriod's pull and jump his way into deep space.

- A more careful but just as unlucky miner could bound with an upward velocity that lands him back on the asteroid after his air supply has been exhausted.

Your job is (1) to determine how high a miner will fly if he jumps straight up with an initial velocity of v_0, and then (2) to determine the maximum safe upward velocity (i.e., the miner will still have oxygen in his pack when he comes back down). Some important details are provided below.

1. This particular miner has been out working for a while and has only 11 minutes of oxygen left in his pack.

2. The mass of the asteroid is 3.8934×10^{13} kilograms, and it has an average radius of 23,000 meters. (We'll assume that the miner starts at this distance from the center of the asteroid.)

3. We will be interested only in vertical position. In particular, we are interested in the miner's distance from the center of the asteroid, which we will denote by x.

4. If the miner's mass is denoted by m, M is the mass of the asteroid, and G the gravitational constant ($G = 6.67 \times 10^{-11} \frac{\text{N·m}^2}{\text{kg}^2}$), the gravitational force exerted on a miner is $F = -\frac{GmM}{x^2}$.

5. When mass is constant, as in this case because the miner and his respiratory gases are encapsulated in his suit, Newton's Second Law reduces to the famous $F = ma$.

6. Because there's no atmosphere to exert drag force, it takes the same length of time to go up as it does to come down.

▷ An initial push (for you, not the miner)

Since the only force the astronaut experiences during each bound is the asteroid's gravity, Newton's Second Law says

$$\underbrace{-\frac{GmM}{x^2}}_{\text{net force}} = \underbrace{mx''}_{\text{(mass)} \times \text{(acceleration)}}.$$

After we cancel the factor of m and multiply by x', we can integrate both sides with respect to time:

$$\int -\frac{GM}{x^2} \, x' \, dt = \int x' \, x'' \, dt$$
$$\frac{GM}{x} + C = \frac{1}{2}\left(x'\right)^2.$$

If we use the letter y to denote the miner's velocity, instead of x', this equation is

$$C = \frac{1}{2}y^2 - \frac{GM}{x}.$$

This equation allows you find the maximum height of a bound, x_f, once you know the miner's initial velocity (which determines the value of C). We know that the miner's initial distance from the asteroid's center is $x_i = 23000$. If we had a formula for time as a function of x, we could calculate

$$\text{the duration of bound} = 2\Big(t(x_f) - t(x_i)\Big).$$

So that's the bulk of your job—develop a formula for time as a function of x. *(Hint: you already have a relationship between y and x.)*

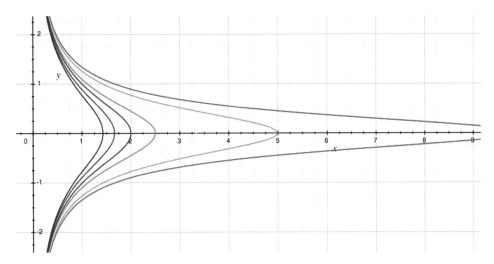

Figure 10.3: The graphs of $C = \frac{1}{2}y^2 - \frac{GM}{x}$ for several values of C (which depends on the miner's initial velocity) when $GM = 1$.

✧ Stopping on a Dime

As mentioned in the previous project, there's no source of friction in space to slow you down. So if you want to stop motion, you have to rely on other forces. Consider the problem of moving a satellite using only the following commands:

1. Turn on the main thruster.

2. Turn off the main thruster.

3. Rotate 180°.

How fast can you make the satellite traverse a linear distance of exactly 1500 kilometers (starting and stopping with a velocity of zero)? When answering, you should consider the following information:

- Fully fueled with hydrazine monopropellant, the satellite's mass is 2420 kg.

- The main thruster produces a force of 96.8 newtons and consumes the hydrazine at a rate of 1.6 grams per second.

- The rotational thrusters consume 0.8 grams per second.

- The satellite requires 400 seconds to complete a rotation of 180°.

- Unlike the previous project, we cannot use $F = ma$ in this scenario because the satellite's thrusters accelerate it by ejecting mass. Instead, you should use the more general $F = \frac{dp}{dt}$, where p denotes momentum (the product of mass and velocity, both of which depend on time).

- Because the mass of the satellite at the beginning of the trip is not the same as the mass toward the end, the duration of the first and last legs of this 1500-km trek will not be the same.

Chapter 8
Sequences and Series

8.1 Sequences

At therapeutic concentrations, the rate at which the human body metabolizes most medicines is proportional to the amount present. If $a(t)$ is the amount of some drug in the human body, this idea is expressed by the equation

$$a'(t) = -\eta\, a(t), \tag{1.1}$$

where $\eta > 0$ depends on the particular drug and patient. On p. 283 we established that only exponential functions (and scaled versions of them) are described by equation (1.1), so

$$a(t) = a_0 e^{-\eta t}, \tag{1.2}$$

where a_0 is the initial amount of the drug that's present. This formula describes the amount of medicine in the body when a single dose is administered, but many medications are taken on a regular basis in order to manage chronic conditions, so let's talk about the effect of repeated dosing on medication levels in the body.

Because $a(t)$ is an exponential function, the time required to metabolize half of the drug is always the same, no matter how much is present initially (check!). We call that length of time the *half-life* of the drug, and we'll denote it here by τ. For the sake of discussion, let's suppose that a dose of 100 milligrams is administered every τ hours. Then the amount of medicine in the patient immediately before and after each dose will be as listed in Table 1.1 and depicted in Figure 1.1.

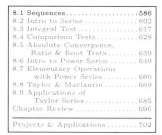

Figure 1.1: Levels of medication when 100 mg doses are given every 5 hours, and the half-life of the medication is also 5 hours. Complementary data is provided in Table 1.1.

Dose	pre-dose (mg)	post-dose (mg)
1	0	100
2	50	150
3	75	175
4	87.5	187.5
⋮	⋮	⋮
15	99.9939	199.9939
16	99.9969	199.9969
⋮	⋮	⋮

Table 1.1: Levels of medicine immediately before and immediately after repeated dosage of 100 mg at regular intervals of τ hours. A complementary graph is provided in Figure 1.1.

Table 1.1 seems to indicate that the medication levels are closing in on a maximum 200 mg. On p. 590 we'll prove that this is indeed true, but first we need to develop the relevant vocabulary.

When a set of numbers has a specific order, such as those in the columns of Table 1.1 (ordered top to bottom), we say it is a **sequence**. When calculating

586

the kth number in a sequence requires us to know the number that came before it, we say the sequence is defined **recursively** and refer to its defining equation as a **recurrence relation** or a **difference equation**. For example, if we denote the amount of medicine (in mg) immediately following the kth dose by a_k, the sequence of post-dosage levels in Table 1.1 is described by the difference equation

$$a_k = \frac{1}{2}a_{k-1} + 100 \qquad (1.3)$$

Don't just read it. Check it for yourself!

when $k \geq 1$. By contrast, when we have a formula that tells us how to calculate the kth number in the sequence directly, based only on k, we say the sequence is defined **explicitly**. For example,

$$a_k = \frac{k}{k+1} \quad \text{defines the sequence} \quad \underbrace{\frac{1}{2}}_{k=1}, \; \underbrace{\frac{2}{3}}_{k=2}, \; \underbrace{\frac{3}{4}}_{k=3}, \; \underbrace{\frac{4}{5}}_{k=4} \cdots . \qquad (1.4)$$

In many situations, such as the scenario of repeated dosage discussed above, we're interested in what happens to a_k when k gets very large. Specifically, does it tend toward a particular value (e.g., a *steady-state* level of medication) or not? This is exactly the idea of a limit, which is the central topic of this section.

✦ Limits of Sequences

You can understand limits of sequences in much the same way that you understand $\lim_{t \to \infty} f(t)$. In short, we're asking whether the numbers of the sequence get closer and closer to a particular value.

Introduction to limits of sequences

Example 1.1. Discuss the behavior of (a) $a_k = (-1)^{k+1}$ and (b) $a_k = \frac{k}{k+1}$ as k gets larger and larger.

Solution: First let's note that both of these sequences are defined explicitly.

(a) The numbers in this sequence are $\overbrace{1}^{k=1}, \overbrace{-1}^{k=2}, \overbrace{1}^{k=3}, \overbrace{-1}^{k=4} \ldots$ which alternate between 1 and -1, and never approach any particular value in the long run.

(b) The first few numbers in this sequence are shown in equation (1.4). They appear to be getting closer and closer to 1, but we've only calculated the first four terms so making a firm conclusion seems premature. However, a common way to understand the long-term behavior of a sequence is to think of its numbers as function values. In this example, we have

$$a_k = f(k) \quad \text{where } f \text{ is the function} \quad f(t) = \frac{t}{t+1}.$$

Since $\lim_{t \to \infty} f(t) = 1$, the numbers a_k must get closer and closer to 1 as $k \to \infty$ (see Figure 1.2). ∎

In Example 1.1 we saw that one sequence of numbers approached a particular value as k grew, while the other did not. This leads us directly to the idea of a limit in the context of sequences: when the value of a_k approaches the number L as $k \to \infty$, we write

$$\lim_{k \to \infty} a_k = L.$$

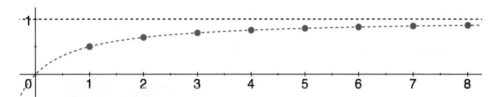

Figure 1.2: The sequence $a_k = \frac{k}{k+1}$ depicted on the graph of $f(t) = \frac{t}{t+1}$, which approaches $y = 1$ in the limit.

While correct in spirit, we're using the word "approaches" to mean "heads toward," and that lacks precision. For example, in Example 1.1 we saw that subsequent values of $a_k = k/(k+1)$ rose toward 1. However, they were also rising toward 2, and we don't mean to say that the limit value of that sequence is 2. This issue is remedied by the formal definition of the limit, in which the number ε is used to say that the distance between a_k and L becomes small.

Limit of a Sequence: The notation $\lim_{k \to \infty} a_k = L$ means that for each $\varepsilon > 0$ there is an N so that

$$\underbrace{|a_k - L| < \varepsilon}_{\substack{\text{this says the distance between } a_k \text{ and } L \\ \text{is small, where "small" is quantified by } \varepsilon}} \quad \text{whenever} \quad \underbrace{k \geq N.}_{\substack{\text{this is a quantitative way} \\ \text{of saying "eventually"}}}$$

Note: Since ε can be any positive number, $\lim_{k \to \infty} a_k = L$ means that a_k will eventually get as close to L as you want it to. The statement "whenever $k \geq N$" means that *all* of the numbers in the sequence beyond the Nth are at least that close to L.

When we want to denote a whole sequence, rather than a particular number in a sequence, it's common to write $\{a_k\}$. When $\lim_{k \to \infty} a_k$ exists, we say that $\{a_k\}$ is a **convergent** sequence, or that it **converges**. Otherwise, we say that it's **divergent** or that it **diverges**.

> In everyday usage, the word "diverge" means "to separate," or "to get farther apart," but in this context it simply means that a limit does not exist, for whatever reason.

The limit of an explicitly defined sequence

Example 1.2. Suppose $a_k = \frac{3k^2+4k+7}{5k^2-k+10}$ when $k \geq 1$. Determine the limit of the sequence $\{a_k\}$.

Solution: In this case, the number a_k is just the function $f(t) = \frac{3t^2+4t+7}{5t^2-t+10}$ evaluated at the integer $t = k$, and we know from L'Hôpital's Rule that

$$\lim_{t \to \infty} f(t) = \lim_{t \to \infty} \frac{3t^2 + 4t + 7}{5t^2 - t + 10} = \lim_{t \to \infty} \frac{6t + 4}{10t - 1} = \lim_{t \to \infty} \frac{6}{10} = \frac{3}{5}.$$

Since $f(t)$ converges to $3/5$ as $t \to \infty$, and (k, a_k) is a point on the graph of f, it must be that $\lim_{k \to \infty} a_k = \lim_{k \to \infty} f(k) = 3/5$. ■

Try It Yourself: The limit of an explicitly defined sequence

Example 1.3. Suppose $a_k = k \sin\left(\frac{1}{k}\right)$ when $k \geq 1$. Determine $\lim_{k \to \infty} a_k$.

Answer: 1. ■

Full Solution On-line

In Examples 1.1–1.3 we calculated limit values by embedding the sequence in the graph of a function. This is such an important technique that we pause to highlight it:

> **Calculating the Limit of a Sequence:** Suppose that $a_k = f(k)$, and $\lim_{t \to \infty} f(t) = L$. Then $\lim_{k \to \infty} a_k = L$ also.

Sometimes it's impractical or impossible to embed a sequence in the graph of a function. The Riemann sums from Chapter 5 are a great example.

The limit of an explicitly defined sequence

Example 1.4. Determine $\lim_{n \to \infty} a_n$ when $a_n = \sum_{k=1}^{n} 2 \sin\left(\frac{k\pi}{n}\right) \frac{\pi}{n}$.

Solution: For each integer $n \geq 1$, we can understand the number a_n as a Riemann sum (with $\Delta t = \pi/n$). More specifically, these Riemann sums approximate the area between the graph of $y = 2\sin(t)$ and the horizontal axis over $[0, \pi]$. Since $\sin(t)$ is continuous over this interval,

$$\lim_{n \to \infty} a_n = \lim_{n \to \infty} \sum_{k=1}^{n} 2 \sin\left(\frac{k\pi}{n}\right) \frac{\pi}{n} = \int_0^\pi 2\sin(t) \, dt = 4. \qquad \blacksquare$$

> We didn't refer to the Riemann sums as a sequence, but each time you increase n you get a new value. That's a sequence.

❖ Cobwebbing

We've been using graphs of functions to help us determine limits of explicitly defined sequences. We also use graphs to assist us with recursively defined sequences, but in a different way. In Figure 1.3 you see something called a *cobwebbing diagram* that's associated with the repeated dosing scenario discussed earlier. It allows us to conclude that $\lim_{k \to \infty} u_k = 200$ in that case. The following discussion shows you how to generate such diagrams and use them to calculate limits of recursively defined sequences.

In equation (1.3) we said that the amount of medication in the patient immediately following the kth dose was

$$a_k = \frac{1}{2} a_{k-1} + 100.$$

This leads us to consider the line

$$y = \frac{1}{2} x + 100. \qquad (1.5)$$

More generally, the relationship $a_{k+1} = f(a_k)$ is depicted by the curve $y = f(x)$. We refer to f as the **updating function**, and we use its graph to understand the long-term behavior of a recursively defined sequence.

In the medical context of our current discussion, you can think of x as the level of medication after the "previous" dose, and y as the level after "this" dose. Because the patient starts with none of the drug, we begin by looking at $x = 0$, and the amount of medication present in the patient's bloodstream immediately after the first dose is

$$\underbrace{y}_{\text{the next level of medication}} = f(\underbrace{0}_{\text{the initial level of medication}}) = 100.$$

We can make this calculation graphically by locating $x = 0$ on the x-axis and traveling vertically to the graph of f, a shown in Figure 1.4.

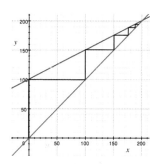

Figure 1.3: An example of a cobwebbing diagram.

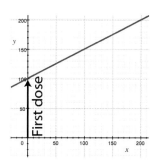

Figure 1.4: Calculating $a_1 = 100$ from $a_0 = 0$ using the graph of the updating function.

After the first dose, the patient's blood stream carries 100 mg of medication. To find the level after the second dose, we calculate

$$y \quad = \quad f(100) = 150.$$

$\underbrace{}$ the next level of medication $\qquad$ $\underbrace{}$ the current level of medication

Working graphically, we find $x = 100$ on the x-axis and then move vertically to the graph of the updating function, as seen in the left-hand image of Figure 1.5. In the middle image of that figure, you see that we can make the same calculation by moving horizontally to the $y = x$ line and then vertically to the graph of the updating function. Repeating the process allows us to determine the medication level after the third dose, the fourth, the fifth, and so on.

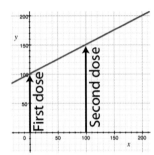 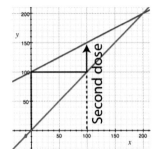 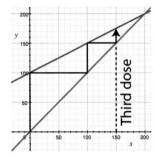 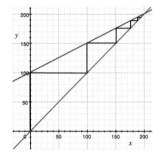

Figure 1.5: (first) Using the updating function to calculate medication levels after the first and second doses; (second) making the same calculation; (third) calculating the third term in the sequence; (fourth) repeating the method allows us to see the limit.

The graph that results from this process is called a **cobwebbing diagram**. In this case, it shows us that a_k converges to 200 in the limit. Note that, had we started with $a_0 = 200$, the recursion would have generated the sequence

$$a_0 = 200 \quad \Rightarrow \quad a_1 = 200 \quad \Rightarrow \quad a_2 = 200 \quad \Rightarrow \quad a_3 = 200 \ldots .$$

For this reason, we say that $a^* = 200$ is an **equilibrium** of the difference equation.

> The word "equilibrium" is meant to communicate the idea of "no change."

> **Drawing a cobwebbing diagram:** To draw a cobwebbing diagram, graph $y = f(x)$ and $y = x$ on the same axes. Then starting at $x = a_0$,
>
> 1. move vertically to the graph of the updating function;
>
> 2. move horizontally to the graph of $y = x$;
>
> 3. repeat ad infinitum.

Cobwebbing

Example 1.5. Suppose $a_k = 1/\sqrt{a_{k-1}}$, and $a_0 = 1.5$. (a) Construct a cobwebbing diagram for this sequence and use it to make a conjecture about whether the sequence converges, and (b) verify that $a_0 = 1$ is an equilibrium.

Solution: The updating function for this sequence is $f(x) = 1/\sqrt{x}$, and we begin by drawing its graph and the line $y = x$ together. Then starting at $x = 1.5$ we draw line segments according to the instructions above, moving vertically to the graph of f and then horizontally to the line $y = x$. The first five steps of the process are shown in Figure 1.6, and you can see that the cobwebbing seems to be spiraling toward the intersection of $y = x$ and $y = 1/\sqrt{x}$. That leads us to conjecture that the sequence converges and that its limit value is 1.

(b) When we start at $a_0 = 1$, the subsequent values of this sequence are $a_1 = 1$, $a_2 = 1$, $a_3 = 1 \ldots$. Since there is no change, $a_0 = 1$ is an equilibrium. ∎

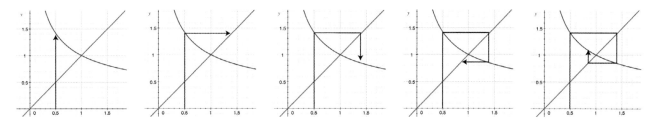

Figure 1.6: The first five steps of the cobwebbing process in Example 1.5.

Try It Yourself: Cobwebbing

Example 1.6. Suppose $b_{k+1} = 0.3b_k + 20$, and $b_0 = 300$. (a) Use a cobwebbing diagram to make a conjecture about $\lim_{k \to \infty} b_k$, and (b) find an equilibrium of the difference equation.

Answer: (a) the sequence appears to converge to $200/7$; (b) $200/7$. ∎

Full Solution
On-line

A recursively defined sequence that diverges because it surpasses all finite values

Example 1.7. Suppose $p_0 = 1000$ and $p_{k+1} = 1.5p_k$. Show that $\{p_k\}$ is divergent.

Solution: Numerically, the equation $p_{k+1} - 1.5p_k$ tells us that each number in this sequence is 50% larger than the one before it. This behavior can be seen in the cobwebbing diagram in Figure 1.7. Since these numbers grow forever and surpass every finite value, no particular value is approached in the limit. So this sequence diverges. ∎

We say that a sequence follows a **geometric progression** when the ratio of consecutive terms is always the same (e.g., p_{k+1}/p_k is always 1.5 in Example 1.7), and the number p_{k+1}/p_k is called the **common ratio**. On the associated cobwebbing diagram, this number is the slope of the (linear) graph of the updating function (e.g., see Figure 1.7).

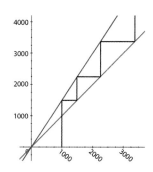

Figure 1.7: The cobwebbing diagram associated with Example 1.7.

In Example 1.1 we saw an explicitly defined sequence that diverged because it alternated between two specific values. This can happen in recursively defined sequences also, as you'll see in Example 1.8.

A divergent sequence of voltages in the human heart

Example 1.8. The human heart has a natural pacemaker, but also a natural "trigger lock" that prevents it from beating if it's not had enough time to rest since the last contraction. This "trigger lock" is called the *atrioventricular* (AV) node, and it uses electrical voltage to determine whether the heart can beat.

Suppose the heart has just beat, and the voltage on its AV node is V_k. By the time the *next* signal from the pacemaker arrives at the AV node, its voltage will have decayed to some fraction of V_k, which we'll write as mV_k where $m \in (0, 1)$. If mV_k is less than a *critical voltage,* the heart will beat and increase the voltage on the AV node by b volts. Otherwise the heart continues to relax. Because there are two possible events (*beat* and *not beat*), the relationship between V_k and V_{k+1} is described with a piecewise defined function:

$$V_{k+1} = \begin{cases} b + mV_k & \text{if } mV_k \leq C \\ mV_k & \text{if } mV_k > C \end{cases}$$

where $C > 0$ is the critical voltage and $b > 0$. Suppose $m = \frac{2}{3}$, $b = 1$ and $C = 1$. If $V_0 = \frac{9}{5}$, show that $\{V_k\}$ diverges.

Solution: From a computation standpoint, we can check that

$$mV_0 = \left(\frac{2}{3}\right)\left(\frac{9}{5}\right) = \frac{6}{5} > C,$$

so the heart does not beat and $V_1 = mV_0 = \frac{6}{5}$. However,

$$mV_1 = \left(\frac{2}{3}\right)\left(\frac{6}{5}\right) = \frac{4}{5} \leq C,$$

so the heart beats when the next signal from the pacemaker arrives, and

$$V_2 = mV_1 + b = \frac{4}{5} + 1 = \frac{9}{5},$$

which is where we started. This sequence will continue to alternate between $\frac{4}{5}$ and $\frac{9}{5}$ forever, so $\lim_{k\to\infty} V_k$ does not exist. That is, the sequence $\{V_k\}$ diverges. This is shown graphically in Figure 1.8. ∎

| The medical condition in which the heart beats at every other signal from the pacemaker is called *2:1 AV block.* |

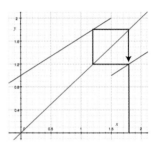

Figure 1.8: Cobwebbing diagram for Example 1.8.

✧ Monotone Sequences

The sequence $\{p_k\}$ from Example 1.7 is said to be **increasing** because $p_{k+1} > p_k$. By contrast, the sequence $\{b_k\}$ from Example 1.6 is said to be **decreasing** because $b_{k+1} < b_k$. We say that a sequence is **monotonic** if it's either increasing or decreasing.

Note that both $\{b_k\}$ and $\{p_k\}$ are monotonic, but $\{p_k\}$ diverges while $\{b_k\}$ converges. In short, this happens because $\{p_k\}$ surpasses every finite value but $\{b_k\}$ is caught inside a closed interval. Said formally:

| **Bounded Monotone Convergence Theorem:** Suppose $\{a_k\}$ is monotonic and there are numbers m and M such that each $a_k \in [m, M]$. Then $\{a_k\}$ is convergent. |

Note: When $\{a_k\}$ is an increasing sequence, we can use $m = a_0$, so all we need to find is an appropriate value of M, called an **upper bound** of the sequence. Similarly, when $\{a_k\}$ is a decreasing sequence, we can use $M = a_0$, so all we need to find is an appropriate value of m, called a **lower bound** of the sequence. When a sequence has both an upper bound and a lower bound, we say that it's **bounded**.

| Example 1.8 demonstrates that a sequence can diverge, even if it's bounded. Boundedness, by itself, does not imply convergence. |

Proving convergence

Example 1.9. Suppose $s_1 = \sqrt{2}$ and $s_{k+1} = \sqrt{2 + s_k}$. Use the Bounded Monotone Convergence Theorem to prove that $\{s_k\}$ is a convergent sequence.

Solution: First we show that the sequence is increasing. According to the definition,

$$s_2 = \sqrt{2 + s_1} = \sqrt{2 + \sqrt{2}}.$$

Replacing the $\sqrt{2}$ with zero would make the radicand smaller, so

$$s_2 = \sqrt{2 + \sqrt{2}} > \sqrt{2 + 0} = \sqrt{2} = s_1.$$

Similarly,

$$s_3 = \sqrt{2 + \sqrt{2 + \sqrt{2}}} > \sqrt{2 + \sqrt{2 + 0}} = \sqrt{2 + \sqrt{2}} = s_2$$

$$s_4 = \sqrt{2 + \sqrt{2 + \sqrt{2 + \sqrt{2}}}} > \sqrt{2 + \sqrt{2 + \sqrt{2 + 0}}} = \sqrt{2 + \sqrt{2 + \sqrt{2}}} = s_3$$

$$\vdots \qquad \vdots$$

$$s_{k+1} > s_k.$$

Next we show that the sequence is bounded. Each term is clearly non-negative, so we can use $m = 0$, and need only establish an upper bound. It's a convenient fact that $s_1 = \sqrt{2} < 7$, because it allows us to conclude that

$$s_2 = \sqrt{2 + s_1} < \sqrt{2 + 7} = 3,$$

which is less than 7. Then because $s_2 < 7$, we know

$$s_3 = \sqrt{2 + s_2} < \sqrt{2 + 7} = 3,$$

which is less than 7. Then because $s_3 < 7$, we know

$$s_4 = \sqrt{2 + s_3} < \sqrt{2 + 7} = 3$$

$$\vdots \qquad \vdots$$

$$s_{k+1} = \sqrt{2 + s_k} \leq \sqrt{2 + 7} = 3,$$

so s_k never exceeds $M = 3$. Since the sequence is bounded and increasing, it converges. In this case, the cobwebbing technique discussed earlier can be used to show that the limit value is 2 (see Figure 1.9). ∎

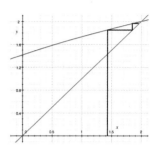

Figure 1.9: Cobwebbing diagram for Example 1.9.

> We said that this sequence is bounded above by 3. That doesn't mean it will ever reach 3, much in the same way that your height is bounded above by 3 meters, but you'll never grow to be 3 meters tall.

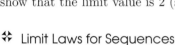 **Limit Laws for Sequences**

As with functions, limits of sequences often behave the way you want them to. For example, suppose the value of a_k gets closer and closer to 4 as $k \to \infty$, and the value of b_k gets closer and closer to 5 at the same time. Then—intuitively speaking—the value of the sum $a_k + b_k$ *should* converge to $4 + 5 = 9$ as $k \to \infty$. That's exactly what happens, and all of the limit laws can be understood in this way.

The Limit Laws: Suppose $\lim_{k\to\infty} a_k = L$ and $\lim_{k\to\infty} b_k = M$, that p and q are positive integers, and that β is any fixed scalar. Then,

1. $\lim_{k\to\infty} \left(\beta a_k \right) = \beta L$ 3. $\lim_{k\to\infty} \left(a_k b_k \right) = LM$

2. $\lim_{k\to\infty} \left(a_k \pm b_k \right) = L \pm M$ 4. $\lim_{k\to\infty} \left(\frac{a_k}{b_k} \right) = \frac{L}{M}$ when $M \neq 0$

and lastly,

5. $\lim_{k\to\infty} \left(a_k \right)^{p/q} = L^{p/q}$, provided that $L > 0$ if q is even.

> **Said in English:**
>
> Limit Law 2 says *the limit of a sum is the sum of the limits.*
>
> Limit Law 3 says *the limit of a product is the product of the limits.*
>
> Limit Law 4 says *the limit of a quotient is the quotient of the limits.*

The Squeeze Theorem also has an analog in the context of sequences.

Squeeze Theorem for Sequences: Suppose that there is a number N for which $a_k \leq b_k \leq c_k$ whenever $\underbrace{k > N}$.

$$\text{This is a precise way of saying "eventually"}$$

Further, suppose that

$$\lim_{k \to \infty} a_k = L = \lim_{k \to \infty} c_k.$$

Then $\lim_{k \to \infty} b_k$ exists, and its value is also L.

> The sequences $\{a_k\}, \{b_k\}$ and $\{c_k\}$ might not satisfy the order relation at first, but they do eventually. The word "eventually" is quantified by N in any given case.

Using the Squeeze Theorem

Example 1.10. Suppose that $\lim_{k \to \infty} |b_k| = 0$. Show that $\lim_{k \to \infty} b_k = 0$.

Solution: In this case, we set $c_k = |b_k|$, and $a_k = -|b_k|$. Then $a_k \leq b_k \leq c_k$, and Limit Law 1 allows us to write

$$\lim_{k \to \infty} a_k = \lim_{k \to \infty} \left(-|b_k| \right) = (-1)(0) = 0.$$

Since both $\lim_{k \to \infty} a_k = 0$ and $\lim_{k \to \infty} c_k = 0$, and b_k is caught in between, $\lim_{k \to \infty} b_k = 0$ also. ∎

Using the Squeeze Theorem

Example 1.11. Suppose $b_k = (-1)^k / k$. Show that $\lim_{k \to \infty} b_k = 0$.

Solution: This follows directly from Example 1.10. ∎

> This example shows that monotonicity is not required for convergence.

Using the Squeeze Theorem

Example 1.12. Suppose $a_k = 3^k / (k!)$ when $k \geq 1$, where $k!$ (read **"k factorial"**) denotes the product $k(k-1)(k-2) \cdots (3)(2)(1)$. Determine $\lim_{k \to \infty} a_k$.

Solution: The first three terms of the sequence are

$$a_1 = \frac{3}{1}, \quad a_2 = \frac{3^2}{(2)(1)} = \frac{9}{2}, \quad \text{and} \quad a_3 = \frac{3^3}{(3)(2)(1)} = \frac{27}{6}.$$

Starting at the fourth, we see an important pattern:

$$a_4 = \frac{3^4}{(4)(3)(2)(1)} = \left(\frac{3}{4} \right) \frac{3^3}{(3)(2)(1)} = \left(\frac{3}{4} \right) a_3.$$

Similarly,

$$a_5 = \frac{3^5}{(5)(4)(3)(2)(1)} = \left(\frac{3}{5} \right)\left(\frac{3}{4} \right) \frac{3^3}{(3)(2)(1)} < \left(\frac{3}{4} \right)\left(\frac{3}{4} \right) a_3$$

$$a_6 = \frac{3^6}{(6)(5)(4)(3)(2)(1)} = \left(\frac{3}{6} \right)\left(\frac{3}{5} \right)\left(\frac{3}{4} \right) \frac{3^3}{(3)(2)(1)} < \left(\frac{3}{4} \right)^3 a_3$$

$$a_7 = \frac{3^7}{(7)(6)(5)(4)(3)(2)(1)} = \left(\frac{3}{7} \right)\left(\frac{3}{6} \right)\left(\frac{3}{5} \right)\left(\frac{3}{4} \right) \frac{3^3}{(3)(2)(1)} < \left(\frac{3}{4} \right)^4 a_3.$$

In general, we see $0 \leq a_k \leq \left(\frac{3}{4} \right)^{k-3} a_3$ when $k \geq 4$. Since $(3/4)^{k-3}$ converges to zero as $k \to \infty$, the Squeeze Theorem tells us that $\lim_{k \to \infty} a_k = 0$. ∎

> Alternatively, if you worked through the project called *An Introduction to Stirling's Approximation* (on p. 400) you know that $k! \geq \left(\frac{k}{e} \right)^k$, so
>
> $$0 \leq \frac{3^k}{k!} \leq \frac{3^k}{(k/e)^k} = \left(\frac{3e}{k} \right)^k.$$
>
> Once $k > 3e$, this is a fraction that's less than one put to a large power. As k grows, the fraction gets smaller and the power gets bigger, both of which drive its value to zero. Since $3^k / k!$ is caught between this number and zero, it must also converge to zero in the limit.

You should know

- the terms *sequence, explicit, recursive, recurrence relation, difference equation, equilibrium, limit (of a sequence), convergent, divergent, updating function, cobwebbing, increasing, decreasing, monotonic, upper bound, lower bound, bounded, factorial,* and the notation $\{a_k\}$;

- the Bounded Monotone Convergence Theorem, the Squeeze Theorem for Sequences, and the Limit Laws;

- not all sequences have limits.

You should be able to

- use the cobwebbing technique;

- determine the limit of a sequence when it's expressed in the form $a_n = f(n)$;

- determine the limit of a sequence when it's defined recursively (in simple situations).

✤ 8.1 Skill Exercises

Determine whether each of the sequences in #1–12 converges and, if so, determine its limit.

1. $s_n = \frac{3+5n}{n+2}$

2. $s_n = (-1)^n \sin(1/n)$

3. $s_n = n2^{-n}$

4. $s_n = \frac{1+\sqrt{n}}{\sqrt{n}}$

5. $s_n = n!/2^n$

6. $s_n = (1+3n)^{1/n}$

7. $s_n = 1/(3 - s_{n-1})$ where $s_1 = 2$

8. $s_n = \frac{s_{n-1}+5}{2}$ where $s_1 = 1$

9. $s_n = \sqrt{3 + s_{n-1}}$ where $s_1 = 4$

10. $s_n = -1.5s_{n-1} + s_{n-1}^2$ where $s_1 = \frac{5}{2}$

11. $s_n = -1.5s_{n-1} + s_{n-1}^2$ where $s_1 = \frac{1}{5}$

12. $s_{n+1}s_n = 3$ where $s_1 = 5$

Denote by $[\![x]\!]_k$ the kth digit in the decimal expansion of x. For example, $\pi = 3.14159\ldots$ so $[\![\pi]\!]_1 = 3$ and $[\![\pi]\!]_2 = 1$. In exercises #13–16 determine whether the sequence $\{[\![x]\!]_k\}$ converges and, if so, to what.

13. $x = \frac{14}{9}$ 14. $x = \frac{3895}{900}$ 15. $x = \frac{1}{33}$ 16. $x = \sqrt{2}$

Calculate $\lim_{n\to\infty} r_n$ in each of #17–20 by interpreting r_n as a Riemann sum.

17. $r_n = \sum_{k=1}^{n} \cos\left(\frac{2\pi k}{n}\right) \frac{2\pi}{n}$

18. $r_n = \sum_{k=1}^{n} \left(-1 + \frac{2k}{n}\right)^{26} \frac{2}{n}$

19. $r_n = \sum_{k=1}^{n} \ln\left(1 + \frac{4k}{n}\right) \frac{4}{n}$

20. $r_n = \sum_{k=1}^{n} \frac{k}{n^2} e^{k/n}$

21. Suppose $a_0 = 1$, and $a_{k+1} = m\, a_k$. Use a cobwebbing diagram to show that $\lim_{k\to\infty} a_k = \infty$ when $m > 1$, and $\lim_{k\to\infty} a_k = 0$ when $m \in (0,1)$.

22. Suppose $a_0 = 1$, and $a_{k+1} = b + a_k$. Use a cobwebbing diagram to show that $\lim_{k\to\infty} a_k = \infty$ when $b > 0$, and $\lim_{k\to\infty} a_k = \infty$ when $b < 0$.

✧ 8.1 Concept and Application Exercises

23. Suppose p_k is the cumulative amount of product, in grams, produced from a chemical reaction after k minutes. Use limits to write an equation that says, "In the long run, the reaction produces 13 grams of product."

24. Suppose a_k and b_k indicate the remaining mass (in grams) of reactants A and B after k minutes of a reaction. Explain what the equations $\lim_{k\to\infty} a_k = 0$ and $\lim_{k\to\infty} b_k = 4$ tell us about the reaction.

25. Is every bounded sequence convergent? If so, why? If not, give an example.

26. Is every convergent sequence bounded? If so, why? If not, give an example.

27. Is every monotone sequence convergent? If so, why? If not, give an example.

28. Is every convergent sequence monotone? If so, why? If not, give an example.

29. Suppose $\lim_{n\to\infty} s_{2n} = 9$.

    (a) Can we conclude that $\{s_n\}$ converges? If so, why? If not, explain why not.

    (b) If we also know that $\lim_{n\to\infty} s_{2n+1} = 9$, can we conclude that $\{s_n\}$ converges? If so, why? If not, explain why not.

30. Find sequences $\{a_k\}$ and $\{b_k\}$ that both diverge, but for which the sequence $\{a_k - b_k\}$ converges.

31. Suppose a ball is dropped from a height of 1 meter, and it returns to 75% of its previous altitude with each bounce. Write the first 10 numbers in the sequence of altitudes, $a_1 = 1, a_2 = 0.75\ldots$

                               — Recreation

32. The human genetic code is written with an alphabet of four letters $\{A, C, G, T\}$. For the purposes of this exercise, a string of these letters will be called a *word*.

                               — Genetics

    (a) Determine the number of words with k letters for each of $k = 1$, $k = 2$, $k = 3$ and $k = 4$.

    (b) Determine a general formula for the number of words with k letters.

    (c) We'll say that a word is *legitimate* if it contains an even number of As (possibly none). If a_k denotes the number of legitimate words of length k, write a formula for the number of *illegitimate* words with k letters. *(Hint: use your answer from part (b).)*

    (d) Use the following two facts to write a formula that expresses a_{k+1} in terms of a_k:

       • If a legitimate word of length $k + 1$ starts with an A, the remaining k letters must have an odd number of As.

       • If the first letter of a legitimate word of length $k + 1$ is C, G, or T, the remaining k letters must have an even number of As.

    (e) Explain why $a_1 = 3$, and use your recursion formula from part (c) to determine $a_2, a_3, \ldots, a_{10}$.

33. (a) Write an equation that says, "The amount of a drug in the human body decreases at a rate that's proportional to the amount present." Then (b) show that the body takes the same length of time to metabolize 25% of a drug, regardless of how much drug is present initially.

                               — Medical Sciences

34. (a) Write an equation that says, "The amount of a drug in the human body decreases at a rate that's proportional to the amount present." Then (b) show that the body takes the same length of time to metabolize $p\%$ of a drug, regardless of how much drug is present initially.

35. Suppose an initial dose of 300 mg of a drug is administered to a patient, and an additional 150 mg is administered every 6 hours thereafter. If the patient's body neutralizes 25% of the drug in 6 hours, and a_k is the amount of medication in the patient's body immediately after the kth dose, determine $\lim_{k \to \infty} a_k$ and explain what it means to us.

36. Suppose an initial dose of 300 mg of a drug is administered to a patient, an additional 150 mg is administered every 6 hours thereafter, and the fraction of medication neutralized by the patient's body in that time is $1 - p$, where $p \in (0, 1)$.

    (a) What does the number p mean to us?

    (b) Determine the equilibrium level of medication in the patient's body as a function of p.

    (c) Graph the equilibrium level as a function of p.

    (d) Your graph from part (c) has a vertical asymptote at $p = 1$. Explain its meaning in the medical context of this exercise.

Exercises 37–40 are about the sequence V_k, defined by

$$V_{k+1} = \begin{cases} b + mV_k & \text{if } mV_k \leq C \\ mV_k & \text{if } mV_k > C, \end{cases}$$

which was introduced on p. 591.

37. Suppose $C = 1$, $b = 1$, $m = 0.1$, and $V_0 = 0$. Determine $\lim_{k \to \infty} V_k$.

38. Suppose $C = 1$, $b = 1$, and $V_0 = 0$. The number $V = \lim_{k \to \infty} V_k$ depends on the number m.

    (a) Find a formula for $V(m)$.

    (b) For what values of m is your formula valid, and why?

    (c) Sketch a graph of $V(m)$.

    (d) Verify that $V'(m) < 0$, and based on the graph of the updating function, explain why that makes sense.

39. The medical condition in which the heart misses every n^{th} beat is called the *Wenkebach phenomenon*.

    (a) Use a spreadsheet program to calculate $V_1, \dots, V_{30}$ when $C = 0.6$, $b = 0.3$, $m = 0.69$ and $V_0 = 0.75$, and verify that the heart misses every 5th beat. (You can use an `if` command to implement the piecewise definition of this recursion equation.)

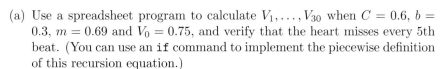

    (b) Graph $V_1, \dots, V_{30}$ as a scatter plot, and identify where in the graph we see the heart miss a beat.

40. Determine the relationship among m, b, and c necessary to guarantee the existence of an equilibrium (i.e., a number V_0 for which $V_{k+1} = V_k$).

In #21 you showed that the sequence $a_{k+1} = m\, a_k$ exhibits geometric growth when $m > 1$. In population modeling we often want to describe growth, but *bounded* growth. The **discrete logistic** model achieves this by replacing the constant m with $m = (r)\left(1 - \frac{a_k}{K}\right)$, where $r > 0$ is called the *intrinsic growth rate* and $K > 0$ is called the *carrying capacity*. So

$$a_{k+1} = r\left(1 - \frac{a_k}{K}\right) a_k.$$

Exercises 41–47 address sequences of this form.

41. (a) Determine a formula for the ratio a_{k+1}/a_k, and (b) explain what it says about the relationship between the current population level and the percentage of growth.

42. Show that the discrete logistic equation has an equilibrium at $a_* = 0$, and explain why that makes sense in the context of population modeling.

43. Find a formula for the nonzero equilibrium of the discrete logistic model, $a_* \neq 0$ (it depends on r and K).

44. Show that the only biologically relevant equilibrium is $a_* = 0$ when $r \in (0, 1)$.

45. Suppose $K = 1$.

   (a) Sketch a graph that depicts the relationship between a_* and $r > 0$ (in population models, only $a_* \geq 0$ makes sense).

   (b) Explain the effect of r on the graph of the updating function.

   (c) Determine $\lim_{r \to \infty} a_*(r)$ and explain your answer in light of your answer to (b).

46. Suppose $K = 1$, $r = 2.5$, and $a_0 = 0.25$.

   (a) Use a spreadsheet to calculate $a_1, \ldots, a_{50}$ and plot the points as a scatter plot. Then describe the behavior in full sentences.

   (b) Use a graphing utility to plot the updating function, print it, and draw the cobwebbing diagram that illustrates the calculation of the first several numbers in $\{a_k\}$.

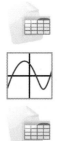

47. Suppose $K = 1$ and $a_0 = 0.25$. (Read parts (a), (b) and (c) before making any calculations.)

   (a) Use a spreadsheet to calculate $a_1, \ldots, a_{50}$ and plot the points as a scatter plot for $r = 2.5, 2.6, 2.7, \ldots, 3.3$.

   (b) Use the formula from #43 to locate the nonzero equilibrium, a_*, and calculate the derivative of the updating function at a_* for each value of r in the list above.

   (c) In full sentences, describe how the behavior of the sequence $\{a_k\}$ appears to depend on $f'(a_*)$.

   (d) In full sentences, describe how the behavior of the sequence $\{a_k\}$ appears to depend on r.

Exercises #48–51 have to deal with gas exchange in the lungs. The term *tidal volume* refers to the volume of air exhaled and replaced with each breath.

Biology

Biology

48. Suppose there's a gas in the atmosphere that your body does *not* absorb, although you breathe it in and out. Further suppose your lung has a volume of 5 L, your tidal volume is 0.4 L, and the concentration of this mystery gas in the local atmosphere is $\gamma = 8 \times 10^5$ molecules per liter. We'll denote by c_n the concentration of the gas in your lung immediately following the nth breath.

    (a) What do the numbers $5c_n$, $0.4c_n$, and $5c_{n+1}$ tell us in this context? *(Hint: see where the numbers 5 and 0.4 occur in the description above.)*

    (b) Explain what the equation $5c_{n+1} = 5c_n - 0.4c_n + 0.4\gamma$ means to us in this context.

    (c) Find the equilibrium value of this recursion relation, and explain what it tells us.

49. Suppose your lung volume is V liters, your tidal volume is T liters, c_n denotes the concentration (number per liter) of the mystery gas from #48 that's in your lung immediately following the nth breath, and γ is its concentration in the local atmosphere (in molecules per liter).

    (a) As in #48 part (a), write a recurrence relation between c_{n+1} and c_n.

    (b) If $c_1 = 0$, calculate $\lim_{n \to \infty} c_n$.

    (c) Explain what your answer from part (b) means in this context, and determine whether it seems reasonable to you.

50. Suppose there's a gas in the atmosphere whose concentration is $\gamma = 9 \times 10^{13}$ molecules per liter, and the fraction of this gas in your lung that's absorbed by your body during each breath is $\alpha = 1/500$. Further suppose your lung has a volume of 5.2 L, your tidal volume is 0.45 L, and c_n is the concentration of the gas in your lung immediately following the nth breath.

    (a) What do the numbers $5.2c_n$, $\alpha(5.2c_n)$ and $(1-\alpha)5.2c_n$ mean in this context?

    (b) What do the numbers $(1-\alpha)c_n$ and $0.45(1-\alpha)c_n$ mean in this context?

    (c) Explain what the equation $5.2c_{n+1} = (1-\alpha)5.2c_n - 0.45(1-\alpha)c_n + 0.45\gamma$ means to us in this context.

    (d) If $c_1 = 0$, determine $\lim_{n \to \infty} c_n$.

    (e) Explain what your answer from part (d) means in this context, and determine whether it seems reasonable to you.

51. Suppose your lung volume is V liters, your tidal volume is T liters, c_n denotes the concentration (number per liter) of a gas that's in your lung immediately following the nth breath, α is the fraction of it that's absorbed during each respiration cycle, and γ is the concentration of the gas in the local atmosphere. We'll denote by c_* the equilibrium value of c_n.

    (a) As in #50, write a recurrence relation between c_{n+1} and c_n.

    (b) With all other parameters held fixed, how do you think an increase in γ would affect c_*?

    (c) With all other parameters held fixed, how do you think an increase in α would affect c_*?

    (d) Determine a formula for c_* in terms of α, γ, and $q = \frac{T}{V}$.

    (e) Use your formula from part (d) to check your answers from (b) and (c).

(f) The numbers α and γ might be fixed, but you can increase q by increasing T. Derive a formula for $c'_*(q)$ and use it to determine the effect of increasing q.

52. Leonardo of Pisa, posthumously named Fibonacci, is perhaps most famous for introducing the Hindu-Arabic number system to Western culture in his book *Liber Abaci* in 1202. Among other applications of the new system demonstrated in that book, Fibonacci addressed a question about a hypothetical population of rabbits. Suppose a rabbit of this hypothetical species is sexually immature for 1 month after it's born, but every pair of rabbits begets a new pair in each of the following months. We'll denote by f_n the number of pairs of rabbits after n months, and say the first pair of rabbits is born in January, so $f_1 = 1$. Since those rabbits are initially immature, no new rabbits are born in February, so $f_2 = 1$. However, the rabbits beget a new pair in March, so $f_3 = 2$. This new generation is still immature come April, but the first generation produces another pair, so $f_4 = 3$, and so on. In general,

$$f_n = \underbrace{f_{n-1}}_{\substack{\text{all rabbits from the previ-}\\\text{ous month survive}}} + \underbrace{f_{n-2}}_{\substack{\text{this accounts for the progeny of each}\\\text{pair that was alive 2 months prior}}}.$$

(a) Use this recursion relation to calculate f_4 though f_{10}.

(b) Explain why the Fibonacci sequence diverges.

(c) Denote by q_n the factor by which the population grows in month n, so $q_n = f_n/f_{n-1}$. Write a recurrence relation that relates q_n to q_{n-1}. (*Hint: divide the recurrence relation above by f_{n-1}.*)

(d) Based on your answer to part (b), sketch a cobwebbing diagram that illustrates the convergence of $\{q_n\}$.

(e) Calculate $\lim_{n\to\infty} q_n$.

53. Suppose we send messages using only two signals, $\{a, b\}$. Signal a takes one unit of time and signal b takes two units of time. We will denote by w_k the number of different code words whose transmission takes at most k units of time. For example, only the words b, ba, a, aa, ab, and aaa can all be transmitted in 4 units of time (or less), so $w_3 = 6$.

| Computer Science |

(a) By making a list like the one above, calculate w_4.

(b) Use the following ideas to write a recursion formula for w_k:

- If a word of length $k + 1$ starts with the letter a, the remaining portion of the word must be transmitted in k units of time.
- If a word of length $k + 1$ starts with the letter b, the remaining portion of the word must be transmitted in $k - 1$ units of time.

(c) Use your formula from part (a) to calculate $w_5, w_6, \ldots, w_{10}$.

The "zeroth" row of **Pascal's triangle** is the single number 1. Each number in subsequent rows is the sum of the numbers above it, as seen in Figure 1.10 (you can think of each 1 along the edges as having an unwritten 0 above it). Exercises #54–55 address properties of this famous triangle.

54. Denote by s_n the sum of the entries in the nth row of Pascal's triangle, as shown in Figure 1.10, so $s_0 = 1$, $s_1 = 2$, etc. (a) Calculate s_3 through s_7, and (b) determine a general formula for s_n.

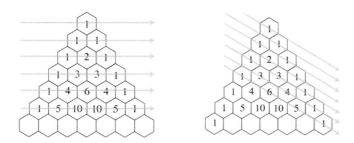

Figure 1.10: (left) For #54; (right) for #55.

55. Denote by f_n the sum of the nth diagonal, so $f_1 = 1$, $f_2 = 1$, and $f_3 = 2$, as shown in Figure 1.10. Calculate f_4 through f_{10} and compare them to the answer you found in part (a) of #52.

56. The sequence of *Lucas numbers*, named for Edouard Lucas (1842-1891), has the same recursive pattern as the Fibonacci numbers, but uses different initial values: $L_n = L_{n-1} + L_{n-2}$ with $L_0 = 2$ and $L_1 = 1$.

    (a) Calculate the first 10 Lucas numbers.

    (b) Suppose $q_n = L_n/L_{n-1}$. Determine $\lim_{n\to\infty} q_n$. *(Hint: see #52.)*

    (c) Show that Lucas numbers are related to Fibonacci numbers by $5f_n = L_{n-1} + L_{n+1}$ when $n \geq 1$.

    (d) Determine an equation that relates f_n to L_{n-2} and L_{n+2} when $n \geq 2$.

8.2   Introduction to Series

The left-hand image of Figure 2.1 shows an equilateral triangle whose area is 1. It has been subdivided into four equal parts by an inverted equilateral triangle at its center. Let's pick two of these parts, shade one, and subdivide the other into four equal parts in the same manner. Then we do it again: choose one of these new parts to shade, and another to subdivide.

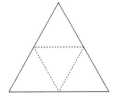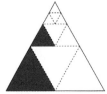

Figure 2.1: (left to right) Shading smaller and smaller fractions of a triangle.

As we continue this process, the amount of shaded area continues to grow, and we can say exactly how. Let's denote by s_n the cumulative amount of shaded area after the nth time we shade a triangle. Then

$$s_1 = \frac{1}{4}$$

$$s_2 = \frac{1}{4} + \left(\frac{1}{4}\right)^2$$

$$s_3 = \frac{1}{4} + \left(\frac{1}{4}\right)^2 + \left(\frac{1}{4}\right)^3$$

$$s_4 = \frac{1}{4} + \left(\frac{1}{4}\right)^2 + \left(\frac{1}{4}\right)^3 + \left(\frac{1}{4}\right)^4,$$

and in general,

$$s_n = \sum_{k=1}^{n} \left(\frac{1}{4}\right)^k.$$

Notice that $\{s_n\}$ is an increasing sequence of numbers (since we shade a little more in each step), but that the shaded area never exceeds 1 (the area of the original triangle). Since the sequence is increasing but bounded above, the Bounded Monotone Convergence Theorem (p. 592) guarantees that it has a limit. But what is it?

Figure 2.2 shows how the inverted equilateral triangles partition the initial triangle into horizontal "levels." Since 1/3 of each level is shaded, in the limit we'll have 1/3 of the entire triangle shaded. So $\lim_{n\to\infty} s_n = 1/3$. Said differently,

$$\lim_{n\to\infty} \sum_{k=1}^{n} \left(\frac{1}{4}\right)^k = \frac{1}{3},$$

which is commonly written in shorthand as $\sum_{k=1}^{\infty} \left(\frac{1}{4}\right)^k = \frac{1}{3}$. More generally,

$$\text{by writing } \sum_{k=1}^{\infty} a_k \quad \text{we mean} \quad \lim_{n\to\infty} \sum_{k=1}^{n} a_k. \tag{2.1}$$

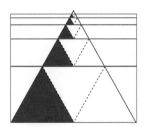

Figure 2.2: The inverted equilateral triangles used in our procedure subdivide the triangle into levels.

This is reminiscent of the way we defined

$$\int_{1}^{\infty} f(t)\, dt$$

as the *limit* of definite integrals.

The expression $\sum_{k=1}^{\infty} a_k$ is called a **series**, or **infinite series**. When the limit in equation (2.1) exists, we say the series **converges**, or that it's a **convergent series**. Otherwise we say the series **diverges**, or that it's a **divergent series**.

When talking about shading the triangle we wrote a formula for s_n, which denotes the cumulative total after adding the first n summands. Since there are many more summands, we refer to each s_n as a **partial sum**. In general, we say that

$$s_n = \sum_{k=1}^{n} a_k$$

is a partial sum of the series $\sum_{k=1}^{\infty} a_k$. If the series converges, its **value** (or **sum**) is the limit of its partial sums (compare with equation (2.1)).

Partial sums show that the Harmonic Series diverges

Example 2.1. The **Harmonic Series** is $\sum_{k=1}^{\infty} \frac{1}{k}$. (a) Write the first four partial sums of this series, and (b) show that the series diverges.

Solution: (a) The first four partial sums are

$$s_1 = 1$$
$$s_2 = 1 + \tfrac{1}{2}$$
$$s_3 = 1 + \tfrac{1}{2} + \tfrac{1}{3}$$
$$s_4 = 1 + \tfrac{1}{2} + \tfrac{1}{3} + \tfrac{1}{4}.$$

(b) For brevity, we'll skip to the particular partial sums that illuminate why this series diverges. For instance, look to the expression for s_4 written above. Since $\frac{1}{3} > \frac{1}{4}$, we know that $s_4 > 1 + \frac{1}{2} + \frac{1}{4} + \frac{1}{4} = 1 + \frac{2}{2}$. Similarly,

> In Section 8.3 we'll use an improper integral to demonstrate that this series diverges.

$$s_8 = 1 + \frac{1}{2} + \overbrace{\frac{1}{3} + \frac{1}{4}}^{\text{both} \geq 1/4} + \overbrace{\frac{1}{5} + \frac{1}{6} + \frac{1}{7} + \frac{1}{8}}^{\text{all} \geq 1/8} > 1 + \frac{1}{2} + \frac{2}{4} + \frac{4}{8} = 1 + \frac{3}{2}.$$

We see an important pattern when we line up these partial sums:

$$2 = 2^1 \quad \text{and} \quad s_2 \geq 1 + \tfrac{1}{2}$$
$$4 = 2^2 \quad \text{and} \quad s_4 \geq 1 + \tfrac{2}{2}$$
$$8 = 2^3 \quad \text{and} \quad s_8 \geq 1 + \tfrac{3}{2}.$$

In general, we find that after adding 2^n summands,

$$s_{2^n} \geq 1 + \frac{n}{2}.$$

Therefore $\lim_{n \to \infty} s_n = \infty$, so the series does not converge. $\blacksquare$

The Harmonic Series is a good "landmark" to keep in mind, because it demonstrates that *a series might not converge, even if its summands tend to zero in the limit.* The next theorem addresses the other side of the coin, as it were: if a series converges, what do we know about the summands?

Theorem: Suppose $\sum_{k=1}^{\infty} a_k$ converges. Then $\lim_{k \to \infty} a_k = 0$.

Proof. By saying that $\sum_{k=1}^{\infty} a_k$ converges, we mean that the sequence of partial sums has a limit, so let's write $\lim_{n\to\infty} s_n = L$. And since

$$s_{n+1} = a_1 + a_2 + \cdots a_n + a_{n+1} = s_n + a_{n+1},$$

we can write $a_{n+1} = s_{n+1} - s_n$, so

$$\lim_{n\to\infty} a_{n+1} = \lim_{n\to\infty}\left(s_{n+1} - s_n\right) = \lim_{n\to\infty} s_{n+1} - \lim_{n\to\infty} s_n = L - L = 0. \qquad \blacksquare$$

Testing for divergence

Example 2.2. Show that $\sum_{k=1}^{\infty} \frac{3k}{5k+1}$ diverges.

Solution: If this series converged, its summands would tend to zero in the limit. Since

$$\lim_{k\to\infty} \frac{3k}{5k+1} = \frac{3}{5} \neq 0,$$

this series cannot converge. $\qquad \blacksquare$

❖ Geometric Series

The series that we examined when shading the triangle is called a **geometric series** because its summands exhibit a geometric progression. In general, a geometric series takes the form

$$a + ar + ar^2 + ar^3 + ar^4 + ar^5 + \cdots .$$

Whenever we discuss geometric series, we'll use the letter a to denote the first summand. Note that the ratio of any two consecutive summands is always the same. For example,

$$\frac{\text{summand \#4}}{\text{summand \#3}} = \frac{ar^3}{ar^2} = r.$$

For this reason, we refer to r as the **common ratio** of the series.

Calculating the common ratio

Example 2.3. Determine the common ratio of the series $\sum_{k=7}^{\infty} 3^{3k-2} 8^{1-7k}$.

Solution: To find the common ratio, we consider the ratio of any two consecutive summands, say $k = 9$ and $k = 10$:

$$\frac{\text{summand with } k=10}{\text{summand with } k=9} = \frac{3^{28} 8^{-69}}{3^{25} 8^{-62}} = \frac{3^3}{8^7} = \frac{27}{2097152}. \qquad \blacksquare$$

Try It Yourself: Calculating the common ratio

Example 2.4. Determine the common ratio of the series $\frac{1}{3} + \frac{1}{15} + \frac{1}{75} + \cdots$

Answer: $1/5$. $\qquad \blacksquare$

Full Solution On-line

Geometric series are important in the larger scheme of things, but also make a nice introduction to the subject of infinite series because it's easy to tell when they converge, and to what. In short, a geometric series converges when the magnitude of its common ratio is less than 1. To understand this intuitively, look back to our discussion of shading triangles on p. 602, where each summand is $1/4$ the size of the one before it.

Geometric Series Theorem: The series $\sum_{k=0}^{\infty} ar^k$ converges if and only if $|r| < 1$, in which case

$$\sum_{k=0}^{\infty} ar^k = \frac{a}{1-r}.$$

Proof. A series converges when $\lim_{n \to \infty} s_n$ exists, so let's begin by listing some partial sums:

$$
\begin{aligned}
s_1 &= a \\
s_2 &= a + ar \\
s_3 &= a + ar + ar^2 \\
&\;\;\vdots \qquad\qquad \vdots \\
s_n &= a + ar + ar^2 + \cdots + ar^{n-1} \\
s_{n+1} &= a + ar + ar^2 + \cdots + ar^{n-1} + ar^n.
\end{aligned}
$$

We typically think of generating s_{n+1} by adding one more term to s_n:

$$s_{n+1} = s_n + ar^n.$$

However, because the summands follow a geometric progression, we can also generate s_{n+1} by scaling s_n and adding a:

$$s_{n+1} = a + r s_n.$$

That is,

$$s_n + ar^n = a + r s_n.$$

As long as $r \neq 1$ we can rewrite this as

$$s_n = \frac{a}{1-r} - \left(\frac{a}{1-r} \right) r^n, \qquad\qquad (2.2)$$

which allows us to calculate the limit of the partial sums. When $|r| < 1$ we have

$$\lim_{n \to \infty} s_n = \lim_{n \to \infty} \frac{a}{1-r} - \left(\frac{a}{1-r} \right) r^n = \frac{a}{1-r},$$

and when $|r| > 1$ the limit doesn't exist. ∎

> When $r > 1$, the number r^n surpasses every finite value as n grows. When $r < -1$, the number r^n is alternately positive and negative as n increases, and its magnitude surpasses every finite value. In both cases, the numbers s_n don't approach any particular value, so the limit does not exist.

Value of a geometric series

Example 2.5. Determine whether each of the following series converges, and if so, calculate its value: (a) $\sum_{k=1}^{\infty} 7 \left(\frac{1}{10} \right)^k$, and (b) $\sum_{k=4}^{\infty} \left(\frac{3}{2} \right)^k$.

Solution: (a) This series is

$$\frac{7}{10} + \frac{7}{10^2} + \frac{7}{10^3} + \frac{7}{10^4} + \cdots .$$

The first term of this geometric series is $a = \frac{7}{10}$, so let's rewrite it as

$$a + a \left(\frac{1}{10} \right) + a \left(\frac{1}{10} \right)^2 + a \left(\frac{1}{10} \right)^3 + \cdots ,$$

which allows us to see that the common ratio is $r = \frac{1}{10}$. Since $|r| < 1$ the series converges, and we can use our formula for geometric series to write

$$\sum_{k=1}^{\infty} 7\left(\frac{1}{10}\right)^k = \frac{\frac{7}{10}}{1 - \frac{1}{10}} = \frac{7}{9}.$$

(b) In this case we have

$$\left(\frac{3}{2}\right)^4 + \left(\frac{3}{2}\right)^5 + \left(\frac{3}{2}\right)^6 + \left(\frac{3}{2}\right)^7 + \cdots$$

in which $a = \left(\frac{3}{2}\right)^4$ and $r = \frac{3}{2}$. Since $r > 1$, the partial sums grow forever and eventually exceed every finite value. Therefore, $\lim_{n\to\infty} s_n$ does not exist. This series diverges. ∎

Try It Yourself: Value of a geometric series

Example 2.6. Identify the numbers a and r, and find the value of each series if it converges: (a) $\sum_{k=3}^{\infty} \left(\frac{2}{3}\right)^k$, and (b) $\sum_{k=2}^{\infty} 2\left(-1\right)^k$.

Answer: (a) $a = (2/3)^3$, $r = 2/3$, and $\lim_{n\to\infty} s_n = 8/9$; (b) $a = 2$, $r = -1$, and the series diverges. ∎

Full Solution
On-line

Approximation and error

Example 2.7. Approximate the number $\sum_{k=0}^{9999999} 3\left(\frac{1}{2}\right)^k$.

Solution: We have ten *million* summands. Instead of adding them directly, let's treat this quantity as a partial sum of a geometric series, s_n where $n = 10^7$, $a = 3$ and $r = 1/2$. Then we know from equation (2.2) that

$$s_n = \frac{a}{1-r} - \frac{a}{1-r}r^n = 6 - 6\left(\frac{1}{2}\right)^{10000000}.$$

So if we say the value is 6, we're off the mark by only $6(0.5)^{10000000}$. ∎

❖ Alternating Series

Suppose you shade a disk of radius 1. Then your friend paints a white disk of radius $1/2$ over it. You shade a disk of radius $1/3$ on top of that, and your friend paints a white disk of radius $1/4$ on top of that ... As you continue, the shaded portion looks more and more like a target, as seen in Figure 2.3.

Figure 2.3: Painting a target by shading concentric disks blue and then white, ad infinitum.

If you could continue this forever (with *really* small paintbrushes!), how much of the circle would be shaded *in the limit?* As we did in the triangle example at the beginning of this section, let's calculate the shaded area in each step of the process:

$$s_1 = \pi$$

$$s_2 = \pi - \pi\left(\frac{1}{2}\right)^2 = \pi\left(1 - \frac{1}{2^2}\right)$$

$$s_3 = \pi - \pi\left(\frac{1}{2}\right)^2 + \pi\left(\frac{1}{3}\right)^2 = \pi\left(1 - \frac{1}{2^2} + \frac{1}{3^2}\right)$$

$$s_4 = \pi - \pi\left(\frac{1}{2}\right)^2 + \pi\left(\frac{1}{3}\right)^2 - \pi\left(\frac{1}{4}\right)^2 = \pi\left(1 - \frac{1}{2^2} + \frac{1}{3^2} - \frac{1}{4^2}\right)$$

$$\vdots \qquad \vdots$$

and in general we have

$$s_n = \pi \sum_{k=1}^{n} \frac{(-1)^{k+1}}{k^2}.$$

> Note that these summands do not follow a geometric progression (the ratio of one summand to the next is not constant).

Since the shaded area is between 0 and π at every step, and each turn alters the amount of shaded area by less than the turn before it, we're tempted to say that the limit of these partial sums must exist. In the exercise set, you'll prove that this is, in fact, correct. So the series converges, but to what? Answering that question *exactly* is very difficult, but with a little thought we can *approximate* its value.

Approximating the value of an alternating series

Example 2.8. By how much does s_2 differ from $\sum_{k=1}^{\infty}(-1)^{k+1}\left(\frac{\pi}{k^2}\right)$?

Solution: You began by painting a circle whose area was π, and your friend took some of that shading away, after which the total shaded area was $s_2 = \pi - \pi/4 = 3\pi/4$. All of the shaded area that's added in the remaining steps is contained inside your *next* disk, and that has an area of $\pi/9$. So if we were to approximate the limit value as $\approx s_2$, we wouldn't miss the mark by any more than $\pi/9$. ∎

Approximating the value of an alternating series

Example 2.9. By how much does s_3 differ from $\sum_{k=1}^{\infty}(-1)^{k+1}\left(\frac{\pi}{k^2}\right)$?

Solution: In the third step of the scenario described above, you shade a disk of radius $1/3$ (see the middle image of Figure 2.3). Some of that shading will be removed as the process continues, but no more than the area of the white disk that's painted in the *next* step, which has an area of $\pi/16$. So s_3 differs from the actual limit value by no more than $\pi/16$. ∎

Approximating the value of an alternating series

Example 2.10. How many summands are needed to approximate $\sum_{k=1}^{\infty}(-1)^{k+1}\left(\frac{\pi}{k^2}\right)$ to within 0.0001 of its actual value?

Solution: As in Example 2.8 and Example 2.9, if we stop our calculation after the nth term, our number will be within $a_{n+1} = \pi/(n+1)^2$ of the limit value. Consequently, if we choose n large enough that $\frac{\pi}{(n+1)^2} \leq 0.0001$, we're sure that s_n (the nth partial sum, which tells us the amount of shading after the nth step) is within 0.0001 of the limit value. That happens when $n = 177.2453$, from which we conclude that we need 178 of the summands. ∎

> In Example 2.8 we said that s_2 differs from the limit value by no more than the area of the next disk, $a_3 = \frac{\pi}{9}$. And in Example 2.9 we used the same reasoning to assert that s_3 differs from the limit value by no more than $a_4 = \frac{\pi}{16}$.

In general, a series in which the summands alternate sign is called—you guessed it—an **alternating series**, and it's easy to tell when such a series converges.

Note that $|a_{k+1}|$ is the size of the first summand that's unused by the partial sum.

> **Alternating Series Theorem:** Suppose $\{a_k\}$ is a sequence of positive numbers for which
>
> $$a_{k+1} \leq a_k \quad \text{and} \quad \lim_{k \to \infty} a_k = 0.$$
>
> Then the alternating series $\sum_{k=1}^{\infty}(-1)^{k+1}a_k$ converges, and its nth partial sum differs from the actual value of the series by no more than $|a_{n+1}|$. Written technically,
>
> $$\left| s_n - \sum_{k=1}^{\infty}(-1)^{k+1}a_k \right| \leq |a_{n+1}|.$$

We've written $(-1)^{k+1}$ instead of $(-1)^k$ in this alternating series so that the first summand is positive. That's not necessary to the theorem, but fits well with the example of shading the disk that we discussed above.

Alternatively, we could write an alternating series whose first term is positive as

$$\sum_{k=0}^{\infty}(-1)^k a_k,$$

starting with an index of 0 instead of 1.

In Examples 2.8–2.10 we motivated this error estimate using areas, but we can also understand it in terms of lengths. Figure 2.4 depicts the location of the partial sums of a convergent alternating series whose first term is positive. Think of starting at zero and stepping right, then left, then right ... in ever-decreasing amounts. When you arrive at s_3 after three steps, how far from the limiting position are you? Because of the back-and-forth nature of your motion, you know the limiting position is somewhere between s_3 and s_4, so you're no farther away from it than $|s_4 - s_3| = |a_4|$. For the same reason, when you're at s_n, you're no farther away from the limiting position than $|s_{n+1} - s_n| = |a_{n+1}|$.

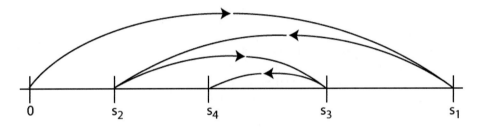

Figure 2.4: The partial sums of a convergent alternating series whose first summand is positive.

A divergent alternating series

Example 2.11. Determine whether the alternating series $\sum_{k=1}^{\infty}(-1)^{k+1}$ converges, and if so, approximate it to within 10^{-3} of its actual value.

Solution: This is an alternating series, but the summands don't converge to zero. This results in the partials sums

$$s_n = \begin{cases} 1 & \text{if } n = \text{even} \\ 0 & \text{if } n = \text{odd}. \end{cases}$$

which don't converge to a particular number in the limit, so this series diverges. ∎

Approximating the value of an alternating series

Example 2.12. Determine whether the series $\sum_{k=1}^{\infty}(-1)^{k+1}\left(\frac{1}{k}\right)$ converges, and if so, approximate it to within 10^{-2} of its actual value.

Solution: This is a famous series called the **Alternating Harmonic Series**. Its summands clearly decrease as we increment $k = 1, 2, 3, \ldots$, and $\lim_{k \to \infty}(1/k) = 0$, so this alternating series converges. The error bound above tells us that the nth

partial sum is no farther away from the limit value than $|a_{n+1}|$ (the size of the *next* step). So we determine n by solving the inequality

$$\frac{1}{n+1} \leq \frac{1}{100} \quad \Rightarrow \quad n \geq 99.$$

Therefore s_{99} is within 0.01 of the actual limit value. A spreadsheet or computer algebra system can be used to quickly add these summands, and we get $s_{99} = 0.6981721793$.　■

To be exact, the partial sum s_{99} is a fraction whose numerator and denominator are 97353652632905 8233802478942580320423 1637 and 1394407504594249543290 6761787062460711360, respectively.

Try It Yourself: Approximating the value of an alternating series

Example 2.13. Determine whether the alternating series $\sum_{k=1}^{\infty}(-1)^{k+1}\left(\frac{1}{k^4+89}\right)$ converges, and if so, how many summands are required to approximate the value of the series to within 10^{-5} of its actual value.

Answer: It converges; $n = 17$.　■

Full Solution On-line

✧ Arithmetic of Series

Because the value of a convergent series is defined by a limit, the basic arithmetic of series is tied to the limit laws.

Arithmetic of Series: Suppose that $\sum_{k=1}^{\infty} a_k$ and $\sum_{k=1}^{\infty} b_k$ are convergent series, and β is any constant. Then

$$\sum_{k=1}^{\infty}(a_k \pm b_k) = \sum_{k=1}^{\infty} a_k \pm \sum_{k=1}^{\infty} b_k \quad \text{and} \quad \sum_{k=1}^{\infty} \beta a_k = \beta \sum_{k=1}^{\infty} a_k.$$

The first of these equalities says simply that *the limit of a sum is the sum of the limits* (when both limits exist independently).

Note: In practice, we often omit parentheses unless a situation requires delimiters for added clarity. So instead of writing $\sum(a_k + b_k)$ we just write $\sum a_k + b_k$.

This is like writing

$$\int 2t + 4 \, dt$$

instead of

$$\int (2t + 4) \, dt.$$

Arithmetic of series

Example 2.14. Determine the value of $\sum_{k=1}^{\infty} \frac{2}{3^k} + \frac{5^k}{6^k}$.

Solution: The summands of this series are comprised of two convergent geometric series:

$$\sum_{k=1}^{\infty} \frac{2}{3^k} = \frac{2}{3} + \frac{2}{3^2} + \frac{2}{3^3} + \frac{2}{3^4} + \cdots$$

is a geometric series with $a = 2/3$ and $r = 1/3$. Similarly,

$$\sum_{k=1}^{\infty} \left(\frac{5}{6}\right)^k = \frac{5}{6} + \left(\frac{5}{6}\right)^2 + \left(\frac{5}{6}\right)^3 + \left(\frac{5}{6}\right)^4 + \cdots$$

is a geometric series with $a = 5/6$ and $r = 5/6$. Since both series converge independently,

$$\sum_{k=1}^{\infty} \frac{2}{3^k} + \frac{5^k}{6^k} = \sum_{k=1}^{\infty} \frac{2}{3^k} + \sum_{k=1}^{\infty} \frac{5^k}{6^k} = \frac{2/3}{1 - \frac{1}{3}} + \frac{5/6}{1 - \frac{5}{6}} = 1 + 5 = 6.$$　■

Arithmetic of series

Example 2.15. Determine the value of $\sum_{k=1}^{\infty} \frac{1}{k} - \frac{1}{k+1}$.

Solution: It's important to note that this series is *not* the difference of two convergence series, so we *cannot separate the series into the sum of two others* (recall that $\sum_{k=0}^{\infty} 1/k$ is the Harmonic Series, which diverges—see Example 2.1). Instead, we go back to the definition: the value of a series is the limit of its partial sums. So we compute . . .

$$\begin{aligned} s_1 &= 1 - \tfrac{1}{2} \\ s_2 &= 1 - \tfrac{1}{2} + \underbrace{\tfrac{1}{2} - \tfrac{1}{3}}_{} = 1 - \tfrac{1}{3}. \end{aligned}$$

$$\underbrace{}_{k=1} \quad \underbrace{}_{k=2}$$

Similarly, the positive and negative terms in the "interior" of s_3 sum to zero:

$$s_3 = \underbrace{1 - \frac{1}{2}}_{k=1} + \underbrace{\frac{1}{2} - \frac{1}{3}}_{k=2} + \underbrace{\frac{1}{3} - \frac{1}{4}}_{k=3} = 1 + \left(-\frac{1}{2} + \frac{1}{2} - \frac{1}{3} + \frac{1}{3} \right) - \frac{1}{4} = 1 - \frac{1}{4}.$$

Because the "interior" summands will always sum to zero like this, collapsing like an old sea captain's looking glass, we say that this is a **telescoping series**. For this particular telescoping series we have $s_n = 1 - \frac{1}{n+1}$, so

$$\sum_{k=1}^{\infty} \frac{1}{k} - \frac{1}{k+1} = \lim_{n \to \infty} s_n = 1. \qquad \blacksquare$$

Arithmetic of series

Example 2.16. Determine the value of $\sum_{k=1}^{\infty} \frac{5}{4^k} + \frac{10}{k^2+4k+3}$.

Solution: Our strategy will be to write

$$\sum_{k=1}^{\infty} \frac{5}{4^k} + \frac{10}{k^2 + 4k + 3} = \sum_{k=1}^{\infty} \frac{5}{4^k} + \sum_{k=1}^{\infty} \frac{10}{k^2 + 4k + 3}.$$

We can do it only if both series on the right-hand side converge independently, and it's useful only if we can calculate their values, but they do and we can. The first series is geometric with $a = 5/4$ and $r = 1/4$, so it converges to

$$\sum_{k=1}^{\infty} \frac{5}{4^k} = \frac{\frac{5}{4}}{1 - \frac{1}{4}} = \frac{5}{3}.$$

The second series is not familiar, but you know from your study of partial fractions decompositions that we can rewrite it as

$$\sum_{k=1}^{\infty} \frac{10}{k^2 + 4k + 3} = \sum_{k=1}^{\infty} \frac{5}{k+1} - \frac{5}{k+3},$$

after which it looks like a telescoping series. It is, and we can determine its value by writing its partial sums.

$$\begin{aligned} s_1 &= \tfrac{5}{2} - \tfrac{5}{4} \\ s_2 &= \tfrac{5}{2} - \tfrac{5}{4} + \tfrac{5}{3} - \tfrac{5}{5} \\ s_3 &= \tfrac{5}{2} - \tfrac{5}{4} + \tfrac{5}{3} - \tfrac{5}{5} + \tfrac{5}{4} - \tfrac{5}{6} = \tfrac{5}{2} + \tfrac{5}{3} - \tfrac{5}{5} - \tfrac{5}{6} \\ s_4 &= \underbrace{\tfrac{5}{2} - \tfrac{5}{4}}_{k=1} + \underbrace{\tfrac{5}{3} - \tfrac{5}{5}}_{k=2} + \underbrace{\tfrac{5}{4} - \tfrac{5}{6}}_{k=3} + \underbrace{\tfrac{5}{5} - \tfrac{5}{7}}_{k=4} = \tfrac{5}{2} + \tfrac{5}{3} - \tfrac{5}{6} - \tfrac{5}{7}. \end{aligned}$$

In Example 2.15 the telescoping characteristic of the series was manifest as early as s_2. In this example, we have to wait until s_3. The "lag time" before telescoping begins depends on the difference in the denominators of the partial fractions decomposition. In Example 2.15 the denominators differ by 1, so we begin to see telescoping in $s_{1+1} = s_2$. In this example, the denominators differ by 2, so we begin to see telescoping in $s_{1+2} = s_3$.

Here we see "interior" terms that sum to zero in s_3 and in s_4. This continues to happen in subsequent partial sums, resulting in

$$s_n = \frac{5}{2} + \frac{5}{3} - \frac{5}{n+2} - \frac{5}{n+3}.$$

Therefore $\lim_{n\to\infty} s_n = \frac{5}{2} + \frac{5}{3} = \frac{25}{6}$, and

$$\sum_{k=1}^{\infty} \frac{5}{4^k} + \frac{10}{k^2+4k+3} = \sum_{k=1}^{\infty} \frac{5}{4^k} + \sum_{k=1}^{\infty} \frac{10}{k^2+4k+3} . = \frac{5}{3} + \frac{25}{6} = \frac{35}{6}. \qquad \blacksquare$$

❖ Common Difficulties with Series

Many people mistakenly think that a series must converge if its summands tend to zero, but the Harmonic Series is a counterexample. You have to know more before you can say that a series converges (e.g., that the terms tend to zero *and* do so geometrically, or that they tend to zero monotonically *and* are alternating).

The topic of series is one in which it's very handy to have "landmark" examples against which to test ideas. The Harmonic Series and the Alternating Harmonic Series are good landmarks to keep in mind.

> Sections 8.3–8.5 introduce five different tests that help us determine whether $\sum_{k=0}^{\infty} a_k$ converges. We wouldn't need any of them if $\lim_{k\to\infty} a_k = 0$ were sufficient.

❖ In Closing

In general, we typically ask two different questions about series:

- Does $\sum_{k=1}^{\infty} a_k$ represent a number? (i.e., does it converge?)

- If so, what number is it? (i.e., to what?)

When you're wondering whether a particular series converges, ask yourself the following questions ...

1. Do the summands tend to zero in the limit? If not, the partial sums cannot converge.

2. Is it an alternating series? If so, it's relatively easy to tell whether it converges (see p. 608).

> **IMPORTANT:** A series can diverge even when its summands tend to zero in the limit (e.g., the Harmonic Series).
>
> In order to be sure that a series converges, you need to know (1) that the terms go to zero *and* (2) some *other* piece of information (for example, that it's alternating or geometric or ...).

Once you know that a series converges, the next question is, "To what?" With the exception of geometric and telescoping series, it's unlikely that you'll be able to find out exactly, but you can always approximate. For alternating series, we know that the nth partial sum differs from the actual value of the series by no more than the magnitude of the first unused summand (see p. 608).

You should know

- the terms *partial sum, (infinite) series, geometric series, converge, diverge, alternating series, Harmonic Series,* and *telescoping series;*

- that the nth partial sum is the cumulative total after adding the first n summands;

- that a series cannot converge unless its summands tend to zero in the limit;

- that the Harmonic Series diverges, even though its summands tend to zero in the limit.

You should be able to

- determine the numbers a and r for a geometric series;

- calculate the value of a convergent geometric series;

- use a geometric series to estimate the value of a finite sum, and determine the amount of error in the approximation;

- determine the amount of error in approximating an alternating series by its nth partial sum;

- determine the value of a series that's the sum of other convergent series.

✥ 8.2 Skill Exercises

In #1–8 (a) determine whether the series is geometric, and if so (b) determine its common ratio.

1. $\displaystyle\sum_{k=0}^{\infty} 3^k$

2. $\displaystyle\sum_{k=1}^{\infty} \frac{1}{7^k}$

3. $\displaystyle\sum_{k=2}^{\infty} \frac{(-2)^{k+1}}{5^k}$

4. $\displaystyle\sum_{k=3}^{\infty} \frac{8^k}{(-3)^{k+2}}$

5. $\displaystyle\sum_{k=4}^{\infty} \frac{2^k}{k}$

6. $\displaystyle\sum_{k=5}^{\infty} \frac{\sqrt{k}}{(-3)^k}$

7. $\displaystyle\sum_{k=6}^{\infty} \frac{8k}{(-3)^k}$

8. $\displaystyle\sum_{k=7}^{\infty} \frac{k^2}{(k+3)^k}$

In #9–24 (a) determine whether the geometric series converges, and if so (b) determine its value.

9. $\displaystyle\sum_{k=0}^{\infty} \left(\frac{1}{4}\right)^k$

10. $\displaystyle\sum_{k=1}^{\infty} \left(\frac{1}{5}\right)^k$

11. $\displaystyle\sum_{k=2}^{\infty} \frac{3}{5^k}$

12. $\displaystyle\sum_{k=3}^{\infty} \frac{8}{3^k}$

13. $\displaystyle\sum_{k=4}^{\infty} \frac{7^k}{4^k}$

14. $\displaystyle\sum_{k=0}^{\infty} (0.5)^k \frac{5^k}{3^k}$

15. $\displaystyle\sum_{k=1}^{\infty} \frac{(-1)^k}{4^k}$

16. $\displaystyle\sum_{k=2}^{\infty} \frac{(-9)^k}{4^k}$

17. $\displaystyle\sum_{k=3}^{\infty} \frac{(-1)^k}{4^k}$

18. $\displaystyle\sum_{k=0}^{\infty} \frac{(-2)^k}{5^k}$

19. $\displaystyle\sum_{k=4}^{\infty} 9\left(-\frac{1}{5}\right)^k$

20. $\displaystyle\sum_{k=1}^{\infty} \frac{7}{(-3)^k}$

21. $\displaystyle\sum_{k=1}^{\infty} 7^{k-1} 6^{3-2k}$

22. $\displaystyle\sum_{k=1}^{\infty} (-2)^{3k+1} 3^{1-2k}$

23. $\displaystyle\sum_{k=0}^{\infty} x^k$

24. $\displaystyle\sum_{k=0}^{\infty} x^{2k}$

In #25–28 (a) estimate the value of each sum by treating it as an infinite geometric series, and (b) determine the amount of error in this approximation.

25. $\displaystyle\sum_{k=0}^{1000} (0.1)^k$

26. $\displaystyle\sum_{k=1}^{20} (0.02)^k$

27. $\displaystyle\sum_{k=1}^{10^6} \left(\frac{2}{3}\right)^k$

28. $\displaystyle\sum_{k=1}^{10^9} \left(\frac{1}{3}\right)^k$

You've probably heard that every repeating decimal can be expressed as a fraction. Here's how: suppose we want to write $4.1\overline{7}$ as a fraction. We begin by writing it as

$$\frac{41}{10} + \frac{7}{100} + \frac{7}{1000} + \frac{7}{10000} + \cdots = \frac{41}{10} + \frac{7}{100} \sum_{k=0}^{\infty} \frac{1}{10^k}.$$

Since we know how to sum a geometric series, we can write this expression as a fraction. In #29–34 write each number as a fraction.

29. $5.\overline{11}$

31. $7.31\overline{44}$

33. $0.\overline{99}$

30. $6.8\overline{22}$

32. $2.925\overline{33}$

34. $0.\overline{xx}$

In #35–38 determine the value of each alternating series to within ε of its actual value.

35. $\sum_{k=1}^{\infty} \frac{(-1)^k}{k^2}$, $\varepsilon = 0.01$

37. $\sum_{k=2}^{\infty} \frac{(-1)^k}{k \ln(k)}$, $\varepsilon = 0.1$

36. $\sum_{k=2}^{\infty} \frac{(-1)^k}{k^3}$, $\varepsilon = 0.01$

38. $\sum_{k=1}^{\infty} \frac{(-1)^k}{\sqrt{k}}$, $\varepsilon = 0.5$

In each of #39–42 (a) use a partial fractions decomposition to rewrite the summand, and (b) calculate the value of the series as we did in Example 2.15.

39. $\sum_{k=1}^{\infty} \frac{1}{k^2 + 2k}$

40. $\sum_{k=1}^{\infty} \frac{1}{k^2 + 3k}$

41. $\sum_{k=1}^{\infty} \frac{3}{k^2 + 5k + 4}$

42. $\sum_{k=1}^{\infty} \frac{4}{k^2 + 7k + 10}$

Calculate the value of the series in #43–46.

43. $\sum_{k=1}^{\infty} \left(\frac{1}{4}\right)^k + \frac{(-1)^k}{3^k}$

45. $\sum_{k=1}^{\infty} \frac{3}{4^k} + \frac{1}{k^2 + k}$

44. $\sum_{k=1}^{\infty} (0.12)^k + 6^{k-1} 2^{5-3k}$

46. $\sum_{k=1}^{\infty} \frac{2}{k^2 + 8k + 15} + \frac{7}{k^2 + 4k}$

❖ 8.2 Concept and Application Exercises

47. Can a series be both alternating *and* geometric? If so give an example. If not, explain why not.

48. Suppose s_n denotes the nth partial sum of $\sum_{k=1}^{\infty} a_k$, which is a series that converges to 5. Determine $\lim_{n \to \infty} s_n$.

49. Suppose a_k is defined for all integers k, and $\sum_{k=100}^{\infty} a_k$ converges. Must $\sum_{k=1}^{\infty} a_k$ also converge? If so, explain why? If not, provide a counterexample.

50. Suppose you know the first 10^6 summands of a series.

    (a) What percentage of the summands do you know?

    (b) How reliable would a converge/diverge decision be, based on that amount of data?

51. Suppose that $a_k = 1 + \frac{1}{k}$.

    (a) Verify that $0 \le a_{k+1} < a_k$

    (b) Calculate the first 10 partial sums of $\sum_{k=1}^{\infty} (-1)^k a_k$.

    (c) Explain why this alternating series diverges, and how this fact does not contradict the Alternating Series Theorem.

52. Assume $x > 0$. Since 1 is not greater than itself, the computation

$$1 < (1 + x + x^2 + x^3 + \cdots) + \left(1 + \frac{1}{x} + \frac{1}{x^2} + \frac{1}{x^3} + \cdots \right) = \frac{1}{1-x} - \frac{x}{1-x} = 1$$

must be wrong. Find the error.

53. Since $1 \neq 0$, the computation

$$0 = (1-1)+(1-1)+(1-1)+\cdots = \sum_{k=0}^{\infty}(-1)^k = 1+(-1+1)+(-1+1)+\cdots = 1+0+0+\cdots = 1$$

must contain an error. Find it.

54. You know that the Harmonic Series diverges, but also that $1 \neq 0$, so

$$1 = \sum_{k=1}^{\infty} \frac{1}{k} - \frac{1}{k+1} = \sum_{k=1}^{\infty} \frac{1}{k} - \sum_{k=1}^{\infty} \frac{1}{k+1} = \infty - \infty = 0$$

must contain an error. Find it.

55. In each part below, determine whether $\sum_{k=1}^{\infty} a_k$ converges.

(a) Figure 2.5 depicts the summands a_k.

(b) Figure 2.5 depicts the partial sums s_n.

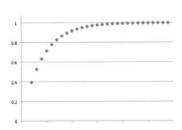

Figure 2.5: For #55.

56. Suppose a ball is dropped from a height of 1 meter and it returns to 75% of its previous altitude with each bounce.

(a) If the bouncing were to continue forever, what would be the cumulative distance traveled by the ball?

(b) If the ball bounces 20 times, and comes to a stop when it meets the ground for the 21st time, what's the cumulative distance traveled by the ball?

(c) How different are your answers in parts (a) and (b)?

57. Suppose the large square shown in Figure 2.6 has a side length of 1. If the shaded brackets continue ad infinitum into the corners, what's the area of the white space (ignoring the gray guidelines)?

58. Suppose the equilateral triangle shown in Figure 2.6 has a side length of 1. If the disks continue ad infinitum into the corner, what's the area of the shaded region?

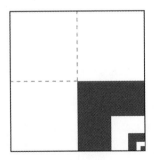 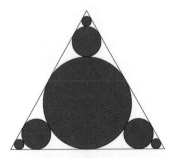

Figure 2.6: (left) For #57; (right) for #58.

59. Suppose we form a right triangle in the first quadrant by using a line segment to connect the points $(a, 0)$ and $(0, 1)$. Then we make a right-triangle in the second quadrant using a line segment perpendicular to our first one, and a right-triangle in the third quadrant using a line segment perpendicular to that, and so on (see Figure 2.7).

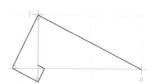

Figure 2.7: For #59.

(a) Supposing we continue the process ad infinitum, find a formula for the length of the resulting path in terms of a.

(b) Determine which values of a will result in a finite length, and explain why some do while others don't.

60. The isosceles triangle with angles $36°$, $72°$, and $72°$, shown in Figure 2.8, is called the *golden triangle*.

(a) Verify that bisecting one of the $72°$ angles with a line segment results in another, smaller golden triangle.

(b) If the side length of the original golden triangle is L, what is the length of the line segment that bisects the $72°$ as shown in Figure 2.8?

(c) Because bisecting the $72°$ angle results in another golden triangle, we can bisect a $72°$ angle of this smaller triangle and get an even smaller one, ad infinitum (as indicated by Figure 2.8). Denote by b_k the length of the bisector in the kth step of this process. Determine an explicit formula for b_k.

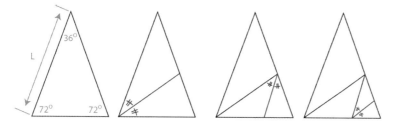

Figure 2.8: Golden triangles for #60.

61. Suppose we draw the graph of $\sin(x)$ over $x \in [0, \pi]$. Separately, we also draw the graph of $\sin(2x)$ over $[0, \pi/2]$, the graph of $\sin(4x)$ over $[0, \pi/4]$, ...the graph of $\sin(2^n x)$ over $[0, \pi/2^n]$, etc. Suppose further that we append these graphs in order, one after the other as indicated in Figure 2.9. Determine the area between the resulting curve and the horizontal axis.

Figure 2.9: For #61.

62. In exercise #60 you developed a formula for the length of the kth bisector in an iterative process that involved the golden triangle, b_k. If we were to set all those bisectors end to end, what would be their total length?

63. In this exercise you will answer the question, "How much space on the number line is taken up by all of the rational numbers?"

(a) Verify that the table in Figure 2.10 includes all the rational numbers.

(b) By traversing the table along the suggested path, show that the rational numbers can be indexed by the positive integers. For the rest of this exercise, we'll denote by r_k the rational number indexed by k.

(c) Suppose $\varepsilon > 0$, and around each number r_k on the number line we put an interval of length $\varepsilon/2^k$. What is the total length of those intervals? (If you think of $(r_k - \varepsilon/2^{k+1}, r_k + \varepsilon/2^{k+1})$ as a "blanket" that we lay over r_k, how much "cloth" would we need to cover all the rational numbers?)

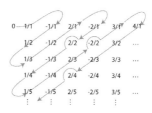

Figure 2.10: Indexing the rational numbers, avoiding repeats (for #63).

(d) The fact that we didn't specify ε beforehand allows us some flexibility now. Based on your answer to part (c) and the fact that we can now choose ε to be as small as we want, how much space is taken up by all of the rational numbers on the number line?

(e) If you throw a mathematical "dart" that hits exactly one point on the number line, what is the probability that you'll hit a rational number?

Checking the details

64. In this exercise you'll prove that $\sum_{k=1}^{\infty}(-1)^k a_k$ converges when $0 < a_{k+1} \le a_k$ and $\lim_{k\to\infty} a_k = 0$ (i.e., this is the claim made on p. 608).

(a) Show that the *even* partial sums, $s_{2n} = \sum_{k=1}^{2n}(-1)^k a_k$, constitute a monotonically decreasing sequence. *(Hint: compare s_2 and s_4, using what you know about the numbers a_k.)*

(b) Show that $\{s_{2n}\}$ is contained in $[-a_1, 0]$.

(c) Show that the *odd* partial sums, $s_{2n+1} = \sum_{k=1}^{2n+1}(-1)^k a_k$, constitute a monotonically increasing sequence that's contained in $[-a_1, 0]$.

(d) The Bounded Monotone Convergence Theorem tells us that each of these sequences converges. Let's define $L_{\text{even}} = \lim_{n\to\infty} s_{2n}$ and $L_{\text{odd}} = \lim_{n\to\infty} s_{2n}$. Show that $L_{\text{odd}} = L_{\text{even}}$ by showing that their difference is zero. *(Hint: the difference of limits is the limit of the difference, and $s_{2n+1} - s_{2n} = a_{2n+1}$.)*

(e) Explain why $\lim_{n\to\infty} s_n$ exists.

(f) Where in this argument did you rely on the fact that the summands are alternating?

65. The Alternating Series Theorem (see p. 608) requires not only $\lim_{k\to\infty} a_k = 0$, but the summands must be monotonically decreasing. In this exercise, we provide an example of why both hypotheses must be satisfied in order to guarantee convergence. Consider the series $\sum_{k=2}^{\infty}(-1)^k a_k$ where

$$a_k = \begin{cases} 4/k & \text{if } k \text{ is even} \\ 2/(k-1) & \text{if } k \text{ is odd.} \end{cases}$$

(a) Use the Squeeze Theorem to show that $\lim_{k\to\infty} a_k = 0$.

(b) Verify that $|a_{2k}| > |a_{2k+1}|$ (i.e., each even-indexed summand has a larger magnitude than the next odd-indexed summand).

(c) Calculate the first five partial sums, s_1, s_2, s_3, s_4 and s_5.

(d) Explain why, in general, $s_{2n+2} = s_{2n} + \frac{1}{n+1}$.

(e) Based on part (d), explain why this series cannot converge.

8.3   The Integral Test and Estimating Sums

In Section 8.2 we used geometric examples to introduce the idea of series, and the fact that an infinite series can be a finite number. Series arise in many other ways too. For example, the series

$$\sum_{k=1}^{\infty} \frac{1}{k^2} = 1 + \frac{1}{2^2} + \frac{1}{3^2} + \frac{1}{4^2} + \cdots$$

arises when calculating how much heat is required to raise the temperature of a Fermi gas (i.e., when determining its *heat capacity*). This is very much like the series we used to determine the shaded area of the "target" on p. 606 except that it's not alternating, so the fact that the summands tend to zero in the limit is not sufficient to guarantee convergence. In this section, we'll use improper integrals to determine when series like this converge, and to estimate the value of those that do.

> Recall that the Harmonic Series diverges, even though its summands tend to zero in the limit.

❖ Integral Test

When all the summands of $\sum_{k=1}^{\infty} a_k$ are non-negative, we can think of each of them as contributing area, but unlike the triangle example on p. 602 there's generally not a specific geometric context, so we develop our own:

a_1 = area of a rectangle that's 1 unit wide, and a_1 units high

a_2 = area of a rectangle that's 1 unit wide, and a_2 units high

   ⋮   ⋮

Figure 3.1 shows these rectangles side by side on the number line, with the right-hand edge of the kth rectangle set at k. It also shows the graph of a decreasing function f for which $f(k) = a_k$. When we know such a function, we can say

$$s_3 \;=\; a_1 + a_2 + a_3 \leq a_1 + \int_1^3 f(t)\,dt$$

$$s_4 \;=\; a_1 + a_2 + a_3 + a_4 \leq a_1 + \int_1^4 f(t)\,dt$$

   ⋮   ⋮

And in general,

$$s_n = a_1 + a_2 + \cdots + a_n \leq a_1 + \int_1^n f(t)\,dt. \tag{3.1}$$

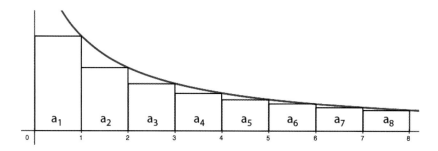

Figure 3.1: Each a_k is the area of a rectangle.

This leads us to the following conclusion.

Integral Test (Part 1): Suppose that f is a continuous, decreasing function for which $f(k) = a_k \geq 0$. If

$$\int_1^\infty f(t)\, dt \quad \text{converges, so does} \quad \sum_{k=1}^\infty a_k.$$

Proof. The sequence of partial sums is increasing since all $a_k \geq 0$, and equation (3.1) tells us that

$$s_n \leq a_1 + \int_1^n f(t)\, dt \leq a_1 + \int_1^\infty f(t)\, dt.$$

When the improper integral converges, the right-hand side of this inequality is a particular finite number. Since the sequence $\{s_n\}$ is increasing (recall that all $a_k \geq 0$) and bounded above, the Bounded Monotone Converge Theorem (p. 592) guarantees that $\lim_{n\to\infty} s_n$ exists. ∎

Using the Integral Test

Example 3.1. Use the Integral Test to show that $\sum_{k=1}^\infty k e^{-k}$ converges.

Solution: In order to use the Integral Test, we need a continuous, decreasing function that generates the summands at the integers. Based on our series, a natural choice is $f(t) = te^{-t}$. This function is clearly continuous, and

$$f'(t) = (1 - t)e^{-t}, \quad \text{which is negative when } t > 1,$$

so it's decreasing over $[1, \infty)$. Using an integration by parts, we see that

$$\int_1^\infty te^{-t}\, dt = \lim_{T\to\infty} \int_1^T te^{-t}\, dt = \lim_{T\to\infty} \frac{2}{e} - (T+1)e^{-T} = \frac{2}{e}.$$

Since the improper integral converges, the Integral Test tells us that the series does too. ∎

> This is the essential idea of the Integral Test: instead of working with series (which are new and unfamiliar) let's work with integrals (which are familiar).

Now let's turn our attention to Figure 3.2, which shows the same rectangles as Figure 3.1, except that they've been shifted to the right by 1 unit, so that the left-hand edge of the kth rectangle is at k. It also shows the graph of the same decreasing function f whose value at $t = k$ is a_k.

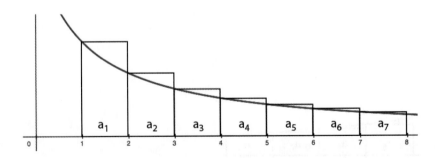

Figure 3.2: Each a_k is the area of a rectangle.

With the rectangles set like this, we see

$$s_3 = a_1 + a_2 + a_3 \geq \int_1^4 f(t)\, dt$$

$$s_4 = a_1 + a_2 + a_3 + a_4 \geq \int_1^5 f(t)\, dt$$

$$\vdots \qquad \vdots$$

$$s_n = a_1 + a_2 + a_3 + \cdots + a_n \geq \int_1^{n+1} f(t)\, dt. \qquad (3.2)$$

This leads us to the following conclusion.

Integral Test (Part 2): Suppose that f is a continuous, decreasing function for which $f(k) = a_k \geq 0$. If

$$\int_1^\infty f(t)\, dt \quad \text{diverges, so does} \quad \sum_{k=1}^\infty a_k.$$

Proof. The integral $\int_1^{n+1} f(t)\, dt$ increases with n since $f(t) \geq 0$. And since it diverges, it surpasses each finite number as $n \to \infty$. Since the integral is smaller than s_n, the sequence of partial sums must also surpass each finite number, so it cannot converge. ∎

Using the Integral Test to prove convergence

Example 3.2. Prove that $\sum_{k=1}^\infty 1/k^p$ converges if and only if $p > 1$.

Solution: On p. 461 we showed that $\int_1^\infty 1/t^p\, dt$ converged if and only if $p > 1$. Our conclusion follows directly from the Integral Test. ∎

Series of the form $\sum_{k=1}^\infty 1/k^p$, as seen in Example 3.2, are called p-**series**. The Harmonic Series is a p-series with $p = 1$. On p. 603 we saw that it diverges, even though the summands tend to zero in the limit. In essence, this happens because the summands don't head toward zero "fast enough." Example 3.2 tempts us to define "fast enough" to mean that the summands tend toward zero "faster than $1/k$," and while that's a good rule of thumb, there are exceptions. For example …

Using the Integral Test to prove divergence

Example 3.3. Use the Integral Test to prove that $\sum_{k=2}^\infty \frac{1}{k \ln(k)}$ diverges.

Solution: In this case we use the function $f(t) = \frac{1}{t \ln(t)}$. Using the substitution $x = \ln(t)$, $dx = \frac{1}{t}\, dt$ we see that

$$\int_2^\infty \frac{1}{t \ln(t)}\, dt = \lim_{L \to \infty} \int_2^L \frac{1}{t \ln(t)}\, dt = \lim_{L \to \infty} \int_{\ln(2)}^{\ln(L)} \frac{1}{x}\, dx = \int_{\ln(2)}^\infty \frac{1}{x}\, dx,$$

which diverges (see p. 461). Since the improper integral diverges, the Integral Test guarantees that the series does too. ∎

❖ Error Estimates

In Section 8.2 we approximated the value of an alternating series with one of its partial sums, and we calculated the error in such an approximation. When the summands of a convergent series are all non-negative, the Integral Test can often be used to make the same kind of calculation.

For the sake of discussion, suppose we know that $\sum_{k=1}^{\infty} a_k$ converges, but we don't know the limiting value, so we approximate $\sum_{k=1}^{\infty} a_k \approx a_1 + a_2 + a_3$. There's error in this approximation, of course, because we've omitted $a_4 + a_5 + a_6 + \cdots$ However, Figure 3.1 (p. 617) shows that

$$a_4 + a_5 + a_6 + \cdots \leq \int_3^{\infty} f(t)\, dt,$$

and Figure 3.2 (p. 618) shows us that

$$\int_4^{\infty} f(t)\, dt \leq a_4 + a_5 + a_6 + \cdots .$$

We can write these inequalities simultaneously as

$$\int_4^{\infty} f(t)\, dt \leq a_4 + a_5 + a_6 + \cdots \leq \int_3^{\infty} f(t)\, dt.$$

More generally,

$$\int_{n+1}^{\infty} f(t)\, dt \leq \sum_{k=n+1}^{\infty} a_k \leq \int_n^{\infty} f(t)\, dt. \qquad (3.3)$$

A direct consequence of this fact is the following.

> **Error Bounds for Series of Positive Terms:** Suppose $\sum_{k=1}^{\infty} a_k$ is a convergent series and f is a continuous, decreasing function for which $f(k) = a_k \geq 0$. Then the difference between the value of $\sum_{k=1}^{\infty} a_k$ and its nth partial sum is
>
> $$\sum_{k=1}^{\infty} a_k - \sum_{k=1}^{n} a_k = \sum_{k=n+1}^{\infty} a_k,$$
>
> which is at least $\int_{n+1}^{\infty} f(t)\, dt$ but no more than $\int_n^{\infty} f(t)\, dt.$

Estimating error

Example 3.4. Suppose we approximate $\sum_{k=1}^{\infty} 1/k^{1.2}$ by adding the first 100 summands. How much error is in our approximation?

Solution: We know that

$$\sum_{k=1}^{\infty} \frac{1}{k^{1.2}} = \sum_{k=1}^{100} \frac{1}{k^{1.2}} + \sum_{k=101}^{\infty} \frac{1}{k^{1.2}}.$$

If we only use the first hundred terms, we're neglecting $\sum_{k=101}^{\infty} 1/k^{1.2}$, and according to the error bounds shown above,

$$\int_{101}^{\infty} \frac{1}{t^{1.2}}\, dt \leq \sum_{k=101}^{\infty} \frac{1}{k^{1.2}} \leq \int_{100}^{\infty} \frac{1}{t^{1.2}}\, dt.$$

You know how to find the exact value of convergent geometric and telescoping series. For almost all other convergent series, you will not be able to calculate an exact value (even though there is one), so you have to approximate. Part of making a good approximation is understanding the magnitude of the error in it.

Figure 3.3: Both of the bounding integrals in (3.3) are improper, but the integral on the left-hand side begins at a later value of t (at $n + 1$ instead of n). Since it has less of the real axis along which to accumulate area, its value is smaller.

So when we approximate $\sum_{k=1}^{\infty} \frac{1}{k^{1.2}}$ by its 100th partial sum, our error is at least

$$\int_{101}^{\infty} \frac{1}{t^{1.2}}\, dt = \lim_{L\to\infty} \int_{101}^{L} \frac{1}{t^{1.2}}\, dt = \lim_{L\to\infty} \frac{-5}{t^{0.2}}\Big|_{101}^{L} = \frac{5}{101^{0.2}} \approx 1.98657,$$

but is no more than

$$\int_{100}^{\infty} \frac{1}{t^{1.2}}\, dt = \lim_{L\to\infty} \int_{100}^{L} \frac{1}{t^{1.2}}\, dt = \lim_{L\to\infty} \frac{-5}{t^{0.2}}\Big|_{100}^{L} = \frac{5}{100^{0.2}} \approx 1.99054. \qquad \blacksquare$$

Try It Yourself: Estimating error

Example 3.5. Suppose we approximate $\sum_{k=1}^{\infty} k/(k^2+1)^2$ by adding the first 50 summands. How much error is in our approximation?

Full Solution On-line

Answer: $\frac{1}{5204} \le \sum_{k=51}^{\infty} k/(k^2+1)^2 \le \frac{1}{5002}$. $\qquad \blacksquare$

Achieving accuracy

Example 3.6. How many summands are required to approximate $\sum_{k=1}^{\infty} 1/k^{1.2}$ to within 0.001 of its actual value using s_n?

Solution: We know that

$$\sum_{k=1}^{\infty} \frac{1}{k^{1.2}} = \sum_{k=1}^{n} \frac{1}{k^{1.2}} + \sum_{k=n+1}^{\infty} \frac{1}{k^{1.2}}.$$

So s_n will be within 0.001 units of the actual series value when $\sum_{k=n+1}^{\infty} 1/k^{1.2}$ is less than $1/1000$. Said differently, n must be large enough that

$$\sum_{k=n+1}^{\infty} \frac{1}{k^{1.2}} \le \frac{1}{1000}.$$

We know from our discussion of error bounds that

$$\sum_{k=n+1}^{\infty} \frac{1}{k^{1.2}} \le \int_{n}^{\infty} \frac{1}{t^{1.2}}\, dt = \frac{5}{n^{0.2}},$$

so we can achieve our goal by choosing n sufficiently large that

$$\frac{5}{n^{0.2}} \le \frac{1}{1000} \quad \Rightarrow \quad 5 \times 10^3 \le n^{1/5} \quad \Rightarrow \quad 2.5 \times 10^{16} \le n.$$

That is, we need 25 million *billion* summands! $\qquad \blacksquare$

> In Examples 3.4 and 3.5 the number of summands was specified, and you were asked about the amount of error in the approximation. By contrast, this example asks you how many summands you need to use so that s_n is "close enough" to the actual series value (where "close enough" is quantified by the specified tolerable error).

Don't just read it. Check it for yourself!

> You'll see a much more efficient way to make an approximation on p. 623.

$$s_n \qquad\qquad\qquad\qquad\qquad s_n + \int_{n}^{\infty} f(t)\, dt$$

Figure 3.4: Our error bounds guarantee that the series value lies somewhere in this interval. In Examples 3.6 and 3.7 we guarantee that s_n is close to the actual value of the series by choosing n large enough that the value of the improper integral (and so the length of this interval) is small.

Try It Yourself: Achieving accuracy

Example 3.7. How many summands are required to approximate $\sum_{k=1}^{\infty} k/(k^2+1)^2$ to within 0.001 of its actual value?

Full Solution On-line

Answer: $n = 23$. $\qquad \blacksquare$

❖ Approximating with Comparison

In both Examples 3.6 and 3.7 (see p. 621), we relied on the fact that $\int_n^\infty f(t)\,dt$ overestimates the error incurred by using a partial sum to approximate a series. This requires us to calculate that improper integral, but integration isn't always so easy. In the following examples, the integral is too difficult to compute, so we approximate it by something larger that we *can* compute.

Approximating with comparison

Example 3.8. Use the technique demonstrated in Example 3.6 (on p. 621) to determine the number of summands that are required to approximate $\sum_{k=1}^\infty \frac{k \tan^{-1}(k)}{5k^4 - 3k + 4}$ to within 0.01 of its actual value.

Solution: Because $f(t) = \frac{t \tan^{-1}(t)}{5t^4 - 3t + 4}$ is continuous and decreasing when $t \geq 1$, we know that

$$\sum_{k=n+1}^\infty \frac{k \tan^{-1}(k)}{5k^4 - 3k + 4} \leq \int_n^\infty \frac{t \tan^{-1}(t)}{5t^4 - 3t + 4}\,dt, \tag{3.4}$$

> You can check that $f(t)$ is decreasing when $t \geq 1$ by verifying that $f'(t) < 0$.

but finding an antiderivative by which to evaluate this integral seems unlikely. Since we're currently in the business of estimation, let's continue by approximating this integral with one that we *can* integrate. *It's important that any algebraic changes we make during our approximation increase the value of the integrand, so that we're always sure to overestimate the actual amount of error.*

One way to increase the size of a fraction is to increase the size of its numerator. Since the arctangent never exceeds $\pi/2$, it follows from equation (3.4) that

$$\sum_{k=n+1}^\infty \frac{k \tan^{-1}(k)}{5k^4 - 3k + 4} \leq \int_n^\infty \frac{\frac{\pi}{2}t}{5t^4 - 3t + 4}\,dt. \tag{3.5}$$

Another way to increase the size of a fraction is to decrease the size of its denominator. In this case, we can do that by omitting the "+4" from the denominator of the integrand. That is, it follows from equation (3.5) that

$$\sum_{k=n+1}^\infty \frac{k \tan^{-1}(k)}{5k^4 - 3k + 4} \leq \int_n^\infty \frac{\frac{\pi}{2}t}{5t^4 - 3t}\,dt. \tag{3.6}$$

Since the lower bound of integration is at least 1, we know that $3t \leq 3t^4$, so we can reduce the size of the denominator even further by replacing the "$-3t$" with "$-3t^4$" (so that we're subtracting *more* from $5t^4$). It follows from (3.6) that

$$\sum_{k=n+1}^\infty \frac{k \tan^{-1}(k)}{5k^4 - 3k + 4} \leq \int_n^\infty \frac{\frac{\pi}{2}t}{5t^4 - 3t^4}\,dt = \int_n^\infty \frac{\pi}{4t^3}\,dt = \frac{\pi}{8n^2}.$$

Because $\frac{\pi}{8n^2} < 0.01$ when $n \approx 6.3$, we know that s_7 is within 0.01 of the series' actual value. ∎

Try It Yourself: Approximating with comparison

Example 3.9. Use the technique demonstrated in Example 3.8 to determine the number of summands that are required to approximate $\sum_{k=1}^\infty \frac{k}{13k^3 - \cos(k)}$ to within 0.01 of its actual value. *(Hint: replace the cosine.)*

Full Solution On-line

Answer: $n = 9$. ∎

❖ Approximating with Midpoints

You know that

$$\sum_{k=1}^{\infty} \frac{1}{k^{1.2}} = s_{100} + \sum_{k=101}^{\infty} \frac{1}{k^{1.2}} = 3.60303 + \sum_{k=101}^{\infty} \frac{1}{k^{1.2}}.$$

The value of s_{100} is easily determined with a spreadsheet.

In Example 3.4 (on p. 620) we determined that the value of the remaining sum is at least 1.98657 but no more than 1.99054. So we know that

$$3.60303 + 1.98657 \leq \sum_{k=1}^{\infty} \frac{1}{k^{1.2}} \leq 3.60303 + 1.99054.$$

That is, the actual value of the series is some number in [5.5896, 5.59357]. Since this interval has a length of 0.00397 units, its midpoint is no more than $\frac{0.00397}{2} = 0.001985$ units away from the actual value of the series (though we don't know whether the midpoint is above or below the actual value). This is an important idea, so let's pause for a moment. If we approximate the value of the series using s_{100} our error is at least 1.98657. If we use the midpoint of [5.5896, 5.59357] instead, our error is at most 0.001985. That's roughly a 99.9% improvement in accuracy using the same 100 summands!

The number 0.001985 is less than one tenth of one percent of 1.98657.

In general, since

$$\sum_{k=1}^{\infty} a_k = s_n + \sum_{k=n+1}^{\infty} a_k,$$

the bounds on $\sum_{k=n+1}^{\infty} a_k$ that we established on p. 620 guarantee that $\sum_{k=1}^{\infty} a_k$ is some number in the interval

$$\left[s_n + \int_{n+1}^{\infty} f(t)\, dt, \ s_n + \int_{n}^{\infty} f(t)\, dt \right]. \tag{3.7}$$

If we use its midpoint as our approximation of the series value (as depicted in Figure 3.5), we'll be off the mark by no more than half the length of the interval.

> **Midpoint Approximation of Series Value:** Suppose $\sum_{k=1}^{\infty} a_k$ is a convergent series, f is a continuous, decreasing function for which $f(k) = a_k \geq 0$, and
>
> $$m_n = s_n + \frac{1}{2}\left(\int_{n+1}^{\infty} f(t)\, dt + \int_{n}^{\infty} f(t)\, dt \right).$$
>
> Then m_n differs from the series value by no more than $\frac{1}{2}\int_{n}^{n+1} f(t)\, dt.$

Don't just read it. Check for yourself that m_n is the midpoint of the interval in (3.7) and

$$\frac{1}{2}\int_{n}^{n+1} f(t)\, dt$$

is half the length.

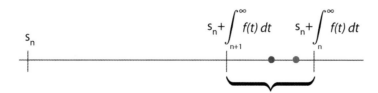

Figure 3.5: Error bounds tells us that the actual value of an infinite series is in *this* interval (e.g., it might be the red dot, •). By estimating its value with the midpoint of the interval, here indicated by the blue dot •, we get a very good approximation with relatively small n. In Examples 3.6 and 3.7, the lower bound of our interval was s_n, so our uncertainty about the actual value was greater.

Using midpoints to improve efficiency

Example 3.10. Use the idea of midpoints discussed above to reduce the number n in Example 3.6 (see p. 621).

Solution: In Example 3.6 we want to approximate $\sum_{k=1}^{\infty} 1/k^{1.2}$ to within 0.001 of its actual value. We can achieve that goal with the midpoint method by choosing n sufficiently large that

$$\frac{1}{2} \int_n^{n+1} f(t)\, dt = \frac{5}{2} \left(\frac{1}{n^{0.2}} - \frac{1}{(n+1)^{0.2}} \right) < \frac{1}{1000}.$$

Newton's method quickly finds that $\frac{5}{2} \left(\frac{1}{n^{0.2}} - \frac{1}{(n+1)^{0.2}} \right) = 0.001$ when $n = 157.9603$. So the desired accuracy is attained when $n = 158$. Calculating $s_{158} = 3.776217357$ is quick and easy with a spreadsheet program, and our improper integrals are

> Our $n = 158$ is a *huge* gain over the billions of summands that we thought we'd need in Example 3.6.

$$\int_{159}^{\infty} \frac{1}{t^{1.2}}\, dt = \frac{5}{\sqrt[5]{159}} \quad \text{and} \quad \int_{158}^{\infty} \frac{1}{t^{1.2}}\, dt = \frac{5}{\sqrt[5]{158}},$$

from which we conclude that the actual value of the series lies somewhere in

$$\left[3.776217357 + \frac{5}{\sqrt[5]{159}}, 3.776217357 + \frac{5}{\sqrt[5]{158}} \right].$$

By approximating with the midpoint,

$$\sum_{k=1}^{\infty} \frac{1}{k^{1.2}} \approx 3.776217357 + \frac{1}{2} \left(\frac{5}{\sqrt[5]{158}} + \frac{5}{\sqrt[5]{159}} \right),$$

we incur no more than 0.001 units of error. ∎

Try It Yourself: Using midpoints to improve efficiency

Example 3.11. Use the idea of midpoints discussed above to (a) reduce the number n in Example 3.7 (see p. 621), and (b) approximate the actual value of the series as the midpoint of some interval.

Answer: (a) $n = 8$, and (b) $\sum_{k=1}^{\infty} \frac{k}{(k^2+1)^2} \approx 0.39721$, which is the midpoint of $\left[s_8 + \frac{1}{162}, s_8 + \frac{1}{130} \right]$. ∎

Full Solution On-line

❖ Common Difficulties with the Integral Test

Some people mistakenly think that the Integral Test tells us the value of a series. It does not. It tells us whether a series converges in the first place, and if so, we can use it to determine an interval that contains the series value (see Example 3.4). With the methods discussed so far in this text, you can only determine the exact value of geometric and telescoping series.

❖ Strategy for Testing Series

When you're wondering whether a particular series converges, ask yourself the following questions:

1. Do the summands tend to zero in the limit? If not, the partial sums cannot converge.

2. Is it an alternating series? If so, it's relatively easy to tell whether it converges (see p. 608).

> **IMPORTANT:** A series can diverge even when its summands tend to zero in the limit (e.g., the Harmonic Series).
>
> In order to be sure that a series converges, you need to know (1) that the terms go to zero *and* (2) some *other* piece of information (for example, that it's alternating or geometric or . . .).

3. If the summands have the form $f(k)$, can you integrate $f(x)$? If so, consider using the Integral Test.

Once you know that a series converges, the next question is, "To what?" With the exception of geometric and telescoping series, it's unlikely that you'll be able to find out exactly, but you can always approximate. For a series of positive terms of the form $a_k = f(k)$, where f is continuous and decreasing, we found that improper integrals bound the difference between the nth partial sum and the actual value of the series:

$$\int_{n+1}^{\infty} f(x) \, dx \leq \sum_{k=n+1}^{\infty} a_k \leq \int_{n}^{\infty} f(x) \, dx.$$

You should know

- the terms *Integral Test* and *p-series*.

You should be able to

- use the Integral Test to determine whether a given series converges;

- determine how many summands are needed to approximate the value of a convergent series to within a specified error tolerance of its actual value;

- determine how close/far a specified partial sum is to the actual value of the series.

✧ 8.3 Skill Exercises

For each of the *p*-series in #1–4 (a) determine the value of p, and (b) classify the series as either convergent or divergent.

1. $\displaystyle\sum_{k=1}^{\infty} \frac{1}{k^6}$

2. $\displaystyle\sum_{k=1}^{\infty} \frac{1}{\sqrt[3]{k^7}}$

3. $\displaystyle\sum_{k=1}^{\infty} \frac{1}{\sqrt{k}}$

4. $\displaystyle\sum_{k=1}^{\infty} k^{-0.1}$

In #5–16 use the Integral Test to determine whether the series converges.

5. $\displaystyle\sum_{k=1}^{\infty} ke^{-k^2}$

9. $\displaystyle\sum_{k=1}^{\infty} \frac{8k}{9k^2 + 1}$

13. $\displaystyle\sum_{k=3}^{\infty} \frac{\ln(k)}{k^7}$

6. $\displaystyle\sum_{k=1}^{\infty} \frac{1}{k^2} e^{-1/k}$

10. $\displaystyle\sum_{k=1}^{\infty} \frac{k^3}{\sqrt{k^4 + 2}}$

14. $\displaystyle\sum_{k=1}^{\infty} \frac{1}{k^2 - 4k + 5}$

7. $\displaystyle\sum_{k=1}^{\infty} \frac{\arctan(k)}{k^2 + 1}$

11. $\displaystyle\sum_{k=2}^{\infty} \frac{1}{k \ln(k)^2}$

15. $\displaystyle\sum_{k=1}^{\infty} \frac{1}{k + 5}$

8. $\displaystyle\sum_{k=1}^{\infty} \frac{k}{(3 + 7k^2)^3}$

12. $\displaystyle\sum_{k=2}^{\infty} \frac{1}{k\sqrt{\ln(k)}}$

16. $\displaystyle\sum_{k=1}^{\infty} \frac{1}{k + 5} - \frac{1}{k + 6}$

In #17–20 use the Integral Test to determine all values of p for which the series converges.

17. $\displaystyle\sum_{k=1}^{\infty} k(1+k^2)^p$ 18. $\displaystyle\sum_{k=2}^{\infty} \frac{\ln(k)}{k^p}$ 19. $\displaystyle\sum_{k=2}^{\infty} \frac{1}{k^p \ln(k)}$ 20. $\displaystyle\sum_{k=1}^{\infty} k^p e^{-k}$

Each of the series in #21–24 diverges. Use equation (3.2) from p. 619 to determine the number of terms needed to raise the value of the series above N.

21. $\displaystyle\sum_{k=1}^{\infty} \frac{1}{k}$, $N = 10$ 23. $\displaystyle\sum_{k=1}^{\infty} \frac{k}{k^2+4}$, $N = 100$

22. $\displaystyle\sum_{k=1}^{\infty} \frac{1}{\sqrt{k}}$, $N = 2$ 24. $\displaystyle\sum_{k=2}^{\infty} \frac{1}{k\ln(k)}$, $N = 1000$

In #25–30 use the technique demonstrated in Examples 3.6 and 3.7 (see p. 621) to determine the number of summands required to approximate the series to within ε of its actual value using s_n.

25. $\displaystyle\sum_{k=1}^{\infty} \frac{10}{k^2}$, $\varepsilon = 10^{-5}$ 27. $\displaystyle\sum_{k=1}^{\infty} e^{-k/10}$, $\varepsilon = 10^{-6}$ 29. $\displaystyle\sum_{k=1}^{\infty} \frac{3}{k^2+5k+4}$, $\varepsilon = 0.1$

26. $\displaystyle\sum_{k=1}^{\infty} \frac{13}{k^{1.7}}$, $\varepsilon = 10^{-4}$ 28. $\displaystyle\sum_{k=2}^{\infty} \frac{\ln(k)}{k^2}$, $\varepsilon = 0.01$ 30. $\displaystyle\sum_{k=1}^{\infty} \frac{4}{k^2+8k+15}$, $\varepsilon = 10^{-3}$

In #31–36 use the comparison techniques demonstrated on p. 622 to determine a value of N so that the nth partial sum of the series is within 0.1 of the actual value.

31. $\displaystyle\sum_{k=1}^{\infty} \frac{k}{k^4+5}$ 33. $\displaystyle\sum_{k=1}^{\infty} \frac{1}{12k^2-\sin(k)}$ 35. $\displaystyle\sum_{k=1}^{\infty} \frac{2k^2-\sin^2(k)}{k^4}$

32. $\displaystyle\sum_{k=1}^{\infty} \frac{k\sin^2(k)}{k^5+1}$ 34. $\displaystyle\sum_{k=1}^{\infty} \frac{k}{6k^5-\arctan(k)}$ 36. $\displaystyle\sum_{k=1}^{\infty} \frac{3k^2-\arcsin(1/k)}{k^5+7}$

In #37–42 use the midpoint technique discussed on p. 623 to approximate the series to within ε of its actual value.

37. $\displaystyle\sum_{k=1}^{\infty} \frac{1}{k^2}$, $\varepsilon = 0.001$ 39. $\displaystyle\sum_{k=1}^{\infty} k^{-9/4}$, $\varepsilon = 0.002$ 41. $\displaystyle\sum_{k=1}^{\infty} ke^{-k}$, $\varepsilon = 10^{-7}$

38. $\displaystyle\sum_{k=1}^{\infty} \frac{1}{k^3}$, $\varepsilon = 0.0001$ 40. $\displaystyle\sum_{k=1}^{\infty} k^{-5/4}$, $\varepsilon = 0.0016$ 42. $\displaystyle\sum_{k=1}^{\infty} \frac{4}{k^2+9k+20}$, $\varepsilon = 10^{-3}$

Mixed Practice: The goal of this set of exercises is to help you learn to determine which convergence test is appropriate for a given series. Determine whether the series in #43–50 converge by using the ideas from either this section **or** Section 8.2.

43. $\displaystyle\sum_{k=1}^{\infty} 2^{1/k}$ 45. $\displaystyle\sum_{k=1}^{\infty} \frac{3^{2k}}{4^k}$ 47. $\displaystyle\sum_{k=1}^{\infty} \frac{\cos(k\pi)}{k}$ 49. $\displaystyle\sum_{k=1}^{\infty} \frac{1}{k^2+6k+5}$

44. $\displaystyle\sum_{k=2}^{\infty} \frac{1}{k(\ln(k))^3}$ 46. $\displaystyle\sum_{k=1}^{\infty} \frac{(-1)^k}{\sqrt{k}}$ 48. $\displaystyle\sum_{k=1}^{\infty} (-1)^k \frac{4k+1}{9k-2}$ 50. $\displaystyle\sum_{k=1}^{\infty} \frac{k^2}{4k^3+7}$

❖ 8.3 Concept and Application Exercises

51. For what values of p does $\sum_{k=1}^{\infty} \frac{1}{k^{2p}}$ converge while $\sum_{k=1}^{\infty} \frac{1}{k^p}$ diverges?

52. Suppose f is a positive, decreasing function, and $a_k = f(k)$ at each integer $k \geq 1$. Put the numbers $\sum_{k=1}^{n-1} a_k$, $\sum_{k=2}^{n} a_k$, and $\int_1^n f(t)\,dt$ in increasing order and support your answer with a sketch.

53. Suppose that $b > 0$.

    (a) Use the Integral Test to determine the values of b for which $\sum_{k=1}^{\infty} b^{\sqrt{k}}$ converges.

    (b) How does your answer change if we consider other roots of k in the exponent (e.g., $\sqrt[3]{k}$, $\sqrt[4]{k}$, ...)?

54. Suppose that $b > 0$.

    (a) Use the Integral Test to determine the values of b for which $\sum_{k=1}^{\infty} b^{\ln(k)}$ converges.

    (b) How does your answer change if we consider other logarithms of k in the exponent (e.g., $\log_3(k)$, $\log_4(k)$, ...)?

55. Suppose $g_n(x)$ is the function defined by $g_n(x) = \begin{cases} \frac{2^n x^n}{n} & \text{if } x \in [0, 0.5] \\ \frac{2^n (1-x)^n}{n} & \text{if } x \in [0.5, 1]. \end{cases}$

    (a) Calculate $g_n(0.5)$.

    (b) Calculate $\int_0^1 g(x)\,dx$.

    (c) Sketch graphs of $g_1(x - 0.5)$, $g_2(x - 1.5)$, $g_3(x - 2.5)$ and $g_4(x - 3.5)$.

    (d) Suppose we define $g(x) = g_n(x - n + 0.5)$ when $x \in [n - 0.5, n + 0.5]$, whose graph looks like a row of tents, with smaller tents to the right. Calculate $g(n)$.

    (e) Verify that $\int_1^{\infty} g(x)\,dx$ is finite, even though $\sum_{n=1}^{\infty} \frac{1}{n}$ diverges, and explain why this does not contradict the assertion of the Integral Test.

56. The *Riemann zeta function*, denoted by ζ, is used to study the distribution of prime numbers. It's defined by $\zeta(x) = \sum_{k=1}^{\infty} \frac{1}{k^x}$. For what real numbers x does $\zeta(x)$ converge?

In Section 8.8 you'll see that many functions can be written in terms of series, and some have to be. In #57–60 determine whether (a) $x = 1$, and (b) $x = -1$ are in the domain of f.

57. $f(x) = \displaystyle\sum_{k=1}^{\infty} \frac{x^k}{k\sqrt{k}}$

58. $f(x) = \displaystyle\sum_{k=1}^{\infty} \frac{x^k}{k^2 + 3k + 2}$

59. $f(x) = \displaystyle\sum_{k=1}^{\infty} \frac{x^k}{\sqrt[3]{k}}$

60. $f(x) = \displaystyle\sum_{k=1}^{\infty} \frac{\sqrt{k+7}}{\sqrt{k+5}} x^k$

8.4　Comparison Tests

In the discussion below you'll see that we can often avoid using the Integral Test by answering the question, "How does *this* series compare to those whose convergence I already know?" By "compare" we could mean a direct summand-by-summand comparison, or we could mean a comparison of the rate at which the summands tend to zero in the limit. We begin with the former.

❖　Direct Comparison

The basic idea of a comparison test is simple. Suppose you have two series whose summands are positive, and the larger of the two series converges. Then the smaller must also converge. Said formally,

---

Direct Comparison Theorem for Series (Part 1): Suppose $0 \leq a_k \leq b_k$, and $\sum_{k=1}^{\infty} b_k$ converges. Then $\sum_{k=1}^{\infty} a_k$ does too.

> This is extremely similar to the Direct Comparison Theorem for Improper Integrals (see p. 463)

Note: At an intuitive level, we're saying that if b_k tends to zero "fast enough" for a series to converge, so does a_k (because it's always smaller). A more formal proof is given below.

Proof. Let's begin by noting that the partial sums of $\sum_{k=1}^{\infty} a_k$ are increasing, since all the summands are positive. And because $0 \leq a_k \leq b_k$, we know

$$\sum_{k=1}^{n} a_k \leq \sum_{k=1}^{n} b_k \leq \sum_{k=1}^{\infty} b_k.$$

The right-hand side is finite since $\sum_{k=1}^{\infty} b_k$ converges, so the partial sums $s_n = \sum_{k=1}^{n} a_k$ are bounded above. Since the sequence of partial sums is monotone and bounded, the Bounded Monotone Convergence Theorem (p. 592) guarantees that it converges. ∎

> The number $\sum_{k=1}^{\infty} b_k$ acts as a kind of "ceiling," above which the partial sums of $\sum_{k=1}^{\infty} a_k$ cannot rise.

Proving convergence with Direct Comparison

Example 4.1. Use the Direct Comparison Theorem to prove that $\sum_{k=4}^{\infty} \frac{3^k}{k!}$ converges.

Solution: On p. 594 we established that $\frac{3^k}{k!} < \left(\frac{3}{4}\right)^{k-3} \frac{27}{6}$ when $k \geq 4$. So

$$s_n = \sum_{k=4}^{n} \frac{3^k}{k!} \; < \; \sum_{k=4}^{n} \left(\frac{3}{4}\right)^{k-3} \frac{27}{6} \; < \; \frac{27}{6} \sum_{k=4}^{\infty} \left(\frac{3}{4}\right)^{k-3} .$$

> The first term of the series is $a = 3/4$, so we have
> $$\sum_{k=4}^{\infty} \left(\frac{3}{4}\right)^{k-3} = \frac{\frac{3}{4}}{1 - \frac{3}{4}} = 3.$$

Since the series on the right-hand side is geometric, with $r = \frac{3}{4}$, it converges. Its value is irrelevant. The point is that s_n is increasing (since all summands are positive) and bounded above, so the Bounded Monotone Convergence Theorem (p. 592) guarantees that the sequence converges. ∎

　　Example 4.1 is not intended to show that a particular series converges, but to demonstrate the *way* we use the Direct Comparison Theorem. Typically, we have a series that we *think* converges. To prove it definitively, we have to find or design a series whose terms are larger that we *know* converges. (That is, we are given a_k and need to find or design b_k.)

Proving convergence with Direct Comparison

Example 4.2. Use the Direct Comparison Theorem to prove that $\sum_{k=7}^{\infty} \frac{k}{k^3+1}$ converges.

And the inequality $a_k < (L + \varepsilon)b_k$ guarantees that if

$$\sum_{k=N}^{\infty} b_k \ \text{converges, so does} \ \sum_{k=N}^{\infty} a_k.$$

In brief, if one series converges, the other must also.   ■

Using the Limit Comparison Theorem

Example 4.6. Use the Limit Comparison Theorem to determine whether $\sum_{k=1}^{\infty} \frac{9k^2+1}{k^4+k+2}$ converges.

Solution: When k is very large, we know that

$$\frac{9k^2+1}{k^4+k+2} \approx \frac{9k^2}{k^4} = \frac{9}{k^2},$$

and $\sum_{k=1}^{\infty} \frac{9}{k^2}$ converges since it's a scaled p-series with $p > 1$ (see p. 619). That leads us to believe that the series of this example does too. We check this intuition by performing a limit comparison with $a_k = \frac{9k^2+1}{k^4+k+2}$ and $b_k = \frac{9}{k^2}$ ...

> We try to pick out the dominant behavior of the summands, and let that define b_k for us. Note that b_k is (a) simple and (b) familiar to us.

$$\lim_{k\to\infty} \frac{a_k}{b_k} = \lim_{k\to\infty} \frac{\frac{9k^2+1}{k^4+k+2}}{\frac{9}{k^2}} = \lim_{k\to\infty} \frac{9k^4+k^2}{9k^4+9k+18}.$$

After factoring k^4 out of the numerator and denominator of this fraction, we see

$$\lim_{k\to\infty} \frac{9 + \frac{1}{k^2}}{9 + \frac{9}{k^3} + \frac{18}{k^4}} = \frac{9+0}{9+0+0} = 1.$$

Since this limit exists and is positive, the Limit Comparison Theorem tells us that both $\sum a_k$ and $\sum b_k$ behave in the same way. Therefore, since we know that $\sum b_k$ converges, so does our series.   ■

> Alternatively, you could think about the numbers
>
> $$\frac{9k^4 + k^2}{9k^4 + 9k + 18}$$
>
> as $f(k)$ where
>
> $$f(x) = \frac{9x^4 + x^2}{9x^4 + 9x + 18},$$
>
> and then use L'Hôpital's Rule to calculate
>
> $$\lim_{x\to\infty} \frac{9x^4 + x^2}{9x^4 + 9x + 18} = 1.$$

Try It Yourself: Using the Limit Comparison Theorem

Example 4.7. Use the Limit Comparison Theorem to determine whether $\sum_{k=1}^{\infty} \frac{3k+2}{(k+1)^2}$ converges.

Answer: It diverges.   ■

Full Solution On-line

Though we've motivated the Limit Comparison Theorem with series whose summands are rational functions of k, it can also be used with more complicated series.

Using the Limit Comparison Theorem

Example 4.8. Determine whether $\sum_{k=1}^{\infty} \left(\frac{k+1}{4k+3}\right)^k$ converges by using the Limit Comparison Theorem.

Solution: When k is very large, we know that $\frac{k+1}{4k+3} \approx \frac{1}{4}$, and $\sum_{k=1}^{\infty} \left(\frac{1}{4}\right)^k$ is a geometric series with a common ratio of $1/4$, so it converges. That makes us think that the series of this example does too. Let's check by using the Limit Comparison Theorem with $a_k = \left(\frac{k+1}{4k+3}\right)^k$ and $b_k = (1/4)^k$:

> As in the last example, we try to pick out the dominant behavior of the summands, and let that define b_k for us. Note that b_k is (a) simple and (b) familiar to us.

$$\lim_{k\to\infty} \frac{a_k}{b_k} = \lim_{k\to\infty} \frac{\left(\frac{k+1}{4k+3}\right)^k}{\left(\frac{1}{4}\right)^k} = \lim_{k\to\infty} \left(\frac{4k+4}{4k+3}\right)^k.$$

As in Example 4.6, let's think about $\left(\frac{4k+4}{4k+3}\right)^k$ as $f(k)$, where $f(x) = \left(\frac{4x+4}{4x+3}\right)^x$. Then $\lim_{x \to \infty} f(x)$ results in a 1^∞ indeterminate form, but we have techniques for dealing with such limits. As we did on p. 288, we'll use the fact that

$$f(x) = e^{\ln(f(x))} \quad \text{when } f(x) > 0$$

(which it is when $x > 0$). Then, since the exponential function is continuous,

$$\lim_{x \to \infty} f(x) = \lim_{x \to \infty} e^{\ln(f(x))} = \exp\left(\lim_{x \to \infty} \ln(f(x))\right)$$

> Recall that $\exp(u)$ is another way of writing e^u.

when $\lim_{x \to \infty} \ln(f(x))$ exists. So our question has been reduced to calculating that limit. Note that

$$\ln(f(x)) = \ln\left(\frac{4x+4}{4x+3}\right)^x = x \ln\left(\frac{4x+4}{4x+3}\right) = \frac{\ln\left(\frac{4x+4}{4x+3}\right)}{1/x},$$

so

$$\lim_{x \to \infty} \ln(f(x)) = \lim_{x \to \infty} \frac{\ln\left(\frac{4x+4}{4x+3}\right)}{1/x} = \frac{0}{0}.$$

The $0/0$ indeterminate form allows us to use L'Hôpital's Rule in our calculation:

$$\lim_{x \to \infty} \frac{\ln\left(\frac{4x+4}{4x+3}\right)}{1/x} = \lim_{x \to \infty} \frac{-\frac{4}{(4x+3)(4x+4)}}{-\frac{1}{x^2}} = \lim_{x \to \infty} \frac{4x^2}{(4x+3)(4x+4)} = \frac{1}{4}.$$

That is, we have determined $\lim_{x \to \infty} e^{\ln(f(x))} = e^{1/4}$, from which it follows that

$$\lim_{k \to \infty} \frac{a_k}{b_k} = e^{1/4} > 0.$$

Since the limit value is finite and positive, either both series converge or both series diverge. Because $\sum_{k=1}^{\infty} b_k$ converges, so does our series. ∎

Part 1 of the Limit Comparison Theorem requires a positive limit value, but consider the following comparison between $\sum_{k=1}^{\infty} \left(\frac{1}{3}\right)^k$ and $\sum_{k=1}^{\infty} \frac{1}{k^2}$, both of which converge. With $a_k = (1/3)^k$ and $b_k = \frac{1}{k^2}$ we see

$$\lim_{k \to \infty} \frac{a_k}{b_k} = \lim_{k \to \infty} \frac{(1/3)^k}{1/k^2} = \lim_{k \to \infty} \frac{k^2}{3^k} = 0.$$

> This limit value is zero because the geometric growth of 3^k is so much faster than the polynomial growth of k^2. You can verify this claim by using L'Hôpital's Rule (twice).

In short, the limit value is zero because although both kinds of summands tend to zero in the limit, the numbers a_k head toward zero *much faster*. This is the basic idea behind Part 2 of the Limit Comparison Theorem.

Limit Comparison Theorem for Series (Part 2): Suppose $\sum_{k=1}^{\infty} a_k$ and $\sum_{k=1}^{\infty} b_k$ are series with positive summands for which $\lim_{k \to \infty} \frac{a_k}{b_k} = 0$.

- If $\sum_{k=1}^{n} b_k$ converges, so does $\sum_{k=1}^{n} a_k$.

- If $\sum_{k=1}^{n} a_k$ diverges, so does $\sum_{k=1}^{n} b_k$.

Note: You'll prove this in the exercise set using a technique very similar to the proof of Part 1.

Using Part 2 of the Limit Comparison Theorem

Example 4.9. Determine whether $\sum_{k=1}^{\infty} \frac{\ln(k)}{k^{3/2}}$ converges.

Solution: Let's begin by talking about the rate at which the logarithm grows. Remember that

$$\sum_{k=1}^{\infty} \frac{1}{k^{1.25}} \quad \text{converges but} \quad \sum_{k=1}^{\infty} \frac{1}{k \ln(k)}$$

does not (see Example 3.3, p. 619). That's because $\ln(k)$ is outpaced by $k^{0.25}$ as k gets large. With this in mind, we expect that

$$\frac{\ln(k)}{k^{3/2}} \leq \frac{k^{1/4}}{k^{3/2}} = \frac{1}{k^{5/4}},$$

when k gets large, so we think that our series will converge. We validate our conjecture by using the Limit Comparison Theorem with $a_k = \ln(k)/k^{3/2}$ and $b_k = 1/k^{5/4}$ ...

$$\lim_{k \to \infty} \frac{\frac{\ln(k)}{k^{3/2}}}{\frac{1}{k^{5/4}}} = \lim_{k \to \infty} \frac{\ln(k)}{k^{3/2}} \frac{k^{5/4}}{1} = \lim_{k \to \infty} \frac{\ln(k)}{k^{1/4}},$$

which results in an ∞/∞ indeterminate form. This allows us to apply L'Hôpital's Rule, and see that the limit value is

$$\lim_{k \to \infty} \frac{\ln(k)}{k^{1/4}} = \lim_{k \to \infty} \frac{1/k}{0.25 k^{-3/4}} = \lim_{k \to \infty} \frac{4}{k^{1/4}} = 0.$$

Since $\lim_{k \to \infty} \frac{a_k}{b_k} = 0$ and $\sum_{k=1}^{\infty} b_k$ converges, so does our series. ∎

> In fact, k^{ε} will eventually surpass $\ln(k)$ as $k \to \infty$ for *any* $\varepsilon > 0$.

Try It Yourself: Using Part 2 of the Limit Comparison Theorem

Example 4.10. Determine whether $\sum_{k=2}^{\infty} \left(\frac{1}{\ln(k)} \right)^2$ converges.

Answer: It diverges. ∎

Full Solution On-line

❖ Common Difficulties with Direct Comparison

When trying to find a comparable series to use with the Direct Comparison Theorem, people often make algebraic changes that are "too big," and so move a series from the convergent to the divergent category, or vice versa. For example ...

• Suppose we're trying to show that $\sum_{k=7}^{\infty} \frac{k}{k^3+1}$ converges (it does) by designing a series whose summands are larger, so we increase the numerator from k to k^2, and decrease the denominator by replacing "+1" with "+0." Then

$$\frac{k}{k^3+1} \leq \frac{k^2}{k^3+1} \leq \frac{k^2}{k^3+0} = \frac{1}{k},$$

from which it follows that

$$s_n = \sum_{k=7}^{n} \frac{k}{k^3+1} \leq \sum_{k=7}^{n} \frac{1}{k}. \tag{4.1}$$

While true, this fact is unhelpful because the right-hand side is the Harmonic Series, which diverges as $n \to \infty$. That is, equation (4.1) implies that

$$s_n = \sum_{k=7}^{n} \frac{k}{k^3+1} \leq \sum_{k=7}^{n} \frac{1}{k} < \sum_{k=7}^{\infty} \frac{1}{k} = \infty. \tag{4.2}$$

The strict inequality in equation (4.2) tells us that $s_n < \infty$—i.e., each partial sum is finite. But that's *always* true, regardless of whether the partial sums converge or diverge. So based only on equation (4.2) we cannot make a conclusion about whether the partial sums have a limit.

- Suppose we're trying to show that $\sum_{k=1}^{\infty} \frac{2k}{4+3k^2}$ diverges (it does) by finding a series with smaller summands. Since $k \geq 1$,

$$\frac{2k}{4+3k^2} \geq \frac{2}{4+3k^2} \geq \frac{2}{4k^2+3k^2} = \frac{2}{7k^2}.$$

Therefore,

$$\lim_{n\to\infty} \sum_{k=1}^{n} \frac{2k}{4+3k^2} \geq \lim_{n\to\infty} \sum_{k=1}^{n} \frac{2}{7k^2}. \tag{4.3}$$

Since $\sum_{k=1}^{\infty} \frac{2}{7k^2}$ is a scaled p-series with $p=2$ it converges, so equation (4.3) says

$$\lim_{n\to\infty} \sum_{k=1}^{n} \frac{2k}{4+3k^2} \geq \text{a finite number,}$$

which is useless information. From this mathematics alone, we still don't now whether the increasing sequence of partial sums, $s_n = \sum_{k=1}^{n} \frac{2k}{4+3k^2}$, converges to a finite value or surpasses every finite value in the limit.

There's a certain art to making the algebraic adjustments by which we arrive at a known "landmark" series (such as a p-series) while preserving the character of the series in which we're actually interested, and it comes with practice. The first step in learning that art is recognizing when our work results in a useless statement, such as $\lim_{n\to\infty} s_n \leq \infty$ or $\lim_{n\to\infty} s_n \geq$ a finite number.

✢ Strategy for Testing Series

The margin notes alongside Examples 4.6–4.8 mention that the dominant aspects of a series' summands can often be used to determine b_k. We saw polynomials in those examples, and you already know that polynomials are dominated by their leading term when k is large. As we continue our study, we'll see other kinds of terms also. As a rule of thumb,

> factorial terms (e.g., $k!$) dominate geometric terms (e.g., 3^k)
> geometric terms dominate power terms (e.g., k^2)
> power terms dominate logarithmic terms (e.g., $\ln(k)$).

We could expand this list of relative dominance by including products such as (power)×(exponential) or (power)×(log)×(log), but even this short list can often give you a sense of whether a given series converges or diverges. Once you know what you believe, you have to prove that you're right, so consider the following:

1. Do the summands tend to zero in the limit? If not, the partial sums cannot converge.

2. Is it an alternating series? If so, it's relatively easy to tell whether it converges (see p. 608).

3. Are the summands larger (or smaller) than those from a series whose convergence you already know, such as a p-series? If so, consider using the Direct Comparison Theorem.

The project called *An Introduction to Stirling's Approximation* (on p. 400) leads you through a proof that

$$k! \geq \left(\frac{k}{e}\right)^k.$$

Both the base *and* the exponent are growing on the right-hand side, which is why the factorial grows faster than a geometric term such as 2^k, in which the base is fixed.

IMPORTANT: A series can diverge even when its summands tend to zero in the limit (e.g., the Harmonic Series).

In order to be sure that a series converges, you need to know (1) that the terms go to zero *and* (2) some *other* piece of information (for example, that it's alternating or geometric or ...).

When using a comparison technique, we typically compare to p-series or geometric series. Those two serve as "landmarks" for us.

4. When k is large, do the summands behave like a series whose convergence you know? If so, consider using the Limit Comparison Theorem. Try to compare to a series that's (a) simple and (b) familiar.

5. If the summands have the form $f(k)$, can you integrate $f(x)$? If so, consider using the Integral Test.

> We've moved the Integral Test to the end of the list because it's often more difficult to use than other convergence tests.

You should know

- the Direct Comparison Theorem (parts 1 and 2);

- the Limit Comparison Theorem (parts 1 and 2);

- when designing/finding a series to which you can make a comparison, make small algebraic changes in order to avoid changing the fundamental character of the summands.

You should be able to

- establish the convergence or divergence of a given series using the Direct Comparison Theorem;

- establish the convergence or divergence of a given series using the Limit Comparison Theorem.

❖ 8.4 Skill Exercises

Show that each of the series in #1–6 converges by using the Direct Comparison Theorem.

1. $\displaystyle\sum_{k=1}^{\infty} \frac{k}{k^5 + 1}$

2. $\displaystyle\sum_{k=2}^{\infty} \frac{(0.75)^k}{k + 1}$

3. $\displaystyle\sum_{k=1}^{\infty} \frac{1}{3^{k^2}}$

4. $\displaystyle\sum_{k=3}^{\infty} \frac{1}{k^2 + 3k + 17}$

5. $\displaystyle\sum_{k=1}^{\infty} \frac{\sin^2(k)}{k^7 + 1}$

6. $\displaystyle\sum_{k=1}^{\infty} \frac{k}{2k^3 + \cos^2(k)}$

Show that each of the series in #7–12 diverges by using the Direct Comparison Theorem.

7. $\displaystyle\sum_{k=4}^{\infty} \frac{k^3}{k^2 + 3k + 15}$

8. $\displaystyle\sum_{k=5}^{\infty} \frac{\sqrt{k}}{k + 1}$

9. $\displaystyle\sum_{k=6}^{\infty} \frac{k!}{5^k}$

10. $\displaystyle\sum_{k=1}^{\infty} \frac{4 - k^{-k}}{k}$

11. $\displaystyle\sum_{k=1}^{\infty} \frac{k^2 - k}{k^3 + 4}$

12. $\displaystyle\sum_{k=1}^{\infty} \frac{k^2 + 1}{k^3 - \cos(k)}$

In #13–20 (a) determine whether you think the series converges, and (b) use the Direct Comparison Theorem to prove that you're right.

13. $\displaystyle\sum_{k=1}^{\infty} \frac{\sqrt{k}}{k+1}$

15. $\displaystyle\sum_{k=1}^{\infty} \frac{3\sqrt{k}}{\sqrt{k+\sqrt{5k^7+2}}}$

17. $\displaystyle\sum_{k=2}^{\infty} \frac{\ln(k)}{k}$

19. $\displaystyle\sum_{k=1}^{\infty} \frac{k}{2k^3-\cos^2(k)}$

14. $\displaystyle\sum_{k=1}^{\infty} \frac{3+k}{k^2+\sqrt{k}}$

16. $\displaystyle\sum_{k=1}^{\infty} \frac{\arctan(k)\sqrt{k}}{k^6+1}$

18. $\displaystyle\sum_{k=1}^{\infty} \frac{\sin^2(k)}{2k^7-1}$

20. $\displaystyle\sum_{k=1}^{\infty} \frac{4+k^{-k}}{k}$

In #21–28 (a) determine whether you think the series converges, and (b) use the Limit Comparison Theorem to prove that you're right.

21. $\displaystyle\sum_{k=1}^{\infty} \frac{3}{k+0.25}$

23. $\displaystyle\sum_{k=1}^{\infty} \frac{k-1}{k^2+1}$

25. $\displaystyle\sum_{k=1}^{\infty} \frac{\sqrt{k}}{k^2+k+8}$

27. $\displaystyle\sum_{k=1}^{\infty} \frac{k^2-k}{k^3+4}$

22. $\displaystyle\sum_{k=1}^{\infty} \frac{k-7}{k^8+k^2+1}$

24. $\displaystyle\sum_{k=1}^{\infty} \frac{k^6+7k+1}{k^7+11}$

26. $\displaystyle\sum_{k=1}^{\infty} \frac{\sqrt{k}}{(1+\sqrt{k})^3}$

28. $\displaystyle\sum_{k=1}^{\infty} \frac{\ln(k)}{k^{1.1}}$

Mixed Practice: The goal of this set of exercises is to help you learn how to determine which convergence test is appropriate for a given series. In #29–44 determine whether the given series converges (you might need to use ideas from previous sections).

29. $\displaystyle\sum_{k=1}^{\infty} \frac{6\sqrt[5]{k}}{\sqrt[3]{k+\sqrt{9k^8-2}}}$

33. $\displaystyle\sum_{k=1}^{\infty} \frac{1}{2k^3-k^{1/3}}$

37. $\displaystyle\sum_{k=1}^{\infty} \frac{1}{\sqrt{k}\ln(k+1)}$

41. $\displaystyle\sum_{k=10}^{\infty} \left(\frac{1}{\ln(k)}\right)^2$

30. $\displaystyle\sum_{k=1}^{\infty} \frac{\ln(k)}{\sqrt{k}}$

34. $\displaystyle\sum_{k=1}^{\infty} \frac{5+|\sin(k)|}{(k+1)\sqrt{k}}$

38. $\displaystyle\sum_{k=1}^{\infty} \frac{3k+k^{1/k}}{4k-1}$

42. $\displaystyle\sum_{k=100}^{\infty} \left(\frac{1}{\ln(k)}\right)^3$

31. $\displaystyle\sum_{k=1}^{\infty} 11^k 3^{-5-2k}$

35. $\displaystyle\sum_{k=1}^{\infty} \frac{k^2+\sqrt{k}}{4+\sin(k)+k^6}$

39. $\displaystyle\sum_{k=1}^{\infty} \frac{k^2}{e^k}$

43. $\displaystyle\sum_{k=10}^{\infty} \frac{1}{k\,(\ln(k))^2}$

32. $\displaystyle\sum_{k=1}^{\infty} \frac{3+(-1)^k}{k}$

36. $\displaystyle\sum_{k=1}^{\infty} \frac{\ln(k)}{k^{1.2}}$

40. $\displaystyle\sum_{k=1}^{\infty} (-1)^k \frac{k}{k+1}$

44. $\displaystyle\sum_{k=100}^{\infty} \frac{1}{k\ln(k)}$

❖ 8.4 Concept and Application Exercises

45. Suppose a friend of yours is trying to determine whether $\sum_{k=1}^{\infty} \frac{\sqrt{k}}{k^2+1}$ converges. She writes

$$\frac{\sqrt{k}}{k^2+1} \le \frac{k}{k^2+1} \le \frac{k}{k^2} = \frac{1}{k},$$

and since the Harmonic Series diverges, concludes that $\sum_{k=1}^{\infty} \frac{\sqrt{k}}{k^2+1}$ does too. Determine whether her answer is correct and supported by the evidence. If not, provide assistance.

46. Suppose a friend of yours is trying to determine whether $\sum_{k=1}^{\infty} \frac{\sqrt{k}}{k+1}$ converges. He writes

$$\frac{\sqrt{k}}{k+1} \le \frac{\sqrt{k}}{k} = \frac{1}{\sqrt{k}},$$

and, since $\sum_{k=1}^{\infty} \frac{1}{k^{1/2}}$ diverges (it's a p-series with $p < 1$), concludes that $\sum_{k=1}^{\infty} \frac{\sqrt{k}}{k+1}$ does too. Determine whether his answer is correct and supported by the evidence. If not, provide assistance.

47. Suppose a friend is trying to determine whether $\sum_{k=1}^{\infty} \frac{\ln(k)}{k^2}$ converges. Using the Limit Comparison Theorem, he writes

$$\lim_{k \to \infty} \frac{\frac{\ln(k)}{k^2}}{1/k} = \underbrace{\lim_{k \to \infty} \frac{\ln(k)}{k}}_{\text{indeterminate form}} = \underbrace{\lim_{k \to \infty} \frac{1/k}{1} = 0}_{\text{according to L'Hôpital's Rule}}$$

and since the Harmonic Series diverges, concludes that $\sum_{k=1}^{\infty} \frac{\ln(k)}{k^2}$ does too. Determine whether his answer and reasoning are correct and, if not, provide assistance.

48. Suppose a friend is trying to determine whether $\sum_{k=1}^{\infty} \frac{\ln(k)}{k^2}$ converges. Using the Limit Comparison Theorem, she writes

$$\lim_{k \to \infty} \frac{\frac{\ln(k)}{k^2}}{1/k^2} = \lim_{k \to \infty} \ln(k) = \infty$$

and concludes that $\sum_{k=1}^{\infty} \frac{\ln(k)}{k^2}$ diverges. Determine whether her answer and reasoning are correct and, if not, provide assistance.

In Section 8.8 you'll see that many functions can be written in terms of series, and some have to be. In #49–52 use a comparison test to determine whether x_0 is in the domain of f.

49. $f(x) = \sum_{k=1}^{\infty} \frac{x^k}{3^k + \sqrt{k+1}}$, $x_0 = 2$

50. $f(x) = \sum_{k=2}^{\infty} \frac{k\,x^k}{4^k + \sqrt{k+1}}$, $x_0 = 5$

51. $f(x) = \sum_{k=4}^{\infty} \frac{\sqrt{k+1}}{3k + 2\sin(k)} x^k$, $x_0 = 1$

52. $f(x) = \sum_{k=3}^{\infty} \frac{4x^k}{k + k^{1/k}}$, $x_0 = 1$

Exercises #53 and #54 refer to the following figure, in which the blue circles represent the summands of $\sum_{k=1}^{\infty} a_k$.

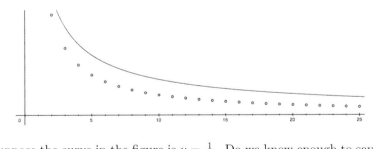

53. Suppose the curve in the figure is $y = \frac{1}{t^2}$. Do we know enough to say whether the series $\sum_{k=1}^{\infty} a_k$ is convergent or divergent? If so, classify the series and explain how you know. If not, why not?

54. Suppose the curve in the figure is $y = \frac{1}{\sqrt{t}}$. Do we know enough to say whether the series $\sum_{k=1}^{\infty} a_k$ is convergent or divergent? If so, classify the series and explain how you know. If not, why not?

Checking the details

55. Suppose that $\sum_{k=1}^{\infty} a_k$ and $\sum_{k=1}^{\infty} b_k$ are series with positive summands, and $\lim_{k \to \infty} \frac{a_k}{b_k} = 0$.

(a) Explain why, for every $\varepsilon > 0$, there is some index N so that $\frac{a_k}{b_k} \in [0, \varepsilon]$ whenever $k \geq N$.

(b) Based on part (a), you know that $a_k \leq \varepsilon b_k$ whenever $k \geq N$. Explain why, if $\sum_{k=1}^{\infty} b_k$ converges, $\sum_{k=1}^{\infty} a_k$ must also.

(c) Explain why, if $\sum_{k=1}^{\infty} a_k$ diverges, $\sum_{k=1}^{\infty} b_k$ must also.

56. We've only asserted that the Limit Comparison tests work for series with non-negative summands. Here's an example of why, first introduced by the mathematician Lejeune Dirichlet (1805–1859). Consider the series

$$\underbrace{\sum_{k=2}^{\infty} \frac{(-1)^k}{\sqrt{k}}}_{\text{series } (i)} \quad \text{and} \quad \underbrace{\sum_{k=2}^{\infty} \frac{(-1)^k}{\sqrt{k}} \left(1 + \frac{(-1)^k}{\sqrt{k}}\right)}_{\text{series } (ii)}.$$

(a) Verify that the summands of both series alternate sign.

(b) Show that adding consecutive odd-indexed and even-indexed terms in series (ii), say $k = 2\ell + 1$ and $k = 2\ell + 2$, results in a net increase of more than $\frac{1}{2\ell+2}$ in the partial sums. That is, the partial sums of series (ii) obey the inequality $s_{2\ell+2} \geq s_{2\ell} + \frac{1}{2\ell+2}$.

(c) Use part (b) to explain why the series (ii) diverges.

(d) Explain how we know that series (i) converges.

(e) Show that the limit value in the Limit Comparison Theorem is 1.

8.5   Absolute Convergence, Ratio and Root Tests

The Integral Test and comparison theorems help us to determine whether a series of positive terms converges, but many series include negative summands (e.g., alternating series, see p. 606). In this section, we introduce a convergence test that can be applied to series with the occasional negative summand, and separate such series into two categories.

❖ Conditional and Absolute Convergence

It's easy to see that the Alternating Harmonic Series,

$$\sum_{k=1}^{\infty} \frac{(-1)^{k+1}}{k} = 1 - \frac{1}{2} + \frac{1}{3} - \frac{1}{4} + - \cdots$$

converges to some number in $[0, 1]$, but something interesting happens when we rearrange the order of the summands.

Rearranging the Alternating Harmonic Series

Example 5.1. Show that we can make the Alternating Harmonic Series converge to any number we want by rearranging the order of its summands.

Solution: In this solution we'll rearrange the summands so that the series converges to -1.2, which was chosen arbitrarily. (Using the same technique, we can make the series converge to any other number instead.) We'll begin by adding negative summands from the Alternating Harmonic Series until the first time that we fall below the target value . . .

$$-\frac{1}{2} - \frac{1}{4} - \frac{1}{6} - \frac{1}{8} - \frac{1}{10} - \frac{1}{12} = -1.225.$$

Now that we're below the target value of -1.2, let's add a positive summand of our series in order to step above again. The first positive summand is a 1, whose addition brings us up to

$$-\frac{1}{2} - \frac{1}{4} - \frac{1}{6} - \frac{1}{8} - \frac{1}{10} - \frac{1}{12} + 1 = -0.225.$$

Now we add more negative terms until the sum is less than -1.2 again. This will take longer than it did before because the negative summands are decreasing in size, but it happens eventually. In fact, it takes 40 more negative summands before we finally pass the target value of -1.2:

$$\underbrace{-\frac{1}{2} - \frac{1}{4} - \cdots - \frac{1}{12}}_{\text{step 1}} \underbrace{+ 1}_{\text{step 2}} \underbrace{- \frac{1}{14} - \frac{1}{16} - \cdots - \frac{1}{92}}_{\text{step 3 of rearranging}} = -1.2083436.$$

The next positive term in the Alternating Harmonic Series is $1/3$, whose addition brings us above -1.2 again. Then the addition of more negative terms brings us below -1.2 again . . . We continue like this forever, getting closer and closer to -1.2 by using negative terms to slip below it and positive summands to step us above it, always by smaller and smaller amounts. All the summands are used in the limit, and the partial sums converge to -1.2. ∎

Example 5.1 confuses many people, because never in their experience with arithmetic has the order of summands mattered, and here we find a case in which it does!

So now we're faced with the question, "Does the value of an infinite series always depend on the arrangement of its summands?" No, not always, and that leads to some vocabulary: when the value of a convergent series depends on the order in which we add its summands, we say that it's **conditionally convergent**. Otherwise, we say it is **unconditionally convergent**.

The word "conditional" refers to the value of the series, not whether it converges in the first place.

To get a sense of why the Alternating Harmonic Series is conditionally convergent, consider its negative summands:

$$a_2 = -\frac{1}{2}, \quad a_4 = -\frac{1}{4}, \quad a_6 = -\frac{1}{6}, \quad a_8 = -\frac{1}{8}, \quad a_{10} = -\frac{1}{10} \cdots$$

If we were to add only these negative terms, we'd see

$$-\frac{1}{2} - \frac{1}{4} - \frac{1}{6} - \frac{1}{8} - \frac{1}{10} - \cdots = \sum_{k=1}^{\infty} -\frac{1}{2k} = -\frac{1}{2} \sum_{k=1}^{\infty} \frac{1}{k},$$

which diverges. Loosely said, the Alternating Harmonic Series contains an "infinite amount of negative," which is why we're sure that we can push the partial sums below -1.2 over and over again in Example 5.1. Similarly, the accumulation of the positive summands (only) in the Alternating Harmonic Series also diverges, which is why we know that we'll always be able to step above -1.2 again.

By contrast, the series $\sum_{k=1}^{\infty} (-1)^{k+1}/k^2$ has a "finite amount" of negative since

$$\sum_{k=\text{even}} \frac{(-1)^{k+1}}{k^2} = -\sum_{\ell=1}^{\infty} \frac{1}{(2\ell)^2} = -\frac{1}{4} \sum_{\ell=1}^{\infty} \frac{1}{\ell^2}$$

converges. With only a finite amount of negative value available, we cannot rearrange the series so that its partial sums decrease below -1.2 over and over again (as happened in Example 5.1) because eventually we don't have enough negative value remaining to overcome the contribution from the next positive term in the series.

Since we're not omitting any summands of the series, but only rearranging them, every positive summand appears in the series eventually.

Recall that the number -1.2 was chosen arbitrarily in Example 5.1. There's nothing particularly special about it in the present discussion of $\sum_{k=1}^{\infty} (-1)^{k+1}/k^2$. Rather, the important fact is that the negative terms of the series sum to a finite amount, as do the positive terms. We test for this condition by looking for something called *absolute convergence*. Simply put, we say the series $\sum_{k=1}^{\infty} a_k$ is **absolutely convergent** when $\sum_{k=1}^{\infty} |a_k|$ converges.

Many people ask whether a series that's *absolutely convergent* is also *convergent*. The answer is, *yes,* and here's one way to understand why:

In the discussion above we've tried to emphasize that *absolute convergence* is a special kind of *convergence*. The proof below establishes the fact algebraically.

Theorem: Suppose the series $\sum_{k=1}^{\infty} a_k$ is *absolutely* convergent. Then $\sum_{k=1}^{\infty} a_k$ converges.

Proof. Most of the facts that we've established about series apply to those with non-negative summands, so let's note that $0 \leq a_k + |a_k|$. Our strategy will be to show that $\sum_{k=1}^{\infty} (a_k + |a_k|)$ converges. Then because the difference of convergent series is a convergent series, we'll know that

$$\sum_{k=1}^{\infty} (a_k + |a_k|) - \sum_{k=1}^{\infty} |a_k| = \sum_{k=1}^{\infty} (a_k + |a_k| - |a_k|) = \sum_{k=1}^{\infty} a_k$$

converges. To show that $\sum_{k=1}^{\infty} (a_k + |a_k|)$ converges, note that $a_k + |a_k| \leq 2|a_k|$. Since $\sum_{k=1}^{\infty} a_k$ is absolutely convergent, we know $\sum_{k=1}^{\infty} 2|a_k|$ converges, and the convergence of $\sum_{k=1}^{\infty}(a_k + |a_k|)$ follows from the Direct Comparison Theorem. ■

Now let's formally connect the idea of absolute convergence to the discussion of conditional convergence that started this section.

> **Theorem:** A convergent series of real numbers is either absolutely convergent or conditionally convergent, but not both.

Note: As a mnemonic, consider this: a series is absolutely convergent when it absolutely always converges to the same number, regardless of the order in which we add the summands.

> This theorem is equivalent to saying that a series of real numbers is unconditionally convergent if and only if it's absolutely convergent. The proof is beyond the scope of this book, but see [42, p. 78].

Verifying absolute convergence

Example 5.2. Show that $\sum_{k=1}^{\infty}(-1)^{k+1}/k^2$ is absolutely convergent.

Solution: The series is absolutely convergent because

$$\sum_{k=1}^{\infty}\left|\frac{(-1)^{k+1}}{k^2}\right| = \sum_{k=1}^{\infty}\frac{1}{k^2}$$

converges (it's a p-series with $p > 1$). ■

Checking for absolute convergence

Example 5.3. Determine whether $\sum_{k=1}^{\infty}(-1)^k/\sqrt{k^2+1}$ is absolutely convergent, conditionally convergent, or neither.

Solution: The summands of this alternating series decrease term by term, and converge to zero in the limit, so it converges (see p. 608). However, the absolute value of the summands is $1/\sqrt{k^2+1}$, and

$$\frac{1}{\sqrt{k^2+1}} \geq \frac{1}{\sqrt{k^2+k^2}} = \frac{1}{\sqrt{2}\,k}.$$

> Alternatively, you could perform a limit comparison to $1/k$.

Since $\sum_{k=1}^{\infty}1/k$ diverges, so does $\sum_{k=1}^{\infty}\frac{1}{\sqrt{k^2+1}}$, which means that this series is *not* absolutely convergent. Since the series converges but is not absolutely convergent, it's conditionally convergent. ■

Try It Yourself: Checking for absolute convergence

Example 5.4. Determine whether $\sum_{k=1}^{\infty}(-1)^k/\sqrt{k^2+k^k}$ is absolutely convergent, conditionally convergent, or neither.

Answer: Absolutely convergent. ■

Full Solution On-line

✥ Ratio Test

The Ratio Test is designed to answer the question, "Does this series behave like a geometric series? And if so, what is (effectively) the common ratio?" Recall that the common ratio of a geometric series,

$$\underbrace{a}_{a_1} + \underbrace{ar}_{a_2} + \underbrace{ar^2}_{a_3} + \underbrace{ar^3}_{a_4} + \underbrace{ar^4}_{a_5} + \cdots$$

can be determined by calculating the ratio of any two consecutive terms:

$$\frac{a_2}{a_1} = r, \qquad \frac{a_3}{a_2} = r, \qquad \frac{a_4}{a_3} = r \ldots.$$

If a series isn't *actually* geometric, but its summands are dominated by geometric terms when k is large, we'll see the analog of a common ratio emerge in the limit:

$$\lim_{k \to \infty} \left| \frac{a_{k+1}}{a_k} \right|,$$

where we've employed absolute values to suppress the sign because, from the standpoint of determining convergence, we only care about the magnitude of the common ratio. Intuitively, if this analog of the common ratio is less than 1, we expect the series to converge (just like a geometric series with $|r| < 1$). Similarly, if it's larger than 1, we expect the series to diverge. That intuition is correct, and is called . . .

> At a technical level, the absolute value also prevents random flip-flops in the sign of the summands from causing the limit to diverge.

Ratio Test: Suppose $\lim_{k \to \infty} \left| \frac{a_{k+1}}{a_k} \right| = R$.

(a) If $R < 1$, the series $\sum a_k$ converges absolutely.

(b) If $R > 1$ (including $R = \infty$), the series $\sum a_k$ diverges.

(c) A limit of $R = 1$ is inconclusive.

> The fraction
> $$\left| \frac{a_{k+1}}{a_k} \right|$$
> is exactly the expression that we use to determine the common ratio of geometric series.

Proof. You'll see why part (c) is true in Example 5.7. We'll prove part (a) here, and leave part (b) to the exercise set. Like the proof of the Limit Comparison Theorem on p. 630, our proof will rely on the fact that the first N terms of a series (any finite number of them) affect its value, but not whether it converges in the first place.

(a) In the case when $R < 1$, let's choose an $r \in (R, 1)$. Then $\lim_{k \to \infty} \left| \frac{a_{k+1}}{a_k} \right| = R$ implies that there is some index n past which

$$\left| \frac{a_{k+1}}{a_k} \right| \leq r.$$

> The value of r affects the size of n, but not the fact that there *is* an index past which this inequality holds.

Said differently, $|a_{k+1}| \leq r|a_k|$ when $k \geq n$. It follows that

$$
\begin{aligned}
|a_{n+1}| &\leq r|a_n| \\
|a_{n+2}| &\leq r|a_{n+1}| \leq r^2|a_n| \\
|a_{n+3}| &\leq r|a_{n+2}| \leq r^3|a_n| \\
&\vdots \qquad \vdots \\
|a_{n+m}| &\leq r^m|a_n|.
\end{aligned}
$$

Since a_n is a fixed number and $r < 1$, this says that $|a_{n+m}|$ is bounded above by the summands of a convergent geometric series. So the Direct Comparison Theorem guarantees that $\sum_{k=n}^{\infty} |a_k|$ converges. And since the first $n - 1$ summands don't affect convergence, that means $\sum_{k=1}^{\infty} |a_k|$ converges. ∎

Note: The Ratio Test is most effective when the ratio of summands results in cancellation.

Checking absolute convergence with the Ratio Test

Example 5.5. Prove that $\sum_{k=4}^{\infty} \frac{3^k}{k!}$ converges using the Ratio Test.

Solution: In Example 4.1 we used the Direct Comparison Theorem to determine that this series converges. It's significantly easier to do this with the Ratio Test.

We begin by looking at the ratio of consecutive terms:

$$\frac{a_{k+1}}{a_k} = \frac{\frac{3^{k+1}}{(k+1)!}}{\frac{3^k}{k!}} = \frac{3^{k+1}}{(k+1)!}\frac{k!}{3^k} = 3\frac{(k)(k-1)(k-2)\cdots(3)(2)(1)}{(k+1)(k)(k-1)(k-2)\cdots(3)(2)(1)} = \frac{3}{k+1}.$$

The absolute value required by the Ratio Test is irrelevant in this case, since a_{k+1}/a_k is always positive, but we include it as a reminder:

$$\lim_{k\to\infty}\left|\frac{a_{k+1}}{a_k}\right| = \lim_{k\to\infty}\frac{3}{k+1} = 0.$$

Since the limit value is less than 1, the Ratio Test tells us that the series is absolutely convergent. (In fact, the limit value of 0 is telling us that these terms tend to zero *faster* than those of a convergent geometric series.) ∎

Try It Yourself: Checking absolute convergence with the Ratio Test

Example 5.6. Use the Ratio Test to determine whether $\sum_{k=1}^{\infty}\frac{3k^4+1}{(-5)^k}$ converges absolutely.

Full Solution On-line

Answer: $R = 1/5$, so it does. ∎

Inconclusive results with the Ratio Test

Example 5.7. Show that $R = 1$ for both $\sum_{k=1}^{\infty}1/k$ and $\sum_{k=1}^{\infty}1/k^2$, and discuss the implications of that fact.

Solution: When we consider $a_k = 1/k$, the ratio of consecutive terms is

$$\left|\frac{a_{k+1}}{a_k}\right| = \frac{\frac{1}{k+1}}{\frac{1}{k}} = \frac{k}{k+1}.$$

And when $a_k = 1/k^2$, the ratio of consecutive terms is

$$\left|\frac{a_{k+1}}{a_k}\right| = \frac{\frac{1}{(k+1)^2}}{\frac{1}{k^2}} = \frac{k^2}{k^2+2k+1}.$$

In both cases we see $\lim_{k\to\infty}\left|\frac{a_{k+1}}{a_k}\right| = 1$. Because one of these series converges but the other does not, we learn from this example that we *cannot make a conclusion about absolute convergence when $R = 1$*. ∎

✧ Root Test

The Root Test is another way to tell whether a given series behaves like a geometric series. Let's note that, in addition to using the ratio of consecutive terms, the common ratio of a geometric series can be calculated as the kth root of the kth term, in the limit:

$$r = \lim_{k\to\infty}\left(ar^{k-1}\right)^{1/k} = \lim_{k\to\infty}a^{1/k}r^{1-1/k} = \left(\lim_{k\to\infty}a^{1/k}\right)\left(\lim_{k\to\infty}r^{1-1/k}\right) = r\,,$$

since $\lim_{k\to\infty}a^{1/k} = 1$ when $a > 0$. When a series isn't *actually* geometric, but its summands are dominated by geometric terms when k is large, we expect to see the analog of a common ratio emerge in the limit:

$$\lim_{k\to\infty}|a_k|^{1/k}.$$

The Root Test articulates this idea precisely.

Root Test: Suppose $\lim_{k\to\infty} |a_k|^{1/k} = R$.

(a) If $R < 1$, the series $\sum a_k$ converges absolutely.

(b) If $R > 1$ (including $R = \infty$), the series $\sum a_k$ diverges.

(c) A limit of $R = 1$ is inconclusive.

> The conclusions of the Root Test are identical to those of the Ratio Test because they answer the same basic question: does the given series behave geometrically when k is large? So if one of these tests yields $R = 1$, you should expect that the other will also.

Proof. We'll prove part (a) here, leave part (b) to the exercise set, and demonstrate part (c) in an example. When $r \in (R, 1)$, as in our proof of the Ratio Test on p. 642, there is an index n beyond which $|a_k|^{1/k} \leq r$. This implies that $|a_k| \leq r^k$ when $k \geq n$. Since $r < 1$, this says all but a finite number of $|a_k|$ are bounded above by the summands of a convergent geometric series, and the convergence of $\sum |a_k|$ follows from the Direct Comparison Theorem. ∎

Note: The Root Test is most effective when whole summands are raised to a power.

Using the Root Test

Example 5.8. Use the Root Test to determine whether $\sum_{k=1}^{\infty} \left(\frac{k+1}{4k+3} \right)^k$ converges.

Solution: We showed that this series converged in Example 4.8 by using the Limit Comparison Theorem. It's much easier to do it with the Root Test. We begin by calculating the kth root of the kth term. Since all the summands are positive,

$$|a_k|^{1/k} = \left| \left(\frac{k+1}{4k+3} \right)^k \right|^{1/k} = \frac{k+1}{4k+3}.$$

Since $\lim_{k\to\infty} |a_k|^{1/k} = 1/4$, which is less than one, the Root Test tells us that this series converges absolutely. ∎

Try It Yourself: Using the Root Test

Example 5.9. Use the Root Test to determine whether $\sum_{k=1}^{\infty} \frac{(2k-3)^k}{(3k-2)^k}$ converges absolutely.

Answer: It does ($R = 2/3$). ∎

Full Solution On-line

Inconclusive results with the Root Test

Example 5.10. Show that $R = 1$ for both $\sum_{k=1}^{\infty} 1/k$ and $\sum_{k=1}^{\infty} 1/k^2$ when we use the Root Test, and discuss the implications of that fact.

Solution: This result relies on the fact that $\lim_{k\to\infty} k^{1/k} = 1$, which is established using L'Hôpital's Rule. When $a_k = 1/k$ we have

$$\lim_{k\to\infty} |a_k|^{1/k} = \lim_{k\to\infty} \frac{1}{k^{1/k}} = \frac{1}{\lim_{k\to\infty} k^{1/k}} = 1.$$

Similarly, when $a_k = 1/k^2$ we have

$$\lim_{k\to\infty} |a_k|^{1/k} = \lim_{k\to\infty} \frac{1}{k^{2/k}} = \lim_{k\to\infty} \left(\frac{1}{k^{1/k}} \right)^2 = \left(\frac{1}{\lim_{k\to\infty} k^{1/k}} \right)^2 = 1.$$

As before, this is an important example because it shows us that $R = 1$ can happen with both convergent and divergent series. ∎

❖ In Closing

When using the Ratio Test or Root Test, you're asking the question, "Are the summands of this series dominated by geometric behavior?" One of the possible answers is, "No," which is how you should interpret a limit of $R = 1$ (see Example 5.7). The other possible answer is, "Yes, and the analog of the common ratio is R," from which you can determine whether the series converges or diverges.

❖ Strategy for Testing Series

When you're wondering whether a particular series converges, ask yourself the following questions:

1. Do the summands tend to zero in the limit? If not, the partial sums cannot converge.

2. Is it an alternating series? If so, it's relatively easy to tell whether it converges (see p. 608).

3. Do the summands have factors that cancel in a a_{k+1}/a_k ratio, such as powers or factorials? If so, consider using the Ratio Test.

4. Are the summands raised to a power? If so, consider using the Root Test.

5. Are the summands larger (or smaller) than those from a series whose convergence you already know, such as a p-series? If so, consider using the Direct Comparison Theorem.

6. When k is large, do the summands behave like a series whose convergence you know? If so, consider using the Limit Comparison Theorem. Try to compare to a series that's (a) simple and (b) familiar.

7. If the summands have the form $f(k)$, can you integrate $f(x)$? If so, consider using the Integral Test.

> **IMPORTANT:** A series can diverge even when its summands tend to zero in the limit (e.g., the Harmonic Series).
>
> In order to be sure that a series converges, you need to know (1) that the terms go to zero *and* (2) some *other* piece of information (for example, that it's alternating or geometric or ...).

> The Ratio Test is particularly useful when dealing with *power series*, which is the subject of the next section.

> When using the comparison tests, we typically compare to *p*-series or geometric series. Those two serve as "landmarks" for us.

You should know

- the terms *conditionally convergent* and *absolutely convergent;*

- the Ratio Test and Root Test examine the extent to which a series appears to be geometric when k is large;

- the Ratio Test and Root Test are inconclusive when $R = 1$.

You should be able to

- determine whether an alternating series is absolutely or conditionally convergent;

- use the Ratio Test and the Root Test to determine whether a series is absolutely convergent.

❖ 8.5 Skill Exercises

Show that the series in #1–4 converge absolutely.

1. $\displaystyle\sum_{k=1}^{\infty} \frac{1}{k^{2.1}}$
2. $\displaystyle\sum_{k=1}^{\infty} \frac{k^2}{2^k}$
3. $\displaystyle\sum_{k=1}^{\infty} (-1)^k \frac{\sqrt{k}}{k!}$
4. $\displaystyle\sum_{k=1}^{\infty} (-1)^k \frac{3^k}{k!}$

Show that the series in #5–8 converge conditionally.

5. $\displaystyle\sum_{k=2}^{\infty} \frac{(-1)^k}{\sqrt{k}}$
6. $\displaystyle\sum_{k=2}^{\infty} \frac{(-1)^k}{k \ln(k)}$
7. $\displaystyle\sum_{k=2}^{\infty} (-1)^k \frac{k^{3/2}}{k^2 + 1}$
8. $\displaystyle\sum_{k=2}^{\infty} (-1)^k \frac{k \ln(k)}{k^2 + 1}$

Apply the Ratio Test to the series in #9–14 and report whether it identifies the series as absolutely convergent or divergent, or is inconclusive.

9. $\displaystyle\sum_{k=2}^{\infty} \frac{8k^2 + 7}{2^k}$
11. $\displaystyle\sum_{k=2}^{\infty} \frac{31}{7k + 8}$
13. $\displaystyle\sum_{k=2}^{\infty} \frac{(-1)^k}{k(0.3)^k}$

10. $\displaystyle\sum_{k=2}^{\infty} \frac{0.7^k}{k^2 + k + 1}$
12. $\displaystyle\sum_{k=2}^{\infty} \frac{10\sqrt{k}}{k^2 + k + 21}$
14. $\displaystyle\sum_{k=2}^{\infty} \frac{(-1.2)^k}{\ln(k)}$

Apply the Root Test to the series in #15–20 and report whether it identifies the series as absolutely convergent or divergent, or is inconclusive.

15. $\displaystyle\sum_{k=2}^{\infty} \left(\frac{2k + 1}{8k + 7} \right)^k$
17. $\displaystyle\sum_{k=2}^{\infty} \left(\frac{3}{7k} \right)^k$
19. $\displaystyle\sum_{k=2}^{\infty} \left(\frac{-\sqrt{k} - 9k}{9k + \ln(k)} \right)^k$

16. $\displaystyle\sum_{k=2}^{\infty} \left(\frac{9k^2 + k + 1}{14k - 2} \right)^k$
18. $\displaystyle\sum_{k=2}^{\infty} \left(\frac{10k + \sin(k)}{10k + \sqrt{k}} \right)^k$
20. $\displaystyle\sum_{k=2}^{\infty} \left(\frac{-1}{\ln(k)} \right)^{k!}$

Determine whether the series in #21–32 converge absolutely, conditionally, or not at all.

21. $\displaystyle\sum_{k=1}^{\infty} \frac{(-1)^k}{k^{1/3}}$
25. $\displaystyle\sum_{k=1}^{\infty} \frac{k}{k^3 + 2}$
29. $\displaystyle\sum_{k=1}^{\infty} (-1)^k \left(1 + \frac{k}{k^2 - 3} \right)^k$

22. $\displaystyle\sum_{k=2}^{\infty} \frac{(-1)^k \ln(k)}{k}$
26. $\displaystyle\sum_{k=4}^{\infty} \frac{\sqrt{k}}{2 + \ln(k)}$
30. $\displaystyle\sum_{k=6}^{\infty} e^{(-1)^k (k^2 + 1)}$

23. $\displaystyle\sum_{k=1}^{\infty} (-1)^k k e^{-k}$
27. $\displaystyle\sum_{k=1}^{\infty} (-1)^k \ln\left(\frac{k}{2k + 7} \right)$
31. $\displaystyle\sum_{k=1}^{\infty} (-1)^k \tan\left(\frac{1}{k} \right)$

24. $\displaystyle\sum_{k=3}^{\infty} \sin(k) e^{-k}$
28. $\displaystyle\sum_{k=5}^{\infty} \frac{(-1)^k}{k} \sin\left(\frac{1}{k} \right)$
32. $\displaystyle\sum_{k=7}^{\infty} (-1)^k \frac{k \ln(k)}{k^2 + 1}$

In #33–38 determine the values of p (if any) for which the series is conditionally convergent.

33. $\displaystyle\sum_{k=1}^{\infty}(-1)^{k}\frac{\sqrt{k}}{(k+1)^{p}}$

34. $\displaystyle\sum_{k=1}^{\infty}(-1)^{k}\frac{k^{2p}}{k^{4}+5k+1}$

35. $\displaystyle\sum_{k=1}^{\infty}(-1)^{k}\frac{k^{p}}{2^{k}}$

36. $\displaystyle\sum_{k=1}^{\infty}\frac{4}{k^{2}+5k^{p}}$

37. $\displaystyle\sum_{k=1}^{\infty}(-1)^{k}k^{3+2p-p^{2}}$

38. $\displaystyle\sum_{k=1}^{\infty}(-1)^{k}\frac{k^{\ln(p)}}{k^{2}+1}$

In #39–44 determine the values of x for which the Ratio Test (a) guarantees convergence, (b) guarantees divergence, and (c) is inconclusive.

39. $\displaystyle\sum_{k=1}^{\infty}5^{k+1}x^{k}$

40. $\displaystyle\sum_{k=1}^{\infty}(0.125)^{k}x^{3k}$

41. $\displaystyle\sum_{k=1}^{\infty}10^{k}(x-2)^{k}$

42. $\displaystyle\sum_{k=1}^{\infty}4^{k}(x+3)^{k}$

43. $\displaystyle\sum_{k=1}^{\infty}\frac{2k^{3}+4}{(k-1)!}x^{k}$

44. $\displaystyle\sum_{k=1}^{\infty}\frac{(x-1)^{k}}{(k+1)!}$

In #45–48 suppose that $a_{1}=1$, and other values of a_{k} are determined by the given recurrence relation. (a) Use a cobwebbing diagram to show that $\lim_{k\to\infty}a_{k}=0$. (b) What does the Ratio Test tell you about the series $\sum_{k=1}^{\infty}a_{k}$?

45. $a_{k+1}=e^{0.5a_{k}}-1$

46. $a_{k+1}=1-0.5^{a_{k}}$

47. $a_{k+1}=\sin(a_{k})$

48. $a_{k+1}=\ln(1+a_{k})$

Mixed Practice: The goal of this set of exercises is to help you learn how to determine which convergence test is appropriate for a given series. Exercises 49–64 require the ideas from either this section **or** previous sections.

49. $\displaystyle\sum_{k=1}^{\infty}\frac{1}{2k^{2}-3k+4}$

50. $\displaystyle\sum_{k=1}^{\infty}\frac{\sqrt{k}}{\sqrt{k+\sqrt{k^{8}+13}}}$

51. $\displaystyle\sum_{k=2}^{\infty}\sqrt{k}3^{-k}$

52. $\displaystyle\sum_{k=1}^{\infty}(0.3)^{k}$

53. $\displaystyle\sum_{k=3}^{\infty}\frac{\ln(k+1)-\ln(k)}{\ln(k)-\ln(k-1)}$

54. $\displaystyle\sum_{k=4}^{\infty}\sin\left(\frac{1}{k}\right)$

55. $\displaystyle\sum_{k=5}^{\infty}\frac{4^{k}+k^{-7}}{\sqrt{k^{k}}}$

56. $\displaystyle\sum_{k=1}^{\infty}\left(\frac{-4}{k}\right)^{k}$

57. $\displaystyle\sum_{k=2}^{\infty}\frac{k^{k}}{k!}$

58. $\displaystyle\sum_{k=3}^{\infty}\left(\frac{3k+4}{6k+7}\right)^{2k}$

59. $\displaystyle\sum_{k=4}^{\infty}\frac{k^{2k+1}}{(2k+1)^{k}}$

60. $\displaystyle\sum_{k=5}^{\infty}(-1)^{k}\frac{\cos(\pi k)}{k+3}$

61. $\displaystyle\sum_{k=6}^{\infty}\frac{4+k^{2}}{8^{k-2}}$

62. $\displaystyle\sum_{k=5}^{\infty}\left(\frac{8k+1}{3+8k}\right)^{k}$

63. $\displaystyle\sum_{k=4}^{\infty}(1-2^{-k})^{k}$

64. $\displaystyle\sum_{k=3}^{\infty}\frac{(k+1)^{2}5^{k}}{(k-1)!}$

❖ 8.5 Concept and Application Exercises

65. Suppose a friend asks you to look at his work. He's trying to determine whether $\sum_{k\geq 1}k^{2}e^{-k}$ converges, so he calculates the ratio

$$\left|\frac{a_{2}}{a_{1}}\right|=\frac{4/e^{2}}{1/e}=\frac{4}{e}>1.$$

Since the ratio is larger than 1, your friend concludes that the series diverges. Is he right? If so, explain why. If not, explain his error to him.

66. Suppose that when using the Root Test to determine whether a particular series converges, you see a limit of "1." What will happen if you use the Ratio Test (and why)?

67. Suppose $\sum_{k=1}^{\infty} a_k$ converges absolutely. Do we know whether $\sum_{k=1}^{\infty} a_k^2$ converges absolutely? If so, prove it. If not, why not?

68. Suppose $\sum_{k=1}^{\infty} a_k^2$ converges absolutely. Do we know whether $\sum_{k=1}^{\infty} a_k$ converges absolutely? If so, prove it. If not, why not?

69. Suppose $\sum_{k=1}^{\infty} a_k$ converges conditionally. Do we know whether $\sum_{k=1}^{\infty} a_k^2$ converges? If so, prove it. If not, why not?

70. Suppose $\sum_{k=1}^{\infty} a_k$ and $\sum_{k=1}^{\infty} b_k$ are absolutely convergent series. Is $\sum_{k=1}^{\infty} a_k + b_k$ also? If so, prove it. If not, why not?

71. Suppose $\sum_{k=1}^{\infty}(a_k + b_k)$ is an absolutely convergent series. Are $\sum_{k=1}^{\infty} a_k$ and $\sum_{k=1}^{\infty} b_k$ also? If so, prove it. If not, why not?

Exercises #72 and #73 refer to the following figure, in which the blue circles represent the summands of $\sum_{k=1}^{\infty}(-1)^k a_k$.

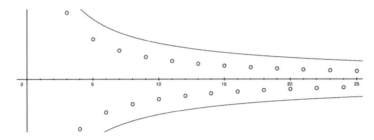

72. Suppose the curves in the figure are $y = \pm\frac{1}{t^2}$. Do we know enough to say whether the series $\sum_{k=1}^{\infty}(-1)^k a_k$ is divergent, conditionally convergent, or absolutely convergent? If so, classify the series and explain how you know. If not, why not?

73. Suppose the curves in the figure are $y = \pm\frac{1}{\sqrt{t}}$. Do we know enough to say whether the series $\sum_{k=1}^{\infty}(-1)^k a_k$ is divergent, conditionally convergent, or absolutely convergent? If so, classify the series and explain how you know. If not, why not?

8.6   Introduction to Power Series

We can write numbers in many different forms. For example, you know that

$$\frac{6}{3}, \quad \sqrt[3]{8}, \quad \log(100), \quad \lim_{\phi \to 0} \frac{\sin(2\phi)}{\phi}, \quad \text{and} \quad \int_1^2 \frac{8}{15} x^3 \, dx \qquad (6.1)$$

are all ways of writing the number 2. In this section we introduce a new way to write *functions*. Specifically, you'll begin learning how to write them as series. The simplest examples arise in the context of geometric series. You know that a geometric series with a common ratio of $|t| < 1$ converges, and

$$\sum_{k=0}^{\infty} t^k = \frac{1}{1-t}.$$

Reading from right to left, this equality says that the function $f(t) = 1/(1-t)$ generates the same values as a geometric series when $|t| < 1$. That is, we have both

$$f(t) = \frac{1}{1-t} \qquad (6.2)$$

$$f(t) = \sum_{k=0}^{\infty} t^k \qquad (6.3)$$

which we understand—much in the spirit of (6.1)—as two different ways of writing down the same function. We say that the familiar expression in equation (6.2) is the **closed form** for $f(t)$, and because the expression in equation (6.3) is a series of powers of t, we say it's a **power series expansion** of $f(t)$.

> The Geometric Series Theorem on p. 605 tells us that
> $$\sum_{k=0}^{\infty} ar^k = \frac{a}{1-r}$$
> when $|r| < 1$. Recall that a is the first summand of the series, whatever that happens to be.

Transitioning from power series to closed form

Example 6.1. Suppose $f(t) = \sum_{k=5}^{\infty} t^k$. Determine the domain of f and a closed form for the function.

Solution: In this case, $f(t)$ is defined to be the value of a geometric series with a common ratio of t, so only $|t| < 1$ is admissible. Therefore, the domain of f is $(-1, 1)$. For values of t inside this interval, the geometric series formula

$$\sum_{k=5}^{\infty} t^k = \frac{t^5}{1-t}.$$

> The first term of this series is $a = t^5$ because k starts at 5.

gives us a closed form for $f(t)$.   ∎

Transitioning from power series to closed form

Example 6.2. Suppose $f(t) = \sum_{k=0}^{\infty} \frac{(-1)^k}{2^{k+1}} t^k$. Determine the domain of f and a closed form for the function.

Solution: Again, we have $f(t)$ defined to be the value of a geometric series. In this case, the common ratio is $-t/2$, so we must have $|t/2| < 1$ in order for the function value to exist. That is, the domain of f is $(-2, 2)$. Again, for values of t inside this interval, the geometric series formula

$$\sum_{k=0}^{\infty} \frac{(-1)^k}{2^{k+1}} t^k = \frac{1/2}{1 - \left(\frac{-t}{2}\right)} = \frac{1}{2+t}.$$

gives us a closed form for $f(t)$.   ∎

Try It Yourself: Transitioning from power series to closed form

Example 6.3. Suppose $f(t) = \sum_{k=1}^{\infty} \frac{3}{5^k} t^{2k}$. Determine (a) the domain of f, and (b) a closed form for $f(t)$ inside its domain.

Answer: (a) $(-\sqrt{5}, \sqrt{5})$, (b) $3t^2/(5 - t^2)$.   ∎

Transitioning from closed form to power series

Example 6.4. Suppose $f(t) = \frac{7}{5+t}$. Find a power series expansion of f and determine the interval over which it converges.

Solution: The key to making this transition from a closed form to a power series is the fact that

$$\frac{1}{1-u} = \sum_{k=0}^{\infty} u^k \quad \text{when } |u| < 1. \tag{6.4}$$

We can rewrite $f(t)$ in the form of $1/(1-u)$ by factoring out both the 7 and the 5 from our formula:

$$f(t) = \frac{7}{5} \frac{1}{1 + \frac{t}{5}} = \frac{7}{5} \frac{1}{1 - \left(-\frac{t}{5}\right)} = \frac{7}{5} \frac{1}{1-u} \quad \text{where} \quad u = -\frac{t}{5}.$$

Therefore, when $|-t/5| < 1$ we have

$$f(t) = \frac{7}{5} \sum_{k=0}^{\infty} \left(-\frac{t}{5}\right)^k = \sum_{k=0}^{\infty} (-1)^k \frac{7}{5^{k+1}} t^k.$$

The requirement that $|-t/5| < 1$ is the same as saying that this power series expansion for $f(t)$ converges when $t \in (-5, 5)$; and for any t in this interval, the power series expansion of $f(t)$ and the closed form for $f(t)$ are equivalent ways of expressing the function.   ∎

Transitioning from closed form to power series

Example 6.5. Suppose $f(t) = \frac{3t}{8-t^2}$. Find a power series expansion of f and determine the interval over which it converges.

Solution: As in Example 6.4 we begin by factoring out terms from $f(t)$ in order to rewrite the formula with a $1/(1-u)$ form:

$$f(t) = \frac{3t}{8} \frac{1}{1 - \frac{t^2}{8}} = \frac{3t}{8} \frac{1}{1-u} \quad \text{where} \quad u = \frac{t^2}{8}.$$

Therefore, when $|t^2/8| < 1$ we have

$$f(t) = \frac{3t}{8} \sum_{k=0}^{\infty} \left(\frac{t^2}{8}\right)^k = \frac{3t}{8} \sum_{k=0}^{\infty} \frac{1}{8^k} t^{2k} = \sum_{k=0}^{\infty} \frac{3}{8^{k+1}} t^{2k+1}.$$

The requirement that $|t^2/8| < 1$ is the same as saying that this power series expansion for $f(t)$ converges when $t \in (-\sqrt{8}, \sqrt{8})$.   ∎

Try It Yourself: Transitioning from closed form to power series

Example 6.6. Suppose $f(t) = \frac{2t^5}{4+t^3}$. Find a power series expansion of f and determine the interval over which it converges.

Answer: $\sum_{k=0}^{\infty} \frac{2(-1)^k}{4^{k+1}} t^{3k+5}$ converges when $t \in (-\sqrt[3]{4}, \sqrt[3]{4})$.   ∎

Full Solution On-line

Full Solution On-line

❖ Expansion Points

When a series is composed of powers of $(t - t_0)$ we say that t_0 is the **expansion point**, or that the power series is **expanded at** t_0. Until now, we've been dealing with $t_0 = 0$, in which case we refer to the power series as a **Maclaurin series**. When $t_0 \neq 0$, we refer to the power series as a **Taylor series** at (or about) t_0.

> You can think of the names *Maclaurin* and *Taylor* as adjectives that tell you whether a power series includes powers of t or powers of $t - t_0$.

Identifying expansion points

Example 6.7. Determine the expansion points for the following series:

$$\text{(a)} \sum_{k=3}^{\infty} 5^k (t - 17)^k, \quad \text{(b)} \sum_{k=4}^{\infty} \frac{1}{k^2}(t + 8)^k, \quad \text{and (c)} \sum_{k=0}^{\infty} (3t - 5)^k.$$

Solution: We look for powers of $(t - t_0)$. In part (a), we see powers of $(t - 17)$, so $t_0 = 17$ is the expansion point of that Taylor series. In part (b) we see powers of $(t + 8) = (t - (-8))$, so $t_0 = -8$ is the expansion point of that Taylor series. In part (c), the series isn't written in the standard form, with powers of $(t - t_0)$, so we have to begin by rewriting it as

$$\sum_{k=0}^{\infty} (3t - 5)^k = \sum_{k=0}^{\infty} \left(3\left(t - \frac{5}{3}\right)\right)^k = \sum_{k=0}^{\infty} 3^k \left(t - \frac{5}{3}\right)^k.$$

Now we see powers of $(t - 5/3)$, so the expansion point is $t_0 = 5/3$.   ∎

Transitioning from Taylor series to closed form

Example 6.8. Suppose $f(t) = \sum_{k=0}^{\infty} \frac{(-1)^k}{9^{k+1}}(t - 7)^k$. Determine (a) the expansion point, (b) the domain of f, and (c) a closed form for $f(t)$.

Solution: (a) Since we see powers of $(t - 7)$, the expansion point is $t_0 = 7$. (b) In this case $f(t)$ is defined to be the value of a geometric series with $a = 1/9$ and $r = -(t - 7)/9$, so it only converges when

> Check that this is the common ratio by calculating the ratio of any two consecutive terms.

$$\left| \frac{t - 7}{9} \right| < 1 \quad \Rightarrow \quad |t - 7| < 9.$$

You should read this inequality as saying "the distance from t to $t_0 = 7$ must be less than 9." That is, the admissible values of t extend all the way down to $7 - 9 = -2$, and all the way up to $7 + 9 = 16$, though neither -2 nor 16 are close enough to $t_0 = 7$ to be in the domain (because the distance to t_0 is *equal to*, not less than 9). Therefore, the domain of this function is $t \in (-2, 16)$. As for part (c) we can get a closed form by using our formula for geometric series:

$$f(t) = \sum_{k=0}^{\infty} \frac{(-1)^k}{9^{k+1}}(t - 7)^k = \frac{1/9}{1 - \left(-\frac{(t-7)}{9}\right)} = \frac{1}{2 + t}. \qquad ∎$$

Transitioning from Taylor series to closed form

Example 6.9. Suppose $f(t) = \sum_{k=0}^{\infty} \frac{5^{k+2}}{3^{k+1}}(9t - 8)^k$. Determine (a) the expansion point, (b) the domain of f, and (c) a closed form for $f(t)$.

Solution: (a) To determine the expansion point, we rewrite this series in standard form using the technique demonstrated in part (c) of Example 6.7:

$$f(t) = \sum_{k=0}^{\infty} \frac{5^{k+2} 9^k}{3^{k+1}} \left(t - \frac{8}{9}\right)^k,$$

from which we conclude that the expansion point is $t_0 = 8/9$. (b) This function is defined to be the value of a geometric series with $a = 25/3$ and $r = \frac{5}{3}(9t - 8)$. The series converges if and only if

$$\left| \frac{5}{3}(9t - 8) \right| < 1 \quad \Rightarrow \quad |9t - 8| < \frac{3}{5} \quad \Rightarrow \quad \left| t - \frac{8}{9} \right| < \frac{1}{15}.$$

You should read this inequality as saying "the distance from t to $t_0 = \frac{8}{9}$ must be less than $\frac{1}{15}$." That is, the admissible values of t extend all the way down to $\frac{8}{9} - \frac{1}{15} = \frac{37}{45}$, and all the way up to $\frac{8}{9} + \frac{1}{15} = \frac{43}{45}$, though neither $37/45$ nor $43/45$ are close enough to $t_0 = 8/9$ for the series to converge (because the distance to t_0 is *equal to*, not less than $1/15$). Therefore the domain of this function is $(\frac{37}{45}, \frac{43}{45})$. As for part (c) we can get a closed form by using our formula for geometric series:

$$f(t) = \frac{\frac{25}{3}}{1 - \frac{5}{3}(9t - 8)} = \frac{25}{3 - 5(9t - 8)} = \frac{25}{43 - 45t}. \qquad \blacksquare$$

Try It Yourself: Transitioning from Taylor series to closed form

Example 6.10. Suppose $f(t) = \sum_{k=0}^{\infty} \frac{1}{7^{k+2}}(2t - 10)^k$. Determine (a) the expansion point, (b) the domain of f, and (c) a closed form for $f(t)$.

Answer: (a) $t_0 = 5$, (b) $t \in (\frac{3}{2}, \frac{17}{2})$, and (c) $f(t) = 1/(119 - 14t)$. $\qquad \blacksquare$

Full Solution On-line

Transitioning from closed form to Taylor series

Example 6.11. Suppose $f(t) = \frac{1}{2+t}$. Determine (a) the Taylor series for f about $t_0 = 10$, and (b) the interval in which that series converges.

Solution: (a) As before, the key to this transition is equation (6.4), but this time we want u to involve the term $(t - 10)$. We can make that happen by adding 0 (in the form of $-10 + 10$) to the denominator of our fraction:

$$f(t) = \frac{1}{2 + t - 10 + 10} = \frac{1}{12 + (t - 10)} .$$

Now the technique demonstrated earlier can be used to obtain the Taylor series. We begin by factoring the 12 out of the denominator,

$$f(t) = \frac{1}{12} \frac{1}{1 + \frac{(t-10)}{12}} = \frac{1}{12} \frac{1}{1 - \left(-\frac{(t-10)}{12} \right)} = \frac{1}{12} \frac{1}{1 - u} \quad \text{where} \quad u = -\frac{(t - 10)}{12}.$$

So the geometric series formula tells us that

$$f(t) = \frac{1}{12} \sum_{k=0}^{\infty} \left(-\frac{t - 10}{12} \right)^k = \frac{1}{12} \sum_{k=0}^{\infty} \frac{(-1)^k}{12^k} (t - 10)^k = \sum_{k=0}^{\infty} \frac{(-1)^k}{12^{k+1}} (t - 10)^k.$$

(b) Our Taylor series for $f(t)$ is a geometric series with $r = -(t - 10)/12$, so it converges if and only if

$$\left| \frac{t - 10}{12} \right| < 1 \quad \Rightarrow \quad |t - 10| < 12.$$

That is, the distance between t and 10 must be less than 12. So the set of admissible t stretches all the way down to $10 - 12 = -2$, and all the way up to $10 + 12 = 22$, though neither -2 nor 22 is close enough to t_0 in order for the series to converge (the distance to t_0 is *equal* to, not less than 12 at those points). So our series converges for $t \in (-2, 22)$. $\qquad \blacksquare$

Transitioning from closed form to Taylor series

Example 6.12. Suppose $f(t) = \frac{1}{5+3t}$. Determine (a) the Taylor series for f about $t_0 = -4$, and (b) the interval in which that series converges.

Solution: (a) Since $t_0 = -4$, we want our Taylor series to be composed of powers of $(t - t_0) = (t + 4)$. We make this happen by adding 0 (in the form of $-4 + 4$) to t:

$$f(t) = \frac{1}{5 + 3(t+4-4)} = \frac{1}{5 + 3(t+4) - 12} = \frac{1}{-7 + 3(t+4)}$$

Now factoring out the -7 and using the geometric series equation with $u = \frac{3}{7}(t+4)$, we get

$$f(t) = -\frac{1}{7} \sum_{k=0}^{\infty} \left(\frac{3}{7}(t+4)\right)^k = \sum_{k=0}^{\infty} -\frac{3^k}{7^{k+1}}(t+4)^k.$$

(b) This is a geometric series with $r = 3(t+4)/7$ so it converges when

$$\left|\frac{3}{7}(t+4)\right| < 1 \quad \Rightarrow \quad |t+4| < \frac{7}{3}.$$

When this inequality is rewritten as $|t - t_0| < 7/3$, where $t_0 = -4$, we can understand it as saying that admissible values of t must be closer to $t_0 = -4$ than $7/3$ of a unit. That is, admissible t are larger than $-4 - \frac{7}{3} = -\frac{19}{3}$ and smaller than $-4 + \frac{7}{3} = -\frac{5}{3}$, so the interval in which this series converges is $\left(-\frac{19}{3}, -\frac{5}{3}\right)$. ∎

Try It Yourself: Transitioning from closed form to Taylor series

Example 6.13. Suppose $f(t) = \frac{3}{2+7t}$. Determine (a) the Taylor series for f about $t_0 = 1$, and (b) the interval in which that series converges.

Answer: (a) $\sum_{k=0}^{\infty} \frac{(-7)^k}{3^{2k+1}}(t-1)^k$, (b) $t \in \left(-\frac{2}{7}, \frac{16}{7}\right)$. ∎

Full Solution On-line

Transitioning from closed form to Taylor series

Example 6.14. Suppose $f(t) = \frac{2t}{1+t}$. Determine (a) the Taylor series for f about $t_0 = 8$, and (b) the interval in which that series converges.

Solution: (a) Because $t_0 = 8$, we want powers of $(t-8)$ in our Taylor series. So as in previous examples we'll add zero to all occurrences of t in the form of $0 = -8 + 8$, but not immediately. Because $f(t)$ is not written in lowest terms, we begin by using polynomial division to rewrite $f(t)$ as $2 - \frac{2}{1+t}$. After this preparatory step, we can proceed with the add-zero technique:

Appendix B includes a review of polynomial division.

$$f(t) = 2 - \frac{2}{1+t} = 2 - \frac{2}{1 + (t-8+8)} = 2 - \frac{2}{9 + (t-8)}.$$

Next we write the faction in the form $1/(1-u)$ by factoring out the 2 and the 9:

$$2 - \frac{2}{9 + (t-8)} = 2 - \frac{2}{9} \frac{1}{1 + \frac{(t-8)}{9}} = 2 - \frac{2}{9} \frac{1}{1-u} \quad \text{where} \quad u = -\frac{(t-8)}{9}.$$

So the geometric series formula tells us that

$$f(t) = 2 - \frac{2}{9} \sum_{k=0}^{\infty} \left(-\frac{(t-8)}{9}\right)^k = 2 - \frac{2}{9} \sum_{k=0}^{\infty} \frac{(-1)^k}{9^k}(t-8)^k = 2 - \sum_{k=0}^{\infty} \frac{2(-1)^k}{9^{k+1}}(t-8)^k.$$

(b) This geometric series has a common ratio of $-(t-8)/9$, so it converges when

$$\left|\frac{t-8}{9}\right| < 1 \quad \Rightarrow \quad |t-8| < 9 \quad \Rightarrow \quad t \in (-1, 17).$$ ∎

You might also find it tempting to factor the $2t$ out of the numerator and then use the add-zero technique on what remains, but that leads to a series whose summands have the form $t(t-8)^k$. This is not a Taylor series because it involves powers of t as well as powers of $(t-8)$. Taylor series only comprise scaled powers of $(t-t_0)$.

Note: Starting with the add-zero technique in part (a), above, would have led to

$$f(t) = \frac{2(t-8+8)}{1+(t-8+8)} = \frac{2(t-8)+16}{9+(t-8)} = \frac{2(t-8)}{9+(t-8)} + \frac{16}{9+(t-8)}.$$

Then we would have rewritten each term using the geometric series formula, and merged the results into a single series. That would have worked, but the step of merging the series requires a *re-indexing* technique that many people find confusing, so we started with polynomial division instead.

Try It Yourself: Transitioning from closed form to Taylor series

Example 6.15. Suppose $f(t) = \frac{3t}{5+t}$. Determine (a) the Taylor series for f about $t_0 = 1$, and (b) the interval in which that series converges.

Full Solution On-line

Answer: (a) $\frac{1}{2} + \sum_{k=0}^{\infty} \frac{5(-1)^k}{(2)6^{k+1}}(t-1)^{k+1}$, (b) $(-5,7)$. ∎

❖ Radius of Convergence and Intervals of Convergence

In all of the examples so far, our work with power series has led us to a statement of the form $|t - t_0| < R$, from which we have concluded that the series converges when $t \in (t_0 - R, t_0 + R)$. This interval is called the **interval of convergence**, and while it's been an open interval until now, it doesn't have to be. The number R, which is how far you step away from the expansion point in either direction to find the endpoints of that interval, is called the **radius of convergence**.

Radius of convergence

Example 6.16. Determine the radii of convergence in Examples 6.8–6.15.

Solution: This information is displayed in the table below.

Example	R	Expansion point = t_0	τ_0 = Location of discontinuity	Distance between t_0 and τ_0
6.8	7	9	-2	7
6.9	$\frac{1}{15}$	$\frac{8}{9}$	$\frac{43}{45}$	$\frac{1}{15}$
6.10	$\frac{7}{2}$	5	$\frac{119}{14}$	$\frac{7}{2}$
6.11	12	10	-2	12
6.12	$\frac{7}{3}$	-4	$-\frac{5}{3}$	$\frac{7}{3}$
6.13	$\frac{9}{7}$	1	$-\frac{2}{7}$	$\frac{9}{7}$
6.14	9	8	-1	9
6.15	6	1	-5	6

∎

After comparing the second and last columns of the table above, you might conjecture that a Taylor series' radius of convergence is the distance from the expansion point to the closest discontinuity of the associated closed form. That's correct in the current context of closed forms whose Taylor expansions are geometric series, but in general the relationship between radius of convergence and discontinuity is more subtle. We'll discuss it further in Section 8.8.

When power series become more complicated, as in the next example, we rely heavily on the Ratio Test to determine the radius of convergence.

The interval of convergence

Example 6.17. Determine the interval of convergence for $f(t) = \sum_{k=0}^{\infty} \frac{1}{\sqrt{k}}(t-4)^k$.

Solution: This series is *not* geometric, but the powers of $(t-4)$ dominate the summands when k is large, so we expect it to behave like a geometric series. When we use the Ratio Test, we see

$$\lim_{k\to\infty} \left| \frac{a_{k+1}}{a_k} \right| = \lim_{k\to\infty} \left| \frac{\frac{1}{\sqrt{k+1}}(t-4)^{k+1}}{\frac{1}{\sqrt{k}}(t-4)^k} \right| = \lim_{k\to\infty} \left| \sqrt{\frac{k}{k+1}}(t-4) \right| = |t-4|.$$

The Ratio Test guarantees that this series converges absolutely when $|t-4| < 1$, which happens when $t \in (3,5)$; but our limit value is 1 when $t = 3$ and $t = 5$, so the Ratio Test is inconclusive there. We have to check the convergence at these values of t by other means. When $t = 5$, we see

$$f(5) = \sum_{k=0}^{\infty} \frac{1}{\sqrt{k}},$$

which diverges (it's a p-series with $p < 1$). But when $t = 3$ we have

$$f(3) = \sum_{k=0}^{\infty} \frac{(-1)^k}{\sqrt{k}},$$

which converges because it's alternating, and the summands are monotonically decreasing in size, approaching zero in the limit. Therefore, the interval of convergence is $[3,5)$, and the radius of convergence is 1 (the distance from the expansion point to an endpoint of the interval). ∎

The interval of convergence

Example 6.18. Determine the interval of convergence for $f(t) = \sum_{k=0}^{\infty} \frac{2^k}{k^3}(t-1)^k$.

Solution: As in Example 6.17, we use the Ratio Test to establish the radius of convergence:

$$\lim_{k\to\infty} \left| \frac{a_{k+1}}{a_k} \right| = \lim_{k\to\infty} \left| \frac{\frac{2^{k+1}}{(k+1)^3}(t-1)^{k+1}}{\frac{2^k}{k^3}(t-1)^k} \right| = \lim_{k\to\infty} \left| \frac{2k^3}{(k+1)^3}(t-1) \right| = 2|t-1|.$$

The Ratio Test guarantees that the series converges absolutely when this limit is less than one, and

$$2|t-1| < 1 \quad \Rightarrow \quad |t-1| < \frac{1}{2} \quad \Rightarrow \quad t \in \left(\frac{1}{2}, \frac{3}{2}\right).$$

However, the Ratio Test is inconclusive at the endpoints of the interval, where $2|t-1| = 1$, so we check them separately:

$$f\left(\frac{1}{2}\right) = \sum_{k=0}^{\infty} \frac{(-1)^k}{k^3} \quad \text{and} \quad f\left(\frac{3}{2}\right) = \sum_{k=0}^{\infty} \frac{1}{k^3},$$

both of which converge. So the interval of convergence is $\left[\frac{1}{2}, \frac{3}{2}\right]$ and the radius of convergence is $1/2$. ∎

Try It Yourself: The interval of convergence

Example 6.19. Determine (a) the radius of convergence, and (b) the interval of convergence for $f(t) = \sum_{k=0}^{\infty} \frac{(-0.1)^k}{k^{1/3}}(t+7)^k$

Answer: (a) the radius is 10, and (b) the interval is $(-17, 3]$. ∎

Full Solution
On-line

When a power series converges at all values of t, we say that its **radius of convergence is infinite**.

An infinite interval of convergence

Example 6.20. Show that $f(t) = \sum_{k=0}^{\infty} \frac{1}{k!}t^k$ has an infinite radius of convergence.

Solution: This follows directly from the Ratio Test since

$$\lim_{k \to \infty} \left| \frac{a_{k+1}}{a_k} \right| = \lim_{k \to \infty} \left| \frac{\frac{1}{(k+1)!}t^{k+1}}{\frac{1}{k!}t^k} \right| = \lim_{k \to \infty} \left| \frac{t}{k+1} \right| = 0.$$

Since the series converges absolutely at all t, its radius of convergence is infinite. ∎

At the other end of the spectrum, a power series might converge at *only* its expansion point, in which case we say that its radius of convergence is zero.

A trivial interval of convergence

Example 6.21. Show that $f(t) = \sum_{k=0}^{\infty} k!t^k$ converges *only* when $t = 0$.

Solution: Using the Ratio Test, we see that $\lim_{k \to \infty} \left| \frac{a_{k+1}}{a_k} \right| = \lim_{k \to \infty} k|t|$, which exists only when $t = 0$. ∎

The following theorem summarizes what we've learned about the radius of convergence.

Theorem: When a power series is expanded about t_0, either

1. there is a positive number R such that the series converges absolutely when $|t - t_0| < R$ and diverges when $|t - t_0| > R$, or

2. the power series converges for all t (in which case we say that $R = \infty$), or

3. the power series converges only at t_0 (in which case we say that $R = 0$).

You should know

- the terms *closed form, power series, expansion point, Taylor series, Maclaurin series, radius of convergence,* and *interval of convergence;*

- inside the radius of convergence, a geometric power series and its associated closed form are just two different ways of writing the same function;

- when a power series is not geometric, we use the Ratio Test to determine its radius of convergence.

You should be able to

- use the Ratio Test to determine the radius of convergence of a power series;

- determine the interval of convergence of a power series;

- rewrite a power series representation of a function in closed form;

- rewrite the closed form representation of a function as a power series.

❖ 8.6 Skill Exercises

In #1–8 determine a power series representation of the function about $t_0 = 0$.

1. $f(t) = \dfrac{1}{1 - 6t}$

3. $f(t) = \dfrac{3}{5 + 9t}$

5. $f(t) = \dfrac{8t}{4 - 3t}$

7. $f(t) = \dfrac{5t^3}{3 + 7t^2}$

2. $f(t) = \dfrac{2}{1 + 8t}$

4. $f(t) = \dfrac{4}{2 - 11t}$

6. $f(t) = \dfrac{4t}{6 + 13t}$

8. $f(t) = \dfrac{at^\alpha}{b + ct^\beta}$

In #9–14 determine (a) a power series representation of $f(t)$ about the specified t_0, and (b) the interval in which the series converges.

9. $f(t) = \dfrac{3}{4 - t}$; $t_0 = 1$

11. $f(t) = \dfrac{7}{5 + 8t}$; $t_0 = 3$

13. $f(t) = \dfrac{4t}{1 - t}$; $t_0 = 2$

10. $f(t) = \dfrac{3}{4 - t}$; $t_0 = -2$

12. $f(t) = \dfrac{2}{5 + 6t}$; $t_0 = 11$

14. $f(t) = \dfrac{8t}{2 + 5t}$; $t_0 = -1$

In #15–18 determine (a) a closed form for the (geometric) power series that's given, and (b) the value of the series at the specified value of t_0.

15. $\displaystyle\sum_{k=3}^{\infty} 4^k t^k$, $t_0 = \frac{1}{12}$

17. $\displaystyle\sum_{k=1}^{\infty} 8^k (t-3)^k$, $t_0 = 2.9$

16. $\displaystyle\sum_{k=2}^{\infty} \dfrac{2^{k-1}}{3^k} t^k$, $t_0 = 1$

18. $\displaystyle\sum_{k=2}^{\infty} \dfrac{7^{2k+1}}{11^k} (t+2)^k$, $t_0 = 2.1$

In #19–22 (a) use a partial fractions decomposition to rewrite $f(t)$, (b) find a power series expanded at $t = t_0$ for each term in the PFD and add them together to get a series expansion for f, then (c) determine the interval of convergence for that series. (Hint: each of the constituent series must converge.)

No re-indexing of series is necessary in these exercises.

19. $f(t) = \dfrac{1}{t^2 + 2t - 3}$; $t_0 = 5$

21. $f(t) = \dfrac{1}{t^3 + 7t^2 + 12t}$; $t_0 = -1$

20. $f(t) = \dfrac{1}{t^2 + 2t - 3}$; $t_0 = -4$

22. $f(t) = \dfrac{1}{t^3 + 7t^2 + 12t}$; $t_0 = 2$

In #23–30 (a) use the Ratio Test to determine the radius of convergence, and (b) determine the interval of convergence.

23. $\displaystyle\sum_{k=1}^{\infty} \dfrac{t^k}{2^{k+1}}$

25. $\displaystyle\sum_{k=1}^{\infty} \dfrac{(-3)^k}{k} t^k$

27. $\displaystyle\sum_{k=1}^{\infty} \dfrac{(t-3)^k}{k^2 + 11k}$

29. $\displaystyle\sum_{k=1}^{\infty} \dfrac{\sqrt{k}}{8^k + 3^k} (t+1)^{2k}$

24. $\displaystyle\sum_{k=1}^{\infty} \dfrac{k2^{k+1}}{9^{k-1}} t^k$

26. $\displaystyle\sum_{k=1}^{\infty} \dfrac{7^k}{3^k \sqrt{k}} t^k$

28. $\displaystyle\sum_{k=1}^{\infty} \dfrac{(6t-6)^k}{7k^4 + 3k}$

30. $\displaystyle\sum_{k=1}^{\infty} \dfrac{5^{2k-1} k}{4^{3k+2}} (t+5)^{3k}$

❖ 8.6 Concept and Application Exercises

31. Suppose that $f(x) = \sum_{k=0}^{\infty} a_k x^k$ and $f(4)$ converges. Can you make any conclusions about $f(3)$ or $f(5)$? Explain why or why not.

32. Suppose that $f(x) = \sum_{k=0}^{\infty} a_k x^k$ and $f(4)$ converges. Can you make any conclusions about $f(-3)$ or $f(-5)$? Explain why or why not.

In #33–38 design a power series whose interval of convergence is specified.

33. $(-3, 3)$ 35. $[10, 13]$ 37. $[-5, 5)$ 39. $(15, 16)$

34. $[-3, 3]$ 36. $(10, 13)$ 38. $(-5, 5]$ 40. $[-22, 4)$

41. In Section 8.7 (on p. 661) we'll show that the power series $\sum_{k=0}^{\infty} \frac{(-1)^k}{2k+1} t^{2k+1}$ is the same as $\arctan(t)$ where it converges.

    (a) Determine the expansion point of this power series.

    (b) Use a graphing utility to render the graph of $\sum_{k=0}^{n} \frac{(-1)^k}{2k+1} t^{2k+1}$ for $n = 5, 10, 15$, and 20.

    (c) Based on these graphs, make a conjecture about the interval of convergence for this series.

    (d) Use the Ratio Test to determine the radius of convergence for the series.

    (e) Determine the interval of convergence, and check it against your prediction from part (c).

42. In Section 8.7 (on p. 661) we'll show that the power series $\sum_{k=1}^{\infty} \frac{(-1)^{k+1}}{k} (t-1)^k$ is the same as $\ln(t)$ where it converges.

    (a) Determine the expansion point of this power series.

    (b) Use a graphing utility to render the graph of $\sum_{k=1}^{n} \frac{(-1)^{k+1}}{k} (t-1)^k$ for $n = 5, 10, 15$, and 20.

    (c) Based on these graphs, make a conjecture about the interval of convergence for this series.

    (d) Use the Ratio Test to determine the radius of convergence for the series.

    (e) Determine the interval of convergence, and check it against your prediction from part (c).

43. Suppose f is defined to be the power series $f(t) = \sum_{k=0}^{\infty} \frac{t^{2k}}{(2k)!}$, where $m!$ is the product $m(m-1)(m-2)\cdots(3)(2)(1)$.

    (a) Use a graphing utility to render graphs of the polynomials $\sum_{k=0}^{n} \frac{(-1)^k}{(2k)!} t^{2k}$ over $[-20, 20]$ when $n = 5, 10, 15, 20, 25$.

    (b) Based on these graphs, make a conjecture about what familiar function is approached in the limit (i.e., the number $f(t)$ is really ...).

44. Suppose f is defined to be the power series $f(t) = \sum_{k=0}^{\infty} \frac{(-1)^k}{(2k+1)!} t^{2k+1}$, where $m!$ is the product $m(m-1)(m-2)\cdots(3)(2)(1)$.

    (a) Use a graphing utility to render graphs of the polynomials $\sum_{k=0}^{n} \frac{t^{2k+1}}{(2k+1)!}$ over $[-20, 20]$ when $n = 5, 10, 15, 20, 25$.

    (b) Based on these graphs, make a conjecture about what familiar function is approached in the limit. (i.e., the number $f(t)$ is really ...)

In the exercise set for Section 8.1 you worked with recursively defined sequences. With such a definition, calculating a_{100} requires us to know a_{99}, which requires us to know a_{98}, which requires us to know $a_{97}, a_{96}, \ldots, a_1, a_0$. A closed form for calculating a_k would be much faster, and #45–48 give you practice finding one by using something called a *generating function*, which is written as a power series.

45. In #32 on p. 596 you showed that the number of "legitimate" code words of length k, denoted by a_k, is described by $a_{k+1} = 4^k + 2a_k$. Suppose we define $g(x) = \sum_{k=0} a_k x^k$, where $a_0 = 1$ and $a_{k+1} = 4^k + 2a_k$.

[Computer Science]

(a) Verify that $g(x) = 1 + \sum_{k=0} a_{k+1} x^{k+1}$.

(b) Use the fact that $a_{k+1} = 4^k + 2a_k$ to show that $g(x) = 1 + 2xg(x) + \frac{1}{1-4x}$.

(c) By solving that equation for $g(x)$, verify that $g(x) = \frac{1}{1-2x} + \frac{x}{(1-2x)(1-4x)}$.

(d) Use a partial fractions decomposition to find numbers α and β for which $g(x) = \frac{\alpha}{1-2x} + \frac{\beta}{1-4x}$.

(e) After rewriting these rational functions as geometric series, add them together to get a Maclaurin series for $g(x)$.

(f) Based on the Maclaurin series in part (e), write a closed form for a_k (which is the coefficient of x^k in the Maclaurin series).

46. Suppose we send messages using only two signals, $\{a, b\}$. Signal a takes one unit of time, and signal b takes two units of time. We will denote by w_k the number of different code words whose transmission takes at most k units of time. In #53 on p. 600 you derived a recurrence relation for w_k.

[Computer Science]

(a) Set $f(x) = \sum_{k=0} w_k x^k$ where $w_0 = 1$, $w_1 = 1$ and w_k satisfies the recurrence relation from #53 on p. 600, and verify that $f(x) = 1 + x + \sum_{k=0} w_{k+2} x^{k+2}$

(b) Use the recurrence relation to express w_{k+2} in terms of previous coefficients, and solve the resulting equation for $f(x)$ as we did in parts (b) and (c) of #45.

(c) Use a partial fractions decomposition to rewrite $f(x)$, and convert the summands into geometric series, as in parts (d) and (e) of #45.

(d) Extract a closed form expression for w_k from the series in part (c).

47. In #48 on p. 599 you denoted by c_k the concentration of a gas in your lung immediately following the kth breath and, under the assumption that none of it was absorbed, showed that $c_{k+1} = 0.92c_k + 0.08\gamma$ where γ is the concentration of the gas in the local atmosphere. Given that $c_0 = 0$, use the technique modeled in #45 to derive a closed form for calculating c_k. Your answer will be in terms of γ. (*Hint: assuming that $|x| < 1$ when working with the generating function will allow you to use what you know about geometric series.*)

[Biology]

48. In #50 on p. 599 you denoted by c_k the concentration of the gas in your lung immediately following the kth breath and, under the assumption that $1/500^{\text{th}}$ of the gas is absorbed during each cycle of respiration, showed that $c_{k+1} = 0.9116c_k + 0.0865\gamma$ where γ is the concentration of the gas in the local atmosphere. Given that $c_0 = 0$, use the technique modeled in #45 to derive a closed form for calculating c_k. Your answer will be in terms of γ. (*Hint: assuming that $|x| < 1$ when working with the generating function will allow you to use what you know about geometric series.*)

[Biology]

8.7   Elementary Operations with Power Series

In Section 8.5 we saw that absolutely convergent series behave like finite sums in the sense that the order of addition is irrelevant. Then in Section 8.6 the Ratio Test showed us that a power series converges absolutely inside its radius of convergence. Consequently, the order in which we add the summands of a power series is irrelevant inside its interval of convergence. In that sense, it's like a polynomial. In this section, we discuss other ways that power series behave like polynomials inside their intervals of convergence. In particular, we study the differentiation, integration, multiplication and division of power series.

❖ Differentiation and Integration of Power Series

As indicated above, a power series behaves like a polynomial inside its interval of convergence. So in the same way that we differentiate and integrate polynomials term-by-term, you might expect that we can differentiate and integrate power series term-by-term. That's exactly right, as stated formally below.

Differentiation and Integration of Power Series: Suppose the radius of convergence of $f(t) = \sum_{k=0}^{\infty} a_k (t - t_0)^k$ is $R > 0$ (including the possibility of $R = \infty$). Then

$$f'(t) = \sum_{k=1}^{\infty} k \, a_k (t - t_0)^{k-1} \quad \text{and} \quad \int f(t) \, dt = C + \sum_{k=0}^{n} \frac{a_k}{k + 1} (t - t_0)^{k+1}$$

when $|t - t_0| < R$, and the radius of convergence is R for both series.

Calculating the first derivative of a Maclaurin series

Example 7.1. Suppose $f(t) = \sum_{k=0}^{\infty} (k + 1)^2 t^k$, whose radius of convergence is $R = 1$. Determine $f'(0)$.

Solution: The function f is

$$f(t) = \sum_{k=0}^{\infty} (k + 1)^2 t^k = \underbrace{1}_{k=0} + \underbrace{4t}_{k=1} + \underbrace{9t^2}_{k=2} + \underbrace{16t^3}_{k=3} + \cdots .$$

When we differentiate this series directly, we have

$$f'(t) = \sum_{k=1}^{\infty} k(k + 1)^2 t^{k-1} = \underbrace{4}_{k=1} + \underbrace{18t}_{k=2} + \underbrace{48t^2}_{k=3} + \cdots ,$$

from which we conclude that $f'(0) = 4$. ∎

Try It Yourself: Calculating the second derivative of a Maclaurin series

Example 7.2. Suppose $f(t) = \sum_{k=0}^{\infty} \frac{1}{(k+1)6^k} t^k$, whose radius of convergence is $R = 6$. Determine $f''(0)$.

Answer: $f''(0) = 1/54$. ∎

Full Solution On-line

Using the first derivative of a Taylor series

Example 7.3. Suppose $f(t) = \sum_{k=0}^{\infty} (-1)^k \frac{3}{5^k} (t - 2)^k$. Determine whether f is increasing or decreasing at $t = 2$.

Solution: The series that defines $f(t)$ has a positive radius of convergence (specifically, $R = 5$), and we can differentiate directly within that distance of the expansion point:

$$f'(t) = \sum_{k=1}^{\infty}(-1)^k\frac{3k}{5^k}(t-2)^{k-1} = \underbrace{-\frac{3}{5}}_{k=1} + \underbrace{\frac{6}{25}(t-2)}_{k=2} + \underbrace{\frac{9}{125}(t-2)^2}_{k=3} + \cdots,$$

from which we conclude that $f'(2) = -3/5$. Since this is negative, $f(t)$ is decreasing at $t = 2$. ∎

Try It Yourself: Using the second derivative of a Taylor series

Example 7.4. Suppose $f(t) = \sum_{k=0}^{\infty}(-1)^k\frac{3}{5^k}(t-2)^k$, whose radius of convergence is $R = 5$. Determine whether the graph of f is concave up at $t = 2$.

Full Solution On-line

Answer: Yes, the graph is concave up. ∎

Calculating the integral of a Maclaurin series

Example 7.5. Derive a Maclaurin series for $\tan^{-1}(t)$ and calculate its radius of convergence.

Solution: We know that

$$\frac{d}{dt}\tan^{-1}(t) = \frac{1}{1+t^2} = \frac{1}{1-(-t^2)} = \sum_{k=0}^{\infty}(-1)^k t^{2k}$$

when $|t| < 1$. So

$$\tan^{-1}(t) = \int \sum_{k=0}^{\infty}(-1)^k t^{2k}\, dt = C + \sum_{k=0}^{\infty}\frac{(-1)^k}{2k+1}t^{2k+1}$$

for some value of C. In fact, $C = \tan^{-1}(0) = 0$. Since integration doesn't change the radius of convergence (mentioned above), the radius of convergence for this series is also 1. ∎

Try It Yourself: Calculating the integral of a Maclaurin series

Example 7.6. Derive a Maclaurin series for $\int \tan^{-1}(t)\, dt$ and calculate its radius of convergence.

Full Solution On-line

Answer: $C + \sum_{k=0}^{\infty}\frac{(-1)^k}{(2k+1)(2k+2)}t^{2k+2}$, with $R = 1$. ∎

Calculating the integral of a Taylor series

Example 7.7. Derive a Taylor series for $\ln(t)$ about $t_0 = 1$, and calculate its radius of convergence.

Solution: We know that

$$\frac{d}{dt}\ln(t) = \frac{1}{t} = \frac{1}{1+(t-1)} = \frac{1}{1-\left(-(t-1)\right)} = \sum_{k=0}^{\infty}(-1)^k(t-1)^k$$

when $|t-1| < 1$. So

$$\ln(t) = \int \sum_{k=0}^{\infty}(-1)^k(t-1)^k\, dt = C + \sum_{k=0}^{\infty}\frac{(-1)^k}{k+1}(t-1)^{k+1}$$

for some value of C. In fact, $C = \ln(1) = 0$. Since integration doesn't change the radius of convergence (mentioned above), the radius of convergence for this series is also 1. ∎

❖ Multiplication and Division of Power Series

We can multiply and divide power series in the same way that we do polynomials. Let's take a moment to work with polynomials so that we recall the method. The product of

$$
\begin{aligned}
f(t) &= a_0 + a_1 t + a_2 t^2 + a_3 t^3 \text{ and} \\
g(t) &= b_0 + b_1 t + b_2 t^2 + b_3 t^3
\end{aligned}
$$

is found by collecting like powers of t. For example, the only way to get a constant (i.e., no powers of t) is to multiply $a_0 b_0$, so that's the constant term of the product:

$$ f(t)g(t) = a_0 b_0 + \cdots . $$

Similarly, we can produce a term with a single power of t by multiplying either a_0 and $b_1 t$, or b_0 and $a_1 t$. So

$$ f(t)g(t) = a_0 b_0 + (a_0 b_1 + a_1 b_0)t + \cdots . $$

We can produce a term with t^2 by multiplying either the two linear terms or a constant and quadratic term. So

$$ f(t)g(t) = a_0 b_0 + (a_0 b_1 + a_1 b_0)t + (a_0 b_2 + a_1 b_1 + a_2 b_0)t^2 + \cdots . $$

And lastly, we produce terms with t^3 by multiplying either a constant and a cubic term, or a linear and a quadratic term. So

$$ f(t)g(t) = a_0 b_0 + (a_0 b_1 + a_1 b_0)t + (a_0 b_2 + a_1 b_1 + a_2 b_0)t^2 + (a_0 b_3 + a_1 b_2 + a_2 b_1 + a_3 b_0)t^3. $$

Notice that the subscripts of the coefficients always add to the power of t. This is exactly what happens with power series.

Multiplying Power Series: Suppose $R > 0$, and $f(t) = \sum_{k=0}^{\infty} a_k (t - t_0)^k$ and $g(t) = \sum_{k=0}^{\infty} b_k (t - t_0)^k$ both converge absolutely when $|t - t_0| < R$. Then

$$ f(t)g(t) = \sum_{k=0}^{\infty} c_k (t - t_0)^k $$

when $|t - t_0| < R$, where

$$ c_k = \sum_{n=0}^{k} a_n b_{k-n}. $$

Multiplication of power series

Example 7.8. Suppose $f(t) = \frac{1}{1-t}$ and $g(t) = \frac{1}{1+t}$. Find a Maclaurin series for $f(t)g(t)$ that's valid when $|t| < 1$.

Solution: Let's write the Maclaurin series for each function (valid when $|t| < 1$):

$$
\begin{aligned}
f(t) &= 1 + 1t + 1t^2 + 1t^3 + 1t^4 + \cdots \\
g(t) &= 1 - 1t + 1t^2 - 1t^3 + 1t^4 - \cdots
\end{aligned}
$$

When we multiply these power series, we collect terms according to how many powers of t are in them:

$$
\begin{aligned}
f(t)g(t) = & \overbrace{(1)(1)}^{\text{constants}} + \overbrace{(1t)(1) + (1)(-1t)}^{\text{linear}} + \overbrace{(1t^2)(1) + (1t)(-1t) + (1)(1t^2)}^{\text{quadratic}} \\
& + \underbrace{(1t^3)(1) + (1t^2)(-1t) + (1t)(1t^2) + (1)(-1t^3)}_{\text{cubic}} + \cdots = 1 + t^2 + t^4 + \cdots .
\end{aligned}
$$

This pattern, in which odd powers of t have a coefficient of 0 and even powers of t have a coefficient of 1, has to do with the number of negative terms contributed by the power series for $g(t)$. We can validate our calculation by doing it differently. In this case, because f and g are so simple, we can multiply their closed forms together and then pass to a power series:

$$f(t)g(t) = \left(\frac{1}{1-t}\right)\left(\frac{1}{1+t}\right) = \overbrace{\left(\frac{1}{1-t^2}\right) = \sum_{k=0}^{\infty}(t^2)^k}^{\text{using the formula for geometric series with a common ratio of } t^2} = 1 + t^2 + t^4 + t^6 + \cdots \quad \blacksquare$$

In the same way that you first learned elementary division facts in terms of multiplication, you can understand the division of power series in terms of multiplication. For example, based on the results of Example 7.8 we can say that

$$\frac{\sum_{k=0}^{\infty} t^{2k}}{\sum_{k=0}^{\infty} t^k} = \sum_{k=0}^{\infty}(-1)^k t^k \quad \text{because} \quad \sum_{k=0}^{\infty} t^{2k} = \left(\sum_{k=0}^{\infty} t^k\right)\left(\sum_{k=0}^{\infty}(-1)^k t^k\right).$$

This leads us directly to an algorithm for calculating quotients of power series.

| When you first learned that $\frac{12}{4} = 3$, you understood it based on the knowledge that $12 = (4)(3)$. |

Division of power series

Example 7.9. Calculate the first four summands of the Maclaurin series for $f(t)/g(t)$, where f and g are defined as in Example 7.8.

Solution: The Maclaurin series for $f(t)/g(t)$ has the form $\sum_{k=0}^{\infty} a_k t^k$, so

$$\frac{f(t)}{g(t)} = \sum_{k=0}^{\infty} a_k t^k \quad \Rightarrow \quad f(t) = g(t)\left(\sum_{k=0}^{\infty} a_k t^k\right).$$

That is,

$$1 + t + t^2 + t^3 + t^4 + t^5 + \cdots = \left(1 - t + t^2 - t^3 + t^4 - t^5 + \cdots\right)\left(a_0 + a_1 t + a_2 t^2 + a_3 t^3 + a_4 t^5 + \cdots\right)$$

Now we find the product on the right-hand side, and compare it to what we see on the left-hand side:

- The constant term on the right-hand side is $(1)(a_0)$, and the constant term on the left is 1. These have to be the same, so $1 = a_0$.

- The linear term on the right-hand side is $(a_1 - a_0)t$. The linear term on the left-hand side of the equation is $1t$. Since these have to be the same, $a_1 - a_0 = 1$. We already know the value of a_0, so

$$a_1 - a_0 = 1 \quad \Rightarrow \quad a_1 = 1 + a_0 = 2.$$

- The quadratic term on the right-hand side is $(a_2 - a_1 + a_0)t^2$. The quadratic term on the left-hand side is $1t^2$. Since these have to be the same, $a_2 - a_1 + a_0 = 1$. We already know a_0 and a_1, so this allows us to solve for a_2.

$$a_2 - a_1 + a_0 = 1 \quad \Rightarrow \quad a_2 = 1 + a_1 - a_0 = 2.$$

- The cubic term on the right-hand side is $(a_3 - a_2 + a_1 - a_0)t^3$. The cubic term on the left-hand side is $1t^3$, so

$$a_3 - a_2 + a_1 - a_0 = 1 \quad \Rightarrow \quad a_3 = 1 + a_2 - a_1 + a_0 = 2.$$

We could continue, but we're only out to find the first four terms of the power series, and we have. So let's write

$$\frac{f(t)}{g(t)} = 1 + 2t + 2t^2 + 2t^3 + \cdots .$$

Again, because $f(t)$ and $g(t)$ are such simple functions, we can validate this work by calculating in a different manner and seeing that we get the same result:

$$
\begin{aligned}
\frac{f(t)}{g(t)} &= \frac{1+t}{1-t} = \frac{1}{1-t} + \frac{t}{1-t} = \sum_{k=0}^{\infty} t^k + t \sum_{k=0}^{\infty} t^k \\
&= \left(1 + t + t^2 + t^3 + t^4 + \cdots \right) + t \left(1 + t + t^2 + t^3 + t^4 + \cdots \right) \\
&= \left(1 + t + t^2 + t^3 + t^4 + \cdots \right) + \left(t + t^2 + t^3 + t^4 + t^5 + \cdots \right) \\
&= 1 + 2t + 2t^2 + 2t^3 + 2t^4 + \cdots ,
\end{aligned}
$$

which explains why we saw a string of 2s in our calculation. ∎

You should know

- that you can differentiate and integrate a power series in a direct fashion within its radius of convergence;

- that $f'(t)$ and $\int f(t)\, dt$ have the same radius of convergence as $f(t)$;

- that we multiply power series in the same way that we multiply polynomials;

- that the division of power series can be accomplished by calculating a product.

You should be able to

- differentiate and integrate power series;

- multiply and divide power series.

❖ 8.7 Skill Exercises

In #1–4 (a) write out the first five terms of each power series, and (b) determine the first five terms of their product.

1. $\displaystyle\sum_{k=1}^{\infty} kt^k$ and $\displaystyle\sum_{k=1}^{\infty} \frac{t^k}{k}$

3. $\displaystyle\sum_{k=3}^{\infty} (2-k)t^k$ and $\displaystyle\sum_{k=2}^{\infty} (2+k)t^k$

2. $\displaystyle\sum_{k=1}^{\infty} (2t)^k$ and $\displaystyle\sum_{k=1}^{\infty} \frac{(-t)^k}{k}$

4. $\displaystyle\sum_{k=3}^{\infty} \frac{1}{k+4} t^k$ and $\displaystyle\sum_{k=1}^{\infty} 5^k t^k$

In #5–8 (a) write power series for $f(t)$ and $g(t)$ expanded about $t = 0$, and (b) write the first three terms in the series expansion of their product, $f(t)g(t)$.

5. $f(t) = \dfrac{1}{1-t}$, $g(t) = \dfrac{3}{4-t}$ 7. $f(t) = \dfrac{t}{3+t^2}$, $g(t) = \dfrac{t^2}{15+t}$

6. $f(t) = \dfrac{1}{2+t}$, $g(t) = \dfrac{4}{5+t}$ 8. $f(t) = \dfrac{t^3}{12+t^2}$, $g(t) = \dfrac{3t^2}{7-t^4}$

9. Determine the first four terms in the power series of $f(t) = \frac{1}{t^2+7t+12}$ about $t = 0$. (*Hint: factor the denominator and treat $f(t)$ as a product.*)

10. Determine the first four terms in the power series of $f(t) = \frac{1}{3t^2-10t-8}$ about $t = 0$. (*Hint: factor the denominator and treat $f(t)$ as a product.*)

In #11–14 (a) write out the first five terms of each power series, and (b) determine the first three terms of the quotient (first series)/(second series).

11. $\displaystyle\sum_{k=1}^{\infty} kt^k$ and $\displaystyle\sum_{k=1}^{\infty} \dfrac{t^k}{k}$ 13. $\displaystyle\sum_{k=3}^{\infty} (2-k)t^k$ and $\displaystyle\sum_{k=2}^{\infty} (2+k)t^k$

12. $\displaystyle\sum_{k=1}^{\infty} (2t)^k$ and $\displaystyle\sum_{k=1}^{\infty} \dfrac{(-t)^k}{k}$ 14. $\displaystyle\sum_{k=3}^{\infty} \dfrac{1}{k+4}t^k$ and $\displaystyle\sum_{k=1}^{\infty} 5^k t^k$

In #15–18 (a) write power series for $f(t)$ and $g(t)$ expanded about $t = 0$, and (b) write the first three terms in the series expansion of the quotient $f(t)/g(t)$.

15. $f(t) = \dfrac{1}{1-t}$, $g(t) = \dfrac{3}{4-t}$ 17. $f(t) = \dfrac{t^2}{3+t^2}$, $g(t) = \dfrac{t}{15+t}$

16. $f(t) = \dfrac{1}{2+t}$, $g(t) = \dfrac{4}{5+t}$ 18. $f(t) = \dfrac{t^3}{12+t^2}$, $g(t) = \dfrac{3t^2}{7-t^4}$

In Section 8.8 we'll develop power series that allow us to calculate the values of sine and cosine. Specifically,

$$\sin(t) = \sum_{k=0}^{\infty} \dfrac{(-1)^k}{(2k+1)!} t^{2k+1} \quad \text{and} \quad \cos(t) = \sum_{k=0}^{\infty} \dfrac{(-1)^k}{(2k)!} t^{2k},$$

where $m!$ denotes the product $m(m-1)(m-2)\cdots(2)(1)$. Use division to determine the first four terms of the series (expanded at $t_0 = 0$) for the functions in #19–22.

19. $\sec(t)$ 20. $\csc(t)$ 21. $\tan(t)$ 22. $\cot(t)$

In #23–26 (a) determine a power series for f about t_0, (b) differentiate that power series to determine a power series for $f'(t)$, and (c) use the Ratio Test to verify that the series for $f(t)$ and $f'(t)$ have the same radius of convergence.

23. $f(t) = \dfrac{3}{2+t}$; $t_0 = 0$ 25. $f(t) = \dfrac{9}{3-t}$; $t_0 = 5$

24. $f(t) = \dfrac{8t}{3-t}$; $t_0 = 0$ 26. $f(t) = \dfrac{11}{2+t}$; $t_0 = 6$

27. Verify that differentiating the series cited above for sine results in the series cited above for cosine.

28. Consider the function $f(t) = \sum_{k=0}^{\infty} \frac{1}{k!} t^k$.

(a) Show that the interval of convergence for f is $(-\infty, \infty)$.

(b) Determine a power series for $f'(t)$, and compare it to the power series for $f(t)$. *(Hint: try writing out the first several terms of each.)*

(c) Based on what you saw in part (b), what familiar function is $f(t)$?

29. Suppose $f(t) = \sum_{k=0}^{\infty} \frac{(-1)^k}{7-4k} t^k$.

    (a) Determine whether f is increasing or decreasing at $t = 0$.

    (b) Determine whether the graph of f is concave up or concave down at $t = 0$.

30. Suppose $f(t) = \sum_{k=0}^{\infty} \frac{1}{k^2+k-3} t^k$.

    (a) Determine whether f is increasing or decreasing at $t = 0$.

    (b) Determine whether the graph of f is concave up or concave down at $t = 0$.

In #31-34 determine the integral as a power series.

31. $\displaystyle\int \sum_{k=1}^{\infty} \frac{3\sqrt{k+1}}{k!} t^k \, dt$

32. $\displaystyle\int \sum_{k=2}^{\infty} \frac{(-1)^k}{k^2+1} t^k \, dt$

33. $\displaystyle\int_0^1 \sum_{k=1}^{\infty} k3^{-k} t^{k-1} \, dt$

34. $\displaystyle\int_1^2 \sum_{k=1}^{\infty} (0.3)^{k-1} t^k \, dt$

35. Suppose $f(t) = \frac{t^2}{t^5+2}$.

    (a) Find a Maclaurin series representation of f. *(Hint: use geometric series techniques.)*

    (b) Determine the radius of convergence, R, for the series from part (a).

    (c) Determine a formula for an antiderivative of $f(t)$ in $(-R, R)$.

36. Repeat #35 using $f(t) = \frac{t}{t^7-3}$.

✤ 8.7 Concept and Application Exercises

37. Suppose $f(x) = \sum_{k=0}^{\infty} a_k x^k$ and $f(4)$ converges. Can you make any conclusions about $f'(-5)$, $f'(-3.7)$ or $f'(1.2)$?

38. We can find a closed form for $\sum_{k=1}^{\infty} \frac{1}{k+1}(x+1)^k$ by relating it to $\sum_{k=1}^{\infty}(x+1)^k$ by way of integration. Consider the following:

$$\sum_{k=1}^{\infty} \frac{1}{k+1}(x+1)^k = \frac{1}{x+1}\sum_{k=1}^{\infty}\frac{1}{k+1}(x+1)^{k+1} = \frac{1}{x+1}\int\sum_{k=1}^{\infty}(x+1)^k \, dx$$

$$= \frac{1}{x+1}\int\frac{x+1}{1-(x+1)} \, dx = \frac{1}{x+1}\int -1-\frac{1}{x} \, dx$$

$$= -\frac{1}{x+1}\left(x+\ln|x|\right).$$

If the left-hand side and right-hand side of this equation were really the same function, they would have the same function value at every value of x inside the series' radius of convergence. However, when we check at $x = -1$ the series yields a value of zero while the closed form is undefined. There must be an error somewhere. Find it.

39. Suppose the radius of convergence of $f(t) = \sum_{k=0}^{\infty} a_k(t-2)^k$ is $R = 6$. Determine whether the left-hand image in Figure 7.1 could be a clipping from the graph of f. Whether your answer is "yes" or "no," identify aspects of the curve that lead you to your conclusion.

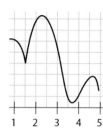 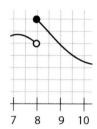

Figure 7.1: (left) Image for #39; (right) image for #40.

40. Suppose the radius of convergence of $f(t) = \sum_{k=0}^{\infty} a_k(t-9)^k$ is $R = 4$. Determine whether the curve in the right-hand image of Figure 7.1 could be a clipping from the graph of f. Whether your answer is "yes" or "no," identify aspects of the curve that lead you to your conclusion.

In #41–46 you see a power series for a function that's evaluated at t_0. (a) Determine a closed form for the function, and (b) use it to determine the value of the series.

41. $\displaystyle\sum_{k=0}^{\infty} \frac{1}{3^k}$, $t_0 = \frac{1}{3}$

42. $\displaystyle\sum_{k=0}^{\infty} \frac{5^k}{27^k}$, $t_0 = \frac{5}{9}$

43. $\displaystyle\sum_{k=0}^{\infty} \frac{k}{2^k}$, $t_0 = \frac{1}{2}$

44. $\displaystyle\sum_{k=1}^{\infty} \frac{1}{k\,2^k}$, $t_0 = \frac{1}{2}$

45. $\displaystyle\sum_{k=0}^{\infty} \frac{1}{4^k(k+1)}$, $t_0 = \frac{1}{4}$

46. $\displaystyle\sum_{k=0}^{\infty} \frac{k}{4^k(k+1)}$, $t_0 = \frac{1}{4}$

47. Suppose that when an airborne rescuer glimpses in the direction of a lost skier (who's stationary, not skiing), the probability of detecting the skier is x, so the probability of *failing* to find the skier is $1 - x$ per glimpse. Of course, the rescuer will continue to look! Consider the situation in which the following three events occur together. (1) the rescuer misses the skier on the first glimpse *and* (2) misses on the second glimpse *and* (3) sees the skier on the third glimpse (yay!). If we treat these events as independent, the probability of all three happening together is $(1-x)(1-x)x$. More generally, the probability that the skier is first seen on the nth glimpse is $(1-x)^{n-1}x$. To determine how many times the rescuer should expect to glimpse in the skier's direction before seeing the skier, we calculate the weighted average

$$\sum_{n=1}^{\infty} \overbrace{n}^{\text{number of glimpses required to find the skier}} \underbrace{(1-x)^{n-1}x}_{\text{probability of that event}} = x\sum_{n=1}^{\infty} n(1-x)^{n-1}$$

By treating this sum as the derivative of a (familiar) power series, determine a closed form for this expression and answer the following question: if the probability of seeing the skier is 0.25, how many glimpses should we expect before locating the skier?

48. Suppose that two types of molecules, x and y, are in an aqueous solution. If these surfactants combine to make long tube-like structures whose end caps contain at least ℓ_0 molecules of species x, the number density of type x is given by

Physical Chemistry

overall number density of species x

$$X \;=\; \overbrace{X_1}^{} + \underbrace{X_1^{\ell_0}\mathcal{E}_e\left(\ell_0\sum_{k=0}^{\infty}(X_1\mathcal{E}_c)^k + \sum_{k=0}^{\infty}k(X_1\mathcal{E}_c)^k\right)}_{\text{number density of singletons of species } x}$$

where $\mathcal{E}_e$ and $\mathcal{E}_c$ are parameters that depend on the change in free energy due to variations in the number of molecules in the end caps and cylindrical wall of a tube, respectively. Assuming that $|X_1\mathcal{E}_c| < 1$ write X as a rational function of X_1.

49. Consider the integral $\int_0^1 \frac{1-x^n}{1-x}\,dx$, where $n > 1$ is an integer.

   (a) By writing the integrand as a power series, show that its value is $\sum_{k=1}^{\infty}\frac{1}{k(k+n)}$.

   (b) Determine the value of the integral (as a function of n) by calculating the value of this telescoping series.

Checking the details

50. Suppose R is the radius of convergence for $f(t) = \sum_{k=0}^{\infty} a_k(t-t_0)^k$.

   (a) Show that $\lim_{k\to\infty}\left|\frac{a_{k+1}}{a_k}\right|$ exists, and determine its value.

   (b) Verify that $\lim_{k\to\infty}\left|\frac{(k+1)\,a_{k+1}}{k\,a_k}\right|$ also exists, and has the same value.

   (c) What does this tell us about the series for $f'(t)$?

51. Suppose R is the radius of convergence for $f(t) = \sum_{k=0}^{\infty} a_k(t-t_0)^k$.

   (a) Show that $\lim_{k\to\infty}\left|\frac{a_{k+1}}{a_k}\right|$ exists, and determine its value.

   (b) Verify that $\lim_{k\to\infty}\left|\frac{a_{k+1}/(k+2)}{a_k/(k+1)}\right|$ also exists, and has the same value.

   (c) What does this tell us about the series for $\int f(t)\,dt$?

8.8   Taylor and Maclaurin Series

Modern computers are very fast, but at a computational level they can only add, multiply, subtract, and divide. So if we want a computer to calculate $\arctan(0.5)$, for example, we have to give it an algorithm that involves only those four arithmetic operations. The Maclaurin series for the arctangent that was developed on p. 661 (Example 7.5) using our knowledge of geometric series does exactly that. In this section you'll learn how to generate Maclaurin (and Taylor) series for functions that aren't related to the geometric series, such as $\sin(t)$ and e^t.

Let's begin by assuming that there *is* a Taylor series,

$$f(t) \;=\; a_0 + a_1(t - t_0) + a_2(t - t_0)^2 + a_3(t - t_0)^3 + a_4(t - t_0)^4 + a_5(t - t_0)^5 + \cdots .$$

Then for each t inside its radius of convergence, we have

$$f'(t) \;=\; a_1 + 2a_2(t - t_0) + 3a_3(t - t_0)^2 + 4a_4(t - t_0)^3 + \cdots$$

$$f''(t) \;=\; 2a_2 + (3)(2)a_3(t - t_0) + (4)(3)a_4(t - t_0)^2 + (5)(4)a_5(t - t_0)^3 + \cdots$$

$$f'''(t) \;=\; (3)(2)a_3 + (4)(3)(2)a_4(t - t_0) + (5)(4)(3)a_5(t - t_0)^2 + (6)(5)(4)(t - t_0)^3 + \cdots$$

$$f''''(t) \;=\; (4)(3)(2)a_4 + (5)(4)(3)(2)a_5(t - t_0) + (6)(5)(4)(3)(t - t_0)^2 + (7)(6)(5)(4)(t - t_0)^3 + \cdots$$

$$\vdots \qquad \vdots$$

Calculating these derivatives when $t = t_0$, we see that

$$f(t_0) = a_0, \quad f'(t_0) = (1!)a_1, \quad f''(t_0) = (2!)a_2, \quad f'''(t_0) = (3!)a_3, \quad f''''(t_0) = (4!)a_4, \; \ldots .$$

This leads us to conclude that the coefficients of the series are

$$a_k = \frac{f^{(k)}(t_0)}{k!},$$

where $f^{(k)}(t_0)$ denotes the kth derivative of f at t_0.

> $k = 0$ indicates that you should take *no* derivatives. That is, $f^{(0)}(t_0)$ is the same as $f(t_0)$.

Taylor and Maclaurin Series: Suppose f is a function with derivatives of all orders throughout an open interval that contains t_0. Then the **Taylor series** generated by f at t_0 is

$$\sum_{k=0}^{\infty} \frac{f^{(k)}(t_0)}{k!}\,(t - t_0)^k. \tag{8.1}$$

We refer to this series as the **Maclaurin series** generated by f when $t_0 = 0$.

In the remainder of this section, you will see several examples in which we generate power series using equation (8.1). We'll also introduce the subtle fact that the Taylor series generated by f *might not* tell us $f(t)$.

❖ Examples of Generating Maclaurin Series

Let's begin with a pair of familiar examples.

Generating a familiar Maclaurin series

Example 8.1. Suppose $f(t) = 3 + 5t + 8t^2$. Use equation (8.1) to generate a Maclaurin series from f.

Solution: Since $t_0 = 0$ in a Maclaurin series, we have $a_0 = f(0) = 3$. Then to use (8.1) we need to know the derivative of f at $t_0 = 0$, so let's begin by differentiating:

$$\begin{aligned} f'(t) &= 5 + 16t \\ f''(t) &= 16 \\ f'''(t) &= 0. \end{aligned}$$

Now we can use equation (8.1) to calculate

$$a_1 = \frac{f'(0)}{1!} = 5, \qquad a_2 = \frac{f''(0)}{2!} = 8, \qquad a_3 = \frac{f'''(0)}{3!} = 0, \qquad a_4 = \frac{f''''(0)}{4!} = 0 \dots .$$

So the Maclaurin series generated by f is $3 + 5t + 8t^2 + 0t^2 + 0t^3 + \cdots$. In this sense, you can think of polynomials as Maclaurin series with a finite number of nonzero coefficients. $\blacksquare$

Generating a familiar Maclaurin series

Example 8.2. Suppose $f(t) = 1/(1-t)$. Use equation (8.1) to generate a Maclaurin series from f.

Solution: Since $t_0 = 0$ in a Maclaurin series, we have $a_0 = f(0) = 1$. Then

$$\begin{aligned} f'(t) &= (1-t)^{-2} & &\Rightarrow & a_1 &= \tfrac{f'(0)}{1!} = 1 \\ f''(t) &= 2(1-t)^{-3} & &\Rightarrow & a_2 &= \tfrac{f''(0)}{2!} = 1 \\ f'''(t) &= (3!)(1-t)^{-4} & &\Rightarrow & a_3 &= \tfrac{f'''(0)}{3!} = 1 \\ &\;\;\vdots & \vdots & & &\;\;\vdots \\ f^{(k)}(t) &= (k!)(1-t)^{-(k+1)} & &\Rightarrow & a_k &= \tfrac{f^{(k)}(0)}{k!} = 1. \end{aligned}$$

So the Maclaurin series generated by f is the familiar series

$$1 + 1t + 1t^2 + 1t^3 + 1t^4 + 1t^5 + \cdots = \sum_{k=0}^{\infty} t^k. \qquad \blacksquare$$

> The geometric series methods of previous sections generate the same series as equation (8.1) but are often faster to use.

Now that we've generated some familiar series, let's apply this technique to generate some new ones.

Generating a Maclaurin series from the Euler exponential function

Example 8.3. Generate a Maclaurin series from $f(t) = e^t$ using equation (8.1), and determine its radius of convergence.

Solution: Because the derivative of e^t is itself, we know that $f^{(k)}(t) = e^t$, so equation (8.1) tells us that $a_k = 1/(k!)$. That is, the Maclaurin series is

$$\sum_{k=0}^{\infty} \frac{1}{k!} t^k,$$

where $0! = 1$. In Example 6.20 (p. 656) we showed that this series has an infinite radius of convergence (i.e., it converges absolutely for all values of t). $\blacksquare$

Generating a Maclaurin from the sine

Example 8.4. Use equation (8.1) to generate a Maclaurin series from $\sin(t)$, and determine its radius of convergence.

Solution: We know $a_0 = 0$. Other coefficients depend on the derivatives of $\sin(t)$.

$$\frac{d}{dt}\sin(t)\Big|_{t=0} = \cos(0) = 1 \qquad \Rightarrow \quad f'(0) = 1$$

$$\frac{d^2}{dt^2}\sin(t)\Big|_{t=0} = -\sin(0) = 0 \qquad \Rightarrow \quad f''(0) = 0$$

$$\frac{d^3}{dt^3}\sin(t)\Big|_{t=0} = -\cos(0) = -1 \quad \Rightarrow \quad f'''(0) = -1$$

$$\frac{d^4}{dt^4}\sin(t)\Big|_{t=0} = \sin(0) = 0 \qquad \Rightarrow \quad f^{(4)}(0) = 0$$

$$\vdots \qquad\qquad\qquad\qquad \vdots$$

Since all the even derivatives are zero at $t = 0$, equation (8.1) tells us that $a_k = 0$ when k is even. That is, our power series will have only odd powers of t. Specifically, equation (8.1) gives us

$$t - \frac{1}{3!}t^3 + \frac{1}{5!}t^5 - \frac{1}{7!}t^7 + \cdots = \sum_{k=0}^{\infty} \frac{(-1)^k}{(2k+1)!}t^{2k+1}.$$

To determine the radius of convergence, we turn to the Ratio Test. When we take limit of the ratio of consecutive terms, we see

$$\lim_{k\to\infty} \left| \frac{\frac{(-1)^{k+1}}{(2(k+1)+1)!}t^{2(k+1)+1}}{\frac{(-1)^k}{(2k+1)!}t^{2k+1}} \right| = \lim_{k\to\infty} \left| \frac{t^2}{(2k+3)(2k+2)} \right| = 0.$$

Since this is less than 1 for all values of t, this series also has an infinite radius of convergence. ■

Try It Yourself: Generating a Maclaurin series from the cosine

Example 8.5. (a) Use equation (8.1) to generate a Maclaurin series from $\cos(t)$, and (b) determine its radius of convergence.

Answer: (a) $\sum_{k=0}^{\infty} \frac{(-1)^k}{(2k)!}t^{2k}$ where $0! = 1$, and (b) $R = \infty$. ■

Full Solution
On-line

The Binomial Series

Example 8.6. Use equation (8.1) to generate a Maclaurin series for $f(t) = (1+t)^p$, and determine its radius of convergence.

Solution: As before, we begin by determining a general formula for $f^{(k)}(t)$.

$$\begin{aligned}
f'(t) &= p(1+t)^{p-1} \\
f''(t) &= p(p-1)(1+t)^{p-2} \\
f'''(t) &= p(p-1)(p-2)(1+t)^{p-3} \\
f''''(t) &= p(p-1)(p-2)(p-3)(1+t)^{p-4} \\
&\vdots \qquad \vdots \\
f^{(k)}(t) &= p(p-1)\cdots(p-k+1)(1+t)^{p-k},
\end{aligned}$$

from which we conclude that the coefficients of a Maclaurin series for $f(t)$ are

$$\frac{f^{(k)}(0)}{k!} = \frac{p(p-1)\cdots(p-k+1)}{k!}, \text{ which we denote by } \binom{p}{k}. \qquad (8.2)$$

The notation $\binom{p}{k}$ tells you that the product in the numerator starts with the number p, and has k factors.

Using this notation, the Maclaurin series generated by $(1+t)^p$ is

$$\sum_{k=0}^{\infty} \binom{p}{k} t^k.$$

> When p is a positive integer, this series terminates after a finite number of summands, but $\binom{p}{k}$ is never zero if p is either a non-integer, a negative number, or both.

This is called the **Binomial Series**, and is particularly handy when p is not an integer. We use the Ratio Test to calculate the radius of convergence. In the limit, the ratio of consecutive summands in this series has a magnitude of

$$\lim_{k \to \infty} \left| \frac{\frac{p(p-1)\cdots(p-k)}{(k+1)!} t^{k+1}}{\frac{p(p-1)\cdots(p-k+1)}{k!} t^k} \right| = \lim_{k \to \infty} \left| \frac{p-k}{k+1} t \right| = |t|,$$

since p is a fixed number. Therefore, the series converges absolutely when $|t| < 1$, and $R = 1$ is the radius of convergence. ∎

> See Appendix C for more about the Binomial coefficients, including their relation to Pascal's Triangle, and their utility as a quick counting tool.

The coefficients of the Maclaurin series that we generated in Example 8.6 are called the **Binomial coefficients**. If p is an integer, the numbers a_k are exactly the entries of Pascal's Triangle and $a_k = 0$ when $k > p$.

❖ Examples of Generating Taylor Series

As we did with Maclaurin series, let's begin with some familiar examples.

Generating a familiar Taylor series

Example 8.7. Use $f(t) = t^2$ to generate a Taylor series about $t_0 = 5$.

Solution: The Taylor series for f about $t_0 = 5$ will have the form $\sum_{k=0}^{\infty} a_k (t-5)^k$. All we need to do is use equation (8.1) to determine the numbers a_k, and for that we need to know the derivatives of f at $t_0 = 5$. So let's calculate …

$$
\begin{array}{lll}
f(t) = t^2 & f(5) = 25 & a_0 = \frac{25}{0!} = 25 \\
f'(t) = 2t & f'(5) = 10 & a_1 = \frac{10}{1!} = 10 \\
f''(t) = 2 & f''(5) = 2 & a_2 = \frac{2}{2!} = 1 \\
f'''(t) = 0 & f'''(5) = 0 & a_3 = \frac{0}{3!} = 0 \\
\quad \vdots & \quad \vdots & \quad \vdots \\
f^{(k)}(t) = 0 & f^{(k)}(5) = 0 & a_k = \frac{0}{k!} = 0 \, .
\end{array}
$$

So the Taylor series for f about $t_0 = 5$ terminates after three summands:

$$25 + 10(t-5) + (t-5)^2. \qquad\qquad ∎$$

> Note: If you expand $(t-5)^2$ and collect powers of t in the polynomial above, you see that it's just a rewritten version of t^2. Try it!

Try It Yourself: Generating a familiar Taylor series

Example 8.8. Use $f(t) = 2t^3 - 4t + 9$ to generate a Taylor series about $t_0 = 2$.

Answer: $17 + 20(t-2) + 12(t-2)^2 + 2(t-2)^3$. ∎

Full Solution On-line

Using the natural logarithm to generate a Taylor series about $t_0 = 1$

Example 8.9. Use $f(t) = \ln(t)$ to generate a Taylor series about $t_0 = 1$, and determine its radius of convergence.

Solution: We begin by calculating the derivatives of f at $t_0 = 1$.

$$
\begin{aligned}
f(1) &= \ln(1) = 0 \\
f'(t) &= 1/t \;\Rightarrow\; f'(1) = 1 \\
f''(t) &= -t^{-2} \;\Rightarrow\; f''(1) = -1 \\
f'''(t) &= 2t^{-3} \;\Rightarrow\; f'''(1) = (2!) \\
f''''(t) &= -(3)(2)t^{-4} \;\Rightarrow\; f'''(1) = -(3!) \\
f'''''(t) &= (4)(3)(2)t^{-5} \;\Rightarrow\; f'''(1) = (4!) \\
&\;\;\vdots \qquad\qquad \vdots
\end{aligned}
$$

In general, we have $f^{(k)}(t) = (-1)^{k-1}(k-1)!t^{-k}$, so equation (8.1) tells us that the coefficients of our power series will be

$$
a_k = \frac{(-1)^{k-1}(k-1)!}{k!} = \frac{(-1)^{k-1}}{k},
$$

when $k > 0$, and $a_0 = f(0) = 0$. That is, the Taylor series generated by $ln(t)$ about $t_0 = 1$ is

$$
\sum_{k=1}^{\infty} \frac{(-1)^{k-1}}{k}(t-1)^k.
$$

If we re-index this series to start at $k = 0$, we have

$$
\sum_{k=0}^{\infty} \frac{(-1)^k}{k+1}(t-1)^{k+1},
$$

which is exactly what we found in Example 7.7 (p. 661). We noted in that example that the radius of convergence is $R = 1$. ∎

> Note that there's one unit between the expansion point and the discontinuity of $\ln(t)$ at $t = 0$.

Generating a Taylor series for the sine at $t_0 = \pi/4$

Example 8.10. Use $f(t) = \sin(t)$ to generate a Taylor series about $t_0 = \pi/4$, and determine its radius of convergence.

Solution: In this case, we see that

$$
f\left(\frac{\pi}{4}\right) = \frac{\sqrt{2}}{2}, \qquad f'\left(\frac{\pi}{4}\right) = \frac{\sqrt{2}}{2}, \qquad f''\left(\frac{\pi}{4}\right) = -\frac{\sqrt{2}}{2}, \qquad f'''\left(\frac{\pi}{4}\right) = -\frac{\sqrt{2}}{2},
$$

and then the pattern repeats because $f''''(t) = f(t)$. Notice that this sequence of numbers has two positives in a row, followed by two negatives. We can describe this kind of behavior using the **greatest integer** function (also called the **floor** function), which is defined by

$$
\lfloor x \rfloor = \text{the greatest integer that's less than or equal to } x.
$$

Since

$$
\left\lfloor \frac{0}{2} \right\rfloor = 0, \qquad \left\lfloor \frac{1}{2} \right\rfloor = 0, \qquad \left\lfloor \frac{2}{2} \right\rfloor = 1, \qquad \left\lfloor \frac{3}{2} \right\rfloor = 1,
$$

we have

$$
(-1)^{\lfloor 0/2 \rfloor} = 1, \qquad (-1)^{\lfloor 1/2 \rfloor} = 1, \qquad (-1)^{\lfloor 2/2 \rfloor} = -1, \qquad (-1)^{\lfloor 3/2 \rfloor} = -1,
$$

which exhibits the same two-positives, then two-negatives pattern that we see in our derivatives. This allows us to write our Taylor series as

$$
\sum_{k=0}^{\infty} \frac{f^{(k)}\left(\frac{\pi}{4}\right)}{k!}\left(t - \frac{\pi}{4}\right)^k = \sum_{k=0}^{\infty} \frac{(-1)^{\lfloor k/2 \rfloor}}{k!}\frac{\sqrt{2}}{2}\left(t - \frac{\pi}{4}\right)^k \qquad\qquad ∎
$$

> For example,
>
> $$\lfloor 5 \rfloor = 5$$
> $$\lfloor \pi \rfloor = 3$$
> $$\lfloor e \rfloor = 2$$
> $$\lfloor -1 \rfloor = -1$$
> $$\lfloor -\pi \rfloor = -4.$$

❖ Taylor Polynomials

Think about using $f(t)$ to generate a Taylor about t_0, one term at a time.

Step 1: a_0

Step 2: $a_0 + a_1(t - t_0)$

Step 3: $a_0 + a_1(t - t_0) + a_2(t - t_0)^2$

$\vdots$ $\vdots$

Step $(n+1)$: $a_0 + a_1(t - t_0) + a_2(t - t_0)^2 + \cdots + a_n(t - t_0)^n$.

At each step in the process, we have a polynomial. After $n + 1$ steps we see a polynomial of degree n, which is called the nth**-degree Taylor polynomial**, and is denoted by $T_n(t)$. Each Taylor polynomial is a partial sum of the Taylor series, so they play a central role in understanding Taylor series.

> The subscript of n tells you the degree of the polynomial.

Taylor polynomials

Example 8.11. Suppose $f(t) = 2t^3 - 4t + 9$. Determine the second-degree Taylor polynomial generated by f about $t_0 = 4$.

Solution: In Example 8.8 we used this function to generate a Taylor series. The second-degree Taylor polynomial comprises the first three terms (constant, linear, quadratic) of that series. So $T_2(t) = 17 + 20(t - 2) + 12(t - 4)^2$.   ∎

Taylor polynomials

Example 8.12. Suppose $f(t) = e^t$. Determine the third-degree Taylor polynomial generated by f about $t_0 = 0$.

Solution: In Example 8.3 we used the exponential function to generate the Maclaurin series $1 + t + \frac{1}{2}t^2 + \frac{1}{3!}t^3 + \frac{1}{4!}t^4 + \cdots$. The third-degree Taylor polynomial is just the first four terms of this series (constant through cubic),

$$T_3(t) = 1 + t + \frac{1}{2}t^2 + \frac{1}{3!}t^3.$$ ∎

Try It Yourself: Taylor polynomials

Example 8.13. Suppose $f(t) = t^2$. Determine the first-degree Taylor polynomial generated by f about $t_0 = 5$.

Answer: $T_1(t) = 25 + 10(t - 5)$.   ∎

Full Solution
On-line

❖ Taylor Series Don't Always Converge to $f(t)$

We started this section by discussing the ability of modern computers to calculate values of transcendental functions, and used arctan(0.5) as an example. That led us to generate a Taylor series using the value of f and its derivatives at t_0. Since f and the Taylor series have the same function value and derivatives at t_0, it's tempting to assume that they're the same, but on p. 669 we mentioned that, in general, the Taylor series generated by f *might not* tell us the value of $f(t)$. Here we provide a pair of examples to illustrate that point.

A Taylor series that converges everywhere but equals $f(t)$ only sometimes

Example 8.14. Suppose $f(t) = |t|$. (a) Use $f(t)$ to generate a Taylor series about $t_0 = 3$, and determine its radius of convergence, and (b) compare the function value at $t = -10$ to the value generated by the Taylor series at $t = -10$.

Solution: (a) When $t > 0$ we have $f(t) = t$. So $f(3) = 3$, $f'(3) = 1$, and $f^{(k)}(3) = 0$ for all $k > 1$. This means that the Taylor series generated by this function is $3 + 1(t - 3)$. Since it has a finite number of terms, this series converges for all t, and we have an infinite radius of convergence.

(b) We have $f(-10) = 10$, but the value of the Taylor series is -10 at $t = -10$.   ∎

A Maclaurin series that equals $f(t)$ at only one point

Example 8.15. Suppose $f(t)$ is the function defined below. Show that $f(t)$ is equal to the Maclaurin series generated by f at only one point.

$$f(t) = \begin{cases} e^{-1/t^2} & \text{when } t \neq 0 \\ 0 & \text{when } t = 0. \end{cases}$$

Solution: It can be shown (though not easily) that f is differentiable, and all its derivatives are 0 at $t_0 = 0$. Since $f^{(k)}(0) = 0$ for all integer $k \geq 0$, the Maclaurin series is just

$$0 + 0t + 0t^2 + 0t^3 + \cdots = 0.$$

But $f(t) = 0$ at only the point $t_0 - 0$.   ∎

You should understand Examples 8.14 and 8.15 as cautionary in nature, meant to indicate the need for further investigation and deeper understanding. As we'll prove in just a moment, however, in the most important cases the value of a Taylor series agrees with the function used to generate it, enabling us to calculate values of exponential and logarithmic functions, sine, and cosine using only the four arithmetic operations from elementary school.

✣ Taylor's Theorem

A moment ago we said that the Taylor series generated by f is equal to $f(t)$ in the most important cases, and now we prove that assertion. The first step in our proof will be using integration by parts to develop the formula

$$f(t) = f(0) + f'(0)t + \frac{f''(0)}{2!}t^2 + \cdots + \frac{f^{(n)}(0)}{n!}t^n + (-1)^n \int_0^t \frac{(x - t)^n}{n!} f^{(n+1)}(x) \, dx,$$

in which you see $f(t)$ expressed as the sum of a definite integral and the nth-degree Taylor polynomial. As $n \to \infty$ the Taylor polynomial becomes the Taylor series, so the conclusion that $f(t)$ equals its Taylor series will rest on whether the integral term vanishes in the limit.

▷ Integrating by parts to generate a series

We know from the Fundamental Theorem of Calculus that

$$f(t) = f(0) + \int_0^t f'(x) \, dx \tag{8.3}$$

provided that f' is continuous over $[0, t]$, which we'll assume for the sake of discussion. A moment ago we indicated that we would integrate by parts, which we do by writing the integrand as $1 \cdot f'(x)$. Then we set $u = f'(x)$ and $v' = 1$. This would typically lead us to write $v = x$, which is the difference between x and the lower bound of integration in this case ($v = x - 0$); but $v' = 1$ also allows us to define v as the difference between x and the upper bound of integration, $v = x - t$, and this turns out to be a better choice. With this choice of u and v,

$$\int_0^t f'(x)\, dx = \int_0^t u(x)v'(x)\, dx$$

$$= u(x)v(x)\Big|_{x=0}^{x=t} - \int_0^t v(x)u'(x)\, dx$$

$$= f'(0)t - \int_0^t (x - t)f''(x)\, dx.$$

This allows us to rewrite equation (8.3) as

$$f(t) = f(0) + f'(0)t - \int_0^t (x - t)f''(x)\, dx, \qquad (8.4)$$

whose first two terms constitute the first-degree Taylor polynomial generated by f at $t_0 = 0$. When we integrate by parts a second time, we'll see the second-degree Taylor polynomial emerge: setting $u(x) = f''(x)$ and $v'(x) = (x - t)$, we have $v(x) = \frac{1}{2}(x - t)^2$ so

$$\int_0^t (x - t)f''(x)\, dx = \int_0^t u(x)v'(x)\, dx$$

$$= u(x)v(x)\Big|_{x=0}^{x=t} - \int_0^t v(x)u'(x)\, dx$$

$$= -\frac{f''(0)}{2}t^2 - \int_0^t \frac{1}{2}(x - t)^2 f'''(x)\, dx.$$

Substituting this into equation (8.4) yields

$$f(t) = f(0) + f'(0)t + \frac{f''(0)}{2}t^2 + \int_0^t \frac{1}{2}(x - t)^2 f'''(x)\, dx, \qquad (8.5)$$

whose first three terms are precisely $T_2(t)$. Let's integrate by parts one more time together. Setting $u(x) = f'''(x)$ and $v'(x) = \frac{1}{2}(x - t)^2$, we have $v(x) = \frac{1}{3!}(x - t)^3$ so

$$\int_0^t \frac{1}{2}(x - t)^2 f'''(x)\, dx = \int_0^t u(x)v'(x)\, dx$$

$$= u(x)v(x)\Big|_{x=0}^{x=t} - \int_0^t v(x)u'(x)\, dx$$

$$= \frac{f'''(0)}{3!}t^3 - \int_0^t \frac{1}{3!}(x - t)^3 f''''(x)\, dx.$$

When we substitute this into equation (8.5) we have

$$f(t) = f(0) + f'(0)t + \frac{f''(0)}{2}t^2 + \frac{f'''(0)}{3!}t^3 - \int_0^t \frac{1}{3!}(x - t)^3 f''''(x)\, dx. \qquad (8.6)$$

Trying this once for yourself will help you understand it better.

Don't just read it. Work it out!

Don't just read it. Work it out!

Don't just read it. Work it out!

Let's assume that e is a rational number, and see what follows. If e is rational, we can write it as $\frac{a}{b}$, where a and b have no common factors. Let's remark that whenever N is an integer larger than b, the number b is a factor in $N!$, so

$$N!e = N!\left(\frac{a}{b}\right)$$

is an integer. The number

$$N!\sum_{k=0}^{N}\frac{1}{k!}$$

is also an integer, since the factorial cancels the denominator of each fraction. Further, since

$$e = \sum_{k=0}^{\infty}\frac{1}{k!} > \sum_{k=0}^{N}\frac{1}{k!}$$

the difference of integers

$$N!e - N!\sum_{k=0}^{N}\frac{1}{k!}$$

is positive. Using equation (9.2) we can rewrite this positive integer as

$$N!\sum_{k=0}^{\infty}\frac{1}{k!} - N!\sum_{k=0}^{N}\frac{1}{k!} = N!\sum_{k=N+1}^{\infty}\frac{1}{k!}. \qquad (9.3)$$

The right-hand side of equation (9.3) can be rewritten as

$$
\begin{aligned}
N!\sum_{k=N+1}^{\infty}\frac{1}{k!} &= \frac{1}{N+1} + \frac{1}{(N+1)(N+2)} + \frac{1}{(N+1)(N+2)(N+3)} + \cdots \\
&\leq \frac{1}{N+1} + \frac{1}{(N+1)^2} + \frac{1}{(N+1)^3} + \cdots \\
&= \sum_{k=1}^{\infty}\left(\frac{1}{N+1}\right)^k = \frac{1}{N}.
\end{aligned}
$$

That is, the assumption that e is rational leads to the conclusion that the number

$$N!e - N!\sum_{k=1}^{N}\frac{1}{k!}$$

is both (1) a positive integer, and (2) smaller than $1/N$, where N is an arbitrarily large number (remember, it just had to be larger than b for this to work). These things cannot be true simultaneously, so e is not rational. ∎

You should know

- the terms *Gaussian Error Function (erf)*, *differential equation*, *first order*, and *second order*;

- that the number e is irrational.

You should be able to

- use known Maclaurin series to generate others;

- use series to approximate values of functions to a specified accuracy;

- use series to work with indefinite integrals and definite integrals;

- determine the Maclaurin series of a function that's described by a differential equation.

❖ 8.9 Concept and Application Exercises

In #1–6 use the Binomial Series to determine a Taylor series for $f(t)$ at t_0 as demonstrated in Example 9.2.

1. $f(t) = \sqrt{t}$, $t_0 = 16$ 3. $f(t) = t^{5/2}$, $t_0 = 9$ 5. $f(t) = t^{-1/2}$, $t_0 = 25$

2. $f(t) = \sqrt[3]{t}$, $t_0 = 8$ 4. $f(t) = t^{7/3}$, $t_0 = 125$ 6. $f(t) = t^{-4/3}$, $t_0 = 27$

In each of #7–12 use the appropriate series to determine the value of the limit.

7. $\lim_{t \to 0} \frac{\cos(t) - e^t}{5t}$ 9. $\lim_{\theta \to 0} \frac{15 \tan(\theta) - 15\theta - 2\theta^5}{13\theta^7}$ 11. $\lim_{t \to 0} \frac{\sinh(t) - t}{4t^3(t + 3)}$

8. $\lim_{\theta \to 0} \frac{\sin(4\theta) - 4\theta + 32\theta^3}{9\theta^3}$ 10. $\lim_{\theta \to 0} \frac{2\cos(\theta^2) - 2 - \theta^4}{12\theta^5}$ 12. $\lim_{t \to 0} \frac{\cosh(t) - 1 - 0.5t^2}{t^5 + 9t^4}$

In #13–16 determine the coefficient a_k from the Maclaurin series of the specified function.

13. e^{2x}; $k = 10$ 14. $e^{x/3}$; $k = 100$ 15. $\cos(x^2)$; $k = 36$ 16. $\sin(x^3)$; $k = 35$

In #17–20 determine the Maclaurin series for $f(x)$.

17. $f(x) = \sin(x^5)$ 18. $f(x) = \tan(x^3)$ 19. $f(x) = e^{x^7}$ 20. $f(x) = \text{erf}(x^2)$

In #21–24 use your knowledge of the Maclaurin series for sine, cosine, and the Euler exponential function to determine the 4th-degree Taylor polynomial at $t_0 = 0$ for $f(x)$

21. $f(x) = \sin(\sin(x))$ 22. $f(x) = \cos(\sin(x))$ 23. $f(x) = \cos(e^x - 1)$ 24. $f(x) = e^{\cos(x) - 1}$

25. The *imaginary error function* is similar to erf(x), but is defined as

$$\text{erfi}(x) = \frac{2}{\sqrt{\pi}} \int_0^x e^{t^2} \, dt \quad \text{when } x \in \mathbb{R}.$$

(a) Determine a Maclaurin series for erfi(x), and (b) use your answer from part (a) to approximate erfi(0.3) to within 0.001 of its actual value.

26. The *complementary error function* is defined as erfc(x) $= 1 - \text{erf}(x)$. (a) Determine a Maclaurin series for erfc(x), and (b) use your answer from part (a) to approximate erfi(0.3) to within 0.001 of its actual value.

27. Use a change of variable to verify that

$$\text{erf}(x) = \int_{-\sqrt{2}x}^{\sqrt{2}x} \frac{1}{\sqrt{2\pi}} e^{-t^2/2} \, dt \quad \text{when } x \geq 0.$$

28. (a) Verify that the Gaussian Error Function is odd, and (b) determine $\text{erf}'(0)$.

In each of #29–34 (a) express the number as the sum of an alternating (Maclaurin) series, (b) determine the number of summands needed to approximate the given number to within 0.001 of its actual value using the alternating series, and (c) use that number of summands to approximate the value.

29. $\sin(0.9)$

30. $\cos(1.5)$

31. $e^{-1.5}$

32. $e^{-0.7}$

33. $\text{erf}(0.5)$

34. $\text{erf}(0.7)$

In #35–38 (a) use the Maclaurin series for $\ln(1+x)$ to write each number as an alternating series, (b) determine the number of summands needed to approximate the given number to within 0.001 of its actual value using the alternating series, and (c) use that number of summands to approximate the value.

35. $\ln(1.2)$

36. $\ln(1.3)$

37. $\ln(3) - \ln(2)$

38. $\ln(0.25) + \ln(5)$

Since $12^{1/3} = (8+4)^{1/3} = (8(1+0.5))^{1/3} = 2(1+0.5)^{1/3}$, we can use the Binomial Series to approximate the value of $\sqrt[3]{12}$. Use this technique to approximate each of the numbers in exercises #39–42 to within 0.001 of its actual value.

39. $\sqrt[3]{12}$; $(12 = 8+4)$

40. $\sqrt{18}$; $(18 = 16+2)$

41. $\sqrt[4]{17}$; $(17 = 16+1)$

42. $\sqrt[3]{29}$; $(29 = 27+2)$

In #43–46 (a) write the specified value as an integral of the form $\int_0^x f(t)\,dt$, (b) use Taylor's Inequality to determine a value of n so that the integral of the nth-degree Taylor polynomial about $t_0 = 0$, $\int_0^x T_n(t)\,dt$, is within 0.001 of $\int_0^x f(t)\,dt$, and (c) calculate $\int_0^x T_n(t)\,dt$.

43. $e^{0.5}$; $f(t) = e^t$

44. $e^{1.3}$; $f(t) = e^t$

45. $\sin(0.9)$; $f(t) = \cos(t)$

46. $\cos(1.1)$; $f(t) = \sin(t)$

47. $\ln(0.9)$; $f(t) = \frac{1}{1+t}$

48. $\cosh(0.5)$; $f(t) = 2 + \sinh(t)$

In #49–54 write the indefinite integral as a Maclaurin series.

49. $\int \frac{\sin(x)}{x}\,dx$

50. $\int x\tan(x)\,dx$

51. $\int \frac{1}{1+t^5}\,dt$

52. $\int \frac{1}{1-t^6}\,dt$

53. $\int \sqrt{1+t^7}\,dt$

54. $\int \frac{1}{(1+t^7)^{1/3}}\,dt$

In #55–60 (a) write the integrand as a Maclaurin series, (b) integrate the series term by term, and (c) approximate the resulting series to within 0.01 of its actual value.

55. $\int_0^1 \sin(t^2)\,dt$

56. $\int_0^1 \cos(t^3)\,dt$

57. $\int_0^{1/2} \frac{e^x - 1}{x}\,dx$

58. $\int_0^{1/2} \frac{2\cos(x) - 2}{x}\,dx$

59. $\int_0^{3/5} \frac{t}{1+t^4}\,dt$

60. $\int_0^{2/7} \frac{t^2}{1+t^6}\,dt$

61. Einstein's theory of special relativity tells us that when an object whose rest mass is m travels at velocity v, its kinetic energy is

$$K = mc^2\left(\gamma - 1\right) \quad \text{where} \quad \gamma = \left(1 - \left(\frac{v}{c}\right)^2\right)^{-1/2},$$

in which c is the speed of light. (a) Use the Binomial Series to write K as a series in powers of $\left(\frac{v}{c}\right)$, (b) show that when the fraction v/c is small, Einstein's theory predicts that the kinetic energy is $\frac{1}{2}mv^2$, and (c) determine how fast the object has to travel in order for the first relativistic correction to be 5% of the Newtonian kinetic energy.

62. Suppose an object is moving toward you at a speed v (measured as a fraction of the speed of light) and emits a light signal. The frequency of the signal that you receive will be higher than what was emitted. The fraction by which the observed frequency exceeds the emitted frequency is $Z = \sqrt{1 + \frac{2v}{1-v}}$. Use the Binomial Series to write a Maclaurin series for Z in terms of $x = \frac{2v}{1-v}$.

63. Owing to the effect of gravity on its angular momentum, the period of a simple pendulum is proportional to

$$\int_0^{\pi/2} \frac{1}{\sqrt{1 - \varepsilon^2 \sin^2(\theta)}} \, d\theta,$$

where $\varepsilon \in [0, 1)$ depends on the angle off vertical at which the pendulum is released. This integral is called an *elliptic integral of the first kind*. In this exercise, we'll take $\varepsilon = 1/2$.

(a) Use the Binomial Series to verify that

$$\int_0^{\pi/2} \frac{1}{\sqrt{1 - \varepsilon^2 \sin^2(\theta)}} \, d\theta = \sum_{k=0}^{\infty} (-1)^k \binom{-1/2}{k} \varepsilon^{2k} \int_0^{\pi/2} \sin^{2k}(\theta) \, d\theta.$$

(b) Explain why we know that $\int_0^{\pi/2} \sin^{2k}(\theta) \, d\theta < \pi/2$ for every integer $k \geq 0$.

(c) Because this series is alternating, we know that

$$\sum_{k=0}^{n} (-1)^k \binom{-1/2}{k} \varepsilon^{2k} \int_0^{\pi/2} \sin^{2k}(\theta) \, d\theta$$

is within 0.001 of the series' value when

$$\left| \binom{-1/2}{n+1} \varepsilon^{2(n+1)} \int_0^{\pi/2} \sin^{2(n+1)}(\theta) \, d\theta \right| < \frac{1}{1000}.$$

Use the result from part (b) to determine the smallest integer $n > 0$ for which this is true.

(d) In order to approximate the value of the series, we need to know $\int_0^{\pi/2} \sin^{2k}(\theta) \, d\theta$ for several values of k. Calculate the value of this integral when $k = 1$.

(e) Integrate by parts using $u = \sin^{2k-2}(\theta)$ and $v' = \sin^2(\theta)$ to show that

$$\int_0^{\pi/2} \sin^{2k}(\theta) \, d\theta = \left(1 - \frac{1}{2k}\right) \int_0^{\pi/2} \sin^{2k-2}(\theta) \, d\theta.$$

(f) Use the recursion formula from part (e) to determine the value of $\int_0^{\pi/2} \sin^{2k}(\theta) \, d\theta$ when $k = 2, 3, 4, 5, 6$ and 7.

(g) Approximate the value of this integral to within 0.001 of its actual value.

64. Suppose $a > b > 0$. In Chapter 9 you'll see that the arc length of the ellipse $(x/a)^2 + (y/b)^2 = 1$ is

$$L = 4b \int_0^{\pi/2} \sqrt{1 + \varepsilon^2 \sin^2(\theta)} \, d\theta,$$

where $\varepsilon^2 = (a^2 - b^2)/b^2$. This integral is called an *elliptic integral of the second kind*. In this exercise, we'll assume that $a = 5$ and $b = 4$.

(a) Use the Binomial Series to verify that

$$L = \sum_{k=0}^{\infty} 4b \binom{1/2}{k} \varepsilon^{2k} \int_0^{\pi/2} \sin^{2k}(\theta) \, d\theta.$$

(b) Because this series is alternating, we know that its nth partial sum is within 0.001 of L when

$$\left| 4b \binom{1/2}{n+1} \varepsilon^{2(n+1)} \int_0^{\pi/2} \sin^{2(n+1)}(\theta) \, d\theta \right| < 0.001.$$

Determine the smallest integer $n > 0$ for which this happens (the recursion formula from #63 (part e) is useful here).

(c) Approximate the value of L to within 0.001 of its actual value.

In #65–70 (a) determine the Maclaurin series of the function $y(t)$ that's described by the differential equation.

65. $y' = 4y$, $y(0) = 1$
66. $y' = -2y$, $y(0) = 3$

67. $y' = 5t^2 y$, $y(0) = 4$
68. $y' = t^3 y$, $y(0) = 2$

69. $ty' = ty + y$, $y(1) = 2$
70. $ty' = t^2 y + y$, $y(1) = 3$

71. Consider the equation $y'' + 4y = 0$, $y(0) = 1$, $y'(0) = 2$. (a) Determine the Maclaurin series of the function $y(t)$ described by this differential equation, (b) rewrite your solution as a sum of two series, one with only even powers of t (including t^0) and the other with exclusively odd powers of t, and (c) rewrite these series in more familiar form.

72. Consider the equation $y'' + 9y = 0$, $y(0) = -1$, $y'(0) = 3$. (a) Determine the Maclaurin series of the function $y(t)$ described by this differential equation, (b) rewrite your solution as a sum of two series, one with only even powers of t (including t^0) and the other with exclusively odd powers of t, and (c) rewrite these series in more familiar form.

In each of #73–76 you see a second-order differential equation with specified $y(0)$ and $y'(0)$. You should (a) determine a recurrence relation between the coefficients of the series solution, $y(t) = \sum_{k=0}^{\infty} a_k t^k$, (b) use your answer from part (a) to write down the specified Taylor polynomial approximation of $y(t)$, and (c) use a graphing utility to render a plot of that polynomial and verify that has the correct function value and derivative at $t = 0$.

73. $y'' + 2y' + y = 0$
    $y(0) = 1$, $y'(0) = -1$; $T_5(t)$

74. $y'' - 2y' + y = 0$
    $y(0) = 1$, $y'(0) = 1$; $T_6(t)$

75. $y'' + 8y' + 15y = 0$
    $y(0) = 0$, $y'(0) = 4$; $T_7(t)$

76. $y'' + 3y' + 7y = 0$
    $y(0) = -2$, $y'(0) = 0$; $T_8(t)$

Chapter 8 Review

❖ True or False

1. Suppose $\{s_n\}$ is a monotonically increasing sequence. Then $\lim_{n\to\infty} s_n$ exists.

2. Suppose $\{s_n\}$ is a monotonically decreasing sequence, and $s_n > 0$. Then $\lim_{n\to\infty} s_n = 0$.

3. The value of a series is defined to be the limit of its partial sums.

4. The Integral Test tells you the value of a series.

5. If a series converges, its summands tend to zero.

6. If the summands of a series tend to zero, it converges.

7. If a geometric series is alternating, it converges.

8. If $\sum_{k=1}^{\infty} a_k$ converges and $a_k \geq 0$, the series is absolutely convergent.

9. If $\sum_{k=1}^{\infty} a_k$ is absolutely convergent, it's also convergent.

10. If $a_1 = 1$ and $a_{k+1} = 3/(a_k^2 + 1)$, the sequence $\{a_k\}$ converges.

11. If $a_1 = 1$ and $a_{k+1} = 3/(a_k^2 + 1)$, the series $\sum_{k=1}^{\infty} a_k$ converges.

12. The series $\sum_{k=1}^{\infty} \frac{1}{k^2} = \frac{1}{1 - \frac{1}{k^2}}$.

13. If $\sum_{k=1}^{\infty} |a_k|$ diverges, so does $\sum_{k=1}^{\infty} a_k$.

14. If $\sum_{k=32}^{\infty} a_k$ converges, so does $\sum_{k=1}^{\infty} a_k$.

15. If $\sum_{k=1}^{\infty} a_k$ converges, so does $\sum_{k=47}^{\infty} a_k$.

16. Suppose $f(x) = \sum_{k=1}^{\infty} a_k x_k$ and $g(x) = \sum_{k=8}^{\infty} a_k x_k$ share the same coefficients. Then the interval of convergence for f is the same as the interval of convergence for g.

17. Suppose $f(x) = \sum_{k=1}^{\infty} a_k x^k$ and $g(x) = \sum_{k=8}^{\infty} a_k x^k$ share the same coefficients, and $x = 1$ is in the domain of both functions. Then $g(1) = f(1)$.

Suppose we know that $0 \leq |a_k| < b_k$, and $\sum_{k=1}^{\infty} b_k$ converges. Questions #18–20 make statements about $\sum_{k=1}^{\infty} a_k$. Determine whether they are true or false.

18. The series $\sum_{k=1}^{\infty} |a_k|$ also converges, and $\sum_{k=1}^{\infty} |a_k| < \sum_{k=1}^{\infty} b_k$.

19. The series $\sum_{k=1}^{\infty} a_k$ also converges, and $\sum_{k=1}^{\infty} a_k < \sum_{k=1}^{\infty} b_k$.

20. The series $\sum_{k=1}^{\infty} a_k$ also converges, and $\left|\sum_{k=1}^{\infty} a_k\right| < \sum_{k=1}^{\infty} b_k$.

❖ Multiple Choice

21. Suppose each $a_k \in [1, 2]$. Then the sequence $\{a_k\}$ . . .

   (a) converges to 2.
   (b) converges to 1.
   (c) converges to some point in $(1, 2)$.
   (d) bounces back and forth between 1 and 2.
   (e) might do none of the above.

22. A geometric series ...

   (a) converges when its common ratio is $r < 1$.

   (b) converges when its common ratio is r, where $|r| \leq 1$.

   (c) converges when its common ratio is r, where $-1 < r < 1$.

   (d) diverges.

   (e) None of the above.

23. Suppose $\sum_{k=1}^{\infty} a_k$ is conditionally convergent. Then ...

   (a) $\sum_{k=1}^{\infty} a_k$ is absolutely convergent.

   (b) $\sum_{k=1}^{\infty} |a_k|$ diverges.

   (c) we don't know whether $\sum_{k=1}^{\infty} a_k$ converges.

   (d) we don't know whether $\sum_{k=1}^{\infty} |a_k|$ converges.

   (e) None of the above.

24. Suppose $\sum_{k=1}^{\infty} a_k$ is absolutely convergent. Then ...

   (a) $\sum_{k=1}^{\infty} |a_k|$ converges.

   (b) $\sum_{k=1}^{\infty} a_k$ might or might not converge.

   (c) $\sum_{k=1}^{\infty} a_k^2$ converges.

   (d) All of the above.

   (e) None of the above.

25. Suppose $|a_k| < b_k$ and $\sum_{k=1}^{\infty} b_k$ diverges. Then ...

   (a) $\sum_{k=1}^{\infty} a_k$ diverges.

   (b) $\sum_{k=1}^{\infty} a_k$ converges.

   (c) $\sum_{k=1}^{\infty} |a_k|$ diverges.

   (d) $\sum_{k=1}^{\infty} |a_k|$ converges.

   (e) we don't have enough information to decide.

26. Consider the series (i) $\sum_{k=4}^{\infty} 2^k x^k$, (ii) $\sum_{k=5}^{\infty} (0.5) 2^k x^{k-1}$, and (iii) $\sum_{k=3}^{\infty} 2^k x^{k+1}$.

   (a) Only (i) and (ii) are the same.

   (b) Only (ii) and (iii) are the same.

   (c) Only (i) and (iii) are the same.

   (d) All of these are the same.

   (e) None of these are the same.

27. Consider the series (i) $\sum_{k=7}^{\infty} \frac{k^2+1}{3^k}$, (ii) $\sum_{k=5}^{\infty} \frac{k^2+4k+5}{(9)(3^k)}$, and (iii) $\sum_{k=10}^{\infty} \frac{k^2-6k+10}{3^{k-3}}$.

   (a) Only (i) and (ii) are the same.

   (b) Only (ii) and (iii) are the same.

   (c) Only (i) and (iii) are the same.

   (d) All of these are the same.

   (e) None of these are the same.

28. Consider the power series (i) $\sum_{k=0}^{\infty} \frac{x^k}{3^k}$, (ii) $\sum_{k=3}^{\infty} \frac{x^k}{3^k}$, and (iii) $\sum_{k=0}^{\infty} \frac{(x-4)^k}{3^k}$.

   (a) Only (i) is a Maclaurin series.

(b) Only (i) and (ii) are Maclaurin series.

(c) Only (i) and (iii) are Maclaurin series.

(d) All of these are Maclaurin series.

(e) None of these are Maclaurin series.

Exercises #29 and #30 refer to the following figure, in which the blue circles represent the summands of $\sum_{k=1}^{\infty}(-1)^k a_k$. Though difficult to see in the figure, it's true that $0 < a_{k+1} < a_k$ and $\lim_{k\to\infty} a_k = 0$.

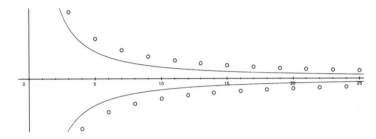

29. Suppose the curves in the figure are $y = \pm\frac{1}{t}$. Based on the figure ...

    (a) the series is divergent.

    (b) the series is conditionally convergent.

    (c) the series is absolutely convergent.

    (d) the series is either absolutely or conditionally convergent, but we cannot tell which.

    (e) we don't have enough information to tell whether the series converges or not.

30. Suppose the curves in the figure are $y = \pm\frac{1}{t^2}$. Based on the figure ...

    (a) the series is divergent.

    (b) the series is conditionally convergent.

    (c) the series is absolutely convergent.

    (d) the series is either absolutely or conditionally convergent, but we cannot tell which.

    (e) we don't have enough information to tell whether the series converges or not.

Exercises #31 and #32 refer to the following figure, in which the blue circles represent the summands of $\sum_{k=1}^{\infty} a_k$. As indicated by the figure, it's true that $\lim_{k\to\infty} a_k = 0$.

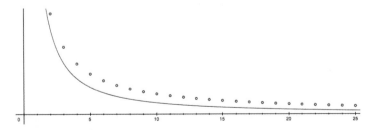

31. Suppose the curve in the figure is $y = \frac{1}{\sqrt{t}}$. Based on the figure ...

    (a) the series is divergent.

(b) the series is conditionally convergent.

(c) the series is absolutely convergent.

(d) the series is either absolutely or conditionally convergent, but we cannot tell which.

(e) we don't have enough information to tell whether the series converges or not.

32. Suppose the curves in the figure are $y = \frac{1}{t^2}$. Based on the figure ...

(a) the series is divergent.

(b) the series is conditionally convergent.

(c) the series is absolutely convergent.

(d) the series is either absolutely or conditionally convergent, but we cannot tell which.

(e) we don't have enough information to tell whether the series converges or not.

❖ Exercises

33. Suppose $a_1 \in (0, 1)$ and $a_{k+1} = -\frac{1}{a_k^2 - b}$, where $b \geq 2$ is an integer.

(a) Show that $\lim_{k \to \infty} a_k$ exists, and its value is a root of $f(x) = x^3 - bx + 1$. (Hint: use a cobwebbing diagram.)

(b) The linear approximation of $f(x)$ about $x = 0$ is $y = 1 - bx$. Show that the x-intercept of $y = 1 - bx$ is less than $\lim_{k \to \infty} a_k$ by verifying that $f(1/b) > 0$ and $f'(1/b) < 0$.

(c) If we cut the slope of the linear approximation in half, we can get an upper bound on $\lim_{k \to \infty} a_k$. Show that $x = 2/b$ is larger than $\lim_{k \to \infty} a_k$ by verifying that $f(2/b) < 0$.

(d) Based on the results of parts (b) and (c), you know that $\lim_{k \to \infty} a_k \in (\frac{1}{b}, \frac{2}{b})$. What does this tell you about the series $\sum_{k=1}^{\infty} a_k$?

34. Suppose \$100,000 of counterfeit money is introduced into circulation. If 25% of the money is identified as counterfeit and confiscated in each transaction, but the rest remains in circulation to be used again, determine the total value of all fraudulent purchases made with the counterfeit money.

Business

35. Suppose a ground station sends a signal to a satellite. Due to interference there's no guarantee that a signal will reach the satellite intact, but if it does, the satellite responds with a "ping." Let's denote by $p \in (0, 1)$ the probability of the signal reaching the satellite intact, so the probability of the signal arriving fragmented or not at all is $1 - p$. Then the probability that the ground station has to send the signal exactly k times before it hears the satellite's ping is $(1 - p)^{k-1}p$, and the series

Telecommunications

$$\sum_{k=m}^{\infty} (1 - p)^{k-1} p$$

tells us the probability that the satellite responds with a "ping" on or after the ground station's m^{th} attempt.

(a) Find a closed form for this series, which we'll denote by $G(m)$.

(b) Verify that $G(1) = 1$ and explain what that means in this context.

(c) The ratio $G(m+n)/G(m)$ is understood as the probability that "if the station has not heard a ping by the mth try, it will not hear a ping for another n attempts." Show that this ratio does not depend on m, and explain what that fact means in this context.

36. Suppose we approximate $\sum_{k=0}^{10^6} \left(\frac{1}{4}\right)^k \approx \frac{4}{3}$. How much error have we made?

37. When $x < 0$ the series for e^x is an alternating series. How many terms are required to calculate e^{-1} to within 10^{-7} of its actual value?

Determine the value of each series in #38–41.

38. $\sum_{k=0}^{\infty} \frac{3}{2^k}$

39. $\sum_{k=2}^{\infty} \frac{7^{k+1}}{11^k}$

40. $\sum_{k=0}^{\infty} \frac{1}{k+2} - \frac{1}{k+4}$

41. $\sum_{k=0}^{\infty} \frac{1}{k^2+7k+10}$

42. Explain why $\sum_{k=1}^{\infty} \frac{1}{2k^2+6k+4}$ is a telescoping series but $\sum_{k=1}^{\infty} \frac{1}{2k^2+5k+2}$ is not.

In #43–54 (a) determine whether each of the following series converges, and (b) prove you're right by using one of the convergence tests.

43. $\sum_{k=1}^{\infty} (-2)^k$

44. $\sum_{k=1}^{\infty} (-0.9)^{k+3}$

45. $\sum_{k=1}^{\infty} \frac{1}{k\sqrt{\pi}}$

46. $\sum_{k=1}^{\infty} \frac{1}{k\sqrt{0.5}}$

47. $\sum_{k=1}^{\infty} \cos\left(\frac{1}{k^3}\right)$

48. $\sum_{k=1}^{\infty} \sin\left(\frac{1}{k^2}\right)$

49. $\sum_{k=1}^{\infty} k^3 e^{-k}$

50. $\sum_{k=1}^{\infty} \frac{\ln(k)}{k\sqrt{\ln(k)+1}}$

51. $\sum_{k=1}^{\infty} \frac{1}{2k^2 - k + 1}$

52. $\sum_{k=1}^{\infty} \frac{k\ln(k)}{k^3 - 0.5k + 2}$

53. $\sum_{k=1}^{\infty} \ln\left(1 + \frac{1}{k}\right)$

54. $\sum_{k=1}^{\infty} \ln\left(1 + \frac{(-1)^k}{k}\right)$

55. Suppose $\sum_{k=1}^{\infty} a_k$ is a convergent series of positive terms. Based on that information alone, do we know whether $\sum_{k=1}^{\infty} \sin(a_k)$ converges?

56. Suppose $\{a_k\}$ is a sequence of positive terms, and $\sum_{k=1}^{\infty} \sin(a_k)$ converges. Based on that information alone, do we know whether $\sum_{k=1}^{\infty} a_k$ converges?

Determine the interval of convergence for each power series in #57–60.

57. $\sum_{k=1}^{\infty} \frac{2^k x^k}{1 + \sqrt{k}}$

58. $\sum_{k=1}^{\infty} \frac{k^2 x^k}{4^k + k^6}$

59. $\sum_{k=1}^{\infty} \left(\frac{8k+1}{8k+9\sqrt{k}}\right)^k (x-3)^k$

60. $\sum_{k=1}^{\infty} \frac{k^2 + \sqrt{k}}{4^k}(x-1)^k$

61. Denote by $\pi[k]$ the kth digit in the decimal expansion of π. For example, $\pi[1] = 3$, $\pi[2] = 1$ and $\pi[3] = 4$.

(a) Determine whether the series $\sum_{k=1}^{\infty} \frac{1}{k^2}(x-1)^{\pi[k]}$ converges for any values of x other than 1.

(b) If not, why not? If so, how do you know, and what is its radius of convergence?

62. Determine the values of x for which $\sum_{k=1}^{\infty} a_k x^k$ converges when $a_k = \left(\frac{4k-2}{\pi k+5}\right)^k$.

63. Determine a closed form for $\sum_{k=1} 3^{k+1} x^k$.

64. Determine a closed form for $\sum_{k=1} k 2^{k-3} x^{k-1}$.

65. Write the Maclaurin series for $f(t) = \sin(t^4)$.

66. Write the 5th-degree Taylor polynomial for $f(t) = e^{\sin(t)}$ about $t = 0$.

67. Write $\int \cos(\sqrt{\theta})\, d\theta$ as a Maclaurin series.

68. Write $\int \tan(\theta^6)\, d\theta$ as a Maclaurin series.

69. Use the Maclaurin series for $\ln(1+t)$ to approximate $\ln(0.9)$ to within 0.01 of its actual value.

70. Use the Taylor series for $\ln(t)$ about $t_0 = 1$ to approximate $\ln(1.1)$ to within 0.01 of its actual value.

Chapter 8 Projects and Applications

❖ Revisiting Stirling's Approximation

In the project called *The Gamma Function* on p. 482 you saw that

$$n! = \int_0^\infty t^n e^{-t}\, dt.$$

(A graph of the integrand is shown in the Figure 11.1.) In this brief project, based on an exposition in [11], we use this characterization of the factorial to refine the approximation that we established on p. 400.

1. Use the First Derivative Test to verify that, when restricted to the interval of integration, the integrand has a global maximum at $t = n$.

2. Verify that the integrand can be written as $\exp(n\ln(t) - t)$

3. Calculate the second-order Taylor polynomial for $n\ln(t) - t$ about $t_0 = n$.

4. Use your answer from #3 to explain the approximation

$$t^n e^{-t} \approx \left(\frac{n}{e}\right)^n e^{-(t-n)^2/2n}.$$

The graphs of $t^n e^{-t}$ and $\left(\frac{n}{e}\right)^n e^{-(t-n)^2/2n}$ are shown in Figure 11.2. Notice that $\left(\frac{n}{e}\right)^n e^{-(t-n)^2/2n}$ overestimates $t^n e^{-t}$ when $t \in (0, n)$ and underestimates it when $t > n$.

The approximation that you established in #4 leads us to write

$$\int_0^\infty t^n e^{-t}\, dt \approx \int_0^\infty \left(\frac{n}{e}\right)^n e^{-(t-n)^2/2n}\, dt.$$

Since $\left(\frac{n}{e}\right)^n$ is constant and $e^{-(t-n)^2/2n}$ is small when $t < 0$, we can further approximate by writing

$$\int_0^\infty t^n e^{-t}\, dt \approx \left(\frac{n}{e}\right)^n \int_{-\infty}^\infty e^{-(t-n)^2/2n}\, dt.$$

The substitution $x = \frac{t-n}{\sqrt{n}} \Rightarrow dt = \sqrt{n}\, dx$, which allows us to rewrite this approximation as

$$\int_0^\infty t^n e^{-t}\, dt \approx \left(\frac{n}{e}\right)^n \sqrt{n} \int_{-\infty}^\infty e^{-x^2/2}\, dx.$$

During our discussion of probability density functions on p. 567, we asserted that

$$\frac{1}{\sqrt{2\pi}} \int_{-\infty}^\infty e^{-x^2/2}\, dx = 1.$$

Substituting this into the approximation above, we have

$$n! = \int_0^\infty t^n e^{-t}\, dt \approx \left(\frac{n}{e}\right)^n \sqrt{2\pi n}\,,$$

which is an important fact known as **Stirling's Approximation**.

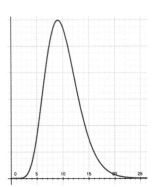

Figure 11.1: The graph of $y = t^n e^{-t}$ has a global maximum at $t = n$.

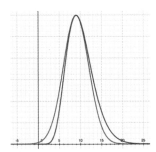

Figure 11.2: Graphing the curves $y = t^n e^{-t}$ and $y = (n/e)^n e^{-(t-n)^2/2n}$ together.

✤ Weighing Exploitation Against Retaliation

Robert Axelrod's famous book *The Evolution of Cooperation* details an experiment in which various computer algorithms were entered into a competition. After each round of the competition, algorithms that did well were duplicated while those that did not were eliminated in Darwinistic fashion. The game they played against each other is called the *iterated prisoner's dilemma,* which goes like this:

- There are two players, whom we'll call Blue and Red.

- In each round, both players decide whether to *cooperate with* or *exploit* the other (C or E).

- The payoff to Blue depends on the choice made by both players, and is often represented by a 2×2 matrix of numbers such as:

$$
\begin{array}{cc}
 & \text{Red's choice} \\
 & \begin{array}{cc} C & E \end{array} \\
\text{Blue's choice} \left\{ \begin{array}{c} C \\ E \end{array} \right. &
\begin{array}{|c|c|}
\hline
7 & 1 \\
\hline
11 & 3 \\
\hline
\end{array}
\end{array}
$$

meaning that if both players choose to exploit the other, Blue gets a payoff of 3, but the payoff to Blue is only 1 if Blue chooses to cooperate while Red chooses to exploit.

- The payoff matrix for Red is similar in structure:

$$
\begin{array}{cc}
 & \text{Red's choice} \\
 & \begin{array}{cc} C & E \end{array} \\
\text{Blue's choice} \left\{ \begin{array}{c} C \\ E \end{array} \right. &
\begin{array}{|c|c|}
\hline
7 & 11 \\
\hline
1 & 3 \\
\hline
\end{array}
\end{array}
$$

- The players cannot communicate except to reveal their choices (i.e., no coordination is possible).

- After a certain number of rounds, which is unknown to the players, the game is stopped and the player with the most points wins.

This last point is important, because it means that the players always have some doubt about whether the next round will be played. So next round's payoff is less important than the payoff in the current move. This leads to the idea of a *discount parameter,* $\omega \in (0, 1)$, that's used to scale down the value of future payoffs. Specifically, a payoff of P that's expected k turns in the future is said to have a *present value* of $\omega^k P$. Using the payoff matrix above, the present value continued mutual cooperation is

$$
\underbrace{7}_{\text{this round}} + \underbrace{7\omega}_{} + \underbrace{7\omega^2}_{\text{two rounds away}} + \cdots + \underbrace{7\omega^n}_{n \text{ rounds away}} = \sum_{k=0}^{n} 7\omega^k.
$$

where the "next round" brace is over 7ω.

Of course, we don't know the value of n so we often consider the limiting case (a game that continues forever!).

1. What does a high value of ω say about the players' expectations, contrasted with low values of ω?

2. Suppose that the discount parameter is $\omega = 0.95$. Calculate the present value of mutual cooperation that continues forever.

3. If Red were to successfully exploit Blue, he'd earn 11 points in this round, but might earn only 3 in subsequent rounds (if Blue were to hold a grudge and play E forever). Calculate the present value of this scenario for both Blue and Red when $\omega = 0.95$.

4. For what values of ω does the present value calculation indicate that it's better for Red to exploit Blue and face the continuing consequences rather than continuing to cooperate forever?

5. A moment ago we said that Blue would play E forever. This strategy (cooperate until the other player betrays you and then never trust him again) is called *permanent retaliation*. Another strategy is called *tit-for-tat* and can be summarized this way:

   Begin by cooperating. Thereafter, do what your opponent did on the previous turn.

   Suppose that Red plays tit-for-tat and Blue *always* chooses E. Calculate the present value of the game to both players using the discount parameter $\omega = 0.75$. What do you think this says about a single tit-for-tat player in a community of others who *always* choose to exploit?

6. Calculate the present value of the game (using $\omega = 0.75$) when both players are playing the tit-for-tat strategy.

7. Calculate the present value of a game (using $\omega = 0.75$) if both players are playing the "all-E" strategy. Verify that it's less than your answer from #6, and explain why that makes sense.

Until this point, we've been talking about a single game between two players. Now we talk about a *round*, in which each player is matched against every other player in a round-robin tournament. At the beginning of the round, each player chooses to use either the tit-for-tat strategy or the "all-E" strategy for the entire round, and *you're* included as a player.

8. Suppose there are 10000 players, N of whom employ the tit-for-tat strategy. If you always play tit-for-tat, and $\omega = 0.75$ as before, what is the present value of this round to you? (Your answer will be in terms of N.)

9. To calculate the *expected value* of a game, divide your answer from #8 by 10000 (the number of players). This will result in an expression that involves the fraction $\frac{N}{10000}$, which is the fraction of players who employ the tit-for-tat strategy. Henceforth, we refer to this fraction as α.

10. Now suppose that you choose to play the "all-E" strategy. Much as you did in #9, calculate the expected value of a game. Your answer will be in terms of α.

11. What fraction of players must be playing tit-for-tat in order for that strategy and the "all-E" strategy to have the same expected value?

12. Based on the evolutionary structure of the game, what do you think happens to the community of tit-for-tat players when they make up a larger fraction of the population than you calculated in #11?

13. Much as you did in #11, determine a general formula for α as a function of ω so that we can calculate the minimal fraction of the population that must play tit-for-tat in order for its expected payoff to be greater than or equal to that of an "all-E" player.

14. Since α is a fraction of a population, it must be in $[0, 1]$. Consequently, the number ω must be in some interval $[a, b]$ in order for the expected payoff of a tit-for-tat player to be greater than or equal to that of an "all-E" player. Determine the numbers a and b.

15. Determine a formula for $\frac{d\alpha}{d\omega}$ and show that it's always negative when $\omega \in (a, b)$. Given the meanings of α and ω, how do we interpret this fact?

Chapter 9
Parametric Curves, Polar Coordinates, and Complex Numbers

Planets moving around the sun, the flow velocity in an artery, the electromagnetic field around a wire, and many other phenomena have a naturally circular structure to them, so we describe them using *polar coordinates*. And when a situation requires several variables for its description, any or all of which change with time, we visualize it as a *parametric curve*. In this chapter we begin a discussion of these topics and combine them with the ideas and techniques of calculus. Additionally, we bring together the ideas of polar coordinates and infinite series in an introduction to *complex numbers*.

9.1 Parametric Curves

Imagine tracking yourself on a GPS device as you drive. Your coordinates change with time as you move, so let's write them as $x(t)$ and $y(t)$. While the particulars of $x(t)$ and $y(t)$ will depend on when you speed up and slow down, the relationship between them is governed by the road. When a point moves in the plane, the path along which it travels (e.g. the road) is called a **parametric curve**, and the independent variable is called the **parameter**.

The word *parametric* is pronounced "para-MET-rik," and the word *parameter* is pronounced "puh-RAM-it-er."

Eliminating the parameter in a simple case

Example 1.1. Describe the motion of the object whose coordinates change with time as follows:

$$x(t) = 4t + 7 \tag{1.1}$$
$$y(t) = -3t + 9. \tag{1.2}$$

Solution: Since $x'(t) = 4$ is constant, the value of x is *increasing* at a constant rate (think "headed east in the Cartesian plane"); and since $y'(t) = -3$ is constant, the value of y is *decreasing* at a constant rate (think "headed south in the Cartesian plane"). So it seems that $(x(t), y(t))$ is moving along a line with negative slope. We can check that intuition by determining the relationship between x and y, independent of time.

In a simple case like this, many people find it easiest to solve $x = 4t + 7$ for t, and then substitute the result into the equation describing y:

$$x = 4t + 7 \quad \Rightarrow \quad \frac{x-7}{4} = t \quad \Rightarrow \quad y = -3\left(\frac{x-7}{4}\right) + 9,$$

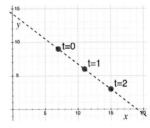

Figure 1.1: The location of $(x(t), y(t))$ at times $t = 1, 2, 3$ as calculated from equations (1.1) and (1.2).

which looks more familiar when we write it as

$$y = -\frac{3}{4}x + \frac{57}{4}. \tag{1.3}$$

As we guessed, this describes motion on a line with negative slope. ∎

In Example 1.1, equations (1.1) and (1.2) tell us explicitly how x and y depend on the parameter, t, so we refer to them as **parametric equations.** By contrast, equation (1.3) tells us about the relationship between x and y in general, regardless of t. We call this the **Cartesian equation** of the parametric curve. In general, the technique of passing from a pair of parametric equations to a single Cartesian equation is called **eliminating the parameter.**

Try It Yourself: Eliminating the parameter in a simple case

Example 1.2. Suppose an object's coordinates at time t are $x(t) = -8t + 11$ and $y(t) = 6t+6$. (a) Describe the motion, and (b) verify your description by eliminating the parameter to find a Cartesian equation for the curve.

Answer: (a) motion along a line, always moving in a northwest direction; (b) $y = -\frac{3}{4}x + \frac{57}{4}$. ∎

Full Solution
On-line

The point $(x(t), y(t))$ moved along the line $y = -\frac{3}{4}x + \frac{57}{4}$ in both Example 1.1 and Example 1.2, but it happened in different directions and at different rates (we'll address the calculation of speed later in this section). We say that these are different **parameterizations** of the same line. Returning to the discussion of driving that led off the section, you can think of each particular parameterization as putting you in the driver's seat—you can calculate your location and direction of travel at any moment. By contrast, the Cartesian description of the parametric curve suppresses that instantaneous information but shows you the whole stretch of road at once.

> Like the word parameter, the word parameterization should be emphasized on the "ram:" puh-RAM-itter-iz-AY-shun.

We've discussed linear motion so far, but we can also investigate other kinds of motion. Circular motion, for example, is often described by the sine and cosine. Whenever trigonometric functions are involved in a parameterization, we employ the trigonometric identities to eliminate the parameter rather than working algebraically.

When trigonometric functions are involved

Example 1.3. Suppose an object's position is $(x(t), y(t))$ where $x(t) = \cos(6\pi t)$ and $y(t) = \sin(6\pi t)$. (a) Describe the horizontal and vertical motion of the object, and (b) eliminate the parameter to find a Cartesian equation for the curve described by these equations.

Solution: (a) Since $x(t) = \cos(t)$ our point moves as far left as $x = -1$, as far right as $x = 1$, and continues to travel back and forth forever. Similarly, the y-coordinate of our point oscillates between $y = -1$ and $y = 1$ forever. This could happen in a variety of ways, so the Cartesian equation of the curve will be very helpful in understanding the way that x and y vary together. (b) Let's make use of the fact that $\cos^2(\theta) + \sin^2(\theta) = 1$ for all values of θ. When $\theta = 6\pi t$, this identity tells us that

$$x^2 + y^2 = \cos^2(6\pi t) + \sin^2(6\pi t) = \cos^2(\theta) + \sin^2(\theta) = 1,$$

so this point is moving along the unit circle. ∎

Try It Yourself: When trigonometric functions are involved

Example 1.4. Suppose $x(t) = \sin(2\pi t)$ and $y(t) = \cos(2\pi t)$. Eliminate the parameter to find a Cartesian equation for the curve described by these equations.

Answer: $x^2 + y^2 = 1$.   ∎

Full Solution On-line

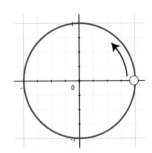

Comparing parameterizations

Example 1.5. Discuss the differences between the parameterizations in Example 1.3 and Example 1.4.

Solution: Both parameterizations traverse the unit circle, but in different ways. Specifically:

Position: When $t = 0$, the point in Example 1.3 is at the location $(1, 0)$, but the point in Example 1.4 starts at $(0, 1)$.

Speed: In Example 1.3 we have $y(t) = \sin(6\pi t)$. So by the time $t = 1$, the argument of the sine is 6π, which means the y coordinate has completed 3 whole cycles—i.e., we've moved around the circle 3 times in 1 second. By contrast, in Example 1.4 we have $y(t) = \cos(2\pi t)$, which completes only 1 cycle by the time $t = 1$, so we've moved around the circle only 1 time in 1 second.

Direction: In Example 1.3 the point is at $(1, 0)$ at time $t = 0$; and since $y'(0) > 0$, the value of y initially increases, so the point travels "northward" along the circle, in a counterclockwise direction. By contrast, in Example 1.4 the point begins at $(0, 1)$, and $x'(0) > 0$, so the point moves "east"—to the right along the unit circle, which means a clockwise direction.   ∎

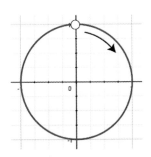

Figure 1.2: The motion described in Example 1.3 (top) and Example 1.4 (bottom).

Examining a parameterization

Example 1.6. Suppose $x(t) = -3\cos(9\pi t)$ and $y(t) = 3\sin(9\pi t)$. Discuss (a) the Cartesian curve that's traversed, (b) the position at time $t = 0$, (c) the direction of travel, and (d) the rate of travel,.

Solution: Because $x^2 + y^2 = 3^2$, we know the point $(x(t), y(t))$ traverses the circle of radius 3 that's centered at the origin; (b) the point starts at $(x(0), y(0)) = (-3, 0)$, which is the leftmost point of the circle; (c) since the point is initially at the leftmost point of the circle and $y'(0) > 0$, it's traveling clockwise around the circle; (d) the point $(x(t), y(t))$ completes 4.5 cycles of the circle "per second" (i.e., by the time that $t = 1$).   ∎

> We know from part (b) that the point is as far left as it can get. In order to determine the direction of travel, we need to know whether it's moving "up" or "down" at $t = 0$, which is why we check $y'(0)$.

Try It Yourself: Designing a parameterization

Example 1.7. Based on what you learned in Examples 1.5 and 1.6, design a parameterization the circle of radius 5 (centered at the origin) that starts at the bottom of the circle, moves counterclockwise, and completes 7 cycles per second.

Answer: $x(t) = 5\sin(14\pi t)$ and $y(t) = -5\cos(14\pi t)$.   ∎

Full Solution On-line

The following example shows that trigonometric functions don't always parameterize a circle.

Parameterizing an ellipse

Example 1.8. Suppose $x(t) = 3\cos(t)$ and $y(t) = 4\sin(t)$. Determine what curve is parameterized by these equations.

Solution: Based on our knowledge of the cosine and sine functions, these parametric equations suggest a circle-like shape, but note that x ranges from -3 to 3, while y ranges from -4 to 4. So while the curve is circle-like, it's wider in one direction than another, which makes us think of an ellipse. That's exactly right, and we can show it by eliminating the parameter. Solving our parametric equations for sine and cosine, we see that $\left(\frac{x}{3}\right)^2 + \left(\frac{y}{4}\right)^2 = \cos^2(t) + \sin^2(t) = 1$.   ■

Combining "overview" and "details"

Example 1.9. Suppose $x(t) = 1 + \cos(t)$ and $y(t) = 3 + \sin^2(t)$. Eliminate the parameter to find a Cartesian equation for the curve parameterized by these equations.

Solution: We'd like to use the Pythagorean identity, $\sin^2(\theta) + \cos^2(\theta) = 1$. To do so, we need to isolate the trigonometric function in each of our parametric equations:

$$x = 1 + \cos(t) \quad \Rightarrow \quad \cos(t) = x - 1$$
$$y = 3 + \sin^2(t) \quad \Rightarrow \quad \sin^2(t) = y - 3.$$

Substituting these into the Pythagorean identity, we get

$$\cos^2(\theta) + \sin^2(\theta) = 1$$
$$(x - 1)^2 + y - 3 = 1.$$

That is, our parametric equations describe the motion of a point along the parabola $y = -(x - 1)^2 + 4$, whose graph is shown in Figure 1.3. As discussed above, though eliminating the parameter has provided us with a familiar description of the path that's followed by $(x(t), y(t))$, the equation $y = -(x - 1)^2 + 4$ doesn't tell the whole story. Specifically, notice that our equation for $y(t)$ looks like

$$y(t) = 3 + something\ that's\ never\ negative.$$

So the y-coordinate of our point is never less than 3, so most of the parabola is *never visited.*

Also note that the x-coordinate of our point depends on the cosine, which achieves its maximum when $t = 0$ and its minimum when $t = \pi$. If we start keeping track of position at $t = 0$, our point begins at $(2, 3)$ and then moves to the left (since $x'(0) < 0$). It moves along the parabola $y = -(x - 1)^2 + 4$ until time $t = \pi$, when it reaches the point $(0, 3)$. Then it turns around, and travels back along the parabola to $(2, 3)$. It continues to "bounce" back and forth along the upper arch of the parabola forever.   ■

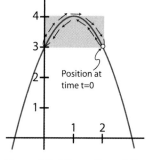

Figure 1.3: The motion of the point in Example 1.9.

Distinguishing between intersection points and collision points

Example 1.10. Suppose the point P moves in the plane according to $x_p(t) = -\cos(t)$ and $y_p(t) = -\sin(t)$, and the point Q moves according to $x_q(t) = t/\pi$ and $y_q(t) = -1.5 + 3t/2\pi$. (a) Show that the paths of P and Q intersect twice, and (b) determine whether the points collide.

Solution: (a) The point P travels around the unit circle, $x^2 + y^2 = 1$, and the point Q travels along the line $y = 1.5x - 1.5$. Graphing these curves shows two intersection points (see Figure 1.4). To find them, we substitute $y = 1.5x - 1.5$ into the equation of the circle, and the quadratic formula tells us that $x = 5/13$ and $x = 1$.

(b) It's easy to see that $x_q(5\pi/13) = 5/13$, but $x_p(5\pi/13) = 0.3546$ at that time. Since these x-coordinates are not the same, the points P and Q don't arrive at the left-hand point of intersection simultaneously, so they don't collide. However, both points arrive at the right-hand point of intersection when $t = \pi$, which we check by calculating $(x_q(\pi), y_q(\pi)) = (1, 0)$ and $(x_p(\pi), y_p(\pi)) = (1, 0)$, so the points P and Q *do* collide there. In more complicated cases, graphing the distance between P and Q as a function of time can be helpful:

$$\text{Distance between } P \text{ and } Q = \sqrt{(x_1(t) - x_2(t))^2 + (y_1(t) - y_2(t))^2}.$$

The right-hand image of Figure 1.4 shows the distance between P and Q in this example. Notice that it's zero at time $t = \pi$. ■

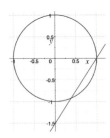
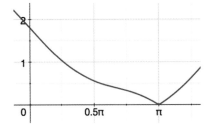

Figure 1.4: (left) The path of P and the path of Q in Example 1.10; (right) the distance between points P and Q as a function of time.

In all of the examples so far, the parameter has been *time*, but it doesn't have to be. For example . . .

Dilution paths (an example in which the parameter is not time)

Example 1.11. Suppose we have two kinds of molecules that don't combine chemically, say type A and type B, and we put them into a container of water. After everything has had time to diffuse thoroughly, we start taking samples of the solution. On average, what fraction of the molecules in a sample will be type A, what fraction will be type B, and how do these fractions change when we add more water?

Solution: The answer to this question depends on three quantities:

$$a \; = \; \text{the number of molecules of type } A \text{ in the container,}$$
$$b \; = \; \text{the number of molecules of type } B \text{ in the container, and}$$
$$w \; = \; \text{the number of water molecules in the container.}$$

With this notation, the total number of molecules in the solution is $N = a + b + w$, and the fraction of them that are of type A is

$$x = \frac{a}{N} = \frac{a}{a + b + w}. \tag{1.4}$$

Similarly, the fraction of molecules in the container that are of type B is

$$y = \frac{b}{N} = \frac{b}{a + b + w}. \tag{1.5}$$

Both x and y decrease when we add more water (since a and b remain fixed but w increases), so the point $(x(w), y(w))$ moves toward the origin. Note that

> In this case x and y change with the number of water molecules, so w is the parameter.

$$\frac{y}{x} = \frac{\frac{b}{N}}{\frac{a}{N}} = \frac{b}{a} \tag{1.6}$$

is always the same, so $(x(w), y(w))$ moves toward the origin along a line (called the *dilution path*) as w increases. Specifically, the line $y = \frac{b}{a}x$. ■

❖ The Cycloid

Suppose you bring your car to a stop, right on a spot of wet white paint. Now you have a white dot on the bottom of your tire. When you start up again, assuming that the tire doesn't slip, what path does that white dot follow? Let's make the question simpler by focusing on an idealized scenario: a unit circle sits atop the x-axis and the point P is at the bottom of the circle. When that circle rolls forward, the point P moves too.

  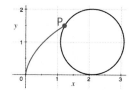 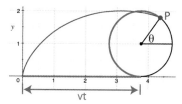

Figure 1.5: (left to right) Tracking the position of the point P.

For the sake of discussion, suppose the center of the circle moves forward at a constant velocity, v. Then after t seconds, the center has moved vt units to the right, and its height is always 1, so its coordinates are $(vt, 1)$. By designating θ as the angle depicted in the right-hand image of Figure 1.5, we can write the coordinates of P as

$$x(t) = vt + \cos(\theta) \quad \text{and} \quad y(t) = 1 + \sin(\theta).$$

The last step is writing θ in terms of time. To do it, we'll make an argument about distance and length. Specifically, because the wheel is not slipping, the length of arc that has touched the road is the same as the distance traveled by the center, vt. Additionally, the length of arc between points P and Q is θ when we measure in radians (as seen in the top image of Figure 1.6). So on the one hand, we can express the length of arc between the road and the point Q (measured clockwise) as $vt + \theta$. On the other hand, we know that length of arc is $3\pi/2$ since we're talking about a unit circle. Therefore, $vt + \theta = \frac{3\pi}{2}$, so

$$\cos(\theta) = \cos\left(\frac{3\pi}{2} - vt\right) = -\sin(vt), \quad \text{and} \quad \sin(\theta) = \sin\left(\frac{3\pi}{2} - vt\right) = -\cos(vt).$$

This allows to write the coordinates of the point P as

$$x(t) = vt - \sin(vt) \quad \text{and} \quad y(t) = 1 - \cos(vt). \tag{1.7}$$

You'll extend this derivation to a circle of radius R in the exercise set, in which case the parametric equations describing the position of P are

$$x(t) = vt - R\sin\left(\frac{vt}{R}\right) \quad \text{and} \quad y(t) = R - R\cos\left(\frac{vt}{R}\right). \tag{1.8}$$

The curve that's parameterized by these equations is called the **cycloid**, and though we won't show it here, it's the solution of two famous problems. For the sake of discussion, suppose the point P_0 is above and to the left of the point P_1.

The **Tautochrone** Problem: Is there a curve from P_0 to P_1 down which a bead will slide (under the force of gravity), arriving at P_1 in the same amount of time no matter where it starts? If we neglect friction, yes, and that curve is an inverted (i.e., cupped-up) cycloid.

The **Brachistochrone** Problem: What curve minimizes the time it takes a bead to slide from P_0 to P_1 under the force of gravity? An inverted cycloid.

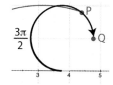

Figure 1.6: Two ways of calculating the distance between the road and the point Q (measured clockwise around the circle).

The word *tautochrone* is pronounced "TOTE-oh-krone."

The word *brachistochrone* is pronounced "brah-KEEST-oh-krone."

❖ Common Difficulties

Many people have trouble eliminating the parameter from a pair of equations that involve the trigonometric functions. As a simple example, let's take $x(t) = \cos(t)$ and $y(t) = \sin(t)$. Some people try to solve one of the equations for t and substitute into the other:

$$t = \cos^{-1}(x) \quad \Rightarrow \quad y = \sin(\cos^{-1}(x)). \tag{1.9}$$

The problem here is that $t = \cos^{-1}(x)$ is only true when $t \in [0, \pi]$, because that's the range of the arccosine. So equation (1.9) says

$$y = \sin(\textit{some angle between } 0 \textit{ and } \pi) \geq 0,$$

which only describes the upper half of the unit circle, even though the point $(x(t), y(t))$ travels all the way around. To avoid this problem, try to work with trigonometric identities whenever parametric equations include trigonometric functions.

You should know

- the terms *parametric curve, parametric equations, parameter, Cartesian equation,* and *cycloid;*

- that parametric equations describe the position of a point in the plane, and how it moves as the parameter changes.

You should be able to

- eliminate the parameter to arrive at a Cartesian equation for a parametric curve, using both algebraic techniques and trigonometric identities;

- discuss the motion of $(x(t), y(t))$ on based on x' and y';

- parameterize linear, circular, and elliptical motion.

❖ 9.1 Skill Exercises

In #1–8 (a) eliminate the parameter to find the Cartesian equation that describes the curve parameterized by $x(t)$ and $y(t)$, (b) sketch the curve, indicating the position at $t = 0$ and the direction of motion (as t increases from zero), and (c) determine what portion of the curve is visited by $(x(t), y(t))$ as t increases from zero.

1. $x(t) = -4t + 1$, $y(t) = 8t - 7$ 5. $x(t) = -\cos^2(t)$, $y(t) = -\sin(t)$

2. $x(t) = -9t - 6$, $y(t) = -2t + 4$ 6. $x(t) = \cos(t)$, $y(t) = -\sin^2(t)$

3. $x(t) = t^3 + 4$, $y(t) = 5t + 1$ 7. $x(t) = \tan(t)$, $y(t) = \sec(t)$

4. $x(t) = t + 4$, $y(t) = 3t^5 + 1$ 8. $x(t) = \tan^2(t)$, $y(t) = \sec^2(t)$

In #9–12 (a) complete the square in order to solve for t in terms of x or y, and (b) eliminate the parameter to find the Cartesian equation that describes the curve parameterized by $x(t)$ and $y(t)$, (c) sketch the curve, indicating the position at

$t = 0$ and the direction of motion (as t increases from zero), and (d) determine what portion of the curve is visited by $(x(t), y(t))$ as t increases from zero.

9. $x(t) = t^2 + 2t + 1$, $y(t) = 3t^2 + 19$

10. $x(t) = t^4 + 5t - 1$, $y(t) = 3t^2 + 18t + 27$

11. $x(t) = t^2 + 2t + 12$, $y(t) = t^4 - t^2 + 19$

12. $x(t) = t^6 + t^2 + 10$, $y(t) = t^2 + 8t + 20$

In #13–16 the graphs of $x(t)$ and $y(t)$ are shown together. Combine the information they provide to sketch a graph of the curve parameterized by $(x(t), y(t))$ when $t \in [0, 1]$.

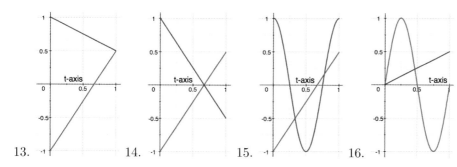

13.   14.   15.   16.

In #17–22 you'll work with the circle of radius 7 that's centered at the origin. Based on what you saw in Examples 1.5 and 1.7, write a parameterization whose position at time $t = 0$ is the specified point, moves in the specified direction (cw=clockwise, ccw=counterclockwise) at the given speed (Hz=cycles (around the circle) per second).

17. $(0, 7)$, ccw, 13 Hz

18. $(0, -7)$, ccw, 14 Hz

19. $(0, 7)$, cw, 1.8 Hz

20. $(0, -7)$, cw, 9.7 Hz

21. $\left(\frac{7\sqrt{2}}{2}, \frac{7\sqrt{2}}{2}\right)$, ccw, 4 Hz

22. $\left(\frac{7}{2}, \frac{21\sqrt{2}}{2}\right)$, cw, 5 Hz

In exercises #23–28 you'll work with the circle of radius 4 that's centered at the point (5,11). Write a parameterization whose position at time $t = 0$ is the specified point, moves in the specified direction (cw=clockwise, ccw=counterclockwise) at the given speed (Hz=cycles (around the circle) per second).

23. $(9, 11)$, ccw, 21 Hz

24. $(1, 11)$, ccw, 22 Hz

25. $(5, 15)$, cw, 2 Hz

26. $(5, 7)$, cw, 3 Hz

27. $(5 - 2\sqrt{2}, 11 + 2\sqrt{2})$, cw, 2 Hz

28. $(5 + 2\sqrt{3}, 13)$, cw, 3 Hz

29. Suppose $x(t) = 8\sin(t)$ and $y(t) = 2\cos(t)$. (a) Sketch a graph of the ellipse parameterized by $(x(t), y(t))$, (b) locate the point $(x(0), y(0))$ and indicate the direction of motion when $t > 0$.

30. Suppose $x(t) = -13\cos(8t)$ and $y(t) = 4\sin(8t)$. (a) Sketch a graph of the ellipse parameterized by $(x(t), y(t))$, (b) locate the point $(x(0), y(0))$ and indicate the direction of motion when $t > 0$.

31. Determine equations for $x(t)$ and $y(t)$ so that $(x(0), y(0)) = (0, 0.5)$ and $(x(t), y(t))$ proceeds around the ellipse $9x^2 + 4y^2 = 1$ in a counterclockwise direction, completing one cycle in 3 seconds.

32. Determine equations for $x(t)$ and $y(t)$ so that $(x(0), y(0)) = (0.1, 0)$ and $(x(t), y(t))$ proceeds around the ellipse $100x^2 + 16y^2 = 1$ in a clockwise direction, completing one cycle in 9 seconds.

❖ 9.1 Concept and Application Exercises

33. Suppose P moves in the plane according to $x_p(t) = 13t$ and $y_p(t) = 7t + 4$, and Q moves in the plane according to $x_q(t) = 8ct$ and $y_q(t) = -1 + 16ct$, where $c > 0$.

    (a) By eliminating the parameter, show that P and Q move along intersecting lines.

    (b) Determine a value of c for which a collision happens at that intersection point.

34. Suppose P moves in the plane according to $x_p(t) = 4t$ and $y_p(t) = 4t - 5$, and Q moves in the plane according to $x_q(t) = 8t$ and $y_q(t) = b - 8t$, where $b > 0$.

    (a) By eliminating the parameter, show that P and Q move along intersecting lines.

    (b) Show the intersection point is never a collision point, regardless of b.

35. Suppose P moves in the plane according to $x_p(t) = 2\cos(t)$ and $y_p(t) = 2\sin(t)$, and Q moves in the plane according to $x_q(t) = 2 + 3\cos(\omega t)$ and $y_q(t) = 3\sin(\omega t)$, where $\omega > 0$.

    (a) By eliminating the parameter, show that P and Q move along intersecting circles.

    (b) Determine the x-coordinate of the intersection points.

    (c) Determine a value of ω for which the intersection point in the 2nd quadrant is not a point of collision, but the intersection point in the 3rd quadrant is. (It's an important fact that P moves at a constant speed of 2 units per second, and Q moves at a constant speed of 3ω units per second.)

36. Suppose P moves in the plane according to $x_p(t) = 3\cos(t)$ and $y_p(t) = 3\sin(t)$, and Q moves in the plane according to $x_q(t) = 3 + \cos(\omega t)$ and $y_q(t) = \sin(\omega t)$, where $\omega > 0$.

    (a) By eliminating the parameter, show that P and Q move along intersecting circles.

    (b) Determine the x-coordinate of the intersection points.

    (c) Determine a value of ω for which the intersection point in the 1st quadrant is not a point of collision, but the intersection point in the 4th quadrant is.

Exercises #37–39 address the bifold door shown in Figure 1.7, where the sliding pin closes along the track according to $x_s(t) = 100 - t$, and $y_s(t) = 0$. Treat the panels as being 50 cm wide and the pins as being points.

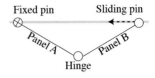

Figure 1.7: For #37–39.

37. Parameterize the path followed by the hinge, using time as the parameter. *(Hint: the hinge is attached rigidly to the sliding pin by Panel B.)* What kind of curve is parameterized by these equations?

38. Parameterize the path followed by the midpoint of Panel A, using time as the parameter. What kind of curve is parameterized by these equations?

39. Parameterize the path followed by the midpoint of Panel B, using time as the parameter. What kind of curve is parameterized by these equations?

40. A *linear* **Bézier** curve between control points $P_0 = (x_0, y_0)$ and $P_1 = (x_1, y_1)$ is defined by $x(t) = x_0(1-t) + tx_1$ and $y(t) = y_0(1-t) + ty_1$.

    (a) Verify that $(x(0), y(0)) = P_0$ and $(x(1), y(1)) = P_1$.

    (b) Eliminate the parameter to show that $x(t)$ and $y(t)$ parameterize a line.

    (c) Parameterize the line segment connecting the points $P_0 = (2, 4)$ and $P_1 = (3, 7)$ with a linear Bézier curve.

    (d) How does the linear Bézier curve change if we swap the control points, making $P_0 = (3, 7)$ and $P_1 = (2, 4)$?

41. The parameterization of the cycloid described by equation (1.7) assumes that the wheel is not slipping.

    (a) Derive equations for $x(t)$ and $y(t)$ in the case when the wheel (radius= 1 foot) is spinning clockwise at 3π radians per second, and its center is moving to the right at 2 feet per second.

    (b) Use a graphing utility to graph $(x(t), y(t))$ when $t \in [0, 5]$.

42. The parameterization of the cycloid described by equation (1.7) assumes that the wheel is not slipping.

    (a) Derive equations for $x(t)$ and $y(t)$ in the case when the wheel (radius= 1 foot) is spinning clockwise at 5π radians per second, and its center is moving to the left at 4 feet per second.

    (b) Use a graphing utility to graph $(x(t), y(t))$ when $t \in [0, 5]$.

Exercises #43–46 ask you to graph the position and velocity of an object simultaneously by treating them as the x and y coordinates of a point in the xy-plane. When we do this, we refer to the plane as the *phase plane*.

43. Suppose the displacement of an object from its equilibrium is $D(t) = 4\sin(3t)$. Graph the curve parameterized by $x = D(t)$ and $y = D'(t)$.

44. The *harmonic oscillator equation* predicts that the displacement from equilibrium of atoms in a diatomic molecule is $D(t) = A\cos\left(\sqrt{k/m}\ t\right)$, where $k > 0$ depends on the electrostatic properties of the atoms, and m is their (common) mass. (a) Show that the curve parameterized by $x = D(t)$ and $y = D'(t)$ is an ellipse, and (b) explain how the numbers A, k, and m change its shape.

45. Suppose a point starts at rest at the origin and then moves along the x-axis with an acceleration described by $a(t) = t\sin(\pi t)$.

    (a) Find the functions describing the point's velocity $v(t)$ and position $x(t)$ as functions of time.

    (b) Use a graphing utility to render the parametric curve defined by $x = x(t)$ and $y = v(t)$.

    (c) Based on your graph in part (b), what can you say about the relationship between position and velocity?

46. Suppose a point starts at rest at the origin and then moves along the x-axis according to $x(t) = e^{-t}\left(2\cos(2t) + \sin(2t)\right)$.

    (a) Find the function describing the point's velocity, $v(t)$.

    (b) Use a graphing utility to render the parametric curve defined by $x = x(t)$ and $y = v(t)$.

(c) Based on your graph in part (b), what can you say about the relationship between position and velocity?

In Example 1.11 you saw that the parameter might not be time. In #47–54 you see other such examples based on topics from Chapter 7.

47. Consider a rod that stretches across the interval $[0, b]$, with density $\rho(D) = D + 1$ kg/m, where D represents the distance from the left end of the rod in meters.

   (a) Find the mass of the rod, $m(b)$.

   (b) Find the first moment of the rod with respect to the y-axis, $M(b)$.

   (c) Use a graphing utility to render the parametric curve defined by $x = m(b)$ and $y = M(b)$.

   (d) Based on the curve you graphed in part (c), explain how the mass and moment change together as b grows.

48. Consider a rod that stretches across the interval $[0, b]$, with density $\rho(D) = D + 1$ kg/m, where D represents the distance from the left end of the rod in meters.

   (a) Find the mass of the rod, $m(b)$.

   (b) Find the centroid of the rod, $\bar{x}(b)$.

   (c) Use a graphing utility to render the parametric curve defined by $x = m(b)$ and $y = \bar{x}(b)$.

   (d) What does it mean in this context when the parametric curve from part (c) lies *below* the line $y = x$? What about when it is above the line $y = x$?

49. Consider a rod of length b and density $\rho(D) = 8e^{-D/3}$ kg/m, where D represents the distance from the left end of the rod in meters.

   (a) Find the mass of the rod $m(b)$.

   (b) Find the first moment of the rod with respect to the y-axis, $M(b)$.

   (c) Use a graphing utility to render the parametric curve defined by $x = m(b)$ and $y = M(b)$.

   (d) What does the graph from part (c) indicate about $\lim_{b \to \infty} m(b)$ and $\lim_{b \to \infty} M(b)$?

   (e) Calculate $\lim_{b \to \infty} m(b)$ and $\lim_{b \to \infty} M(b)$, and compare them to your answer from part (d).

50. Consider a rod of length b and density $\rho(D) = 5e^{-D/2}$ kg/m, where D represents the distance from the left end of the rod in meters.

   (a) Find the mass of the rod, $m(b)$.

   (b) Find the centroid of the rod, $\bar{x}(b)$.

   (c) Use a graphing utility to render the parametric curve defined by $x = m(b)$ and $y = \bar{x}(b)$.

   (d) What does the graph from part (c) indicate about $\lim_{b \to \infty} m(b)$ and $\lim_{b \to \infty} \bar{x}(b)$?

   (e) Calculate $\lim_{b \to \infty} m(b)$ and $\lim_{b \to \infty} \bar{x}(b)$, and compare them to your answer from part (d).

51. Consider the curve $y = x^{3/2}$ on the interval $[0, b]$ and the surface generated by rotating the curve about the y-axis.

   (a) Find the arc length $L(b)$ and the surface area $S(b)$.

   (b) Use a graphing utility to render the parametric curve defined by $x = L(b)$ and $y = S(b)$.

   (c) Based on the curve you graphed in part (b), explain how arclength and surface area change together as b grows.

52. Suppose $\mathcal{R}$ is the region between the curve $y = x^{3/2}$ and the x-axis when $0 \le x \le b$.

   (a) Find the surface area, $S(b)$, and the volume, $V(b)$, of the solid generated by revolving $\mathcal{R}$ about the x-axis.

   (b) Use a graphing utility to render the parametric curve defined by $x = S(b)$ and $y = V(b)$.

   (c) Based on the curve you graphed in part (b), explain how surface area and volume change together as b grows.

53. Suppose a rectangular tank with a length of 1 meter, a width of 0.75 meters and a height of 0.5 meters contains water that is D meters deep. (Recall that the mass density of water is 1000 kg/m^3, and take g to be 9.8 m/sec^2.)

   (a) Determine the amount of work $W(D)$ done by gravity in draining the tank from the bottom.

   (b) Determine the amount of force $F(D)$ from the water pushing against one of the 1×0.5 meter faces.

   (c) Use a graphing utility to render the parametric curve defined by $x = W(D)$ and $y = F(D)$.

   (d) Based on your graph in part (c), explain how the work done by gravity and the hydrostatic force change together as D grows.

54. Suppose a trough is 3 meters long and 0.5 meters high, and its ends are in the shape of equilateral triangles (points down). Suppose the trough contains water that is D meters deep. (Recall that the mass density of water is 1000 kg/m^3.)

   (a) Determine the amount of work $W(D)$ done by gravity in draining the trough from the bottom.

   (b) Determine the amount of force $F(D)$ from the water pushing against one of the triangular end-faces.

   (c) Use a graphing utility to render the parametric curve defined by $x = W(D)$ and $y = F(D)$.

   (d) Based on your graph in part (c), explain how the work done by gravity and the hydrostatic force change together as D grows.

Checking the details

55. Verify the parameterization of the cycloid suggested in equation (1.8).

9.2   Calculus with Parametric Curves

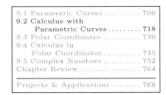
In Example 1.1 on p. 706 we saw that $y'(t) = -3$ and $x'(t) = 4$ were constant, and we established that the point $(x(t), y(t))$ moved along a line. More specifically, over the course of Δt seconds we see $\Delta y = -3\Delta t$ and $\Delta x = 4\Delta t$, from which we conclude that the slope of the line is

$$\frac{\Delta y}{\Delta x} = \frac{-3\Delta t}{4\Delta t} = \frac{-3}{4} = \frac{y'(t)}{x'(t)}. \tag{2.1}$$

In this section, we'll see how the ideas embodied by this equation apply to non-linear parametric curves. We'll also discuss arc length and area in this context of parametric curves.

❖ Tangent Lines to Parametric Curves

Equation (2.1) extends naturally to nonlinear parametric curves through the Chain Rule. Specifically, as long as $x'(t) \neq 0$,

$$\frac{dy}{dt} = \frac{dy}{dx}\frac{dx}{dt}, \quad \text{which is the same as writing} \quad y'(t) = \frac{dy}{dx}x'(t).$$

Solving this equation for $\frac{dy}{dx}$ leads us to the following assertion.

> **Tangent lines to parametric curves:** Suppose $\mathcal{C}$ is a curve parameterized by $x(t)$ and $y(t)$, both of which are differentiable at time t_0, and y is a differentiable function of x at the point $P_0 = (x(t_0), y(t_0))$. Then the slope of the tangent line to $\mathcal{C}$ at P_0 is
> $$\frac{dy}{dx} = \frac{y'(t_0)}{x'(t_0)}, \tag{2.2}$$
> provided that $x'(t_0) \neq 0$.

> Algebraically, the assertion that $x'(t) \neq 0$ allows us to divide by $x'(t)$. At an intuitive level, $x'(t) \neq 0$ means that the curve is wandering to the right or left, so we can think of y as a function of x, and talk about dy/dx.

Calculating slopes of parametric curves

Example 2.1. Suppose $\mathcal{C}$ is the curve parameterized by $x(t) = 7t^4 + \sin(\pi t)$ and $y(t) = \ln(t^2 + 3)$ for $t \in [0.5, 2]$. Determine the equation of the tangent line to $\mathcal{C}$ when $t = 1$.

Solution: The point on $\mathcal{C}$ visited by $(x(t), y(t))$ at time $t = 1$ is $(x(1), y(1)) = (7, \ln(4))$. To find the slope of the tangent line to $\mathcal{C}$ at that point, we need to know

$$x'(t) = 28t^3 + \pi \cos(\pi t) \quad \text{and} \quad y'(t) = \frac{2t}{t^2 + 3}.$$

With those formulas, we can calculate that

$$\frac{dy}{dx}\bigg|_{t=1} = \frac{y'(1)}{x'(1)} = \frac{1/2}{28 - \pi} = \frac{1}{56 - 2\pi}.$$

So the equation of the tangent line is

$$y - \ln(4) = \left(\frac{1}{56 - 2\pi}\right)(x - 7). \qquad \blacksquare$$

Calculating slopes of parametric curves

Example 2.2. The parametric equations $x(t) = \cos(t) + 2\sin(t)$ and $y(t) = 2\sin(t) - \cos(t)$ parameterize a rotated ellipse (see Figure 2.1). What are the coordinates of its lowest and highest points?

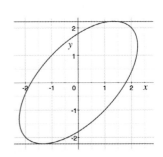

Figure 2.1: The rotated ellipse in Example 2.2.

Solution: The tangent line will be horizontal at the lowest and highest points, so let's look for its slope to be zero. The slope of the tangent line at $(x(t), y(t))$ is

$$\frac{dy}{dx} = \frac{y'(t)}{x'(t)} = \frac{2\cos(t) + \sin(t)}{2\cos(t) - \sin(t)},$$

which is zero when $2\cos(t) + \sin(t) = 0$. Solving this equation for t yields $t = \arctan(-2)$, at which time $x = -3/\sqrt{5}$ and $y = -\sqrt{5}$. It also happens when $t = \arctan(-2) + \pi$, at which time $x = 3/\sqrt{5}$ and $y = \sqrt{5}$. The former is the lowest point, and the latter is the highest.   ∎

Try It Yourself: Calculating slopes of parametric curves

Example 2.3. Suppose $\mathcal{C}$ is the curve parameterized by $x(t) = t + \sin(\pi t)$ and $y(t) = 0.5t + \cos(\pi t)$. Determine the equation of the tangent line to $\mathcal{C}$ when $t = 0.5$.

Answer: $y - \frac{1}{4} = \left(\frac{1}{2} - \pi\right)\left(x - \frac{3}{2}\right)$ ∎

Full Solution On-line

We cannot always use equation (2.2)

Example 2.4. Suppose $\mathcal{C}$ is the loop depicted in Figure 2.2, which is parameterized by $x(t) = t + 1.5\cos(t)$ and $y(t) = 1 + 1.5\sin(t)$. Identify points at which we cannot use equation (2.2) to calculate the slope of the tangent line.

Solution: At points P_0 and P_1 you see that the tangent line is vertical, which happens because $x'(t) = 0$ but $y'(t)$ is not, so dy/dx is undefined. At point P_2 you see that y is not a differentiable function of x because the curve self-intersects. Consequently, the loop doesn't have a well defined tangent line at P_2.   ∎

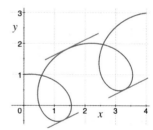
Figure 2.2: This parametric loop does not have a well-defined tangent line at P_2.

The curve parameterized in Example 2.3 is shown in Figure 2.3. Note that it begins at $(x(0), y(0)) = (0, 1)$, and at time $t = 4$ the parameterization arrives at $(x(4), y(4)) = (4, 3)$. Since this smooth curve rises a net distance of 2 units while it meanders a net distance of 4 units to the right, we're led to conjecture that there is some point where its tangent line has a slope of rise/run $= 2/4 = 0.5$. Cauchy's extension of the Mean Value Theorem, which we first saw on p. 284, guarantees that we're right.

Cauchy's Mean Value Theorem: Suppose $x(t)$ and $y(t)$ are both continuous when $a \leq t \leq b$, and differentiable when $a < t < b$. Then there is a number $t^* \in (a, b)$ at which

$$\Big(x(b) - x(a)\Big)y'(t^*) = \Big(y(b) - y(a)\Big)x'(t^*).$$

Note: You proved this fact on p. 290.

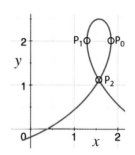
Figure 2.3: The curve $\mathcal{C}$ in Example 2.3.

On p. 284 we motivated Cauchy's MVT with a story about running a race, and then used it to prove L'Hôpital's Rule. Though we indicated that there is a geometric interpretation of the formula, we had to wait until now to see it.

Using Cauchy's Mean Value Theorem

Example 2.5. Suppose $\mathcal{C}$ is the curve parameterized in Example 2.3. Use Cauchy's Mean Value Theorem to prove that there is a point on $\mathcal{C}$ at which the tangent line has a slope of $1/2$.

Solution: At time $t = 0$ the parameterization in Example 2.3 gives $x(0) = 0$ and $y(0) = 1$. At time $t = 2$, it tells us that $x(4) = 4$ and $y(4) = 3$. Since $x(t)$ and $y(t)$ are continuous on $[0, 4]$ and differentiable on $(0, 4)$, Cauchy's Mean Value Theorem tells us that there is a time $t^* \in (0, 4)$ at which

$$\Big(x(4) - x(0)\Big)y'(t^*) = \Big(y(4) - y(0)\Big)x'(t^*)$$
$$4y'(t^*) = 2x'(t^*).$$

But this is the same as saying that

$$\frac{2}{4} = \frac{y'(t^*)}{x'(t^*)} = \frac{dy}{dx}\bigg|_{t=t^*}.$$ ∎

❖ Concavity of Parametric Curves

When $\mathcal{C}$ is the graph of a function $y(x)$, we use $\frac{d^2y}{dx^2} = \frac{d}{dx}\left(\frac{dy}{dx}\right)$ to determine its concavity at each point. As we did with slopes of tangent lines, we can extend this idea to parametric curves by using the Chain Rule. For notational convenience in the following calculation, let's temporarily denote $\frac{dy}{dx}$ by y_x. Then

$$\frac{dy_x}{dt} = \frac{dy_x}{dx}\frac{dx}{dt} \quad\Rightarrow\quad \frac{dy_x}{dx} = \frac{\frac{dy_x}{dt}}{x'(t)}$$

when $x'(t) \neq 0$. Since y_x is $y'(t)/x'(t)$, the numerator on the right-hand side of this equation is really

$$\frac{d}{dt}\left(\frac{y'(t)}{x'(t)}\right) = \frac{x'(t)y''(t) - y'(t)x''(t)}{(x'(t))^2},$$

so our equation is

$$\frac{dy_x}{dx} = \frac{x'(t)y''(t) - y'(t)x''(t)}{(x'(t))^3}.$$

And since the left-hand side of the equation is

$$\frac{dy_x}{dx} = \frac{d}{dx}\left(\frac{dy}{dx}\right) = \frac{d^2y}{dx^2},$$

we have the following:

Concavity of parametric curves: Suppose $\mathcal{C}$ is a curve parameterized by $x(t)$ and $y(t)$, both of which are twice differentiable at time t_0, and y is a twice-differentiable function of x at the point $P_0 = (x(t_0), y(t_0))$. Then the concavity of $\mathcal{C}$ at P_0 is

$$\frac{d^2y}{dx^2}\bigg|_{P_0} = \frac{x'(t_0)y''(t_0) - y'(t_0)x''(t_0)}{(x'(t_0))^3}, \tag{2.3}$$

provided that $x'(t_0) \neq 0$.

Calculating concavity of a parametric curve

Example 2.6. Suppose $\mathcal{C}$ is the curve parameterized by $x(t) = 3t$ and $y(t) = 1/t$ when $t > 0$. Determine the concavity of $\mathcal{C}$ at $(x(\frac{1}{3}), y(\frac{1}{3}))$.

Solution: We begin by calculating the necessary derivatives of $x(t)$ and $y(t)$:

$$\begin{aligned} x'(t) &= 3 & y'(t) &= -1/t^2 \\ x''(t) &= 0 & y''(t) &= 2/t^3. \end{aligned}$$

Upon substituting these into equation (2.3), we see that the concavity of $\mathcal{C}$ is

$$\frac{d^2y}{dx^2}\bigg|_{(x(t),y(t))} = \frac{2}{9t^3}$$

at each $t > 0$. At time $t = 1/3$ this tells us $d^2y/dx^2 = 6 > 0$, so the graph is concave up. By eliminating the parameter, we see that $\mathcal{C}$ is the familiar curve $y = 3/x$, so we can check our answer at $x(1/3) = 1$ using the methods of Chapter 4:

Check that $y = 3/x$ by eliminating the parameter for yourself!

$$\frac{d^2y}{dx^2} = \frac{6}{x^3} \quad \text{which is 6 when } x = 1.$$ ∎

Try It Yourself: Calculating concavity of a parametric curve

Example 2.7. Suppose $x(t) = 12t^2$ and $y(t) = 4t^2 + 8$ when $t > 0$. (a) Determine the d^2y/dx^2 at $(x(2), y(2))$, and (b) explain why the result makes sense.

Full Solution
On-line

Answer: (a) 0 (b) See the on-line solution. ∎

❖ Arc Length of Parametric Curves

Suppose $(x(t), y(t))$ traverses the curve $\mathcal{C}$ as t increases from $t = a$ to $t = b$. We can use this parameterization of $\mathcal{C}$ to determine its length by way of a definite integral. As we develop this integral below, we'll assume that $x'(t)$ and $y'(t)$ are continuous over (a, b).

As usual, we begin with an approximation. Let's partition the interval $[a, b]$ into n subintervals of equal length, $a = t_0 < t_1 < t_2 < \cdots < t_n = b$. Then each $P_k = (x(t_k), y(t_k))$ is a point on $\mathcal{C}$ (see Figure 2.4).

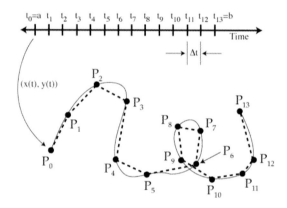

Figure 2.4: Partition points on $\mathcal{C}$.

The line segment connecting P_{k-1} to P_k has a length of

$$\left|\overline{P_{k-1}P_k}\right| = \sqrt{\Big(x(t_k) - x(t_{k-1})\Big)^2 + \Big(y(t_k) - y(t_{k-1})\Big)^2}.$$

Since $y(t)$ is continuous over $[t_{k-1}, t_k]$ and differentiable over (t_{k-1}, t_k), the Mean Value Theorem guarantees a time $\tau_k \in (t_{k-1}, t_k)$ at which

$$\frac{y(t_k) - y(t_{k-1})}{\Delta t} = y'(\tau_k) \quad \Rightarrow \quad y(t_k) - y(t_{k-1}) = y'(\tau_k)\Delta t. \tag{2.4}$$

When Δt is very small, τ_k is very near the midpoint of $[t_{k-1}, t_k]$, which we'll denote by t_k^*. Since $y'(t)$ is continuous we know that $y'(\tau_k)$ converges to $y'(t_k^*)$ as $n \to \infty$, so let's approximate $y'(\tau_k) \approx y'(t_k^*)$ in equation (2.4), and write

$$y(t_k) - y(t_{k-1}) \approx y'(t_k^*)\Delta t.$$

Similarly,

$$x(t_k) - x(t_{k-1}) \approx x'(t_k^*)\Delta t.$$

So we can rewrite the length of $\overline{P_{k-1}P_k}$ as

$$\left|\overline{P_{k-1}P_k}\right| \approx \sqrt{\left(x'(t_k^*)\Delta t\right)^2 + \left(y'(t_k^*)\Delta t\right)^2} = \sqrt{\left(x'(t_k^*)\right)^2 + \left(y'(t_k^*)\right)^2} \; \Delta t.$$

The sum of these lengths approximates the length of $\mathcal{C}$, denoted by $|\mathcal{C}|$. That is,

$$|\mathcal{C}| \approx \sum_{k=1}^{n} \sqrt{\left(x'(t_k^*)\right)^2 + \left(y'(t_k^*)\right)^2} \; \Delta t.$$

We pass to an exact calculation in the limit, and so define the length of $\mathcal{C}$ as

$$|\mathcal{C}| = \lim_{n \to \infty} \sum_{k=1}^{n} \sqrt{\left(x'(t_k^*)\right)^2 + \left(y'(t_k^*)\right)^2} \; \Delta t.$$

Arc length of a parameterized curve: Suppose $\mathcal{C}$ is a curve parameterized by $(x(t), y(t))$ when $a \leq t \leq b$, where $x'(t)$ and $y'(t)$ are continuous over $[a, b]$, and each point on $\mathcal{C}$ is visited exactly once as t increases from $t = a$ to $t = b$. Then the length of $\mathcal{C}$ is

$$|\mathcal{C}| = \int_a^b \sqrt{\left(x'(t)\right)^2 + \left(y'(t)\right)^2} \; dt. \qquad (2.5)$$

Calculating Arc Length

Example 2.8. Suppose $\mathcal{C}$ is parameterized by $x(t) = e^t + e^{-t}$ and $y(t) = -7 + 2t$ when $1 \leq t \leq 3$. Calculate the length of $\mathcal{C}$.

Solution: Since $x'(t) = e^t - e^{-t}$ and $y'(t) = 2$, the length of $\mathcal{C}$ is

$$|\mathcal{C}| = \int_1^3 \sqrt{\left(e^t - e^{-t}\right)^2 + \left(2\right)^2} \; dt = \int_1^3 \sqrt{\left(e^t + e^{-t}\right)^2} \; dt$$

$$= \int_1^3 e^t + e^{-t} \; dt = e^t - e^{-t} \Big|_{t=1}^{t=3} = \left(e^3 - e^{-3}\right) - \left(e^1 - e^{-1}\right) \approx 17.68534. \quad \blacksquare$$

Note: If we measure distance in meters and time in seconds, $x'(t)$ and $y'(t)$ both have units of $\frac{\text{m}}{\text{sec}}$, so

$$\left(x'(t)\right)^2 + \left(y'(t)\right)^2 \quad \text{has units of} \quad \left(\frac{\text{m}}{\text{sec}}\right)^2$$

and

$$\sqrt{\left(x'(t)\right)^2 + \left(y'(t)\right)^2} \quad \text{has units of} \quad \frac{\text{m}}{\text{sec}}.$$

That means

$$|\mathcal{C}| = \int_a^b \underbrace{\sqrt{\left(x'(t)\right)^2 + \left(y'(t)\right)^2}}_{\text{m/sec}} \underbrace{dt}_{\text{sec}} \quad \text{has units of meters.}$$

So the integrand of the arc length integral can be understood as a speed.

✤ Surface Area

Suppose $\mathcal{C}$ is the graph of the function f over the interval $[A, B]$. On p. 494 we said that the surface generated by revolving $\mathcal{C}$ about the x-axis has an area of

$$\int_A^B \overbrace{2\pi f(x)}^{\text{circumference}} \underbrace{\sqrt{1 + \left(f'(x)\right)^2}}_{\text{this is the arc length element}} dx$$

when $f(x)$ remains nonnegative. In much the same way that we developed the formula for arc length, we can extend this formula to the parametric setting.

Surface area: Suppose $\mathcal{C}$ is a curve parameterized by $(x(t), y(t))$, where $x'(t)$ and $y'(t)$ are continuous, and each point on $\mathcal{C}$ is visited exactly once as t increases from $t = a$ to $t = b$.

- If $y(t)$ doesn't change sign when $t \in [a, b]$, the surface generated by revolving $\mathcal{C}$ about the x-axis has an area of

$$\int_a^b \overbrace{2\pi |y(t)|}^{\text{circumference}} \underbrace{\sqrt{\left(x'(t)\right)^2 + \left(y'(t)\right)^2}}_{\text{this is the arc length element}} dt.$$

- If $x(t)$ doesn't change sign when $t \in [a, b]$, the surface generated by revolving $\mathcal{C}$ about the y-axis has an area of

$$\int_a^b \overbrace{2\pi |x(t)|}^{\text{circumference}} \underbrace{\sqrt{\left(x'(t)\right)^2 + \left(y'(t)\right)^2}}_{\text{this is the arc length element}} dt.$$

> The number $|y(t)|$ is acting as a radius in this formula.
>
> The hypothesis that y doesn't change sign is stronger than it needs to be, though see the note below.

Note: The hypothesis that $x(t)$ doesn't change sign prevents "double counting" in certain situations. For example, suppose $\mathcal{C}$ is the portion of the parabola $y = x^2$ that sits over $-1 \le x \le 1$. When we revolve $\mathcal{C}$ about the y-axis we see a parabolic cup called a *paraboloid*. Due to symmetry, the same paraboloid is generated by revolving only the segment of $\mathcal{C}$ that sits over $[-1, 0]$, or only the segment that sits over $[0, 1]$. So when we add

$$\begin{pmatrix}\text{the area swept out by} \\ \text{points on } \mathcal{C} \text{ over } [-1, 0]\end{pmatrix} + \begin{pmatrix}\text{the area swept out by} \\ \text{points on } \mathcal{C} \text{ over } [0, 1]\end{pmatrix}$$

we get *twice* the actual area of the paraboloid.

Area of a sphere

Example 2.9. The functions $x(t) = r\cos(t)$ and $y(t) = r\sin(t)$ parameterize the upper semicircle of radius r when $t \in [0, \pi]$. Use the surface area formula to calculate the area of the sphere of radius r (which is produced by revolving the semicircle about the x-axis).

Solution: Since $x'(t) = -r\sin(t)$ and $y'(t) = r\cos(t)$, the surface area is

$$\text{area} = \int_0^\pi 2\pi r \sin(t) \sqrt{\Big(- r \sin(t)\Big)^2 + \Big(r\cos(t)\Big)^2}\ dt$$

$$= \int_0^\pi 2\pi r \sin(t) \sqrt{r^2 \Big(\sin^2(t) + \cos^2(t)\Big)}\ dt$$

$$= \int_0^\pi 2\pi r^2 \sin(t)\ dt = -2\pi r^2 \cos(t)\Big|_{t=0}^{t=\pi} = -2\pi r^2(-1 - 1) = 4\pi r^2. \qquad \blacksquare$$

Try It Yourself: Area of a cone

Example 2.10. Suppose $x(t) = 3t$ and $y(t) = 2t$ when $t \in [0,1]$. (a) Verify that $(x(t), y(t))$ parameterizes a line segment, and (b) use the surface area formula above to calculate the surface area of the cone that's generated by revolving the line segment about the x-axis.

Full Solution
On-line

Answer: (a) See the on-line solutions, (b) $2\pi\sqrt{13}$. $\qquad \blacksquare$

❖ Areas Bounded by Parametric Curves

Suppose $(x(t), y(t))$ parameterizes a curve that's the graph of a function, f, and that $x(a) < x(b)$. On the one hand, we know the net area between the graph of f and the x-axis over the interval $[x(a), x(b)]$ is

$$\int_{x(a)}^{x(b)} f(x)\ dx.$$

On the other hand, we know that $dx = x'(t)\ dt$, so we can rewrite this integral as

$$\int_a^b f(x(t))\ x'(t)\ dt.$$

And since $(x(t), y(t))$ is on the graph of f, we know $y(t) = f(x(t))$, so we can express the net area as

$$\int_a^b y(t)\ x'(t)\ dt.$$

Areas bounded by parametric curves: Suppose $(x(t), y(t))$ parameterizes the graph of the function f, and $x(a) < x(b)$. Then the net area between the graph of f and the x-axis is

$$\int_a^b y(t)\ x'(t)\ dt$$

when $y(t)$ and $x'(t)$ are continuous functions.

Calculating Area

Example 2.11. Suppose $\mathcal{C}$ is the curve parameterized by $x(t) = 1 + \cos(t)$ and $y(t) = 3 + \sin^2(t)$. Determine the area between $\mathcal{C}$ and the x-axis over $[x(\pi), x(2\pi)]$.

Solution: Since $x(\pi) < x(2\pi)$, and $y(t) = 3 + \sin^2(t)$, our formula says that the net area bounded between the curve and the x-axis is

$$\int_\pi^{2\pi} \Big(3 + \sin^2(t)\Big)\Big(-\sin(t)\Big)\ dt = \int_\pi^{2\pi} \Big(3 + (1 - \cos^2(t))\Big)\Big(-\sin(t)\Big)\ dt$$

$$= \int_\pi^{2\pi} -4\sin(t) + \cos^2(t)\sin(t)\ dt$$

$$= 4\cos(t) - \frac{1}{3}\cos^3(t)\Big|_\pi^{2\pi} = \frac{22}{3}.$$

In Example 1.9 (see p. 709) we showed that $(x(t), y(t))$ parameterizes the parabola $y = -(x-1)^2 + 4$, and $x(\pi) = 0$ and $x(2\pi) = 2$, so we can check this number using the more familiar calculation of area:

$$\int_0^2 -(x-1)^2 + 4 \; dx = -\frac{1}{3}(x-1)^3 + 4x \Big|_{x=0}^{x=2} = \frac{22}{3}. \qquad \blacksquare$$

Try It Yourself: Calculating Area

Example 2.12. Suppose $\mathcal{C}$ is the curve parameterized by $x(t) = \cos(6\pi t)$ and $y(t) = \sin(6\pi t)$ when $t \in [1/6, 1/3]$. Determine the net area between $\mathcal{C}$ and the x-axis.

Full Solution
On-line

Answer: $-\pi/2$. $\qquad \blacksquare$

❖ Common Difficulties

Many people have trouble remembering what the integral formulas in this section mean. You can check with a quick dimensional analysis:

$$\int_a^b \overbrace{y(t)}^{\text{m}} \; \overbrace{x'(t)}^{\text{m/sec}} \; \overbrace{dt}^{\text{sec}} \qquad \int_a^b \overbrace{\sqrt{(x'(t))^2 + (y'(t))^2}}^{\text{m/sec}} \; \overbrace{dt}^{\text{sec}}$$

$$\underbrace{}_{\text{m}^2} \qquad\qquad \underbrace{\phantom{\sqrt{(x'(t))^2 + (y'(t))^2} \; dt}}_{\text{m}}$$

so the first integral is an *area*, and the second is a *length*.

┌─── You should know ───┐

- Cauchy's Mean Value Theorem and its implications in the context of parametric curves.

┌─── You should be able to ───┐

- determine the equation of the tangent line to a parametric curve at a specified time;

- determine the concavity of a parametric curve at a specified time;

- calculate the length of a parametric curve;

- calculate the area of a surface that's generated by revolving a parametric curve about either the x-axis or y-axis;

- calculate the area bounded by a parametric curve that's the graph of a function.

❖ 9.2 Skill Exercises

In #1–6 determine the equation of the tangent line to the parameterized curve at the specified time.

1. $x(t) = t^3 + 3t + 4$, $y(t) = 8t + 2$; $t_0 = 2$

2. $x(t) = 13t - 7$, $y(t) = 2t^5 - t + 2$; $t_0 = 1$

3. $x(t) = 2t^2 + 3t$, $y(t) = t - \sin(2t)$; $t_0 = 0$

4. $x(t) = \sin(t)$, $y(t) = t + \sin(2t)$; $t_0 = 0$

5. $x(t) = \ln(t)$, $y(t) = 3 + 2t$; $t_0 = 1$

6. $x(t) = e^t$, $y(t) = e^{-t}$; $t_0 = \ln(7)$

In #7–10 the curve $\mathcal{C}$ is parameterized by $x(t)$ and $y(t)$ when $t \in [0, 2\pi]$. Determine the highest and lowest points on $\mathcal{C}$ by determining when the tangent line is horizontal.

7. $\begin{cases} x(t) &=& 3\cos(t) - 14\sin(t) \\ y(t) &=& 3\cos(t) + 14\sin(t) \end{cases}$

8. $\begin{cases} x(t) &=& \sin(t) \\ y(t) &=& \sin(2t) \end{cases}$

9. $\begin{cases} x(t) &=& \cos(t) + \cos^2(t) \\ y(t) &=& \sin(t) + 0.5\sin(2t) \end{cases}$

10. $\begin{cases} x(t) &=& e^t \\ y(t) &=& \sin(t) \end{cases}$

In #11–16 determine the concavity of the parameterized curve at the given time.

11. $\begin{cases} x(t) &=& t^3 + t + 1 \\ y(t) &=& 8t^3 + 8t - 5 \end{cases}$; $t_0 = 2$

12. $\begin{cases} x(t) &=& \sin^2(t) \\ y(t) &=& \cos^2(t) \end{cases}$; $t_0 = \pi/4$

13. $\begin{cases} x(t) &=& \sin^2(3t) \\ y(t) &=& \cos(3t) \end{cases}$; $t_0 = \pi/4$

14. $\begin{cases} x(t) &=& e^t \\ y(t) &=& e^{2t} \end{cases}$; $t_0 = 1$

15. $\begin{cases} x(t) &=& t + \ln(t + 7) \\ y(t) &=& t^3 \end{cases}$; $t_0 = -1$

16. $\begin{cases} x(t) &=& t\arctan(t) \\ y(t) &=& 1/(t^2 + 1) \end{cases}$; $t_0 = 1$

In #17–22 determine the arc length of the parametric curve.

17. $x(t) = 8t + 3$, $y(t) = 9t - 1$; $t \in [3, 5]$

18. $x(t) = 3 - \frac{1}{4}t^4$, $y(t) = 9 + \frac{1}{5}t^5$; $t \in [2, 3]$

19. $x(t) = t^2 + 3$, $y(t) = 4t^2 + 8$; $t \in [1, 2]$

20. $x(t) = t - \tanh(t)$, $y(t) = \mathrm{sech}(t)$; $t \in [0, 1]$;

21. $x(t) = e^t$, $y(t) = e^{2t}$; $t \in [0, \ln(2)]$

22. $x(t) = t\cos(t)$, $y(t) = t\sin(t)$; $t \in [0, 1]$

In #23–28 a curve $\mathcal{C}$ is parameterized by $x(t)$ and $y(t)$ when $t \in [a, b]$. Use the specified numerical integration technique (R for right-sampled Riemann sum, M for midpoint-sampled Riemann sum, T for trapezoid rule, and S for Simpson's method) with n subintervals to approximate the arc length of $\mathcal{C}$.

23. $x(t) = t^3 + t$, $y(t) = 9t^7 + 3t^2$, $[0, 1]$, R, $n = 5$

24. $x(t) = \sqrt{t} + t^3$, $y(t) = 8$, $[1, 3]$, R, $n = 7$

25. $x(t) = \ln(t^2 + 1)$, $y(t) = \arctan(9t)$, $[0, 1]$, M, $n = 4$

26. $x(t) = \arcsin(t)$, $y(t) = 9t^3 + 3t$, $[0, 0.5]$, T, $n = 5$

27. $x(t) = \int_0^t \cos(x^2 + \sqrt{x})\, dx$, $y(t) = 10t$, $[0, 1]$, S, $n = 6$

28. $x(t) = t^2 - \sin(t)$, $y(t) = \int_t^9 x\cos(\cos(x))\, dx$, $[0, 1]$, S, $n = 4$

In #29–32 determine the area of the surface that's generated by revolving the parametric curve about the x-axis.

29. $x(t) = 4t^2 + 8$, $y(t) = t^2 + 3$; $t \in [1, 2]$

30. $x(t) = t$, $y(t) = 1/t$; $t \in [1, 10]$

31. $x(t) = t - \tanh(t)$, $y(t) = \mathrm{sech}(t)$; $t \in [0, 1]$;

32. $x(t) = t - \sin(t)$, $y(t) = 1 - \cos(t)$; $t \in [0, \pi]$

In #33–38 a curve $\mathcal{C}$ is parameterized by $x(t)$ and $y(t)$ when $t \in [a, b]$. Use the specified numerical integration technique (L for left-sampled Riemann sum, M for midpoint-sampled Riemann sum, T for trapezoid rule, and S for Simpson's method) with n subintervals to approximate the area of the surface that's generated by revolving $\mathcal{C}$ about the y-axis.

33. $x(t) = 2t$, $y(t) = t\ln(t) - t$, $[1, 2]$, L, $n = 5$ 36. $x(t) = 4t + e^t$, $y(t) = 13t^6$, $[2, 5]$, M, $n = 6$

34. $x(t) = e^t$, $y(t) = t^2$, $[-1, 1]$, L, $n = 7$ 37. $x(t) = 2 + \cos(t)$, $y(t) = 3 - 5t$, $[0, \pi/2]$, S, $n = 6$

35. $x(t) = t^7$, $y(t) = 3t$, $[1, 2]$, T, $n = 5$ 38. $x(t) = 8t + 2$, $y(t) = \sin(t)$, $[2, 5]$, S, $n = 6$

In #39–44 determine the net area bounded between the x-axis and the parametric curve.

39. $x(t) = 2t + 1$, $y(t) = \sin(t)$; $t \in [0, \pi/2]$ 42. $x(t) = e^t$, $y(t) = 1 + 3t$; $t \in [1, e]$

40. $x(t) = \arctan(t)$, $y(t) = 2t$; $t \in [0, 1]$ 43. $x(t) = \ln(t)$, $y(t) = 3 + t$; $t \in [1, e]$

41. $x(t) = 3t^2$, $y(t) = 2 - t^2$; $t \in [1, 2]$ 44. $x(t) = t^2$, $y(t) = \ln(t)$; $t \in [e, e^2]$

❖ 9.2 Concept and Application Exercises

45. Suppose $x(t) = \alpha t + a$ and $y(t) = \beta t + b$. What must be true about the numbers α, β, a and b in order for the line parameterized by $(x(t), y(t))$ to have a positive (a) slope, (b) x-intercept, (c) y-intercept? (Not all at once!)

46. Suppose $\mathcal{C}$ is parameterized by $x(t) = 3t + \cos(\pi t)$ and $y(t) = \sin(\pi t + \pi/2)$. Explain how we know that the tangent line to $\mathcal{C}$ has a slope of -2 at some point between $(1, 1)$ and $(2, -1)$.

47. Suppose $\mathcal{C}$ is the curve parameterized by $x(t) = \cos(8t)$, $y(t) = \sin(8t)$ when $0 \le t \le 2\pi$.

    (a) Eliminate the parameter to show that $\mathcal{C}$ is a circle of radius 1.

    (b) Based on part (a) determine the length of the curve $\mathcal{C}$.

    (c) Calculate $\int_0^{2\pi} \sqrt{(x'(t))^2 + (y'(t))^2} \, dt$.

    (d) Your answers from parts (b) and (c) are not the same. Explain why equation (2.5) does not seem to work in this case.

48. Suppose $\mathcal{C}$ is the curve parameterized by $x(t) = t^3 - 0.2\cos(20\pi t)$, $y(t) = 0$ when $-1 \le t \le 1$.

    (a) Show that $\mathcal{C}$ is the segment of the x-axis that stretches across $[-1.2, 0.98]$.

    (b) Based on part (a) determine the length of the curve $\mathcal{C}$.

    (c) Use a Riemann sum with 10 subintervals to approximate $\int_{-1}^{1} \sqrt{(x'(t))^2 + (y'(t))^2} \, dt$.

    (d) Your answers from parts (b) and (c) are substantially different, which is a problem not with the Riemann-sum approximation but, rather, with the parameterization of $\mathcal{C}$. Graph $x'(t)$ and use it to explain why equation (2.5) does not seem to work in this case.

49. Suppose a cycloid is made by rolling a wheel of radius 1 along the x-axis at unit speed (see p. 711, and take $v = 1$). Show that at each point P on this cycloid, the tangent line to the curve passes through the topmost point of the wheel (as depicted in Figure 2.5).

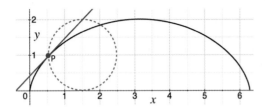

Figure 2.5: The cycloid described in #49.

50. Verify that the result of #49 is independent of the speed, v, at which the wheel rolls, and its radius, R.

Exercises #51–56 address a kind of parametric curve called a **Bézier curve**, named after the French engineer Pierre Bézier (who used them to design car bodies), which are often used by modern animation programs, and vector graphics programs. Bézier curves use so-called *control points* to determine the location at times $t = 0$ and $t = 1$ and the direction it heads in between.

51. A *quadratic* Bézier curve, which is used to render TrueType fonts, uses the control points $P_0 = (x_0, y_0)$, $P_1 = (x_1, y_1)$ and $P_2 = (x_2, y_2)$. Specifically,

$$x(t) = x_0(1-t)^2 + 2t(1-t)x_1 + t^2 x_2 \qquad (2.6)$$
$$y(t) = y_0(1-t)^2 + 2t(1-t)y_1 + t^2 y_2. \qquad (2.7)$$

(a) Verify that $(x(0), y(0)) = P_0$ and $(x(1), y(1)) = P_2$.

(b) Verify that the tangent line to the Bézier curve at time $t = 0$ connects P_0 and P_1. (In this way, P_1 controls the shape of the curve.)

(c) Verify that the tangent line to the Bézier curve at time $t = 1$ connects P_1 and P_2.

(d) Verify that the numbers t^2, $2t(1-t)$ and $(1-t)^2$ sum to 1, so that we can understand equation (2.6) as calculating a weighted average of the control points' x-coordinates (the particular weights change with time, but at each moment we have a weighted average).

52. Suppose that $P_0 = (0,0)$ and $P_2(1, 1)$.

(a) Use a graphing utility to render graphs of the quadratic Bézier curves (see #51) with $P_1 = (1, -2)$, $P_1 = (2, -4)$ and $P_1 = (4, -8)$ on the same axes.

(b) Verify that all three of these Bézier curves have the same initial position and slope, and the same terminal position.

(c) All of the points P_1 listed above lie on the line $y = -2x$. Describe how sliding P_1 along that line changes the shape of the Bézier curve.

53. Design a quadratic Bézier curve that starts at $P_0 = (3, 4)$ where the slope is -3, and ends at $P_2 = (1, 5)$.

54. Design a quadratic Bézier curve that starts at $P_0 = (0, 0)$ where the slope is $1/2$, and ends at $P_2 = (1, 1)$.

55. A *cubic* Bézier curve, which is used to render Postscript fonts, uses the control points $P_0 = (x_0, y_0)$, $P_1 = (x_1, y_1)$, $P_2 = (x_2, y_2)$ and $P_3 = (x_3, y_3)$. Specifically,

$$x(t) = x_0(1-t)^3 + 3t(1-t)^2 x_1 + 3t^2(1-t)x_2 + t^3 x_3 \qquad (2.8)$$
$$y(t) = y_0(1-t)^3 + 3t(1-t)^2 y_1 + 3t^2(1-t)y_2 + t^3 y_3. \qquad (2.9)$$

(a) Verify that $(x(0), y(0)) = P_0$ and $(x(1), y(1)) = P_3$.

(b) Verify that the tangent line to the Bézier curve at time $t = 0$ connects P_0 and P_1. (In this way, P_1 controls the shape of the curve.)

(c) Verify that the tangent line to the Bézier curve at time $t = 1$ connects P_2 and P_3.

(d) Verify that the numbers t^3, $3t^2(1-t)$, $3t(1-t)^2$ and $(1-t)^3$ sum to 1, so that we can understand equation (2.8) as calculating a weighted average of the control points' x-coordinates. *(Hint: $1^3 = (t + (1-t))^3$.)*

56. In general, a Bézier curve with control points $P_0 = (x_0, y_0)$, $P_1 = (x_1, y_1)$, ..., $P_n = (x_n, y_n)$ is defined by

$$x(t) = \sum_{k=0}^{n} \binom{n}{k} t^{n-k}(1-t)^k x_k \quad \text{and} \quad y(t) = \sum_{k=0}^{n} \binom{n}{k} t^{n-k}(1-t)^k y_k$$

where $\binom{n}{k}$ are the binomial coefficients.

(a) Verify that $(x(0), y(0)) = P_0$ and $(x(1), y(1)) = P_n$.

(b) Verify that the tangent line to the Bézier curve at time $t = 0$ connects P_0 and P_1. (In this way, P_1 controls the shape of the curve.)

(c) Verify that the tangent line to the Bézier curve at time $t = 1$ connects P_{n-1} and P_n.

(d) Verify that $\sum_{k=0}^{n} \binom{n}{k} t^{n-k}(1-t)^k = 1$, so that we can understand $x(t)$ and $y(t)$ as weighted averages of the control points' coordinates. *(Hint: $1^n = (t + (1-t))^n$.)*

57. **Cornu's spiral** (also called the **clothoid**) is the curve parameterized by $x(t) = \int_0^t \cos(s^2)\, ds$ and $y(t) = \int_0^t \sin(s^2)\, ds$ (such integrals also arise in Fresnel diffraction).

(a) Determine the arc length of the spiral for $t \in [0, A]$.

(b) Determine a formula for the concavity of the spiral at time t.

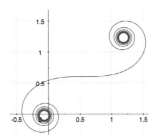

58. Consider the clothoid defined by $x(t) = \int_0^t \cos(\beta s^2)\, ds$ and $y(t) = \int_0^t \sin(\beta s^2)\, ds$.

Figure 2.6: A clothoid when $t \in [-7, 7]$.

(a) Determine the arc length of the curve for $t \in [0, A]$.

(b) Use a graphing utility to plot the resulting curve for several values of β, and then explain how β affects the graph.

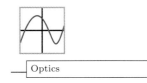

59. The **nephroid** is a curve, often seen as an envelope of reflected light at the bottom of a cup, that's parameterized by $x(t) = 3a\cos(t) - a\cos(3t)$ and $y(t) = 3a\sin(t) - a\sin(3t)$.

(a) Use a graphing utility to render a graph of the nephroid for $0 \le t \le 2\pi$.

(b) Print or reproduce the graph from part (a). Then locate $(x(0), y(0))$ and indicate the initial direction of motion as t increases.

(c) Locate $(x(\pi/2), y(\pi/2))$, and calculate the length of the nephroid curve in the first quadrant (it will depend on a).

(d) Determine the length of the entire nephroid.

9.3   Polar Coordinates

When you're playing catch with a friend, you probably don't lay down a mental grid by which to approximate the difference in your latitudes and longitudes before throwing the ball. Rather, you look in his direction, gauge the distance between you, and throw. In this section, we develop a quantitative way of talking about an object's location in terms of direction and distance, and we discuss the transition between this new system and the familiar Cartesian coordinate system of the plane.

A point's distance from the origin is easily calculated using the Pythagorean Theorem (or a tape measure), and the "direction" to it is quantified by the angle that separates it from a specified reference direction. By convention, we agree to use the positive x-axis as that direction and call the angle θ (see Figure 3.1). This method of locating points is the **polar coordinate** system, and we write the coordinates of each point in the plane as (r, θ).

Starting from the North Pole, you can specify any point on the planet by citing its distance from the pole and the angle that separates it from the Prime Meridian

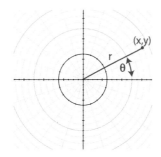

Figure 3.1: Polar coordinates in the plane.

Transitioning into polar coordinates

Example 3.1. Determine the polar coordinates of the point (a) whose Cartesian coordinates are $(\sqrt{3}, 1)$, and (b) whose Cartesian coordinates are $(-\sqrt{3}, -1)$.

Solution: (a) The distance from this point to the origin is

$$r = \sqrt{(0 - \sqrt{3})^2 + (0 - 1)^2} = \sqrt{4} = 2.$$

To find the angle θ, we use the right-triangle trigonometry that you see in Figure 3.2:

$$\tan(\theta) = \frac{1}{\sqrt{3}} \quad \Rightarrow \quad \theta = \arctan\left(\frac{1}{\sqrt{3}}\right) = \frac{\pi}{6}.$$

So the polar coordinates of this point are $(r, \theta) = (2, \pi/6)$.

(b) This part of the example is intended to point out a computational hazard to avoid when finding θ. The point $(-\sqrt{3}, -1)$ is the same distance from the origin as the point in part (a), and its y/x ratio is the same. So if we use the arctangent to try to calculate θ, we'll see $\pi/6$, just as we did in part (a), but this point is in the *third quadrant* (see Figure 3.3). We correct for this fact by adding π to the arctangent value, and seeing that $\theta = 7\pi/6$. That is, the polar coordinates of this point are $(r, \theta) = (2, 7\pi/6)$.   ∎

In the same way that a negative x-coordinates means "step *left* from the origin" and a negative y-coordinate means "move *down* from the origin," we can understand negative r and θ in terms of direction. Specifically, by writing $(-3, \pi/4)$ we mean

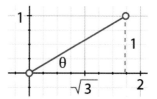

Figure 3.2: Calculating θ in Example 3.1.

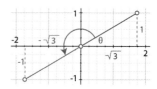

Figure 3.3: The arctangent tells us angles in quadrants I and IV, but our point is in quadrant III.

the point located by standing at the origin, looking in the direction of $\theta = \pi/4$, and taking three steps *backwards* (see Figure 3.4). Similarly, by writing $(4, -\pi/7)$, we mean the point that you find by standing at the origin, turning $\pi/7$ radians *clockwise,* and taking four steps forwards. Further, we can allow θ to grow beyond 2π, meaning that we complete more than one full turn around the origin (see Figure 3.4).

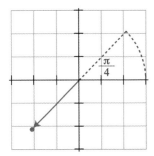 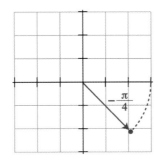 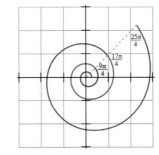

Figure 3.4: (left) A negative value of r means to "back up" from the origin; (middle) negative values of θ mean to move clockwise in the plane; (right) values of $\theta > 2\pi$ indicate to wrap around the origin more than once.

Multiple coordinate representations in polar coordinates

Example 3.2. Suppose the point P has Cartesian coordinates $(\sqrt{3}, 1)$. Find three ways of writing polar coordinates for P.

Solution: We already know from Example 3.1 that we can write the location of this point as $(r, \theta) = (2, \pi/6)$. That's one way. Another would be to loop around the origin and arrive back at the same point, so $\theta = \pi/6 + 2\pi = 13\pi/6$ and $(r, \theta) = (2, 13\pi/6)$. A third way would be to stand at the origin looking toward $(-\sqrt{3}, -1)$ and then step backwards two units. In that case, $\theta = 7\pi/6$ (as we calculated in part (b) of Example 3.1), so our point is at $(r, \theta) = (-2, 7\pi/6)$. ∎

> In a coordinate system, different ordered pairs should refer to different points. As you see in this example, that doesn't happen when we allow $r < 0$ and allow θ outside of $[0, 2\pi)$. Said technically, the transition from (r, θ) to (x, y) is no longer invertible. However, the benefits of allowing $r < 0$ and larger (or negative) angles θ tend to outweigh the drawbacks.

> **Transitioning into polar coordinates:** Suppose the Cartesian coordinates of the point P, are (x, y). Then the polar coordinates of P are (r, θ), where
>
> $$r^2 = x^2 + y^2 \qquad (3.1)$$
>
> and
>
> $$\tan(\theta) = \frac{y}{x}. \qquad (3.2)$$

Now that we can transition from Cartesian to polar coordinates, let's practice going the other way.

Transitioning out of polar coordinates

Example 3.3. Suppose the polar coordinates of the point P are $(10, \pi/3)$. Determine the Cartesian coordinates of P.

Solution: Because $\theta = \pi/3$, we know that P is in the first quadrant (see Figure 3.5). We calculate the x and y coordinates using right-triangle geometry:

$$\frac{x}{10} = \cos\left(\frac{\pi}{3}\right) \quad \Rightarrow \quad x = 10\cos\left(\frac{\pi}{3}\right) = 5.$$

Similarly,

$$\frac{y}{10} = \sin\left(\frac{\pi}{3}\right) \quad \Rightarrow \quad y = 10\sin\left(\frac{\pi}{3}\right) = 5\sqrt{3},$$

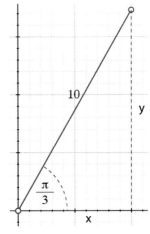

Figure 3.5: The point for Example 3.3.

so the Cartesian coordinates are $(5, 5\sqrt{3})$.   ∎

The right-trangle geometry that we used to calculate Cartesian coordinates in Example 3.3 can be easily generalized.

Transitioning out of polar coordinates: Suppose the polar coordinates of the point P are (r, θ). Then the Cartesian coordinates of P are

$$x \quad = \quad r\cos(\theta) \tag{3.3}$$
$$y \quad = \quad r\sin(\theta). \tag{3.4}$$

Transitioning out of polar coordinates

Example 3.4. Suppose that P is at $(r, \theta) = (-2, \pi/3)$. What are the Cartesian coordinates of P?

Solution: This example is slightly different from Example 3.3 because $r < 0$. However, the transition formulas seen above automatically take care of that for us. When we use them we see

$$x = (-2)\cos(\pi/3) = -\sqrt{3} \quad \text{and} \quad y = (-2)\sin(\pi/3) = -1,$$

which is in the third quadrant.   ∎

❖ Conic Sections in Polar Coordinates

Because each of the conic sections can be discussed in terms of a single line and a central point, called the **directrix** and the **focus**, respectively, it is easy to describe them with polar coordinates. In fact, when the focus is at the origin, all three conic sections are described by the same kind of equation and are distinguished by the value of a single parameter.

> A review of conic sections in Cartesian coordinates, including relevant vocabulary, is provided in Appendix D.

Parabolas, Ellipses, and Hyperbolas: When $a \neq 0$, $\varepsilon > 0$, and θ_0 is any constant, the equation

$$r = \frac{a\varepsilon}{1 + \varepsilon\cos(\theta - \theta_0)}$$

describes an ellipse when $\varepsilon < 1$, a parabola when $\varepsilon = 1$, and a hyperbola when $\varepsilon > 1$. The directrix of the curve is $|a|$ units away from the origin.

We derive this formula in the discussion below, and discuss how to work with it.

▷ Parabolas

A **parabola** is the set of points that are equidistant from the directrix and the focus, which we take to be the origin. Suppose the directrix is the line $x = a > 0$, and the point at (r, θ) is on the parabola (as in Figure 3.6). The x-coordinate of that point is $r\cos(\theta)$, so its distance from the directrix is $a - r\cos(\theta)$. Its distance from the origin is just r, so

$$r = a - r\cos(\theta), \tag{3.5}$$

which we typically rewrite as

$$r = \frac{a}{1 + \cos(\theta)}. \tag{3.6}$$

This equation tells us how a point's distance from the origin depends on its direction, so we think of r as a function of θ. Note how the denominator of $r(\theta)$ tends toward zero as $\theta \to \pi$. Consequently, $\lim_{\theta \to \pi} r(\theta) = \infty$, which tells us that the

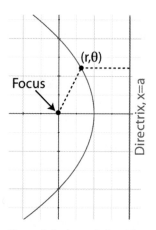

Figure 3.6: A parabola with a vertical directrix whose focus is at the origin.

parabola is unbounded in the direction $\theta = \pi$. Said differently, the parabola's axis of symmetry is in the $\theta = \pi$ direction.

Try It Yourself: Deriving the polar equation of a parabola

Example 3.5. Derive the equation of a parabola whose focus is at the origin and whose directrix is at $y = a > 0$, as seen in Figure 3.7.

Answer: $r = \frac{a}{1+\sin(\theta)}$. ∎

Full Solution On-line

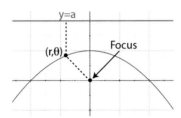

Figure 3.7: The parabola for Example 3.5

The graph that you see in Figure 3.7 is just a rotated version of what you see in Figure 3.6. Specifically, the directrix (and so the axis of symmetry) was rotated by $\pi/2$ radians about the origin. We can use the fact that $\sin(\theta) = \cos(\theta - \pi/2)$ to rewrite our formula so that it displays this rotation explicitly:

$$r = \frac{a}{1 + \cos\left(\theta - \frac{\pi}{2}\right)}.$$

More generally, the equation

$$r = \frac{a}{1 + \cos\left(\theta - \theta_0\right)}$$

describes the parabola in equation (3.6) after it's rotated by θ_0 radians about the origin in the counterclockwise direction.

> Recall that $y = f(t - t_0)$ is just a shifted version of the graph of f. Similarly, the curve $r = f(\theta - \theta_0)$ is just a shifted version of $r = f(\theta)$, except that it's a *shift in angle*—i.e., a rotation.

The polar equation of a rotated parabola

Example 3.6. Write the equation of the parabola whose focus is at the origin and whose directrix is the line $y = 4 - 2x$.

Solution: This parabola and its directrix are shown in Figure 3.8. Our job is to determine the number a and the angle θ_0. We do that by thinking of this parabola as the rotation of Figure 3.6, in which the number a is *the perpendicular distance between the directrix and the origin*.

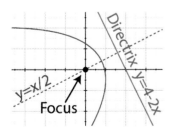 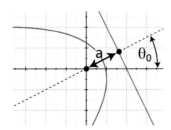

Figure 3.8: (left) A parabola's axis is perpendicular to its directrix; (right) calculating a and θ_0.

The line $y = x/2$ is perpendicular to the directrix and passes through the origin (see Figure 3.8). It intersects the directrix at $(8/5, 4/5)$, which is $4/\sqrt{5}$ units away from the origin, so $a = 4/\sqrt{5}$. And since $\tan(\theta_0) = 1/2$ (see Figure 3.9), we know that $\theta_0 = \arctan(1/2)$. Therefore, the equation of this parabola is

$$r = \frac{4/\sqrt{5}}{1 + \cos(\theta - \arctan(1/2))}.$$

■

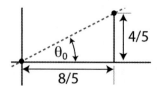

Figure 3.9: Calculating θ_0.

Try It Yourself: The polar equation of a rotated parabola

Example 3.7. Write the equation of the parabola whose focus is at the origin and whose directrix is the line $y = 6 + 3x$.

Answer: $a = \frac{3\sqrt{6}}{5}$, and $\theta_0 = \arctan(-1/3) + \pi$.

■

Full Solution
On-line

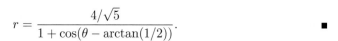

▷ **Ellipses**

Loosely said, points on an ellipse are closer to its focus than they are to its directrix. More specifically, when the focus is at the origin and the directrix is at $x = a > 0$, the set of points P for which

$$\begin{array}{c} \text{distance from } P \text{ to the focus} \rightarrow \\ \text{distance from } P \text{ to the directrix} \rightarrow \end{array} \frac{r}{a - r\cos(\theta)} = \varepsilon \qquad (3.7)$$

is an ellipse when $\varepsilon \in (0, 1)$. The number ε is called the **eccentricity** of the ellipse. Figure 3.10 shows three snapshots of an ellipse being traced out as θ increases, and you can check that the ratio ε, described by equation (3.7), is 0.8 in all of them.

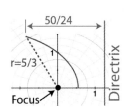
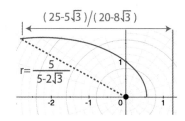
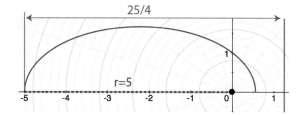

Figure 3.10: (all) Images show snapshots of an ellipse being traced out as θ increases; (left) $\theta = 2\pi/3$; (middle) $\theta = 5\pi/6$; (right) $\theta = \pi$.

We typically solve equation (3.7) for r in terms of θ, and rewrite it as

$$r = \frac{a\varepsilon}{1 + \varepsilon\cos(\theta)}. \qquad (3.8)$$

Compare to equation (3.6) on p. 732.

Since the value of the cosine ranges over $[-1, 1]$, the denominator of this fraction cannot reach zero when $\varepsilon \in (0, 1)$. Consequently the value of r remains bounded, never exceeding $a\varepsilon/(1 - \varepsilon)$ at any angle, which is one way to tell when the equation describes an ellipse.

An ellipse in polar coordinates

Example 3.8. Suppose $\mathcal{C}$ is the ellipse described by $r = \frac{12}{18 + 9\cos(\theta)}$ in polar coordinates. (a) Determine its eccentricity, (b) determine the distance between its focus and directrix, and (c) determine the length of its major axis.

Solution: (a) We can determine the eccentricity when the constant term in the denominator is a 1, so let's factor the 18 out of the denominator and write the equation of this ellipse as

$$r = \frac{1}{18}\frac{12}{1 + 0.5\cos(\theta)} = \frac{2/3}{1 + 0.5\cos(\theta)}.$$

Now we see that $\varepsilon = 0.5$.

(b) The distance between the focus and the directrix is the number a in equation (3.8). Comparing the numerator of our fraction with equation (3.8), we see that $2/3 = a\varepsilon$. We already know that $\varepsilon = 0.5$ from our previous work, so we conclude that $a = 4/3$.

(c) Because the fraction $r(\theta)$ has a constant numerator, its value is minimized where its denominator is largest. That happens when $\theta = 0$, where $r = 4/9$. Similarly, $r(\theta)$ is maximized where its denominator is smallest. That happens when $\theta = \pi$, where $r = 4/3$. The distance between these two points is $4/9 + 4/3 = 16/9$, which is the length of the major axis, as seen in Figure 3.11. ∎

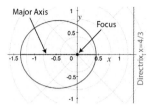

Figure 3.11: The graph of the ellipse $r = 12/(18 + 9\cos(\theta))$ from Example 3.8.

Try It Yourself: An ellipse in polar coordinates

Example 3.9. Suppose C is the ellipse described by $r = \frac{5}{6 + 2\cos(\theta)}$ in polar coordinates. Determine (a) its eccentricity, (b) the distance between its focus and directrix, and (c) the length of its major axis.

Answer: (a) $\varepsilon = 1/3$, (b) $5/2$, (c) $15/8$. ∎

More generally, when $\varepsilon \in (0, 1)$ the equation

$$r = \frac{a\varepsilon}{1 + \varepsilon\cos(\theta - \theta_0)}$$

Full Solution
On-line

describes the ellipse in equation (3.8) after it's rotated by θ_0 radians about the origin in the counterclockwise direction. This rotation changes the direction of the major axis, but not the length-to-width aspect ratio of the ellipse.

> Recall that $y = f(t - t_0)$ is just a shifted version of the graph of f. Similarly, the curve $r = f(\theta - \theta_0)$ is just a shifted version of $r = f(\theta)$, except that it's a *shift in angle*—i.e., a rotation.

A rotated ellipse in polar coordinates

Example 3.10. Suppose C is the ellipse described by $r = \frac{3}{8 + 2\cos(\theta - \pi/6)}$ in polar coordinates. Determine (a) its eccentricity, (b) the distance between its focus and directrix, and (c) the length of its major axis.

Solution: (a) After factoring the 8 out of the denominator we see $r = \frac{0.375}{1 + 0.25\cos(\theta - \pi/6)}$, so $\varepsilon = 0.25$. (b) Since $a\varepsilon = 0.375$ and $\varepsilon = 0.25$, we have $a = 1.5$. (c) The number $r(\theta)$ is minimized when $\theta = \pi/6$ and maximized when $\theta = 7\pi/6$. Adding these respective values, we find that the major axis is 0.8 units long. ∎

Try It Yourself: A rotated ellipse in polar coordinates

Example 3.11. Suppose C is the ellipse described by $r = \frac{7}{8 + 3\cos(\theta - \pi/4)}$ in polar coordinates. Determine (a) its eccentricity, (b) the distance between its focus and directrix, and (c) the length of its major axis.

Full Solution
On-line

Answer: (a) $\varepsilon = 3/8$, (b) $7/3$, (c) $112/55$. ∎

▷ Hyperbolas

In contrast to an ellipse, points on a hyperbola are closer to its directrix than they are to its focus. More specifically, equations (3.7) and (3.8) describe a hyperbola when $\varepsilon > 1$, rather than an ellipse. As before, the number ε is called the hyperbola's **eccentricity.** Figure 3.12 shows several snapshots of a hyperbola being traced out as θ increases. Note that because $\varepsilon > 1$ the denominator in equation (3.8) tends toward zero as θ approaches $\pm\arccos(-1/\varepsilon)$. As with the parabola, this means

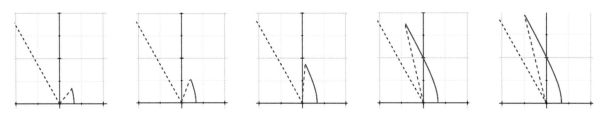

Figure 3.12: (all) The black dashed ray extends in the direction of $\arccos(-1/\varepsilon)$; (left to right) r grows without bound as θ increases toward $\arccos(-1/\varepsilon)$.

that r surpasses all finite values as θ approaches those angles.

Because the cosine is an even function, the same thing happens as θ decreases from zero toward $-\arccos(-1/\varepsilon)$, thereby completing one "branch" of a hyperbola (see Figure 3.13).

As seen in Figure 3.14, the number $r < 0$ when $\varepsilon > 1$ and

$$\arccos\left(-\frac{1}{\varepsilon}\right) < \theta < 2\pi - \arccos\left(-\frac{1}{\varepsilon}\right).$$

Remember that $r < 0$ means we step "backwards" from the origin, so although we look into the 2nd and 3rd quadrants, the point (r, θ) is in the 4th or 1st.

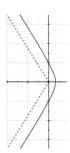

Figure 3.13: One branch of a hyperbola. The directions $\theta = \pm \arccos(-1/\varepsilon)$ are indicated by black dashed lines.

 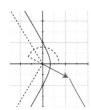 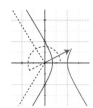 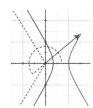 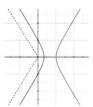

Figure 3.14: (all) The black dashed rays extend in the direction of $\theta = \arccos\left(-\frac{1}{\varepsilon}\right)$ and $\theta = 2\pi - \arccos(1/\varepsilon)$; (left to right) r remains negative while $\theta \in \left(\arccos\left(-\frac{1}{\varepsilon}\right), 2\pi - \arccos\left(-\frac{1}{\varepsilon}\right)\right)$, during which the second branch of the hyperbola is created.

A hyperbola in polar coordinates

Example 3.12. The equation $r = \frac{2}{5 + 13\cos(\theta)}$ describes a hyperbola. Determine (a) the eccentricity of the hyperbola, and (b) how close its branches come together.

Solution: (a) As we did with the ellipse, let's factor the constant summand out of the denominator. Then we see that

$$r = \frac{1}{5} \frac{2}{1 + \frac{13}{5}\cos(\theta)} = \frac{2/5}{1 + \frac{13}{5}\cos(\theta)},$$

from which we conclude that the eccentricity is $\varepsilon = 13/5$. (b) The denominator is maximized when $\theta = 0$, at which we have $r = 1/9$. Similarly, the denominator is minimized at $\theta = \pi$, where $r = -1/4$. That is, when we look in the direction of $\theta = \pi$, the corresponding point on the hyperbola is 0.25 units behind us, at $x = 1/4$ (see Figure 3.15). Therefore, the distance between the points corresponding to $\theta = 0$ and $\theta = \pi$ is $\frac{1}{4} - \frac{1}{9} = \frac{5}{36}$. ∎

More generally, when $\varepsilon > 1$ the equation

$$r = \frac{a\varepsilon}{1 + \varepsilon \cos(\theta - \theta_0)}$$

Recall that $y = f(t - t_0)$ is just a shifted version of the graph of f. Similarly, the curve $r = f(\theta - \theta_0)$ is just a shifted version of $r = f(\theta)$, except that it's a *shift in angle*—i.e., a rotation.

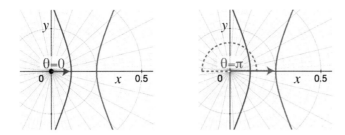

Figure 3.15: The hyperbola from Example 3.12.

describes the hyperbola in equation (3.8) after it's rotated by θ_0 radians about the origin in the counterclockwise direction. As with parabolas and ellipses, this rotation affects the orientation but not the basic structure of the hyperbola.

Try It Yourself:　A rotated hyperbola in polar coordinates

Example 3.13. Consider the hyperbola described by $r = \frac{9}{3+17\cos(\theta-\pi/3)}$. Determine (a) its eccentricity, (b) how close its branches come together.

Answer: (a) $17/3$, (b) $153/140$. ∎

Full Solution On-line

❖ Other Polar Curves

We've seen that parabolas, ellipses, and hyperbolas are all described by an equation of the form $r = f(\theta)$. Many others are also, such as the **cardioid**, which describes the most common sensitivity pattern of microphones (see Figure 3.16), as well as *stress-sum contours* around the tip of a crack.

The cardioid

Example 3.14. Graph the curve described by $r = 1 - \sin(\theta)$.

Solution: The curve $y = 1 - \sin(x)$ is shown in Figure 3.17, where you see that y initially decreases from 1, is zero when $x = \pi/2$, and then rises again. Similarly, the equation $r = 1 - \sin(\theta)$ tells us that $r = 1$ when we look in the direction of $\theta = 0$ (i.e., out the x-axis), and r decreases as we begin to rotate in the counterclockwise direction (i.e., as θ increases away from zero). The point (r, θ) arrives at the origin when $\theta = \pi/2$, since $r(\pi/2) = 0$, and then it moves away again when θ increases beyond $\pi/2$. Note that $r = 1 - \sin(\theta)$ is never negative, so when we look in the θ direction, the point is never behind us. ∎

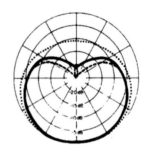

Figure 3.16: The sensitivity pattern for the Shure SM58 Dynamic Microphone. (Copyright Shure, Inc. Used with permission.)

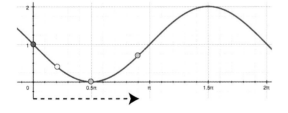 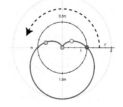

Figure 3.17: (left) The graph $y = 1 - \sin(x)$; (right) the graph of $r = 1 - \sin(\theta)$.

Try It Yourself:　The cardioid

Example 3.15. Graph the curve described by $r = 1 + \cos(\theta)$.

Answer: See the on-line solution.   ∎

The cardioid that you graphed in Example 3.15 is just a rotated version of what you saw in Example 3.14. Specifically, the two cardioids differ by a rotation of $\pi/2$ radians. This rotation can been seen in the formula if we rewrite the equation in Example 3.15 as

$$r = 1 - \sin\left(\theta - \frac{\pi}{2}\right).$$

More generally, the equation $r = 1 - \sin(\theta - \theta_0)$ describes the cardioid in Example 3.14 after it's been rotated through θ_0 radians.

> This is the same kind of thing we saw with parabolas, hyperbolas, and ellipses.

The equation of a circle

Example 3.16. Verify that the equation $r = \sin(\theta)$ describes a circle.

Solution: When we have a computer draw this graph, it sure *looks* like a circle (see Figure 3.18). But looks can be deceiving, so let's check. If we multiply the equation by r we get

$$r^2 = r\sin(\theta).$$

Using equations (3.1) and (3.4) from p. 731, we can rewrite this as

$$x^2 + y^2 = y, \quad \text{which is the same as} \quad x^2 + y^2 - y = 0.$$

When we complete the square, this becomes

> Appendix B includes a review of the completing-the-square technique.

$$x^2 + y^2 - y + \frac{1}{4} = \frac{1}{4}, \quad \text{which is the same as} \quad x^2 + \left(y - \frac{1}{2}\right)^2 = \left(\frac{1}{2}\right)^2,$$

which describes a circle of radius $\frac{1}{2}$ whose center has Cartesian coordinates $(0, \frac{1}{2})$. In Figure 3.18 you see that $r = \sin(\theta)$ traces out the circle when $0 \leq \theta \leq \pi$. While $\theta \in (\pi, 2\pi)$, we're looking into the 3rd and 4th quadrants but $r < 0$, so the point (r, θ) is "behind us," tracing out the circle a second time.   ∎

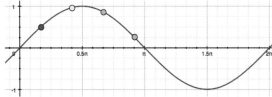

Figure 3.18: (left) The graph of $y = \sin(x)$; (right) the curve $r = \sin(\theta)$. The circle is traced out once as θ increases from $\theta = 0$ to $\theta = \pi$, then again as θ increases from π to 2π, during which $r < 0$.

Try It Yourself: The equation of a circle

Example 3.17. Verify that the equation $r = \cos(\theta)$ describes a circle, and determine the Cartesian coordinates of its center.

Answer: The center is at $(0.5, 0)$.   ∎

If we start with the basic equation of a circle but change the angular frequency, r becomes negative more often, and we end up with something called a **rose**.

The equation of a rose

Example 3.18. Draw the graph of $r = \sin(3\theta)$.

Solution: Generally speaking, it's helpful to know the angles at which $r = 0$ when trying to understand how an equation of the form $r = f(\theta)$ is related to its graph. In this case, we have $r = 0$ when $\theta = \pi/3 + 2k\pi/3$, where $k = 0, 1, 2, 3, 4, 5$. You can see in Figure 3.19 that $r > 0$ as θ ranges over $(0, \pi/3)$, during which the right-hand leaf is created. Then $r < 0$ when $\theta \in (\pi/3, 2\pi/3)$, during which we're tracing the lower leaf. The rose is complete by the time that θ reaches π, and then it is traced a second time as θ increases from π to 2π.   ■

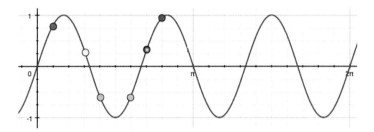 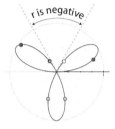

Figure 3.19: (left) The graph of $y = \sin(3x)$; (right) the curve $r = \sin(3\theta)$—the bottom leaf is traced out as θ ranges over $\left(\frac{\pi}{3}, \frac{2\pi}{3}\right)$, during which $r < 0$, and again when $\theta \in \left(\frac{4\pi}{3}, \frac{5\pi}{3}\right)$, during which $r > 0$.

Try It Yourself: The equation of a rose

Example 3.19. The graph of $r = \sin(5\theta)$ is shown in Figure 3.20. Determine which values of θ correspond to points on the vertical petal.

Answer: $\theta \in \left(\frac{2\pi}{5}, \frac{3\pi}{5}\right) \cup \left(\frac{7\pi}{5}, \frac{8\pi}{5}\right)$; full solution on-line.   ■

✤ Intersection of Polar Curves

In Example 1.10 on p. 709 we saw that path intersection doesn't always indicate a collision, because the points moving along those paths can arrive at the intersection at different times. For a similar reason, it's sometimes difficult to locate the intersection of polar curves, as seen in the next example.

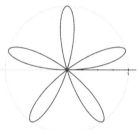

Figure 3.20: The graph of the polar equation in Example 3.19.

Intersection of polar curves

Example 3.20. Locate all intersections of the four-leaved rose $r = \cos(2\theta)$ and the cardioid $r = 1 + \sin(\theta)$.

Solution: Figure 3.21 depicts the region of the plane where these curves intersect. To find the points of intersection, we look for those angles at which $\cos(2\theta)$ and $1 + \sin(\theta)$ are the same. That requires us to use the double-angle formula for the cosine, and the Pythagorean identity between sine and cosine:

$$
\begin{aligned}
\cos(2\theta) &= 1 + \sin(\theta) \\
\cos^2(\theta) - \sin^2(\theta) &= 1 + \sin(\theta) \\
1 - 2\sin^2(\theta) &= 1 + \sin(\theta).
\end{aligned}
$$

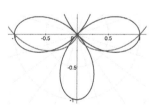

Figure 3.21: The cardioid and the rose from Example 3.20.

After moving all these terms to the right-hand side of the equation, we see

$$
0 = 2\sin^2(\theta) + \sin(\theta) = (2\sin(\theta) + 1)\sin(\theta),
$$

which happens when

$$\sin(\theta) = 0 \Rightarrow \begin{cases} \theta = 0 \\ \theta = \pi \end{cases} , \quad \text{and when} \quad \sin(\theta) = -\frac{1}{2} \Rightarrow \begin{cases} \theta = -\frac{\pi}{6} \\ \theta = \frac{7\pi}{6} \end{cases} .$$

We have found four points of intersection, but a close inspection of Figure 3.21 reveals seven, one of which is the origin. Our method didn't find it because the rose arrives there when θ is an odd multiple of $\pi/4$, but the cardioid arrives there when $\theta = 3\pi/2$. The two remaining points of intersection are more difficult to find because the "southern" petal of the rose is traced out when $\pi/4 < \theta < 3\pi/4$, but the cardioid doesn't reach it until $\theta > \pi$. Technical difficulties such as this often arise when working with polar curves, and the method of handling them depends on the particular problem you're facing. In this case, a phase shift of π radians in the cardioid allows us to locate the remaining points. Specifically, to find the remaining point of intersection in the 3rd quadrant, we look for an angle $\theta \in (\pi/4, \pi/2)$ at which

$$\overbrace{- \cos(2\theta)}^{\text{negative } r \text{ for rose}} = \overbrace{(1 + \sin(\theta + \pi))}^{\text{angle in the 3rd quadrant}} . \qquad (3.9)$$

$$\underbrace{}_{\text{positive}} \qquad \underbrace{}_{\text{positive } r \text{ for cardioid in the 3rd quadrant}}$$

Understanding this equation requires a bit of thought. It equates values of r corresponding to θ in different quadrants and with different signs.

The same formulation, applied to $\theta \in (\pi/2, 3\pi/4)$, allows us to locate the remaining point of intersection in the 4th quadrant. Using trigonometric identities to rewrite $\cos(2\theta)$, as done above, and the fact that $\sin(\theta + \pi) = -\sin(\theta)$, we arrive at the equation

$$2\sin^2(\theta) + \sin(\theta) - 2 = 0,$$

which is quadratic in $\sin(\theta)$. The quadratic formula tells us that $2x^2 + x - 2 = 0$ is solved by $x = \frac{-1 \pm \sqrt{17}}{4}$, but our "$x$" is really $\sin(\theta)$ so we must have $|x| \leq 1$. Therefore, our quadratic equation is solved when

$$\sin(\theta) = \frac{-1 + \sqrt{17}}{4} \Rightarrow \theta = \arcsin\left(\frac{-1 + \sqrt{17}}{4}\right) \quad \text{or} \quad \theta = \pi - \arcsin\left(\frac{-1 + \sqrt{17}}{4}\right).$$

Remember that equation (3.9) refers to angles in the 1st and 2nd quadrants, but the actual points are in the 3rd and 4th. We get there by adding π to our answers. The remaining points of intersection are at

$$\theta = \pi + \arcsin\left(\frac{-1 + \sqrt{17}}{4}\right) \quad \text{and} \quad \theta = 2\pi - \arcsin\left(\frac{-1 + \sqrt{17}}{4}\right). \qquad \blacksquare$$

❖ A Comment on Convention

There is not agreement across disciplines about whether to denote the polar angle by θ or ϕ. Mathematicians tend to use θ, as do many software programs, but physicists and engineers sometimes use ϕ instead.

❖ Common Difficulties

Students sometimes forget how to determine whether a particular version of equation (3.8) describes a parabola, ellipse, or hyperbola. It comes down to asking whether r is ever forced to infinity because the denominator goes to zero. If not, you have an ellipse. If once, you have a parabola. If twice, you have a hyperbola.

You should know

- the terms *polar coordinates, focus, directrix, eccentricity, cardioid,* and *rose.*

You should be able to

- transition from Cartesian to polar coordinates, and back;

- determine whether an equation of the form $r = a/(1+\varepsilon \cos(\theta - \theta_0))$ describes a parabola, ellipse, or hyperbola, based on the value of ε;

- write down the equation of a parabola, ellipse, or hyperbola whose focus is at the origin, and whose directrix is a specified line;

- sketch the graph of $r = f(\theta)$ based on the graph of f.

✧ 9.3 Skill Exercises

In #1–6 determine the polar coordinates of the specified point.

1. $(x, y) = (8\sqrt{3}, 8)$ 3. $(x, y) = (-4, -4)$ 5. $(x, y) = (-4, 6)$

2. $(x, y) = (9, 9\sqrt{3})$ 4. $(x, y) = (-11, -11\sqrt{3})$ 6. $(x, y) = (8, -17)$

In #7–12 determine the Cartesian coordinates of the specified point.

7. $(r, \theta) = (7, \pi/4)$ 9. $(r, \theta) = (-2, 5\pi/6)$ 11. $(r, \theta) = (13, -\pi/3)$

8. $(r, \theta) = (5, 2\pi/3)$ 10. $(r, \theta) = (-4, 7\pi/6)$ 12. $(r, \theta) = (4, -11\pi/6)$

In #13–20 sketch the region described by the polar equation or inequalities.

13. $r = 2$ 15. $1 \le r \le 3$ 17. $\pi/6 \le \theta \le \pi/3$ 19. $\theta \in [0, \pi/2], r \in [1, 2]$

14. $\theta = \pi/4$ 16. $0 < r \le 2$ 18. $\pi < r \le 5\pi/4$ 20. $\theta \in [\pi/4, 4\pi/4], r \in [2, 3]$

In #21–28 (a) determine whether the equation describes a parabola, hyperbola, or an ellipse, and (b) sketch a graph of the curve and its directrix.

21. $r = \frac{4}{1+0.5\cos(\theta)}$ 23. $r = \frac{2}{1+\cos(\theta)}$ 25. $r = \frac{2}{3-1.2\sin(\theta)}$ 27. $r = \frac{8}{1.3-\pi\sin(\theta)}$

22. $r = \frac{3}{1+1.5\cos(\theta)}$ 24. $r = \frac{1}{1-\sin(\theta)}$ 26. $r = \frac{15}{6-6\cos(\theta)}$ 28. $r = \frac{5}{10-9\sin(\theta)}$

In #29–32 determine how close the curve comes to its directrix.

29. $r = \frac{3}{1+\cos(\theta)}$ 30. $r = \frac{4}{7+7\sin(\theta)}$ 31. $r = \frac{8}{5+2\cos(\theta)}$ 32. $r = \frac{5}{3+5\cos(\theta)}$

In #33–40 use polar coordinates to write the equation of the parabola with the specified directrix whose focus is at the origin.

33. $x = 2$ 35. $y = -2$ 37. $y = x - 5$ 39. $y = -2x - 3$

34. $x = -13$ 36. $y = 3$ 38. $y = -x + 7$ 40. $y = 3x + 4$

In #41–46 (a) use polar coordinates to write the equation of the conic section with the specified directrix and eccentricity whose focus is at the origin, (b) identify the curve based on its eccentricity, and (c) use a graphing utility to render a graph of the curve you described in part (a).

41. $x = 4$; $\varepsilon = 0.2$ 43. $y = 4x - 6$; $\varepsilon = 0.9$ 45. $y = 8x + 17$; $\varepsilon = 6$

42. $y = -17$; $\varepsilon = 0.4$ 44. $y = 3x + 2$; $\varepsilon = 5$ 46. $y = -2x + 11$; $\varepsilon = 0.125$

In #47–52 (a) determine the polar equation of the curve, and (b) use a graphing utility to render a graph of the equation you wrote in part (a).

47. $y = 3$ 49. $3y + x = -5$ 51. $y = x^2$

48. $x = -5$ 50. $y + 4x = 5$ 52. $x^4 + y^4 = 1$

In #53–60 (a) sketch a graph of the curve $y = f(t)$, and (b) based on your sketch from part (a), draw the curve $r = f(\theta)$.

53. $f(t) = t$ 55. $f(t) = -\cos(t)$ 57. $f(t) = \cos(3t)$ 59. $f(t) = \sin(\pi t)$

54. $f(t) = |t|$ 56. $f(t) = \cos^2(t)$ 58. $f(t) = \sin(4t)$ 60. $f(t) = \cos(\sqrt{2}t)$

In #61–66 find the intersections of each pair of curves. (Newton's method can be applied in #65 and #66.)

61. $r = \sin(2\theta)$, $r = \sin(\theta)$ 63. $r = \cos^2(\theta)$, $r = \sin^2(\theta)$ 65. $r = 1 + 0.8\cos(\theta)$, $r = 1 - 1.5\sin(\theta)$

62. $r = \sin(2\theta)$, $r = \cos(\theta)$ 64. $r = \sin(2\theta)$, $r = \cos(2\theta)$ 66. $r = \theta + \sin(\theta)$, $r = 1 - \sin(\theta)$

❖ 9.3 Concept and Application Exercises

Exercises #67 and #68 are about **limaçons** (pronounced "lee-mah-SONES"), which are curves defined by the polar equation $r = A + B\sin(\theta)$ or $r = A + B\cos(\theta)$. The cardioid is a special case when $A = B$.

67. Consider the family of limaçons defined by $r = 1 + B\sin(\theta)$. (a) Use a graphing utility to plot representative members of this family by setting $B = -2.5, -2, -1, -0.5, 0, 0.5, 1, 2$ and 2.5. Then (b) explain what the value of B tells us about the shape.

68. Repeat #67 for the limaçons $r = 2 + B\cos(\theta)$, and (c) explain how the shape generated by the cosine differs from the shape generated by the sine.

69. Suppose P_0 is the point with polar coordinates (r_0, θ_0). Based on Figure 3.22, use the Law of Cosines to show that the circle of radius R centered at P_0 is described by $R^2 = r^2 + r_0^2 - 2Rr\cos(\theta - \theta_0)$.

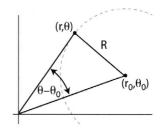

70. Write the equation for the circle centered at $(3, \pi/6)$ of radius 4.

Figure 3.22: Diagram for #69.

The conditions that $f(\theta) \geq 0$ and $0 < \beta - \alpha < 2\pi$ avoid "double counting" by ensuring that we don't trace over the same curve twice while θ increases from $\theta = \alpha$ to $\theta = \beta$.

The area of a rose petal

Example 4.3. The graph of $r = \sin(5\theta)$ is shown in Figure 4.3. Determine the area of one petal.

Solution: In Example 3.19 (p. 739) we determined that the vertical petal corresponded to $\theta \in \left(\frac{2\pi}{5}, \frac{3\pi}{5}\right)$. So the area of one petal is

$$\int_{2\pi/5}^{3\pi/5} \frac{1}{2} \sin^2(5\theta) \, d\theta.$$

The half-angle identity for the sine allows us to rewrite this as

$$\frac{1}{4} \int_{2\pi/5}^{3\pi/5} 1 - \cos(10\theta) \, d\theta = \frac{1}{4} \left(\theta - \frac{1}{10} \sin(10\theta) \right) \Bigg|_{2\pi/5}^{3\pi/5} = \frac{\pi}{20}. \qquad \blacksquare$$

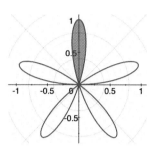

Figure 4.3: The graph of the polar equation in Example 4.3.

Area between cardioids

Example 4.4. Calculate the area between the cardioids $r = 1 - \sin(\theta)$ and $r = 2 - 2\sin(\theta)$, shown as the blue region in Figure 4.4.

Solution: To calculate the area of this region let's determine the area enclosed by each cardioid, respectively, and then subtract the smaller number from the larger:

$$\text{Area} = \int_0^{2\pi} \frac{1}{2} \Big(2 - 2\sin(\theta)\Big)^2 \, d\theta - \int_0^{2\pi} \frac{1}{2} \Big(1 - \sin(\theta)\Big)^2 \, d\theta.$$

When we factor a 2 out of the squared term in the first integral, this becomes

$$\begin{aligned}
\text{Area} &= \int_0^{2\pi} 2 \Big(1 - \sin(\theta)\Big)^2 \, d\theta - \int_0^{2\pi} \frac{1}{2} \Big(1 - \sin(\theta)\Big)^2 \, d\theta \\
&= \frac{3}{2} \int_0^{2\pi} \Big(1 - \sin(\theta)\Big)^2 \, d\theta \\
&= \frac{3}{2} \int_0^{2\pi} 1 - 2\sin(\theta) + \sin^2(\theta) \, d\theta.
\end{aligned}$$

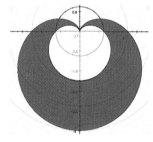

Figure 4.4: The region described in Example 4.4 is shown in blue.

The half-angle identity for the sine allows us to rewrite this as

$$\begin{aligned}
\text{Area} &= \frac{3}{2} \int_0^{2\pi} 1 - 2\sin(\theta) + \frac{1}{2} - \frac{1}{2}\cos(2\theta) \, d\theta \\
&= \frac{3}{2} \left(\frac{3}{2}\theta + 2\cos(\theta) - \frac{1}{4}\sin(2\theta) \right) \Bigg|_0^{2\pi} = \frac{9\pi}{2}. \qquad \blacksquare
\end{aligned}$$

❖ Arc Length of Polar Curves

To calculate the length of a polar curve $r = f(\theta)$, we return to Cartesian coordinates using $x = r\cos(\theta)$ and $y = r\sin(\theta)$. This allows us to think of the curve as being parameterized by $(x(\theta), y(\theta))$, after which we can use the arc length formulation from p. 723. Specifically,

$$\text{length} = \int_\alpha^\beta \sqrt{\Big(x'(\theta)\Big)^2 + \Big(y'(\theta)\Big)^2} \, d\theta.$$

In the exercise set, you'll use the Product Rule to calculate these derivatives and verify that the integrand always reduces to a specific form:

Arc Length of a Polar Curve: Suppose $\mathcal{C}$ is the curve described by $r = f(\theta)$ when $\alpha \leq \theta \leq \beta$. If the curve is traversed exactly once as θ increases from $\theta = \alpha$ to $\theta = \beta$, and f' is continuous, the length of $\mathcal{C}$ is

$$|\mathcal{C}| = \int_\alpha^\beta \sqrt{\left(r(\theta)\right)^2 + \left(r'(\theta)\right)^2}\, d\theta.$$

The arc length of a polar curve

Example 4.5. Suppose $\mathcal{C}$ is the spiral $r = \theta$, $0 \leq \theta \leq \pi$ (see Figure 4.5). Calculate the length of $\mathcal{C}$.

Solution: Since $r'(\theta) = 1$, our formula tells us that

$$|\mathcal{C}| = \int_0^\pi \sqrt{\theta^2 + 1}\, d\theta.$$

This integral requires us to use trigonometric substitution (see p. 437). The result is that

$$|\mathcal{C}| = \frac{1}{2}\left(\theta\sqrt{1+\theta^2} + \ln|\theta + \sqrt{1+\theta^2}|\right)\Big|_{\theta=0}^{\theta=\pi} \approx 6.109919. \qquad \blacksquare$$

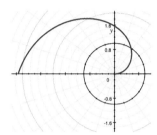

Figure 4.5: The spiral in Example 4.5.

You should know

- that the slope of the tangent line to the graph of $r = f(\theta)$ is $y'(\theta)/x'(\theta)$ when $x'(\theta) \neq 0$;

- the integral formula for calculating the area between a polar curve and the origin;

- the integral formula for calculating the length of a polar curve.

You should be able to

- calculate the slope of a tangent line to a polar curve;

- calculate the area bounded between a polar curve and the rays $\theta = \alpha$ and $\theta = \beta$;

- calculate the length of a polar curve.

❖ 9.4 Skill Exercises

For each of the curves described in #1–4, determine the slope of the tangent line at the specified point.

1. $r = \cos(\theta)$, $\theta = \pi/6$

2. $r = 1 + \sin(\theta)$, $\theta = \pi/4$

3. $r = 3 + \cos(8\theta)$, $\theta = \pi/3$

4. $r = 3\cos(\theta) - \cos(3\theta)$, $\theta = -\pi/6$

5. Determine the y-coordinate of the lowest point on the cardioid $r = 1 + \sin(\theta)$.

6. Determine the x-coordinate of the rightmost point on the cardioid $r = 1 + \sin(\theta)$.

7. Determine the point(s) on the cardioid $r = 1 + \sin(\theta)$ at which the tangent line is parallel to $y = x$.

8. Determine the point(s) on the cardioid $r = 1 + \cos(\theta)$ at which the tangent line is perpendicular to $y = 2x$.

In #9–14 use the integral formula on p. 746 to determine the area of the bounded region enclosed by the curve $r = f(\theta)$ when $\theta \in [a, b]$.

9. $r = 3 + 3\sin(\theta)$; $[0, 2\pi]$　　　　　　　12. $r = 2 + \sin(2\theta)$; $[0, 2\pi]$

10. $r = 4 - 4\cos(\theta)$; $[0, 2\pi]$　　　　　　　13. $r = \theta$; $[0, \pi]$

11. $r = \cos(4\theta)$; $[-\pi/8, \pi/8]$　　　　　　14. $r = \ln(1 + \theta)$; $[0, 2\pi]$

In #15–20 (a) use a graphing utility to render a plot of the specified curve, and (b) use the integral formula on p. 748 to determine its length.

15. $r = \sec(\theta)$; $\theta \in [0, \pi/4]$　　　　　　18. $r = e^{\theta}$; $\theta \in [0, 2\pi]$

16. $r = \csc(\theta)$; $\theta \in [\pi/4, 3\pi/4]$　　　19. $r = 1 + \sin(\theta)$; $\theta \in [0, 2\pi]$

17. $r = \theta^2$; $\theta \in [0, \pi]$　　　　　　　20. $r = \theta^4$; $\theta \in [0, \pi]$

21. Suppose $\mathcal{R}$ is the locus of points that lies inside both circles $r = \cos(\theta)$ and $r = \sin(\theta)$. Determine the area of $\mathcal{R}$.

22. Suppose $\mathcal{R}$ is the locus of points that lies inside the circle $r = \cos(\theta)$ but outside the cardioid $r = 1 - \cos(\theta)$. Determine the area of $\mathcal{R}$.

23. Determine the area of the region that lies inside the circle $r = \cos(\theta)$ but outside the curve $r^2 = \cos(3\theta)$, as seen in Figure 4.6.

24. Determine the area of the region that lies inside both the circle $r = 3\sin(\theta)$ and the cardioid $r = 1 + \sin(\theta)$, as seen in Figure 4.6.

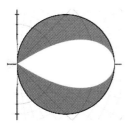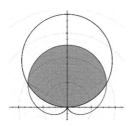

Figure 4.6: (left) The region for exercise #23; (right) the region for exercise #24.

25. Determine the area of the blue-shaded region in the left-hand image of Figure 4.7, which is inside the circle $r = 4$ and outside the curve $r = 3\cos(3(\theta - \pi/2))$.

26. Determine the area of the blue-shaded region in the right-hand image of Figure 4.7, which is bounded inside the circle $r = 1$ and outside the curve $r = 3\cos(3(\theta - \pi/2))$.

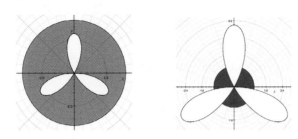

Figure 4.7: (left) The region for exercise #25; (right) the region for exercise #26.

❖ 9.4 Concept and Application Exercises

27. Suppose f is a continuous, positive function with a period of 2π, and $\mathcal{R}$ is the region enclosed by $r = f(\theta)$ for $0 \leq \theta \leq 2\pi$. Use the integral formula on p. 746 to show that dilating $\mathcal{R}$ by a factor of $\lambda > 0$, so that the bounding curve is $r = \lambda f(\theta)$, scales the area of the region by λ^2.

28. Suppose f is a continuous positive function with a period of 2π, and $\mathcal{R}$ is the region enclosed by $r = f(\theta)$ for $0 \leq \theta \leq 2\pi$. Use the integral formula on p. 748 to show that dilating $\mathcal{R}$ by a factor of $\lambda > 0$ scales the length of its perimeter by λ.

29. Suppose an aluminum lamina is made 0.1 cm thick, and its cross-section is the bounded region enclosed by the curve $r = 2 + \sin(2\theta)$, where r is measured in cm. Determine the mass of the lamina. (The density of aluminum is 2.7 g/cm^3.)

30. Suppose a company adopts a new icon for use on its letterhead. The icon is the locus of points that lies inside *both* $r = 2 + \sin(2\theta)$ and $r = 2 + \cos(2\theta)$, as seen in Figure 4.8. If r is measured in millimeters, how much ink (in mm^2) per page is needed to print it?

31. Suppose that after looking at Figure 4.8 the company in #30 decides to use the locus of points that are enclosed by one but not both curves—including all four disjoint regions—as its new icon. If r is measured in millimeters, how much ink (in mm^2) per page is needed to print this design?

32. Suppose f is a differentiable, positive function with a period of 2π, and $\mathcal{C}$ is the curve $r = f(\theta)$ for $\theta \in [0, 2\pi]$. Then there is some point on $\mathcal{C}$ that's closest to the origin, say at $\theta = \theta_0$. Show that the tangent line to $r = f(\theta)$ at $\theta = \theta_0$ is perpendicular to the ray $\theta = \theta_0$.

33. Show for a spiral of the form $r = Ae^{B\theta}$, called a *logarithmic* or *equiangular* spiral, the angle between the radial line and the tangent line is always the same (see Figure 4.9).

34. On p. 738 we verified that $r = \sin(\theta)$ describes a circle of radius $1/2$.

    (a) Using basic geometry, verify that the area enclosed by $r = \sin(\theta)$ is $\pi/4$.

    (b) Calculate $\int_0^{2\pi} \frac{1}{2} r^2 \, d\theta$.

    (c) Explain why your answers from parts (a) and (b) are different (is there a flaw in our formula)?

35. On p. 738 we verified that $r = \sin(\theta)$ describes a circle of radius $1/2$.

    (a) Using basic geometry, verify that the arc length of the curve is π.

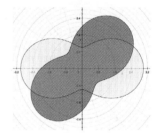

Figure 4.8: For #30.

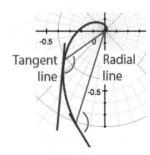

Figure 4.9: The angle between the radial line and the tangent line is always the same in a logarithmic spiral.

(b) Calculate $\int_0^{2\pi} \sqrt{\Big(r(\theta)\Big)^2 + \Big(r'(\theta)\Big)^2}\ d\theta$.

(c) Explain why your answers from parts (a) and (b) are different (is there a flaw in our formula)?

36. The curve $r^2 = \cos(2\theta)$ is called a **lemniscate**.

(a) Use a graphing utility to render a plot of $r^2 = \cos(2\theta)$, and verify that the area enclosed by its loops is not zero.

(b) Determine what's wrong with the area calculation

$$\int_0^{2\pi} \frac{1}{2} r^2\ d\theta = \int_0^{2\pi} \frac{1}{2}\cos(2\theta)\ d\theta = \frac{1}{4}\sin(2\theta)\Big|_0^{2\pi} = 0.$$

(c) Calculate the area enclosed by $r^2 = \cos(2\theta)$.

37. The polar curve $r = A\theta$ is called a **spiral of Archimedes**. (a) Examine the graph for various positive and negative values of A. (b) Find the slope of the tangent line to the curve at the point $(r,\theta) = (A\pi/4, \pi/4)$, and discuss how it depends on the value of A. (c) Find the length of the curve from the origin to the point $(A\pi/4, \pi/4)$. (Your answer will depend on A.)

38. A **lituus** is an Archimedean spiral in which the angle is inversely proportional to the square of the radius, such as $r^2 = 1/\theta$. Use a graphing utility to plot this lituus, and determine its length between $\theta = \pi/6$ and $\theta = \pi/4$.

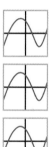

39. The curve described by $r = 2 + \sec(\theta)$ is an example of a **conchoid**. Use a graphing utility to plot this curve, and determine the area of the loop.

40. The polar curve $r = \frac{\cos(2\theta)}{\cos(\theta)}$ is an example of a **strophoid**. (a) Use a graphing utility to plot this curve, (b) find the cartesian equation describing the vertical asymptote, and (c) find the area of the loop.

41. The curve $r = 3\sin(\theta)\cos(\theta)/(\cos^3(\theta)+\sin^3(\theta))$ is an example of a **folium of Descartes**. (a) Use a graphing utility to plot this folium, and (b) approximate the area enclosed by its loop using Simpson's method with 10 subintervals.

42. The curve $r = \sin(\theta)/\theta$ is an example of a **cochleoid**. (a) Determine $\lim_{\theta \to 0} r$, (b) without using a graphing utility, sketch a graph of the cochleoid for $0 \le \theta \le 2\pi$, (c) use the half-angle identity for the sine to verify that $r^2 = (1 - \cos(2\theta))/2\theta^2$, (d) write the Maclaurin series for r^2, and (e) use the 6th-degree Taylor polynomial for r^2 (about $\theta = 0$) to approximate the area of the loop.

Checking the details

43. Suppose $x = r\cos(\theta)$ and $y = r\sin(\theta)$, where r is a function of θ.

(a) Use the Product Rule to determine $x'(\theta)$ and $y'(\theta)$.

(b) Use your answers from part (a) to rewrite $(x'(\theta))^2 + (y'(\theta))^2$ in terms of $r'(\theta), r(\theta), \sin(\theta)$, and $\cos(\theta)$.

(c) Use the Pythagorean Identity $\sin^2(\theta) + \cos^2(\theta) = 1$ to verify that

$$\int_a^b \sqrt{\Big(x'(\theta)\Big)^2 + \Big(y'(\theta)\Big)^2}\ d\theta = \int_a^b \sqrt{\Big(r(\theta)\Big)^2 + \Big(r'(\theta)\Big)^2}\ d\theta.$$

9.5   Introduction to Complex Numbers

In Section 8.6 we saw that

$$\frac{1}{1-t} = \sum_{k=0}^{\infty} t^k \quad \text{when} \quad |t| < 1,$$

and noted that the radius of convergence for this geometric power series is exactly the distance between its expansion point ($t_0 = 0$) and the discontinuity that we see in its closed form (at $t = 1$). In fact, we saw that relationship in many of our examples, but *not* when we found

$$\frac{1}{1+t^2} = \sum_{k=0}^{\infty} (-1)^k t^{2k}.$$

The radius of convergence for the series is still 1, but the closed form doesn't seem to have any discontinuities to speak of. The only way this function could have a discontinuity would be if

$$1 + t^2 = 0 \quad \Longleftrightarrow \quad t = \sqrt{-1} \text{ or } t = -\sqrt{-1}.$$

When people first saw $\sqrt{-1}$ arise in calculations, they weren't sure what to make of it. It's not a real number in the sense that we understand real numbers, so it was saddled with the unfortunate name *imaginary*, and we denote it by i today. In this section we introduce numbers of the form $x + iy$, where $x, y \in \mathbb{R}$, including the basics of their algebra and calculation of their magnitude. You'll find that i is exactly 1 unit away from $t_0 = 0$, which is one way to understand why the radius of convergence for the Maclaurin series of $1/(1+t^2)$ is 1.

> Electrical engineers write $\sqrt{-1}$ as j, since i is used to denote "current" in that profession.

❖ Complex Numbers

A complex number is a quantity that has two components, called its *real part* and its *imaginary part*. One of the first significant steps in understanding them at a rigorous level was made by Jean Robert Argand (1768–1822), who wrote these two parts as an ordered pair (x, y) and explained the algebra of complex numbers in terms of operations on vectors. While this famous work (called the **Argand map**) still pervades the way we think about complex numbers today, many people find it easier to write complex numbers as

$$x + iy.$$

We typically denote a complex variable by z (so $z = x + iy$). We say that the **real part** of z is the number x, written $\operatorname{Re}(z) = x$, and the **imaginary part** of z is the number y, written $\operatorname{Im}(z) = y$. Just as the set of real numbers is denoted by $\mathbb{R}$, we denote the set of complex numbers by $\mathbb{C}$.

Note: Remember that a complex number has two parts. It's irrelevant whether we write it as $x + iy$ or $x + yi$.

Real and imaginary parts of complex numbers

Example 5.1. Determine the real and imaginary parts of the complex numbers (a) $2 + 3i$, (b) 4, and (c) $5i$, and then plot them as points in the plane.

Solution: (a) $\operatorname{Re}(2 + 3i) = 2$ and $\operatorname{Im}(2 + 3i) = 3$; (b) remember that each complex number has two parts. In this case, when we write 4 as $4 + 0i$ we see that $\operatorname{Re}(4) = 4$ and $\operatorname{Im}(4) = 0$; (c) as we did in part (b), let's write $8i$ as $0 + 8i$. Then we see that

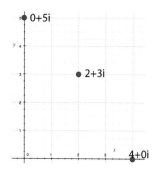

Figure 5.1: Points in the complex plane.

$\text{Re}(5i) = 0$ and $\text{Im}(5i) = 5$. These complex numbers have been plotted in Figure 5.1 by treating $x + iy$ as the point (x, y). ∎

Note: The imaginary part of $2 + 3i$ is *not* $3i$, but 3.

Based on the Argand map, by which we understand $z = x + iy$ as a point in the plane, you might think to define the magnitude of a complex number using the distance formula (i.e., the magnitude of a complex number is its distance from $0 = 0 + 0i$, which is the origin of the complex plane). That's exactly right:

$$|x + iy| = \sqrt{x^2 + y^2}. \tag{5.1}$$

Magnitude of complex numbers

Example 5.2. Determine the magnitudes of the complex numbers (a) $2 + 3i$, (b) 4, and (c) $5i$.

Solution: Using equation (5.1), we see that (a) $|2 + 3i| = \sqrt{13}$, (b) $|4| = 4$, and (c) $|5i| = 5$. ∎

❖ Arithmetic of Complex Numbers

The addition and multiplication of complex numbers works as you might hope, with the commutativity and associativity to which you're accustomed. Specifically,

$$(a + bi) + (c + di) = (a + c) + (b + d)i \tag{5.2}$$

and

$$
\begin{aligned}
(a + bi)(c + di) &= a(c + di) + (bi)(c + di) \\
&= ac + adi + bci + bdi^2 \\
&= (ac - bd) + (ad + bc)i, \tag{5.3}
\end{aligned}
$$

The method of addition and multiplication is akin to the addition and multiplication of binomials, $a + bx$ and $c + dx$, with the additional fact that $x^2 = -1$.

since $i^2 = -1$. In the exercise set, you'll verify that addition and multiplication of complex numbers is both commutative and associative.

Addition and multiplication of complex numbers

Example 5.3. Suppose $z = 2 + 3i$ and $w = 4 - 8i$. Calculate the numbers (a) $z + w$ and (b) zw.

Solution: These calculations can be made in a straightforward manner:

$$(2 + 3i) + (4 - 8i) = (2 + 4) + (3 - 8)i = 6 - 5i,$$

and

$$(2 + 3i)(4 - 8i) = (2)(4 - 8i) + (3i)(4 - 8i) = 8 - 16i + 12i + 24 = 32 - 4i,$$

where we've used the fact that $i^2 = -1$ in our calculation of $(3i)(-8i)$. ∎

In the same sense that $\frac{1}{7}$ is the reciprocal of 7 because their product is 1, we say that w is a reciprocal of z when $zw = 1$ (i.e., $1 + 0i$). All complex numbers except $z = 0$ have a unique reciprocal, and in the exercise set you'll show that

$$(x + iy)^{-1} = \frac{x}{x^2 + y^2} - \frac{y}{x^2 + y^2}i.$$

Try It Yourself: Reciprocals of complex numbers

Example 5.4. Verify that $z = 3 + 4i$ is the reciprocal of $w = \frac{3}{25} - \frac{4}{25}i$ by showing that their product is 1.

Answer: See the on-line solution.   ∎

Full Solution On-line

The last of the basic algebraic operations is called *conjugation*. The **conjugate** of $z = x + iy$ is denoted by $\overline{z}$ and defined by

$$\overline{z} = x - iy.$$

> Conjugation is sometimes denoted by z^*.

Graphically speaking $\overline{z}$ is the reflection of z across the x-axis in the complex plane.

Conjugates of complex numbers

Example 5.5. Determine the conjugate of (a) $2 + 3i$, (b) 4, and (c) $5i$.

Solution: Conjugation negates the imaginary part of a complex number, so

$$\begin{aligned} \overline{2 + 3i} &= 2 - 3i \\ \overline{4} &= \overline{4 + 0i} = 4 - 0i = 4 \end{aligned}$$

> Note that since $4 = 4 + 0i$ sits on the x-axis in the complex plane, its reflection is just itself.

and

$$\overline{5i} = \overline{0 + 5i} = 0 - 5i = -5i.$$ ∎

In the exercise set, you'll verify the following algebraic facts about conjugation:

$$\overline{z + w} = \overline{z} + \overline{w} \qquad\qquad \overline{zw} = \overline{z}\,\overline{w} \qquad\qquad z\overline{z} = |z|^2$$

❖ Polar Form of Complex Numbers

In Section 8.8 we established three extremely important and useful power series, which we now cite as algebraic facts:

$$e^z = \sum_{k=0}^{\infty} \frac{1}{k!}z^k, \qquad \cos(z) = \sum_{k=0}^{\infty} \frac{(-1)^k}{(2k)!}z^{2k}, \qquad \text{and} \qquad \sin(z) = \sum_{k=0}^{\infty} \frac{(-1)^k}{(2k+1)!}z^{2k+1},$$

> As with real numbers, a series of complex numbers is understood as limits of its partial sums.

all of which are absolutely convergent, even when z is a complex number. In his 1748 work titled *Introductio in Analysin Infinitorum*, Leonhard Euler (1707–1783) published a relationship among these three series that is now widely acclaimed as the most remarkable formula in mathematics.

Euler's Formula: $e^{i\theta} = \cos(\theta) + i\sin(\theta)$.

> The famous equation
> $$e^{i\pi} + 1 = 0,$$
> known as *Euler's identity*, is a direct consequence of this fact. Check it!

Proof. If we take $z = i\theta$, the power series for the exponential function becomes

$$e^{i\theta} = \sum_{k=0}^{\infty} \frac{1}{k!}(i\theta)^k.$$

When we separate this into two series, one with even powers of $i\theta$ and the other with odd, we see

$$e^{i\theta} = \sum_{k=0}^{\infty} \frac{i^{2k}}{(2k)!}\theta^{2k} + \sum_{k=0}^{\infty} \frac{i^{2k+1}}{(2k+1)!}\theta^{2k+1}.$$

> Recall that for each integer k, the number $2k$ is even and $2k + 1$ is odd.

Factoring an i out of the second series leaves us with

$$e^{i\theta} = \sum_{k=0}^{\infty} \frac{i^{2k}}{(2k)!}\theta^{2k} + i\sum_{k=0}^{\infty} \frac{i^{2k}}{(2k+1)!}\theta^{2k+1},$$

which simplifies matters because $i^{2k} = (i^2)^k = (-1)^k$. That is,

$$e^{i\theta} = \sum_{k=0}^{\infty} \frac{(-1)^k}{(2k)!}\theta^{2k} + i\sum_{k=0}^{\infty} \frac{(-1)^k}{(2k+1)!}\theta^{2k+1}.$$

These are exactly the series for the cosine and sine! ∎

If we scale Euler's formula by $r > 0$, we see

$$re^{i\theta} = r\cos(\theta) + i\,r\sin(\theta).$$

Writing $x = r\cos(\theta)$ and $y = r\sin(\theta))$, as we did when studying polar coordinates, this becomes

$$re^{i\theta} = x + iy. \tag{5.4}$$

We say that the left-hand side of this equation is the **polar form** of a complex number, and the right-hand side is the **Cartesian form.** We pass back and forth between these forms in exactly the same way that we passed between polar and Cartesian coordinates. In particular, note that

$$|re^{i\theta}| = \sqrt{x^2 + y^2} = \sqrt{r^2\cos^2(\theta) + r^2\sin^2(\theta)} = \sqrt{r^2} = |r| = r$$

since $r > 0$, so we say that r is the **magnitude** of $z = re^{i\theta}$, and write $r = |z|$. The angle θ is called the **argument** of $z = re^{i\theta}$, and is often denoted by $\arg(z)$.

Changing between forms of complex numbers

Example 5.6. Determine (a) the polar form of $\sqrt{3}+1i$, and (b) the Cartesian form of $10e^{(2\pi/3)i}$.

Solution: (a) We need to know r and θ. Since r is the magnitude of the number,

$$r = \sqrt{(\sqrt{3}\,)^2 + (1)^2} = 2,$$

from which it follows that $\sqrt{3} = x = 2\cos(\theta)$ and $1 = y = 2\sin(\theta)$. Since both sine and cosine are positive, we conclude that

$$\frac{\sqrt{3}}{2} = \cos(\theta) \quad \Rightarrow \quad \theta = \cos^{-1}\left(\frac{\sqrt{3}}{2}\right) = \frac{\pi}{6}.$$

That is, $\sqrt{3} + 1i = 2e^{i\pi/6}$. (b) The Cartesian form can be determined by direct substitution:

$$10e^{(2\pi/3)i} = 10\left(\cos\left(\frac{2\pi}{3}\right) + i\,\sin\left(\frac{2\pi}{3}\right)\right) = 5 + 5\sqrt{3}\,i. \qquad ∎$$

Try It Yourself: Changing between forms of complex numbers

Example 5.7. Determine (a) the polar form of $6 + 6i$, and (b) the Cartesian form of $10e^{(-4\pi/3)i}$.

Answer: (a) $6\sqrt{2}e^{i\pi/4}$, (b) $-5 + 5\sqrt{3}\,i$. ∎

Full Solution On-line

In Examples 5.6 and 5.7 you saw that $5 + 5\sqrt{3}\,i$ can be written in polar form with an argument of $2\pi/3$, or an argument of $-4\pi/3$. That happens because these values of θ differ by 2π, which is exactly the period of the sine and cosine. In general, for this very reason,

$$re^{i\theta} = re^{i(\theta + 2\pi k)} \quad \text{when } k \text{ is any integer.}$$

Because of this fact, the argument of a complex number is a set rather than a value. For example,

$$\arg(5 + 5\sqrt{3}\, i) = \left\{ \ldots - \frac{10\pi}{3}, -\frac{4\pi}{3}, \frac{2\pi}{3}, \frac{8\pi}{3}, \frac{14\pi}{3}, \frac{20\pi}{3}, \ldots \right\}.$$

This leads us to define the **principal argument** of a complex number to be the angle in $\arg(z)$ that lies in $(-\pi, \pi]$. The particular angle is denoted by $\text{Arg}(z)$. For example,

$$\text{Arg}(5 + 5\sqrt{3}\, i) = \frac{2\pi}{3}.$$

▷ Multiplication of complex numbers in polar form

Multiplying complex numbers in polar form is very easy. In brief, we can just *multiply the magnitudes* and *add the arguments*. Here's why: when

$$re^{i\theta} = r\cos(\theta) + ir\sin(\theta) \quad \text{and} \quad Re^{i\phi} = R\cos(\phi) + iR\sin(\phi),$$

equation (5.3) tells us that their product is

$$\left(re^{i\theta}\right)\left(Re^{i\phi}\right) = \left(\overbrace{r\cos(\theta)}^{a} + \overbrace{ir\sin(\theta)}^{b} \right)\left(\overbrace{R\cos(\phi)}^{c} + \overbrace{iR\sin(\phi)}^{d} \right)$$

$$= \left(rR\cos(\theta)\cos(\phi) - rR\sin(\theta)\sin(\phi) \right) + i\left(rR\cos(\phi)\sin(\theta) + rR\sin(\phi)\cos(\theta) \right)$$

$$= rR\left(\cos(\theta)\cos(\phi) - \sin(\theta)\sin(\phi) \right) + i\, rR\left(\cos(\phi)\sin(\theta) + \sin(\phi)\cos(\theta) \right).$$

The sum identities for the sine and cosine allow us to rewrite this as

$$\left(re^{i\theta}\right)\left(Re^{i\phi}\right) = rR\cos(\theta + \phi) + i\, rR\sin(\theta + \phi)$$

$$= rR\left(\cos(\theta + \phi) + i\, \sin(\theta + \phi) \right).$$

That is, we've just shown

$$\left(re^{i\theta}\right)\left(Re^{i\phi}\right) = rRe^{i(\theta + \phi)}.$$

Using $r = 1$ and $R = 1$, this is a handy way to remember the addition formulas for the sine and cosine.

The reciprocal in polar form

Example 5.8. Show that the reciprocal of $z = re^{i\theta}$ is $\frac{1}{r}e^{-i\theta}$.

Solution: Since the product of complex numbers is found by multiplying their magnitudes and adding their arguments,

$$\left(re^{i\theta}\right)\left(\frac{1}{r}e^{-i\theta}\right) = \frac{r}{r}e^{i(\theta - \theta)} = 1^{i0} = 1.$$

Since reciprocals are unique, we've shown that $z^{-1} = \frac{1}{r}e^{-i\theta}$. ∎

The 6th roots of unity

Example 5.9. Solve the equation $z^6 = 1$.

Solution: Since multiplication is so easy in polar form, let's write $z = re^{i\theta}$, where $\theta \in (-\pi, \pi]$ is the principal argument. Then

$$z^6 = \left(re^{i\theta}\right)\left(re^{i\theta}\right)\left(re^{i\theta}\right)\left(re^{i\theta}\right)\left(re^{i\theta}\right)\left(re^{i\theta}\right) = r^6 e^{i6\theta},$$

so the equation $z^6 = 1$ can be written as $r^6 e^{i6\theta} = 1$. Since the left and right sides of this equation are the same, they have the same magnitude and the same argument. That is,

$$\overbrace{\begin{array}{c} \\ r^6 = 1 \\ r = 1 \end{array}}^{\text{same magnitude}} \qquad \overbrace{\begin{array}{l} 6\theta = 0 + 2\pi k \quad \text{for some integer } k \\ \theta = \dfrac{\pi}{3}k \quad \text{for some integer } k. \end{array}}^{\text{same argument}}$$

Since $\theta \in (-\pi, \pi]$, there are only six admissible values of k:

$$k = -2 \Rightarrow \theta = -\tfrac{2\pi}{3} \qquad k = 0 \Rightarrow \theta = 0 \qquad k = 2 \Rightarrow \theta = \tfrac{2\pi}{3}$$

$$k = -1 \Rightarrow \theta = -\tfrac{\pi}{3} \qquad k = 1 \Rightarrow \theta = \tfrac{\pi}{3} \qquad k = 3 \Rightarrow \theta = \pi.$$

So our six solutions, called the 6th *roots of unity*, are

$$z_1 = 1e^{-2\pi i/3} \qquad z_3 = 1e^{i0} = 1 \qquad z_5 = 1e^{2\pi i/3}$$

$$z_2 = 1e^{-\pi i/3} \qquad z_4 = 1e^{\pi i/3} \qquad z_6 = 1e^{-\pi i} = -1,$$

which are shown in Figure 5.2.   ■

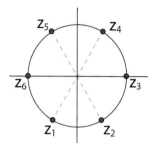

Figure 5.2: The 6th roots of unity are evenly spaced around the unit circle in the complex plane.

Calculating the square root of a complex number

Example 5.10. Find all solutions to the equation $z^2 = \frac{9}{\sqrt{2}} + \frac{9}{\sqrt{2}}i$.

Solution: The first step is to rewrite the equation in its polar form, beginning with the right-hand side. Since the real and imaginary parts of $\frac{9}{\sqrt{2}} + \frac{9}{\sqrt{2}}i$ are equal and positive, we know that its principal argument is $\pi/4$; and it's easy to check that the magnitude of the number is 9. So the right-hand side of this equation can be rewritten as $9e^{i\pi/4}$.

When we set $z = re^{i\theta}$, where $\theta \in (-\pi, \pi]$, we have $z^2 = r^2 e^{i2\theta}$, so

$$z^2 = \frac{9}{\sqrt{2}} + \frac{9}{\sqrt{2}}i \quad \text{can be written as} \quad r^2 e^{i2\theta} = 9e^{i\pi/4}.$$

Since the numbers on the left- and right-hand sides are equal, they have the same magnitude and the same argument. That is,

$$\overbrace{\begin{array}{c} \\ r^2 = 9 \\ r = 3 \end{array}}^{\text{same magnitude}} \qquad \overbrace{\begin{array}{l} 2\theta = \dfrac{\pi}{4} + 2\pi k \quad \text{for some integer } k \\ \theta = \dfrac{\pi}{8} + \pi\,k \quad \text{for some integer } k. \end{array}}^{\text{same argument}}$$

Since $\theta \in (-\pi, \pi]$, there are only two admissible values of k:

$$k = 0 \Rightarrow \theta = \frac{\pi}{8} \qquad z_1 = 3e^{i\pi/8}$$

$$k = -1 \Rightarrow \theta = -\frac{3\pi}{8} \qquad z_2 = 3e^{-i7\pi/8},$$

which are shown in Figure 5.3.   ■

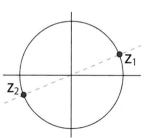

Figure 5.3: The square roots of a complex number evenly spaced around a circle in the complex plane.

❖ The Complex Exponential and Logarithm Functions

Euler's formula also allows us to understand the exponential of a complex number. When $z = x + iy$,

$$e^z = e^{x+iy} = e^x e^{iy}.$$

Calculating the exponential of a complex number

Example 5.11. Suppose $z_1 = \ln(6) + i\frac{\pi}{3}$ and $z_2 = -4 + i\pi$. Calculate (a) e^{z_1} and (b) e^{z_2}.

Solution: (a) Using the algebra outlined above,

$$e^{z_1} = e^{\ln(6)+i\pi/3} = e^{\ln(6)}e^{i\pi/3} = 6e^{i\pi/3} = 6\left(\frac{1}{2} + \frac{\sqrt{3}}{2}i\right) = 3 + 3\sqrt{3}\,i$$

and (b) $e^{z_2} = e^{-4+i\pi} = e^{-4}e^{i\pi} = e^{-4}(-1) = -e^{-4}.$ ■

Try It Yourself: Calculating the exponential of a complex number

Example 5.12. Suppose $z_1 = \ln(6) + i\frac{7\pi}{3}$ and $z_2 = 14 + i8\pi$. Calculate (a) e^{z_1}, and (b) e^{z_2}.

Full Solution On-line

Answer: (a) $3 + 3\sqrt{3}\,i$, and (b) e^{14}.   ■

In Examples 5.11 and 5.12 you saw that both

$$\ln(6) + i\frac{\pi}{3} \quad \text{and} \quad \ln(6) + i\frac{7\pi}{3} \quad \text{solve the equation} \quad e^z = 3 + 3\sqrt{3}\,i,$$

so *the exponential function is not one-to-one when we allow it to act on complex numbers.* However, all the numbers that solve $e^z = 3 + 3\sqrt{3}\,i$ have the same real part, so they lie on a vertical line in the complex plane (see Figure 5.4), and their imaginary parts differ by multiples of 2π. That is, they all have the form $\ln(6) + i\left(\frac{\pi}{3} + 2\pi k\right)$ where k is an integer. So, in accord with our previous usage of the word *logarithm,* we write

$$\log(3 + 3\sqrt{3}\,i) = \left\{ \ln(6) + i\left(\frac{\pi}{3} + 2\pi k\right) \text{ , where } k \text{ is an integer}\right\}.$$

More generally, when $w \neq 0$ the set of solutions to $e^z = w$ is

$$\log(w) = \left\{ \ln|w| + \arg(w)i \right\}.$$

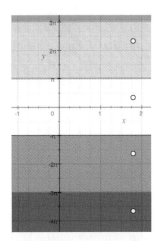

Figure 5.4: Solutions of $e^z = 3 + 3\sqrt{3}\,i$ lie in different "strata" of the complex plane. Note that all solutions have a real part (x-coordinate) of $\ln(6)$.

We can make the exponential function one-to-one by restricting its domain, in much the same way that we were able to invert the sine and cosine by restricting their domains appropriately. Specifically, we can make e^z a one-to-one function by restricting its domain to any horizontal strip of the complex plane that has two properties: (1) its height is 2π, and (2) it includes either its upper or lower boundary, but not both. Some examples are shown as colored strata in Figure 5.4. The different logarithms (inverse functions) corresponding to different restrictions of the exponential function's domain are called different **branches** of the logarithm. The branch corresponding to the white strip in Figure 5.4, wherein $-\pi < \text{Im}(z) \leq \pi$, is called the **principal branch** of the logarithm and is denoted by **Log**(z).

The capital "L" is a convention for identifying the principal branch of the logarithm.

Calculating the principal logarithm of a complex number

Example 5.13. Calculate (a) $\text{Log}(3 + 3\sqrt{3}\,i)$, (b) $\text{Log}(e^{14})$, and (c) $\text{Log}(1 + i)$.

Solution: (a) In Example 5.11 we showed that

$$e^z = 3 + 3\sqrt{3}\,i \quad \text{when} \quad z = \ln(6) + i\frac{\pi}{3}.$$

Since $\text{Im}\left(\ln(6) + i\frac{\pi}{3}\right) = \frac{\pi}{3}$, which is in $(-\pi, \pi]$, this is the value of the principal logarithm. That is, $\text{Log}(3 + 3\sqrt{3}\,i) = \ln(6) + i\frac{\pi}{3}$.

(b) In Example 5.12 we saw that $e^w = e^{14}$ when $w = 14 + i8\pi$, but this is not $\text{Log}(e^{14})$ because its imaginary part is too large. Since all solutions of the equation have the same real part but differ by multiples of 2π in the imaginary part, another solution is $14 + 0i$, whose imaginary part is in $(-\pi, \pi]$. So $\text{Log}(e^{14}) = 14 + 0i = 14$.

(c) When we write $z = x + iy$, and $1 + i = \sqrt{2}e^{i\pi/4}$, the equation $e^z = 1 + i$ becomes $e^x e^{iy} = \sqrt{2}e^{i\pi/4}$. Since the left and right sides are the same, they must have the same magnitude and the same argument. That is,

$$\overbrace{e^x = \sqrt{2}}^{\text{same magnitude}} \qquad \overbrace{y = \frac{\pi}{4} + 2\pi k}^{\text{same argument}} \quad \text{for some integer } k,$$

from which we conclude that $x = \ln(\sqrt{2}) = \frac{1}{2}\ln(2)$ and $y = \frac{\pi}{4}$ (taking $k = 0$ so that our answer is in the correct strip of the complex plane). That is $\text{Log}(1 + i) = \frac{1}{2}\ln(2) + i\frac{\pi}{4}$. ∎

Some solutions of $e^z = e^{14}$ are

$$\vdots$$
$$14 + i10\pi,$$
$$14 + i8\pi,$$
$$14 + i6\pi,$$
$$14 + i4\pi,$$
$$14 + i2\pi,$$
$$14 + i0,$$
$$14 - i2\pi,$$
$$14 - i4\pi,$$
$$\vdots$$

You should know

- the terms *Argand map, real part, imaginary part, conjugate, Euler's formula, Euler's identity, polar form, Cartesian form, magnitude, argument, principal argument, branch* (of the logarithm), and *principal branch* (of the logarithm);

- $e^{i\theta} = \cos(\theta) + i\sin(\theta)$;

- the notation $\text{Re}(z)$, $\text{Im}(z)$, $|z|$, $\bar{z}$, $\arg(z)$, $\text{Arg}(z)$, and $\text{Log}(z)$.

You should be able to

- transition between the Cartesian and polar forms of a complex number;

- add and multiply complex numbers;

- calculate the reciprocal of a complex number;

- determine $|z|$ and $\bar{z}$;

- solve equations of the form $z^n = w$ and $e^z = w$.

❖ 9.5 Skill Exercises

In #1–6 determine (a) the real part of z, (b) the imaginary part of z, (c) the magnitude of z, and then (d) plot z as a point in the Cartesian plane.

1. $z = 2 + 4i$ 3. $z = -2 + 3i$ 5. $z = -5 + 0i$

2. $z = 5 - 8i$ 4. $z = -4 - 3i$ 6. $z = 0 + 3i$

In #7–12 (a) determine $z + w$ and then (b) plot z, w and $z + w$ as points in the Cartesian plane.

7. $z = 3 + 4i$, $w = 4 + 2i$ 9. $z = 8$, $w = 3 + 3i$ 11. $z = 8i$, $w = 1 - 3i$

8. $z = 2 - 8i$, $w = -5 - i$ 10. $z = 1 + 4i$, $w = 7$ 12. $z = 5 + 9i$, $w = 2i$

In #13–18 (a) plot z and $\overline{z}$ as points in the Cartesian plane, then determine (b) $\overline{z} + \overline{w}$, (c) $\overline{z + w}$, (d) $\overline{z}\,\overline{w}$, and (e) $\overline{zw}$.

13. $z = 13 + 2i$, $w = 3 + 6i$ 15. $z = 5$, $w = 1 + i$ 17. $z = 5i$, $w = 1 + 2i$

14. $z = 1 - 9i$, $w = 5 + i$ 16. $z = -3$, $w = 4 - i$ 18. $z = 3 + 9i$, $w = 72i$

In #19–24 (a) plot z, w, and zw as points in the Cartesian plane, (b) determine the principal arguments of z, w and zw, and (c) use your answers from part (b) to verify that $\arg(zw) = \arg(z) + \arg(w)$.

19. $z = 2 + 2i$, $w = 1 - i$ 21. $z = 2 - 2i$, $w = i$ 23. $z = 5 + 5\sqrt{3}\,i$, $w = i$

20. $z = -3 + 3i$, $w = -1 - i$ 22. $z = -2 + 2i$, $w = -i$ 24. $z = 6\sqrt{3} + 6i$, $w = -i$

In #25–30 write the reciprocal of z in its Cartesian form.

25. $z = 2 + 2i$ 27. $z = -\sqrt{3} - i$ 29. $z = 3 + 4i$

26. $z = -3 + 3i$ 28. $z = 4 + 4\sqrt{3}i$ 30. $z = 5 + 6i$

In #31–38 write the specified complex number in polar form using its principal argument.

31. $z = 8 + 8i$ 33. $z = -4\sqrt{3} - 4i$ 35. $z = -9$ 37. $z = -13i$

32. $z = -3 + 3i$ 34. $z = 7 - 7\sqrt{3}i$ 36. $z = 3$ 38. $z = 4i$

In #39–46 (a) write the specified complex number in Cartesian form, and (b) write $\overline{z}$ in polar form.

39. $z = 7e^{(5\pi/6)i}$ 41. $z = 2e^{(17\pi/3)i}$ 43. $z = 4e^{(9\pi/4)i}$ 45. $z = 14e^{(\pi/2)i}$

40. $z = 17e^{(-\pi/6)i}$ 42. $z = 3e^{(11\pi/3)i}$ 44. $z = 5e^{(-\pi/4)i}$ 46. $z = 5e^{(-\pi/2)i}$

In #47–52 write the product zw in polar form.

47. $z = 2e^{(\pi/6)i}$, $w = 3e^{(\pi/4)i}$ 50. $z = 7e^{(-\pi/6)i}$, $w = 3e^{(\pi/4)i}$

48. $z = 7e^{(-\pi/6)i}$, $w = 5e^{(-\pi/3)i}$ 51. $z = 8e^{(-\pi/2)i}$, $w = 13e^{(\pi/2)i}$

49. $z = 2e^{(\pi/6)i}$, $w = 5e^{(-\pi/3)i}$ 52. $z = 6e^{\pi i}$, $w = 4e^{2\pi i}$

In #53–60 find all solutions to the given equation.

53. $z^3 = 1$ 55. $z^4 = 1 + i$ 57. $z^3 = 8\sqrt{3} + 8i$ 59. $z^6 = 2 + 5i$

54. $z^5 = 1$ 56. $z^6 = -2 + 2i$ 58. $z^4 = -6 - 6\sqrt{3}i$ 60. $z^8 = 3 - 6\sqrt[3]{7}i$

In #61–66 write e^z in both its polar and Cartesian representations.

61. $z = \ln(6) + \frac{\pi}{4}i$ 63. $z = 2 + 6\pi i$ 65. $z = 7 - i$

62. $z = \ln(7) - \frac{\pi}{6}i$ 64. $z = -1 + \frac{\pi}{2}i$ 66. $z = -8 + 2i$

In #67–78 determine (a) $\text{Log}(z)$, and (b) $\log(z)$. Then (c) plot a representative sampling as points in the Cartesian plane.

67. $z = 42$ 71. $z = 7 - 7i$ 75. $z = 2 + 2\sqrt{3}i$

68. $z = 19$ 72. $z = -3 - 3i$ 76. $z = -4\sqrt{3} - 4i$

69. $z = -4$ 73. $z = 5 - 5\sqrt{3}i$ 77. $z = 2 + 23i$

70. $z = -9$ 74. $z = -6\sqrt{3} + 6i$ 78. $z = 43 - 5i$

❖ 9.5 Concept and Application Exercises

79. Find all complex numbers z (if any) for which $\overline{z} = iz$.

80. Suppose $f(t) = \sum_{k=0}^{n} a_k t^k$ is a polynomial whose coefficients are real numbers, and $f(z) = 0$. Use the algebraic facts about conjugation listed on p. 754 to show that $\overline{z}$ is also a root of f.

81. Suppose $f(t)$ is a polynomial function with roots at $z = 2+3i$, $z = 4-7i$, and $z = 17$. In light of #80, what is the smallest possible degree of the polynomial $f(t)$?

82. Determine whether the statement $\text{Arg}(zw) = \text{Arg}(z) + \text{Arg}(w)$ is true for all nonzero complex numbers z and w. If so, explain why. If not, give an example where it fails.

83. Suppose the dashed circle in Figure 5.5 is the unit circle in the Cartesian plane (i.e., where $|z| = 1$). Plot the solutions of $z^3 = z_0$ as points in the Cartesian plane.

84. Suppose the solid circle in Figure 5.5 is the unit circle in the Cartesian plane (i.e., where $|z| = 1$). Plot the solutions of $z^3 = z_0$ as points in the Cartesian plane.

Figure 5.5: For #83 and #84.

85. Suppose that $z = re^{i\theta}$. Determine the polar form of $\overline{z}$.

86. Suppose that $z = re^{i\theta}$. Determine the polar form of z^{-1}.

87. Consider the number $z(t) = e^{\lambda t}$, where λ is a complex number and $t \in \mathbb{R}$.

(a) Show that $\lim_{t \to \infty} z(t) = 0$ if $\text{Re}(\lambda) < 0$.

(b) For what values of λ will $z(t)$ travel counterclockwise around the origin as $t \to \infty$?

88. Consider the number $z(t) = e^{\lambda t}$, where λ is a complex number and $t \in \mathbb{R}$.

(a) Show that $\lim_{t\to\infty} |z(t)| = \infty$ if $\mathrm{Re}(\lambda) > 0$.

(b) For what values of λ will $z(t)$ travel clockwise around the origin as $t \to \infty$?

89. Sketch the locus of points e^z, where (a) $z = -1 + yi$, (b) $z = \frac{1}{2} + yi$, and (c) $z = 1 + yi$, where y ranges over all of $\mathbb{R}$.

90. Sketch the locus of points e^z, where (a) $z = x - \frac{\pi}{4}i$, (b) $z = x - \frac{\pi}{6}i$, (c) $z = x + \frac{\pi}{6}i$, and (d) $z = x + \frac{\pi}{3}i$, where x ranges over all of $\mathbb{R}$.

91. The hyperbolic trigonometric functions are $\cosh(t) = \frac{1}{2}\left(e^t + e^{-t}\right)$ and $\sinh(t) = \frac{1}{2}\left(e^t - e^{-t}\right)$. Use Euler's formula to express $\cosh(it)$ and $\sinh(it)$ as familiar functions.

92. Use Euler's formula to write expressions for $e^{i(\theta+\phi)}, e^{i\theta}$, and $e^{i\phi}$. Then multiply the numbers on the right-hand side of the equation $e^{i(\theta+\phi)} = e^{i\theta}e^{i\phi}$ to recover the addition formulas for sine and cosine (since the left-hand and right-hand sides must have the same real and imaginary parts).

93. Use Euler's formula to verify (a) $\frac{1}{2}\left(e^{it} + e^{-it}\right) = \cos(t)$, and (b) $\frac{1}{2i}\left(e^{it} - e^{-it}\right) = \sin(t)$.

We can use the results of #93 above to extend the domain of the sine and cosine to complex numbers by defining

$$\cos(z) = \frac{1}{2}\left(e^{iz} + e^{-iz}\right) \quad \text{and} \quad \sin(z) = \frac{1}{2i}\left(e^{iz} - e^{-iz}\right).$$

Use these definitions in #94–97.

94. Using the formula for $\sin(z)$ cited above and the fact that $e^{ix} = \cos(x) + i\sin(x)$, show that $\sin(x + iy) = \sin(x)\cosh(y) + i\cos(x)\sinh(y)$.

95. For each z, the number $\sin(z)$ is a complex number, so it has a real and an imaginary part. Let's write $\sin(z) = u + iv$.

(a) Based on the formula that you established in #94, show that $\sin(z)$ is a point in the uv-plane that's on the hyperbola

$$\left(\frac{u}{\sin(4)}\right)^2 - \left(\frac{v}{\cos(4)}\right)^2 = 1$$

when $z = 4 + iy$ and y is allowed to range over all of $\mathbb{R}$. (We say that the sine *maps* vertical lines to hyperbolas.)

(b) How does the hyperbola change when we consider different vertical lines (e.g., $z = 5 + iy$, $z = 6 + iy$, etc.)?

96. For each z, the number $\sin(z)$ is a complex number, so it has a real and an imaginary part. Let's write $\sin(z) = u + iv$.

(a) Based on the formula that you established in #94, show that the locus of points $\sin(z)$ is the ellipse

$$\left(\frac{u}{\cosh(4)}\right)^2 + \left(\frac{v}{\sinh(4)}\right)^2 = 1$$

in the uv-plane when $z = x + 4i$ and x is allowed to range over all of $\mathbb{R}$. (We say that the sine *maps* horizontal lines to ellipses.)

(b) How does the ellipse change when we look at different horizontal lines (e.g., $z = x + 5i$, $z = x + 6i$, etc.)?

97. Find all values of z for which $\cos(z) = 17$.

Checking the details

98. Suppose $z = x + yi$ is a particular nonzero complex number (so that x and y are fixed numbers). If $w = a + bi$ is the reciprocal, we know that $zw = 1$.

   (a) Determine a formula for the product zw.

   (b) Write an equation that says the real part of zw is 1.

   (c) Write an equation that says the imaginary part of zw is 0.

   (d) Solve these equations for a and b in terms of x and y.

99. Suppose $z = x + yi$ and $w = a + bi$.

   (a) Verify that $\overline{z} + \overline{w} = \overline{z + w}$.

   (b) Verify that $\overline{z}\,\overline{w} = \overline{zw}$.

   (c) Verify that $z\,\overline{z} = |z|^2$.

Chapter 9 Review

✧ True or False

1. The parameterization $x(t) = -\cos(t)$, $y(t) = \sin(t)$ proceeds counterclockwise around the unit circle.

2. The parameterization $x(t) = t^2 + 7$, $y(t) = 3t^2 - 4$ proceeds along a line.

3. No tangent line to the curve parameterized by $x(t) = 1 + 4t + t^3$, $y(t) = 3t$ has a slope of 1.

4. The equations $x(t) = -\sin(t)$ and $y(t) = \cos(t)$ parameterize the unit circle by arc length.

5. Suppose $x(t)$ and $y(t)$ parameterize Γ, and $\int_0^5 \sqrt{(x'(t))^2 + (y'(t))^2} \, dt = 5$. Then $(x(t), y(t))$ parameterize Γ by arc length.

6. If the polar coordinates of P_1 and P_2 are $(r, \theta) = (7, \pi/2)$ and $(7, 5\pi/2)$ respectively, P_1 and P_2 are the same point.

7. If the polar coordinates of P_1 and P_2 are $(r, \theta) = (7, \pi/2)$ and $(-7, 3\pi/2)$ respectively, P_1 and P_2 are the same point.

8. The point $(r, \theta) = (-3, 7\pi/6)$ is in the first quadrant.

9. The rose $r = \cos(5\theta)$ has 5 petals.

10. The rose $r = \sin(6\theta)$ has 6 petals.

11. The regions $\{(r, \theta) : r \in [2, 3] \text{ and } \theta \in [\pi/6, \pi/3]\}$ and $\{(r, \theta) : r \in [8, 9] \text{ and } \theta \in [\pi/6, \pi/3]\}$ have the same area.

12. The regions $\{(r, \theta) : r \in [2, 3] \text{ and } \theta \in [\pi/6, \pi/3]\}$ and $\{(r, \theta) : r \in [2, 3] \text{ and } \theta \in [5\pi/6, \pi]\}$ have the same area.

13. The area enclosed by $r = 1.5\cos(3\theta)$ in the first quadrant is larger than the area enclosed by $r = \sin(6\theta)$ in the first quadrant.

14. There are exactly two points on the cardioid $r = 1 - \sin(\theta)$ at which the tangent line has a slope of 1.

15. Suppose $f(\theta)$ is continuous and positive, and has a period of 2π. Then $r = f(\theta)$ has at least 2 points where the tangent line has a slope of 1.

16. When z is a complex number whose imaginary part is zero, $\bar{z} = -z$.

17. The product $z\bar{z}$ is always a real number.

18. All complex numbers have a reciprocal.

19. The exponential function e^z is 1-to-1 when z is allowed to be a complex number.

20. All solutions of $e^z = 3 + 5i$ lie on a vertical line in the complex plane.

❖ Multiple Choice

21. The parameterization $x(t) = 3\cos(2\pi t)$, $y = -3\sin(2\pi t)$ ...

    (a) traverses a circle of radius 3 counterclockwise, and completes 2π cycles per second.

    (b) traverses a circle of radius 3 counterclockwise, and completes $1/2\pi$ cycles per second.

    (c) traverses a circle of radius 3 clockwise, and completes 2π cycles per second.

    (d) traverses a circle of radius 3 clockwise, and completes $1/2\pi$ cycles per second.

    (e) none of the above.

22. The functions $x(t) = 4\cos(t)$ and $y(t) = 5\sin^2(t)$ parameterize ...

    (a) a circle.

    (b) an ellipse.

    (c) a line (or segment).

    (d) a parabola.

    (e) none of the above.

23. The functions $x(t) = 4\cos(t)$ and $y(t) = 5\sin(t)$ parameterize ...

    (a) a circle.

    (b) an ellipse.

    (c) a line (or segment).

    (d) a parabola.

    (e) none of the above.

24. The functions $x(t) = 4\cos^2(t)$ and $y(t) = 5\sin^2(t)$ parameterize ...

    (a) a circle.

    (b) an ellipse.

    (c) a line (or segment).

    (d) a parabola.

    (e) none of the above.

25. The equation $r = 2/(3 + \cos(\theta + \pi/6))$ describes ...

    (a) a parabola.

    (b) a hyperbola.

    (c) an ellipse.

    (d) a cardioid.

    (e) none of the above.

26. The equation $r = 0.5/(0.1 + \cos(\theta - \pi/4))$ describes ...

    (a) a parabola.

    (b) a hyperbola.

    (c) an ellipse.

    (d) a cardioid.

    (e) none of the above.

27. The equation $r = 8\cos(\theta)$ describes ...

    (a) a parabola.

    (b) a hyperbola.

    (c) an ellipse.

    (d) a cardioid.

    (e) none of the above.

28. The equation $r = \sin(\theta - \pi/4)$ describes ...

(a) a circle whose center is on the ray $\theta = \pi/4$.

(b) a circle whose center is on the ray $\theta = -\pi/4$.

(c) a circle whose center is on the ray $\theta = 3\pi/4$.

(d) a circle whose center is on the ray $\theta = -3\pi/4$.

(e) none of the above

29. When $z = 4 - 4i$ ...

(a) $\arg(z) = \pi/4$.   (d) $\text{Arg}(z) = -\pi/4$.

(b) $\text{Arg}(z) = \pi/4$.   (e) none of the above.

(c) $\arg(z) = -\pi/4$.

30. Suppose z is a nonzero complex number. Then ...

(a) $z\overline{z}$ is always a real number.   (d) all of the above.

(b) $\overline{\overline{z}} = z$.   (e) none of the above.

(c) z^{-1} exists.

31. Suppose z is a non-zero complex number. Then ...

(a) $\log(z)$ and $\text{Log}(z)$ are the same. (d) $\arg(\text{Log}(z)) \in (-\pi, \pi]$.

(b) $\log(z)$ is a set of points.   (e) none of the above.

(c) $\arg(\text{Log}(z)) \in [0, 2\pi)$.

❖ Exercises

32. Determine the Cartesian equation of the curve that's parameterized by $x(t) = 3t - 4$ and $y(t) = 17t + 22$.

33. In our discussion of the stereographic projection on p. 18, we derived the formulas $x(t) = \frac{4t}{t^2+4}$ and $y(t) = \frac{2t^2}{t^2+4}$. Show that $(x(t), y(t))$ parameterizes all but one point of a circle.

34. Suppose $\mathcal{R}$ is the triangular lamina bounded by $y = 0$, $x = 4$, and the line $y = x$, where all lengths are measured in cm. The mass density in $\mathcal{R}$ is $\rho(x) = kx + 1$ g/cm^2, where $k \geq 0$.

(a) Determine the location of the centroid $(\overline{x}(k), \overline{y}(k))$.

(b) Determine $\lim_{k \to \infty} \overline{x}(k)$ and $\lim_{k \to \infty} \overline{y}(k)$.

(c) Based on your answer to part (b), explain what happens to the centroid of the lamina as k increases.

(d) Use a graphing utility to render the parametric curve defined by $x = \overline{x}(k)$ and $y = \overline{y}(k)$. Does the graph confirm your conclusion from part (c)?

35. Suppose $\mathcal{R}$ is the triangular lamina bounded by $y = 0$, $x = 4$, and the line $y = \frac{1}{4}kx$, where $k \geq 0$ and all lengths are measured in cm. The mass density in $\mathcal{R}$ depends on position according to $\rho(x) = x + 1$ g/cm^2.

(a) Determine the location of the centroid $(\overline{x}(k), \overline{y}(k))$.

(b) Graph the parametric curve defined by $x = \overline{x}(k)$ and $y = \overline{y}(k)$ for $1 \leq k \leq 10$.

36. Suppose $x(t) = 3t^2 + 4t + 1$ and $y(t) = \cos(t)$ parameterize the curve $\mathcal{C}$. Find the equation of the tangent line to $\mathcal{C}$ at the point $(x(0), y(0))$.

37. Suppose $x(t) = t^3$ and $y(t) = 3t + 1$ parameterize the curve $\mathcal{C}$. Determine the area between $\mathcal{C}$ and the x-axis over $[x(0), x(2)]$.

38. Suppose $x(t) = 13t + 4$ and $y(t) = -12t + 2$ parameterize $\mathcal{C}$ when $t \in [2, 4]$. What's the arc length of $\mathcal{C}$?

39. Suppose $x(t) = 11 - 2t$ and $y(t) = 2\cosh(t)$ parameterize $\mathcal{C}$ when $t \in [0, 1]$. What's the arc length of $\mathcal{C}$?

40. Suppose $x(t) = \ln|\sec(t)|$ and $y(t) = t + 14$ parameterize the curve $\mathcal{C}$ when $t \in [0, \pi/4]$. Determine the length of $\mathcal{C}$.

41. Suppose $x(t) = 3t + 4$ and $y(t) = -2t + 1$ parameterize $\mathcal{C}$ when $t \in [1, 5]$. What's the area of the surface generated by revolving $\mathcal{C}$ about the x-axis?

42. Suppose $x(t) = e^{2t}$ and $y(t) = e^t$ parameterize $\mathcal{C}$ when $t \in [0, 1]$. What's the area of the surface generated by revolving $\mathcal{C}$ about the x-axis?

43. Write the equation of the parabola whose focus is at the origin and whose directrix is the line $x = 7$.

44. Write the equation (in polar coordinates) of the parabola whose focus is at the origin and whose directrix is the line $x = -2x + 7$.

45. Write the equation (in polar coordinates) of the ellipse whose focus is at the origin, directrix is the line $x = 3$, and eccentricity is $\varepsilon = 0.5$.

46. Write the equation of the ellipse (in polar coordinates) whose focus is at the origin, directrix is the line $x = 3x - 11$, and eccentricity is $\varepsilon = 0.25$.

47. Write the equation (in polar coordinates) of the hyperbola whose focus is at the origin, directrix is the line $x = 13$, and eccentricity is $\varepsilon = 2.5$.

48. Write the equation (in polar coordinates) of the hyperbola whose focus is at the origin, directrix is the line $x = -3x - 11$, and eccentricity is $\varepsilon = 4$.

49. Determine the location of all points (if any) at which the tangent line to $r = 1 + \cos(\theta)$ has a slope of -1.

50. Determine the area enclosed by the cardioid $r = 3 + 3\sin(\theta)$.

51. Determine the area enclosed by the curve $r = 3\sin(\theta)$.

52. Suppose $z = 5 + 5i$. Plot (a) z, and (b) $\bar{z}$ as points in the Cartesian plane.

53. Suppose $z = 2e^{(\pi/4)i}$. Plot (a) z, and (b) $\bar{z}$ as points in the Cartesian plane.

54. Write $z = 5\sqrt{3} - 5i$ in polar form.

55. Write $z = 7e^{(\pi/2)i}$ in Cartesian form.

56. Determine the magnitude and principal argument of (a) $z = 6 - 6\sqrt{3}i$, and (b) $z = -2 + 4i$.

57. Suppose $z = 3 + 4i$ and $w = 9 - 2i$. Determine (a) $z + w$, (b) zw, and (c) $\bar{z}$.

58. Find all solutions (if any) to the equation $(3\sqrt{3} + 3i)z = 1$.

59. Find all solutions (if any) to the equation $z^3 = 7 + 7\sqrt{3}i$.

60. Find all solutions (if any) to the equation $e^z = -6$.

61. Find all solutions (if any) to the equation $\sin(z) = 32$.

Chapter 9 Projects and Applications

✧ Planetary Orbits

In 1684 when Edmund Halley asked Isaac Newton whether a force that obeyed an *inverse square law* would result in the elliptical orbits that Kepler had described, Newton's ready response was *yes*. In this project, you'll show that he was right.

▷ Conservation of Angular Momentum

The single piece of the puzzle that we haven't yet discussed is a physical quantity called *angular momentum.* When the sun is at the origin and a planet of mass m is at (r, θ), its angular momentum is

$$L = mr^2 \frac{d\theta}{dt}, \tag{7.1}$$

which you can understand as

$$L = \underbrace{r}_{\text{distance from the sun}} \underbrace{\left(m \, r \, \overbrace{\frac{d\theta}{dt}}^{\text{speed}} \right)}_{(\text{mass}) \times (\text{speed}) \, = \, \text{linear momentum}}.$$

Later, in our study of vector-valued functions, we'll verify that angular momentum is constant when the gravitational attraction between the sun and planet is the only force present. We say that angular momentum is *conserved,* and Kepler's Second Law of planetary motion (which he extracted from a meticulous study of data) follows from that fact. Specifically, Kepler's Second Law says that the line segment connecting the sun to a planet sweeps out equal areas in equal times.

1. If the planet's coordinates are (r_1, θ_1) at time t_1, and (r_2, θ_2) at time t_2, verify that the area swept out by the aforementioned line segment is

$$\int_{\theta_1}^{\theta_2} \frac{1}{2} r^2 \, d\theta = \int_{t_1}^{t_2} \frac{1}{2} r^2 \, \frac{d\theta}{dt} \, dt.$$

2. Use equation (7.1) to verify that this integral depends on the length of time, $t_2 - t_1$, but not where the planet is in its orbit.

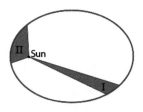

Figure 7.1: In region I the planet is moving slowly, but is farther away from the sun. In region II the planet is moving quickly, but is closer to the sun.

▷ Conservation of Energy

In the context of a story about lifting this book (see p. 525) we showed that the sum of potential and kinetic energy is constant when gravity is the only force present. That was a one-dimensional example (up/down along a line) and we're currently talking about a two-dimensional scenario, but the fact remains true, so let's write down the kinetic energy and potential energy of a planet.

- **Kinetic energy of a planet:** When we discussed the arc length of a parametric curve on p. 722, we said that an object moving according to $(x(t), y(t))$ has a speed of

$$v = \sqrt{\left(x'(t) \right)^2 + \left(y'(t) \right)^2},$$

so its kinetic energy is

$$K = \frac{1}{2} mv^2 = \frac{m}{2} \left(x'(t) \right)^2 + \frac{m}{2} \left(y'(t) \right)^2.$$

where m is the planet's mass.

- **Gravitational potential energy of a planet:** On p. 320 you investigated the relationship between force and the derivative of potential energy with respect to position. In this context, that relationship can be written as

$$-F = \frac{dU}{dr}, \tag{7.2}$$

where U denotes the gravitational potential energy between the planet and the sun. Since the distances involved in our calculation are so large, we cannot assume that the force due to gravity is constant. Rather, we use Halley's suggestion of an inverse square law, and write

$$F = -\frac{GmM}{r^2},$$

> The force is negative because it pulls an object toward the center.

where m is the mass of the planet, M is the mass of the sun, and G is a constant of proportionality called the *gravitation constant*. Combining this with equation (7.2), we see

$$\frac{GmM}{r^2} = \frac{dU}{dr} \quad \Rightarrow \quad U = -\frac{GmM}{r} + E_1,$$

where E_1 is some constant. The particular value of E_1 is irrelevant, since differentiating U results in the same calculation of force, so let's set $E_1 = 0$.

> Setting the number E_1 amounts to deciding what energy level to call "zero." This is very similar to deciding what altitude to call "zero." If we say that Miami is at an altitude of zero, Denver is at an altitude of 1 mile; but if we say that Denver is at an altitude of zero, Miami is at an altitude of −1 mile. What's important is that the *change* in altitude is the same when you drive from Denver to Miami, regardless of the system we choose.

Since the sum of kinetic and potential energy is constant, there is a number E for which

$$E = \underbrace{\frac{m}{2}\left(x'(t)\right)^2 + \frac{m}{2}\left(y'(t)\right)^2}_{(A)} \underbrace{- \frac{GmM}{r}}_{(B)}. \tag{7.3}$$

3. Explain where parts (A) and (B) come from in equation (7.3). What do they mean to us?

▷ **The answer to Halley's question**

Equation (7.3) has a mix of Cartesian and polar coordinates. Let's switch fully into the polar system using $x = r\cos(\theta)$ and $y = r\sin(\theta)$.

4. Using the fact that both r and θ change with time, verify that equation (7.3) can be rewritten as

$$E = (r')^2 + (r\theta')^2 - \frac{2GM}{r}.$$

5. Use the conservation of angular momentum, as expressed by equation (7.1), to rewrite this equation as

$$E = (r')^2 + \left(\frac{H}{r}\right)^2 - \frac{2GM}{r},$$

where $H = L/m$ is the planet's angular momentum per unit mass.

6. Verify that adding $(GM/H)^2$ to both sides of the equation results in

$$E + \left(\frac{GM}{H}\right)^2 = (r')^2 + \left(\frac{H}{r} - \frac{GM}{H}\right)^2. \tag{7.4}$$

7. Use the Chain Rule and equation (7.1) to verify that $\frac{dr}{dt} = \frac{H}{r^2}\frac{dr}{d\theta}$, so that equation (7.4) can be written as

$$E + \left(\frac{GM}{H}\right)^2 = \left(\frac{H}{r^2}\frac{dr}{d\theta}\right)^2 + \left(\frac{H}{r} - \frac{GM}{H}\right)^2.$$

8. Notice that the right-hand side of our equation is the sum of squares, so it's not negative. That means the left-hand side is non-negative, so we can legitimately call it C^2. Verify that our equation is the same as

$$C^2 = \left(\frac{du}{d\theta}\right)^2 + u^2,$$

where $u = \frac{H}{r} - \frac{GM}{H}$ and $C^2 = E + \left(\frac{GM}{H}\right)^2$, so that

$$\frac{du}{d\theta} = \pm\sqrt{C^2 - u^2}. \tag{7.5}$$

We'll deal with the positive version of this equation (the technique for the negative version is the same).

9. After rewriting equation (7.5) as

$$1 = \frac{1}{\sqrt{C^2 - u^2}}\frac{du}{d\theta},$$

integrate both sides with respect to θ (Hint: $du = \frac{du}{d\theta}d\theta$) to show that

$$\theta = \sin^{-1}\left(\frac{u}{C}\right) + \beta, \tag{7.6}$$

where β can be any constant. For the sake of convenience (which will become clear in a moment), let's choose $\beta = -\frac{\pi}{2}$.

> Choosing β amounts to designating a direction to call the x-axis.

10. Solve equation (7.6) for u, and use the fact that $u = \frac{H}{r} - \frac{GM}{H}$ to verify that

$$\frac{H}{r} - \frac{GM}{H} = C\cos(\theta).$$

11. Solve this equation for r to verify that

$$r = \frac{a\varepsilon}{1 + \varepsilon\cos(\theta)}$$

where ε and a depend on H, G, M, and E.

Turning back to p. 732, we see that this equation describes a conic section whose focus is at the origin and whose directrix is the line $x = a$. Of the three kinds—parabola, hyperbola, ellipse—only the *ellipse* remains bounded (as do the orbits of the planets). So we conclude that the planets move around the sun in elliptical orbits.

Note: We made a simplifying assumption that the gravitational force between the sun and a planet is the only force present. Actually, the planets attract each other gravitationally, too.

Chapter 10
Differential Equations

On p. 263 we wanted to describe the motion of a baseball that had been hit by a bat. Based on some facts about drag, we used Newton's Second Law to say that the ball's vertical velocity, v, changes according to

$$m\,v' = -9.8m - \beta\,v^2 \tag{0.1}$$

where m is the mass of the ball and β is a constant that depends the roughness of its surface. This equation tells us how the acceleration experienced by the ball depends on its speed, but doesn't tell us explicitly how velocity changes over time. For that, we need to do some work—we say that we must *solve* equation (0.1). In this chapter we discuss what it means to solve such equations, and see how to do it in some of the most common cases.

10.1 Introduction to Differential Equations

In 2008 the *Los Angeles Times* reported that a music video had been seen by over 2 million people after being posted on the internet for only three days. This phenomenon of rapid dissemination has since been called "going viral" in the media, and in its early stages it can be described with a simple equation. Let's say that $y(t)$ is the number of people who have seen the video after t days. Some of them continually disseminate the link to friends via social networking sites, email, and conversations, thereby increasing the value of y. So the rate of change in y (i.e., the number of new viewers per day) is proportional to the current value of y, which leads us to write

$$y' = \lambda y \tag{1.1}$$

where λ is the constant of proportionality. In the following examples you'll see that functions described by this equation grow in magnitude when $\lambda > 0$, as in the case of the video dissemination, and shrink in magnitude when $\lambda < 0$.

> The constant function $y(t) = 0$ is described by equation (1.1), but we don't mean to include it in this statement.

Continuously compounded interest

Example 1.1. Suppose \$100 is invested into an account that pays interest at 4% (APR), compounded continuously. Then the account value after t years, $A(t)$, is described by the equation $A' = 0.04A$ dollars per year. Determine the account value after 5 years.

Solution: The statement that $A' = 0.04A$ is just equation (1.1) with $\lambda = 0.04$, and on p. 283 we proved that the only nontrivial functions described by equation (1.1) are scaled exponential functions. So $A(t)$ must have the form $Ce^{0.04t}$. Using this formula at $t = 0$ we see that

$$100 = A(0) = Ce^0 = C,$$

771

so $A(t) = 100e^{0.04t}$. Now we can answer the question by calculating

$$A(5) = 100e^{(0.04)(5)} = 122.14 \text{ dollars.} \quad \blacksquare$$

Try It Yourself: Population growth

Example 1.2. Suppose 10000 bacteria are isolated in a nutrient-rich environment. If an average of 2.4 daughter cells are spawned by each bacterium per hour, the number of bacteria after t hours, $b(t)$, is described by the equation $b' = 2.4b$. Determine the size of the bacteria population after 3.1 hours.

Full Solution On-line

Answer: $b(3.1) \approx 17027502$ cells. $\quad \blacksquare$

Radioactive decay

Example 1.3. Potassium-40 is a radioactive element that decays into either argon-40 or calcium-40. Based on its half-life we know the amount of potassium-40 in a given sample, $K(t)$, changes at a rate of $K' = -0.569 \times 10^{-9} K$. Determine the amount of a 1-gram sample that remains after a million years.

Solution: The statement that $K' = -0.569 \times 10^{-9}K$ is just equation (1.1) with $\lambda = -0.569 \times 10^{-9}$, so the function $K(t)$ must have the form $Ce^{-0.569 \times 10^{-9}t}$. Using this formula at time $t = 0$ we find that $C = 1$ gram, so the amount that remains after 10^6 years is

$$K(10^6) = e^{-0.569 \times 10^{-9} \times 10^6} = e^{-0.569 \times 10^{-3}} = 0.999 \text{ grams.} \quad \blacksquare$$

Try It Yourself: Atmospheric pressure

Example 1.4. Suppose $P(a)$ is the atmospheric pressure, measured in pascals, at an altitude of a meters. Then $P(a)$ changes according to $P' = -0.001 \ln(25/3)P$. The pressure at sea level is 101325 pascals. Determine the atmospheric pressure at an altitude of 700 meters above sea level.

Full Solution On-line

Answer: 22969.07 pascals. $\quad \blacksquare$

Equation (1.1) is like (0.1) in the sense that doesn't tell us explicitly how y' depends on time, but rather how y' is related to the current value of y. Equations like these, which describe the relationship between a function and its derivatives, are called **differential equations (DEs)** More specifically, equations (0.1) and (1.1) are called **first-order** differential equations because they involve the first derivative, but no higher derivatives. By contrast, the equation

$$y'' = -\lambda^2 y \quad (1.2)$$

is called a **second-order** differential equation because it involves the second derivative, but no higher derivatives. In general, the **order** of a differential equation refers to the highest-order derivative that appears in it. The majority of this chapter will be devoted to first-order differential equations.

> You saw this kind of an equation on p. 264 when we described the distance between atoms of a diatomic molecule using Newton's Second Law.

While equations (0.1) and (1.1) are similar in that they are both first-order DEs, they differ in an important way.

> A first-order differential equation is said to be **linear** if it can be written in the form $y' + p(t)y = q(t)$. Otherwise we say it is **nonlinear**.

Linearity

Example 1.5. Determine whether equations (0.1) and (1.1) are linear.

Solution: In equation (0.1) we see the term v^2. No amount of rewriting will get rid of that quadratic term, so the DE is nonlinear. By contrast, we can rewrite equation (1.1) as $y' - \lambda y = 0$, in which $p(t) = -\lambda$ and $q(t) = 0$ are both constant functions. So equation (1.1) is a linear DE. ∎

Try It Yourself: Linearity

Example 1.6. Determine which (if either) of the following equations are linear: (a) $5y' + 4t^2 y = 10$, and (b) $y' + t\cos(y) = 8t + 1$.

Answer: (a) linear (b) nonlinear. ∎

Full Solution
On-line

In the same way that an algebraic equation such as $t^3 - t + 6 = 0$ describes a collection of numbers, and our job is typically to find those numbers (e.g., $t = -2$ is a solution), a differential equation describes a family of functions, and our job is typically to figure out which functions are being described. When we find such a function, we say that it **solves** the DE.

Verifying a solution

Example 1.7. Verify that $f(t) = Ct^2 - 1/t$ solves the equation $ty' = 2y + 3/t$ for any value of C.

Solution: Saying that f solves this DE is the same as saying that the DE describes f, so let's calculate $tf'(t)$ and $2f(t) + 3/t$ and see if they're the same:

$$tf'(t) = t\left(2Ct + \frac{1}{t^2}\right) = 2Ct^2 + \frac{1}{t},$$

and

$$2f(t) + \frac{3}{t} = 2\left(Ct^2 - \frac{1}{t}\right) + \frac{3}{t} = \left(2Ct^2 - \frac{2}{t}\right) + \frac{3}{t} = 2Ct^2 + \frac{1}{t}.$$

Since these are the same, this DE *does* describe f. Said differently, the function f solves the differential equation. ∎

Checking a proposed solution

Example 1.8. Determine whether $f(t) = \sin(t)$ solves $y' + 4y = 1$.

Solution: Since $f'(t) = \cos(t)$, we have $f'(t) + 4f(t) = \cos(t) + 4\sin(t)$. While that can be 1 at certain times (e.g., at $t = 0$), it's not *always* equal to 1, so this differential equation does not describe $f(t)$. That is, $f(t)$ is *not* a solution of the differential equation. ∎

Try It Yourself: Verifying a solution

Example 1.9. Verify that $g(t) = 7t + Ce^{-2t}$ solves $3y' + 6y = 21 + 42t$ for any value of C.

Answer: See solution on-line. ∎

Full Solution
On-line

Note that all of the solutions in Example 1.7 have the same form, as do all of the solutions in Example 1.9, which is why we say that a DE describes a *family* of functions. We lock down on a particular member of that family by specifying the function value at a given time. This information is called **initial data**, and when it's included with a DE, we say that we have an **initial value problem (IVP)**.

Finding the solution to an IVP

Example 1.10. Find a solution of $ty' - 2y = 3/t$ for which $y(1) = 3$.

Solution: We know from Example 1.7 that $f(t) = Ct^2 - 1/t$ will solve the equation with any value of C. If we set $C = 4$, this function satisfies the given data. ∎

Try It Yourself: Finding the solution to an IVP

Example 1.11. Find a solution of $3y' + 6y = 21 + 42t^2$ for which $y(0) = 11$. *(Hint: see Example 1.9.)*

Answer: $y(t) = 7t + 11e^{-2t}$. ∎

Full Solution
On-line

❖ Mathematical Modeling with Differential Equations

You first saw equations (0.1) and (1.2) in Chapter 4, when we had something to say about force and acceleration, and we wrote down equation (1.1) because we had something to say about the dissemination of information. Differential equations often arise when we have something to say about the way quantities change, and we close this section with additional examples from various disciplines.

▷ Biology: potassium gates in the cell wall

Nerve cells send signals by controlling the flow of potassium and sodium ions through gates in the cell membrane. Suppose a nerve cell has a total of K potassium gates, and $k(t)$ are open at time t. Then Δt seconds later, the number of open gates is

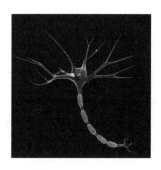

$$\underbrace{\begin{pmatrix}\text{the number of}\\ \text{gates that are}\\ \text{already open}\end{pmatrix}}_{=k(t)} + \begin{pmatrix}\text{gates that open}\\ \text{during that } \Delta t\\ \text{seconds}\end{pmatrix} - \begin{pmatrix}\text{gates that close}\\ \text{during that } \Delta t\\ \text{seconds}\end{pmatrix}. \qquad (1.3)$$

Let's write the *fraction* of gates that close during a short span of Δt seconds as $\beta\Delta t$, where $\beta > 0$ is a number that depends on the cell's environment. Then the *number* of gates that close during a span of Δt seconds is the product

$$\overbrace{k(t)}^{\text{the number of gates that are open}} \underbrace{\beta\Delta t}_{\text{the fraction of them that close}}.$$

Similarly, let's write the fraction of gates that open during a short span of Δt seconds as $\alpha\Delta t$, where $\alpha > 0$ is a number that depends on the cell's environment. Then the *number* of gates that open in that span of time is the product

$$\overbrace{(K - k(t))}^{\text{the number of gates that are closed}} \underbrace{\alpha\Delta t}_{\text{the fraction of them that open}}.$$

Now we're ready to express the ideas in (1.3) with an equation: at time $t + \Delta t$ the number of open gates will be

$$k(t + \Delta t) = k(t) + (K - k(t))\alpha\Delta t - k(t)\beta\Delta t.$$

When we subtract $k(t)$ from both sides and divide by Δt, we see that the number of open gates changes at an average rate of

$$\frac{k(t + \Delta t) - k(t)}{\Delta t} = \alpha K - (\alpha + \beta)k(t).$$

And in the limit as $\Delta t \to 0$,

$$k'(t) = \lambda_1 - \lambda_2 \, k(t) \tag{1.4}$$

where $\lambda_1 = \alpha K$ and $\lambda_2 = \alpha + \beta$. This is a first-order linear differential equation, and you'll learn how to solve it later in this chapter.

> We said that α and β depend on cell's environment. More specifically, they depend on the voltage difference across the cell membrane.

▷ Ecology: logistic population model

Whether we're talking about a community of people, a herd of animals, or a colony of bacteria, population growth is affected by the availability of *resources*. In a socioeconomic context, the resource might be a job. In an ecological context, the resource might be food, or oxygen in water, or metabolic catalysts. The exact relationship between population growth and the availability of resources is complicated, but we can begin to describe it quantitatively by studying very simple situations.

So let's talk about a simple situation: a population of *yeast* and the way its growth depends on the availability of *sugar*. We'll denote by $y(t)$ the number of yeast at time t, and use $s(t)$ to denote the amount of sugar present in the environment. The first thing we want to say is that the *rate* of population growth should increase with population size, and with the availability of sugar. Written in the language of calculus,

$$y'(t) = \alpha \, y(t)s(t), \tag{1.5}$$

where $\alpha > 0$ depends on the strain of yeast. The second important idea is that while the yeast population is increasing, the sugar supply is decreasing. Moreover, if the yeast population is growing quickly, the sugar supply is decreasing quickly. The simplest equation that captures this fact is

$$s'(t) = -\beta y'(t),$$

where $\beta > 0$ also depends on the strain of yeast. Because both sides of this last equation are derivatives, we can integrate it in time and see

> There will be no population growth if there is no sugar (all the yeast will become dormant), and if there are no yeast in the first place we should have $y' = 0$. Equation (1.5) captures both of these qualitative aspects of population growth.

$$\int_0^t s'(\tau) \, d\tau = -\beta \int_0^t y'(\tau) \, d\tau$$

$$s(t) - s(0) = -\beta \Big(y(t) - y(0) \Big)$$

When we solve this equation for $s(t)$ we find that

$$s(t) = s(0) + \beta y(0) - \beta y(t)$$

Substituting this formula for $s(t)$ into equation (1.5) gives us the differential equation

$$y'(t) = r \, y(t) \left(1 - \frac{y(t)}{K} \right),$$

where $r = \alpha(s(0) + \beta y(0))$ is a number called the **intrinsic growth rate** and $K = y(0) + s(0)/\beta$ is a constant called the **carrying capacity**. This is a famous nonlinear DE known as the **logistic equation**, and you'll learn how to solve it later in the chapter.

▷ Engineering: flow rates

Suppose $a(z)$ is the cross-sectional area of a container at a height of z. When the liquid in the container is y meters deep, you know from studying the method of slicing in Chapter 7 that the liquid volume is

$$V = \int_0^y a(z)\,dz.$$

Now suppose the container springs a leak in the bottom. **Torricelli's Law** tells us that the volumetric flow through the hole is $A\sqrt{2gy}$ m^3/sec, where A is the area of the hole, g is the acceleration per unit mass due to gravity, and y is the depth of the liquid that's above the hole. So according to the Fundamental Theorem of Calculus,

$$-A\sqrt{2gy} = \frac{dV}{dt} = \frac{d}{dt}\int_0^y a(z)\,dz = a(y)y'.$$

> Since fluid is *leaving* the tank, dV/dt must be negative.

Dividing this equation by $a(y)$, we see that the height of fluid in the tank is governed by the differential equation

$$y' = -\frac{A\sqrt{2gy}}{a(y)}.$$

You should know

- the terms *order, linear, nonlinear, differential equation (DE), initial data,* and *initial value problem (IVP)*;

- what it means for a function to *solve* a DE.

You should be able to

- determine whether a given function solves a specified DE;

- determine which member of a family of solutions satisfies specified initial data.

✥ 10.1 Skill Exercises

In #1–6 determine whether $y(t)$ solves the DE.

1. $y' = 3y$, $y(t) = 4e^{3t}$

2. $y' = 7y$, $y(t) = 5 + e^{7t}$

3. $ty' + y = 0$, $y(t) = 1/t$

4. $y' - 2/y = 0$, $y(t) = \sqrt{t}$

5. $y' + 2y = 4t + 1$, $y(t) = 3t + 7$

6. $y' = e^{-y}$, $y(t) = \ln(t)$

7. Determine values of a and b for which $y(t) = at + b$ solves the differential equation $y' + 7y = 3t - 9$.

8. Determine values of a, b and c for which $y(t) = at^2 + bt + c$ solves the differential equation $y' + 2y = 2t^2 + 19$.

9. Determine the value(s) of A for which $y(t) = Ae^{5t}$ solves the differential equation $y' - 9y = e^{5t}$.

10. Determine the value(s) of λ for which $y(t) = e^{\lambda t}$ solves the differential equation $y'' + 2y' - 5y = 0$.

In #11–18 classify each DE as linear or nonlinear.

11. $y' = 3ty + 4$ 13. $y' - 34t^2 y = e^t$ 15. $y' + y^2 = 2$ 17. $y' + e^y = 0$

12. $y' - 21y = 4t$ 14. $t^3 y' - 11y = \cos(t)$ 16. $(y')^3 + y = t$ 18. $y' - \cos(y) = 2$

Determine the order of each DE in #19-26.

19. $y'' + y' = 4t$ 21. $\cos(y) + y' = 0$ 23. $y' + y^{2/3} = t^7$ 25. $1 + (y')^2 = 2ty$

20. $y''' + t^2 y = 4t^6$ 22. $y''' + y'' = \sin(y)$ 24. $\sqrt{y'} + t^9 y = 0$ 26. $(y'')^3 - y^3 = y$

✤ 10.1 Concept and Application Exercises

27. We can describe learning as the rate of change in performance $p(t) \in (0,1)$, where $p = 1$ indicates mastery. Determine which of the equations $p' = 0.01(1-p)$ and $p' = 0.01p(1-p)$ you think is a better characterization of (a) learning calculus, (b) learning to drive a car, and (c) children learning language, and explain your reasoning. (Pay special attention to $p \approx 0$.) | Psychology |

28. Suppose we put equal amounts of two chemicals in pure water, say X and Y, which combine to make the compound Z according to $X + Y \to Z$. To describe the rate at which Z is produced, let's denote by x, y, z the respective concentrations of these molecules. (a) Explain what the equation $z' = kxy$ means in this setting, where $k > 0$ is a fixed number called the *rate constant*. (b) Explain why $z' = -x'$, and $y' = x'$. (c) Rewrite the DE entirely in terms of x. | Chemistry |

29. Write a differential equation that says, "The rate of change in the population is proportional to the difference between the environment's carrying capacity, $K > 0$, and the current population size, y." | Biology |

30. A sick student carries a virus to a small school of 1400 people. Write a differential equation that says, "The rate at which the pathogen spreads is proportional to both the number of infected students *and* the number of healthy students." (Note: by using the fact that these two populations sum to 1400, you can write your equation in terms of a single variable.) | Public Health |

31. Explain how you know that $y(t)$ is decreasing at $t = 0$ when (a) $y(0) = 4$ and $y'(t) = -t^2$, (b) $y(0) = 4$ and $y' = -y^2$. | Mathematics |

32. Suppose $y(t)$ solves the DE $y' = -y^2$. Differentiate the equation implicitly (using the Chain Rule on the right-hand side) and explain what the resulting equation tells you about the concavity of $y(t)$ when $y > 0$. | Mathematics |

33. Suppose y' depends on the current value of y according to $y' = y(y^2 - 1)$. Explain how we know the graph of y is continuous, and has neither corners nor cusps.

34. Suppose y' depends on the current value of y according to $y' = y(y^2 - 1)$. Explain why a non-constant solution to this DE cannot have a local extremum.

35. The curves $y = f(t)$ and $y = g(t)$ are shown in the left-hand image of Figure 1.1. Explain why neither f nor g can be a solution of $y' = t^2(y - 1)$. | Mathematics |

36. The curves $y = f(t)$ and $y = g(t)$ are shown in the right-hand image of Figure 1.1. Determine which (if either) of f and g is a solution of $y' = t^2(y - 1)$, and explain how you came to your conclusion. | Mathematics |

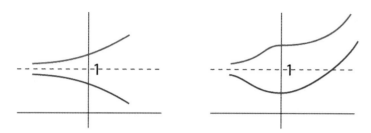

Figure 1.1: (left) Graphs for #35; (right) graphs for #36.

37. In Section 10.5 we'll discuss Newton's Law of Cooling, which says that the temperature of an object varies with time according to $T' = -k(T-A)$, where A is the temperature of the environment (which we'll assume to be constant for now) and k is a number called the *time constant* that depends on the object. (a) Show that $T(t) = A + Ce^{-kt}$ is a solution of this equation for any constant C. Then (b) determine $\lim_{t \to \infty} T(t)$, and (c) explain why your answer from part (b) makes sense.

     Mechanical Engineering

38. When an annuity earns interest at $\gamma\%$/year, compounded continuously, and pays out \$3000 per year continuously, the amount of money in the account varies over time according to $y' = \gamma y - 3000$. Show that $y(t) = \frac{3000}{\gamma} + Ce^{\gamma t}$ is a solution to this equation for every constant C.

     Finance

39. On p. 775 we introduced the logistic population model, which says that population varies with time according to $P' = rP(1 - P/K)$, where $r > 0$ depends on the species and $K > 0$ depends on the environment. (a) Show that the constant functions $P(t) = 0$ and $P(t) = K$ are solutions of this DE, and (b) explain what that means to us in the context of population modeling.

     Biology

40. The charge on the capacitor in an RC circuit varies according to $Rq' + \frac{q}{C} = V$, where R is the resistance of the resistor, C is the capacitance of the capacitor, and V is the impressed voltage (all of which we assume to be constant). Show that the constant function $q(t) = CV$ is a solution.

     Electrical Engineering

41. On p. 775 the conclusion of our discussion about potassium gates was the equation $k'(t) = \lambda_1 - \lambda_2 k(t)$, where λ_1 and λ_2 are constants. Show that $k(t) = \frac{\lambda_1}{\lambda_2} + Ce^{-\lambda_2 t}$ is a solution for any value of C.

     Biology

42. The equation $my'' + \gamma y + ky = 0$ is often used to describe the motion of a mass on a spring. Determine (a) the order of this equation, and (b) whether it's linear. Then (c) determine whether $y(t) = e^{-\gamma t} \sin\left(\sqrt{\frac{k}{m}}\, t\right)$ is a solution.

Exercises #43–44 refer the "sales equation," by which we mean $s' = \rho \alpha - \lambda s$, where $s(t)$ describes a company's sales at time t (measured in dollars per month), α is the advertising expenditure (also measured in dollars per month), and ρ and λ are positive constants.

     Marketing

43. Suppose the company does no advertising. (a) What are the units of s'? (b) Based on your answer to part (a), what are the units of λ? (c) Based on your answer to part (b), what does λ mean to us? (d) Show that $s(t) = \rho \alpha / \lambda + Ce^{-\lambda t}$ solves the sales equation for any value of C, and (e) explain why $C > -\rho \alpha / \lambda$ in this context.

44. Based on the units of s', (a) what are the units of ρ, and (b) what does it mean to us in this context?

10.2 Autonomous Equations and Equilibria

We're often interested in the qualitative behavior of solutions to differential equations, and in long-term behavior rather than moment-by-moment information about function value. In this section, we consider an important kind of differential equation that lends itself to such analysis.

❖ Autonomous Equations

A differential equation is said to be **autonomous** when it describes the derivative in terms of the current function value, but not (explicitly) the independent variable.

Determining whether an equation is autonomous

Example 2.1. Determine whether (a) equation (0.1), and (b) equation (1.1) on p. 771 are autonomous differential equations.

Solution: Both equations are autonomous. (a) Equation (0.1) can be written as

$$v' = -9.81 - \frac{\beta}{m}v^2,$$

in which t does *not* appear explicitly on the right-hand side, so v' depends only on the current value of v. (b) Similarly, equation (1.1) says $y' = \lambda y$, where λ is constant. So y' depends only on the current value of y. ∎

Try It Yourself: Determining whether an equation is autonomous

Example 2.2. Determine whether the equations in Example 1.7 (p. 773) and Example 1.9 (p. 773) are autonomous differential equations.

Answer: Neither is autonomous. ∎

Full Solution
On-line

Because autonomous equations have such a simple structure (i.e., no explicit mention of time in the equation), we can understand the general behavior of their solutions even before we even attempt to solve them, as seen in the next example.

Analyzing behavior of a solution to an autonomous DE

Example 2.3. Suppose $x(t)$ is the position of an object whose motion is governed by the equation $x' = x^2 - 5x + 6$. Discuss the motion of the object when (a) $x(0) = 1.5$, (b) $x(0) = 2.5$, and (c) $x(0) = 3.5$.

Solution: This DE has the form $x' = f(x)$, where $f(x) = x^2 - 5x + 6$. In cases like this, the graph of f can greatly facilitate our analysis, so we begin by sketching it. That's been done in Figure 2.1.

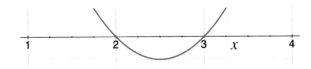

Figure 2.1: The graph of $f(x) = x^2 - 5x + 6$ is a parabola with intercepts at $x = 2$ and $x = 3$.

(a) In Figure 2.1 we see that $x^2 - 5x + 6$ is positive at $x = 1.5$. Consequently, the DE tells us that $x' > 0$ there, so $x(t)$ will increase. That is, when $x(t)$ starts at $x = 1.5$ it moves to the right. In fact, since $x^2 - 5x + 6$ is positive whenever $x < 2$, the object will move to the right as long as $x(t) < 2$.

(b) In Figure 2.1 we see that $x^2 - 5x + 6$ is negative at $x = 2.5$. Consequently, the DE tells us that $x' < 0$ there, so $x(t)$ will decrease. That is, when $x(t)$ starts at $x = 2.5$ it moves to the left. Since $x^2 - 5x + 6 < 0$ whenever $2 < x < 3$, the same thing will happen as long as $x(t) \in (2, 3)$.

(c) In Figure 2.1 we see that $x^2 - 5x + 6$ is positive whenever $x(0) > 3$. Since $x' > 0$ at all such points, an object starting at $x(0) = 3.5$ will move to the right. In fact, it will move to the right whenever $x(t) > 3$. ∎

When we draw arrows onto Figure 2.1 that communicate the increase/decrease facts that we established in Example 2.3, the result is called a **phase line** for the differential equation $y' = F(y)$. (Typically, a phase line contains the axis and arrows, but not the graph of F.)

Figure 2.2: The graph of $x^2 - 5x + 6$ for Example 2.3, and information about the motion of $x(t)$ for various initial data.

> Based on this phase line, it's not hard to convince yourself that if $x(t)$ is a solution of the DE that starts at $x(0) = 2.5$, it will never reach $x = 1$, or $1.25, 1.5 \ldots$ or any $x < 2$, but you might wonder whether $x(t)$ ever reaches $x = 2$. The short answer is "no," and we'll see why in a few moments. It happens in the same way that $x(t) = e^{-t}$ approaches zero but never reaches it.

In Figure 2.2 it seems that $x = 2$ and $x = 3$ separate the phase line into segments over which the behavior of $x(t)$ is always the same. These numbers are also important for another reason.

Analyzing behavior of a solution to an autonomous DE

Example 2.4. Show that the constant functions (a) $x(t) = 2$, and (b) $x(t) = 3$ are solutions of $x' = x^2 - 5x + 6$.

Solution: (a) Because $x(t) = 2$ is constant we know $x'(t) = 0$, and it's easy to check that $2^2 - 5(2) + 6 = 0$. Since this DE describes the function $x(t) = 2$, the constant function is a solution. (b) When $x(t) = 3$ is a constant function, $x'(t) = 0$, and it's easy to check that $3^2 - 5(3) + 6 = 0$. ∎

Because the constant function $x(t) = 2$ is a solution of the DE in Example 2.4 that's *unchanging,* we say that it's an *equilibrium* of the differential equation. More generally, we say that the constant function $x(t) = x_0$ is an **equilibrium** of the equation $x' = F(x)$ when $F(x_0) = 0$.

Locating equilibria

Example 2.5. Locate the equilibria of $x' = x^2(x - 2)$, and draw a phase line for the differential equation.

Solution: The equilibria are the constant functions $x(t) = 0$ and $x(t) = 2$, because those are the values of x that make $x^2(x - 2) = 0$. Note that these are the *only* places at which $x' = x^2(x - 2) = 0$, so if $x(t)$ is at another position on the number line it must be moving right or left. Since $x^2(x - 2) < 0$ when $x \in (-\infty) \cup (0, 2)$, solutions for which $x(0) \in (-\infty) \cup (0, 2)$ will move to the left. Similarly, since $x^2(x - 2) > 0$ when $x > 2$, solutions for which $x(0) > 2$ will move to the right. This information is depicted in Figure 2.3. ∎

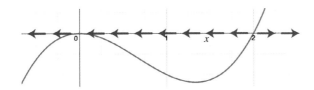

Figure 2.3: The graph of $x^2(x-2)$ for Example 2.5, and information about the motion of $x(t)$ for various initial data.

Toward developing some vocabulary that describes the behavior of solutions near the equilibria, imagine that the blue arrows in Figure 2.2 represent the flow of water. Because the arrows seem to be flowing away from $x = 3$, as if they emerge there, we say the equilibrium at $x = 3$ is a **source**. And since the arrows near $x = 2$ all head toward it, we say that $x = 2$ is a **sink**. In Figure 2.3 you see that the equation $x' = x^2(x-2)$ has a source at $x = 2$, but the equilibrium at $x = 0$ is neither a sink nor a source. Rather, it seems to be half-and-half. We refer to such an equilibrium as a **node**. Using this vocabulary, we can say the following:

Classifying Equilibria: Suppose x_0 is an isolated root of the continuous function F (meaning there's a small neighborhood about x_0 in which F has no other roots). Then $x' = F(x)$ has

- a source at x_0 if $F(x)$ changes from negative to positive at x_0;

- a sink at x_0 if $F(x)$ changes from positive to negative at x_0;

- a node at x_0 if $F(x)$ is doesn't change sign at x_0.

Compare to the phase lines in Figures 2.2 and 2.3.

Instead of using nouns, we often use adjectives to describe equilibria: a sink is said to be **stable** because small perturbations away from it always head back toward the equilibrium, and a source is said to be **unstable** because initial data that's even slightly different from the equilibrium value is pushed away. Because a node exhibits stability on one side but not the other, we describe it as **semi-stable**.

Try It Yourself: Classifying equilibria

Example 2.6. For the equation $x' = x(x^2 - 1)(x^4 - 1)$, (a) locate and classify the equilibria, and (b) draw a phase line.

Answer: (a) nodes at $x = \pm 1$, and a source at $x = 0$, (b) see on-line solution. ∎

We began our discussion of equilibria with autonomous equations of the form $x' = f(x)$ because, frankly, the graph of $f(x)$ is something familiar. Our qualitative analysis of $y' = f(y)$ is exactly the same, but we think about motion along the y-axis instead of horizontal motion.

Full Solution On-line

Phase lines and equilibria

Example 2.7. Consider the equation $y' = y^2 - 5y + 6$. (a) Locate the equilibria, and (b) draw a phase line.

Solution: The letter we use to represent the dependent variable, whether x, y, z, ..., is irrelevant to the mathematics. (a) We're looking for constant functions. As before, these are $y(t) = 2$ and $y(t) = 3$ because 2 and 3 make $y^2 - 5y + 6 = 0$. (b)

Figure 2.4: The phase line for $y' = y^2 - 5y + 6$.

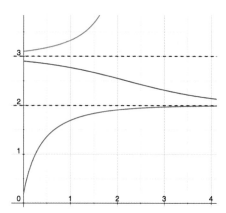

Figure 2.5: Solutions to $y' = y^2 - 5y + 6$ with (a) initial condition $y(0) = 0.2$, (b) initial condition $y(0) = 2.9$, and (c) initial condition $y(0) = 3.1$.

Rotate the phase line in Figure 2.2 by $90°$ counterclockwise (see Figure 2.4).   ∎

Figure 2.5 shows the graphs of three solutions to $y' = y^2 - 5y + 6$ corresponding to initial conditions between and around the equilibria at $y = 2$ and $y = 3$. You can see that their behavior agrees with the qualitative analysis of increase and decrease that led us to the phase line in Figure 2.4. You might wonder whether the red curve in Figure 2.5 will ever reach the horizontal line at $y = 2$. The answer is no, the line is an asymptote, in much the same way that the line $y = 0$ is an asymptote of the curve $y = e^{-t}$. We know this because of the following theorem:

> **Barrier Theorem:** Suppose $F(y)$ is a differentiable function that has isolated roots at $y_1, y_2, \ldots, y_n$, and $y(t)$ is a non-constant solution of the autonomous equation $y' = F(y)$. Then the graph of $y(t)$ will never touch the horizontal lines $y = y_1, y = y_2 \ldots y = y_n$.

Note: This theorem is actually a special case of a much broader fact called the *First-Order Existence and Uniqueness Theorem,* whose statement requires something called a *partial derivative* that we discuss in Chapter 13, and whose proof is beyond the scope of this book.

This theorem is not typically named, so if you mention the *Barrier Theorem* to other people, they probably won't know what you're talking about. However, we're naming it here in hopes of helping you remember it (the horizontal lines at the equilibria act as *barriers* that graphs of other solutions cannot touch).

You should know

- the terms *autonomous, equilibrium, source, sink, node, stable, unstable, semi-stable, phase line,* and *direction field;*

- the Barrier Theorem.

You should be able to

- locate and classify equilibria of $y' = F(y)$ using either the formula or the graph of $F(y)$;

- draw a phase line;

- determine whether a direction field is associated with an autonomous or non-autonomous DE;

- use a direction field to sketch the graph of a solution to an IVP.

❖ 10.2 Skill Exercises

In #1–8 (a) draw a phase line for the DE, (b) locate all equilibria, and (c) classify each equilibrium as a sink, source, or node.

1. $x' = x - 7$

2. $y' = 3 - y$

3. $y' = y^2 - 5y + 6$

4. $x' = x^2 + x - 12$

5. $x' = x^3 - x^2 - 3x$

6. $y' = y^4 + y^3 - 12y^2$

7. $y' = \cos(y)$

8. $x' = \sin^2(x)$

9. The graph of f is shown in Figure 2.6. Use it to (a) locate all equilibria of $x' = f(x)$, (b) classify each equilibrium as a source, sink, or node, then determine $\lim_{t \to \infty} x(t)$ when (c) $x(0) = 1.2$ and (d) $x(0) = 1.9$.

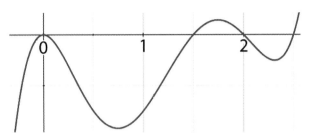
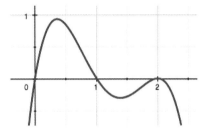

Figure 2.6: (left) The graph of f for #9; (right) the graph of f for #10.

10. The graph of f is shown in Figure 2.6. Use it to (a) locate all equilibria of $y' = f(y)$, (b) classify each equilibrium as a source, sink, or node, then determine $\lim_{t \to \infty} y(t)$ when (c) $y(0) = 0.5$, and (d) $y(0) = 2.1$.

11. Suppose $y' = f(y)$, where $f(y) = y(y - 4)^2$.

   (a) Verify that $y = 0$ and $y = 4$ are equilibria.

   (b) Sketch a graph of f, and use it to classify each equilibria as a source, sink, or node.

   (c) Determine $\lim_{t \to \infty} y(t)$ when $y(0) = -0.001$, when $y(0) = 0.001$, and when $y(0) = 4.001$

12. Suppose a physical system is described by $y' = f(y)$ where $f(y) = y^3 + y^2 - 5y + k$, where the number $k = 3$ is determined experimentally.

   (a) Verify that $y = 1$ is an equilibrium.

   (b) Use a graphing utility to render a plot of f, zoom in on the root at 1, and determine whether $y = 1$ is stable, unstable, or semi-stable.

   (c) Based on your plot in part (b), determine $\lim_{t \to \infty} y(t)$ when $y(0) = 0$.

   (d) Because $k = 3$ was measured in an experiment, and there's always error in experiments, plot f for several values of k in $[2.8, 3.2]$, and then discuss how small changes in k affect $\lim_{t \to \infty} y(t)$ when $y(0) = 0$.

❖ 10.2 Concept and Application Exercises

13. Suppose $x' = f(x)$ where f is a continuous function with two (or more) equilibria. Can they all be nodes? If so, provide an example. Otherwise, explain why this is impossible.

Mathematics

14. Suppose $x' = f(x)$ where f is a continuous function with two (or more) equilibria. Can they all be sinks? If so, provide an example. Otherwise, explain why this is impossible.

     — Mathematics

15. In #43 on p. 778 we introduced the "sales equation," $s' = \rho\alpha - \lambda s$. (a) Locate and classify the equilibria of this equation. (b) What does this model predict about the sales of a company that continually increases its monthly advertising expenditure? (c) Does this model predict an upper limit on sales for the company described in part (b)?

     — Marketing

16. In 1957 **M. Vidale and H. Wolfe** proposed a model of sales response to advertising that's similar to the "sales equation" introduced on p. 778. They suggest that $s' = \rho\alpha(1 - s/m) - \lambda s$, where m is a quantity called the *market saturation*. (a) Assuming constant ρ, α, λ and m, locate and classify the equilibrium of this DE. (Its particular value depends on the parameters, but its (in)stability does not.) (b) What does this equation say about a company's sales (in dollars per month) as the company continually increases its monthly advertising expenditure?

     — Marketing

17. The **Sethi model** of sales response to advertising is an adaptation of the model proposed by Vidale and Wolfe. It asserts $s' = \rho\alpha\sqrt{1 - s/m} - \lambda s$. (a) Assuming constant ρ, α, λ and m, locate and classify the equilibrium of this DE. (The particular value of the equilibrium depends on the parameters, but its (in)stability does not). (b) The equilibrium from part (a) depends on α, which the company can control, so let's write it as $s_*(\alpha)$. Determine $\lim_{\alpha \to \infty} s_*(\alpha)$, and explain what it means. (c) Verify that $s_*''(\alpha) < 0$ and explain what that means in the context of marketing.

     — Marketing

18. The **Allee effect** (also called the *underpopulation effect*) refers to a decline in population growth that occurs at low population densities. Consider the following adaptation of the logistic population model (introduced on p. 775):

$$y' = ry\left(1 - \frac{y}{K_1}\right)\left(1 - \frac{y}{K_2}\right),$$

where r, K_1 and K_2 are positive constants with $K_1 < K_2$. (a) Draw a phase line for this equation. (b) Locate and classify the equilibria (the particular values of the equilibria depend on the parameters, but not their (in)stability). (c) Explain what happens to a population when $y(0) < K_1$, when $K_1 < y(0) < K_2$, and when $K_2 < y(0)$.

     — Ecology

19. In this exercise we investigate how the equilibria of $y' = y^2 - 3y + k$ depend on the number k.

     — Mathematics

    (a) Show that there are two distinct equilibria when $k = 0$, $y_1 < y_2$, and show that y_1 is a source while y_2 is a sink.

    (b) Show that there are two distinct equilibria when $k = 0.1$, $y_1 < y_2$; that they are a little closer than they were in part (a), but their stability has not changed.

    (c) In part (b) you saw y_1 and y_2 moved when we perturbed k. Determine formulas that state explicitly how y_1 and y_2 depend on k.

    (d) Explain why the stability of the equilibria is the same for all $k < 9/4$.

    (e) Sketch graphs of $y_1(k)$ and $y_2(k)$ on the same ky-plane, making the graph of y_1 a dashed curve and the graph of y_2 a solid curve.

20. The graph you made in part (e) of #19, which shows us how the equilibria of a system depend on a parameter, is called a *bifurcation* diagram (the word *bifurcation* refers a change in the number or stability of equilibria). Create bifurcation diagrams for (a) $y' = y^2 + ky + 1$, and (b) $y' = ky^2 + y + 1$.

Mathematics

21. In Section 10.5 we'll discuss Newton's Law of Cooling, which says that the temperature of an object varies with time according to $T' = -k(T - A)$, where A is the temperature of the environment (which we'll assume to be constant for now) and k is a number called the *time constant* that depends on the object. (a) Verify that the constant function $T = A$ is a stable equilibrium, (b) explain what that means to us in this physical context.

Mechanical Engineering

22. On p. 775 we introduced the logistic population model, which says that population varies with time according to $P' = rP(1 - P/K)$, where $r > 0$ depends on the species and $K > 0$ depends on the environment. (a) Show that $P = 0$ is an unstable equilibrium and $P = K$ is a stable equilibrium, and (b) explain what that means to us in the context of population modeling.

Biology

23. When an annuity earns interest at $\gamma\%$/year, compounded continuously, and pays $3000 per year continuously, the amount of money in the account varies over time according to $y' = \gamma y - 3000$. (a) Show that this DE has an unstable equilibrium at $3000/\gamma$, and (b) explain what that means in this context (consider $\lim_{t\to\infty} y(t)$ when $y(0)$ starts below, at, and above $3000/\gamma$).

Finance

24. When fluid creeps up a pipette or capillary, its height varies over time according to an equation of the form $y' = (F - Ay)/(\gamma y)$, where F is a force due to surface tension, the number A depends on quantities such as the cross-sectional area of the capillary, and γ quantifies the friction force at the liquid-solid interface along the capillary wall. (a) Locate all equilibria of this equation and determine whether each is stable or unstable, and (b) determine $\lim_{t\to\infty} y(t)$ when $y(0)$ is near zero, but positive.

Fluid Mechanics

25. The charge on the capacitor in an RC circuit varies according to $Rq' + \frac{q}{C} = V$, where R is the resistance of the resistor, C is the capacitance of the capacitor, and V is the impressed voltage (all of which we assume to be constant). Show that this differential equation has a stable equilibrium at $q = CV$.

Electrical Engineering

26. Suppose an inverted conical tank (point down, like an icecream cone) is 8 meters tall and has a diameter of 2 meters at its open, circular top. We pump water into the top at 1 m^3 minute, but after a short while, the tank springs a leak at its tip, and water begins to drain.

Engineering

    (a) Write the volume of fluid in the tank, V, as a function of its height, y.

    (b) In the moments before the leak, what's dV/dt?

    (c) If the hole has an area of 0.01 m^2, and water exits through it according to Torricelli's Law (see p. 776), write an equation describing dV/dt in terms of y (accounting for both inflow and outflow per second).

    (d) Use the relationship between V and y to rewrite your equation from part (c) entirely in terms of y.

    (e) Based on your answer to part (e), will the water overow, drain entirely, or approach an equilibrium height if we continue to pump water in?

27. Suppose the hole in #26 has an area of A m^2. For what value of A (if any) will the water height come to equilibrium at the top of the tank if we are pumping 0.5 m^3 per minute?

Engineering

28. Like most cells, a neuron at rest has a negative voltage potential across its cell wall (meaning the concentration of positive ions is greater outside the cell than inside). Small perturbations in this voltage are not amplified, but when the neuron is brought above a threshold voltage it opens channels in its cell wall that allow positive sodium ions to flow in, thereby raising the potential. Consider the differential equation $v' = -v(v - \alpha)(v - \beta)$ where $\alpha > \beta > 0$.

   (a) Draw a phase line for this DE.

   (b) Verify that the rest voltage of $v = 0$ is stable, and explain what that means in this biological model.

   (c) Suppose the cell's voltage is shifted to $v = \alpha/2$. Explain what happens next.

   (d) Show that if $v(0) > \alpha$, the cell's voltage never returns to rest (so this equation does not, by itself, model the behavior of a neuron).

The model of a neuron developed in #28 only allowed the cell to amplify a signal once, after which the voltage across its cell wall was stuck at $v = \beta$ (i.e., it could not return to rest, $v = 0$). In fact, neurons can counter the effects of sodium influx by, for example, moving potassium ions through the cell wall. In exercises #29–31 we quantify the strength of such a *blocking* mechanism as the number w.

29. Suppose that $\alpha = 0.1$ and $\beta = 1$.

   (a) Draw the phase lines for $v' = -v(v-\alpha)(v-\beta)-w$ when $w = 0.05, 0.1, 0.15,$ and 0.2.

   (b) Determine the values of w at which the DE has only one stable equilibrium.

   (c) In complete sentences, explain the effect of w on this differential equation.

   (d) When $w = 0.2$ and $v(0) = 0.3$, what is $\lim_{t \to \infty} v(t)$?

30. Suppose we write $v' = f(v)-w$, where $f(v) = -v(v-\alpha)(v-\beta)$ and $\beta > \alpha > 0$.

   (a) Verify that the rightmost critical point of f is at $v_* = \frac{\alpha+\beta+\sqrt{\alpha^2-\alpha\beta+\beta^2}}{3}$.

   (b) What happens to the action threshold of the neuron when w varies between 0 and $f(v_*)$?

   (c) Can the neuron amplify more than one signal (i.e., more than one occurrence of $v(0) > \alpha$) when $0 \le w < f(v_*)$?

   (d) Explain why the neuron doesn't amplify any signal with $v(0) > 0$ when $w \ge f(v_*)$.

31. In #30 we established that a constant value of w cannot fix our model of the neuron (so that it amplifies a signal and then returns to rest, at $v = 0$, ready to amplify another). This leads us to conjecture that w is a function of time. We want to say that $w(0) = 0$ and that w grows when the cell's voltage is positive, but not without bound. To better understand $w(t)$, consider the DE $w' = \varepsilon(v - \gamma w)$, where unlike previous exercises, v is understood to be a *fixed* positive number, and $\varepsilon, \gamma > 0$. (We'll combine the behaviors of v and w on p. 829.)

   (a) Locate all equilibria of the DE and classify each as a source, sink or node.

   (b) Draw a phase line for the DE.

10.3   Slope Fields and Euler's Method

It's an inconvenient fact that many mathematical models of the real world involve differential equations that we cannot solve explicitly. In such situations, we turn to approximation methods, two of which are discussed in this section.

✣ Slope Fields

Even when we cannot solve a first-order differential equation, we can produce a visual way of understanding its solutions by using it to calculate slopes of tangent lines. For the sake of discussion, let's consider the equation

$$y' = \frac{1}{4}(y^2 - 3y - 4). \tag{3.1}$$

If a solution to this equation has an initial value of $y(0) = 2$, the tangent line to its graph has a slope of

$$y' = \frac{1}{4}\left((2)^2 - 3(2) - 4\right) = -\frac{3}{2}$$

at time $t = 0$. We can communicate this fact visually by locating time $t = 0$ in the ty-plane and drawing an arrow or small line segment at a height of $y = 2$ that has a slope of -1.5. Similarly, if $y(0) = 3$, the tangent line to the graph of $y(t)$ has a slope of

$$y' = \frac{1}{4}\left((3)^2 - 3(3) - 4\right) = -1$$

at time $t = 0$. So we draw an arrow at a height of $y = 3$ that has a slope of -1. After doing this for several initial values of y, as seen in Figure 3.1, we have a basic understanding of how various solutions are changing at time zero.

We've been discussing the function value at time $t = 0$, but equation (3.1) describes y' in terms of only the function value, y, not what time it is. So if $y(1) = 2$, the slope of the tangent line will be -1.5 when $t = 1$; and if $y(3) = 2$, the slope of the tangent line will be -1.5 when $t = 3$. For this reason, we replicate the arrows in Figure 3.1 over and over again at different times, as seen in the left-hand image of Figure 3.2. The resulting field of arrows is called a **slope field** or **direction field**.

Figure 3.1: Slopes of tangent lines at time $t = 0$. If you've read Section 10.2, you can think of this as an enhanced phase line, which depicts not only whether a solution is increasing or decreasing, but also the rate at which it happens.

Equation (3.1) is said to be *autonomous*. Such equations were discussed in Section 10.2.

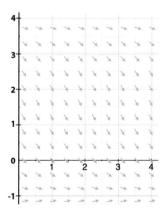 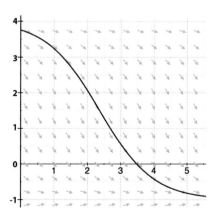

Figure 3.2: (left) We can add information to the phase line by graphing slope instead of simply increase and decrease; (right) the direction field for $y' = y^2 - 5y + 6$.

The slope field visually communicates moment-to-moment information about $y'(t)$. We can integrate that information by drawing a curve that follows the slope field, as shown in the right-hand image of Figure 3.2. If we had arrows to follow at every instant, the resulting curve would be the graph of a solution. With arrows spaced out across a grid, we get an approximation.

Try It Yourself: Using a direction field

Example 3.1. The direction field for $y' = F(y)$ is shown in the left-hand image of Figure 3.3. Use it to approximate the graph of the solution $y(t)$ with (a) $y(0) = 0$, and then (b) $y(0) = 4$.

Answer: See on-line solution.   ∎

Full Solution On-line

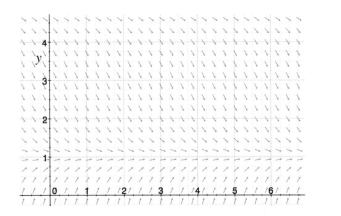
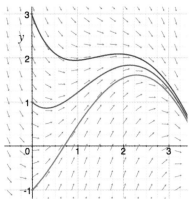

Figure 3.3: (left) Direction field for Example 3.1; (right) a direction field for the non-autonomous differential equation $y' = 3t - t^2 - y$, with solutions whose initial data is $y(0) = 3$, $y(0) = 1$, and $y(0) = -1$.

Direction fields can also help us understand the qualitative behavior of solutions to differential equations that are *not* autonomous—meaning that y' depends explicitly on both y and t. The basic idea is the same, except that you have to calculate slopes at each time t on your grid rather than simply duplicating the arrows from time $t = 0$. For example, the direction field for $y' = 3t - t^2 - y$ is shown with in the right-hand image of Figure 3.3 with the graphs of three solutions.

Using a direction field

Example 3.2. A direction field is shown below. Determine whether it corresponds to an autonomous or non-autonomous differential equation.

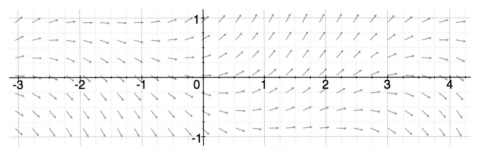

Solution: Let's pick a particular height on the graph, say $y = 1/3$. Looking at the arrows at that height, we see that they are not all the same. Some point up, and others point down, which means that y' depends on time, not just on the value of y. So this differential equation is not autonomous.   ∎

Try It Yourself: Using a direction field

Example 3.3. Suppose $y(t)$ is a solution of the differential equation that corresponds to the direction field from Example 3.2, and for which $y(-2) = 3/4$. Sketch a graph of $y(t)$.

Full Solution
On-line

Answer: See on-line solution.   ∎

✢ Euler's Method

In our discussion above, we said that a slope field contains moment-to-moment information about $y'(t)$, and that we can integrate it by drawing a curve that follows the direction field. This idea leads us to an algorithm for approximating $y(t)$ called **Euler's method**, which is demonstrated below.

Approximating function values for solutions to DEs

Example 3.4. Suppose $y(t)$ is a solution of the DE $y' = y^2 - 5y + 6$ for which $y(0) = 1$. Estimate $y(0.1)$ and $y(0.2)$.

Solution: We know that the definite integral calculates net change, so

$$y(0.1) = \underbrace{y(0) + \int_0^{0.1} y'(t)\ dt}_{\text{initial value + net change}}.$$

We don't know y' as a function of time, so we can't calculate this integral exactly, but we know that $y(0) = 1$, so the DE tells us that the slope of the graph is initially

$$y'(0) = (1)^2 - 5(1) + 6 = 2.$$

We don't expect the value of y' to remain 2, but we're estimating, so let's approximate by taking $y'(t) = 2$ at all times between times $t = 0$ and $t = 0.1$. Then

$$y(0.1) = y(0) + \int_0^{0.1} y'(t)\ dt \approx y(0) + \int_0^{0.1} 2\ dt = 1 + 2(0.1) = 1.2.$$

Similarly, we know that

$$y(0.2) = y(0.1) + \int_{0.1}^{0.2} y'(t)\ dt,$$

which we can't calculate exactly because we don't have a formula for $y'(t)$. However, we know that $y(0.1) \approx 1.2$, so

$$y'(0.1) \approx (1.2)^2 - 5(1.2) + 6 = 1.44.$$

This allows us to estimate $y(0.2)$ by approximating $y'(t) = 1.44$ at all $t \in [0.1, 0.2]$. So doing yields

$$y(0.2) = y(0.1) + \int_{0.1}^{0.2} y'(t)\ dt \approx 1.2 + \int_{0.1}^{0.2} 1.44\ dt = 1.2 + 1.44(0.1) = 1.344. \quad ∎$$

In Example 3.4 we stepped from $t = 0$ to $t = 0.1$ and then from $t = 0.1$ to $t = 0.2$, using the differential equation to approximate the slope of the graph in each step. In essence, this algorithm is a numerical way of "following" a direction field.

Try It Yourself: Using Euler's method

Example 3.5. Suppose $y(t)$ is the function from Example 3.4. Use Euler's method to estimate $y(0.3)$ and $y(0.4)$.

Answer: Starting with $y(0.2) \approx 1.344$, $y(0.3) \approx 1.453$ and $y(0.4) \approx 1.537$. ■

Full Solution On-line

Applying Euler's method when the DE includes explicit mention of time

Example 3.6. Suppose $y(t)$ is a solution of $y' = 3t - 0.5y$ for which $y(0) = 2$. Use Euler's method to estimate $y(0.25)$ and $y(0.5)$.

Solution: The definite integral calculates net change, so

$$y(0.25) = y(0) + \int_0^{0.25} y'(t)\, dt.$$

We don't know y' as a function of time, so we can't use this equation to calculate $y(0.25)$ exactly; but we know that $y' = 3t - 0.5y$ and $y = 2$ at time $t = 0$, so

$$y'(0) = 3\,\underbrace{(0)}_{\text{current time}} - 0.5\,\underbrace{(2)}_{\text{height of the graph}} = -1.$$

> Suppose y' has units of distance/time. In order for this equation to make sense, the coefficient of 3 must have units of distance/time^2, and the coefficient of 0.5 must have units of 1/time. Otherwise, we're equating two entirely different kinds of quantities.

This allows us to approximate

$$y(0.25) = y(0) + \int_0^{0.25} y'(t)\, dt \approx y(0) + \int_0^{0.25} -1\, dt = 2 + (-1)(0.25) = 1.75.$$

Similarly,

$$y(0.5) = y(0.25) + \int_{0.25}^{0.5} y'(t)\, dt. \tag{3.2}$$

We know that $y \approx 1.75$ at time $t = 0.25$, and $y' = 3t - 0.5y$, so

$$y'(0.25) \approx 3\underbrace{(0.25)}_{\text{current time}} - 0.5\underbrace{(1.75)}_{\text{height of the graph}} = -0.125.$$

This allows us to approximate equation (3.2) as

$$y(0.5) \approx 1.75 + \int_{0.25}^{0.5} (-0.125)\, dt = 1.75 + (-0.125)(0.25) = 1.71875. \quad ■$$

Try It Yourself: Applying Euler's method when the DE includes time

Example 3.7. Suppose $y(t)$ is a solution of $y' = 1 + t^2 + 4y$ for which $y(0) = -0.5$. Use Euler's method to estimate $y(0.15)$ and $y(0.3)$.

Answer: $y(0.15) \approx -0.65$ and $y(0.3) \approx -0.8867$. ■

Full Solution On-line

In Examples 3.4 and 3.5 we calculated $y(t)$ at $t = 0.1, 0.2, 0.3$, and 0.4. These times are uniformly separated by $\Delta t = 0.1$, which we call the **step size** of Euler's method. Similarly, the step size in Example 3.6 was $\Delta t = 0.25$, and the step size in Example 3.7 was $\Delta t = 0.15$.

Figure 3.4 shows the graph of the solution to the IVP in Example 3.6, and the approximations that we calculated with Euler's method (shown in red). There are two important things to note:

1. Our approximations have been interpolated with a piecewise linear graph. This is a visual reminder of our assumption that the derivative is constant over small intervals of time.

2. Euler's method produced the correct qualitative behavior, but you can see error beginning to accumulate. We can reduce its effect by taking smaller steps (e.g. use a step size of $\Delta t = 0.05$ instead of $\Delta t = 0.25$), but that means more calculations. In practice, a modern computers can take thousands of steps forward in the blink of an eye, so let's take a moment to talk about doing that.

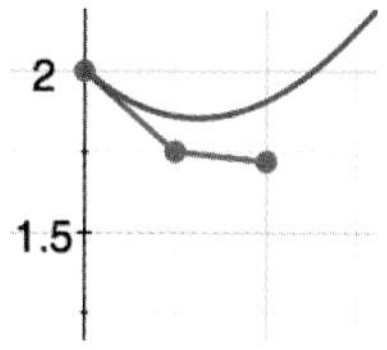

Figure 3.4: The actual solution to Example 3.6, and the iterates of Euler's method.

✣ Streamlining Euler's Method (Spreadsheet Calculation)

When we know $y(t_k)$ and want to know the function value a moment later, at time $t_{k+1} = t_k + \Delta t$, we calculate

$$y(t_{k+1}) = y(t_k) + \int_{t_k}^{t_{k+1}} y'(t) \ dt.$$

In Euler's method we approximate $y'(t)$ as being the constant $y'(t_k)$ across the entire interval of integration, so

$$y(t_{k+1}) \quad = \quad \overbrace{y(t_k)}^{\text{current function value}} + \underbrace{y'(t_k)}_{} \underbrace{\Delta t}_{}. \tag{3.3}$$

$$\underbrace{\phantom{y(t_{k+1})}}_{\text{next function value}} \qquad \underbrace{}_{\text{(current slope)} \times \text{(time step)}}$$

The specified differential equation tells how $y'(t_k)$ depends on t_k and $y(t_k)$, which was estimated in the previous step, so equation (3.3) amounts to multiplication and addition. That's easily done with a spreadsheet.

Implementing Euler's method on a spreadsheet

Example 3.8. Suppose $y(t)$ is a solution of $y' = y^2 - t^2$ for which $y(0) = 0.5$. Use Euler's method to estimate the value of $y(t)$ every tenth of a second between $t = 0$ and $t = 1$.

Solution: The table below was generated using a spreadsheet program. After entering the value of Δt in cell **A1**, and using cells in row **2** to label columns according to their content, initial values of t and y were entered into cells **A3** and **B3** respectively. The value of $y'(0)$ depends on both $t = 0$ and $y(0)$ (found in cells **A3** and **B3** respectively) according to $y' = y^2 - t^2$, so in cell **C3** we typed

$$= \textbf{B3\textasciicircum 2 - A3\textasciicircum 2}$$

Then because the change in y is $\Delta y = y'(t_k)\Delta t$, we entered the code

$$= \textbf{C3*A\$1}$$

into cell **D3**. At this point, the initial data have been entered, and Δy calculated, so the spreadsheet appears as follows:

> The \$ in this equation tells the computer that we are referring to a fixed, not relative position in the spreadsheet. So when we "drag" the equations down the spreadsheet, which we'll do in a moment, every cell in column **D** will refer to **A1**.

	A	**B**	**C**	**D**
1	0.1			
2	t	y	dy/dt	Δy
3	0	0.5	0.25	0.025
4				

To make our first step forward, we increment time by typing $=$**A3**$+$**A\$1** into cell **A4**, and then use equation (3.3) to update the function value by typing $=$ **B3**$+$**D3** into cell **B4**. This yields:

	A	B	C	D
1	0.1			
2	t	y	dy/dt	Δy
3	0	0.5	0.25	0.025
4	0.1	0.525		
5				

Now we're only two quick mouse-actions away from finishing. When we highlight cells **C3** and **D3**, a small box appears in the lower-right corner. Grabbing this box, dragging down one row, and releasing tells the computer to copy the formulas in these cells while preserving the relative location of data. This yields:

	A	B	C	D
1	0.1			
2	t	y	dy/dt	Δy
3	0	0.5	0.25	0.025
4	0.1	0.525	0.265625	0.0265625
5				

If you click on cell **C4** at this stage, you'll see that it uses the same formula as **C3**, but that it refers to data in row 4 rather than row 3. This is the sense in which the spreadsheet preserves the relative location of data.

And finally, after highlighting cells **A4** through **D4**, the same drag-and-release action tells the computer to repeat the calculation across multiple rows. In this case, dragging down to row **13** yields

	A	B	C	D
1	0.1			
2	t	y	dy/dt	Δy
3	0	0.5	0.25	0.025
4	0.1	0.525	0.265625	0.0265625
5	0.2	0.5515625	0.26422119140625	0.026422119140625
6	0.3	0.57798461914063	0.24406621996313	0.024406621996313
7	0.4	0.60239124113694	0.2028752073985	0.02028752073985
8	0.5	0.62267876187679	0.13772884049241	0.013772884049241
9	0.6	0.63645164592603	0.045070697601952	0.004507069760195
10	0.7	0.64095871568622	-0.079171924785865	-0.007917192478587
11	0.8	0.63304152320764	-0.23925842989495	-0.023925842989495
12	0.9	0.60911568021814	-0.43897808811239	-0.043897808811239
13	1	0.5652178714069	-0.68052875784225	-0.068052875784225

where we see that the function at a time $t = 1$ is $y(1) \approx 0.5652178714069$.   ■

❖ A Closing Remark About the Step Size in Euler's Method

Suppose $y(t)$ solves $y' = f(t, y)$, and we know $y(0)$ but need to know $y(100)$. In Euler's method we approximate the value by using short intervals to step our way from $t = 0$ to $t = 100$, assuming that y' is constant over each. Smaller steps result in less error because we check y' more frequently, but that means we have to make more calculations. Consider the following comparison:

1. Using $\Delta t = 50$, we take only two steps, so the approximation method is quick. However we've assumed a constant derivative over *very* long periods of time, which makes us worry about the accuracy of our estimate.

2. Using $\Delta t = 0.0001$, we are much more confident in our accuracy, but we have to make a *million* steps, so it takes a lot longer to do, even with a computer program or spreadsheet.

So what step size should we use? That depends on how much error you can tolerate in your final estimate. You can see a rigorous treatment of error in a course called *numerical analysis,* but in this introduction, the step size (or the number of steps) will always be specified for you.

You should know

- the terms *slope field, direction field, Euler's method,* and *step size.*

You should be able to

- sketch a slope field;

- determine whether a slope field corresponds to an autonomous equation;

- implement Euler's method.

❖ 10.3 Skill Exercises

In #1–6 use Euler's method with the specified step size to approximate $y(1)$.

1. $y' = (t+1)y^2 - y$; $y(0) = 2$; $\Delta t = 0.2$ 4. $y' = y/(t^2+1)$; $y(0) = 1$; $\Delta t = 0.1$

2. $y' = t^2/y$; $y(0) = 1$; $\Delta t = 0.2$ 5. $y' = y^2 - t^2$; $y(0) = 0$; $\Delta t = 0.25$

3. $y' = ty$; $y(0) = 3$; $\Delta t = 0.1$ 6. $y' = y$; $y(0) = 0$; $\Delta t = 0.01$

In #7–12 use Euler's method with n steps to approximate the specified function value when $y(t)$ is a solution of the differential equation and $y(0) = 10$.

7. $n = 4$, find $y(1)$ when $y' = \sqrt{y^2 + 1}$ 10. $n = 8$, find $y(2)$ when $y' = \cos(t - y)$

8. $n = 5$, find $y(2)$ when $y' = t - y$ 11. $n = 10$, find $y(1)$ when $y' = y^{1+t^2}$

9. $n = 6$, find $y(3)$ when $y' = \sin(t + y)$ 12. $n = 12$, find $y(3)$ when $y' = 2^{-yt}$

❖ 10.3 Concept and Application Exercises

Exercises 13 and 14 require that you've read Section 10.2.

13. Consider the differential equation $y' = 20 - 10y$.

    (a) Verify that $y = 2$ is an equilibrium and plot its graph.

    (b) Calculate the first 6 iterations of Euler's method with a step size of $\Delta t = 0.2$ when $y(0) = 4$.

    (c) Plot the points calculated in part (b) on the graph from part (a).

    (d) You know from the Barrier Theorem (p. 782) that solutions to autonomous DE cannot cross equilibria. Explain how to resolve this fact with your plot from part (c).

14. Consider the differential equation $y' = y(y - 2)^2$.

    (a) Verify that $y = 2$ is an equilibrium, and plot its graph.

    (b) Calculate the first 6 iterations of Euler's method with a step size of $\Delta t = 1$ when $y(0) = 1$.

    (c) Plot the points calculated in part (b) on the graph from part (a).

    (d) You know from the Barrier Theorem (p. 782) that the graphs of non-constant solutions to this autonomous DE cannot touch the line $y = 2$. Explain how to resolve this fact with your plot from part (c).

In #15–16 you see the curve $y = f(t)$ in blue, and the curve $y = g(t)$ in red. Determine which (if either) of f and g are solutions of the differential equation that generated the slope field.

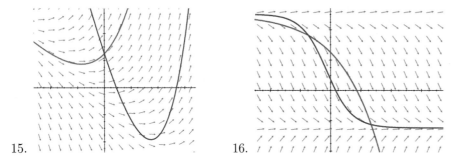

15.   16.

In #17–18 (a) determine whether the direction field corresponds to an autonomous DE. Then draw the graph of the solution whose initial data is (b) at the point **A**, and (c) at the point **B**.

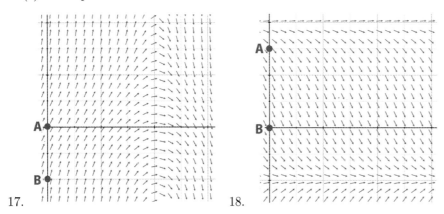

17.   18.

Exercises #19–20 address the depth of fluid in a leaking tank, based on the discussion of flow rates on p. 776.

Engineering

19. Suppose the base of a right circular cylinder has radius of 1 m, and the cylinder is 5 m tall. It contains fluid that's 4 meters deep when a hole of area 0.01 m^2 appears in the base. Use 30 steps of Euler's method to predict the height of the fluid after one minute.

20. Suppose the container from #20 is on its side, so the circular faces are vertical. The fluid it contains is 0.8 meters deep when a hole of area 0.01 m^2 appears in one of the circular faces, 0.1 meters from the bottom. Use 20 steps of Euler's method to predict the height of the fluid after 2 minutes.

Exercises #21–22 address populations that are governed by the logistic growth model discussed on p. 775.

Biology

21. Suppose a bacterial culture has an initial population of $p(0) = 1000$ cells and its intrinsic growth rate is $r = 1.001/\text{minute}$. If the carrying capacity is $K = 10000$, and the population is governed by the logistic growth model, use Euler's method with 20 steps to approximate the population an hour later.

22. Suppose a bacterial culture has an initial population of $p(0) = 1000$ cells and its intrinsic growth rate is $r = 1.01/\text{minute}$. If the carrying capacity is $K = 1000000$, and the population is governed by the logistic growth model, use Euler's method with 30 steps to approximate the population an hour later.

10.4 Separable Equations

On p. 775 we introduced the logistic model in the context of population growth with limited resources. It's also used to describe the spread of rumors, the rate of autocatalytic chemical reactions, and many other things. Mathematically, it's a prime example of something called a *separable* differential equation, defined as follows:

> **Separable Equations:** A differential equation is said to be **separable** if it can be written in the form $p(y)y' = q(t)$. (The functions p and q *can* be constant.)

Note: It's important to note that y' occurs as a *factor* on the left-hand side of this equation. In a few moments, you'll se that this is essential to solving such equations.

Determining whether a DE is separable

Example 4.1. Determine which of the following differential equations are separable: (a) $y' = \frac{t}{y}$, (b) $y' = 2 + 2y + t + ty$, (c) $y' = ty + y^2$.

Solution: (a) We can multiply this equation by y and rewrite it as $yy' = t$. Note that y and t occur on opposite sides of this equation, and that y' occurs as a factor on the left-hand side, so this equation *is* separable.

(b) After factoring the right-hand side, this equation becomes $y' = (2+t)(1+y)$. Then dividing by $1 + y$ yields

$$\frac{1}{1+y}y' = 2 + t.$$

Note that y and t occur on opposite sides of this equation, and that y' occurs as a factor on the left-hand side, so this equation *is* separable.

(c) If we subtract y^2 and then divide by y, this equation becomes

$$\frac{y' - y^2}{y} = t.$$

Though y and t occur on opposite sides of the equation, the y' does not occur as a *factor* on the left-hand side. This equation is *not* separable. ∎

The standard method for solving separable equations is often summarized in a memorable sound bite: *separate and integrate*. This is demonstrated below.

Solving a logistic equation

Example 4.2. Solve the logistic equation $y' = (1 - y)y$.

Solution: The constant functions $y(t) = 0$ and $y(t) = 1$ solve this DE. The discussion that follows addresses solutions for which $y(0) \neq 0$ and $y(0) \neq 1$. We begin by dividing this equation by $(1 - y)y$, and rewriting it as

$$\frac{1}{(1-y)y}\,y' = 1.$$

Now y and t have been separated, and y' occurs as a factor on the left-hand side, so we can perform the second step of the process: *integrate*.

$$\int \frac{1}{(1-y)y}\,y'\,dt = \int 1\,dt.$$

Since $y'\, dt = \frac{dy}{dt}\, dt = dy$, this equation is really

This is why it's important that y' occur as a factor.

$$\int \frac{1}{(1-y)y}\, dy = t + C. \tag{4.1}$$

The remaining integral is accomplished by a partial fractions decomposition:

$$\int \frac{1}{(1-y)y}\, dy = \int \frac{1}{y} + \frac{1}{1-y}\, dy = \ln|y| - \ln|1-y| + B,$$

so equation (4.1) is really

$$\ln|y| - \ln|1-y| + B \;=\; t + C$$
$$\ln\left|\frac{y}{1-y}\right| \;=\; t + D,$$

where $D = C - B$ can be any constant. After exponentiating both sides, this becomes

$$\left|\frac{y}{1-y}\right| = e^{t+D} = e^t e^D. \tag{4.2}$$

It would be a lot easier to simplify this expression algebraically if we could omit the absolute values, so let's take a moment to examine the ratio they delimit. Because the right-hand side of equation (4.2) is never zero, the ratio on the left is never zero. And we know it's continuous because y is a differentiable function of time, so the ratio is either always positive or always negative. If it's always positive, we can omit the absolute value and nothing else changes. If it's always negative, $y/(1-y)$ and $e^t e^D$ are opposites. Because no initial data was specified in this example, we don't know which case we have, so let's just write the equation as

The number A is either e^D or its opposite, and we don't know which at this stage.

$$\frac{y}{1-y} = Ae^t. \tag{4.3}$$

Now we continue our algebraic simplification by cross-multiplying to rewrite the equation as

$$\frac{1}{Ae^t} y = 1 - y \quad \Rightarrow \quad \frac{1}{Ae^t} y + y = 1.$$

Factoring a y out of the left-hand side yields

$$\left(\frac{1 + Ae^t}{Ae^t}\right) y = 1 \quad \Rightarrow \quad y = \frac{Ae^t}{1 + Ae^t}.$$

We began with the assumption that $y(0) \neq 0$, so $A \neq 0$, from which it follows that $y(t)$ is *never* zero. Similarly, we began with the assumption that $y(0) \neq 1$. This formula tells us that $y(t)$ is a fraction whose denominator is always 1 larger than its numerator, so $y(t)$ will *never* be 1. ∎

Try It Yourself: Solving a logistic equation

Example 4.3. Solve the logistic equation $y' = 3(4-y)y$, when $y(0) = 1$.

Answer: $y(t) = \frac{4}{3e^{-12t}+4}$. ∎

Full Solution On-line

Solving a separable equation

Example 4.4. Solve the equation $t^2 y' + 4y = 7$ when $t > 0$.

Solution: If we subtract $4y$ from both sides, the equation becomes $t^2 y' = 7 - 4y$. Note that even though this equation is not autonomous, it has an equilibrium solution at $y = 7/4$ (since $t \neq 0$, y' must be zero when $y = 7/4$). For the rest of this solution, let's assume that we're talking about non-constant solutions, so $7 - 4y \neq 0$. Then dividing both sides of our equation by t^2 and by $7 - 4y$ results in

$$\frac{1}{7 - 4y} y' = \frac{1}{t^2}.$$

Integrating this equation in time, we see

$$\int \frac{1}{7 - 4y} y' \, dt = \int t^{-2} \, dt.$$

The right-hand side is easy to integrate, and on the left-hand side we use the fact that $dy = y' \, dt$ to write

$$\int \frac{1}{7 - 4y} \, dy = -t^{-1} + C.$$

After using the substitution $x = 7 - 4y$ to integrate the left-hand side, we see

$$-\frac{1}{4} \ln|7 - 4y| = -\frac{1}{t} + C,$$

which we rewrite as

$$\ln|7 - 4y| = \frac{4}{t} - 4C.$$

Now exponentiating both sides and solving for y, we have

$$|7 - 4y| = e^{4/t} e^{-4C}.$$

Because $7 - 4y$ is never zero (this was our initial assumption), it's either always positive or always negative; so the same basic argument that led us from equation (4.2) to (4.3) on p. 796 allows us omit the absolute value from our equation and write it as

$$7 - 4y = A e^{4/t} \quad \Rightarrow \quad y = \frac{7}{4} + A e^{4/t}. \qquad \blacksquare$$

Solving a separable equation

Example 4.5. Solve the equation $y' = \frac{t^2 y}{2 \ln(y)}$.

Solution: We can isolate t by multiplying, and rewrite the equation as

$$2 \ln(y) \frac{1}{y} y' = t^2.$$

Then integrating in time, we have

$$\int 2 \ln(y) \frac{1}{y} y' \, dt = \int t^2 \, dt.$$

The right-hand side is easy to integrate, and on the left we use the fact that $dy = y' \, dt$ to write

$$\int 2 \ln(y) \frac{1}{y} \, dy = \frac{1}{3} t^3 + C.$$

The remaining integral is accomplished with the substitution $x = \ln(y)$, which leads us to

$$\left(\ln(y) \right)^2 = \frac{1}{3} t^3 + C. \qquad (4.4)$$

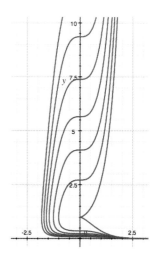

Figure 4.1: Solution curves from Example 4.5 with $C = 0, 1, 2, 3, 4, 5,$ and $C = 6$.

In contrast to Examples 4.2–4.4, in which the differential equation described y as a function of t, equation (4.4) can be solved for t as a function of y:

$$t = \sqrt[3]{\left(\ln(y)\right)^2 - C}.$$

The graphs of several such curves in the ty-plane are shown in Figure 4.1.　　■

Solving a separable equation

Example 4.6. Use the "separate and integrate" technique from Example 4.2 to solve the equation $t + 2y\sqrt{t^2 + 1}\, y' = 0$.

Solution: We can separate the t and y variables by rewriting the equation as

$$2y\, y' = -\frac{t}{\sqrt{t^2 + 1}}.$$

Now when we integrate in time,

$$\int 2y\, y'\, dt = -\int \frac{t}{\sqrt{t^2 + 1}}\, dt.$$

The integral on the right-hand side is done with the substitution $u = t^2 + 1$, and on the left-hand side we use the fact that $dy = y'\, dt$ to write the equation as

$$\int 2y\, dy \;=\; -\frac{1}{2}\sqrt{t^2 + 1} + C$$

$$y^2 \;=\; -\frac{1}{2}\sqrt{t^2 + 1} + C.$$

Several such curves are shown in Figure 4.2.　　■

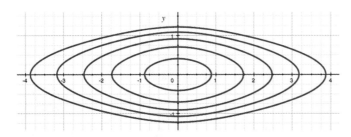

Figure 4.2: Several solution curves for $t + 2y\sqrt{t^2 + 1}\, y' = 0$.

You might feel dissatisfied by the conclusion of Example 4.6, but we can neither write y as a function of t, nor t as a function of y, so there's little point in continuing. However, we did arrive at a relationship between t and y that is free of derivatives, and each value of $C > 0$ corresponds to a curve in the ty-plane. So although we don't have an explicit formula for y as a function of t, we say that we have found an **implicit solution** to the DE.

> We could use the square root to write this as
> $$y = \pm\sqrt{C - \frac{1}{2}\sqrt{t^2 + 1}}$$
> but the $\pm$ we need in order to do it clearly shows that *two* values of y are associated with each admissible t, so we might as well leave the equation alone.

✥ Orthogonal Trajectories

Suppose a topographic map showed the summit of a hill at $(0,0)$ surrounded by circles (each curve on a topographic map indicates a path of constant altitude). Then a ball that's set slightly away from the origin will roll away, and its trajectory will be a line. We say that each line $y = mx$ is an *orthogonal trajectory* of the

> The word *orthogonal* means "perpendicular."

family of circles $x^2 + y^2 = r^2$ because it intersects each circle at a right angle. More generally, suppose that $\mathcal{F}$ is some family of disjoint curves (such as the family of circles, above). We say that a curve $\mathcal{C}$ is an orthogonal trajectory of $\mathcal{F}$ if every intersection with a member of the family happens at a right angle.

> By "disjoint" we mean that no two curves from this family share a common point.

Orthogonal trajectories

Example 4.7. Determine the orthogonal trajectories of the family of curves described by $x^4 + y^4 = k$, which are sometimes called *squircles* due to their shape (see Figure 4.3).

Solution: When an orthogonal trajectory intersects a squircle, they meet at a right angle. Said differently, if the tangent line to the squircle has a slope of m, the tangent line to the orthogonal trajectory is $-1/m$. Of course, the number m will vary from point to point on the squircle, and we can use implicit differentiation to say exactly how:

$$x^4 + y^4 = k \quad \Rightarrow \quad 4x^3 + 4y^3 y' = 0 \quad \Rightarrow \quad y' = -\frac{x^3}{y^3}.$$

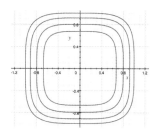

Figure 4.3: The curves described by $x^4 + y^4 = k$ for $k = 0.2, 0.4, 0.7, 1$.

Therefore, the tangent line to the orthogonal trajectory has a slope of

$$y' = \frac{y^3}{x^3}. \tag{4.5}$$

Equation (4.5) describes the trajectory by telling us about the slope of its tangent line at each point. When we separate variables we see

$$\frac{1}{y^3} y' = \frac{1}{x^3},$$

and integrating with respect to x yields

$$\int \frac{1}{y^3} y' \, dx = \int \frac{1}{x^3} \, dx.$$

Since $y' \, dx = \frac{dy}{dx} \, dx = dy$, this becomes

$$\int \frac{1}{y^3} \, dy = \int \frac{1}{x^3} \, dx$$
$$-\frac{1}{2y^2} = -\frac{1}{2x^2} + C.$$

Because both x and y appear in even powers, we cannot recover an explicit formula for one in terms of the other, but our equation expresses the relationship between them without using derivatives. That is, we've found a family of implicit solutions, several of which are shown in Figure 4.4. ∎

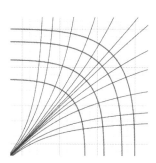

Figure 4.4: Several orthogonal trajectories to the family of squircles, shown in the first quadrant.

Orthogonal trajectories

Example 4.8. Determine the orthogonal trajectories of the family of curves described by $x^2 = ky^3$.

Solution: Like we did in Example 4.7, let's begin by using implicit differentiation to find a formula for the slope of the tangent line to one of the curves in the specified family:

$$x^2 = ky^3 \quad \Rightarrow \quad 2x = 3ky^2 y' \quad \Rightarrow \quad y' = \frac{2x}{3ky^2}.$$

So the tangent line to the orthogonal trajectory through (x, y) has a slope of

$$y' = -\frac{3ky^2}{2x}. \tag{4.6}$$

This differs from Example 4.8 in that the differential equation includes the parameter k. This parameter tells us which member of the family the orthogonal trajectory is crossing at a given point. As the orthogonal trajectory crosses the curves in the family, the number k will change, and we can use the original equation to say exactly how:

$$x^2 = ky^3 \quad \Rightarrow \quad k = \frac{x^2}{y^3}.$$

Substituting this into equation (4.6), we see that the orthogonal trajectory is described by

$$y' = -\frac{3x}{2y} \quad \Rightarrow \quad y\, y' = -\frac{3}{2}x.$$

Now integrating with respect to x, and using the fact that $\frac{dy}{dx}\, dx = dy$, we see

$$\int y\, y'\, dx = -\int \frac{3}{2}x\, dx \quad \Rightarrow \quad \frac{1}{2}y^2 = -\frac{3}{4}x^2 + C \quad \Rightarrow \quad \frac{1}{2}y^2 + \frac{3}{4}x^2 = C,$$

which is the family of ellipses you see in Figure 4.5. (Note that we have arrived at an implicit solution.)   ∎

Note: We began our discussion of orthogonal trajectories by thinking about a ball rolling down a perfectly circular hill. This analogy gets the right idea across, but breaks down after a point. An actual ball rolling down a real hill has momentum, so generally speaking, it crosses the topographic curves at oblique angles as it rolls. By contrast, orthogonal trajectories always cross at right angles.

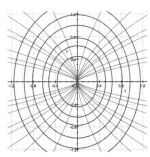

Figure 4.5: The family of curves $x^2 = ky^3$ and their orthogonal trajectories $0.5y^2 + 0.75x^2 = C$.

You should know

- the terms *separable, implicit solution,* and *orthogonal trajectory;*

- the phrase "separate and integrate."

You should be able to

- determine whether a DE is separable;

- use the "separate and integrate" technique to solve a separable equation;

- determine the orthogonal trajectories to a family of curves.

✤ 10.4 Skill Exercises

Use the separate-and-integrate technique to find a general solution to each of the differential equations in #1–6.

1. $y' = -6t\, y$

2. $y' = 24t - 8t\, y$

3. $e^{t+y} + y' = 0$

4. $2\ln(t) + t\, y' = 0$

5. $4y^{1/2}\sin(t) + y' = 2y^{1/2}\cos(t)$

6. $t\, y' = 7 - \dfrac{24}{t+6}$

In #7–14 (a) determine a solution to the DE (the solution may be implicit), and (b) use a graphing utility to render a plot of the relationship between y and t.

7. $y' = 3\,ty - 7y$; $y(2) = 5$

8. $y' + y\,t^2 = 0$; $y(1) = -1$

9. $(ty^2 + t)\,y' = \ln(t)$; $y(1) = 1$

10. $y' = -\dfrac{4}{t^2\,y}$; $y(3) = 5$

11. $2t\,y^2 + 3t\,y = 4y\,y'$; $y(2) = 4$

12. $2t\cos(y)\,y' - \sin(y) = t\sin(y)$; $y(1) = 7$

13. $y\tan(t) + \cos^2(t)\,y' = 0$; $y(0) = 1$

14. $\cos(y)\,y' - 1 = 6\cos(t) + y\sin(y)\,y'$; $y(0) = 0$

Exercises #15–20 contain both differential equations whose solutions are explicit, and also DEs whose solutions are implicit. In either case, (a) solve the DE, and (b) use a graphing utility to render a plots of solution curves that pass through $(1, k)$ for several different values of k.

15. $y' = 5 + y^2$

16. $ye^t = y'(1 - e^t)$

17. $\dfrac{1}{3}y' + 6 = 2y$

18. $1 - y' = 4y$

19. $\sin(y)\sec(y)\,y' = \sin(t)\cos(t)\cos(y)$

20. $y^2 + 1 = y'\sqrt{1 - x^2}$

❖ 10.4 Concept and Application Exercises

In #21–30 determine the orthogonal trajectories to the specified family of curves.

21. $y = kx$

22. $5y^2 + 5x^2 = k$

23. $y^2 + 4x^2 = k$

24. $3y^2 - x^2 = k$

25. $y^2 = -kx^4$

26. $y = kxe^x$

27. $y^2 - kx^3 = 9$

28. $y^2 = \frac{1}{x^2 + k}$

29. $x^2 - ky^2 = 1$

30. $y = \arcsin(kx)$

31. Suppose there are initially a molecules of species A, and b molecules of species B in a 1 L aqueous solution, and they react chemically to form a product, species C, according to $2A + 3B \to 1C$. If $c(t)$ is the number of molecules of species C that have been created after t minutes ...

    Chemistry

    (a) what does the number $a - 2c$ tell us?

    (b) what does the number $b - 3c$ tell us?

    (c) what does the equation $c' = \gamma(a - 2c)(b - 3c)$ mean in this context, where $\gamma > 0$, and why does it make sense? (*Hint: think about what happens when $2c$ approaches a.*)

32. Solve the DE in #31 when $a = 8$ mols, $b = 9$ mols, and $\gamma = 0.1$, and determine the amount of product present after 3 minutes.

    Chemistry

33. Suppose that there are 8 mols of A present initially in #31, and $\gamma = 0.2$. How many mols of B are needed if we want to produce 3 mols of C in 2 minutes?

    Chemistry

34. A virus is unknowingly carried into a school of 15000 students, and after three days, seven people have been infected.

    Public Health

(a) If $n(t)$ is the number of sick students after t days, explain what the equation $n' = \gamma(n)(15000 - n)$ says about the rate at which the virus spreads, where $\gamma > 0$ is constant (consider $n \approx 0$ and $n \approx 1400$).

(b) When is the virus spreading the fastest among the population?

(c) Solve the equation in terms of γ, then use the data provided above to determine γ.

(d) If no countermeasures are taken, how long will it be until 50% of the campus has been infected?

35. On p. 785 (see #24) we said that when fluid creeps up a pipette, or capillary, its height varies according to an equation of the form $y' = (F - Ay)/(\gamma y)$. Solve this equation when $F, A,$ and γ are all 1. | Fluid Mechanics |

36. On p. 785 (see #24) we said that when fluid creeps up a pipette, or capillary, its height varies over time according to an equation of the form $y' = (F - Ay)/(\gamma y)$ where F, A and γ are all positive. Solve this equation in general (i.e., leave the coefficients as $F, A,$ and γ). | Fluid Mechanics |

37. On p. 786 (see #28) we suggested that the voltage across a neuron's cell wall obeys the equation $v' = -v(v - \alpha)(v - \beta)$. (a) Determine implicit solutions to this equation corresponding to $v(0) = 0.25$, $v(0) = 0.75$ and $v(0) = 2.1$ when $\alpha = 1$ and $\beta = 2$, and (b) use a graphing utility to plot the curves from part (a) on the same axes. | Biology |

38. On p. 786 (see #28) we suggested that the voltage across a neuron's cell wall obeys the equation $v' = -v(v - \alpha)(v - \beta)$. | Biology |

(a) Determine implicit solutions to this equation in terms of α and β when $v(0) = 1$.

(b) Verify that your implicit solution from part (a) does not allow for $v(0) = \alpha$ or $v(0) = \beta$, and explain why that happens.

39. A parachutist with a mass of 60 kg is falling at a rate of 54 $\frac{m}{sec}$ at the moment his parachute opens. If the air resists the fall with a force that is equal to $\frac{1}{4}v^2$ N, (a) find the velocity as a function of time, and (b) determine the terminal velocity of the parachutist. | Recreation |

40. When $y = f(x)$ is a chain or rope hanging (at rest) under its own weight, an analysis of the tension along it leads directly to the equation $g\rho\sqrt{1 + (f')^2} = cf''$, where g is the acceleration per unit mass due to gravity, ρ is the mass density of the line, and c is the horizontal component of the force due to tension (which is the same at all values of x). | Engineering |

(a) Treat this equation as a first-order separable equation describing f', and find a formula for $f'(x)$ in terms of ρ, g and c.

(b) Integrate your formula from part (a) to arrive at a formulation of $f(x)$.

(c) Set $g = 9.81$, choose positive values of ρ and c, and use a graphing utility to render a plot of $y = f(x)$. (Provided that your integrations in parts (b) and (c) were done correctly, this so-called *catenary* will look like a hanging cable.)

41. Stefan's Law of Radiation tells us that the rate of change in an object's temperature (kelvins) with respect to time (seconds) is described by | Thermodynamics |

$$\frac{dT}{dt} = \gamma(T^4 - A^4), \qquad (4.7)$$

where T is the temperature of the object, A is the ambient temperature (also in kelvins), and γ is a constant that depends on the material properties of the object in question.

  (a) By factoring the right-hand side of (4.7), show that the rate of change in T with respect to time is roughly proportional to $T - A$ when T is near A (in this sense, Stefan's Law acts as a refinement of Newton's).

  (b) Solve equation (4.7) to find relationship between temperature and time.

  (c) Suppose the temperature of molten metal is 1800 K when it is first poured into a vat that is sitting in a room that measures 295 K. If the metal has cooled to 1750 one hour later, use Stefan's Law to determine how long it will take for the metal to cool to 400 K.

42. Monod kinetics is a description of reactions that are catalyzed by enzymes. One of the important equations describes the concentration of substrate (i.e., nutrients) by $s' = -\frac{\nu s}{\kappa + s}$, where ν and κ are positive constants. (The number ν is the maximum rate of consumption, and because $s' = \nu/2$ when $s = \kappa$, the number $\kappa > 0$ is called the 1/2-saturation coefficient.) Solve this differential equation for s.

| Biochemistry |

Exercises #43–46 address the depth of fluid in a leaking tank, based on the discussion of flow rates (Torricelli's Law) on p. 776.

| Engineering |

43. Suppose the base of a right circular cylinder has radius of 1.5 m, and the cylinder is 7 m tall. It contains liquid that's 6 meters deep when a hole of area 0.001 m^2 appears in the base. Determine the height of liquid in the tank as a function of time.

44. Suppose a tank is in the shape of an inverted cone (i.e., point down, like an ice cream cone). Its circular top has radius of 1 m, and the tank is 5 m tall. It contains fluid that's 4 meters deep when a hole of area 0.01 m^2 appears in the tip. Determine the height of fluid in the tank as an implicit function of time.

45. Suppose a tank is in the shape of an inverted cone (i.e., point down, like an ice cream cone). Its circular top has radius of 1 m, and the tank is 5 m tall. It contains fluid that's 4 meters deep when a hole of area 0.01 m^2 appears in the side, 0.25 meters from the tip. Determine the height of fluid in the tank as an implicit function of time.

46. Suppose that a tank is 4 meters tall, and when seen from the side, its outline is the parabola $y = x^2$. When the liquid in the tank is 2.5 meters deep, a hole of area 0.03 m^2 appears in the side, 0.25 meters from the floor. Determine the height of fluid in the tank as a function of time.

47. On p. 784 we introduced the **Vidale and Wolfe** model of sales response to advertising: $s' = \rho\alpha(1 - s/m) - \lambda s$, where ρ, α and λ are positive constants. Solve this separable differential equation (your answer will involve ρ, α and λ, and the constant of integration that depends on initial sales volume).

| Marketing |

48. On p. 784 we introduced the **Sethi model** of sales response to advertising, which is a separable differential equation of the form $s' = \sqrt{1 - s} - s$. Solve this differential equation (without given initial data, your answer will involve an unspecified constant of integration).

| Marketing |

10.5 Linear Equations

On p. 774 we suggested that the number of open potassium gates in the cell wall is governed by a first-order linear differential equation. Such equations model a wide variety of phenomena, and in this section we'll see them occur in descriptions of freefall, temperature change, and mixing problems. As a reminder, recall:

> A first-order differential equation is said to be **linear** if it can be written as
>
> $$y' + p(t)y = q(t).$$

Our principal tool for solving linear equations is called the *method of integrating factors,* and is discussed below.

❖ Integrating Factors

We're going to develop the standard solution method in the context of **Newton's Law of Cooling**, which asserts that the rate at which an object heats up (or cools down) is proportional to the difference between its own temperature and the temperature of its surroundings (called the *ambient* temperature). This is commonly written as

$$\frac{dT}{dt} = -k(T - A), \tag{5.1}$$

where T is the temperature of the object, A is the ambient temperature, and $k > 0$ is a number called the **time constant** that depends on the particular object and environment. For the sake of discussion, let's rewrite equation (5.1) as

$$T' + kT = kA. \tag{5.2}$$

On the left-hand side we see T' appear in one summand, and T in the other. *That's exactly the kind of thing we see when we use the Product Rule.* With that said, we should be clear that the left-hand side of equation (5.2) is *not* the result of differentiating with the Product Rule because there's not a second function. Said differently, it's lacking a factor, and that leads to an idea: if it's lacking a factor, let's multiply one onto the equation. In fact, that idea pays off.

A hot cup of coffee cools off and a cold can of soda warms up. Newton's Law of Cooling describes the rate at which it happens.

The standard technique for solving equations like (5.2) is to multiply by a second function, called an **integrating factor**, which is chosen so that the left-hand side becomes the derivative of a product. For example, when we multiply a function onto equation (5.2), say $u(t)$, it becomes

$$uT' + kuT = kuA. \tag{5.3}$$

Now keep our goal in mind: the left-hand side of this equation is supposed to be the result of differentiating a product with the Product Rule. Were that the case, it would have the form $uT' + u'T$. Comparing this to the left-hand side of equation (5.3), we conclude that $u(t)$ must be a function for which

$$u'(t) = ku(t).$$

On p. 283 we proved that only exponential functions are described by this equation (when k is a nonzero constant), so the integrating factor must be $u(t) = Ce^{kt}$ for some number C. In fact, any number C will do, so we often choose $C = 1$. With this choice of $u(t)$, equation (5.3) becomes

$$e^{kt}T'(t) + ke^{kt}T(t) = kAe^{kt}$$

$$\frac{d}{dt}\left(e^{kt}T(t)\right) = kAe^{kt}.$$

We finish the solution process by integrating in time and solving for $T(t)$, as demonstrated in the next example.

Practice with integrating factors

Example 5.1. Use an integrating factor to solve the equation $T' = -6(T - 100)$ when $T(0) = 75$.

Solution: As we did above, let's gather all occurrences of T on the left-hand side of this equation. Then we have

$$T' + 6T = 600.$$

When we multiply this equation by $u(t)$ we see

$$uT' + 6u\,T = 600u. \tag{5.4}$$

In order for the left-hand side of this equation to be the result of differentiating with the Product Rule, we need

$$6u(t) = u'(t),$$

Don't just read it. Check it for yourself!

from which we conclude that $u(t) = Ce^{6t}$ for some constant C. For convenience, we choose $C = 1$, and rewrite equation (5.4) as

$$e^{6t}T' + 6e^{6t}T = 600e^{6t}$$

in which the left-hand side is exactly the derivative of a product. Specifically, the equation has become

$$\frac{d}{dt}\left(e^{6t}T\right) = 600e^{6t}.$$

Now integrating with respect to time yields

$$\int \frac{d}{dt}\left(e^{6t}T\right)\,dt \;=\; \int 600e^{6t}\,dt$$
$$e^{6t}T \;=\; 100e^{6t} + C.$$

Don't just read it. Check it for yourself!

Finally, we isolate T by dividing both sides of the equation by e^{6t}, and see

$$T(t) = 100 + Ce^{-6t}.$$

We can determine the number C by using the initial data:

$$\overbrace{75 = T(0)}^{\text{given data}} \underbrace{= 100 + C}_{\text{using our formula for } T} \quad \Rightarrow \quad C = -25 \quad \Rightarrow \quad T(t) = 100 - 25e^{-6t}. \quad \blacksquare$$

Try It Yourself: Practice with integrating factors

Example 5.2. Solve the equation $T' = -5(T + 23)$, with $T(0) = 5$.

Answer: $T(t) = -23 + 28e^{-5t}$. $\blacksquare$

Full Solution On-line

The numbers k and A were constant in Examples 5.1 and 5.2, but the method of integrating factors works even when they vary over time.

Practice with integrating factors (non-constant ambient temperature)

Example 5.3. Suppose a particular building has a time constant of $k = 0.1$. If the outside temperature varies according to $A(t) = 70 + 10\sin\left(\frac{\pi}{12}\, t\right)$, and the building temperature is $70°$ at time $t = 0$, find a formula that describes the temperature of the building.

Solution: Using this $A(t)$, Newton's law of cooling becomes

$$T' = -0.1\left(T - 70 - 10\sin\left(\frac{\pi}{12}\, t\right)\right).$$

When we gather all terms with T on one side of the equation, we see

$$T' + 0.1T = 7 + \sin\left(\frac{\pi}{12}\, t\right),$$

and multiplying this equation by $u(t)$ yields

$$uT' + 0.1uT = 7u + u\,\sin\left(\frac{\pi}{12}\, t\right). \tag{5.5}$$

In order for the left-hand side of this equation to be the result of differentiating with the Product Rule, we need

$$0.1u(t) = u'(t),$$

from which we conclude that $u(t) = Ce^{0.1t}$. Taking $C = 1$, equation (5.5) becomes

$$e^{0.1t}T' + 0.1e^{0.1t}T = 7e^{0.1t} + e^{0.1t}\sin\left(\frac{\pi}{12}\, t\right),$$

in which the left-hand side is exactly the derivative of a product. Specifically, the equation has become

$$\frac{d}{dt}\left(e^{0.1t}T\right) = 7e^{0.1t} + e^{0.1t}\sin\left(\frac{\pi}{12}\, t\right).$$

Now integrating with respect to time gives us

$$\int \frac{d}{dt}\left(e^{0.1t}T\right)\,dt = \int 7e^{0.1t} + e^{0.1t}\sin\left(\frac{\pi}{12}\, t\right)\,dt$$

$$e^{0.1t}T = 70e^{0.1t} + \frac{144}{1.44 + \pi^2}\left(0.1\sin\left(\frac{\pi}{12}t\right) - \frac{\pi}{12}\cos\left(\frac{\pi}{12}t\right)\right)e^{0.1t} + C.$$

After dividing this by $e^{0.1t}$ in order to isolate $T(t)$, we see

$$T(t) = 70 + \frac{144}{1.44 + \pi^2}\left(0.1\sin\left(\frac{\pi}{12}t\right) - \frac{\pi}{12}\cos\left(\frac{\pi}{12}t\right)\right) + Ce^{-0.1t}.$$

Lastly, the number C is determined using the initial data:

$$70 = T(0) = 70 - \left(\frac{144}{1.44 + \pi^2}\right)\frac{\pi}{12} + C \quad \Rightarrow \quad C = \frac{12\pi}{1.44 + \pi^2}.$$

The graphs of $T(t)$ and $A(t)$ are show in Figure 5.1.   ∎

Don't just read it. Check it for yourself!

Integrating the product of

requires an integration by parts twice (it's cyclic).

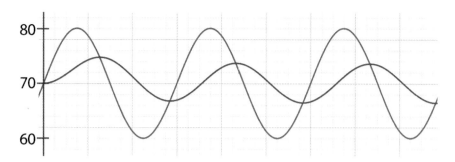

Figure 5.1: Graphs of the building temperature and the ambient temperature from Example 5.3.

Try It Yourself: Practice with integrating factors

Full Solution On-line

Example 5.4. Solve the equation $T' + 7T = e^{-2t}$, when $T(0) = 1$.

Answer: $T(t) = 0.2e^{-2t} + 0.8e^{-7t}$. ∎

We've been working with equations of the form $T' + kT = \ldots$ in which k is constant. The following table summarizes our work, and highlights the relationship between the exponent of the integrating factor and the number k.

Example	$T' + kT$	$u(t)$
5.1	$T' + 6T$	e^{6t}
5.2	$T' + 5T$	e^{5t}
5.3	$T' + 0.1T$	$e^{0.1t}$
5.4	$T' + 7T$	e^{7t}

Notice that $6t$ is an antiderivative of 6, that $5t$ is an antiderivative of 5, and so on …

There are situations in which k is not constant (e.g., mixing problems that you'll see on p. 810), and integrating factors work even then, but finding them is slightly more involved.

> **Integrating factors:** The function $e^{K(t)}$ is an integrating factor for
>
> $$y'(t) + k(t)y(t) = q(t)$$
>
> when $K(t)$ is an antiderivative of $k(t)$.

It's important that the coefficient of y' is 1. If it's not, begin by dividing.

Note that when k is constant, $K(t) = kt$ is an antiderivative. So each of the integrating factors in Examples 5.1–5.4 have the form shown here.

Proof. An integrating factor is a function, $u(t)$, that we design to do the following job: when we multiply it onto the equation $y' + ky = q$, the left-hand side becomes the derivative of the product $u(t)y(t)$. That is,

$$uy' + ku\,y \quad = \quad qu$$

is the same as

$$uy' + u'y \quad = \quad qu.$$

This means that $u' = ku$. We know that only exponential functions are described by this equation when k is constant, which makes it easy for us to determine $u(t)$, but when k is a function of time we have more work to do. In general, the key to finding $u(t)$ is recognizing that $u' = ku$ is a separable equation:

$$u'(t) = k(t)u(t) \quad \Rightarrow \quad \frac{1}{u}u' = k(t).$$

Since $du = u'\, dt$, integrating with respect to time yields

$$\int \frac{1}{u} u'\, dt = \int k(t)\, dt$$

$$\int \frac{1}{u}\, du = \int k(t)\, dt$$

$$\ln|u| = K(t)$$

where K is any antiderivative of k (recall that all antiderivatives differ by an additive constant). Now exponentiating gives us $|u| = e^{K(t)}$, so $u(t) = \pm e^{K(t)}$. Since we're going to multiply it onto both sides of an equation, the sign of $u(t)$ is irrelevant, so we typically choose $u(t) = e^{K(t)}$.   ■

The argument above not only proves the theorem but also provides a way of determining the integrating factor: use the "separate and integrate" technique to solve $u' = ku$. With that said, the whole point of the theorem is that $u(t)$ always has the same basic form, so in practice we don't need to solve $u' = ku$ every time. Rather, we can just find an antiderivative of k, and use it to define $u(t) = e^{K(t)}$. As you'll see in Example 5.5, the formula for $u(t)$ occasionally simplifies so that you don't see an exponential function in the end.

Practice with integrating factors

Example 5.5. Solve the equation $y' + \frac{1}{1+2t} y = 7$ for $t \geq 0$, when $y(0) = 19$.

Solution: In this case we have $k(t) = \frac{1}{1+2t}$, and

$$\int k(t)\, dt = \int \frac{1}{1+2t}\, dt = \frac{1}{2}\ln|1+2t| + C = \ln|1+2t|^{1/2} + C.$$

Since we just need *an* antiderivative of $k(t)$, we'll choose $C = 0$ as a matter of convenience, and write our integrating factor as

$$e^{\ln|1+2t|^{1/2}} = |1+2t|^{1/2}.$$

And because $t \geq 0$ we know that $1 + 2t$ is positive, so we can drop the absolute value and designate $u(t) = (1+2t)^{1/2}$ as the integrating factor. When we multiply it onto the differential equation, we see

$$(1+2t)^{1/2}\, y' + (1+2t)^{-1/2} y = 7(1+2t)^{1/2},$$

in which the left-hand side is exactly the derivative of a product. Specifically, the equation has become

$$\frac{d}{dt}\left((1+2t)^{1/2}\, y(t)\right) = 7(1+2t)^{1/2}.$$

Now integrating with respect to time, we get

$$\int \frac{d}{dt}\left((1+2t)^{1/2}\, y(t)\right) dt = \int 7(1+2t)^{1/2}\, dt$$

$$(1+2t)^{1/2}\, y(t) = \frac{7}{3}(1+2t)^{3/2} + \underbrace{B}.\qquad (5.6)$$

we're using B to denote the constant of integration because we've already used the letter C in this problem

Don't just read it. Check it for yourself!

At time $t = 0$, when $y = 19$, this equation tells us that

$$(19) = \left(\frac{7}{3}\right) + B \quad \Rightarrow \quad B = \frac{50}{3}.$$

Substituting this into equation (5.6) and then dividing the equation by $(1 + 2t)^{1/2}$ in order to isolate $y(t)$, we see

$$y(t) = \frac{7}{3}(1 + 2t) + \frac{50}{3\sqrt{1 + 2t}}. \qquad \blacksquare$$

Try It Yourself: Practice with integrating factors

Example 5.6. Solve the equation $y'(t) + 6ty(t) = 5t$, when $y(0) = 0.5$.

Answer: $y(t) = \frac{5}{6} - \frac{1}{3}e^{-3t^2}$. $\qquad \blacksquare$

Full Solution
On-line

✦ Linear Drag

The term **linear drag** (also called **Newtonian drag**) is used to mean that fluid resistance is proportional to velocity. In such cases, Newton's Second Law becomes a first-order linear differential equation for velocity, $v(t)$.

Newtonian Drag

Example 5.7. A parachutist whose mass is 100 kg drops from a plane that's 3000 m above the ground. Assume the force due to air resistance is proportional to the velocity of the parachutist when her chute is open, with the proportionality constant $b = 200$ N-sec/m. If the parachutist opens the chute immediately after leaving the plane, what is her velocity after 5 seconds?

Solution: Newton's Second Law of motion tells us that net force is related to momentum according to $F_{\text{net}} = \frac{dp}{dt}$, where $p = mv$ is the parachutist's momentum. Since mass is constant at $m = 100$ kg in this case, this reduces to $F_{\text{net}} = mv'$.

To determine the net force acting on the parachutist, we consider two principal forces: (1) gravity, and (2) drag. The force due to gravity is

$$F_g = \left(-9.8\frac{\text{m}}{\text{sec}^2}\right)(100 \text{ kg}) = -980 \text{ N}.$$

> F_g is negative because it pulls *down.*

The drag due to the parachute is

$$F_{\text{drag}} = -200v \text{ N}.$$

> F_{drag} is written with a negative sign so that velocity and F_{drag} have opposite signs. This way, when an object falls the drag force pushes upward.

Therefore, the net force acting on the parachutist is

$$F_{\text{net}} = F_g + F_{\text{drag}} = -980 - 200v \text{ N}.$$

With this information in hand we can write Newton's Second Law as

$$-980 - 200v = 100v'.$$

After gathering all the velocity terms on one side, we see

$$100v' + 200v = -980.$$

In preparation for the development of an integrating factor, let's divide this equation by 100 so that the coefficient on v' is 1. Then we have

$$v' + 2v = -9.8$$

When we multiply the integrating factor e^{2t} onto this equation, it becomes

$$\frac{d}{dt}\left(e^{2t}v\right) = -9.8e^{2t}$$

$$\int \frac{d}{dt}\left(e^{2t}v\right)\,dt = \int -9.8e^{2t}\,dt$$

$$e^{2t}v = -\frac{9.8}{2}e^{2t} + C.$$

Dividing by the exponential term allows us to isolate the velocity function, $v(t) = -4.9 + Ce^{-2t}$. And since the vertical velocity of the parachutist was initially zero,

$$0 = v(0) = -4.9 + C \quad \Rightarrow \quad C = 4.9 \quad \Rightarrow \quad v(t) = -4.9 + 4.9e^{-2t}.$$

Lastly, we answer the question by calculating $v(5) = -4.9 + 4.9e^{-10} \approx -4.8997775 \approx -4.9$ meters per second.   ∎

❖ Mixing Problems

Toward the end of 2003 the United States issued a new \$20 bill that includes several features designed to prevent counterfeiting, such as a thin strip that glows under a black light. Over time, the old bills are removed from circulation, and only the new remain. With a little bit of information, we can predict how long that will take. In order to keep our mathematical description of this situation simple, let's suppose that the banks issue only new bills, and remove old ones from circulation as they're deposited. Further, we'll assume bank customers don't care whether they're depositing old or new currency. Because of the latter assumption, we expect a mix of new and old bills to be deposited (on average); that mix will change over time since the number of new bills in circulation, though small at first, grows.

Let's denote by $p(t)$ the number of new bills in people's pockets after t days. Then p changes according to

$$\frac{dp}{dt} = \begin{pmatrix} \text{new bills that enter cir-} \\ \text{culation because they are} \\ \text{withdrawn from banks} \end{pmatrix} - \begin{pmatrix} \text{new bills that leave} \\ \text{circulation because} \\ \text{they are deposited} \end{pmatrix}.$$

We'll use n_w to denote the average number of \$20 bills that are withdrawn from banks per day (all of which are new), so our equation becomes

$$\frac{dp}{dt} = n_w - \begin{pmatrix} \text{new bills that leave} \\ \text{circulation because} \\ \text{they are deposited} \end{pmatrix}.$$

> We're using the subscript d for "deposit," and w for "withdraw"

Similarly, we'll use n_d to denote the average number of \$20 bills that are deposited in banks per day, *some* of which are new bills. More specifically, if the total number of \$20 bills in circulation is denoted by N, the ratio p/N is the fraction of them that are newly issued bills.

- If one hundred \$20 bills were deposited each day, we'd expect $\left(\frac{p}{N}\right)100$ of them to be new.

- If two hundred \$20 bills were deposited each day, we'd expect $\left(\frac{p}{N}\right)200$ of them to be new bills.

Similarly, since n_d is the average number of \$20 bills deposited per day, we expect $\left(\frac{p}{N}\right)n_d$ of them to be new bills. With this in mind, we can finish our description of the change in currency by writing

$$\frac{dp}{dt} = \underbrace{n_w}_{\substack{\text{the number of new bills enter-} \\ \text{ing circulation each day}}} - \underbrace{\left(\frac{p}{N}\right)n_d}_{\substack{\text{the number of new bills leaving circulation} \\ \text{each day because they are deposited}}}.$$

This is more commonly written as

$$p' = n_w - \left(\frac{n_d}{N}\right) p,$$

which is a first-order linear differential equation describing p. Scenarios like this are typically called *mixing* problems. In the monetary example above, new currency was "mixed" with old, so that some fraction of what exited circulation was the new currency. The example of potassium gates on p. 774 is another example, in that gates enter and exit the *open* state. Whether we're talking about transitions between old and new systems, ventilation of buildings, blood transfusions or a host of other scenarios, the basic setup will always be the same: "stuff" enters and exits, and because of mixing, some fraction of what exits is the "stuff" you care about.

Our first example will address the simple scenario of salt water in a tank. Before proceeding, we note that the term **well mixed** is used to mean that the concentration of salt is assumed to be same throughout. This allows us to use the overall concentration when calculating the quantity of salt that's carried out of the tank by each unit of fluid.

E.g., to determine the number of new bills that were deposited each day, we calculated their average concentration in the "tank" of all $20 bills in circulation.

A mixing problem with constant volume

Example 5.8. Suppose a tank has 400 L of water in which 30 kg of salt are suspended. New fluid enters the tank at a rate of 6 L/min, and fluid is removed at the same rate through an overflow valve. If each liter of new fluid carries 1/3 kg of salt, and the tank is well mixed, determine a formula for the mass of salt in the tank after t minutes.

Solution: In order to talk about how the amount of salt in the tank changes over time, we need to know how much enters and how much exits each minute.

Each liter of new fluid carries 1/3 kg of salt, so when 6 L enters the tank (which happens every minute) it brings with it

$$(\text{volume}) \times (\text{concentration}) = (6 \text{ L})\left(\frac{1}{3} \frac{\text{kg}}{\text{L}}\right) = 2 \text{ kg of salt.}$$

Fluid also leaves the tank at 6 L/min in this scenario, but the amount of salt that goes with it depends on how much salt is in the tank at the moment. Let's denote the mass of salt in the tank by $m(t)$. Since the tank is well mixed (meaning that the salt is distributed evenly across the 400 L), the fluid in the tank has a uniform concentration of $\frac{m(t)}{400}$ kilograms per liter. So when 6 L of fluid leaves the tank each minute, it carries with it

$$(\text{volume}) \times (\text{concentration}) = \underbrace{6}_{\text{L}} \underbrace{\left(\frac{m(t)}{400}\right)}_{\text{kg/L}} \text{ kg of salt.}$$

Now we can talk quantitatively about the rate at which $m(t)$ changes:

$$m'(t) = \underbrace{2}_{\substack{\text{the amount of salt} \\ \text{entering per minute}}} - \underbrace{(6)\left(\frac{m(t)}{400}\right)}_{\substack{\text{the amount of salt} \\ \text{exiting per minute}}}.$$

After rewriting this as

$$m' + \frac{3}{200}m = 2,$$

we multiply it by the integrating factor $u(t) = e^{3t/200}$ so that it becomes

$$\frac{d}{dt}\left(e^{3t/200}s\right) = 2e^{3t/200}.$$

Then integrating with respect to time and isolating $m(t)$ yields

$$m(t) = \frac{400}{3} + Ce^{-3t/200}.$$

We finish our work by using initial data to determine the value of C. Since there are initially 30 kg of salt,

$$30 = m(0) = \frac{400}{3} + C \quad\Rightarrow\quad C = -\frac{310}{3} \quad\Rightarrow\quad m(t) = \frac{400}{3} - \frac{310}{3}e^{-3t/200}. \quad\blacksquare$$

Although setting up a mixing problem can be complicated, the long-term effect of mixing "new" with "old" is often intuitive (especially in the case of constant volume), and you should use it to check your work. For example, since each liter of new fluid in Example 5.8 carries 1/3 kg of salt, we expect the 400 L tank to have $(1/3)(400) = 400/3$ kg of salt in the long run. So we check our formula for $m(t)$ by calculating $\lim_{t\to\infty} m(t) = 400/3$ kg.

> If the limit value were anything else, we'd conclude that there's a mistake in our work.

Try It Yourself: A mixing problem with constant volume

Example 5.9. Suppose a tank has 200 L of water in which 10 g of salt are suspended. New fluid enters the tank at a rate of 2 L/hr, and fluid is removed at the same rate through an overflow valve. If the new fluid carries 0.75 g/L, and the tank is well mixed, determine how long it will take until the salt in the tank has a concentration of 0.7 g/L.

Answer: $100\ln(14)$ hours ≈ 11 days. $\quad\blacksquare$

Full Solution On-line

A mixing problem with non-constant volume

Example 5.10. A patient undergoes an operation during which she loses blood at an average rate of 0.25 L/hr, and is given a whole blood transfusion at a rate of 0.2 L/hr. If her blood volume is initially 5 L, 38% of which is red blood cells (RBCs), and the whole blood she receives during the operation is 48% RBCs by volume, write the patient's volume of RBCs as a function of time.

Solution: Let's denote by $r(t)$ the patient's volume of RBCs after t hours of the operation. Then r changes at a rate of

$$\frac{dr}{dt} = \begin{pmatrix}\text{the volume of red} \\ \text{blood cells in-} \\ \text{fused per hour}\end{pmatrix} - \begin{pmatrix}\text{the volume of red} \\ \text{blood cells lost to} \\ \text{bleeding per hour}\end{pmatrix}. \qquad (5.7)$$

Since 0.2 L of whole blood enters per hour and 48% of its volume is RBCs, the first term on the right-hand side of this equation is

$$(\text{flow}) \times (\text{concentration}) = \left(0.2\frac{\text{L whole blood}}{\text{hour}}\right)\left(0.48\frac{\text{L of RBC}}{\text{L of whole blood}}\right) = 0.096\frac{\text{L of RBC}}{\text{hr}}$$

The second term on the right-hand side of equation (5.7) is also calculated as $(\text{flow}) \times (\text{concentration})$, but the concentration changes with time. More precisely, when the patient's total blood volume is v, the fraction of whole blood that is RBCs is

$$\begin{array}{ccc}\text{volume of RBCs} & \rightarrow & r \\ \text{volume of whole blood} & \rightarrow & v \end{array},$$

so RBCs are lost at a rate of

$$(\text{flow})\times(\text{concentration}) = \left(0.25\frac{\text{L whole blood}}{\text{hour}}\right)\left(\frac{r}{v}\frac{\text{L of RBC}}{\text{L of whole blood}}\right) = 0.25\left(\frac{r}{v}\right)\frac{\text{L of RBC}}{\text{hr}}$$

due to bleeding. We've written v for blood volume instead of using the number 5 because the patient is losing blood more quickly than it's being infused. Specifically, her blood volume changes at a rate of

$$v'(t) = 0.2\frac{\text{L}}{\text{hr}} - 0.25\frac{\text{L}}{\text{hr}} = -0.05\frac{\text{L}}{\text{hr}} \quad \Rightarrow \quad v(t) = C - 0.05t.$$

And because she begins with 5 L of blood, we know that $C = 5$. Now substituting these terms into equation (5.7), we see

$$r' = 0.096 - \frac{0.25}{5 - 0.05t}r,$$

which we rewrite as

$$r' + \frac{5}{100 - t}r = 0.096.$$

In this case $k(t) = 5/(100 - t)$ so the exponent of the integrating factor will be

$$\int \frac{5}{100 - t}\,dt = -5\ln|100 - t| + C = \ln|100 - t|^{-5} + C.$$

Since we just need *an* antiderivative of $k(t)$ to make an integrating factor, let's choose $C = 0$ as a matter of convenience. Then the integrating factor for our differential equation is

$$u(t) = e^{\ln|100-t|^{-5}} = |100 - t|^{-5}.$$

Since $t \in (0, 100)$, we can drop the absolute values and write $u(t) = (100 - t)^{-5}$. When we multiply this function onto our differential equation we see

$$(100 - t)^{-5}r' + 5(100 - t)^{-6}r = 0.096(100 - t)^{-5},$$

in which the left-hand side is exactly the derivative of a product. Specifically, the equation has become

$$\frac{d}{dt}\left((100 - t)^{-5}\,r(t)\right) = 0.096(100 - t)^{-5}.$$

Now integrating with respect to time gives us

$$\int \frac{d}{dt}\left((100 - t)^{-5}\,r(t)\right) dt = \int 0.096(100 - t)^{-5}\,dt$$
$$(100 - t)^{-5}\,r(t) = 0.024(100 - t)^{-4} + B.$$

At time $t = 0$ we know that the patient's volume of RBCs is $(0.38)(5) = 1.9$ L. Substituting this fact into our equation tells us that $B = -5 \times 10^{-11}$. Now we multiply both sides of our equation by $(100 - t)^5$ in order to isolate r, and find that

$$r(t) = 2.4 - 0.024t - \frac{5}{10^{11}}(100 - t)^5. \qquad \blacksquare$$

> Based on our formula for v, the patient would have $v(100) = 0$ liters of blood at time $t = 100$, by which time the operation isn't going to help her.

Don't just read it. Check it for yourself!

> We're using B as the constant of integration because we've already used C once during our work.

Try It Yourself: A mixing problem with non-constant volume

Example 5.11. Suppose that fluid is removed from the tank at 3 L/hr in Example 5.9. What's the concentration of salt in the tank after a day?

Answer: ≈ 0.2079 g/L. $\qquad \blacksquare$

Full Solution On-line

❖ Common Difficulties

Students tend to focus on the left-hand side of $y' + k(t)y = q(t)$. That's good, since the integrating factor is determined by $k(t)$. But many beginners lose sight of the *whole equation,* and erroneously multiply the integrating factor onto the left-hand side *only!*

Many people have trouble writing down the appropriate differential equation for a mixing problem. Dimensional analysis good way to check for errors. If the left-hand side of your equation has units of kg/sec, the right-hand side must also—if not, your equation is certainly wrong!

You should know

- the terms *time constant, integrating factor,* and *linear (Newtonian) drag;*
- Newton's law of heating/cooling.

You should be able to

- determine the integrating factor for a first-order DE;
- use an integrating factor to solve a first-order DE;
- solve heating/cooling, linear drag, and mixing problems.

❖ 10.5 Skill Exercises

In #1–6 determine the appropriate integrating factor in simplest terms.

1. $y' + 3y = 7$ 3. $5y' + \frac{y}{t} = t^2$ 5. $ty' + y = e^{-3t}$

2. $y' + 9ty = t$ 4. $t^2 y' + y = 1$ 6. $\cot(t)y' + y = \sin(t)$

Solve each of the differential equations in #7–22 using an integrating factor.

7. $y' - 2y = 6$ 11. $y' + 3y = e^{-3t}$ 15. $y' + 2y = \sin(t)$ 19. $(t^2 + 1)y' - y = 4$

8. $y' + 3y = 7$ 12. $y' + 5y = e^{7t}$ 16. $5y' + \frac{y}{t} = t^2$ 20. $ty' + y = e^{-3t}$

9. $y' = 5 - 6y$ 13. $y' + 9ty = t$ 17. $ty' - y = t\ln(t)$ 21. $y'\sin(t) + y = \tan(t)$

10. $4y = y' + 10$ 14. $y' + t^2 y = 6t^2$ 18. $t^2 y' + y = 1$ 22. $y' + t^4 y = (t - 8)e^{-t^3}$

❖ 10.5 Concept and Application Exercises

23. Suppose a thermos of boiling water (212 °F) is placed in a freezer whose temperature is held at 10 °F, and after 10 minutes the water has a temperature of 209 °F. Determine how long it will take for the water in the kettle to reach the freezing point (32 °F).

         | Newtonian Cooling |

24. If a 130°F cup of warm apple cider has a time constant of $k = 0.25$, and it is placed in a room whose temperature is varies according to $A(t) = 65 + 15e^{-0.2t}$, find a formula that describes the temperature of the cider as a function of time.

         | Newtonian Cooling |

25. Suppose the thermos of boiling water from #23 is placed outside on a spring morning. The ambient temperature is changing according to $A(t) = 60 - 20\cos\left(\frac{\pi}{12}t\right)$, where t is measured in hours, and $t = 0$ corresponds to 6 AM. Determine the length of time it takes for the kettle of water to reach 58 °F. *(Note: exercise #23 used minutes, but this exercise is in terms of hours. You need to choose one or the other and write everything in terms of it, including the time constant k.)*

26. On p. 784 we introduced the **Vidale and Wolfe** model of sales response to advertising: $s' = \rho\alpha(1 - s/m) - \lambda s$, where ρ, α and λ are positive constants. Solve this linear differential equation (your answer will involve ρ, α and λ, and the constant of integration that depends on initial sales volume).

27. In #31 on p. 786 we described the blocking mechanism of a neuron by the equation $w' = \varepsilon(v - \gamma w)$, where $\varepsilon, \gamma > 0$ and v is understood to be a particular fixed number.

   (a) Solve this linear equation for $w(t)$.

   (b) Show that ε does not affect $\lim_{t\to\infty} w(t)$, but controls the rate at which the limit value is approached.

   (c) Suppose we solve the DE three times, with $0 < \varepsilon_1 < \varepsilon_2 < \varepsilon_3$ and $w(0) = 0$, and we use the same values of γ and v in all three cases. Which solution will converge to the equilibrium most quickly?

28. A parachutist whose mass is 120 kg drops from a plane that's 2500 m above the ground. Assume the force due to air resistance is $F_{\text{air}} = -250v$, where v is her velocity. If she opens the chute immediately after leaving the plane, how fast is she falling after 6 seconds? What is the parachutist's terminal velocity?

29. A parachutist of mass m is falling under a deployed parachute that has a drag coefficient k. Assume the force due to air resistance is proportional to the velocity of the parachutist, $F_{\text{air}} = -kv$.

   (a) Find a formula for the parachutist's velocity as a function of time.

   (b) What is the parachutist's terminal velocity?

   (c) Suppose you operate a store that caters to parachute enthusiasts, and two people come in wanting to purchase parachutes. One person has a mass of 125 kg, while the other has a mass of only 47 kg. You know that, for a safe landing, their terminal velocity should be no more than $5 \frac{\text{m}}{\text{sec}}$. What will the drag coefficient be on the chutes you sell to each person?

30. A small motorboat with mass 200 kg is towed at a velocity of 5 meters per second across a still lake. The tow rope is cast off just as the motor boat's engine starts. If the engine exerts a force of 8 newtons and the water exerts a drag force that is twice the velocity, (a) determine a formula that describes the boat's velocity as a function of time, and (b) determine the boat's terminal velocity.

31. In #30 you saw that the boat slowed down after the tow rope was cast off. Show that the boat speeds up when $v_0 < f_e/\gamma$, where v_0 is the boat's velocity at the moment the rope is cast off, f_e is the force supplied by the boat's engine, and γ is the drag coefficient from the water (which was 2 in #30).

32. Suppose a tank has 800 L of water in which 20 kg of salt are suspended. New fluid enters the tank at a rate of 10 L/min, and fluid is removed at the same

rate through an overow valve. If the new fluid carries 0.4 kg/L of salt and the tank is well mixed, determine a formula for the amount of salt in the tank after t minutes.

33. A sugar solution contains 40 kg of sugar suspended in 50 L of water. New solution is mixed in at a rate of 16 L/min, and an equal amount is removed at the same rate. If the new fluid carries $2 \frac{\text{kg}}{\text{L}}$ of sugar and the container is well mixed, (a) determine a formula for the amount of sugar in the tank, and (b) determine when the solution will be the consistency of heavy syrup (approximately $4/3 \frac{\text{kg}}{\text{L}}$ of sugar). | Mixing |

34. Suppose a cylindrical tank has a capacity of 10 gallons. We pump water into it at 6 gallons per minute, but after 3 gallons have been pumped in, the tank springs a leak at its base and water begins to drain at a rate of 4.7 gallons per minute. If the pump is left running and water exits the tank at a rate that is proportional to the volume, (a) determine a formula for volume as a function of time, and (b) whether the water overows (if so, when?), drains entirely (if so, when?), or approaches an equilibrium height (if so, what is it?). | Mixing |

35. The carbon dioxide (CO_2) concentration in a certain room is 2%, which is dangerously high. The volume of the room is 204 cubic meters. The outside air tests at 0.0387% CO_2. Assume that the air in the room is well mixed by fans, and that the amount of air in the room is held constant. If outside air is pumped into the room at a constant rate of 25 m^3 per minute, how long will it take to reduce the CO_2 levels to 0.05%? | Mixing |

36. Suppose the available outside air in #35 is pumped in at a constant rate. What's the minimum volumetric flow (m^3/min) that will reduce the concentration to 0.05% within 15 minutes? | Mixing |

Exercises #37–40 address RC circuits, which contain only a resistor, a capacitor and a voltage source. The charge on the capacitor of an RC circuit is often denoted by $q(t)$, measured in coulombs, and is governed by the differential equation

$$Rq' + \frac{1}{C}q = E(t),$$

where $E(t)$ describes the voltage source, the number R describes the strength of the resistor (in ohms), and the number C is a measure of the capacitor's ability to store electrons (measured in farads).

37. Suppose a circuit contains a 4-ohm resistor, a 0.01-farad capacitor and a constant voltage source of 10 volts. If the capacitor initially held 2 coulombs of charge, determine the amount of charge on it after 10 seconds. | RC circuit |

38. Consider a circuit containing a 4 microfarad capacitor, a 450-ohm resistor and a 20-millivolt voltage source. If the initial charge on the circuit is 20 nano-coulombs, determine the charge after 0.01 seconds. | RC circuit |

39. A 5-ohm resistor and a 0.1-farad capacitor are hooked up to a variable voltage source modeled by $E(t) = 5t - t^2$ volts (for $0 \leq t \leq 5$). If there is no charge on the capacitor initially, (a) find the charge on the capacitor as a function of time and (b) graph $q(t)$ and $E(t)$ together over $[0, 5]$. | RC circuit |

40. Over time, a resistor degrades according to $R = R_0 + \alpha e^{-kt}$. If the resistor is placed in a circuit containing a 1-farad capacitor and a voltage source $E(t)$, (a) write down the differential equation describing the charge on the capacitor, and (b) find the integrating factor used to solve this DE. | RC circuit |

Exercises #41–44 address LR circuits, which contain only a resistor (which resists current), an inductor (which resists *changes* in current), and a voltage source. The current in an LR circuit, denoted by $I(t)$ and measured in amperes (amps), is governed by the equation

$$LI' + RI = E(t),$$

where the number R describes the strength of the resistor (in ohms), the function $E(t)$ describes the voltage source, and L quantifies the ability of the inductor to resist changes in the current (measured in henrys).

41. A circuit contains a 4-ohm resistor, a 2-henry inductor and a variable voltage source given by $E(t) = 5 + 2e^{-0.4t}$. If initially there is no current flowing, determine the amount of current 5 seconds after the voltage is switched on.

    LR circuit

42. A circuit contains a 2.0-h inductor and a 30-ohm resistor. Initially, the current is 0.02 amps and the voltage is switched *off* at $t = 0$. Find a formula to describe the current as a function of time

    LR circuit

43. Suppose you are wiring up a circuit with a 6-henry inductor and an alternating voltage source described by $E(t) = 12\sin(120\pi t)$ volts. You accidentally touch two of the live wires, thereby completing the circuit and, in effect, becoming a resistor with a resistance of 17 kilo-ohms. Find a formula that describes the current passing through your body as a function of time.

    LR circuit

44. Solve the general form of the LR circuit equation

    $$LI' + RI = E$$

    where L, R, and I are constant. Then examine your general solution and discuss the effect that different values of L have on the behavior of the current over time.

    LR circuit

45. Recall from p. 778 (#38) that $y' = \gamma y - 3000$ describes the amount of money in an annuity that earns $\gamma\%$ interest per year, compounded continuously.

    Finance

    (a) Since y' is measured in \$/year, interpret (based on its units) the number 3000 on the right-hand side of the equation.

    (b) Since y' is measured in \$/year, what are the units of γ?

    (c) Determine the amount of money in the annuity after 5 years if a sum of \$35000 was deposited initially, and $\gamma = 0.04$.

46. Suppose an annuity earns interest at $\gamma\%$/year, compounded continuously, and pays \$K per year continuously.

    Finance

    (a) If $y(t)$ is the amount of money in the annuity after t years, describe y' in terms of y, K and γ.

    (b) Solve the differential equation from part (a), assuming an initial value of P for the annuity. (*Hint: the equation from #45 does not account for any payout.*)

    (c) Under what conditions will the annuity (i) decrease in value over time, (ii) increase in value over time, (iii) remain constant over time.

47. Suppose you earn 7% annual interest on your retirement savings, which compounds continuously, and starting at age 20 you deposit \$800 continuously per year until age 65, at which point you retire. Then the amount of money in your retirement account after t years is described by $y' = 0.07y + 800$.

    Finance

(a) Ignoring for a moment the "+800" on the right-hand side of the differential equation, explain what $y' = 0.07y$ tells us about the money in the retirement account.

(b) What role is played by the number 800 in this equation (i.e., what aspect of this scenario requires it), and what are the units associated with it?

(c) Determine a formula for the amount of money in your retirement savings as a function of time.

(d) How much money do you have in your retirement savings when you retire?

(e) If you started depositing money at age 30 (instead of age 20) and you invested $1500 continuously per year, when could you retire if you wanted to have the same amount of money determined in part (d)?

(f) You would prefer to retire at age 60, but not start investing until age 30. How much money would you need to deposit yearly if you wanted to acquire the same amount of money as in part (d)?]

48. Suppose that you deposit money into your retirement savings starting at age 20 (see #47), and each year you deposit a little more money than the year before so that the value of the account is described by $y' = 0.07y + (50t + 300)$, where t is measured in years.

> Finance

(a) Determine the units associated with the number 300, and what it says about your contributions to the account.

(b) Determine the units associated with the number 50, and what it says about your contributions to the account.

(c) How much money will you have by the time you retire at age 65?

49. Suppose that at age 20 you begin making deposits (continuously) into the retirement account from #48, so that the value of the account changes according to $y' = 0.07y + (2250 - 50t)$.

> Finance

(a) What part of this equation describes your contributions, and what part comes from the interest earned by the account?

(b) What does the equation say is happening to your contributions over time?

(c) What is the continuous interest rate?

(d) How much money will there be in the account when you retire at age 65?

(e) Compare your results with those obtained from question #48, and determine which is the better investment strategy.

Checking the details

50. When calculating the integrating factor, we have always chosen the constant of integration to be $C = 0$ as a matter of convenience (e.g., see Example 5.6 on p. 809).

(a) Re-solve the equation in Example 5.6, but this time choose $C = 4$. Show that the solution is the same.

(b) Show that e^{5t-7} and e^{5t+1} both work as integrating factors for the equation $y' + 5y = 3$.

(c) Show algebraically that the constant of integration $(+C)$ that arises in the determination of an integrating factor for $y' + p(t)y = q(t)$ does not affect the final answer.

10.6   Introduction to Dynamical Systems

In this section we introduce the topic of dynamical systems by way of two examples. In brief, a continuous dynamical system is a combination of two other topics you already know about: differential equations and parametric curves. A point (x, y) moves around in the plane as time evolves (parametric curves), but instead of having explicit formulas for $x(t)$ and $y(t)$, the motion is governed by differential equations. There are two basic questions to answer in the context of dynamical systems:

1. Are there any equilibria?

2. What happens when you're not at an equilibrium?

As it has throughout this chapter, the word equilibrium means "no change" in the context of dynamical systems. So we can rephrase the first question as, "Are there any points (x, y) at which x' and y' remain 0?"

❖ Damped Linear Oscillator

Suppose an object of mass m is anchored to a wall by a spring, as depicted in Figure 6.1. Based on our experience with springs, we know that the spring will pull when it's extended and push when it's compressed, so the object will move back and forth about the spring's natural length. We can make this qualitative knowledge quantitatively precise by using Newton's Second Law. For the sake of simplicity, we'll address the case when there are two dominant forces:

1. *The force from the spring:* we'll assume the spring is governed by Hooke's Law, meaning the force it exerts is proportional to its extension ($x > 0$) or compression ($x < 0$). Written as an equation, $F_s = -kx$, where $k > 0$ is the *spring constant*, which depends on the material properties of the spring.

2. *Fluid resistance:* we're assuming that this motion is happening in a fluid (e.g., air), and we'll model the fluid resistance as a force that always acts in the direction opposite the object's velocity (so when the object moves to the right, the force pushes left, and vice versa). Written as an equation, $F_d = -\gamma x'$, where $\gamma > 0$ is called the **damping coefficient** and depends on the fluid.

> When used as a verb, the word *damp* means "to take energy out." The word "dampen" *can* mean the same thing, but is more commonly used in the sense of "to make wet."

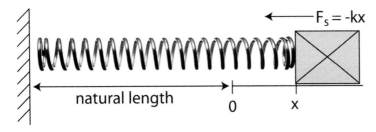

Figure 6.1: The scenario of "a spring anchored to a wall by a spring" acts as a stand-in for any scenario in which there is a restoring force and a force that inhibits motion.

Newton's Second Law says that our object's momentum changes instantaneously in response to the net force it experiences. More specifically, since momentum is (mass)×(velocity) = mx',

$$\underbrace{\frac{d}{dt}\left(mx'\right)}_{\text{instantaneous rate of change in momentum}} = \underbrace{-kx - \gamma x'}_{\text{the sum of all the forces}}$$

or, since the mass is constant,

$$mx'' + \gamma x' + kx = 0. \tag{6.1}$$

Equation (6.1) is called the **linear spring-mass** equation. We can turn this second-order equation into a first-order equation by introducing a "helper variable." Let's define $y = x'$. Then equation (6.1) becomes

$$
\begin{aligned}
x' &= y \\
my' + \gamma y + kx &= 0,
\end{aligned}
$$

which is often rewritten as

$$x' = y \tag{6.2}$$

$$y' = -\frac{k}{m}x - \frac{\gamma}{m}y. \tag{6.3}$$

This pair of equations is an example of a continuous **dynamical system**.

Looking for equilibria

Example 6.1. Determine (a) whether there are any equilibria for the dynamical system described by equations (6.2) and (6.3), and (b) what happens if $(x(0), y(0))$ is not an equilibrium, assuming that γ, m, and k are all positive.

Solution: (a) Only when $x = 0$ and $y = 0$ are both x' and y' zero simultaneously, so the origin is the only equilibrium.

(b) When $(x(t), y(t))$ is in the first quadrant, equations (6.2) and (6.3) show us that $y' < 0$ and $x' > 0$, so the point heads down and to the right; and when $(x(t), y(t))$ is in the fourth quadrant, the equations dictate that $y' < 0$ and $x' < 0$, so the point is traveling down and to the left. The qualitative analyses for quadrants two and three are similar, and the behavior in all four quadrants is depicted by the direction field in Figure 6.2. ∎

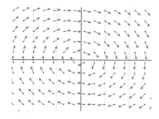

Figure 6.2: An example of a direction field to accompany Example 6.1. The direction of an arrow is determined by the location of the black point at its tail.

We can perform a much more comprehensive analysis of the behavior described by equations (6.2) and (6.3) if we recall what we know about mechanical energy. On p. 531 we established that the potential energy in a linear spring is $0.5kx^2$, and earlier we defined the kinetic energy of an object to be $0.5my^2$, where $y =$ velocity, as it does here. The sum of these is called the mechanical energy, and equations (6.2) and (6.3) allows us to verify that its derivative is

> Actually, we showed that $0.5x^2$ is the *change* in potential energy. This statement is true if we define the potential energy to be zero when $x = 0$.

$$
\begin{aligned}
\frac{d}{dt}\left(\frac{k}{2}x^2 + \frac{m}{2}y^2\right) &= \frac{k}{2}xx' + \frac{m}{2}yy' \\
&= \frac{k}{2}x(y) + \frac{m}{2}y\left(-\frac{k}{m}x - \frac{\gamma}{m}y\right) = -\gamma y^2. \tag{6.4}
\end{aligned}
$$

In Example 6.1 we assumed that $\gamma > 0$, but mathematically there's no reason to require that. So let's address three cases:

Case 1 ($\gamma = 0$): In this case, equation (6.4) says that the derivative of mechanical energy with respect to time is always zero, so it must be constant. That is,

$$\frac{k}{2}x^2 + \frac{m}{2}y^2 = C.$$

So when $(x(0), y(0))$ is not at the origin, the point $(x(t), y(t))$ stays on an ellipse as it moves through the xy-plane (see Figure 6.3). Interpreted physically, $\gamma = 0$ happens in a vacuum, and equation (6.4) says that the mechanical energy is constant in such a scenario.

Case 2 $(\gamma > 0)$**:** Mathematically speaking, equation (6.4) tells us that $0.5kx^2 + 0.5my^2$ is decreasing from moment to moment, provided that $y \neq 0$. So at time zero, the quantity $0.5kx^2 + 0.5my^2$ is some number

$$\frac{k}{2}x^2 + \frac{m}{2}y^2 = C_0, \tag{6.5}$$

> While y can be zero, equation (6.3) guarantees that it won't *stay* zero unless $x = 0$ also (i.e., we're at the equilibrium).

and at time $t = 1$ it's another number

$$\frac{k}{2}x^2 + \frac{m}{2}y^2 = C_1 \tag{6.6}$$

where $C_1 < C_0$. These equations both describe ellipses in the xy-plane, both centered at the origin and with the same aspect ratio, but the ellipse described by equation (6.6) is smaller because $C_1 < C_0$. So in short, as time evolves forward, the point (x, y) moves through the plane around an ever-shrinking ellipse—i.e., it spirals toward the origin, as seen in Figure 6.3.

Case 3 $(\gamma < 0)$**:** As in Case 2, the point (x, y) is traveling through the plane on an ellipse of ever-changing size, but in this case the ellipse is expanding—i.e., the point $(x(t), y(t))$ spirals away from the origin. Physically speaking, equation (6.4) would say that the mechanical energy is increasing—so the fluid resistance is *adding* energy to the motion, which is the stuff of science fiction. ∎

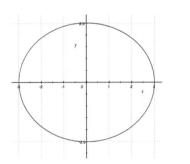 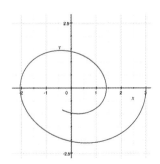

Figure 6.3: Starting at $x(0) = 3$ and $y(0) = 0$, the curve described by equations (6.2) and (6.3) when $m = 1$, $k = 0.7$, and (left) $\gamma = 0$; (right) $\gamma = 0.2$; (both) $0 \leq t \leq 10$.

✥ Competing Species

On p. 775 we developed the logistic population model,

$$y' = ry\left(1 - \frac{y}{K}\right),$$

which describes the growth of a population when resources are limited. The number $r > 0$ is called the intrinsic growth rate, and $K > 0$ is called the carrying capacity. You can think of the factor $(1 - y/K)$ as accounting for the effects of crowding, which changes if there is a second species that uses the same resource (e.g., the same food supply). The members of that second species, say species X, must be

> As the population $y(t)$ approaches K, the number $(1 - y/K) \rightarrow 0$, so population growth slows.

counted against the carrying capacity for species Y, and vice versa. So we write

$$x + y = \text{the total population consuming the resource}$$

$$x' = r_1 x \left(1 - \overbrace{\frac{x+y}{K_1}} \right)$$

$$y' = r_2 y \left(1 - \frac{x+y}{K_2} \right),$$

allowing for $K_1 \neq K_2$ since one species might be more sensitive to crowding than the other, and allowing for $r_1 \neq r_2$ since the species might have different intrinsic growth rates. This dynamical system is called the **competing species model**

> Said in the language of Darwin, one species might be less "fit" than the other.

Like before, equilibria occur where both $x' = 0$ and $y' = 0$ simultaneously. Finding such points can be tricky, so we often break the problem into two simpler tasks: (1) find the points where $x' = 0$ (regardless of y'), and then (b) find the points where $y' = 0$ (regardless of x'). These sets of points typically form lines or curves that we call **nullclines**.

Nullclines of the competing species model

Example 6.2. Determine (a) the nullclines for x in the competing species model, and (b) describe what happens when (x, y) is not at such a point.

Solution: Because $x(t)$ and $y(t)$ are population numbers, we'll restrict our discussion to points in the Cartesian plane at which $x \geq 0$ and $y \geq 0$.

(a) The differential equation that governs x' indicates that $x' = 0$ when

$$r_1 x \left(K_1 - x - y \right) = 0,$$

which happens when $y = K_1 - x$ or when $x = 0$ (i.e., there are no members of species X present). This pair of lines are the nullclines for x.

(b) The point (x, y) is either above or below the line $y = K_1 - x$. These cases are addressed separately below.

Case 1: When (x, y) is a point in the plane that's below the nullcline $y = K_1 - x$ we have $y < K_1 - x$, which is the same as saying that $0 < K_1 - x - y$. Consequently,

$$x' = \underbrace{r_1 x}_{>0} \underbrace{\left(K_1 - x - y \right)}_{>0} > 0.$$

Therefore $x(t)$ is increasing. That is, the point $(x(t), y(t))$ is moving to the right.

Case 2: When (x, y) is a point in the plane that's above the $y = K_1 - x$ nullcline we see $y > K_1 - x$, which is the same as saying that $0 > K_1 - x - y$. So

$$x' = \underbrace{r_1 x}_{>0} \underbrace{\left(K_1 - x - y \right)}_{<0} < 0.$$

Therefore $x(t)$ is decreasing. That is, the point $(x(t), y(t))$ is moving to the left.

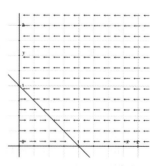

Figure 6.4: The point (x, y) moves toward the nullcline (shown with $r_1 = 1$ and $K_1 = 1$).

The two cases are combined graphically in Figure 6.4, where we see that $(x(t), y(t))$ will always move toward the nullcline. (We've only depicted the horizontal aspect of the movement, but the trajectory isn't actually horizontal.)   ■

Try It Yourself: Nullclines of the competing species model

Example 6.3. Determine (a) nullclines for y in the competing species model, and (b) what happens when (x, y) is not at such a point.

Answer: (a) $y = 0$ and $y = K_2 - x$, (b) see Figure 6.5 .   ■

Full Solution On-line

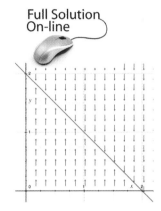

The results of Examples 6.2 and 6.3 are overlaid in the left-hand image of Figure 6.6, called a **motion diagram**. There you see ...

- when (x, y) is below the $x' = 0$ nullcline, it moves up and to the right (i.e., in a "northeasterly" direction);

- when (x, y) is between the nullclines, it moves up and to the left (i.e., in a "northwesterly" direction);

- when (x, y) is above the $y' = 0$ nullcline, it moves down and to the left (i.e., in a "southwesterly" direction).

Figure 6.5: The change in y depends on whether (x, y) is above or below the nullcline.

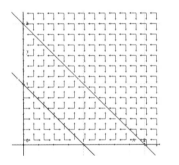 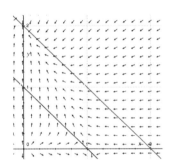

Figure 6.6: (left) An overlay of Figures 6.5 and 6.4; (right) arrows with slope $dy/dx = y'/x'$.

Since the slope of a parametric curve is

$$\frac{dy}{dx} = \frac{y'(t)}{x'(t)},$$

we can refine our graphical analysis of the competing species model by drawing an arrow whose slope is

$$\frac{y'(t)}{x'(t)} = \frac{r_2 y(K_2 - y - x)}{r_1 x(K_1 - x - y)}$$

at each point (x, y), using the bullet points above to decide which end has the arrow (see the right-hand image of Figure 6.6). The resulting direction field shows us clearly that trajectories $(x(t), y(t))$ will move toward the equilibrium at which the less-fit species is extinct, as seen in Figure 6.7.

When trajectories tend toward an equilibrium, we say it is **stable** and refer to it as a **sink** or an **attractor**. When trajectories are pushed away from an equilibrium, we say it is **unstable** and refer to it as a **source** or a **repeller**. If trajectories orbit the equilibrium (as we saw in the spring-mass model when $\gamma = 0$) we say that the equilibrium is a **center**.

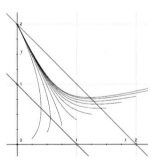

Figure 6.7: When $K_1 = 1$, $r_1 = 1$, $K_2 = 2$, and $r_2 = 1$, trajectories of the competing species model are pushed away from the equilibrium at $(1, 0)$ and toward the equilibrium at $(0, 2)$, meaning that species X becomes extinct.

Note: There is a precise quantitative method for classifying equilibria called *linear stability analysis,* but it's beyond the scope of this text.

Try It Yourself: Classifying equilibria of a dynamical system

Example 6.4. Suppose the parameters of the competing species model are $K_1 = 1$, $r_1 = 1$, $K_2 = 2$ and $r_2 = 1$. (a) Verify that $(1, 0)$ and $(0, 2)$ are equilibria of the system, and (b) classify each as stable or unstable.

Answer: (b) $(1,0)$ is unstable, and $(0, 2)$ is stable. ∎

Full Solution On-line

Now suppose that species X and Y each require several resources, only some of which are consumed by both. In that case, it takes several members of species Y to exert the same population pressure on X as a single member of X. We account for this by writing the competing species model as

$$x' = r_1 x \left(1 - \frac{x + \omega_1 y}{K_1}\right) \quad \text{and} \quad y' = r_2 y \left(1 - \frac{\omega_2 x + y}{K_2}\right),$$

where ω_1 and ω_2 are both numbers in $(0, 1)$. These scale factors reduce the effect of y in the equation governing x', and lessen the impact of x in the equation that governs y'.

Coexistence of competing species

Example 6.5. Suppose $r_1 = 1$, $r_2 = 1$, $K_1 = 5000$, $K_2 = 8000$, $\omega_1 = 0.4$ and $\omega_2 = 0.7$. Show that species X and Y can coexist, and determine the equilibrium population of each species.

Solution: With these parameters, the population numbers of species X and Y change according to

$$x' = x \left(1 - \frac{x + 0.4y}{5000}\right) \quad \text{and} \quad y' = y \left(1 - \frac{0.7x + y}{8000}\right).$$

The nullclines $x = 0$ and $y = 0$ are not of interest to us in our current discussion, so we consider only

$$\underbrace{0 = 1 - \frac{x + 0.4y}{5000}}_{x'=0} \quad \text{and} \quad \underbrace{0 = 1 - \frac{0.7x + y}{8000}}_{y'=0},$$

which we rewrite as

$$y = 12500 - 2.5x \quad \text{and} \quad y = 8000 - 0.7x.$$

These nullclines intersect when $x = 2500$ and $y = 6250$. Since x' and y' are both zero at $(2500, 6250)$, this is a point of equilibrium. Figure 6.8 shows the intersection of the nullclines. Figure 6.9 shows several trajectories that have been started at points $(x(0), y(0))$ near the equilibrium. Based on the motion indicated in the figure, we conclude that $(2500, 6250)$ is a stable equilibrium. ∎

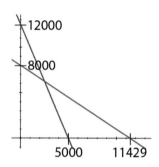

Figure 6.8: The nullclines in Example 6.5 intersect.

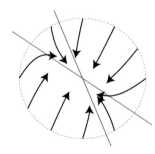

Figure 6.9: Zoomed in on the intersection of nullclines shown in Figure 6.8, with several trajectories that start at points near by.

The competing species model becomes more intricate still if we allow for predation. If species Y tends to eat X, encounters between the populations will be good for Y and bad for X. Since the probability of such an encounter rises with greater x and with greater y, our first approximation of the effect of predation is

$$x' = r_1 x \left(1 - \frac{x + \omega_1 y}{K_1}\right) - \alpha xy \quad \text{and} \quad y' = r_2 y \left(1 - \frac{\omega_2 x + y}{K_2}\right) + \beta xy,$$

where α and β can be different numbers because the direct benefit to the survival of species Y is not equal to the direct impact of predation on the population of X.

Mathematically speaking, these models (with and without predation) are the same. Each of them can be written as

$$x' = Ax - Bxy - Cx^2 \quad \text{and} \quad y' = Dy - Exy + Fy^2.$$

The existence of an equilibrium and how trajectories behave near it depend on the relationship between the coefficients. For example, the system

$$
\begin{aligned}
x' &= 35x - 3xy - x^2 \\
y' &= 14y - 0.7xy - y^2
\end{aligned}
$$

has an equilibrium at $(6.3636, 9.5455)$. In this case, as you can see in Figure 6.10, some trajectories are initially draw toward the equilibrium, while others are pushed away. When this happens we say that the equilibrium is **unstable** and refer to it as a **saddle point** of the dynamical system.

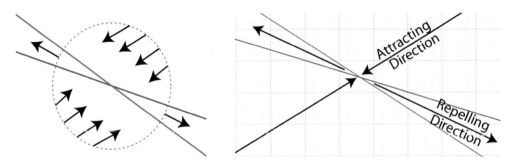

Figure 6.10: (left) Some trajectories initially approach the equilibrium while others are initially pushed away; (right) a saddle point has a corresponding line along which it attracts, and another along which it repels.

You should know

- the terms *dynamical system, equilibrium, nullcline, motion diagram, attractor (sink), repeller (source), center,* and *saddle point.*

You should be able to

- draw the nullclines of a dynamical system;

- draw a motion diagram;

- classify an equilibrium as an attractor, repeller, center or saddle.

❖ 10.6 Skill Exercises

In #1–4, sketch the nullclines and find all (any) equilibria of the dynamical system.

1. $\begin{cases} x' &= y + 4x \\ y' &= 1 - 2y + 5x \end{cases}$

2. $\begin{cases} x' &= 3 - y - 7x \\ y' &= 5 + 6y - x \end{cases}$

3. $\begin{cases} x' &= x^3 - x - y \\ y' &= x^2 - y \end{cases}$

4. $\begin{cases} x' &= y - \sin(x) \\ y' &= 1 - y - x^2 \end{cases}$

5. Determine whether the equilibrium in the left-hand image of Figure 6.11, shown as a red dot, is an attractor, repeller, saddle or center.

6. Determine whether the equilibrium in the right-hand image of Figure 6.11, shown as a red dot, is an attractor, repeller, saddle or center.

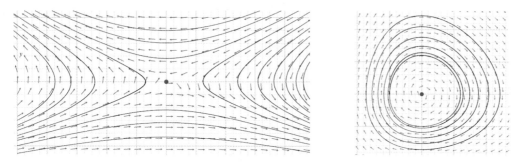

Figure 6.11: (both) There is an equilibrium at the red dot; (left) for #5; (right) for #6.

Instead of working in Cartesian coordinates, we can discuss motion in the plane in polar coordinates. For example, exercises #7–10 refer to the dynamical system

$$r' = \beta r - r^2 \tag{6.7}$$
$$\phi' = 1 \tag{6.8}$$

where β is a fixed number, and r is understood to be non-negative (although we allow ϕ to take values beyond $[0, 2\pi]$).

7. In this exercise we consider the case when $\beta \leq 0$.

   (a) Show that the only equilibrium occurs at $r = 0$.

   (b) Show that when $\beta \leq 0$ and $r(0) > 0$, the trajectory $(r(t), \phi(t))$ spirals toward the origin.

   (c) Does the trajectory in part (b) proceed clockwise or counterclockwise around the origin?

   (d) Determine whether the origin is an attractor, repeller, saddle or center.

8. In this exercise we consider the case when $\beta > 0$.

   (a) Show that the only equilibrium occurs at $r = 0$.

   (b) Show that points on the circle $r = \beta$ stay on the circle.

   (c) Show that points that are not on the circle (whether inside or outside) spiral toward it.

   (d) Determine whether the origin is an attractor, repeller, saddle or center.

9. In this exercise, we take $\beta > 0$ but change equation (6.7) so that it reads $r' = r^2 - \beta r$.

   (a) Show that the only equilibrium occurs at $r = 0$.

   (b) Show that points on the circle $r = \beta$ stay on the circle.

   (c) Show that points that are not on the circle (whether inside or outside) spiral away from it.

   (d) Determine whether the origin is an attractor, repeller, saddle or center.

10. In this exercise, we take $\beta > 0$ but change equation (6.8) so that it reads $\phi' = \beta r - r^2$. (Note: this exercise is independent of #9.)

    (a) Locate all equilibria of this dynamical system.

(b) Choose a value of $\beta > 0$, and sketch the trajectory whose initial position is $(r_0, \phi_0) = (0.9\beta, 0)$.

(c) For the same value of $\beta > 0$, sketch the trajectory whose initial position is $(r_0, \phi_0) = (1.1\beta, 0)$.

(d) Compare and contrast the behavior of trajectories in this example to the behavior exhibited by trajectories in #8.

✤ 10.6 Concept and Application Exercises

11. In our introduction to Newton's Law of Cooling, we assumed that the temperature of the environment did not change due to heat transfer to/from the object. That's a reasonable assumption when a small object is placed in a large room, or its temperature is close to the ambient temperature, but if we put a very hot rock into a small vial of water, the rock will cool and the water will warm. In order to describe the dynamics of this situation, we apply Newton's Law of Cooling to both the rock, and the water. If we denote by w the temperature of the water, and by r the temperature of the rock, Newton's Law of Cooling tells us

$$\begin{aligned} w'(t) &= -\alpha(w - r), \\ r'(t) &= -\beta(r - w), \end{aligned}$$

where α and β are positive constants.

(a) Draw the nullclines of the dynamical system and a motion diagram.

(b) Show that the system is at equilibrium only when $r = w$.

(c) Suppose the initial temperature of the rock and water are r_0 and w_0 respectively, where $r_0 > w_0$. If T_* denotes the equilibrium temperature of the system, use your motion diagram from part (a) to show that $T_* < r_0$ and $T_* > w_0$.

(d) Show that $w(t)$ and $r(t)$ parameterize a line with negative slope in the wr-plane by deriving a formula for dw/dr. (Hint: you have formulas for $w'(t)$ and $r'(t)$.)

(e) Suppose $\alpha = 2$, $\beta = 1/3$, $w_0 = 40$ and $r_0 = 200$. Determine T_*.

(f) Derive a formula for T_* in terms of α, β, r_0, and w_0.

Suppose $I(t)$ is the number of people who are infected by a pathogen at time t and $S(t)$ is the rest of the population, who are susceptible but not infected. The rate at which the pathogen spreads through the population (i.e., the rate at which $I(t)$ changes) depends on how many interactions occur between infected and susceptible people each day. So larger values of I should correspond to larger $I'(t)$ because a larger group of infected people typically means there are more interactions with the susceptible population. Similarly, larger values of S should correspond to larger $I'(t)$ because a larger susceptible population affords infected individuals more opportunities to come in contact with healthy people. Lastly, the pathogen will not spread if either $I = 0$ (nobody's sick) or $S = 0$ (everybody's sick). We can capture these ideas by writing

$$I' = \alpha I S.$$

However, this equation doesn't account for those infected individuals who recover, or those who die. If the fraction of infected people who recover per day is $\rho \in (0, 1)$, and the fraction who die per day is $\mu \in (0, 1)$, we have

$$I'(t) = \alpha I S - \rho I - \mu I \tag{6.9}$$

If the fraction of susceptible people who die per day is $\delta \in (0, 1)$, and $\beta(S + I)$ is the number of babies born into the population per day, the susceptible population is described by

$$S'(t) = -\alpha IS + \rho I - \delta S + \beta(S + I). \qquad (6.10)$$

We've assumed that babies cannot be infected in utero (so they enter S but not I, even if born to an infected parent). Exercises #12–15 focus on this population model.

12. This question focuses on reading equations (6.9) and (6.10) as statements of epidemiological ideas.

   (a) Determine the units on the left-hand side of (6.9) and (6.10).

   (b) Based on your answer to part (a), what are the units associated with β, ρ, δ and μ?

   (c) Do you expect $\delta > \mu$ or $\mu > \delta$, and why?

   (d) Why does ρI enter as a negative quantity in equation (6.9) but a positive quantity in (6.10)?

   (e) Why does $\beta(S + I)$ appears in (6.10) but not (6.9)?

   (f) Based on your answer to part (a), what are the units associated with αIS, and how can we use that to understand the term role of αI?

13. In this exercise we'll assume that $\mu > \beta > \delta$.

   (a) Explain what the inequalities $\beta > \delta$ and $\mu > \delta$ mean in this context.

   (b) Sketch the nullclines of this dynamical system.

   (c) Based on your sketch from part (b) draw a motion diagram for this dynamical system.

   (d) Determine all equilibrium values of S and I when $\alpha = 0.001$, $\beta = 0.016$, $\delta = 0.007$, $\mu = 0.019$, and $\rho = 0.15$.

   (e) Let's define $I_* = \lim_{t \to \infty} I(t)$. Determine a formula for I_* in terms of $\alpha, \beta, \rho, \delta$ and μ, assuming that $\mu > \beta > \delta$.

   (f) Use your formula from part (e) to show that $dI_*/d\alpha < 0$. Then explain what this fact means and why it makes sense (or doesn't, in which case we must have missed an important detail).

14. Suppose that $\beta > \mu > \delta$ in #13.

   (a) Explain what $\beta > \mu$ means in this context.

   (b) Show that $\lim_{t \to \infty} S(t)$ exists when $\beta > \mu > \delta$, $S(0) > 0$ and $I(0) > 0$.

   (c) Show that $\lim_{t \to \infty} I(t)$ does not exist when $\beta > \mu > \delta$, $S(0) > 0$ and $I(0) > 0$, and explain how that can happen.

15. Let's embellish our description of the population in #13 by asserting that it obeys a logistic growth model, so that instead of including $\beta(S + I)$ in equation (6.10) we account for births with $\beta(S + I)(K - (S + I))$, where K is the carrying capacity.

   (a) Determine formulas for the nullclines in terms of $\alpha, \beta, \rho, \delta, \mu$ and K.

   (b) Use a graphing utility to render plots of the nullclines when $\alpha = 0.001$, $\beta = 0.016$, $\delta = 0.007$, $\mu = 0.019$, $\rho = 0.15$, and $K = 1000$.

(c) Use the graph from part (b) to sketch a motion diagram for the dynamical system.

(d) Compare and contrast the behavior of this system to the one described in #13.

16. In a classical **predator-prey** system, the population of predators declines in the absence of prey, and grows when the number of prey is sufficiently large. The dynamics of the prey are just the opposite. Consider the dynamical system

$$\begin{aligned} x'(t) &= -0.01x + 0.0002xy \\ y'(t) &= 4y - 0.2xy. \end{aligned}$$

(a) Based on the discussion above, which of $x(t)$ and $y(t)$ denote the population of predators?

(b) What part of the system says that the population of predators grows in the presence of prey?

(c) What part of the system says that encountering predators is bad for prey?

(d) Sketch the nullclines of this dynamical system and draw a motion diagram.

(e) Locate the equilibrium (x_*, y_*) that's in the first quadrant.

(f) Based on your sketch in part (d), do you think that (x_*, y_*) is an attractor, repeller, saddle or center?

17. Consider the dynamical system

$$\begin{aligned} x'(t) &= \alpha x - \beta xy \\ y'(t) &= \gamma y - \delta xy \end{aligned}$$

where $x(t)$ and $y(t)$ denote the number of animals of species X and Y respectively, and $\alpha, \beta, \gamma, \delta > 0$. More specifically, suppose that $\beta > \delta$ and $\alpha > \gamma$.

(a) Which population grows faster in the absence of the other?

(b) Are encounters between the populations good for either species, and if so, which? How can you tell?

(c) Determine formulas for the nullclines in terms of α, β, γ and δ.

(d) Sketch a motion diagram for this system when $\alpha = 4$, $\beta = 0.2$, $\gamma = 0.01$ and $\delta = 0.0002$.

The **Fitzhugh-Nagumo** equations are often used to understand the dominant characteristics of excitable systems, such as neurons or heart muscle. They are

$$\begin{aligned} v'(t) &= -v(v - \alpha)(v - \beta) - w & (6.11) \\ w'(t) &= \varepsilon(v - \gamma w), & (6.12) \end{aligned}$$

where $\alpha, \beta, \varepsilon$ and γ are positive constants. In equation (6.11) we see that w acts to reduce v', thereby inhibiting the growth of v. This is the element that was missing from our model of the neuron on p. 786 (#28). The strength of the inhibition is controlled by equation (6.12), which was introduced on p. 786 (#31). Exercises #18–20 refer to these equations.

18. Suppose $\alpha = 0.2$, $\beta = 1$, and $\varepsilon = 0.1$.

(a) Find a value of $\gamma > 0$ so that the nullclines intersect only at the origin.

(b) Find a value of $\gamma > 0$ so that the nullclines intersect exactly twice.

(c) Find a value of $\gamma > 0$ so that the nullclines intersect exactly three times.

19. Suppose $\alpha = 0.2$, $\beta = 1$, $\gamma = 4$, and $\varepsilon = 0.1$.

   (a) Starting with $w = 0$ and $v(0) = 0.5$, approximate the solutions of (6.11) and (6.12) for $0 \le t \le 28$ using Euler's method with a step size of $\Delta t = 0.5$. *(Hint: use a spreadsheet! In each step of Euler's method you'll need to update both v and w, but otherwise this is done in the same way as you've done it before.)*

   (b) Use your data from part (a) to graph v and w together as functions of time.

   (c) Repeat parts (a) and (b) for $v(0) = 0.1, 0.2, 0.3, 0.4., \dots, 0.9$.

   (d) Based on your graphs from part (c), which values of $v(0)$ are amplified, and which are immediately attenuated?

20. We can include external conditions in our model (e.g., the influence of a nearby neuron) by appending a so-called *driving* term to equation (6.11). Specifically, let's write $v' = -v(v - \alpha)(v - \beta) - w + d$, where d can depend on time but will be constant $d = 0.1$ for the purposes of this exercise. We'll also fix $\alpha = 0.2$, $\beta = 1$, $\varepsilon = 0.05$, $\gamma = 2$.

   (a) Starting with $w = 0$ and $v(0) = 1.5$, approximate the solutions of (6.11) and (6.12) for $0 \le t \le 250$ using Euler's method with a step size of $\Delta t = 0.5$. *(Hint: use a spreadsheet! In each step of Euler's method you'll need to update both v and w, but otherwise this is done in the same way as you've done it before.)*

   (b) Use your data from part (a) to graph v and w together as functions of time.

   (c) Repeat parts (a) and (b) for $v(0) = 0$ and $v(0) = 0.5$, and describe the qualitative and/or quantitative changes that you see.

   (d) Based on your graphs from part (c), what does it seem that a neuron can do with a constant applied current?

Chapter 10 Review

❖ True or False

1. The function $y(t) = 3t + 4$ solves the equation $y' + 2y = 6t + 10$.

2. The equation $ty' = y^3 + y$ is linear.

3. The equation $ty' = yt + t$ is separable.

4. The equation $ty' = y^3 + y$ is autonomous.

5. All autonomous equations are separable.

6. All linear equations are separable.

7. A direction field comes from an autonomous DE if all the arrows along each horizontal line have the same slope.

8. If $y(t)$ is a solution of $y' = (y-2)^2$ and $y(0) = 1$, the limit $\lim_{t \to \infty} y(t) = \infty$.

9. Suppose $f(y_0) = 0$ and $f'(y_0) > 0$. Then y_0 is a stable equilibrium of $y' = f(y)$.

10. Suppose $y(t)$ solves $y' = 4(3 - y)$ when $y(0) = 18$. Then $y(t)$ is within 10^{-4} of the equilibrium by time $t = 3$.

11. Suppose that $y(t)$ solves the equation $y' = 17(6 - y)$ and $y(8) = 6$. Then $y(9) = 6$ also.

12. Suppose that $y(t)$ solves the equation $y' = 17(6 - y)$ and $y(8) = 6$. Then $y(7) = 6$ also.

13. Suppose f is a continuous function whose only roots are at 1 and 4. If $y(t)$ is a solution of $y' = f(y)$ with $y(0) = 3$, the graph of $y(t)$ is either always increasing or always decreasing.

14. Suppose f is a continuous function whose only roots are at 1 and 4. If $y(t)$ is a solution of $y' = f(y)$ with $y(0) = 7$, the graph of $y(t)$ is either always increasing or always decreasing.

❖ Multiple Choice

15. The equation $y' = (y-2)^2(y-3)$ has ...

    (a) a semi-stable equilibrium at $y = 2$
    (b) a stable equilibrium at $y = 2$
    (c) an semi-stable equilibrium at $y = 3$
    (d) a stable equilibrium at $y = 3$
    (e) none of the above

16. The equation $y' = (y+1)^3(y-4)$ has ...

    (a) a source at $y = -1$
    (b) a node equilibrium at $y = -1$
    (c) an sink equilibrium at $y = 4$
    (d) a node equilibrium at $y = 4$
    (e) none of the above

17. The function $y(t) = e^{-t}$ solves ...

(a) $y' + y = 0$

(b) $y'' - y = 0$

(c) $y'' + 2y' + y = 0$

(d) all of the above

(e) none of the above

18. The function $y(t) = \ln(t)$ solves ...

(a) $e^y y' = 1$

(b) $y' = \frac{1}{2y}$

(c) $y' = \frac{1}{\sqrt{y}}$

(d) all of the above

(e) none of the above

19. The integrating factor for the equation $6y' + 12y = 24$ is ...

(a) e^{24t}

(b) e^{12t}

(c) e^{6t}

(d) e^{4t}

(e) none of the above

20. The integrating factor for the equation $4y' + \frac{2y}{t} = 16t$ is ...

(a) $e^{2 \ln |t|}$

(b) $2|t|$

(c) t^2

(d) $\sqrt{t}$

(e) none of the above

21. The equation $y' + 3y = 4$ is ...

(a) linear

(b) autonomous

(c) separable

(d) all of the above

(e) none of the above

22. The equation $y' + t^2 \sqrt{y} = 2$ is ...

(a) linear

(b) autonomous

(c) separable

(d) all of the above

(e) none of the above

23. Based on the motion diagram associated with the dynamical system $x' = f(x, y)$, $y' = g(x, y)$, which is depicted in Figure 7.1, the equilibrium at point **A** is ...

(a) an attractor

(b) a repeller

(c) a saddle

(d) a center

(e) none of the above

24. Based on the motion diagram associated with the dynamical system $x' = f(x, y)$, $y' = g(x, y)$, which is depicted in Figure 7.1, the equilibrium at point **B** is ...

(a) an attractor

(b) a repeller

(c) a saddle

(d) a center

(e) none of the above

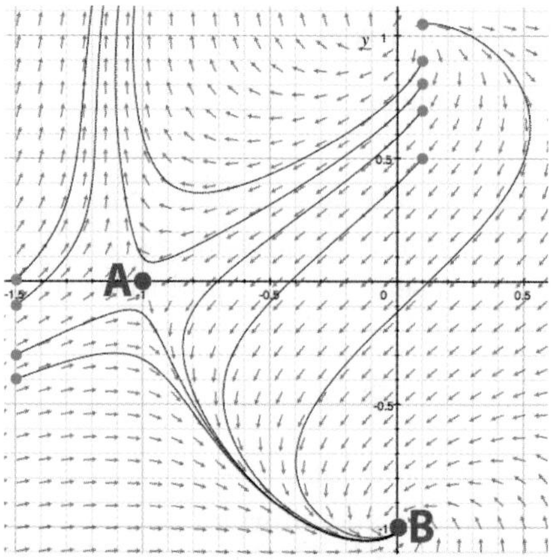

Figure 7.1: A motion diagram for #23 and #24, overlaid with trajectories that start at the green dots, •.

❖ Exercises

25. Locate the equilibria of $p' = -0.8p(10 - p)$ and classify each as a sink, source, or node.

26. Locate the equilibria of $p' = -0.8p^2(1 - p^2)^2$ and classify each as a sink, source, or node.

27. Suppose $y(t)$ solves $y' = 8t - 4y^2$ with $y(1) = 0$. Use Euler's method with a step size of $\Delta t = 0.25$ to estimate $y(2)$.

28. Suppose $y(t)$ solves $y' = 3ty^2$ with $y(0) = 4$. Use Euler's method with a step size of $\Delta t = 0.1$ to estimate $y(1)$.

Solve the differential equations in #29–36 subject to the specified initial data.

29. $y' = ty - 8t$; $y(0) = 10$

30. $y' = ty - 4y + 2t - 8$; $y(0) = 1$

31. $y' + y/(1000 + 3t) = 5$; $y(0) = 2$

32. $y' + 4y/(1 + t^2) = 5/(1 + t^2)$; $y(0) = 7$

33. $y' = 1 + y^2$; $y(0) = 4$

34. $y' = \sec(y)$; $y(0) = 0$

35. $y' = y^2 - 5y$; $y(0) = 8$

36. $ty' = y^3 + y$; $y(1) = 3$

37. Solve the equation $7y' = 4 + y$ by treating it as (a) a linear equation, and (b) a separable equation.

38. Find the orthogonal trajectories to the family of curves $y^2 = 1/(x^2 + k)$.

39. Find the orthogonal trajectories to the family of curves $x^2 - ky^2 = 1$.

40. Suppose an object that's 200 °F is put into a bath of water that's held at 40 °F, and one minute later the object's temperature is 195 °F. If we assume that heat flows out of the object according to Newton's Law of Cooling, how many minutes will pass before the object has reached a temperature of 60 °F? | Newtonian Cooling |

41. Suppose a tank contains 200 m³ of water in which salt is suspended. The concentration of salt is $0.25\frac{\text{kg}}{\text{m}^3}$. A salt solution with $0.1\frac{\text{kg}}{\text{m}^3}$ is pumped into the tank at a rate of $10\frac{\text{m}^3}{\text{sec}}$, and water flows out through an overflow valve at the same rate as it is pumped in. Assuming that the tank is well mixed, determine the mass of salt in the tank as a function of time. | Mixing |

42. A particular room has a volume of 300 ft³ and 14 grams of ozone are initially suspended in the air. Suppose a gas with a concentration of $0.05 \frac{\text{g}}{\text{ft}^3}$ ozone is blown into the room through a duct at a rate of $10 \frac{\text{ft}^3}{\text{min}}$. The air in the room is well mixed by a fan and leaves through an exit vent at the same rate that new air enters. Write down the appropriate mixing equation, and use it to determine a function that describes the amount of ozone in the room after t minutes. | Mixing |

43. In Example 5.10 on p. 812 we argued that the volume of red blood cells in a patient's body, $r(t)$, changed according to $r' = 0.12 - \frac{0.25}{5 - 0.05t}r$ during the course of an operation. What really matters to patient health is *hematocrit*, which is the fraction of blood volume that is red blood cells: $h(t) = r(t)/v(t)$, where v is the patient's blood volume t hours into the operation. Based on the DE for $r(t)$, (a) develop a DE for $h(t)$ and (b) solve it, assuming that $h(0) = 0.35$. | Mixing |

44. Suppose a 0.01-g pellet is shot into a liquid that offers resistance proportional to the object's velocity (see Figure 7.2). If the constant of proportionality is $\gamma = 800 \frac{\text{kg}}{\text{sec}}$, Newton's Second Law says that $(1 \times 10^{-5})v' = -800v$. If the pellet's velocity is $v(0) = 0.03\frac{\text{m}}{\text{sec}}$ when it enters the fluid, how far does it penetrate?

Penetration Distance

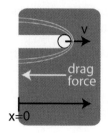

45. Suppose a 0.01-g pellet is shot into a liquid that offers resistance proportional to the object's velocity. If the constant of proportionality is $\gamma\frac{\text{kg}}{\text{sec}}$, Newton's Second Law says that $mv' = -\gamma v$. If the pellet's velocity is $v_0\frac{\text{m}}{\text{sec}}$ when it enters the fluid, determine a formula in terms of m, v_0 and γ that describes how far the pellet will penetrate the fluid.

Figure 7.2: For #44–48.

46. Suppose a 0.01-g pellet is shot into a liquid that offers resistance that's proportional to the *square of the object's velocity* when $v \geq 0.01$ and proportional to the velocity itself when $v \in [0, 0.01)$. If the constant of proportionality is always $\gamma = 700 \frac{\text{kg}}{\text{sec}}$, Newton's Second Law says that $(1 \times 10^{-5})v' = -800v^2$ when $v \geq 0.01$ and $(1 \times 10^{-5})v' = -800v$ when $v \in [0, 0.01)$. If the pellet's velocity is $v(0) = 0.03\frac{\text{m}}{\text{sec}}$ when it enters the fluid, how far does it penetrate?

Penetration Distance

47. Suppose the fluid in #46 always provided resistance proportional to the square of the velocity (i.e., there's no transition to linear drag at low speeds). (a) Show that a pellet with mass m whose velocity is $v_0 > 0$ when it enters the fluid will travel through the fluid forever (i.e., the penetration distance is ∞), and (b) explain why that happens, mathematically speaking.

Penetration Distance

48. In #46 the drag force changed suddenly from quadratic to linear at a particular velocity. Here we'll continue to use the data regarding the pellet's mass and initial velocity, but we'll make a gradual change in the fluid resistance by using the function $f(v) = 2 - e^{-0.05v}$.

Penetration Distance

   (a) Verify that $\lim_{v \to \infty} f(v) = 2$ and $\lim_{v \to 0^+} f(v) = 1$.

   (b) In complete sentences, explain what the equation $mv' = -\gamma v^{f(v)}$ means to us when $v > 0$.

   (c) Use Euler's method with a step size of $\Delta t = 0.2$ to estimate $v(t)$ for $t \in [0, 2]$.

   (d) Use your data from part (c) and the trapezoid rule to estimate $x(2)$.

49. Suppose that a population grows according to the logistic equation, $P' = 0.001P(3 \times 10^6 - P)$, where time is measured in hours. If $P(0) = 1000$, how long does it take the population to reach 95% of the carrying capacity?

Biology

50. Suppose $x' = x^2 + xy + y^2 - 1$ and $y' = x^2 - xy + y^2 - 1$.(a) Use a graphing utility to render a plot of the nullclines, and (b) determine the number of equilibria.

51. Suppose $x' = x^2 + xy + y^2 - 1$ and $y' = x^2 - xy + y^2 - k$.(a) Use a graphing utility to render a plot of the nullclines for several values of k between 1 and 5, and (b) determine the value of k for which this dynamical system has exactly two equilibria.

Chapter 10 Projects and Applications

❖ Crosswind Landing

In this project, we're going to use the term *air speed* to mean the rate at which air flows over an aircraft's wing, and *ground speed* to mean the rate at which its position in the Cartesian plane changes. For example, suppose a wind is blowing north at 10 kilometers per hour:

- If the plane is flying due north at an airspeed of 100 kph, it travels through 100 km of air in an hour. The air itself travels 10 km north in that time, so the plane's ground speed is 110 kph.

- If the plane is flying due south at an airspeed of 100 kph, its ground speed is 90 kph.

More generally, suppose that the plane is at the point (x,y), traveling at an air speed of v kph, and wind is blowing due north at w kph.

- If the plane is traveling northward, $y' = w + v$ and $x' = 0$.
- If the plane is traveling southward, $y' = w - v$ and $x' = 0$.
- If the plane is traveling eastward, $y' = w$ and $x' = v$.
- If the plane is traveling westward, $y' = w$ and $x' = -v$.

We're making the assumption here that the wind moves the plane unless the plane acts to counter it.

Notice that wind affects y' when it blows due north, but not x'. This happens even when the plane is flying in an oblique direction. For example, suppose the plane has an air speed of v kph, and its flight path makes an angle of θ with lines of latitude, as seen in Figure 8.1. Then in still air it travels v kilometers in one hour, and its x-coordinate changes by $\Delta x = -v\cos(\theta)$. Similarly, it travels $v\Delta t$ kilometers in Δt hours, so its x-coordinate changes by $\Delta x = -v\Delta t\cos(\theta)$. That is,

$$\frac{\Delta x}{\Delta t} = -v\cos(\theta) \quad \Rightarrow \quad x'(t) = \lim_{\Delta t\to 0}\frac{\Delta x}{\Delta t} = -v\cos(\theta).$$

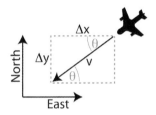

Figure 8.1: An airplane traveling in still air for one hour at an air speed of v kph.

Even if the wind is blowing due north, we get the same formula for x'. However, the formula for y' requires us to account for the wind.

1. Develop a formula for y' when the plane is flying with an air speed of v kph in still air, and its flight path makes an angle of θ with lines of latitude, as seen in Figure 8.1.

2. Adjust your formula for y' by adding the effect of wind that blows due north at w kph.

3. Use the equations for y' and x' to write an expression for $\frac{dy}{dx}$ in terms of v, w, and θ.

Now suppose the a plane departs from a point that's x_0 km due east of the origin, and heads toward the origin with a constant air speed of v kilometers per hour, but it's flying in a wind that blows due north at w kph.

4. Because the plane always points at the origin, the angle θ can be understood as a polar angle (see Figure 8.2), which we saw in Chapter 9. On p. 745 we derived the equation

$$\frac{dy}{dx} = \frac{r'\sin(\theta) + r\cos(\theta)}{r'\cos(\theta) - r\sin(\theta)}$$

where r' denotes $\frac{dr}{d\theta}$. Equate this expression for $\frac{dy}{dx}$ to the one you developed in #3, and solve the resulting separable differential equation for r as a function of θ (your answer will involve the parameters v and w, too).

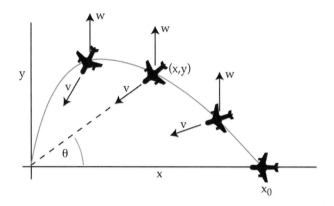

Figure 8.2: The path of the plane.

5. Use a graphing utility to graph $r(\theta)$ for various values of v and w. Explain what happens when $v < w$, and why that makes sense.

❖ Pursuit Curve

Consider two aircraft flying at constant altitude over the Cartesian plane. One starts at the origin and heads due north at $b \frac{\text{km}}{\text{hr}}$. The other is x_0 kilometers to the east, heading due west. This second aircraft is hostile, and fires a missile at the first. The missile travels at a constant speed of $v \frac{\text{km}}{\text{hr}}$ and always points at the target aircraft (see Figure 8.3). In this project, you'll find the path that's followed by the missile, called the **pursuit curve**

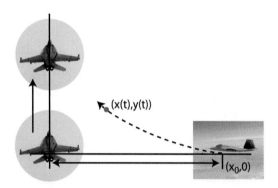

Figure 8.3: The curve along which the missile travels is called a pursuit curve.

After t seconds, the missile is at (x, y) and the target aircraft is at $(0, bt)$. Since the missile always points at its target, the curve it's following has a slope of

$$\frac{dy}{dx} = \frac{y - bt}{x} \tag{8.1}$$

at (x, y). As it stands, the t in this equation prevents us from thinking of (8.1) as a differential equation that relates y to x. However, we've not used all the information we're given. On the one hand, we know the missile has traveled a distance of vt in these t hours. On the other, if we think of y as a function of x, we can calculate this same distance as an arc length. That is,

$$vt = \int_x^{x_0} \sqrt{1 + (y'(X))^2}\, dX \qquad (8.2)$$

The X on the right-hand side of this equation is a dummy variable for the definite integral, and x_0 is constant. The only part that varies is the lower bound of integration, x.

where y' denotes $\frac{dy}{dx}$.

1. Solve equation (8.2) for t and substitute it into equation (8.1).

2. Multiply the resulting equation by x in order to remove the fraction.

While we no longer have t in our equation, we have an integral. We get around this problem by differentiating.

3. Differentiate both sides of your equation with respect to x (using the Fundamental Theorem of Calculus to differentiate the integral).

4. After making the substitution $u = y'$, $u' = y''$, this equation becomes a first-order separable DE. Solve this equation, and verify that

$$\sqrt{1 + u^2} + u = Ax^{b/v}, \qquad (8.3)$$

where $A > 0$ is constant.

5. Solve equation (8.3) for u as a function of x, and then undo your previous substitution by writing the resulting equation as $y' = \cdots$

6. Since the missile was launched horizontally, $y'(x_0) = 0$. Use this fact to determine the value of A in terms of x_0.

7. Now integrate your equation $y' = \cdots$ to find y as a function of x. (You'll have two cases to consider: one when $b = v$, and one when $b \neq v$.)

8. Use the fact that the missile's trjectory started at $(x_0, 0)$ to determine the constant of integration.

9. Set $x_0 = 10$, and use a graphing utility to plot trajectories for several values of b and v.

10. Explain what happens when $b > v$, and why it makes sense.

11. Explain what happens when $b \leq v$, and why it makes sense.

Appendix A

Trigonometry

This appendix assumes that you've already been exposed to trigonometry and know many of the basic facts (e.g., the interior angles of a triangle in the plane sum to $180°$), but you either need a review, are looking for some mnemonics, or want to see the subject from a different point of view.

❖ The Pythagorean Theorem

While we expect that you already know the Pythagorean Theorem, it's extremely important, and this review of trigonometry would be decidedly lacking without it.

> **Pythagorean Theorem:** Suppose the legs of a right triangle have lengths a and b, and the length of the hypotenuse is c. Then
>
> $$a^2 + b^2 = c^2.$$

Proof. Consider taking a pole of length a and joining it to another of length b, making one long pole whose length is $a+b$. In fact, suppose you do that four times, and then lay the poles in the shape of a square, as seen in Figure A.1 (notice that the poles are always oriented in the same direction as you walk around the square).

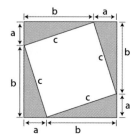

Figure A.1: Making a square of side length $a + b$.

If we connect the points on the sides of this square where the short and long pole are joined, we make a smaller square contained in the first, as seen in Figure A.2. We'll denote by c the side length of that interior square, and now we're ready to prove the theorem. We'll do it by calculating the area of the inner square in two different ways.

- On the one hand, the area of this inner square is just c^2.

- On the other hand, we can calculate its area by subtracting the areas of the shaded triangles from the area of the larger square. The larger square has an area of $(a+b)^2$, and each of the four triangles has an area of $\frac{1}{2}ab$, so the area of the inner square is $(a+b)^2 - 4(\frac{1}{2}ab)$.

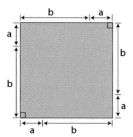

Figure A.2: Making an inner square of side length c.

Since both of these expressions represent the area of the inner square, they must be equal:

$$\begin{aligned} (a+b)^2 - 2ab &= c^2 \\ a^2 + 2ab + b^2 - 2ab &= c^2, \end{aligned}$$

which proves the result. ∎

Calculating the side lengths of a $45°, 45°, 90°$ triangle

Example A.1. Consider the right triangle that's made by cutting a square along its diagonal (so the interior angles of the triangle are $45°, 45°, 90°$). If the diagonal has a length of 1, what are the lengths of the legs of this right triangle?

Solution: Since we cut a square across its diagonal, the legs of this right triangle have the same length. Let's call it x (see Figure A.3). Then the Pythagorean Theorem tells us that

$$1^2 = x^2 + x^2 \quad \Rightarrow \quad 1 = 2x^2 \quad \Rightarrow \quad x = \frac{1}{\sqrt{2}},$$

which is often rewritten as $x = \sqrt{2}/2$. ∎

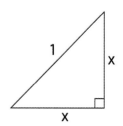

Figure A.3: A right triangle whose legs are the same length.

Calculating the side lengths of a $30°, 60°, 90°$ triangle

Example A.2. Consider the right triangle that's made by dropping an altitude from one vertex of an equilateral triangle (so the interior angles of our triangle are $30°, 60°, 90°$). If the equilateral triangle has a side length of 1, what are the lengths of the legs of this right triangle?

Solution: Since this triangle is equilateral, the altitude bisects the angle at the vertex from which it originates, and also bisects the opposite side (see Figure A.4), resulting in a pair of triangles whose interior angles are $30°, 60°, 90°$. The only unknown length is that of the altitude, but we can calculate it using the Pythagorean Theorem:

$$\left(\frac{1}{2}\right)^2 + y^2 = 1^2 \quad \Rightarrow \quad y^2 = \frac{3}{4} \quad \Rightarrow \quad y = \frac{\sqrt{3}}{2}.$$ ∎

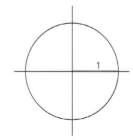

Figure A.4: Dropping an altitude from one vertex of the triangle creates a pair of triangles whose interior angles are $30°, 60°, 90°$.

✧ The Trigonometric Functions

Suppose you draw a circle in the sand, and to provide an orientation for our discussion, you also draw a pair of perpendicular lines through the center—one traveling from north to south, and the other from east to west. We're going to talk about lengths in a moment so we need a unit of some kind, and since the circle cuts the orientation lines so nicely, it seems natural to use those segments somehow. Whether we use the radius or the diameter of the circle is irrelevant in the larger scheme of things, so we'll just choose the radius of the circle as our unit. By virtue of this choice, the radius of the circle is 1.

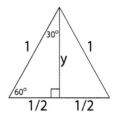

Figure A.5: The unit circle, with north-south and east-west lines providing orientation.

With that said, suppose you draw a line that intersects the circle at exactly one point, somewhere in the northeast quadrant of the circle, as seen in Figure A.6. Such a line is called a *tangent line,* and is perpendicular to the line segment that joins the point of its intersection to the center of the circle. This right angle can be understood as part of a right triangle whose hypotenuse is the east-west axis, as seen in the left-hand image of Figure A.7. The lengths of the legs and hypotenuse of this triangle depend on exactly where the tangent line intersects the circle. We'll keep track of that location using the interior angle of our triangle at the origin, which we'll denote by the Greek letter ϕ (pronounced "fee").

- The length of this triangle's altitude is called the **sine** of ϕ, and is denoted by $\sin(\phi)$.

- One leg of our right triangle has a length of 1.

- The other leg of this right triangle lies along the tangent line. Its length is called the **tangent** of ϕ, and is denoted by $\tan(\phi)$.

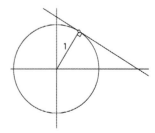

Figure A.6: The radial line connecting the center of the circle to the point of intersection is perpendicular to the tangent line.

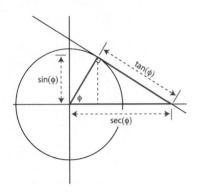
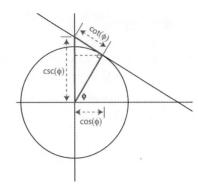

Figure A.7: (left) The lengths of the triangle's legs and hypotenuse depend on the point of intersection, which we quantify using the angle ϕ; (right) analogous lengths on the "co"-triangle.

- The hypotenuse of this right triangle is on a line that cuts through the circle (such a line is called secant line). Its length is called the **secant** of ϕ, and is denoted by $\sec(\phi)$.

- According to the Pythagorean Theorem $\Big(\tan(\phi) \Big)^2 + 1 = \Big(\sec(\phi) \Big)^2$, which is usually written as
$$\tan^2(\phi) + 1 = \sec^2(\phi).$$

You might ask why we made a triangle whose hypotenuse was on the east-west axis, rather than one whose hypotenuse is on the north-south axis instead. Couldn't we measure the corresponding lengths of that other—let us say "co"-triangle? Yes. See Figure A.7.

- The length of this co-triangle's altitude is called the **cosine** of ϕ, and is denoted by $\cos(\phi)$.

- One leg of the co-triangle has a length of 1.

- The other leg of the co-triangle sits on the tangent line. Its length is called the **cotangent** of ϕ, and is denoted by $\cot(\phi)$.

- The hypotenuse of the co-triangle is on a line that cuts through the circle (such a line is called secant line). Its length is called the **cosecant** of ϕ, and is denoted by $\csc(\phi)$.

- According to the Pythagorean Theorem $\Big(\cot(\phi) \Big)^2 + 1 = \Big(\csc(\phi) \Big)^2$, which is usually written as
$$\cot^2(\phi) + 1 = \csc^2(\phi).$$

When we combine the information in the left and right images of Figure A.7, we see that $\cos(\phi)$ and $\sin(\phi)$ are the side lengths of a right triangle whose hypotenuse has a length of 1 (see Figure A.8), so the Pythagorean Theorem tells us that

$$\Big(\cos(\phi) \Big)^2 + \Big(\sin(\phi) \Big)^2 = 1,$$

which is usually written as

$$\cos^2(\phi) + \sin^2(\phi) = 1.$$

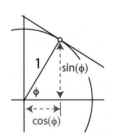

Figure A.8: A right triangle whose side lengths are $\sin(\phi)$ and $\cos(\phi)$.

Lastly, because the ratio of side lengths is the same in similar triangles, we find that

$$\frac{\tan(\phi)}{1} = \frac{\sin(\phi)}{\cos(\phi)} \tag{a.1}$$

$$\frac{\sec(\phi)}{1} = \frac{1}{\cos(\phi)} \tag{a.2}$$

$$\frac{\csc(\phi)}{1} = \frac{1}{\sin(\phi)}. \tag{a.3}$$

> Many people have trouble remembering whether these equations say that secant is the reciprocal of the sine or the cosine. It might help to note that when you read equations (a.2) and (a.3) you hear the phoneme "co" exactly one time.

❖ Radian Measure

Have you ever wondered why there are $360°$ in a circle? We inherited this way of measuring angle from the ancient Babylonians, who used a sexagesimal instead of a decimal number system (i.e., base 60 instead of base 10). We could measure angle in other ways, too, and the one that we use in calculus is called *radian* measure. This way of measuring angle, which we'll discuss in a moment, has the practical benefit of simplifying discussions about the rate at which the sine and cosine change as the angle ϕ varies. (This discussion happens in Chapter 3.)

> Imagine that you're a Babylonian astronomer, and you want to communicate the location of stars. So you divide the arc of the sky into three equal segments: one directly overhead, one to the west, and one to the east. Then you use the number system—i.e., base 60—to refine the segments. You have now divided the semi-circle into $3 \times 60 = 180$ equal parts.

So what are radians? On p. 839 we drew a circle, and used its radius as our unit of length. Figure A.9 shows the portion of the circle's circumference that you'd cross were you to start at the easternmost point and walk 1, 2, or 3 units around it. If you walk 1 unit around the edge of this circle, we say that the angle subtended by your path is $\phi = 1$ **radian**. If you walk 2 units around it, we say that the angle subtended by your path is $\phi = 2$ radians, and an angle of $\phi = 3$ radians means that you walked 3 units around the edge of the circle.

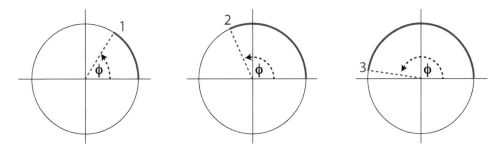

Figure A.9: (left to right) The bold blue arc shows how much of the circumference you would cover by walking 1, 2, or 3 units around the edge of the circle whose radius is 1.

You can see from Figure A.9 that the upper half of the unit circle is a little more than 3 units long. In fact, its length is the irrational number $\pi \approx 3.14159\ldots$ so we say that the upper semicircle subtends an angle of π radians, and there are 2π radians of angle in the whole circle. We can use this fact to convert back and forth between radians and degrees.

Converting degrees to radians

Example A.3. Convert the angles $30°, 60°$ and $90°$ into radians.

Solution: There are $360°$ of angle in a circle, so an angle of $30°$ corresponds to

$$\frac{30°}{360°} = \frac{1}{12} \quad \text{of the the total angle in the circle.}$$

Therefore, that same angle is $\frac{1}{12}(2\pi) = \frac{\pi}{6}$ radians. Similarly, we can determine the radian measure of $60°$ by calculating

$$2\pi\left(\frac{60°}{360°}\right) = 2\pi\left(\frac{1}{6}\right) = \frac{\pi}{3}.$$

And the radian measure of $90°$ is

$$2\pi\left(\frac{90°}{360°}\right) = 2\pi\left(\frac{1}{4}\right) = \frac{\pi}{2}. \qquad \blacksquare$$

Try It Yourself: Converting degrees to radians

Example A.4. Convert the angles (a) $45°$, and (b) $120°$ into radians.

Answer: (a) $\pi/4$, (b) $2\pi/3$. $\qquad \blacksquare$

Full Solution On-line

Converting radians to degrees

Example A.5. Convert the angles (a) $\frac{3\pi}{4}$ and (b) $\frac{7\pi}{6}$ to degree measure.

Solution: (a) Since there are π radians of angle in the upper semicircle, a radian measure of $\frac{3\pi}{4} = \frac{3}{4}\pi$ corresponds to $\frac{3}{4}$ of the angle in a semicircle. Since there are $180°$ degrees in a semicircle, we conclude that

$$\frac{3\pi}{4} \text{ radians is the same angle as } \frac{3}{4}(180°) = 135°.$$

> Most people have to think for a moment before realizing that $135°$ corresponds to 3/4 of a semicircle, but that fact appears explicitly in the radian measure of the angle.

(b) Since $\frac{7\pi}{6} > \pi$, this angle sweeps over the upper semicircle and then some. Specifically, since $\frac{7\pi}{6} = \frac{7}{6}\pi = \pi + \frac{1}{6}\pi$, we know that it sweeps across the upper semicircle and then, starting from the westernmost point of the circle, $\frac{1}{6}$ of the lower semicircle. Since each semicircle corresponds to $180°$,

$$\frac{7\pi}{6} \text{ radians is the same angle as } 180° + \frac{1}{6}(180°) = 210°. \qquad \blacksquare$$

Try It Yourself: Converting radians to degrees

Example A.6. Convert the angles (a) $\frac{5\pi}{4}$, (b) $\frac{11\pi}{6}$ and (c) $\frac{3\pi}{2}$ to degree measure.

Answer: (a) $225°$, (b) $330°$, (c) $270°$. $\qquad \blacksquare$

Full Solution On-line

❖ Trigonometric Functions and the Cartesian Plane

The perpendicular axes and circle that we've drawn can be used to define a Cartesian plane: the center of the circle will be the reference point to which we relate all locations (i.e., the origin), the radius of the circle will be what we call the unit of length (as it has been heretofore), and we'll say that a point has coordinates (x, y) if you can arrive there from the origin by walking x units east and y units north ($x < 0$ means to travel west, and $y < 0$ means to travel south). For example, the easternmost point on the unit circle has coordinates $(1, 0)$; and the northernmost point has coordinates $(0, 1)$.

Now suppose P is some point on the northeast part of the unit circle, and it's separated from $(1, 0)$ by ϕ units of length along the circle. Based on our work so far, the coordinates of P are the cosine and sine, as depicted in Figure A.10:

$$\begin{aligned} x &= \cos(\phi) \\ y &= \sin(\phi). \end{aligned}$$

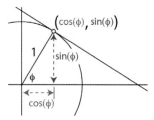

Figure A.10: From Figure A.8 we see that the cosine tells us the x-coordinate of a point on the unit circle, and the sine tells us its y-coordinate.

Values of the trigonometric functions at standard reference angles

Example A.7. Determine the values of sine and cosine at the standard reference angles, $\phi = 0, \frac{\pi}{6}, \frac{\pi}{4}, \frac{\pi}{3}, \frac{\pi}{2}$.

Solution: We discussed the side lengths of some particular right triangles in Examples A.1 and A.2. In the latter, we saw that the legs have lengths $\frac{\sqrt{3}}{2}$ and $\frac{1}{2}$.

- If we lay that triangle with one vertex at the origin and its long side on the x-axis, as shown in the left-hand image of Figure A.11, the hypotenuse reaches to the point $(\frac{\sqrt{3}}{2}, \frac{1}{2})$ and makes a $\pi/6$ angle with the horizontal. So $\cos\left(\frac{\pi}{6}\right) = \frac{\sqrt{3}}{2}$ and $\sin\left(\frac{\pi}{6}\right) = \frac{1}{2}$.

- If we flip the triangle so that its short side is on the x-axis, as shown in the right-hand image of Figure A.11, the hypotenuse reaches to the point $(\frac{1}{2}, \frac{\sqrt{3}}{2})$ and makes a $\pi/3$ angle with the horizontal. So $\cos\left(\frac{\pi}{3}\right) = \frac{1}{2}$ and $\sin\left(\frac{\pi}{3}\right) = \frac{\sqrt{3}}{2}$.

Similarly, if we lay the triangle from Examples A.1 on the plane with one vertex at the origin and one leg on the x-axis (see the middle image of Figure A.11), the hypotenuse reaches to the point $(\frac{\sqrt{2}}{2}, \frac{\sqrt{2}}{2})$ and makes a $\pi/4$ angle with the horizontal. So $\cos\left(\frac{\pi}{4}\right) = \frac{\sqrt{2}}{2}$ and $\sin\left(\frac{\pi}{4}\right) = \frac{\sqrt{2}}{2}$.

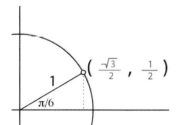 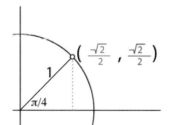 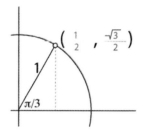

Figure A.11: (left to right) Using familiar triangles to determine the sine and cosine of $\pi/6$, $\pi/4$ and $\pi/3$.

As seen in Figure A.12, an angle of $\phi = 0$ means that we've not moved from the easternmost point at all, whose coordinates are $(1, 0)$. So thinking of the cosine and sine as x- and y-coordinates of that point, respectively, we say that $\cos(0) = 1$ and $\sin(0) = 0$.

And lastly, an angle of $\phi = \frac{\pi}{2}$ corresponds to the northernmost point of the circle, whose coordinates are $(0, 1)$. So thinking of the cosine and sine as x- and y-coordinates of that point, respectively, we say $\cos\left(\frac{\pi}{2}\right) = 0$ and $\sin\left(\frac{\pi}{2}\right) = 1$. ∎

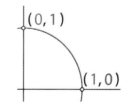

Figure A.12: Angles of 0 and $\pi/2$ correspond to the easternmost and northern-most points of the circle, whose coordinates we already know.

When we combine the information from Example A.7 into a single diagram, we see Figure A.13. We've written 1 as $\frac{\sqrt{4}}{2}$, and 0 as $\frac{\sqrt{0}}{2}$ in that figure because it makes a pattern in the numbers particularly clear. As we move counterclockwise around the circle, increasing ϕ from zero to $\frac{\pi}{2}$ we see that . . .

- the y-coordinate increases. Specifically, the y-coordinate (the sine) at the standard reference angles is $\frac{\sqrt{0}}{2}, \frac{\sqrt{1}}{2}, \frac{\sqrt{2}}{2}, \frac{\sqrt{3}}{2}, \frac{\sqrt{4}}{2}$, in which the number under the radical increments by 1 each time we increase ϕ from one of the standard reference angles to the next.

- the x-coordinate decreases. Specifically, the x-coordinate (the cosine) at the standard reference angles is $\frac{\sqrt{4}}{2}, \frac{\sqrt{3}}{2}, \frac{\sqrt{2}}{2}, \frac{\sqrt{1}}{2}, \frac{\sqrt{0}}{2}$, in which the number under the radical is reduced by 1 each time we increase ϕ from one of the standard reference angles to the next.

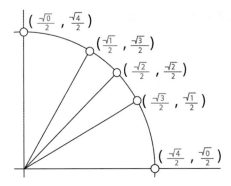

Figure A.13: Combining the information from Example A.7 into a single diagram.

Alternatively, it might help you to remember the values of the sine and cosine if you note that they can only produce the numbers $1, \frac{\sqrt{3}}{2}, \frac{\sqrt{2}}{2}, \frac{1}{2}, 0$ at the standard reference angles. If you think of this list as a numerical teeter-totter whose fulcrum is the number $\frac{\sqrt{2}}{2}$, the numbers $\cos(\phi)$ and $\sin(\phi)$ always "balance."

▷ Negative values of sine and cosine

When we introduced the sine and cosine, we said that they were lengths. While that motivation is true to their geometric origin, at this time we want to extend the idea that $(\cos(\phi), \sin(\phi))$ are the coordinates of a point on the unit circle by addressing points that do not lie in the northeast quadrant of our Cartesian plane.

- When a point on the unit circle is in the left half-plane, its x-coordinate (the cosine) will be negative (see Figure A.14).

- When a point on the unit circle is in the lower half-plane, its y-coordinate (the sine) will be negative (see Figure A.14).

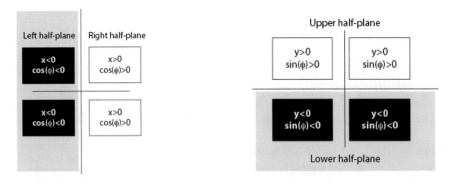

Figure A.14: (left) $\cos(\phi) < 0$ when the angle ϕ identifies a point on the unit circle that's in the left half-plane; (right) $\sin(\phi) < 0$ when the angle ϕ identifies a point on the unit circle that's in the lower half-plane.

Negative values of sine and/or cosine

Example A.8. Determine the sine and cosine of (a) the angle $\phi = \pi$, (b) the angle $\phi = \frac{5\pi}{6}$, and (c) the angle $\phi = \frac{5\pi}{4}$.

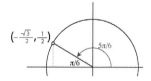

Solution: (a) The radian angle $\phi = \pi$ corresponds to the westernmost point of the circle, whose coordinates are $(-1, 0)$. So we say that $\cos(\pi) = -1$ and $\sin(\pi) = 0$.

Figure A.15: The cosine is negative in part (b) of Example A.8 because the corresponding point on the unit circle is in the left half-plane.

(b) As mentioned in part (a), the westernmost point of the circle is found when $\phi = \pi$, so the point on the unit circle corresponding to $\phi = \frac{5}{6}\pi$ is just shy of that—i.e., it's in the northwest (second) quadrant. We can locate it precisely by laying the long leg of the triangle from Example A.2 (whose interior angles are $\pi/6, \pi/3$ and $\pi/2$) on the negative-x axis, as seen in Figure A.15. Since we know the lengths of that triangle's legs, we know that the point's coordinates are $(-\frac{\sqrt{3}}{2}, \frac{1}{2})$. That is, $\cos\left(\frac{5\pi}{6}\right) = -\frac{\sqrt{3}}{2}$ and $\sin\left(\frac{5\pi}{6}\right) = \frac{1}{2}$.

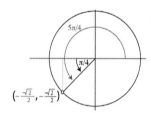

(c) Similarly, since $\frac{5}{4}\pi = \pi + \frac{1}{4}\pi$, we have to traverse more than just the upper semicircle in order to locate the point on the unit circle corresponding to an angle of $\phi = \frac{5\pi}{4}$. It's found in the southwest (third) quadrant. We can locate it precisely by laying a leg edge of the triangle from Example A.1 along the negative-x axis, as seen in Figure A.16. Since we know that both lengths of that triangle have a length of $\frac{\sqrt{2}}{2}$, we know that the point's coordinates are $(-\frac{\sqrt{2}}{2}, -\frac{\sqrt{2}}{2})$. That is, $\cos\left(\frac{5\pi}{4}\right) = -\frac{\sqrt{2}}{2}$ and $\sin\left(\frac{5\pi}{4}\right) = -\frac{\sqrt{2}}{2}$. ∎

Figure A.16: The cosine is negative in part (c) of Example A.8 because the corresponding point on the unit circle is in the left half-plane, and the sine is negative because it lies in the lower half-plane.

Try It Yourself: Negative values of sine and/or cosine

Example A.9. Determine the sine and cosine of (a) the angle $\phi = 2\pi$, (b) the angle $\phi = \frac{4\pi}{3}$ and (c) the angle $\phi = \frac{5\pi}{3}$.

Answer: (a) $(1, 0)$, (b) $\left(-\frac{1}{2}, -\frac{\sqrt{3}}{2}\right)$, (c) $\left(\frac{1}{2}, -\frac{\sqrt{3}}{2}\right)$. ∎

Full Solution On-line

✦ Radian Angles Beyond $[0, 2\pi]$

When a radian angle is negative, it simply means to start at $(1, 0)$ and proceed *clockwise* around the circle, rather than counterclockwise, and radian angles larger than 2π mean simply that you have to complete more that one full circuit of the circle in order to find the corresponding point on the circle. For example, the radian angles of $-\pi/6$, $11\pi/6$ and $23\pi/6$ all locate the same point on the circle. Since the x-coordinate of the point is always the same, regardless of how we arrived at it, we have

$$\cos\left(-\frac{\pi}{6}\right) = \cos\left(\frac{11\pi}{6}\right) = \cos\left(\frac{23\pi}{6}\right).$$

Similarly, since the y-coordinate of the point is always the same, regardless of how we arrived at it, we have

$$\sin\left(-\frac{\pi}{6}\right) = \sin\left(\frac{11\pi}{6}\right) = \sin\left(\frac{23\pi}{6}\right)$$

More generally, since moving from a point to itself by traversing the unit circle any number of times will always leave us with the same x and y coordinates, we see that

$$\sin(\phi) = \sin(\phi + 2\pi m) \quad \text{and} \quad \cos(\phi) = \cos(\phi + 2\pi m)$$

for any integer m. For this reason, we say that the sine and cosine are **periodic** functions and that their period is 2π. This periodicity is explicitly apparent in the graphs of sine and cosine (as functions of ϕ) shown in Figure A.17. Beginners sometimes have difficulty remembering which graph corresponds to which function. You can pin it down by remembering that $\sin(0) = 0$, so its graph will pass through the origin.

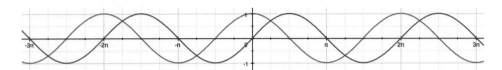

Figure A.17: The periodicity of $\cos(\phi)$ and $\sin(\phi)$ is manifest in their graphs.

❖ Area and Circumference

We said earlier that the upper half of the circle has a length of π units (where "unit" means the length of the radius). Amazingly, the area enclosed by the unit circle is also π. To see why, suppose we slice the circle into many equal sectors—much like cutting a pizza for a large group of friends who are all equally hungry. If we line up those slices, alternating "tip up" and "tip down" (as in Figure A.18) we see an almost rectangular region. As we make the slices thinner, their cumulative area is always the same, but the sides of the "rectangle" become more vertical while its top and bottom become less bumpy—i.e., it becomes more and more like a rectangle. So as we use more and more slices, we're led to the fact that

$$\text{the area of the unit circle } = (1)\pi = \pi.$$

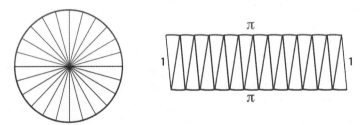

Figure A.18: By reorienting the slices of a circle, we see a rectangular shape, which allows us to relate circumference to area.

❖ When the Radius is Not 1

On p. 839 we began our discussion of the trigonometric functions by imagining that you drew a circle in the sand. We declared the radius of the circle to be our "unit," meaning that all lengths have been expressed in terms of it. Now suppose a friend comes along with a tape measure, and tells you that your radius is actually ...

- 5 feet. Since you know that the circumference is 2π lengths of radius, and each radius is 5 feet, the circumference of the circle is $2\pi \times 5$ feet long.

- 60 inches. Since you know that the circumference is 2π lengths of radius, and each radius is 60 inches, the circumference of the circle is $2\pi \times 60$ inches long.

- 152.4 centimeters. Since you know that the circumference is 2π lengths of radius, and each radius is 152.4 centimeters, the circumference of the circle is $2\pi \times 152.4$ centimeters long.

In short, if the length of the radius is r, the circumference of the circle is $2\pi r$. And if some arc of the circle subtends an angle of ϕ radians, its length is ϕr, which you can think of as

$$\phi r = (\text{number of lengths of radius}) \times (\underbrace{\text{ units per radius}})$$
$$\text{ft, in, cm} \dots$$

or as

$$\phi r = \left(\frac{\phi}{2\pi}\right)(2\pi r) = \begin{pmatrix}\text{fraction of the cir-}\\\text{cle's total angle sub-}\\\text{tended by the arc}\end{pmatrix} \times (\text{circumference}).$$

Changing scale affects our measurement of area, much in the same way that it affected our calculation of circumference. Since each half of the circle has a length of πr, and the radius has a length of r, the technique of slicing the circle into wedges and making a "rectangle" (as seen in Figure A.18) leads directly to the conclusion that the area of the circle is $(\pi r) \times r = \pi r^2$.

Similarly, we've said that a point on the unit circle that's separated from $(1,0)$ by ϕ radians has an x-coordinate of $\cos(\phi)$ and a y-coordinate of $\sin(\phi)$. That is, its horizontal displacement from the origin is $\cos(\phi)$ units, and its vertical displacement is $\sin(\phi)$ units. If 1 "unit" is actually ...

- 5 feet, the horizontal displacement is $5\cos(\phi)$ feet.

- 60 inches, the vertical displacement is $60\sin(\phi)$ inches.

In general, when the radius of the circle is r, the point has coordinates

$$\begin{aligned} x &= r\cos(\phi) \\ y &= r\sin(\phi), \end{aligned}$$

whose consequences we'll study in greater detail when we discuss *polar coordinates* in Chapter 9. For now let's note that, when x and y are nonnegative, the first of these equations allows us to express

$$\cos(\phi) = \frac{x}{r} = \frac{\text{length of leg adjacent to } \phi}{\text{length of hypotenuse}}, \quad \text{or in brief} \quad \cos(\phi) = \frac{\text{adjacent}}{\text{hypotenuse}}.$$

Similarly,

$$\sin(\phi) = \frac{y}{r} = \frac{\text{length of leg opposite from } \phi}{\text{length of hypotenuse}}, \quad \text{or in brief} \quad \sin(\phi) = \frac{\text{opposite}}{\text{hypotenuse}}.$$

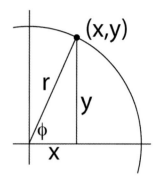

Figure A.19: A point on the circle of radius r.

❖ Dimensional Analysis

The idea that ϕ radians is a *number of lengths* of radius is helpful when using dimensional analysis to check that an equation makes sense. For example, a moment ago we said that the circumference of a circle is

$$\underbrace{C}_{\text{measured in, e.g., centimeters}} = 2\pi \underbrace{r}_{\text{measured in the same unit, e.g., centimeters}}.$$

An equation that says mass = length is nonsensical. Mass and length are entirely different characteristics of an object. Similarly, an equation that says time = color is meaningless. In order for an equation to make sense, both sides must quantify the same *kind* of thing.

In order for both sides of this equation to have units of *length*, the number 2π must not affect the dimensions (i.e., the *kind* of quantity being described) on the right-hand side. More generally, when an arc of the circle subtends an angle of ϕ radians, we said that its length, L is

$$\underbrace{L}_{\text{length of arc}} = \phi \underbrace{r}_{\text{length of radius}}.$$

For a more detailed discussion of what's meant by the words "dimensions" and "units," see Appendix E.

Here you see that ϕ is acting as a constant of proportionality which, in order for this statement to make sense, must have units of "length of arc length *per* length of radius." That is, the dimensions of ϕ are

$$\frac{\text{length of arc}}{\text{length of radius}}.$$

You can see this at work in Figure A.20, which shows several circular arcs that all subtend an angle of $\pi/6$ radians. Each time the radius is incremented by 1, the circular arc gets longer by ϕ units. Because ϕ has units of "(length)/(length)," we say that radians are a **dimensionless** quantity.

For a more detailed discussion of *dimensionless* quantities, see Appendix E.

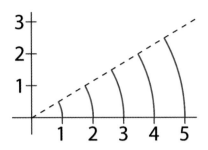

radius of circle	angle subtended	length of arc
1	$\pi/6$	$\pi/6$
2	$\pi/6$	$2(\pi/6)$
3	$\pi/6$	$3(\pi/6)$
4	$\pi/6$	$4(\pi/6)$
5	$\pi/6$	$5(\pi/6)$

Figure A.20: The length of a circular arc that subtends ϕ radians is proportional to the radius of the circle.

❖ Law of Cosines

The Law of Cosines is an extension of the Pythagorean Theorem that allows us to address non-right triangles. For example, an acute triangle with side lengths A, B and C is depicted in the left-hand image of Figure A.21.

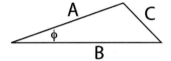 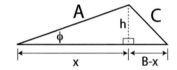

Figure A.21: (left) An acute triangle; (right) introducing "helper" variables x and h.

Since we know a lot about right triangles, let's introduce them into the diagram by drawing an altitude of the triangle, as indicated by the right-hand image of Figure A.21. We'll refer to the length of the altitude as h, and say that it separates the horizontal leg of the triangle into segments of length x and $B - x$. Now using the Pythagorean Theorem on these right triangles, we find

$$h^2 + x^2 = A^2 \quad \text{and} \quad h^2 + (B - x)^2 = C^2.$$

Solving both equations for h^2 leads us to write

$$\begin{aligned} C^2 - (B - x)^2 &= A^2 - x^2 \\ C^2 - (B^2 - 2Bx + x^2) &= A^2 - x^2 \\ C^2 &= A^2 + B^2 - 2Bx. \end{aligned}$$

If we think of our triangle as sitting in the plane, with its left corner at the origin, the topmost point of the altitude has coordinates (x, h) and is A units away from the origin, so $x = A\cos(\phi)$. Therefore,

$$C^2 = A^2 + B^2 - 2AB\cos(\phi), \tag{a.4}$$

which is called the **Law of Cosines**. Note that equation (a.4) reduces to the Pythagorean Theorem when $\phi = \pi/2$ because $\cos(\pi/2) = 0$.

❖ Sum Formulas for Sine and Cosine

Suppose P and Q are points on the unit circle at $(\cos(\phi), \sin(\phi))$ and $(\cos(\theta), \sin(\theta))$ respectively, as shown in the left-hand image of Figure A.22. Then the distance between them is

$$\sqrt{(\cos(\theta) - \cos(\phi))^2 + (\sin(\theta) - \sin(\phi))^2}.$$

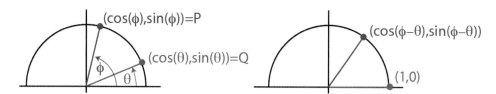

Figure A.22: (left) A pair of points on the unit circle; (right) the same pair of points after rotating the circle.

If we rotate the circle clockwise by θ radians, the point Q arrives at $(1, 0)$, and the point P ends up at $(\cos(\phi - \theta), \sin(\phi - \theta))$, as seen in the right-hand image of Figure A.22. The distance between them is

$$\sqrt{(1 - \cos(\phi - \theta))^2 + (0 - \sin(\phi - \theta))^2}.$$

Since rotating the circle changes the position of our points but not the distance between them, these numbers are the same:

$$\sqrt{(1 - \cos(\phi - \theta))^2 + (0 - \sin(\phi - \theta))^2} = \sqrt{(\cos(\theta) - \cos(\phi))^2 + (\sin(\theta) - \sin(\phi))^2}.$$

In the following steps, we'll solve this equation for $\cos(\phi - \theta)$. We begin by squaring both sides:

$$(1 - \cos(\phi - \theta))^2 + (0 - \sin(\phi - \theta))^2 = (\cos(\theta) - \cos(\phi))^2 + (\sin(\theta) - \sin(\phi))^2. \qquad (a.5)$$

When we expand the left-hand side of this equation, we see

$$1 - 2\cos(\phi - \theta) + \underbrace{\cos^2(\phi - \theta) + \sin^2(\phi - \theta)}_{=1\ \text{(Pythagorean Thereom)}} = 2 - 2\cos(\phi - \theta).$$

The right-hand side of equation (a.5) is

$$
\begin{aligned}
(\cos(\theta) - \cos(\phi))^2 + (\sin(\theta) - \sin(\phi))^2 &= \cos^2(\theta) - 2\cos(\theta)\cos(\phi) + \cos^2(\phi) \\
&\quad + \sin^2(\theta) - 2\sin(\theta)\sin(\phi) + \sin^2(\phi) \\
&= 2 - 2\cos(\phi)\cos(\theta) - 2\sin(\phi)\sin(\theta).
\end{aligned}
$$

This equation also relies on the Pythagorean Theorem.

Using these expansions, we rewrite equation (a.5) as

$$2 - 2\cos(\phi - \theta) = 2 - 2\cos(\phi)\cos(\theta) - 2\sin(\phi)\sin(\theta),$$

after which it's easy to see that

$$\cos(\phi - \theta) = \cos(\phi)\cos(\theta) + \sin(\phi)\sin(\theta), \qquad (a.6)$$

which is called the **difference formula** for the cosine. When $\theta = -\beta$ it becomes

$$\cos(\phi + \beta) = \cos(\phi)\cos(-\beta) + \sin(\phi)\sin(-\beta).$$

Since $\cos(-\beta) = \cos(\beta)$ and $\sin(-\beta) = -\sin(\beta)$, which follow from their interpretations as x and y-coordinates (also see Figure A.17), we can rewrite this as

$$\cos(\phi + \beta) = \cos(\phi)\cos(\beta) - \sin(\phi)\sin(\beta), \tag{a.7}$$

which is called the **sum formula** for the cosine. Similarly, when we have $\theta = -\phi$ in equation (a.6), we see

$$\cos(2\phi) = \cos^2(\phi) - \sin^2(\phi), \tag{a.8}$$

which is called the **double angle** formula for the cosine. When we use the Pythagorean Theorem to write $\sin^2(\phi) = 1 - \cos^2(\phi)$, we can rearrange the equation to read

$$\cos^2(\phi) = \frac{1}{2} + \frac{1}{2}\cos(2\phi),$$

which is sometimes called the **half-angle formula** for the cosine. Similarly, when we use the Pythagorean Theorem to write $\cos^2(\phi) = 1 - \sin^2(\phi)$, equation (a.8) can be rewritten as

$$\sin^2(\phi) = \frac{1}{2} - \frac{1}{2}\cos(2\phi),$$

which is called the **half-angle formula** for the sine. Both of these formulas are particularly helpful in certain techniques of integral calculus.

> Note that adding the half-angle formulas for sine and cosine returns us to the Pythagorean Theorem:
> $$\cos^2(\phi) + \sin^2(\phi) = 1.$$

For the sake of completeness, we note that when $\theta = \beta - \frac{\pi}{2}$, the difference formula for the cosine becomes

$$\cos\left(\phi - \beta + \frac{\pi}{2}\right) = \cos(\phi)\cos\left(\beta - \frac{\pi}{2}\right) + \sin(\phi)\sin\left(\beta - \frac{\pi}{2}\right). \tag{a.9}$$

Since for all angles ω it's true that

$$\cos\left(\omega + \frac{\pi}{2}\right) = -\sin(\omega)$$
$$\cos\left(\omega - \frac{\pi}{2}\right) = \sin(\omega)$$
$$\sin\left(\omega + \frac{\pi}{2}\right) = \cos(\omega),$$

> You can convince yourself of this fact by understanding the sine and cosine as coordinates, or by looking at their graphs in Figure A.17.

equation (a.9) reduces to

$$\sin(\phi - \beta) = \cos(\phi)\sin(\beta) - \sin(\phi)\cos(\beta),$$

which is called the **difference formula** for the sine. When $\beta = -\theta$ this relationship becomes

$$\sin(\phi + \theta) = \cos(\phi)\sin(\theta) + \sin(\phi)\cos(\theta),$$

> We're using the fact that $\sin(-\theta) = -\sin(\theta)$.

which is called the **sum formula** for the sine. When $\theta = \phi$, this becomes the **double angle** formula for the sine:

$$\sin(2\phi) = 2\sin(\phi)\cos(\phi).$$

❖ An Application of Radian Measure to Sighting Theory

In the mid-1940s Bernard Koopman developed the modern theory of sighting based the experiences of fighter pilots searching for ships in the open ocean. His basic premise was simple: you're more likely to see something if it takes up a larger fraction of your field of vision. More specifically, he said that the likelihood of seeing an object is directly proportional the *solid angle* it subtends. The term *solid*

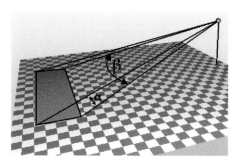 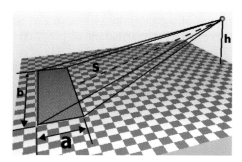

Figure A.23: (both) The point denotes a plane that's searching the water, and the rectangle approximates the object that's in the water.

angle refers to the product of the angles α and β that are shown in the left-hand image of Figure A.23.

By using a little geometry, we can rephrase Koopman's idea in terms of distance, altitude, and area rather than solid angle. Consider the right-hand image of Figure A.23, in which the altitude of a plane is h, the object in the ocean is approximated by an $a \times b$ rectangle, and the distance between them is s. If we were to look at this image from above, we would see Figure A.24. There you see that the width of the object is about equal to the length of a circular arc. When β is measured in radians, that arc has a length of $\beta s b$ so $b \approx \beta s$, from which we conclude that $\beta \approx \frac{b}{s}$.

Deriving a formula for α takes a little more work, but relies on the same ideas. Figure A.25 depicts our scenario as seen from the side, and Figure A.26 zooms in on the object in the water.

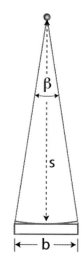

Figure A.24: Looking at the right-hand image of Figure A.23 from above.

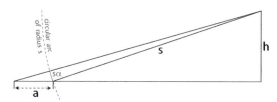

Figure A.25: The altitude of an aircraft is h, its direct distance from the object is s, and the width of the object is a.

The leg of a right triangle has been superimposed on Figure A.26, and you can see that its length is very close to αs (the length of the circular arc) when the width of the object is much smaller than its distance to the pilot. Using the Pythagorean Theorem, we can approximate the length of this right triangle's other leg as $\sqrt{a^2 - (s\alpha)^2}$, so that the large right triangle in Figure A.25 has a hypotenuse whose length is

$$s + \sqrt{a^2 - (s\alpha)^2}.$$

With real objects in real oceans, the number a is much smaller than s, and αs is even smaller than that. So let's approximate

$$s + \sqrt{a^2 - (s\alpha)^2} \approx s + \sqrt{a^2} = s + a \approx s.$$

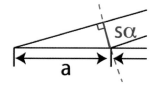

Figure A.26: Zooming in on the lower-left portion of Figure A.25; note that the circular arc nearly forms a right triangle when α is small.

What this approximation loses in accuracy it makes up for in simplicity. Now using the similarity of the large right triangle in Figure A.25 and the small right triangle highlighted by Figure A.26, we have

$$\frac{\alpha s}{a} \approx \frac{h}{s} \quad \Rightarrow \quad \alpha \approx \frac{ha}{s^2}.$$

Now recall our premise: the likelihood of sighting the object in the ocean, which we'll denote by L, is proportional the $\alpha\beta$. If the constant of proportionality is k, we can write this as

$$L = k \; \alpha\beta = kh\frac{ab}{s^3}.$$

Note how this formula, called *Koopman's inverse cube law of sighting*, includes the altitude of the plane, the area of the object that's in the water, and the distance between them.

✤ Exercises

In exercises #1–4 the lengths of a right triangle's legs are given. Use the Pythagorean Theorem to determine the length of the hypotenuse.

1. $a = 3, b = 4$ 2. $a = 5, b = 12$ 3. $a = 2, b = 7$ 4. $a = 3, b = x$

In exercises #5–8 the length of a right triangle's hypotenuse and the length of one of its legs are given. Use the Pythagorean Theorem to determine the length of the remaining leg.

5. $a = 1, c = 4$ 6. $a = 2, c = 15$ 7. $a = 3, c = 8$ 8. $a = 4, c = x$

In exercises #9–12 convert the given angle from degrees to radians.

9. $240°$ 10. $120°$ 11. $300°$ 12. $350°$

In exercises #13–16 convert the given angle from radians to degrees.

13. $4\pi/3$ 14. $8\pi/3$ 15. $3\pi/2$ 16. $11\pi/6$

In exercises #17–24 locate the point on the unit circle whose coordinates are $(\cos(\phi), \sin(\phi))$.

17. $\phi = 2\pi/3$ 19. $\phi = -3\pi/2$ 21. $\phi = 15\pi/4$ 23. $\phi = -34\pi/6$

18. $\phi = 4\pi/3$ 20. $\phi = -7\pi/6$ 22. $\phi = 25\pi/3$ 24. $\phi = -41\pi/2$

Answer #25–28 based on the geometric discussion of the trigonometric functions on p. 839.

25. Explain what happens to the number $\tan(\phi)$ as ϕ increases toward $\pi/2$.

26. Explain what happens to the number $\cot(\phi)$ as ϕ decreases toward 0.

27. Explain what happens to the number $\cos(\phi)$ as ϕ increases from $\phi = -1$ to $\phi = 1$.

28. Explain what happens to the number $\sin(\phi)$ as ϕ increases from $\phi = 1.4\pi$ to $\phi = 1.6\pi$.

Appendix B

Review of Algebraic Techniques

In this appendix we review some algebraic techniques that often come in handy.

❖ Completing the Square

The term **completing the square** means answering questions such as, "What constant would I append to $t^2 + 2t$ in order to make it a perfect square?" In this case we would append "+1" to make a perfect square since $t^2 + 2t + 1 = (t + 1)^2$. Similarly, appending "+9" to the end of the expression $t^2 + 6t$ makes a perfect square since $t^2 + 6t + 9 = (t + 3)^2$, but it's more difficult to figure out what we should append to $t^2 + \sqrt{7}\,t$. So let's look at the general structure of a perfect square:

$$(t + b)^2 = t^2 + 2bt + b^2. \tag{b.1}$$

What we're talking about are cases in which we know the first two terms, $t^2 + 2bt$, but have to figure out the third. Consider the two examples that we've discussed above. Comparing to equation (b.1), we see

$$t^2 + 2t \quad \Rightarrow \quad 2b = 2 \quad \Rightarrow \quad b = 1,$$

so we append "+1^2" to make a perfect square (which is what we did). Similarly,

$$t^2 + 6t \quad \Rightarrow \quad 2b = 6 \quad \Rightarrow \quad b = 3,$$

so we append "+3^2" to make a perfect square (which is what we did).

Completing the square

Example B.1. Determine the constant that would you append to $t^2 + \sqrt{7}\,t$ in order to make it a perfect square.

Solution: In this case we have $2b = \sqrt{7}$, so $b = \frac{\sqrt{7}}{2}$. That means we should append $+\left(\frac{\sqrt{7}}{2}\right)^2$ to the end of the expression in order to make it a perfect square. We can check by expanding

$$\left(t + \frac{\sqrt{7}}{2}\right)^2 = t^2 + 2\left(\frac{\sqrt{7}}{2}\right)t + \left(\frac{\sqrt{7}}{2}\right)^2 = t^2 + \sqrt{7}\,t + \frac{7}{4}. \qquad \blacksquare$$

We often complete the square in order to rewrite expressions in simpler or more compact form, but we want to avoid actually changing them.

Example B.2. Rewrite the expression $t^2 + 14t + 3$ in the form $(t + b)^2 + c$.

Solution: Start by looking at the quadratic and linear terms, as we had above. With $t^2 + 14t$ we have $2b = 14 \Rightarrow b = 7$, so we should append "$+7^2$" to this expression in order to make it a perfect square. Now that we know what we would like to see, we insert it by adding zero ...

$$t^2 + 14t + 3 = t^2 + 14t + 49 - 49 + 3$$
$$= (t^2 + 14t + 49) - 49 + 3 = (t + 7)^2 - 46.$$

So $b = 7$ and $c = -46$.   ∎

Now consider the perfect square in which you see a difference, rather than a sum:

$$(t - b)^2 = t^2 - 2bt + b^2. \tag{b.2}$$

Comparing this to equation (b.1), see that a positive linear term in the expression indicates a *sum* in the perfect square, and a negative linear term indicates a difference.

Example B.3. Rewrite the expression $t^2 - 3t + 10$ in the form $(t - b)^2 + c$.

Solution: As before, we begin by focusing on the quadratic and linear terms. With $t^2 - 3t$ we have $2b = 3 \Rightarrow b = \frac{3}{2}$, so we should append "$+ \left(\frac{3}{2}\right)^2$" to this expression in order to make a perfect square. So let's write

$$t^2 - 3t + 10 = t^2 - 3t + \frac{9}{4} - \frac{9}{4} + 10 = \left(t - \frac{3}{2}\right)^2 + \frac{31}{4}.$$

So $b = 3/2$ and $c = 31/4$.   ∎

Lastly, note that both equations (b.1) and (b.2) address quadratic expressions whose leading coefficient is 1. If that's not true, you have to begin by factoring our the leading coefficient.

Example B.4. Rewrite the expression $5t^2 - 3t + 20$ in the form $a(t - b)^2 + c$.

Solution: We begin by factoring out the 5:

$$5t^2 - 3t + 20 = 5\left(t^2 - \frac{3}{5}t + 4\right).$$

Now let's work with the expression in parentheses, in which $2b = 3/5$, so $b = 3/10$ and we can make a perfect square by adding $9/100$:

$$5\left(t^2 - \frac{3}{5}t + 4\right) = 5\left(t^2 - \frac{3}{5}t + \frac{9}{100} - \frac{9}{100} + 4\right)$$
$$= 5\left(\left(t - \frac{3}{10}\right)^2 + \frac{391}{400}\right).$$

Lastly, we redistribute the factor of 5 to arrive at the desired form:

$$5\left(t - \frac{3}{10}\right)^2 + \frac{391}{80},$$

Now we see that $a = 5$, $b = -3/10$ and $c = 391/80$.   ∎

❖ Polynomial Division

The key to understanding polynomial division is the division algorithm that you learned in elementary school. It's been a while since most people have thought about it any detail, so we're going to begin by using an example to review. Simple as it might seem, read Example B.5 carefully because it presents the first principles on which the entire method is based.

Principles that are central to the division algorithm

Example B.5. Use the division algorithm to calculate $\frac{852}{4}$.

Solution: The first important fact to recognize is that $4\left(\frac{852}{4}\right) = 852$. So we're looking for a number that we can multiply by 4 to get 852. The key to finding it is *place value*. Recall that

$$852 = 800 + 50 + 2 = 8(10^2) + 5(10^1) + 2.$$

Similarly, the number we're calculating could include hundreds, tens and units. So let's write

$$
\begin{aligned}
8(10^2) + 5(10^1) + 2 &= 4\Big(a(10^2) + b(10^1) + c \Big) &\text{(b.3)}\\
&= 4a(10^2) + 4b(10^1) + 4c.
\end{aligned}
$$

Since the two sides of this equation are the same, they have the same number of hundreds in them. That is, $8 = 4a$, from which we gather that $a = 2$. So

$$852 = 800 + 4b(10^1) + 4c.$$

By designating $a = 2$ we have "built" 800 of the 852, so $4b(10^1) + 4c$ must account for the remaining 52. Written as an equation,

$$52 = 4b(10^1) + 4c.$$

Now recall that we're only dealing with integers. If $b = 2$, the right-hand side would have 8 tens, but the left-hand side has only 5. So let's try $b = 1$. In that case, our equation reduces to

$$52 = 40 + 4c.$$

By designating $b = 1$, we've accounted for 40 of the 52, so $4c$ must account for the remaining 12. Written as an equation,

$$12 = 4c \quad \Rightarrow \quad c = 3.$$

Now that we know a, b and c, we can rewrite equation (b.3) as

$$852 = 4(2(10^2) + 1(10^1) + 3) = 4(213).$$

Dividing both sides of the equation by 4 brings us to the answer: $\frac{852}{4} = 213$.   ∎

In essence, the process demonstrated above deconstructs a number by using the idea of *place value*—the 8 in 852 means 8 hundreds, and the 5 means 5 tens. The technique is often done in a shorthand notation. Most recently, it was encoded using the ⌐ symbol, which we use to organize the relevant information by putting the divisor on the left, the dividend "inside," and the quotient "atop." For example, writing

$$\frac{852}{4} = 213 \quad \text{is the same as writing} \quad \frac{213}{4\,\big|\,852}$$

We mention it here because the same notation is used to encode polynomial division. Because understanding the notation is important to using it, Example B.6 will show the calculation of 852/4 using the ⌐‾‾ shorthand. It's important to note that the steps we perform below are exactly the ones we discussed above, and the ⌐‾ notation is acting solely as a bookkeeping device.

Principles that are central to the division algorithm

Example B.6. Use the division algorithm to calculate $\frac{852}{4}$.

Solution: Before beginning, let's use the idea of place value to rewrite

$$4\,\overline{|\,852} \quad \text{as} \quad 4\,\overline{|\,8(10^2) + 5(10^1) + 2}\,.$$

Now we're ready to begin the process. First, since the quotient can have 2 hundreds in it but no more, we place a 2 in the hundreds column:

$$\begin{array}{r} 2(10^2) \\ \hline 4\,|\,8(10^2) \quad + \quad 5(10^1) \quad + \quad 2(10^1) \end{array}$$

Since $4 \times 2(10^2) = 8(10^2)$, we have "built" 800 of the 852. To find out how much remains unaccounted for, we subtract 800 from 852:

$$\begin{array}{r} 2(10^2) \\ \hline 4\,|\ \ 8(10^2) \quad + \quad 5(10^1) \quad + \quad 2(10^1) \\ -8(10)^2 \\ \hline 5(10^1) \quad + \quad 2(10^1) \end{array}$$

So we have to "build" an additional $5(10^1) + 2(10^1) = 52$. Since the remaining 52 units can accommodate 4 tens but not 8, we add 10 onto our quotient and account for its contribution by writing

$$\begin{array}{r} 2(10^2) \quad + \quad 1(10^1) \\ \hline 4\,|\ \ 8(10^2) \quad + \quad 5(10^1) \quad + \quad 2(10^1) \\ -8(10)^2 \\ \hline 5(10^1) \quad + \quad 2(10^1) \\ -4(10^1) \\ \hline 1(10^1) \quad + \quad 2(10^1) \end{array}$$

$$\boxed{4 \times 1(10^1) = 4(10^1)}$$

meaning that we have accounted for all but $1(10^1) + 2(10^1) = 12$ of the original 852. At this point, we've already added hundreds and tens into our quotient, so all that remains is units. Since $(4)(3) = 12$, we add 3 units onto our quotient and see ...

$$\begin{array}{r} 2(10^2) \quad + \quad 1(10^1) \quad + \quad 3 \\ \hline 4\,|\ \ 8(10^2) \quad + \quad 5(10^1) \quad + \quad 2(10^1) \\ -8(10)^2 \\ \hline 5(10^1) \quad + \quad 2(10^1) \\ -4(10^1) \\ \hline 1(10^1) \quad + \quad 2(10^1) \\ -(1(10^1) \quad + \quad 2(10^1)) \\ \hline 0 \end{array}$$

> This is often written in a more compact form as
>
> $$\begin{array}{r} 213 \\ \hline 4\,|\ 852 \\ -800 \\ \hline 52 \\ -52 \\ \hline 0 \end{array}$$
>
> We're using the verbose form because our actual goal is to prepare for polynomial division.

The zero at the bottom of our calculation indicates that we have successfully divided 852 into 4, with no remainder, and we see in our notation that the quotient is

$$2(10^2) + 1(10^1) + 3 = 213.$$ ∎

Practice with the division algorithm

Example B.7. Use the division algorithm to calculate $3225/75$.

Solution: Since $3225 = 3(10^3) + 2(10^2) + 2(10^1) + 5$ and $75 = 7(10^1) + 5$, let's begin by writing

$$7(10^1) + 5 \,\overline{\big)\, 3(10^3) \quad + \quad 2(10^2) \quad + \quad 2(10^1) \quad + \quad 5}$$

It might seem strange to write 75 as $7(10^1) + 5$, but remember that our real objective to prepare for polynomial division, in which the divisor might have the form $7x + 5$.

We know the quotient cannot have any thousands since $75 \times 1000 = 75000$, which is far greater than 3225. Similarly, the quotient cannot have any hundreds since $75 \times 100 = 7500$, which is larger than 3225. However, it can have tens. Since $75 \times 50 = 3750$ (which is too large) but $75 \times 40 = 3(10^3)$, let's add 4 tens into our quotient and account for their contribution by writing

$$
\begin{array}{r}
4(10^1) \\
\hline
7(10^1) + 5 \,\big)\, 3(10^3) \quad + \quad 2(10^2) \quad + \quad 2(10^1) \quad + \quad 5 \\
-3(10^3) \\
\hline
2(10^2) \quad + \quad 2(10^1) \quad + \quad 5
\end{array}
$$

which means that we still have to account for $2(10^2) + 2(10^1) + 5 = 225$ of the original 3225. Since $75 \times 3 = 225$, we add 3 units onto the quotient and write

$$
\begin{array}{r}
4(10^1) \quad + \quad 3 \\
\hline
7(10^1) + 5 \,\big)\, 3(10^3) \quad + \quad 2(10^2) \quad + \quad 2(10^1) \quad + \quad 5 \\
-3(10^3) \\
\hline
2(10^2) \quad + \quad 2(10^1) \quad + \quad 5 \\
-(\,2(10^2) \quad + \quad 2(10^1) \quad + \quad 5\,) \\
\hline
0
\end{array}
$$

This tells us that $3225 = 75 \times (4(10^1) + 3)$, so $\frac{3225}{75} = 4(10^1) + 3 = 43$. ∎

Practice with the division algorithm

Example B.8. Use the division algorithm to calculate $3228/75$.

Solution: In Example B.7 we established that 3225 is a multiple of 75, so we know that 3228 is not. In this division we will have a remainder of of 3. This is seen in the last step of the algorithm . . .

$$
\begin{array}{r}
4(10^1) \quad + \quad 3 \\
\hline
7(10^1) + 5 \,\big)\, 3(10^3) \quad + \quad 2(10^2) \quad + \quad 2(10^1) \quad + \quad 8 \\
-3(10^3) \\
\hline
2(10^2) \quad + \quad 2(10^1) \quad + \quad 8 \\
-(\,2(10^2) \quad + \quad 2(10^1) \quad + \quad 5\,) \\
\hline
3
\end{array}
$$

We conclude from this calculation that $3228 = (75)(43) + 3$, so $\frac{3228}{75} = 43 + \frac{3}{75}$. ∎

The point of spending so much time on elementary arithmetic and notation is that the factors of 10 were entirely inert in our calculation. All that mattered to us during the calculations was that, for example, $(10^2) \times (10^1) = 10^3$, but that's a matter of exponents and has nothing to do with the fact that the base is 10. It might as well have been 8, or 18, or . . . you guessed it, x.

Polynomial division with no remainder

Example B.9. Determine $(6x^3 + 30x^2 + 38x + 4)/(2x + 4)$.

Solution: The process for polynomial division is *exactly* the same as what you saw with integer division a moment ago. We begin by writing

$$2x + 4 \overline{\big)\,6x^3 \;+\; 30x^2 \;+\; 38x \;+\; 4}$$

The quotient cannot have any x^3 terms, since $(2x + 4)(x^3) = 2x^4 + 4x^3$ and our dividend has no powers of x^4. However, the quotient *can* include quadratic factors. Specifically, since

$$(2x + 4)(3x^2) = 6x^3 + 12x^2$$

has the same leading term as $6x^3 + 30x^2 + 38x + 4$, we begin our quotient with $3x^2$. Note that $6x^3 + 12x^2$ differs from our dividend by $18x^2 + 38x + 4$, which we'll account for in later steps of the algorithm using lower-degree terms in the quotient. Written in the notation of the division algorithm,

> We chose *three* x^2 precisely because its product with $2x + 4$ yields the first term in the dividend.

$$
\begin{array}{r}
3x^2 \\
2x + 4 \overline{\big)\,6x^3 \;+\; 30x^2 \;+\; 38x \;+\; 4} \\
\underline{-(6x^3 \;+\; 12x^2)} \\
18x^2 \;+\; 38x \;+\; 4
\end{array}
$$

We can produce the $18x^2$ of the remaining polynomial by adding $9x$ onto the quotient. Specifically,

$$(2x + 4)(3x^2 + 9x) = 6x^3 + 12x^2 + 18x^2 + 36x = 6x^3 + 30x^2 + 36x,$$

which is another step closer to $6x^3 + 30x^2 + 38x + 4$. In fact, our product now differs from $6x^3 + 30x^2 + 38x + 4$ by exactly $2x + 4$. Written in the notation of the division algorithm,

$$
\begin{array}{r}
3x^2 \;+\; 9x \\
2x + 4 \overline{\big)\,6x^3 \;+\; 30x^2 \;+\; 38x \;+\; 4} \\
\underline{-(6x^3 \;+\; 12x^2)} \\
18x^2 \;+\; 38x \;+\; 4 \\
\underline{-(18x^2 \;+\; 36x)} \\
2x \;+\; 4
\end{array}
$$

Now adding 1 onto the quotient, we arrive at the conclusion of the division algorithm:

$$
\begin{array}{r}
3x^2 \;+\; 9x \;+\; 1 \\
2x + 4 \overline{\big)\,6x^3 \;+\; 30x^2 \;+\; 38x \;+\; 4} \\
\underline{-(6x^3 \;+\; 12x^2)} \\
18x^2 \;+\; 38x \;+\; 4 \\
\underline{-(18x^2 \;+\; 36x)} \\
2x \;+\; 4 \\
\underline{-(2x \;+\; 4\,)} \\
0
\end{array}
$$

This tells us that (check!)

$$6x^3 + 30x^2 + 38x + 4 = (2x + 4)(3x^2 + 9x + 1).$$

Now dividing both sides by $2x + 4$ yields our answer:

$$\frac{6x^3 + 30x^2 + 38x + 4}{2x + 4} = 3x^2 + 9x + 1.$$ ∎

Polynomial division when there's a remainder

Example B.10. Calculate $(5x^3 + 2x^2 + 4x + 1)/(8x + 2)$ using the division algorithm.

Solution: We begin by writing

$$8x + 2 \overline{\smash{\big)}\ 5x^3 \ + \ 2x^2 \ + \ 4x \ + \ 1}$$

The quotient cannot have any x^3 terms, since $(8x + 2)(x^3) = 8x^4 + 2x^3$, and our dividend has no x^4 terms in it. However, the quotient can include quadratic terms. Specifically, since

$$(8x + 2)\left(\frac{5}{8}x^2\right) = 5x^3 + \frac{5}{4}x^2,$$

which has the same leading term as $5x^3 + 2x^2 + 4x + 1$, we begin our quotient with $\frac{5}{8}x^2$. Note that $5x^3 + \frac{5}{4}x^2$ differs from our dividend by $\frac{3}{4}x^2 + 4x + 1$. Written in the notation of the division algorithm,

We chose a coefficient of $\frac{5}{8}$ precisely so that this product would yield the leading term of the dividend.

$$
\begin{array}{r}
\frac{5}{8}x^2 \qquad\qquad\qquad\quad \\
8x + 2 \overline{\smash{\big)}\ 5x^3 \ + \ 2x^2 \ + \ 4x \ + \ 1} \\
-(8x^3 \ + \ \tfrac{5}{4}x^2) \qquad\qquad\quad \\
\hline
\tfrac{3}{4}x^2 \ + \ 4x \ + \ 1
\end{array}
$$

Next we add linear terms into the quotient. Specifically, since we need an additional $\frac{3}{4}$ of an x^2, we'll add $\frac{3}{32}x$ onto the quotient. Then we have

$$(8x + 2)\left(\frac{5}{8}x^2 + \frac{3}{32}x\right) = 5x^3 + \frac{5}{4}x^2 + \frac{3}{4}x^2 + \frac{3}{16}x = 5x^3 + 2x^2 + \frac{3}{16}x,$$

whose first and second terms are the same as $5x^3 + 2x^2 + 4x + 1$. Note that $5x^3 + 2x^2 + \frac{3}{16}x$ differs from our dividend by $\frac{61}{16}x + 1$. Written in the notation of the division algorithm,

$$
\begin{array}{r}
\frac{5}{8}x^2 \ + \ \frac{3}{32}x \qquad\qquad\quad \\
8x + 2 \overline{\smash{\big)}\ 5x^3 \ + \ 2x^2 \ + \ 4x \ + \ 1} \\
-(8x^3 \ + \ \tfrac{5}{4}x^2) \qquad\qquad\quad \\
\hline
\tfrac{3}{4}x^2 \ + \ 4x \ + \ 1 \\
-(\tfrac{3}{4}x^2 \ + \ \tfrac{3}{16}x \ ) \quad \\
\hline
\tfrac{61}{16}x \ + \ 1
\end{array}
$$

Next we add constant terms into the quotient. Specifically, since we need an additional $\frac{61}{16}$ of an x, we'll add $\frac{61}{128}$ onto the quotient. Then we have

$$(8x+2)\left(\frac{5}{8}x^2 + \frac{3}{32}x + \frac{61}{128}\right) = 5x^3 + \frac{5}{4}x^2 + \frac{3}{4}x^2 + \frac{3}{16}x + \frac{61}{16}x + \frac{61}{64} = 5x^3 + 2x^2 + 4x + \frac{61}{64}.$$

This differs from $5x^3 + 2x^2 + 4x + 1$ by only $\frac{3}{64}$. Written in the notation of the division algorithm,

$$
\begin{array}{r}
\frac{5}{8}x^2 \quad + \quad \frac{3}{32}x \quad + \quad \frac{61}{128} \\[4pt]
8x+2 \enclose{longdiv}{5x^3 \quad + \quad 2x^2 \quad + \quad 4x \quad + \quad 1} \\[2pt]
-(8x^3 \quad + \quad \frac{5}{4}x^2) \\[2pt]
\hline
\frac{3}{4}x^2 \quad + \quad 4x \quad + \quad 1 \\[2pt]
-(\frac{3}{4}x^2 \quad + \quad \frac{3}{16}x \quad) \\[2pt]
\hline
\frac{61}{16}x \quad + \quad 1 \\[2pt]
-(\frac{61}{16}x \quad + \quad \frac{61}{64}) \\[2pt]
\hline
\frac{3}{64}
\end{array}
$$

The algorithm stops at this point, just like it would in the case of elementary arithmetic, and we write

$$5x^3 + 2x^2 + 4x + 1 = (8x+2)\left(\frac{5}{8}x^2 + \frac{3}{32}x + \frac{61}{128}\right) + \frac{3}{64}.$$

After dividing both sides of this equation by $8x + 2$ we have our answer:

$$\frac{5x^3 + 2x^2 + 4x + 1}{8x + 2} = \frac{5}{8}x^2 + \frac{3}{32}x + \frac{61}{128} + \frac{3/64}{8x + 2}. \quad \blacksquare$$

The division algorithm can be expedited by modifications in special cases, but all of them come back to the same basic ideas shown here.

> For example, the technique called *synthetic division* is particularly handy when the divisor has the form $x - a$, where a is any number.

❖ Exercises

In #1–16 rewrite the expression in the form $a(t+b)^2+c$ (b and c might be negative).

1. $t^2 + 12t + 100$
2. $t^2 + 20t + 321$
3. $t^2 + 18t + 70$
4. $t^2 + 16t + 50$

5. $t^2 - 8t + 22$
6. $t^2 - 60t + 1$
7. $t^2 - 7t + 10$
8. $t^2 + 9t + 13$

9. $2t^2 + 12t + 100$
10. $4t^2 + 20t + 88$
11. $5t^2 + 18t + 70$
12. $7t^2 + 16t + 50$

13. $2t^2 - 8t + 22$
14. $3t^2 - 60t + 12$
15. $3t^2 - 7t + 10$
16. $5t^2 + 9t + 13$

In #17–24 use polynomial division to write the quotient as $q(x) + r(x)$ where $q(x)$ is a polynomial and r is a rational function whose numerator has a lower degree than its denominator (we say that such a rational function is in *lowest terms*).

17. $\frac{3x^2+4x+7}{x+2}$
18. $\frac{5x^2+8x+17}{x+5}$

19. $\frac{16x^2+7x+9}{2x+1}$
20. $\frac{10x^2+6x+2}{3x+5}$

21. $\frac{7x^3+5x^2-x+19}{x^2+1}$
22. $\frac{2x^3+15x^2+2x+1}{x^2-3}$

23. $\frac{4x^5+8x+1}{7x^2+10}$
24. $\frac{8x^4+3x^3+x}{2x^2+7x+1}$

In #25–26 suppose that $p(x)$ is a polynomial whose degree is 2 or more.

25. Explain why the remainder of $\frac{p(x)}{x-a}$ must be constant.

26. Based on #25 we know that $p(x)$ can be written as $q(x)(x - a) + r$ where q is a polynomial function and r is a constant. Explain why $r = 0$ if and *only if* a is a root of p.

In #27–30 use the result of #26 to check whether $x - a$ is a factor of $p(x)$ by calculating $p(a)$.

27. $p(x) = 7x^4 + 2x^2 + x - 10$, $a = 1$

28. $p(x) = 13x^4 - x^2 + 3x - 9$, $a = -1$

29. $p(x) = x^5 - x^4 + 8x - 30$, $a = 2$

30. $p(x) = 2x^4 - 15x^2 - 14x + 3$, $a = -2$

In Chapter 6 you'll see a technique called *partial fractions decomposition*. The presentation there focuses on its usage in *integration*, but doesn't push far into the algebra that underlies the general formulation. In exercises #31–34 we use polynomial division to provide you with a motivation for the general form that you see in Chapter 6.

31. In this exercise you're going to use polynomial division as a tool to rewrite

$$\frac{3x^5 + 9x^4 + 8x^3 + 17x^2 + 6x + 15}{(x^2 + 1)^3}$$

as the sum of three simpler fractions.

(a) Use polynomial division to verify that

$$\frac{3x^5 + 9x^4 + 8x^3 + 17x^2 + 6x + 15}{x^2 + 1} = 3x^3 + 9x^2 + 5x + 8 + \frac{x + 7}{x^2 + 1}.$$

(b) Use polynomial division to verify that

$$\frac{3x^3 + 9x^2 + 5x + 8}{x^2 + 1} = 3x + 9 + \frac{2x - 1}{x^2 + 1}.$$

(c) Combine parts (a) and (b) to verify that

$$\frac{3x^5 + 9x^4 + 8x^3 + 17x^2 + 6x + 15}{(x^2 + 1)^2} = 3x + 9 + \frac{2x - 1}{x^2 + 1} + \frac{x + 7}{(x^2 + 1)^2}.$$

(d) Based on part (c), verify that

$$\frac{3x^5 + 9x^4 + 8x^3 + 17x^2 + 6x + 15}{(x^2 + 1)^3} = \frac{3x + 9}{x^2 + 1} + \frac{2x - 1}{(x^2 + 1)^2} + \frac{x + 7}{(x^2 + 1)^3}.$$

(e) Based on the method used above, explain why all the numerators are linear (it's not coincidence).

By following the procedure demonstrated in #31 rewrite the following rational functions as the sum of simpler fractions.

32. $\dfrac{2\,x^5 + 3\,x^4 + 21\,x^3 + 25\,x^2 + 56\,x}{(x^2 + 4)^3}$

33. $\dfrac{x^5 + 3\,x^4 + 19\,x^3 + 25\,x^2 + 67\,x + 14}{(x^2 + x + 5)^3}$

34. $\dfrac{x^5 + 12\,x^4 + 47\,x^3 + 153\,x^2 + 233\,x + 304}{(x^2 + 2x + 6)^3}$

Appendix C

Binomial Coefficients

The prefix "bi" means 2, as in America's *bi*cameral legislature, or your *bi*weekly paycheck. The expression $(a + b)$ is called a *bi*nomial because it has 2 parts, a and b (*nomos* is the Greek word for *part*).

The **binomial coefficients** are numbers that arise when we expand expressions of the form $(a + b)^n$, but they are also extremely useful in a variety of other situations. It's easiest to get a feel for them by looking at an example, so consider

$$(a + b)^3 = (a + b)(a + b)(a + b) \qquad (c.1)$$
$$= (aa + ab + ba + bb)(a + b)$$
$$= aaa + aba + baa + bba + aab + abb + bab + bbb. \qquad (c.2)$$

When we collect terms according to how many factors of a and b they include, we get the standard equation

$$(a + b)^3 = a^3 + 3a^2b + 3ab^2 + b^3. \qquad (c.3)$$

There are two important points to notice here:

Notice that $(a+b)^3$ is written as (blue)(green)(red) on the right-hand side of (c.1), and that each summand on the right-hand side of (c.2) has one blue, one red, and one green factor.

- Each factor of $(a + b)$ in $(a + b)^3$ contributes either a factor of a or a factor of b (but not both) to each summand on the right-hand side of (c.2).

- The coefficients in equation (c.3) *count the number of ways* to get each combination of a and b. For example, aba, baa, and aab are all equal to a^2b, so the coefficient of a^2b is 3. Similarly, there are three ways to get ab^2, but there's only one way to get a^3 (all a), and only one way to get b^3 (all b).

When we combine these ideas, we get a powerful tool for making calculations. For example, suppose we want to know the coefficient of a^6b^4 in $(a + b)^{10}$. If we printed each of the 10 factors of $(a + b)$ with a different color, much as we did on the right-hand side of (c.1), each summand in the expanded form of $(a + b)^{10}$ would have 10 *differently* colored factors, much as you see on the right-hand side of (c.2). We're interested in finding out how many of them have exactly four factors of b, so let's ask ourselves, "How many ways can four different colors (indicating which factors are b) be selected from a group of 10 distinct colors?"

We're *really* selecting which specific copies of $(a + b)$ contribute a factor of b to the product.

1. Without knowing the particular colors that are used or the order in which they appears, let's just say that the first factor of b could be any of the 10 colors.

2. For each of those possibilities, there are 9 possibilities for the color of the second factor of b.

3. For each of *those* possibilities, there are 8 possibilities for the color of the third factor of b.

4. For each of *those* possibilities, there are only 7 possibilities for the color of the fourth factor of b.

So our initial estimate is that there are $10 \cdot 9 \cdot 8 \cdot 7 = 5040$ ways to chose four distinct colors from a group of 10. That's a good beginning, but it's not right because we've overcounted. We said, "There are 10 possibilities for making the first choice, and for each of those ..." so we've counted $bbbba^6$ as being different from $bbbba^6$; but the summand whose factors of b are blue, green, red, and black occurs only once in the expanded form of $(a + b)^{10}$.

So how many times did we overcount it? To answer, think about using blue, green, red and black to color the factors of b in $bbbba^6$. How many ways can we color the first factor? Four, since there are four colors. How many ways can we color the second? Three, since only three colors remain to be used. The third factor? Two, then the color that goes in the last slot is determined. There are $4 \cdot 3 \cdot 2 \cdot 1 = 24$ ways to order the four colors, so we've counted the summand $bbbb a^6$ a total of 24 times. To fix the overcounting problem, we divide our original answer by 24, and conclude that the total number of ways to choose four colors from 10 is $5040/24 = 210$. Therefore, the coefficient of $a^6 b^4$ is 210. We refer to this binomial coefficient as **10 choose 4** and write it as $\binom{10}{4}$.

In calculating $\binom{10}{4}$ we saw products of consecutive decreasing integers: $10 \cdot 9 \cdot 8 \cdot 7$ and $4 \cdot 3 \cdot 2 \cdot 1$. When that decreasing string of integers is allowed to decrease all the way to 1, we refer to the product as a **factorial** and denote it with an exclamation mark. For example,

$$
\begin{aligned}
4! &= 4 \cdot 3 \cdot 2 \cdot 1 \\
5! &= 5 \cdot 4 \cdot 3 \cdot 2 \cdot 1 \\
6! &= 6 \cdot 5 \cdot 4 \cdot 3 \cdot 2 \cdot 1.
\end{aligned}
$$

Notice that

$$
\binom{10}{4} = \frac{10 \cdot 9 \cdot 8 \cdot 7}{4!} = \frac{(10 \cdot 9 \cdot 8 \cdot 7)6!}{6!4!} = \frac{10!}{6!4!}.
$$

More generally, after expanding $(a + b)^n$ and collecting like terms, the coefficient of $a^{n-k}b^k$ is

> k consecutive integers in the numerator, starting at n and decreasing

$$
\binom{n}{k} = \frac{n \cdot (n-1) \cdot (n-(k-1))}{k!} = \frac{(n \cdot (n-1) \cdot (n-(k-1))(n-k)!}{(n-k)!k!} = \frac{n!}{(n-k)!k!}.
$$

This formula is reminiscent of the color discussion above, but ...

... *this* formulation is often easier for people to remember because of its symmetry.

Note that, after expanding $(a + b)^n$ and collecting like terms, the coefficient of a^n is 1, and so is the coefficient of b^n. That's one way to remember that

$$
\binom{n}{n} = 1 \quad \text{and} \quad \binom{n}{0} = 1.
$$

Alternatively, you can apply the general formula if you remember that $0!$ is the number 1. (You can think of $0! = 1$ as a convention for now, but it comes from the relationship between the factorial and something called the *Gamma function* which is defined by way of a *improper integral* in Chapter 6.)

Calculating binomial coeffiecients

Example C.1. Calculate $\binom{7}{4}$, $\binom{7}{3}$, and $\binom{5}{0}$.

Just as the summand whose factors of b are blue and green occurs only once on the right-hand side of (c.2).

Solution: Using the formula above, we see that

$$\binom{7}{4} = \frac{7!}{4!3!} = 35, \quad \binom{7}{3} = \frac{7!}{3!4!} = 35, \quad \text{and} \quad \binom{5}{0} = \frac{5!}{0!5!} = \frac{5!}{(1)(5!)} = 1. \quad \blacksquare$$

Expanding a binomial

Example C.2. Verify that $(a+b)^3 = \binom{3}{0}a^3b^0 + \binom{3}{1}a^2b^1 + \binom{3}{2}a^1b^2 + \binom{3}{3}a^0b^3$.

Solution: Using the formula $\binom{3}{k} = \frac{3!}{k!(3-k)!}$ we calculate

$$\binom{3}{0} = 1, \quad \binom{3}{1} = 3, \quad \binom{3}{2} = 3, \quad \text{and} \quad \binom{3}{3} = 1,$$

so the coefficients are exactly what we see in equation (c.3). $\blacksquare$

In summary, when n is a positive integer ...

$$(a+b)^n = \sum_{k=0}^{n} \binom{n}{k} a^{n-k}b^k. \tag{c.4}$$

Note: This formula is extended to non-integer n in Chapter 8.

Expanding a binomial

Example C.3. Use equation (c.4) to expand $(1+u)^5$.

Solution: When $a = 1$ and $b = u$, equation (c.4) tells us that

$$(1+u)^5 = \binom{5}{0}1^5u^0 + \binom{5}{1}1^4u^1 + \binom{5}{2}1^3u^2 + \binom{5}{3}1^2u^3 + \binom{5}{4}1^1u^4 + \binom{5}{5}1^0u^5$$

$$= 1 + 5u + 10u^2 + 10u^3 + 5u^4 + u^5.$$ $\blacksquare$

> Notice the symmetry in the pattern of coefficients.

❖ Binomial coefficients and counting

We've introduced the binomial coefficients as numbers that occur in the expansion of a binomial, but they are also extremely useful in other ways, some of which are introduced in the following examples.

Using binomial coefficients for quick counting

Example C.4. You walk up to a fruit tray at a friend's wedding, but all that's left is 1 strawberry, 1 chunk of pineapple, 1 slice of mango, 1 chunk of watermelon, and 1 slice of a strange green fruit that you've never seen before. How many different ways are there for you to select three of them?

Solution: You might put (in order) the strawberry, the pineapple, and the strange green fruit on your plate. On the other hand, you might put (in order) the pineapple, the strange green fruit, and the pineapple chunk on your plate. Either way, you have the same selection of fruit on your plate when you leave. *The order in which you choose them doesn't matter,* and that's the kind of counting the binomial coefficients are good for. So we can say that the number of ways is "five choose three," or $\binom{5}{3} = 10$. Alternatively, since choosing three is the same as leaving two, we could also say that the answer is "five choose two" (where we're choosing which fruits to leave), or $\binom{5}{2} = 10$. $\blacksquare$

In Example C.4 we said that choosing 3 of the 5 fruit slices is the same as leaving 2 of them, and verified that $\binom{5}{3} = \binom{5}{2}$ by calculating both numbers independently. And in Example C.1, you saw that $\binom{7}{4} = \binom{7}{3}$, which makes sense in the context of expanding $(a + b)^7$. Much as we saw in equation (c.2) on p. 862, each summand of the expanded $(a + b)^7$ comprises 7 factors. Choosing 4 of them to be b is the same as choosing 3 of them to be a. More generally,

$$\binom{n}{k} = \binom{n}{n - k}.$$

> If n things are available, selecting k of them is mathematically equivalent to excluding $n - k$ of them.

Using binomial coefficients for quick counting

Example C.5. Suppose that five cars exit the turnpike and head toward three toll booths, numbered 1 through 3. From the point of view of the workers in the booths, the cars are identical (it doesn't matter *which* cars line up at your booth, only the number of them). Determine the number of ways the cars could line up at the booths.

Solution: It's tempting to start by setting the cars into different configurations (one of which is depicted in Figure C.1). That would work *eventually,* but it involves a lot of bookkeeping and takes a lot of time. Instead, let's draw 8 slots, from left to right, and place either a car or a booth into each slot. We'll place the booths in order, so that the leftmost is #1 and the rightmost is #3. If a car is at booth #j, it is placed to the *right* of booth #j in our diagram. For example, the configuration that you see in Figure C.1 is depicted in Figure C.2.

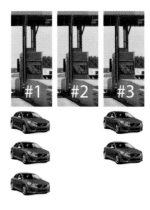

Figure C.1: One way that 5 cars could line up at 3 booths.

Each car is at *some* booth, so it must be to the *right* of some booth. Therefore, the leftmost slot in our diagram will never be a car. It will always be booth #1. The other 7 slots are up for grabs. The number of ways to choose 2 of the 7 slots (selecting them for the booths) is $\binom{7}{2} = 21$, so that's the number of different ways that the cars could line up at the booths. ∎

Figure C.2: Determining the number of ways that 5 identical cars can line up at 3 distinct toll booths.

Each particular way that the cars could line up at the booths in Example C.5 is called a *microstate* of the system. For example, the particular configuration depicted in Figure C.1 is a microstate. Another microstate would be 5 cars at booth #1 and no cars at the other booths. We found that there are 21 different microstates when there are 5 cars and 3 booths. Of course, that number would change if there were a different number of cars or booths. The particular numbers of cars and booths is called a *macrostate* of the system.

The terms *microstate* and *macrostate* are used in thermal physics to describe the distribution of identical energy packets across a body with n atoms (the energy packets are like the cars, and the atoms are like the booths). The number of atoms in a given object is fixed, so the macrostate tells us the number of energy packets that are in the object. A microstate tells us the particular fashion in which those energy packets are distributed across the atoms. In the exercises at the end of this appendix, you'll use the technique that was demonstrated in Example C.5 to show that the number of ways that q identical packets of energy can arrange themselves across n atoms, denoted by $\Omega(q)$, is

$$\Omega(q) = \binom{q + n - 1}{q}.$$

Now suppose there are *two* bodies that share the q packets of thermal energy, the first with n_1 atoms, and the second with n_2 atoms. There could be 0 in the first and q in the second, or 1 in the first and $q - 1$ in the second, or 2 in the first and $q - 2$ in the second, ... or q in the first and 0 in the second. Let's consider the case when there are k packets of energy in the first body and $q - k$ in the second. The number of possible microstates in the first body is $\Omega_1(k) = \binom{k+n_1-1}{k}$, and for each of those arrangements the packets in the second body have $\Omega_2(q - k) = \binom{q-k+n_2-1}{q_2}$ ways to distribute themselves, so the total number of ways that the packets of energy could arrange themselves in this configuration is

$$\Omega_1(k)\Omega_2(q-k) = \binom{k + n_1 - 1}{k}\binom{q - k + n_2 - 1}{q - k},$$

which we denote by $\Omega(k, q - k)$. In virtually any situation that concerns us, from cells to cell phones, the number of atoms involved is *astonishingly large*, so the number Ω is larger than you could possibly comprehend. We cut it down to a manageable, but still amazingly big number by asking how many powers of e are in it. Specifically, we define the *thermal entropy* of the system to be $S = k_B \ln(\Omega)$. With this vocabulary we can state ...

The Second Law of Thermodynamics: Any spontaneous change in the distribution of thermal energy will increase the entropy of a system.

Because Ω (and so S) counts the number of ways to rearrange energy packets, the Second Law of Thermodynamics says that Nature makes spontaneous changes in order to provide more thermodynamic flexibility.

❖ Pascal's Triangle

A famous array of numbers known as **Pascal's triangle** is commonly described as follows: the so-called 0th row contains only the number 1. The nth row $(n \geq 1)$ begins and ends with the number 1, and comprises a total of $n + 1$ numbers. Each "interior" entry is the sum of the entries that are found in the previous row, to its upper left and upper right. The first few rows of Pascal's triangle are shown below:

$n = 0$............................... 1
$n = 1$............................. 1   1
$n = 2$........................... 1   2   1
$n = 3$......................... 1   3   3   1
$n = 4$....................... 1   4   6   4   1
$n = 5$..................... 1   5   10   10   5   1

Notice that the numbers $\binom{3}{k}$ compose the third row. Similarly, the numbers $\binom{5}{k}$ compose the fifth row. This happens in general: the nth row of Pascal's triangle contains the numbers $\binom{n}{k}$, and so gives us the coefficients in the expanded form of $(a + b)^n$.

Side notes:

When you have q packets and n atoms, the number of different microstates is Ω.

The function ln, called the *natural logarithm*, is introduced in Chapter 1.

The number k_B is Boltzmann's constant. Its value is $1.3806503 \times 10^{-23}$, and carries units of joules per kelvin.

The numbers Ω and $\ln(\Omega)$ are pure numbers, so entropy has units of J/K (i.e., the units associated with k_B).

This happens because the binomial coefficients satisfy the same recursion relationship that defines Pascal's triangle. Specifically:

Suppose n and k are positive integers. Then

$$\binom{n}{k} = \binom{n-1}{k} + \binom{n-1}{k-1}. \tag{c.5}$$

> The number $\binom{n}{k}$ is in the nth row of Pascal's triangle. The numbers $\binom{n-1}{k}$ and $\binom{n-1}{k-1}$ are in row $(n-1)$. Because the columns are slightly offset from row to row, columns k and $k-1$ of row $n-1$ are on the right and left of column k in row n.

Proof. Imagine that we have a collection of n items, and want to know how many different groups of k items can be made by selecting from it. The key to understanding equation (c.5) is recognizing that any such group must either (a) *exclude* the first item in the collection or (b) *include* the first item in the collection.

(a) How many groups of k items (selected from our set of n) *exclude* the first item of the collection? In this case, we have k choices to make, but there are only $n-1$ items to choose from (since we cannot choose item #1). Therefore, there are a total of $\binom{n-1}{k}$ groups of k items (selected from a set of n) that exclude item #1.

(b) How many groups of k items (selected from our set of n) *include* the first item of the collection? Since item #1 *has* to be in the group of k, each group is completed by making $k-1$ other choices. None of those choices can be item #1 (since it has already been selected), so there are $n-1$ items to choose from. Therefore, there are a total of $\binom{n-1}{k-1}$ groups of k items (selected from our set of n) that include item #1.

Adding these numbers together gives us (c.5).   ∎

❖ Exercises

1. How many ways are there to choose 0 items from a group of 9?

2. How many ways are there to choose 9 items from a group of 9?

3. In complete sentences, using the ideas of "choosing" and "leaving," explain why it *should* be true that $\binom{n}{k} = \binom{n}{n-k}$. Then verify that the two numbers are the same by using the formula for $\binom{n}{k}$ that's provided on p. 863.

4. Suppose there are five (different) DVDs next to your computer. How many ways are there for you to choose three of them?

5. Suppose there are 360 (different) mp3's on your computer. How many ways can you make a playlist of 10 songs? (Suppose your iPod is set to shuffle the playlist, so the order in which you select the songs is irrelevant.)

In #6–8 write the expanded form of the given expression.

6. $(1+x)^7$ 7. $(2+x)^5$ 8. $(2x+4)^6$ 9. $(3x+2)^5$

10. Suppose that 14 cars exit the turnpike and head toward five toll booths, numbered 1 through 5. From the point of view of the workers in the booths, the cars are identical (it doesn't matter *which* cars line up at your booth, only the number of them). How many different ways could the cars could line up at the booths?

11. Suppose that, at the same time the cars are exiting the turnpike in #10, there are four cars that are about to *enter* the turnpike, and are approaching another set of five booths. How many different ways are there for the 18 cars to line up at the 10 booths, given the division of the cars and booths? *(For each way that the* entering *cars could line up at the booths, how many ways are there for the* exiting *cars to line up?)*

12. Suppose that q identical "packets" of thermal energy are in a body with n atoms, numbered 1 through n. By mimicking the reasoning in Example C.5, show that there are $\binom{q+n-1}{q}$ different ways that the packets of energy could distribute themselves across the atoms.

13. Suppose there are q packets of energy shared between two bodies. Then there might be 0 in the first and q in the second, or 1 in the first and $q-1$ in the second, or 2 in the the first and $q-2$ in the second .... When there are k in the first and $q-k$ in the second, the number of microstates is $\Omega(k, q-k)$, so the total number of microstates is

$$\Omega_T = \sum_{k=0}^{q} \Omega(k, q-k). \tag{c.6}$$

> The Σ-notation for summation is introduced at the beginning of Chapter 5.

(a) Find the total number of microstates when $n_1 = 10$, $n_2 = 8$, and $q = 20$.

(b) When all microstates are equally likely, the probability of seeing $q_1 = k$ and $q_2 = 20 - k$ is
$$P(k) = \frac{\Omega(k, 20-k)}{\Omega_T}.$$
Calculate $P(k)$ for each k between 0 and 20.

(c) Explain why we expect $\sum_{j=0}^{20} P(k) = 1$, then verify that it's true.

(d) Plot these probabilities as a function of k (i.e., plot the points $(k, P(k))$).

(e) What is the most likely macrostate (what are q_1 and q_2)?

(f) How does your answer from part (e) correspond to the values of n_1 and n_2?

(g) Complete parts (a)–(f) with $n_1 = 30$, $n_2 = 50$ and $q = 100$.

14. The **Laguerre polynomials** help us determine the *radial wave functions* for an electron in a hydrogen atom (see [8]). They are defined as follows:

> The name *Laguerre* is pronounced "Log-AIR."

$$L_0(x) \;=\; 1 \tag{c.7}$$
$$L_1(x) \;=\; 1 - x \tag{c.8}$$
$$L_2(x) \;=\; 1 - 2x + \frac{1}{2}x^2. \tag{c.9}$$

In general,

$$L_n(x) = \sum_{k=0}^{n} (-1)^k \binom{n}{k} \frac{x^k}{k!}. \tag{c.10}$$

(a) Verify that $L_1(x)$ and $L_2(x)$ are described by (c.10).

(b) Use (c.10) to write down formulas for $L_3(x)$ and $L_4(x)$ like those you see in (c.7)–(c.9).

Appendix D

Conic Sections

We refer to parabolas, ellipses, and hyperbolas as conic sections because they can be formed by intersecting a plane with a cone, as seen in Figure D.1.

Figure D.1: (all) Intersecting a plane with a cone generates the conic sections; (left) the intersection of the cone with a plane that's parallel to its edge but offset from its tip creates a parabola; (middle) the intersection of the cone with a plane that's parallel to its central axis creates a hyperbola; (right) planes at other angles create ellipses (or circles) when intersected with the cone.

The conic sections can also be described in terms of distances, which is how we proceed below. As with our presentation of trigonometry on p. 838, we assume that this appendix is review material, not presented to you for the first time, but used as a refresher.

❖ Ellipses

Suppose you want to send a signal from one radio tower to another. Because transmission occurs in all directions, some of the signal can hit the ground between the towers and reflect to the receiver; but since such signals have to travel a larger distance than the direct signal, they arrive later. The net effect is that the receiver hears many copies of the same message but time-delayed and overlapped. This makes it difficult to read the signal at the receiving end.

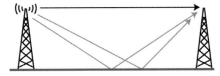

Figure D.2: Sending from one tower to another; reflections from the ground cause interference.

In the discussion below we'll show that when the curve of the ground is an ellipse rather than a line, all of the reflected signals travel the same distance, and so arrive at the receiver simultaneously. So rather than having multiple overlapping signals, the receiver hears only two.

Suppose the transceivers at the top of the radio towers are at $(\pm c, 0)$ in the Cartesian plane, where $c > 0$ (i.e., the x-axis of the plane runs through the transceivers). Then all the points on the y-axis are equidistant from them. For example, the origin is c units from both, and the point $(0, -2)$ is $\sqrt{4 + c^2}$ from both (using the Pythagorean Theorem). More generally, the point $(0, -b)$ is $\sqrt{b^2 + c^2}$ from each transceiver. We'll call this distance a, as depicted in Figure D.3. That is,

$$a^2 = b^2 + c^2. \tag{d.1}$$

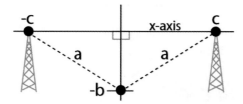

Figure D.3: Points on the y-axis are equidistant from $(\pm c, 0)$.

If (x, y) is any other point of reflection for which the signal travels a path of length $2a$,

$$\underbrace{\sqrt{(x + c)^2 + (y - 0)^2}}_{\text{distance from } (-c,0) \text{ to } (x,y)} + \underbrace{\sqrt{(x - c)^2 + (y - 0)^2}}_{\text{distance from } (x,y) \text{ to } (c,0)} = 2a.$$

We can remove the roots from this equation by isolating one radical at a time and then squaring both sides, which takes two steps. Step 1 is to write the equation as

$$\sqrt{(x + c)^2 + y^2} = 2a - \sqrt{(x - c)^2 + y^2}$$

and square both sides. This yields

$$(x + c)^2 + y^2 = (2a)^2 - 2(2a)\sqrt{(x - c)^2 + y^2} + (x - c)^2 + y^2,$$

which simplifies to

$$a\sqrt{(x - c)^2 + y^2} = a^2 - xc.$$

We eliminate the remaining radical by squaring again, thereby arriving at

$$a^2\left((x - c)^2 + y^2\right) = a^4 - 2a^2cx + c^2x^2,$$

which reduces to

$$(a^2 - c^2)x^2 + a^2y^2 = a^2(a^2 - c^2).$$

Looking back to equation (d.1) we see that $a^2 - c^2 = b^2$, so we can rewrite this equation as

$$b^2x^2 + a^2y^2 = a^2b^2,$$

or after dividing by a^2b^2,

$$\frac{x^2}{a^2} + \frac{y^2}{b^2} = 1, \quad \text{where } a \geq b > 0. \tag{d.2}$$

The curve described by equation (d.2) is an **ellipse**. Notice that $|y|$ cannot exceed b since the left-hand side sums to 1 and x^2/a^2 cannot be negative, and the points $(0, \pm b)$ are on the curve. Similarly, $|x|$ cannot exceed a, and the points $(\pm a, 0)$ are on the ellipse. We say that $(\pm c, 0)$ are the **foci** of the ellipse, and we refer to $(\pm a, 0)$ as its **vertices**. (Some people also refer to $(0, \pm b)$ as vertices of the ellipse.)

> The word foci (pronounced "FOE-sigh") is the plural of *focus*, and the word vertices (pronounced "VUR-tuh-sees") is the plural of the word *vertex*.

If the foci had been at $(0, \pm c)$ instead, the same work would have resulted in the equation

$$\frac{x^2}{b^2} + \frac{y^2}{a^2} = 1 \quad \text{where} \quad a \geq b > 0, \tag{d.3}$$

and a and b are related to c by equation (d.1). In this case, the vertices are the points $(0, \pm a)$.

You can tell whether the foci are displaced vertically or horizontally from the ellipse's center by the denominators of (d.2) and (d.3)—horizontal if the denominator associated with x^2 is larger, and vertical otherwise. In either case, the line that passes through the foci is called the **focal axis** of the ellipse, and the segment of it that connects the vertices is called the **major axis** of the ellipse. The **conjugate axis** is the line that passes through the center of the ellipse and is perpendicular to the focal axis, and the **minor axis** of the ellipse is the segment of the conjugate axis enclosed by the ellipse. We often use the prefix *semi-* to mean *half*, so a **semi-major axis** is half of the major axis (starting at the center and heading toward a vertex), and a **semi-minor axis** is half of the minor axis (starting at the center and heading toward the edge of the ellipse).

Major axes, vertices and foci of an ellipse

Example D.1. Determine (a) the major axis, (b) the location of the vertices, and (c) the location of the foci for the ellipse $8x^2 + 6y^2 = 48$.

Solution: (a) If we divide both sides of this equation by 48, it becomes

$$\frac{x^2}{6} + \frac{y^2}{8} = 1.$$

This equation allows $|y|$ to get as large as $\sqrt{8}$, but $|x|$ can only grow to $\sqrt{6}$, so this ellipse is taller than it is wide. That is, the vertical axis of this ellipse is its major axis. (b) Since the major axis is vertical, and $|y|$ can get as large as $\sqrt{8}$, the vertices are at $(0, \pm\sqrt{8})$. (c) Our equation indicates that $a^2 = 8$ and $b^2 = 6$ (a^2 is always the larger denominator), so $c^2 = 8 - 6 = 2$. That is, the foci are located $c = \sqrt{2}$ units from the origin. Since they're on the major axis, their coordinates are $(0 \pm \sqrt{2})$. ∎

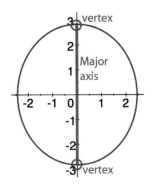

Figure D.4: The ellipse in Example D.1.

Try It Yourself: Major axes, vertices and foci of an ellipse

Example D.2. Determine (a) the major axis, (b) the location of the vertices, and (c) the location of the foci for the ellipse $7x^2 + 13y^2 = 91$.

Answer: (a) the segment $[-\sqrt{13}, \sqrt{13}]$ on the x-axis, (b) the points $(\pm\sqrt{13}, 0)$, and (c) at $(\sqrt{6}, 0)$. ∎

Full Solution On-line

A slight modification changes the center of the ellipse, but not its width or height. Specifically, consider the equation

$$\frac{(x-h)^2}{a^2} + \frac{(y-k)^2}{b^2} = 1 \tag{d.4}$$

where $a, b > 0$. This equation requires $|x - h| \le a$, so $h - a \le x \le h + a$. Similarly, $|y - k| \le b$ means that $k - b \le y \le k + b$, so this ellipse is centered at (h, k).

An ellipse that's not at the origin

Example D.3. Determine (a) the center, (b) the location of the vertices, and (c) the location of the foci for the ellipse $8x^2 + 16x + 10y^2 - 60y = 0$.

Solution: We begin by completing the square in both x and y, writing

$$
\begin{aligned}
8(x^2 + 2x) + 10(y^2 - 6y) &= 0 \\
8(x^2 + 2x + 1) + 10(y^2 - 6y + 9) &= 0 + 8 + 90 \\
8(x + 1)^2 + 10(y - 3)^2 &= 98.
\end{aligned}
$$

Dividing both sides by 98 yields

$$
\frac{4(x + 1)^2}{49} + \frac{5(y - 3)^2}{49} = 1,
$$

which is the same as

$$
\frac{(x + 1)^2}{\frac{49}{4}} + \frac{(y - 3)^2}{\frac{49}{5}} = 1.
$$

Note that $49/4 > 49/5$, so this is a version of equation (d.4) with $a^2 = 49/4$, $b^2 = 49/5$, $h = -1$ and $k = 3$. Now we can read off our answers: (a) The center of the ellipse is at $(-1, 3)$. (b) Since this ellipse is wider than it is tall, its vertices are found by starting at the center and moving $a = 7/2$ units to the left or right. That is, the vertices are at $(-1 \pm 7/2, 3)$. (c) To locate the vertices, we need to know c, which we find by calculating

$$
c^2 = a^2 - b^2 = \frac{49}{4} - \frac{49}{5} = \frac{49}{20} \quad \Rightarrow \quad c = \frac{7\sqrt{5}}{10}.
$$

Since this ellipse is wider than it is tall, the foci are found by starting at the center and moving c units to the left or right. That is, the foci are at $(-1 \pm 0.7\sqrt{5}, 3)$. ∎

Try It Yourself: An ellipse not at the origin

Example D.4. Determine (a) the center, (b) the location of the vertices, and (c) the location of the foci for the ellipse $4x^2 - 80x + 5y^2 + 30y = 0$.

Answer: (a) center at $(10, -3)$, (b) vertices at $(-0.5475, -3)$ and $(20.5475, -3)$, and (c) foci at $(5.283, -3)$ and $(14.717, -3)$. ∎

Full Solution
On-line

As we saw in equation (d.3), when the numbers a^2 and b^2 in equation (d.4) are swapped, the ellipse that's described is taller than it is wide. In summary:

Suppose $a > b > 0$. Then the equation

$$
\frac{(x - h)^2}{a^2} + \frac{(y - k)^2}{b^2} = 1 \tag{d.5}
$$

describes an ellipse centered at the point (h, k) whose vertices are at $(h \pm a, k)$. The foci of the ellipse are at $(h \pm c, k)$, where $c^2 = a^2 - b^2$. Similarly, the equation

$$
\frac{(x - h)^2}{b^2} + \frac{(y - k)^2}{a^2} = 1 \tag{d.6}
$$

describes an ellipse centered at the point (h, k) whose vertices are at $(h, k \pm a)$. The foci of the ellipse are at $(h, k \pm c)$, where $c^2 = a^2 - b^2$.

▷ Eccentricity of an ellipse

When the foci of an ellipse are closer to its center than to its vertices, the curve appears almost circular. By contrast, when the foci are closer to the vertices than to the center, the ellipse is more elongated (see Figure D.5). We quantify this structural characteristic with a number called the **eccentricity.** Specifically, the eccentricity of an ellipse described by (d.5) or (d.6) is

$$\varepsilon = \frac{c}{a}. \tag{d.7}$$

Since the semi-major axis has a length of a, the number ε tells us how far out the semi-major axis (as a fraction of its total length) we find the foci.

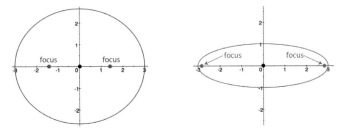

Figure D.5: (left) An ellipse whose eccentricity is $\varepsilon \approx 0.5$; (right) an ellipse whose eccentricity is near 1.

Calculating eccentricity of an ellipse

Example D.5. Determine the eccentricity of the ellipse $x^2 + 9y^2 = 9$, which is shown in the right-hand image of Figure D.5.

Solution: When we divide this equation by 9, we see

$$\frac{x^2}{9} + \frac{y^2}{1} = 1,$$

from which we conclude that $a = 3$ and $b = 1$. Therefore, $c = \sqrt{8}$, and $\varepsilon = c/a = 0.94$. That is, the foci can be found by traveling 94% of a semi-major axis away from the center of the ellipse.   ∎

Try It Yourself: Calculating eccentricity of an ellipse

Example D.6. Determine the eccentricity of the ellipse $x^2 + 9y^2 = 9$, which is shown in the left-hand image of Figure D.5.

Answer: $\varepsilon = 47/100$.   ∎

Full Solution
On-line

The relationship between eccentricity and the shape of an ellipse becomes apparent when we consider the fact that $c^2 = a^2 - b^2$, which allows us to see that

$$\varepsilon^2 = \left(\frac{c}{a}\right)^2 = \frac{a^2 - b^2}{a^2} = 1 - \left(\frac{b}{a}\right)^2 \quad \Rightarrow \quad \varepsilon = \sqrt{1 - \left(\frac{b}{a}\right)^2}.$$

An almost-circular ellipse has $b \approx a$, so that $\varepsilon \approx 0$. Looking back to equation (d.7), we conclude that the foci are very close to the center. Similarly, the number a is much larger than b in a very elongated ellipse, so $b/a \approx 0 \Rightarrow \varepsilon \approx 1$. Looking back to equation (d.7), we conclude that the foci are very close to the vertices.

❖ Hyperbolas

Suppose you touch the surface of still water at two points, called *touchpoints*, one after the other, and the leading edge of each disturbance travels at the same speed. The wavefronts will eventually meet, and after that time the point of intersection will proceed away from both touchpoints (see Figure D.6).

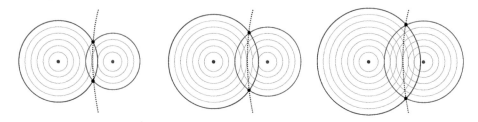

Figure D.6: When the wavefronts travel at the same speed, they meet on the edge of a hyperbolic curve (the first touchpoint is marked with ●, and the second with ●).

Let's denote the point were the wavefronts meet by P, and the touchpoints by F_1 and F_2. Since the wavefronts are traveling at the same speed, the lengths $|PF_1|$ and $|PF_2|$ are increasing at the same rate. Consequently, their difference is constant. That is,

$$|PF_1| - |PF_2| = \text{constant}.$$

This equation describes a hyperbola, and as we did with ellipses, we can rewrite it in Cartesian coordinates. Suppose the touchpoints are at $(\pm c, 0)$, and the wavefronts first meet at $(a, 0)$, so $0 < a < c$. If P is at (x, y),

$$\underbrace{\sqrt{(x + c)^2 + (y - 0)^2}}_{\text{distance from } (-c, 0) \text{ to } (x, y)} - \underbrace{\sqrt{(x - c)^2 + (y - 0)^2}}_{\text{distance from } (x, y) \text{ to } (c, 0)} = 2a.$$

> The point $(a, 0)$ is $a + c$ units away from one touchpoint, and $c - a$ units away from the other. The difference of these distances is
>
> $$(a + c) - (c - a) = 2a.$$

By using the same "isolate-and-square" technique as we did with the ellipse, we arrive at the following:

$$\frac{x^2}{a^2} - \frac{y^2}{b^2} = 1, \quad \text{where} \quad b^2 = c^2 - a^2. \tag{d.8}$$

As with the ellipse, we can derive a similar equation when the touchpoints—called the **foci** of the hyperbola—are at $(0, \pm c)$, and when the whole enterprise is re-centered on (h, k). The first point where the wavefronts meet, $(a, 0)$ in our discussion above, and its reflection across the center of the hyperbola are called the **vertices** of the hyperbola.

Suppose $a, b > 0$. Then the equation

$$\frac{(x - h)^2}{a^2} - \frac{(y - k)^2}{b^2} = 1 \tag{d.9}$$

describes a hyperbola whose vertices are at $(h \pm a, k)$ and whose foci are at $(h \pm c, k)$ where $c^2 = a^2 + b^2$. Similarly, the equation

$$\frac{(y - k)^2}{a^2} - \frac{(x - h)^2}{b^2} = 1 \tag{d.10}$$

describes a hyperbola whose vertices are at $(h, k \pm a)$ and whose foci are at $(h, k \pm c)$ where $c^2 = a^2 + b^2$.

> Equation (d.9) prohibits $x = h$, so the hyperbola has two *branches*, one on each side of that vertical line.
>
> Similarly, equation (d.10) prohibits $y = h$, so the hyperbola has two branches that are separated by that horizontal line.

As with ellipses, the line connecting the foci of a hyperbola is called its **focal axis,** and the perpendicular line through the center is called the **conjugate axis.**

Locating the center, vertices, and foci of a hyperbola

Example D.7. Determine the location of (a) the center, (b) the vertices, and (c) the foci of the hyperbola $4x^2 + 24x - 3y^2 + 12y = 0$.

Solution: As in Example D.3, we begin by completing the square in both x and y, writing our equation as

$$4(x^2 + 6x) - 3(y^2 - 4y) = 0$$
$$4(x + 3)^2 - 3(y - 2)^2 = 24.$$

Dividing both sides by 24 yields

$$\frac{(x+3)^2}{6} - \frac{(y-2)^2}{8} = 1.$$

This is a version of equation (d.9) in which $h = -3$, $k = 2$, $a = \sqrt{8}$ and $b = \sqrt{6}$. Now we can answer the questions: (a) the center is at $(-3, 2)$, (b) the vertices are at $(-3 \pm \sqrt{8}, 2)$, and (c) the number $c = \sqrt{14}$ so the foci are at $(-3 \pm \sqrt{14}, 2)$. These facts are shown in Figure D.7. ∎

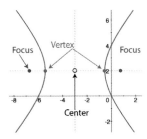

Figure D.7: The hyperbola in Example D.7.

Try It Yourself: Locating the center, vertices, and foci of a hyperbola

Example D.8. Determine the location of (a) the center, (b) the vertices, and (c) the foci of the hyperbola $3y^2 + 24y - 7x^2 + 28x = 0$.

Answer: (a) The center is at $(2, -4)$, (b) the vertices are at $\left(2, -4 \pm \frac{2\sqrt{15}}{3}\right)$, (c) the foci are at $\left(2, -4 \pm \frac{10\sqrt{42}}{21}\right)$. ∎

Full Solution On-line

▷ Asymptotes of a Hyperbola

We can rewrite equation (d.8) as

$$y^2 = \frac{b^2}{a^2}x^2 - b^2 \quad \Rightarrow \quad y = \pm\sqrt{\frac{b^2}{a^2}x^2 - b^2}.$$

After factoring x^2 out of the radicand, this becomes

$$y = \pm\sqrt{\left(\frac{b^2}{a^2} - \frac{b^2}{x^2}\right)x^2} = \pm\sqrt{\frac{b^2}{a^2} - \frac{b^2}{x^2}}\ x.$$

When x is much larger than b, the number $(b/x)^2 \approx 0$ and this reduces to

$$y = \pm\frac{b}{a}x.$$

That is, points on a hyperbola described by equation (d.8) approach the lines $y = \pm\frac{b}{a}x$ when x is large. We say that these lines are **asymptotes** of the hyperbola. More generally:

Suppose $a, b > 0$. Then the asymptotes of the hyperbola described by equation (d.9) are $y - k = \pm\frac{b}{a}(x - h)$.

Similarly, the asymptotes of the hyperbola described by equation (d.10) are $y - k = \pm\frac{a}{b}(x - h)$.

This information is often used to make quick sketches of hyperbolas.

Asymptotes of a hyperbola

Example D.9. Determine the center, vertices, and asymptotes of the hyperbola $x^2 - 7y^2 = 21$, and sketch a graph of it.

Solution: After dividing both sides of this equation by 21, it becomes

$$\frac{x^2}{21} - \frac{y^2}{3} = 1.$$

This is an example of equation (d.9) where $h = 0$, $k = 0$, $a = \sqrt{21}$ and $b = \sqrt{3}$. So this hyperbola is centered at the origin, and its vertices are at $(\pm\sqrt{21}, 0)$. Based on our work above, we know that the asymptotes are the lines $y = \pm\sqrt{\frac{3}{21}}\, x$. After plotting these points and asymptotes (which we draw dashed to indicate their role as guidelines), we can sketch a graph of the hyperbola by passing a curve through the vertices that's curved away from the center, and approaches the asymptotes as $|x|$ grows (see Figure D.8). ∎

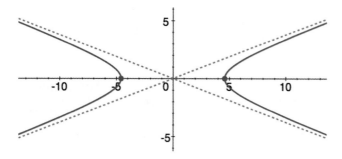

Figure D.8: The hyperbola for Example D.9.

Try It Yourself: Asymptotes of a hyperbola

Example D.10. Determine the center, vertices, and asymptotes of the hyperbola $3y^2 + 48y - 2x^2 = 0$, and sketch a graph of it.

Full Solution On-line

Answer: The center is at $(0, -8)$, the vertices at $(0, 0)$ and $(0, -16)$, and the asymptotes are $y + 8 = \pm\frac{8}{9}x$. ∎

▷ Eccentricity of a hyperbola

As with an ellipse, we define the **eccentricity** of a hyperbola to be

$$\varepsilon = \frac{c}{a} = \frac{\text{distance between the foci}}{\text{distance between the vertices}}.$$

Note: Since $c > a$, the eccentricity of a hyperbola is always > 1.

Eccentricity of a hyperbola

Example D.11. Determine the eccentricity of the hyperbola described by $(x + 3)^2 - 6(y - 2)^2 = 6$.

Solution: Dividing both sides of this equation by 6 yields

$$\frac{(x+3)^2}{6} - \frac{(y-2)^2}{1} = 1,$$

from which we conclude that $c^2 = a^2 + b^2 = 6 + 1 = 7 \Rightarrow c = \sqrt{7}$. Therefore, the eccentricity is $\varepsilon = c/a = \sqrt{7}/6$. ∎

Try It Yourself: Eccentricity of a hyperbola

Example D.12. Determine the eccentricity of the hyperbola in Example D.8.

Answer: $\varepsilon = \sqrt{10/7}$. ∎

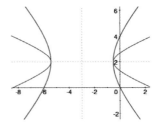

Figure D.9: The hyperbolas from Example D.11 and Example D.7.

The hyperbolas in Examples D.11 and D.7 are closely related. They have the same center and vertices, but the eccentricity of the hyperbola in Example D.7 is larger ($\varepsilon = \sqrt{14}/6$). The pair of hyperbolas are graphed in Figure D.9, where you can see that an eccentricity closer to 1 results in greater curvature at the vertices. As before, we can understand this by using the fact that $c^2 = a^2 + b^2$ to write

$$\varepsilon^2 = \frac{c^2}{a^2} = \frac{a^2 + b^2}{a^2} = 1 + \left(\frac{b}{a}\right)^2 \quad \Rightarrow \quad \varepsilon = \sqrt{1 + \left(\frac{b}{a}\right)^2}.$$

When the focal axis of a hyperbola is horizontal, the slopes of its asymptotes are $\pm b/a$. When ε is near 1, that number must be small, so the angle between the asymptotes is small. Consequently, there is greater curvature at the vertices.

❖ Parabolas

Unlike hyperbolas and ellipses, in which points are related to a pair of foci, a parabola is the set of points that are equidistant from a specified point, called its **focus**, and a given line, called its **directrix**. For example, if the focus is at $(0, c)$ and the directrix is the line $y = -c$, the origin is on the parabola because it's c units from both. More generally, the point at (x, y) is on the parabola if

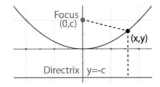

Figure D.10: Example of a parabola.

$$\underbrace{\sqrt{(x-0)^2 + (y-c)^2}}_{\text{distance from } (x,y) \text{ to the focus at } (0,c)} = \underbrace{y + c}_{\text{distance from } (x,y) \text{ to the directrix } y = -c.}$$

Squaring both sides of this equation and solving the resulting expression for y yields

$$y = \frac{1}{4c}x^2.$$

Similarly, when the focus is at $(c, 0)$ and the directrix at $x = -c$, as seen in Figure D.11, points on the parabola are described by

$$\sqrt{(x-c)^2 + y^2} = x + c \quad \Rightarrow \quad x = \frac{1}{4c}y^2.$$

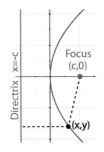

Figure D.11: Example of a parabola.

While we have drawn $c > 0$ in Figures D.10 and D.11, the mathematics is the same when $c < 0$. The sign of c determine whether the parabola cups up or down, right or left. In any event, the line through the focus that's perpendicular to the directrix is called the **axis** of the parabola, and its **vertex** is the point of the parabola on its axis (half way between the focus and the directrix). As before, a slight change in the formula shifts the vertex of the parabola.

Suppose $c \neq 0$. The equation

$$y = k + \frac{1}{4c}(x - h)^2 \qquad (d.11)$$

> No matter the value of x, the number $(x - h)^2 > 0$. So when $c > 0$, this equation says that $y = k + \text{more}$, allowing y to be any number $\geq k$. That's one way to understand why $c > 0$ makes the parabola open up.

describes a parabola whose focus is the point $(h, k + c)$, and whose directrix is the horizontal line $y = k - c$. The parabola's vertex is at (h, k), it opens upward when $c > 0$, and it opens downward when $c < 0$. Similarly, the equation

$$x = h + \frac{1}{4c}(y - k)^2 \qquad (d.12)$$

> No matter the value of y, the number $(y - k)^2 > 0$. So when $c > 0$, this equation says that $x = h + \text{more}$, so x cannot be less than h. That's one way to understand why $c > 0$ makes the parabola open to the right.

describes a parabola whose focus is the point $(h + c, k)$, and whose directrix is the vertical line $x = h - c$. The parabola's vertex is at (h, k), it opens to the right when $c > 0$, and it opens to the left when $c < 0$.

Determining a parabola

Example D.13. Determine the focus, directrix, and vertex of the parabola $13x + 26y^2 = 7$.

Solution: When we solve this equation for the linear variable, we see

$$x = \frac{7}{13} - 2y^2.$$

Since x cannot be larger than $7/13$, this parabola opens to the left. Its vertex is its rightmost point, which is $(7/13, 0)$. Comparing this with equation (d.12) yields

$$\frac{1}{4c} = -2 \quad \Rightarrow \quad c = -\frac{1}{8},$$

which means the focus and directrix are $1/8$ of a unit away from the focus. Since this parabola opens to the left, the focus is left of the vertex and the directrix is to the right. Specifically, the directrix is the line $x = \frac{7}{13} + \frac{1}{8} = \frac{57}{104}$, and the focus is the point $\left(\frac{55}{104}, 0\right)$. ∎

Try It Yourself: Determining a parabola

Example D.14. Determine (a) the focus, (b) the directrix, and (c) the vertex of the parabola $3x^2 - 5y = 2$.

Answer: (a) The point $(0, \frac{1}{60})$, (b) the line $y = -\frac{49}{60}$, (c) the point $(0, -\frac{2}{5})$. ∎

Full Solution
On-line

▷ Eccentricity of a Parabola

Earlier in our discussion, eccentricity told us the extent to which an ellipse was elongated rather than nearly circular, and whether the two sides of a hyperbola raced away from each other or lazily meandered apart. Whereas ellipses have eccentricity $\varepsilon < 1$, and hyperbolas have $\varepsilon > 1$, the eccentricity of a parabola is always $\varepsilon = 1$, so we cannot use it to quantify structure. However, the number c tells us how close the focus is to the vertex, which is at the heart of the idea. Small values of c indicate that the focus and vertex are very close, and the result is greater curvature at the vertex. By contrast, larger values of c result parabolas that appear more flat (see Figure D.12).

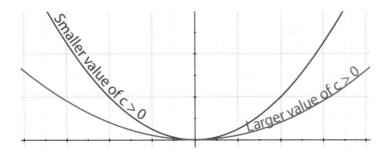

Figure D.12: Larger values of c mean the vertex and focus are farther apart, so the parabola appears "flatter."

❖ Reflective Properties of Conic Sections

We started this discussion by showing that signals that are emitted from one focus of an ellipse and arrive at the other after reflecting off the curve all travel the same distance, but we didn't show that a reflected signal is actually reflected toward that second focus. That can be done by showing that the line segments PF_1 and PF_2 make equal angles with the tangent line at each point P on the ellipse (see Figure D.13), and calculus allows us to determine the slope of the tangent line at P.

In a similar fashion, a parabola reflects signals to its focus that arrive on trajectories parallel to its axis. This fact, which is widely used with parabolic dishes, can be shown by calculating the slope of the tangent line and relevant angles.

Lastly, the tangent line to a hyperbola at the point P bisects the angle F_1PF_2, so a signal directed toward F_2 will reflect off the hyperbola and head toward F_1. This fact is used, for example, in the construction of Cassegrain reflecting telescopes.

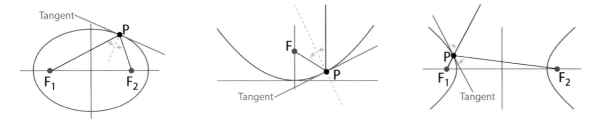

Figure D.13: Reflective properties of (left) ellipses; (middle) hyperbolas; (right) parabolas.

❖ Skill Exercises

For questions #1-4, determine (a) the length of the major axis, (b) the location of the vertices, and (c) the location of the foci of the given ellipse.

1. $4x^2 + 9y^2 = 36$ 3. $16x^2 + 9y^2 = 144$

2. $6x^2 + 10y^2 = 30$ 4. $5^2 + 8y^2 = 12$

For questions #5-8, determine (a) the center, (b) the location of the vertices, and (c) the location of the foci of the given ellipse.

5. $x^2 + 4x + 8y^2 = 1$ 7. $3x^2 + 2y^2 = 8x - 4y$

6. $2x^2 + 12x + y^2 - 6y = 0$ 8. $9x^2 - 4x = 7y - 3y^2$

For questions #9-12 (a) determine the eccentricity of the given ellipse, and (b) sketch the ellipse.

9. $x^2 + 5y^2 = 1$ 11. $x^2 + 4x + y^2 + 2y = 0$

10. $8x^2 + 3y^2 = 1$ 12. $2x^2 + 12x = 4y^2 + 9y$

For questions #13-16 determine the equation of the ellipse.

13. Centered at the origin with a major axis of 4 units and an eccentricity of $1/2$.

14. Centered at $(2, 4)$ with an eccentricity of 0.3 and a semi-minor axis of 6 units.

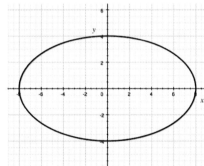

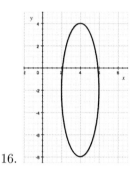

16.

15.

For questions #17-20 determine (a) the vertices, (b) the foci, and (c) the equations of the asymptotes of the hyperbola.

17. $4x^2 - 9y^2 = 36$ 19. $5y^2 - 9x^2 = 45$

18. $25x^2 - 4y^2 = 100$ 20. $10y^2 - 6x^2 = 12$

For questions #21-24 determine (a) the vertices, (b) the foci, and (c) the equations of the asymptotes of the hyperbola.

21. $x^2 - y^2 + 6y = 0$ 23. $6x^2 + 2x = y^2 + 8y$

22. $2x^2 - 8x + 18y - 9y^2 = 0$ 24. $10x^2 - 6y^2 = 40x - 30y$

For questions #25-28 (a) determine the eccentricity, and (b) sketch the hyperbola.

25. $x^2 - 8y^2 = 48$ 27. $2x^2 - 10y^2 = 4y + 3x$

26. $9x^2 - 4y^2 = 72$ 28. $8x^2 + 6x = 7y^2 - 11y$

For questions #29-34 use the foci F, vertices V, point P on the curve, or eccentricity E to find the equation of the hyperbola.

29. F: $(\pm 10, 0)$, V: $(\pm 4, 0)$ 31. V: $(\pm 2, 0)$, P: $(6, 4)$ 33. F: $(2, 3)$ and $(5, 3)$, E: $3/2$

30. F: $(0, \pm 6)$, V: $(0, \pm 5)$ 32. V: $(0, \pm 1)$, P: $(-2, 3)$ 34. F: $(-1, 2)$ and $(-1, -8)$, E: 4

For questions #35-42 (a) find the vertex, focus and directrix of the parabola, then (b) sketch its graph, showing the focus and directrix.

35. $7y = x^2$

36. $4x = 3y^2$

37. $8x = -5y^2$

38. $10y = -x^2$

39. $3 = x^2 + y$

40. $y = 4 + 7x - 3x^2$

41. $y^2 + 2y = 3x$

42. $6x + 4 = 3 - 8y^2$

For questions #43-50 use the focus F, directrix D, vertex V, point P, or orientation O to find the equation of the parabola that satisfies the given conditions.

43. F: $(4,0)$, $D : x = 1$

44. F: $(2,3)$, D: $x = 8$

45. F: $(0,6)$, D: $y = 1$

46. F: $(-2,5)$, D: $y = 5$

47. V: $(1,0)$, D: $x = -1$

48. V: $(4,-3)$, D: $x = -8$

49. V: $(0,10)$, D: $y = 1$

50. V: $(-2,-5)$, D: $y = 3/2$

51. V: $(7,0)$, P: $(8,2)$, O: horizontal

52. V: $(2,9)$, P: $(1,1)$, O: horizontal

53. V: $(0,-5)$, P: $(-2,-8)$, O: vertical

54. V: $(4,12)$, P: $(2,13)$, O: vertical

✧ Concept and Application Exercises

55. The moon's orbit about the earth is elliptical with an eccentricity of 0.0549 and a major axis of 769496 km. The earth is located at one of the foci, which we'll take as the origin of the Cartesian plane. Determine (a) the equation of the ellipse describing the moon's orbit, and (b) the maximum and minimum distance between the moon the earth.

56. Lithotripsy is a procedure used to break up kidney stones using the reflective properties of the ellipse. High-energy waves are generated at one focus of an elliptical reflector (called a *lithotripter*), which directs them to the second focus (see Figure D.14). With careful positioning, the second focus is centered on the kidney stone. A standard lithotripter has a semi-major axis of 13.8 cm and a minor axis of 15.5 cm. (a) Determine the equation of the ellipse described by the lithotripter dimensions. (b) If the kidney stone is 140 mm below the skin, how far from the body should the lithotripter's vertex be located?

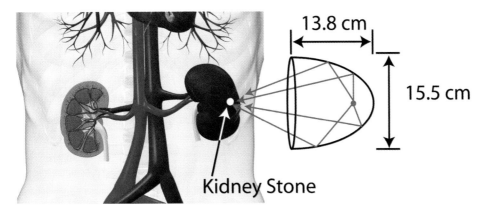

Figure D.14: Diagram of a lithotripter in action (not to scale).

57. On an ellipse, the line segment perpendicular to the major axis, through one of the foci, and with endpoints on the ellipse is called a **focal chord.** Show that the length of an ellipse's focal chord is $2b^2/a$.

58. In the LORAN navigation system, synchronized radio signals from three navigation stations (whose locations are known) are detected by ships. Since the signals travel at the same speed, the difference in their arrival time allows us to calculate the difference in the distances they have traveled.

    Suppose three navigation stations are situated along a coastline so that Station 1 is six kilometers due west of Station 2 and Station 3 is eight kilometers further east (see Figure D.15). As a ship approaches from the west, it detects signals from each of the stations and determines that the signal from Station 2 traveled four kilometers farther than the signal from Station 1, and the signal from Station 3 traveled six kilometers farther than the signal from Station 2. (a) Place Station 2 at the origin, and then determine the equation for the hyperbola whose foci are at Stations 1 and 2 and for which $2a = 2$ (i.e. the difference in the distance their signals traveled). (b) Keeping Station 2 at the origin, determine the equation for the hyperbola whose foci are at Stations 2 and 3, and for which $2a = 6$. (c) Determine the exact location of the ship, which sits on both hyperbolas from parts (a) and (b).

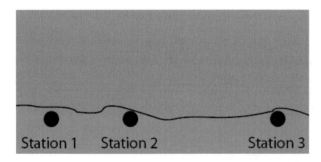

Figure D.15: The distribution of LORAN stations for #58.

59. The Cassegrain reflecting telescope uses a double-mirror system such as the one shown in Figure D.16. The primary mirror is parabolic and reflects the incoming light rays up towards the parabola's focus, F_1. The secondary mirror is hyperbolic and one of its foci is also at F_1. It reflects light rays downwards to the hyperbola's second focus at F_2.

    Consider the Cassegrain reflecting telescope in Figure D.16. It has a tube length of 1 meter and a primary focal length of 1.4 meters. Determine the equation of the hyperbola describing the secondary mirror.

60. What kind of curve is represented by

$$\frac{x^2}{p} + \frac{y^2}{p-9} = 1$$

for (a) $p > 9$, (b) $0 < p < 9$, and (c) $p < 0$?

61. The Photon Energy Transformation and Astrophysics Laboratory (PETAL) in Sede Boqer, Israel is one of the world's largest solar energy concentrators. It is a parabolic dish that is 24 m across and has a focal length (the distance

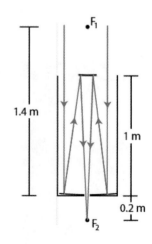

Figure D.16: Diagram of light rays in a Cassegrain reflecting telescope.

from the vertex to the focus) of approximately 13 m (see Figure D.17). (a) Determine the equation of the parabola that runs from edge to edge through the center, and (b) determine the depth, d, of the dish.

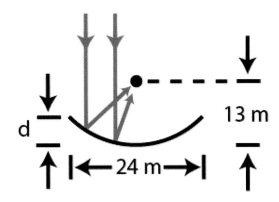

Figure D.17: Diagram of the parabolic reflector for #61.

62. A bridge spanning a two-lane highway is designed with a parabolic arch. The standard width for a two-lane highway is 36 feet (including both lanes and the shoulders). Each lane is 12 feet wide and the the parabolic arch must be 14 feet high on the outside edge of the lanes to accommodate tractor-trailers (see Figure D.18). Determine the height above ground of the highest point of the arch.

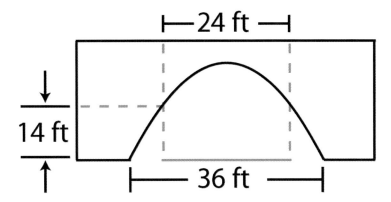

Figure D.18: Diagram of the bridge for #62.

63. Consider the parabola $y = \frac{1}{4p}x^2$, where $p > 0$. An ellipse shares the focus at $(0, p)$, and the endpoints of its minor axis lie on the parabola. Which curve has a vertex closest to the shared focus?

64. Consider the ellipse with a fixed focus at $(0, c)$ and centered at $(0, c + k)$. Show that as k grows without bound the equation of the ellipse reduces to the parabola $x^2 = 4cy$.

Appendix E

The World We Live In

The vast majority of your life experience has involved something we call *contact force*. When you *flip* open the cover of this book, *push* on the keys of a computer, *pull* up your covers at night, *turn* the steering wheel of your car, *twist* a door knob, *tug* open a refrigerator door, *kick* a soccer ball, or *lift* your backpack, you act on another object by making contact with it. This is the kind of force that Newton, Hooke, Halley and the others had experienced in their lives, too, so the suggestion that the sun could exert a force on the planets even though it doesn't touch them seemed very strange. Newton called the idea "action at a distance," and he was uncomfortable with it.

Since we've grown up knowing about gravity, electricity, and magnetism, it's hard for many people to empathize with Newton, but here's a modern example that elicits the same intellectual discomfort in many people: When NASA wants to send a probe into the outer reaches of the solar system, it uses the planets to "slingshot" the vehicle forward along its trajectory. In short, when the probe gets "close enough" to the planet, the two exchange something called *momentum*, which is discussed below, and the result is roughly equivalent to a fast-swung bat (the planet) colliding with a golf ball (the probe). On the one hand, that's pretty cool. On the other, it's also a little spooky. How was momentum or anything else exchanged since, unlike the bat and ball, the probe and the planet were never in contact?

After a few minutes thinking about that example, or about gravity, or about electrostatics—all of which involve "action at a distance," you might begin to wonder whether the word "force" is being used to gloss over things that we don't fully comprehend, but in today's world we have a very good understanding of force. The essence of that understanding is encapsulated in Newton's Laws of Motion, which focus on the quantity called *momentum*.

> Electrostatic force and magnetic force were not well understood in 1684, when Newton and his peers were struggling with the idea of gravity, so Newton couldn't look to them (as we can today) as other examples of "action at a distance."
>
> In fact, it wasn't until 1733 that Charles François du Fay discovered that electricity comes in two kinds, which he called resinous(-) and vitreous (+); and it wasn't until 1750 that John Michell discovered that the two poles of a magnet are equal in strength and that the force associated with an individual pole obeys an inverse square law.

> You might guess that the probe picks up speed because it falls into the planet's gravity well, but it also has to climb *out* of the well! What's happening is a bit more complicated.

❖ Momentum

Informally, you might think of an object's *momentum*, which is often denoted by p, as quantifying how much it will hurt when it hits you. Objects that are more massive will hurt more, as will objects that are traveling faster. From that standpoint, **momentum** should be (and is) defined as $p = mv$, where v is the object's velocity.

Technically speaking, an object's velocity has both a magnitude (speed) and a direction, so we call it a *vector*. Consequently momentum (which inherits the directional information included in the object's velocity) is also a vector. Vectors are typically denoted by bold characters, so we write $\mathbf{p} = m\mathbf{v}$. Note that the units associated with momentum are $(\text{kg})\left(\frac{\text{m}}{\text{sec}}\right)$.

> See Chapter 11 for a discussion of vectors.

❖ Newton's Laws of Motion

In short, Newton's Laws describe the relationship between momentum and force. We've already talked a little bit about momentum, but what's force? That's a fair question, but it's also the wrong one to ask—at least inititally. The first job of the physical sciences is to say what things *do,* not what they are. So let's address action instead of essence, and ask, "What does a force *do?*" Newton's answer is that force changes an object's momentum. In fact, Newton's First Law of motion tells us that force is the *only* thing that changes an object's momentum.

> **Newton's First Law of Motion:** Consider a body that's free of external forces. If the body is in motion, it will stay in motion (maintaining both its direction and speed). If the body is at rest, it will stay at rest.

But what about a body that's *not* free of external forces? How does its momentum change? In order to simplify the answer, let's restrict our discussion to cases in which the object's mass is constant (unlike a jet or rocket, which is throwing fuel out the back end). When m is constant, momentum can only change when there's a change in velocity, which we call an *acceleration* and often denote by **a**. Of course, you know from experience that when you apply the same force to different objects, those with more mass experience less acceleration (smaller Δp). This is the essence of Newton's Second Law of motion.

Notice that the word *acceleration* is not used only when an object "speeds up." It means only that there is a *change* in velocity. That could mean that the object speeds up, slows down, or changes direction.

> **Newton's Second Law of Motion:** When a force, **F**, acts on a body whose mass is m kilograms (constant), the body's velocity changes by $\mathbf{a} = \mathbf{F}/m$ meters per second each second.

If we multiply that equation by m, we have the famous formula

$$\mathbf{F} = m\mathbf{a}. \tag{e.1}$$

Since mass is measured in kilograms (kg), and acceleration is measured as $\frac{\text{m/sec}}{\text{sec}}$, which we often write at $\frac{\text{m}}{\text{sec}^2}$, equation (e.1) leads us to say that force has units of $\frac{\text{kg m}}{\text{sec}^2}$. The term *kilogram-meters-per-second-squared* is cumbersome, so we just say that force has units of **newtons**, which we denote by N.

When we read (e.1) aloud, we read the equals sign as the word "is," so (e.1) says that force is the product of mass and acceleration. Many people find the equation $a = F/m$ to be more closely in line with their intuition.

Now let's return to the question that we posed initially, "What *is* a force?" Newton's answer was that force is an *interaction between two objects.* This is an extremely important idea, so let's spend a moment with it.

Consider the following simple question: when you jump, do you push on the planet or does the planet push on you? Based on symmetry, you might be tempted to answer, "both," and that's exactly right. It's not that one object acts on another but, rather, that two objects act on *each other.*

You may have heard it said that "forces come in pairs," which expresses the spirit of the Third Law of Motion.

> **Newton's Third Law of Motion:** Suppose that Object #1 acts on Object #2 with a force of **F**. Then simultaneously, Object #2 acts on Object #1 with a force of $-\mathbf{F}$.

The negative sign indicates an opposite *direction* of force.

For example, when a Hummer hits a Mini Cooper, delivering a force **F**, the Mini Cooper delivers the same amount of force to the Hummer, but in the opposite direction. That assertion is often met with some degree of sarcasm. "*Sure* it does. Why don't we test that out? *I'll* drive the Hummer, and *you* sit in the Mini Cooper." Of course the driver of the Mini will be much worse off in a collision, but that's because the Mini experiences a much greater acceleration than the Hummer, not a greater force.

Figure E.1: A Hummer is much more massive than a Mini Cooper.

When we look at Newton's Second Law, we see

$$\mathbf{a}_{\text{hummer}} = \frac{\mathbf{F}}{m_{\text{hummer}}} \quad \text{and} \quad \mathbf{a}_{\text{mini}} = -\frac{\mathbf{F}}{m_{\text{mini}}}.$$

The negative sign in a_{mini} indicates an opposite direction.

Since m_{hummer} is so much larger than m_{mini}, the Hummer experiences much less acceleration. The same kind of thing happens in the example of jumping. You apply a force to the Earth, and the Earth applies the same amount of force to you. Since the Earth is so much more massive than you, the acceleration it experiences because of this force is much, *much* smaller than the acceleration you experience.

> **In Summary:** You can understand Newton's Laws of Motion as telling us three things: (1) *without* external force, momentum is constant, (2) *with* external force, momentum changes, and (3) forces occur in pairs.

The ideas of momentum and force, and the way they relate (according to Newton's Laws of Motion), are extremely helpful as we try to understand the world we live in, but they don't tell us everything. The next section introduces an important concept—*energy*—that allows us to understand and predict a much wider variety of phenomena.

✤ Energy

When you slow your car by applying the brakes, you generate a lot of heat from friction, and afterwards you're traveling more slowly. So the statement that

$$(\text{initial momentum}) - (\text{heat from brakes}) = (\text{final momentum})$$

might appeal to you. While the spirit of this equation is right, it doesn't actually make sense. To see why, rewrite it as

$$-(\text{heat from brakes}) = (\text{final momentum}) - (\text{initial momentum})$$

We talk about how much heat is generated, but you never hear someone say, "Wow, that stove has a lot of westerly heat!" Though heat can be carried by a flow, heat itself does not have a direction.

On the right-hand side you see the difference in the car's momentum, $\Delta\mathbf{p}$, which accounts for changes in both direction and magnitude, but heat is described only in terms of magnitude. That is, on the right-hand side we see a *vector* quantity, but on the left-hand side we see a *scalar* quantity, so our statement equates two quantities that are fundamentally different, much as if we had equated mass and color, or length and time.

This is not to say that heat and motion are unrelated in our example, just that the equation we wrote is nonsensical at a technical level. We fix this problem by relating both heat and momentum to something called **energy,** which then allows us to relate them to each other:

$$-\begin{pmatrix}\text{the energy generated} \\ \text{from the friction of} \\ \text{the brakes}\end{pmatrix} = \begin{pmatrix}\text{the final energy} \\ \text{associated with} \\ \text{motion}\end{pmatrix} - \begin{pmatrix}\text{the initial en-} \\ \text{ergy associated} \\ \text{with motion}\end{pmatrix}. \quad (e.2)$$

The energy associated with motion is called **kinetic energy**, and is often denoted by K. In Chapter 7 you'll see kinetic energy arise naturally from Newton's Second Law by way of something called the *definite integral.* For now, let's just cite the conclusions that are developed there: when a body has a mass of m and a velocity of $\mathbf{v}$, its kinetic energy is $K = 0.5m\|\mathbf{v}\|^2$, where $\|\mathbf{v}\|$ denotes the object's speed. Therefore, the units of kinetic energy are

$$(\text{kg})\left(\frac{\text{m}}{\text{sec}}\right)^2 = \underbrace{\left(\frac{\text{kg m}}{\text{sec}^2}\right)}_{\substack{\text{remember that we} \\ \text{call this a newton}}}(\text{m}) = Nm,$$

> The notation $\|\mathbf{v}\|$ is used to mean the magnitude of the vector $\mathbf{v}$, much as $|t|$ means the magnitude of the number t.

which we read as **newton-meters.** This unit is also called a **joule**, and is denoted by J. In fact, the joule is used not only for *kinetic* energy, but for *all* kinds of energy.

The idea of energy and its consequences are quite a bit more slippery than the idea of force. In [28, p.4-2], Richard Feynman remarks, "... in physics today, we have no knowledge of what energy *is*. We do not have a picture that energy comes in little blobs of a definite amount. It is not that way. [And yet there is a quantity, which we call *energy,* that] does not change when something happens. It is not a description of a mechanism, or anything concrete; it is just a strange fact that we can calculate some number and when we finish watching nature go through her tricks and calculate the number again, it is the same." This fact is called the **Conservation of Energy** and it's one of the most important things that we know about the world we live in.

> In Chapter 7, you'll see simple cases in which the Conservation of Energy can be seen in the equations.

Another extremely important fact about energy is that it can change forms. For example, when you jump, your body converts stored *chemical* energy into *kinetic* energy (you move), which is then converted into *potential* energy (you rise). When all of the kinetic energy has been converted into potential energy, you stop rising. As you fall back to Earth, the potential energy is converted back into kinetic energy, which (when you land) is converted into *thermal* energy. So while the total amount of energy doesn't change, the place that we find it—the way we experience it—the *form* it takes does change.

In closing we should mention that, aside from kinetic energy, the most important form of energy is called **potential energy,** which is often denoted by U. Potential energy is important because, when mechanical systems change spontaneously, they do so in a way that reduces potential energy.

> You'll see potential energy arise naturally from Newton's Second Law in Chapter 7.

✦ Dimensions and Units

In the physical sciences, the word **dimension** refers to the *kind* of thing being measured, and the word **unit** refers to the *way* we measure. For example, the physical characteristic of *length* is a dimension, and it can be measured in units of *inches, meters, miles,* etc. Similarly, *time, mass, temperature, velocity, force,* and *energy* are physical dimensions that are measured in units of *seconds, kilograms, kelvins, meters per second, newtons,* and *joules.*

Other settings make use of different dimensions, depending on what's important. For example, *money* is a dimension in business and economics, and can be measured in units of *dollars, cents, euros,* or *yen.* And *population* is a dimension in the social and biological sciences that's measured in units of *cells, individuals,* or *family groups* as appropriate.

> There are seven *base dimensions* in the SI system of measurement, including *time, mass* and *length.* Dimensions such as *velocity, force* and energy are called *derived dimensions* because, as you might guess, they are derived from the base dimensions. For example, velocity is calculated as length per unit time. Force and energy are also derived dimensions.

A basic principle of applied mathematics is that an equation can only be meaningful if both sides have the same dimensions. For example, an equation that says length = mass is nonsensical because length and mass are fundamentally different characteristics—i.e., different physical *dimensions* of an object—so cannot be equated. The technique of using quantities' units to check that an equation makes sense is called **dimensional analysis**.

Dimensional analysis

Example E.1. Suppose that if you travel for t minutes, you cross x meters, where $x = 3t$. Determine the dimensions of the "3."

Solution: Since the left-hand side of $x = 3t$ has dimensions of *length*, the right-hand side must also. The variable t has dimensions of *time*, so in order for the equation to make sense, the coefficient of 3 must have dimensions of *length per time*. Dimensionally speaking, the equation

$$x = 3t \quad \text{looks like} \quad (\text{length}) = \left(\frac{\text{length}}{\text{time}}\right)(\text{time}).$$

More specifically, since x is measured in meters and t is measured in seconds,

$$x = 3t \quad \text{looks like} \quad (\text{meters}) = \left(\frac{\text{meters}}{\text{second}}\right)(\text{second}),$$

in which the units of seconds on the right-hand side appear to cancel, like common factors in a fraction, leaving us with (meters) = (meters). ∎

Here we should also mention **dimensionless** quantities, which often occur as ratios of other quantities that each have associated dimensions. Dimensionless quantities often act as scale factors in equations where the two sides have—at least at first glance—the same dimensions but different units.

Dimensionless quantities

Example E.2. When converting from miles, m, to yards, y, we have $y = 1760m$. Show that 1760 is a dimensionless quantity.

Solution: In order for the equation $y = 1760m$ to make sense, both sides must have the same dimensions. Since y has dimensions of length, the quantity $1760m$ must also have dimensions of length—and m already does—so 1760 cannot have dimensions.

This is not to say that 1760 is without *units,* and that's often where people get confused so let's take a moment to look closer. Since m has units of "miles," and y has units of "yards," the number 1760 must have (and *does* have) units of "yards per mile." Written as a fraction, the units of of the scale factor are

$$\frac{1760}{1} \frac{\text{yards}}{\text{mile}}.$$

So dimensionally speaking, the equation

$$y = 1760m \quad \text{looks like} \quad (\text{length}) = \left(\frac{\text{length}}{\text{length}}\right)(\text{length})$$

in which the dimensions of (length) on the right-hand side appear to cancel, like common factors in a fraction.

Because the scale factor has dimensions of length per length, it doesn't change the *kind* of thing on the right-hand side of $y = 1760m$. *This is the sense in which 1760 is dimensionless.* ∎

Dimensional analysis

Example E.3. Suppose you invest P dollars in an account that pays $r\%$ APR, compounded continually. If you leave the account alone for t years, the amount of money in the account, A, is related to the initial investment by the equation $A = Pe^{rt}$. Perform a dimensional analysis of this equation.

Solution: In order for $A = Pe^{rt}$ to be meaningful, both sides must have the same dimensions. Since the left-hand side has dimensions of *money*, the quantity Pe^{rt} must also—and P already does—so the factor e^{rt} must be dimensionless. ∎

Dimensional analysis

Example E.4. Suppose the altitude of an oscillating mass (in centimeters) is $y = \sin(\omega t)$, where ω is constant and t is measured in seconds. Perform a dimensional analysis of this equation.

Solution: The argument of a trigonometric function, in this case ωt, should be in radians. Since t is measured in seconds, the number ω must have units of radians per second. Note that the left-hand side of this equation is measured in centimeters, but the output of a trigonometric function is dimensionless. That might make you think that the equation is dimensionally inconsistent. However, remember that we often neglect to write a coefficient of 1, and that's what has happened here. The equation is actually $y = 1\sin(\omega t)$, and the "1" has units of centimeters. ∎

> Recall from Appendix A that the sine is a ratio of lengths: (opposite length)/(hypotenuse length). Since its dimensions are length/length, it's dimensionless.

> We typically omit a coefficient of 1, and write $y = x + 4$ rather than $y = 1x + 4$.

We close by remarking that, at least for the purposes of this book, the most important dimensionless quantity is the *radian,* which is discussed in Appendix A.

Appendix F

End Notes

❖ Chapter 1

In Chapter 1 we asserted that Koopman's Sighting Law could be derived from some simple geometry, though we didn't show it at the time. This section is where we take care of that omission.

▷ Koopman's sighting formula

On p. 27 we introduced a simplified version of Koopman's Sighting Law, which describes the likelihood that a pilot will see an object in the water. The formula was "simplified" by suppressing information such as the altitude of the plane, h, and the area of the object in the water, A; but in the derivation of Koopman's Sighting Law on p. 850 you see those quantities enter the formula as follows:

$$L = kh \, \frac{A}{s^3},$$

where $k > 0$ is a constant of proportionality, and s is the direct line-of-sight distance between the plane and the object. By using the Pythagorean Theorem to write $s = \sqrt{h^2 + r^2}$, where r is the across-the-water distance between the plane and the object (see Figure F.1), we arrive at

$$L = kh \, \frac{A}{(h^2 + r^2)^{3/2}},$$

which has the structure cited on p. 27.

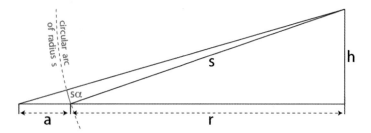

Figure F.1: The across-the-water distance between the plane and the object in the water is r (this figure shows a minor change in Figure A.25, on p. 851).

❖ Chapter 10

In various places through the book we cite Torricelli's Law for fluid flow, and in Chapter 10 we use it extensively. Here we provide a brief derivation of the law based on the ideas of force, work, and energy from Chapter 7.

▷ Torricelli's Law

Suppose a small hole forms in the side of a large tank, and the liquid it holds begins to drain. For the sake of discussion, let's assume the hole is circular, and consider a small "coin" of water next to the hole that's pushed out because of the pressure in the tank (see Figure F.2). Let's denote by A the area of coin's face, and denote the coin's thickness by Δx.

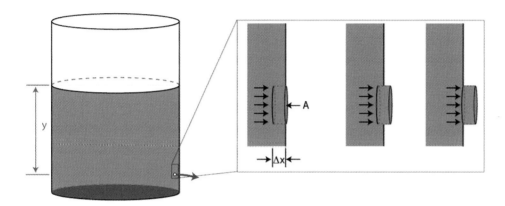

Figure F.2: A small "coin" of water is pushed out of a hole due to pressure in the tank.

We know from Chapter 7 that when there are y meters of liquid above the hole, the hydrostatic pressure next to the hole is $P = g\rho y$, where g is the acceleration per units mass due to gravity and ρ is the density of the liquid. So the force that acts to push the coin through the hole is $F = PA = g\rho yA$. Because the coin of water has a thickness of Δx, the hydrostatic force moves it a distance of Δx in order to expel it from the tank. We know from Chapter 7 that (force)×(distance)=(work), and that work changes kinetic energy. Assuming that the coin of water had no horizontal velocity before it was expelled, we get

$$\underbrace{F\Delta x}_{\text{work}} = \underbrace{\frac{1}{2}mv^2 - 0}_{\Delta K}$$

> We're assuming that the hole is small enough, relative to the size of the tank, that it's reasonable to approximate by saying that every point of the hole has the same amount of water above it.

where m is the mass of the "coin." Since $m =$(density)×(volume)$= \rho A\Delta x$, we can rewrite this equation as

$$g\rho yA\Delta x = \frac{1}{2}\rho A\Delta x\, v^2 \quad \Rightarrow \quad v = \sqrt{2gy}.$$

Selected Answers

Section 1.1:

1 black graph **3** blue graph **5** Both function values are nonnegative. The graph of f increases away from $y = 0$ and curves up as $|x|$ increases. The graph of g moves towards $y = 0$ as $|x|$ increases. **7** $(-\infty, 0)$ is outside the domain of both functions, and $g(x)$ is not defined for $x = 0$. The graph of f moves away from $y = 0$ curving up, and the graph of g moves towards $y = 0$. **9** $f(x)$ & $g(x)$ both move away from $y = 0$, curving downward on the right. $f(x)$ is non-negative, but $g(x)$ has a range of all real numbers. **11** $y = t^2$ is blue. **17** m/s **19** N m^2/c^2 **21** (a) $F(r)$, (b) $F(r)$

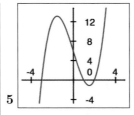

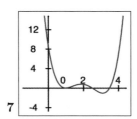

9 The graph "crashes through" the axis at the linear root, "bounces off" at the quadratic root, and "snakes through the axis" at the cubic root

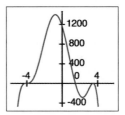

Section 1.2:

1 The function f behaves like $3t^5$ (an odd power) when $|t|$ is large, so $f(t) \to \infty$ as $t \to \infty$ and $f(t) \to -\infty$ as $t \to -\infty$.

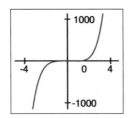

3 The function k behaves like $-8t^6$ (an even power) when $|t|$ is large, so $k(t) \to -\infty$ as $t \to \infty$ and as $t \to -\infty$.

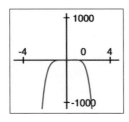

11 (a) 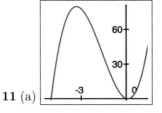 (b)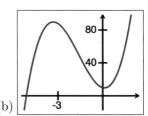

13 (a) The graph "bounces off" the axis at $t = 6$, and "crashes through" the axis at $t = 0$; (b) Shift the graph from part (a) down by 2 units.

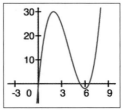

15 $y = t^3(t-2)^2$ **17** $y = (t-1)^2(t-2)^3$
21 (a) x-intercept: $(500, 0)$, y-intercept: $(0, 75)$;

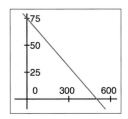

(b) $\frac{\$}{\#\ \text{of items sold}}$; (c) the slope; (d) the first price at which nobody would purchase the product; (e) the number of item that could be "sold" for \$0; (f) $0 \le x \le 500$

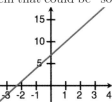

23 (a) (b) The units of 7 are feet. The units of 3 are feet per second.

25 $[0, 8.5911]$ **27** (c) $P_4(x) = \frac{1}{8}(35x^4 - 30x^2 + 3)$, $P_5(x) = \frac{1}{8}(63x^5 - 70x^3 + 15x)$ **31** the image is inverted, so you see the negative   **33** the image becomes (much) darker because all pixels with $x < 0.7$ in the original image have $f(x) < 0.5$ in the new

Section 1.3:

1 $f(t) \to \infty$ as $t \to \infty$ and $f(t) \to -\infty$ as $t \to -\infty$ **3** $f(t) \to \infty$ **5** $f(t) \to 1$

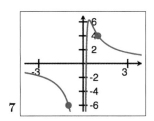   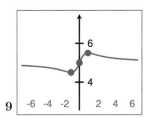

7   **9**

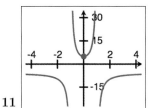

11   **13** (a) $y = 2$; (b) slope is 0   **15** (a) oblique asymptote; (b) slope = 5   **17** (a) $y = \frac{1}{6}$; (b) slope = 0   **19** (a) oblique asymptote; (b) slope = -1 **21** (a) no asymptote   **23** (a) $y = 0$; (b) slope = 0   **29** one   **31 (e)** below

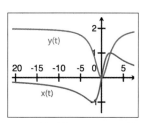

33 (a) $\frac{25{,}000}{n}$; (b) $\frac{12{,}500}{n}$; (c) $\frac{37{,}500}{n}$; (d) $40n$; (e) $C_T(n) = \frac{37{,}500}{n} + 40n$ **35** (a) as the two atoms get far away, their potential drops to zero, (b) as the two atoms get close, their potential increases toward ∞.

Section 1.4:

1 (a) $16t+2$; (b) $16t+31$ **3** (a) $\frac{8}{\sqrt{3t+4}}$; (b) $\sqrt{\frac{24}{t} + 4}$ **5** (a) $\frac{3t^2+14}{9t^4+78t^2+171}$; (b) $3\frac{t^2+2t+1}{t^4+4t^2+4} + 13$ **7** $t^6 + 2$ **9** $\sqrt[6]{3t^4 + 1}$ **11** $\frac{t+1}{t+2}$ **13** $f(t) = \frac{81}{t^4-6t^2+13}$

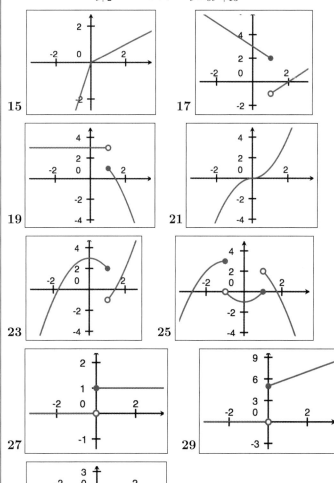

15   **17**

19   **21**

23   **25**

27   **29**

31   **33** (a) 3, (b) 1, (c) 7

35 (a) 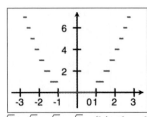 breaks at $t = \sqrt{1}, \sqrt{2}, \sqrt{3}, \sqrt{4}, \sqrt{5}, \sqrt{6}$; (b) the distances are shorter due to the increasing climb of x^2. **37**

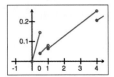 **39 (a)**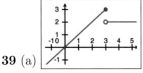

43 $g(7)$ is in the domain of f **45** (a) $t \neq \frac{-1 \pm \sqrt{41}}{2}$; (b) $t \neq 0$ **47** (a) $(-\infty, \infty)$; (b) $|t| \leq 1$ **49** pixels at less than 50% intensity are left unchanged, while those in the upper 50% are made dimmer (the intensity of bright pixels ranges over $[0.5, 1]$ in the original image, but only from $[0.5, 0.8]$ after the filter, so the variation in the intensity of the bright pixels has been reduced by 40%) **51** pixels that are either very dark or very light are left unchanged, but all pixels with mid-range intensity are brightened

Section 1.5:

1 (a) $f(x+2) = \cos(5x+10)$; (b) $f(x-5) = \cos(5x-25)$; (c) $f(x + 4t) = \cos(5x + 20t)$; (d) $f(x + h) = \cos(5x + 5h)$ **3** (a) $f(x + 2) = \tan(7x + 14)$; (b) $f(x - 5) = \tan(7x - 35)$; (c) $f(x+4t) = \tan(7x + 28t)$; (d) $f(x+h) = \tan(7 + 7h)$ **5** (a) $f(x+2) = 3x + 13$; (b) $f(x - 5) = 3x - 8$; (c) $f(x + 4t) = 3x + 12t + 7$; (d) $f(x + h) = 3x + 3h + 7$ **7** (a) $f(x + 2) = x^2 + 7x + 18$, (b) $f(x - 5) = x^2 - 7x + 18$; (c) $f(x + 4t) = x^2 + 8xt + 16t^2 + 3x + 12t + 8$; (d) $x^2 + 2xh + h^2 + 3x + 3h + 8$ **9** (a) $f(x+2) = \sqrt{x + 2}$; (b) $f(x - 5) = \sqrt{x - 5}$; (c) $f(x + 4t) = \sqrt{x + 4t}$; (d) $f(x + h) = \sqrt{x + h}$ **11** (a) $f(x + 2) = \frac{x-1}{2x+9}$; (b) $f(x - 5) = \frac{x-8}{2x-5}$; (c) $f(x + 4t) = \frac{x+4t-3}{2x+8t+5}$; (d) $f(x + h) = \frac{x+h-3}{2x+2h+5}$

13 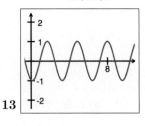 **15**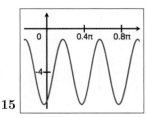

17 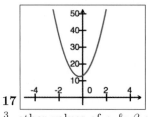 **19** $\alpha = \frac{1}{2}, \beta = 1, h = 2, k = \frac{3}{2}$, other values of α & β as long as $\alpha\beta^2 = \frac{1}{2}$ **21** $\alpha = 2, \beta = 3\pi, h = 0, k = 2$

23 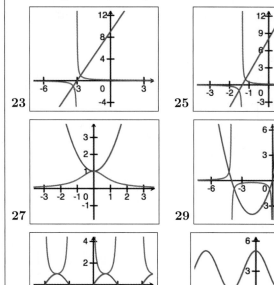 **25**

27 **29**

31 **33**

35 $4\frac{\text{rad}}{\text{sec}}$ **37** $\pi\frac{\text{rad}}{\text{sec}}$ **39** $\frac{4\pi}{5}\frac{\text{rad}}{\text{sec}}$ **41** $\frac{\pi}{4}$ meters **43** $\frac{1}{2000}$ meters **45** $\frac{16}{5}$ meters **47** (a) $w = \frac{1}{3}$, (b) $w = 8$, (c) $w = 7\pi$ **49** $\lambda = \frac{\pi}{2}, w = 19, s = \frac{19}{4}$ **51** $\lambda = 2, w = 3, s = \frac{3}{\pi}$ **53** $\lambda = \frac{\pi}{\sqrt{2}}, w = \sqrt{21}, s = \sqrt{\frac{21}{8}}$

55 (a)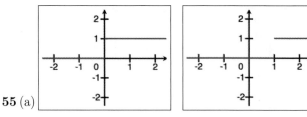

(b)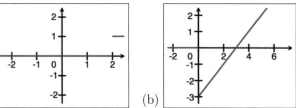

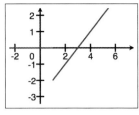

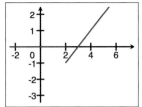

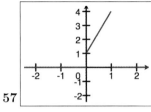

57

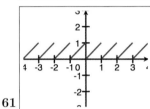

59

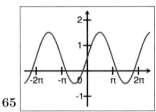

61

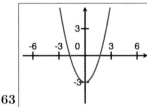

63

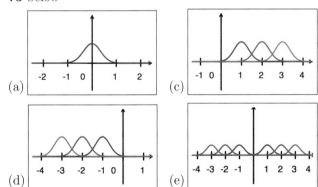

65   **67** odd   **69** $h(2) = -17$: "After 2 seconds, the object was 17 feet below sea level."   **71** 3

73 below

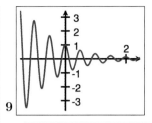

9

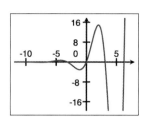

11

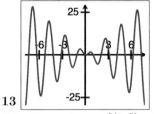

13   **15** (a) $f(t+2) = e^{5(t+2)}$; (b) $f(t-5) = e^{5(t-5)}$; (c) $f(t+\sqrt{2}) = e^{5(t+\sqrt{2})}$; (d) $f(t+h) = e^{5(t+h)}$ **17** (a) $f(t+2) = e^{12(t+2)^2-7(t+2)+1}$; (b) $f(t-5) = e^{12(t-5)^2-7(t-5)+1}$; (c) $f(t+\sqrt{2}) = e^{12(t+\sqrt{2})^2-7(t+\sqrt{2})+1}$; (d) $f(t+h) = e^{12(t+h)^2-7(t+h)+1}$ **19** (a) $f(t+2) = e^{7(t+2)}\cos(8(t+2))$; (b) $f(t-5) = e^{7(t-5)}\cos(8(t-5))$; (c) $f(t+\sqrt{2}) = e^{7(t+\sqrt{2})}\cos(8(t+\sqrt{2}))$; (d) $f(t+h) = e^{7(t+h)}\cos(8(t+h))$ **21** $f(t) = 0.4^t$ **27** (a) red graph, (b) larger than 1. The exponential's bases are greater than one because they increase as x increases.   **31** below

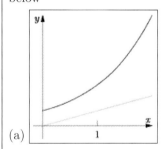

35 (a) $100(1.1527)^3 \approx 153.16$; (b) $100(1.1527)^{10} \approx 414.16$; (c) $V = 100(1.1527)^n$ **37** (a) decrease; (b) $\frac{3}{4}$; (c) $\frac{3}{4^{11}}$; (d) bigger

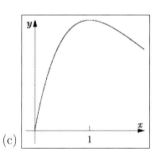

75 (a) $\frac{\pi}{3}, \frac{2\pi}{5}$; (b) 2π; (c) $6, 5$ **77** no   **79** cosine is even, sine is odd   **81** 6   **85** $\frac{299{,}792{,}458}{475 \times 10^{-9}}$ Hz

Section 1.6:

1 $A(t) = \left(\frac{1}{32}\right)^t$ **3** $A(t) = 14(64)^t$ **5** $A(t) = 5^{5t}$ **7** $A(t) = 14\left(\frac{10}{3}\right)^{-t}$

Section 1.7:

1 6   **3** 10   **5** 3   **7** -3 **9** $\frac{3}{2}$ **11** $\frac{2}{3}$ **13** $3^x = 26$ **15** $\frac{1}{3}$ **17** 1.4854   **19** 2.0959   **21** 0   **23** $\frac{\ln(5)}{\ln\left(\frac{27}{64}\right)}$ **25** $x = 0, x = \frac{9\ln(3)}{\ln(4)}$ **27** $x = \frac{9\ln(3) \pm \sqrt{81\ln^2(3)+4\ln(4)\ln(15)}}{2\ln(4)}$ **29** (a) $A(t) = 3\left(10^{t\log(5)}\right), B(t) = 4(10^{-t\log(3)})$; (b) $A(t) = 3e^{t\ln(5)}, B(t) = 4e^{-t\ln(3)}$ **31** Addition doesn't distribute through the logarithm.   **33** $\ln(0.5)$ is negative, so $3\ln(0.5) < 2\ln(0.5)$ **37** $g(t)$ **39** (a) 7; (b) 4.78% probability that the queue currently holds 7 items   **41** (a) $n = \frac{\ln(1000)}{2\ln(1.06)} \approx 59.2748$ years; (b)

$n = \frac{\ln(1000)}{4\ln(1.03)} \approx 58.4238$ years

Section 1.8:

1 7   **3** $f^{-1}(t) = \sqrt[7]{t+15}$ **5** The range of $\sin^{-1}$ is restricted to $\left[-\frac{\pi}{2}, \frac{\pi}{2}\right]$, and $\frac{2\pi}{3}$ is outside that range.   **7** (a) a,b equal to two adjacent minima/maxima; (b) x_0 equal to a minima/maxima   **9** below

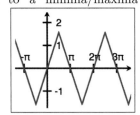

11 $\frac{\pi}{6}$ **13** $\frac{1}{\sqrt{3}}$ **15** $-\frac{\pi}{4}$ **17** $-\frac{\pi}{2}$ **19** $\frac{2\pi}{3}$ **21** $\frac{\sqrt{3}}{2}$ **23** $\frac{\sqrt{7}}{4}$ **25** $\frac{x}{\sqrt{x^2+1}}$ **27** $\frac{5x}{\sqrt{1-25x^2}}$ **29** $\sqrt{x+1}$ **31** $\sqrt{1-x^2}$ **33** $\frac{\sqrt{9-25x^2}}{5x}$ **35** $\frac{\sqrt{x^2+36}}{x}$ **37** $-\frac{1}{2}\sqrt{4-x^2}$ **41** $\frac{t-8}{17}$ **43** $\frac{20t+1}{9t-3}$ **45** $\frac{t^3-3}{t^3-1}$ **47** $\sqrt[3]{\frac{t-8}{37}}$ **49** $\log_3(\log_2(t-8))$ **51** $\sqrt[3]{\frac{\ln t - 84}{5}}$ **53** $\frac{1}{e^t-7}$ **57**

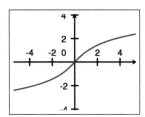

65 $\sqrt[3]{\frac{2}{3}}$ **67** The domain is $\left[-\frac{3}{8}, \infty\right)$ and the function is $f^{-1}(t) = \frac{-3+\sqrt{16t+265}}{8}$.   Other answers are possible.   **69** (a) Yes (b) below

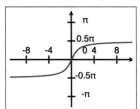

(c) Domain: all real numbers, Range: $\left(-\frac{\pi}{2}, \frac{\pi}{2}\right)$

Chapter 1 Review:

1 T   **3** T   **5** F   **7** F   **9** F   **11** F   **13** F   **15** F   **17** T   **19** F   **21** T   **23** T   **25** c   **27** b   **29** b   **31** d   **33** d   **35** a   **37** $F = k\,y(t)$ **39** (b) 10.409   **41** (b) APR $= 100\left(\left(1+\frac{r}{100m}\right)^m - 1\right)$; (c) APR $= 100\left(e^{\frac{r}{100}} - 1\right)$ **47** $T_{28}(x) \approx \cos(x)$ **49** $\sin\left(\frac{2x}{\pi} + 238t\right)$ **51** $t^2 + 2th + h^2 + 17t + 17h + 1$ **53**

$2\sqrt{2}$ **57** $\left(-\frac{\pi}{2}, \frac{\pi}{2}\right)$ **59** (a) $6\log_4 x + 5\log_4 y - 5\log_4(x+y)$; (b) $\log\left(\frac{x^3}{yz^7}\right)$; (c) $\ln\left(\frac{x^3}{y^{1/\log 3} \cdot z^{7/\log 2}}\right)$ **61** $11,460,000$

Section 2.1:

1 $f(2) = 2, \lim_{t\to2} f(t) = 3$ **3** $f(1.5) = 2.5, \lim_{t\to1.5} f(t) = 2.5$ **5** $\lim_{t\to-1^-} f(t) = 2, \lim_{t\to-1^+} f(t) = -1$ **7** $f(1) = 2$, and $\lim_{t\to1} f(t)$ does not exist   **9** $f(15) = 1/2$, and $\lim_{t\to1.5} f(t) = 1/2$ **11** $\lim_{t\to-1^-} f(t) = 1$, and $\lim_{t\to-1^+} f(t) = 3$ **13** 0   **15** 20   **17** 1   **19** $\frac{1}{2}$ **21** ∞ **23** nothing   **25** $f =$ amount of fuel, $p =$ purchase price, $\lim_{t\to15^-} p = 75$ **27** $s =$ speed, $I =$ inertia, $\lim_{s\to c^-} I = \infty$ **33** $[A]$ cannot approach 0 from the left   **41** below

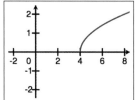

Section 2.2:

1 4   **3** $\sqrt{3}$ **5** 9   **7** 15   **9** 0   **11** 18   **13** 0   **15** $\sqrt[5]{47}$ **17** $\frac{29}{36}$ **19** $\frac{5}{3}$ **21** does not exist   **23** $\frac{5}{2}\left(\frac{25}{4}\right)^{\frac{1}{3}}$ **25** $\sqrt[4]{\frac{1}{10}}$ **27** 8   **29** $\frac{1}{6}$ **31** $\frac{1}{32}$ **33** $\frac{8}{7}$ **35** $\frac{8}{7}$ **37** 0   **39** $\frac{64}{49}$ **41** $\frac{3}{17}$ **43** $\frac{3}{4}$ **45** 12   **47** (a) $m(h) = \frac{\sqrt{4+h}-2}{h}$, (b) $\frac{1}{4}$ **49** (a) $B(a) = \frac{1}{2a}$; (b) $\lim_{a\to0^+} B(a) = \infty$ **63** below

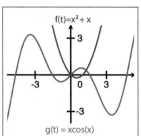

g(t) = xcos(x)

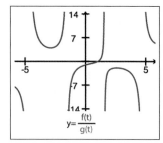

$y = \frac{f(t)}{g(t)}$

65 (a) no; (b) no   **67** below

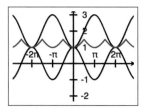

71 (b) ∞; (c) $A(\theta)$

Section 2.3:

1 $\frac{3}{8}$ **3** $\frac{\sqrt{5}}{18}$ **5** 12   **7** 0   **9** $\frac{4}{5}$ **11** $\frac{15}{2}$ **13** ∞ **15** ∞ **17** $\frac{4}{5}$ **19** ∞ **21** 7   **23** 7   **25** ∞ **27** $\frac{1}{8}$ **29** 9   **31** $\frac{7}{8}$ **33** -9

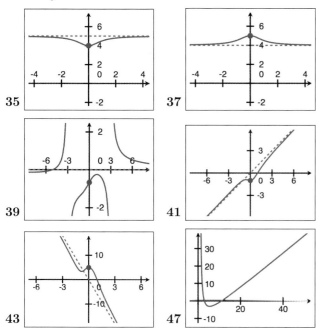

35   **37**   **39**   **41**   **43**   **47**

49 0   **51** 0   **53** 3   **55** 3   **57** 0
59 $\lim_{t \to \infty} a(t) = 100$ **61** $\lim t \to \infty P(t) = 0$
69 (a) $f(n)$; (b) $g(n)$; (c) $g(n)$; (d) $f(n)$

Section 2.4:

1 $\delta = 0.1$ **3** $\delta = 0.1$ **5** $\delta = 0.5$ **7** $\delta = 0.07$ **9** $\delta = 0.05$ **11** $\delta = 0.05$ **13** $(t-4)^2 + 5(t-4) + 16$ **15** $6(t+3)^2 - 23(t+3) + 8$ **17** $(t-1)^2 + 3(t-1)^2 + 5(t-1) + 8$ **19** $(t-5)^2 + 2(t-5) + 1$ **21** $-2(t - \frac{7}{6})^2 - \frac{35}{3}(t - \frac{7}{6}) - \frac{152}{9}$ **23** $-(t - \sqrt{6})^2 + (1 - 2\sqrt{6})(t - \sqrt{6}) + 11 + \sqrt{6}$ **25** $\delta = \min\{1, \frac{\varepsilon}{12}\}$ **27** $\delta = \min\{1, \frac{\varepsilon}{24}\}$ **29** $\delta = \min\{1, \frac{\varepsilon}{6}\}$ **31** $\delta = \min\{1, \frac{\varepsilon}{4}\}$ **33** $\delta = \min\{1, \frac{\varepsilon}{8}\}$ **35** $\delta = \min\{1, \frac{\varepsilon}{6}\}$ **37** $\delta = \log(1 + \varepsilon)$ **39** $\delta = 1 + \log(1 + \varepsilon/10)$ **41** $\delta = 2 + \log(1 + \varepsilon/100)$ **43** $M = \sqrt{\frac{3}{8\varepsilon} - \frac{1}{8}}$ **45** $M = \ln(\varepsilon)$ **47** $M = \frac{3}{(1 + \varepsilon/4)^2 - 1}$ **49** $\delta = \frac{1}{N}$ **51** $\delta = \frac{1}{36N}$ **53** $\delta = \sqrt[4]{7/N}$ **55** $\delta = \frac{27}{6N}$ **57** $\delta = \frac{24}{N}$ **59** $\delta = 32 - \left(\frac{3}{N} + 2\right)^5$ **61** (a) 4; (b) 50; (c) 50; (d) $y(6)$ **63** We say $\lim_{t \to \infty} f(t) = -\infty$ if each number $N > 0$ is associated with a number $M > 0$ such that $f(t) > N$ whenever $t < M$.   **65** $\delta = \frac{\varepsilon}{2}$ **67** $\delta = \sqrt{\varepsilon}$ **69** (a) $\delta(1) = \sqrt{1 + \varepsilon} - 1$; (c) $\delta(t) = -t + \sqrt{t^2 + \varepsilon}$; (d) below

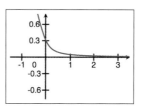

71 Nothing   **73** $93 - \sqrt{0.24} \le f(7.24) \le 93 + \sqrt{0.24}$
75 We say $\lim_{t \to t_0^+} f(t) = L$ when every $\varepsilon > 0$ is associated with a distance $\delta > 0$ such that $|f(t) - L| < \varepsilon$ whenever $t_0 < t < t_0 + \delta$.

Section 2.5:

1 (a) discontinuous at $t = -2.5, \lim_{t \to -2.5^-} f(t) = 2, \lim_{t \to -2.5^+} f(t) = -2$, and discontinuous at $t = 2, \lim_{t \to 2^-} f(t) = 2.5, \lim_{t \to 2^+} f(t) = 1$; (b) at $t = 1$ set $f(1) = 1$; (c) $t = 2$, (d) none   **3** (a) discontinuous at $t = -2, \lim_{t \to -2^-} f(t) = 2, \lim_{t \to -2^+} f(t) = 0.5$, and discontinuous at $t = 1, \lim_{t \to 1^-} f(t) = \lim_{t \to 1^+} f(t) = 1$, and discontinuous at $t = 2, \lim_{t \to 2^-} f(t) = 3, \lim_{t \to 2^+} f(t) = 1.5$; (b) at $t = 1$ set $f(1) = 1$; (c) $t = -2$, (d) $t = 2$ **5** $f(t_0) = \frac{8}{5}$ **7** $f(t_0) = 24$ **9** $f(t_0) = 14$ **11** $\lim_{t \to -} f(t) = \lim_{t \to 0^+} f(t) = k, k = 0$ or 1   **13** $\lim_{t \to 0^-} f(t) = \lim_{t \to 0^+} f(t) = \frac{1}{k}, k = 2 \pm \sqrt{5}$ **15** $\lim_{t \to 0^-} f(t) = 15k - 2$ and $\lim_{t \to 0^+} f(t) = 3k^2 + 10k - 10$; $\lim_{t \to 1^-} f(t) = 2k^2 + 13k + 5$ and $\lim_{t \to 1^+} f(t) = k - 5$; $k = -1$ **17** all $t \in \mathbb{R}$ **19** $(-\infty, -4) \cup (-4, -3) \cup (-3, \infty)$ **21** $(-\infty, -4) \cup (-4, -3) \cup (-3, \infty)$ **23** 112   **25** $(-\infty, \frac{-5 - \sqrt{73}}{2}] \cup [\frac{-5 + \sqrt{73}}{2}, \infty)$ **27** $\ldots \cup (\frac{4\pi}{3}, \frac{8\pi}{3}) \cup (\frac{10\pi}{3}, \frac{14\pi}{3}) \cup \ldots$ **29** -7 **31** 1   **33** 2   **35** 1.5   **37** $\frac{5}{16}$ **39** a, c, d, e   **41** (a) $\lim_{t \to b^-} p(t) = 80$; (b) $\lim_{t \to b^+} p(t) = 100$

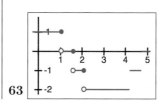

49

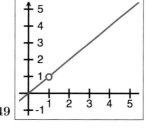

51

57 19   **59** $f(t) = 1/t$ isn't continuous between -1 and 1   **61** (b) $\frac{887}{1024}$

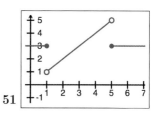

63

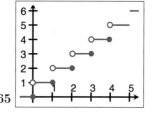

65

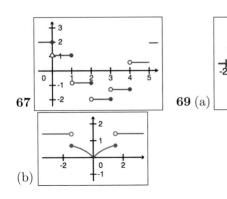

67

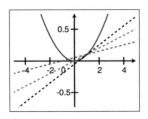

69 (a)

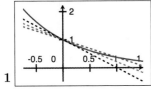

(b)

(b) greater; (c) below, above   **35** 3   **37** 3   **39**
1/2   **41** 13   **43** $\frac{2}{e^2} - \frac{1}{e}$ **45** (a) barrens; (b) both
have $\frac{1}{50}\frac{\text{mm}}{\text{day}}$ during the first 50 days. The barrens grows
at $\frac{0.75}{100}\frac{\text{mm}}{\text{day}}$ and the kelp bed urchins grow at $\frac{2.25}{100}\frac{\text{mm}}{\text{day}}$;
(c) barrens $\approx 0.005\frac{\text{mm}}{\text{day}}$, kelp bed $\approx 0.025\frac{\text{mm}}{\text{day}}$ **49**
$y = 4 - 3(x - 2)$ **51** see below

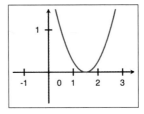

Chapter 2 Review:

1 T   **3** T   **5** F   **7** F   **9** F   **11** T   **13** T   **15**
T   **17** d   **19** e   **21** d   **23** d   **25** a   **27** 18   **29**
16   **31** $\frac{1}{3}$ **33** $-\infty$ **35** 0   **37** $\frac{25}{6}$ **39** 9   **41** does
not exist   **47** (a) 100; (b) 250

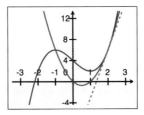

53 (a) $4\pi r^2$; (b) $24\pi < 40\pi$ **55** (a) $-6\frac{\text{m}}{\text{s}}$; (b) $-3\frac{\text{m}}{\text{s}}$ **57**
(a) $\frac{\text{m}}{\text{sL}}, \frac{\text{m}}{\text{L}}, \frac{\text{m}}{\text{L}}, \frac{\text{L}}{\text{ms}}$ **59** (a) $\frac{\text{Pa}}{\text{s}}$; (b) $\frac{Nk_B}{V}$ **61** (a) $\frac{°\text{C}}{\text{s}}$ **63**
$\frac{\text{sn}}{\text{min}}$

Section 3.1:

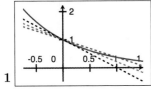

1

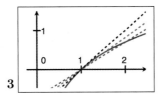

3

5 2   **7** -64 **9** 13.5   **11** (a) 13; (b) $13(t-1)$; (c)
below

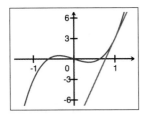

13 (a) 0; (b) -80; (c) below

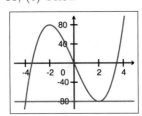

15 (a) 32; (b) 8   **17** (a) 2; (b) 3   **19** (a) 21; (b)
-4 **21** (a) 9; (b) $9 + 6\Delta h + \Delta h^2$; (c) 6   **23** (a) -8;
(b) $-8 + 12\Delta h - 6\Delta h^2 + \Delta h^3$; (c) -6 **25** (a) 9; (b)
$9 + 14\Delta h + \Delta h^2$; (c) 14   **27** 7   **29** 0   **31** b   **33** (a)
see below

Section 3.2:

1 $f'(x) = 6x - 8$ **3** $f'(x) = 3x^2$ **5** $f'(x) =$
$-\frac{3}{(3x+1)^2}$ **7** $f'(x) = \frac{3}{2\sqrt{3x+1}}$ **9** $f'(x) = -\frac{1}{2x^{\frac{3}{2}}}$ **11**
$f'(x) = -\frac{\beta}{(\beta x+\gamma)^2}$ **13** $(f+g)'(1) \approx \frac{5}{6}$ **15**
$(2f+3g)'(1.4) \approx 0$ **17** $f'(x) = 8 + 42x^6 + 3x^{-4}$ **19**
$f'(x) = 5.4 - 24x^3$ **21** $f'(x) = \frac{1}{2\sqrt{x}} + \frac{1}{3x^{\frac{2}{3}}}$ **23**
$f'(x) = \frac{5}{2}x^{3/2}$ **25** 15   **27** $-\frac{7}{2^{13}}$ **29** $\frac{106}{27}$

37 (a) & (c) below; (b) they cause a vertical shift; (d)
they are parallel;

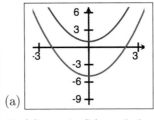

(a)

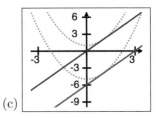

(c)

39 (a) $x = 2$, (b) see below

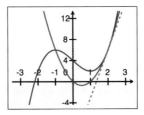

41 (b) $f'(4) = -3$; (a) & (c) see below

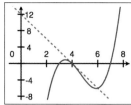

43 (b) $f'(1) = 8$; (a) & (c) see below

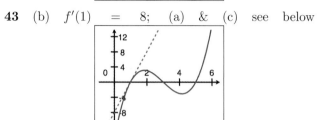

47 (a) $b = at^2 + \frac{1}{2a}$, (b) $R = \sqrt{t^2 + \frac{1}{4a^2}}$, (c) $t = \pm\sqrt{R^2 - \frac{1}{4a^2}}$ **49** (a) $v_f = \pi r^2 \sqrt{19.62h}$; (b) $v_f = \frac{r^2\sqrt{\pi}}{R}\sqrt{19.62v}$; (c) $\frac{m^2}{sec}$; (d) $\frac{r^2\sqrt{19.62\pi}}{24\sqrt{v}}v'$ **51** (a) $P'(\rho) = RT + 2RTB\rho$, $\frac{Nm}{mol}$ **53** (a) 4; (b) $I'(R) = -\frac{80}{R^2}$ **55** (a) $v(t) = 2t - 3$; (b) $t < \frac{3}{2}$; (c) see below

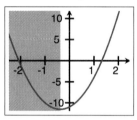

57 (a) $v(t) = -6t + 9$; (b) $t > \frac{3}{2}$; (c) see below

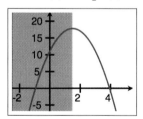

59 (a) $t = 0, \frac{2}{3}$, (b) see below

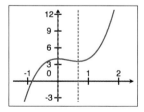

61 (a) $t = -3, 1$ (b) see below

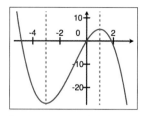

63 $a = -6x10^{-5}, b = -3x10^{-3}, c = 0.4, d = 0$ **65** (a) $-3\beta s^{-4}$; (b) $\frac{likelihood}{meter}$

Section 3.3:

1 $y = 3x + 1$ **3** $y - 16 = 16\ln(2)(x - 4)$ **5** $y - \frac{7}{2} = -\frac{7\sqrt{3}}{2}(x - \frac{\pi}{3})$ **7** $y + \frac{1}{2} = -\sqrt{3}(x - \frac{\pi}{3})$ **9** $y + \frac{7}{2} = -7\sqrt{3}(x - \frac{\pi}{3})$ **11** $y - \beta\cos(\omega t) = -\beta\omega\sin(\omega t)(x - t)$ **13** (a) $f'(x) = -e^{-x}$; (b) $f'(x) = (\frac{1}{e})^x \ln(\frac{1}{e})$ **17** $t = \ln(1 \pm 2\sqrt{2})$ **19** $y = e^{7t}$ **21** $y = e^{3t}$ **23** (a) $x = \frac{3\pi}{4} + n\pi$, for $n \in \mathbb{N}$; (b) $f'(x)$ has a maximum; graph below

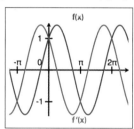

25 (a) $f'(x) = 1 - \sin(x)$; (b) $f'(x) = 0$ at $x = \frac{\pi}{2} + n\pi$, for $n \in \mathbb{N}$; graph below

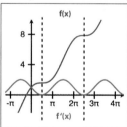

27 (a) $x^2 + 16 - 8x\cos(\theta)$; (b) $dx/d\theta$ has units of millimeters per radian, and tells us the rate at which x decreases when θ increases **29** (a) 625 ft; (b) see below

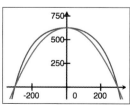

31 (a) $3.54215\cos\left(\frac{\pi t}{6}\right)$; (c) see below

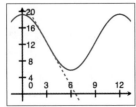

33 (a) $-1.00819\sin\left(\frac{2\pi}{12.421}t\right)$, $H'(5) \approx -0.579$, $H'(6.2105) \approx 0$; (c) see below

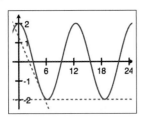

35 (a) $\bar{q} \approx -0.878227\frac{\text{C}}{\text{s}}$; (b) $q'(0.8) \approx -0.852144\frac{\text{C}}{\text{s}}$ **37** (a) 1800   **39** $-0.8e^{-0.4t}$ **41** $c'(t) = \alpha c(t)$, where α is some constant   **43** $t = 6.02\log_2(e)$ **45** (b) $\alpha e^{\beta x} - \alpha$ **47** negative

Section 3.4:

1 $x = -2, 1$ **3** (a) $-\frac{7}{3}$; (b) $x < -\frac{7}{3}$; (c) see below

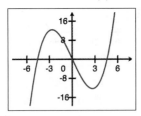

5 (a) $\pm\sqrt{7}$; (b) $-\sqrt{7} < x < \sqrt{7}$; (c) see below

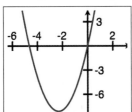

7 (a) 0; (b) $x < 0$; (c) see below

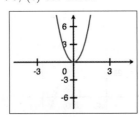

9 (a) $x = -\frac{1}{2}, \frac{3}{4}$; (b) $-\frac{1}{2} < x < \frac{3}{4}$; (c) see below

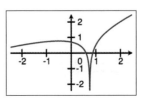

11 (a) $x = \pi n, \pi n - \frac{pi}{2}, n \in \mathbb{Z}$; (b) $\pi n - \frac{pi}{2} < x < \pi n$; (c) see below

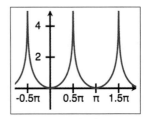

13 (b) no; (c) 3; (d) -7; (a) & (e) see below

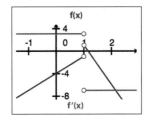

17 (b) no; (c) $-2x$; (d) 2; (a) & (e) see below

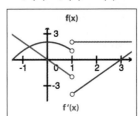

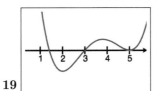 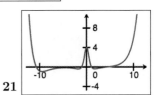

19   **21**

23 (a) 0, 1; (b) $0 < x < 1$; (c) see below

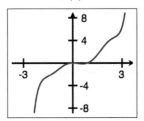

25 (a) $-2, \frac{1}{2}, \frac{3}{4}$; (b) $\frac{1}{2} < x < \frac{3}{4}$; (c) see below

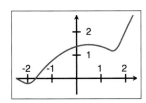

27 $x(3) = 4, x'(3) = 5$ **29** $x(10) = -8, x'(10) = -6$ **31** (a) $3 < t < \frac{3}{2}$; (b) $\frac{3}{2} < t < 4$; (c) $4 < t$; (d) $t < 3$ **33** (a) $\frac{1}{3}(-7 + \sqrt{52}) < t < \frac{1}{2}(-7 + \sqrt{53})$; (b) $t < \frac{1}{2}(-7 - \sqrt{53}), 0 < t < \frac{1}{3}(-7 + \sqrt{52})$; (c) $\frac{1}{2}(-7 - \sqrt{53}) < t < \frac{1}{3}(-7 - \sqrt{52}), \frac{1}{2}(-7 + \sqrt{53}) < t$; (d) $\frac{1}{3}(-7 - \sqrt{52}) < t < 0$ **41** $0, 1, 7, 12$ **43** $g(t)$ is not defined at zero   **45** (a) Yes. If $a > 0$ and $b > 0$; (b) Yes. If $a < 0$ and $b < 0$.   **49** Yes, if a and b are both positive or both negative.   **51** See below (other answers possible)

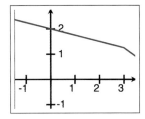   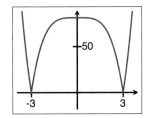

61 (b) a corner; (a) see below

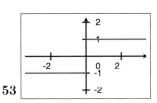

63 negative   **65** positive   **67** $y = a\left(R^2 - \frac{1}{4a^2}\right)$ **69** (a) $\frac{\text{population}}{\text{year}}$ **71** (a) $c'(0)$ is positive; (b) c will increase in the next moment; (c) positive, but smaller than $c'(0)$; (d) 1

Section 3.5:

1 $2(6x - 1) + 6(2x + 4)$ **3** $26(13x + 7)^2$ **5** $\sin(6x) + 6x\cos(6x)$ **7** $21x^2e^{5x} + 36x^3e^{5x}$ **9** $0.1e^{0.1x}\cos(5x) - 5e^{0.1x}\sin(5x)$ **11** $7\cos(7x)\cos(5x) - 5\sin(7x)\sin(5x)$ **13** $-36\csc^2(18x)\cot^2(18x)$ **15** $2x9^x\cos(10x) + \ln(9)x^29^x\cos(10x) - 10x^29^x\sin(10x)$ **17** $15\cos(5x)\sin^2(5x)$ **19** $\frac{4(9x+3) - 9(4x+8)}{(9x+3)^2}$ **21** $\frac{(2x^2-8x-10)(6x+9) - (4x-8)(3x^2+9x+1)}{(2x^2-8x-10)^2}$ **23** $\frac{2xe^x - (x^2+1)e^x}{(x^2+1)^2}$ **25** $\sec(x)\tan(x)$ **27** $\frac{(2x+1)\sin(8x) - 8(x^2+x+1)\cos(8x)}{\sin^2(8x)}$ **29** $\frac{0.5x^{\frac{1}{2}}(4x-1) - 4\sqrt{x}}{(4x-1)^2}$ **31** $\frac{(x^2-4)\sec^2(x) - 2x\tan(x)}{(x^2-4)^2}$

33 $\frac{-9\sin(9x)(1+\sqrt{x}) + \cos(9x)x^{-\frac{1}{2}}}{(1+\sqrt{x})^2}$ **35** $\frac{(6e^{6x}\cos(2x) - 2e^{6x}\sin(x))(x^2+2) - 2xe^{6x}\cos(2x)}{(x^2+2)^2}$ **37** $y = -2e^{-3}(t - 3) + 3e^{-3}$ **39** $y = x$ **41** $y = -\frac{\pi}{2}e^{2t}(x - 1)$ **43** does not exist   **45** $y = -\frac{1}{10}(x - 2) + \frac{12}{100}$ **47** $y = -\frac{10}{9\pi+122}(x - \frac{\pi}{2})$ **49** (a) $2(x + 5)$; (b) $2x + 10$ **51** (a) $2e^x(1 + e^x)$; (b) $2e^x - 2e^{2x}$ **53** (a) $2(e^x + e^{-x})(e^x - e^{-x})$; (b) $2(e^{2x} - e^{-2x})$ **55** $\text{sech}^2(\theta)$ **57** $-e^{-x}\sin(x) - e^{-x}\cos(x)$ **67** (b) 2.2   **69** $\frac{\text{Pa}}{\text{m}^3}$ **77** see below

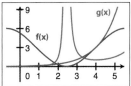

81 $\Delta x > \Delta y$ **85** (a) $V'(t) = I'(t)R(t) + I(t)R'(t)$ **87** (d) $R'(p) = f(p) + pf'(p)$ **89** (a) elastic; (b) increase the price

Section 3.6:

1 $\frac{13}{2\sqrt{13x-4}}$ **3** $28x\sin(7 - 14x^2)$ **5** $e^{\tan(x)}\sec^2(x)$ **7** $-9\sin(3x + 9)\cos^2(3x + 9)$ **9** $48\tan^2(2x)\sec^2(2x) - 112\tan(7x)\sec^4(7x)$ **11** $-5\sin(\sin(5x))\cos(5x)$ **13** $48x^2(8x^3 - 4)(3x^2 + 7)^6 + 36x(8x^3 - 4)^2(3x^2 + 7)^5$ **15** $\frac{2x^2}{\sqrt{4x+9}} + 2x\sqrt{4x + 9}$ **17** $3x^27^{\cos(x)} - x^3\ln(7)\sin(x)7^{\cos(x)}$ **19** $-\left(\frac{24x}{19x^3-3x^2} - \frac{(12x^2-5)(57x^2-6x)}{(19x^3-3x^2)}\right)\cot\left(\frac{12x^2-5}{19x^3-3x^2}\right) \times \csc\left(\frac{12x^2-5}{19x^3-3x^2}\right)$ **21** $\frac{10(2x+5)^4}{(9x-2)^5} - \frac{45(2x+5)^5}{(9x-2)^6}$ **23** $\frac{12e^{3x}(e^{3x}+1)^4}{(1-e^{3x})^5} + \frac{12e^{3x}(e^{3x}+1)^3}{(1-e^{3x})^4}$ **25** $\frac{9(3x-15)^2}{(19x-2)^7} - \frac{133(3x-15)^3}{(19x-2)^8}$ **27** $\frac{e^x}{\sqrt{x^2-9x+1}} - \frac{(e^x+2)(2x+9)}{2(x^2+9x+1)^{\frac{3}{2}}}$ **29** $\frac{1}{\sqrt{\frac{x}{3x+1}}}\left(\frac{1}{(3x+1)^2}\right)\tan\left(2\sqrt{\frac{x}{3x+1}}\right)\sec\left(2\sqrt{\frac{x}{3x+1}}\right)$ **31** $-2\tan(2x)\sec(2x)\sec^2(\cos(\sec(2x)))\sin(\sec(2x))$ **33** $16x^{15} + 112x^{13} + 384x^{11} + 800x^9 + 1104x^7 + 1008x^5 + 576x^3 + 160x$ **35** $\sinh(x^2\sin(7x) + 7)(7x^2\cos(7x) + 2x\sin(7x))$ **37** $n\left(\frac{\alpha x+\beta}{\gamma x+\delta}\right)^{n-1}\left(\frac{\alpha}{\gamma x+\delta} - \frac{\gamma(\alpha x+\beta)}{(\gamma x+\delta)^2}\right)$ **39** $e^{(x/\sigma)^2}\left(\frac{2x}{\sigma^2}\right)$ **41** $\frac{x^4+29x^2-2x-18}{(x^2+9)^2}$ **43** $\frac{7}{\cos(x)+3} + \frac{(7x+1)\sin(x)}{(\cos(x)+3)^2}$ **45** $\frac{\cos(x)}{3\sin^{2/3}(x)}$ **47** $\cos(x)\cos(\sin(x))$ **49** $\frac{24x^7-2}{3(3x^8-2x+1)^{2/3}}$ **51** $\frac{\cos(x)\cos(\sin(x))}{3\sin^{2/3}(\sin(x))}$ **55** (a) $10(5t + 3)$; (b) $5(5t + 3) + 5(5t + 3)$; (c) $50t + 30$ **57** $4e^{4t}$ **59** -28 **61** $\sec^2(g + h)(g' + h')$ **63** $-\frac{4g'}{f^4g^5} - \frac{4f}{f^3g^4}$ **65** $0.1\frac{\text{kg}}{\text{day}}\frac{1}{24}\frac{\text{day}}{\text{hour}}$ **67** up   **69** left   **71** (b) 1   **73** (a) $K(6x^{-7} - 21112.9x^{-13})$; (b) $K \times 10^8(6(x \times 10^8)^{-7} - 21112.9(x \times 10^8)^{-13})$; (c) 3.9×10^{-8} **75** positive   **77** negative   **79** (a) $422\frac{\text{mi}}{\text{hr}}$; (c) $\frac{1}{\text{sec}}$ **81** (a) 12; (b) 1.2×10^7; (c) $\Omega_A(q)\Omega_B(Q - q)$;

(d) $\Omega'_A(q)\Omega_B(Q-q) < \Omega_A(q)\Omega'_B(Q-q)$

Section 3.7:

1 $f'(t) = \frac{4}{1+16t^2}$ **3** $f'(t) = \frac{6t}{\sqrt{1+9t^4}}$ **5** $f'(t) = \frac{\cot(t)}{\sqrt{\sin^2(t)-1}}$ **7** $f'(t) = \csc^{-1}(7t) - \frac{1}{\sqrt{49t^2-1}}$ **9** $f'(t) = \frac{2\cos^{-1}(t)\sqrt{1-t^2}+2t}{(\cos^{-1}(t))^2\sqrt{1-t^2}}$ **11** $f'(t) = 6t\sin^{-1}(3t^2) + \frac{18t^3}{\sqrt{1-9t^4}}$ **13** $f'(t) = \frac{-4}{(1+16t^2)\cot^{-1}(4t)}$ **15** $f'(t) = \frac{2t}{[1+\ln^2(t^2+1)](t^2+1)}$ **17** $f'(t) = \frac{2\sec^2(\sin^{-1}(2t))}{\sqrt{1-4t^2}}$ **19** $f'(t) = \frac{1}{2(1+t^2)\sqrt{\tan^{-1}(t)}}$ **21** $f'(t) = e^{-t}\left(\frac{2}{\sqrt{1-4t^2}} - \sin^{-1}(2t)\right)$ **23** $f'(t) = -\tan(t)$ **25** $f'(t) = \sec(t)$ **27** (a) $f'(t) = \frac{4}{t}$, (b) $f'(t) = \frac{4}{t}$ **29** (a) $f'(t) = -2/t$, (b) $f'(t) = -2/t$ **31** (a) $f'(t) = \frac{-10}{(2t+1)(4t-3)}$, (b) $f'(t) = \frac{2}{2t+1} - \frac{4}{4t-3}$ **33** (a) $f'(t) = \frac{2\sin(t)+t\cos(t)}{t\sin(t)}$, (b) $f'(t) = \frac{2}{t} + \cot(t)$ **35** (a) $f'(t) = \frac{2}{\ln(2)t}$, (b) $f'(t) = \frac{2}{\ln(2)t}$ **37** (a) $f'(t) = \frac{1+t\tan(t)}{t}$, (b) $f'(t) = \frac{1}{t} + \tan(t)$ **39** $f'(t) = 0.5 - \frac{12}{12t+7}$ **41** $f'(t) = \frac{3}{2t} - \frac{1}{2(t-1)}$ **43** $f'(t) = \frac{2}{\ln(5)t} - \frac{2t}{\ln(5)(t^2-1)}$ **45** (a) $f^{-1}(x)' = \frac{1}{3}$, (b) $f^{-1}(x) = \frac{1}{3}$ **49** $(f^{-1})'(3) = \frac{1}{7}$, so the inverse function is increasing. **51** $t'(0.5) = -17.37$; It takes 17.37 hours for the pertechnetate to decay by 1 gm. **53** $D'(6 \times 10^{-11}) = \frac{1.667\times10^{11}}{\ln(10)}$; the sound is increasing at a rate of $\frac{1.667\times10^{11}}{\ln(10)}$ decibels per watt. **55** $t'(2000) = -\frac{30}{1999}$; the population decreases by 1999 every 14 days. **57** $M'(E) = \frac{1}{1.5\ln(10)E}$; the magnitude is increasing at a rate of $M'(30,000) = 9.651 \times 10^{-6}$ per joule of energy. **59** $\theta'(40) = -\frac{1}{82}$; at the current distance, the angle is decreasing by $1/82$ radians per 1 foot of horizontal distance the eagle flies away. **61** (a) $\theta = \sin^{-1}\left(\frac{r\sin(\phi)}{\sqrt{r^2+(r+h)^2-2r(r+h)\cos(\phi)}}\right)$; (b) $\frac{d\theta}{d\phi} = A/B$ where $A = (r^2 + (r+h)^2) - 2r(r+h)\cos(\phi))r\cos(\phi) - r^2(r+h)\sin^2(\phi)$ and $B = (r^2+(r+h)^2 - 2r(r+h)\cos(\phi))^{3/2}\sqrt{1 - \frac{r^2\sin^2(\phi)}{r^2+(r+h)^2-2r(r+h)\cos(\phi)}}$.

Section 3.8:

1 $y' = -\frac{x}{y}$ **3** $y' = -\frac{2x+2}{3y^2}$ **5** $y' = -\frac{8x+9y}{9x-4y}$ **7** $y' = -\tan(y)$ **9** $y' = \frac{2x^2-y^2}{2xy\ln(x)}$ **11** $y' = \frac{1}{y\sqrt{1-4x^2}}$ **13** $x = 2$ **15** $y = -4.9x + 5.9$ **17** $y = \frac{2\pi}{3} - x$ **19** $y = -\frac{3}{5}x$ **21** $y = \frac{\pi}{6} - x$ **23** $y = \frac{7\pi}{8} - \frac{3}{4}x$ **25** $x = -0.4725$ **27** $y = -0.5333$ **29** $y' = (2\ln(x)+2)x^{2x}$ **31** $y' = \left(\frac{1}{x}\tan(x) + \sec^2(x)\ln(x)\right)x^{\tan(x)}$

33 $y' = \left(\frac{1}{x}e^{x^2} + 2xe^{x^2}\ln(x)\right)x^{e^{x^2}}$
35 $y' = (5 + 5\ln(x) + 3\cot(3x))x^{5x}\sin(3x)$
37 $y' = \left(\frac{1}{x}\tan(x) + \ln(x) + \sec(x)\tan(x)\right)x^{\sin(x)}$
39 $y' = 2x - (1+\ln(x))x^x$ **41** $y' = \frac{1}{2}\left(\frac{45(4x^3-3)}{x^4-3x} + \frac{13(7x^6-36x^5)}{x^7+6x^6}\right)\sqrt{(x^4-3x)^{45}(x^7+6x^6)^{13}}$
43 $y' = \left(\frac{12}{x} + 6\pi\cot(\pi x) + \frac{x}{x^2-1}\right)9x^{12}\sin(\pi x)\sqrt{x^2-1}$
45 $y' = \left(\frac{4}{x} + \frac{6x^2}{8x^3-5} - 3\sec(3x)\csc(3x)\right)\frac{30x^4\sqrt[4]{8x^3-5}}{\tan(3x)}$
47 $y' = \left(\frac{4}{8x-7} - 0.15 - \frac{2}{x} - \frac{33x-8}{6(11x^2-2x^3)}\right)\sqrt{\frac{(8x-7)e^{-0.3x}}{x^4\sqrt[3]{11x^3-2x^4}}}$
51 $y' = \frac{b^2x}{a^2y}$ **53** $\left(\frac{1}{2\sqrt{2}}, \frac{1}{2\sqrt{2}}\right)$ and $\left(\frac{-1}{2\sqrt{2}}, \frac{-1}{2\sqrt{2}}\right)$ **55** $(1,2)$; $(-1,-2)$ **57** $(-\sqrt{2},0)$, $(\sqrt{2},0)$ **59** $(0.8603, 1.317)$, $(0.8603, 0.359)$ **61** (a) Equatorial Radius: 6378.1370 km, Height: 2641.9108 km; (b) 10269 km. **63** $-\frac{a}{b}f'(ax+by)$

Chapter 3 Review p. 227:

1 F **3** T **5** T **7** T **9** F **11** T **13** T **15** F **17** F **19** F **21** c **23** d **25** e **27** c **29** e **31** $y - 13 = 16(x-2)$ **33** $y = \frac{\pi}{3} + \frac{[\sqrt{3}]}{2} + 1 - x$ **35** $y - 2e^6 = 10e^6(x-3)$ **37** $y - 2 = \frac{3}{2}(x-1)$ **39** $y' = \frac{\sin y + y \sin x}{\cos x - x \cos y}$ **41** $y' = \frac{ye^{xy}-e^{x+y}}{e^{x+y}-xe^{xy}}$ **43** $y' = 2x\cot^2 y$ **45** $y' = \frac{e^y-2x}{3y^2-x}$ **47** (a) $V'(t) = 3[\ell(t)]^2\ell'(t)$; (b) $m^3 = m^2\frac{\text{m}}{\text{s}}$ **49** (a) the right-hand side is at most 1, the left-hand side is > 1 if $x = 0$ is not a solution; (b) $y' = \frac{4\cos 2x\cos 2y - e^{0.5x}}{4\sin 2y}$; (c) roots of $\cos[2x, x = \frac{\pi}{4} + \frac{\pi}{2}n$; (d) 0; (e) $\frac{\pi}{2} + n\pi$, y is free. **51** $-\frac{2(x-2)^5x^3(8x^5-64x^4-5x+4)}{(8x^4+1)^4}$ **53** $(x^4 + 3)^{\sin(8x+1)}\left(8\ln(x^4+3)\cos(8x+1) + \frac{4x^3\sin(8x+1)}{x^4+3}\right)$ **55** (a) $x = \frac{1}{3}(7\pm2\sqrt{7})$; (b) $x < \frac{1}{3}(7-2\sqrt{7}), x > \frac{1}{3}(7+2\sqrt{7})$;

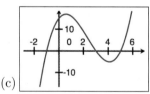

(c) **57**

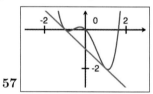

59 (a) $-1 < x < 1, 3 < x$; (b) $x < -1, 1 < x < 3$; (c) $x = -1, 1, 3$ **63** $\frac{\text{g}}{\text{cm}}$ **67** $\frac{\text{cal}}{^\circ\text{K}}$ **69** $V'(P) > 0$ **71** (a) $R'(f) = \tilde{\rho}$ **73** (b) $-0.06e^{-0.06(\tau-30)}$

75 $\frac{m_0}{c^2\sqrt{\left(1-\left(\frac{v}{c}\right)^2\right)^3}} \frac{1}{\text{s}}$

Section 4.1:

1 3.007 **3** 0.0011 **5** 2.09 **7** -0.1 **9** 4.9 **11** -1 **13** 4.1230 **15** 1.1323 **17** $-\frac{3}{4}$ **19** (a) $\alpha^2 - 17 = 0$; (b) $\alpha_0 = 4$; (c) 4.123 **21** (a) $\alpha^4 - 9 = 0$; (b) $\alpha_0 = 2$;

(c) 1.732 **23** $t = 0.53543$ **25** -0.26604 **27**
(a) $f\left(\frac{\pi}{3}\right) = \frac{\sqrt{3}}{2}$; (b) $df = 0.005$ **29** (a) $f(2) = 3$;
(b) $df = \dfrac{1}{150} \approx 0.0067$ **31** (a) $f(1) = 1$; (b)
$0.8 \le f(t) \le 1.2$ **33** (a) $f\left(\frac{\pi}{6}\right) = \sqrt{3}$; (b) $1.3321 \le$
$f(t_0) \le 2.1321$ **35** (a) $f(1 + \Delta t) = 8 + 5\Delta t + (\Delta t)^2$;
(b) $f(1 + \Delta t) \approx 8 + 5\Delta t$ **37** (a) $f(1 + \Delta t) =$
$3 + 4\Delta t + 3(\Delta t)^2 + (\Delta t)^3$; (b) $f(1 + \Delta t) \approx 3 + 4\Delta t$ **45**
$-\frac{1}{2}$ **47** $f(t_0) = f(t_6) = 1$ **51** $|t| \le \sqrt{3}$ **53**
(a) $p(d) \approx L(d) = 1.11135 \times 10^5 + 9810(d - 1)$, where
1.11135×10^5 is in pascals, 9810 is in $\frac{\text{Pascals}}{\text{meter}}$, and $(d-1)$ is
in meters; (b) $\Delta p \approx 9810 \Delta d$ **55** does not makes sense
(it says that volume decreases as height increases) **57**
makes sense (as the asteroids get father away, the gravitational force between them reduces) **59** does not
make sense (this says the amount you pay goes *down* the
longer you talk) **61** $df = -0.01dr$ **63** $dp = m\,dv$ **67**
(a) $\frac{s}{m^2}$ or $\frac{s/m}{m}$, (b) $830\sqrt{95.3}$, (c) $dq = \frac{830}{\sqrt{95.3}} \approx 85$ meters. **69** (a) 5.1044 ft, (b) $5.0407 \le h \le 5.1681$

Section 4.2:

1 12 **3** $\frac{-14}{\sqrt{95}}$ **5** (a) decreasing; (b) increasing **7**
24 **9** $\frac{5}{2\pi}\left(\frac{15}{\pi}\right)^{-2/3}$ **11** $-\frac{2800}{13}\frac{\text{rad}}{\text{sec}}$ **13** $-\frac{3}{13}\frac{\text{rad}}{\text{sec}}$ **15**
$\frac{1}{16}\frac{\text{m}}{\text{min}}$ **17** $\frac{10800}{\sqrt{901}}\frac{\text{km}}{\text{hr}}$ **19** $-90\sqrt{3}/13\frac{\text{radians}}{\text{hr}}$ **21**
(a) $\dfrac{d\theta}{dt} = -\dfrac{720\cos(\phi)}{5 + 24\sin(\phi)}\frac{\text{rad}}{\text{hr}}$ **23** $-31.25\frac{\text{ft}}{\text{min}}$ **25**
$25000\pi\frac{\text{ft}^2}{\text{hr}}$ **27** $-\dfrac{9}{80\sqrt{2-\sqrt{3}}}\frac{\text{m}}{\text{s}}$ **29** $\frac{\sqrt{65}}{8}\frac{\text{m}}{\text{s}}$ **31**
$12000\pi\frac{\text{cm}}{\text{min}}$ **33** $-\frac{1}{25\pi}\frac{\text{ft}}{\text{min}}$ **35** $\frac{dh}{dt} = -\frac{2}{13\pi}\frac{\text{ft}}{\text{hr}}$ **37**
$\dfrac{390 - 10\sqrt{106}}{\sqrt{163876 - 8268\sqrt{106}}}\frac{\text{ft}}{\text{sec}}$ **39** $\frac{24}{5}\frac{\text{ft}}{\text{sec}}$

Section 4.3:

1 0 **3** $8 + 20x^3$ **5** $4x^2 + \frac{12}{5}x^{-5}$ **7** $1.288x^{-0.6} +$
$0.0585x^{-2.3}$ **9** $-\frac{2}{9}x^{-\frac{5}{3}} - 4\left(x^2 + 4\right)^{-\frac{3}{2}}$ **11** $f''(x) =$
$-\frac{36}{(2x-1)^3}$ **13** (a) $f''(\frac{7\pi}{6}) = \frac{\sqrt{3}}{2}$, (b) concave up **15** (a)
$f''(0) = -49$, (b) concave down **17** (a) $f''(1) = 4e^2$,
(b) concave up **19** (a) $f''(1) = -1$, (b) concave
down **21** $f''(\frac{\pi}{4}) = 4$, (b) concave up **23** (a)
$f''(0) = 48$, (b) concave up **25** (a) see below; (b)
critical points: $\frac{\pi}{2} + n\pi$; (c) all are critical points

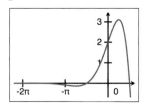

27 (a) see below; (b) critical point at $x = 0$; (c) not an

inflection

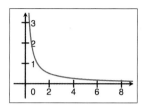

29 (a) see below; (b) no critical points; (c) NA

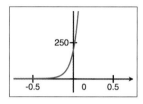

31 (a) see below; (b) no critical points; (c) NA

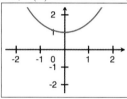

33 (a) see below; (b) $x = 0$; (c) inflection point

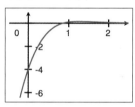

35 $f'(x) = 3x^2$ and $f''(x) = 6x$ **37** 0 **39** $-\cos(x)$ **41**
e^x **43** $\frac{d^k f}{dx^k} = e^x(x + k)$ **45** $f(x)$, $f'(x)$ $f''(x)$ **47**
see below

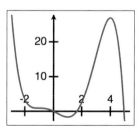

49 $y'' = -1/y^3$ **51** $y'' = \dfrac{(3y^2x^3 - 2x - y^3)(y - x^2)}{(y^3 - x)^3} -$
$\dfrac{3x^2}{y^3 - x}$ **53** $y'' = -2x^2\sec(y) + \frac{4}{7}x^3\tan(y)\sec^3(y)$ **55**
$y'' = \dfrac{2y^2e^x - (e^x - 1)^2}{4y^3}$ **57** $y = \cos(3x)$ **59** $y =$
$\sin(\sqrt{7}x), y = \cos(\sqrt{7}x)$ **61** $y = e^{7x}, y = e^{-7x}$ **63**
$y = e^{\sqrt{2}x}, y = e^{-\sqrt{2}x}$ **65** (a) towards for $t < \frac{5}{3}$; (b)
not speeding up or slowing down **67** (a) towards ori-
gin for $t \in (-\infty, -1 - \sqrt{3}) \cup (-1, -1 + \sqrt{3})$, away from
origin $t \in (-1 - \sqrt{3}, -1) \cup (-1 + \sqrt{3}, \infty)$; (b) slowing down for $t < -1$, speeding up for $t > -1$ **69**
(a) towards for $t \in (-\infty, 6 - \sqrt{41}) \cup (6, 6 + \sqrt{41})$,

away for $t \in (6 - \sqrt{41}, 6) \cup (6 + \sqrt{41}, \infty)$; (b) slowing down for $t < 6$, speeding up for $t > 6$ **71** (a) towards for $t \in \left(-\infty, \frac{9-\sqrt{77}}{2}\right) \cup \left(3 - \sqrt{\frac{26}{3}}, 0\right) \cup$ $\left(3 + \sqrt{\frac{26}{3}}, \frac{9+\sqrt{77}}{2}\right)$, away for $t \in \left(\frac{9-\sqrt{77}}{2}, -\sqrt{\frac{26}{3}}\right) \cup$ $\left(0, 3 + \sqrt{\frac{26}{3}}\right) \cup \left(\frac{9+\sqrt{77}}{2}, \infty\right)$; (b) slowing down when $t \in \left(-\infty, 3 - \sqrt{\frac{26}{3}}\right) \cup \left(0, \frac{9+\sqrt{77}}{2}\right)$, speeding up when $t \in \left(3 - \sqrt{\frac{26}{3}}, 0\right) \cup \left(\frac{9+\sqrt{77}}{2}, \infty\right)$ **73** (a) towards when $t < 0$, away when $t > 0$; (b) speed up $t < 0$, slow down $t > 0$ **75** towards when $t \in (n\pi, \frac{\pi}{2} + n\pi)$, away when $t \in (\frac{\pi}{2} + n\pi, (n+1)\pi)$; (b) speeding up: $t \in (n\pi, \frac{\pi}{4} + \frac{n\pi}{2}) \cup (\frac{\pi}{2} + n\pi, \frac{\pi}{4} + \frac{(n+1)\pi}{2})$, slowing down: $t \in (\frac{\pi}{4} + \frac{n\pi}{2}, \frac{\pi}{2} + n\pi) \cup (\frac{\pi}{4} + \frac{(n+1)\pi}{2}, (n+1)\pi)$ **77** (a) always approaches origin; (b) speeding up when $t \in (0, \sqrt[3]{2/3})$, slowing down when $t \in (-\infty, 0) \cup (\sqrt[3]{2/3}, \infty)$ **79** (a) never approaches origin; (b) speeding up when $t < -3$ or $t > -1$, and slowing down when $t \in (-3, -2)$

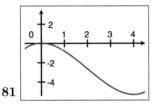
81

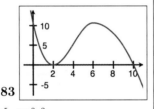
83

85 yes   **87** (b) $x''(r) = \frac{8t^5 - 96t^3 r^2}{(t^2 + 4r^2)^3}$ **95** $f'(4) = 0$, $f''(4) < 0$ **97** $f''(x) < 0$ **99** $x'' = \alpha x'$ **101** $x' > 0, x'' < 0$ **103** (a) $f'(x+a) > f'(x)$, where $a > 0$; (b) $f''(x+a) < f''(x)$, where $a > 0$ **105** $mx'' = -kx - \beta x'$

Section 4.4:

1 (a) $x = \frac{3}{2}$, (b) maximum   **3** (a) no critical values   **5** (a) $x = \frac{5}{8}$, (b) minimum   **7** $x = 0$ is a local minimum, and $x = 2$ is a local maximum   **9** (a) $x = 3$, $x = -\frac{1}{3}$, (b) $x = 3$ minimum, $x = -\frac{1}{3}$ maximum   **11** local max at $x = 0$, local minima at $x = \pm\sqrt{2}$ **13** critical points at $x = k\pi$, local max when k =even, and local min when k =odd   **15** $t = 1.916898769$ is a local min   **17** (a) see below; (b) endpoints

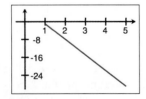

19 (a) see below; (b) endpoints

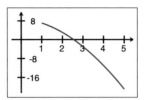

21 (a) see below; (b) minimum is inside, maximum at endpoint

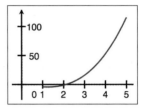

23 local maxima at $x = -16, 4, 12$ and local minima at $x = -4, 8, 16$ **25** $x \in \left[-\frac{7}{3}, 1\right]$ **27** $x \in \left[-\frac{7}{3}, c\right], c \neq 1$ **29** $x \in \left(-\frac{7}{3}, \frac{8}{3}\right)$ **35** (b) only that the tangent line is horizontal; (c) local minimum; (d) see below

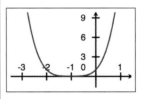

39 (a)
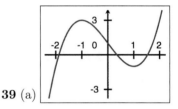

Section 4.5:

1 $t^* = 2$ **3** $t^* = 1.5\pi$ **5** (a) 41, (b) $t^* = \sqrt{13}$ **7** (a) $\Delta f = \frac{e^8 - 1}{4}$, (b) $t^* = \frac{1}{2} \ln \frac{e^8 - 1}{8}$ **9** (a) not applicable, (b) $frac{39}{49}$ **11** (a) applicable, (b) $-\frac{16}{3}$ **13** (a) applicable, (b) $-\frac{1}{4}$ **15** (b) $\lim_{x \to \infty} (x^2 + 3)^{\frac{5}{9}}$, (f) 1   **17** (a) 3, (b) $f(t) = \sin 3t$, (c) $f'(0) = 3$ **19** (a) $\frac{\sqrt{2}}{2}$, (b) $f(t) = \sin t$, (c) $f'(\frac{\pi}{4}) = \frac{\sqrt{2}}{2}$ **21** (a) 1, (b) $f(t) = e^t$, (c) $f'(0) = 1$ **23** (a) ∞^0, (b) $e^{\frac{21}{8}}$ **25** (a) ∞^0, (b) 1   **27** (a) 0^0, (b) 1   **29** (a) 1^∞, (b) 1   **33** $f(1) = \frac{1}{2}$ **35** g is steeper   **37** (c) the function is not differentiable everywhere; (d) at t=4, the person's velocity changes from -5 m/sec to 5 m/sec instantaneously, which requires an infinite acceleration (i.e., an infinite amount of force)   **39** $\alpha = \frac{1}{22}$ **41** (a) $\frac{1}{4}$; (b) $t = 4$ **43** (b) for $A'(t) = m$ at $t = 2$ and $t \approx 6.9$, and $B'(t) = m$ at $t \approx 3.5$ and $t \approx 10.2$ **45** (a) Friend 1 is correct, Friend 2 is incorrect because L'Hôpital's Rule provides a method to calculate a limit that exists, not a method to prove that one does not.   **51** $b = \frac{\beta g(\alpha) - \alpha g(\beta)}{\beta - \alpha}$

Section 4.6:

1 (a) $x > -1$; (b) all x; (c) see below

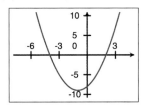

3 (a) $x < -4$ and $x > 2$; (b) $x > -1$; (c) see below

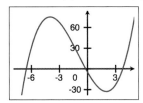

5 (a) $x \in \left(\frac{-3-\sqrt{105}}{2}, 0 \right)$ and $x \in \left(\frac{-3+\sqrt{105}}{2}, \infty \right)$; (b) $x < -4$ and $x > -1$; (c) see below

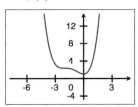

7 (a) $x > 2$; (b) $x \in (-\infty, 0)$ and $x \in \left(\frac{8}{7}, \infty \right)$; (c) see below

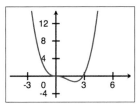

9 (a) $x \in (0, 1)$; (b) $x \in \left(-\frac{1}{\sqrt{2}}, \frac{1}{\sqrt{2}} \right)$; (c) see below

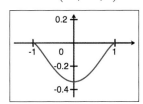

11 (a) never positive; (b) $x \in (-\infty, 0)$ and $x \in (3, \infty)$; (c) see below

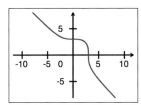

13 (a) $x \in \left(0, \frac{1}{\sqrt{3}} \right)$; (b) $x > \sqrt{\frac{9+4\sqrt{6}}{15}}$; (c) see below

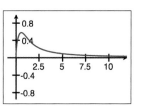

15 (a) $x \in \left(\frac{3-\sqrt{5}}{2}, \frac{3+\sqrt{5}}{2} \right)$; (b) $x < 1$ and $x > 4$; (c) see below

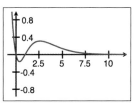

17 (a) $x \in (-1, 0)$ and $x \in (1, \infty)$; (b) $x \in \left(0, \sqrt{2 + \sqrt{5}} \right)$; (c) see below

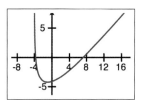

19 (a) $x \in \left(\frac{n\pi}{2}, \frac{(n+1)\pi}{2} \right)$ where n is even; (b) $x \in \left(-\frac{1}{2}\arctan 6 + \frac{n\pi}{2}, -\frac{1}{2}\arctan 6 + \frac{(n+1)\pi}{2} \right)$ for n even; (c) see below

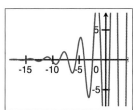

21 (a) $x \in \left(\frac{\pi}{4} + \frac{n\pi}{4}, \frac{\pi}{4} + \frac{(n+1)\pi}{4} \right)$ for $n = 4k$ and $n = 4k + 1$; (b) $x \in \left(n\pi, n\pi + \frac{\pi}{2} \right)$ for $n \in \mathbb{Z}$; (c) see below

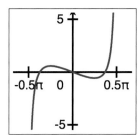

23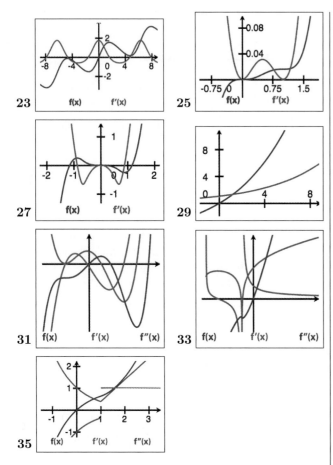

f(x)   f'(x)

25

f(x)   f'(x)

27

f(x)   f'(x)

29

31

f(x)   f'(x)   f''(x)

33

f(x)   f'(x)   f''(x)

35

f(x)   f'(x)   f''(x)

Section 4.7:

1 25

3 (a)–(d) see below; (e) the line shifts up but does not change slope; (f) $\left(\frac{1}{\sqrt{22}}, \frac{5\sqrt{22}}{11}\right)$

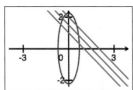

5 $\left(-\frac{1}{\sqrt{53}}, -\sqrt{\frac{300}{31}}\right)$ **7** $\frac{18\sqrt{10}}{5}$ units   **9** $D \approx$ 9.2606   **11** cut at $\frac{40}{\pi+4} \approx 5.601$cm from left, make square form longer piece   **13** cut $\frac{30\sqrt{3}}{2+3\sqrt{3}} \approx$ 7.2207cm from the left, make triangle from longer piece   **15** $\left(6 - \frac{2*\sqrt{21}}{3}\right)^2 \approx 8.6727$cm^2 **17** $(r,h) = \left(\frac{1}{\sqrt[3]{5\pi}}, \sqrt[3]{\frac{25}{\pi}}\right) \approx (0.3993, 1.9965)$ **19** 64.3142 ft   **21** 500 by 125   **23** 70 by 17.5   **25** $6\sqrt{35}$ feet   **31** $w = 2r/\sqrt{3}$ and $h = 2r\sqrt{2/3}$ **33** $y = \frac{6(0.911)}{\sqrt{(6.4)^2-(0.911)^2}}$ **37** $y = \frac{sx}{\sqrt{r^2-s^2}}$ **39** Run only   **42** max area $\sqrt{12}$ **43**

$8\sqrt[3]{\frac{3}{2}}\sqrt{1 + \sqrt[3]{\frac{9}{4}}} + 12\sqrt{1 + \sqrt[3]{\frac{4}{9}}}$ **45** 51.7762 ft (using Newton's method)   **47** $x = 0.29289$ (using Newton's method)

Chapter Review:

1 T   **3** F   **5** T   **7** T   **9** T   **11** F   **13** F   **15** T   **17** F   **19** F ($\geq$)   **21** c   **23** e   **25** b   **27** a   **29** a   **31** e   **33** (b) $t_1 = \frac{7}{3}$, $t_2 = \frac{103}{51}$, $t_3 = \frac{26215}{13107}$ **35** $y(t) = -1$ (constant function)   **37** greater external force $\Rightarrow$ greater density makes sense   **39** $\frac{dr}{dt} = 2(\pi\tan(35°))^{-1/3}(51^{-2/3}) \frac{\text{cm}}{\text{sec}}$ **41** see below

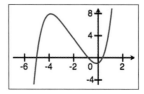

43 when $t \in (-1, 2/3)$ and when $t \in (7/3, \infty)$ **45** a   **47** (a) increasing on $(-\infty, 1]$, decreasing on $[1, \infty)$, concave down on $(-\infty, 2)$, concave down on $(2, \infty)$; (b) the origin; (c) $\lim_{t\to\infty} g(t) = 0$ and $\lim_{t\to-\infty} g(t) = -\infty$; (d) see below

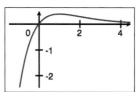

49 $(12 + 8t + t^2)e^t$ **51** f does not have an extremum at $t = 4$ **53** (b) yes (c) nothing   **55** f' is red (curve C), f'' is blue (curve A)   **57** $f'(x) = g'(x)$ at some $x \in (0, 4)$ **59** (b) $f'(t) = -g'(t)$ (c) $t = 0$ or $t = 10/3$ **61** (a) see below; (b) $a = -1 - \sqrt{8}$ and $b = 0$, which are points at which $f(x) = 1$ (other answers possible)

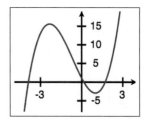

63 1225 ft^2 **65** (b) -1/4   **67** 0   **69** 1/6   **71** ∞ **73** (a) $b(t_0) = 16 - t_0^2$; (b) $b''(t_0) = -2 < 0$

Section 5.1:

1 45   **3** 210   **5** 17,025   **7** −246,705   **9** 128,710   **11** 1,722   **13** $t_k^* = 0 + \frac{4k}{300}$ **15** $t_k^* = 3 + \frac{4(k-1)}{1000}$ **17** $t_k^* = -5 + \frac{7(k-0.5)}{30}$ **19** 90   **21** 50.25   **23** $\frac{61}{3}$ **25** 10.88   **27** 736   **29** −7.34   **31** this is the revenue generated by the entire sales force, across the country   **33** the total time it took him to finish the race   **35** $\sum_{k=1}^{21} d_k$ **37** (a) 58.1 liters (b) underestimate   **39** (a) 56.6 cubic cm (b) 40.2 cubic cm   **41** (a) $\Delta m = \frac{53}{4000}$ and $\sum_{k=1}^{4000} f\left(\frac{53k}{4000}\right)\frac{53}{4000}$; (b) $\Delta m = \frac{53}{3000}$ and $\sum_{k=1}^{3000} f\left(\frac{53(k-1)}{3000}\right)\frac{53}{3000}$; (c) $\Delta m = \frac{53}{2000}$ and $\sum_{k=1}^{2000} f\left(\frac{53(k-0.5)}{2000}\right)\frac{53}{2000}$;   **43** 0.12375 km   **45** (a) depth increased by ≈ 17.31 feet (b) 27.31 feet   **47** 1296 beats   **49** 1284 beats   **51** −75 millivolts   **53** −12.5 millivolts

Section 5.2:

1 $\int_3^8 \sin(t)\,dt$ **3** $\int_1^6 \ln(2t)\,dt$
5 $\lim_{n\to\infty} \sum_{k=1}^n \ln\left(1 + \frac{5k}{n}\right)\frac{5}{n}$
7 $\lim_{n\to\infty} \sum_{k=1}^n \ln\left(4 + \cos\left(1 + \frac{6k}{n}\right)\right)\frac{6}{n}$
9 $\lim_{n\to\infty} \sum_{k=1}^n \tan\left(\frac{4k}{n} - 1\right)\frac{1}{n}$ **11** (a) $\frac{64}{3}$ (b) $\frac{64}{3}$ **13** (a) $\frac{15}{2}$ (b) $\frac{7}{3}$ (c) $\frac{59}{6}$ **15** $\int_1^4 q(t)\,dt = 205$ **17** Since da has units of years and the integral has units of people, the integrand must have units of people/year. The function $f(a)$ tells us how many people are a years old.   **19** $\int_0^{60} b(t)\,dt$ is the distance run by the brown horse, and $\int_0^{60} b(t)\,dt$ is the distance run by the gray horse, so this inequality means simply that the gray horse ran farther.   **21** 0   **23** ≈ $\frac{3}{4}$ (from approximating the curve with a triangle)   **25** 0   **27** 1   **29** 3.5   **31** 268/3   **33** 97/3   **35** (a) see below; (b) 205/4 units²

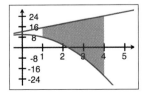

37 (a) see below; (b) 58.5 units²

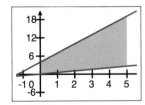

39 (a) see below; (b) 28/3 units²

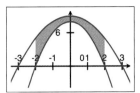

41 4.5π   **43** (a) 135 miles; (b) estimating area using triangles and rectangles   **45** 27.5

Section 5.3:

1 $\frac{85}{2}$ **3** 728.5   **5** 0   **7** 300   **9** $\frac{13130}{3}$ **11** 0   **13** 5.4   **15** $\frac{65}{6}$ **17** 159.25   **19** $n \geq 13867$ **21** $n \geq 368$ **23** $n \geq 1$ **25** $n \geq 8266$ **27** (a) $n = 5$ using $M_1 = 1$; (b) $\ln(2) \approx \frac{1879}{2520}$ **29** (a) $n = 7$ using $M_2 = 2$; (b) $\ln(4) \approx \frac{10653567}{7607600}$ **31** (a) using $M_2 = 2 \Rightarrow n = 13$ subintervals, which yields an estimate of 0.7851516152; (b) 3.140606461 which is at most 0.004 from the actual value   **33** f decreasing over $[a, b]$ **35** 3228   **37** 233   **39** 42.5 megabits   **41** $\frac{1796}{15}$ in³   **43** 4 units²   **45** ≈ 93,667 trades

Section 5.4:

1 $4t^2 + 7t + C$ **3** $\frac{8}{7}t^7 + \frac{7}{3}t^3 + C$ **5** $e^t - 8t + C$ **7** $7e^t + C$ **9** $\frac{6^t}{\ln(6)} + C$ **11** $\frac{7(4^t)}{\ln(4)} + C$ **13** $3\ln|t| + 3t^2 + C$ **15** $-2\cos(t) + C$ **17** $-\frac{1}{2}\cos(2t) + C$ **19** $-\frac{3}{8}\cos(8t) + C$ **21** $-\frac{9}{13}\cot(13t) + C$ **23** $\frac{1}{2}t^2 - \ln|t| + C$ **25** $\frac{1}{2}t^2 - \ln|t| + C$ **27** $2t^{1/2} + \frac{2}{7}t^{7/2} + C$ **29** 2   **31** 21   **33** 2   **35** 18   **37** $\frac{7}{11}$ **39** $\frac{2943}{64}$ **43** (a) see below; (b) & (c) 32

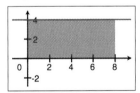

45 (a) see below; (b) & (c) 8

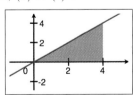

47 130   **49** 5π²/8   **51** $2 - \frac{\pi}{1}$ **92**   **53** 2   **55** ≈ 2.5421494 seconds (found using Newton's Method)   **57** 0.2352 pounds   **59** (a) -4.9876 (b) ≈ 16.09 seconds (found using Newton's Method)   **61** (a) f'' does not change sign at $t = 1$; (b) e.g., $f'(t) = t - t^2 + \frac{1}{3}t^3$; (c) e.g.,

$f(t) = 1 + \frac{1}{2}t^2 - \frac{1}{3}t^3 + \frac{1}{4}t^4$; (d) the graph is "flat" at $t = 1$ (see below)

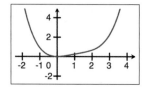

Section 5.5:

1 $A(t) = t^3 - 4t$ **3** $A(t) = 6t^{11} - 8t^3$ **5** $\sin(\cos(t))\sin(t)$ **7** $A(t) = \sin(t) - 2t\sin(t^2)$ **9** $2f(t)$ **11** $f(x) = 12x^5 - 7$ **13** $-7/3$ **15** $t = 0, \frac{-5 \pm \sqrt{193}}{24}$

The graphs for #17–23 shown below are examples of answers. Other answers are possible.

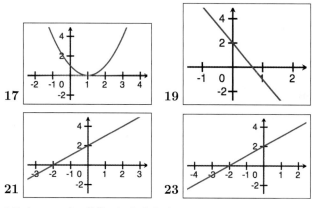

17   **19**

21   **23**

25 the graphs differ in height by some constant   **27** increasing   **29** $t = 1.5$ and $t = 8.5$ **31** $(2,3)\cup(5,6)$ **33**

$t = 0, 3, 6$ **35** $(0, 1.5) \cup (2.5, 2.5) \cup (5, 6.5)$ **37** see below

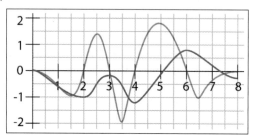

39 $2/3$ **41** $A'(0.5) < A'(8.5) < A'(4)$ **43** (a) always increasing (b) $(-4, 0)$ **45** (a) $A'(5)$ is positive but is small relative to all other values of $A'(t)$, so the graph of $A(t)$ is less steep there than anywhere else, (b) $t = 5$ is a second-order critical point   **47** the graph of f bounds more area below the axis than above when $5 \le x \le 7$.   **49** (d) $\ln(at) = \ln(a) + \ln(t)$

Chapter Review:

1 F   **3** T   **5** T   **7** F   **9** T   **11** F   **13** T   **15** T   **17** d   **19** c   **21** b   **23** b   **25** 84   **27** 595/6   **29** 0   **31** $-\frac{1}{2}t^{-6} + \frac{8}{7}t^7 + C$ **33** $-\frac{8}{5}\cos(5t) + C$ **35** $\frac{5}{6}e^{6t} + C$ **37** 174   **39** 0   **41** 1   **43** 2961 subintervals, using $M_1 = 0.6$ **45** 2 subintervals, using $M_2 = 22$ **47** $[1, 1.25]$ **49** 65/3   **51** 22   **53** 37 °F   **55** $25\pi/4$ lbs (look at the graph)   **57** 38.55 mg   **59** (a) $F_{\text{gravity}} = -0.98$ N; (e) 34.11 meters; (f) $y = 34.11 + 48.2(t - 1) - 4.9(t - 1)^2$; (g) max altitude 152.64 meters (500.78 feet) at $t = 5.93$ seconds after lift-off; (b)-(d) are shown in the table below

for Chapter Review #59

time (sec)	0.1	0.2	0.3	0.4	0.5	0.6	0.7	0.8	0.9	1
net force (N)	2.02	12.02	13.02	5.02	4.02	4.02	4.02	4.02	1.02	−0.98
momentum (kg-m/sec)	0.202	1.404	2.706	3.208	3.61	4.012	4.414	4.816	4.918	4.82
velocity (m/sec)	2.02	14.04	27.06	32.08	36.1	40.12	44.14	48.16	49.18	48.2

Section 6.1:

1 (a) $x = 4t$; (b) $dx = 4\,dt$; (c) $\int \frac{1}{4}\tan(x)\,dx = \frac{1}{4}\ln|\sec(x)| + C$; (d) $\frac{1}{4}\ln|\sec(4t)| + C$ **3** (a) $x = 5t$; (b) $dx = 5\,dt$; (c) $\int \frac{1}{5}\sec(x)\,dx = \frac{1}{5}\ln|\sec(x) + \tan(x)| + C$; (d) $\frac{1}{5}\ln|\sec(5t) + \tan(5t)| + C$ **5** (a) $x = t/4$; (b) $4\,dx = dt$; (c) $\int 24\cos(x)\,dx = 24\sin(x) + C$; (d) $24\sin(t/4) + C$ **7** (a) $x = \pi t/6$; (b) $\frac{6}{\pi}dx = dt$; (c) $\int \frac{24}{\pi}\csc(x)\,dx = -\frac{24}{\pi}\ln|\csc(x) + \cot(x)| + C$; (d) $-\frac{24}{\pi}\ln\left|\csc\left(\frac{\pi t}{6}\right) + \cot\left(\frac{\pi t}{6}\right)\right| + C$ **9** $\int_1^9 \frac{1}{2x}\,dx$ **11**

$\int_4^{84} \frac{2}{x}\,dx$ **13** $\int_1^2 \frac{1}{x}\,dx$ **15** $\int x^{11}\,dx$, $x = t^2 + 7$ **17** $\int \frac{3}{2}\sqrt{x}\,dx$, $x = t^2 + 1$ **19** $\int \frac{1}{2}e^x\,dx$, $x = t^2$ **21** $\int e^x\,dx$, $x = \tan(t)$ **23** $\int \frac{5}{x^2}\,dx$, $x = t^2 - 8$ **25** $\int x^{12}\,dx$, $x = \sin(t)$ **27** $\int -x^9\,dx$, $x = \cot(t)$ **29** $\int -\frac{1}{3}x^{4/3}\,dx$, $x = \cos(3t)$ **31** $\frac{3}{2}\arctan(2t) + C$ **33** $\frac{7\sqrt{5}}{5}\arctan(t/\sqrt{5}) + C$ **35** $\frac{\sqrt{7}}{7}\arctan\left(\frac{t+1}{\sqrt{7}}\right) + C$ **37** $\frac{4\sqrt{19}}{19}\arctan\left(\frac{2t+3}{\sqrt{19}}\right) + C$ **39** $\frac{12\sqrt{39}}{39}\arctan\left(\frac{4t+5}{\sqrt{39}}\right) + C$ **41** $\ln(t^2 + 1) + 6\arctan(t) +$

C **43** $2\ln(t^2 + 11) + \frac{12\sqrt{11}}{11}\arctan(t/\sqrt{11}) + C$ **45**
$\ln(t^2 + 4t + 10) + \frac{\sqrt{6}}{6}\arctan\left((t+2)/\sqrt{6}\right) + C$ **47**
$2\ln(t^2 + 8t + 101) + \frac{4\sqrt{85}}{85}\arctan\left((t+4)/\sqrt{85}\right) + C$
49 $\frac{1}{90}(3t+4)^{30} + C$ **51** $-\frac{1}{3}\cos(3t+7) + C$ **53**
$\frac{1}{9}\tan(9t-5) + C$ **55** $\frac{1}{405}(5\sin(t)+16)^{81} + C$ **57**
$4\sqrt{e^t+5}+C$ **59** $-2\cos(\sqrt{t})+C$ **61** $\frac{1}{2}(\ln(t))^2+C$ **63**
$-\cos(\ln(t)) + C$ **65** $\sin(t^3 + 2t) + C$ **67**
$\frac{4}{3}(t^3+8)^{9/4} - \frac{96}{5}(t^3+8)^{5/4} + C$ **69** $\frac{2}{25}(t^5+13)^{5/2} -$
$\frac{26}{15}(t^5+13)^{3/2}+C$ **71** $\frac{1}{6}(2t-1)^{3/2}-\frac{1}{2}(2t-1)^{1/2}+C$ **73**
$\frac{1}{5}\cos^5(t) - \frac{1}{3}\cos^3(t)+C$ **75** $\frac{1}{10}\cos^{10}(t) - \frac{1}{12}\cos^{12}(t) -$
$\frac{1}{8}\cos^8(t) + C$ **77** $\frac{1}{10}\tan^{10}(t) + \frac{1}{8}\tan^8(t) + C$ **79**
$\frac{1}{3}\sec^3(t)-\sec(t)+C$ **81** $\frac{1}{5}\csc^5(t)-\frac{1}{7}\csc^7(t)+C$ **83**
$\frac{t}{2}+\frac{1}{16}\sin(8t)+C$ **85** $\frac{3}{56}(204^7-8^7)$ **87** $\frac{1}{35}(59^7+1)$ **89**
$\frac{2}{11}$ **91** $\frac{20}{3}(5^{3/2} - 2^{3/2}) - \frac{4}{5}(5^{5/2} - 2^{5/2})$ **93**
$x = t + 13$ **95** $x = \ln(t)$ **97** (a) 2; (b) $\int_0^2 3f(x)\,dx$
is three times larger than $\int_0^2 f(x)\,dx$; (c) $\int_0^6 f(x/3)\,dx$
is three times larger than $\int_0^2 f(x)\,dx$; (d) they're equal,
(e) $\int_0^6 f(x/3)\,dx = \int_0^2 f(u)\,du$, where $u = x/3$.   **99**
$\int_{-a}^a f(t)\,dt = 0$ **101** use $u = x/a$, $dx = a\,du$ **103**
(a) see below; (b) 1

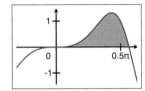

105 (a) see below; (b) $10(e^3 - 3)$

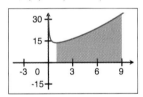

107 (a) $\frac{3}{2}\ln(5) - \frac{4}{3}$ cm^2, (b) $\frac{81}{200}\ln(5) - \frac{108}{300}$ grams   **109**
(a) $(e - 1) + \pi^2 - \frac{\pi^3}{6}$, cm^2; (b) $(162(e - 1) + 162\pi^2 - 27\pi^3)/600$ grams   **111** (a) see below; (b)
$x \approx -2.356152$, $x \approx 0.785398$; (c) 3.070093

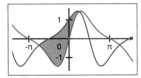

113 $\frac{17}{3}$ cm   **115** (a) $t = 10 - \left(10^{31/10} - 620\right)^{10/31}$; (b)
$\left(\frac{50}{31}\right)10^{31/10}$ **117** (a) $\sum_{k=1}^n \frac{1}{4\pi\varepsilon_0} 2\pi\sigma y \frac{r_k^*}{(y^2+(r_k^*)^2)^{3/2}}\,\Delta r$;
(b) $\frac{2\pi\sigma y}{4\pi\varepsilon_0}\int_0^R \frac{r}{y^2+r^2}\,dr$; (c) $\frac{1}{4\pi\varepsilon_0}\left(2\pi\sigma - \frac{2\pi\sigma y}{\sqrt{y^2+R^2}}\right)$

Section 6.2:

1 $\frac{1}{4}\sin(2t)-\frac{t}{2}\cos(2t)+C$ **3** $t\tan(t)-\ln|\sec(t)|+C$ **5**
$\frac{3}{2}te^{6t} - \frac{1}{4}e^{6t} + C$ **7** $\frac{3t}{\ln(25)}25^t - \frac{3}{(\ln(25))^2}25^t + C$ **9**
$\frac{9}{4}t^2\sin(4t)-\frac{9}{32}\sin(4t)+\frac{9}{8}t\cos(4t)+C$ **11** $-5(50+10t+t^2)e^{-t/5} + C$ **13** $2(t^3 - 6t^2 + 24t - 48)e^{t/2} + C$ **15**
$\frac{3}{10}\sin(t)e^{3t} - \frac{1}{10}\cos(t)e^{3t} + C$ **17** $\frac{2}{5}t^2\sin(\ln(t)) -$
$\frac{1}{5}t^2\cos(\ln(t))+C$ **19** $\frac{t}{a}e^{at} - \frac{1}{a^2}e^{at} + C$ **21** $\frac{8}{3}\ln(4) -$
$\frac{1}{3}\ln(2) - \frac{7}{9}$ **23** $1/8$ **25** $\frac{486}{5}\ln(3) - \frac{64}{5}\ln(2) - \frac{422}{25}$ **27**
$2(\sqrt{t}-1)e^{\sqrt{t}} + C$ **29** $2\sqrt{t}\ln(\sqrt{t}) - 2\sqrt{t} + C$ **31**
$\frac{\pi-2}{3}$ **33** $\frac{1}{\sqrt{3}}(\ln(\sqrt{3})+1-\sqrt{3})$ **35** $\frac{a}{a^2+b^2}e^{at}\sin(bt) -$
$\frac{b}{a^2+b^2}e^{at}\cos(bt)$ **37** $\sec(\theta) - \ln(\csc(\theta) + \cot(\theta))$
39 $0.5\theta - 0.5\sin(\theta)\cos(\theta)$ **41** $\frac{7}{45}\cos(2\theta)\sin(7\theta) -$
$\frac{2}{45}\sin(2\theta)\cos(7\theta)$ **43** $\frac{5}{16}\sin(3\theta)\sin(5\theta) +$
$\frac{3}{16}\cos(3\theta)\cos(5\theta)$ **45** $5t\sin(t) + 5\cos(t) + C$ **47**
$-\frac{t}{2}e^{-2t} - \frac{1}{4}e^{-2t} + C$ **49** $-\frac{1}{8}e^{-4t^2} + C$ **51**
$3\arctan(t-4)+C$ **53** $\frac{3}{34}e^{3t}\cos(5t)+\frac{5}{34}e^{3t}\sin(5t)+C$
55 $\frac{1}{2}(1+t^2)\arctan(t) - \frac{t}{2} + C$ **57** (a) see below; (b)
$\pi/3$

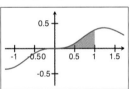

59 (a) see below; (b) $\frac{1}{2} - \frac{1}{e}$

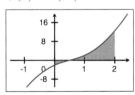

61 (a) see below; (b) $3\ln(64) - 6$

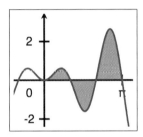

63 (a) see below; (b) $t_1 = 1.412406$, $t_2 = 3.057236$; (c)
net area$= 0.822436$

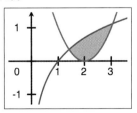

65 (a) see below; (b) $t_1 = 0$, $t_2 = 1.125358$; (c) net
area$= 0.165275$

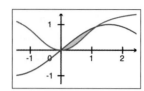

67 (a) $400 - \frac{800}{e}$; (b) 9.6129 seconds   **69** $q(t) = \frac{231}{226} - \frac{5}{226}\cos(3t)e^{-t/5} + \frac{75}{226}e^{-t/5}\sin(3t)$ **71** $(b-a)f(b) - \int_a^b f(x)\,dx$ = (area of rectangle) − (area under graph of f)   **73** (a) $\hat{f}_k = \frac{4\sqrt{2}(1-\cos(k\pi))}{(k\pi)^3}$

Section 6.3:

1 $\sin(\theta)-\sin^3(\theta)+\frac{3}{5}\sin^5(\theta)-\frac{1}{7}\sin^7(\theta)+C$ **3** $\frac{1}{5}\cos^5(\theta)-\frac{1}{3}\cos^3(\theta)+C$ **5** $\frac{1}{20}\sin^3(5\theta)-\frac{1}{30}\sin^6(\theta)+C$ **7** $\frac{3}{8}x-\frac{1}{32}\sin^3(8x)\cos(8x)-\frac{3}{64}\cos(8x)\sin(8x)+C$ **9** $\frac{1}{8}x-\frac{1}{4}\sin(x)\cos^3(x)+\frac{1}{8}\cos(x)\sin(x)+C$ **11** $\frac{5}{2048}x-(\frac{1}{56}\sin^7(4x)+\frac{1}{96}\sin^5(4x)+\frac{1}{192}\sin^3(4x)+\frac{1}{512}\sin(4x))\cos^7(4x)+\frac{1}{3072}\cos^5(4x)\sin(4x)+\frac{5}{12288}\sin(4x)\cos^3(4x)+\frac{5}{8192}\cos(4x)\sin(4x)+C$ **13** $\frac{1}{5}\tan^5(\theta)+C$ **15** $\frac{1}{6}\cot^3(\theta)-\frac{1}{5}\cot^5(\theta)+C$ **17** $\theta+\frac{2}{3}\tan^3(\theta)+\frac{1}{5}\tan^5(\theta)+C$ **19** $\frac{1}{3}\sec^3(\theta)-\sec(\theta)+C$ **21** $-\frac{1}{8}\cot^8(\theta)-\frac{1}{6}\cot^6(\theta)+C$ **23** $\frac{1}{3}\tan^3(\theta)-\tan(\theta)+\theta+C$ **25** $\frac{3}{16}\ln|\csc(2\theta)-\cot(2\theta)|-\frac{1}{8}\cot(2\theta)\csc^3(2\theta)-\frac{3}{16}\cot(2\theta)\csc(2\theta)+C$ **27** $\frac{1}{4}\tan(\theta)\sec^3(\theta)-\frac{5}{8}\tan(\theta)\sec(\theta)-\frac{3}{8}\ln|\sec(\theta)+\tan(\theta)|+C$ **29** $\frac{1}{2}\sec(\theta)\tan(\theta)-\frac{1}{2}\ln|\sec(\theta)+\tan(\theta)|+C$ **31** $\frac{1}{16}\cos(\theta)+\frac{1}{48}\cos^3(\theta)+\frac{1}{16}\ln|\csc(\theta)-\cot(\theta)|-\frac{1}{6}\cot^5(\theta)\csc(\theta)-\frac{1}{24}\cos(\theta)\cot^4(\theta)+\frac{1}{48}\cot^2(\theta)\cos^3(\theta)+C$ **33** $\frac{1}{2}\sec^2(\theta)+C$ **35** $\frac{1}{3}\tan^3(\theta)+C$ **37** $2\cos(x)+2x\sin(x)-x^2\cos(x)+C$ **39** $e^{\tan(t)}+C$ **41** $\frac{1}{3}\sin^3(t)-\frac{1}{5}\sin^5(t)+C$ **43** $j\frac{1}{5}\tan^5(t)+\frac{2}{3}\tan^3(t)+\tan(t)+C$ **45** $\frac{1}{2}\sin(2t)+C$ **47** $\frac{1}{6}\cot^6(t)-\frac{1}{4}\cot^4(t)+C$ **49** (a) $\frac{1}{4}\tan^4(\theta)$; (b) $\frac{1}{4}\sec^4(\theta)-\frac{1}{2}\sec^2(\theta)$ **51** (a) $\frac{1}{4}\sin^4(\theta)$; (b) $\frac{1}{4}\cos^4(\theta)-\frac{1}{2}\cos^2(\theta)$ **55** $\sqrt{2}$ [5 pt]

Section 6.4:

Integrals are valid over an interval: **1** $\frac{t}{2}\sqrt{9-t^2}+\frac{9}{2}\arcsin(\frac{t}{3})+C$ **3** $\frac{t}{2}\sqrt{5-4t^2}+\frac{5}{4}\arcsin(\frac{2\sqrt{5}}{5}t)+C$ **5** $\frac{\sqrt{5}}{5}\arcsin(\frac{\sqrt{5}}{\sqrt{12}}t)+C$ **7** $-\frac{1}{5}\ln|5+\sqrt{25-49t^2}|-\frac{1}{5}\ln|7t|+C$ **9** $-\frac{t}{18}\sqrt{64-9t^2}+\frac{32}{27}\arcsin(\frac{3}{8}t)+C$ **11** $\frac{t}{2}\sqrt{t^2-81}-\frac{81}{2}\ln(t+\sqrt{t^2-81})+C_{\ell,r}$ **13** $\sqrt{5}\ln|\sqrt{5}\ t+\sqrt{5t^2-36}|-\frac{5t^2-36}{t}+C_{\ell,r}$ **15** $\frac{\sqrt{3}}{3}\sec^{-1}(\frac{2\sqrt{3}}{3}t)+C$ **17** $\frac{t}{2}\sqrt{t^2+12}+6\ln|\sqrt{t^2+12}+t|+C$ **19** $\frac{\sqrt{6}}{6}\ln|\sqrt{6t^2+1}+\sqrt{6}\ t|+C$ **21** $\frac{t+4}{2}\sqrt{t^2+8t}-8\ln|t+4+\sqrt{t^2+8t}|+C_{\ell,r}$ **23** $\frac{(3-t)}{2}\sqrt{6t-t^2}-\frac{9}{2}\ln|3-t+\sqrt{6t-t^2}|+C_{\ell,r}$ **25** $\frac{\sqrt{2}}{32}(4t+3)\sqrt{16t^2+24t}-\frac{9\sqrt{2}}{32}\ln|4t+3+\sqrt{16t^2+24t}|+C$

27 $\frac{1}{9}\sqrt{9t^2-4}+C_{\ell,r}$ **29** $t(\ln(t))^2-2t\ln(t)+2t+C$ **31** $4\ln|t^2+4|+\frac{1}{2}\arctan(\frac{t}{2})+C$ **33** $\frac{1}{54}(6t+5)^{3/2}-\frac{11}{18}\sqrt{6t+5}+C$ **35** $\frac{8}{\pi}t^2\sin(\frac{\pi t}{8})+2(\frac{8}{\pi})^2t\cos(\frac{\pi t}{8})-2(\frac{8}{\pi})^3\sin(\frac{\pi t}{8})+C$ **37** $\arctan(t)+\frac{t}{t^2+1}+C$ **39** $-\frac{3}{2}\frac{1}{t^2+1}+C$ **41** $\frac{1}{2}t^2\arcsin(t)+\frac{t}{4}\sqrt{1-t^2}-\frac{1}{4}\arcsin(t)+C$ **43** $k(\pi+\sin(x-1)+\frac{1}{2}(x-1)\sqrt{1-(x-1)^2})$ **45** cut at $x=\pm 0.553295R$, where R is the radius of the pizza   **47** $2\arcsin(\frac{1}{2}\sqrt{4-H^2})+\frac{1}{4}H(H-2)\sqrt{4-H^2}$

Section 6.5:

1 $\frac{1}{2}\ln\left|\frac{t-2}{t+2}\right|+C$ **3** $\frac{1}{3}\ln\left|\frac{t-4}{t-1}\right|+C$ **5** $\frac{1}{5}\ln\left|\frac{(t+3)^{10}}{(t+2)^7}\right|+C$ **7** $\frac{11}{4}\ln|2t-1|-\frac{21}{16}\ln|4t-1|+C$ **9** $\frac{1}{3}\ln|t|+\frac{37}{6}\ln|t-3|-\frac{5}{2}\ln|t-1|+C$ **11** $\frac{10}{9}\ln|t-8|+\frac{5}{36}\ln|t+1|+C$ **13** $\frac{2}{\sqrt{7}}\arctan(\frac{\sqrt{7}}{3}t)+C$ **15** $\frac{5}{\sqrt{21}}\arctan(\sqrt{\frac{7}{3}}t)-\frac{2}{7}\ln|3+7t^2|+C$ **17** $8\arctan(t+2)+C$ **19** $\frac{1}{10}\ln|5t^2+6t+10|-\frac{23}{5\sqrt{41}}\arctan(\frac{5\sqrt{41}}{41}(t+\frac{3}{5}))+C$ **21** $8\ln|4t-1|-\frac{2}{t}-8\ln|t|+C$ **23** $\frac{3}{16}\ln\left|\frac{t}{t+4}\right|+\frac{3}{4}\frac{1}{t+4}+C$ **25** $\frac{1}{4}\frac{1}{t+2}+\frac{1}{8}\ln\left|\frac{t}{t+2}\right|-\frac{11}{4}\frac{1}{(t+2)^2}+C$ **27** $\frac{6}{25}\ln|t^2+1|-\frac{12}{25}\ln|t-2|-\frac{8}{5}\frac{1}{t-2}-\frac{16}{25}\arctan(t)+C$ **29** $\ln|t|+\frac{1}{2}\ln|t^2+1|+\frac{1}{t^2+1}+C$ **31** $\frac{1}{2}t^2+9t+54\ln|t-5|+C$ **33** $3t^2+12t-\frac{8}{t-1}+18\ln|t-1|+C$ **35** $\frac{9}{2}t^2-18t-\frac{41}{2}\ln|t^2+2t+8|+\frac{369\sqrt{7}}{14}\arctan(\frac{1}{\sqrt{7}}(t+1))+C$ **37** (a) $\int 1-\frac{1}{t-2}\,dt$; (b) $t-\ln|t+2|+C$ **39** (a) $\int 2-\frac{11}{t+5}\,dt$; (b) $2t-11\ln|t+5|+C$ **41** $\int -\frac{1}{4}+\frac{61}{4}\frac{1}{4t+1}\,dt$; (b) $-\frac{t}{4}+\frac{61}{16}\ln|4t+1|+C$ **43** $t+(a-1)\ln|t+1|+C$ **45** $\frac{t}{c}+\frac{3c-d}{c^2}\ln|ct+d|+C$ **47** $\frac{7}{5}\ln|t+14|+\frac{1}{10}\ln|t-1|+C$ **49** $\frac{13\sqrt{3}}{18}\arctan(\frac{\sqrt{3}}{2}t)+C$ **51** $\frac{1}{2}t^2+3t+9\ln|t-2|-2\ln|t-1|+C$ **53** $\frac{3}{169}\ln|13t-16|-\frac{t}{13}+C$ **55** $8\arctan(t)+4\ln|t-1|-4\ln|t+1|+C$ **57** $\frac{44}{25}\ln\left|\frac{t-5}{t}\right|-\frac{1}{5t}+C$ **59** $x=\sqrt{t+9}$ leads to $\ln|\sqrt{t+9}-3|-\ln|\sqrt{t+9}+3|+C$ **61** $x=\sqrt{t}$ leads to $\frac{2}{3}\ln|t^{3/2}-1|+C$ **63** $x=\sqrt{t+6}$ leads to $\frac{96}{7}\ln|6+\sqrt{t+6}|+\frac{1}{7}\ln|1-\sqrt{t+6}|+C$ **65** $t-11\ln|t-6|+C$ **67** $\frac{1}{2}\ln|t^2-3t+4|-\frac{5\sqrt{7}}{7}\arctan(\frac{2}{\sqrt{7}}(t-\frac{3}{2}))+C$ **69** $\frac{7}{2}\ln|t^2+8|-\frac{6\sqrt{8}}{8}\arctan(\frac{t}{\sqrt{8}})+C$ **71** $\frac{1}{16}\ln\left|\frac{(t+4)^{17}}{t}\right|-\frac{1}{4t}+C$ **73** $(t-1.5)\ln(t^2-3t+17)-2t+\sqrt{59}\arctan(\frac{2}{\sqrt{59}}(t-1.5))+C$ **75** $\frac{3}{2\sqrt{7}}\arctan(\sqrt{7}t)+\frac{3}{2}\frac{t}{7t^2+1}+C$ **77** $b^2>4a$ **79** $b^2=4c$ **81** (a) $100-10\sqrt{20}\arctan(\sqrt{5})$ meters; (b) $T\approx 15.79$ seconds   **83** (a) 3; (b) 2; (c) denominator 4, numerator 3   **85** (a) $\frac{1}{s^2}+\frac{1}{s^2+1}$; (b) $t+\sin(t)$ **87** (a) $\frac{s}{s^2+1}+\frac{4}{s}$; (b) $4t+\cos(\sqrt{5}\ t)$

Section 6.6:

1 $1/7$ **3** $1/3$ **5** $4/\sqrt{e}$ **7** $17/e^3$ **9** $6\ln\left|\frac{\sqrt{2}+1}{\sqrt{2}-1}\right|$

11 $1/2$ **13** -1 **15** $\frac{10}{3}\arctan(5/3)+\frac{10\pi}{6}$ **17** $\frac{1}{2}\ln(5)$

19 converge **21** diverge **23** (a) $\Delta t=\infty$ for a Riemann sum; (b) converges; (c) $\frac{1}{4t^2+9}<\frac{1}{t^2}$ when $t>1$, and $\int_1^\infty 1/t^2\,dt$ converges **25** (a) there are singularities inside the domain of integration; (b) diverges; (c) the partial-fraction decomposition of this integrand has the form $\frac{a}{t-2}+\frac{b}{2t+1}$, from which direct calculation shows divergence **27** (a) $\Delta t=\infty$ for a Riemann sum; (b) diverges; (c) Direct Comparison: $\frac{10t}{t^2+8}>\frac{10}{9t}$ when $t>1$, and $\int_4^\infty\frac{10}{9t}\,dt$ diverges (calculation) **29** (a) $\Delta t=\infty$ for a Riemann sum; (b) diverges; (c) Limit Comparison: $\frac{1}{\sqrt{t+1}}\approx\frac{1}{\sqrt{t}}$ when t is large, and $\int_1^\infty\frac{1}{\sqrt{t}}\,dt$ diverges. **31** (a) singularity at $t=1$; (b) diverges; (c) Direct Comparison: $\frac{t^2}{t^4-1}=\frac{t^2}{(t^2+1)(t+1)(t-1)}=\frac{t^2}{(t^2+1)(t+1)}\frac{1}{t-1}>\frac{1}{4}\frac{1}{t-1}$ when $t>1$ is close to 1, and $\int_1^2\frac{0.25}{t-1}\,dt$ diverges (direct calculation) **33** (a) $\Delta t=\infty$ for a Riemann sum; (b) diverges; (c) Limit Comparison to $\frac{2}{t}$ **35** (a) singularity at $t=\pi/2$; (b) diverges; (c) Direct Comparison: $\frac{1}{(t-\pi/2)\cos(t)}>\frac{1}{t-\pi/2}$ when $t\in(0,\pi/2)$ and $\int_0^{\pi/2}\frac{1}{t-\pi/2}\,dt$ diverges (direct calculation) **37** (a) $\Delta t=\infty$ for a Riemann sum; (b) converges; (c) direct calculation **39** (a) $\Delta t=\infty$ for a Riemann sum; (b) diverges; (c) Direct Comparison: the integrand is larger than $\frac{2}{3}$ when $t>4$, and $\int_4^\infty\frac{2}{3}\,dt$ diverges **41** (a) converges; (b) 4π **43** (a) converges; (b) $e^{\pi/2}-e^{-\pi/2}$ **45** (a) diverges **47** (a) converges; (b) 6 **49** (a) diverges **51** (a) diverges **53** (a) diverges **55** (a) converges; (b) $30\sqrt[5]{4}$ **57** (a) converges; (b) $\ln(4/3)$ **59** $\ell\geq\ln(100)$ from using $\frac{1}{1+t^6}e^{-t}\leq e^{-t}$ **61** $\ell\geq\sqrt[9]{\frac{40000}{81}}$ from using $\frac{\sqrt{t}}{\sqrt{t}+t^6}\leq t^{-11/2}$ **63** $\sqrt{21}-\sqrt{5}$ **65** diverges **67** diverges **69** $\frac{1}{2}\ln\left|\frac{2\sqrt{3}+2\sqrt{2}}{\sqrt{6}+\sqrt{2}}\right|$ **71** Student #2 is wrong. Check with $f(t)=1/\sqrt{t}$. **77** $2\sigma q/y$ **79** $p(\chi)=1-e^{-2m/(h^2+\chi^2)}$ **81** (b) 1.9128 (using $\ell=6.5$) **83** (b) 3.9952 (using $\ell=11$) **85** $F(s)=1/s^2$ **87** $F(s)=1/(s-3)$ **89** $F(s)=1/(s^2+1)$ **91** $f(t)=s/(s^2+1)$

Chapter 6 Review:

1 F **3** T **5** T **7** F **9** F **11** F **13** F **15** c **17** a **19** d **21** b **23** b **25** $-\ln|e^{-t}-1|+C$ **27** $(2(\sqrt{t}-1)e^{\sqrt{t}}+C$ **29** $\frac{3}{10}(t-1)^{10/3}+\frac{6}{7}(t-1)^{7/3}+\frac{3}{4}(t-1)^{4/3}+C$ **31**

$-\frac{5}{2}\frac{1}{t^2-6}+C$ **33** $\frac{1}{108}\ln\left|\frac{3t+1}{3t-1}\right|-\frac{1}{54}\arctan(3t)+C$ **35** $-4\ln|\csc(t/4)|+C$ **37** $\frac{1}{4}e^{t^4}+C$ **39** $-\frac{1}{5}\ln|5t-8|+C$ **41** $t-16\ln|t+12|+C$ **43** $\frac{1}{5}\ln\left|\frac{t-4}{t+1}\right|+C$ **45** $t^2+5-5\ln|t^2+5|+C$ **47** $0.01\sin(10t)-0.1t\cos(10t)+C$ **49** $\frac{3t}{32}\cos(4t)-\frac{t^3}{4}\cos(4t)+\frac{3t^2}{16}\sin(4t)-\frac{3}{128}\sin(4t)+C$ **51** $t\arcsin(t/4)+4\sqrt{1-\left(\frac{t}{4}\right)^2}+C$ **53** $\frac{t}{2}\sqrt{t^2+4}+2\ln|t+\sqrt{t^2+4}|+C$ **55** $\sqrt{9t^2-1}-\arccos\left(\frac{1}{3t}\right)+C$ **57** converges **59** converges **61** diverges **63** diverges **65** $\frac{432}{15}$ **67** $\frac{1}{3}\arctan\left(\frac{2}{3}t\right)-\frac{7}{8}\ln|4t^2+9|+C$ **69** $100(20201e^{-0.01}-20609e^{-0.03})$ **71** $\frac{1}{2}\ln|e^{2t}-3|+C$ **73** $\arcsin(1/\sqrt{7})$ **75** $\pi/4$ **77** $\frac{3}{\pi}\ln\left|\sec\left(\frac{\pi}{3}t\right)\right|+C$ **79** $\left(\frac{5}{226}\cos(3t)+\frac{75}{226}\sin(3t)\right)e^{t/5}+C$ **81** 2 **83** $\cos(t)-\frac{1}{3}\cos^3(t)+C$ **85** $\pi/2$ **87** $\ln\left|\frac{2t+1}{t}\right|-\frac{5}{2t+1}+C$ **89** $\frac{1}{4}\sqrt{4t^2+4t+10}-\frac{1}{4}\ln|2t+1+\sqrt{4t^2+4t+10}|+C$ **91** $\frac{1}{3}t^3-t^2+5t-17\ln|t+2|+C$ **93** $\frac{1}{4}\tan^4(t)-\frac{1}{2}\tan^2(t)+C$ **95** $\frac{1}{2}(t^2-1)e^{t^2}+C$

Section 7.1:

1 $4\sqrt{10}$ **3** $\frac{2}{243}\left(163^{\frac{3}{2}}-82^{\frac{3}{2}}\right)$ **5** $\frac{18085}{2592}$ **7** $\ln\left(\sqrt{2}+1\right)$ **9** $\ln\left|\frac{1+\sqrt{1+e^{-10}}}{e^{-3}}\right|+\sqrt{2}-\ln\left|1+\sqrt{2}\right|-\sqrt{1+e^{-10}}$ **11** $\frac{1}{2}\ln(\sqrt{3})$ **13** $3\pi\sqrt{10}$ **15** $(\sqrt{21^3}-1)\frac{\pi}{6}$ **17** $\left(\frac{[3\sqrt{2}-\ln(\sqrt{2}+1)]}{8}\right)\pi$ **19** $3\pi\sqrt{26}$ **21** $\frac{\pi}{54}\left(145^{1.5}-1\right)$ **23** $33\pi/2$ **25** 4.1629 **27** 5.0315 **29** 4.6936 **31** 14664.196 **33** 1.2764 **35** 59.0257 **37** 2786.07197 **39** 70.8616 **41** 2.0285 **43** $n=895$ using $M_1=1.79$ **45** $n=52$ using $M_2=1/4$ **47** $n=6$ using $M_2=0.71$ **51** $1.2\pi+\frac{\pi}{6}\left(5^{\frac{3}{2}}-1\right)$ **55** (a) Area= 1; (b) $\int_0^1\sqrt{1+\frac{1}{4x^3}}\,dx$

Section 7.2:

1 36π **3** $16\pi/3$ **5** $56\pi/3$ **7** (a) see below; (b) $\frac{1296\pi}{5}$

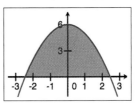

9 (a) see below; (b) $\frac{4\pi}{5}$

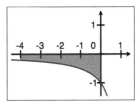

11 (a) see below; (b) $\frac{32\sqrt{2}\pi}{15}$

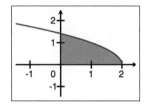

13 (a) see below; (b) π

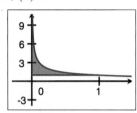

15 (a) see below; (b) $\frac{3544\pi}{105}$

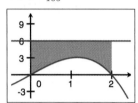

17 (a) see below; (b) $\frac{536\pi}{3}$

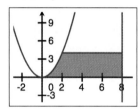

19 (a) see below; (b) $\frac{500\pi}{3}$

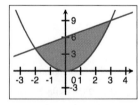

21 (a) see below; (b) $\left(16\arctan\left(\frac{23}{10}\right) - \frac{1}{2}\left(\frac{23}{10}\right)^2 \right)\pi$

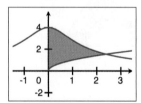

23 (a) see below; (b) $\frac{664\pi}{15}$

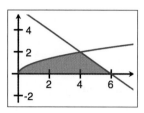

25 (a) see below; (b) 4π

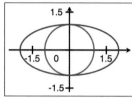

27 (a) see below; (b) $\frac{1096\pi}{15}$

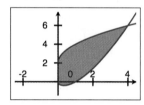

29 (a) see below; (b) $32\pi^2$

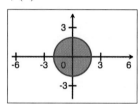

31 (a) see below; (b) 8π

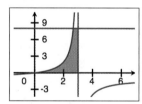

33 (a) see below; (b) $8\pi\sqrt{2}$

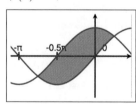

35 (a) see below; (b) $\frac{28\pi}{3}$

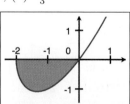

37 (a) $\frac{9}{4}(150 - y)^2\Delta y$; (b) $\sum_{k=1}^{n} \frac{9}{4}(150 - y_k)^2\Delta y$; (c)

$\int_0^{150} \frac{9}{4}(150 - y)^2 \, dy$; (d) 2531250 m^3 **39** 8π **41** 0.7823 cm^3 **43** (a) $\sum_{k=1}^n (x_k + 1) \times \pi(5x_k - x_k^2)^2 \Delta x$; (b) $4375\pi/12$ **45** (a) $62.42796\pi(r^2 h - \frac{1}{3}(r^2 - (r - h)^3))$ lbs; (b) $0.7556r$ (starting Newton's method at $h/r = 0.5$)   **47** $\sqrt{\frac{8}{5\pi}}$ meters   **49** (a) $\frac{3\pi}{4}\left(\frac{24}{27\pi}\right)^{2/3}$ mm^2 **51** 32π mm^3

Section 7.3:

1 (a) see below; (b) 18π

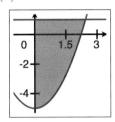

3 (a) see below; (b) $2\pi^2$

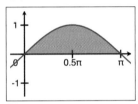

5 (a) see below; (b) $41\pi/42$

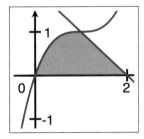

7 (a) see below; (b) $184\pi/3$

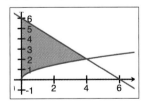

9 (a) see below; (b) $\pi(\pi - \ln(4))/4$

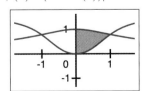

11 (a) see below; (b) $2\pi(12\ln(2) - 7 - (\ln(2))^2 + \ln(4))$

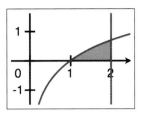

13 (a) shells; (c) $\int_{-5/2}^0 \frac{\pi}{3}(0 - x)(12 - 4x - 4x^2 - 3\sin(\pi x)) \, dx$; (d) $\frac{1}{\pi} + \frac{925}{144}\pi$ **15** (a) shells; (c) $\int_0^{3/2} 2\pi x(x + \frac{1}{2}\sin(\pi x)) \, dx + \int_{3/2}^2 2\pi x(2 + \frac{1}{3}x - \frac{2}{3}x^2) \, dx$; (d) $\frac{127}{144}\pi - \frac{4 - 9\pi^2}{4\pi}$ **17** (a) washers; (c) $\int_{3/2}^2 \pi(x + \frac{1}{2}\sin(\pi x))^2 - \pi(2 + \frac{1}{3}x - \frac{2}{3}x^2)^2 \, dx$; (d) $\frac{1537\pi^2 - 2160\pi + 1080}{1080\pi}$ **19** (a) disks; (c) $\int_0^2 \pi(x + \frac{1}{2}\sin(\pi x))^2 \, dx$; (d) $\frac{35\pi - 24}{12}$ **21** (a) shells; (c) $\int_{-5/2}^{3/2} 2\pi(2 - x)(2 - \frac{2}{3}x - \frac{2}{3}x^2 - \frac{1}{2}\sin(\pi x)) \, dx$; (d) $230\pi/9$ **23** (a) washers; (c) $\int_{1/2}^2 \pi(1 - \frac{1}{3}x + \frac{2}{3}x^2)^2 - \pi \, dx$; (d) $289\pi/80$ **25** (a) see below; (b) $13\pi/6$

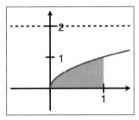

27 (a) see below; (b) $2\pi^2(\pi - 1)$

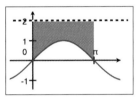

29 (a) see below; (b) $288\pi/5$

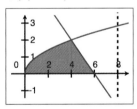

31 1.5 **33** 18.452 **35** 41.966 **37** $n = 67,222$ using $M_1 = 43.36\pi$ **39** $n = 35$ using $M_2 = 5.74\pi$ **41** $n = 29$ using $M_2 = (20 + 4e)\pi$

Section 7.4:

1 8575 J   **3** $980 - 196e^{6/5}$ J   **5** 441 J   **7** $19600/3$ J   **9** $30625\pi/6$ J   **11** $1254400/3$ J   **13** 46.8 J   **15** $27/700$ J   **17** $\sqrt{2}/50$ meters   **19** $558/121$ N   **21** 6.93×10^{-18} J   **23** 4.2733×10^{-18} J   **25** 747 J   **27** 49367.5 J   **29** 372.4 J   **31** 0.016 J

Section 7.5:

1 115/36 **3** $7 - 3/\ln(2)$ **5** $a = 1, b = 2$ **7** length $= (3 + \sqrt{105})/2$, mass $= 17.59$ kg **9** $581728\pi^2/375$ J **11** $1.1172\pi^2$ J **13** (a) see below; (b) $5\sqrt{6/7}$ meters

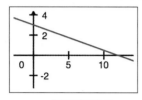

15 (a) see below; (b) $\sqrt{24308/665}$ meters

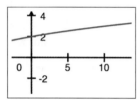

17 (a) see below; (b) $\overline{x} = 1/(2e-4), \overline{y} = (e^2-3)/(4e-8)$

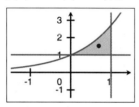

19 (a) see below; (b) $\overline{x} = \frac{1}{\pi} + \frac{3}{2}\pi, \overline{y} = \frac{9}{8}$

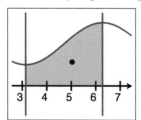

21 $\overline{x} = 32/55, \overline{y} = 161/22$ **23** $\overline{x} = 295/56, \overline{y} = 315/256$ **25** $m = 12 - 5\ln(5), M_y = 15\ln(5) - \frac{76}{3}, M_x = \frac{25}{2}\ln(5) - 14, \overline{x} \approx -0.301498, \overline{y} \approx 1.547753$ **27** $\overline{x} = \frac{12}{23}\left(4 - \left(\frac{1}{\pi}\right)^2\right), \overline{y} = \frac{53}{46}$ **29** (a) $\frac{2247}{640}\pi^2$ J; (b) $\sqrt{749/170}$ m **31** (a) $1215\pi^4 - 972\pi^2$ J; (b) $\sqrt{7290\pi^4 - 5832\pi^2}/(54\pi)$ m **33** (a) & (c) see below; (b) 80.4441 nm; (c) 787.4342572 nm

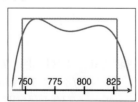

Section 7.6:

1 235,200,000 N **3** 235200 N **5** 352,800 N **7** 1,452,068,940 N **9** 19,600,000/3 N **11** 3,845,260.58 N **13** 2,975,334.022 N **15** 13,747.3995 N **17** 1379840/3 N **19** 108959956.8 N **21** $2116800\sqrt{2}$ N **23** (a) 24500 N (b) 392000 N (c) 1151500/3 N **25** $\frac{245}{3}(6 - t)^2$ **27** 3/70 m^3 **29** 13/20 m^3 (water depth 1/2 meter)

Section 7.7:

1 160/3 **3** $2/\pi$ **5** $20/\pi$ **7** $2-22e^{-3}$ **9** $2-\frac{1}{8}\ln(5)$ **11** $\sqrt{7/3}$ **13** $\arccos\left(\frac{2}{3\pi}\right)$ **15** $\ln(e - 1)$ **17** 0 **19** $\pi/2$ **21** $2/\sqrt{3}$ **23** see below

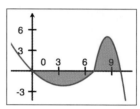

27 1 m/sec **29** 7.1593 m **31** (a) \$110/3; (b) \$33 **33** (a) 1418.3444; (b) 1513.0695 **35** 4/3 **37** $4/\pi$ **39** 5.0930 **41** 7.5250

Section 7.8:

1 (a) $1/\pi$; (b) 0 **3** (a) $\arcsin(0.5)$; (b) 0 **5** (a) 1/3; (b) 5/4 **7** (b) 1/18 **9** (b) 1/5 **11** (b)13/48 **13** (a) kg; (b) kg^2; (c) kg **17** (b) $(a + b)/2$ **19** (a) $\lambda = 1/5$; (b) $1 - e^{-8/5}$; (c) e^{-3}; (d) 5; (e) $1 - e^{-2}$ **21** $\int_{10}^{17.25} \frac{1}{7.25\sqrt{2\pi}} \exp\left(-\frac{(x-24.5)^2}{105.125}\right) dx$ **23** (a) 5.65 mg/L; (b) 0.525 mg/L; (c) this is the probability that a person selected from the general population has zinc levels between 6 and 6.25 mg/L **25** 0.9545 **27** 0.001884

Chapter Review:

1 T **3** F **5** F **7** F **9** T **11** T **13** T **15** T **17** d **19** d **21** b **23** a **25** 3.1905 **27** $\left(\frac{2\sqrt{2}-\sqrt{3}}{2} - \ln\left|\frac{1+\sqrt{3}}{2+\sqrt{2}}\right|\right)\pi$ **29** $8 + \frac{224}{3}\pi$ **31** $\frac{1}{14\pi}(3 + 28\pi)$ **33** $537\pi/2$ **35** 2871/2980375 **37** $\pi/2$ mm^3 **39** 51.3π **41** $\sqrt{2783/175}$ **43** $\frac{19600}{3\sqrt{2}}(1.3\sqrt{3}t)^{3/2}$ **45** 6860 J **47** $8 - \sqrt{60}$ **49** 200/3 **51** (a) 0.6321; (b) 0.2231; (c) 0.1723; (d) $\sigma = 20$

Section 8.1:

1 converges to 5 **3** converges to 0 **5** diverges   **7** converges to $\frac{3-\sqrt{5}}{2}$ **9** converges to $\frac{1+\sqrt{13}}{2}$ **11** diverges   **13** converges to 5   **15** diverges   **17** 0   **19** $5\ln(5) - 4$ **23** $\lim_{k\to\infty} p_k = 13$ **25** no   **27** no   **29** (a); no (b) yes   **31** $1, \frac{3}{4}, \left(\frac{3}{4}\right)^2, \left(\frac{3}{4}\right)^3, \left(\frac{3}{4}\right)^4, \left(\frac{3}{4}\right)^5, \left(\frac{3}{4}\right)^6, \left(\frac{3}{4}\right)^7, \left(\frac{3}{4}\right)^8, \left(\frac{3}{4}\right)^9$ **33** (a) $A'(t) = -k\,A(t)$ **35** $\lim_{k\to\infty} a_k = 600$ **37** $\frac{10}{9}$ **39** (b) see below

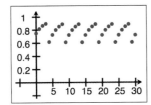

41 (a) $\frac{a_{k+1}}{a_k} = r\left(1 - \frac{a_k}{K}\right)$; (b) the percentage growth (which is $\frac{a_{k+1}}{a_k}$) is very small when relatively large ($\approx rK$) when a_k is small, but the percentage growth is small when $a_k/K \approx 1$ **43** $a_* = K - \frac{K}{r}$ **45** (b) larger r correspond to parabolas that are taller, but the graph of the updating function always has roots at 0 and 1; (c) $\lim_{r\to\infty} a_*(r) = 1$ because as the parabolic graph of the updating function gets taller (steeper) its intersection with the line $y = x$ moves to the right, though never beyond the right-hand root.   **47** (b) $f'(a_*) = 2 - r$; (c) $\lim_{k\to\infty} a_k = a_*$ when $-1 < f'(a_*) < 0$, but not when $f'(a_*) < -1$ (c) part (b) translates directly to a statement about r since $f'(a_*) = 2 - r$ when $K = 1$ **49** (a) $Vc_{n+1} = Vc_n - Tc_n + \gamma T$; (b) γ; (c) over time, your lungs equilibrate with the atmosphere   **51** (a) $Vc_{n+1} = V(1-\alpha)c_n - T(1-\alpha)c_n + \gamma T$; (b) higher atmospheric concentration should increase c_*; (c) higher absorption rate should decrease resident amounts of the gas, and so decrease c_*; (d) $c_* = \frac{\gamma q}{\alpha + q(1-\alpha)}$; (e) thinking of all parameters except γ as fixed constants, $dc/d\gamma = q/(\alpha + q(1-\alpha))^2 > 0$ when $\alpha \in (0,1)$, thinking of all parameters except α as fixed constants, $dc/d\alpha = -\gamma q(1-q)/(\alpha + q(1-\alpha))^2 < 0$ since $q \in (0,1)$ **53** (a) 11 (b) $w_{k+1} = w_k + w_{k-1}$ **55** they're Fibonacci numbers!

Section 8.2:

1 (b) 3   **3** (b) $-2/5$ **5** not geometric   **7** not geometric **9** $\frac{4}{3}$ **11** $\frac{3}{20}$ **13** diverges   **15** $-\frac{1}{5}$ **17** $-\frac{1}{80}$ **19** $\frac{3}{250}$ **21** $\frac{216}{29}$ **23** $\frac{1}{1-x}$ **25** (a) $\frac{10}{9}$ (b) $\frac{1}{9} \times 10^{-1000}$ **27** (a) 2 (b) $2\left(\frac{2}{3}\right)^{1000000}$ **29** $\frac{46}{9}$ **31** $\frac{6583}{900}$ **33** 1   **35** $\sum_{k=1}^{9} \frac{(-1)^k}{k^2} = -\frac{5257891}{6350400} \approx -0.8279622$ **37**

$\sum_{k=2}^{5} \frac{(-1)^k}{k\ln(k)} \approx 0.4740043383$ **39** $\frac{3}{4}$ **41** $\frac{13}{12}$ **43** $\frac{1}{12}$ **45** 2   **47** Yes, $\sum_{k=1}^{\infty} \left(\frac{-1}{2}\right)^k$ **49** yes   **51** (a) $\frac{1}{k+1} < \frac{1}{k}$; (b) $s_1 = -2$, $s_2 = -\frac{1}{2}$, $s_3 = -\frac{11}{6}$, $s_4 = -\frac{7}{12}\ldots s_{10} = -\frac{1627}{2520}$; (c) this has to do with the fact that $|a_k| \to 1$ (remember that a series is defined to be the limit of its partial sums)   **55** (a) diverge (b) converge (remember that a series is defined to be the limit of its partial sums)   **57** $\frac{4}{5}$ **59** (a) $\frac{a\sqrt{1+a^2}}{a-1}$; (b) $a > 1$ results in finite length   **61** 4   **63** (c) ε; (d) zero; (e) zero

Section 8.3:

1 (a) 6; (b) converge   **3** (a) $\frac{1}{2}$; (b) diverge   **5** converges   **7** converges   **9** diverges   **11** converges **13** converges   **15** diverges   **17** $p < -1$ **19** $p > 1$ **21** $n \geq e^{10} - 1 \Rightarrow n = 22026$ **23** $n \geq \sqrt{5e^{200} - 4} - 1 \Rightarrow n \approx 6.0108 \times 10^{43}$ **25** $n = 10^6$ **27** $n \geq 70\ln(10) \Rightarrow n = 162$ **29** $(4 - e^{0.1})/(e^{0.1} - 1) \leq n \Rightarrow n = 28$ **31** using $\frac{t}{t^4+5} < \frac{t}{t^4}$ when $t > 0$: $5 \leq n^2 \Rightarrow n = 3$ **33** using $\frac{1}{12t^2 - \sin(t)} < \frac{1}{12t^2 - t^2}$ when $t \geq 1$, $n = 1$ **35** using $\frac{2t^2 - \sin^2(t)}{t^4} \leq \frac{2}{t^2}$ when $t \geq 1$, $n \geq 20$ **37** 1.64496333616 using $n = 22$ **39** 1.460253304 using $n = 16$ **41** 0.9206625274 using $n = 13$ **43** diverge   **45** diverge   **47** converge   **49** converge   **51** $p \in (0.5, 1]$ **53** (a) $b \in (0,1)$; (b) no change   **55** (a) $\frac{1}{n}$; (b) $\frac{1}{n(n+1)}$; (c) see below; (d) $\frac{1}{n}$

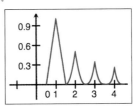

57 (a) yes; (b) yes   **59** (a) no; (b) yes

Section 8.4:

13 diverges   **15** converges   **17** diverges   **19** converges   **21** diverges   **23** diverges   **25** converges   **27** diverges   **29** converges   **31** diverges   **33** converges **35** converges   **37** diverges   **39** converges   **41** diverges   **43** converges   **45** incorrect conclusion   **47** incorrect conclusion   **49** yes   **51** no   **53** convergent

Section 8.5:

9 absolutely convergent   **11** inconclusive   **13** divergent   **15** absolutely convergent   **17** absolutely convergent   **19** inconclusive   **21** conditionally convergent   **23** absolutely convergent   **25** absolutely convergent   **27** divergent   **29** divergent   **31** conditionally convergent   **33** $p \in (0.5, 1.5)$ **35** none   **37** $p \in [1-\sqrt{5}, -1) \cup (3, 1+\sqrt{5}]$ **39** (a) $|x| < \frac{1}{5}$ (b) $|x| > \frac{1}{5}$ (c) $|x| = \frac{1}{5}$ **41** (a) $|x - 2| < \frac{1}{10} \Rightarrow x \in (1.9, 2.1)$, (b) $|x - 2| > \frac{1}{10} \Rightarrow x \in (-\infty, 1.9) \cup (2.1, \infty)$ (c) $x = 1.9$ and $x = 2.1$ **43** (a) all values of x (b) no values of x (c) no values of x **45** (b) absolutely convergent   **47** (b) inconclusive   **49** converges   **51** converges   **53** diverges   **55** converges   **57** diverges   **59** diverges   **61** converges   **63** diverges   **65** incorrect assertion   **67** it does   **69** we don't know   **71** not necessarily   **73** the series converges, but we don't know enough to tell whether its *absolutely* or *conditionally*

Section 8.6:

1 $\sum_{k=0}^{\infty} 6^k t^k$ **3** $\sum_{k=0}^{\infty}(-1)^k \frac{3^{2k+1}}{5^{k+1}} t^k$ **5** $\sum_{k=0}^{\infty} \frac{3^k}{2^{2k-1}} t^{k+1}$ **7** $\sum_{k=0}^{\infty} 5(-1)^k \frac{7^k}{3^{k+1}} t^{2k+3}$ **9** $\sum_{k=0}^{\infty} \frac{1}{3^k}(t-1)^k$ **11** $\sum_{k=0}^{\infty} 7\frac{(-8)^k}{29^{k+1}}(t-3)^k$ **13** $-8 + \sum_{k=1}^{\infty}(-1)^{k+1}(4)(t-2)^k$ **15** (a) $\frac{(4t)^3}{1-4t}$ (b) $\frac{1}{18}$ **17** (a) $\frac{8(t-3)}{1-8(t-3)}$ (b) $-\frac{4}{9}$ **19** (a) $\frac{1}{4}\frac{1}{t-1} - \frac{1}{4}\frac{1}{t+3}$, (b) $\sum_{k=0}^{\infty}(-1)^k\left(\frac{1}{4^{k+2}} - \frac{1}{2^{3k+5}}\right)(t-5)^k$, (c) $(1, 9)$ **21** (a) $\frac{1}{12t} + \frac{1}{4}\frac{1}{t+4} - \frac{1}{3}\frac{1}{t+3}$, (b) $\sum_{k=0}^{\infty} -\frac{1}{12}\left(1 + \frac{(-1)^k}{3^k} + \frac{(-1)^k}{2^{k-1}}\right)(t+1)^k$, (c) $(-2, 0)$ **23** (a) $R = 2$ (b) $(-2, 2)$ **25** (a) $R = \frac{1}{3}$ (b) $\left(-\frac{1}{3}, \frac{1}{3}\right)$ **27** (a) $R = 1$ (b) $[2, 4]$ **29** (a) $R = \sqrt{8}$ (b) $(-1-\sqrt{8}, -1+\sqrt{8})$ **31** $f(3)$ also converges   **41** (a) $t_0 = 0$; (b) see below; (c) $[-1, 1]$; (d) $R = 1$

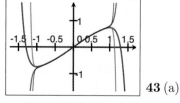

43 (a)

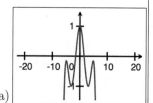

(b)

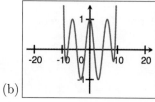

(c)

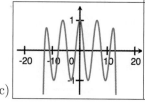

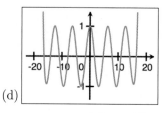

(d)

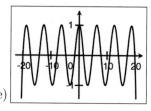

(e)

45 (d) $g(x) = \frac{1/2}{1-2x} + \frac{1/2}{1-4x}$; (e) $\sum_{k=0}^{\infty}\left(\frac{2^k + 4^k}{2}\right)x^k$ **47** $c_k = \left(1 - \left(\frac{23}{25}\right)^k\right)\gamma$

Section 8.7:

1 (b) $t^2 + \frac{5}{2}t^3 + \frac{13}{3}t^4 + \frac{77}{12}t^5 + \frac{87}{10}t^6 + \cdots$ **3** (b) $-4t^5 - 13t^6 - 28t^7 - 50t^8 - 80t^9 + \cdots$ **5** (b) $\frac{3}{4} + \frac{15}{16}t + \frac{63}{64}t^2 + \cdots$ **7** (b) $\frac{1}{45}t^3 - \frac{1}{675}t^4 - \frac{74}{10125}t^5 + \cdots$ **9** $\frac{1}{12} - \frac{7}{144}t + \frac{37}{1728}t^2 - \frac{175}{20736}t^3 + \cdots$ **11** (b) $\frac{2}{3} + \frac{4}{3}t + \frac{50}{27}t^2 + \cdots$ **13** (b) $-\frac{1}{4}t - \frac{3}{16}t^2 - \frac{9}{64}t^3 - \cdots$ **15** (b) $\frac{4}{3} + t + t^2 + \cdots$ **17** (b) $5t + \frac{1}{3}t^2 - \frac{5}{3}t^3 - \frac{1}{9}t^4 + \cdots$ **19** $1 + \frac{1}{2}t^2 + \frac{3}{24}t^4 + \frac{61}{720}t^6 + \cdots$ **21** $t + \frac{1}{3}t^3 + \frac{2}{15}t^5 + \frac{17}{315}t^7 + \cdots$ **23** (b) $\sum_{k=1}^{\infty}(-1)^k\frac{3k}{2^{k+1}}t^{k-1}$ **25** (b) $\sum_{k=1}^{\infty}(-1)^{k+1}\frac{9k}{2^k}(t-5)^{k-1}$ **29** (a) decreasing; (b) down   **31** $C + \sum_{k=1}^{\infty}\frac{3\sqrt{k+1}}{(k+1)!}t^{k+1}$ **33** $\frac{1}{2}$ **35** (a) $\sum_{k=0}^{\infty}\frac{(-1)^k}{2^{k+1}}t^{5k+2}$; (b) $\sqrt[5]{2}$ (c) $\sum_{k=0}^{\infty}\frac{(-1)^k}{(5k+3)2^{k+1}}t^{5k+3}$ **37** the series for $f'(t)$ converges at $t = -3.7$ and $t = 1.2$ **41** 3/2   **43** 2   **45** $4\ln(3/4)$ **47** 4   **49** $\frac{1}{n}\left(1 + \frac{1}{2} + \frac{1}{3} + \cdots + \frac{1}{n-1}\right)$

Section 8.8:

1 (a) $1 - 2t$; (b) see below

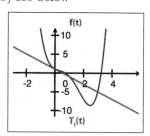

3 (a) $-1 - 4(t-1) - 3(t-1)^2$; (b) see below

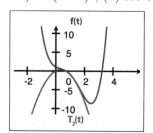

5 (a) $11 - 24(t+1) + 21(t+1)^2 - 8(t+1)^3$; (b) see below

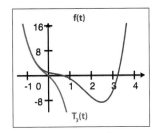

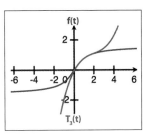

7 (a) $-5+16\,(t-3)+21\,(t-3)^2+8\,(t-3)^3+(t-3)^4$; (b) see below

17 (a) $t-\frac{1}{3}t^3$; (b) see below

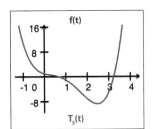

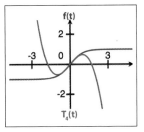

9 (a) $3+\frac{1}{6}(t-9)-\frac{1}{216}(t-9)^2+\frac{1}{3888}(t-9)^3+\frac{5}{279936}(t-9)^4$; (b) see below

19 (a) $1-\frac{7}{3}t+\frac{14}{9}t^2-\frac{14}{81}t^3-\frac{7}{243}t^4$; (b) see below

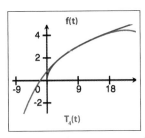

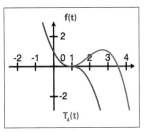

11 (a) $-(t-\pi)+\frac{1}{6}\,(t-\pi)^3-\frac{1}{120}\,(t-\pi)^5$; (b) see below

21 (a) $1-\frac{5}{9}t-\frac{100}{81}t^2-\frac{8500}{2187}t^3-\frac{276250}{19683}t^4$; (b) see below

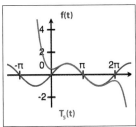

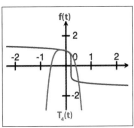

13 (a) $(t-1)-\frac{1}{2}(t-1)^2$; (b) see below

23 (a) $3+\frac{1}{6}t-\frac{1}{216}t^2+\frac{1}{3888}t^3-\frac{5}{279936}t^4$; (b) see below

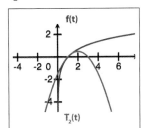

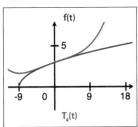

15 (a) $\frac{\pi}{4}+\frac{1}{2}\,(t-1)-\frac{1}{4}\,(t-1)^2+\frac{1}{12}\,(t-1)^3$; (b) see below

25 (a) $\frac{1}{3}-\frac{1}{243}t+\frac{2}{19683}t^2-\frac{14}{4782969}t^3+\frac{35}{387420489}t^4$; (b) see below

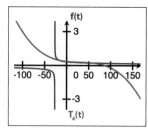

27 $1+2t+\frac{5}{2}t^2+\frac{8}{3}t^3+\frac{65}{24}t^4$ **29** t^3 **31** $\sum_{k=0}^{\infty}\frac{1/2}{(2k)!}t^{2k}$ **33**

$\sum_{k=0}^{\infty}\frac{(-3)^k}{2^{k+1}}t^k$ **35** $\sum_{k=0}^{\infty}\frac{(-1)^{\lfloor k/2\rfloor}}{k!}t^k$ where $\lfloor\ \rfloor$ is the

greatest integer (floor) function **37** d **39** c **41**

$t\in(6.5,7.5)$ **43** see below

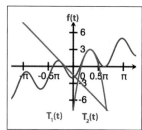

45 (a) 2^k (b) $2^{23}(23!)$ **47** $2^{50}(50^2-51)$ **51**

$f(t)=2+\ln(1-t)$ **53** $t\in[4,6]$ **55** all $t\in\mathbb{R}$ **57** (a)

$-3+5(t-1)+(t-1)^2$; (b) $-\frac{3}{(t-1)^3}+\frac{5}{(t-1)^2}+\frac{1}{t-1}$; (c)

$\frac{3}{2(t-1)^2}-\frac{5}{t-1}+\ln|t-1|+C$ **59** (a) $-18+12(t+3)-$

$5(t+3)^2+(t+3)^3$; (b) $-\frac{18}{(t+3)^4}+\frac{12}{(t+3)^3}-\frac{5}{(t+3)^2}+\frac{1}{t+3}$;

(c) $\frac{6}{(t+3)^3}-\frac{6}{(t+3)^2}+\frac{5}{t+3}+\ln|t+3|+C$ **61** (a)

$3705-2991(t+5)+900(t+5)^2-120(t+5)^3+6(t+5)^4$;

(b) $\frac{3705}{t+5}-2991+900(t+5)-120(t+5)^2+6(t+5)^3$;

(c) $3705\ln|t+5|-2991t+450(t+5)^2-40(t+5)^3+$

$\frac{3}{2}(t+5)^4+C$ **63** $T_3(x)=x$

Section 8.9:

1 $\sum_{k=0}^{\infty}\frac{4}{16^k}\binom{1/2}{k}(t-16)^k$ **3** $\sum_{k=0}^{\infty}\frac{3^5}{9^k}\binom{5/2}{k}(t-$

$9)^k$ **5** $\sum_{k=0}^{\infty}\frac{1}{5^{2k+1}}\binom{-1/2}{k}(t-25)^k$ **7** $-1/5$ **9**

$17/273$ **11** $1/72$ **13** $2^{10}/(10!)$ **15** $-1/(18!)$ **17**

$\sum_{k=0}^{\infty}\frac{(-1)^k}{(2k+1)!}x^{10k+5}$ **19** $\sum_{k=0}^{\infty}\frac{1}{k!}x^{7k}$ **21** $x-\frac{1}{3}x^3$ **23**

$1-\frac{1}{2}x^2-\frac{1}{2}x^3-\frac{1}{4}x^4$ **25** (a) $\sum_{k=0}^{\infty}\frac{2}{k!\sqrt{k}(2k+1)}x^{2k+1}$; (b)

$\sum_{k=0}^{2}\frac{2}{k!\sqrt{k}(2k+1)}(0.3)^{2k+1}=\frac{309243}{500000}\approx0.6184860$ **29**

(a) $\sum_{k=0}^{\infty}\frac{(-1)^k}{(2k+1)!}(0.9)^k$; (b) 4; (c) $\sum_{k=0}^{3}\frac{(-1)^k}{2k+1)!}(0.9)^k=$

$\frac{479699}{560000}\approx0.8566053571$ **31** (a) $\sum_{k=0}^{\infty}\frac{(-1)^k}{k!}(1.5)^k$;

(b) 8; (c) $\sum_{k=0}^{\infty}\frac{(-1)^k}{k!}(1.5)^k=\frac{3191}{14336}\approx0.2225864955$

33 (a) $\sum_{k=0}^{\infty}\frac{(-1)^k}{(k!)(2k+1)\sqrt{\pi}}\left(\frac{1}{2}\right)^{2k}$; (b) 3; (c)

$\sum_{k=0}^{2}\frac{(-1)^k}{(k!)(2k+1)\sqrt{\pi}}\left(\frac{1}{2}\right)^{2k}=\frac{443}{480}\approx0.9229166667$ **35**

(a) $\sum_{k=0}^{\infty}\frac{(-1)^k}{k+1}\left(\frac{1}{5}\right)^k$; (b) 4; (c) $\sum_{k=0}^{3}\frac{(-1)^k}{k+1}\left(\frac{1}{5}\right)^k=$

$\frac{1367}{1500}\approx0.91133$ **37** (a) $\sum_{k=0}^{\infty}\frac{(-1)^k}{k+1}\left(\frac{1}{2}\right)^k$; (b) 7;

(c) $\sum_{k=0}^{6}\frac{(-1)^k}{k+1}\left(\frac{1}{2}\right)^k=\frac{909}{1120}\approx0.8116071429$ **39**

$2\sum_{k=0}^{5}\binom{1/3}{k}\left(\frac{1}{2}\right)^k=\frac{35615}{15552}\approx2.290059156$ **41**

$2\sum_{k=0}^{1}\binom{1/4}{k}\left(\frac{1}{16}\right)^k=\frac{65}{32}\approx2.03125$ **43** (a) $\sqrt{e}=$

$1+\int_0^{1/2}e^x\,dx$; (b) $n=4$; (c) $\frac{6331}{3840}\approx1.648697917$ **45**

(a) $\sin(0.9)=\int_0^{0.9}\cos(x)\,dx$; (b) $n=4$; (c) $\frac{3133683}{4000000}\approx$

0.78342075 **47** (a) $\ln(0.9)=\int_0^{-0.1}1/(1+x)\,dx$;

(b) $n=2$; (c) $-\frac{79}{750}\approx-0.105333$ **49** $C+$

$\sum_{k=0}^{\infty}\frac{(-1)^k}{(2k+1)!}\frac{x^{2k+1}}{2k+1}$ **51** $C+\sum_{k=0}^{\infty}\frac{(-1)^k}{(5k+1)}t^{5k+1}$ **53** $C+$

$\sum_{k=0}^{\infty}\binom{1/2}{k}\frac{t^{7k+1}}{7k+1}$ **55** (a) $\sin(t^2)=\sum_{k=0}^{\infty}\frac{(-1)^k}{(2k+1)!}t^{4k+2}$;

(b) $\sum_{k=0}^{\infty}\frac{(-1)^k}{(2k+1)!}\frac{1}{4k+3}$; (c) $13/42$ **57** (a) $\frac{e^x-1}{x}=$

$\sum_{k=1}^{\infty}\frac{1}{k!}x^{k-1}$; (b) $\sum_{k=1}^{\infty}\frac{1}{(k)(k!)}\left(\frac{1}{2}\right)^k$; (c) $9/16$ **59**

(a) $\frac{t}{1+t^4}=\sum_{k=0}^{\infty}(-1)^kt^{4k+1}$; (b) $\sum_{k=0}^{\infty}\frac{(-1)^k}{4k+2}\left(\frac{3}{5}\right)^{4k+2}$;

(c) $9/50$ **61** (a) $\sum_{k=1}^{\infty}\binom{-1/2}{k}(-1)^kmc^2\left(\frac{v}{c}\right)^{2k}$; (c)

$v=\sqrt{\frac{5}{75}}c\approx25.82\%$ of c **63** (c) $n=3$; (d) $\pi/4$; (f) $\frac{3\pi}{16}$,

$\frac{5\pi}{32},\frac{35\pi}{256},\frac{63\pi}{512},\frac{231\pi}{2048},\frac{429\pi}{4096}$; (g) $\frac{17577}{32768}\pi$ **65** $\sum_{k=0}^{\infty}\frac{4^k}{k!}t^k$ **67**

$4\sum_{k=0}^{\infty}\frac{5^k}{3^k\,k!}t^{3k}$ **69** $\frac{2}{\sqrt{e}}\sum_{k=0}^{\infty}\frac{1}{2^k\,k!}t^{2k+1}$ **71** (a) $1+t-$

$\frac{1}{2}t^2-\frac{1}{3!}t^3+\frac{1}{4!}t^4+\frac{1}{5!}t^5--++\cdots$; (b) $\sum_{k=0}^{\infty}\frac{(-1)^k}{(2k)!}t^{2k}+$

$\sum_{k=0}^{\infty}\frac{(-1)^k}{(2k+1)!}t^{2k+1}$; (c) $\cos(t)+\sin(t)$ **73** (a) see below;

(b) $T_5(t)=1-t+\frac{1}{2}t^2-\frac{1}{6}t^3+\frac{1}{24}t^4-\frac{1}{120}t^5$

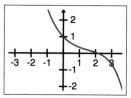

75 (a) see below; (b) $T_7(t)=4t-16t^2+\frac{122}{3}t^3-\frac{232}{3}t^4+$

$\frac{3529}{30}t^5-\frac{1342}{9}t^6+\frac{204133}{1260}t^7$

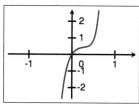

75 (a) see below; (b) $T_7(t)=4t-16t^2+\frac{122}{3}t^3-\frac{232}{3}t^4+$

$\frac{3529}{30}t^5-\frac{1342}{9}t^6+\frac{204133}{1260}t^7$

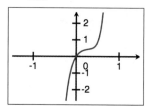

Chapter Review:

1 F **3** T **5** T **7** F **9** T **11** F **13** F

15 T **17** F **19** T **21** e **23** b **25** e **27**

d   **29** b   **31** a   **33** (d) the series diverges   **35** (a) $G(m) = (1-p)^{m-1}$ **37** 10   **39** $\frac{7^3}{44}$ **41** $\frac{47}{180}$ **43** diverges   **45** converges   **47** diverges   **49** converges   **51** converges   **53** diverges   **55** it converges also   **57** $\left(-\frac{1}{2}, \frac{1}{2}\right)$ **59** $(2, 4)$ **61** (a) Yes (b) It's a polynomial of degree 9, so $R = \infty$ **63** $\frac{9x}{1-3x}$ **65** $\sum_{k=0}^{\infty} \frac{(-1)^k}{(2k+1)!} t^{8k+4}$ **67** $C + \sum_{k=0}^{\infty} \frac{1}{(2k)!\,(k+1)} \theta^{k+1}$ **69** 23/150 (using the first three terms of the series)

Section 9.1:

1 (a) $y = -2x - 5$; (b) starting at $(1, -7)$ and moving in the direction of decreasing x (graph below); (c) all of the line to the left of the starting point is visited

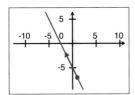

3 (a) $x = \frac{1}{125}(y - 1)^3 + 4$; (b) starting at $(4, 1)$ and moving in the direction of increasing x (graph below); (c) all of the curve to the right of the starting point is visited

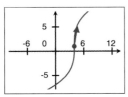

5 (a) $x = y^2 - 1$; (b) starting at $(-1, 0)$ and moving in the direction of decreasing y (graph below); (c) all of (and only) the part of the parabola with $|y| \leq 1$ is visited

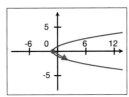

7 (a) $x^2 = y^2 - 1$ (b) starting at $(0, 1)$ and moving in the direction of increasing x (graph below); (c) due to jump discontinuities in the formulas for x and y, all of the hyperbola is visited

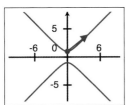

9 (a) $y = 3(\sqrt{x} - 1)^2 + 19$; (b) starting at $(1, 19)$ and moving in the direction of increasing x; (c) all points on the curve with $x \geq 1$ (graph below)

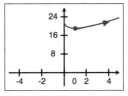

11 (a) $y = (\sqrt{x - 11} - 1)^4 - (\sqrt{x - 11} - 1)^2 + 19$ (b) starting at $(12, 19)$ and moving in the direction of increasing x (c) all points on the curve with $x \geq 12$ (graph below)

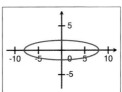

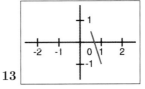
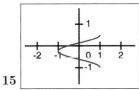

13   **15**

17 $x(t) = -7\sin(26\pi t)$, $y(t) = 7\cos(26\pi t)$ **19** $x(t) = 7\sin\left(\frac{18\pi}{5} t\right)$, $y(t) = 7\cos\left(\frac{18\pi}{5} t\right)$ **21** $x(t) = 7\cos\left(8\pi\left(t + \frac{1}{32}\right)\right)$, $y(t) = 7\sin\left(8\pi\left(t + \frac{1}{32}\right)\right)$ **23** $x(t) = 5 + 4\cos(42\pi t)$, $y(t) = 11 + 4\sin(42\pi t)$ **25** $x(t) = 5 + 4\sin(4\pi t)$, $y(t) = 11 + 4\cos(4\pi t)$ **27** $x(t) = 5 + 4\cos\left(4\pi\left(t - \frac{3}{16}\right)\right)$, $y(t) = 11 + -4\sin\left(4\pi\left(t - \frac{3}{16}\right)\right)$ **29** (a) see below; (b) $(0, 2)$; (c) moving clockwise

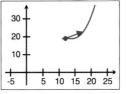

31 $x(t) = -\frac{1}{3}\sin\left(\frac{2\pi}{3} t\right)$, $y(t) = \frac{1}{2}\cos\left(\frac{2\pi}{3} t\right)$ **33** (b) $v = 13/8$ **35** (b) $x = -1/4$, (c) $v = \frac{\pi + \arcsin(\sqrt{63}/12)}{\pi + \arcsin(\sqrt{63}/8)}$ **37** $x_H(t) = 50 - \frac{t}{2}$ and $y_H(t) = -\sqrt{2500 - 0.25\left(50 - \frac{t}{2}\right)^2}$ parameterize a circle   **39** $x_{m,b}(t) = 75 - \frac{3t}{4}$ and $y_{m,b}(t) = -\frac{1}{2}\sqrt{2500 - 0.25\left(50 - \frac{t}{2}\right)^2}$ parameterize an ellipse   **41** (a) $x(t) = 2t - \sin(3\pi t)$ and $y(t) = 1 - \cos(3\pi t)$; (b) see below

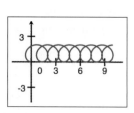

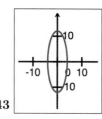

43

45 (a) $v(t) = \frac{1}{\pi^2}\sin(\pi t) - \frac{1}{\pi}t\cos(\pi t)$ and $x(t) = -\frac{2}{\pi^3}\cos(\pi t) - \frac{1}{\pi^2}t\sin(\pi t)$ **47** (a) $\frac{1}{2}b^2 + b$; (b) $\frac{1}{3}b^3 + \frac{1}{2}b^2$; (c) see below

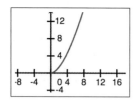

49 (a) $24(1 - e^{-b/3})$; (b) $72 - (72 + 24b)e^{-b/3}$; (c) see below

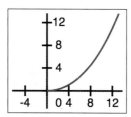

51 (a) $L = \frac{2}{3}(\sqrt{1 + 2.25b} - 1)$ and $S = 60\pi\left(3(1 + 2.25b)^{2.5} - 5(1 + 2.25b)^{1.5} + 2\right)$; (b) see below

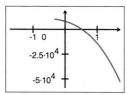

53 (a) $W = 3675D^2$; (b) $F = 4900D^2$; (c) see below

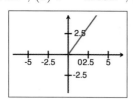

Section 9.2:

1 $y = \frac{8}{15}x + \frac{42}{5}$ **3** $y = -\frac{1}{3}x$ **5** $y = 2x + 5$ **7** $\left(\pm\frac{187\sqrt{205}}{205}, \mp\sqrt{205}\right)$ **9** $(0.75, \pm 0.75\sqrt{3})$ **11** $240/169$ (concave up)   **13** $1/\sqrt{2}$ (concave up)   **15** $-83/12$ (concave down)   **17** $2\sqrt{145}$ **19** $\sqrt{68}$ **21** $0.5\sqrt{17} + 0.125\ln|4 + \sqrt{17}|$ **23** 20.4357 **25** 1.7253 **27** 10.01989 **29** $\frac{33\pi\sqrt{17}}{10}$ **31** $2\pi(1 - \text{sech}(1))$ **33** 35.9114 **35** 59111.4455 **37** 131.3008 **39** 2 **41**

$-\frac{9}{2}$ **43** $2 + e$ **45** (a) $\beta/\alpha > 0$ (b) $a - \frac{\alpha b}{\beta} > 0$ (c) $b - \frac{\beta a}{\alpha} > 0$ **53** any point on the line $y = 13 - 3x$ with $x > 3$ can serve as P_1 **57** (a) A, (b) $2t\sec^2(t^2)$ **59** (a) initial position is $(2a, 0)$, headed counterclockwise (graph below); (b) position at $t = \pi/2$ is $(0, 2a)$, and the length is $6a$; (c) $24a$

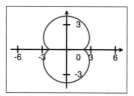

Section 9.3:

1 $(r, \phi) = (16, \pi/6)$ **3** $(r, \phi) = (\sqrt{32}, 5\pi/4)$ **5** $(r, \phi) = \sqrt{52}, \pi + \arctan(-3/2))$ **7** $(x, y) = (\frac{7\sqrt{2}}{2}, \frac{7\sqrt{2}}{2})$ **9** $(x, y) = (\sqrt{3}, 1)$ **11** $(x, y) = (\frac{13}{2}, -\frac{13\sqrt{3}}{2})$

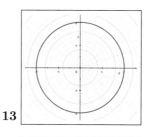

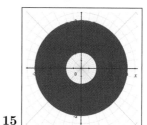

13   **15**

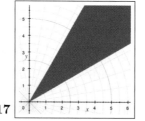

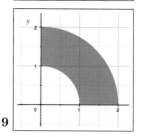

17   **19**

21 (a) ellipse; (b) see below

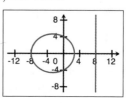

23 (a) parabola; (b) see below

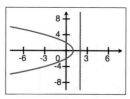

25 (a) ellipse; (b) see below

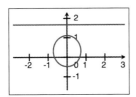

27 (a) hyperbola; (b) see below

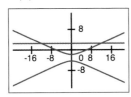

29 1.5 units   **31** 20/7 units   **33** $r = \frac{2}{1+\cos(\phi)}$

35 $r = \frac{2}{1-\sin(\phi)}$ **37** $r = \frac{5/\sqrt{2}}{1+\cos(\phi+\pi/4)}$ **39**

$r = \frac{3/\sqrt{5}}{1+\cos(\phi-\arctan(1/2)-\pi)}$ **41** (a) $r = \frac{0.8}{1+0.2\cos(\phi)}$;

(b) ellipse; (c) see below

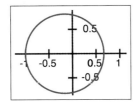

43 (a) $r = \frac{54\sqrt{17}/170}{1+0.9\cos(\phi+\arctan(-1/4))}$; (b) ellipse; (c) see

below

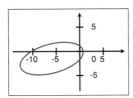

45 (a) $r = \frac{102/\sqrt{65}}{1+6\cos(\phi-\pi-\arctan(-1/8))}$; (b) hyperbola; (c)

see below

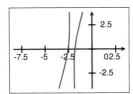

47 (a) $r = 3\csc(\phi)$; (b) see below

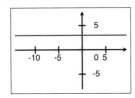

49 (a) $r = -\frac{5}{3\sin(\phi)+\cos(\phi)}$; (b) see below

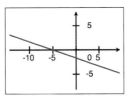

51 (a) $r = \sec(\phi)\tan(\phi)$; (b) see below

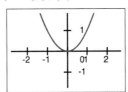

53
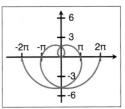

55
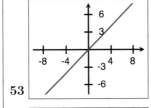
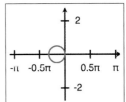

57
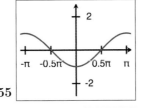
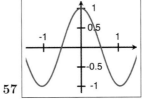

59
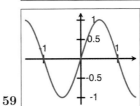
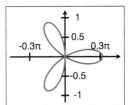
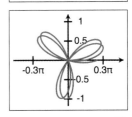

61 origin, and points (r, θ) where $r = \sqrt{3}/2$ and $\theta = \pi/3, 2\pi/3$ **63** origin, and points (r, θ) where $r = 1/2$ and $\theta = \pi/4, 3\pi/4, 5\pi/4, 7\pi/4$ **65** $(r, \theta) = (29/17, \arctan(-8/15))$ and $(r, \theta) = (5/17, \pi + \arctan(-8/15))$ **67** (a) see below; (b) the magnitude of B determines the "magnitude" of the loop (e.g., whether it's a loop or a dent) and the sign of B determine which side of the shape has the loop

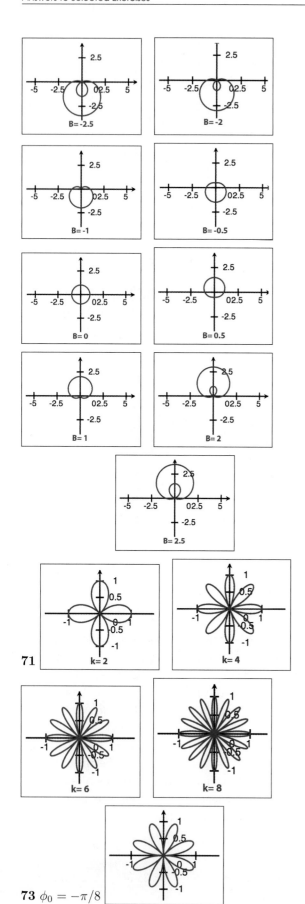

71

73 $\phi_0 = -\pi/8$

75 (a) see below; (b) $7\pi/24$; (c) $\frac{(2R\Delta r+(\Delta r)^2)\pi}{24}$; (d) $\frac{dA}{dR} = \frac{\pi}{12}\Delta r$

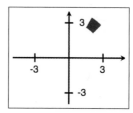

Section 9.4:

1 $-\frac{\sqrt{3}}{3}$ **3** $\frac{19\sqrt{3}}{39}$ **5** $y = -1/4$ **7** $(r,\phi) = (1,0)$ and $(r,\phi) = (\frac{2+\sqrt{3}}{2}, \frac{2\pi}{3})$ **9** $\frac{27\pi}{2}$ **11** $\frac{\pi}{16}$ **13** $\frac{\pi^3}{6}$ **15** (a) see below; (b) 1

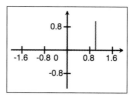

17 (a) see below; (b) $\frac{(4+\pi^2)^{3/2}-8}{3}$

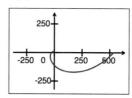

19 (a) see below; (b) 8

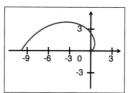

21 $\frac{\pi-1}{8}$ **23** $\frac{3\pi-4}{12}$ **25** $\frac{55\pi}{4}$ **29** 3.815 grams (using $\rho = 2.7\frac{\text{g}}{\text{cm}^3}$)   **31** $16\sqrt{2}$ mm^2 **35** (c) allowing ϕ to range over $[0, 2\pi]$ traces the circle twice   **37** (b) $(4+\pi)/(4-\pi)$ (c) $0.43A$

39 (a) 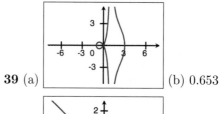 (b) 0.653

41 (a) 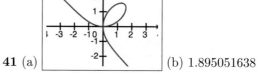 (b) 1.895051638

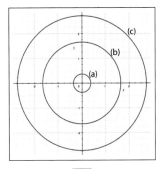

Section 9.5:

1 (a) 2; (b) 4; (c) $2\sqrt{5}$; (d) the point $(2,4)$ **3** (a) -2; (b) 3; (c) $\sqrt{13}$; (d) the point $(-2,3)$ **5** (a) -5; (b) 0; (c) 5; (d) the point $(-5,0)$ **7** (a) $7+6i$ **9** (a) $11+3i$ **11** (a) $1+5i$ **13** (b) $16-8i$; (c) $16-8i$; (d) $27-84i$; (e) $27-84i$ **15** (b) $6-i$; (c) $6-i$; (d) $5-5i$; (e) $5-5i$ **17** (b) $1-7i$; (c) $1-7i$; (d) $-10-5i$; (e) $-10-5i$ **19** (b) $\text{Arg}(z)=-\frac{\pi}{4}$, $\text{Arg}(w)=\frac{\pi}{4}$, $\text{Arg}(z+w)=0$ **21** (b) $\text{Arg}(z)=-\frac{\pi}{4}$, $\text{Arg}(w)=\frac{\pi}{2}$, $\text{Arg}(z+w)=\frac{\pi}{4}$ **23** (b) $\text{Arg}(z)=\frac{\pi}{3}$, $\text{Arg}(w)=\frac{\pi}{2}$, $\text{Arg}(z+w)=\frac{5\pi}{6}$ **25** $\frac{\sqrt{16}}{16}-\frac{\sqrt{16}}{16}i$ **27** $\frac{\sqrt{3}}{4}+\frac{1}{4}i$ **29** $\frac{3}{5}-\frac{4}{5}i$ **31** $8\sqrt{2}e^{(\pi/4)i}$ **33** $8e^{(-5\pi/6)i}$ **35** $9e^{\pi i}$ **37** $13e^{(-\pi/2)i}$ **39** (a) $-\frac{7\sqrt{3}}{2}+\frac{7}{2}i$; (b) $\frac{1}{7}e^{(-5\pi/6)i}$ **41** (a) $1-\sqrt{3}\,i$; (b) $\frac{1}{2}e^{(\pi/3)i}$ **43** (a) $2\sqrt{2}+2\sqrt{2}\,i$; (b) $\frac{1}{4}c^{(-\pi/4)i}$ **45** (a) $14i$; (b) $\frac{1}{14}e^{(-\pi/2)i}$ **47** $6e^{(5\pi/12)i}$ **49** $10e^{(-\pi/6)i}$ **51** 104   **53** $1e^{\frac{2\pi k}{3}i}$ where $k\in\{0,1,2\}$ **55** $\sqrt[8]{2}\exp\left(\left(\frac{\pi}{16}+\frac{2\pi}{4}k\right)i\right)$, where $k\in\{0,1,2,3\}$ **57** $2\sqrt[3]{2}\exp\left(\left(\frac{\pi}{18}+\frac{2\pi}{3}k\right)i\right)$, where $k\in\{0,1,2\}$ **59** $\sqrt[12]{29}\exp\left(\left(\frac{\arctan(5/2)}{6}+\frac{2\pi}{6}k\right)i\right)$, where $k\in\{0,1,2,3,4,5\}$ **61** $6e^{(\pi/4)i}=3\sqrt{2}+3\sqrt{2}\,i$ **63** $e^2=e^2+0i=e^2e^{2\pi i}$ **65** $e^7e^{-1i}=e^7\cos(-1)+e^7\sin(-1)i$ **67** (a) $\ln(42)+0i$ (b) $\ln(42)+(0+2\pi k)i$ where k is any integer   **69** (a) $\ln(4)+\pi i$ (b) $\ln(4)+(\pi+2\pi k)i$ where k is any integer   **71** (a) $\ln(\sqrt{98})-\frac{\pi}{4}i$; (b) $\ln(\sqrt{98})+(\frac{\pi}{4}+2\pi k)i$ where k is any integer   **73** (a) $\ln(10)-\frac{\pi}{3}i$; (b) $\ln(10)+(-\frac{\pi}{3}+2\pi k)i$ where k is any integer   **75** (a) $\ln(4)+\frac{\pi}{3}i$; (b) $\ln(4)+(\frac{\pi}{3}+2\pi k)i$ where k is any integer   **77** (a) $\ln(\sqrt{533})+\arctan(23/2)i$; (b) $\ln(\sqrt{533})+(\arctan(23/2)+2\pi k)i$ where k is any integer   **79** $z=re^{(3\pi/4)i}$ for any $r\in\mathbb{R}$ (why?)   **81** 5

83 see below

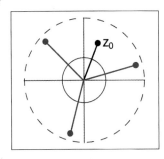

85 $\bar{z}=re^{-i\phi}$ **87** (b) $\text{Im}(\lambda)>0$ **89** see below

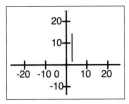

97 $z=2\pi k-i\ln\left(17\pm\sqrt{288}\right)$, where k is any integer

Chapter 9 Review:

1 F   **3** T   **5** F   **7** T   **9** T   **11** F   **13** F   **15** F (think of a star shape)   **17** T   **19** F   **21** e   **23** b   **25** c   **27** e   **29** d   **31** b   **35** (a) $(32/11, 16k/11)$ (b) see below

37 44   **39** $e+\frac{1}{e}$ **41** $40\pi\sqrt{13}$ **43** $r=\frac{7}{1+\cos(\phi)}$ **45** $r=\frac{1.5}{1+0.5\cos(\phi)}$ **47** $r=\frac{32.5}{1+2.5\cos(\phi)}$ **49** $(r,\phi)=\left(-\frac{\pi}{2},1\right)$ and $(r,\phi)=\left(\frac{\pi}{6},\frac{2+\sqrt{3}}{2}\right)$ **51** $9\pi/4$ **53** plot z as $(2,2)$, and $\bar{z}$ as $(2,-2)$ **55** $z=7i$ **57** (a) $12+2i$ (b) $35+30i$ (c) $3-4i$ **59** $z=\sqrt[3]{14}e^{i\phi}$ where $\phi=\frac{\pi}{18}+2\pi k$ where $k\in\{0,1,2\}$ **61** $z=\frac{\pi}{2}+2\pi k-(32\pm\sqrt{1023})i$ for any integer k

Section 10.1:

1 yes   **3** yes   **5** no   **7** $a=3/7$ and $b=-60/7$ **9** $A=-1/4$ **11** linear   **13** linear   **15** nonlinear   **17** nonlinear   **19** 2   **21** 1   **23** 1   **25** 1   **37** (b) A

Section 10.2:

1 (a) see below; (b) $x=7$; (c) source

3 (a) see below; (b) $y=2$ and $y=3$; (c) $y=2$ is a sink, and $y=3$ is a source

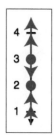

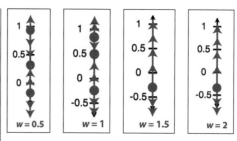

31 (a) $w = v/\gamma$ is a sink (b) see below

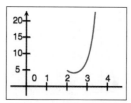

5 (a) see below; (b) $x = 0$ and $x = \frac{1}{2} \pm \frac{\sqrt{13}}{2}$; (c) $x = 0$ is a sink, the others are sources

7 (a) see below; (b) $y = (2k + 1)\frac{\pi}{2}$; (c) the sinks and sources are interdigitated along the number line, with a sink at $y = \pi/2$, a source at $y = 3\pi/2$, a sink at $y = 5\pi/2$ etc.

Section 10.3:

1 92.2711   **3** 4.6413   **5** −0.2172   **7** 24.47036   **9** 7.6267   **11** 403.2596   **13** (a) see below; (b) 4, 0, 4, 0, 4, 0

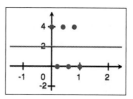

15 the red curve   **17** (a) nonautonomous; (b) & (c) see below

9 (a) $x = 0, 1.5, 2, 2.5$ (b) node, source, sink, source (from left to right) (c) 0 (d) 2   **11** (b) source at $y = 0$, node at $y = 4$ (graph below); (c) $y(0) = -0.001 \Rightarrow \lim_{t\to\infty} y(t) = -\infty$, $y(0) = 0.001 \Rightarrow \lim_{t\to\infty} y(t) = 4$, $y(0) = 4.001 \Rightarrow \lim_{t\to\infty} y(t) = \infty$

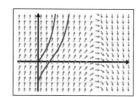

19 2.4824 m   **21** 11327 cells

Section 10.4:

1 $y(t) = Ae^{-3t^2}$ **3** $y(t) = -\ln(e^t + C)$ **5** $y(t) = (2\cos(t)+\sin(t)+C)^2$ **7** (a) $y(t) = 5e^{1.5t^2-7t+8}$; (b) see below

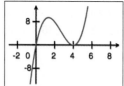

15 (a) $s = \frac{\rho\alpha/\lambda}{}$ **17** (a) $s = -\frac{\rho^2\alpha^2}{2\lambda^2 m} + \frac{\rho\alpha}{2m\lambda}\sqrt{\left(\frac{\rho\alpha}{\lambda}\right)^2 + 4m^2}$ (b) stable   (c) $\lim_{\alpha\to\infty} s_*(\alpha) = m$ **19** (e) see below

29 (a) see below; (b) $w = 0.1262$; (d) 0

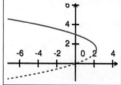

9 (a) $\frac{1}{3}y^3 + y = (\ln(t))^2 + \frac{4}{3}$; (b) see below

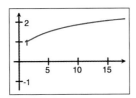

11 (a) $y(t) = -\frac{3}{2} + \frac{11}{2e}e^{0.25t^2}$; (b) see below

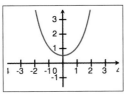

13 (a) $y(t) = e^{0.5 - 0.5\sec^2(t)}$; (b) see below

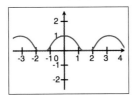

15 (a) $y(t) = \frac{1}{\sqrt{0.2}}\tan(0.2t + \arctan(\sqrt{0.2}\,k)) - 0.2$; (b) see below

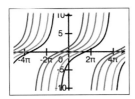

17 (a) $y(t) = 3 + (k-3)e^{6t-6}$; (b) see below

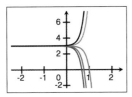

19 (a) $\sec(y) = -\frac{1}{2}\sin^2(t) + \sec(k) + \frac{1}{2}\sin^2(1)$; (b) see below

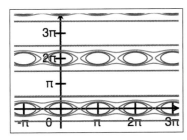

21 $x^2 + y^2 = r^2$ (circles)   **23** $y^4 = A|x|$ **25** $x^2 + 2y^2 = C$ (ellipses)   **27** $\frac{1}{2}y^2 - 9\ln|y| = -\frac{1}{3}x^2 + C$ **29** $\frac{1}{2}y^2 = \frac{1}{2}x^2 - \ln|x| + C$ **33** ≈ 3.001683025 mols (using Newton's method)   **35** $y - \ln|1 - y| = -t + C$ **37** (a) implicit solutions are $v(v-2) = A(v-1)^2 e^{-2t}$ where $A = -\frac{7}{9}, -15, \frac{21}{121}$ respectively; (b) see below

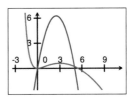

39 (a) $v(t) = \sqrt{4mg}\,\dfrac{1 + A\exp(\sqrt{g/m}\,t)}{1 - A\exp(\sqrt{g/m}\,t)}$ where $g = 9.8$, $m = 60$, and $A = (-54 - \sqrt{4mg})/(-54 + \sqrt{4mg}) \approx 18.627166$; (b) $-\sqrt{4mg}$ **41** (b) $\ln\left|\dfrac{T-A}{T+A}\right| + \dfrac{2(A-2)}{A}\arctan(T/A) = 4A^2\gamma t + C$; (c) 1.3392 hours   **43** $y(t) = \left(-\dfrac{2\sqrt{2g}}{9000\pi}t + \sqrt{6}\right)^2$ where $g = 9.8$ **45** $\frac{2}{5}\left(y - \frac{1}{4}\right)^{5/2} + \frac{1}{3}\left(y - \frac{1}{4}\right)^{3/2} + \frac{1}{8}\left(y - \frac{1}{4}\right)^{1/2} = -\dfrac{\sqrt{2g}}{4\pi}t + C$ where $g = 9.8$ and $C = \frac{2}{5}(23/4)^{5/2} + \frac{1}{3}(23/4)^{3/2} + \frac{1}{8}(23/4)^{1/2} \approx 36.6082$ **47** $s = \dfrac{m\alpha\rho}{\alpha\rho + m\lambda} + Ae^{-(\alpha\rho + m\lambda)t/m}$ where A depends on initial data

Section 10.5:

1 $u(t) = e^{3t}$ **3** $u(t) = t^{1/5}$ **5** $u(t) = 1$ (it is already in the right form)   **7** $y(t) = -3 + Ce^{2t}$ **9** $y(t) = \frac{5}{6} + Ce^{-6t}$ **11** $y(t) = te^{-3t} + Ce^{-3t}$ **13** $y(t) = \frac{1}{9} + Ce^{-4.5t^2}$ **15** $y(t) = \frac{2}{5}\sin(t) - \frac{1}{5}\cos(t) + Ce^{-2t}$ **17** $y(t) = t(\ln(t))^2 + Ct$ **19** $y(t) = 4\arctan(t)\ln(\arctan(t)) + C\arctan(t)$ **21** $y(t) = (\csc(t) + \cot(t))(\ln\left|\dfrac{1 + \cos(t)}{\cos(t)}\right| + C)$ **23** $10\ln\left(\frac{202}{22}\right)/\ln\left(\frac{202}{199}\right) \approx 1481.81$ minutes ≈ 24.69 hours   **25** 49.1753 hours (using Newton's method)   **27** (a) $w(t) = \frac{v}{\gamma} + Ce^{-\varepsilon\gamma t}$; (c) ε_3 **29** (a) $v(t) = -\frac{gm}{k} + Ce^{-kt/m}$; (b) $-\frac{gm}{k}$; (c) $k = \frac{(9.8)(47)}{5} = 92.12$ for the lighter person, and $k = \frac{(9.8)(125)}{5} = 245$ for the heavier person   **33** (a) $m(t) = 100 - 60e^{-8t/25}$; (b) $\frac{25}{8}\ln(9/5) \approx 1.836$ minutes   **35** 42.0775 minutes   **37** $0.1 + 1.9e^{-250}$ coulombs   **39** (a) $q(t) = -0.1t^2 + 0.6t - 0.3 + 0.3e^{-2t}$; (b) see below

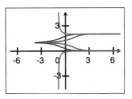

41 1.334499 amps   **43** $I(t) = \frac{2k}{k^2 + b^2}\sin(bt) - \frac{2b}{k^2 + b^2}\cos(bt) + \frac{2b}{k^2 + b^2}e^{-kt}$ where $b = 120\pi$ and $k = \frac{17}{4} \times 10^3$ **45** (a) continuous withdrawal at a rate of \$3000 per year; (b) per year; (c) \$26143.89   **47** (c) $y(t) = -\frac{800}{0.07} + \frac{800}{0.07}e^{0.07t}$; (d) \$255,269.31; (e) 66.5 years old; (f) \$1949.44   **49** (d) \$522168.76

Section 10.6:

1 equilibrium at $\left(-\frac{1}{13}, \frac{4}{13}\right)$, graph below

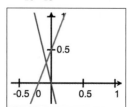

3 equilibria at $(0,0)$ and $\left(\frac{1+\sqrt{5}}{2}, \left(\frac{1+\sqrt{5}}{2}\right)^2\right)$, and $\left(\frac{1-\sqrt{5}}{2}, \left(\frac{1-\sqrt{5}}{2}\right)^2\right)$, graph below

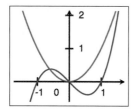

5 repeller **7** (b) $r'(t) < 0$ whenever $r > 0$, and $\phi'(t) > 0$ so the point at (r, ϕ) is orbiting the origin as it falls in; (c) counterclockwise since ϕ is increasing; (d) attractor **9** (a) ϕ is always increasing, so only the origin could be an equilibrium; (b) $r' = 0$ when $r = \beta$ so points on the circle don't change their distance to the origin; (d) $r' < 0$ when $r \in (0, \beta)$ so points inside the circle head toward the origin; and $r' > 0$ when $r > \beta$, so points outside the circle are pushed farther away from the origin; (e) attractor **11** (a) see below; (d) $\frac{dw}{dr} = -\frac{\alpha}{\beta}$ is constant; (e) $T_* = \frac{1240}{7}$

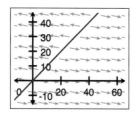

13 (a) $\beta > \delta$ means that the birth rate is larger than the death rate of healthy individuals, and $\mu > \delta$ means that the death rate is higher for infected individuals;

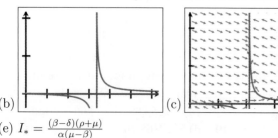

(e) $I_* = \frac{(\beta-\delta)(\rho+\mu)}{\alpha(\mu-\beta)}$

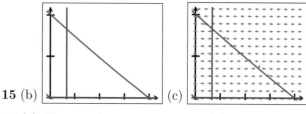

15 (b) (c)

17 (a) X grows faster since $\alpha > \gamma$; (b) no; (c) $x' = 0$ when $y = \alpha/\beta$ or $x = 0$, and $y' = 0$ when $x = \gamma/\delta$ or $y = 0$; (d) see below

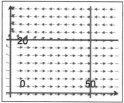

19 (b) & (c) graphs below

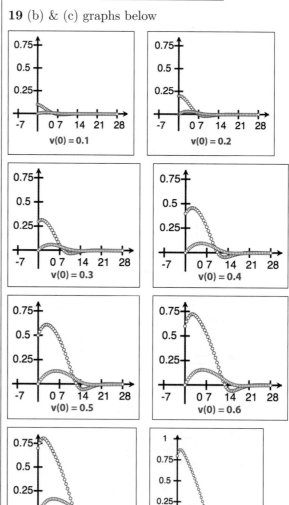

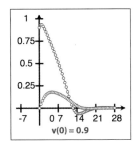

v(0) = 0.9

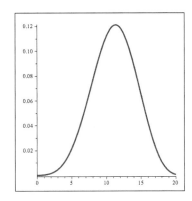

Chapter Review:

1 F **3** T **5** T **7** T **9** F **11** T **13** T **15** a **17** c **19** e **21** c **23** c **25** $p = 0$ is a sink, $p = 10$ is a source **27** 7.279356956 **29** $y(t) = 8 + 2e^{0.5t^2}$ **31** $y(t) = \frac{5}{4}(1000 + 3t) - 12480(1000 + 3t)^{-1/3}$ **33** $y(t) = \tan(t + \arctan(4))$ **35** $y(t) = 40/(8 - 3e^{5t})$ **37** $y(t) = -4 + Ce^{t/7}$ **39** $y^2 + x^2 = \ln(x^2) + C$ **41** $m(t) = 20 + 30e^{-t/20}$ grams **43** (a) $h' + \frac{4}{100-t}h = \frac{12}{5}\frac{1}{100-t}$; (b) $h(t) = 0.6 - 2.5 \times 10^{-9}(100 - t)^4$ **45** mv_0/γ **47** $x(t) = \frac{m}{\gamma}\ln(m + \gamma v_0 t) - \frac{m}{\gamma}\ln(m)$ **49** $\ln(5698.1)$ hours **51** (a) see below; (b) $k = 3$ and $k = 1/3$

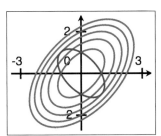

Appendix A: 1 5 **3** $\sqrt{53}$ **5** $\sqrt{15}$ **7** $\sqrt{55}$ **9** $4\pi/3$ **11** $15\pi/9$ **13** $240°$ **15** $270°$ **17** $(-\frac{1}{2}, \frac{\sqrt{3}}{2})$ **19** $(0, 1)$ **21** $(-\frac{\sqrt{2}}{2}, \frac{\sqrt{2}}{2})$ **23** $(\frac{\sqrt{3}}{2}, \frac{1}{2})$ **25** it continues to grow, and eventually surpasses each finite value

Appendix B: 1 $(t+6)^2 + 44$ **3** $(t+9)^2 - 11$ **5** $(t-4)^2 + 6$ **7** $\left(t - \frac{7}{2}\right)^2 - \frac{9}{4}$ **9** $2(t+3)^2 + 82$ **11** $5(t+9)^2 + \frac{269}{5}$ **13** $2(t-2)^2 + 14$ **15** $3\left(t - \frac{7}{6}\right)^2 + \frac{71}{12}$ **17** $3x - 2 + \frac{11}{x+2}$ **19** $8x - \frac{1}{2} + \frac{19/2}{2x+1}$ **21** $7x + 5 + \frac{14-8x}{x^2+1}$ **23** $\frac{4}{7}x^3 - \frac{40}{49}x + \left(1 + \frac{792}{49}x\right)/(7x^2 + 10)$ **27** yes **29** no

Appendix C: 1 1 **5** 8881600973913295056 **7** $32 + 80x + 80x^2 + 40x^3 + 10x^4 + x^5$ **9** $32 + 240x + 720x^2 + 1080x^3 + 810x^4 + 243x^5$ **11** 214200 **13** (a) $\Omega_T = 15905368710$; (b) see below

Appendix D: 1 (a) 6; (b) $(\pm 3, 0)$; (c) $(\pm\sqrt{5}, 0)$ **3** (a) 8; (b) $(0, \pm 4)$; (c) $(0, \pm\sqrt{7})$ **5** (a) $(-2, 0)$; (b) $(-2 \pm \sqrt{5}, 0)$; (c) $(-2 \pm \frac{\sqrt{70}}{4}, 0)$ **7** (a) $(\frac{4}{3}, -1)$; (b) $(\frac{4}{3}, -1 \pm \frac{\sqrt{33}}{3})$; (c) $(\frac{4}{3}, -1 \pm \frac{\sqrt{11}}{3})$ **9** (a) $\varepsilon = \frac{4\sqrt{5}}{5}$; (b) see below

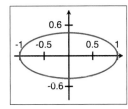

11 (a) $\varepsilon = 0$ (b) see below

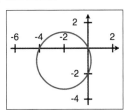

13 $\frac{x^2}{4} + \frac{y^2}{3} = 1$ **15** $\frac{x^2}{64} + \frac{y^2}{16} = 1$ **17** (a) $(\pm 3, 0)$; (b) $(\pm\sqrt{13}, 0)$; (c) $y = \pm\frac{2}{3}x$ **19** (a) $(0, \pm 3)$; (b) $(0, \pm\sqrt{14})$; (c) $y = \pm\frac{3\sqrt{5}}{5}x$ **21** (a) $(0, 3 \pm 3)$; (b) $(0, 3 \pm 3\sqrt{2})$; (c) $y = 3 \pm x$ **23** (a) $(-\frac{1}{6}, -4 \pm \frac{\sqrt{95}}{6})$; (b) $(-\frac{1}{6}, -4 \pm \frac{\sqrt{665}}{6})$; (c) $y = -4 \pm 6(x + \frac{1}{6})$ **25** (a) $27\sqrt{6}/12$; (b) see below

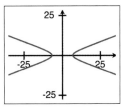

27 (a) $\sqrt{30}/5$; (b) see below

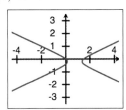

29 $\frac{x^2}{16} - \frac{y^2}{84} = 1$ **31** $\frac{x^2}{4} - \frac{y^2}{2} = 1$ **33** $\frac{x^2}{1} - \frac{4y^2}{5} = 1$ **35** (a) vertex: origin, focus: $(0,7/4)$, directrix: $y = -7/4$; (b) see below

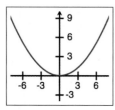

37 (a) vertex: origin, focus: $(0,-2/5)$, directrix: $x = 2/5$; (b) see below

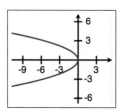

39 (a) vertex: $(0,3)$, focus: $(0,2.75)$, directrix: $y = 3.25$; (b) see below

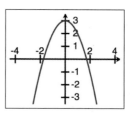

41 (a) vertex: $(-1/3,-1)$, focus: $(5/12,-1)$, directrix: $x = -13/12$; (b) see below

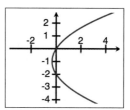

43 $x = \frac{5}{2} + \frac{1}{6}y^2$ **45** $y = \frac{7}{2} + \frac{1}{10}x^2$ **47** $x = 1 + \frac{1}{8}y^2$ **49** $y = 10 + \frac{1}{36}x^2$ **51** $x = 7 + \frac{1}{4}y^2$ **53** $y = -5 - \frac{3}{4}x^2$ **55** $\frac{(x+21122.6652)^2}{(384748)^2} + \frac{y^2}{0.99698599(384748)^2} = 1$ **59** $\frac{(y-0.6)^2}{0.16} - \frac{x^2}{0.2496} = 1$ **61** (a) $y = \frac{1}{52}x^2$ (b) $\frac{152}{44}$ m   **63** the parabola

Bibliography

[1] Simoson, A. Finding spring on planet x. *PRIMUS*, XVII(3):228–238, 2007.

[2] Anderson, B. *The Physics of Sailing Explained*. Sheridan House, New York, 2003.

[3] Hayes, B. Gauss's day of reckoning. *American Scientist*, 94:200–205, May-June 2006.

[4] Conselice, C. The universe's invisible hand. *Scientific American*, pages 34–41, February 2007.

[5] Elder, A., Gelein, R., Finkelstein, J., Cox, C., and Oberdörsterl, G. Pulmonary inflammatory response to inhaled ultrafine particles is modified by age, ozone exposure, and bacterial toxin. *Inhal. Toxicol.*, (12 (Suppl. 4)):227–246, 2000.

[6] Falbo, C. `http://www.sonoma.edu/Math/faculty/falbo/gauss.html` .

[7] Lutzer, C., and Ross, D. The dynamics of embedded-charge microenergy harvesting. *Journal of Computational and Nonlinear Dynamics*, 5(2), 2010.

[8] Griffiths, D. *Introduction to Quantum Mechanics*. Pearson/Prentice Hall, Upper Saddle River, 2005.

[9] Halliday, D. and Resnick, R. *Fundamentals of Physics*. Wiley, Hoboken, NJ, 3d edition, 1988.

[10] Lide, D., editor. *Handbook of Chemistry and Physics*. CRC Press, New York, 86th edition, 2005.

[11] Schroeder, D. *An Introduction to Thermal Physics*. Addison Wesley Longman, San Francisco, 2000.

[12] Watson, D. G-loc, could it happen to you? *AOPA (Australia) Magazine*, 43(8), 1990.

[13] Vanhaeren, M., d'Errico, F., Stringer, C., James, S., Todd, J. and Mienis, H. Middle paleolithic shell beads in Israel and Algeria. *Science*, 312(5781):1785–1788, June 2006.

[14] Purcell, E., and Varberg, D. *Calculus with Analytic Geomety*. Prentice Hall, Engelwood Cliffs, 1987.

[15] Adler, F. *Modeling the Dynamics of Life*. Brooks Cole, Belmont, 2005.

[16] Hoppensteadt, F. and Peskin, C. *Modeling and Simulation in Medicine and the Life Sciences*. Springer, New York, 2002.

[17] Anton, H. and Busby, R. *Contemporary Linear Algebra*. Wiley, Hoboken, NJ, 2003.

[18] Anton, H., Bivens, I., and Davis, S. *Calculus*. Wiley, Hoboken, NJ, 2005.

[19] Rogawski, J. *Calculus*. W. H. Freeman, New York, 2008.

[20] Stewart, J. *Calculus, Early Transcendentals*. Brooks Cole, Belmont, 2008.

[21] Zweibel, K., Mason, J., and Fthenakis, V. A solar grand plan. *Scientific American*, pages 64–73, January 2008.

[22] Krane, K. *Modern Physics*. Wiley, Hoboken, NJ, 1983.

[23] Combs, L. http://stern.kennesaw.edu/inter/in01004.htm .

[24] Boas, M. *Mathematical Methods in the Physical Sciences*. Wiley, Hoboken, NJ, 2006.

[25] Zeilik, M. and Gustad, J. *Astronomy—The Cosmic Perspective*. Wiley, Hoboken, NJ, 2nd edition, 1990.

[26] Ko, M., Sze, N., and Prather, M. Better protection of the ozone layer. *Nature*, (367):505–508, February 1994.

[27] Lax, P. *Hyperbolic Systems of Conservation Laws and the Mathematical Theory of Shock Waves*. Society for Industrial and Applied Mathematics, Philadelphia, 1973.

[28] Feynman, R., Leighton, R., and Sands, M. *The Feynman Lectures on Physics, Definitive Edition*. Addison-Wesley, San Francisco, 2006.

[29] Hatch,   R.   http://web.clas.ufl.edu/users/rhatch/pages/01-Courses/current-courses/08sr-newton.htm .

[30] Jarmakani, J., Graham, T., Benson, D., Canent, R., and Greenfield, J. In vivo pressure-radius relationships of the pulmonary artery in children with congenital heart disease. *Circulation*, 43:585–592, 1971.

[31] Silbey, R., Alberty, R., and Bawendi, M. *Physical Chemistry*. Wiley, Hoboken, NJ, 2005.

[32] Chaboud, A., Chernenko, S., and Wright, J. Trading activity and exchange rates in high-frequency ebs data. International Finance Discussion Papers, September 2007.

[33] Pennings, T. Do dogs know calculus? *College Math. J.*, 34(3):178–182, 2003.

[34] URL. http://www.geophys.washington.edu/tsunami/general/physics/characteristics.html .

[35] URL. http://www.mos.org/oceans/planet/features.html .

[36] URL. http://sgs.nozzle-network.com/en/sample/pdqgraph.jsp .

[37] URL. http://math.furman.edu/~mwoodard/mqs/mquot.shtml .

[38] URL. http://quickfacts.census.gov/qfd/states/36/3663000lk.html .

[39] URL. http://www.madsci.org/posts/archives/2000-07/963425573.Ch.r.html .

[40] URL. http://www.simetric.co.uk/si_liquids.htm .

[41] URL. http://www.everyscience.com/Chemistry/Physical/Mixtures/a.1265.php .

[42] Rudin, W. *Principles of Mathematical Analysis*. McGraw-Hill, New York, 1976.

[43] Strauss, W. *Partial Differential Equations, An Introduction*. Wiley, Hoboken, NJ, 1992.

Index